Microbiome, Immunity, Digestive Health and Nutrition

Epidemiology, Pathophysiology, Prevention and Treatment

Microbiome, Immunity, Digestive Health and Nutrition

Epidemiology, Pathophysiology, Prevention and Treatment

Edited by

Debasis Bagchi
College of Pharmacy and Health Sciences, Texas Southern University, Houston, TX, United States; Department of Biology, Adelphi University, Garden City, NY, United States; Department of R&D, Victory Nutrition International, Inc., Bonita Springs, FL, United States

Bernard William Downs
Division of Precision Nutrition, Victory Nutrition International, Inc., Bonita Springs, FL, United States

ELSEVIER

ACADEMIC PRESS
An imprint of Elsevier

Academic Press is an imprint of Elsevier
125 London Wall, London EC2Y 5AS, United Kingdom
525 B Street, Suite 1650, San Diego, CA 92101, United States
50 Hampshire Street, 5th Floor, Cambridge, MA 02139, United States
The Boulevard, Langford Lane, Kidlington, Oxford OX5 1GB, United Kingdom

Notices
Knowledge and best practice in this field are constantly changing. As new research and experience broaden our understanding, changes in research methods, professional practices, or medical treatment may become necessary.

Practitioners and researchers must always rely on their own experience and knowledge in evaluating and using any information, methods, compounds, or experiments described herein. In using such information or methods they should be mindful of their own safety and the safety of others, including parties for whom they have a professional responsibility.

To the fullest extent of the law, neither the Publisher nor the authors, contributors, or editors, assume any liability for any injury and/or damage to persons or property as a matter of products liability, negligence or otherwise, or from any use or operation of any methods, products, instructions, or ideas contained in the material herein.

ISBN: 978-0-12-822238-6

For Information on all Academic Press publications
visit our website at https://www.elsevier.com/books-and-journals

Publisher: Nikki P. Levy
Acquisitions Editor: Megan R. Ball
Editorial Project Manager: Lindsay Lawrence
Production Project Manager: Bharatwaj Varatharajan
Cover Designer: Miles Hitchen

Typeset by MPS Limited, Chennai, India

Dedication

Dedicated this legendary book to Mr. Tapan Kumar Pain, my beloved and respected guide, philosopher, and friend for his esteemed contribution, encouragement, and inspiration in my life.

—Debasis Bagchi

This book is dedicated with love and awe-struck admiration to Kim, my wife. She is the most amazing woman I know and a source of constant inspiration to me. She achieves extraordinary accomplishments and makes them seem simply like routine occurrences.

—Bernard William Downs

Contents

List of contributors xix
Preface xxv

Section I
Microbiome and human health: an introduction

1. Oral microbiome: a gateway to your health 3

Na-Young Song, Se-Young Park, Won-Yoon Chung, Young-Joon Surh, Kyung-Soo Chun and Kwang-Kyun Park

1.1 Introduction 3
1.2 Oral microbiome and oral disease 3
1.2.1 Caries 3
1.2.2 Periodontal disease 4
1.2.3 Oral cancer 4
1.3 Oral microbiome and systemic disease 5
1.3.1 Alzheimer's disease 5
1.3.2 Cardiovascular disease 6
1.3.3 Diabetes 6
1.4 Concluding remarks 7
Acknowledgments 8
References 8

2. Influence of microbiome in shaping the newborn immune system: an overview 11

Manoj Kumar Kingsley and B. Vishnu Bhat

2.1 Introduction 11
2.2 How the microbiota shapes the development of immunity in early life 11
2.3 Influence of the microbiota on the development and function of specific immune cell subsets in early life 13
2.3.1 Myeloid cells 13
2.3.2 Lymphoid cells 14
2.3.3 Epithelial cells 20
2.4 Conclusion 21
Author disclosure statement 21
References 21

3. Impact of the gut microbiome on human health and diseases 25

Samudra Prosad Banik, Shalini Sehgal, Rituparna Banik Ghosh and Ahana Das

3.1 Introduction 25
3.1.1 Definition of microbiome 25
3.1.2 Uneven distribution of microbiota across various sites and organs of the body: reasons and implications 25
3.2 The gut as the most significant niche of the human microbiome 26
3.2.1 Establishment of the gut microbiome: journey from womb to adulthood 26
3.2.2 Heritable component of the gut microbiome 26
3.2.3 Role of diet and other environmental factors in shaping the gut microbiome 27
3.3 Physiological roles of the gut microbiome 27
3.3.1 Gut Microbiota in metabolism 27
3.3.2 Gut microbiota in boosting immune response: role of short-chain fatty acids and cross-talk with diet 28
3.3.3 The role of the gut microbiome in the development of gut-associated lymphoid tissue 29
3.3.4 Gut microbes as a defense mechanism against pathogens 30
3.3.5 Role in cognitive development: the gut-brain axis 30
3.3.6 The gut microbiome as a human organ 32
3.4 Dysbiosis of normal flora: implications in human diseases and intervention strategies 32
3.4.1 Gut microbiome in human ailments 32
3.4.2 How to keep the gut happy: prebiotics, probiotics, postbiotics, and synbiotics 34
3.5 Concluding remarks 35
Acknowledgments 36
References 36

4. The gut microbiome, human nutrition, and immunity: visualizing the future 41

Soching Luikham and Jhimli Bhattacharyya

4.1 Introduction 41
4.2 Studying the gut microbiome and its importance 42
4.3 The role of the microbiome in nutrition 44
4.3.1 Digestion of breast milk 45
4.3.2 Digestion of Fiber 45
4.3.3 Enhancement of immune competence 45
4.3.4 Promotion of brain health 45
4.3.5 Boosting of cardiovascular health 45
4.3.6 Potentiation of insulin sensitization and decrease in the risk of diabetes 45
4.3.7 Prebiotic foods and dietary fiber 45
4.3.8 Probiotic foods 46
4.4 The role of the microbiome in immune system development 46
4.5 Various ways to improve our gut microbiome 47
4.5.1 Eat a diverse diet 47
4.5.2 Consider fermented foods 47
4.5.3 Avoid artificial sweeteners 47
4.5.4 Consume prebiotic foods 47
4.5.5 Breastfeed for six months 47
4.5.6 Include whole grains 48
4.5.7 Include plant-based foods 48
4.5.8 Consume polyphenol-containing food 48
4.5.9 Take antibiotics when required 48
4.6 Future areas of research 48
4.7 Conclusion 48
Acknowledgments 48
References 49

5. Individual microbiota correction and human health: programming and reprogramming of systemic and local immune response 53

Tamara Meleshko, Oleksandra Pallah and Nadiya Boyko

5.1 Introduction 53
5.1.1 What is known and what is new? Beneficial microbes: a new concept? 53
5.1.2 The concept of a healthy human microbiome 53
5.1.3 The basic concept of a healthy normal microbiota and concepts of early formation of the microbiome 54
5.1.4 Microbiome formation and early colonization 54
5.1.5 The endocannabinoid axis: the microbiome as a mechanism by which lifestyle affects various aspects of metabolism, which in turn can lead to metabolic syndrome and metabolic disorders 54
5.2 Brief methodology overview 55
5.3 Biomarker validation: metabolic syndrome and cardiovascular diseases 55
5.3.1 Limited clinical trial: person-oriented treatment of type 2 diabetes 55
5.4 Results and discussion 55
5.4.1 Patterns of clinically validated biomarkers in the gastrointestinal microbiome for patients with metabolic syndrome, obesity, type 2 diabetes, and cardiovascular diseases 55
5.4.2 Data from human trials: treatment by individual nutrition and biopreparation with pharmabiotics 56
5.4.3 IT tools for implementation of 3P medicine and individual nutrition 58
5.4.4 Databases 58
5.4.5 IT tools 59
5.4.6 Mechanisms involved and prospective research 59
5.4.7 Biological active compounds as potential modern prebiotics 59
5.4.8 Early health programming 62
5.5 Conclusion 65
References 66

Section II
Microbiome and digestive health

6. The etiology of gut dysbiosis and its role in chronic disease 71

Alvin H. Danenberg

6.1 Introduction 71
6.2 First things first 71
6.3 Diet and the gut 72
6.4 Short-chain fatty acids 72
6.5 Fasting and the gut 72
6.6 The effect of unhealthy diets on the gut 72
6.7 The benefit of healthy diets for the gut 73
6.7.1 Mediterranean diet 74
6.7.2 Low-carbohydrate diets 74
6.7.3 Ketogenic diets 74

6.7.4 Paleolithic diet 74
6.8 Inflammation 74
6.8.1 Acute inflammation 75
6.8.2 Chronic inflammation 75
6.9 Gluten 75
6.10 Gliadin 75
6.11 Leaky gut 76
6.12 Other irritants to the gut 76
6.13 A closer look into the inner tube of the gut 77
6.13.1 The gut microbiome and infections 78
6.13.2 Cross-talk between microbes 78
6.14 Chronic inflammatory and autoimmune diseases 79
6.14.1 Type 1 diabetes 79
6.14.2 Rheumatoid arthritis 79
6.14.3 Celiac disease 79
6.14.4 Inflammatory bowel disease 80
6.14.5 Allergic diseases 80
6.14.6 Systemic lupus erythematosus 80
6.14.7 Skin-related autoimmune pathologies 80
6.14.8 Neurological inflammatory diseases 80
6.14.9 Periodontal disease 80
6.14.10 Cancer 81
6.15 Unraveling the mystery of treatment 82
6.15.1 Treatment for active infections 82
6.15.2 Understanding periodontal disease 82
6.15.3 Repairing and restoring the gut 82
6.15.4 Spore-based bacillus probiotics 83
6.15.5 Prebiotics 83
6.15.6 The mucosal layer 83
6.15.7 The importance of the mouth 84
6.15.8 Personal oral hygiene protocol 84
6.16 Concluding remarks 85
References 85

7. Role of the microbiome in the function and diseases of the digestive system 93

F A Dain Md Opo, Addisu Demeke Teklemariam, Rokeya Pervin and Md. Akil Hossain

7.1 Introduction 93
7.2 Microbes in gastrointestinal tract 93
7.3 Development of the gut flora 94
7.4 Physiological functions of gut microbes 95
7.4.1 Metabolism 95
7.4.2 Immune system development 96
7.4.3 Central nervous system development 96
7.5 Contribution of human intestinal microbes in causing diseases 96
7.5.1 *Clostridium difficile* infection 97
7.5.2 Inflammatory bowel disease 97
7.5.3 Colorectal cancer 97
7.5.4 Rheumatoid arthritis 98
7.5.5 Obesity 98
7.5.6 Esophageal cancer 98
7.6 Microorganism that can prevent human disease 99
7.6.1 Colorectal cancer 99
7.6.2 Gastric cancer 99
7.6.3 Inflammatory bowel diseases 100
7.6.4 Obesity 100
7.6.5 Hepatic encephalopathy 100
7.7 Dietary modulation of the gut microbiota 100
7.8 Conclusion 102
References 102

Section III
Microbiome and metabolic syndrome

8. The beneficial role of healthy microbiome in metabolic syndrome and cardiovascular health 109

Smriti Arora, Nitika Bhambri, Samikshha Puri, Anamika, Nandini Dasgupta, Shradha Khater and Gautam Das

8.1 Introduction 109
8.2 Gut microbiota and metabolic syndrome 110
8.2.1 Obesity 111
8.2.2 Gut microbiota and hyperglycemia and diabetes 112
8.2.3 Gut microbiota and dyslipidemia 112
8.2.4 Gut microbiota and hypertension 113
8.2.5 Gut microbiota and cardiovascular disease 113
8.2.6 Early gut microbiota and MetS 114
8.3 Mechanism at the molecular level 114
8.3.1 The gut barrier 114
8.3.2 Energy metabolites 115
8.4 Metabolic syndrome therapy by managing gut microbiota 117
8.4.1 Influence of diet 117
8.4.2 Influence of prebiotics 118
8.4.3 Influence of probiotics 118
8.4.4 Influence of exercise 119
8.5 Conclusion 119
References 119

9. The role of the intestinal microbiota in weight loss in overweight and obese humans 125

Isabel Sospedra López, Gema Alfonso Sigüenza, Estela González Rodríguez, Mar Lozano Casanova and José Miguel Martínez-Sanz

9.1 Causes of overweight and obesity 125
9.2 The microbiota and obesity 125
9.3 The bacterial composition of the microbiota according to weight status 126
9.4 Effect of dietary interventions on the microbiota 127
9.5 Conclusions 133
References 133

10. Role of microbial metabolites in cardiovascular and human health 137

Abhilasha Singh

Abbreviations 137
10.1 Introduction 137
10.2 Gut microbial metabolites and cardiovascular disease 138
10.2.1 Short-chain fatty acids 139
10.2.2 Other metabolites of clinical significance 143
10.3 Conclusions 145
References 145

11. Beneficial role of gut microbiome in metabolic syndrome, obesity, and cardiovascular diseases 149

Pradipta Banerjee, Sriya Choudhury, Komal Jalan, Amitava Das, Nandini Ghosh and Debasis Bagchi

11.1 Introduction 149
11.1.1 Gut microbiome 149
11.1.2 Dietary intake for reducing the risk factor of metabolic syndrome, obesity, and cardiovascular disease 149
11.2 Metabolic syndrome 150
11.3 Composition of a healthy gut microbiome 150
11.4 Beneficial role of a healthy microbiome in preventing metabolic syndrome 154
11.4.1 How the microbiome is related to health and diseases 154
11.4.2 Gut microbiome and metabolic syndrome 154
11.4.3 Gut microbiome: a new strategy to treat obesity 155
11.4.4 Intestinal microbiota in cardiovascular diseases: the heart-gut axis 156
11.5 Influence of dietary components in boosting a healthy microbiome and reducing the risk of metabolic syndrome 157
11.5.1 Carbohydrates 158
11.5.2 Proteins 158
11.5.3 Fats 159
11.5.4 Fatty acids 159
11.5.5 Polyphenols 160
11.5.6 Spices 160
11.5.7 Prebiotics 160
11.5.8 Probiotics 161
11.5.9 Selective diets 161
11.6 Conclusion 162
References 162
Further reading 165

12. Restoring gut biome balance for weight loss: clinical applications 167

Naras Bhat, Jyoti Bhat, Kusum Bhat, Imran Junaid, Anita Bhat and Joy Bhat

12.1 Introduction 167
12.2 Gut microbiome resiliency 167
12.3 Obesogenic features of the gut microbiome 168
12.3.1 Obesogenic gut microbiome is transferrable 168
12.3.2 Mother-to-child transmission of obesity 168
12.4 Dysbiosis as obesogenic signature 168
12.4.1 Decreased diversity 168
12.4.2 Firmicutes/bacteroidetes ratio 168
12.4.3 Keystone bacteria 168
12.4.4 Gut barrier permeability 169
12.4.5 Concept of prebiotics, probiotics, and postbiotics 169
12.5 Obesogenic metabolic endotoxemia and insulin resistance 170
12.5.1 Diet-induced lipopolysaccharide endotoxemia 170
12.5.2 Insulin resistance and inflammation 172
12.5.3 Stress and metabolic endotoxemia 172
12.5.4 Immune activation 172
12.6 Effect of dietary components on metabolic endotoxemia 172
12.6.1 Effect of fasting on metabolic endotoxemia 172

12.6.2 Exercise and metabolic endotoxemia 172
12.6.3 How long does it take to change the gut biome? 172
12.7 Microbial regulation of host energy balance 173
12.7.1 Microbiota-associated energy harvest 173
12.7.2 Cross-talk between the gut microbiota and the host metabolism 173
12.7.3 Gut-brain axis of eating behavior 173
12.7.4 Gut-muscle axis and energy output 173
12.7.5 Gut-adipose-liver axis and energy storage 173
12.7.6 Gut biome and circadian rhythm 174
12.7.7 Fasting and autophagy 174
12.8 Western lifestyle as a menace to the gut biome 174
12.9 Pragmatic mitigation of western lifestyle effects on the gut biome 176
12.10 Eating and fasting 176
12.10.1 What to eat? 176
12.10.2 Clinical protocol for microbiome-centered eating 178
12.10.3 When to eat? Intermittent fasting 178
12.10.4 Clinical protocol for fasting-mimicking time-restricted eating 179
12.11 Physical activity versus physical inactivity 179
12.11.1 Exercise versus nonexercise 179
12.11.2 Sedentary versus sitting disease 179
12.11.3 Clinical protocol for physical activity 180
12.12 Wake-sleep circadian rhythm 180
12.12.1 Clinical protocol for circadian rhythm balance 180
12.13 Conclusion 181
References 181
Further reading 184

13. Harnessing gut friendly microbiomes to combat metabolic syndrome 185

Chitra Selvan and Satinath Mukhopadhyay

13.1 Introduction 185
13.2 Gut microbiome 185
13.3 Gut microbiota and metabolic disorders: mechanisms 187
13.3.1 Gut microbiome and energy homeostasis 187
13.4 Dysbiosis 187
13.5 Gut microbiome and inflammation 188
13.6 Future: harnessing the gut microbiota to treat metabolic disorders 188
13.7 Intestinal transplantation 188
13.8 Oral butyrate 189
13.8.1 Probiotics 189
13.9 Heat-killed probiotics 190
13.10 Prebiotics 190
13.11 Synbiotics 190
13.12 Postbiotics 190
13.13 Summary and concluding remarks 190
References 191

14. The effect of resveratrol-mediated gut microbiota remodeling on metabolic disorders 193

Marzena Wojcik and Lucyna Wozniak

14.1 Introduction 193
14.2 Gut microbiota and obesity 194
14.2.1 Resveratrol, the gut microbiota, and obesity 196
14.3 Gut microbiota and type 2 diabetes mellitus 199
14.3.1 Resveratrol, the gut microbiota, and type 2 diabetes mellitus 200
Acknowledgments 200
References 200

Section IV
Microbiome and immune health

15. The microbiome, immunity, anaerobism, and inflammatory conditions: a multifaceted systems biology intervention 205

Bernard William Downs, Samudra Prosad Banik, Manashi Bagchi, Rituparna Banik Ghosh, Steve Kushner and Debasis Bagchi

15.1 Introduction 205
15.2 Restoring the ideal biological environment and optimal homeostatic equilibrium 205
15.3 Digestion: the first step in metabolism 206
15.4 Antioxidant therapy: firing at the wrong target 206
15.5 Role of the microbiome 207
15.6 Epigenetic influences 209
15.7 Microbiome functional competence 209

15.8 Inflammatory events 210
15.9 Multifaceted systems biology intervention: conceptual overview 210
15.10 Microbial dysbiosis 211
15.11 Functions of microbial strains 212
15.12 Microbial restoration 212
15.13 Butyric acid 213
15.14 Conclusion 213
References 214
Further reading 216

16. Preclinical data support the strong specificity and advertising ability to control the immune reactions at mucosal sites 217

Tamara Meleshko, Oleksandra Pallah and Nadiya Boyko

16.1 Introduction 217
16.2 Microbiota detection tools 218
16.2.1 Technological approaches 218
16.2.2 Mice models 218
16.2.3 Schaedler's *E. coli* induces diminished translocation of *Lactobacillus salivarius* into the lung: the role of surfactant protein D 221
16.3 Conclusion 225
References 225

Section V
Microbiome and cognitive health

17. Achieving dopamine homeostasis to combat brain-gut functional impairment: behavioral and neurogenetic correlates of reward deficiency syndrome 229

Kenneth Blum, Bernard William Downs, Debasis Bagchi, Gene-Jack Wang, Panyotis K. Thanos, Igor Elman, Rehan Jalali, Abdalla Bowirrat, David Baron, Diwanshu Soni, Rajendra D. Badgaiyan and Mark S. Gold

17.1 Introduction 229
17.2 Understanding endorphin deficiency syndrome and opioid deficiency syndrome 232
17.3 Understanding obesity as a hypodopaminergia 233
17.4 Gut-hypothalamic homeostatic energy regulation 234
17.4.1 The set point dilemma 235
17.4.2 Homeostatic systems 235
17.4.3 Feeding behavior regulation by dopaminergic systems 235
17.4.4 Food and drug addiction identification of common neurobiological mechanisms 236
17.5 Summary 237
17.6 Conclusion 239
References 239

18. Influence of gut microbial flora in body's serotonin turnover and associated diseases 245

Sabyasachi Choudhuri, Jyotirmoy Panda and Sayantan Maitra

List of abbreviation 245
18.1 Introduction 245
18.2 The gut microbiota 246
18.3 Role of the gut microbiota in maintaining the body's normal physiological homeostasis 249
18.4 Influence of neurotransmitters on the gut microbiome population through the gut-brain axis 251
18.4.1 Role of biogenic amines in the gut microbiota 252
18.4.2 Role of esters in the gut microbiota 256
18.4.3 Role of miscellaneous neurotransmitters in the gut microbiota 257
18.5 Association between the microbiome-gut-brain axis and serotonin turnover 257
18.6 Dysregulation in the occurrence of gut microbiota and its implications in diverse pathological conditions 259
18.7 Conclusion 262
References 262

19. The connection between diet, gut microbes, and cognitive decline 265

Abhai Kumar, Smita Singh and Rameshwar Nath Chaurasia

19.1 Introduction 265
19.2 Microbiome composition and evolution 265
19.3 Microbiota and brain functioning 266

19.4 Microbiome-gut-brain axis 267
19.5 Microbiome and cognitive decline 268
19.6 Diet and cognition 268
19.7 Conclusion 269
References 269

20. Role of the gut microbiome in Rett syndrome 273

Carly Albright, Amanda Friend, Benjamin Peters, Miranda Reed, Vishnu Suppiramaniam and Subhrajit Bhattacharya

20.1 Introduction to the gut microbiome 273
20.1.1 Digestion and the gut microbiome 273
20.1.2 Absorption, distribution, metabolism and excretion 274
20.1.3 Gut microbiome altering medication efficacy 274
20.1.4 Medication-induced gut microbiome dysbiosis 276
20.2 Diseases of the gut 276
20.2.1 Diseases central to the gut 276
20.2.2 Diseases distant from the gut: psychiatric and neurological disorders 277
20.3 The gut microbiome and Rett syndrome 280
20.3.1 Background and history of Rett syndrome 280
20.3.2 Pathology, symptoms, and progression 281
20.3.3 Genetic links 281
20.3.4 DNA methylation in Rett syndrome 282
20.3.5 Biogenic amine systems in Rett syndrome 282
20.3.6 Rett syndrome in males 282
20.3.7 Role of different excitatory and inhibitory receptor systems: GABA, NMDAR, and AMPAR 283
20.3.8 Treatment paradigm 283
20.4 Relationship between gut microbiome and Rett syndrome 284
20.4.1 Effect on gut microbiome diversity and dysbiosis 285
20.4.2 Recent and past studies 286
20.4.3 Rett syndrome and other neurological and psychiatric disorders that share gut microbiome alterations 287
20.4.4 Intestinal inflammation in Rett syndrome 289
20.4.5 Rett syndrome therapeutics and the gut microbiome 289
20.5 Summary 290
References 291

Section VI
Microbiome, dermal health and wound healing

21. Skin microbiota and its role in health and disease with an emphasis on wound healing and chronic wound development 297

J.H. Kim, A. Ahamed, K. Chen, E.G. Lebig, B. Petros, S. Saeed and M. Martins-Green

21.1 Introduction 297
21.2 Skin structure and function 298
21.3 Skin microbiome diversity in different parts of the body 299
21.4 Microbial dysbiosis in skin-related diseases 300
21.4.1 Atopic dermatitis 300
21.4.2 Acne vulgaris 300
21.4.3 Rosacea 301
21.4.4 Psoriasis 301
21.4.5 Skin cancers 302
21.5 The microbiome and cutaneous wound healing 302
21.5.1 The microbiome and cutaneous chronic wound development 303
21.5.2 Specific chronic wounds and their wound microbiome 304
21.6 Common techniques for surveying the skin microbiota 306
21.7 Manipulation of the skin microbiota to prevent or treat skin-associated ailments 307
21.7.1 Probiotics 307
21.7.2 Skin microbiota transplant 307
21.7.3 Prebiotics 307
21.8 Perspectives 308
References 308

Section VII
Microbiome and cancer

22. Healthy gut microbiome in the prevention of colorectal cancer 315

Marufa Rumman, Misako Nagasaka, Bayan Al-Share, Asfar S. Azmi and Md. Hafiz Uddin

22.1 Introduction 315

22.2 The gut microbiome in the progression of colorectal cancer 315
22.3 Factors affecting the gut microbiome 317
22.3.1 Dietary fats 317
22.3.2 Red and processed meats 317
22.3.3 Alcohol consumption 318
22.3.4 Smoking 318
22.3.5 Obesity 318
22.3.6 Physical activity 319
22.4 The gut microbiome in the protection against colorectal cancer 319
22.5 Prevention of colorectal cancer by enhancing healthy gut flora 320
22.5.1 Prebiotics 320
22.5.2 Probiotics 321
22.5.3 Synbiotics 322
22.6 Conclusion and future perspectives 322
Conflict of interest 322
References 322

Section VIII
Microbiome, arthritis and multiple sclerosis

23. Regulatory roles of the microbiome in arthritis, fibromyalgia, and multiple sclerosis 331

Liz J. Vázquez-Chinea, Savannah N. Johnson and Elizabeth A. Lipski

23.1 Introduction 331
23.2 Rheumatoid arthritis 331
23.2.1 Microbiome profile of individuals with rheumatoid arthritis 332
23.2.2 Probiotic supplementation in individuals with rheumatoid arthritis 332
23.3 Fibromyalgia 334
23.3.1 Pain and fibromyalgia 334
23.3.2 Gut microbiota and fibromyalgia 334
23.4 Multiple sclerosis 335
23.4.1 Gut microbiota and multiple sclerosis 335
23.4.2 Multiple sclerosis and modifiable factors 336
23.5 Discussion 338
23.6 Conclusion 340
References 340

Section IX
Environmental pollutants and gut microbiome

24. Microplastic toxicity and the gut microbiome 345

Muhammad Manjurul Karim, Pooja Shivappa, Nadiya Dileep, Tania Akter Jhuma and Ashfaque Hossain

24.1 Introduction 345
24.2 History of microplastics 345
24.3 Microplastics pollution and the threat to habitats 346
24.4 Gut microbiome 347
24.4.1 Dysbiosis and eubiosis 348
24.4.2 Short-chain fatty acids and branched-chain fatty acids 349
24.4.3 Virome and microplastics 349
24.4.4 Akkermansia muciniphila 351
24.5 Toxic effects of microplastics on human systems 351
24.5.1 Microplastic modulation of cellular pathways 352
24.5.2 Microplastics-induced reactive oxygen species 352
24.5.3 Accumulation of microplastic particles in different tissues 353
24.6 Plastisphere: biofilm on microplastics 353
24.7 Microplastics as pathogen carriers 354
24.8 Conclusion and perspectives 354
References 354

Section X
Molecular and immunological mechanisms associated with healthy gut microbiome functions

25. Role of the gut microbiome and probiotics for prevention and management of tuberculosis 361

Tejaswini Baral, Shilia Jacob Kurian, Sonal Sekhar M., Murali Munisamy, Chandrashekar Udyavara Kudru, Bidita Khandelwal, Mithu Banerjee, Chiranjay Mukhopadhyay, Kavitha Saravu, Jitendra Singh, Sarman Singh and Mahadev Rao

25.1 Introduction 361
25.2 Gut-lung axis 362
25.3 Gut-liver axis 362

25.4 Gut–immune system axis 363
25.5 Role of the human gut microbiome in tuberculosis 363
25.6 Role of probiotics in tuberculosis 363
25.6.1 Probiotics as immunomodulator 365
25.6.2 Probiotics as an antiinfective mediator 366
25.6.3 Preclinical and clinical evidence of the effects of probiotics in managing tuberculosis 366
25.7 Role of probiotics in preventing anti-TB drug-induced adverse effects 367
25.7.1 Amelioration of adverse gastrointestinal effects 367
25.7.2 Amelioration of drug-induced hepatotoxicity 367
25.8 Probiotics in the improvement of nutritional status in tuberculosis 368
25.9 Dose and dosage of probiotic formulations for dietary supplements 368
25.10 Conclusion 369
References 369

26. Lactic acid bacteria–based beverages in the promotion of gastrointestinal tract health 373

Luís Cláudio Lima de Jesus, Tales Fernando da Silva, Rafael de Assis Glória, Andria dos Santos Freitas, Monique Ferrary Américo, Lucas Jorge da Silva Fernandes, Gabriela Munis Campos, Gabriel Camargos Gomes, Rhayane Cristina Viegas Santos, Rodrigo Dias de Oliveira Carvalho, Debmalya Barh and Vasco Azevedo

26.1 Introduction 373
26.2 Lactic Acid Bacteria 373
26.3 LAB-derived bioactive compounds 374
26.4 The LAB fermentation process 375
26.5 Organoleptic characteristics of LAB-fermented foods 375
26.6 Commercial products from LAB fermented foods 376
26.7 Application of LAB-fermented beverages in gastrointestinal tract–related diseases 376
26.7.1 Antibiotic-associated diarrhea 378
26.7.2 Constipation 378
26.7.3 Inflammatory Bowel Diseases 380
26.7.4 Intestinal mucositis 381
26.7.5 Colorectal cancer 382
26.7.6 Enteric infections 382
26.8 Final considerations 383
References 383

27. Prospective role of prebiotics and probiotics in gut immunity 387

Saptadip Samanta

27.1 Introduction 387
27.2 The gut microbiota and its beneficial effects 388
27.2.1 Shape of the gut microbiota 388
27.2.2 Role of the microbiota in the synthesis of essential metabolites and bioactive components 388
27.3 Impacts of dysbiosis 390
27.3.1 Inflammatory bowel diseases 391
27.3.2 *Clostridium difficile* infection 391
27.3.3 Metabolic disorders 391
27.3.4 Role in cancer progression 391
27.3.5 Gastric cancer 392
27.3.6 Colorectal cancer 392
27.3.7 Hepatocellular carcinoma 392
27.3.8 Other cancers 392
27.4 Prebiotics 393
27.5 Probiotics 394
27.6 Role of prebiotics and probiotics in immunomodulation 394
27.6.1 Prebiotics and immune-boosting effects 394
27.6.2 Production of short-chain fatty acids and their immunomodulatory effects 394
27.6.3 Protective role against inflammation 396
27.6.4 Role of bile acids 396
27.6.5 Effect on development of gut-associated lymphoid tissue and expansion of gut immunity 396
27.6.6 Protective role of probiotic organisms against pathogenesis 397
27.6.7 Role in innate immunity 398
27.6.8 Probiotic bacteria and modulation of cytokine profile 398
27.6.9 Role in enhanced secretion of IgA 398
27.6.10 Role in immunosurveillance-mediated cancer prevention 399
27.7 Additional benefits of probiotics 399
27.8 Concluding remarks 399
References 399

28. Gut–brain communication: a novel application of probiotics 405

Sujit Das, Trupti J. Undhad and Subrota Hati

28.1 Introduction 405

28.2 The gut-brain axis: a concept 405
28.3 Factors regulating the microbiota-gut-brain axis 407
28.3.1 Genetics and epigenetics 407
28.3.2 Mode of delivery at birth 407
28.3.3 Fats 407
28.3.4 Medications and the microbiome 407
28.4 Interactions within the gut-brain axis 407
28.4.1 Neuronal pathway 408
28.4.2 Endocrine pathway 408
28.4.3 Humoral/metabolic pathway 408
28.4.4 Immune pathway 409
28.5 Function of the microbiota in the gut-brain-axis 409
28.5.1 From the gut microbiota to the brain 409
28.5.2 From the brain to the gut microbiota 410
28.6 The gut-brain axis in the regulation of food intake 411
28.7 Behavior and the microbiota gut-brain axis 411
28.7.1 Animal studies 411
28.7.2 Human studies 413
28.8 Conclusions and future perspectives 413
References 413

29. Flaxseed has a pronounced effect on gut microbiota 417

M.N. Sharath Kumar and S. Devaraja

29.1 Introduction 417
29.2 Flaxseed oil prevents alcoholic liver disease by modulating the gut microbiota through its antiinflammatory property 418
29.3 Flaxseed oil reduces the severity of type 2 diabetes mellitus by modulating the gut microbiota via its antiinflammation effect 419
29.4 By altering the composition of the gut microbiota, flaxseed oil has beneficial immunological and metabolic effects 421
29.5 Flaxseed polysaccharide modulates the composition and structure of the gut microbiota by altering the firmicutes-bacteroidetes ratio 423
29.6 Flaxseed oil ameliorated polycystic ovary syndrome through the sex steroid hormone–microbiota-inflammation axis 424
References 427

30. Role of mushroom polysaccharides in improving gut health and associated diseases 431

Divya Yadav and Pradeep Singh Negi

30.1 Introduction 431
30.2 The gut microbiome 432
30.3 Mushroom polysaccharides 432
30.3.1 D-glucan 433
30.4 Proteoglycan 437
30.4.1 Glycoprotein 438
30.5 Prebiotic efficiency of mushroom polysaccharides 438
30.6 Hydrolysis of mushroom polysaccharides by intestinal bacterial enzymes 439
30.7 Impact of mushroom polysaccharides on the gut microbiome 439
30.8 Prevention of various diseases by prebiotic mushroom polysaccharides through modulation of the gut microbiota 440
30.8.1 Inflammatory bowel disease 440
30.8.2 Antitumor effect 444
30.8.3 Immunomodulatory effect 445
30.9 Conclusion and future perspectives 445
References 445

Section XI
Microbiome and immunomodulatory peptides

31. The role of the microbiome in sports nutrition 451

Estela González-Rodríguez, Laura Albero Montava, Isabel Sospedra López and José Miguel Martínez-Sanz

31.1 Introduction 451
31.2 The microbiome and sport 451
31.3 Research related to the microbiome and athletic performance 452
31.4 Conclusion 457
References 457

Section XII
Study design and statistical interpretation

32. Food-derived immunomodulatory peptides: science to applications 461

Chopada Kevalkumar Dineshbhai, Pratik Shukla, Amar Sakure, Brij Pal Singh, Sujit Das, Sreeja V. and Subrota Hati

32.1 Introduction 461
32.2 Immunomodulation and immunomodulatory peptides 462
32.3 Mechanism of immunomodulatory peptide 463
32.4 Sources of some immunomodulatory peptides 463
32.4.1 Milk and milk products 464
32.4.2 Fish 467
32.4.3 Marine organisms 468
32.4.4 Eggs 468
32.4.5 Plants 468
32.5 Bioavailability of bioactive peptides in foods 470
32.5.1 Absorption 471
32.5.2 Distribution 471
32.5.3 Metabolism 472
32.5.4 Elimination 472
32.6 Methods used to assay immunomodulatory activity 472
32.7 Clinical studies 474
32.7.1 Related to COVID-19 474
32.7.2 Impact of probiotics strains on T cell response 474
32.8 Applications 474
32.9 Conclusion 475
References 475

33. Clinical intervention, study design, and statistical methodology: statistical approaches for pre-post studies 481

Kohsuke Hayamizu

33.1 Introduction 481
33.2 Pre-post design 481
33.3 Outcome measure 482
33.4 Sample data: hydroxycitric acid study 482
33.5 Method for analyzing the intervention effect without adjusting for the pre-value effect 482
33.6 Method for analyzing an intervention effect, adjusting for the effect of the pre-value: analysis of covariance 485
33.7 Change point regression model 489
References 492

Commentary from the Editor's Desk 495

A treatise on a healthy microbiome: contribution to human health and disease prevention 497

Debasis Bagchi, Samudra Prosad Banik and Bernard William Downs

A.1 Microbiome terminology: the origin 497
A.2 Microbiota, microorganisms, and microbiome 497
A.3 Microbiome: human health and disease pathology 498
A.4 Microbiome and metabolic syndrome 499
A.5 Microbiome and obesity 500
A.6 Microbiome and diabetes 500
A.7 Microbiome and cancer 500
A.8 Microbiome, brain health, and advancing age 500
A.9 Summary and conclusions 501
References 501

Index 503

List of contributors

A. Ahamed Department of Molecular, Cell and Systems Biology, University of California, Riverside, CA, United States

Laura Albero Montava Nursing Department, Faculty of Health Sciences, University of Alicante, Alicante, Spain

Carly Albright Department of Drug Discovery and Development, Auburn University, Auburn, AL, United States

Bayan Al-Share Department of Oncology, Karmanos Cancer Institute, Wayne State University School of Medicine, Detroit, MI, United States

Monique Ferrary Américo Laboratório de Genética Celular e Molecular (LGCM), Instituto de Ciências Biológicas, Departamento de Genética, Ecologia e Evolução, Universidade Federal de Minas Gerais (UFMG), Belo Horizonte, Minas Gerais, Brazil

Anamika Department of Allied Health Sciences, University of Petroleum and Energy Studies (UPES), Dehradun, Uttarakhand, India

Smriti Arora Department of Allied Health Sciences, University of Petroleum and Energy Studies (UPES), Dehradun, Uttarakhand, India

Vasco Azevedo Laboratório de Genética Celular e Molecular (LGCM), Instituto de Ciências Biológicas, Departamento de Genética, Ecologia e Evolução, Universidade Federal de Minas Gerais (UFMG), Belo Horizonte, Minas Gerais, Brazil

Asfar S. Azmi Department of Oncology, Karmanos Cancer Institute, Wayne State University School of Medicine, Detroit, MI, United States

Rajendra D. Badgaiyan Department of Psychiatry, South Texas Veteran Health Care System, Audie L. Murphy Memorial VA Hospital, and Long School of Medicine, University of Texas Health Science Center, San Antonio, TX, United States; Department of Psychiatry, MT. Sinai School of Medicine, New York, NY, United States

Debasis Bagchi College of Pharmacy and Health Sciences, Texas Southern University, Houston, TX, United States; Department of Biology, Adelphi University, Garden City, NY, United States; Department of R&D, Victory Nutrition International, Inc., Bonita Springs, FL, United States

Manashi Bagchi Department of R&D, Dr. Herbs LLC, Concord, CA, United States

Mithu Banerjee Department of Biochemistry, All India Institute of Medical Sciences, Jodhpur, Rajasthan, India

Pradipta Banerjee Department of Biochemistry and Plant Physiology, Centurion University of Technology and Management, Parlakhemundi, Odisha, India; Department of Surgery, Indiana University School of Medicine, Indianapolis, IN, United States

Samudra Prosad Banik Department of Microbiology, Maulana Azad College, Kolkata, West Bengal, India

Tejaswini Baral Department of Pharmacy Practice, Manipal College of Pharmaceutical Sciences, Manipal Academy of Higher Education, Manipal, Karnataka, India

Debmalya Barh Laboratório de Genética Celular e Molecular (LGCM), Instituto de Ciências Biológicas, Departamento de Genética, Ecologia e Evolução, Universidade Federal de Minas Gerais (UFMG), Belo Horizonte, Minas Gerais, Brazil; Centre for Genomics and Applied Gene Technology, Institute of Integrative Omics and Applied Biotechnology (IIOAB), Nonakuri, West Bengal, India

David Baron Department of Psychiatry, College of Osteopathic Medicine of the Pacific, Western University of Health Sciences, Pomona, CA, United States

Nitika Bhambri Department of Allied Health Sciences, University of Petroleum and Energy Studies (UPES), Dehradun, Uttarakhand, India

Anita Bhat Jivahealth Clinics, Concord, CA, United States

B. Vishnu Bhat Director-Medical Research, Professor of Pediatrics and Neonatology, Aarupadaiveedu Medical College and Hospital, Vinayaka Mission's Research Foundation-DU, Puducherry, India

Joy Bhat Jivahealth Clinics, Concord, CA, United States

Jyoti Bhat Jivahealth Clinics, Concord, CA, United States

Kusum Bhat Jivahealth Clinics, Concord, CA, United States

Naras Bhat Jivahealth Clinics, Concord, CA, United States

Subhrajit Bhattacharya School of Pharmacy and Health Sciences, Keck Graduate Institute, Claremont, CA, United States

Jhimli Bhattacharyya Department of Chemistry, National Institute of Technology Nagaland, Dimapur, Nagaland, India

Kenneth Blum Department of Psychiatry, College of Osteopathic Medicine of the Pacific, Western University of Health Sciences, Pomona, CA, United States

Abdalla Bowirrat Department of Molecular Biology, Adelson School of Medicine, Ariel University, Ariel, Israel

Nadiya Boyko Department of Clinical Laboratory Diagnostics and Pharmacology, Faculty of Dentistry, Uzhhorod National University, Uzhhorod, Ukraine; Research Development and Educational Centre of Molecular Microbiology and Mucosal Immunology, Uzhhorod National University, Uzhhorod, Ukraine; Ediens LLC, Velyki Lasy, Ukraine

Gabriela Munis Campos Laboratório de Genética Celular e Molecular (LGCM), Instituto de Ciências Biológicas, Departamento de Genética, Ecologia e Evolução, Universidade Federal de Minas Gerais (UFMG), Belo Horizonte, Minas Gerais, Brazil

Rodrigo Dias de Oliveira Carvalho Laboratório de Genética Celular e Molecular (LGCM), Instituto de Ciências Biológicas, Departamento de Genética, Ecologia e Evolução, Universidade Federal de Minas Gerais (UFMG), Belo Horizonte, Minas Gerais, Brazil

Mar Lozano Casanova Nursing Department, Faculty of Health Sciences, University of Alicante, Alicante, Spain

Rameshwar Nath Chaurasia Department of Neurology, Institute of Medical Sciences, Banaras Hindu University, Varanasi, India

K. Chen Department of Molecular, Cell and Systems Biology, University of California, Riverside, CA, United States

Sabyasachi Choudhuri Department of Pharmaceutical Science and Technology, Maulana Abul Kalam Azad University of Technology, Haringhata, West Bengal, India

Sriya Choudhury Department of Biochemistry, Institute of Biomedical Sciences, The West Bengal University of Health Sciences, Kolkata, West Bengal, India

Kyung-Soo Chun College of Pharmacy, Keimyung University, Daegu, South Korea

Won-Yoon Chung Department of Oral Biology, Yonsei University College of Dentistry, Seoul, South Korea; Department of Applied Life Science, The Graduate School, Yonsei University, Seoul, South Korea; BK21 Four Project, Yonsei University College of Dentistry, Seoul, South Korea

Tales Fernando da Silva Laboratório de Genética Celular e Molecular (LGCM), Instituto de Ciências Biológicas, Departamento de Genética, Ecologia e Evolução, Universidade Federal de Minas Gerais (UFMG), Belo Horizonte, Minas Gerais, Brazil

Alvin H. Danenberg Balanced Metabolic Coaching, Periodontist, Certified Functional Medicine Practitioner, Certified Primal Health Coach, Charleston, SC, United States

Ahana Das Department of Microbiology, Maulana Azad College, Kolkata, West Bengal, India

Amitava Das Department of Surgery, Indiana University School of Medicine, Indianapolis, IN, United States

Gautam Das miBiome Therapeutics LLP, Mumbai, Maharashtra, India

Sujit Das Department of Rural Development and Agricultural Production, North-Eastern Hill University, Tura, Meghalaya, India

Nandini Dasgupta miBiome Therapeutics LLP, Mumbai, Maharashtra, India

Luís Cláudio Lima de Jesus Laboratório de Genética Celular e Molecular (LGCM), Instituto de Ciências Biológicas, Departamento de Genética, Ecologia e Evolução, Universidade Federal de Minas Gerais (UFMG), Belo Horizonte, Minas Gerais, Brazil

S. Devaraja Department of Studies and Research in Biochemistry, Tumkur University, Tumkur, Karnataka, India

Nadiya Dileep Department of Medical Microbiology and Immunology, RAK Medical and Health Sciences University, Ras Al Khaimah, United Arab Emirates

Chopada Kevalkumar Dineshbhai Department of Dairy Microbiology, SMC College of Dairy Science, Kamdhenu University, Anand, Gujarat, India

Bernard William Downs Division of Precision Nutrition, Victory Nutrition International, Inc., Bonita Springs, FL, United States

Igor Elman Behavioral Neuropharmacology and Neuroimaging Laboratory on Addictions, Clinical Research Institute on Addictions, Department of Pharmacology and Toxicology, Jacobs School of Medicine and Biosciences, State University of New York at Buffalo, Buffalo, NY, United States

Lucas Jorge da Silva Fernandes Laboratório de Genética Celular e Molecular (LGCM), Instituto de Ciências Biológicas, Departamento de Genética, Ecologia e Evolução, Universidade Federal de Minas Gerais (UFMG), Belo Horizonte, Minas Gerais, Brazil

Andria dos Santos Freitas Laboratório de Genética Celular e Molecular (LGCM), Instituto de Ciências Biológicas, Departamento de Genética, Ecologia e Evolução, Universidade Federal de Minas Gerais (UFMG), Belo Horizonte, Minas Gerais, Brazil

Amanda Friend Department of Drug Discovery and Development, Auburn University, Auburn, AL, United States

Nandini Ghosh Department of Surgery, Indiana University School of Medicine, Indianapolis, IN, United States

Rituparna Banik Ghosh Department of Physiology, Bhairab Ganguly College, Kolkata, West Bengal, India

Rafael de Assis Glória Laboratório de Genética Celular e Molecular (LGCM), Instituto de Ciências Biológicas, Departamento de Genética, Ecologia e Evolução, Universidade Federal de Minas Gerais (UFMG), Belo Horizonte, Minas Gerais, Brazil

Mark S. Gold Department of Psychiatry, Washington University School of Medicine, St. Louis, MO, United States

Gabriel Camargos Gomes Laboratório de Genética Celular e Molecular (LGCM), Instituto de Ciências Biológicas, Departamento de Genética, Ecologia e Evolução, Universidade Federal de Minas Gerais (UFMG), Belo Horizonte, Minas Gerais, Brazil

Estela González-Rodríguez Nursing Department, Faculty of Health Sciences, University of Alicante, Alicante, Spain

Subrota Hati Department of Dairy Microbiology, SMC College of Dairy Science, Kamdhenu University, Anand, Gujarat, India

Kohsuke Hayamizu Yokohama University of Pharmacy, Yokohama, Japan

Ashfaque Hossain Department of Medical Microbiology and Immunology, RAK Medical and Health Sciences University, Ras Al Khaimah, United Arab Emirates

Md. Akil Hossain Department of Pharmacology and Experimental Therapeutics, School of Medicine, Boston University, MA, United States

Rehan Jalali Division of Nutrigenomics, The Kenneth Blum Behavioral Neurogenetic Institute, Austin, TX, United States

Komal Jalan Department of Agriculture, University of Calcutta, Kolkata, West Bengal, India

Tania Akter Jhuma Department of Microbiology, University of Dhaka, Dhaka, Bangladesh

Savannah N. Johnson Department of Clinical Nutrition, Maryland University of Integrative Health, Laurel, MD, United States

Imran Junaid Jivahealth Clinics, Concord, CA, United States

Muhammad Manjurul Karim Department of Microbiology, University of Dhaka, Dhaka, Bangladesh

Bidita Khandelwal Department of General Medicine, Sikkim Manipal Institute of Medical Sciences, Sikkim Manipal University, Gangtok, Sikkim, India

Shradha Khater miBiome Therapeutics LLP, Mumbai, Maharashtra, India

J.H. Kim Department of Molecular, Cell and Systems Biology, University of California, Riverside, CA, United States

Manoj Kumar Kingsley Department of Neonatology, JIPMER, Puducherry, India

Chandrashekar Udyavara Kudru Department of Medicine, Kasturba Medical College and Hospital, Manipal Academy of Higher Education, Manipal, India

Abhai Kumar Department of Neurology, Institute of Medical Sciences, Banaras Hindu University, Varanasi, India

Shilia Jacob Kurian Department of Pharmacy Practice, Manipal College of Pharmaceutical Sciences, Manipal Academy of Higher Education, Manipal, Karnataka, India

Steve Kushner Department of R&D, ALM R&D, Oldsmar, FL, United States

E.G. Lebig Department of Molecular, Cell and Systems Biology, University of California, Riverside, CA, United States

Elizabeth A. Lipski Faculty Advisor, Department of Clinical Nutrition, Maryland University of Integrative Health, Laurel, MD, United States

Isabel Sospedra López Nursing Department, Faculty of Health Sciences, University of Alicante, Alicante, Spain

Soching Luikham Department of Chemistry, National Institute of Technology Nagaland, Dimapur, Nagaland, India

Sayantan Maitra Institute of Pharmacy, Jalpaiguri, Government of West Bengal, Department of Health and Family Welfare, West Bengal, India

José Miguel Martínez-Sanz Nursing Department, Faculty of Health Sciences, University of Alicante, Alicante, Spain

M. Martins-Green Department of Molecular, Cell and Systems Biology, University of California, Riverside, CA, United States

Tamara Meleshko Department of Clinical Laboratory Diagnostics and Pharmacology, Faculty of Dentistry, Uzhhorod National University, Uzhhorod, Ukraine; Research Development and Educational Centre of Molecular Microbiology and Mucosal Immunology, Uzhhorod National University, Uzhhorod, Ukraine

Chiranjay Mukhopadhyay Department of Microbiology, Coordinator, Center for Emerging and Tropical Diseases, Kasturba Medical College and Hospital, Manipal Academy of Higher Education, Manipal, Karnataka, India

Satinath Mukhopadhyay Department of Endocrinology, Institute of Postgraduate Medical Education & Research, Kolkata, West Bengal, India

Murali Munisamy Department of Translational Medicine, All India Institute of Medical Sciences, Bhopal, Madhya Pradesh, India

Misako Nagasaka Department of Oncology, Karmanos Cancer Institute, Wayne State University School of Medicine, Detroit, MI, United States

Pradeep Singh Negi Department of Fruit and Vegetables Technology, CSIR-Central Food Technological Research Institute, Mysuru, India; Academy of Scientific and Innovative Research (AcSIR), Ghaziabad, India

F A Dain Md Opo Department of Biological Sciences, King Abdulaziz University, Jeddah, Saudi Arabia

Oleksandra Pallah Department of Clinical Laboratory Diagnostics and Pharmacology, Faculty of Dentistry, Uzhhorod National University, Uzhhorod, Ukraine; Research Development and Educational Centre of Molecular Microbiology and Mucosal Immunology, Uzhhorod National University, Uzhhorod, Ukraine

Jyotirmoy Panda Department of Pharmaceutical Science and Technology, Maulana Abul Kalam Azad University of Technology, Haringhata, West Bengal, India

Kwang-Kyun Park Department of Oral Biology, Yonsei University College of Dentistry, Seoul, South Korea

Se-Young Park Department of Oral Biology, Yonsei University College of Dentistry, Seoul, South Korea; Department of Applied Life Science, The Graduate School, Yonsei University, Seoul, South Korea; BK21 Four Project, Yonsei University College of Dentistry, Seoul, South Korea

Rokeya Pervin Foot and Mouth Disease Division, Animal and Plant Quarantine Agency, Gimcheon-si, South Korea

Benjamin Peters Department of Drug Discovery and Development, Auburn University, Auburn, AL, United States

B. Petros Department of Molecular, Cell and Systems Biology, University of California, Riverside, CA, United States

Samikshha Puri Department of Allied Health Sciences, University of Petroleum and Energy Studies (UPES), Dehradun, Uttarakhand, India

Mahadev Rao Department of Pharmacy Practice, Coordinator, Center for Translational Research, Manipal College of Pharmaceutical Sciences, Manipal Academy of Higher Education, Manipal, Karnataka, India

Miranda Reed Department of Drug Discovery and Development, Auburn University, Auburn, AL, United States; Center for Neuroscience Initiative, Auburn University, Auburn, AL, United States

Estela González Rodríguez Nursing Department, Faculty of Health Sciences, University of Alicante, Alicante, Spain

Marufa Rumman Department of Microbiology, National Center for Toxicological Research, US Food and Drug Administration, Redfield, AR, United States

S. Saeed Department of Molecular, Cell and Systems Biology, University of California, Riverside, CA, United States

Amar Sakure Deparment of Agriculture Biotechnology, Anand Agricultural University, Anand, Gujarat, India

Saptadip Samanta Department of Physiology, Midnapore College, Midnapore, West Bengal, India

Rhayane Cristina Viegas Santos Laboratório de Genética Celular e Molecular (LGCM), Instituto de Ciências Biológicas, Departamento de Genética, Ecologia e Evolução, Universidade Federal de Minas Gerais (UFMG), Belo Horizonte, Minas Gerais, Brazil

Kavitha Saravu Department of Infectious Diseases, Coordinator, Manipal Center for Infectious Diseases, Kasturba Medical College and Hospital, Manipal Academy of Higher Education, Manipal, Karnataka, India

Shalini Sehgal Department of Food Technology, Bhaskaracharya College of Applied Sciences, University of Delhi, New Delhi, India

Sonal Sekhar M. Department of Pharmacy Practice, Manipal College of Pharmaceutical Sciences, Manipal Academy of Higher Education, Manipal, Karnataka, India

Chitra Selvan Department of Endocrinology, Ramaiah Medical College, Bengaluru, Karnataka, India

M.N. Sharath Kumar Department of Studies and Research in Biochemistry, Tumkur University, Tumkur, Karnataka, India

Pooja Shivappa Department of Basic Sciences, RAK Medical and Health Sciences University, Ras Al Khaimah, United Arab Emirates

Pratik Shukla Department of Dairy Microbiology, SMC College of Dairy Science, Kamdhenu University, Anand, Gujarat, India

Gema Alfonso Sigüenza Nursing Department, Faculty of Health Sciences, University of Alicante, Alicante, Spain

Abhilasha Singh Cardiovascular Research Centre, Massachusetts General Hospital, Harvard Medical School, Boston, MA, United States

Brij Pal Singh Deparment of Microbiology, Central University of Haryana, Mahendragarh, Haryana, India

Jitendra Singh Department of Translational Medicine, All India Institute of Medical Sciences, Bhopal, Madhya Pradesh, India

Sarman Singh All India Institute of Medical Sciences, Bhopal, Madhya Pradesh, India

Smita Singh Department of Geriatric Medicine, Institute of Medical Sciences, Banaras Hindu University, Varanasi, India

Na-Young Song Department of Oral Biology, Yonsei University College of Dentistry, Seoul, South Korea; Department of Applied Life Science, The Graduate School, Yonsei University, Seoul, South Korea; BK21 Four Project, Yonsei University College of Dentistry, Seoul, South Korea

Diwanshu Soni Department of Psychiatry, College of Osteopathic Medicine of the Pacific, Western University of Health Sciences, Pomona, CA, United States

Vishnu Suppiramaniam Department of Drug Discovery and Development, Auburn University, Auburn, AL, United States; Center for Neuroscience Initiative, Auburn University, Auburn, AL, United States

Young-Joon Surh College of Pharmacy, Seoul National University, Seoul, South Korea; Department of Molecular Medicine and Biopharmaceutical Sciences, Graduate School of Convergence Science and Technology, Seoul National University, Seoul, South Korea; Cancer Research Institute, Seoul National University, Seoul, South Korea

Addisu Demeke Teklemariam Department of Biological Sciences, King Abdulaziz University, Jeddah, Saudi Arabia

Panyotis K. Thanos Department of Psychology and Behavioral Neuropharmacology and Neuroimaging Laboratory on Addictions (BNNLA), Research Institute on Addictions, University at Buffalo, Buffalo, NY, United States

Md. Hafiz Uddin Department of Oncology, Karmanos Cancer Institute, Wayne State University School of Medicine, Detroit, MI, United States

Trupti J. Undhad Department of Dairy Microbiology, SMC College of Dairy Science, Kamdhenu University, Anand, Gujarat, India

Sreeja V. Department of Dairy Microbiology, SMC College of Dairy Science, Kamdhenu University, Anand, Gujarat, India

Liz J. Vázquez-Chinea Department of Clinical Nutrition, Maryland University of Integrative Health, Laurel, MD, United States

Gene-Jack Wang Laboratory of Neuroimaging, National Institute of Alcohol Abuse and Alcoholism, National Institutes of Health, Bethesda, MD, United States

Marzena Wojcik Department of Structural Biology, Medical University of Lodz, Lodz, Poland

Lucyna Wozniak Department of Structural Biology, Medical University of Lodz, Lodz, Poland

Divya Yadav Department of Fruit and Vegetables Technology, CSIR-Central Food Technological Research Institute, Mysuru, India; Academy of Scientific and Innovative Research (AcSIR), Ghaziabad, India

Preface

This book explores a wide range of topics related to the role of nutrition in boosting a healthy microbiome in chapters written by leading experts in the field. To provide a comprehensive approach to the emerging field of nutrition, gut health, food recommendations, and the human microbiome, this book covers a wide spectrum of agents including food, minerals, vitamins, dietary fibers, prebiotics and probiotics, nutraceutical supplements, and phytonutrients that boost overall human health and performance by improving the health of the microbiome. This book specifically addresses the causes and pathological manifestations of microbiome dysbiosis and the molecular and cellular mechanisms and pathways involved by which nutritional components contribute to the physiology and functionality of a healthy microbiome and consequential human health. In addition to an extensive review of a variety of functional ingredients and dietary supplements, this book addresses information regarding nutritional guidelines and healthy lifestyle that are important to correct, boost, and maintain a healthy gut microbiome for optimal human health.

Digestion and metabolism are the most important steps in human physiology. It is important to note that 70%–80% of the lymphatic tissue in humans resides in the gastrointestinal tract. Moreover, there is more neurotransmitter activity in the gut than in the brain, prompting labels of the gastrointestinal tract as "the second brain." Proper nutrition is essential for optimal human health, daily activities, and performance. The microbiota that resides in the gastrointestinal tract performs functions that are essential to overall health, including the synthesis and/or metabolization of certain vitamins, fatty acids, fibers, and amino acids. The health of the GI tract is ultimately foundational to maintaining good health and appropriately regulating immune function and maintaining metabolic homeostasis. Otherwise, unhealthy lifestyle choices and disruptions of health can cause drastic alterations of these gut microbial communities, which ultimately leads to dysbiosis and the potential for a range of disorders, including but not limited to immune dysregulation, mental dysfunction, and even autoimmune disorders. Therefore there is a growing need for a comprehensive appraisal of the proper nutritional benefits in boosting the human microbial ecosystem, which ultimately promotes optimal gastrointestinal health, immunity, and mental function.

This textbook is divided into 12 major section comprising of 33 chapters. The book starts with, the editorial preface, which was crafted by the editors to provide an overview of this book. Section I "Microbiome and Human Health: An Introduction" is comprised of five chapters. The first chapter, "Oral Microbiome: A Gateway to Your Health," demonstrates the intricate aspects of the oral microbiome and its association with human health, while the second chapter highlights how a healthy microbiome is involved in shaping the newborn immune system. The third chapter demonstrates the impact of diverse gut microbiota on human health and prevention against diverse degenerative diseases, and the fourth and fifth chapters discusses how the individual microbiota participates in programming and reprogramming both systemic and local immune responses to modulate immune health.

Section II, "Microbiome and Digestive Health," is composed of two chapters. The first of these chapters discusses the etiology of gut dysbiosis and its role in chronic diseases, and the second chapter explores the role of microbiome in the function and diseases of the digestive system. Section III, "Microbiome and Metabolic Syndrome," is composed of seven independent chapters. These chapters discuss diverse aspects on the roles of a healthy microbiome in obesity, weight management, and cardiovascular health. The fourteenth chapter demonstrates the beneficial roles of resveratrol in remodeling the gut microbiota to prevent metabolic disorders. Section IV, "Microbiome and Immune Health," is composed of two chapters. These two chapters emphasize the regulatory roles of the microbiome in immune competence, anaerobism, and inflammatory conditions.

Section V, "Microbiome and Cognitive Health," is composed of four chapters. The first of these chapters demonstrates the importance of achieving dopamine homeostasis to combat brain-gut functional impairment and discusses the behavioral and neurogenetic correlation with reward deficiency syndrome. The second chapter discussed the influence of the gut microbial flora in the body's serotonin turnover and its associated diseases. The third chapter explores the connection between diet, gut microbes, and cognitive decline, while the fourth chapter discusses the role of the gut microbiome in Rett syndrome, a rare genetic mutation that affects brain development in some girls. Section VI,

"Microbiome, Dermal Health, and Wound Healing," discusses the roles of the skin microbiota in dermal health, development of chronic wounds, and wound healing. Section VII, "Microbiome and Cancer," is composed of one chapter discussing the roles of a healthy gut microbiome in the prevention of colorectal cancer.

Section VIII, "Microbiome, Arthritis, and Multiple Sclerosis," explains the regulatory roles of the microbiome in arthritis, fibromyalgia, and multiple sclerosis in one independent chapter. Section IX, "Environmental Pollutants and Gut Microbiome," explores the consequences of microplastic exposure on the gut microbiome.

Section X, "Molecular and Immunological Mechanisms Associated With Healthy Gut Microbiome Functions," elaborates on the subject in seven independent chapters. The first three of these chapters discuss the beneficial roles of a healthy gut microbiome, prebiotics, and probiotics, while the second chapter highlights the roles of lactic acid bacteria in diverse foods and beverages in the promotion of gastrointestinal health. The fourth chapter explores demonstrates the novel application of probiotics in enhancing gut-brain communication. The following two chapters demonstrate the beneficial effects of flaxseeds, dietary flavonoids, and mushroom polysaccharides in improving the gastrointestinal microbiome and human health. The last chapter in this section demonstrates the roles and importance of healthy gut microbiome in sports nutrition.

Section XI, "Microbiome and Immunomodulatory Peptides," details the beneficial roles of immunomodulatory peptides in human health. Section XII, "Study Design and Statistical Interpretation," demonstrates the aspects of clinical intervention, study design, and different statistical methodologies for pretest—posttest studies.

The book concludes with a commentary entitled "A Treatise on a Healthy Microbiome," which summarizes a vivid scenario of diverse aspects of a healthy microbiome.

We would like to express our sincere gratitude and thanks to all our eminent contributors as well as the helpful Elsevier/Academic Press editorial team members including Megan R. Ball, Lindsay Lawrence, and Praveen Sachidanandam for their continued support, cooperation, and assistance.

Debasis Bagchi
Bernard William Downs

Section I

Microbiome and Human Health: An Introduction

Chapter 1

Oral microbiome: a gateway to your health

Na-Young Song[1,2,3], Se-Young Park[1,2,3], Won-Yoon Chung[1,2,3], Young-Joon Surh[4,5,6], Kyung-Soo Chun[7] and Kwang-Kyun Park[1]

[1]*Department of Oral Biology, Yonsei University College of Dentistry, Seoul, South Korea,* [2]*Department of Applied Life Science, The Graduate School, Yonsei University, Seoul, South Korea,* [3]*BK21 Four Project, Yonsei University College of Dentistry, Seoul, South Korea,* [4]*College of Pharmacy, Seoul National University, Seoul, South Korea,* [5]*Department of Molecular Medicine and Biopharmaceutical Sciences, Graduate School of Convergence Science and Technology, Seoul National University, Seoul, South Korea,* [6]*Cancer Research Institute, Seoul National University, Seoul, South Korea,* [7]*College of Pharmacy, Keimyung University, Daegu, South Korea*

1.1 Introduction

It is now well recognized that the microbiome plays a crucial role in the human health and pathogenesis (Shreiner et al., 2015). The microbes in and on the human body outnumber the human cells by about ten to one and contribute to maintaining the immune tone by regulating the innate and adaptive immune responses in the body (Kamada and Nunez, 2014). Thus the microbiome is currently considered as a new organ system in the human body, and its dysbiosis is implicated in various types of diseases, from infectious disease to cancers. Conversely, the microbiota dysbiosis-driven disorders can be treated by transferring "normal-like" microorganisms into the patients, such as fecal microbiota transplantation to treat recurrent *Clostridium difficile* infection, which is already used in the clinics worldwide (Gupta et al., 2016).

The gastrointestinal (GI) tract is fully covered by mucosal biofilm, a thick layer of mucus, which creates environments for microorganisms to thrive (Donaldson et al., 2016). Because the GI tract is the largest microbial ecosystem, it is not surprising that most investigations to date have been focusing on the gut microbiome. The cumulative evidence from this intensive research strongly supports the conclusion that gut dysbiosis is associated with the pathogenesis of numerous diseases, from intestinal disorders, such as inflammatory bowel disease and irritable bowel syndrome, to extraintestinal diseases, such as cognitive disorders, cardiopulmonary disease, and obesity (Carding et al., 2015).

The oral cavity possesses the second largest and most complex microbiome after the gut (Zhang et al., 2018). According to the human oral microbiome database (HOMD), the oral cavity contains approximately 700 species of microorganisms, about 54% of which are named and cultivated, 14% of which are not named but cultivated, and 32% of which are known only as uncultivated phylotypes (from the HOMD website: http://www.homd.org). Like the GI tract, the oral cavity is lined by the mucous membrane. Owing to its warm and humid characteristics, the oral cavity provides a proper environment for microorganisms to reside and develop diverse biofilms, like a natural microbial growth medium in the human body (Deo and Deshmukh, 2019). In spite of its abundance, the potential effects of the oral microbiome on human health have been greatly underestimated to date. Here, we will highlight the role of the oral microbiome in human health, particularly in the context of its local effects in the oral cavity and its systemic implications.

1.2 Oral microbiome and oral disease

1.2.1 Caries

Dental caries, the most prevalent oral disease, is caused by polymicrobial infection (Bowen et al., 2018). Dental caries involves tooth destruction as well as pulp and periapical infection, which can reduce the quality of life. Intake of dietary

Microbiome, Immunity, Digestive Health and Nutrition. DOI: https://doi.org/10.1016/B978-0-12-822238-6.00025-X

sugars and carbohydrates is highly associated with dental caries. Sugars are degraded by the bacteria in the dental plaque, generating an acidic environment that is favorable to growth of acidophilic and cariogenic bacteria (Sharma et al., 2018). The acidogenic *Streptococcus mutans* and *Streptococcus sobrinus* are the most well-known species responsible for dental caries (Forssten et al., 2010). Salivary microbiota analysis revealed that individuals with high sugar intake showed the enrichment of acidogenic and acid-tolerant species in *Actinomyces*, *Bifidobacterium*, and *Veillonella* (Esberg et al., 2020). Overall, sucrose consumption promotes an imbalance between acid-producing and alkali-producing bacteria in oral biofilms, facilitating demineralization on enamel blocks in vitro (Du et al., 2020). Demineralization can be corrected by supplements of fluoride. In addition, modulation of oral microbiota could be a more proactive strategy to prevent or inhibit formation of dental caries. In this regard, it has been reported that the use of probiotics containing *Lactobacillus rhamnosus* GG, *Streptococcus salivarius* M18, or *Lactobacillus paracasei* SD1 effectively reduces the bacterial load of *S. mutans* and the risk of dental caries (Sivamaruthi et al., 2020).

1.2.2 Periodontal disease

Periodontal diseases cause destruction of the tooth-supporting periodontium and are divided into gingivitis and periodontitis (Gao et al., 2018). Gingivitis is a reversible inflammatory disease that is limited to the gingiva. However, if not controlled, gingivitis will progress into periodontitis, a chronic, irreversible inflammatory disease with further destruction of the periodontium, including the periodontal ligament and alveolar bone (Zhang et al., 2018). In conjunction with inflammatory responses, oral microbial dysbiosis is involved in the pathogenesis of the periodontal diseases (Van Dyke et al., 2020). Unlike dental caries, the periodontal diseases involve shifts in microbial communities rather than enrichment of single specific species identified as the etiology (Bartold and Van Dyke, 2019).

The subgingival region is one of the distinct microbial niches in the oral cavity. In healthy individuals the subgingival microbiota comprises mostly gram-positive bacteria, such as *Actinomyces naeslundii*, *A. meyeri*, *Streptococcus sanguinis*, *S. oralis*, and *S. gordonii*, and some gram-negative bacteria, including *Capnocytophaga gingivalis* and *Fusobacterium nucleatum* (Curtis et al., 2020). During the evolution of microbial communities from health to gingivitis, the relative abundances of *Rothia dentocariosa*, *Propionibacterium*, and *Stenotrophomonas maltophilia* are reduced (Van Dyke et al., 2020). In contrast, the gingivitis states showed five predominant phyla compared to healthy states: *Actinobacteria*, *Firmicutes*, *Saccharibacteria* (TM7), *Bacteroidetes*, and *Fusobacteria* (Huang et al., 2014). Thus the development of gingivitis is closely associated with an increased microbial load as well as a shift in the subgingival microbiota.

The subgingival microbiota transition is also observed during the progression from gingivitis to periodontitis. Both gingivitis and periodontitis involve a shift from gram-positives to gram-negative bacteria compared to healthy individuals, but the distinct gram-negative bacteria are enriched by each other. It is well known that the so-called red complex is enriched in periodontal microbial communities, consisting of *Porphyromonas gingivalis*, *Tannerella forsythia*, and *Treponema denticola* (Mohanty et al., 2019). Among them, *P. gingivalis* is considered a keystone pathogen in periodontitis, which can change the oral microbiota and provoke leukocyte recruitment (Darveau et al., 2012). In mice, oral inoculation of *P. gingivalis* induced periodontal bone loss, bacterial load, neutrophil recruitment, and serum proinflammatory cytokine levels (Hajishengallis et al., 2011; Maekawa et al., 2014). Concomitantly, the periodontitis-associated microbial community was much more diverse than the healthy microbial ecosystem (Curtis et al., 2020).

While it is obvious that oral microbiota dysbiosis is associated with periodontitis, the interplay between dysbiosis and inflammation has not yet been clarified. Recently, Dutzan et al. have shown that resident memory T helper 17 (Th17) cells expanded in human and murine experimental periodontitis models (Dutzan et al., 2018). In a murine ligature-induced periodontitis model, the oral microbiome composition was altered in comparison to the healthy control group, displaying shifts from *Streptococcus*-predominant to enrichment of *Enterococcus*, *Lactobacillus*, and *Clostridium*. However, systemic treatment with a broad-spectrum antibiotic cocktail suppressed not only the microbial shift but also Th17 expansion, implying a correlation between oral dysbiosis and inflammation in periodontitis. Whether through oral microbiota dysbiosis or not, periodontitis can induce systemic inflammation as well, contributing to extraoral pathogenesis (Hajishengallis, 2015), which will be discussed in later sections.

1.2.3 Oral cancer

Head and neck cancers occur in the oral cavity, salivary gland, pharynx, larynx, nasal cavity, thyroid, and bone, accounting for 5% of all cancers (Heroiu Cataloiu et al., 2013). Oral squamous cell carcinoma (OSCC) is the most prevalent type of the head and neck cancers and its 5-year overall survival rate remains low at approximately 50%

(Kademani, 2007). Smoking and alcohol consumption are the best-known risk factors for developing OSCC (Rivera, 2015). In addition to these environmental factors, human papillomavirus (HPV) is believed to be a major microbial suspect for OSCC. On the basis of metaanalysis data, it is evident that HPV infection remarkably increases the incidence of OSCC (Rapado-Gonzalez et al., 2020; Saulle et al., 2015). In spite of its significance, however, the majority of OSCC cases are not related to HPV infection (Irfan et al., 2020).

Instead, recent investigations have been focusing on changes in the oral bacterial composition in OSCC patients. At the phylum level, OSCC patients showed enrichment of *Bacteroidetes*, *Proteobacteria*, *Firmicutes*, *Fusobacteria*, and *Actinobacteria* (Pushalkar et al., 2012; Schmidt et al., 2014; Zhang et al., 2019a; Zhao et al., 2017). On the basis of the Phylogenetic Investigation of Communities by Reconstruction of Unobserved States (PICRUSt) analysis to predict functional profiles of microbial communities, the OSCC microbiome had a relatively higher abundance of genes associated with proinflammatory bacterial components, such as lipopolysaccharide (LPS) biosynthesis (Zhang et al., 2019a,b). These data suggest a correlation between oral microbiota dysbiosis and inflammation in OSCC.

Chronic inflammation is well known to contribute to cancer development and progression. In this regard, metaanalyses have revealed a strong correlation between periodontitis and OSCC (Javed and Warnakulasuriya, 2016; Ye et al., 2016). Of note, the OSCC microbiome signature is greatly overlapped with the periodontitis-associated communities (Pushalkar et al., 2012; Zhao et al., 2017). Chronic infection of periodontal pathogens promoted tumorigenesis of OSCC in mouse models (Binder Gallimidi et al., 2015; Kamarajan et al., 2020; Lafuente Ibanez de Mendoza et al., 2020). Thus oral microbiome dysbiosis promotes periodontitis, which can lead to OSCC development if not treated properly.

1.3 Oral microbiome and systemic disease

1.3.1 Alzheimer's disease

Alzheimer's disease (AD), the most common type of dementia in older adults, is a neurodegenerative disorder characterized by a progressive decline of cognitive functions. The hallmark lesions of AD are the extracellular deposition of amyloid β (Aβ) aggregates and the intracellular accumulation of neurofibrillary tangles (NFTs) derived from hyperphosphorylated tau protein (Kametani and Hasegawa, 2018). It is well known that inflammation plays a crucial role in AD pathogenesis by exacerbating the accumulation of Aβ and NFTs (Kinney et al., 2018). The blood levels of the proinflammatory markers, such as interleukin-1β (IL-1β), IL-6, and C-reactive protein (CRP), are elevated in patients with AD compared to healthy individuals, which correlates with increased risk for dementia and incidence of cognitive impairment (Alley et al., 2008; Engelhart et al., 2004; Holmes et al., 2003). Recently, the comparative proteomic profiling of plasma and brain tissues revealed elevation of the IL-6 level as well as enrichment of proteins associated with the IL-6 signaling pathway in AD patients compared to healthy subjects, further supporting a major role of systemic inflammation in AD pathogenesis (Chen and Xia, 2020).

Interestingly, a case-control study in Japan showed a correlation between AD risk and premature tooth loss that occurs largely as a result of periodontitis in adults, implying the relationship between dental health and AD (Kondo et al., 1994). In line with this, a longitudinal study has demonstrated that tooth loss is strongly associated with a higher prevalence and incidence of dementia (Stein et al., 2007). Chen et al. have further supported the significant correlation between chronic periodontitis and AD risk by conducting a retrospective matched-cohort study (Chen et al., 2017).

This association between periodontitis and AD might be explained by systemic inflammation as well as oral dysbiosis. Periodontitis is associated with elevated serum levels of the proinflammatory IL-1β, IL-6, and CRP, which are also increased in AD patients (Cheng et al., 2020; D'Aiuto et al., 2004; Noack et al., 2001). The murine ligature-induced periodontitis model showed increased serum IL-6 level and neuroinflammation due to permeability of the blood–brain barrier (BBB) (Furutama et al., 2020). However, the neutralization of IL-6 remarkably suppressed BBB disruption as well as proinflammatory cytokine expression levels in the hippocampus. Taken together, these data support the hypothesis that the periodontitis-induced systemic inflammation is a primary risk factor for AD.

In periodontitis-associated AD pathogenesis, oral microbiota dysbiosis, such as infection, is another crucial axis. As was mentioned in the previous section, *P. gingivalis* is the most distinguished species among the periodontitis signature microbial communities, the red complex. In a transgenic mouse model of AD, oral infection with *P. gingivalis* promoted alveolar bone loss in the oral cavity as well as proinflammatory cytokine release and Aβ accumulation in the brain, resulting in cognitive decline (Ishida et al., 2017). In an experimental chronic periodontitis model, *P. gingivalis* and its product gingipain were detected in the brain tissues along with an accumulation of Aβ and phosphorylated tau (Ilievski et al., 2018). Notably, synthetic small molecule inhibitors targeting gingipain reduced the bacterial load, neuroinflammation, and neurodegeneration in the brain of *P. gingivalis*–infected mice (Dominy et al., 2019). It is noteworthy that Poole et al. demonstrated the coexistence of *P. gingivalis* and its major virulence factor LPS in brain tissues from short-term postmortem AD patients (Poole et al., 2013). In middle-aged wild-type mice, chronic systemic exposure to *P. gingivalis* LPS induced neuronal Aβ accumulation, neuroinflammation, and learning and memory deficits (Wu et al., 2017). Similar to systemic exposure, topical application of *P. gingivalis* LPS induced periodontitis, neuroinflammation, and learning and memory impairment in Sprague-Dawley rats (Hu et al., 2020). These data all together suggest that oral pathogens, such as *P. gingivalis*, may translocate to the brain, and then either the pathogen itself, its products, or both can lead to neuroinflammation and neurodegeneration. Concomitantly, oral pathogen–induced systemic inflammation can further contribute to AD pathogenesis.

1.3.2 Cardiovascular disease

Cardiovascular disease (CVD) is the leading cause of death worldwide, including coronary heart disease (CHD), stroke, and myocardial infarction. The traditional risk factors of CVD have been well established, such as hyperlipidemia, obesity, hypertension, and smoking (Mahmood et al., 2014). Regardless of the type of risk factors, atherosclerosis is the most common pathological process that triggers CVDs (Ministrini et al., 2020). This can restrict the blood flow and oxygen supply due to the thickening and/or stiffening of the walls of arteries. Atherosclerotic lesions are characterized by the accumulation of lipids and infiltration of immune cells, mostly macrophages (Hansson and Hermansson, 2011). Thus atherosclerosis is thought to be a chronic inflammatory state (Galkina and Ley, 2009).

As was mentioned earlier, periodontitis involves systemic inflammation, not just local infection. Notably, periodontitis and atherosclerosis show similar patterns of systemic inflammatory mediators, including elevated serum levels of CRP, IL-6, and IL-18 (Schenkein et al., 2020). Based on the immunohistological analyses, the atherosclerotic lesions concomitantly had infiltration of inflammatory cells and periodontal pathogens, such as *P. gingivalis* (Ford et al., 2006). Furthermore, Spahr et al. demonstrated that the periodontal pathogen burden is correlated with the presence of CHD (Spahr et al., 2006). Metaanalysis data also showed that patients with periodontal diseases had an increased risk of CHD, myocardial infarction, and stroke (Blaizot et al., 2009; Janket et al., 2003; Khader et al., 2004). Given these investigations together, it is evident that periodontitis increases the incidence of CVDs through systemic inflammation and infection with periodontal pathogens.

In addition to periodontal pathogens, oral commensal microbiota can play an important role in CVD progression by regulating nitrogen oxide (NO), a potent vasodilator. NO controls vascular homeostasis and cardiac contractility, conferring preventive and therapeutic effects against CVDs (Blekkenhorst et al., 2018; Lundberg et al., 2015). Of note, the oral commensal bacteria, particularly facultative anaerobes, can reduce nitrate in the diet into nitrite, which is further reduced to NO (Duncan et al., 1995). In the oral microbiome a higher abundance of these nitrate-reducing bacteria, such as species of *Firmicutes* and *Actinobacteria*, was potentially associated with lower systolic blood pressure (Goh et al., 2019; Koch et al., 2017). In healthy adults, 10-day supplementation with nitrate increased plasma levels of nitrite and NO while reducing blood pressure compared to the placebo group, potentially through enrichment of nitrate-responsive species, such as *F. nucleatum* subsp. *nucleatum*, *Prevotella melaninogenica*, and *Campylobacter concisus* (Vanhatalo et al., 2018). Thus modulation of the oral microbiome can be a promising strategy to prevent and treat CVDs by regulating NO levels.

1.3.3 Diabetes

Diabetes mellitus (DM) is a chronic metabolic disease characterized by high levels of blood glucose. According to the pathogeny of diabetes, DM can be classified into type 1 (T1DM) and type 2 (T2DM). T1DM is also known as insulin-dependent diabetes, since it results from the inability of the pancreas to produce insulin. In contrast, peripheral insulin resistance is the key feature of T2DM, the more common type of diabetes.

Both types of DM are linked with periodontitis in a bidirectional way, which means that diabetes can aggravate periodontitis and vice versa (Mealey, 2006; Taylor, 2001). Patients with severe T1DM had increased gingival inflammation, dental plaque, and periodontal bone loss (Jindal et al., 2015; Tervonen et al., 2000). In case of T2DM, a population-based cross-sectional study demonstrated that patients with poorly controlled diabetes had a significantly higher prevalence of periodontitis compared to individuals without diabetes (Tsai et al., 2002). These results strongly support DM as a risk factor for periodontitis.

The underlying biological evidence for DM as a risk factor for periodontitis has not yet been fully clarified. However, it is noteworthy that diabetes can affect the oral microbiome. Compared to non-DM subjects, poorly controlled DM patients retained gingival enrichment of the periodontal pathogens, also known as the red complex, such as *P. gingivalis*, *T. denticola*, and *T. forsythia* (Aemaimanan et al., 2013). Furthermore, individuals with T1DM displayed alterations in the oral microbial composition, such as higher abundance of taxa belonging to the phyla *Actinobacteria* and *Firmicutes*, while *Bacteroidetes* and *Proteobacteria* phyla were enriched in controls (de Groot et al., 2017). The whole metagenomics shotgun sequencing of the subgingival microbiome showed that T2DM was associated with less diverse microbial composition and higher relative abundance of *Anaerolineaceae* bacterium oral taxon 439 (Farina et al., 2019). Still, these alteration of the oral microbiota cannot explain how DM increases the risk of periodontitis. In spite of the limitations of mouse model studies, Xiao et al. have provided more direct evidence to solve this question. Diabetic *db/db* mice had increased periodontal inflammation and bone loss as well as enrichment of periodontitis-associated bacteria in the oral cavity (Xiao et al., 2017). Moreover, local injection of anti-IL-17 into the gingiva antibody markedly reduced the oral microbiome–induced periodontal pathogenicity in diabetic *db/db* mice, implying that DM can increase severity of periodontal disease through periodontal inflammation and subsequent alteration of the oral microbiome.

Conversely, periodontitis can affect the pathogenesis of DM as well. Metaanalysis studies showed that periodontitis can exacerbate glycemic control and increase the risk of developing diabetes in healthy individuals (Borgnakke et al., 2013; Graziani et al., 2018). Even in subjects with diabetes, periodontitis exerted an adverse effect on glycemic control and diabetic complications (Borgnakke et al., 2013; Graziani et al., 2018). In a wider perspective, it has been reported that there is a correlation between the severity of periodontitis and metabolic syndrome, including hypertension, obesity, dyslipidemia, and DM (Gomes-Filho et al., 2020). Interestingly, the periodontal pathogens have been detected in the oral cavity of both T1DM and T2DM patients (Aoyama et al., 2018; Schara et al., 2013). In high-fat diet–induced diabetic mice, *P. gingivalis* infection exacerbated periodontitis, systemic inflammation, glucose intolerance, and insulin resistance, simultaneously displaying the oral microbiota alteration (Blasco-Baque et al., 2017). Taken all together, these data suggest a two-way relationship between the oral microbiota and diabetes.

1.4 Concluding remarks

There is a growing body of evidence to support the theory that the oral microbiome is closely associated with human health, not limited to the oral cavity, but also related to diverse systemic physiological functions. As depicted in Fig. 1.1, in certain pathogenic circumstances the oral microbiota can translocate into the other organs and directly facilitate pathogenesis. In rather indirect manners, the oral microbiota can regulate systemic inflammatory and immune responses, which is considered a main crossroad between oral dysbiosis and systemic diseases. Moreover, recent investigations suggest that oral microbiota dysbiosis can change the microbial ecosystem in distal organs, such as the gut microbiome, suggesting an oral-gut axis (Kitamoto et al., 2020; Olsen and Yamazaki, 2019). This means that it is

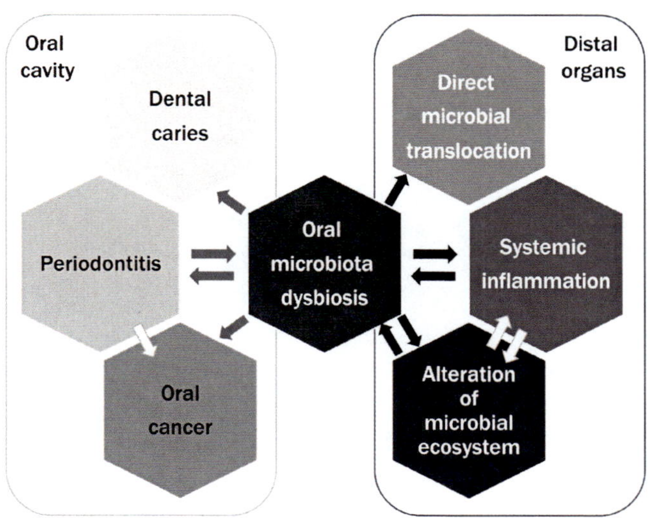

FIGURE 1.1 Oral microbiota dysbiosis in oral and systemic diseases. The oral microbiome can affect pathogenesis locally in the oral cavity, including dental caries, periodontal disease, and oral cancer. In addition to the local effects, oral dysbiosis can regulate systemic diseases, such as Alzheimer's disease, cardiovascular disease, and diabetes, possibly through provoking direct translocation of certain types of oral pathogens to the distal organs, sustaining systemic inflammation, and/or changing the microbial ecosystem in other organs.

possible to regulate the majority of the gut microbiota–associated human diseases by modulating the oral microbiota. Therefore it is not an overstatement to say that oral health is a gateway to maintaining whole-body health. In this regard, further research on the role of the oral microbiome in human health will enable the development of new strategies for diagnosis and treatment of oral diseases as well as systemic diseases.

Acknowledgments

This research was supported by National Research Foundation of Korea (NRF) Grants funded by the Korean Government (grant numbers NRF-2020R1C1C1003338, NRF-2016R1A5A2008630, and NRF-2022M3A9F3016364 to N.Y.S.) and by the Yonsei University Research Fund of 2021 (Yonsei Signature Research Cluster Program 2021-22-0017).

References

Aemaimanan, P., Amimanan, P., Taweechaisupapong, S., 2013. Quantification of key periodontal pathogens in insulin-dependent type 2 diabetic and non-diabetic patients with generalized chronic periodontitis. Anaerobe 22, 64–68.

Alley, D.E., Crimmins, E.M., Karlamangla, A., Hu, P., Seeman, T.E., 2008. Inflammation and rate of cognitive change in high-functioning older adults. J. Gerontol. A Biol. Sci. Med. Sci. 63, 50–55.

Aoyama, N., Suzuki, J.I., Kobayashi, N., Hanatani, T., Ashigaki, N., Yoshida, A., et al., 2018. Increased oral *Porphyromonas gingivalis* prevalence in cardiovascular patients with uncontrolled diabetes mellitus. Int. Heart J. 59, 802–807.

Bartold, P.M., Van Dyke, T.E., 2019. An appraisal of the role of specific bacteria in the initial pathogenesis of periodontitis. J. Clin. Periodontol. 46, 6–11.

Binder Gallimidi, A., Fischman, S., Revach, B., Bulvik, R., Maliutina, A., Rubinstein, A.M., et al., 2015. Periodontal pathogens *Porphyromonas gingivalis* and *Fusobacterium nucleatum* promote tumor progression in an oral-specific chemical carcinogenesis model. Oncotarget 6, 22613–22623.

Blaizot, A., Vergnes, J.N., Nuwwareh, S., Amar, J., Sixou, M., 2009. Periodontal diseases and cardiovascular events: *meta*-analysis of observational studies. Int. Dent. J. 59, 197–209.

Blasco-Baque, V., Garidou, L., Pomie, C., Escoula, Q., Loubieres, P., Le Gall-David, S., et al., 2017. Periodontitis induced by *Porphyromonas gingivalis* drives periodontal microbiota dysbiosis and insulin resistance via an impaired adaptive immune response. Gut 66, 872–885.

Blekkenhorst, L.C., Bondonno, N.P., Liu, A.H., Ward, N.C., Prince, R.L., Lewis, J.R., et al., 2018. Nitrate, the oral microbiome, and cardiovascular health: a systematic literature review of human and animal studies. Am. J. Clin. Nutr. 107, 504–522.

Borgnakke, W.S., Ylostalo, P.V., Taylor, G.W., Genco, R.J., 2013. Effect of periodontal disease on diabetes: systematic review of epidemiologic observational evidence. J. Clin. Periodontol. 40 (Suppl 14), S135–S152.

Bowen, W.H., Burne, R.A., Wu, H., Koo, H., 2018. Oral biofilms: pathogens, matrix, and polymicrobial interactions in microenvironments. Trends Microbiol. 26, 229–242.

Carding, S., Verbeke, K., Vipond, D.T., Corfe, B.M., Owen, L.J., 2015. Dysbiosis of the gut microbiota in disease. Microb. Ecol. Health Dis. 26, 26191.

Chen, C.K., Wu, Y.T., Chang, Y.C., 2017. Association between chronic periodontitis and the risk of Alzheimer's disease: a retrospective, population-based, matched-cohort study. Alzheimers Res. Ther. 9, 56.

Chen, M., Xia, W., 2020. Proteomic profiling of plasma and brain tissue from Alzheimer's disease patients reveals candidate network of plasma biomarkers. J. Alzheimers. Dis. 76, 349–368.

Cheng, R., Wu, Z., Li, M., Shao, M., Hu, T., 2020. Interleukin-1beta is a potential therapeutic target for periodontitis: a narrative review. Int. J. Oral. Sci. 12, 2.

Curtis, M.A., Diaz, P.I., Van Dyke, T.E., 2020. The role of the microbiota in periodontal disease. Periodontol. 2000 83, 14–25.

D'Aiuto, F., Parkar, M., Andreou, G., Suvan, J., Brett, P.M., Ready, D., et al., 2004. Periodontitis and systemic inflammation: control of the local infection is associated with a reduction in serum inflammatory markers. J. Dent. Res. 83, 156–160.

Darveau, R.P., Hajishengallis, G., Curtis, M.A., 2012. *Porphyromonas gingivalis* as a potential community activist for disease. J. Dent. Res. 91, 816–820.

de Groot, P.F., Belzer, C., Aydin, O., Levin, E., Levels, J.H., Aalvink, S., et al., 2017. Distinct fecal and oral microbiota composition in human type 1 diabetes, an observational study. PLoS One 12, e0188475.

Deo, P.N., Deshmukh, R., 2019. Oral microbiome: unveiling the fundamentals. J. Oral. Maxillofac. Pathol. 23, 122–128.

Dominy, S.S., Lynch, C., Ermini, F., Benedyk, M., Marczyk, A., Konradi, A., et al., 2019. *Porphyromonas gingivalis* in Alzheimer's disease brains: evidence for disease causation and treatment with small-molecule inhibitors. Sci. Adv. 5, eaau3333.

Donaldson, G.P., Lee, S.M., Mazmanian, S.K., 2016. Gut biogeography of the bacterial microbiota. Nat. Rev. Microbiol. 14, 20–32.

Du, Q., Fu, M., Zhou, Y., Cao, Y., Guo, T., Zhou, Z., et al., 2020. Sucrose promotes caries progression by disrupting the microecological balance in oral biofilms: an in vitro study. Sci. Rep. 10, 2961.

Duncan, C., Dougall, H., Johnston, P., Green, S., Brogan, R., Leifert, C., et al., 1995. Chemical generation of nitric oxide in the mouth from the enterosalivary circulation of dietary nitrate. Nat. Med. 1, 546–551.

Dutzan, N., Kajikawa, T., Abusleme, L., Greenwell-Wild, T., Zuazo, C.E., Ikeuchi, T., et al., 2018. A dysbiotic microbiome triggers TH17 cells to mediate oral mucosal immunopathology in mice and humans. Sci. Transl. Med. 10.

Engelhart, M.J., Geerlings, M.I., Meijer, J., Kiliaan, A., Ruitenberg, A., van Swieten, J.C., et al., 2004. Inflammatory proteins in plasma and the risk of dementia: the rotterdam study. Arch. Neurol. 61, 668–672.

Esberg, A., Haworth, S., Hasslof, P., Lif Holgerson, P., Johansson, I., 2020. Oral microbiota profile associates with sugar intake and taste preference genes. Nutrients 12.

Farina, R., Severi, M., Carrieri, A., Miotto, E., Sabbioni, S., Trombelli, L., et al., 2019. Whole metagenomic shotgun sequencing of the subgingival microbiome of diabetics and non-diabetics with different periodontal conditions. Arch. Oral Biol. 104, 13–23.

Ford, P.J., Gemmell, E., Chan, A., Carter, C.L., Walker, P.J., Bird, P.S., et al., 2006. Inflammation, heat shock proteins and periodontal pathogens in atherosclerosis: an immunohistologic study. Oral Microbiol. Immunol. 21, 206–211.

Forssten, S.D., Bjorklund, M., Ouwehand, A.C., 2010. *Streptococcus mutans*, caries and simulation models. Nutrients 2, 290–298.

Furutama, D., Matsuda, S., Yamawaki, Y., Hatano, S., Okanobu, A., Memida, T., et al., 2020. IL-6 induced by periodontal inflammation causes neuroinflammation and disrupts the blood-brain barrier. Brain Sci. 10.

Galkina, E., Ley, K., 2009. Immune and inflammatory mechanisms of atherosclerosis. Annu. Rev. Immunol. 27, 165–197.

Gao, L., Xu, T., Huang, G., Jiang, S., Gu, Y., Chen, F., 2018. Oral microbiomes: more and more importance in oral cavity and whole body. Protein Cell 9, 488–500.

Goh, C.E., Trinh, P., Colombo, P.C., Genkinger, J.M., Mathema, B., Uhlemann, A.C., et al., 2019. Association between nitrate-reducing oral bacteria and cardiometabolic outcomes: results from ORIGINS. J. Am. Heart Assoc. 8, e013324.

Gomes-Filho, I.S., Balinha, I., da Cruz, S.S., Trindade, S.C., Cerqueira, E.M.M., Passos-Soares, J.S., et al., 2020. Moderate and severe periodontitis are positively associated with metabolic syndrome. Clin. Oral Investig.

Graziani, F., Gennai, S., Solini, A., Petrini, M., 2018. A systematic review and *meta*-analysis of epidemiologic observational evidence on the effect of periodontitis on diabetes. An update of the EFP-AAP review. J. Clin. Periodontol. 45, 167–187.

Gupta, S., Allen-Vercoe, E., Petrof, E.O., 2016. Fecal microbiota transplantation: in perspective. Therap. Adv. Gastroenterol. 9, 229–239.

Hajishengallis, G., 2015. Periodontitis: from microbial immune subversion to systemic inflammation. Nat. Rev. Immunol. 15, 30–44.

Hajishengallis, G., Liang, S., Payne, M.A., Hashim, A., Jotwani, R., Eskan, M.A., et al., 2011. Low-abundance biofilm species orchestrates inflammatory periodontal disease through the commensal microbiota and complement. Cell Host Microbe 10, 497–506.

Hansson, G.K., Hermansson, A., 2011. The immune system in atherosclerosis. Nat. Immunol. 12, 204–212.

Heroiu Cataloiu, A.D., Danciu, C.E., Popescu, C.R., 2013. Multiple cancers of the head and neck. Maedica (Bucur) 8, 80–85.

Holmes, C., El-Okl, M., Williams, A.L., Cunningham, C., Wilcockson, D., Perry, V.H., 2003. Systemic infection, interleukin 1beta, and cognitive decline in Alzheimer's disease. J. Neurol. Neurosurg. Psychiatry 74, 788–789.

Hu, Y., Li, H., Zhang, J., Zhang, X., Xia, X., Qiu, C., et al., 2020. Periodontitis induced by *P. gingivalis*-LPS is associated with neuroinflammation and learning and memory impairment in Sprague-Dawley rats. Front. Neurosci. 14, 658.

Huang, S., Li, R., Zeng, X., He, T., Zhao, H., Chang, A., et al., 2014. Predictive modeling of gingivitis severity and susceptibility via oral microbiota. ISME J. 8, 1768–1780

Ilievski, V., Zuchowska, P.K., Green, S.J., Toth, P.T., Ragozzino, M.E., Le, K., et al., 2018. Chronic oral application of a periodontal pathogen results in brain inflammation, neurodegeneration and amyloid beta production in wild type mice. PLoS One 13, e0204941.

Irfan, M., Delgado, R.Z.R., Frias-Lopez, J., 2020. The oral microbiome and cancer. Front. Immunol. 11, 591088.

Ishida, N., Ishihara, Y., Ishida, K., Tada, H., Funaki-Kato, Y., Hagiwara, M., et al., 2017. Periodontitis induced by bacterial infection exacerbates features of Alzheimer's disease in transgenic mice. NPJ Aging Mech. Dis. 3, 15.

Janket, S.J., Baird, A.E., Chuang, S.K., Jones, J.A., 2003. *Meta*-analysis of periodontal disease and risk of coronary heart disease and stroke. Oral Surg. Oral Med. Oral Pathol. Oral Radiol. Endod. 95, 559–569.

Javed, F., Warnakulasuriya, S., 2016. Is there a relationship between periodontal disease and oral cancer? A systematic review of currently available evidence. Crit. Rev. Oncol. Hematol. 97, 197–205.

Jindal, A., Parihar, A.S., Sood, M., Singh, P., Singh, N., 2015. Relationship between severity of periodontal disease and control of diabetes (glycated hemoglobin) in patients with type 1 diabetes mellitus. J. Int. Oral Health 7, 17–20.

Kademani, D., 2007. Oral cancer. Mayo Clin. Proc. 82, 878–887.

Kamada, N., Nunez, G., 2014. Regulation of the immune system by the resident intestinal bacteria. Gastroenterology 146, 1477–1488.

Kamarajan, P., Ateia, I., Shin, J.M., Fenno, J.C., Le, C., Zhan, L., et al., 2020. Periodontal pathogens promote cancer aggressivity via TLR/MyD88 triggered activation of Integrin/FAK signaling that is therapeutically reversible by a probiotic bacteriocin. PLoS Pathog. 16, e1008881.

Kametani, F., Hasegawa, M., 2018. Reconsideration of amyloid hypothesis and tau hypothesis in Alzheimer's disease. Front. Neurosci. 12, 25.

Khader, Y.S., Albashaireh, Z.S., Alomari, M.A., 2004. Periodontal diseases and the risk of coronary heart and cerebrovascular diseases: a *meta*-analysis. J. Periodontol. 75, 1046–1053.

Kinney, J.W., Bemiller, S.M., Murtishaw, A.S., Leisgang, A.M., Salazar, A.M., Lamb, B.T., 2018. Inflammation as a central mechanism in Alzheimer's disease. Alzheimers Dement (NY) 4, 575–590.

Kitamoto, S., Nagao-Kitamoto, H., Jiao, Y., Gillilland 3rd, M.G., Hayashi, A., Imai, J., et al., 2020. The intermucosal connection between the mouth and gut in commensal pathobiont-driven colitis. Cell 182, 447–462. e414.

Koch, C.D., Gladwin, M.T., Freeman, B.A., Lundberg, J.O., Weitzberg, E., Morris, A., 2017. Enterosalivary nitrate metabolism and the microbiome: intersection of microbial metabolism, nitric oxide and diet in cardiac and pulmonary vascular health. Free Radic. Biol. Med. 105, 48–67.

Kondo, K., Niino, M., Shido, K., 1994. A case-control study of Alzheimer's disease in Japan—significance of life-styles. Dementia 5, 314–326.

Lafuente Ibanez de Mendoza, I., Maritxalar Mendia, X., Garcia de la Fuente, A.M., Quindos Andres, G., Aguirre Urizar, J.M., 2020. Role of *Porphyromonas gingivalis* in oral squamous cell carcinoma development: a systematic review. J. Periodontal. Res. 55, 13–22.

Lundberg, J.O., Gladwin, M.T., Weitzberg, E., 2015. Strategies to increase nitric oxide signalling in cardiovascular disease. Nat. Rev. Drug Discov. 14, 623–641.

Maekawa, T., Krauss, J.L., Abe, T., Jotwani, R., Triantafilou, M., Triantafilou, K., et al., 2014. *Porphyromonas gingivalis* manipulates complement and TLR signaling to uncouple bacterial clearance from inflammation and promote dysbiosis. Cell Host Microbe 15, 768–778.

Mahmood, S.S., Levy, D., Vasan, R.S., Wang, T.J., 2014. The Framingham Heart Study and the epidemiology of cardiovascular disease: a historical perspective. Lancet 383, 999–1008.

Mealey, B.L., 2006. Periodontal disease and diabetes. A two-way street. J. Am. Dent. Assoc. 137 (Suppl), 26S–31S.

Ministrini, S., Carbone, F., Montecucco, F., 2020. Updating concepts on atherosclerotic inflammation: from pathophysiology to treatment. Eur. J. Clin. Invest. e13467.

Mohanty, R., Asopa, S.J., Joseph, M.D., Singh, B., Rajguru, J.P., Saidath, K., et al., 2019. Red complex: polymicrobial conglomerate in oral flora: a review. J. Family Med. Prim. Care 8, 3480–3486.

Noack, B., Genco, R.J., Trevisan, M., Grossi, S., Zambon, J.J., De Nardin, E., 2001. Periodontal infections contribute to elevated systemic C-reactive protein level. J. Periodontol. 72, 1221–1227.

Olsen, I., Yamazaki, K., 2019. Can oral bacteria affect the microbiome of the gut? J. Oral Microbiol. 11, 1586422.

Poole, S., Singhrao, S.K., Kesavalu, L., Curtis, M.A., Crean, S., 2013. Determining the presence of periodontopathic virulence factors in short-term postmortem Alzheimer's disease brain tissue. J. Alzheimers Dis. 36, 665–677.

Pushalkar, S., Ji, X., Li, Y., Estilo, C., Yegnanarayana, R., Singh, B., et al., 2012. Comparison of oral microbiota in tumor and non-tumor tissues of patients with oral squamous cell carcinoma. BMC Microbiol. 12, 144.

Rapado-Gonzalez, O., Martinez-Reglero, C., Salgado-Barreira, A., Rodriguez-Fernandez, A., Aguin-Losada, S., Leon-Mateos, L., et al., 2020. Association of salivary human papillomavirus infection and oral and oropharyngeal cancer: a *meta*-analysis. J. Clin. Med. 9.

Rivera, C., 2015. Essentials of oral cancer. Int. J. Clin. Exp. Pathol. 8, 11884–11894.

Saulle, R., Semyonov, L., Mannocci, A., Careri, A., Saburri, F., Ottolenghi, L., et al., 2015. Human papillomavirus and cancerous diseases of the head and neck: a systematic review and *meta*-analysis. Oral Dis. 21, 417–431.

Schara, R., Skaleric, E., Seme, K., Skaleric, U., 2013. Prevalence of periodontal pathogens and metabolic control of type 1 diabetes patients. J. Int. Acad. Periodontol. 15, 29–34.

Schenkein, H.A., Papapanou, P.N., Genco, R., Sanz, M., 2020. Mechanisms underlying the association between periodontitis and atherosclerotic disease. Periodontol. 2000 83, 90–106.

Schmidt, B.L., Kuczynski, J., Bhattacharya, A., Huey, B., Corby, P.M., Queiroz, E.L., et al., 2014. Changes in abundance of oral microbiota associated with oral cancer. PLoS One 9, e98741.

Sharma, N., Bhatia, S., Sodhi, A.S., Batra, N., 2018. Oral microbiome and health. AIMS Microbiol. 4, 42–66.

Shreiner, A.B., Kao, J.Y., Young, V.B., 2015. The gut microbiome in health and in disease. Curr. Opin. Gastroenterol. 31, 69–75.

Sivamaruthi, B.S., Kesika, P., Chaiyasut, C., 2020. A review of the role of probiotic supplementation in dental caries. Probiotics Antimicrob. Proteins 12, 1300–1309.

Spahr, A., Klein, E., Khuseyinova, N., Boeckh, C., Muche, R., Kunze, M., et al., 2006. Periodontal infections and coronary heart disease—role of periodontal bacteria and importance of total pathogen burden in the coronary event and periodontal disease (CORODONT) study. Arch. Intern. Med. 166, 554–559.

Stein, P.S., Desrosiers, M., Donegan, S.J., Yepes, J.F., Kryscio, R.J., 2007. Tooth loss, dementia and neuropathology in the Nun study. J. Am. Dent. Assoc. 138, 1314–1322. quiz 1381–1312.

Taylor, G.W., 2001. Bidirectional interrelationships between diabetes and periodontal diseases: an epidemiologic perspective. Ann. Periodontol. 6, 99–112.

Tervonen, T., Karjalainen, K., Knuuttila, M., Huumonen, S., 2000. Alveolar bone loss in type 1 diabetic subjects. J. Clin. Periodontol. 27, 567–571.

Tsai, C., Hayes, C., Taylor, G.W., 2002. Glycemic control of type 2 diabetes and severe periodontal disease in the US adult population. Community Dent. Oral Epidemiol. 30, 182–192.

Van Dyke, T.E., Bartold, P.M., Reynolds, E.C., 2020. The nexus between periodontal inflammation and dysbiosis. Front. Immunol. 11, 511.

Vanhatalo, A., Blackwell, J.R., L'Heureux, J.E., Williams, D.W., Smith, A., van der Giezen, M., et al., 2018. Nitrate-responsive oral microbiome modulates nitric oxide homeostasis and blood pressure in humans. Free Radic. Biol. Med. 124, 21–30.

Wu, Z., Ni, J., Liu, Y., Teeling, J.L., Takayama, F., Collcutt, A., et al., 2017. Cathepsin B plays a critical role in inducing Alzheimer's disease-like phenotypes following chronic systemic exposure to lipopolysaccharide from *Porphyromonas gingivalis* in mice. Brain Behav. Immun. 65, 350–361.

Xiao, E., Mattos, M., Vieira, G.H.A., Chen, S., Correa, J.D., Wu, Y., et al., 2017. Diabetes enhances IL-17 expression and alters the oral microbiome to increase its pathogenicity. Cell Host Microbe 22, 120–128. e124.

Ye, L., Jiang, Y., Liu, W., Tao, H., 2016. Correlation between periodontal disease and oral cancer risk: a *meta*-analysis. J. Cancer Res. Ther. 12, C237–C240.

Zhang, L., Liu, Y., Zheng, H.J., Zhang, C.P., 2019a. The oral microbiota may have influence on oral cancer. Front. Cell Infect. Microbiol. 9, 476.

Zhang, Y., Wang, X., Li, H., Ni, C., Du, Z., Yan, F., 2018. Human oral microbiota and its modulation for oral health. Biomed. Pharmacother. 99, 883–893.

Zhang, Z., Yang, J., Feng, Q., Chen, B., Li, M., Liang, C., et al., 2019b. Compositional and functional analysis of the microbiome in tissue and saliva of oral squamous cell carcinoma. Front. Microbiol. 10, 1439.

Zhao, H., Chu, M., Huang, Z., Yang, X., Ran, S., Hu, B., et al., 2017. Variations in oral microbiota associated with oral cancer. Sci. Rep. 7, 11773.

Chapter 2

Influence of microbiome in shaping the newborn immune system: an overview

Manoj Kumar Kingsley[1] and B. Vishnu Bhat[2]

[1]*Department of Neonatology, JIPMER, Puducherry, India,* [2]*Director-Medical Research, Professor of Pediatrics and Neonatology, Aarupadaiveedu Medical College and Hospital, Vinayaka Mission's Research Foundation-DU, Puducherry, India*

2.1 Introduction

The microbiome is the community of microorganisms that dwell in and on the body. During birth, newborns are subject to colonization by foreign microorganisms that inhabit most exposed surfaces, such as the skin, mouth, vagina, and gut (Round and Mazmanian, 2009). Most of the human microbiome is present in the gastrointestinal tract, and a major portion of these microorganisms are present in the colon. The human intestinal microbiota comprises 10^{13} to 10^{14} microorganisms that have evolved alongside the host immune system (Gill et al., 2006). These mainly consist of 500 species of bacteria (Brown et al., 2013). Though the microbiome is predominantly bacterial it also contains other microbes such as fungi, yeast and viruses (Salzman, 2014). Recent studies have shown that the mammalian gut contains a rich fungal community that interacts strongly with the immune system (Iliev et al., 2012).

The commensal microbiota colonizes various locations in the host and strongly influences the development of the immune system in early life (Gensollen et al., 2016). Humans and other mammals maintain a symbiotic relationship with this microbial ecosystem. As long as this host-microbiota relationship is operating optimally, it results in protective immune responses to pathogens and maintenance of tolerance to innocuous antigens. When this symbiosis is affected, it leads to a rise in autoimmune and inflammatory disorders, as has been seen in several high-income countries (Belkaid and Hand, 2014). It is evident from a number of studies that early life colonization coincides with a limited time period during which the immune system is permissive to instructions from the microbiota. Furthermore, the immune influences that occur during this time may have long-lasting effects on one's health and immunity, including proper (or improper) development of specific immune cell subsets and resistance (or susceptibility) to various diseases later in life (Gensollen et al., 2016). When appropriate colonization does not happen during this time window, aberrant outcomes may occur.

2.2 How the microbiota shapes the development of immunity in early life

The microbiome influences the development of the immune system. Studies have shown that 45%−50% of the genes that are induced by the microbiota are related to immune response pathways (Gaboriau-Routhiau et al., 2009). The microbiota promotes and shapes the development of both innate and adaptive immunity. The postnatal period is an important time when early life microbial exposures shape the morphological and functional development of the immune system (Gensollen et al., 2016). Moreover, epidemiological data reveal a critical time window in early life during which the microbiota shapes the development of the neonatal immune system (Holt and Jones, 2000). For instance, the exposure of mice to microbiota-induced T regulatory cells during a critical time window during the weaning reaction reduced the susceptibility to several inflammatory pathologies in adult life. When microbiota exposure happens outside this time window, these positive outcomes could not be replicated (Nabhani et al., 2019). Similarly, the microbial antigen encounter from certain gut bacteria has to take place within a specific preweaning interval for developing tolerance to the gut bacteria (Knoop et al., 2017). Likewise, toll-like receptor 5 (TLR5) mediated counter selection of colonizing flagellated bacteria is restricted to a critical time window in the neonatal life and has a significant influence in shaping the gut microbiome

Microbiome, Immunity, Digestive Health and Nutrition. DOI: https://doi.org/10.1016/B978-0-12-822238-6.00035-2

(Fulde et al., 2018). Thus it is evident that microbial colonization during early life is required to take place within a "window of opportunity" during which the gut microbiome establishes the host's mucosal immune homeostasis in a way that cannot be accomplished in later life. The immune development was found to be substantially different in mice that were conventionalized in adulthood despite the achievement of immune homeostasis (Aidy et al., 2013).

Commensals have a dual role in both training the immune system and functional tuning, thus acting as adjuvants to the immune system. The studies on germ-free (GF) mice and colonized mice have revealed the significance of the microbiome in the development of the immune system. GF animals have no commensal microflora, and their immune responses have been found not to be affected by the molecules of pathogenic and beneficial microorganisms. Specific pathogen-free (SPF) animals of the same strain have simple flora. The mucosal immune system GF animals is underdeveloped (Macpherson and Harris, 2004). They have severe defects in the development of gut-associated lymphoid tissues (GALT) (Round and Mazmanian, 2009) and antibody production. They have hypoplastic Peyer's patches (PP) with fewer germinal centers and reduced number of IgA-secreting plasma cells and $CD4^+$ T cells. Apart from the underdeveloped mucosal immune system, their spleen and lymph nodes are structureless with poorly formed B and T cell zones (Fig. 2.1) (Macpherson and Harris, 2004). Likewise, in immunocompetent mouse pups that had not been exposed to milk IgA, the FDC network and primary follicles in the spleen were not properly developed (Rosado et al., 2018).

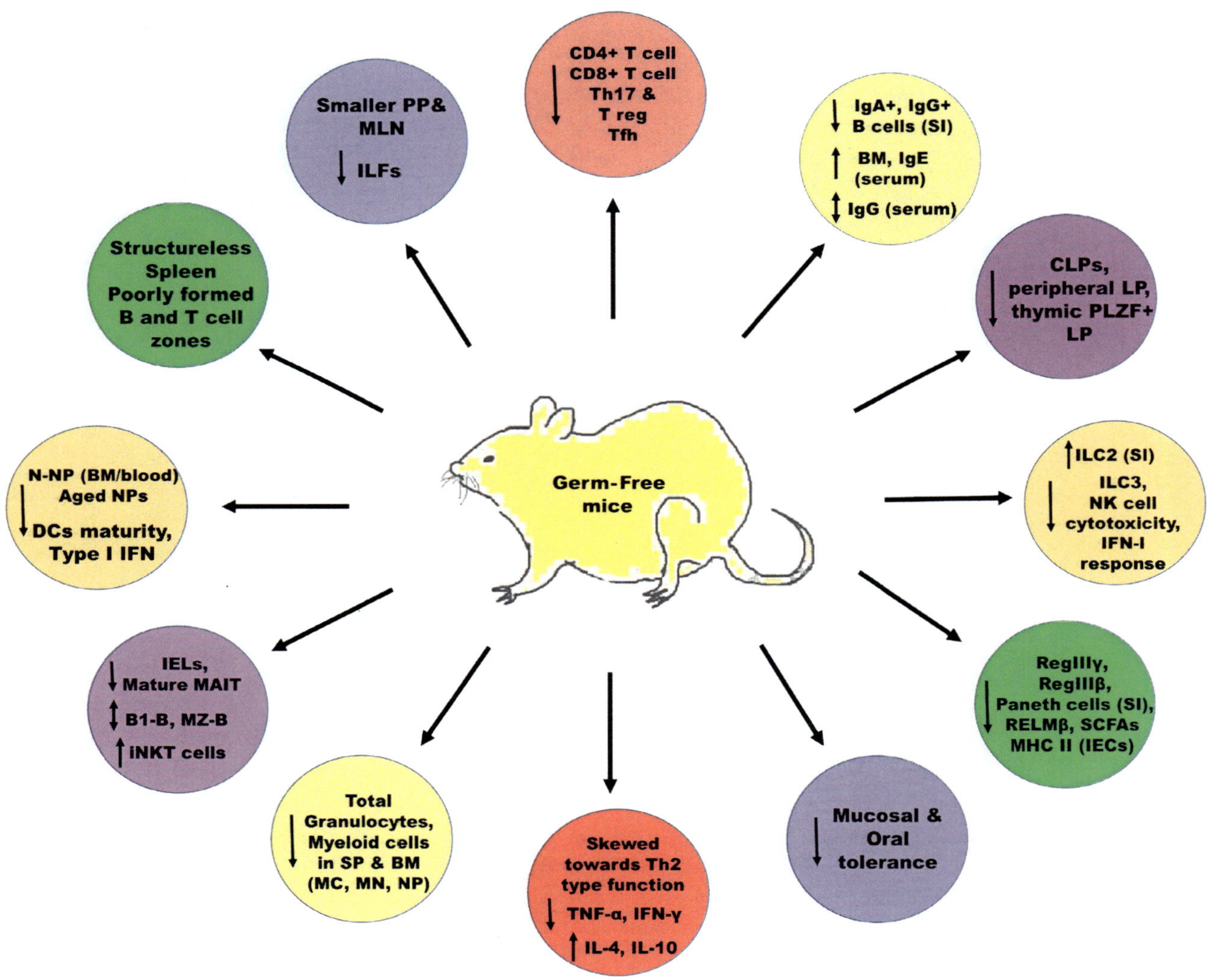

FIGURE 2.1 Characteristics of germ-free mice. *BM*, bone marrow; *CLP*, common lymphoid progenitor; *DC*, dendritic cells; *IEC*, intestinal epithelial cells; *IEL*, intraepithelial lymphocytes; *ILC*, innate lymphoid cells; *ILF*, isolated lymphoid follicles; *iNKT*, invariant natural killer T cells; *LP*, lymphocytes; *MAIT*, mucosa-associated invariant T cells; *MC*, macrophages; *MHC II*, major histocompatibility complex class II; *MLN*, mesenteric lymph nodes; *MN*, monocytes; *MZ*, marginal zone; *NK*, natural killer cell; *N-NP*, neonatal neutrophils; *NP*, neutrophils; *PP*, Peyer's patches; *RELM-β*, resistin-like molecule β; *SCFAs*, short-chain fatty acids; *SI*, small intestine; *SP*, spleen; *Treg*, T regulatory cells; *TCR*, T cell receptor; *Tfh*, T follicular cells.

These defects seemed to be rectified by colonizing GF animals with commensal bacteria (Round and Mazmanian, 2009). This can be done just by keeping an SPF mouse in a cage along with GF animals (Macpherson and Harris, 2004). The induction of commensal bacteria results in the development of organized GALT, an increase in $CD4^+$ T cell numbers, and the generation of secretory IgA (Macpherson and Harris, 2004). Another study showed that spleens from GF mice that were colonized with *Bacteroides fragilis* appeared normal along with well-developed lymphocyte zones (Mazmanian et al., 2005). Although most abnormalities in GF animals could be corrected by induction of commensal microbes, certain cellular defects could be rectified only within a short time period in early life.

2.3 Influence of the microbiota on the development and function of specific immune cell subsets in early life

The microbiota plays a crucial role in the development and function of individual immune cell subsets in early life. Moreover, microbial exposure that occurs within a critical developmental time window in early life is crucial for the proper development of immune system and individual immune cell subsets.

2.3.1 Myeloid cells

Germ-free mice are deficient in resident myeloid cell populations (macrophages, monocytes, and neutrophils) in the spleen and bone marrow. Complex molecular signals from the microbiota enhance the hematopoietic differentiation of myeloid cells (Khosravi et al., 2014). Likewise, antibiotics have shown to suppress hematopoiesis by depleting the intestinal microbiota (Josefsdottir et al., 2017).

Newborn infants are highly susceptible to infection. This is mainly due to the immaturity and hyporesponsiveness of neonatal immune cells. This immaturity of innate immune cells in neonates is due to lack of microbial priming. The infant gut microbiome provides appropriate stimuli for priming of innate immune cells in early life (Yu et al., 2018). The gut bacteria direct the innate immune cell development by promoting hematopoiesis. Depletion of the microbiota in mice by using antibiotic treatment resulted in the reduction of granulocytes in bone marrow and decreased bone marrow cellularity (Josefsdottir et al., 2017).

The microbiota plays an important role in promoting neutrophil homeostasis in early life. The transfer of normal microbiota to neonatal mice born to antibiotic exposed mouse dams induced interleukin 17 (IL-17) production by intestinal innate lymphoid cells (ILCs) and increased plasma granulocyte colony stimulating factor (G-CSF) levels and neutrophil numbers in circulation and bone marrow (Deshmukh et al., 2014). Recent studies have shown that neutrophil ageing, which results in the production of a functionally overly active subset of neutrophils, is induced by the microbiome(Zhang et al., 2015). Furthermore, microbiota depletion inhibits neutrophil extracellular trap (NET) formation (Zhang et al., 2015).

The intestinal microbiota regulates the steady state cellular life span of neutrophils and inflammatory monocytes (Hergott et al., 2016). In microbiota-depleted mice there were significantly reduced numbers of $F4/80^+$ monocytes and macrophages in the intestines (Ekmekciu et al., 2017). The bone marrow monocytes from gut flora–depleted mice using broad-spectrum antibiotic treatment displayed lower migratory capacity and maturation (Emal et al., 2017).

Gestational colonization increased the numbers of early postnatal intestinal $F4/80^+CD11c^+$ mononuclear cells in mice (de Agüero et al., 2016). The gut microbiome drives the influx of monocyte precursors that differentiate into mature macrophages at the time of weaning. This process needs to be continued throughout adult life to maintain a normal intestinal macrophage pool (Bain et al., 2014). Depletion of the microbiota causes defects in the bactericidal capacity of alveolar macrophages (Brown et al., 2017). Likewise, macrophages that were isolated from antibiotic-treated mice displayed reduced expression of genes related to antiviral immunity. They also exhibited a defective response to type I and II interferons (IFNs) and impaired ability to control viral replication (Abt et al., 2012).

Following antibiotic treatment, the various dendritic cell (DC) subsets have been found to be reduced in mucosal and systemic sites (Ekmekciu et al., 2017; Thackray et al., 2018). Some *Lactobacillus* species have been found to enhance the priming and maturation of DCs, and a decrease in this can result in delayed immune response in neonates (Sanidad and Zeng, 2020). Likewise, the first exposure to the microbiota after birth induces tumor necrosis factor (TNF) secretion by monocytes and macrophages. This initiates the differentiation and functional maturation of the neonatal splenic type I preconventional DC (pre-cDC1) compartment (Köhler et al., 2020).

2.3.1.1 *Erythrocytes in the spleen*

$CD71^+$ erythroid cells are enriched in the spleen from birth to 2 weeks of age but are reduced in adult life. Their numbers in this neonatal stage are not influenced by the microbiota. Neonatal $CD71^+$ cells exhibit immunosuppressive properties and suppress the aberrant inflammation that is associated with microbial colonization in early life, and their elimination alleviates these protective benefits. In particular, the intestinal $CD11b^+$ and $CD11c^+$ cells from $CD71^+$ cell–depleted neonatal mice produced increased levels of TNF-α and costimulatory molecule expression. The removal of commensal bacteria, in turn, corrected the elevated TNF-α production and costimulatory molecule expression induced by $CD71^+$ cell depletion (Elahi et al., 2013). This information shows the reciprocal interactions between the immune system and the microbiota in early life.

2.3.1.2 *Myeloid-derived suppressor cells*

Myeloid-derived suppressor cells (MDSCs) are a cell population having T cell–suppressive properties. They are rarely found in healthy adults but are found to be elevated in neonates (Gervassi et al., 2014). The elevated levels of MDSCs in the first few weeks of life may be a mechanism to enhance tolerance during microbial colonization among neonates (Fike et al., 2019; He et al., 2018).

2.3.2 Lymphoid cells

The depletion of the microbiota using antibiotic treatment results in severe reduction in the common lymphoid progenitors (CLPs) in the bone marrow along with reduction in the peripheral lymphocytes (Josefsdottir et al., 2017; Thackray et al., 2018). GF mice also display reduced number of CLPs (Iwamura et al., 2017). This shows that the microbiome has an important role in the differentiation and function of lymphoid cells.

2.3.2.1 *T cell repertoire*

The intestinal T cell receptor (TCR) repertoire is shaped during infancy. In normal SPF mice at birth, the repertoire is polyclonal, which changes toward a more restrictive oligoclonal repertoire after weaning and into adulthood. This may be due to the selective clonal expansion that happens. The T cell clones that react to commensal and food antigens are deleted, and those that react to pathogenic antigens are selected and expanded. Thus aberrant responses to commensal flora and food antigens are avoided. Here, bacteria drive the clonal selection and expansions after weaning either independently or along with food antigens. Likewise, in GF mice the TCR repertoire was oligoclonal at birth and remains oligoclonal at and after weaning. Nevertheless, in GF mice, because microbial colonization is absent, there are far fewer intestinal epithelial lymphocytes, and it is unlikely that any specific clone would have undergone selection and expansion. However, in SPF mice the repertoire is oligoclonal, since there are numerous intraepithelial lymphocytes (IELs), specific clones would have undergone selection and expansion (Probert et al., 2007).

Furthermore, intestinal microbes have also been found to influence the thymic T cell repertoire. The impaired thymic development of the transcription factor PLZF expressing innate lymphocytes in GF neonatal mice is corrected by colonization with *B. fragilis*. The intestinal bacteria regulate the thymic distribution of $PLZF^+$ cells through plasmacytoid dendritic cells, which migrate from the colon to the thymus (Ennamorati et al., 2020).

The gut immune maturation is based on colonization with host-specific bacterial species that induces the expansion of T cell numbers in intestinal secondary lymphoid organs such as PPs and mesenteric lymph nodes (MLNs) and epithelial compartment (IEL and lamina propria) in mice. Segmented filamentous bacteria (SFB) could partially restore intestinal T cell numbers and were not fully protective against *Salmonella* infection, suggesting that other microbes are needed for complete immune maturation and host defense (Chung et al., 2012).

2.3.2.2 *$CD4^+$ T cells*

Recent studies have shown that neonatal gut microbiome dysbiosis enhances $CD4^+$ T cell dysfunction (Fujimura et al., 2016). The depletion of the microbiota resulted in the reduction of $CD4^+$ T cells in the peripheral blood, small intestinal, and colonic lamina propria of mice. Reconstitution with microbiota was able to reverse these effects. In contrast, bone marrow $CD4^+$ T cells increased, and splenic $CD4^+$ T cells were not affected by microbiota depletion (Ekmekciu et al., 2017; Josefsdottir et al., 2017).

Commensal gut flora has been shown to drive the expansion of $CD4^+$ T cells in the intestines of adult mice (Niess et al., 2008). The expansion of $CD4^+$ T cells was observed in the colonic lamina propria of conventionalized adult mice. Specifically, SFB was able to induce the differentiation of a wide range of proinflammatory and regulatory $CD4^+$ T cells in the intestines (Gaboriau-Routhiau et al., 2009). However, the situation in neonatal intestines is quite different. The $CD4^+$ T lymphocytes exhibited an immature phenotype throughout the postnatal period in spite of the rapid establishment of the neonatal gut microbiome. The maternal secretory IgA and neonatal T regulatory cells act together to impede the postnatal maturation of $CD4^+$ T cells. This reduced maturity of $CD4^+$ T cells in the neonatal intestines may protect from the microbiota-induced maturation of cross-reactive T lymphocytes and autoimmunity in the postnatal period. This might also provide a time window in early life to reinstate self-tolerance and sustain a broad TCR repertoire throughout infancy to adulthood (Torow et al., 2015). Likewise, the absence of gut microbial stimuli resulted in the reduction of $CD4^+$ memory cells in the intestines and spleen. This can be reversed by recolonization with complex murine intestinal microbiota (Ekmekciu et al., 2017).

Likewise, the development and maintenance of germinal center T follicular cells (Tfh) within the gut are influenced by signals from the commensal microbiota (Kubinak et al., 2015). The gut commensal SFB induces the differentiation of Tfh cells within the PPs and drives their migration into systemic sites such as the spleen (Teng et al., 2016).

2.3.2.3 Th1 and Th2 Cells

Effector $CD4^+$ T cells are of two general subtypes: T helper 1 (Th1) and T helper 2 (Th2), which have distinct and opposite roles. The balance between Th1 and Th2 immune responses is important to maintaining the individual's overall health (Mazmanian et al., 2005). Commensal bacteria have been found to play an important role in maintaining this equilibrium. Likewise, immunity in the newborn is poised toward rapid Th2 function and minimal Th1 function. Because Th1 lymphocytes are crucial in countering microbes and Th2 lymphocytes favor allergies, the newborn is susceptible to microbial infections and allergic disorders (Zaghouani et al., 2009). GF mice also are found to be skewed toward Th2-type function (Gensollen et al., 2016). In antibiotic-treated mice the IFN-γ-producing $CD4^+$ T cells were reduced in the small and large intestines but could be restored by reconstitution with microbiota (Ekmekciu et al., 2017).

B. fragilis is an important commensal bacterium within the human gut that has been found to be transmitted from the maternal gut to the offspring (Bjerke et al., 2011). Monocolonization with *B. fragilis* was sufficient to correct the $CD4^+$ T cell deficiency in the spleen of GF mice and the production of appropriate cytokines. *B. fragilis* produces a microbial molecule polysaccharide A (PSA), which is required for suppressing the levels of Th2-type cytokine IL-4 and upregulating the levels of Th1 cytokines such as IFN-γ during microbial colonization in GF mice (Fig. 2.2). This enables the maintenance of the Th1/Th2 balance during early life (Mazmanian et al., 2005).

Another study using newborn rat pups showed that the oral administration of a bacterial extract overcomes the Th2 bias through the selective upregulation of Th1-dependent IgG subclass production (Bowman and Holt, 2001). The Th1 immune responses are also induced by other microbes, such as yeast and fungi that are present in the gut (Romani, 1997).

2.3.2.4 Th17 cells

Individual commensal species have been responsible for the development of various subsets of the lamina propria T lymphocytes. T helper 17 (Th17) cells are a subset of T helper cells that produce effector molecules such as IL-17A, IL-17F, IL-22, and IL-26 that have crucial roles in protecting the host from bacterial and fungal infection mainly at the mucosal surfaces. Depletion of the microbiota resulted in the reduction of IL-17-expressing $CD4^+$ cells in the intestines, MLNs, and spleen, but this could be reversed following reconstitution with microbiota (Ekmekciu et al., 2017). The gut microbiome has been found to promote the development of Th17 cells in the lamina propria of the intestines (Atarashi et al., 2008; Ivanov et al., 2008). Among the commensals, SFB are the most effective inducers of Th17 cells. Mice that are colonized with SFB alone show robust accumulation of Th17 cells in the lamina propria of the small intestines (Gaboriau-Routhiau et al., 2009; Ivanov et al., 2009). Likewise, colonization of GF mice with human infant microbiota associated with allergies leads to enhanced intestinal and systemic Th17 responses (Petursdottir et al., 2017).

2.3.2.5 T regulatory cells

T regulatory (Treg) cells play a crucial role in establishing and maintaining immune homeostasis in peripheral tissues, especially at various barrier sites. When this homeostasis is affected, it leads to the loss of mucosal tolerance and impaired effector responses in the gut. The depletion of the microbiota resulted in the significant reduction of $FoxP3^+$ Treg cells in the intestines, MLNs, and spleen of mice. Reconstitution with microbiota could reverse these effects (Ekmekciu et al., 2017; Ochoa-Repáraz et al., 2009).

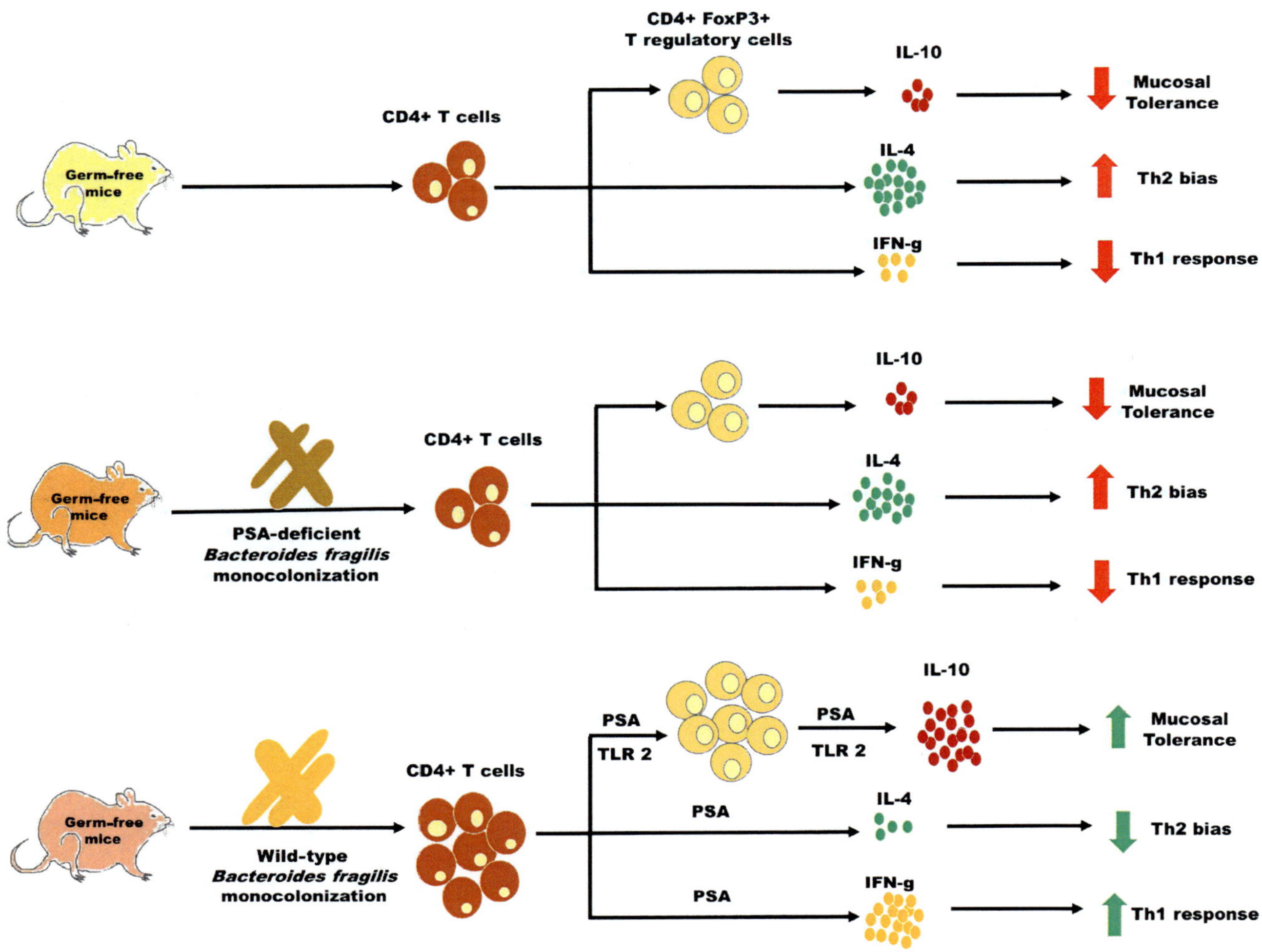

FIGURE 2.2 Monocolonization of germ-free mice with *Bacteroides fragilis* during early life expands the $CD4^+$ T cell pool, corrects any Th1/Th2 imbalance, and induces IL-10-producing $Foxp3^+$ Tregs. Their effect is mediated by the immunomodulatory molecule polysaccharide A (PSA).

Microbial exposure during a specific time window in neonatal life influences the induction of Treg cells. Microbial exposure to neonatal mice caused an increase in Treg cells in lungs at day 8 after birth, and this was not observed in GF mice (Gollwitzer et al., 2014). *Bifidobacterium* spp. are predominant members of the infant gut microbiome and have a crucial role as modulators of the immune response (Turroni et al., 2012). The oral administration of *Bifidobacterium breve* resulted in increased numbers of $FoxP3^+$ Treg cells in mice only if administered before the pre-weaning stage (Lyons et al., 2010).

Given the important role of Treg cells in maintaining mucosal homeostasis, it should be that a large sector of any microbiota rather than specific commensals has evolved to enhance the induction of Treg cells in early life. Recently, it has been found that a range of microorganisms and bacterial strains have been shown to induce the accumulation of Tregs in the colon (Belkaid and Hand, 2014; Faith et al., 2014). These data show the crucial role of the microbiota in shaping Treg cell development and expansion in early life; thus it indirectly influences the regulation of systemic control of immune responses and maintenance of host-microbiota mutualism.

2.3.2.6 $CD8^+$ T cells

The $CD8^+$ T cells are important for the response against intracellular pathogens and viruses but in neonates there is altered and attenuated $CD8^+$ T cell response (Fike et al., 2019; He et al., 2018). In many neonatal infection models, elevated levels of the antiinflammatory cytokine TGF-β was found, which may promote the suppression of $CD8^+$ T cell response during acute infection or microbial colonization (Fike et al., 2019). Likewise, MDSCs are present in high levels in neonates, and they suppress both $CD4^+$ and $CD8^+$ T cell responses during early life (He et al., 2018).

The depletion of the microbiota resulted in the reduction of $CD8^+$ cytotoxic T cells in the peripheral blood, small intestinal, and colonic lamina propria of mice. Reconstitution with microbiota was able to reverse these effects. Likewise, treatment with antibiotics resulted in the strong reduction of $CD8^+$ memory/effector cells in intestinal and other systemic sites such as the spleen of mice. Reconstitution with complex intestinal microbiota significantly improved these cell populations (Ekmekciu et al., 2017). Recently, it has been found that microbiota-derived SCFA butyrate drives the transition of antigen-activated $CD8^+$ T cells into memory cells, which could not be observed in GF mice (Bachem et al., 2019).

Likewise, maternal antibiotic treatment during pregnancy and lactation resulted in marked changes in the intestinal microbiome composition of both mothers and infants. Following viral infection the antigen-specific IFN-γ-producing $CD8^+$ T cells were reduced in mice born to mothers that had been treated with antibiotics during pregnancy, resulting in suppressed antiviral immunity (Gonzalez-Perez et al., 2016).

Exposure to microbiome very early in life promotes pre-cDC1 to induce protective $CD8^+$ T cell responses during neonatal *L. monocytogenes* infection by modulating their ability to produce IL-10 and IL-12p40 (Köhler et al., 2020).

2.3.2.7 Innate-like lymphocytes

Innate-like lymphocytes are a diverse group of cells, such as CD8αα + IELs, γδ T cells, mucosa-associated invariant T cells (MAIT), B-1 B cells, and invariant NKT cells, which are found at barrier sites such as skin, liver, lungs, or gut and in the lymph nodes that drain these barrier tissues (Ataíde and Kastenmüller, 2016). Owing to their tissue localization, they are in close vicinity to the microbiota. They are crucial in shaping the microbiota in early life, as they are found abundantly in mucosal tissues during neonatal colonization. Conversely, the microbiota influences the development and function of many IELs (Constantinides, 2018; Gensollen et al., 2016).

IELs are a distinct population of T cells that are present in-between the epithelial cells. In GF mice, IELs are reduced in numbers with diminished cytotoxicity. Furthermore, early life microbial colonization is important for the proper development of diet-induced CD8αβ intraepithelial T cells during and after weaning. More important, delayed supplementation of microbial and dietary antigens was ineffective in the induction of CD8αβ IELs (Jung et al., 2019). The $CD4^+$ $CD8\alpha\alpha^+$ double-positive IELs (DP-IELs) exhibit immunoregulatory functions in the small intestinal epithelium but are absent in GF mice and diminished in antibiotic-treated mice. The commensal *Lactobacillus reuteri* promotes the generation of DP-IELs (Cervantes-Barragan et al., 2017).

Likewise, MAIT are not present in GF mice, suggesting that commensal microbes are needed for their expansion (Treiner et al., 2003). Interestingly, the presence of immature thymic MAIT cells in GF mice is not affected, whereas the number of mature MAIT cells are reduced, suggesting the importance of the microbiota in the development of MAIT cells (Koay et al., 2016).

B-1 B cells and marginal zone B cells exhibit features of innate-like lymphocytes, such as quick primary response and tissue localization in peritoneal cavities or marginal sinuses in the spleen (Constantinides, 2018). Their frequencies were unaffected in GF mice, which suggests the lack of requirement of commensals for their development (Lino et al., 2013). However, most recent studies have shown that exposure to commensal bacteria–derived antigens in neonatal life directs polysaccharide-specific B-1 B cell repertoire development. The early life exposure to microbial antigens determines the clonality of the B-1 B cell repertoire and the subsequent antibody response (New et al., 2020).

2.3.2.7.1 Invariant natural killer T cells

The invariant natural killer T cells (iNKT cells) constitute an innate-like T cell subset and express an invariant TCR-α chain. They play a crucial role during various immune responses and pathological conditions (Crosby and Kronenberg, 2018). In GF mice the iNKT cells have a reduced mature phenotype and are hyporesponsive to lipid antigen stimulation. Exposure of GF mice to bacteria expressing iNKT cell antigens completely induced phenotypic maturity of iNKT cells (Wingender et al , 2012).

Likewise, studies have shown that in GF mice, iNKT cells are found to accumulate in the mucosal tissues such as the colon and lung, leading to heightened morbidity in models of asthma and inflammatory bowel disease (IBD). Importantly, colonization of only neonatal (not adult) GF mice with conventional microbiota prevented the iNKT cell accumulation and the susceptibility to IBD and asthma (Olszak et al., 2012). Likewise, symbiotic bacterial sphingolipids inhibit iNKT cell proliferation and accumulation during neonatal development. The exposure of mice to symbiotic sphingolipids has to happen during a critical time window in early life for their iNKT cell numbers to be restricted in adult life (An et al., 2014). It is understood that early life microbial exposure induces long-lasting effects on iNKT cells, without which the later life exposure of these cells to stimulating factors could induce autoinflammatory responses (Olszak et al., 2012) (Fig. 2.3).

2.3.2.8 Innate lymphoid cells

Innate lymphoid cells are a family of immune cells that emulate the phenotype and functions of T cells. However, they do not express antigen-specific receptors or clonal selection and expansion when stimulated. There are three major groups of ILCs: ILC1, ILC2, and ILC3. Apart from these groups, natural killer (NK) cells and lymphoid tissue inducer cells are generally included in the ILC family (Panda and Colonna, 2019).

While most of the innate immune cells show restricted presence and activity during infancy, ILCs show a higher presence and activity in infancy compared to later life. Moreover, ILCs appear and are programmed at the embryonic stages of development, suggesting that they have a crucial role in immunity during early life (Yu et al., 2018). The phenotypic diversity and functional plasticity of ILCs in the intestines are regulated by signals derived from the microbiome (Gury-BenAri et al., 2016).

Although ILC2s occupy the small intestines at birth, the appearance of tuft cells was observed only during weaning in mice. Dietary fibers enabled *Trichomonas* colonization, which promoted ILC2 activation and tuft cell expansion

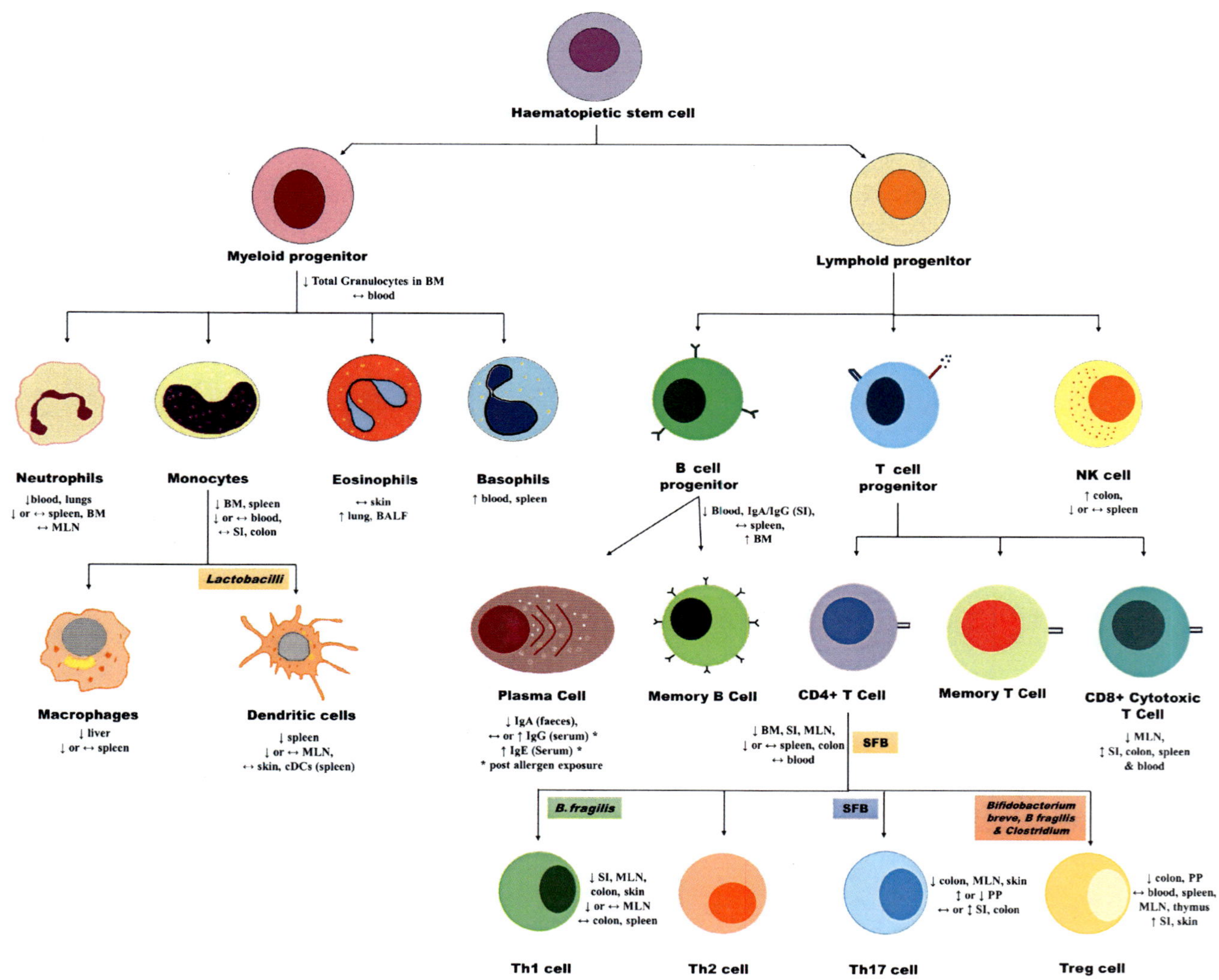

FIGURE 2.3 **Cells of the immune system in germ-free animals.** The populations of different immune cell types are altered in germ-free mice. This shows the importance of the microbiome in the development and function of various immune cell types in early life. *BM*, bone marrow; *cDCs*, conventional dendritic cells; *MLN*, mesenteric lymph nodes; *PP*, Peyer's patches; *SFB*, segmented filamentous bacteria; *SI*, small intestine; ↓, reduced; ↑, increased; ↔, similar.

(Schneider et al., 2018). The ILC-3s (RORγT^{+} ILCs) seemed to be reduced in numbers in GF mice (Brugman et al., 2015; Sonnenberg and Artis, 2012). Among commensals, SFB colonization has been shown to activate ILC-3 and induce IL-22 production (Sano et al., 2015).

Neonatal mice that were transiently exposed to broad-spectrum antibiotics from birth showed a significant decrease in ILC-3 in the lamina propria. Reconstitution with mature microbiota results in an increase in ILC-3 and reduced susceptibility to sepsis (Niu et al., 2020). Likewise, the transfer of normal microbiota to neonatal mice born to antibiotic-treated mouse dams induced the production of IL-17 by ILC-3s in the intestine and restored IL-17-dependent resistance to neonatal sepsis (Deshmukh et al., 2014).

2.3.2.8.1 NK cells

NK cells are involved in host defense and tissue growth. The CD127^{-} CD103^{+} Eomes^{+} NK cells are the major cytotoxic ILC population in the small intestines of infants, and their absolute numbers gradually decrease during the first year of life with a simultaneous increase in Eomes^{+} T cells. Such dynamic changes in the infant intestinal lymphocyte compartment occur in the first year of life and may be due to exposure to large number of microbes and dietary changes (Sagebiel et al., 2019).

In GF mice the type I IFN-dependent priming of NK cells was reduced; therefore they exhibited impaired antiviral immunity. This could be reversed by colonization of GF mice with standard microbiota. The signals from the commensal microbiota primes mononuclear phagocytes for effective IFN-1 production, thus improving NK cell antiviral responses (Ganal et al., 2012).

2.3.2.9 B cells

B cells are one of the major effectors of host defense against infections and produce five different isotypes of functionally distinct antibodies: IgA, IgE, IgD, IgG, and IgM. In microbiota-depleted mice there were significantly reduced numbers of B lymphocytes in the intestines, peripheral blood, and bone marrow. Reconstitution with microbiota could reverse these effects (Ekmekciu et al., 2017; Josefsdottir et al., 2017).

The commensal microbiota has a crucial role in shaping the antibody repertoire in early life. It was found that apart from the bone marrow, postnatal B cell development also occurred in the lamina propria of mouse intestines; this is controlled by extracellular signals from commensal microbiota that influence gut immunoglobulin repertoires (Wesemann et al., 2013). Likewise, the antibacterial IgM^{+} and IgD^{+} B cells in the spleen and intestines increased following exposure to microbial symbionts (Chen et al., 2018).

Shortly after birth, bacterial colonization of the intestine begins, resulting in the induction of protective IgA antibodies. Symbiotic bacteria have been found to shape the production of secretory IgA. GF mice have only a few IgA-expressing B cells in the small intestine. This is mostly due to the reduced numbers of ILFs. Conventionalization of GF mice increases the ILF numbers and rescues IgA production by B cells in the small intestine (Gensollen et al., 2016; Hapfelmeier et al., 2010).

B cell memory is characterized by the persistence of elevated titers of specific antibodies and persistence of memory B cells. The antigen-induced proliferation and differentiation of memory B cells into short-lived plasma cells results in the production of increased levels of protective antibodies (Ochsenbein et al., 2000). But this mechanism required the persistence of antigen for a long time, which is not feasible. The B lymphocyte proliferation and differentiation in the absence of antigen are mediated by polyclonal stimuli, such as those derived from microbial products such as LPS or CpGoligonucleotides (Bernasconi et al., 2002; Krieg, 2002). In response to these microbial stimuli, all memory B cells continuously proliferate and differentiate to antibody-secreting cells. Thus optimal levels of plasma cells and serum antibodies could be maintained throughout a life span (Bernasconi et al., 2002).

IgE antibodies are produced by B cells, and these have a central role in the allergic response and immunity to pathogens (Gensollen et al., 2016). In GF mice the absence of microbial colonization or colonization with low-diversity microbiota in early life led to increased levels of IgE in adult mice. This is because B cells in GF mice undergo IgE class switch recombination at mucosal sites. It is evident that in the absence of commensals the immune regulatory pathways that normally maintain IgE at baseline levels are disrupted. Diverse microbial colonization has to take place during a critical time window in neonatal life in order to restrict the default immune pathways that result in excessive IgE levels (Fig. 2.4) (Cahenzli et al., 2013). These studies show an important role for the microbiota in shaping the B cell, IgA and IgE responses in early life.

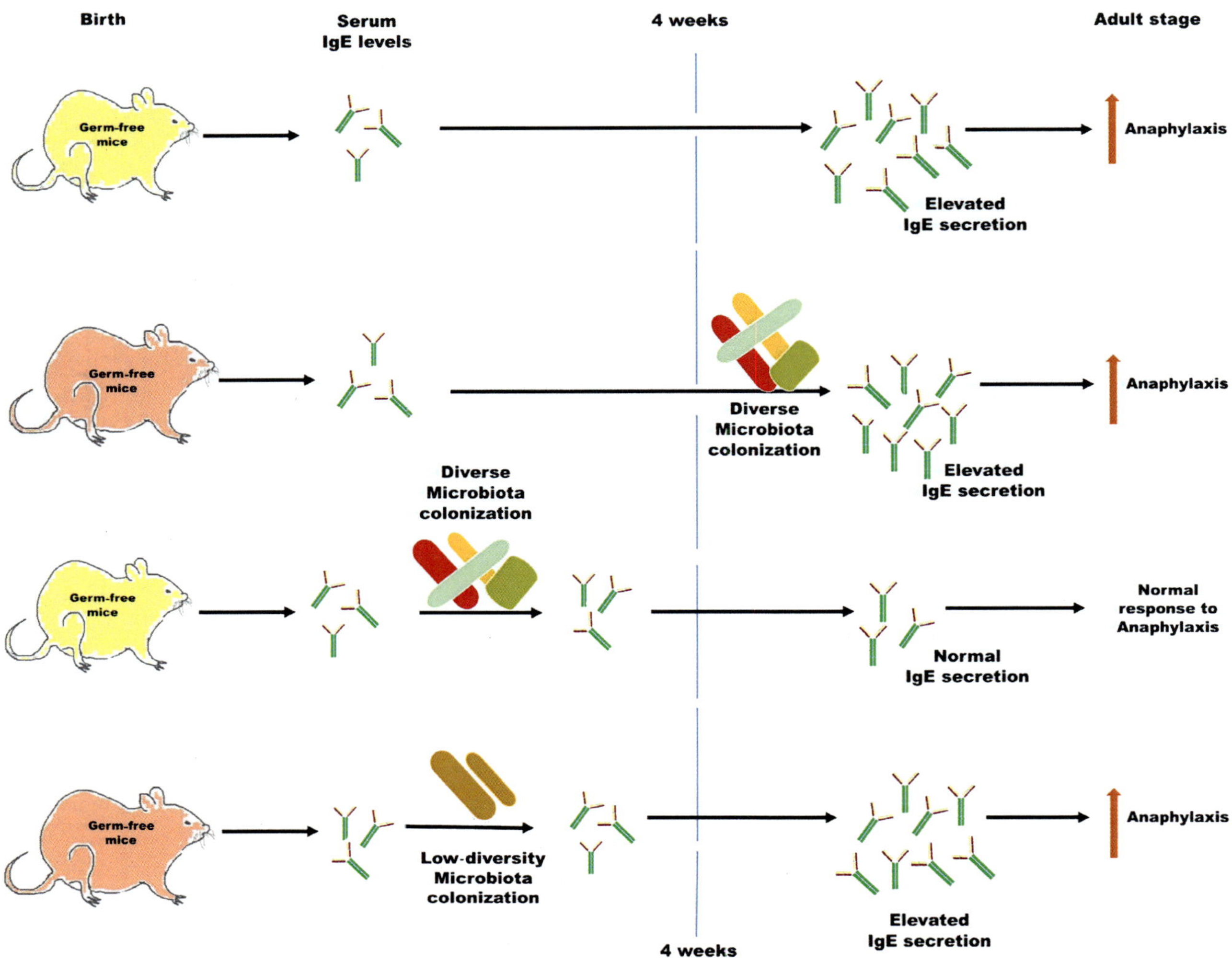

FIGURE 2.4 Exposure to a diversified microbiota within a crucial time window in early life is required to regulate the serum IgE levels.

2.3.3 Epithelial cells

Intestinal epithelial cells (IECs) act as a physical and biochemical barrier that segregates the host from the commensal bacteria and maintains intestinal homeostasis. IECs can detect and respond to microbial signals and promote the coordination of appropriate immune responses. Thus IECs influence the development and homeostasis of mucosal immune cells. The induction of intestinal epithelial tolerance is essential to provide protection from pathogen-induced epithelial damage in neonates. The establishment of tolerance in early life can be regulated by nonimmune cells. The establishment of tolerance is an active condition that is required for attaining intestinal mucosal homeostasis. For instance, TLR signaling is specifically downregulated during the first two weeks of life in mouse IECs but not afterwards (Chassin et al., 2010). This makes the IECs less responsive to TLR stimuli such as LPS. This neonatal tolerance depends on TLR-4 and the mode of delivery. IECs from caesarean-born neonates do not exhibit downregulated TLR signaling and are more susceptible to epithelial damage. This shows the importance of different types of microbial exposure in neonatal life for the establishment of epithelial tolerance (Gensollen et al., 2016).

Because the mammalian intestine houses trillions of bacteria, it is important to understand how symbiotic host-bacterial relationships are maintained without any aberrant immune responses. The secretory IECs maintain their barrier function by the production of mucins and antimicrobial peptides (Peterson and Artis, 2014). Antimicrobial peptides and innate immune effectors produced in the mucosal surfaces play an important role in protecting the host from pathogens and containment of the microbiota. Studies in GF and neonatal mice have shown that a few of these peptides, such as lysozyme and α-defensins, are produced without bacterial colonization. Some are produced in response to bacterial colonization such as angiogenins and RegIIIγ (a secreted C type lectin) (Salzman, 2014). RegIIIγ is an antibacterial lectin that binds peptidoglycan and has direct antimicrobial activity. It is required to maintain a 50-μm zone between the microbiota and the intestinal epithelium (Vaishnava et al., 2011). Microbial colonization of GF mice with standard microbiota elicits the epithelial expression of RegIIIγ. Monocolonization with *Bacteroides thetaiotaomicron* also resulted in the upregulated expression of RegIIIγ in the small intestines and colon (Cash et al., 2006). RegIIIγ expression can also be induced by LPS or flagellin (Kinnebrew et al., 2010; Vaishnava et al., 2008).

Resistin-like molecule β (RELM-β) is a recently described intestinal goblet-specific protein. The secretion of RELM-β is drastically suppressed in GF mice. The induction of GF mice into a conventional environment resulted in the upregulated expression and increased secretion of RELM-β (He et al., 2003).

2.4 Conclusion

The maturation of the immune system in early life is accomplished through bidirectional approaches in which the host immune system determines the microbial composition while at the same time the microbiota promotes the development and maturation of the neonatal immune system. The regulatory nature of the neonatal immune system and the effect of commensals in shaping the development of the immune system coordinately establish a stable host-microbiota relationship. These initial interactions have a long-term influence on host immunity and health. Disruptions in these interactions could lead to several abnormalities in human health. However, these abnormalities do not arise as a result of a single microbial product but rather result from the shifted balance state of an entire group. Maintenance of the host-microbial mutualism is of high importance in establishing proper health.

The maintenance of homeostasis and tolerance to environmental exposures during early life is determined by a number of immune mechanisms that occur during a short time frame in early life. Moreover, a wide array of studies have shown that there exists a window of opportunity in early life during which microbial exposures take place that have a lasting impact on the immunity of an individual throughout the lifetime. In fact, gestational colonization has been found to shape the postnatal immune development and to promote tolerance in early life. Most of the information on the role of the microbiota in immune development comes from animal model studies. Nevertheless, standardization of the microbiota within the various strains is not done. Apart from bacteria, other elements, such as bacteriophage, viruses, and fungi, are also part of the gut microbiota. These components and their role in the development of immunity should be explored further.

Likewise, more studies needs to be performed that focus on the role of the microbiota in immune development in the crucial neonatal period. Although the immediate effects of lactation and breastfeeding have been studied, their long-term consequences on the microbiota and immune development and the specific microbes and factors in breast milk that are responsible for them have not been explored enough. A healthier and more diverse microbiome will definitely lead to enhanced development of immunity and health.

Author disclosure statement

Conflict of interest: The authors have no conflict of interest to declare.
Contribution of authors: Both the authors were involved in the drafting, editing and reviewing of the manuscript.

References

Abt, M., Osborne, L., Monticelli, L., Doering, T., Alenghat, T., Sonnenberg, G., et al., 2012. Commensal bacteria calibrate the activation threshold of innate antiviral immunity. Immunity 37, 158–170.

Agüero, M., de, Ganal-Vonarburg, S., Fuhrer, T., Rupp, S., Uchimura, Y., Li, H., et al., 2016. The maternal microbiota drives early postnatal innate immune development. Science 351, 1296–1302.

Aidy, S., Hooiveld, G., Tremaroli, V., Bäckhed, F., Kleerebezem, M., 2013. The gut microbiota and mucosal homeostasis: colonized at birth or at adulthood, does it matter? Gut Microbes 4, 118–124.

An, D., Oh, S., Olszak, T., Neves, J., Avci, F., Erturk-Hasdemir, D., et al., 2014. Sphingolipids from a symbiotic microbe regulate homeostasis of host intestinal natural killer T cells. Cell 156, 123–133.

Ataíde, M., Kastenmüller, W., 2016. Innate like lymphocytes: moving at the frontline. Elife 5, e20516.

Atarashi, K., Nishimura, J., Shima, T., Umesaki, Y., Yamamoto, M., Onoue, M., et al., 2008. ATP drives lamina propria TH17 cell differentiation. Nature 455, 808–812.

Bachem, A., Makhlouf, C., Binger, K., Souza, D., de, Tull, D., Hochheiser, K., et al., 2019. Microbiota-derived short-chain fatty acids promote the memory potential of antigen-activated CD8 + T cells. Immunity 51, 285–297. e5.

Bain, C., Bravo-Blas, A., Scott, C., Perdiguero, E., Geissmann, F., Henri, S., et al., 2014. Constant replenishment from circulating monocytes maintains the macrophage pool in the intestine of adult mice. Nat. Immunol. 15, 929–937.

Belkaid, Y., Hand, T.W., 2014. Role of the microbiota in immunity and inflammation. Cell 157, 121–141.

Bernasconi, N., Traggiai, E., Lanzavecchia, A., 2002. Maintenance of serological memory by polyclonal activation of human memory B cells. Science 298, 2199–2202.

Bjerke, G.A., Wilson, R., Storrø, O., Øyen, T., Johnsen, R., Rudi, K., 2011. Mother-to-child transmission of and multiple-strain colonization by Bacteroides fragilis in a cohort of mothers and their children. Appl. Environ. Microb 77, 8318–8324.

Bowman, L.M., Holt, P.G., 2001. Selective enhancement of systemic Th1 immunity in immunologically immature rats with an orally administered bacterial extract. Infect. Immun. 69, 3719–3727.

Brown, E., Sadarangani, M., Finlay, B., 2013. The role of the immune system in governing host-microbe interactions in the intestine. Nat. Immunol. 14, 660–667.

Brown, R., Sequeira, R., Clarke, T., 2017. The microbiota protects against respiratory infection via GM-CSF signaling. Nat. Commun. 8, 1512.

Brugman, S., Perdijk, O., Neerven, J., van, Savelkoul, H., 2015. Mucosal Immune Development in Early Life: Setting the Stage. Arch. Immunol. Ther. Exp. (Warsz) 63, 251–268.

Cahenzli, J., Köller, Y., Wyss, M., Geuking, M., McCoy, K., 2013. Intestinal microbial diversity during early-life colonization shapes long-term IgE levels. Cell Host Microbe 14, 559–570.

Cash, H., Whitham, C., Behrendt, C., Hooper, L., 2006. Symbiotic bacteria direct expression of an intestinal bactericidal lectin. Science 313, 1126–1130.

Cervantes-Barragan, L., Chai, J., Tianero, M., Luccia, B., Ahern, P., Merriman, J., et al., 2017. *Lactobacillus reuteri* induces gut intraepithelial CD4 + CD8αα + T cells. Science 357, 806–810.

Chassin, C., Kocur, M., Pott, J., Duerr, C.U., Gütle, D., Lotz, M., et al., 2010. miR-146a mediates protective innate immune tolerance in the neonate intestine. Cell Host Microbe 8, 358–368.

Chen, Y., Chaudhary, N., Yang, N., Granato, A., Turner, J., Howard, S., et al., 2018. Microbial symbionts regulate the primary Ig repertoire. J. Exp. Med. 215, 1397–1415.

Chung, H., Pamp, S., Hill, J., Surana, N., Edelman, S., Troy, E., et al., 2012. Gut immune maturation depends on colonization with a host-specific microbiota. Cell 149, 1578–1593.

Constantinides, M.G., 2018. Interactions between the microbiota and innate and innate-like lymphocytes. J. Leukoc. Biol. 103, 409–419.

Crosby, C., Kronenberg, M., 2018. Tissue-specific functions of invariant natural killer T cells. Nat. Rev. Immunol. 18, 559–574.

Deshmukh, H., Liu, Y., Menkiti, O., Mei, J., Dai, N., O'Leary, C., et al., 2014. The microbiota regulates neutrophil homeostasis and host resistance to Escherichia coli K1 sepsis in neonatal mice. Nat. Med. 20, 524–530.

Ekmekciu, I., Klitzing, E.V., Fiebiger, U., Escher, U., Neumann, C., Bacher, P., et al., 2017. Immune responses to broad-spectrum antibiotic treatment and fecal microbiota transplantation in mice. Front. Immunol. 8, 397.

Elahi, S., Ertelt, J., Kinder, J., Jiang, T., Zhang, X., Xin, L., et al., 2013. Immunosuppressive CD71 + erythroid cells compromise neonatal host defence against infection. Nature 504, 158–162.

Emal, D., Rampanelli, E., Stroo, I., Butter, L.M., Teske, G.J., Claessen, N., et al., 2017. Depletion of gut microbiota protects against renal ischemia-reperfusion injury. J. Am. Soc. Nephrol. 28, 1450–1461.

Ennamorati, M., Vasudevan, C., Clerkin, K., Halvorsen, S., Verma, S., Ibrahim, S., et al., 2020. Intestinal microbes influence development of thymic lymphocytes in early life. Proc. Nat. Acad Sci. 117, 2570–2578.

Faith, J.J., Ahern, P.P., Ridaura, V.K., Cheng, J., Gordon, J.I., 2014. Identifying gut microbe-host phenotype relationships using combinatorial communities in gnotobiotic mice. Sci. Transl. Med. 6, 220ra11.

Fike, A.J., Kumova, O.K., Carey, A.J., 2019. Dissecting the defects in the neonatal CD8 + T-cell response. J. Leukoc. Biol. 106, 1051–1061.

Fujimura, K., Sitarik, A., Havstad, S., Lin, D., Levan, S., Fadrosh, D., et al., 2016. Neonatal gut microbiota associates with childhood multisensitized atopy and T cell differentiation. Nat. Med. 22, 1187–1191.

Fulde, M., Sommer, F., Chassaing, B., Vorst, K., Dupont, A., Hensel, M., et al., 2018. Neonatal selection by Toll-like receptor 5 influences long-term gut microbiota composition. Nature 560, 489–493.

Gaboriau-Routhiau, V., Rakotobe, S., Lécuyer, E., Mulder, I., Lan, A., Bridonneau, C., et al., 2009. The key role of segmented filamentous bacteria in the coordinated maturation of gut helper T cell responses. Immunity 31, 677–689.

Ganal, S., Sanos, S., Kallfass, C., Oberle, K., Johner, C., Kirschning, C., et al., 2012. Priming of natural killer cells by nonmucosal mononuclear phagocytes requires instructive signals from commensal microbiota. Immunity 37, 171–186.

Gensollen, T., Iyer, S., Kasper, D., Blumberg, R., 2016. How colonization by microbiota in early life shapes the immune system. Science 352, 539–544.

Gervassi, A., Lejarcegui, N., Dross, S., Jacobson, A., Itaya, G., Kidzeru, E., et al., 2014. Myeloid derived suppressor cells are present at high frequency in neonates and suppress in vitro T cell responses. PLoS One 9, e107816.

Gill, S., Pop, M., DeBoy, R., Eckburg, P., Turnbaugh, P., Samuel, B., et al., 2006. Metagenomic analysis of the human distal gut microbiome. Science 312, 1355–1359.

Gollwitzer, E., Saglani, S., Trompette, A., Yadava, K., Sherburn, R., McCoy, K., et al., 2014. Lung microbiota promotes tolerance to allergens in neonates via PD-L1. Nat. Med. 20, 642–647.

Gonzalez-Perez, G., Hicks, A., Tekieli, T., Radens, C., Williams, B., Lamousé-Smith, E., 2016. Maternal antibiotic treatment impacts development of the neonatal intestinal microbiome and antiviral immunity. J. Immunol. 196, 3768–3779.

Gury-BenAri, M., Thaiss, C., Serafini, N., Winter, D., Giladi, A., Lara-Astiaso, D., et al., 2016. The spectrum and regulatory landscape of intestinal innate lymphoid cells are shaped by the microbiome. Cell 166 (1231–1246), e13.

Hapfelmeier, S., Lawson, M., Slack, E., Kirundi, J., Stoel, M., Heikenwalder, M., et al., 2010. Reversible microbial colonization of germ-free mice reveals the dynamics of IgA immune responses. Science 328, 1705–1709.

He, W., Wang, M., Jiang, H., Steppan, C., Shin, M., Thurnheer, C., et al., 2003. Bacterial colonization leads to the colonic secretion of RELMβ/FIZZ2, a novel goblet cell-specific protein. Gastroenterology 125, 1388–1397.

He, Y.-M., Li, X., Perego, M., Nefedova, Y., Kossenkov, A., Jensen, E., et al., 2018. Transitory presence of myeloid-derived suppressor cells in neonates is critical for control of inflammation. Nat Med 24, 224–231.

Hergott, C., Roche, A., Tamashiro, E., Clarke, T., Bailey, A., Laughlin, A., et al., 2016. Peptidoglycan from the gut microbiota governs the lifespan of circulating phagocytes at homeostasis. Blood 127, 2460–2471.

Holt, P.G., Jones, C.A., 2000. The development of the immune system during pregnancy and early life. Allergy 55, 688–697.

Iliev, I., Funari, V., Taylor, K., Nguyen, Q., Reyes, C., Strom, S., et al., 2012. Interactions between commensal fungi and the C-type lectin receptor Dectin-1 influence colitis. Science 336, 1314–1317.

Ivanov, I., Atarashi, K., Manel, N., Brodie, E., Shima, T., Karaoz, U., et al., 2009. Induction of intestinal Th17 cells by segmented filamentous bacteria. Cell 139, 485–498.

Ivanov, I., Frutos, R., de, Manel, N., Yoshinaga, K., Rifkin, D., Sartor, B., et al., 2008. Specific microbiota direct the differentiation of IL-17-producing T-helper cells in the mucosa of the small intestine. Cell Host Microbe 4, 337–349.

Iwamura, C., Bouladoux, N., Belkaid, Y., Sher, A., Jankovic, D., 2017. Sensing of the microbiota by NOD1 in mesenchymal stromal cells regulates murine hematopoiesis. Blood 129, 171–176.

Josefsdottir, K., Baldridge, M., Kadmon, C., King, K., 2017. Antibiotics impair murine hematopoiesis by depleting the intestinal microbiota. Blood 129, 729–739.

Jung, J., Surh, C., Lee, Y., 2019. Microbial colonization at early life promotes the development of diet-induced CD8αβ intraepithelial T Cells. Mol. Cells 42, 313–320.

Khosravi, A., Yáñez, A., Price, J., Chow, A., Merad, M., Goodridge, H., et al., 2014. Gut microbiota promote hematopoiesis to control bacterial infection. Cell Host Microbe 15, 374–381.

Kinnebrew, M., Ubeda, C., Zenewicz, L., Smith, N., Flavell, R., Pamer, E., 2010. Bacterial flagellin stimulates Toll-like receptor 5-dependent defense against vancomycin-resistant Enterococcus infection. J Infect Dis 201, 534–543.

Knoop, K., Gustafsson, J., McDonald, K., Kulkarni, D., Coughlin, P., McCrate, S., et al., 2017. Microbial antigen encounter during a preweaning interval is critical for tolerance to gut bacteria. Sci. Immunol. 2, eaao1314.

Koay, H.F., Gherardin, N., Enders, A., Loh, L., Mackay, L., Almeida, C., et al., 2016. A three-stage intrathymic development pathway for the mucosal-associated invariant T cell lineage. Nat. Immunol. 17, 1300–1311.

Köhler, A., Delbauve, S., Smout, J., Torres, D., Flamand, V., 2020. Very early-life exposure to microbiota-induced TNF drives the maturation of neonatal pre-cDC1. Gut gutjnl 2019–319700.

Krieg, A., 2002. CpG motifs in bacterial DNA and their immune effects. Annu. Rev. Immunol. 20, 709–760.

Kubinak, J., Petersen, C., Stephens, Z., Soto, R., Bake, E., O'Connell, R., et al., 2015. MyD88 signaling in T cells directs IgA-mediated control of the microbiota to promote health. Cell Host Microbe 17, 153–163.

Lino, A., Mohr, E., Demengeot, J., 2013. Naturally secreted immunoglobulins limit B1 and MZ B-cell numbers through a microbiota-independent mechanism. Blood 122, 209–218.

Lyons, A., O'Mahony, D., O'Brien, F., MacSharry, J., Sheil, B., Ceddia, M., et al., 2010. Bacterial strain-specific induction of Foxp3 + T regulatory cells is protective in murine allergy models. Clin. Exp. Allergy 40, 811–819.

Macpherson, A., Harris, N., 2004. Interactions between commensal intestinal bacteria and the immune system. Nat. Rev. Immunol. 4, 478–485.

Mazmanian, S., Liu, C., Tzianabos, A., Kasper, D., 2005. An immunomodulatory molecule of symbiotic bacteria directs maturation of the host immune system. Cell 122, 107–118.

Nabhani, Z., Dulauroy, S., Marques, R., Cousu, C., Bounny, S., Déjardin, F., et al., 2019. A weaning reaction to microbiota is required for resistance to immunopathologies in the adult. Immunity 50, 1276–1288. e5.

New, S., Dizon, B., Fucile, C., Rosenberg, A., Kearney, J., King, G., 2020. Neonatal exposure to commensal-bacteria-derived antigens directs polysaccharide-specific B-1 B cell repertoire development. Immunity 53, 172–186. e6.

Niess, J., Leithäuser, F., Adler, G., Reimann, J., 2008. Commensal gut flora drives the expansion of proinflammatory CD4 T cells in the colonic lamina propria under normal and inflammatory conditions. J. Immunol. 180, 559–568.

Niu, X., Daniel, S., Kumar, D., Ding, E., Savani, R., Koh, A., et al., 2020. Transient neonatal antibiotic exposure increases susceptibility to late-onset sepsis driven by microbiota-dependent suppression of type 3 innate lymphoid cells. Sci. Rep. 10, 12974.

Ochoa-Repáraz, J., Mielcarz, D., Ditrio, L., Burroughs, A., Foureau, D., Haque-Begum, S., et al., 2009. Role of gut commensal microflora in the development of experimental autoimmune encephalomyelitis. J. Immunol. 183, 6041–6050.

Ochsenbein, A., Pinschewer, D., Sierro, S., Horvath, E., Hengartner, H., Zinkernagel, R., 2000. Protective long-term antibody memory by antigen-driven and T help-dependent differentiation of long-lived memory B cells to short-lived plasma cells independent of secondary lymphoid organs. Proc. Natl. Acad. Sci. U. S. A. 97, 13263–13268.

Olszak, T., An, D., Zeissig, S., Vera, M., Richter, J., Franke, A., et al., 2012. Microbial exposure during early life has persistent effects on natural killer T cell function. Science 336, 489–493.

Panda, S., Colonna, M., 2019. Innate lymphoid cells in mucosal immunity. Front Immunol. 10, 861.

Peterson, L.W., Artis, D., 2014. Intestinal epithelial cells: regulators of barrier function and immune homeostasis. Nat. Rev. Immunol. 14, 141–153.

Petursdottir, D., Nordlander, S., Qazi, K., Carvalho-Queiroz, C., Osman, O., Hell, E., et al., 2017. Early-life human microbiota associated with childhood allergy promotes the T helper 17 axis in mice. Front. Immunol. 8, 1699.

Probert, C., Williams, A., Stepankova, R., Tlaskalova-Hogenova, H., Phillips, A., Bland, P., 2007. The effect of weaning on the clonality of αβ T-cell receptor T cells in the intestine of GF and SPF mice. Dev. Comp. Immunol. 31, 606–617.

Romani, L., 1997. The T cell response against fungal infections. Curr. Opin. Immunol. 9, 484–490.

Rosado, M., Aranburu, A., Scarsella, M., Cascioli, S., Giorda, E., Chierico, F., et al., 2018. Spleen development is modulated by neonatal gut microbiota. Immunol. Lett. 199, 1–15.

Round, J., Mazmanian, S., 2009. The gut microbiota shapes intestinal immune responses during health and disease. Nat. Rev. Immunol. 9, 313–323.

Sagebiel, A., Steinert, F., Lunemann, S., Körner, C., Schreurs, R., Altfeld, M., et al., 2019. Tissue-resident Eomes + NK cells are the major innate lymphoid cell population in human infant intestine. Nat. Commun. 10, 975.

Salzman, N.H., 2014. The role of the microbiome in immune cell development. Ann. Allergy Asthma Immunol 113, 593–598.

Sanidad, K., Zeng, M., 2020. Neonatal gut microbiome and immunity. Curr. Opin. Microbiol. 56, 30–37.

Sano, T., Huang, W., Hall, J., Yang, Y., Chen, A., Gavzy, S., et al., 2015. An IL-23R/IL-22 circuit regulates epithelial serum amyloid a to promote local effector Th17 responses. Cell 163, 381–393.

Schneider, C., O'Leary, C., Moltke, J., von, Liang, H.-E., Ang, Q., Turnbaugh, P., et al., 2018. A metabolite-triggered tuft. cell-ILC2 circuit drives small intestinal remodeling. Cell 174, 271–284. e14.

Sonnenberg, G., Artis, D., 2012. Innate lymphoid cell interactions with microbiota: implications for intestinal health and disease. Immunity 37, 601–610.

Teng, F., Klinger, C., Felix, K., Bradley, P., Wu, E., Tran, N., et al., 2016. Gut microbiota drive autoimmune arthritis by promoting differentiation and migration of peyer's patch T follicular helper cells. Immunity 44, 875–888.

Thackray, L., Handley, S., Gorman, M., Poddar, S., Bagadia, P., Briseño, C., et al., 2018. Oral antibiotic treatment of mice exacerbates the disease severity of multiple flavivirus infections. Cell Rep 22, 3440–3453. e6.

Torow, N., Yu, K., Hassani, K., Freitag, J., Schulz, O., Basic, M., et al., 2015. Active suppression of intestinal CD4 + TCRαβ + T-lymphocyte maturation during the postnatal period. Nat. Commun. 6, 7725.

Treiner, E., Duban, L., Bahram, S., Radosavljevic, M., Wanner, V., Tilloy, F., et al., 2003. Selection of evolutionarily conserved mucosal-associated invariant T cells by MR1. Nature 422, 164–169.

Turroni, F., Peano, C., Pass, D., Foroni, E., Severgnini, M., Claesson, M., et al., 2012. Diversity of bifidobacteria within the infant gut microbiota. PLoS One 7, e36957.

Vaishnava, S., Behrendt, C.L., Ismail, A.S., Eckmann, L., Hooper, L.V., 2008. Paneth cells directly sense gut commensals and maintain homeostasis at the intestinal host-microbial interface. Proc. Natl. Acad. Sci. U S A. 105, 20858–20863.

Vaishnava, S., Yamamoto, M., Severson, K., Ruhn, K., Yu, X., Koren, O., et al., 2011. The antibacterial lectin RegIIIgamma promotes the spatial segregation of microbiota and host in the intestine. Science 334, 255–258.

Wesemann, D., Portuguese, A., Meyers, R., Gallagher, M., Cluff-Jones, K., Magee, J., et al., 2013. Microbial colonization influences early B-lineage development in the gut lamina propria. Nature 501, 112–115.

Wingender, G., Stepniak, D., Krebs, P., Lin, L., McBride, S., Wei, B., et al., 2012. Intestinal microbes affect phenotypes and functions of invariant natural killer T cells in mice. Gastroenterology 143, 418–428.

Yu, J., Khodadadi, H., Malik, A., Davidson, B., Salles, É., Bhatia, J., et al., 2018. Innate immunity of neonates and infants. Front. Immunol. 9, 1759.

Zaghouani, H., Hoeman, C., Adkins, B., 2009. Neonatal immunity: faulty T-helpers and the shortcomings of dendritic cells. Trends Immunol 30, 585–591.

Zhang, D., Chen, G., Manwani, D., Mortha, A., Xu, C., Faith, J., et al., 2015. Neutrophil ageing is regulated by the microbiome. Nature 525, 528–532.

Chapter 3

Impact of the gut microbiome on human health and diseases

Samudra Prosad Banik[1], Shalini Sehgal[2], Rituparna Banik Ghosh[3] and Ahana Das[1]

[1]*Department of Microbiology, Maulana Azad College, Kolkata, West Bengal, India,* [2]*Department of Food Technology, Bhaskaracharya College of Applied Sciences, University of Delhi, New Delhi, India,* [3]*Department of Physiology, Bhairab Ganguly College, Kolkata, West Bengal, India*

3.1 Introduction

3.1.1 Definition of microbiome

The entire pool of microorganisms that inhabit the human body, including bacteria, fungi, yeast, viruses, and archaea, constitute the human microbiome. About 10,000 microbial species occupy diverse niches in the human body. However, bacteria far outnumber all other microbes in terms of both diversity and cumulative number (Qin et al., 2010). Over the last few decades, concerted efforts by scientific communities have unearthed the diverse facets of symbiotic association of the bacterial life with human body. This has provided a gargantuan volume of information pertaining to the microbial genome and proteome (Peterson et al., 2009) with simultaneous inflation of the corresponding databases (Qin et al., 2010). In spite of so much collective effort, deciphering the microbiota has been a daunting task for the scientific community. This has been attributable primarily to the difficulty in sampling or culturing of these tiny life forms from diverse microenvironments of the body. However, the introduction of phylogenetics-based culture-independent approaches has enabled partially bypassing or overcoming most of the technological bottlenecks (Pace, 1997). There has been no denying that the human gene catalog and the diversity of the human genome have been rendered insignificant in comparison to the microbial genome treasure trove. The entire microbiota of human body consists of about 1%–3% of total body mass, which is close to the mass of the brain. However, in spite of rapid advances in technology, we do not yet know or understand a significant part of the bacterial diversity that remains deep seated within the human body.

3.1.2 Uneven distribution of microbiota across various sites and organs of the body: reasons and implications

Microbes are ubiquitous throughout the human body. They are found even in regions that were previously believed to be sterile, such as the circulatory system (Proal et al., 2014). However, their distribution is significantly higher on the linings of external and internal surfaces, namely, the gastrointestinal tract, skin, saliva, oral mucosa, and conjunctiva. Bacterial density is particularly high in the mouth cavity, including the saliva and dental plaques. However, owing to the smaller volumes of these two microniches, the effective number of bacteria is negligible (about 1%) in comparison to the gut. The stomach and the anterior small intestine have relatively low bacterial densities (10^3–10^4 bacteria per gram or per milliliter of luminal fluid in mice), owing to the acidic pH of the stomach coupled with faster unabated flow of fluid and nutrients (Tannock, 1995). These regions are dominated by acid-tolerant *Lactobacilli* and *Streptococci*. The distal portion of the small intestine or the ileum has higher bacterial density (10^8 cells per gram) with commensurately greater bacterial diversity. However, the region of highest abundance of microbial cells, which essentially constitutes the microbial treasure trove of the gut, is the colon, which hosts more than 400 different known species with a probably still greater number of unknown residents. The colon has a staggering bacterial density of about 10^{13} cells/mL (Sender et al., 2016). Concentration of anaerobes is highest in the small intestine with distribution of species such as *Bacteroides*, *Bifidobacteria*, *Fusobacteria*, and *Peptostreptococci*, each group being present at an

Microbiome, Immunity, Digestive Health and Nutrition. DOI: https://doi.org/10.1016/B978-0-12-822238-6.00026-1

approximate density of 10^9 cells/g. In contrast, aerobes and facultative anaerobes such as *Lactobacilli* and *Enterobacter* are present at moderate concentrations of 10^6-10^8 cells/g.

3.2 The gut as the most significant niche of the human microbiome

The gut outpopulates all other organs and tissue spaces of the body in terms of both number and diversity of microbial cells, encompassing the three domains of life with at least trillions of cells. Hence it is known as the microbiodiversity hotspot of the body. Most of the studies that have been carried out on the diversity of gut microbiota have indicated a long-term stability in their taxonomic composition (David et al., 2014a). The major determinants of the gut microflora are the autochthonous microbes that seem to reside on the epithelium of colonic mucosa, unlike the allochthonous microbes, which transiently pass the lumen as part of the digesta (Schnabl and Brenner, 2014). Although a significant part of the gut microbiota is inherited, other environmental factors related to diet, drugs, and anthropometric measures are larger determinants of microbiota composition (Goodrich et al., 2014; Rothschild et al., 2018), and these differ according to the locations, ages, sexes, races, and diets of the hosts (Arumugam et al., 2011; Wu et al., 2011; Hollister et al., 2014). Additionally, the gut microbes have experienced quite a few changes in their genome due to transfer and acquisition of genes from the accessory genome pool of other intruding bacteria. The gut microflora of a healthy adult individual is chiefly dominated by the genera *Bacteroidetes*, *Firmicutes*, *Actinobacteria*, *Proteobacteria*, *Prevotella*, *Ruminococcus*, *Bifidobacterium*, *Streptococcus*, *Enterobacteriaceae*, *Enterococcus*, *Clostridium*, *Lactobacillus*, and *Akkermansia* (Pilmis et al., 2020; Ley et al., 2008; Tap et al., 2009). Among these, *Bacteroides*, *Bifidobacterium*, *Streptococcus*, *Enterobacteriaceae*, *Enterococcus*, *Clostridium*, *Lactobacillus*, and *Ruminococcus* are mostly lumen associated and therefore can be detected in the stool; others are predominant members of the mucosa (Swidsinski et al., 2005). Although *Escherichia coli* is a normal resident of the gut, an increase in the relative abundance of *Proteobacteria* such as *E. coli* and *Klebsiella* sp. is taken as a cardinal sign of dysbiosis in the body.

3.2.1 Establishment of the gut microbiome: journey from womb to adulthood

Most of the current available reports indicate that a developing womb is free from microbial cells until birth. Therefore it is reasonable to expect that neonates are generally sterile until birth. However, contrasting viewpoints suggest that the amniotic fluid of mothers in preterm labor is enriched with a large and diverse spectrum of bacterial rDNA (DiGiulio et al., 2008). Therefore considering the consumption of 400–500 mL of amniotic fluid per day by a term baby, it is quite unlikely that the baby does not get his or her first share of microbes until delivery, while passing through the vagina or from direct skin contact with the mother (if the delivery is by caesarean section) (Fig. 3.1). The first few genera to colonize the baby's intestine are probably *Lactobacilli*, *Prevotella*, and *Sneathia* (Dominguez-Bello et al., 2010; Reid et al., 2011), which live in the mother's vagina and settle in the baby's intestine to help in digestion. If the baby delivered by caesarean section, *Lactobacilli* will not immediately become part of the baby's microbiota, which will be made up mostly by bacteria from the mother's skin and the baby's environment. In fact, if delivered via caesarean section, the baby receives a significantly smaller share of the mother's microbiota. This results in the baby being in an immunocompromised state (Bach, 2002), owing to the early dependence of immune system development in newborns on the interaction with gut microbiota (Eberl and Lochner, 2009). These differences in a baby's microbiota, resulting from the type of birth that the baby experienced, remains until 12–24 months of age. Newborns also acquire bacteria from the skin of the nurses and medical doctors and the environment in which they live. As soon as external feeding starts, the newborns also start receiving microbes from their food source. During the first few days, the type of the microorganisms that colonize their intestines will be different, depending on whether the baby is consuming breast milk or formula. Prebiotics that are present in breast milk will support increased proliferation of *Bifidobacteria* and *Lactobacilli*, whereas infants receiving formula milk will be colonized more by *Enterococci* and *Enterobacteria*. The acquisition of microbes continues exponentially as babies gradually grow, depending upon the solid food they start ingesting and their daily activities, including crawling on floor, licking toys, and many other sources. It has been postulated that that the gut microflora stays highly dynamic till about 3 years of age, after which it becomes more or less stable.

3.2.2 Heritable component of the gut microbiome

Although gut microbiomes can differ markedly in diversity across adults (Qin et al., 2010), individuals in the same family often have similar profiles of gut microbiota (Tims et al., 2013). This similarity owes largely to the genetic similarity of members of the same family (Spor et al., 2011; Goodrich et al., 2014) and partly to similarity in external contributors

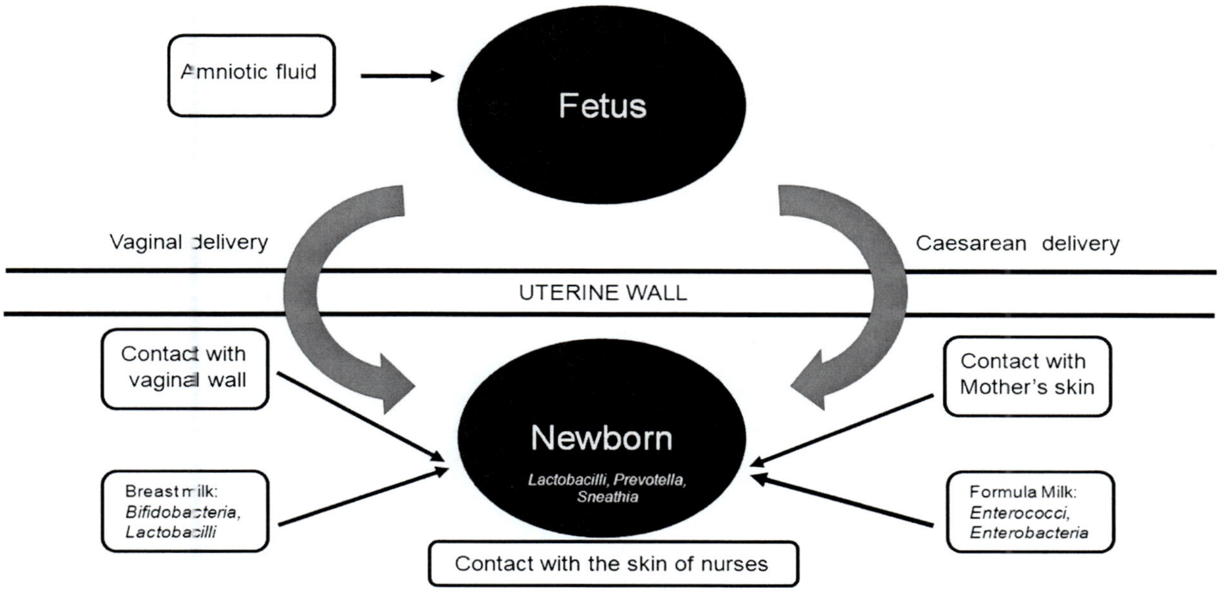

FIGURE 3.1 Chief sources of acquisition of microbes by the fetus and the newborn.

such as food habits (David et al., 2014b). The strong dependence of the gut microbiota constitution on the genetic makeup of individuals has also been shown by a study involving monozygotic twins (Zoetendal et al., 2001). Identification of this heritable component has been advantageous particularly for early prognosis and detection of risk factors associated with certain cardiometabolic diseases arising out of excessive visceral fat accumulation (Beaumont et al., 2016).

3.2.3 Role of diet and other environmental factors in shaping the gut microbiome

Dietary habit is one of the key factors in governing the composition of the microbiota. Most of the studies that have been carried out on the modulating factors of gut microbiota have been focused on diets, since this has been the most promising approach for therapeutic intervention (Flint et al., 2015). Until lately, a major part of the gut microbial community was unknown; however, the advent of state-of-the-art sequencing technologies has immensely facilitated the fathoming of the gut microbial communities and how our diet influences their colonization. Dietary habits play a key role in determining the gut microbiota starting from early childhood. Food components interact with microbes, directly promoting or inhibiting their growth and consequently altering the colon's microenvironment. The food source is also responsible for indirectly mediating the development of host's immune response. A gross imbalance in dietary constituents is responsible in most cases of disruption of the healthy host-microbiome interface and triggers the immune system to initiate an inflammatory response. Studies have revealed significant differences in gut microbial populations between subjects fed with a high percentage of meat (a typical Western diet) with those fed with a diet that is rich in fibers (Yatsunenko et al., 2012). Depending on the diet, growth of a specific strain of microbes may be promoted, which may alter the conditions of the intestinal lumen, including the pH. This in turn can lead to colonization by a pathogenic flora. Similarly, diets that are rich in fat support the development of proinflammatory gut microbiota and consequently increased levels of lipopolysaccharides in circulation (Bibbò et al., 2016). Thus a thorough understanding of the dependence of the gut microflora on the food consumption pattern of individuals can help us immensely in predicting their cardiovascular health for timely therapeutic intervention.

3.3 Physiological roles of the gut microbiome

3.3.1 Gut Microbiota in metabolism

A diverse flora of bacteria is established in the gut of a person right from birth throughout the entire period of growth (Rogier et al., 2014). Most of these eventually develop beneficial relationships with the host. These organisms engage in a lot of functions such as metabolism of certain sugars which human cells cannot degrade, synthesis of certain essential nutrients, building up a defense mechanism against assist in the development of the intestinal architecture (Round and Mazmanian, 2009). Often, dietary fibers contain sugar components which cannot be handled by cells of the stomach and small intestine. Xyloglucans, which are commonly present in vegetables, can be metabolized by a specific *Bacteroides* species (Larsbrink et al., 2014). Other nondigestible fibers such as fructooligosaccharides are taken care of by beneficial microbes, such as *Lactobacillus* and *Bifidobacterium* (Goh and Klaenhammer, 2015). Mucous cells of the intestine are normally capable of digesting a small amount of lactose. However, certain individuals stop producing this lactase after 4 years of age and are said to be lactose intolerant. Irrespective of lactose intolerance, the bulk of lactose passes on to the large intestine, where it is metabolized by certain members of *Lactobacillus*, such as *L. bulgaricus*. However, fermentation of lactose by these bacteria often leads to flatulence and in more extreme cases to diarrhea. More important, in general, fermentation of nondigestible carbohydrates arising from dietary fibers leads to the production of short chain fatty acids (SCFAs).

Gut microbiota play a significant role in lipid and protein homeostasis and synthesize a few key vitamins that are essential for the body (Morowitz et al., 2011). The healthy gut microbiota produces around 50–100 mmol per liter per day of SCFAs, including acetic, propionic, and butyric acids. These compounds, which are synthesized ubiquitously by most of the resident microbes but not the human cells, protect the host from a plethora of threats by virtue of their anticarcinogenic and antiinflammatory properties. Butyrates bring about apoptosis of colon cancer cells and also cause gluconeogenesis in the cells of the small intestine, thus helping in energy homeostasis (De Vadder et al., 2014). It also significantly increases oxygen usage of the epithelial cells by promoting beta-oxidation in them. This leads to a state of hypoxia and regulates the oxygen balance in the gut, preventing dysbiosis of the gut microbiota (Byndloss et al., 2017). But most significantly, these molecules serve as the primary source of fuel for the colonic cells (Duncan et al., 2009).

Several bacterial species belonging to the phylum *Firmicutes* and the order *Clostridiales* produce butyrate in significant amounts in the colon (Louis and Flint, 2017). Butyrate production needs thiamine and riboflavin. Precursor synthesis of these two vitamins significantly burdens the cellular energy reserve; therefore the gut colonizers themselves synthesize and share these precursors for optimal production and utilization of the vitamins (Belzer et al., 2017). Other vitamin B types are also produced by several lactic acid bacteria species such as *Lactococcus lactis*, *Lactobacillus gasseri*, and *Lactobacillus reuteri* as well as a few members of the genus *Bifidobacterium* (e.g., *B. adolescentis*). A few species such as *Enterobacter agglomerans*, *Serratia marcescens*, and *Enterococcus faecium* also produce vitamin K, chiefly in the form of menaquinone (vitamin K_2) (Cooke et al., 2006). In fact, the gut microbiota has been shown to synthesize and deliver a multitude of vitamins to the human host, including folates, vitamin K, biotin, riboflavin (vitamin B2), and cobalamin (vitamin B12).

Other important SCFAs that are produced by gut microbes are propionate and acetate. Propionate regulates gluconeogenesis in the liver and interacts with the gut fatty acid receptors to control satiety signaling (De Vadder et al., 2014). Acetate, the most abundant SCFA, is indispensable for the growth of other bacteria. After synthesis it is transported to the peripheral tissues and deployed for cholesterol metabolism and lipogenesis. Studies suggest that it also has a significant role in the regulation of appetite (Frost et al., 2014). Results fromrandomized controlled trials have suggested that individuals who have higher production of SCFAs have lower incidences of diet-induced obesity (Lin et al., 2012) and reduced insulin resistance (Zhao et al., 2018). The enzymes that are secreted by the gut community produce unconjugated and secondary bile acids from bile acid metabolism. These are in turn involved in important host signaling pathways. SCFAs are also involved in a plethora of immune responses (Table 3.1).

3.3.2 Gut microbiota in boosting immune response: role of short-chain fatty acids and cross-talk with diet

E. coli and other gram-negative bacteria produce lipopolysaccharides in the gut (Balagopal et al., 2008; Brenchley et al., 2006; Hofer et al., 2010); they are carried with chylomicrons (Ghoshal et al., 2009) to the blood for delivery across various cell types. Lipopolysaccharides are known to promote innate immunity across the entire animal kingdom. The notion has been supported by the observation that completely germ-free animals have significant defects in the formation of lymphoid structures, including gut-associated lymphoid tissue (GALT) (Madsen et al., 1999), with underdeveloped spleen and lymph nodes as well as abnormal cytokine levels (Macpherson and Harris, 2004). These animals also showed lower levels of antibody titer when challenged with a particular antigen. Therefore the gut microbiota is indispensable for the maturation and normal functioning of immune cells (Bouskra et al., 2008). Deeper insights have revealed that the hematopoietic and nonhematopoietic cells of the innate immune system can recognize the microbial metabolites and convert them into appropriate physiological responses (Thaiss et al., 2016) to promote the normal development of humoral and cellular mucosal immune systems (Cebra, 1999). Butyrate and propionate are known to induce the differentiation of regulatory T cells by inhibiting histone deacetylase. In addition, butyrate also prevents expression of ICAM-1 and other lymphocyte associated antigens thus suppressing the activation of antigen-presenting cells, cytokine production, stimulate chemotaxis of neutrophils, immune cell migration and COX-2 upregulation. Collectively, these lead to elicitation of the antiinflammatory pathway. Thus butyrate mediates colorectal cancer suppression. SCFAs also mediate the immune response via activation of G protein–coupled receptors on immune cells and regulate production of cytokines and prostaglandins. In addition, the SCFAs help the body to achieve a balanced helper T cell response, which gives protection against the onset of various critical illnesses, such as *Helicobacter pylori*–induced gastritis, Crohn's diseases, and other atopic diseases. However, not all microbiota have health benefits. Some induce inflammation under certain conditions. Studies have revealed that legume-rich foods induce butyric acid production and hence promote an antiinflammatory response; on the other hand, lipid-rich foods lead to elevated levels of inflammatory response mediators such as monocyte chemoattractant protein 1. This in turn causes infiltration of adipose tissue by macrophages and subsequent rise in levels of proinflammatory markers such as the C-regulatory protein and tumor necrosis factor alpha.

3.3.3 The role of the gut microbiome in the development of gut-associated lymphoid tissue

The GALT, comprising the lamina propria, the intraepithelial lymphocytes, the mesenteric lymph nodes, the Peyer's patches (PPs), and the isolated lymphoid follicles, is the largest mass of lymphoid cells. It constitutes a second layer of mechanical barrier along with the intestinal epithelial wall against invasion by external factors. The GALT is associated with a host of critical immune functions and is the mainstay of the intestinal immune response. Secretory-immunoglobulin A (S-IgA) constitutes the main arsenal in this regard (Brandtzaeg, 2010). S-IgA is a dedicated surveillance molecule that sequesters pathogens and prevents them from adsorbing to the lumen epithelial cells. S-IgA antibody also helps in colonization of the gut microflora. PPs, the major type of GALT, are found chiefly in the mucosa

TABLE 3.1 Physiological functions of different short-chain fatty acids produced by gut microbiota.

SCFA	Producer bacteria	Physiological role	References
Butyric acid	*Clostridium* sp.	Promotes Treg development in the colon	Atarashi et al. (2011)
Propionic acid	*Akkermansia muciniphila*	Key propionate producing mucin-degrading organisms	Derrien et al. (2004)
Acetate	*Bifidobacteria* sp.	Protects the host from entero pathogenic infection via carbohydrate transporters	Fukuda et al. (2012)
Formic acid	*Clostridium ljungdahlii*	Formic acid formation at elevated pressures of carbon dioxide and hydrogen	Oswald et al. (2018)
Valeric acid	*Mobilicoccus massiliensis*	Valeric acid protects against radiation injuries	Li et al. (2020)
I-caproic acid	*Clostridium difficile*	Causes allergy in infants	Böttcher et al. (2000)

and submucosa of the small intestine. In humans there are hundreds of such regions. Each PP is composed of several B cell–rich follicles surrounded by a meshlike structure consisting of T cells known as the interfollicular region. The interfollicular regions constitute the site of initiation of the S-IgA response where intestinal IgA responses are initiated and regulated by cytokines such as interleukin 4 (IL-4), IL-6, IL-21, and transforming growth factor beta (TGF-β) as well as cell-cell interactions between dendritic cells, T cells, and B cells. As has been stated already, the gut commensals play a significant role in the development of functioning of the lymphoid-associated immune system. Thus they are also essential for the expansion of the gut lymphoid tissue. When bacteria belonging to the gut microflora are introduced into germ-free animals, the GALT is seen to expand at a prolific rate. A recent study in this regard reported that *Bacteroides fragilis*, a common colonic gram-negative organism, can mediate the expansion of the GALT via a specific bacterial polysaccharide. Another unique member of the GALT community is the *Alcaligenes* that live inside the PPs. It is thought that *Alcaligenes* promotes the secretion of IgA via the upregulation of specific cytokines from dendritic cells (Kunisawa and Kiyono, 2012).

3.3.4 Gut microbes as a defense mechanism against pathogens

The members of a healthy intestinal microflora together put up a stiff barrier against invasion by pathogens through a concerted mechanism known as colonization resistance (Pilmis et al., 2020). The involvement of gut microbiota in fighting enteric pathogens was first described in 1954 by Bohnhoff et al. (1954), who showed that oral administration of streptomycin could disrupt the gut microbiota of mice and make them susceptible to *Salmonella enterica* infection. The direct mechanisms of colonization resistance involve competition for common resources such as mucins and dietary complex carbohydrates, elimination of pathogens by production of bacteriocins, or cell wall active bactericidal polypeptides. Additionally, type VI secretion systems possessed by certain genera can help to transport bacterial toxins into the body of pathogens (Russell et al., 2011). Evidence has also suggested that microbes can collectively fend off pathogens with enhanced vigor after the initial attack, thus resembling immune memory. In a very recent study, Stacy et al. deciphered the basis of this incredible response by the gut bacteria (Stacy et al., 2021). Bile acids are used by certain gut bacteria, resulting in the production of taurine, which can serve as energy sources for other bacteria. The metabolism of taurine in turn releases hydrogen sulfide, which inhibits cytochrome oxidase and arrests aerobic respiration. Stacy and her team found out that there is an increased expression of genes for taurine metabolism in certain gut microbes after initial attack by the pathogen; the enhanced production of hydrogen sulfide subsequently arrests the growth and respiration of the invading pathogens.

3.3.5 Role in cognitive development: the gut-brain axis

The gut-brain axis (GBA) is a unique and complex network of interactions between the central nervous system and the peripheral intestinal cells (Rhee et al., 2009). Not surprisingly, the gut microbiome is involved in a majority of these two-way communication systems (Fig. 3.2). Signaling between the gut microbiota and the CNS is mediated by a culmination of neural, endocrine, immune, and humoral routes and also involves the autonomic nervous system, the enteric nervous system (ENS), and the hypothalamic pituitary adrenal (HPA) axis. The sympathetic and parasympathetic limbs relay the afferent signals from the lumen to the CNS via the enteric, spinal, and vagal pathways as well as the efferent signals from CNS back to the intestinal wall. The HPA axis is a part of the limbic system that is responsible for mediating adaptive responses to any environmental stress by regulating the release of cortisol hormone from the adrenal gland. Recent findings suggest that the enteric microbiota interacts not only with the intestinal cells but also with cells of the CNS through neuroendocrine and metabolic pathways. An absence of gut microbial colonization has been shown to be associated with abnormality in neurotransmitter expression in both the CNS and the ENS. In germ-free animals, significantly decreased emotional behavior along with memory dysfunction has been observed (Gareau et al., 2011). This has been primarily attributed to different concentrations of a small molecule known as brain-derived neurotrophic factor at various regions of the brain. In addition, the body serotonin levels are altered, suggesting direct involvement of the colonic microflora in the serotoninergic system. Administration of *Lactobacillus rhamnosus* has been found to result in reduced stress-induced secretion of cortisol as well as decreased levels of anxiety and depression. Studies indicate that the gut microbiota communicates with the CNS primarily through the vagus nerve (Bravo et al., 2011). Direct evidence in support of this notion has been obtained in many cases; administration of *Bifidobacterium longum* just before induction of chronic colitis in an animal model resulted in partial alleviation of colitis-induced anxiety, but similar results were not obtained in a vagotomized animal (Bercik et al., 2011). The microbes may also communicate with the brain through the ENS, which is a target of bacterial metabolites (primarily the SCFAs) and influences the memory and learning process (Stefanko et al., 2009). The enteroendocrine cells of the gut play a crucial role in sensing the concentration of nutrients and convey the same to the brain via the secretion of peptide hormones. This also represents a crucial access point for the microbes to enter the GBA via modulation of nutrient availability. Given the ability of the gut microbiota to alter nutrient availability and the close relationship between nutrient sensing and peptide secretion by enteroendocrine cells, the interaction of microbiota and the GBA might also occur through the release of biologically active peptides from enteroendocrine cells that can affect the GBA (Uribe et al., 1994). The microbiota can also convey information to the CNS by a host of other mechanisms, such as modulation of the afferent sensory nerves, by producing molecules that can act as local neurotransmitters or less commonly, through nitric oxide–mediated signaling (Sobko et al., 2006).

In the afferent pathway the brain controls the gut flora, mediated chiefly by the neurotransmitters. Interaction of these molecules with the cognate receptors that are present on the surface of many intestinal bacteria results in significant modulation of their activity, including heightened sensitivity to inflammation and infection (Hughes and Sperandio, 2008). For example, *Pseudomonas fluorescens* possesses a high-affinity receptor for GABA and similar binding characteristics (Guthrie and Nicholson-Guthrie, 1989), and *E. coli* O157:H7 possesses a receptor for host adrenaline/noradrenaline that can be blocked efficiently by adrenergic antagonists (Clarke et al., 2006). Thus behaviors of many members of the gut microbiota are controlled by the brain through neurotransmitters. The brain also plays a major role in controlling other physiological functions of the gut, such as secretion and motility, regulating the net fluid turnover by the intestine as well as the luminal immune response. All of these factors are critical for maintenance of a healthy biofilm on the luminal wall (Macfarlane and Dillon, 2007) and fending off invading pathogens by a coordinated attack together with the host immune cells. However, stress perturbs the healthy gut microbiota population and allows establishment of opportunistic pathogens. Therefore the GBA works seamlessly to ensure our constant well-being along with that of our microbial friends.

New frontiers in gut-brain microbiota research have revealed how the intestinal flora can modulate the progression of neurological disorders in human and animals. A certain *E. coli* protein known as curli possesses amyloidogenic behavior and aggregates to form amyloid-like fibrils much like the human protein alpha-synuclein. Studies in animal models have indicated that aggregated curli can travel through the GBA via the vagus nerve and accelerate the aggregation of alpha-synuclein (Sampson et al., 2020). In another mice model experiment, infection of a pregnant mouse by the

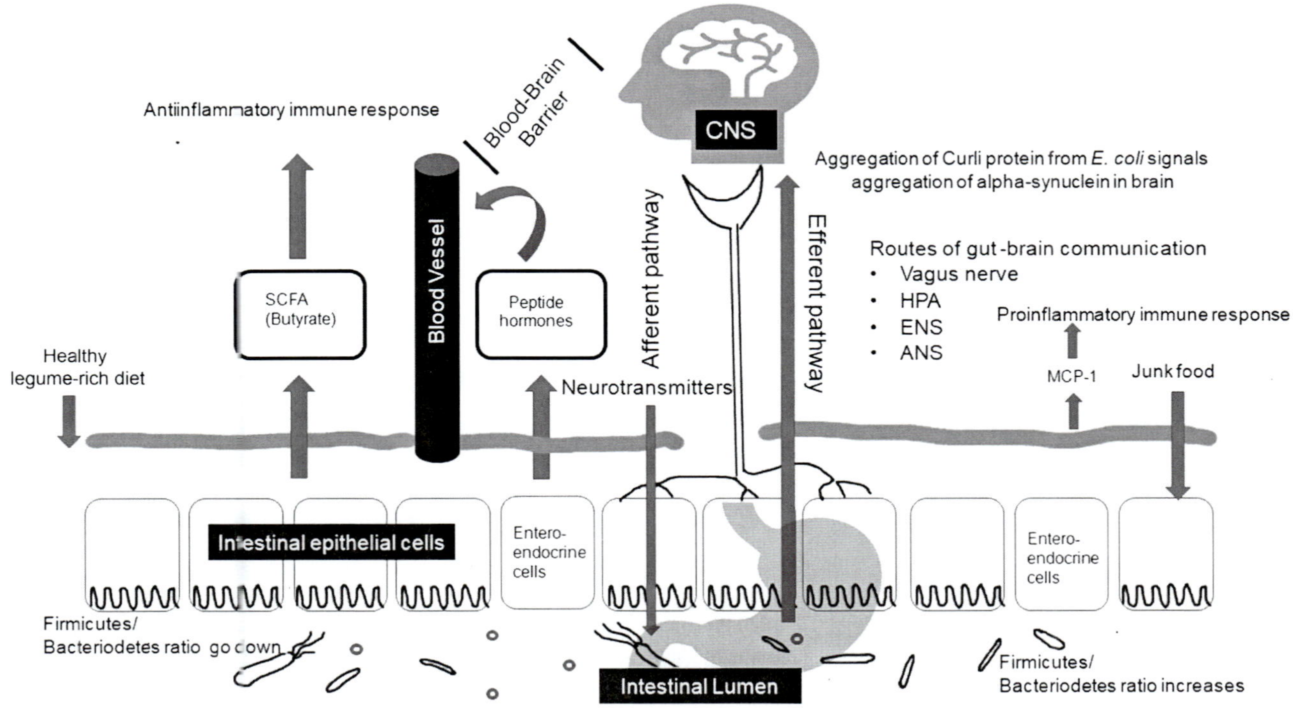

FIGURE 3.2 Implication of gut microbiota in the gut-brain axis and other physiological functions, including metabolic, immune, and neuroendocrine aspects. *SCFA*, Short-chain fatty acid; *CNS*, central nervous system; *HPA*, hypothalamic pituitary adrenal; *ENS*, enteric nervous system; *ANS*, autonomic nervous system; *MCP-1*, monocyte chemoattractant protein 1.

segmented filamentous bacteria caused secretion of $CD4^+$ Th17 immune cells in the gut, which were transmitted to the brain of the developing fetus, inducing autistic changes in the pups (Kim et al., 2017).

3.3.6 The gut microbiome as a human organ

The human microbiota has evolved to get vertically transmitted through generations in almost as precise a manner as the genetic material itself. Thus a common core of the intestinal flora is maintained consistently within specific kin of human individuals along with our genetic material (Gonzalez et al., 2011). How this extremely intricate network formed of thousands of different genera is faithfully inherited remains a mystery. However, some insight has been gained lately in this respect, courtesy of social psychologist Stanley Milgram's "small world" model. According to this model, the gradual colonization and establishment of the gut microflora is mediated by many founder species, each of which represents a node or hub. A network then ensues from each of these hubs through complex interactions between different microbial species that maintain at the same time perfect harmony with the growing human cells. As cells divide synchronously to give rise to a full-fledged organ, it has been tempting to propose the gut microflora as a human organ with indispensable functions and roles to play. Very recently, it has been possible to engraft one person's microflora partially or completely into another person, much like organ transplantations, with limited side effects (Khoruts and Sadowsky, 2011; Gough et al., 2011). This novel approach using state-of-the-art technology can potentially bring about a revolution in the treatment of microbial diseases with tests already conducted in *Clostridium difficile*—associated pathologies, inflammatory bowel diseases, metabolic syndrome, obesity, neurodegenerative diseases, and autoimmune and allergic diseases (Reid et al., 2011; Borody and Khoruts, 2012; Brandt et al., 2012). Current advances in this field include successful preservation of the microbiome equivalent from a person's body (Hamilton et al., 2012) and experimental animal models (Pang et al., 2007).

3.4 Dysbiosis of normal flora: implications in human diseases and intervention strategies

The gut microflora exists in perfect harmony with the host cells and coordinates innumerable body functions. However, any disturbance or deviation from this delicate balance affects the microbial population as it affects the host cells. Therefore it is clear that the dysbiosis of the intestinal bacterial population is intricately involved in many common human ailments. The following subsection gives a brief account of some common diseases and the involvement of the gut microbiome in these ailments.

3.4.1 Gut microbiome in human ailments

3.4.1.1 Obesity

Obesity is a chronic systemic disease that results from an overaccumulation of fat. Over the last few decades, it has become one of the significant lifestyle hazards globally. Adults with a body mass index of 30 and above are said to be obese, and youths who are above the 95th percentile for their respective age groups are said to be in the risk zone (Bischoff et al., 2017; Yanovski and Yanovski, 2018). Studies have revealed a substantial involvement of the gut microbiome in the development of obesity, the host genetic makeup and diet consumed being key regulating factors (Le Chatelier et al., 2013; Sonnenburg and Bäckhed, 2016). Therefore many therapeutic procedures to treat obesity have involved novel dietary interventions to modulate the gut microbiota. Natural compounds such as epigallocatechin 3-O-(3-O-methyl) gallate (EGCG3″Me), a polyphenol that is present in oolong tea, have been implicated in decreasing the overall Firmicutes/Bacteroidetes ratio in the gut and thus helping to prevent the development of obesity (Cheng et al., 2017). Bariatric surgery is practiced frequently for dealing with morbid obesity. This procedure causes a significant restoration in the obesity-induced decrease of population of *Bacteroides thetaiotaomicron* with a corresponding decrease in serum glutamate levels (Liu et al., 2017).

Development of infectious diseases depends on the outcome of the conflict between the gut microflora and the invading pathogens. Pathogens induce a strong inflammatory response, which in turn leads to translocation of the intestinal bacteria (Brenchley and Douek, 2012; Hand et al., 2012). Dysbiosis of the microflora has been successfully demonstrated for the opportunistic pathogen *C. difficile* infection (CDI) (Gu et al., 2016; Ling et al., 2014), HIV (Ling et al., 2016), hepatitis B virus (HBV) (Xu et al., 2012), and a host of other diseases (Hu et al., 2016; Ling et al., 2010). *C. difficile* is an anaerobic, gram-positive, spore-forming bacillus living inside human body whose population is generally kept under check by the intestinal flora. Overuse of antibiotics perturbs the balance of the intestinal mucosa, thus allowing the progression of CDI. Following the administration of an antibiotic, there is a significant drop in the population of butyrate-producing anaerobic bacteria with a simultaneous increase in the number of opportunistic pathogens, all of which are potentially capable of releasing endotoxins. Bacterial vaginosis is a vivid example of the consequence of an altered vaginal flora (Ling et al., 2010).

3.4.1.2 Diabetes

The World Health Organization has identified diabetes as the third major disease after cancer and cardiovascular disorders. It is also the most significant lifestyle-associated disease of the century. The role of the intestinal microflora in regulating the immune response was discussed earlier in the chapter; it is possibly through this involvement that they also play a major part in the commitment of an individual to type I diabetes. An altered flora with a drop in the number of *Clostridium* genus and *Akkermansia muciniphila* has been found to be a good marker of prediabetic conditions in the human body. However, actual development of the disease has been found to be controlled more by the cross-talk among the flora rather by than an overall shift in the balance of the flora itself (Hänninen et al., 2018).

3.4.1.3 Hypertension

A gross imbalance in the Firmicutes/Bacteroidetes ratio as well as dysbiosis in the diversity of gut flora has been linked to development of hypertension in both animal and human studies (Yang et al., 2015). The gut microbiota–derived metabolites indole and indoxyl sulfate have been shown to influence arterial blood pressure via peripheral and central mechanisms utilizing serotonin-based signaling mechanisms.

3.4.1.4 Cancer

The human microbiome has been found to be involved in 20% of all cancer cases globally (Wong et al., 2019). The evidence of involvement of the gut microflora has been the strongest in gastrointestinal malignancy, colorectal cancer, and esophageal cancer.

In patients with prostate cancer, the number of *Bacteroides massiliensis* rose significantly, whereas that of *Faecalibacterium prausnitzii* and *Eubacterium rectale* decreased (Golombos et al., 2018). In a similar study, *B. fragilis* toxin elicited activation of a proinflammatory pathway in colonic epithelial cells that was eventually carcinogenic (Chung et al., 2018). Members of other genera such as *B. fragilis*, *Fusobacterium nucleatum*, and *Peptostreptococcus anaerobius*, were also found to be associated with colorectal cancer (Wong et al., 2017). *H. pylori* is a well-known diagnostic marker of gastric cancer. However, along with *H. pylori*, *Clostridium* and *Fusobacterium* have been found to increase in this disease, indicating their association (Hsieh et al., 2018). Particularly, the role of *F. nucleatum* has been deciphered. This bacterium promotes the growth of cancer cells by suppressing the host immune response and upregulating the cellular proliferation rate. Consumption of food that is rich in whole grains and dietary fiber has been found to reduce the risk of colorectal cancer by promoting the growth of *F. nucleatum*. Oral administration of *Bifidobacterium* sp. has resulted in enhanced antitumor immunity in animal models of melanoma (Sivan et al., 2015). Thus the association of many members of the intestinal community with diet has unearthed many potential therapeutic approaches with minimal side effects (Mehta et al., 2017).

3.4.1.5 Inflammatory bowel disease

Inflammatory bowel disease (IBD) is associated with a selective rise in a subpopulation of the gut flora that has adapted to tolerate high oxidative stress. This often involves a rise in the number of facultative anaerobes, chiefly from the Enterobacteriaceae family and a few others, such as *Ruminococcus gnavus* (Hall et al., 2017). Consumption of antibiotics during pregnancy also increases susceptibility to IBD and may be attributed to changes in the gut microbiome (Örtqvist et al., 2019). Certain probiotics, such as *L. gasseri* SF1138 strain probiotic, have shown promising results against the progression of IBD by secretion of certain antiinflammatory molecules without disrupting the resident microflora (Di Luccia et al., 2018).

3.4.1.6 Gout

Gout is an inherited or acquired defect of metabolism that occurs as a result of enhanced rates of purine metabolism leading to abnormal increases in the levels of serum uric acid (Zhou et al., 2014). In gout patients, the gut microbiome is severely perturbed with an increase in the abundance of opportunistic pathogens (Shao et al., 2017). Along with these pathogens, there is an increase in the Bacteroides/Firmicutes ratio. A new Microbial Index of Gout, based on the relative abundance of 17 bacterial markers, has been framed to diagnose gout with a higher accuracy in comparison to conventional blood uric acid tests (Guo et al., 2016).

3.4.1.7 Depression

There is a popular saying that we are happy if our tiny friends that live in the gut are happy. Indeed, the gut microbiome has been strongly implicated in the emotional and psychological well-being of human beings, especially with regard to states of depression (Bastiaanssen et al., 2019). Accordingly, many studies have been directed toward exploring the efficacy of a number of probiotic strains as therapeutics to treat psychiatric illnesses and boost mental health. In this respect, *B. longum* has shown promising results in alleviating depression and as a treatment for irritable bowel syndrome (Pinto-Sanchez et al., 2017). A combination of *Lactobacillus helveticus* and *B. longum* had been able to reduce depression scores in patients diagnosed with major depressive disorder (Kazemi et al., 2019).

3.4.1.8 Rheumatoid arthritis

Rheumatoid arthritis is a chronic inflammatory autoimmune disorder that involves damage of bones and cartilages in joints mediated by autoantibodies (Gibofsky, 2014; Smolen et al., 2016). Although we have yet to gain in-depth knowledge of the etiology of this disease, the intestinal microflora is believed to play some role (Scher and Abramson, 2011). Just before the onset of arthritis, there is a change in the gut microbiome. Scientists believe that it is possible to arrest the progression of the disease if the gut microbiome can be eliminated by antibiotics at the right juncture. The number of intestinal Th17 immune cells bears a direct correlation with the severity of the disease, which can be curbed by getting rid of the gut microbiota (Rogier et al., 2017; Maeda and Takeda, 2017). However, this approach is bound to have severe side effects on the modulation and functioning of the entire immune system as well as many other indispensable functions that are mediated by the intestinal microflora.

3.4.1.9 Infant health

As was stated earlier in the chapter, the mode of birth plays a very important role in determining the infant's gut microflora, at least through the first year. Milk allergy, a common problem in infants and toddlers, is generally cured by the age of 8 years by an augmented population of Clostridia and Firmicutes. Therefore these can be used as potential probiotics for treating milk allergies (Bunyavanich et al., 2016). After a child attains an age of 1 year, the microbiota, especially the *Ruminococcaceae*, starts to participate in the child's cognitive development along with other functions (Carlson et al., 2018). Infants with an underdeveloped intestinal flora are given *L. rhamnosus* GG as an effective probiotic. The plasticity of the gut microflora is therefore highest during the initial years of growth. This also explains why most probiotic therapies are effective only in infants.

3.4.2 How to keep the gut happy: prebiotics, probiotics, postbiotics, and synbiotics

To keep the gut flora healthy and happy, it sometimes becomes necessary to introduce probiotics from outside and feed them with specialized food sources. These dietary fibers, which can be digested only by the gut bacteria and not by human cells, are known as prebiotics. Compounds that are classified as prebiotics must also not be degraded by stomach acids and enzymes that are present in the human GI tract, according to the guidelines for probiotics and prebiotics laid down by the International Scientific Association. Plant-derived polysaccharides, such as the fructans and galactans, are the major dietary fibers that are classified as prebiotics (Lamsal, 2012). Most of the studies on prebiotics have focused on the two significant probiotic genera, *Lactobacillus* and *Bifidobacteria*. While members of *Lactobacilli* prefer only inulin and fructooligosaccharides, *Bifidobacteria* can ferment a broader range of dietary fibers, including inulin, fructooligosaccharides, xylooligosaccharides, and galactooligosaccharides (Wilson and Whelan, 2017). A good endogenous source of prebiotics is breast milk. The milk oligosaccharides have been shown to promote the growth of certain *Bifidobacterium* species (Enam and Mansell, 2019). They also boost the development of the baby's immune system.

Probiotics are generally administered as a combination of beneficial bacteria and yeasts to revive the damaged indigenous flora. Besides imparting vigor to friendly resident microbes, probiotics modulate a host of immune functions and play a crucial role in restoring the synchronized homeostasis of the host and the gut microflora (Table 3.2). They have been shown to possess antidiabetic, antiallergic, anticarcinogenic, antiinflammatory, and angiogenic activities (Kerry et al., 2018). Most if not all of these functions depend on the SCFAs, unique metabolites that are released by fermentative metabolism of these bacteria. As has been stated already, the most significant probiotics that are used in

TABLE 3.2 Gut microbes dysbiosis leading to different human aliments and the therapeutic probiotic involved therein.

Disease	Gut microbe implicated	Suggested probiotic intervention	References
Obesity	*Bacteroides thetaiotao micron*	*Lactobacillus plantarum*	Aoun et al. (2020)
Diabetes	*Akkermansia muciniphila*	*Lactobacillus* sp.	Kocsis et al. (2020)
Hypertension	*Firmicutes/ bacteroidetes*	*Lactobacillus* sp.	Qi et al. (2020)
Cancer	*Faecali bacterium prausnitzii*	*Bifidobacterium bifidum*	Górska et al. (2019)
Inflammatory bowel disease	*Ruminococcus gnavus*	*Lactobacillus gasseri*	Di Luccia et al. (2018)
Depression	*Bifidobacterium* sp.	*Bifidobacterium longum*	Pinto-Sanchez et al. (2017)

therapeutics currently are the *Lactobacilli*, *Lactococci*, and *Bifidobacteria* and the yeast *Sachharomyces boulardii*. Other important genera whose potential as probiotics are being realized are *Propionibacterium*, *Peptostreptococcus*, *Bacillus*, *Lactococcus*, *Pediococcus*, *Akkermansia*, and *Bacteroides*.

Most of the metabolites that are released by the probiotic species are inconsequential to themselves but continue to offer benefits and protection to the human host. These are collectively termed the postbiotics (Patel and Denning, 2013). The postbiotics must be nontoxic and resistant to host enzyme hydrolysis and commonly comprise bacterial metabolic byproducts, such as bacteriocins, organic acids, ethanol, diacetyl, acetaldehydes, and hydrogen peroxide. Although the practice of gut microbiome transplantation has been in vogue, too few cases have been successful lately, mostly owing to the poor survival rate of the implanted species. As a potential remedy to this, a new therapeutic approach has been invented, known as the synbiotic. A synbiotic is a fusion of a probiotic and its corresponding prebiotic that can be administered together in the body to augment the health benefits offered by the probiotic species (Tufarelli and Laudacio, 2016). Synbiotics have the potential to revolutionize probiotic therapy and increase significantly the rate of success of gut-microbiome transplantation in near future.

3.5 Concluding remarks

The gut community is complex and intricate, and its synchronized management is a challenging feat both for the microbes themselves and for the harboring human cells. Cooperative interaction within this unique microenvironment therefore forms the basis of the well-being of the intestinal flora (Rakoff-Nahoum et al., 2016). Microbes secrete "public goods" molecules, such as metabolites, enzymes, signaling molecules, and toxins, for the greater benefit of their community, ensuring continuous perfect harmony with the host cells. Thanks to recent unprecedented advances in rapid genome-sequencing technologies and state-of-the-art *in silico* modeling tools, most of the gut microbial community has been unearthed. To cope with this gargantuan database of the genome, proteome, and metabolome of our gut friends, scientists have come up with the term "holobiont" (Thaiss et al., 2016) to holistically account for the microbial community as well as the human cells with which they live in harmony. The GBA of microbial communication has particularly emerged as a target of therapeutic interest to treat many neurological disorders for which conventional medicines frequently fail to work. In particular, how microbes control our emotional well-being is an exciting new frontier in gut science. However, the rampant and injudicious use of antibiotics has jeopardized most of these new explorations and brought humankind to a precarious situation in which the threat of antimicrobial resistance is looming large. If we can successfully overcome the threat of antimicrobial resistance and do not destroy our gut community, the day is not far away when nutritionists will prepare our diet charts and doctors will prescribe medicines based upon our unique gut microbiome signatures.

Acknowledgments

The authors express their sincere acknowledgment to the faculty members of the Department of Microbiology, Maulana Azad College, for critically proofreading the chapter to make it as free from errors as possible.

References

Aoun, A., Darwish, F., Hamod, N., 2020. The influence of the gut microbiome on obesity in adults and the role of probiotics, prebiotics, and synbiotics for weight loss. Prev. Nutr. Food Sci. 25, 113.

Arumugam, M., Raes, J., Pelletier, E., Le Paslier, D., Yamada, T., Mende, D.R., et al., 2011. Erratum: enterotypes of the human gut microbiome. Nature 473, 174–180. nature 474.

Atarashi, K., Tanoue, T., Shima, T., Imaoka, A., Kuwahara, T., Momose, Y., et al., 2011. Induction of colonic regulatory T cells by indigenous Clostridium species. Science 331, 337–341.

Bach, J.-F., 2002. The effect of infections on susceptibility to autoimmune and allergic diseases. N. Engl. J. Med. 347, 911–920.

Balagopal, A., Philp, F.H., Astemborski, J., Block, T.M., Mehta, A., Long, R., et al., 2008. Human immunodeficiency virus-related microbial translocation and progression of hepatitis C. Gastroenterology 135, 226–233.

Bastiaanssen, T.F., Cowan, C.S., Claesson, M.J., Dinan, T.G., Cryan, J.F., 2019. Making sense of... the microbiome in psychiatry. Int. J. Neuropsychopharmacol. 22, 37–52.

Beaumont, M., Goodrich, J.K., Jackson, M.A., Yet, I., Davenport, E.R., Vieira-Silva, S., et al., 2016. Heritable components of the human fecal microbiome are associated with visceral fat. Genome Biol. 17, 1–19.

Belzer, C., Chia, L.W., Aalvink, S., Chamlagain, B., Piironen, V., Knol, J., et al., 2017. Microbial metabolic networks at the mucus layer lead to diet-independent butyrate and vitamin B12 production by intestinal symbionts. MBio 8.

Bercik, P., Park, A., Sinclair, D., Khoshdel, A., Lu, J., Huang, X., et al., 2011. The anxiolytic effect of Bifidobacterium longum NCC3001 involves vagal pathways for gut–brain communication. Neurogastroenterol. Motil. 23, 1132–1139.

Bibbò, S., Ianiro, G., Giorgio, V., Scaldaferri, F., Masucci, L., Gasbarrini, A., et al., 2016. The role of diet on gut microbiota composition. Eur. Rev. Med. Pharmacol. Sci. 20, 4742–4749.

Bischoff, S.C., Boirie, Y., Cederholm, T., Chourdakis, M., Cuerda, C., Delzenne, N.M., et al., 2017. Towards a multidisciplinary approach to understand and manage obesity and related diseases. Clin. Nutr. 36, 917–938.

Bohnhoff, M., Drake, B.L., Miller, C.P., 1954. Effect of streptomycin on susceptibility of intestinal tract to experimental Salmonella infection. Proc. Soc. Exp. Biol. Med. 86, 132–137.

Borody, T.J., Khoruts, A., 2012. Fecal microbiota transplantation and emerging applications. Nat. Rev. Gastroenterol. Hepatol. 9, 88–96.

Böttcher, M., Nordin, E., Sandin, A., Midtvedt, T., Björkstén, B., 2000. Microflora-associated characteristics in faeces from allergic and nonallergic infants. Clin. Exp. Allergy 30, 1591–1596.

Bouskra, D., Brézillon, C., Bérard, M., Werts, C., Varona, R., Boneca, I.G., et al., 2008. Lymphoid tissue genesis induced by commensals through NOD1 regulates intestinal homeostasis. Nature 456, 507–510.

Brandt, L.J., Aroniadis, O.C., Mellow, M., Kanatzar, A., Kelly, C., Park, T., et al., 2012. Long-term follow-up of colonoscopic fecal microbiota transplant for recurrent clostridium difficileinfection. Am. J. Gastroenterol. 107, 1079–1087.

Brandtzaeg, P., 2010. Function of mucosa-associated lymphoid tissue in antibody formation. Immunol. Invest. 39, 303–355.

Bravo, J.A., Forsythe, P., Chew, M.V., Escaravage, E., Savignac, H.M., Dinan, T.G., et al., 2011. Ingestion of Lactobacillus strain regulates emotional behavior and central GABA receptor expression in a mouse via the vagus nerve. Proc. Natl. Acad. Sci. 108, 16050–16055.

Brenchley, J.M., Douek, D.C., 2012. Microbial translocation across the GI tract. Annu. Rev. Immunol. 30, 149–173.

Brenchley, J.M., Price, D.A., Schacker, T.W., Asher, T.E., Silvestri, G., Rao, S., et al., 2006. Microbial translocation is a cause of systemic immune activation in chronic HIV infection. Nat. Med. 12, 1365–1371.

Bunyavanich, S., Shen, N., Grishin, A., Wood, R., Burks, W., Dawson, P., et al., 2016. Early-life gut microbiome composition and milk allergy resolution. J. Allergy Clin. Immunol. 138, 1122–1130.

Byndloss, M.X., Olsan, E.E., Rivera-Chávez, F., Tiffany, C.R., Cevallos, S.A., Lokken, K.L., et al., 2017. Microbiota-activated PPAR-γ signaling inhibits dysbiotic Enterobacteriaceae expansion. Science 357, 570–575.

Carlson, A.L., Xia, K., Azcarate-Peril, M.A., Goldman, B.D., Ahn, M., Styner, M.A., et al., 2018. Infant gut microbiome associated with cognitive development. Biol. Psychiat. 83, 148–159.

Cebra, J.J., 1999. Influences of microbiota on intestinal immune system development. Am. J. Clin. Nutr. 69, 1046s–1051s.

Cheng, M., Zhang, X., Miao, Y., Cao, J., Wu, Z., Weng, P., 2017. The modulatory effect of (-)-epigallocatechin 3-O-(3-O-methyl) gallate (EGCG3 ″Me) on intestinal microbiota of high fat diet-induced obesity mice model. Food Res. Int. 92, 9–16.

Chung, L., Orberg, E.T., Geis, A.L., Chan, J.L., Fu, K., Shields, C.E.D., et al., 2018. Bacteroides fragilis toxin coordinates a pro-carcinogenic inflammatory cascade via targeting of colonic epithelial cells. Cell Host Microbe 23, 203–214. e205.

Clarke, M.B., Hughes, D.T., Zhu, C., Boedeker, E.C., Sperandio, V., 2006. The QseC sensor kinase: a bacterial adrenergic receptor. Proc. Natl. Acad. Sci. 103, 10420–10425.

Cooke, G., Behan, J., Costello, M., 2006. Newly identified vitamin K-producing bacteria isolated from the neonatal faecal flora. Microb. Ecol. Health Dis. 18, 133–138.

David, L.A., Materna, A.C., Friedman, J., Campos-Baptista, M.I., Blackburn, M.C., Perrotta, A., et al., 2014a. Host lifestyle affects human microbiota on daily timescales. Genome Biol. 15, 1–15.

David, L.A., Maurice, C.F., Carmody, R.N., Gootenberg, D.B., Button, J.E., Wolfe, B.E., et al., 2014b. Diet rapidly and reproducibly alters the human gut microbiome. Nature 505, 559–563.

De Vadder, F., Kovatcheva-Datchary, P., Goncalves, D., Vinera, J., Zitoun, C., Duchampt, A., et al., 2014. Microbiota-generated metabolites promote metabolic benefits via gut-brain neural circuits. Cell 156, 84–96.

Derrien, M., Vaughan, E.E., Plugge, C.M., de Vos, W.M., 2004. Akkermansia muciniphila gen. nov., sp. nov., a human intestinal mucin-degrading bacterium. Int. J. Syst. Evol. Microbiol. 54, 1469–1476.

Di Lucca, B., Mazzoli, A., Cancelliere, R., Crescenzo, R., Ferrandino, I., Monaco, A., et al., 2018. Lactobacillus gasseri SF1183 protects the intestinal epithelium and prevents colitis symptoms in vivo. J. Funct. Foods 42, 195–202.

DiGiulio, D.B., Romero, R., Amogan, H.P., Kusanovic, J.P., Bik, E.M., Gotsch, F., et al., 2008. Microbial prevalence, diversity and abundance in amniotic fluid during preterm labor: a molecular and culture-based investigation. PLoS One 3, e3056.

Dominguez-Bello, M.G., Costello, E.K., Contreras, M., Magris, M., Hidalgo, G., Fierer, N., et al., 2010. Delivery mode shapes the acquisition and structure of the initial microbiota across multiple body habitats in newborns. Proc. Natl. Acad. Sci. 107, 11971–11975.

Duncan, S.H., Louis, P., Thomson, J.M., Flint, H.J., 2009. The role of pH in determining the species composition of the human colonic microbiota. Environ. Microbiol. 11 2112–2122.

Eberl, G., Lochner, M., 2009. The development of intestinal lymphoid tissues at the interface of self and microbiota. Mucosal Immunol. 2, 478–485.

Enam, F., Mansell, T.J., 2019. Prebiotics: tools to manipulate the gut microbiome and metabolome. J. Ind. Microbiol. Biotechnol. 46, 1445–1459.

Flint, H.J., Duncan, S.H., Scott, K.P., Louis, P., 2015. Links between diet, gut microbiota composition and gut metabolism. Proc. Nutr. Soc. 74, 13–22.

Frost, G., Sleeth, M.L., Sahuri-Arisoylu, M., Lizarbe, B., Cerdan, S., Brody, L., et al., 2014. The short-chain fatty acid acetate reduces appetite via a central homeostatic mechanism. Nat. Commun. 5, 1–11.

Fukuda, S., Toh, H., Taylor, T.D., Ohno, H., Hattori, M., 2012. Acetate-producing bifidobacteria protect the host from enteropathogenic infection via carbohydrate transporters. Gut Microbes 3, 449–454.

Gareau, M.G., Wine, E., Rodrigues, D.M., Cho, J.H., Whary, M.T., Philpott, D.J., et al., 2011. Bacterial infection causes stress-induced memory dysfunction in mice. Gut 60, 307–317.

Ghoshal, S., Witta, J., Zhong, J., De Villiers, W., Eckhardt, E., 2009. Chylomicrons promote intestinal absorption of lipopolysaccharides. J. Lipid Res. 50, 90–97.

Gibofsky, A., 2014. Epidemiology, pathophysiology, and diagnosis of rheumatoid arthritis: A synopsis. Am. J. Manag. Care 20, S128–S135.

Goh, Y.J., Klaenhammer, T.R., 2015. Genetic mechanisms of prebiotic oligosaccharide metabolism in probiotic microbes. Annu. Rev. Food Sci. Technol. 6, 137–156.

Golombos, D.M., Ayangbesan, A., O'Malley, P., Lewicki, P., Barlow, L., Barbieri, C.E., et al., 2018. The role of gut microbiome in the pathogenesis of prostate cancer: a prospective, pilot study. Urology 111, 122–128.

Gonzalez, A., Stombaugh, J., Lozupone, C., Turnbaugh, P.J., Gordon, J.I., Knight, R., 2011. The mind-body-microbial continuum. Dialogues Clin. Neurosci. 13, 55.

Goodrich, J., Waters, J., Poole, A., Sutter, J., Koren, O., Blekhman, R., 2014. Human genetics shape the gut microbiome. Cell [Internet]. Elsevier Inc.

Górska, A., Przystupski, D., Niemczura, M.J., Kulbacka, J., 2019. Probiotic bacteria: a promising tool in cancer prevention and therapy. Curr. Microbiol. 76, 939–949.

Gough, E., Shaikh, H., Manges, A.R., 2011. Systematic review of intestinal microbiota transplantation (fecal bacteriotherapy) for recurrent Clostridium difficile infection. Clin. Infect. Dis. 53, 994–1002.

Gu, S., Chen, Y., Zhang, X., Lu, H., Lv, T., Shen, P., et al., 2016. Identification of key taxa that favor intestinal colonization of Clostridium difficile in an adult Chinese population. Microbes Infect. 18, 30–38.

Guo, Z., Zhang, J., Wang, Z., Ang, K.Y., Huang, S., Hou, Q., et al., 2016. Intestinal microbiota distinguish gout patients from healthy humans. Sci. Rep. 6, 1–10.

Guthrie, G.D., Nicholson-Guthrie, C.S., 1989. gamma-Aminobutyric acid uptake by a bacterial system with neurotransmitter binding characteristics. Proc Natl. Acad. Sci. 86, 7378–7381.

Hall, A.B., Yassour, M., Sauk, J., Garner, A., Jiang, X., Arthur, T., et al., 2017. A novel Ruminococcus gnavus clade enriched in inflammatory bowel disease patients. Genome Med. 9, 1–12.

Hamilton, M.J., Weingarden, A.R., Sadowsky, M.J., Khoruts, A., 2012. Standardized frozen preparation for transplantation of fecal microbiota for recurrentClostridium difficileinfection. Off. J. Am. Coll. Gastroenterol. | ACG 107, 761–767.

Hand, T.W., Dos Santos, L.M., Bouladoux, N., Molloy, M.J., Pagán, A.J., Pepper, M., et al., 2012. Acute gastrointestinal infection induces long-lived microbiota-specific T cell responses. Science 337, 1553–1556.

Hänninen, A., Toivonen, R. Pöysti, S., Belzer, C., Plovier, H., Ouwerkerk, J.P., et al., 2018. Akkermansia muciniphila induces gut microbiota remodelling and controls islet autoimmunity in NOD mice. Gut 67, 1445–1453.

Hofer, U., Schlaepfer, E., Baenziger, S., Nischang, M., Regenass, S., Schwendener, R., et al., 2010. Inadequate clearance of translocated bacterial products in HIV-infected humanized mice. PLoS Pathog 6, e1000867.

Hollister, E.B., Gao, C., Versalovic, J., 2014. Compositional and functional features of the gastrointestinal microbiome and their effects on human health. Gastroenterology 146, 1449–1458.

Hsieh, Y.-Y., Tung, S.-Y., Pan, H.-Y., Yen, C.-W., Xu, H.-W., Lin, Y.-J., et al., 2018. Increased abundance of Clostridium and Fusobacterium in gastric microbiota of patients with gastric cancer in Taiwan. Sci. Rep. 8, 1–11.

Hu, Z., Zhang, Y., Li, Z., Yu, Y., Kang, W., Han, Y., et al., 2016. Effect of Helicobacter pylori infection on chronic periodontitis by the change of microecology and inflammation. Oncotarget 7, 66700.

Hughes, D.T., Sperandio, V., 2008. Inter-kingdom signalling: communication between bacteria and their hosts. Nat. Rev. Microbiol. 6, 111–120.

Kazemi, A., Noorbala, A.A., Azam, K., Eskandari, M.H., Djafarian, K., 2019. Effect of probiotic and prebiotic vs placebo on psychological outcomes in patients with major depressive disorder: a randomized clinical trial. Clin. Nutr. 38, 522–528.

Kerry, R.G., Patra, J.K., Gouda, S., Park, Y., Shin, H.-S., Das, G., 2018. Benefaction of probiotics for human health: a review. J. Food Drug Anal. 26, 927–939.

Khoruts, A., Sadowsky, M., 2011. Therapeutic transplantation of the distal gut microbiota. Mucosal Immunol. 4, 4–7.

Kim, S., Kim, H., Yim, Y.S., Ha, S., Atarashi, K., Tan, T.G., et al., 2017. Maternal gut bacteria promote neurodevelopmental abnormalities in mouse offspring. Nature 549, 528–532.

Kocsis, T., Molnár, B., Németh, D., Hegyi, P., Szakács, Z., Bálint, A., et al., 2020. Probiotics have beneficial metabolic effects in patients with type 2 diabetes mellitus: a *meta*-analysis of randomized clinical trials. Sci. Rep. 10, 1–14.

Kunisawa, J., Kiyono, H., 2012. Alcaligenes is commensal bacteria habituating in the gut-associated lymphoid tissue for the regulation of intestinal IgA responses. Front. Immunol. 3, 65.

Lamsal, B.P., 2012. Production, health aspects and potential food uses of dairy prebiotic galactooligosaccharides. J. Sci. Food Agric. 92, 2020–2028.

Larsbrink, J., Rogers, T.E., Hemsworth, G.R., McKee, L.S., Tauzin, A.S., Spadiut, O., et al., 2014. A discrete genetic locus confers xyloglucan metabolism in select human gut Bacteroidetes. Nature 506, 498–502.

Le Chatelier, E., Nielsen, T., Qin, J., Prifti, E., Hildebrand, F., Falony, G., et al., 2013. Richness of human gut microbiome correlates with metabolic markers. Nature 500, 541–546.

Ley, R.E., Hamady, M., Lozupone, C., Turnbaugh, P.J., Ramey, R.R., Bircher, J.S., et al., 2008. Evolution of mammals and their gut microbes. Science 320, 1647–1651.

Li, Y., Dong, J., Xiao, H., Zhang, S., Wang, B., Cui, M., et al., 2020. Gut commensal derived-valeric acid protects against radiation injuries. Gut Microbes 11, 789–806.

Lin, H.V., Frassetto, A., Kowalik Jr, E.J., Nawrocki, A.R., Lu, M.M., Kosinski, J.R., et al., 2012. Butyrate and propionate protect against diet-induced obesity and regulate gut hormones via free fatty acid receptor 3-independent mechanisms. PLoS One 7, e35240.

Ling, Z., Jin, C., Xie, T., Cheng, Y., Li, L., Wu, N., 2016. Alterations in the fecal microbiota of patients with HIV-1 infection: an observational study in a Chinese population. Sci. Rep. 6, 1–12.

Ling, Z., Kong, J., Liu, F., Zhu, H., Chen, X., Wang, Y., et al., 2010. Molecular analysis of the diversity of vaginal microbiota associated with bacterial vaginosis. BMC Genomics 11, 1–16.

Ling, Z., Liu, X., Jia, X., Cheng, Y., Luo, Y., Yuan, L., et al., 2014. Impacts of infection with different toxigenic Clostridium difficile strains on faecal microbiota in children. Sci. Rep. 4, 1–10.

Liu, R., Hong, J., Xu, X., Feng, Q., Zhang, D., Gu, Y., et al., 2017. Gut microbiome and serum metabolome alterations in obesity and after weight-loss intervention. Nat. Med. 23, 859.

Louis, P., Flint, H.J., 2017. Formation of propionate and butyrate by the human colonic microbiota. Environ. Microbiol. 19, 29–41.

Macfarlane, S., Dillon, J., 2007. Microbial biofilms in the human gastrointestinal tract. J. Appl. Microbiol. 102, 1187–1196.

Macpherson, A.J., Harris, N.L., 2004. Interactions between commensal intestinal bacteria and the immune system. Nat. Rev. Immunol. 4, 478–485.

Madsen, K.L., Doyle, J.S., Jewell, L.D., Tavernini, M.M., Fedorak, R.N., 1999. Lactobacillus species prevents colitis in interleukin 10 gene–deficient mice. Gastroenterology 116, 1107–1114.

Maeda, Y., Takeda, K., 2017. Role of gut microbiota in rheumatoid arthritis. J. Clin. Med. 6, 60.

Mehta, R.S., Nishihara, R., Cao, Y., Song, M., Mima, K., Qian, Z.R., et al., 2017. Association of dietary patterns with risk of colorectal cancer subtypes classified by Fusobacterium nucleatum in tumor tissue. JAMA Oncol. 3, 921–927.

Morowitz, M.J., Carlisle, E.M., Alverdy, J.C., 2011. Contributions of intestinal bacteria to nutrition and metabolism in the critically ill. Surg. Clin. 91, 771–785.

Örtqvist, A.K., Lundholm, C., Halfvarson, J., Ludvigsson, J.F., Almqvist, C., 2019. Fetal and early life antibiotics exposure and very early onset inflammatory bowel disease: a population-based study. Gut 68, 218–225.

Oswald, F., Stoll, I.K., Zwick, M., Herbig, S., Sauer, J., Boukis, N., et al., 2018. Formic acid formation by Clostridium ljungdahlii at elevated pressures of carbon dioxide and hydrogen. Front. Bioeng. Biotechnol. 6, 6.

Pace, N.R., 1997. A molecular view of microbial diversity and the biosphere. Science 276, 734–740.

Pang, X., Hua, X., Yang, Q., Ding, D., Che, C., Cui, L., et al., 2007. Inter-species transplantation of gut microbiota from human to pigs. ISME J. 1, 156–162.

Patel, R.M., Denning, P.W., 2013. Therapeutic use of prebiotics, probiotics, and postbiotics to prevent necrotizing enterocolitis: what is the current evidence? Clin. Perinatol. 40, 11–25.

Peterson, J., Garges, S., Giovanni, M., McInnes, P., Wang, L., Schloss, J.A., et al., 2009. The NIH human microbiome project. Genome Res. 19, 2317–2323.

Pilmis, B., Monnier, Le, Zahar, J.-R., A., 2020. Gut microbiota, antibiotic therapy and antimicrobial resistance: a narrative review. Microorganisms 8, 269.

Pinto-Sanchez, M.I., Hall, G.B., Ghajar, K., Nardelli, A., Bolino, C., Lau, J.T., et al., 2017. Probiotic Bifidobacterium longum NCC3001 reduces depression scores and alters brain activity: a pilot study in patients with irritable bowel syndrome. Gastroenterology 153, 448–459. e448.

Proal, A.D., Albert, P.J., Marshall, T.G., 2014. Inflammatory disease and the human microbiome. Dis. Med. 257–265.

Qi, D., Nie, X.-L., Zhang, J.-J., 2020. The effect of probiotics supplementation on blood pressure: a systemic review and *meta*-analysis. Lipids Health Dis. 19, 1–11.

Qin, J., Li, R., Raes, J., Arumugam, M., Burgdorf, K.S., Manichanh, C., et al., 2010. A human gut microbial gene catalogue established by metagenomic sequencing. Nature 464, 59–65.

Rakoff-Nahoum, S., Foster K.R., Comstock, L.E., 2016. The evolution of cooperation within the gut microbiota. Nature 533, 255–259.

Reid, G., Younes, J.A., Van der Mei, H.C., Gloor, G.B., Knight, R., Busscher, H.J., 2011. Microbiota restoration: natural and supplemented recovery of human microbial communities. Nat. Rev. Microbiol. 9, 27–38.

Rhee, S.H., Pothoulakis, C., Mayer, E.A., 2009. Principles and clinical implications of the brain–gut–enteric microbiota axis. Nat. Rev. Gastroenterol. Hepatol 6, 306.

Rogier, E.W., Frantz, A.L. Bruno, M.E., Wedlund, L., Cohen, D.A., Stromberg, A.J., et al., 2014. Lessons from mother: long-term impact of antibodies in breast milk on the gut microbiota and intestinal immune system of breastfed offspring. Gut Microbes 5, 663–668.

Rogier, R., Evans-Marin, H., Manasson, J., van der Kraan, P.M., Walgreen, B., Helsen, M.M., et al., 2017. Alteration of the intestinal microbiome characterizes preclinical inflammatory arthritis in mice and its modulation attenuates established arthritis. Sci. Rep. 7, 1–12.

Rothschild, D., Weissbrod, O., Barkan, E., Kurilshikov, A., Korem, T., Zeevi, D., et al., 2018. Environment dominates over host genetics in shaping human gut microbiota. Nature 555, 210–215.

Round, J.L., Mazmanian, S.K., 2009. The gut microbiota shapes intestinal immune responses during health and disease. Nat. Rev. Immunol. 9, 313–323.

Russell, A.B., Hood, R.D., Bui, N.K., LeRoux, M., Vollmer, W., Mougous, J.D., 2011. Type VI secretion delivers bacteriolytic effectors to target cells. Nature 475, 343–347.

Sampson, T.R., Challis, C., Jain, N., Moiseyenko, A., Ladinsky, M.S., Shastri, G.G., et al., 2020. A gut bacterial amyloid promotes α-synuclein aggregation and motor impairment in mice. Elife 9, e53111.

Scher, J.U., Abramson, S.B., 2011. The microbiome and rheumatoid arthritis. Nat. Rev. Rheumatol. 7, 569.

Schnabl, B., Brenner, D.A., 2014. Interactions between the intestinal microbiome and liver diseases. Gastroenterology 146, 1513–1524.

Sender, R., Fuchs, S., Milo R., 2016. Revised estimates for the number of human and bacteria cells in the body. PLoS Biol. 14, e1002533.

Shao, T., Shao, L., Li, H., Xie, Z., He, Z., Wen, C., 2017. Combined signature of the fecal microbiome and metabolome in patients with gout. Front. Microbiol. 8, 268.

Sivan, A., Corrales, L., Hubert, N., Williams, J.B., Aquino-Michaels, K., Earley, Z.M., et al., 2015. Commensal Bifidobacterium promotes antitumor immunity and facilitates anti–PD-L1 efficacy. Science 350, 1084–1089.

Smolen, J.S., Breedveld, F.C., Burmester, G.R., Bykerk, V., Dougados, M., Emery, P., et al., 2016. Treating rheumatoid arthritis to target: 2014 update of the recommendations of an international task force. Ann. Rheumatic Dis. 75, 3–15.

Sobko, T., Huang, L., Midtvedt, T., Norin, E., Gustafsson, L.E., Norman, M., et al., 2006. Generation of NO by probiotic bacteria in the gastrointestinal tract. Free Radical Biol. Med. 41, 985–991.

Sonnenburg, J.L., Bäckhed, F., 2016. Diet–microbiota interactions as moderators of human metabolism. Nature 535, 56–64.

Spor, A., Koren, O., Ley, R., 2011. Unravelling the effects of the environment and host genotype on the gut microbiome. Nat. Rev. Microbiol. 9, 279–290.

Stacy, A., Andrade-Oliveira, V., McCulloch, J.A., Hild, B., Oh, J.H., Perez-Chaparro, P.J., et al., 2021. Infection trains the host for microbiota-enhanced resistance to pathogens. Cell 184, 615–627. e617.

Stefanko, D.P., Barrett, R.M., Ly, A.R., Reolon, G.K., Wood, M.A., 2009. Modulation of long-term memory for object recognition via HDAC inhibition. Proc. Nat. Acad. Sci. 106, 9447–9452.

Swidsinski, A., Loening-Baucke, V., Lochs, H., Hale, L.P., 2005. Spatial organization of bacterial flora in normal and inflamed intestine: a fluorescence in situ hybridization study in mice. World J. Gastroenterol.: WJG 11, 1131.

Tannock G.W., 1995. Normal Microflora: An Introduction to Microbes Inhabiting the Human Body. Springer Science & Business Media.

Tap, J., Mondot, S., Levenez, F., Pelletier, E., Caron, C., Furet, J.P., et al., 2009. Towards the human intestinal microbiota phylogenetic core. Environ. Microbiol. 11, 2574–2584.

Thaiss, C.A., Zmora, N., Levy, M., Elinav, E., 2016. The microbiome and innate immunity. Nature 535, 65–74.

Tims, S. Derom, C., Jonkers, D.M., Vlietinck, R., Saris, W.H., Kleerebezem, M., et al., 2013. Microbiota conservation and BMI signatures in adult monozygotic twins. ISME J. 7, 707–717.

Tufarelli V., Laudadio, V., 2016. An overview on the functional food concept: prospectives and applied researches in probiotics, prebiotics and synbiotics. J. Exp. Biol. Agric. Sci. 4, 273–278.

Uribe, A., Alam, M., Johansson, O., Midtvedt, T., Theodorsson, E., 1994. Microflora modulates endocrine cells in the gastrointestinal mucosa of the rat. Gastroenterology 107, 1259–1269.

Wilson, B., Whelan, K., 2017. Prebiotic inulin-type fructans and galacto-oligosaccharides: definition, specificity, function, and application in gastrointestinal disorders. J. Gastroenterol. Hepatol. 32, 64–68.

Wong, S H., Kwong, T.N., Wu, C.-Y., Yu, J., 2019. Clinical applications of gut microbiota in cancer biology. Seminars in Cancer Biology. Elsevier, pp. 28–36.

Wong, S.H., Zhao, L., Zhang, X., Nakatsu, G., Han, J., Xu, W., et al., 2017. Gavage of fecal samples from patients with colorectal cancer promotes intestinal carcinogenesis in germ-free and conventional mice. Gastroenterology 153, 1621–1633. e1626.

Wu, G.D., Chen, J., Hoffmann, C., Bittinger, K., Chen, Y.-Y., Keilbaugh, S.A., et al., 2011. Linking long-term dietary patterns with gut microbial enterotypes. Science 334, 105–108.

Xu, M., Wang, B., Fu, Y., Chen, Y., Yang, F., Lu, H., et al., 2012. Changes of fecal Bifidobacterium species in adult patients with hepatitis B virus-induced chronic liver disease. Microbial Ecol. 63, 304–313.

Yang, T., Santisteban, M.M., Rodriguez, V., Li, E., Ahmari, N., Carvajal, J.M., et al., 2015. Gut dysbiosis is linked to hypertension. Hypertension 65, 1331–1340.

Yanovski, S.Z., Yanovski, J.A., 2018. Toward precision approaches for the prevention and treatment of obesity. JAMA 319, 223–224.

Yatsunenko, T., Rey, F.E., Manary, M.J., Trehan, I., Dominguez-Bello, M.G., Contreras, M., et al., 2012. Human gut microbiome viewed across age and geography. Nature 486, 222–227.

Zhao, L., Zhang, F., Ding, X., Wu, G., Lam, Y.Y., Wang, X., et al., 2018. Gut bacteria selectively promoted by dietary fibers alleviate type 2 diabetes. Science 359, 1151–1156.

Zhou, D., Liu, Y., Zhang, X., Gu, X., Wang, H., Luo, X., et al., 2014. Functional polymorphisms of the ABCG2 gene are associated with gout disease in the Chinese Han male population. Int. J. Mol. Sci. 15, 9149–9159.

Zoetendal, E.G., Akkermans, A.D., Akkermans-van Vliet, W.M., de Visser, J.A.G., de Vos, W.M., 2001. The host genotype affects the bacterial community in the human gastronintestinal tract. Microbial Ecol. Health Dis. 13, 129–134.

Chapter 4

The gut microbiome, human nutrition, and immunity: visualizing the future

Soching Luikham and Jhimli Bhattacharyya
Department of Chemistry, National Institute of Technology Nagaland, Dimapur, Nagaland, India

4.1 Introduction

The human body, which includes the gut (intestines), skin, and various other mucosal parts, is populated by a huge number of different microorganisms, which together are termed the microbiome (Fig. 4.1) (Sender et al., 2016). The gut microbiome, which also refers to the pooled genomes of microorganisms in a specific environment, known as the gut microbiota, is the bacteria that populate our gastrointestinal tract and are well known to play a key role in maintaining host health and metabolic balance. Roughly 100 trillion microorganisms (fungi, viruses, protozoa, bacteria, yeasts, and parasites) live in the gastrointestinal tract of humans (Jovel et al., 2018; Rath and Dorrestein, 2012). Most of the microorganisms that are present in the intestines are in a "pocket" called the cecum, which is part of the large intestine. This collection of microorganisms is also known as the gut microbiota. Every individual has a unique microbiota that is determined by the person's DNA. An individual is first exposed to microbes during birth and through the mother's breast milk (Ursell et al., 2012). Exactly which microbes the newborn is introduced to directly depend on the species that are found in the mother's birth canal. After birth, the diet and environmental exposures can alter the microbiota; this can be beneficial to individual health, or it can cause the person to be at greater risk for various kinds of disease. Even though there are various types of microorganisms inside the body, the most studied ones are the bacteria. The latest findings have unveiled the importance of the gut microbiome in the regulation of host immunity, in interactions between the innate and adaptive immune systems, and in turn in the development of certain diseases. In the human gut

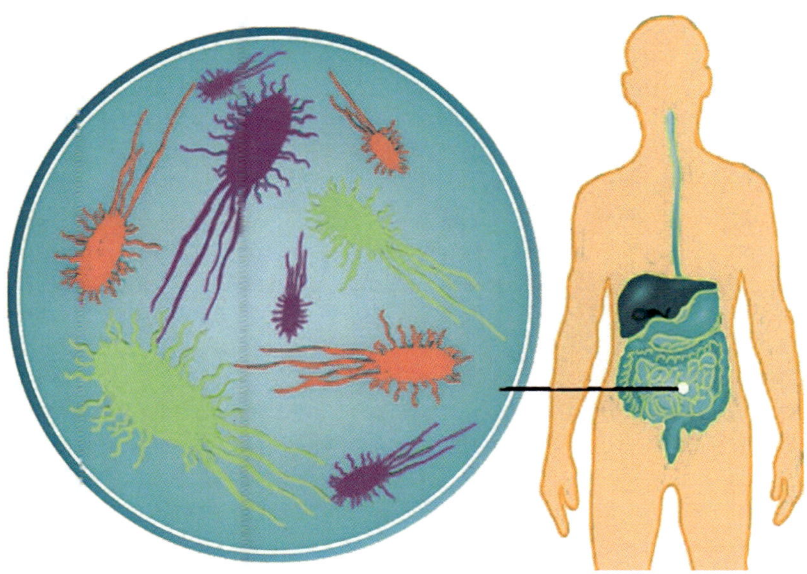

FIGURE 4.1 Human gut microbiome.

Microbiome, Immunity, Digestive Health and Nutrition. DOI: https://doi.org/10.1016/B978-0-12-822238-6.00029-7

microbiome there are approximately 1000 species of bacteria, and every one has a different role to play in the human body. They are very important for health of an individual, and some may cause multiple diseases (Integrative, 2014).

The bacteria that reside in the gut are major inhabitants of the body. They are important in various aspects of the host's health, including better digestion, production of vitamins, detoxifying harmful components, shielding against different pathogenic organisms, and modulation of the immune system. Other benefits include improving heart health, serving to alleviate the symptoms of asthma and allergies, balancing the blood sugar level, and improving mood. The research behind what is known as the hygiene hypothesis has shown that an imbalance within the gut will impair the gut barrier and increase the risk to gut health, thereby increasing the risk for developing various diseases. An unhealthy diet, a lack of exercise, a sedentary lifestyle, binge eating, and persistent stress can all contribute to microbiome imbalance. According to recent research, high levels of dietary fat and fructose disrupt the gastrointestinal barrier, resulting in fatty liver disease and causing inflammation. Positive dietary modifications can help the person to avoid major ailments, including obesity, diabetes, metabolic syndrome, allergies, and cancer. Research has demonstrated that lifestyle and diet play key roles in maintaining gut health and prevention of gastrointestinal disorders, which include irritable bowel syndrome, hyperacidity, acid reflux, diarrhea, and food allergies (Zanteson, 2021). Conventional ancient grains, in contrast to today's processed grains and canned foods, have a large impact on gut microbiome function, which helps in improving and preventing various diseases. A diet that includes large amounts of fruits and vegetables, which are full of nutrients, including vitamins, minerals, and dietary fiber, and have a low glycemic index, will have a positive impact on the gut flora (Zanteson, 2021). It is important to understand more about microbial differences and the sources that can improve lifestyle and hence boost the formation of gut microbial communities in adults and children. Furthermore, the gut microbiota can help to boost the immune systems during the life span and rectify the components of the microbiota that affect the immune system. These findings raise the topic of how the gut microbial community's metabolism of food affects immune mechanisms. It is important to emphasize the need to better understand the relationship between nutritional status, diet, microbial ecology and the immune system in humans at different stages of life in distinct cultural and socioeconomic settings (Kau et al., 2011).

4.2 Studying the gut microbiome and its importance

Although there is a inherited component to the gut microbiota, two studies have revealed that environmental factors, such as nutrition, probiotics, dietary supplements, medicines, and anthropometric measurements, are more important predictors of microbiome composition (Goodrich et al., 2014; Rothschild et al., 2018). Gut microbes are key to many features of human health, including the immune system, (Zhang et al., 2015) metabolism (Rothschild et al., 2018), and neurobehavioral traits (Levy et al., 2017; Yarur et al., 2011). Different levels of evidence support the role of the gut microbiota in human health from animal models (De Palma et al., 2017; Wiley et al., 2017) and human studies (Goodrich et al., 2014). The formation of the gut microbiota is usually measured by using DNA-based approaches, such as next-generation sequencing of 16S RNA or whole-genome sequencing, which allow microbiota interference (Karthik et al., 2014; Ranjan et al., 2016). The microbiota boosts the immune system, potentially breaking down harmful bacterial ingredient, and enhancing the biosynthesis of multiple nutrients, including vitamins and amino acids (Den Besten et al., 2013) such as K vitamins and vitamin B. The main enzymes that are required to make vitamin B12, for example, are present exclusively in bacteria, not in plants or mammals (Morowitz et al., 2011). Sugars, such as table sugar and milk sugar (lactose), are quickly utilized in the upper section of the small intestine, while more complex carbohydrates, such as fibers and starches, are more difficult to digest and may end up in the lower intestine. There, the microbiota is required to disintegrate these compounds with digestive enzymes. Short-chain fatty acids (SCFAs) are produced when indigestible fibers are fermented, and they can be utilized by the body as a food source as well as playing a role in muscular function and possibly preventing chronic diseases, including bowel disorders and cancer. Studies have shown that SCFAs may be helpful for the treatment of ulcerative colitis, Crohn's disease, and antibiotic-associated diarrhea (Den Besten et al., 2013).

Ruminococcus, Prevotella, Firmicutes, and Bacteroides are all large families of bacteria that are present in the human gut. The anaerobic bacteria *Bifidobacterium*, *Peptostreptococcus*, *Clostridium*, and *Lactobacillus* live in the colon, which is low in oxygen (Arumugam et al., 2011). These microbes may prevent harmful bacteria from multiplying by competing for important nutrients and attachment sites on the gut mucus membrane, which play a major part in the immune system and antimicrobial protein production (Canny and McCormick, 2008; Jandhyala et al., 2015). Some beneficial and harmful bacteria are shown in Fig. 4.2. Early postnatal environmental exposures play a crucial role in shaping the overall phylogenetic structure of the adult human gut microbiome, according to metagenomic analyses of the human gut microbiome. During the first initial three years of life (Koenig et al., 2011), the microbiota is assembled into

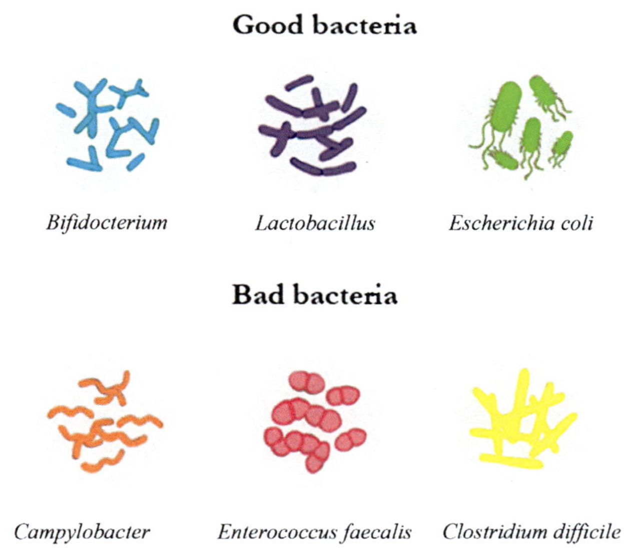

FIGURE 4.2 Types of bacteria in the gut microbiome.

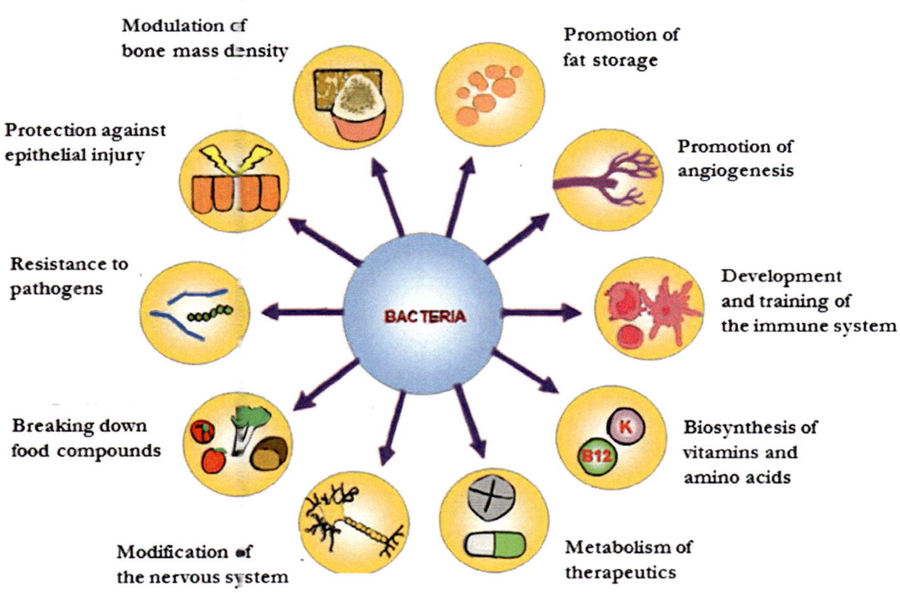

FIGURE 4.3 Various functions of gut bacteria in the body.

an adult configuration; properties of the organism and gene gut communities are shared across family members and passed down through generations (Turnbaugh et al., 2009). There are thousands of different varieties of bacteria in the intestines, the majority of which are beneficial to human health (Fig. 4.3) and perform various functions that are beneficial for health. On the other hand, the presence of harmful microorganisms can lead to disease. Gut dysbiosis is a condition in which the balance between good and bad microorganisms is disturbed, and it can potentially lead to gaining of weight (Patterson et al., 2016). Several studies have shown that the gut microbiotas of identical twins, one who is fat while the other is lean, differ dramatically. This substantiates the hypothesis that those changes in the microbiome are not due to genetic differences (Ridaura et al., 2013; Turnbaugh et al., 2009). Although both groups ate the same foods, the twin who was obese had a microbiota that was different from the microbiota of the lean twin (Ridaura et al., 2013). This study shows that weight gain may be due to microbiome dysbiosis.

Certain foods and dietary patterns can have an impact on the quantity of different types of bacteria in the gut microbiome. High-intensity sweeteners are frequently used as sugar substitutes, since they are multiple times sweeter than sugar while containing only a few calories. Sugar alternatives have been shown to have harmful effects on the gut microbiome (Nettleton et al., 2016). Saccharin, sucralose, and aspartame have all been shown to affect the balance and variety of the gut microbiome. The microbiomes of animals have been demonstrated to be affected by dietary additives such as emulsifiers, which are widespread in processed foods. Mice that were given low doses of two popularly used emulsifiers, namely, carboxymethyl cellulose and polysorbate-80, had less microbial diversity than mice that were not given emulsifiers. Inflammation-promoting Proteobacteria linked with mucus were enriched, while numbers of Verrucomicrobia and Bacteroides were decreased (Chassaing et al., 2015).

4.3 The role of the microbiome in nutrition

For millions of years, humans have evolved to coexist with bacteria. Microbes have come to play critical roles in the body during this time. The gut microbiome has an impact on the body from the minute a person are born. A baby is first subjected to microbes during passage through the mother's birth canal. However, new data reveal that babies may encounter a microorganism while still in the womb (Bäckhed et al., 2015). As one gets older, the person's microbiome starts to evolve, which means that it contains a wider range of bacterial species. Gut flora diversity is thought to be beneficial to health (Koenig et al., 2011). Similarly, nutrition influences the function and structure of the gut microbial population, and the microbiota adjusts in ways that facilitate nutrient digestion; the microbiota's ability to process a certain diet influences the nutrient and energy values of such a diet (Kau et al., 2011). Similarly, nutrition influences the function and structure of the gut microbial population, and the microbiota adjusts in ways that facilitate nutrient digestion; the microbiota's ability to process a certain diet influences the nutrient and energy values of such a diet (Zanteson, 2021). The gut microbiome may partake in the risk and pathogenesis of undernutrition (malnutrition) through effects on nutrient metabolism and immune function (Fig. 4.4). We also know that dietary habits influence the structure of human genes. People who eat high-starch diets, for example, have more copies of the salivary amylase gene (*AMY1*) than people who eat low-starch diets (Perry et al., 2007). We need to know what people consume and how their diets are changing. That also helps us to understand the factors that are associated with nutritional status. Beneficial microorganisms known as probiotic bacteria, such as *Bifidobacteria* and lactic acid bacteria, predominate in the stomach and other human organs, ensuring good health and preventing diseases. The types of bacteria that live in the colon are determined largely by the diet (Jandhyala et al., 2015). All of these elements combine to form a microbiome that is unique to each individual. The kind and abundance of microbiomes in the intestine are influenced by a high-fiber diet in specific. Only enzymes from the microbiome in the colon can break down and ferment dietary fiber. Fermentation results in the release of SCFAs. This decreases the pH of the colon, affecting the bacteria there and their ability to thrive in this acidic medium. Some dangerous microorganisms, such as *Clostridium difficile*, are inhibited by the lower pH. SCFA studies are expanding to look into a variety of health conditions, such as increasing immune cell function and sustaining blood glucose as well as cholesterol levels. The gut microbes ferment nondigestible compounds such as food fibers and endogenous gastrointestinal mucus, which supplies critical functions. The development of specialized microorganisms that create SCFAs as well as gases is facilitated by this fermentation (Julia et al., 2006). Acetate, butyrate, and propionate are the most common SCFAs that are generated. Butyrate is the primary source of energy for human intestinal mucosa, and it has the ability to induce apoptosis in colon cancer cells and activates intestinal glucose production, all of which have positive impacts on glucose and energy (De Vadder et al., 2014). Butyrate is required for epithelial cells to use a substantial supply of oxygen via oxidation, resulting in a state of hypoxia in the gut that maintains the oxygen balance and prevents dysbiosis of the intestinal microbiota (Byndloss et al., 2017). Propionate is transported to the liver,

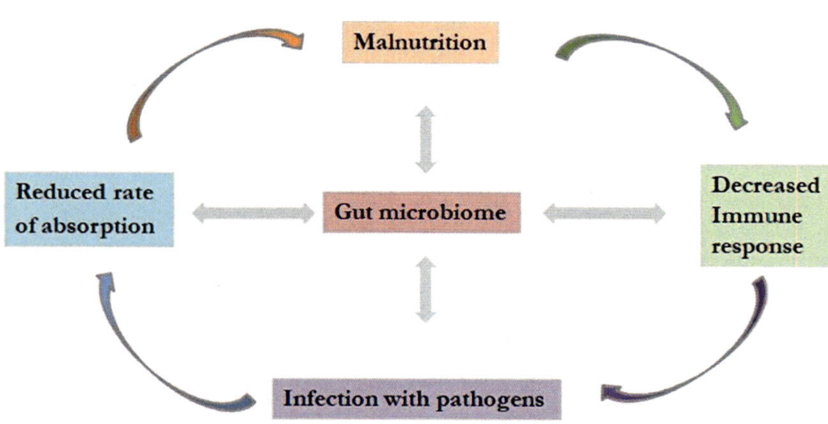

FIGURE 4.4 Schematic representation of the relationship between the gut microbiome, the immune system, and diet.

where it interacts with gastrointestinal fatty acid sensors to control glucose production and signaling (De Vadder et al., 2014). Acetate, the most common SCFA and a necessary molecule for the development of other microorganisms, reaches peripheral tissues and is employed in cholesterol metabolism and lipid metabolism as well as perhaps playing a role in central regulation (Frost et al., 2014). Propionate and butyrate but not acetate appear to regulate gut enzymes and reduce food intake (Lin et al., 2012).

Dietitians would do well to counsel clients to eat diets that help to promote the growth and abundance of beneficial bacteria. Probiotic bacteria utilize the complex carbohydrates, that is, dietary fiber, and harmful bacteria use dietary proteins and fats to produce toxins, which can damage good health. Good sources of complex carbohydrates include fruits, vegetables, legumes, and cereals. Foods that are rich sources of antioxidants include fruits and vegetables. Interestingly the food that we consume may be affecting the microbial diversity of the gut bacteria. Some of the benefit, that the microbiome provides through nutrition are discussed in the following subsections.

4.3.1 Digestion of breast milk

Several of the bacteria that are first present in the infant's intestines are known as *Bifidobacteria*. These bacteria usually digest the healthy sugars, which are present in breast milk and are an important element for growth (Arboleya et al., 2016).

4.3.2 Digestion of Fiber

Specific bacteria digest the fiber that is present in the gut by producing special SCFAs, which are significant for the health of the gut microbiota. Fiber may help to prevent diabetes, heart disease, cancer, and weight gain (Slavin, 2013).

4.3.3 Enhancement of immune competence

The gut microbiota, which controls the immune system and works by interacting with immune cells that are present in the body, is responsible for responding to various infections (Rooks and Garrett, 2016). Overall, a healthy gut microbiome boosts immune competence.

4.3.4 Promotion of brain health

The microbiota may have an impact on the central nervous system, which regulates brain activity. Some species of bacteria can aid in the production of neurotransmitters, which are chemicals produced in the brain. Serotonin, for example, is an antidepressant neurotransmitter that is produced primarily in the gut. As a result, it may have an impact on brain function by assisting in the management of information supplied to the brain via millions of nerves (Cryan and Dinan, 2012). When compared with healthy persons, individuals with different psychological problems have distinct species of bacteria in their stomachs. It is possible that the gut microbiome can help to govern brain health.

4.3.5 Boosting of cardiovascular health

The gut microbiota was found to play a significant role in the promotion of "good" triglycerides and high-density lipoprotein cholesterol (Fu et al., 2015). When consumed as a probiotic, certain bacteria, especially *Lactobacilli*, can lower the cholesterol level (Shimizu et al., 2015), benefiting cardiovascular health.

4.3.6 Potentiation of insulin sensitization and decrease in the risk of diabetes

The gut bacteria aids in the regulation of blood sugar, which may influence the risk of type 1 and type 2 diabetes. Even when patients eat identical foods, their blood glucose levels may differ substantially. This could be related to their gut microbiota (Zeevi et al., 2015).

4.3.7 Prebiotic foods and dietary fiber

Dietary protein is defined as digestible carbohydrate polymers containing three or even more monomeric units that are resilient to endogenous enzymes and hence are not hydrolyzed or digested. Dietary fiber have soluble sources, which imply that they act as growth substrates for microorganisms in the distal colon. Several nondigestible carbohydrates are known as prebiotics, which are described as food components or ingredients that are not digested by humans but particularly or exclusively sustain helpful colonic microbes (Bindels et al., 2015).

4.3.8 Probiotic foods

Probiotics are living microorganisms that provide a nutritional benefit to the host when given in sufficient concentrations (Hill et al., 2014). Probiotics can be found in a number of food products, nutritional supplements, and pharmaceuticals. Probiotics can have an impact on the host's wellness without affecting the gut microbiota. Probiotic supplements have been explored for their therapeutic benefits in a variety of disorders. Probiotic supplements have been shown to help prevent diarrhea, cellulitis, chronic lung conditions in affected children, and dermatitis in children. In patients with type 2 diabetes mellitus, probiotics appear to enhance cardiometabolic markers and reduce serum levels of reactive protein. Novel microorganisms and combinations and mixtures of probiotics and prebiotics are popular as probiotic supplements (symbiotics) (Plovier et al., 2017). Probiotics, which are crucial gut health elements, must be supplemented for proper function of the gut microbiota. The synthesis of SCFAs is the principal function of microbial fermentation of probiotics in the lower intestine. The nutritional value of food is affected in part by the functions and organization of the consumer's gut microbiome, and food, in turn, influences the microbiome and its huge collection of bacterial genes, according to a growing amount of data (Kau et al., 2011).

4.4 The role of the microbiome in immune system development

The gut microflora exerts its efficacy via several mechanisms, primarily by metabolizing the dietary constituents either to detoxify them or to activate them into different forms. A presence of pathogenic bacteria requires a well-functioning and strong immune system to prevent infections. The structure and content of the gut microbiota can thus have an impact on the immune response (Zanteson, 2021). The human immune system is made up of a complex network of innate and adaptive components found throughout the body that play a critical function in defending the host against a variety of potentially damaging foreign agents. Exposure to toxic environments, antibiotic overdose, unhealthy diets, or shifts in geographic location may impair the functionalities of the host microbiome or immune system modifications. This can also induce systemic propagation of commensal microbes, reactions to pathogenic incursion, and abnormal immune responses by disrupting the gut microbiota. Microbiome-immune relationships are observed in a variety of noncommunicable gastrointestinal disorders, such as inflammatory bowel disease (Zhang et al., 2017), and increase the host's vulnerability to pathogenic invasion and abnormal immunological responses to infections, and enable microbial colonization and proliferation (Valitutti et al., 2019). Gut dysbiosis may also be a factor in additional intestinal disorders and rheumatoid arthritis (Maeda and Takeda, 2019), metabolic syndrome (Belizário et al., 2018), neurodegenerative disorders (Main and Minter, 2017), and malignancy (Gopalakrishnan et al., 2018). The interactions between the gut microbiota and host immunity are complex, dynamic, and context dependent. Significant host immunological microbiome interaction generally occurs in childhood, which may have a long-term effect on the immune system, leading to immune disruption and exposure to infectious and inflammatory diseases in advancing age (Zheng et al., 2020a).

Healthy probiotic bacteria aid in maintaining excellent health and immunity in a variety of ways. The most fundamental purpose is to cleanse and facilitate the removal of toxic foreign chemicals that enter the human body. Probiotics can limit the growth of pathogenic bacteria, which live and proliferate in a neutral pH by producing organic acids such as lactic and acetic acids, which decrease the pH of the intestinal tract. This decreased pH also keeps cholesterol and bile acid products out of the intestines. Because cholesterol and bile metabolites are cancer-causing chemicals, probiotic bacteria can aid in the prevention of malignancies of the gastrointestinal tract as well as other organs. By inhibiting the action of an enzyme system in cholesterol manufacturing, these beneficial bacteria could even decrease serum cholesterol levels and reduce the risk of heart disease (Zanteson, 2021). Gut health is determined by the microbiome, often known as the gut flora, and the gut barriers. A trillion live bacteria dwell within the gut, promoting appropriate gastrointestinal action, protecting the body from infection, and regulating metabolic and mucosal immune responses. These bacteria make up more than 75% of the immune response and help in the maintenance of the gastrointestinal barrier.

A healthy gut bacterium maintains gut health, whereas disturbances in the microbiome affect the body's defense systems and therefore can result in leaky gut syndrome, which could compromise the gut flora and lead to inflammation, cancer, chronic fatigue syndrome, depression, and obesity (Zanteson, 2021). It is worth mentioning that certain foods that support increased levels of SCFA are indigestible carbohydrates and fibers such as inulin, resistant starches, gums, pectins, and fructooligosaccharides. Prebiotics are often referred as fibers, which are beneficial as feed for microorganisms. Commercial prebiotic fiber capsules are available, but prebiotics are also available naturally in several nutritious foods. The largest concentrations are available in onions, raw garlic, asparagus, artichokes, leeks, bananas, dandelion greens, and seaweed. Prebiotic fibers are also available in many different types of foods, including vegetables, fruits, legumes, and whole grains such as barley, oats, and wheat.

People with irritable bowel syndrome (Manichanh et al., 2006), psoriatic arthritis (Scher et al., 2015), type 1 diabetes (De Goffau et al., 2013), dermal eczema (Wang et al., 2008), celiac disease (Schippa et al., 2010), obesity or fat deposition (Turnbaugh et al., 2009), type 2 diabetes mellitus (Duff Putu et al., 2015), and arterial blockage (Menni et al., 2018) have reduced microbial abundance in comparison to healthy volunteers. Smokers with Crohn's disease have a much smaller gut microbiome (Opstelten et al., 2016). The abundance of a variety of beneficial microorganisms in the gut appears to be a good predictor of a "healthy gut" (Valdes et al., 2018).

The most crucial events in the development of host defenses may occur within the first years of life, when the microbial community exhibits the greatest intervariability and intravariability until stabilizing at the age of 3 years (Bäckhed et al., 2015). Furthermore, because of the resulting moment of opportunity, newborns will be more resistant to various effects on the microbiome, which could have long-term implications for their immune response (Russell et al., 2012). Many current studies have not shown evidence of repeatable microbial colonization in pregnancy (Wang et al., 2018), and perhaps the majority of colonization occurs after birth, with the majority of microbes coming from mother's microbiome (De Agüero et al., 2016). It is generally known that maternal antibodies supplied by breast milk provide important passive protection against infections in newborns (Caballero-Flores et al., 2019). A recent study found that the commensal microbiomes of pregnant mice generates antibody-mediated protective immunity via lactation (Zheng et al., 2020b).

4.5 Various ways to improve our gut microbiome

The key to a healthy gut microbiota and its beneficial effects requires an optimal balance between the beneficial and pathogenic microorganisms. A healthy balanced diet is essential to provide nutritional health and support a healthy microbiome. There are different ways to improve the gut microbiome, some of which are discussed in the following subsections.

4.5.1 Eat a diverse diet

Eating diverse foods, including fruits, vegetables, and fibers can result in a more diversified microbiota that will promote excellent gut microflora and ecology. Legumes, beans, and fruit are high in fiber, which can encourage the growth of beneficial *Bifidobacteria* (Heiman and Greenway, 2016).

4.5.2 Consider fermented foods

Fermented foods such as yogurt, sauerkraut, and kefir contain beneficial bacteria, mainly *Lactobacilli*, which help to decrease the number of harmful bacteria in the stomach (Alvaro et al., 2007).

4.5.3 Avoid artificial sweeteners

Artificial sweeteners such as aspartame have been demonstrated to raise blood sugar levels by encouraging the growth of pathogenic bacteria such as Enterabacteriaceae in the gut microbiome (Dewulf et al., 2013).

4.5.4 Consume prebiotic foods

Prebiotics are a type of fiber that stimulates the growth of healthy bacteria. Artichokes, asparagus, bananas, apples, and oats are all prebiotic-rich foods (Dewulf et al., 2013).

4.5.5 Breastfeed for six months

Breastfeeding has a critical role in the formation of the gut bacteria. Breast-fed infants have far healthier *Bifidobacteria* than bottle-fed infants for at least 6 months (Madan et al., 2016).

4.5.6 Include whole grains

Whole grains are high in fiber and healthy carbohydrates such as beta-glucan, which are absorbed by gut microbes and may help in weight loss, retarding the progression of cancer, and control of blood sugar and metabolic disorders (Costabile et al., 2008).

4.5.7 Include plant-based foods

Microorganisms such as harmful *Escherichia coli* may induce inflammation and increase cholesterol levels, which may be counteracted by regular consumption of a vegan diet (Kim et al., 2013).

4.5.8 Consume polyphenol-containing food

Red wine, dark chocolate, green tea, whole grains, and olive oil contain polyphenolic antioxidants, which are natural antioxidants. The microbiota breaks them down to boost healthy bacterial development (Cardona et al., 2013).

4.5.9 Take antibiotics when required

Many bad and good bacteria in the gut microbiome are killed by antibiotics, which may contribute to obesity and antibiotic resistance. As a result, use antibiotics only when they are therapeutically essential (Bailey et al., 2014).

4.6 Future areas of research

The microbiome is a living dynamic environment in which the relative abundance of species may fluctuate daily, weekly, and monthly depending on diet, medication, exercise, and a host of other environmental factors. Scientists are still in the initial stages of understanding the broad role of the microbiome in health and the extent of the problems that can occur from an interruption in the normal interactions between the microbiome and its host (Proctor, 2011). Because people's gut microbiota differs, an individual's diet may need to be adapted to suit the person's gut microbiota (Zeevi et al., 2015). Researchers collected a multidimensional microbial profile from 900 patients for a week and tracked their food intake, physical activity, and continuous blood glucose levels. The researchers developed a machine learning algorithm based on clinical and gut microbiome data to predict individual glucose responses after meals and demonstrated that it outperformed alternatives, such as assessing the carbohydrate level or glycemic index. A randomized crossover trial for one-week-long dietary treatments revealed potential interpersonal changes in the glycemic response to various kinds of bread. Purely on the basis of microbiome data that were collected prior to the intervention, the type of bread that generated the lowest glycemic response in each participant could be predicted (Korem et al., 2017). More research studies are required to determine whether tailored diets such as this are beneficial to human health (Valdes et al., 2018). There is a need to develop translational medicine to define more rigorously the value of nutritious food that humans consume daily.

4.7 Conclusion

In the future, we must modify our food habits and intake help to build a healthy gut that is equipped with a beneficial microbiome. It is important to note that dietary fiber is an important component for a healthy gut microbiota. Moreover, refined sugar and fatty foods should be avoided. The adverse effects of pharmaceuticals and processed food items on the microbiome can no longer be overlooked. Controlled randomized, placebo-controlled clinical trials are essential to determine the effects of regular consumption of prebiotics and probiotics as well as fecal microbiota transplantation technology, which may bring differences in gut microbiome composition and health benefits.

Acknowledgments

JB acknowledges the financial assistances from UGC-DAE, Mumbai Centre, BARC, Mumbai (grant no. CRS-M-266); DBT Govt. of India (sanction no. BT/PR25026/NER/95/963/2017); and TEQIP-III Seed Grant, NIT Nagaland.

References

Alvaro, E., Andrieux, C., Rochet, V., Rigottier-Gois, L., Lepercq, P., Sutren, M., et al., 2007. Composition and metabolism of the intestinal microbiota in consumers and non-consumers of yogurt. Br. J. Nutr. 97, 126–133. Available from: https://doi.org/10.1017/S0007114507243065.

Arboleya, S., Watkins, C., Stanton, C., Ross, R.P., 2016. Gut bifidobacteria populations in human health and aging. Front. Microbiol. 7, 1–9. Available from: https://doi.org/10.3389/fmicb.2016.01204.

Arumugam, M., Raes, J., Pelletier, E., Le Paslier, D., Yamada, T., Mende, D.R., et al., 2011. Enterotypes of the human gut microbiome. Nature 474, 666. Available from: https://doi.org/10.1038/nature10187.

Bailey, L.C., Forrest, C.B., Zhang, P., Richards, T.M., Livshits, A., DeRusso, P.A., 2014. Association of antibiotics in infancy with early childhood obesity. JAMA Pediatr. 168, 1063–1069. Available from: https://doi.org/10.1001/jamapediatrics.2014.1539.

Belizário, J.E., Faintuch, J., Garay-Malpartida, M., 2018. Gut microbiome dysbiosis and immunometabolism: new frontiers for treatment of metabolic diseases. Mediators Inflamm. 2018, 1–12. Available from: https://doi.org/10.1155/2018/2037838.

Bindels, L.B., Delzenne, N.M., Cani, P.D., Walter, J., 2015. Opinion: towards a more comprehensive concept for prebiotics. Nat. Rev. Gastroenterol. Hepatol. 12, 303–310. Available from: https://doi.org/10.1038/nrgastro.2015.47.

Byndloss, M.X., Olsan, E.E., Rivera-Chávez, F., Tiffany, C.R., Cevallos, S.A., Lokken, K.L., et al., 2017. Microbiota-activated PPAR-γ signaling inhibits dysbiotic Enterobacteriaceae expansion. Science (80-.) 357, 570–575. Available from: https://doi.org/10.1126/science.aam9949.

Bäckhed, F., Roswall, J., Peng, Y., Feng, Q., Jia, H., Kovatcheva-Datchary, P., et al., 2015. Dynamics and stabilization of the human gut microbiome during the first year of life. Cell Host Microbe 17, 690–703. Available from: https://doi.org/10.1016/j.chom.2015.040.004.

Caballero-Flores, G., Sakamoto, K., Zeng, M.Y., Wang, Y., Hakim, J., Matus-Acuña, V., et al., 2019. Maternal immunization confers protection to the offspring against an attaching and effacing pathogen through delivery of IgG in breast milk. Cell Host Microbe 25, 313–323. e4. Available from: https://doi.org/10.1016/j.chom.2018.120.015.

Canny, G.O., McCormick, B.A., 2008. Bacteria in the intestine, helpful residents or enemies from within? Infect. Immun. 76, 3360–3373. Available from: https://doi.org/10.1128/IAI.00187-08.

Cardona, F., Andrés-Lacueva, C., Tulipani, S., Tinahones, F.J., Queipo-Ortuño, M.I., 2013. Benefits of polyphenols on gut microbiota and implications in human health. J. Nutr. Biochem. 24, 1415–1422. Available from: https://doi.org/10.1016/j.jnutbio.2013.050.001.

Chassaing, B., Koren, O., Goodrich, J.K., Poole, A.C., Srinivasan, S., Ley, R.E., et al., 2015. Dietary emulsifiers impact the mouse gut microbiota promoting colitis and metabolic syndrome. Nature 519, 92–96. Available from: https://doi.org/10.1038/nature14232.

Costabile, A., Klinder, A., Fava, F., Napolitano, A., Fogliano, V., Leonard, C., et al., 2008. Whole-grain wheat breakfast cereal has a prebiotic effect on the human gut microbiota: a double-blind, placebo-controlled, crossover study. Br. J. Nutr. 99, 110–120. Available from: https://doi.org/10.1017/S0007114507793923.

Cryan, J.F., Dinan, T.G., 2012. Mind-altering microorganisms: the impact of the gut microbiota on brain and behaviour. Nat. Rev. Neurosci. 13, 701–712. Available from: https://doi.org/10.1038/nrn3346.

De Agüero, M.G., Ganal-Vonarburg, S.C., Fuhrer, T., Rupp, S., Uchimura, Y., Li, H., et al., 2016. The maternal microbiota drives early postnatal innate immune development. Science (80-.) 351, 1296–1302. Available from: https://doi.org/10.1126/science.aad2571.

De Goffau, M.C., Luopajärvi, K., Knip, M., Ilonen, J., Ruohtula, T., Härkönen, T., et al., 2013. Fecal microbiota composition differs between children with β-cell autoimmunity and those without. Diabetes 62, 1238–1244. Available from: https://doi.org/10.2337/db12-0526.

De Palma, G., Lynch, M.D.J., Lu, J., Dang, V.T., Deng, Y., Jury, J., et al., 2017. Transplantation of fecal microbiota from patients with irritable bowel syndrome alters gut function and behavior in recipient mice. Sci. Transl. Med. 9, 1–15. Available from: https://doi.org/10.1126/scitranslmed.aaf6397.

De Vadder, F., Kovatcheva-Datchary, P., Goncalves, D., Vinera, J., Zitoun, C., Duchampt, A., et al., 2014. Microbiota-generated metabolites promote metabolic benefits via gut-brain neural circuits. Cell 156, 84–96. Available from: https://doi.org/10.1016/j.cell.2013.120.016.

Den Besten, G., Van Eunen, K., Groen, A.K., Venema, K., Reijngoud, D.J., Bakker, B.M., 2013. The role of short-chain fatty acids in the interplay between diet, gut microbiota, and host energy metabolism. J. Lipid Res. 54, 2325–2340. Available from: https://doi.org/10.1194/jlr.R036012.

Dewulf, E.M., Cani, P.D., Claus, S.P., Fuentes, S., Puylaert, P.G.B., Neyrinck, A.M., et al., 2013. Insight into the prebiotic concept: lessons from an exploratory, double blind intervention study with inulin-type fructans in obese women. Gut 62, 1112–1121. Available from: https://doi.org/10.1136/gutjnl-2012-303304.

Duff Putu, G.O., Shoveller, J., Montaner, J., Feng, C., Nicoletti, R., Shannon, K., 2015. Composition, diversity and abundance of gut microbiome in prediabetes and type 2 diabetes. Physiol. Behav. 2, 1–7. Available from: https://doi.org/10.15436/2376-0949.15.031.Composition.

Frost, G., Sleeth, M.L., Sahuri-Arisoylu, M., Lizarbe, B., Cerdan, S., Brody, L., et al., 2014. The short-chain fatty acid acetate reduces appetite via a central homeostatic mechanism. Nat. Commun. 5, 1–11. Available from: https://doi.org/10.1038/ncomms4611.

Fu, J., Bonder, M.J., Cenit, M.C., Tigchelaar, E.F., Maatman, A., Dekens, J.A.M., et al., 2015. The gut microbiome contributes to a substantial proportion of the variation in blood lipids. Circ. Res. 117, 817–824. Available from: https://doi.org/10.1161/CIRCRESAHA.115.306807.

Goodrich, J.K., Waters, J.L., Poole, A.C., Sutter, J.L., Koren, O., Blekhman, R., et al., 2014. Human genetics shape the gut microbiome. Cell 159, 789–799. Available from: https://doi.org/10.1016/j.cell.2014.090.053.

Gopalakrishnan, V., Helmink, B.A., Spencer, C.N., Reuben, A., Wargo, J.A., 2018. The influence of the gut microbiome on cancer, immunity, and cancer immunotherapy. Cancer Cell 33, 570–580. Available from: https://doi.org/10.1016/j.ccell.2018.030.015.

Heiman, M.L., Greenway, F.L., 2016. A healthy gastrointestinal microbiome is dependent on dietary diversity. Mol. Metab. 5, 317–320. Available from: https://doi.org/10.1016/j.molmet.2016.020.005.

Hill, C., Guarner, F., Reid, G., Gibson, G.R., Merenstein, D.J., Pot, B., et al., 2014. Expert consensus document: the international scientific association for probiotics and prebiotics consensus statement on the scope and appropriate use of the term probiotic. Nat. Rev. Gastroenterol. Hepatol. 11, 506–514. Available from: https://doi.org/10.1038/nrgastro.2014.66.

Integrative, T., 2014. The integrative human microbiome project: dynamic analysis of microbiome-host omics profiles during periods of human health and disease corresponding author. Cell Host Microbe 16, 276–289. Available from: https://doi.org/10.1016/j.chom.2014.080.014.

Jandhyala, S.M., Talukdar, R., Subramanyam, C., Vuyyuru, H., Sasikala, M., Reddy, D.N., 2015. Role of the normal gut microbiota. World J. Gastroenterol. 21, 8836–8847. Available from: https://doi.org/10.3748/wjg.v21.i29.8787.

Jovel, J., Dieleman, L.A., Kao, D., Mason, A.L., Wine, E., 2018. The human gut microbiome in health and disease, metagenomics perspect. Methods Appl. 197–213. Available from: https://doi.org/10.1016/B978-0-08-102268-9.00010-0.

Julia, M.W.W., de Souza, R., Kendall, C.W.C., Emam, A., Jenkins, D.J.A., 2006. Colonic health: fermentation and short chain fatty acids. J. Clin. Gastroenterol. 40, 235–243. Available from: https://doi.org/10.12989/sem.2011.37.20.253.

Karthik, L., Kumar, G., Keswani, T., Bhattacharyya, A., Sarath Chandar, S., Bhaskara Rao, K.V., 2014. Protease inhibitors from marine actinobacteria as a potential source for antimalarial compound. PLoS One 9, e90972. Available from: https://doi.org/10.1371/journal.pone.0090972.

Kau, A.L., Ahern, P.P., Griffin, N.W., Goodman, A.L., Gordon, J.I., 2011. Human nutrition, the gut microbiome and the immune system. Nature 474, 327–336. Available from: https://doi.org/10.1038/nature10213.

Kim, M.S., Hwang, S.S., Park, E.J., Bae, J.W., 2013. Strict vegetarian diet improves the risk factors associated with metabolic diseases by modulating gut microbiota and reducing intestinal inflammation. Environ. Microbiol. Rep. 5, 765–775. Available from: https://doi.org/10.1111/1758-2229.12079.

Koenig, J.E., Spor, A., Scalfone, N., Fricker, A.D., Stombaugh, J., Knight, R., et al., 2011. Succession of microbial consortia in the developing infant gut microbiome. Proc. Natl. Acad. Sci. U. S. A. 108, 4578–4585. Available from: https://doi.org/10.1073/pnas.1000081107.

Korem, T., Zeevi, D., Zmora, N., Weissbrod, O., Bar, N., Lotan-Pompan, M., et al., 2017. Bread affects clinical parameters and induces gut microbiome-associated personal glycemic responses. Cell Metab. 25, 1243–1253. e5. Available from: https://doi.org/10.1016/j.cmet.2017.05.002.

Levy, M., Kolodziejczyk, A.A., Thaiss, C.A., Elinav, E., 2017. Dysbiosis and the immune system. Nat. Rev. Immunol. 17, 219–232. Available from: https://doi.org/10.1038/nri.2017.7.

Lin, H.V., Frassetto, A., Kowalik, E.J., Nawrocki, A.R., Lu, M.M., Kosinski, J.R., et al., 2012. Butyrate and propionate protect against diet-induced obesity and regulate gut hormones via free fatty acid receptor 3-independent mechanisms. PLoS One 7, 1–9. Available from: https://doi.org/10.1371/journal.pone.0035240.

Madan, J.C., Hoen, A.G., Lundgren, S.N., Farzan, S.F., Cottingham, K.L., Morrison, H.G., et al., 2016. Association of cesarean delivery and formula supplementation with the intestinal microbiome of 6-week-old infants. JAMA Pediatr. 170, 212–219. Available from: https://doi.org/10.1001/jamapediatrics.2015.3732.

Maeda, Y., Takeda, K., 2019. Host–microbiota interactions in rheumatoid arthritis. Exp. Mol. Med. 51. Available from: https://doi.org/10.1038/s12276-019-0283-6.

Main, B.S., Minter, M.R., 2017. Microbial immuno-communication in neurodegenerative diseases. Front. Neurosci. 11, 1–8. Available from: https://doi.org/10.3389/fnins.2017.00151.

Manichanh, C., Rigottier-Gois, L., Bonnaud, E., Gloux, K., Pelletier, E., Frangeul, L., et al., 2006. Reduced diversity of faecal microbiota in Crohn's disease revealed by a metagenomic approach. Gut 55, 205–211. Available from: https://doi.org/10.1136/gut.2005.073817.

Menni, C., Lin, C., Cecelja, M., Mangino, M., Matey-Hernandez, M.L., Keehn, L., et al., 2018. Gut microbial diversity is associated with lower arterial stiffness in women. Eur. Heart J. 39, 2390a–2397a. Available from: https://doi.org/10.1093/eurheartj/ehy226.

Morowitz, M.J., Carlisle, E.M., Alverdy, J.C., 2011. Contributions of intestinal bacteria to nutrition and metabolism in the critically ill. Surg. Clin. North Am. 91, 771–785. Available from: https://doi.org/10.1016/j.suc.2011.050.001.

Nettleton, J.E., Reimer, R.A., Shearer, J., 2016. Reshaping the gut microbiota: impact of low calorie sweeteners and the link to insulin resistance? Physiol. Behav. 164, 488–493. Available from: https://doi.org/10.1016/j.physbeh.2016.040.029.

Opstelten, J.L., Plassais, J., Van Mil, S.W.C., Achouri, E., Pichaud, M., Siersema, P.D., et al., 2016. Gut microbial diversity is reduced in smokers with Crohn's disease. Inflamm. Bowel Dis. 22, 2070–2077. Available from: https://doi.org/10.1097/MIB.0000000000000875.

Patterson, E.E., Ryan, P.M., Cryan, J.F., Dinan, T.G., Paul Ross, R., Fitzgerald, G.F., et al., 2016. Gut microbiota, obesity and diabetes. Postgrad. Med. J. 92, 286–300. Available from: https://doi.org/10.1136/postgradmedj-2015-133285.

Perry, G.H., Dominy, N.J., Claw, K.G., Lee, A.S., Fiegler, H., Redon, R., et al., 2007. Diet and the evolution of human amylase gene copy number variation. Nat. Genet. 39, 1256–1260. Available from: https://doi.org/10.1038/ng2123.

Plovier, H., Everard, A., Druart, C., Depommier, C., Van Hul, M., Geurts, L., et al., 2017. A purified membrane protein from Akkermansia muciniphila or the pasteurized bacterium improves metabolism in obese and diabetic mice. Nat. Med. 23, 107–113. Available from: https://doi.org/10.1038/nm.4236.

Proctor, L.M., 2011. The human microbiome project in 2011 and beyond. Cell Host Microbe 10, 287–291. Available from: https://doi.org/10.1016/j.chom.2011.100.001.

Ranjan, R., Rani, A., Metwally, A., McGee, H.S., Perkins, D.L., 2016. Analysis of the Microbiome: Advantages of Whole Genome Shotgun vs 16S Amplicon Sequencing. Elsevier Ltd., https://doi.org/10.1016/j.bbrc.2015.120.083.

Rath, C.M., Dorrestein, P.C., 2012. The bacterial chemical repertoire mediates metabolic exchange within gut microbiomes. Curr. Opin. Microbiol. 15, 147–154. Available from: https://doi.org/10.1016/j.mib.2011.120.009.

Ridaura, V.K., Faith, J.J., Rey, F.E., Cheng, J., Duncan, A.E., Kau, A.L., et al., 2013. Gut microbiota from twins discordant for obesity modulate metabolism in mice. Science (80-.) 341, 1–9. Available from: https://doi.org/10.1126/science.1241214.

Rooks, M.G., Garrett, W.S., 2016. Gut microbiota, metabolites and host immunity. Nat. Rev. Immunol. 16, 341–352. Available from: https://doi.org/10.1038/nri.2016.42.

Rothschild, D., Weissbrod, O., Barkan, E., Kurilshikov, A., Korem, T., Zeevi, D., et al., 2018. Environment dominates over host genetics in shaping human gut microbiota. Nature 555, 210–215. Available from: https://doi.org/10.1038/nature25973.

Russell, S.L., Gold, M.J., Hartmann, M., Willing, B.P., Thorson, L., Wlodarska, M., et al., 2012. Early life antibiotic-driven changes in microbiota enhance susceptibility to allergic asthma. EMBO Rep. 13, 440–447. Available from: https://doi.org/10.1038/embor.2012.32.

Scher, J.U., Ubeda, C., Artacho, A., Attur, M., Isaac, S., Reddy, S.M., et al., 2015. Decreased bacterial diversity characterizes the altered gut microbiota in patients with psoriatic arthritis, resembling dysbiosis in inflammatory bowel disease. Arthritis Rheumatol. 67, 128–139. Available from: https://doi.org/10.1002/art.38892.

Schippa, S., Iebba, V., Barbato, M., Di Nardo, G., Totino, V., Checchi, M.P., et al., 2010. A distinctive "microbial signature" in celiac pediatric patients. BMC Microbiol. 10. Available from: https://doi.org/10.1186/1471-2180-10-175.

Sender, R., Fuchs, S., Milo, R., 2016. Are we really vastly outnumbered? Revisiting the ratio of bacterial to host cells in humans. Cell 164, 337–340. Available from: https://doi.org/10.1016/j.cell.2016.010.013.

Shimizu, M., Hashiguchi, M., Shiga, T., Tamura, H.O., Mochizuki, M., 2015. *Meta*-analysis: effects of probiotic supplementation on lipid profiles in normal to mildly hypercholesterolemic individuals. PLoS One 10, 1–16. Available from: https://doi.org/10.1371/journal.pone.0139795.

Slavin, J., 2013. Fiber and prebiotics: mechanisms and health benefits. Nutrients 5, 1417–1435. Available from: https://doi.org/10.3390/nu5041417.

Turnbaugh, P.J., Hamady, M., Yatsunenko, T., Cantarel, B.L., Duncan, A., Ley, R.E., et al., 2009. A core gut microbiome in obese and lean twins. Nature 457, 480–484. Available from: https://doi.org/10.1038/nature07540.

Ursell, L.K., Metcalf, J.L., Parfrey, L.W., Knight, R., 2012. Defining the human microbiome. Nutr. Rev. 70, 38–44. Available from: https://doi.org/10.1111/j.1753-4887.2012.00493.x.

Valdes, A.M., Walter, J., Segal, E., Spector, T.D., 2018. Role of the gut microbiota in nutrition and health. BMJ 361, 36–44. Available from: https://doi.org/10.1136/bmj.k2179.

Valitutti, F., Cucchiara, S., Fasano, A., 2019. Celiac disease and the microbiome. Nutrients 11, 1–19. Available from: https://doi.org/10.3390/nu11102403.

Wang, J., Zheng, J., Shi, W., Du, N., Xu, X., Zhang, Y., et al., 2018. Dysbiosis of maternal and neonatal microbiota associated with gestational diabetes mellitus. Gut Microbiota. 67, 1614–1625. Available from: https://doi.org/10.1136/gutjnl-2018-315988.

Wang, M., Karlsson, C., Olsson, C., Adlerberth, I., Wold, A.E., Strachan, D.P., et al., 2008. Reduced diversity in the early fecal microbiota of infants with atopic eczema. J. Allergy Clin. Immunol. 121, 129–134. Available from: https://doi.org/10.1016/j.jaci.2007.090.011.

Wiley, N.C., Dinan, T.G., Ross, R.P., Stanton, C., Clarke, G., Cryan, J.F., 2017. The microbiota-gut-brain axis as a key regulator of neural function and the stress response: implications for human and animal health. J. Anim. Sci. 95, 3225–3246. Available from: https://doi.org/10.2527/jas2016.1256.

Yarur, A.J., Strobel, S.G., Deshpande, A.R., Abreu, M.T., 2011. Predictors of aggressive inflammatory bowel disease. Gastroenterol. Hepatol. 7, 652–659.

Zanteson, L., 2021. Gut health and immunity-its all about the good bacteria that can help fight disease. Today's Dietit. 14, 58–61.

Zeevi, D., Korem, T., Zmora, N., Israeli, D., Rothschild, D., Weinberger, A., et al., 2015. Personalized nutrition by prediction of glycemic responses. Cell 163, 1079–1094. Available from: https://doi.org/10.1016/j.cell.2015.110.001.

Zhang, H., Sparks, J.B., Karyala, S.V., Settlage, R., Luo, X.M., 2015. Host adaptive immunity alters gut microbiota. ISME J. 9, 770–781. Available from: https://doi.org/10.1038/ismej.20140.165.

Zhang, M., Sun, K., Wu, Y., Yang, Y., Tso, P., Wu, Z., 2017. Interactions between Intestinal microbiota and host immune response in inflammatory bowel disease. Front. Immunol. 8, 1–13. Available from: https://doi.org/10.3389/fimmu.2017.00942.

Zheng, D., Liwinski, T., Elinav, E., 2020a. Interaction between microbiota and immunity in health and disease. Cell Res. 30, 492–506. Available from: https://doi.org/10.1038/s41422-020-0332-7.

Zheng, W., Zhao, W., Wu, M., Song, X., Caro, F., Sun, X., et al., 2020b. Microbiota-targeted maternal antibodies protect neonates from enteric infection. Nature 577, 543–548. Available from: https://doi.org/10.1038/s41586-019-1898-4.

Chapter 5

Individual microbiota correction and human health: programming and reprogramming of systemic and local immune response

Tamara Meleshko[1,2], Oleksandra Pallah[1,2] and Nadiya Boyko[1,2,3]

[1]Department of Clinical Laboratory Diagnostics and Pharmacology, Faculty of Dentistry, Uzhhorod National University, Uzhhorod, Ukraine, [2]Research Development and Educational Centre of Molecular Microbiology and Mucosal Immunology, Uzhhorod National University, Uzhhorod, Ukraine [3]Ediens LLC, Velyki Lasy, Ukraine

5.1 Introduction

5.1.1 What is known and what is new? Beneficial microbes: a new concept?

According to classical concepts, cultured human microorganisms are divided into a resident (autochthonous) microbiota, which is represented by microorganisms that are permanently present in the human body, and a temporary (allochthonous) microbiota, which involves microorganisms that are brought in from other host biotopes or other biotopes of the environment (Nava and Stappenbeck, 2011). However, new, more detailed information about the human microbiome and its changes in various diseases and underlying inflammatory processes became known with the development of genetic research methods, in particular 16S rRNA gene sequencing. Thus, according to different authors (Zhou et al., 2004; Bik et al., 2006), from 20% to 60% of the representatives of the human microbiome are unculturable, which led to an underestimation of the diversity of the microbiome.

The term "microbiome" was first proposed by Joshua Lederberg in 2001 (Lederberg et al., 2001). The terms "microbiome" and "microbiota" are generally interchangeable, but they characterize different concepts. In particular, the microbiota should be understood as a complex consisting of the totality of all commensal, symbiotic, and opportunistic microorganisms, while the microbiome is a complex consisting of the totality of genes (Ursell et al., 2012).

5.1.2 The concept of a healthy human microbiome

Usually, the term "healthy" is used to refer to the absence of overt disease (Aagaard et al., 2013). The first studies of the microbiome sought to identify a core set of microbial taxa that are common in healthy people under the hypothesis that the absence of such microorganisms would indicate dysbacteriosis (Turnbaugh et al., 2007). However, studies of ecological diversity among healthy individuals have found a marked difference in the taxonomic composition of the microbiome, which made such a hypothesis unlikely. Therefore designating a "healthy" microbiome as an ideal set of specific microorganisms is currently insufficient, since the measure of health of a microbiome should be about its function (Shafquat et al., 2014).

An alternative point of view is the novel understanding of healthy as a functional core (functiome), that is, a set of metabolic and other molecular functions that are performed by the microbiome within a particular environment but not necessarily provided by the same organisms in different individuals (Shafquat et al., 2014). Such a functional core provides processes (functions) that are not performed by human cells and therefore represent a potential basis for a symbiotic relationship between humans and the microbiome.

Microbiome, Immunity, Digestive Health and Nutrition. DOI: https://doi.org/10.1016/B978-0-12-822238-6.00015-7

Although there are great individual differences in the taxonomic composition of the microbiome at all sites of the body, the large number of metabolic pathways that occur at a given site is similar in different individuals (Turnbaugh and Gordon, 2009). This suggests that the core of a healthy microbiome may include specific combinations of microbial genes, metabolic blocks, and regulatory pathways that together contribute to a stable microbial ecology (Xu et al., 2014; Martiny et al., 2015).

5.1.3 The basic concept of a healthy normal microbiota and concepts of early formation of the microbiome

The first large-scale study of the human microbiome was carried out as part of the Human Microbiome Project (HMP), which lasted from 2007 to 2016 and took place in two phases. The goal of the first phase of the project (HMP1) was to identify the main representatives of a healthy human microbiome. The most tested niches were the nasal and oral cavities, the skin, the gastrointestinal tract, and the genitourinary system.

One of the main conclusions of HMP1 substantiates the provisions, the human phenotype is closely related to the taxonomic composition of the microbiome and is predicted by the molecular functions of dominant microorganisms, or the personalized strain-specific composition (Huttenhower et al., 2012). This conclusion became the basis for the development of the second phase of HMP - Integrative HMP (HMP2). The aim of HMP2 was to use omics technologies to investigate the role of the pathogenic microbiota in inflammatory bowel disease, the onset of type 2 diabetes (T2D) and in pregnancy and preterm birth (Integrative et al., 2019).

5.1.4 Microbiome formation and early colonization

Factors that influence the dynamics of the microbiota at an early age are important predictors of a healthy microbiome. Enrichment of the intestinal microbiome of infants with bacteria such as *Bacteroides*, *Parabacteroides*, *Clostridium*, *Lactobacillus*, *Bifidobacterium*, and *Faecalibacterium prausnitzii* provide several determinants of a healthy microbiome. This anaerobic component of the microbiome provides synthesis of short-chain fatty acids, which have immunomodulatory (Smith et al., 2013; Lazarenko et al., 2017) and antitumor properties (Chang et al., 2014; Tang et al., 2011).

Childbirth and the way in which infants are fed, namely, with breast milk or formula, can affect the formation of the microbiota at an early age. Babies who were born via natural childbirth have an intestinal microbiome that is similar to their mothers' vaginal microbiome and is composed predominantly of *Lactobacillus*, *Prevotella*, and *Sneathia*. The microbiome of a baby who was born by caesarean section is more similar to the mother's skin microbiome and is dominated by *Corynebacterium* and *Propionibacterium* (Dominguez-Bello et al., 2010) and is enriched with conditionally pathogenic microorganisms, particularly *Haemophilus* spp., *Enterobacter cancerogenus*, *E. hormaechei*, *Veillonella dispar*, *V. parvula* (Bäckhed et al., 2015), and *Staphylococcus* (Smith et al., 2013). Such microorganisms acquired from the mother's skin microbiome can persist during the first year of the baby's life (Bäckhed et al., 2015) and, under certain conditions, can cause some diseases.

5.1.5 The endocannabinoid axis: the microbiome as a mechanism by which lifestyle affects various aspects of metabolism, which in turn can lead to metabolic syndrome and metabolic disorders

The endocannabinoid (eCB) system and its role in many physiological and pathological conditions, including cardiovascular disorders, is well described and accepted. However, the eCB system has been expanded into an "-ome": the endocannabinoidome, which includes endocannabinoid-related mediators, their protein targets, and metabolic enzymes, many of which significantly affect cardiometabolic health (Di Marzo and Silvestri, 2019).

According to recent studies, each human being has a unique microbiome, and the ratio of commensals covering the mucosal sites is highly flexible in biodiversity and function. Still, it remains unclear how to use these data to correct the balance between health and disease. It might be an important issue to understand all the mechanisms by which the microbiota modulates or affects our health and apply them at least in two closely related areas of personalized approaches: in the treatment of patients and for selection of the correct individual methods of prevention of various non-communicable diseases that are connected with the human microbiome disturbances. This will also provide valuable data for practical implementation in all other aspects of predictive, preventive and personalized (3P) medicine (Golubnitschaja et al., 2013).

5.2 Brief methodology overview

This chapter incorporates data that were obtained from clinical trials, cohort studies, databases, and IT tools for novel functional food development. Preliminarily, we were able to analyze in vitro the effects of individual commensal bacteria on monocyte-derived human dendritic cell (moDC)−mediated inflammation and effector T lymphocyte priming conditions mimicking the unique intestinal microenvironment. Human moDCs expressing PPARγ also regulate the cell surface expression of type I and II CD1 glycoprotein receptors as well as mucosa-associated CD103 protein differently in the absence or presence of all-trans retinoic acid (ATRA), when ATRA is providing a tolerogenic effect. In other words, they make possible (enabling) the proinflammatory and antiinflammatory reprogramming of this population of immune cells (Bene et al., 2017).

In another in vitro cell model we show that polyphenols were able to modulate inducible nitric oxide synthase activation of human umbilical vein endothelial cells and regulate the inflammation in a dose-dependent manner. The ability of local traditional plant-based food components, that is, bioactives, to modulate gut microbiota and local immune responses has been detected in fragment culture supernatant from the small intestine (duodenum, jejunum, and ileum). Mesenteric lymph nodes, Payer Patches by profiles of secreted IL (1b, 2, 6, 8, 10, 12, and 17), and secretory IgA (SIgA) have been also noted (Konic-Ristic et al., 2013; Hollands et al., 2013; Woodcock et al., 2013).

5.3 Biomarker validation: metabolic syndrome and cardiovascular diseases

Three groups of patients were formed. Group N.1 included 30 patients with metabolic syndrome (Ms) and T2D; group N.2 included 42 patients with cardiovascular diseases (CVDs); and group N.3 included 15 healthy individuals.

Gut microbiota profiles, biochemical panels (C-reactive protein, uric acid, triglycerides, glycosylated hemoglobin, and cholesterol), and immune parameters (as total antibodies to *Helicobacter pylori*, total immunoglobulin A (IgA) in serum, secretory IgA (SigA) in coprofiltrate, tumor necrosis factor alpha (TNF-α), interleukin-1β (IL-1β), interleukin-10 (IL-10), and interleukin-12 (IL-12)) parameters were measured in all groups of patients.

5.3.1 Limited clinical trial: person-oriented treatment of type 2 diabetes

A randomized controlled trial was conducted in two parallel groups. Group I (experimental) included 35 patients who followed an 18-day personalized diet, which included individually selected products that were rich in biologically active substances (BAS) and yogurts with unique microbial starters. Group II (control) involved 21 patients who, for 18 days, ate berries and yogurt prepared without microbial starters in the morning. Patients were not instructed to do additional physical exercise. A personalized diet was developed for each participant in the experiment in accordance with the state of the person's gut microbiota and immune and biochemical parameters. The diets included products that contained functioning groups of BAS, such as polyphenols, anthocyanins, and flavonoids, as well as unique microbial starters for fermentation (*Lactobacillus casei* IMB B-7412, *Lactobacillus plantarum* IMB B-7414, and *L. plantarum* IMB B-7413). Before and after the diet course we measured parameters such as gut and oral microbiota, biochemical status, immune status (IL-1β, Il-10, TNF-α, IgA, SIgA), and physical parameters (measurement of body weight and circumference of waist, thighs, and upper thighs). Gut and oral microbiota studies were performed in dynamics on days 1, 7, 11, and 18 of the experiment. All biochemical parameters were assayed by using Cobas (Roche/Hitachi) Switzerland. Immune parameters were measured by using indicator immunosorbent systems Vector-Best (Russian Federation); results were read at a wavelength of 450 nm using a plate immunosorbent assay BioTek Elx800.

The first attempt to prioritize biomarkers was made for the patients with oral pathology and patients with T2D (Horzov et al., 2020; Meleshko et al., 2021a,b,c).

5.4 Results and discussion

5.4.1 Patterns of clinically validated biomarkers in the gastrointestinal microbiome for patients with metabolic syndrome, obesity, type 2 diabetes, and cardiovascular diseases

The key changes in the gut microbiota of patients with metabolic disorders include an increase in the total content of *Enterococcus faecalis* and *E. faecium* to 10^9 CFU/g, levels of conditionally pathogenic enterobacteria, while the total amount of *Lactobacillus acidophilus* and *L. salivarius* decreased to 10^5 CFU/g, and number of lactose-positive strains of *Escherichia coli* was reduced up to $(4.0 \pm 0.11) \times 10^5$ CFU/g. The dominant conditionally pathogenic enterobacteria included *Klebsiella pneumoniae* and *Proteus vulgaris*, as well as *Clostridium tertium*, the content of which reached 10^8 CFU/g.

All examined patients with metabolic syndrome demonstrated an increase in the synthesis of total serum IgA, IL-1β and TNF-α levels, as well as a decrease in the synthesis of SIgA in the coprofiltrate, an increase in the amount of IL-12 in the serum, and the appearance of elevated population levels of C-reactive protein and total antibodies to *H. pylori*.

The connection between the content of typical representatives of the commensal microbiota—*Bifidobacterium bifidum*—and the level of cholesterol in the serum was established. Patients with abdominal obesity, overweight, and high cholesterol ($>8.24\pm1.2$ mmol/L) were found to have *B. bifidum* levels of up to $(4.7\pm0.38)\times10^{8}$ CFU/g. At the same time, there was a decrease in the number of *L. acidophilus* ($2.2\pm0.05\times10^{7}$ CFU/g) and lactose-positive *E. coli* ($4.0\pm0.11\times10^{5}$ CFU/g) in individuals with a high content of glycosylated hemoglobin ($>7.4\%\pm2.4\%$) in the serum.

According to our quantitative data, the key changes in the gut microbiota of patients with CVD included a reduction in the total number of representatives of the normal microbiota: *Lactobacillus* spp. (*L. acidophilus*, *L. salivarius* subsp. *salicinius*), and *E. coli* with normal enzymatic properties along with a significant increase in the number of lactose-negative strains of *E. coli*, *E. faecalis*, *Staphylococcus aureus*, *S. saprophyticus*, *Candida albicans*, and *C. krusei*. Exceptions involved bifidobacteria and *C. tertium*, the content of which in the intestines of patients did not undergo significant changes and were in high concentrations (up to $10^{8}-10^{9}$ CFU/mL). A decrease in the synthesis of SIgA in the coprofiltrate and a decrease in the production of all cytokines in the serum were also found.

The level of secretory SIgA in the coprofiltrate significantly correlated only with the amount of *P. vulgaris* and the number of bifidobacteria. TNF-α was interrelated with the levels of both proinflammatory IL-1β and IL-12 and antiinflammatory cytokines IL-10 as well as with the amount of *C. tertium* in the large intestine.

Therefore along with the classic markers of early diagnosis of dyslipidemia, metabolic syndrome, and obesity, generally, such markers as total IgA in the blood and SIgA secreted in the coprofiltrate as well as levels of such representatives of gut microbiota such as *Bifidobacterium* spp., *Lactobacillus*, *Enterococcus* spp., the ratio of lactose-positive to lactose-negative *E. coli* can be used. The bacteria *P. vulgaris* ($7.0\pm0.33\times10^{5}$ CFU/g) should be considered as an indicator of a primary disturbance in a body's lipid-hydrocarbon balance.

The identified stable correlations between species and quantitative composition of intestinal microbiota, immune parameters, and levels of glycosylated hemoglobin, cholesterol, overweight, and impaired glucose tolerance have become the basis for the proposed use of indicators of gut microbiota and immune markers of nonspecific subclinical inflammation for early diagnosis of CVDs and metabolic syndrome (Chendey et al., 2014). The results that were obtained during the study can also be used for the development of targeted microbiota correction approaches for the treatment and prevention of these diseases (Meleshko et al., 2020; Petrov and Boyko, 2014).

5.4.2 Data from human trials: treatment by individual nutrition and biopreparation with pharmabiotics

5.4.2.1 Treatment by personalized nutrition patients with type 2 diabetes

T2D is one of the most common noncommunicable diseases in the world. The aim of our study was to investigate the possibility of correcting the lipid metabolism of patients with T2D by using a personalized diet based on the most important microbial, biochemical, and immunological biomarkers of chronic inflammation.

After 18 days of the experiment, no statistically significant changes in parameters were observed in the control group, but there were changes in blood and physical parameters and microbiota composition in the experimental group.

After adherence to a personalized diet, in the experimental group patients there was a statistically significant reduction in the following parameters: glucose, thymol test, very low-density lipoprotein (VLDL), urea, creatinine, sIgA, and TNF-α, as well as all physical parameters. There was a statistically significant increase in such biochemical parameters as uric acid, sodium, and magnesium levels. Regarding intestinal microbiota indicators, there were statistically significant changes observed in *Enterococcus faecalis*, *Escherichia coli* (lac +), *E. coli* (lac^{-}), *Lactobacillus* spp., and *Candida* spp. The oral microbiota was characterized by a decrease in the number of all representatives except lactobacilli. Such microorganisms as *Enterococcus faecalis*, *Lactobacillus* spp., *Pseudomonas aeruginosa*, and *Candida* spp. demonstrated statistically significant changes.

Statistically significant changes in the concentration of microorganisms (in dynamics) were observed in both the intestinal microbiota (*Enterococcus faecalis*, *Escherichia coli* (lac +), *E. coli* (lac^{-}), *Lactobacillus* spp., and *Candida* spp.) and the oral microbiota (*Enterococcus faecalis*, *Lactobacillus* spp., *P. aeruginosa*, and *Candida* spp.). All these changes indicate an improvement in the patients' condition in the experimental group compared to the control group. The algorithm that we used for the development of personalized diets for patients with T2D demonstrated clinical efficacy in its implementation.

The issue of treatment of T2D is still relevant. The emergence of a number of new markers greatly simplifies and increases the accuracy of the disease diagnosis, but medical personnel still mostly use long-tested, "classic" markers of diabetes, in particular because of their availability for analysis (DeFronzo et al., 2015). In our work, we used a classic set of such markers as well as a number of other indicators, including intestinal and oral microbiota as recent publications emphasize its significant role in the development of T2D and human health in general (Gurung et al., 2020; Sharma and Tripathi, 2019). According to the results of changes in indicators during the experiment, there are improvements in a number of markers, such as VLDL, glucose, creatinine, urea, magnesium, sodium, thymol test, and uric acid.

5.4.2.2 Data of the effective treatment by novel pharmabiotics

Periodontal tissue disease and dental caries are among the most common oral diseases in children and adolescents. Associated pathologies of periodontal tissue inflammation in children are chronic gastritis and duodenitis, which in turn are associated with a violation of the microbial cenosis of the intestine.

In our study, this problem was solved by using the *Lactobacillus acidophilus 042* (IMB B-7413). *L. casei 021* (IMB B-7412), and *Lactobacillus delbrueckii 0–22* (IMB B-7414) strains as probiotic components together with plant-based carriers as prebiotic components in a composite biopreparation for treatment of periodontal tissue inflammation in children and correction of associated gastroduodenal intestinal disorders.

Based on examination by dentists and gastroenterologists of 80 randomly selected children, three groups were formed: (1) a control group of children who had no signs of dental caries, periodontal tissue inflammation, or any registered gastrointestinal diseases; (2) the first experimental group, which consisted of patients diagnosed with dental caries and periodontal tissue inflammation but with no symptoms of gastroduodenal intestinal disorders; and (3) the second experimental group, which consisted of patients who were under dispensary registration with a gastroenterologist and had clearly pronounced signs of dental caries and inflammation of periodontal tissues. The oral and intestinal microbiota were studied in all the examined children.

The oral and gut microbiomes of children with inflammatory processes were characterized by an increased number of microorganisms of the genus *Candida*. *Enterococcus faecalis* in high titers was isolated only in the saliva of children with gastroduodenal intestinal disorders. Representatives of the genus *Lactobacillus* were completely absent in saliva and gingival sulcus samples and found in insignificant amounts in the contents of the large intestine. Thus the indicator microorganisms of the listed oral inflammatory processes include bacteria of *Streptococcus* genus, while the markers of gastroduodenitis are increased titers of *Escherichia coli* (including with altered enzymatic activity), increased frequency of other conditionally pathogenic enterobacteria (primarily *Klebsiella* spp., *Citrobacter* spp., and *Proteus* spp.), appearance of *S. aureus* and *Candida* spp. strains in significant titers, and a decrease in lactobacilli and enterococci, which to some extent clarifies the available literature.

A study of all species of microorganisms that were isolated from children with signs of dental caries, periodontal disease, and gastroduodenal disease and their relation to antibiotics showed that the strains of microorganisms that were most sensitive to antibiotics were strains of lactobacilli. The tested strains of *S. aureus* and *C. albicans* were characterized by resistance to a number of tested antibiotics.

From fermented traditional products and the gastrointestinal contents of healthy individuals we isolated three strains of lactobacilli: *L. casei* IMB B-7412, *L. delbrueckii* IMB B-7414, and *L. acidophilus* IMB-7413. These strains were characterized by highly selective antagonistic activity against cultures of streptococci, *S. aureus*, *P. aeruginosa*, *C. albicans*, and conditionally pathogenic enterobacteria that were isolated from sick children.

We studied the effects of extracts of a number of edible plants (Boyko, 2013) on different groups of microorganisms, including representatives of commensal and opportunistic microbiota, as well as the effect of apple pectin (15% and 30%) on the growth of *L. casei* and *L. delbrueckii*. The studies made it possible to select extracts that not only inhibit targeted pathogenic and conditionally pathogenic bacteria, but also have a stimulating effect on strains of *L. casei*, *L. delbrueckii*, and *L. acidophilus*, which is a probiotic component of synbiotics.

By the method of coculture, such an effect was found in the aqueous extracts of cherries (*Prunus cerasus*) and blueberries (*Vaccinium myrtillus L.*) and vegetable juice from peppers, carrots, and tomatoes. In particular, we showed that cherry extract inhibits the growth of *Enterococcus faecalis*, *E. faecium*, *K. pneumoniae subsp. oxytoca*, *P. morganii*, and *L. salivarius* but stimulates the growth of *L. delbrueckii*, whereas blueberry extract inhibits *E. faecalis* and at the same time stimulates *L. acidophilus*. The tested vegetable juice inhibits the growth of lactose-negative strains of *Escherichia coli* and stimulates the growth of *L. casei* and *L. delbrueckii*. The 15% apple pectin showed a stimulating effect on *L. delbrueckii*, and the 30% apple pectin showed such an effect on *L. casei* and *L. delbrueckii*.

The composite biopreparation was tested in a limited clinical trial on 40 volunteer children of school age (12–15 years). Of these, 28 were clinically diagnosed with dental caries and periodontal tissue disease and associated gastroduodenitis, and 12 formed a control group.

During the first seven days, children of the control group rinsed their mouths with *L. delbrueckii* suspension, and during the next seven days they rinsed with a suspension of *L. casei*. For this purpose, lyophilized strains of lactobacilli (concentration: 10^8 CFU/mL) were dissolved immediately before consumption in boiled water that had been cooled to room temperature. The children in the main group were given a cherry extract with suspended bacteria as a drink for 14 days.

At the beginning of the experiment and at the end of the first and second weeks, the children were examined by a dentist and a gastroenterologist. Oral and gut microbiota were analyzed in all patients, and the levels of secretory IgA in saliva and coprofiltrate before and after they took the biopreparation were determined.

As a result of limited clinical trials, gram-positive cocci, in particular the heterogeneous group of streptococci (*S. salivarius, S. mitis*), have been shown to be dominant among all identified species of aerobic microorganisms that were isolated from dental plaque in sick children.

Several species of enterobacteria (*K. oxytoca, lactose-negative strains of Escherichia coli, Enterococcus faecalis*) and microscopic fungi *Candida albicans*, *Cryptococcus hymiloca*, and *Trichosporon* spp. were isolated from the oral cavity. Microbial analysis of gingival sulcus content showed similar results in terms of the species composition of isolated aerobic microorganisms, but in higher titers (10^5 or 10^6 CFU/mL).

After two weeks of using the biopreparation, changes in the microbiota of the mucous membranes of the oral cavity and intestines were observed, indicating the therapeutic effect of its use.

The most significant changes were in the microbial cenosis at the tooth cervix, where in most cases there was a significant decrease in titers of streptococci (up to 10^3 CFU/mL), which are the main cause of plaque formation on teeth, inflammation of periodontal tissues, and dental caries formation.

In sick children before the start of the clinical trial, an almost complete absence of lactobacilli was registered both in the contents of the gingival sulcus and in saliva samples. After the two-week application of the biopreparation, both in the children's saliva and in the neck of the tooth, both of the lactobacilli that were contained in the biopreparation were detected. The titers of bifidobacteria (10^6) represented by the genus *Bifidobacterium breve* remained unchanged during the study.

In most children, before the use of the biopreparation, conditionally pathogenic microorganisms dominated the intestinal contents. Their numbers after two weeks of use of the biopreparation were lower than the corresponding values in the control group. It should be noted that the content of staphylococci decreased to the limit of determination (up to 10^1).

Thus the use of a composite biopreparation based on lactobacilli provides restoration of oral mucous membrane microbiota and normalization of intestinal microbiota.

To confirm the effectiveness of the biopreparation, we tested the activity of local immunity of oral mucous membranes and intestines by determining changes in the levels of secretory IgA in children before and after using the biopreparation in saliva and coprofilters.

The levels of SIgA in the saliva of most of the examined children who had been diagnosed with dental caries, periodontal disease, and concomitant gastroduodenitis were much higher, indicating the course of their inflammatory processes.

Studies of the levels of secretory IgA in the saliva of children who had taken a composite biopreparation with lactobacilli showed that after a week of its use, the level of SIgA in 22 patients decreased to 40 mg/L; the level increased from 1559.6 to 2000 mg/L in only four cases. After a week of its use, the level of SIgA increased and averaged about 200 mg/L in most subjects, which is within normal limits (115.3–299.7 mg/L) for adults.

In the control group of children (12 subjects) the level of secretory SIgA ranged from 66.76 to 165.19 mg/L.

5.4.3 IT tools for implementation of 3P medicine and individual nutrition

There have been very few attempts to use microbiota indices particularly to connect with foods on the basis of their glycemic index. The algorithm that we used for the formulation of personalized diets for patients with T2D demonstrated clinical efficacy in its implementation. The same approach might be applied for prevention and treatment of chronic diseases connected with inflammation.

5.4.4 Databases

Implementation of these ideas is impossible without the creation of a database with diverse information about the properties, distribution, and ecological safety of selected species. It is quite obvious that this database should be relational and that its development should take into account the need for using the massive amount of data that have been collected about edible plants, their extracts, and biologically active compounds (BAC) in a number of existing databases.

5.4.5 IT tools

For the accessibility of this system, plants should be presented not only biologically, but also geographically, that is, as objects, a particular plant species taken from a particular area (by definition, plant area, known as areal, is part of a specific territory where certain plant species most commonly occur and thrive). Geographical objects are characterized by the so-called spatial data that allow playback on an electronic map location and by the size and shape of the objects.

Our results, in line with similar attempts by others, substantiate the opportunity for proper maintenance of the human microbiome and its direct and targeted regulation in order to prevent or effectively treat a variety of noncommunicable diseases and their comorbidities. These are particularly connected to low-grade inflammation depending on the individual's nutrition.

Finally, the solution for the maintenance of the human microbiome by personalized and healthy nutrition is also connected to local farmers' products in the short supply food chain by actors that may be interested in this approach to food safety and sustainability (Delicato et al., 2019).

5.4.6 Mechanisms involved and prospective research

The already known mechanisms of host-bacterial interaction, such as via selectively induced expression of REG III g/b leptins (Keilbaugh et al., 2005), modulation of immune response by local SIgA production (Boyko et al., 2007), antagonistic properties, controlling of chronic inflammation, and Ag presenting cells reprogramming, can be updated here by a few recently discovered and demonstrated pathways of specific immune response regulation after targeted correction of human microbiota involving the individual composition of mucosal sites.

Among the most fresh, interesting, and very specific signs of cell-to-cell communications are the bacterial cell wall structures and elasticity connections with specific immune responses and specificity of actions and the influence of different plant BACs on the modulation of both microbiota composition and corresponding host immune responsiveness.

The probiotics (pharmabiotics, immunobiotics) that are often presented by lactic acid bacteria (LAB) also have a strong potential to be used in "healthy diets" in an integrated approach to immunity-related disease management (Kobyliak et al., 2018; Bubnov et al., 2015). The cell wall of probiotic bacteria is a dynamic entity, dependent on many components, and plays an essential role in modulating the immune response. We have shown according to atomic force microscopy (AFM) data that strains with a more elastic cell wall demonstrated higher resistance to intracellular digestion in macrophages and higher levels of their activation. All tested LAB strains had a stimulating effect on the functional activity of macrophages and the ability to produce NO/NO_2 in vitro. AFM might be considered to be a fast and accurate method to assess parameters of probiotic strains' cell walls to predict their immunomodulatory properties (Mokrozub et al., 2015).

5.4.7 Biological active compounds as potential modern prebiotics

Anthocyanins are pigments that are responsible for the quality of fruit and are markers of ripening, as most fruits accumulate anthocyanins only in the ripening phase (Roate et al., 2011). Anthocyanins, which are glycosides of anthocyanidins, are derivatives of phenolic compounds and are natural plant pigments that give color to fruit from red to purple. Previous studies showed that the alkaloids that are present in berries have antidiarrheal, antimicrobial, and antiinflammatory properties (Wrolstad et al., 2005).

When choosing berries and fruits, we considered their availability on the market, their use in the food industry, and their ability to suppress opportunistic pathogens. The second stage of the study involved obtaining methanolic extracts of alycha, plum, red and black currants, blueberry, sweet cherry, and jostaberry, followed by distillation of the solvent and a quantitative determination of BAC content by means of thin-layer chromatography. In liquid extracts of berries the contents of biologically active substances were determined that belonged to the class of anthocyanins and polyphenols, and their quantitative content in each investigated extract was measured. Black currant and jostaberry fruits were characterized by the same qualitative composition of BAC (delphinidin-3-glucoside, delphinidin-3-rutinoside, cyanidin-3-glucoside, cyanidin-3-rutinoside) but with different quantitative content of anthocyanidins. These extracts stimulated the growth of *L. casei*, and jostaberry extract additionally stimulated the growth of *L. fermentum* and *L. catenaformis* strains (Table 5.1). The black currant extract (*Ribes nigrum*) contained the following biologically active substances:

TABLE 5.1 Biological influence of fruit and berry extracts on the growth of lactic acid bacteria in dynamics.

Tested extracts	*L. acidophilus* (CFU/mL)		*L. catenaformis* (CFU/mL)		*L. casei* (CFU/mL)		*L. fermentum* (CFU/mL)	
	14 h	**24 h**	**14 h**	**24 h**	**14 h**	**24 h**	**14 h**	**24 h**
Prunus cerasifera	$1 \times 10^3 \pm 0.5$	$1 \times 10^3 \pm 0.5$	$1 \times 10^4 \pm 1.4$	$1 \times 10^3 \pm 0.79$	$1 \times 10^3 \pm 1.7$	$1 \times 10^3 \pm 0.19$	$0.5 \times 10^6 \pm 0.25$	$1 \times 10^3 \pm 0.17$
Ribes x nidigrolaria	$6 \times 10^4 \pm 1.1$	$2 \times 10^5 \pm 1.1$	$3 \times 10^4 \pm 0.5$	$2.5 \times 10^8 \pm 0.5$	$0.5 \times 10^6 \pm 0.35$	$3 \times 10^8 \pm 0.3$	$0.5 \times 10^6 \pm 0.2$	$1 \times 10^9 \pm 0.3$
Prunus avium	$0.5 \times 10^8 \pm 0.79$	$5.5 \times 10^6 \pm 0.5$	$1 \times 10^5 \pm 1.1$	$0.5 \times 10^8 \pm 0.26$	$7.5 \times 10^4 \pm 0.3$	$3 \times 10^8 \pm 0.29$	$1 \times 10^6 \pm 0.42$	$0.5 \times 10^8 \pm 0.4$
Ribes nigrum	$3 \times 10^4 \pm 0.5$	$3 \times 10^5 \pm 0.76$	$3 \times 10^4 \pm 0.5$	$2.5 \times 10^8 \pm 0.5$	$0.5 \times 10^6 \pm 0.35$	$1 \times 10^9 \pm 0.28$	$0.5 \times 10 \pm 0.38$	$1 \times 10^6 \pm 0.57$
Ribes rubrum	$4 \times 10^4 \pm 0.5$	$2.5 \times 10^4 \pm 1.2$	$0.5 \times 10^8 \pm 0.26$	$4.5 \times 10^8 \pm 0.5$	$0.5 \times 10^4 \pm 0.5$	$1 \times 10^4 \pm 0.17$	$0.5 \times 10^8 \pm 0.38$	$1 \times 10^3 \pm 0.45$
Vaccinium myrtillus	$0.5 \times 10^6 \pm 0.83$	$7.5 \times 10^8 \pm 0.5$	$1 \times 10^5 \pm 0.29$	$0.5 \times 10^8 \pm 0.29$	$0.5 \times 10^6 \pm 0.35$	$1 \times 10^9 \pm 0.5$	$0.5 \times 10^6 \pm 1.0$	$1 \times 10^9 \pm 0.92$
Prunus domestica	$1 \times 10^7 \pm 0.2$	$7.5 \times 10^8 \pm 0.5$	$1 \times 10^8 \pm 0.26$	$0.5 \times 10^8 \pm 0.29$	$0.5 \times 10^8 \pm 0.5$	$1 \times 10^8 \pm 0.5$	$1 \times 10^8 \pm 0.78$	$0.5 \times 10^8 \pm 0.38$

delphinidin-3-glucoside (20.4%), delphinidin-3-rutinoside (45%), cyanidin-3-glucoside (5.6%), and cyanidin-3-rutinoside (29%). The black currant extract stimulated the growth of *L. casei*. The red currant fruit (*R. rubrum*) contained the following anthocyanidins: cyanidin-3-sambubioside (15%), cyanidin3-xylosylrutinoside (66%), and cyaniding-3-rutinoside (19%). The fruit of sweet cherry had the ability to stimulate the growth *of L. casei, L. catenaformis*, and *L. fermentum*. However, in contrast to the extracts of black currant and jostaberry, only two types of anthocyanidins were identified in the fruit of sweet cherry: cyanidin-3-glucoside (0.7%) and cyanidin-3-rutinoside (99.3%). Plum fruits (*Prunus domestica*) contained three types of anthocyanins in their composition: cyanidin glycoside (8.6%), cyanidin-3-rutinoside (50%), and peonidin3-rutinoside (41.4%). As to the effect of the plum extract on strains of lactobacilli, we observed the stimulation of growth of all tested LAB and *B. subtilis* 090 strains.

Blueberries (*Vaccinium myrtillus*) contained the largest gross anthocyanin content in their composition. The extract contained 15 compounds of anthocyanins: delphinidin-3-galactoside (13.7%), delphinidin-3-glucoside (15%), cyanidin-3-galactoside (8.9%), delphinidin-3-arabinoside (9%), cyanidin-3-glucoside (11%), petunidin3-galactoside (4%), cyanidin-3-arabinoside (6.4%), petunidin-3-glucoside (9.8%), peonidin-3-galactoside (1.0%), petunidin3-arabinoside (2.4%), peonidin-3-glucoside (4.3%), malvidin-3-galactoside (3.1%), malvidin-3-glucoside (9.6%), and malvidin-3-arabinoside (1.8%) (Table 5.2). We noticed the ability of the blueberries to stimulate the growth of all LAB strains that we selected.

In analyzing the obtained data of the quantitative study of extracts composition, we noted that the number of compounds of anthocyanins and polyphenols is significantly lower in the extracts of the *Prunus* genus than in the extracts of the *Ribes* genus, but according to the experimental data that we obtained, plum and sweet cherry extracts were better in stimulating the growth of probiotic strains of microorganisms.

In analyzing the obtained chromatogram of the plum extract and comparing it with literature sources based on chromatographic analysis of extracts of different varieties of plums, we observed that the substances of the first priority here were cyanidin-3-xyloside and cyanidin-3-glycoside in various amounts (González-García et al., 2015). However, in examining our extract of *P. domestica*, we saw that the peak on the chromatogram corresponded to the cyanidin-3-glycoside substance. This can be explained by the fact that, depending on the geographical location, variety, and the year of life of the plant, the quantitative composition of compounds of the anthocyanins may change.

According to the data that we obtained, the fruits of alycha did not contain anthocyanins in their composition. This was confirmed by the fact that, owing to the presence of anthocyanins in fruits, the fruits were red to violet in color, and the alycha fruit was yellow. In analyzing the obtained chromatogram (Fig. 5.1), we saw that it had a very small amount of flavonoids and a small amount of coumarin. We also observed the content of polar phenols, and according to the analysis of literature data, they could be phenolic acids, such as catechin (Poonam et al., 2011).

The total content of polyphenols was 219 μg/mL. The alycha extract inhibited the growth of all the beneficial microorganisms that we selected; therefore when it is used as a prebiotic component of pharmabiotics, such undesirable effects on the LAB that we tested should be taken into account (see Table 5.1).

TABLE 5.2 **Biological influence of fruit and berry extracts on growth of *Escherichia coli 058*, *Enterococcus faecalis*, and *Bacillus subtilis 090* in dynamics.**

Tested extracts	*E. coli* (CFU/mL)		*E. faecalis* (CFU/mL)		*B. subtilis* 090 (CFU/mL)	
	14 h	**24 h**	**14 h**	**24 h**	**14 h**	**24 h**
Prunus cerasifera	$0.5 \times 10^2 \pm 0.29$	—	$0.5 \times 10^2 \pm 0.5$	—	$0.5 \times 10^5 \pm 0.78$	$1 \times 10^8 \pm 0.83$
Ribes × nidigrolaria	$0.5 \times 10^2 \pm 0.29$	$0.5 \times 10^2 \pm 0.79$	$0.5 \times 10^2 \pm 0.5$	$0.5 \times 10^2 \pm 0.5$	$0.5 \times 10^8 \pm 0.5$	$0.5 \times 10 \pm 0.28$
Prunus avium	$2 \times 10^2 \pm 1.1$	$0.5 \times 10^2 \pm 0.29$	$1 \times 10^2 \pm 0.29$	—	$0.5 \times 10^8 \pm 0.85$	$1 \times 10^3 \pm 0.74$
Ribes nigrum	$0.5 \times 10^4 \pm 0.5$	—	$0.5 \times 10^2 \pm 0.5$	—	$0.5 \times 10^5 \pm 0.3$	$0.5 \times 10^5 \pm 0.57$
Ribes rubrum	$5 \times 10^2 \pm 0.5$	$0.5 \times 10^2 \pm 0.5$	$0.5 \times 10^4 \pm 0.29$	$0.5 \times 10^4 \pm 0.5$	$0.5 \times 10^8 \pm 0.87$	$1 \times 10^3 \pm 1.14$
Prunus domestica	$0.5 \times 10^4 \pm 0.79$	$0.5 \times 10^2 \pm 0.79$	$0.5 \times 10^4 \pm 0.29$	$0.5 \times 10^4 \pm 0.79$	$0.5 \times 10^8 \pm 1.4$	$1 \times 10^8 \pm 0.44$
Vaccinium myrtillus	$1 \times 10^5 \pm 0.2$	$0.5 \times 10^5 \pm 0.5$	$0.5 \times 10^4 \pm 0.29$	$0.5 \times 10^2 \pm 0.29$	$1 \times 10^3 \pm 0.93$	$0.5 \times 10^2 \pm 0.83$

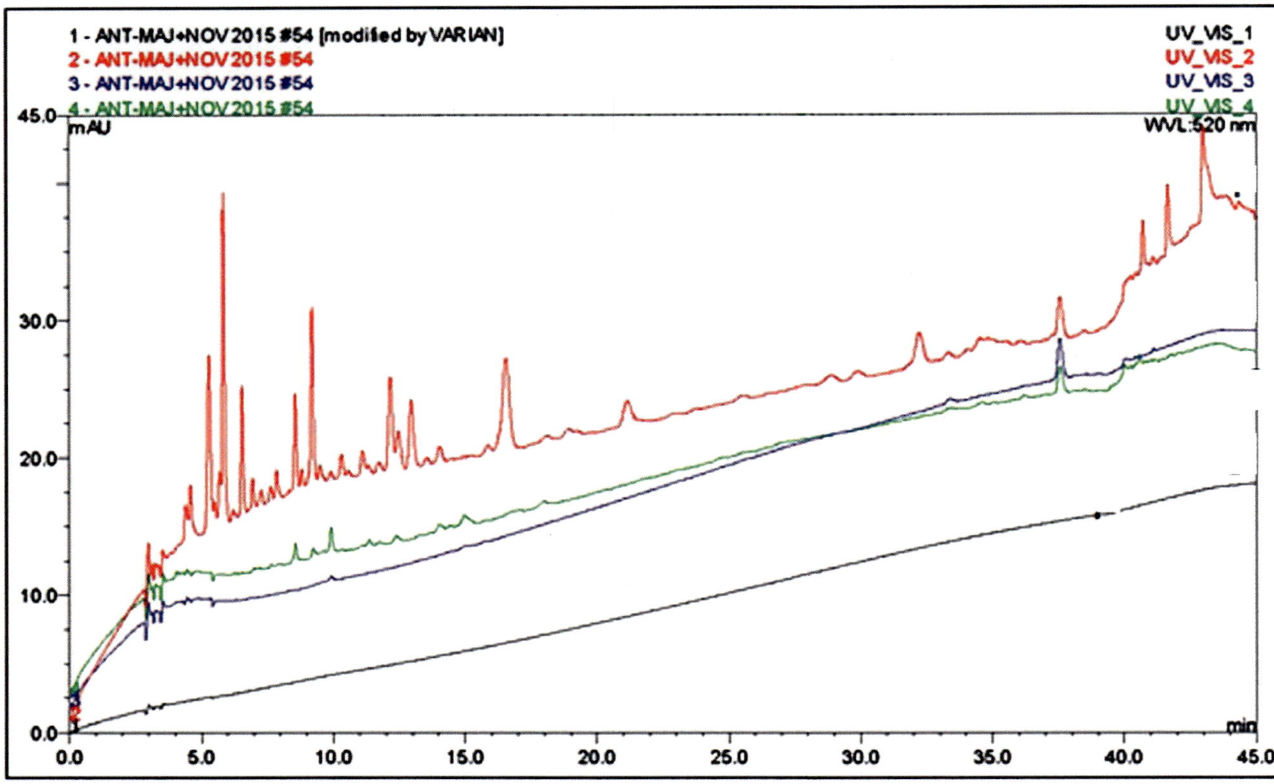

FIGURE 5.1 Chromatography of alycha.

In analyzing the quantitative and qualitative composition of BAC, included in the extracts that we tested, no anthocyanins whatsoever were detected in the alycha extract. We determined the contents of anthocyanins and polyphenols in extracts of fruits and berries and compared them with ePlant LIBRA, a database of BAC content in plant extracts (Plumb et al., 2016).

According to analysis of the content of BAC in the extracts of *Ribes rubrum*, *R. nigrum*, *R. nidigrolaria*, *V. myrtillus*, and *Prunus avium*, which were included in the ePlant LIBRA database, the cyanidin-3-rutinoside compound was not identified (Wu et al., 2004; Veberic et al., 2015), but it was found in the extracts of the fruits and berries that we studied. There was also a difference in a quantitative content of anthocyanins and polyphenols.

When we compared the BAC content of the *P. domestica* and *Prunus cerasifera* extracts that we studied with the ePlant LIBRA database, we noticed a difference, not in the qualitative composition of extracts, but in their quantitative indices (Jakobek et al., 2009). While investigating the probacterial properties of extracts in relation to beneficial strains of microorganisms, we found that extracts of red and black currants, jostaberry, plum, and blueberry exhibited such properties. In other words, these are the extracts that contain anthocyanins and polyphenols in their composition, and therefore we can conclude that the prebiotic properties are the properties of anthocyanins. This hypothesis is confirmed by the experimental data that we obtained; since the extract of alycha had no anthocyanins in its composition, it had an inhibitory action on the strains of lactic bacteria.

5.4.8 Early health programming

Since the available options for obesity treatment are far less than satisfactory, the primary focus in combating the obesity epidemic is on developing and implementing new innovative cost-effective prevention strategies (Koletzko et al., 2017). The recently published World Health Organization (WHO) report "Ending Childhood Obesity," which was developed by an international scientific expert group with contributions from 100 WHO member states, concluded that particular opportunities for effective prevention of obesity and associated noncommunicable diseases exist during pregnancy and before pregnancy, in infancy and early childhood, and in adolescence, which is considered a prepregnancy period in females (Commission-on-Ending-Childhood-Obesity, 2016).

On the basis of this vision and our own strategy that in line with the principles of P4 medicine (predictive, preventative, personalized, participatory), we have designed, initiated, and completed the first, to our knowledge, attempt for targeted individualized correction of disturbed vaginal microbiota of pregnant women. This was accomplished by the application of the multicomponent biopreparation synbiotic Bifiten in order to maintain newborn gut health and its beneficial effects.

The 50 women who were involved in our study were at different stages of pregnancy and were randomly divided into two groups. All the pregnant women were registered at the dispensary in our relevant regional unit and were given a preliminary examination in accordance with the regulations of the Ministry of Health of Ukraine on pregnancy.

The pregnant women were selected because of common complaints of experiencing profuse vaginal discharge (secretions) with an unpleasant odor. To eliminate these symptoms, the first group (25 pregnant women) were treated with the complex synbiotic Bifiten at a dose of capsule once a day orally for 10 days. The second group (25 pregnant women) received routinely prescribed vaginal suppositories with antiinflammatory action against bacteria, which were etiologically proven agents for these vaginal microbiome changes (Tsmur et al., 2016).

Patients in both study groups had comparable demographic indicators in terms of menarche, age, marital status, age at first sexual contact, number of sexual partners, and presence of negative habits (i.e., smoking, alcohol usage).

Before the study, all the pregnant women were informed in detail about the composition of the microbiotic (11 selected strains of live beneficial bacteria that are part of Bifiten and are considered safe and approved by European Food Safety Authority and the U.S. Food and Drug Administration), consisting of six species of lactobacilli (*Lactobacillus rhamnosus*, *L. casei*, *L. plantarum*, *L. acidophilus*, and *L. salivarius*); three species of bifidobacteria (*Bifidobacterium longum*, *B. bifidum*, and *B. lactis*); lactococcal bacteria (*Lactococcus lactis*); thermophilic streptococcus (*Streptococcus thermophilus*); and the prebiotic oligofructose. With the scheme of the application and the need for research, the women correspondingly signed an information acceptance letter (agreement), and after completion of Bifiten, the expected results were detailed, and they signed the informed consent form to participate in the study.

After application of Bifiten, significant improvement in the well-being of the women in the first group, in contrast to those in the second group, was observed, and this effect was stable until the end of the pregnancy.

All the women in the first group demonstrated good tolerability of Bifiten; there were no side effects of the microbial treatment, and the women showed improved intestinal motility, regulation of bowel movements, and reduction of flatulence, which did not happen with the use of typical antiinflammatory vaginal drugs.

Seventeen women from both groups experienced full-term pregnancies, while the pregnancies of 33 women were terminated. Perinatal complications were more common in the second group of patients receiving traditional antiinflammatory medication (Table 5.3).

The biggest, most complex, and most often detected complications were observed in the group of women with the worst indicators of vaginal microbiome.

TABLE 5.3 Variability in the frequency of complications of pregnancy and childbirth in women of the first and second groups.

Type of complications	First group ($n = 25$)	Second group ($n = 25$)
Gestational pyelonephritis	0	1
Threat of premature birth	2	3
Delay intrauterine development	0	1
Polyhydramnios	0	3
Dehydration	1	2
Premature rupture of the fruit shells	1	6
Anomalies of labor	2	7
Premature birth	1	1
Fetal distress	1	1
Delayed childbirth	1	2

The postpartum period in women of the first group proceeded without complications. Complications were observed mainly in the second group, and their course required medical correction (subinvolution of the uterus).

In children born to women in the first group, the early and late neonatal periods passed without complications. In the second group, in children born to women with an unrestored (not corrected) vaginal microbiome (intermediate type of smear), there were minor transient disorders of the digestive tract, namely, sluggish sucking, frequent vomiting, and intestinal cramps, which, however, did not require medical correction.

The health conditions were even more complicated (difficult) in newborn children of women who were diagnosed with combined intestinal dysbiosis and bacterial vaginosis (aerobic bacterial vaginosis). In the early neonatal period, in addition to the above-mentioned symptoms, these children showed decreased motoric (peristaltic) activity, decreased muscle tone, early and prolonged conjugated jaundice, significant weight loss, and prolonged recovery. Cyanosis and respiratory disorders were also observed in these children in the first days of life.

Both groups of pregnant women demonstrated a violation of the vaginal microbiome, including a significant reduction in the lactobacilli component and, in some cases, even its complete absence. At the same time, there was a significant predominance in the microbial cenosis of aerobic and anaerobic representatives of conventionally (potentially) pathogenic microorganisms, which are known agents of opportunistic vagina infections.

Only in the six women's vaginal secretions were lactobacilli in amounts similar to those in healthy individuals found before the study began (Fig. 5.2).

A potentially pathogenic microbiota was represented by different aerobic and anaerobic components. Among the aerobic and facultative strains, the anaerobic representatives enterobacteria (*Escherichia coli*), staphylococci, and *Candida* were most often found. Streptococci and enterococci were isolated less frequently. Among the anaerobic microorganisms, the groups *Gardnerella vaginalis*, *Prevotella bivia*, *Porphyromonas* spp., *Leptotrichia* spp., *Clostridium* spp., and *Atopobium vaginae* were found in all women with symptoms of anaerobic vaginosis. *Mycoplasma genitalium* was not detected in the majority of the pregnant women.

Bifiten application led to increased levels of lactobacilli in vaginal secretions. This indicator ranged from 11% to 16% of the total microbial count, which meant an almost twofold increase in their levels. Interestingly, in patients with a high initial total microbial count (TMC) its levels decreased after Bifiten oral administration, owing to a decrease in the number of obligate anaerobic microorganisms, while the number of lactobacilli in relation to other microorganisms increased significantly (up to 20%). If the TMC in patients was at the lower limit of normal ($10^{6.1}$), then they observed an increase in the number of microorganisms in general (e.g., to $10^{6.9}$ genetic copies), owing to the increase in the number of lactobacilli, while their percentage remained unchanged.

Thus, we found that Bifiten has the ability to restore the vaginal microbiome.

All patients showed a decrease in microscopic fungi of the genus *Candida* two weeks after administration of Bifiten. The fungi were found at levels of no more than 10^1-10^2 CFU/mL of vaginal secretions or were not detected at all.

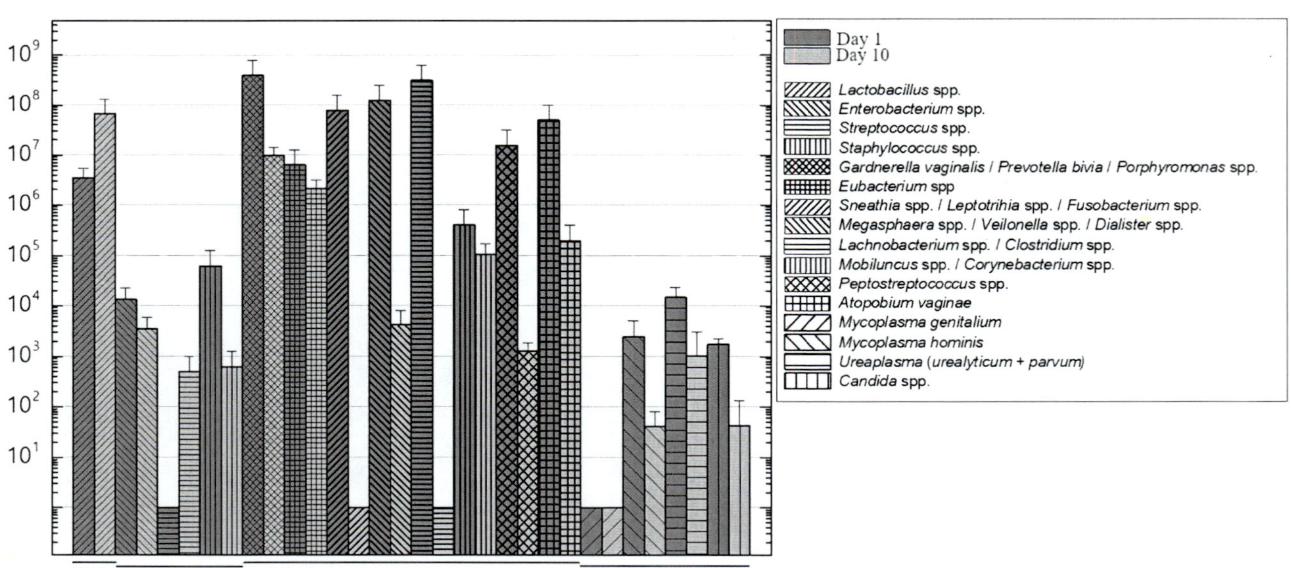

FIGURE 5.2 Number of microorganisms (bacterial DNA). Genome equivalent/sample.

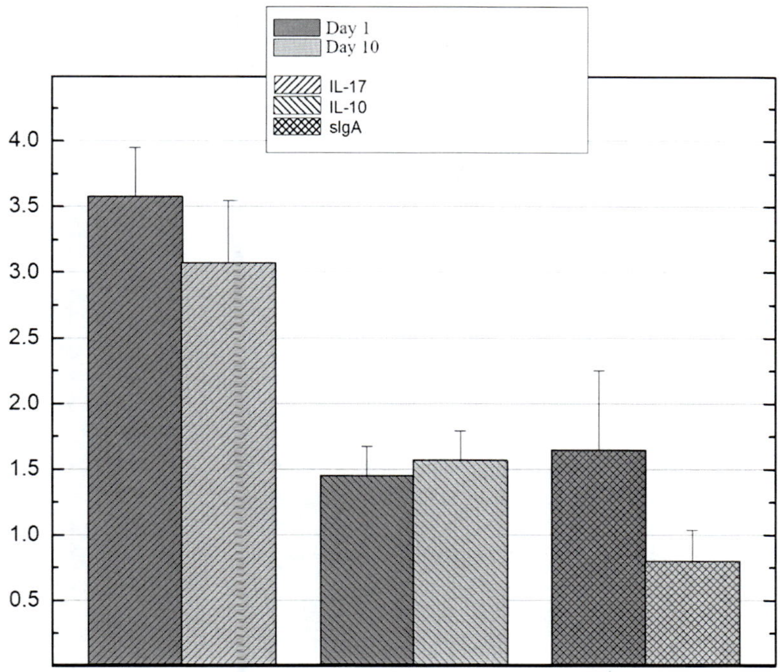

FIGURE 5.3 Changes in the amount of proinflammatory (IL-17) and antiinflammatory (IL-10) cytokines and SIgA in the secretions of pregnant women due to the use of the biological product Bifiten.

The number of anaerobic microorganisms (especially *Enterobacter* spp. and *Staphylococcus* spp.) also decreased, but in some cases they were replaced by bacteria of the genus *Streptococcus* spp. (*S. viridans* and *S. epidermidis*).

With regard to the group of obligate anaerobic microorganisms, the most significant antimicrobial effect was found for representatives of *Gardnerella vaginalis*, *Prevotella bivia*, and *Porphyromonas* spp. Both the number of copies found and their percentage decreased.

An important highly specific marker for the diagnosis of bacterial vaginosis is the presence of the bacterium *A. vaginae*; especially important is its detection in asymptomatic vaginosis. The use of the biological product usually led to a decrease in the level of *A. vaginae*, but in two patients the content increased. The data that we obtained suggest that this biological product, as well as other therapies, should be administered in a personalized manner, possibly in combination with antibiotics.

The majority of patients were found to have a group of gram-positive cocci of *Peptostreptococcus* spp. after administration of the Bifiten, though in minimal amounts (<0.1%). *Peptostreptococcus* spp. reproduced well at pH below 5.5, so it is important to monitor the acid balance of the urogenital tract in women who are being treated with Bifiten.

The study of changes in the levels of vaginal cytokines allows us to assess the effectiveness of Bifiten. In the vaginal secretions of women taking Bifiten, it was found that the level of proinflammatory cytokine IL-17 decreased in more than 60% of patients, while the level of antiinflammatory cytokine IL-10 increased or remained unchanged (Fig. 5.3). The level of SIgA, due to the use of the biological product Bifiten in the vaginal secretions of pregnant women decreased sharply.

It is possible to conclude that the use of Bifiten leads to a rapid and reliable creation of a normal habitat of the vaginal mucosa, with long-term preservation of the effect, preventing recurrence of vaginal infections, reducing perinatal complications and improving the beneficial outcomes of childbirth.

5.5 Conclusion

Trends in the development of national health systems in economically developed countries are centered on the achievements of the actively developed area, which is called predictive, preventive, and personalized medicine or 3P medicine. Approaches utilizing 3P medicine contribute to effective screening of the health of the population, early prevention, stratification of patients by risk groups (phenotypic biomarkers) for optimal treatment planning, disease prediction, and creation of individual patient profiles. Assessment of the course of disease, prognosis, and selection of the correct treatment tactics to provide further improvement in the patient's quality of life cannot be done without precise diagnostics.

The clinical cases that we have described indicate that to perform precise diagnostics, it is necessary to apply a systematic approach, the integral stages of which are the creation of the patient's individual profile and verification of biomarkers in cohort studies.

References

Aagaard, K., Petrosino, J., Keitel, W., et al., 2013. The human microbiome project strategy for comprehensive sampling of the human microbiome and why it matters. FASEB J. 27 (3), 1012–1022.

Bäckhed, F., Roswall, J., Peng, Y., Feng, Q., Jia, H., Kovatcheva-Datchary, P., et al., 2015. Dynamics and stabilization of the human gut microbiome during the first year of life. Cell Host Microbe 17 (5), 690–703.

Bene, K., Varga, Z., Petrov, V.O., Boyko, N., Rajnavolgyi, E., 2017. Gut microbiota species can provoke both inflammatory and tolerogenic immune responses in human dendritic cells mediated by retinoic acid receptor alpha ligation. Front. Immunol. 8, 427.

Bik, E.M., Eckburg, P.B., Gill, S.R., Nelson, K.E., Purdom, E.A., Francois, F., et al., 2006. Molecular analysis of the bacterial microbiota in the human stomach. Proc. Natl. Acad. Sci. 103 (3), 732–737.

Boyko, N.V., 2013. Edible plants specifically modulate systemic and local immune response. In: International Scientific Conference on Probiotics and Prebiotics, Kosice, 11–12 June 2013.

Boyko, N., Kulja, A., Szabo, J., Petrov, V., Kalapos, I., Koval, H., 2007. Endoscopy and imaging modalities/basic science of gastrointestinal disorders/miscellaneous: host/bacteria interaction and influence of beneficial bacteria on mucosal immune responses against opportunistic/enteric pathogens: possible cellular-molecular me. SIDDS 9, 248–249.

Bubnov, R.V., Spivak, M.Y., Lazarenko, L.M., Bomba, A., Boyko, N.V., 2015. Probiotics and immunity: provisional role for personalized diets and disease prevention. EPMA J. 6 (1), 1–11.

Chang, P.V., Hao, L., Offermanns, S., Medzhitov, R., 2014. The microbial metabolite butyrate regulates intestinal macrophage function via histone deacetylase inhibition. Proc. Natl. Acad. Sci. 111 (6), 2247–2252.

Chendey, T.V., Rishko, M.V., Boyko, N.V., Kroon, P.A., 2014. Six weeks ingestion of polyphenol-rich Urtica dioica and Sideritis scardica does not influence endothelial function, blood pressure or lipid profile in patients with coronary artery disease or at high cardiovascular risk: a randomised controlled trial. Український медичний часопис (1), 132–136.

Commission-on-Ending-Childhood-Obesity, 2016. Report of the Commission on Ending Childhood Obesity. Geneva, World Health Organisation.

DeFronzo, R.A., Ferrannini, E., Groop, L., Henry, R.R., Herman, W.H., Holst, J.J., et al., 2015. Type 2 diabetes mellitus. Nat. Rev. Dis. Prim. 1, 15019.

Delicato, C., Collison, M., Myronyuk, I., Symochko, T., Boyko, N., 2019. Is local better?. Consumer value in food purchasing and the role of short food supply chains. Stud. Agric. Econ. 121, 75–83.

Di Marzo, V., Silvestri, C., 2019. Lifestyle and metabolic syndrome: contribution of the endocannabinoidome. Nutrients 11 (8), 1956.

Dominguez-Bello, M.G., Costello, E.K., Contreras, M., Magris, M., Hidalgo, G., Fierer, N., et al., 2010. Delivery mode shapes the acquisition and structure of the initial microbiota across multiple body habitats in newborns. Proc. Natl. Acad. Sci. 107 (26), 11971–11975.

Golubnitschaja, O., Watson, I.D., Topic, E., Sandberg, S., Ferrari, M., Costigliola, V., 2013. Position paper of the EPMA and EFLM: a global vision of the consolidated promotion of an integrative medical approach to advance health care. EPMA J 4 (1), 12.

González-García, E., Puchalska, P., Marina, M.L., García, M.C., 2015. Fractionation and identification of antioxidant and angiotensin-converting enzyme-inhibitory peptides obtained from plum (Prunus domestica L.) stones. J. Funct. Foods 19, 376–384.

Gurung, M., Li, Z., You, H., Rodrigues, R., Jump, D.,B., Morgun, A., et al., 2020. Role of gutmicrobiota in type 2 diabetes pathophysiology. EBioMedicine 51, 102590.

Hollands, W.J., Saha, S., Hayran, O., Boyko, N., Glibetic, M., Konic-Ristic, A., et al., 2013. Lack of effect of bioactive-rich extracts of pomegranate, persimmon, nettle, dill, kale and Sideritis and isolated bioactives on platelet function. J. Sci. Food Agric. 93, 3588–3594.

Horzov, L., Kryvanych, V., Melnyk, V., Drobnych, V., Boyko, N., 2020. Microbial markers of chronic catarrhal gingivitis in treatment of children with fixed orthodontic apparatus. Georgian medical news. 303, 125–134.

Huttenhower, C., Gevers, D., Knight, R., 2012. Structure, function and diversity of the healthy human microbiome. Nature 486 (7402), 207.

Integrative, H.M.P., Proctor, L.M., Creasy, H.H., et al., 2019. The integrative human microbiome project. Nature 569 (7758), 641–648.

Jakobek, L., Šeruga, M., Šeruga, B., Novak, I., Medvidović-Kosanović, M., 2009. Phenolic compound composition and antioxidant activity of fruits of Rubus and Prunus species from Croatia. Int. J. Food Sci. Technol. 44 (4), 860–868.

Keilbaugh, S.A., Shin, M.E., Banchereau, R.F., McVay, L.D., Boyko, N., Artis, D., et al., 2005. Activation of RegIIIβ/γ and interferon γ expression in the intestinal tract of SCID mice: an innate response to bacterial colonisation of the gut. Gut 54 (5), 623–629.

Kobyliak, N., Falalyeyeva, T., Boyko, N., Tsyryuk, O., Beregova, T., Ostapchenko, L., 2018. Probiotics and nutraceuticals as a new frontier in obesity prevention and management. Diabetes Res. Clin. Pract. 141, 190–199.

Koletzko, B., Brands, B., Grote, V., Kirchberg, F.F., Prell, C., Rzehak, P., et al., 2017. Long-term health impact of early nutrition: the power of programming. Ann Nutr Metab 70, 161–169.

Konic-Ristic, A., Srdic-Rajic, T., Kardum, N., Aleksić-Veličković, V., Kroon, P., Hollands, W., et al., 2013. Effects of bioactive-rich extracts of pomegranate, persimmon, nettle, dill, kale and Sideritis and isolated bioactives on arachidonic acid induced markers of platelet activation and aggregation. J. Sci. Food Agric. 93, 3581–3587.

Lazarenko, L.M., Babenko, L.P., Bubnov, R.V., Demchenko, O.M., Zotsenko, V.M., Boyko, N.V., et al., 2017. Immunobiotics are the novel biotech drugs with antibacterial and immunomodulatory properties. Microbiol. J. 79 (1), 66–75.

Lederberg, J., McCray, A.T., 2001. Ome sweetomics—a genealogical treasury of words. Scientist 15 (7), 8–8.

Martiny, J.B., Jones, S.E., Lennon, J.T., Martiny, A.C., 2015. Microbiomes in light of traits: a phylogenetic perspective. Science 350 (6261).

Meleshko, T., Rukavchuk, R., Drobnych, V., Boyko, N., 2020. Personalized pharmabiotics and individual nutrition for nosology specific correction of microbiota and local immune system biomarkers EPMA. 10th Anniversary of the European Association for Predictive, Preventive and Personalised (3P) Medicine - EPMA World Congress Supplement 2020. EPMA J. 11 (Suppl 1), 1–133.

Meleshko, T., Petrov, V., Falalyeyeva, T., Kobyliak, N., Boyko, N., 2021a. Minerva Biotechnol. Biomol. Res. 33 (2), 109–116.

Meleshko, T., Rukavchuk, R., Buhyna, L., Pallah, O., Sukharev, S., Drobnych, V., et al., 2021b. Biologically active substance content in edible plants of zakarpattia and their elemental composition model. Biol. Trace Elem. Res. 199 (6), 2387–2398.

Meleshko, T., Rukavchuk, R., Levchuk, O., Boyko, N., 2021c. Personalized nutrition for microbiota correction and metabolism restore in type 2 diabetes mellitus patients. Advances in Experimental Medicine and Biology. Springer, New York.

Mokrozub, V.V., Lazarenko, L.M., Sichel, L.M., Babenko, L.P., Lytvyn, P.M., Demchenko, O.M., et al., 2015. The role of beneficial bacteria wall elasticity in regulating innate immune response. EPMA J. 6 (1), 1–15.

Nava, G.M., Stappenbeck, T.S., 2011. Diversity of the autochthonous colonic microbiota. Gut Microbes 2 (2), 99–104.

Petrov, V.O., Boyko, N.V. Patent number 90788 Early diagnostic markers of obesity, diabetes and metabolic syndrome. Appl. 08/01/2014, publ. 10.06.2014 Bull. No. 1.

Plumb, J., Lyons, J., Nørby, K., Thomas, M., Nørby, E., Poms, R., et al., 2016. ePlantLIBRA: a composition and biological activity database for bioactive compounds in plant food supplements. Food Chem. 193, 121–127.

Poonam, V., Kumar, G., Reddy L, C.S., Jain, R., Sharma, S.K., Prasad, A.K., et al., 2011. Chemical constituents of the genus Prunus and their medicinal properties. Curr. Med. Chem. 18 (25), 3758–3824.

Roate, A., Jangid, A., Tale, R.V., 2011. Liquid chromatography—tandem mass spectrometric method for simultaneous determination of rutin and quercetin from leaves of Artocarpus Lakoocha Roxb. Int. J. Pharma Biosci. 2 (1), 848–853.

Shafquat, A., Joice, R., Simmons, S.L., Huttenhower, C., 2014. Functional and phylogenetic assembly of microbial communities in the human microbiome. Trends Microbiol. 22 (5), 261–266.

Sharma, S., Tripathi, P., 2019. Gut microbiome and type2 diabetes: where we are and where to go? J. Nutr. Biochem. 63, 101–108.

Smith, P.M., Howitt, M.R., Panikov, N., Michaud, M., Gallini, C.A., Bohlooly-y, M., et al., 2013. The microbial metabolites, short-chain fatty acids, regulate colonic Treg cell homeostasis. Science 341 (6145), 569–573.

Tang, Y., Chen, Y., Jiang, H., Robbins, G.T., Nie, D., 2011. G-protein-coupled receptor for short-chain fatty acids suppresses colon cancer. Int. J. Cancer 128 (4), 847–856.

Tsmur, O.,V., Levchuk, O.B., Lyashina, K.V., Boyko, N.V., 2016. The results of the use of the domestic synbiotic Bifiten for the treatment of bacterial pregnancies in pregnant women. Women's Health 6, 66–72.

Turnbaugh, P.J., Gordon, J.I., 2009. The core gut microbiome, energy balance and obesity. J. Physiol. 587 (17), 4153–4158.

Turnbaugh, P.J., Ley, R.E., Hamady, M., Fraser-Liggett, C.M., Knight, R., Gordon, J.I., 2007. The human microbiome project. Nature 449 (7164), 804–810.

Ursell, L.K., Metcalf, J.L., Parfrey, L.W., Knight, R., 2012. Defining the human microbiome. Nutr. Rev. 70 (Suppl. 1), S38–S44.

Veberic, R., Slatnar, A., Bizjak, J., Stampar, F., Mikulic-Petkovsek, M., 2015. Anthocyanin composition of different wild and cultivated berry species. LWT-Food Sci. Technol. 60 (1), 509–517.

Woodcock, M.E., Hollands, W.J., Konic-Ristic, A., Glibetic, M., Boyko, N., Koçaoglu, B., et al., 2013. Bioactive-rich extracts of persimmon, but not nettle, Sideritis, dill or kale, increase eNOS activation and NO bioavailability and decrease endothelin-1 secretion by human vascular endothelial cells. J. Sci. Food Agric. 93 (14), 3574–3580.

Wrolstad, R.E., Acree, T.E., Decker, E.A., Penner, M.H., Reid, D.S., Schwartz, S.J., et al., (Eds.), 2005. Handbook of Food Analytical Chemistry, Volume 1: Water, Proteins, Enzymes, Lipids, and Carbohydrates. John Wiley & Sons.

Wu, X., Gu, L., Prior, R.L., McKay, S., 2004. Characterization of anthocyanins and proanthocyanidins in some cultivars of Ribes, Aronia, and Sambucus and their antioxidant capacity. J. Agric. Food Chem. 52 (26), 7846–7856.

Xu, Z., Malmer, D., Langille, M.G., Way, S.F., Knight, R., 2014. Which is more important for classifying microbial communities: who's there or what they can do? ISME J. 8 (12), 2357–2359.

Zhou, X., Bent, S.J., Schneider, M.G., Davis, C.C., Islam, M., Forney, L. J., R., 2004. Characterization of vaginal microbial communities in adult healthy women using cultivation-independent methods. Microbiology 150 (8), 2565–2573.

Section II

Microbiome and Digestive Health

Chapter 6

The etiology of gut dysbiosis and its role in chronic disease

Alvin H. Danenberg
Balanced Metabolic Coaching, Periodontist, Certified Functional Medicine Practitioner, Certified Primal Health Coach, Charleston, SC, United States

6.1 Introduction

Implementing a unique starting point for chronic disease is highly recommended. As with the center of a wheel, different paths (like the spokes of a wheel) spread out from a central point, progressing throughout the body. This central point is the gut. The spokes represent various forms of systemic dissemination leading to organ systems. The mechanism of dissemination is based on some compelling medical research involving the gastrointestinal microbiome (GIMB), which is the gateway to the systemic circulation that can be compromised (Fasano, 2012a,b).

The end result, depending on the individual's weaknesses or genetic predispositions, is the manifestation of various chronic diseases. Well before a medical practitioner might diagnose any of these chronic diseases, vicious cycles are already in place to complicate their treatment.

The basis of this author's notion begins with changes in the gut. Changes include (1) an increase in unhealthy microbes resulting in dysbiosis in the gut and (2) changes in the permeability of the gut lining, resulting in leakage of toxic substances into the blood system (Thursby and Juge, 2017; Hawrelak and Stephen, 2004; Cenit et al., 2015; Rosenfeld, 2015). If these changes are persistent and chronic, they could become pathogenic (Kodukula et al., 2017). Then a sequela of cascading events could and does occur. In turn, these changes can set off many differing paths to chronic disease.

6.2 First things first

Our primal ancestors rarely developed chronic disease (Cordain et al., 2005). Skeletal remains of our distant relatives suggest that degenerative diseases were not as prevalent as they are today. In fact, the few primal societies that exist today in remote areas of the world rarely develop degenerative chronic diseases.

Recent human studies have shown that diet and lifestyle are determinants of metabolic syndrome (Manheimer et al., 2015). Medical research has shown that metabolic syndrome is a precursor to many chronic diseases. Metabolic syndrome is the name for a group of risk factors, which include:

- Large waistline
- High triglyceride level
- Low high-density lipoprotein cholesterol level
- High blood pressure
- Elevated blood sugar
- Insulin resistance

In addition, published papers demonstrate that when humans consume fresh nutrient-dense foods, avoid processed foods, obtain restorative sleep, exercise efficiently, and manage stresses of all types, all the risks of metabolic syndrome decrease significantly. Dental health also improves significantly (Baumgartner et al., 2009; Woelber et al., 2016; Sasaki et al., 2016).

Microbiome, Immunity, Digestive Health and Nutrition. DOI: https://doi.org/10.1016/B978-0-12-822238-6.00020-0

6.3 Diet and the gut

Our diet (the foods that we eat and the foods that we avoid eating) has a major impact on our quality of life, health, and longevity (Klement and Pazienza, 2019). The food that we eat directly affects the gut microbiome. The gut microbiome is composed of approximately 3.8×10^{13} microorganisms (Sender et al., 2016), which include over 2100 different species of known bacteria and more than three million genes (Hugon et al., 2015). This garden of microbes is involved in several biological functions, such as carbohydrate digestion, reduction of harmful microbes (by competitive exclusion), vitamin synthesis, immune system activity, and drug metabolism (Jandhyala et al., 2015; Panebianco et al., 2018a,b).

The cells of the body and the gut microbiome talk back and forth. This cross-talk allows for homeostasis between the bacteria and the host (Panebianco et al., 2018a,b). When there is an imbalance of microbes in the gut, it is known as dysbiosis. In a healthy gut, Bacteroidetes and Firmicutes are the dominant bacterial phyla, making up approximately 90% of the total bacterial community. The remaining 10% is made up of Proteobacteria, Verrucomicrobia, and Actinobacteria (Petersen and Round, 2014).

Dysbiosis manifests because of one or more of the following categories: (1) reduction in beneficial microorganisms, (2) overgrowth of harmful microorganisms, and/or (3) reduced diversity of microorganisms (DeGruttola et al., 2016), all of which potentiate an increase in inflammation and GIMB infections.

6.4 Short-chain fatty acids

When functioning in a healthy manner, the gut microbiome produces various short-chain fatty acids (SCFAs). The production of SCFAs in sufficient quantities is the distinctive feature of healthy diets. A major source of SCFAs in the gut is the anaerobic fermentation of polysaccharides that are indigestible by the host. These polysaccharides are the fibers in the diet. A major SCFA that is produced in the gut is butyrate, which provides significant benefits.

Butyrate is the most important energy source for healthy colonocytes, which become impaired in pathological conditions of colitis and colon cancer. Butyrate has been shown to inhibit proliferation and to promote differentiation and apoptosis in different cancers (Canani et al., 2011). Butyrate also inhibits proinflammatory nuclear factor-κB signaling and tumorigenesis. Interestingly, butyrate is a fatty acid in butter, which gave butyrate its name.

6.5 Fasting and the gut

Fasting is a way of not eating. It is the restriction of food so that the body can rest. Fasting includes intermittent fasting, 24-hour fasting, and multiday fasting. Intermittent fasting is eating during a window of 8 hours or less and fasting for about 16 hours or more over the course of a 24-hour day. For example, a person could stop eating all food at 8:00 p.m. in the evening and not eat food until noon the next day. In this example the window of eating is from noon to 8:00 p.m.

Fasting has been associated with numerous health benefits (Mattson et al., 2014). One of the primary mechanisms through which fasting induces metabolic improvements is the positive effects on the gut microbiome (Li et al., 2017). Restricting food also decreases the growth of pathogenic Proteobacteria while increasing the levels of beneficial *Akkermansia muciniphila* (Zheng et al., 2018), a gram-negative anaerobe that uses mucin as its source of carbon and nitrogen, resulting in mucin degradation, which promotes antiinflammatory effects. Another benefit of fasting is related to cancer treatment. Fasting promotes apoptosis in colon cancer models and induces an anti-Warburg effect (Liberti and Locasale, 2016) that increases oxygen consumption but reduces adenosine triphosphate (ATP) synthesis, indicating an increase in mitochondrial uncoupling (Bianchi et al., 2015).

6.6 The effect of unhealthy diets on the gut

The standard American diet (also known as the Western diet) is highly associated with obesity and related metabolic diseases (Zinöcker and Lindseth, 2018). The Western diet promotes inflammation, which is caused by unhealthy changes in the gut microbiome. This diet is loaded with overprocessed foods, including grain products, sugars, vegetable and seed oils, fats from animals that were fed grain products and raised with the use of antibiotics and hormones, produce grown with the use of agrichemicals to ensure maximum harvest potential, preservatives to maximize commercial viability, and other chemical residues from the processing of packaged foods.

6.7 The benefit of healthy diets for the gut

A recent rallying cry is "Eat Organic." To understand healthy food, you must understand the term "organic." The Organic Seal is a certification that can be conferred only by the U.S. Department of Agriculture (USDA) on foods and food products.

For a product to be certified organic, it is required to meet the following strict standards:

- Organic crops cannot be grown with synthetic fertilizers, synthetic pesticides, or sewage sludge.
- Organic crops cannot be genetically engineered or irradiated.
- Animals must eat only organically grown feed (without animal byproducts) and cannot be treated with synthetic hormones or antibiotics.
- Animals must have access to the outdoors, and ruminants (hoofed animals, including cows) must have access to pasture.
- Animals cannot be cloned.

In the practical world, some farms are unable to meet the financial expenses to gain the Organic Seal. However, their products may be grown and raised according to the standards outlined above. If you can ascertain that your local farmer is compliant with the requirements to be "organic" but has not been able to afford the costs to receive the Organic Seal, the farm's products are perfectly fine to consume if you want to eat organic.

You should eat organic fruits, vegetables, nuts, and seeds whenever possible. This explanation also relates to the plants that animals consume for their normal, unadulterated supply of food. As much as possible, animals should be either 100% wild caught or 100% pasture raised. They should not be conventionally raised and confined in cages or feeding lots with inorganic and unnatural foods on which to feed. Healthy animal meat and organs support a healthy gut microbiome, which can produce metabolites that may prevent cancer (Bernstein et al., 2015).

Some compelling reasons to eat organic

- Plants manufacture phytonutrients inside their cells. These phytonutrients are like the immune cells in the body, and there are more than 25,000 different phytonutrients found in plants (Phytonutrients), with more being discovered all the time. These nutrients help the plants to fight infections and maintain their health. The body benefits significantly from many of these phytonutrients when we eat plants. Phytonutrients have health-promoting properties, including antioxidant, antiinflammatory, and other biological activities. The deep colors in plants represent the intensity of the phytonutrients that are available to us.
- If plants are exposed to many insults from their environment, they develop more intense phytonutrients for self-protection. If plants are not exposed to these insults, they develop fewer phytonutrients. When plants are sprayed with insecticides and other chemicals to kill off predators and microbes, these plants not only have a chemical residue that is not healthy for humans, but also produce fewer phytonutrients. Eating organic not only reduces your toxic exposure from the sprayed chemicals but also increases the quantity and quality of the phytonutrients you consume. Locally grown food products are better than food products that are grown thousands of miles away because locally grown food is fresher. Locally grown produce, for example, is also picked at the time closest to optimal ripeness, when nutrient contents are at their highest, naturally protective plant chemicals (lectins, phytoestrogens, etc.) are at their lowest, and flavor is at its best. Protective plant chemicals such as lectins and phytoestrogens act as pest repellents to protect the plant from pests during its development toward maturity, maximum nutritional value, and full ripeness. For the purposes of not spoiling over long shipping distances, plants that are grown many miles from market are picked well before their peak ripeness, have reduced nutrient contents and benefits, and have higher levels of protective plant chemicals, which contribute to adverse health consequences, including damage to the GIMB.
- Plants grow and thrive when exposed to the influence of rain, the nutrients and microbes in the soil, and the sun. When plants have been sprayed with various chemicals, these chemicals can penetrate the soil. They can destroy soil microbes that are required to produce healthy nutrients in the soil. Therefore nonorganic plants could be deficient in necessary nutrients because of contaminated soils.

Several diets have been shown to benefit the gut microbiome. They include the Mediterranean diet, low-carbohydrate diets, ketogenic diets, and the paleolithic diet.

6.7.1 Mediterranean diet

The Mediterranean diet (Del Chierico et al., 2014) consists of eating (1) high levels of vegetables, fruits, cereals (mostly whole grains), nuts, and legumes; (2) low levels of saturated fat, sweets, and meat; (3) high levels of unsaturated fat (mainly olive oil); (4) medium-high levels of fish; (5) moderate levels of wine; and (6) medium-low levels of dairy products (mainly yogurt and cheese). Eating this way is associated with reduced inflammation and increased levels of SCFA production.

6.7.2 Low-carbohydrate diets

Restricting processed carbohydrates and sugars avoids hyperglycemia and hyperinsulinemia and improves the diversity of beneficial gut bacteria (Payne et al., 2012; Segata, 2015). When digestible carbohydrates are replaced with resistant starch, the microbial profile is altered to promote the growth of antiinflammatory microorganisms along with decreasing the proinflammatory ones (Panebianco et al., 2017).

6.7.3 Ketogenic diets

Ketogenic diets represent an extremely low-carbohydrate diet. A ketogenic diet reduces carbohydrate intake to less than 50 g/day. At this level, insulin is kept to lower levels, and cortisol levels are slightly elevated. This will induce the production of ketone bodies in the liver. A ketogenic diet will promote metabolic health and protect against cancer and other noncommunicable chronic diseases.

The mechanism of action of ketogenic diets appears to be (1) lowering insulin levels, (2) enhancing mitochondrial function; and (3) specific antioxidative and antiinflammatory effects (Miller et al., 2018). A high intake of monounsaturated fatty acids and $n-3$ PUFAs as well as nonstarchy vegetables would be beneficial for promoting gut health. Ketogenic diets have been shown to inhibit cancer cell glycolysis and proliferation (Klement, 2017; Weber et al., 2018).

Recent medical research has shown that a ketogenic diet can improve the gut microbiome and reverse dysbiosis in patients with autism, multiple sclerosis, and epilepsy (Newell et al., 2016; Swidsinski et al., 2017; Xie et al., 2017).

6.7.4 Paleolithic diet

A paleolithic diet today, often referred to as a paleo diet, mimics the diet of our ancestors during the Stone Age (the Paleolithic era). This style of eating was prevalent during the course of human existence (Challa and Uppaluri, 2019). The modern paleo diet consists of (1) high consumption of fruits, vegetables, and various herbs and spices; (2) moderate to high consumption of meats, organs, fish, and eggs; (3) moderate consumption of nuts and seeds; and (4) exclusion of all processed foods, legumes, grains, dairy products, and processed vegetable and seed oils (except olive oil and coconut oil). The paleo diet has been shown to be superior to other forms of eating styles (Spreadbury, 2012; Manheimer et al., 2015). The elimination of grains, processed sugars, and legumes has been critical to the success of the paleo diet, since these substances have been shown to damage the intestinal barrier epithelium and to promote chronic autoimmune and inflammatory diseases (Spreadbury, 2012; Kopp, 2019; de Punder and Pruimboom, 2013). The paleo diet also provides the fibers that are critical for the microbiome to function and improve diversity (Wahls, 2017).

It is important to note that the loss of diversity in the gut microbiome with resulting dysbiosis in Western societies in which large quantities of grain products, legumes, sugars, and processed seed and vegetable oils are consumed can be counteracted by returning to a modern paleo diet composed of unprocessed foods (Barone et al., 2019).

6.8 Inflammation

Inflammation by itself is normal and is necessary for a healthy body to function and heal. Inflammation is a response to a pathogenic event. Relentless inflammation is a serious problem in that the cause of the inflammation often remains elusive and the inflammation is persistent and destructive. So when holes in the gut lining occur regularly from continuous reintroduction of unhealthy food choices as well as other irritants that could further damage the gut lining, inflammation becomes constant. Constant inflammation is stressful and destructive and can affect the entire body, capable of inducing damage in any organ or tissue anywhere in the body.

6.8.1 Acute inflammation

Acute inflammation is the response to various types of trauma. It involves the vascular system, the immune system, and the cells that were injured.

Various inflammatory mediators cause vasodilation, allowing blood to flow to the injury site. The blood vessels become more permeable, allowing the plasma and leukocytes to flow through the vessel walls and into the injured tissue to begin a healing process. The movement of plasma into tissue causes fluid buildup and swelling. All of this is necessary for proper healing. The acute inflammatory response should be short and specific. The process breaks down the tissue, targets the damaged tissue and invading pathogens, and then builds the tissue back up.

6.8.2 Chronic inflammation

If the triggers that caused an acute inflammatory reaction are not removed, the inflammation becomes chronic and destructive, also ultimately degrading the gut microbiome. When the gut microbiome becomes unhealthy, the immune system is hyperactivated and hypersensitive, as the gut houses the bulk of the human immune system. The toxic substances that trigger gut dysbiosis can also trigger the abnormal release of zonulin, a protein that modulates the permeability of tight junctions between cells of the wall of the digestive tract (Vanuytsel et al., 2013). Overproduction of zonulin will cause the epithelial lining of the gut to be increasingly permeable (i.e., leaky gut). The result is the spread of chronic systemic inflammation.

Through a complex system that is not completely understood, although it apparently involves immune hyperactivation and hypersensitivity, the immune system may go haywire. In other words, the immune system could begin destroying what appear to be healthy tissues in the body in addition to the foreign invaders. The result is known as autoimmune disease (Wang et al., 2015).

6.9 Gluten

Gluten, which contains the lectins that were mentioned previously, is a family of proteins that is present in wheat, rye, barley, and some other types of grains. Gluten imparts greater pliability to the texture of foods and is used to improve mouth feel. Products that have a high gluten content or extra added gluten have improved texture for more pleasurable mouth feel, like bagels, high-gluten bread, and pasta products. As a matter of fact, gluten could be an ingredient in many processed foods, cosmetics, prescription medications, and nutritional supplements. The human body cannot completely digest gluten. One of the remnants of the incomplete digestion of gluten is gliadin.

6.10 Gliadin

Gliadin, which also contains lectins, has the potential to slowly and deliberately destroy various tissues of the body. Its effects are cumulative, producing chronic inflammation over time. It causes dysbiosis and opens holes in the gut lining, allowing substances, especially undigested proteins, to leak into the bloodstream that should never be there. Researchers have proved that this damage to the gut lining occurs in every human whenever gliadin is present in the intestines (Hollon et al., 2015).

Gliadin causes a chemical reaction in the intestinal wall. It increases the reactivity of zonulin. Zonulin is a protein causes the cells of the intestinal wall to open and close normally to allow digested nutrients to pass into the circulatory system. However, excessive stimulation of zonulin will cause the gut lining cells to separate, opening up large portals to the bloodstream. As cells in the one-cell-layer-thick lining of the gut separate too widely and for too long a period of time, undigested particles of food, bacteria, and toxic substances can leak through the openings into the bloodstream (Miranda-Ribera et al., 2019).

However, the human body is resilient. The cells of the intestinal wall repair, and they replace themselves every couple of days. If gluten or other irritants are not reintroduced into the diet, there will likely be no permanent damage from a onetime occurrence. But the problem is a process of repetition, constantly consuming of gluten, as well exposure to other irritants.

6.11 Leaky gut

Unhealthy leakage from the intestines into the bloodstream is known as increased intestinal permeability (i.e., leaky gut). A preponderance of the evidence indicates that gliadin is one of the triggers that can breakdown the gut lining and increase the overgrowth of harmful, mostly anaerobic bacteria. When undigested particles of food or toxic substances get into the bloodstream that never should be there, the immune system becomes activated. The immune system is like an army that is called into battle. The immune system's purpose is to get rid of foreign elements. The process to rid the body of irritating substances usually begins with inflammation.

The interaction between host genetics, immune response, gut microbiome and function, and exposure to environmental triggers is critical in the breakdown of the mucosal immune response leading to the development of chronic inflammation. A recent study has shown that gut epithelial barrier dysfunction plays a critical role in the initiation of inflammation (Shin and Kim, 2018). Increased intestinal permeability is present in the pathogenesis of several chronic diseases, including obesity and diabetes; diseases of the brain and the nervous system, such as Parkinson's disease; multiple sclerosis; and the autism spectrum disorders (Antalis et al., 2007; Wapenaar et al., 2008; Anderson et al., 2016; Buscarinu et al., 2016; de Magistris et al., 2010; Fiorentino et al., 2016).

In 2000, Alessio Fasano (an expert on autoimmune disease and zonulin) published research demonstrating that the development of autoimmune diseases is a result of three factors: (1) genetic predisposition, (2) environment triggers that are mismanaged by the immune system, and (3) breakdown in the intestinal barrier function secondary to the activation of the zonulin pathway by food-derived environmental triggers or changes in gut microbiota. The potential of correcting genetic predispositions has become an important subject of epigenetic research. What is being revealed is that if environmental triggers can be eliminated, if the gut microbiome can be restored to homeostasis, if inflammatory food choices can be avoided, and if the intestinal barrier can be repaired and restored, then autoimmune diseases could be more successfully arrested rather than just medicated (Fasano, 2012a,b; Gustafson and Fasano, 2018; Fasano, 2012a, b; Fasano, 2011). In 2002, Bagchi et al. published research in which a mechanism of autoimmunity in inflammatory disorders was described in greater detail (Bagchi et al., 2002). Essentially, the research demonstrated that chronic inflammatory disorders—rheumatoid arthritis (RA) and osteoarthritis in this research—are initiated by an inability to form the perfect three-dimensional glycoprotein structures of type II collagen in connective tissues and the IgG immune surveillance molecules. The identity of these structures went from being identified as three-dimensionally competent and biologically usable structures to being unwelcome foreign antigenic invaders (Bagchi et al., 2002). Therefore this now correct classification induced an upregulated "out-of-control" immune attack on the connective and immune tissues of the body. Essentially, the autoimmune response was caused by a "manufacturing error" that instigated the excessive immune response. This explanation provides greater mechanistic detail and insight into the second and third factors reported by Dr. Fasano.

6.12 Other irritants to the gut

The gut microbiome, the intestinal mucous layer, and the epithelial lining of the gut—collectively, the GIMB—can become damaged from many different influences. Some of these irritating influences are as follows:

- Stress (including emotional, physical, or chemical) can be the cause of damage (Gao et al., 2018; Li et al., 2013; Karl et al., 2017). A significant chemical stress to the gut is glyphosate herbicide (e.g., Roundup), which damages the DNA in human cells; deranges glycoprotein synthesis, structure, and function; inhibits the growth of healthy bacteria; and directly causes leaky gut (Samsel and Seneff, 2013; California environmental, 2017). Stress to the body also includes heavy metal toxicity (Li et al., 2019); overexercising (Fry et al., 1994; Aschbacher et al., 2014); sleep deprivation and sleep apnea (Ko et al., 2019); continuous exposure to pathogenic (i.e., "dirty") electromagnetic fields (Crabtree et al., 2017); disturbances in circadian rhythms and excessive blue-light exposure, especially in the evening (Voigt et al., 2016); and lack of proper sunlight that is critical for the production of vitamin D_3 (Tabatabaeizadeh et al., 2018), to name just a few offenders.
- Overly processed vegetable and seed oils, hydrogenated and partially hydrogenated oils, processed sugars, or other chemicals in foods can have a harmful effect (Toor et al., 2019).
- Specific medications such as nonsteroidal antiinflammatory drugs, corticosteroids, alcohol, narcotics, antibiotics, chemotherapy drugs, hydrogen peroxide, and birth control pills (to name a few) could result in leaky gut (Le Bastard et al., 2018; Maier et al., 2018; Cussotto et al., 2019; The Journal of the National Institute, 2019; Singh et al., 2018; Akbari et al., 2017; Blaser, 2016).

- Low-dose ionizing radiation has a cumulative effect. In our society, frequent and unnecessary dental and medical X-rays over the course of time can cause damage to the microbiome, its metabolites, and the epithelial lining. The study cited here was published in 2016 and was performed on mice with potential application to the human gut (Shukla et al., 2016).
- Anything that disturbs the delicate balance of microbes in the gut could be the culprit (Leccioli et al., 2017).

6.13 A closer look into the inner tube of the gut

The garden of microbes in the gut consists of a community of bacteria, fungi, archaea, protozoans, and viruses (Lazar et al., 2018). The epithelial barrier of the gut has an area of 400 m^2 and is essentially an external organ on the inside of the body. It is like the hole in a doughnut. The microbial cells are either floating in the intestinal lumen, adhering to the mucous layer, or adhering to the surface of the epithelial cells (Sekirov et al., 2010). The microbiome from the mother's vaginal tract populates in the baby's gut at birth. However, in babies who were delivered by caesarean section delivered, the gut microbiome is similar to the mother's skin microbiome (Dominguez-Bello et al., 2010). During breastfeeding, human milk contributes to the gut microbiome and the development of the immune system, especially during the first few days of life with the contribution of colostrum.

As the active microbial community develops, it will interact with the host, also generating feedback signaling to maintain homeostasis. The microbiome affects normal gut development mainly through the creation of SCFAs by the microbiome.

SCFAs play an antiinflammatory role and support intestinal homeostasis through cellular proliferation, differentiation, and inhibition of the proliferation of cancerous cells. The beneficial microbiome stimulates nonspecific and specific immune system development and inhibits the adherence of various pathogens. It is important to note that a healthy gut microbiome helps to synthesize many vitamins, such as vitamins B_1, B_2, B_5, B_6, B_{12}, and K, folate, and biotin. In addition, it degrades xenobiotics and sterols and causes biliary acid deconjugation (Gérard, 2014).

If the homeostasis of this beneficial microbiome is disturbed and it becomes dysbiotic, the ultimate result is the development of chronic diseases (Clemente et al., 2012). The intestinal epithelial lining, the mucosal immune system, and the diversity and quality of the gut microbes are affected. The body will become more susceptible to infections; cellular structure and functionality could be impaired, resulting in malformed synthesis of three-dimensionally correct molecular structures; and the immune system will then react against the body's own structurally imperfect cells (autoimmunity) (Bagchi et al., 2002).

To help maintain homeostasis, the intestinal mucosal immune system contains various signaling pathways. Specific sensors, such as Peyer's patches, recognize pathogens, commensal bacteria, and potential pathogens. These sensors are part of the host defense system. The immune system will create intestinal inflammation after it recognizes pathogenic or potentially pathogenic microbes or any potentially pathogenic substances in the gut (Rossi et al., 2013; Yang et al., 2011).

There is a give-and-take process in the gut. In a healthy gut, the mucosal layer resists invasion by unhealthy gut bacteria and inhibits the reproduction of pathological species. Additionally, the garden of balanced bacteria in the gut maintains immune tolerance in the intestine (Abraham and Medzhitov, 2011). When there are genetic predispositions in the host's immune system to intestinal pathogens, these can be overridden by changing the diet to a gut-friendly diet, which will restore the humoral immunity and inhibit pathological bacteria (Lamas et al., 2017).

As has already been mentioned, the mucosal immune system constantly surveils the gut microbiome. Any disruption in the gut microbiome will elicit an immune response. In addition, when there are changes in the gut microbiome, the mucosal and epithelial layers can be affected. For example, when the diet is lacking in healthy fibers, which are the food source of the gut bacteria, gut bacteria can begin to damage the mucous layer to find their nourishment. The ultimate damage to the mucous layer promotes greater epithelial access and lethal colitis by various mucosal pathogens (Desai et al., 2016).

In individuals with dysbiosis the gut microbiome communicates with the gut epithelial layer to trigger antimicrobial substances to attack the pathologic invaders (Hooper et al., 2003; Cash et al., 2006). Chronic dysbiosis will encourage exacerbated autoimmune and inflammatory diseases (Sekirov et al., 2010). Ultimately, this will lead to chronic inflammation, mucosal breakdown, and tissue damage.

The immune system builds its innate responses and its ability to regulate autoimmune and inflammatory diseases early in life (Zuo et al., 2018; Abraham and Medzhitov, 2011; De Filippo et al., 2017). When the host experiences various microorganisms in the first few years of life, the intestinal mucosal immune system learns to react appropriately (Rakoff-Nahoum and Medzhitov, 2008; Thaiss et al., 2016).

Toll-like receptors (TLRs) from the membrane of the epithelial and lymphoid cells of the small intestine are involved in this differential recognition, being responsible for the normal development of the intestinal mucosal immune system. TLRs suppress the occurrence of an inflammatory response and promote immunological tolerance to normal microbiota components.

The gut microbiome affects neutrophil movement and function to cause the differentiation of T cells into the various types of helper cells (i.e., Th1, Th2, Th17) and also into regulatory T cells (Tregs) (Francino, 2014).

The Th17 cells secrete many different cytokines (interleukin-22 (IL-22), IL-17A, and IL-17F), which have a significant influence on immune homeostasis and inflammation (Rossi and Bot, 2013). These Th17 cells maintain both a stimulatory and an inhibitory cytokine function (Weaver et al., 2013).

By contrast, regulatory T cells, known as Tregs, actively maintain immune tolerance. They modulate an excessively high inflammatory response (Collins et al., 2011). If Tregs are malfunctioning, the ultimate results are autoimmune disorders (Wu and Wu, 2012).

Secretory IgA (SIgA) is the main antibody found in secretions of the mucous layer (Woof and Kerr, 2006). It is the first line of defense in protecting the intestinal epithelium from enteric toxins and pathogenic microbes. SIgA assists in the removal of antigens and pathogenic microbes from the intestinal lumen by blocking their access to epithelial receptors, entrapping them in mucus, and causing their removal. Also, SIgA functions in mucosal immunity and intestinal homeostasis (Mantis et al., 2011).

6.13.1 The gut microbiome and infections

The microbiome holds a principal role in the initiation and progression of infectious diseases (Harris et al., 2017). A healthy gut microbiome will prevent the invasion of potentially pathogenic microbes by competing with these unfriendly bacteria for adhesion sites and nutrients as well as releasing toxic molecules to counteract them. The microbiome stimulates the secretory IgA response. Also, the signaling molecules that are released by the microbiome trigger the development of granulocyte/monocyte progenitors in the bone marrow, which ultimately promote the host innate response. In the absence of these signaling molecules, there is increased susceptibility to systemic infection.

The interactions between the gut epithelial cells and the gut microbiome are crucial to maintain gut barrier health. Unfortunately, when dysbiosis occurs, the microbiome will lose its antiinfectious barrier quality, and the general circulation can be easily infected with different pathogenic microorganisms from the environment.

Ongoing research shows the specific role between the gut microbiota and the immune system. The research has focused on infectious diseases caused by microbiome manipulation. Since microbiota manipulation could control the balance between health and infectious disease, intestinal microbiota alteration by a pathogen or a pathobiont could lead to chronic diseases (Carvalho et al., 2012; Vayssier-Taussat et al., 2014; Inoue et al., 2018).

6.13.2 Cross-talk between microbes

All bacteria communicate with each other by signaling molecules. This is a form of cross-talk that includes initiation, responsivity, and feedback generation and includes upstream, downstream, and cross-stream communications. Because of this, bacterial cells can sense their environment, monitor the population density, and adjust their gene expression. Thus bacteria gain the advantage to disseminate and survive in highly competitive environments.

Intercellular communication is divided into two categories, based on the quorum-sensing (QS) mechanism. The first type is intraspecific cell-to-cell communication through specific QS molecules, and the second type consists of the interspecific communication based on a universal chemical "language," which provides interspecific signaling between bacteria and host cells. The QS mechanism allows bacteria to regulate colonization of the host by commensal bacteria and to modulate the host response (Davey and O'toole, 2000; Tlaskalova-Hogenova et al., 2005; Rutherford and Bassler, 2012). The QS mechanism is also used by microbes to detect the presence of other similar microbes (Falcão et al., 2004). In addition, specific intestinal hormones can mimic the action of bacterial signaling molecules, increasing the complexity of cross-talk between the bacteria and the host (Sperandio et al., 2002).

The gut microbiome communicates bidirectionally with the central nervous system by producing and sensing neurochemicals that are derived either within the microorganisms themselves or within their host (Lyte, 2014). Also, the gut microbiome can produce neurochemicals with hormonal activities that could go beyond the gut and affect changes in anxiety, depression, cognition, pain, inflammation, autoimmunity, and metabolic diseases (Grenham et al., 2011; Lin and Zhang, 2017; Ochoa-Reparaz et al., 2011; Erny et al., 2015).

6.14 Chronic inflammatory and autoimmune diseases

Chronic inflammation is evident at the beginning of most chronic diseases and autoimmune diseases. Chronic and autoimmune diseases occur when (1) the gut becomes overly permeable on a constant basis, (2) chronic inflammation prevails and permeates the blood system by means of various elements of the inflammatory cascade, (3) a genetic weakness in some remote tissue reacts to this chronic inflammation, and (4) the confused immune system starts to attack what are in fact healthy tissues.

In the case of gluten, gliadin initiates the damage to the gut lining. It is important to note that other irritants from various sources do cause damage to the gut lining and to gut bacteria as well. An abundance of various irritants to the gut increases the potential for chronic inflammation. Once chronic inflammation persists, chronic and autoimmune diseases begin to manifest.

The manifestation of chronic or autoimmune diseases could take years to develop as a result of repeated insults to the body. The person probably would not know that these destructive changes were slowly occurring. As a matter of fact, the person would not need to experience obvious gut symptoms such as diarrhea, constipation, bloating, or pain. A leaky gut frequently does not cause gut symptoms. As subclinical and frequent insults accumulate over time, disease may make itself known only many years after the first insult occurred.

Another factor to be considered is when children under the age of 5 years living in developed countries are not exposed to many environmental microbes, compounds, and antigens. This lack of early immune stimulation could cause malfunctioning of the immune system and lead to hypersensitivity, hyperreactivity, autoimmune disease, or inflammatory diseases later in life.

We next discuss a variety of chronic diseases.

6.14.1 Type 1 diabetes

Type 1 diabetes (T1DM) is caused by the patient's inability to secrete insulin as a result of the autoimmune destruction of the pancreatic beta cells (Aathira and Jain, 2014). The patient could have a genetic predisposition for T1DM (Polychronakos and Li, 2011), or exposure to environmental toxic substances could trigger the production of antibodies against the patient's beta cells in the islets of Langerhans in the pancreas. Also, a low diversity of gut microbes and the overgrowth of unhealthy bacteria could be a factor in the development of T1DM (Alkanani et al., 2015). Dysbiosis increases the permeability of the intestinal epithelium because of a loss of tight barrier function (Vaarala, 2012). Then leakage into the circulatory system of incompletely digested nutrients, especially proteins from the diet, and microbial antigens will trigger inflammation, which could lead to beta cell attack in the pancreas (Vaarala, 2012). In addition, the loss of healthy commensal microbes in the gut will negatively influence the patient's immune response, increasing the potential for inflammation and beta cell attack (Vaarala et al., 2008).

6.14.2 Rheumatoid arthritis

There is a correlation between RA and dysbiosis, suggesting that decreased communication between the host and the gut microbiome is a significant contributing factor to the inflammation of autoimmune diseases (Scher et al., 2013; Liu et al., 2016). Moreover, as was mentioned previously, RA has been identified as an inability to implant the nutritional sugar galactose into both type II collagen and the IgG immune surveillance molecule (Bagchi et al., 2002).

6.14.3 Celiac disease

Modification of the normal gut microbiota may have a role in the development of celiac disease. Some species of bacteria were reduced in patients with celiac disease compared to healthy people. The fucosyltransferase 2 (FUT2) gene regulates the expression of ABH blood group antigens that are glycosphingolipids or glycoproteins in mucus as well as other body secretions that also influence the composition of mucosa-associated bacteria. ABH antigens are carbohydrate structures that are synthesized in a stepwise fashion by glycosyltransferase enzymes, that is, FUT2 in this case, that sequentially add specific monosaccharides to glycoproteins and glycolipids. A mutation in FUT2 gene impairs structural and functional competence and can cause a decrease in bacterial diversity and imbalance in the gut microbiome. This could cause an overgrowth of *Candida albicans*, for example, which is a culprit in the onset of celiac disease (Nagao-Kitamoto et al., 2016).

6.14.4 Inflammatory bowel disease

Inflammatory bowel disease (IBD) involves a group of chronic intestinal diseases characterized by inflammation of the large or small intestine. The most common types of IBD are ulcerative colitis and Crohn's disease. IBD creates endotoxemia (Vrakas et al., 2017), which compromises immune function and contributes to systemic chronic inflammation (Palmer et al., 2016).

The severity of dysbiosis determines the severity of pathogenesis of IBD. In IBD, excessive amounts of bacteria buildup, adhere to the gut mucosa, and invade mucosal epithelial cells, inciting an inflammatory response. Inflammation leads to changes in the composition of the gut microbiome with an increase in aerobic bacteria accompanied by a significant decrease in the fecal levels of butyric and propionic acid. The unbalanced microbiome contributes to the development of an overproduction of hydrogen peroxide and leads to IBD (Strus et al., 2009; Khan et al., 2012).

6.14.5 Allergic diseases

Although some genetic factors affect the development of allergic diseases, environmental factors (i.e., epigenetics) and gut dysbiosis have a significant effect (van den Elsen et al., 2017). Reduced microbial diversity in infancy is correlated with an increased risk for allergies later in life (Cao et al., 2014). Also, a standard American diet containing unhealthy fats can cause pulmonary damage and impair tight junction protein structure and endothelial integrity (Ji et al., 2014).

6.14.6 Systemic lupus erythematosus

Systemic lupus erythematosus (SLE) is a systemic autoimmune disease that is characterized by the presence of hyperactive and aberrant antibody response to nuclear and cytoplasmic antigens (La Paglia et al., 2017). Dysbiosis is associated with SLE (Hevia et al., 2014). For example, in a mouse study, dysbiosis accelerated the development of SLE (Zhang et al., 2014).

6.14.7 Skin-related autoimmune pathologies

Autoimmune skin diseases are linked to dysbiosis. For example, the gut microbiome dysbiosis in patients with psoriatic arthritis was similar to that in IBD patients (Scher et al., 2015). Dysbiosis was also a factor in patients with systemic sclerosis (Volkmann et al., 2017).

6.14.8 Neurological inflammatory diseases

Medical research has proved that the gut microbiome has an impact on brain development and function. Moreover, the gut houses more neurotransmitter receptors than the brain does. Therefore dysbiosis contributes to anxiety-like behaviors and also upregulates the hypothalamic pituitary adrenal system stress reactivity. Neuroactive bacterial metabolites are transported through the bloodstream to the brain and stimulate inflammation (Rabot et al., 2016).

Dysbiosis can lead to inflammation that is associated with Alzheimer's disease (Jiang et al., 2017) as well as multiple sclerosis (Colpitts and Kasper, 2017). These anaerobic pathologies provide opportunistic environments for anaerobes, such as yeasts, worms, parasites, mold, and viruses, particularly the herpes viral family, to take up residence and promote infectious pathologies.

6.14.9 Periodontal disease

Dental plaque is healthy until it is not healthy (Peterson et al., 2014). Periodontal disease develops from unhealthy dental plaque. Unhealthy plaque results when healthy plaque is transformed into unhealthy dental plaque because of an underlying compromised immune system and unhealthy food choices. The compromised immune system has its roots in unhealthy changes in the gut (Qin and Wade, 2018; Cianci et al., 2018).

A compromised immune system and unhealthy food choices could allow the hundreds of bacteria in dental plaque to get out of balance and become unhealthy (Meuric et al., 2017; Herrero et al., 2016). Then unhealthy bacteria can proliferate and cause the progression of advanced gum disease (Hajishengallis and Lamont, 2012). Autoimmunity may play a role in the progression of periodontal disease (Kaur et al., 2016).

The prevalence of this disease is at epidemic proportions. In 2010 a published paper demonstrated that 93.9% of adults in the United States had some form of gingivitis, making it a disease of epidemic magnitude (Li et al., 2010).

In 2012 the Centers for Disease Control and Prevention published their results in the *Journal of Dental Research*. The report was updated in 2015 in the *Journal of Periodontology* (Eke et al., 2015). It showed that the prevalence of periodontitis was estimated to be 47.2% for American adults (approximately 64.7 million people). For adults 65 years old and older, the prevalence jumped to 70.1%. These findings were the result of the most comprehensive periodontal evaluation performed ever in the United States.

It is critical to diagnose and treat periodontal disease efficiently because its infectious and inflammatory components can infiltrate the blood system and spread throughout the body, just as toxic elements of gut dysbiosis can leak into the circulatory system and spread throughout the body.

One of the most virulent bacteria in periodontal disease is *Porphyromonas gingivalis* (Olsen and Yilmaz, 2016; Hajishengallis, 2015). Among other self-protective measures, this bacterium produces a biofilm, which is resistant to the body's immune defenses (Carvalho-Filho et al., 2016). As the body continues to try to fight the resistant *P. gingivalis*, chronic inflammation results. This chronic inflammation can cause the tissues surrounding the infected gum spaces to breakdown.

As was previously mentioned, once periodontal disease is established in the mouth, its pathological byproducts can seep into the bloodstream, lymph fluid, and bone structures to spread infection and inflammation to all areas of the body. This mechanism of seeping into the blood system is similar to the way in which gut dysbiosis causes leakage of toxic elements into the bloodstream (i.e., leaky gut).

The result could be increased chronic diseases throughout the body (Bale et al., 2017; Corrêa et al., 2017; Potempa et al., 2017). This is similar to the damage caused from a leaky gut. In addition to advanced gum disease, infections from abscessed teeth could cause leakage of unhealthy elements into the bloodstream.

Interestingly, three human studies showed that a healthy diet alone can improve the health of the mouth. These studies also determined that removing dental plaque by brushing and flossing was not critical to improving oral health as long as diet was corrected. Specifically, the investigators demonstrated that changing from a diet that is abundant in high-processed-carbohydrate and inflammatory foods to a diet that excludes highly processed carbohydrates and inflammatory foods will decrease signs of gum inflammation (Baumgartner et al., 2009; Woelber et al., 2016; Woelber et al., 2019). However, active periodontal treatment will be necessary once gum inflammation progresses into the bone structures.

In February 2019 a medical research article was published in *Biomedical Journal (*Bui et al., 2019) entitled "Association Between Periodontal Pathogens and Systemic Disease." The authors describe the correlation between periodontal disease and various chronic diseases and outcomes, such as cardiovascular disease, gastrointestinal and colorectal cancer, diabetes and insulin resistance, Alzheimer's disease, respiratory tract infections, and adverse pregnancy outcomes. The authors go on to state that there are conflicting studies that have tried to prove causal relationships. However, there is significant research to show a strong correlation.

This author's research suggests that periodontal disease is not the seed of all systemic disease. The evidence indicates that periodontal disease is just one of many chronic diseases on the continuum of the spread of systemic disease that starts in the gut, as has been described in this chapter. Since the mouth is visible and easy to examine, the mouth may be the first clinical area where disease is diagnosed. Once systemic disease spreads, a vicious cycle begins because all tissues affect all other tissues in the human body. In the case of periodontal disease, treatment for cascading chronic diseases must include healing both the unhealthy gut and the unhealthy mouth. But for the most part, the starting point of mouth disease is in the gut before the disease becomes visible in the mouth and other areas of the body.

As was previously mentioned, tissues use cross-talk to communicate with other tissues (Tulic et al., 2016; Figueredo et al., 2017; Yap and Mariño, 2018). Unhealthy bacteria in the mouth in turn can interact further with unhealthy bacteria in the gut and vice versa (Lira-Junior and Figueredo, 2016).

6.14.10 Cancer

Dysbiosis influences not only the inflammatory response, but also digestion and other vital functions. In dysbiosis a loss of cellular aerobic glycolysis induces anaerobic glycolysis, also promoting a greater anaerobic bioenvironmental state. In the increasing anaerobic environment, normal commensal bacteria change into pathogenic bacteria, and the homeostatic balance is disrupted. These can sensitize the gut to cancerous lesions (i.e., anaerobic events) by causing cell proliferation and mutation, leading to tumor formation (Spurgeon et al., 2017). There is an increase in the permeability of the mucosal surfaces, which in turn leads to interactions of the gut microbiome and the immune cells. All this increases inflammation, which is generally a response to the overgrowth of anaerobic pathogens or oxygen deprivation, which is otherwise instigated and which further sensitizes the host for cancer progression. The anaerobic pathological process further progresses. Then neighboring cells could become affected, as they share the same anaerobic biological environment.

An anaerobic environment, by default, generates an increase in reactive oxygen species (ROS). Activation of M1 macrophages during dysbiosis causes an even greater increase in the toxic production of ROS and reactive nitrogen intermediates, which can lead to the creation of cancerous lesions (Sheflin et al., 2014). In addition, dysbiosis can change the blood estrogen level and contribute to tumor development (Chen and Madak-Erdogan, 2016). Many pathogenic bacteria and viruses can alter cell cycle progression and inhibit apoptosis, thus increasing the production of cancerous cells (Gagnaire et al., 2017).

Interestingly, studies have shown that increasing beneficial gram-positive bacteria in the gut can enhance Th1 immune responses and offer therapeutic benefits by promoting anticancer immune responses (Viaud et al., 2013; Schwabe and Jobin, 2013). Also, a healthy gut microbiome can produce large amounts of folate, which plays a crucial role in regulating DNA synthesis, and other antioxidant-involved substances, such as selenium, vitamin C, and vitamin A, which help to prevent DNA lesions. In addition, there are antitumoral effects from lactic acid–producing bacteria as well as butyrate-producing bacteria (Goldin et al., 1980; Daïen et al., 2017).

6.15 Unraveling the mystery of treatment

6.15.1 Treatment for active infections

The science is compelling to unravel the mystery of treatment. If there are active inflammation and infections that have affected other areas of the body, these must be treated first. One example is active periodontal disease because it has become a source for leakage of toxic elements into the bloodstream via the tissues surrounding the infected teeth.

Next, it is critical to avoid a shift to anaerobic glycolysis, which stokes the flames of inflammation and infection. Nourish the body with restorative protective resources, but remove the irritant.

If infection is active in the mouth, then the mouth must be treated appropriately. In addition, causes of dysbiosis in the gut and damage to the gut lining must be corrected and subsequently avoided.

6.15.2 Understanding periodontal disease

In 2016, Li and others published their medical research about periodontal disease (Li et al., 2016). They used healthy human gum tissue cells for their experiment. These tissue cells were exposed to lipopolysaccharides (LPS), which are present in the pathogenic bacteria of periodontal disease. The results of the research showed that mitochondria in the gum tissue cells exposed to LPS created excess ROS. Following the production of excess ROS, the gum cells produced excess cytokines, promoting inflammation and inducing periodontal destruction. However, when the mitochondria of these gum tissue cells were treated to reduce excess ROS production, chronic cytokine production also was reduced even in the presence of LPS. It appeared that treating the mitochondrial dysfunction stopped the progression of periodontal disease regardless of the presence of LPS.

Vos and others published a paper in 2012 using an animal model (Vos et al., 2012). In summary, they demonstrated that vitamin K_2 was able to rescue damaged mitochondria, which had been altered at the beginning of the research to represent damaged cells of Parkinson's disease.

In 2018, Ivanova and others described how vitamin K_2 transports out of the liver and then disseminates throughout the body to assist in various biological functions including the prevention of mitochondrial dysfunction (Ivanova et al., 2018).

These three papers suggest that vitamin K_2 could be an important nutrient in decreasing the progression of periodontal disease by improving mitochondrial function.

6.15.3 Repairing and restoring the gut

Repair will consist of eating nutrient-dense, antiinflammatory foods. Several healthy diets were suggested at the beginning of this chapter along with the importance of organic, wild caught, and naturally pastured animal products. A paleo-type diet appears to be best suited to accomplish overall health. In addition, the author of this chapter strongly advocates taking clinically validated dietary supplements, especially prebiotics, probiotics, and digestive enzymes.

The disruption in the healthy balance of bacteria in the gut must be restored and the gut must be repaired. The protocol to treat the gut includes repair of the epithelial lining, support of the mucous layer, and germination and restoration of healthy bacteria in the lumen using proven probiotics and prebiotics.

6.15.4 Spore-based bacillus probiotics

McFarlin and other researchers published a paper in 2017 (McFarlin et al., 2017). The investigators selected 28 participants whose blood tests demonstrated significant endotoxemia after consuming a high-fat, high-carbohydrate meal. This group of individuals were divided into two groups. Both groups took two capsules of a daily supplement for 4 weeks. One group took placebo capsules; the other took capsules containing five different spore-based bacillus probiotics. At the end of the trial, the participants ate a high-fat, high-carbohydrate meal. Their blood was tested before the meal and retested after the meal. Five hours after the meal, the results showed an average 42% decrease in endotoxemia in the group that had taken the probiotic capsules. The group that had taken the placebo had a 36% increase in endotoxemia. The authors suggested that the positive results might be improved significantly if the spore-based probiotics were taken for several more months.

The benefits of spore-based bacillus bacteria are significant for the gut. One study looked at 35 conventional probiotics that primarily contained *Lactobacillus* sp. and *Bifidobacterium* sp. When these probiotics were subjected to a simulated normal stomach acid of pH 1, all 35 brands were inactivated by the stomach acid. At a stomach acid of pH 2–3, only 18 of the 35 brands had 50% survival. However, all the spore-based probiotics survived at a stomach acid of pH 1.3 (Gibson and Rouzaud, 2019).

In another study, when conventional probiotics were dead, their metabolites still entered the intestines and provided a positive biological response. This is referred to as "metabolic response modifiers." However, since the bacteria were dead, they could not germinate in the gut microbiome (Adams, 2010).

Therefore the ideal probiotic to take to repair the gut is a spore-based bacillus probiotic. It was able to create metabolites that functioned as metabolic response modifiers in the gut and to enter the intestines unscathed to germinate into new beneficial bacteria (Lefevre et al., 2015; Gong et al., 2016; Mandel et al., 2010; Bernardeau et al., 2017; Urdaci et al., 2004).

6.15.5 Prebiotics

Prebiotics are nondigestible fibers that feed the bacteria in the gut. Most prebiotics on the market can feed both harmful and beneficial gut bacteria, and this can exacerbate digestive issues. However, oligosaccharides are specific fibers that can increase microbial diversity and selectively feed beneficial keystone bacteria such as *A. muciniphila*, *Faecalibacterium prausnitzii*, and *Bifidobacteria*. Oligosaccharides would be the ideal prebiotic to take as a supplement to assist in repairing the gut (Blatchford et al., 2017; Yang et al., 2015; Vulevic et al., 2013; Chelakkot et al., 2018).

6.15.6 The mucosal layer

The mucosal system is a very important part of the human immune system (Yue et al., 2019). It is the main interface between the human body and the outside world. The mucosal system contains 150 times more surface area than skin, which makes it one of the most important immune barriers. The health of the mucosa determines how the body interacts with antigens; therefore the integrity of the intestinal mucosa can dictate overall immune function (König et al., 2016).

The mucosal layer can be repaired and supported by specific immunoglobulins, amino acids, and citrus polyphenols.

- Immunoglobulins (IgG, IgA, and IgM) will support healthy digestion, neutralize environmental toxins, and help to maintain healthy gut barrier function (Jasion and Burnett, 2015).
- The key amino acids that play an important role in the healthy mucosal layer are L-proline, L-serine, L-cysteine, and L-threonine. These four amino acids have been shown to increase mucin 2 production and to stimulate mucin synthesis in the colon, resulting in a thicker and healthier mucosal barrier (Faure et al., 2006).
- Citrus polyphenols support digestive health and healthy gut barrier function by increasing the integrity of connective tissues and promoting SCFA composition (Romier et al., 2009).

6.15.7 The importance of the mouth

To stop periodontal disease and prevent this infection from entering the systemic circulation, the infection must be treated efficiently. Treatment may consist of a dentist, hygienist, or periodontist removing irritants that have become lodged under the gum tissues and initiating inflammation and infection. In more advanced stages, surgical procedures may be necessary to arrest this disease. Whatever treatment is necessary, an efficient oral hygiene program should be instituted at a frequency based on the patient's ability to take care of his or her mouth. The individual also must have a personal oral hygiene protocol to maintain a healthy mouth.

6.15.8 Personal oral hygiene protocol

Removing the amounts of unhealthy bacteria from around the tooth is the goal of flossing and brushing. The goal is not to kill all the bacteria in the mouth, since many of the bacteria in the mouth are good bacteria. An effective method to clean your mouth is to use (1) something to clean between the teeth, (2) a soft toothbrush to clean the other surfaces of the teeth, and (3) methods to clean your tongue. Perform these methods twice a day: first thing in the morning and last thing at night.

1. Use dental floss to floss between your teeth. Think about sliding up and down a pole. That is how the floss wraps around the tooth and slides up and down to scrape away food particles that could get caught between the contacts of the teeth. Also, always use a small brush that is designed to clean between teeth like a pipe cleaner (one brand is called TePe EasyPick, another is GUM Soft Pick). Imagine the small bristles of this tiny brush scrubbing away the overgrown bacterial film as it is pushed in and out between the teeth at the gum line. These small brushes are the best way to remove unhealthy plaque buildup at the base of the tooth and gum margin.
2. Use an electric toothbrush such as the Sonicare or the Oral B/Braun. Electric brushes are more efficient, and people tend to be lazy. You do not need to use any toothpaste to brush your teeth effectively. Just brush with filtered water GENTLY, angling the bristles into the space where the gums meet the teeth on both the cheek side and the tongue side of all teeth. Brush horizontally but GENTLY.

   If you want to use toothpaste, dip the bristles in a little coconut oil (The author keeps some that stays solid at room temperature in a small jar in the bathroom), and then dip these bristles into a little baking soda (Also keep some in a small jar in the bathroom).

   There is another substance you can use as a toothpaste, but it might surprise you. Research has shown that raw honey is effective at removing potentially harmful bacteria and preventing tooth decay and gum disease (Ramsay et al., 2019). Honey has over 200 known biologically active substances (Rao et al., 2016), and many remain to be discovered in the future. While all raw honey is effective, most of the medical research has shown the superiority of manuka honey because of some of its unique, biologically active compounds (Niaz et al., 2017; Carter et al., 2016).

   Mouthwash need not be used regularly and routinely. Daily use of an antimicrobial mouthwash will kill both bad bacteria and good bacteria. Killing good bacteria daily will compromise the health in your mouth and the rest of your body (Bryan et al., 2017). If you want to use a mouthwash occasionally, use some coconut oil, and swish it around for a minute or so, and then spit it out (this is known as oil pulling). If you use coconut oil as a mouthwash, be sure to spit it out into a napkin or paper towel and throw that in the trash. If you spit coconut oil into your sink, it could clog up the pipes. By the way, raw honey is also an excellent choice for a mouthwash.
3. Most of the odor-forming bacteria are located on the top and back areas of your tongue, closest to your throat. An effective way to remove this overgrown bacteria and food remnants causing odor is to use a teaspoon. Place the inverted teaspoon as far back as is comfortable on the upper side of your tongue. Then, gently glide the teaspoon forward, removing the excess bacterial film and microscopic food particles. Repeat this two to three times, and then wash off the teaspoon.

Some "no-nos"

- Do not floss under the gum tissue. You easily could cut the gum and create a wound. That wound might stay sore and heal like a cleft. Aggressive flossing under the gum also could cause gum recession.
- A water-pick device can be dangerous. It could force food debris and bacteria deeper under the gum tissues if used on a moderate- to high-pressure setting. Also, the force of the water jet could tear gum tissue cells that are trying to heal inside the gum space.
- If you drink very acid drinks, the minerals of the tooth could become "softened" until the acid in the mouth returns to normal. It is best to not brush your teeth right after drinking any acid drink. Research suggests that you should

wait at least an hour before brushing after drinking an acid drink (Choi et al., 2012). It would be a good idea to rinse your mouth with water to help remove the excess acid while your mouth regains its normal acid level.

6.16 Concluding remarks

The human body is a marvelous and an incredibly complex machine. It will repair and replace its tissues on a regular basis. However, when there is a constant and site-specific irritant, the body's reaction is to become chronically inflamed. Once the seed of chronic inflammation is planted and nurtured, chronic degenerative diseases ultimately result. For the most part, chronic diseases find their origin in a gut that is exhibiting dysbiosis. Ultimate successful treatment must eventually include the repair and restoration of a healthy gut microbiome.

If there are toxic components in the mouth, they must be removed. In addition, consider changing lifestyle habits: Start exercising efficiently, sleeping 7–8 hours each night, and practicing stress reduction techniques. If sleep is compromised because of restriction of the airway space, it is critical to have this evaluated and properly treated. If lack of necessary sunlight is a concern, address this. In the absence of adequate sunlight, consuming more vitamin D_3 would be an intelligent supplement choice, combined with a high-quality evidence-based multivitamin, mineral, and phytonutrient formulation. All these efforts and lifestyle changes could prevent chronic disease. They also could improve conditions in your body if you have already been diagnosed with a chronic disease.

We have plenty of evidence that much disease begins in the gut. What you swallow, breathe, and absorb through your skin can affect your gut. In addition, how you sleep, exercise, and deal with stress can affect your gut. An unhealthy gut puts your entire body at risk. It's your gut; it's your health; it's your choice to make healthy adjustments. Take charge of your health!

References

Aathira, R., Jain, V., 2014. Advances in management of type 1 diabetes mellitus. World J. Diabetes 5 (5), 689–696.

Abraham, C., Medzhitov R., 2011. Interactions between the host innate immune system and microbes in inflammatory bowel disease. Gastroenterology 140 (6), 1729–1737.

Adams, C.A., 2010. The probiotic paradox: live and dead cells are biological response modifiers. Nutr. Res. Rev. 23 (1), 37–46.

Akbari, M., Nejat, A.H., Farkhondeh, N., et al., 2017. Does at-home bleaching induce systemic oxidative stress in healthy subjects? Aust. Dent. J. 62 (1), 58–64.

Alkanani, A.K., Hara, N., Gottlieb, P.A., et al., 2015. Alterations in intestinal microbiota correlate with susceptibility to type 1 diabetes. Diabetes 64 (10), 3510–3520.

Anderson, G., Seo, M., Berk, M., Carvalho, A.F., Maes, M., 2016. Gut permeability and microbiota in Parkinson's disease: role of depression, tryptophan catabolites, oxidative and nitrosative stress and melatoninergic pathways. Curr. Pharm. Des. 22 (40), 6142–6151.

Antalis, T.M., Shea-Donohue, T., Vogel, S.N., Sears, C., Fasano, A., 2007. Mechanisms of disease: protease functions in intestinal mucosal pathobiology. Nat. Clin. Pract. Gastroenterol. Hepatol. 4 (7), 393–402.

Aschbacher, K., Kornfeld, S., Picard, M., et al., 2014. Chronic stress increases vulnerability to diet-related abdominal fat, oxidative stress, and metabolic risk. Psychoneuroendocrinology 46, 14–22.

Bagchi, D., Misner, B., Bagchi, M., et al., 2002. Effects of orally administered undenatured type II collagen against arthritic inflammatory diseases: a mechanistic exploration. Int. J. Clin. Pharmacol. Res. 22 (3–4), 101–110.

Bale, B.F., Doneen, A.L., Vigerust, D.J., 2017. High-risk periodontal pathogens contribute to the pathogenesis of atherosclerosis. Postgrad. Med. J. 93 (1098), 215–220.

Barone, M., Turroni, S., Rampelli, S., et al., 2019. Gut microbiome response to a modern Paleolithic diet in a Western lifestyle context. PLoS One 14 (8), e0220619.

Baumgartner, S., Imfeld, T., Schicht, O., et al., 2009. The impact of the stone age diet on gingival conditions in the absence of oral hygiene. J. Periodontol. 80 (5), 759–768.

Bernardeau, M., Lehtinen, M.J., Forssten, S.D., Nurminen, P., 2017. Importance of the gastrointestinal life cycle of *Bacillus* for probiotic functionality. J. Food Sci. Technol. 54 (8), 2570–2584.

Bernstein, A.M., Song, M., Zhang, X., et al., 2015. Processed and unprocessed red meat and risk of colorectal cancer: analysis by tumor location and modification by time. PLoS One 10 (8), e0135959.

Bianchi, G., Martella, R., Ravera, S., et al., 2015. Fasting induces anti-Warburg effect that increases respiration but reduces ATP-synthesis to promote apoptosis in colon cancer models. Oncotarget 6 (14), 11806–11819.

Blaser, M.J., 2016. Antibiotic use and its consequences for the normal microbiome. Science 352 (6285), 544–545.

Blatchford, P., Stoklosinski, H., Eady, S., et al., 2017. Consumption of kiwifruit capsules increases *Faecalibacterium prausnitzii* abundance in functionally constipated individuals: a randomised controlled human trial. J. Nutr. Sci. 6, e52.

Bryan, N.S., Tribble, G., Angelov, N., 2017. Oral microbiome and nitric oxide: the missing link in the management of blood pressure. Curr. Hypertens. Rep. 19 (4), 33.

Bui, F.Q., Almeida-da-Silva, C.L.C., Huynh, B., et al., 2019. Association between periodontal pathogens and systemic disease. Biomed. J. 42 (1), 27–35.

Buscarinu, M.C., Cerasoli, B., Annibali, V., et al., 2016. Altered intestinal permeability in patients with relapsing-remitting multiple sclerosis: a pilot study. Mult. Scler. Available from: http://doi.org/10.1177/1352458516652498.

California environmental protection agencyoffice of environmental health hazard assessmentsafe drinking water and toxic enforcement act of 1986 (Proposition 65) notice to interested parties April 7, 2017 (Posted March 28, 2017) Chemical to be listed as known to the state of California to cause cancer:glyphosate. <https://oehha.ca.gov/media/downloads/crnr/032817tobelistedglyphosate.pdf> (accessed 12.05.19.).

Canani, R.B., Costanzo, M.D., Leone, L., Pedata, M., Meli, R., Calignano, A., 2011. Potential beneficial effects of butyrate in intestinal and extraintestinal diseases. World J. Gastroenterol. 17 (12), 1519–1528.

Cao, S., Feehley, T.J., Nagler, C.R., 2014. The role of commensal bacteria in the regulation of sensitization to food allergens. FEBS Lett. 588 (22), 4258–4266.

Carter, D.A., Blair, S.E., Cokcetin, N.N., et al., 2016. Therapeutic manuka honey: no longer so alternative. Front. Microbiol. 7, 569.

Carvalho, F.A., Koren, O., Goodrich, J.K., et al., 2012. Transient inability to manage proteobacteria promotes chronic gut inflammation in TLR5-deficient mice. Cell Host Microbe 12 (2), 139–152.

Carvalho-Filho, P.C., Gomes-Filho, I.S., Meyer, R., Olczak, T., Xavier, M.T., Trindade, S.C., 2016. Role of *Porphyromonas gingivalis* HmuY in immunopathogenesis of chronic periodontitis. Mediators Inflamm. 2016, 7465852.

Cash, H.L., Whitham, C.V., Behrendt, C.L., Hooper, L.V., 2006. Symbiotic bacteria direct expression of an intestinal bactericidal lectin. Science 313 (5790), 1126–1130.

Cenit, M.C., Olivares, M., Codoñer-Franch, P., Sanz, Y., 2015. Intestinal microbiota and celiac disease: cause, consequence or co-evolution? Nutrients 7 (8), 6900–6923.

Challa, H.J., Uppaluri, K.R., 2019. Paleolithic Diet. StatPearls Publ., Treasure Island, FL.

Chelakkot, C., Choi, Y., Kim, D.K., et al., 2018. *Akkermansia muciniphila*-derived extracellular vesicles influence gut permeability through the regulation of tight junctions. Exp. Mol. Med. 50 (2), e450.

Chen, K.L., Madak-Erdogan, Z., 2016. Estrogen and microbiota crosstalk: should we pay attention? Trends Endocrinol. Metab. 27, 752–755.

Choi, S., Park, K.H., Cheong, Y., et al., 2012. Potential effects of tooth-brushing on human dentin wear following exposure to acidic soft drinks. J. Microsc. 247 (2), 176–185.

Cianci, R., Pagliari, D., Piccirillo, C.A., Fritz, J.H., Gambassi, G., 2018. The microbiota and immune system crosstalk in health and disease. Mediators Inflamm. 2018, 2912539.

Clemente, J.C., Ursell, L.K., Parfrey, L.W., Knight, R., 2012. The impact of the gut microbiota on human health: an integrative view. Cell 148 (6), 1258–1270.

Collins, C.B., Aherne, C.M., Kominsky, D., et al., 2011. Retinoic acid attenuates ileitis by restoring the balance between T-helper 17 and T regulatory cells. Gastroenterology 141 (5), 1821–1831.

Colpitts, S.L., Kasper, L.H., 2017. Influence of the gut microbiome on autoimmunity in the central nervous system. J. Immunol. 198 (2), 596–604.

Cordain, L., Eaton, S.B., Sebastian, A., et al., 2005. Origins and evolution of the Western diet: health implications for the 21st century. Am. J. Clin. Nutr. 81 (2), 341–354.

Corrêa, J.D., Calderaro, D.C., Ferreira, G.A., et al., 2017. Subgingival microbiota dysbiosis in systemic lupus erythematosus: association with periodontal status. Microbiome 5 (1), 34.

Crabtree, D.P.E., Herrera, B.J., Kang, S., 2017. The response of human bacteria to static magnetic field and radiofrequency electromagnetic field. J. Microbiol. 55 (10), 809–815.

Cussotto, S., Clarke, G., Dinan, T.G., Cryan, J.F., 2019. Psychotropics and the microbiome: a chamber of secrets. Psychopharmacology 236 (5), 1411–1432.

Daïen, C.I., Pinget, G.V., Tan, J.K., Macia, L., 2017. Detrimental impact of microbiota-accessible carbohydrate-deprived diet on gut and immune homeostasis: an overview. Front. Immunol. 8, 548.

Davey, M.E., O'toole, G.A., 2000. Microbial biofilms: from ecology to molecular genetics. Microbiol. Mol. Biol. Rev. 64 (4), 847–867.

De Filippo, C., Di Paola, M., Ramazzotti, M., et al., 2017. Diet, environments, and gut microbiota. a preliminary investigation in children living in Rural and Urban Burkina Faso and Italy. Front. Microbiol. 8, 1979.

DeGruttola, A.K., Low, D., Mizoguchi, A., Mizoguchi, E., 2016. Current understanding of dysbiosis in disease in human and animal models. Inflamm. Bowel Dis. 22 (5), 1137–1150.

Del Chierico, F., Vernocchi, P., Dallapiccola, B., Putignani, L., 2014. Mediterranean diet and health: food effects on gut microbiota and disease control. Int. J. Mol. Sci. 15 (7), 11678–11699.

de Magistris, L., Familiari, V., Pascotto, A., et al., 2010. Alterations of the intestinal barrier in patients with autism spectrum disorders and in their first-degree relatives. J. Pediatr. Gastroenterol. Nutr. 51 (4), 418–424.

de Punder, K., Pruimboom, L., 2013. The dietary intake of wheat and other cereal grains and their role in inflammation. Nutrients 5 (3), 771–787.

Desai, M.S., Seekatz, A.M., Koropatkin, N.M., et al., 2016. A dietary fiber-deprived gut microbiota degrades the colonic mucus barrier and enhances pathogen susceptibility. Cell 167 (5), 1339–1353.e21.

Dominguez-Bello, M.G., Costello, E.K., Contreras, M., et al., 2010. Delivery mode shapes the acquisition and structure of the initial microbiota across multiple body habitats in newborns. Proc. Natl. Acad. Sci. USA 107 (26), 11971–11975.

Eke, P.I., Dye, B.A., Wei, L., et al., 2015. Update on prevalence of periodontitis in adults in the United States: NHANES 2009 to 2012. J. Periodontol. 86 (5), 611−622.

Erny, D., Hrabě de Angelis, A.L., Jaitin, D., et al., 2015. Host microbiota constantly control maturation and function of microglia in the CNS. Nat. Neurosci. 18 (7), 965−977.

Falcão, J.P., Sharp, F., Sperandio, V., 2004. Cell-to-cell signaling in intestinal pathogens. Curr. Issues Intest. Microbiol. 5 (1), 9−17.

Fasano, A., 2011. Zonulin and its regulation of intestinal barrier function: the biological door to inflammation, autoimmunity, and cancer. Physiol. Rev. 91 (1), 151−175.

Fasano, A., 2012a. Leaky gut and autoimmune diseases. Clin. Rev. Allergy Immunol. 42 (1), 71−78.

Fasano, A., 2012b. Zonulin, regulation of tight junctions, and autoimmune diseases. Ann. N. Y. Acad. Sci. 1258 (1), 25−33.

Faure, M., Mettraux, C., Moennoz, D., et al., 2006. Specific amino acids increase mucin synthesis and microbiota in dextran sulfate sodium-treated rats. J. Nutr. 136 (6), 1558−1564.

Figueredo, C.M., Martins, A.P., Lira-Junior, R., et al., 2017. Activity of inflammatory bowel disease influences the expression of cytokines in gingival tissue. Cytokine 95, 1−6.

Fiorentino, M., Sapone, A., Senger, S., et al., 2016. Blood-brain barrier and intestinal epithelial barrier alterations in autism spectrum disorders. Mol. Autism. 7, 49.

Francino, M.P., 2014. Early development of the gut microbiota and immune health. Pathogens 3 (3), 769−790.

Fry, R.W., Grove, J.R., Morton, A.R., et al., 1994. Psychological and immunological correlates of acute overtraining. Br. J. Sports Med. 28 (4), 241−246.

Gagnaire, A., Nadel, B., Raoult, et al., 2017. Collateral damage: insights into bacterial mechanisms that predispose host cells to cancer. Nat. Rev. Microbiol.

Gao, X., Cao, Q., Cheng, Y., et al., 2018. Chronic stress promotes colitis by disturbing the gut microbiota and triggering immune system response. Proc. Natl. Acad. Sci. USA 115 (13), E2960−E2969.

Gérard, P., 2014. Metabolism of cholesterol and bile acids by the gut microbiota. Pathogens 3 (1), 14−24.

Gibson, G.R., Rouzaud, C. An evaluation of probiotic effects in the human gut:microbial aspects. <http://citeseerx.ist.psu.edu/viewdoc/download?doi = 10.1.1.526.6176&rep = rep1&type = pdf> (accessed 12.06.19.).

Goldin, B.R., Swenson, L., Dwyer, J., et al., 1980. Effect of diet and Lactobacillus acidophilus supplements on human fecal bacterial enzymes. J. Natl. Cancer Inst. 64 (2), 255−261.

Gong, Y., Li, H., Li, Y., 2016. Effects of *Bacillus subtilis* on epithelial tight junctions of mice with inflammatory bowel disease. J. Interferon Cytokine Res. 36 (2), 75−85.

Grenham, S., Clarke, G., Cryan, J.F., Dinan, T.G., 2011. Brain-gut-microbe communication in health and disease. Front. Physiol. 2, 94.

Gustafson, C., Fasano, A., 2018. How a cholera researcher became an expert on autoimmune disease. Integr. Med. (Encinitas) 17 (2), 20−22.

Hajishengallis, G., 2015. Periodontitis: from microbial immune subversion to systemic inflammation. Nat. Rev. Immunol. 15 (1), 30−44.

Hajishengallis, G., Lamont, R.J., 2012. Beyond the red complex and into more complexity: the polymicrobial synergy and dysbiosis (PSD) model of periodontal disease etiology. Mol. Oral. Microbiol. 27 (6), 409−419.

Harris, V.C., Haak, B.W., Boele van Hensbroek, M., Wiersinga, W.J., 2017. The intestinal microbiome in infectious diseases: the clinical relevance of a rapidly emerging field. Open. Forum Infect. Dis. 4 (3), ofx144.

Hawrelak, J.A., Stephen, P., Myers, S.P., 2004. The causes of intestinal dysbiosis: a review. Altern. Med. Rev. 9 (2), 180−197.

Herrero, E.R., Slomka, V., Boon, N., et al., 2016. Dysbiosis by neutralizing commensal mediated inhibition of pathobionts. Sci. Rep. 6, 38179.

Hevia, A., Milani, C., López, P., et al., 2014. Intestinal dysbiosis associated with systemic lupus erythematosus. MBio 5 (5), e01548-14.

Hollon, J., Puppa, E.L., Greenwald, B., Goldberg, E., Guerrerio, A., Fasano, A., 2015. Effect of gliadin on permeability of intestinal biopsy explants from celiac disease patients and patients with non-celiac gluten sensitivity. Nutrients 7 (3), 1565−1576.

Hooper, L.V., Stappenbeck, T.S., Hong, C.V., Gordon, J.I., 2003. Angiogenins: a new class of microbicidal proteins involved in innate immunity. Nat. Immunol. 4 (3), 269−273.

Hugon, P., Dufour, J.C., ColsonP, Fournier, P.E., Sallah, K., Raoult, D., 2015. A comprehensive repertoire of prokaryotic species identified in human beings. Lancet Infect. Dis. 15 (10), 1211−1219.

Inoue, T., Nakayama, J., Moriya, K., et al., 2018. Gut dysbiosis associated with hepatitis C virus infection. Clin. Infect. Dis.

Ivanova, D., Zhelev, Z., Getsov, P., et al., 2018. Vitamin K: Redox-modulation, prevention of mitochondrial dysfunction and anticancer effect. Redox Biol. 16, 352−358.

Jandhyala, S.M., Talukdar, R., Subramanyam, C., Vuyyuru, H., Sasikala, M., Nageshwar Reddy, D., 2015. Role of the normal gut microbiota. World J. Gastroenterol. 21 (29), 8787−8803.

Jasion, V.S., Burnett, B.P., 2015. Survival and digestibility of orally-administered immunoglobulin preparations containing IgG through the gastrointestinal tract in humans. Nutr. J. 14, 22.

Ji, Y., Sun, S., Goodrich, J.K., et al., 2014. Diet-induced alterations in gut microflora contribute to lethal pulmonary damage in TLR2/TLR4-deficient mice. Cell Rep. 8 (1), 137−149.

Jiang, C., Li, G., Huang, P. et al., 2017. The gut microbiota and Alzheimer's disease. J. Alzheimers Dis.

Karl, J.P., Margolis, L.M., Madslien, E.H., et al., 2017. Changes in intestinal microbiota composition and metabolism coincide with increased intestinal permeability in young adults under prolonged physiological stress. Am. J. Physiol. Gastrointest. Liver Physiol. 312 (6), G559−G571.

Kaur, G. Mohindra, K., Singla, S., 2016. Autoimmunity-basics and link with periodontal disease. Autoimmun. Rev. 16 (1), 64−71.

Khan, M.T., Duncan, S.H., Stams, A.J., et al., 2012. The gut anaerobe *Faecalibacterium prausnitzii* uses an extracellular electron shuttle to grow at oxic-anoxic interphases. ISME J. 6 (8), 1578–1585.

Klement, R.J., 2017. Beneficial effects of ketogenic diets for cancer patients: a realist review with focus on evidence and confirmation. Med. Oncol. 34 (8), 132.

Klement, R.J., Pazienza, V., 2019. Impact of different types of diet on gut microbiota profiles and cancer prevention and treatment. Medicina (Kaunas.) 55 (4), 84.

Ko, C.Y., Liu, Q.Q., Su, H.Z., et al., 2019. Gut microbiota in obstructive sleep apnea-hypopnea syndrome: disease-related dysbiosis and metabolic comorbidities. Clin. Sci. (Lond.) 133 (7), 905–917.

Kodukula, K., Faller, D.V., Harpp, D.N., et al., 2017. Gut microbiota and salivary diagnostics: the mouth is salivating to tell us something. Biores Open. Access. 6 (1), 123–132.

König, J., Wells, J., Cani, P.D., et al., 2016. Human intestinal barrier function in health and disease. Clin. Transl. Gastroenterol. 7 (10), e196.

Kopp, W., 2019. How western diet and lifestyle drive the pandemic of obesity and civilization diseases. Diabetes Metab. Syndr. Obes. 12, 2221–2236.

Lamas, B., Michel, M.L., Waldschmitt, N., et al., 2017. Card9 mediates susceptibility to intestinal pathogens through microbiota modulation and control of bacterial virulence. Gut.

La Paglia, G.M.C., Leone, M.C., Lepri, G., et al., 2017. One year in review 2017: systemic lupus erythematosus. Clin. Exp. Rheumatol. 35 (4), 551–561.

Lazar, V., Ditu, L.M., Pircalabioru, G.G., et al., 2018. Aspects of gut microbiota and immune system interactions in infectious diseases, immunopathology, and cancer. Front. Immunol. 9, 1830.

Le Bastard, Q., Al-Ghalith, G.A., Grégoire, M., et al., 2018. Systematic review: human gut dysbiosis induced by non-antibiotic prescription medications. Aliment. Pharmacol. Ther. 47, 332–345.

Leccioli, V., Oliveri, M., Romeo, M., Berretta, M., Rossi, P., 2017. A new proposal for the pathogenic mechanism of non-coeliac/non-allergic gluten/wheat sensitivity: piecing together the puzzle of recent scientific evidence. Nutrients 9 (11), 1203.

Lefevre, M., Racedo, S.M., Ripert, G., et al., 2015. Probiotic strain *Bacillus subtilis* CU1 stimulates immune system of elderly during common infectious disease period: a randomized, double-blind placebo-controlled study. Immun. Ageing. 12, 24.

Li, Y., Lee, S., Hujoel, P., et al., 2010. Prevalence and severity of gingivitis in American adults. Am. J. Dent. 23 (1), 9–13.

Li, X., Kan, E.M., Lu, J., Cao, Y., Wong, R.K., Keshavarzian, A., et al., 2013. Combat-training increases intestinal permeability, immune activation and gastrointestinal symptoms in soldiers. Aliment. Pharmacol. Ther. 37, 799–809.

Li, X., Wang, X., Zheng, M., Luan, Q.X., 2016. Mitochondrial reactive oxygen species mediate the lipopolysaccharide-induced pro-inflammatory response in human gingival fibroblasts. Cell Res. 347 (1), 212–221.

Li, G., Xie, C., Lu, S., et al., 2017. Intermittent fasting promotes white adipose browning and decreases obesity by shaping the gut microbiota. Cell Metab. 26 (5), 801.

Li, X., Brejnrod, A.K., Ernst, M., et al., 2019. Heavy metal exposure causes changes in the metabolic health-associated gut microbiome and metabolites. Environ. Int. 126, 454–467.

Liberti, M.V., Locasale, J.W., 2016. The Warburg effect: how does it benefit cancer cells? Trends Biochem. Sci. 41 (3), 287.

Lin, L., Zhang, J., 2017. Role of intestinal microbiota and metabolites on gut homeostasis and human diseases. BMC Immunol. 18 (1), 2.

Lira-Junior, R., Figueredo, C.M., 2016. Periodontal and inflammatory bowel diseases: is there evidence of complex pathogenic interactions? World J. Gastroenterol. 22 (35), 7963–7972.

Liu, X., Zeng, B., Zhang, J., et al., 2016. Role of the gut microbiome in modulating arthritis progression in mice. Sci. Rep. 6, 30594.

Lyte, M., 2014. Microbial endocrinology and the microbiota-gut-brain axis. Adv. Exp. Med. Biol. 817, 3–24.

Maier, L., Pruteanu, M., Kuhn, M., et al., 2018. Extensive impact of non-antibiotic drugs on human gut bacteria. Nature 555 (7698), 623–628.

Mandel, D.R., Eichas, K., Holmes, J., 2010. *Bacillus coagulans*: a viable adjunct therapy for relieving symptoms of rheumatoid arthritis according to a randomized, controlled trial. BMC Complement. Altern. Med. 10, 1.

Manheimer, E.W., van Zuuren, E.J., Fedorowicz, Z., Pijl, H., 2015. Paleolithic nutrition for metabolic syndrome: systematic review and *meta*-analysis. Am. J. Clin. Nutr. 102 (4), 922–932.

Mantis, N.J., Rol, N., Corthésy, B., 2011. Secretory IgA's complex roles in immunity and mucosal homeostasis in the gut. Mucosal Immunol. 4 (6), 603–611.

Mattson, M.P., Allison, D.B., Fontana, L., et al., 2014. Meal frequency and timing in health and disease. Proc. Natl. Acad. Sci. USA 111 (47), 16647–16653.

McFarlin, B.K., Henning, A.L., Bowman, E.M., Gary, M.A., Carbajal, K.M., 2017. Oral spore-based probiotic supplementation was associated with reduced incidence of post-prandial dietary endotoxin, triglycerides, and disease risk biomarkers. World J. Gastrointest. Pathophysiol. 8 (3), 117–126.

Meuric, V., Le Gall-David, S., Boyer, E., et al., 2017. Signature of microbial dysbiosis in periodontitis. Appl. Env. Microbiol. 83 (14), e00462-17.

Miller, V.J., Villamena, F.A., Volek, J.S., 2018. Nutritional ketosis and mitohormesis: potential implications for mitochondrial function and human health. J. Nutr. Metab. 2018, 5157645.

Miranda-Ribera, A., Ennamorati, M., Serena, G., et al., 2019. Exploiting the zonulin mouse model to establish the role of primary impaired gut barrier function on microbiota composition and immune profiles. Front. Immunol. 10, 2233.

Nagao-Kitamoto, H., Kitamoto, S., Kuffa, P., Kamada, N., 2016. Pathogenic role of the gut microbiota in gastrointestinal diseases. Intest. Res. 14 (2), 127–138.

Newell C., Bomhof, M.R., Reimer, R.A., Hittel, D.S., Rho, J.M., Shearer, J., 2016. Ketogenic diet modifies the gut microbiota in a murine model of autism spectrum disorder. Mol. Autism. 7 (1), 37.

Niaz, K., Maqbool, F., Bahadar, H., Abdollahi, M., 2017. Health benefits of manuka honey as an essential constituent for tissue regeneration. Curr. Drug. Metab.

Ochoa-Repáraz, J., Mielcarz, D.W., Begum-Haque, S., Kasper, L.H., 2011. Gut, bugs, and brain: role of commensal bacteria in the control of central nervous system disease. Ann. Neurol. 69 (2), 240–247.

Olsen, I., Yilmaz, Ö., 2015. Modulation of inflammasome activity by *Porphyromonas gingivalis* in periodontitis and associated systemic diseases. J. Oral. Microbiol. 8, 30385.

Palmer, C.D., Romero-Tejeda, M., Sirignano, M., et al., 2016. Naturally occurring subclinical endotoxemia in humans alters adaptive and innate immune functions through reduced MAPK and increased STAT1 phosphorylation. J. Immunol. 196 (2), 668–677.

Panebianco, C., Adamberg, K., Adamberg, S., et al., 2017. Engineered resistant-starch (ERS) diet shapes colon microbiota profile in parallel with the retardation of tumor growth in in vitro and in vivo pancreatic cancer models. Nutrients 9 (4), 331.

Panebianco, C., Andriulli, A., Pazienza, V., 2018a. Pharmacomicrobiomics: exploiting the drug-microbiota interactions in anticancer therapies. Microbiome 6 (1), 92.

Panebianco, C., Potenza, A., Andriulli, A., Pazienza, V., 2018b. Exploring the microbiota to better understand gastrointestinal cancers physiology. Clin. Chem. Lab. Med.

Payne, A.N., Chassard, C., Lacroix, C., 2012. Gut microbial adaptation to dietary consumption of fructose, artificial sweeteners and sugar alcohols: implications for host-microbe interactions contributing to obesity. Obes. Rev. 13 (9), 799–809.

Petersen, C., Round, J.L., 2014. Defining dysbiosis and its influence on host immunity and disease. Cell Microbiol. 16 (7), 1024–1033.

Peterson, S.N., Meissner, T., Su, A.I., et al., 2014. Functional expression of dental plaque microbiota. Front. Cell Infect. Microbiol. 4, 108.

Phytonutrients. <http://www.webmd.com/diet/phytonutrients-faq> (accessed WebMed 25.12.19.).

Polychronakos, C., Li, Q., 2011. Understanding type 1 diabetes through genetics: advances and prospects. Nat. Rev. Genet. 12 (11), 781–792.

Potempa, J., Mydel, P., Koziel, J., 2017. The case for periodontitis in the pathogenesis of rheumatoid arthritis. Nat. Rev. Rheumatol.

Qin, Y., Wade, P.A., 2018. Crosstalk between the microbiome and epigenome: messages from bugs. J. Biochem. 163 (2), 105–112.

Rabot, S., Jaglin, M., Daugé, V., Naudon, L., 2016. Impact of the gut microbiota on the neuroendocrine and behavioural responses to stress in rodents. OCL 23 (1), D116.

Rakoff-Nahoum, S., Medzhitov, R., 2008. Innate immune recognition of the indigenous microbial flora. Mucosal Immunol. 1 (Suppl. 1), S10–S14.

Ramsay E.I., Rao, S., Madathil, L., et al., 2019. Honey in oral health and care: a mini review. J. Oral. Biosci. 61 (1), 32–36.

Rao, P.V., Krishnan, K.T., Salleh, N., Gan, S.H., 2016. Biological and therapeutic effects of honey produced by honey bees and stingless bees: a comparative review. Rev. Bras. Faracogn. 26 (5).

Romier, B., Schneider, Y.J., Larondelle, Y., During, A., 2009. Dietary polyphenols can modulate the intestinal inflammatory response. Nutr. Rev. 67 (7), 363–378.

Rosenfeld, C.S., 2015. Special section on drug metabolism and the microbiome—perspective; microbiome disturbances and autism spectrum disorders. Drug Metab. Dispos. 43, 1557–1571.

Rossi, M., Bot, A., 2013. The Th17 cell population and the immune homeostasis of the gastrointestinal tract. Int. Rev. Immunol. 32 (5–6), 471–474.

Rossi, O., van Baarlen, P., Wells, J.M., 2013. Host-recognition of pathogens and commensals in the mammalian intestine. Curr. Top. Microbiol. Immunol. 358, 291–321.

Rutherford, S.T., Bassler, B.L., 2012. Bacterial quorum sensing: its role in virulence and possibilities for its control. Cold Spring Harb. Perspect. Med. 2 (11), a012427.

Samsel, A., Seneff, S., 2013. Glyphosate's suppression of cytochrome P450 enzymes and amino acid biosynthesis by the gut microbiome: pathways to modern diseases. Entropy 15 (4), 1416–1463.

Sasaki, H., Hirai, K., Martins, C.M., Furusho, H., Battaglino, R., Hashimoto, K., 2016. Interrelationship between periapical lesion and systemic metabolic disorders. Curr. Pharm. Des. 22 (15), 2204–2215.

Scher, J.U., Sczesnak, A., Longman, R.S., et al., 2013. Expansion of intestinal *Prevotella copri* correlates with enhanced susceptibility to arthritis. Elife 2, e01202.

Scher, J.U., Ubeda, C., Artacho, A., et al., 2015. Decreased bacterial diversity characterizes the altered gut microbiota in patients with psoriatic arthritis, resembling dysbiosis in inflammatory bowel disease. Arthr. Rheumatol. 67 (1), 128–139.

Schwabe, R.F., Jobin, C., 2013. The microbiome and cancer. Nat. Rev. Cancer 13 (11), 800–812.

Segata, N., 2015. Gut microbiome: westernization and the disappearance of intestinal diversity. Curr. Biol. 25 (14), R611–R613.

Sekirov, I., Russell, S.L., Antunes, L.C.M., Finlay, B.B., 2010. Gut microbiota in health and disease. Physiol. Rev. 90 (3), 859–904.

Sender, R., Fuchs, S., Milo, R., 2016. Revised estimates for the number of human and bacteria cells in the body. PLoS Biol. 14 (8), e1002533.

Sheflin, A.M., Whitney, A.K., Weir, T.L., 2014. Cancer-promoting effects of microbial dysbiosis. Curr. Oncol. Rep. 16 (10), 406.

Shin, W., Kim, H.J., 2013. Intestinal barrier dysfunction orchestrates the onset of inflammatory host-microbiome cross-talk in a human gut inflammation-on-a-chip. Proc. Natl. Acad. Sci. USA 115 (45), E10539–E10547.

Shukla, P.K., Gangwar, R., Manda, B., et al., 2016. Rapid disruption of intestinal epithelial tight junction and barrier dysfunction by ionizing radiation in mouse colon in vivo protection by *N*-acetyl-l-cysteine. Am. J. Physiol. Gastrointest. Liver Physiol. 310 (9), G705–G715.

Singh, A.K., Hertzberger, R.Y., Knaus, U.G., 2018. Hydrogen peroxide production by lactobacilli promotes epithelial restitution during colitis. Redox Biol 16, 11–20.

Sperandio, V., Torres, A.G., Kaper, J.B., 2002. Quorum sensing *Escherichia coli* regulators B and C (QseBC): a novel two-component regulatory system involved in the regulation of flagella and motility by quorum sensing in *E. coli.* Mol. Microbiol. 43 (3), 809–821.

Spreadbury, I., 2012. Comparison with ancestral diets suggests dense acellular carbohydrates promote an inflammatory microbiota, and may be the primary dietary cause of leptin resistance and obesity. Diabetes Metab. Syndr. Obes. 5, 175–189.

Spurgeon, M.E., den Boon, J.A., Horswill, M., et al., 2017. Human papillomavirus oncogenes reprogram the cervical cancer microenvironment independently of and synergistically with estrogen. Proc. Natl. Acad. Sci. USA 114 (43), E9076–E9085.

Strus, M., Gosiewski, T., Fyderek, K., et al., 2009. A role of hydrogen peroxide producing commensal bacteria present in colon of adolescents with inflammatory bowel disease in perpetuation of the inflammatory process. J. Physiol. Pharmacol. 60 (Suppl. 6), 49–54.

Swidsinski, A., Dörffel, Y., Loening-Baucke, V., et al., 2017. Reduced mass and diversity of the colonic microbiome in patients with multiple sclerosis and their improvement with ketogenic diet. Front. Microbiol. 8, 1141.

Tabatabaeizadeh, S.A., Tafazoli, N., Ferns, G.A., et al., 2018. Vitamin D, the gut microbiome and inflammatory bowel disease. J. Res. Med. Sci. 23, 75.

Thaiss, C.A., Zmora, N., Levy, M., Elinav, E., 2016. The microbiome and innate immunity. Nature 535 (7610), 65–74.

The Journal of the National Institute on Alcohol Abuse and Alcoholism; Alcohol and Gut-Derived Inflammation. <https://www.arcr.niaaa.nih.gov/arcr382/article01.htm> (accessed 12.05.19.).

Thursby, E., Juge, N., 2017. Introduction to the human gut microbiota. Biochem. J. 474, 1823–1836.

Tlaskalova-Hogenova, H., Tuckova, L., Mestecky, J., et al., 2005. Interaction of mucosal microbiota with the innate immune system. Scand. J. Immunol. 62 (Suppl. 1), 106–113.

Toor, D., Wsson, M.K., Kumar, P., 2019. Dysbiosis disrupts gut immune homeostasis and promotes gastric diseases. Int. J. Mol. Sci. 20 (10), 2432.

Tulic, M.K., Piche, T., Verhasselt, V., 2016. Lung–gut cross-talk: evidence, mechanisms and implications for the mucosal inflammatory diseases. Clin. Exp. Allergy 46, 519–528.

Urdaci, M.C., Bressollier, P., Pinchuk, I., 2004. *Bacillus clausii* probiotic strains: antimicrobial and immunomodulatory activities. J. Clin. Gastroenterol. 38 (Suppl. 6), S86–S90.

Vaarala, O., 2012. Is the origin of type 1 diabetes in the gut? Immunol. Cell Biol. 90 (3), 271–276.

Vaarala, O., Atkinson, M.A., Neu, J., 2008. The "perfect storm" for type 1 diabetes: the complex interplay between intestinal microbiota, gut permeability, and mucosal immunity. Diabetes 57 (10), 2555–2562.

van den Elsen, L.W., Poyntz, H.C., Weyrich, L.S., Young, W., Forbes-Blom, E.E., 2017. Embracing the gut microbiota: the new frontier for inflammatory and infectious diseases. Clin. Transl. Immunol. 6 (1), e125.

Vanuytsel, T., et al., 2013. The role of Haptoglobin and its related protein, Zonulin, in inflammatory bowel disease. Tissue Barriers 1 (5), e27321. Available from: https://doi.org/10.4161/tisb.27321. PMC 3943850. PMID 24868498.

Vayssier-Taussat, M., Albina, E., Citti, C., et al., 2014. Shifting the paradigm from pathogens to pathobiome: new concepts in the light of *meta*-omics. Front. Cell Infect. Microbiol. 4, 29.

Viaud, S., Saccheri, F., Mignot, G., et al., 2013. The intestinal microbiota modulates the anticancer immune effects of cyclophosphamide. Science 342 (6161), 971–976.

Voigt, R.M., Forsyth, C.B., Green, S.J., et al., 2016. Circadian rhythm and the gut microbiome. Int. Rev. Neurobiol. 131, 193–205.

Volkmann, E.R., Hoffmann-Vold, A.M., Chang, Y.L., et al., 2017. Systemic sclerosis is associated with specific alterations in gastrointestinal microbiota in two independent cohorts. BMJ Open. Gastroenterol. 4 (1), e000134.

Vos, M., Esposito, G., Edirisinghe, J.N., et al., 2012. Vitamin K2 is a mitochondrial electron carrier that rescues pink1 deficiency. Science 336 (6086), 1306–1310.

Vrakas, S., Mountzouris, K.C., Michalopoulos, G., et al., 2017. Intestinal bacteria composition and translocation of bacteria in inflammatory bowel disease. PLoS One 12 (1), e0170034.

Vulevic, J., Juric, A., Tzortzis, G., Gibson, G.R., 2013. A mixture of trans-galactooligosaccharides reduces markers of metabolic syndrome and modulates the fecal microbiota and immune function of overweight adults. J. Nutr. 143 (3), 324–331.

Wahls, T.L., 2017. Feeding your microbiome well. J. Evol. Health 2 (1), Article 5.

Wang, L., Wang, F.-S., Gershwin, M.E., 2015. Human autoimmune diseases: a comprehensive update. J. Intern. Med. 278, 369–395.

Wapenaar, M.C., Monsuur, A.J., van Bodegraven, A.A., Weersma, R.K., Bevova, M.R., Linskens, R.K., et al., 2008. Associations with tight junction genes PARD3 and MAGI2 in Dutch patients point to a common barrier defect for coeliac disease and ulcerative colitis. Gut 57 (4), 463–467.

Weaver, C.T., Elson, C.O., Fouser, L.A., Kolls, J.K., 2013. The Th17 pathway and inflammatory diseases of the intestines, lungs, and skin. Annu. Rev. Pathol. 8, 477–512.

Weber, D.D., Aminazdeh-Gohari, S., Kofler, B., 2018. Ketogenic diet in cancer therapy. Aging (Albany NY.) 10 (2), 164–165.

Woelber, J.P., Bremer, K., Vach, K., et al., 2016. An oral health optimized diet can reduce gingival and periodontal inflammation in humans - a randomized controlled pilot study. BMC Oral. Health 17 (1), 28.

Woelber, J.P., Gärtner, M., Breuninger, L., et al., 2019. The influence of an anti-inflammatory diet on gingivitis. A randomized controlled trial. J. Clin. Periodontol. 46, 481–490.

Woof, J.M., Kerr, M.A., 2006. The function of immunoglobulin A in immunity. J. Pathol. 208 (2), 270–282.

Wu, H.J., Wu, E., 2012. The role of gut microbiota in immune homeostasis and autoimmunity. Gut Microbes 3 (1), 4–14.

Xie, G., Zhou, Q., Qiu, C.Z., et al., 2017. Ketogenic diet poses a significant effect on imbalanced gut microbiota in infants with refractory epilepsy. World J. Gastroenterol. 23 (33), 6164–6171.

Yang, L., Jelsbak, L., Marvig, R.L., et al., 2011. Evolutionary dynamics of bacteria in a human host environment. Proc. Natl. Acad. Sci. USA 108 (18), 7481–7486.

Yang, J., Summanen, P.H., Henning, S.M., et al., 2015. Xylooligosaccharide supplementation alters gut bacteria in both healthy and prediabetic adults: a pilot study. Front. Physiol. 6, 216.

Yap, Y.A., Mariño, E., 2018. An insight into the intestinal web of mucosal immunity, microbiota, and diet in inflammation. Front. Immunol. 9, 2617.

Yue, B., Luo, X., Yu, Z., Mani, S., Wang, Z., Dou, W., 2019. Inflammatory bowel disease: a potential result from the collusion between gut microbiota and mucosal immune system. Microorganisms 7 (10), 440.

Zhang, H., Liao, X., Sparks, J.B., Luo, X.M., 2014. Dynamics of gut microbiota in autoimmune lupus. Appl. Env. Microbiol. 80 (24), 7551–7560.

Zheng, X., Zhou, K., Zhang, Y., Han, X., Zhao, A., Liu, J., et al., 2018. Food withdrawal alters the gut microbiota and metabolome in mice. FASEB J. fj201700614R.

Zinöcker, M.K., Lindseth, I.A., 2018. The western diet-microbiome-host interaction and its role in metabolic disease. Nutrients 10 (3), 365.

Zuo, T., Kamm, M.A., Colombel, J.F., Ng, S.C., 2018. Urbanization and the gut microbiota in health and inflammatory bowel disease. Nat. Rev. Gastroenterol. Hepatol.

Chapter 7

Role of the microbiome in the function and diseases of the digestive system

F A Dain Md Opo[1], Addisu Demeke Teklemariam[1], Rokeya Pervin[2] and Md. Akil Hossain[3]

[1]*Department of Biological Sciences, King Abdulaziz University, Jeddah, Saudi Arabia,* [2]*Foot and Mouth Disease Division, Animal and Plant Quarantine Agency, Gimcheon-si, South Korea,* [3]*Department of Pharmacology and Experimental Therapeutics, School of Medicine, Boston University, MA, United States*

7.1 Introduction

The gastrointestinal tract (GIT) is a complex system that extends from the oral cavity to the anus (Dominguez-Bello et al., 2010). In humans the GIT represents a large microbial ecosystem, housing several trillion microbial cells. An integrated catalog of the human fecal microbial metagenome, based on the data from 1200 people in the United States, China, and Europe, identified an aggregate of 9.9 million microbial genes across these fecal microbiomes (Li et al., 2014; Lynch and Pedersen, 2016). Different microorganisms in the GIT of the human include protozoans, archaea, eukaryotes, the viruses, but the largest proportion are bacteria (Lloyd-Price et al., 2016). The presence of microorganisms among people depends on genetics, age, personal hygiene, disease, use of medication, and dietary habits (O'Hara and Shanahan, 2006). The density of microorganisms and their functions vary from one part of the GIT to others (Dominguez-Bello et al., 2010). More than a billion years of mammalian-microbial coevolution has led to inter dependency. As a result, the intestinal microbiotas play a critical role in the maturation and continued education of the host immune response; provide protection against pathogen overgrowth; influence host-cell proliferation and vascularization; regulate intestinal endocrine functions, neurologic signaling, and bone density; provide a source of energy biogenesis (5%–10% of daily host energy requirements); biosynthesize vitamins, fatty acids, neurotransmitters, and multiple other compounds with as yet unknown targets; metabolize bile salts; react to or modify specific drugs; and eliminate exogenous toxins (Ijssennagger et al., 2015; Lynch and Pedersen, 2016; Neuman et al., 2015; Yano et al., 2015). Therefore the microbiota of the GIT are considered to be an essential organ (Ley et al., 2006). The relevance of these microbial activities to health probably varies across the human population, owing to the variations in the composition and density of microorganisms in different people (Lynch and Pedersen, 2016). Furthermore, the probiotics that contain helpful bacteria and some other microbes are being widely used to treat different diseases. The use of probiotics supplements help to reduce the intensity of diseases; develop the immune system; and work against diarrhea, drug-resistant bacteria, and different cancers in the GIT (Choi et al., 2017; O'Hara and Shanahan, 2006; Tomasik and Tomasik, 2003).

Proper microbial composition and environment is essential for the host to maintain healthy life. An imbalance may cause diseases such as metabolic disorder, gastritis, peptic ulcer, colon cancer, fatty liver diseases, obesity, diabetes, liver cancer, and flatulence in the host (Ley et al., 2006). To maintain a balanced microbiome composition and properly utilize the beneficial effects of gastrointestinal microbiotas for the betterment of human health, it is crucial to understand gastrointestinal microbiotas. In this chapter we present recent information (at the time of this writing) about the microbes in the GIT, how they develop and their physiological functions; and their roles in causing several diseases and in eliminating diseases. Important factors to consider include different dietary types and their effects on the gut microbiome and human health.

7.2 Microbes in gastrointestinal tract

At birth, infants have numerous microorganisms on their skin and in their oral cavity and intestines. Babies who were delivered vaginally have microbiota that is similar to the mother's vaginal flora, such as *Lactobacillus*, *Prevotella*, and

Microbiome, Immunity, Digestive Health and Nutrition. DOI: https://doi.org/10.1016/B978-0-12-822238-6.00021-2

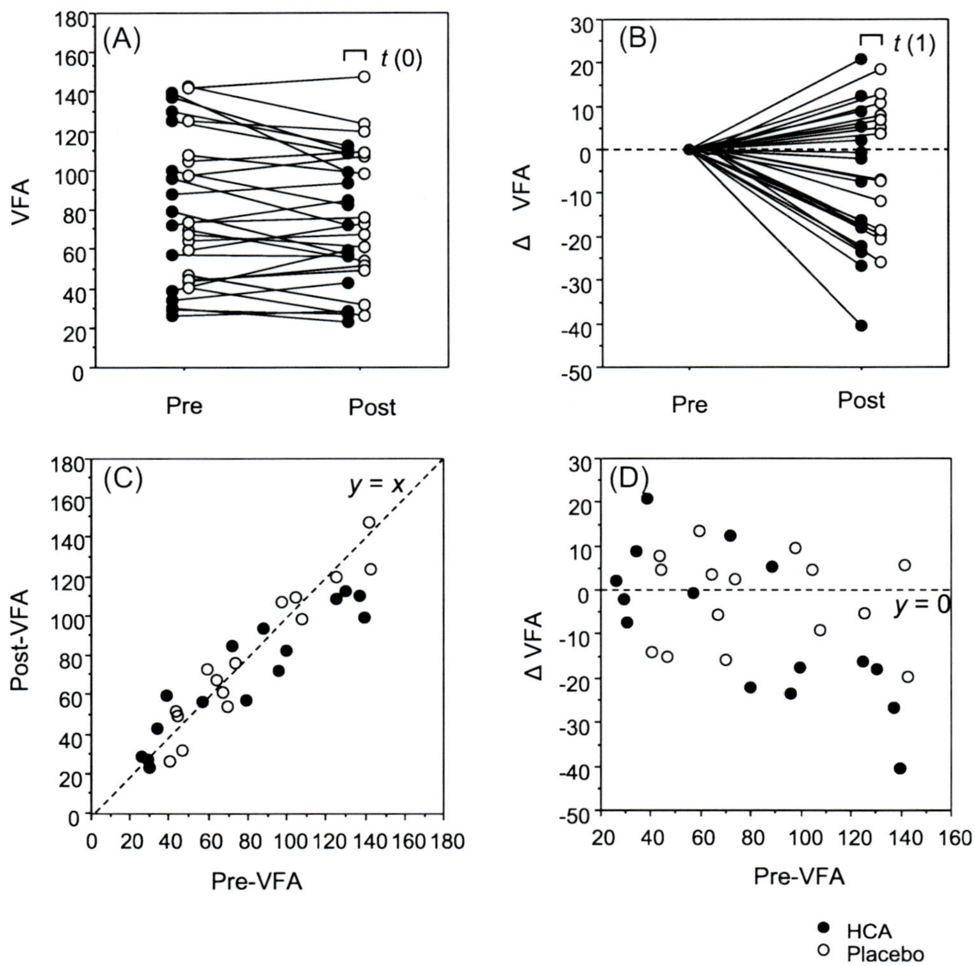

FIGURE 7.1 Distribution of the normal human gut flora.

Sneathia spp. But babies who were delivered by caesarian section have only skin microbiota, including microorganisms such as *Staphylococcus*, *Corynebacterium*, and *Propionibacterium* spp.; these babies ore more prone to certain diseases and allergic reactions until they are 2 years of age. The diverse microorganism community in the human body varies from the esophagus to the colon (Fig. 7.1), and most of the phyla includes Bacteroidetes, Firmicutes, and Proteobacteria (Hillman et al., 2017; Frank et al., 2007). The stomach contains relatively few microorganisms, as the acidic environment prevents the survival of microbiota. But *Helicobacter pylori* is present in the stomach in a low amount that is capable of producing gastritis, ulcers, and cancer (Mackie et al., 1999). The microbe's pattern varies in the gut epithelium and in the gut lumen. Mucin in the gut environment prevents the penetration of microbes. Only a few bacteria can use mucin as a nutrient resource, including *Clostridium*, *Lactobacillus*, and *Enterococcus* (Sekirov et al., 2010). Thus host-microbe interactions in the large intestine are beneficial for energy reabsorption as well as participating in immune system development. *Enterococcus*, *Clostridium*, *Lactobacillus*, *Ruminococcus*, *Bacteroides*, *Bifidobacterium*, *Streptococcus*, and *Enterobacteriaceae* are highly distributed in the feces, and samples are easy to collect for further investigation (Frank et al., 2007).

7.3 Development of the gut flora

The human gut can be colonized by microorganisms at any time starting from birth. The infant's GIT is supposed to be free from microorganisms (normal flora) at birth; nonetheless, the gut is rapidly colonized during and after delivery, and the mode of delivery is one factor that controls the rate and level of colonization. During passage through the birth canal, the neonate acquires microorganism from the mother's vagina (Schloss and Handelsman, 2004). The composition of the human gut microbiome is relatively noncomplex during the first 12 weeks after birth and shows extensive variation between individuals throughout the lifetime (Mackie et al., 1999). It is understood that the primary microbial

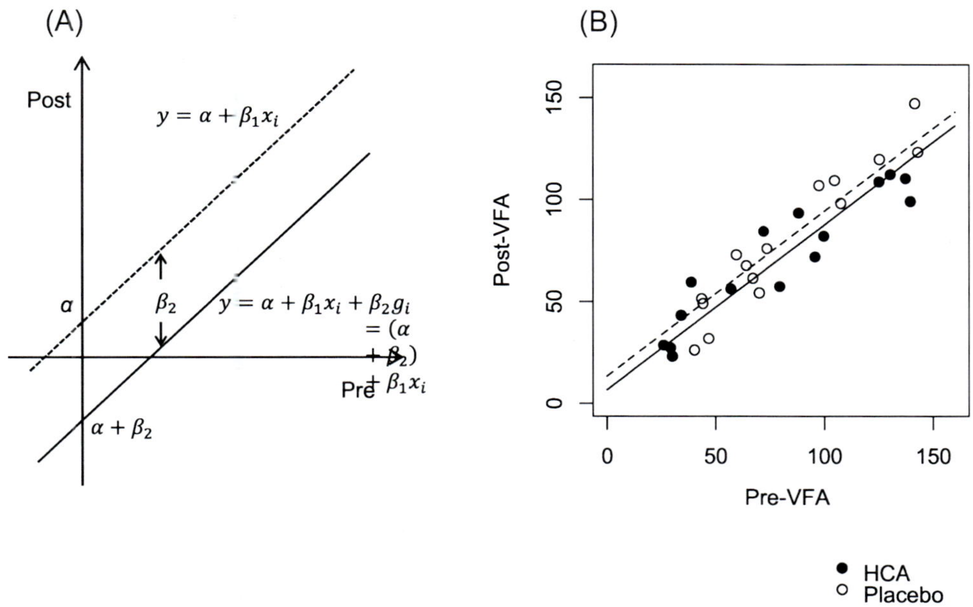

FIGURE 7.2 Schematic diagram showing some of the factors that influence the composition of the human gut microbiota.

colonization of the gut system is influential in shaping the full setup of the adult's gut microbiota and dramatically changes after birth with lactation accompanied by a secondary shift on the feeding of various types of foods. Regardless of the relative similarities of the gut microbiota in mothers and their babies, the level of colonization of gut microbiota is also affected by many internal and external factors (Fig. 7.2).

External factors include, but are not limited to, the type of food eaten and feeding habits and the microbial load of the existing environment coupled with metabolic and temperature-related stresses. Internal factors include host genetics; microbial interactions; gut pH; physiological factors, such as bile acids, peristalsis, intestinal secretions, bacterial GIT receptors, and intrinsic gut immune responses; and antibiotics used to treat gut disease (Adlerberth and Wold, 2009). The types and composition of bacteria are unevenly arranged across the length of the GIT. The number of bacteria found in the GIT can vary from 10^1 to 10^3 bacteria per gram of stomach and duodenal contents, increasing to between 10^4 and 10^7 bacteria per gram in the small intestine, and rising to between 10^{11} and 10^{12} bacteria per gram in the large intestine (Sekirov et al., 2010). The healthy human gut microbiota is largely colonized by strict anaerobes, which outnumber the aerobes and facultative anaerobes by up to 100-fold (Schloss and Handelsman, 2004). The gut microbiota of the adult human is dominated by two phyla of bacteria: the Firmicutes and Bacteroidetes. Early GIT colonizer bacteria include enterococci and enterobacter accompanied by anaerobic organisms such as *Bacteroides* spp., *Bifidobacteria*, anaerobic *Streptococci*, and *clostridia* (Adlerberth and Wold, 2009).

7.4 Physiological functions of gut microbes

More than 1000 microbes are present in the GIT (Rajilić-Stojanović and de Vos, 2014); most of them are Firmicutes and Bacteroidetes (Collins et al., 2012). The cecum and colon are the most abundant sources of microbiota, which perform different functions, based on host physiological demands, including contributing to metabolization of food, development of the immune system, and modulation of nervous system functions (Fabbiano et al., 2018).

7.4.1 Metabolism

Gut microbiota play an important role in digestion and utilization of carbohydrates, amino acids, xenobiotics, and micronutrients in the body and help to produce host energy and increase metabolic proficiency (Gill et al., 2006). For example, *Bacteroides thetaiotaomicron* possesses 260 glycoside hydrolases that are capable of digesting starch and fiber in the human intestine. They readily break down carbohydrates, especially unabsorbed starch and soluble dietary fiber, to short chain fatty acids (SCFAs), acetate, propionate, and butyrate (Topping and Clifton, 2001). The SCFAs are capable of producing energy (70%) via adenosine triphosphate synthesis in the colon after absorption of acetate and distribution into the peripheral tissues. Metabolism of propionate occurs in the liver (Koh et al., 2016), while the majority of butyrate reach has been focused on colonocytes for absorption. The most abundant source of SCFAs derived from bacteria is in the colon; the ileum contains only small amounts (Cuche and Malbert, 1999). Usually, bile acid synthesis occurs from cholesterol in the liver, which is metabolized by intestinal microbiota through the deconjugation and dihydroxylation reaction in the presence of the bacteria-derived bile salt hydrolase, producing secondary bile acids. These acids can be easily absorbed and are hydrophilic in nature, preventing their loss via feces (Jones et al., 2008). Secondary bile acids inhibit microbe growth, maintain mucosal membrane integrity, and prevent the intracellular membrane leakage of content (Nie et al., 2015). Vitamin K_2 helps to prevent several cardiovascular diseases, including atherosclerosis and coronary heart disease, by deceasing cholesterol levels and increasing high-density lipoproteins (HDL). Bacteria that are capable of producing HDL, including *Bacteroides fragilis*, *Eubacterium lentum*, *Enterobacter agglomerans*, *Serratia marcescens*, and *Enterococcus faecium*, generally synthesize vitamin K_2 (Geleijnse et al., 2004). Firmicutes and Actinobacteria mainly participate in this process of metabolism. Intestinal microbiota prevent GIT and neurological disorders and several blood disorders by synthesizing vitamins B_5 and B_{12} and also act as a source for the synthesis of B vitamins (Degnan et al., 2014).

7.4.2 Immune system development

Gastrointestinal microbiota are necessary to mature the immune system (Paust et al., 2004) and maintain homeostatic balance in the gastrointestinal system (Thaiss et al., 2016). The small intestine, having toll-like receptors (TLRs), is capable of maintaining immune response through recognizing several microorganism components, such as bacterial polysaccharide, lipopolysaccharide, and flagella. Interleukin-10 (IL-10) and regulatory T cells (Treg cells) are released by several bacteria, such as *Lactobacillus reuteri* and *Lactobacillus casei*, and act as anti-inflammatory factors that protect the gut barrier. SCFAs that are microbial metabolites also maintain gastrointestinal integrity by working against proinflammatory cytokines (Francino, 2014; Chen et al., 2017). Nucleotide-binding oligomerization domain—like (NOD-like) receptors can identify different microbes and their end products, through which they enhance inflammatory responses. Shortage of the NOD-like receptors, NLPRP6, may induce dysbiosis and intestinal hyperplasia (Elinav et al., 2011). It has been suggested that the bacterium *B. fragilis* is capable of minimizing IL-17 production, protecting the gut barrier from inflammation by bacterial antigen activated $TCD4^+$ lymphocytes that trigger inflammatory response (Mazmanian et al., 2008). Abnormal microbial presence at the time of immune system development may lead to autoimmune disorders and several allergic disorders (Sekirov et al., 2010).

7.4.3 Central nervous system development

Gut microorganisms contribute significantly to the development of the nervous system, as an in vivo study has found. Moreover, a study with mice determined that the intestinal bacteria make a great contribution to enteric brain development and to the regulation of central nervous system functions (McVey Neufeld et al., 2013). In vivo studies have shown that a lack of intestinal microbiota increased nitrergic neurons, which prevent pharmacological blockage of neuronal nitric oxide synthase, and in general decreased myenteric neurons. Lack of intestinal microbiota might be responsible for abnormal growth of the enteric nervous system (Collins et al., 2014). Microorganisms in the ileum are capable of maintaining the proper amount of glial cells in the intestinal membrane of mice (Kabouridis et al., 2015). The intestinal microbiota also contributes to maintaining blood-brain barrier (BBB) permeability in vivo. Administration of *Clostridium tyrobutyricum* is capable of producing butyrate. *B. thetaiotaomicron* mainly produces acetate and propionate and decreases the permeability of the BBB. Intravenous administration of sodium butyrate prevented histone deacetylation and contributed to memory strengthening (Levenson et al., 2004), and the intestinal microbes and sodium butyrate caused histone acetylation of nervous system endothelial cells (Braniste et al., 2014).

7.5 Contribution of human intestinal microbes in causing diseases

The number of microbes in the human intestine is higher than the total number of cells in the human body. This number comprises more than 1200 species of bacteria, archaea, viruses, and eukaryotic microbes, including different phyla of bacteria such as Bacteroides, Firmicutes, Actinomycetes, Proteobacteria, and Verrucomicrobia (Sender et al., 2016). Although they maintain different functions in the human body, an imbalance of microbes leads to the progression of several diseases (Fig. 7.3). Pathogens from outside the human body may colonize the intestinal membrane, leading to progressive inflammatory responses (Hand et al., 2012). A disturbance of the microorganism habitat also causes the formation of human immunodeficiency virus and hepatitis B diseases in the human (Ling et al., 2016). Using

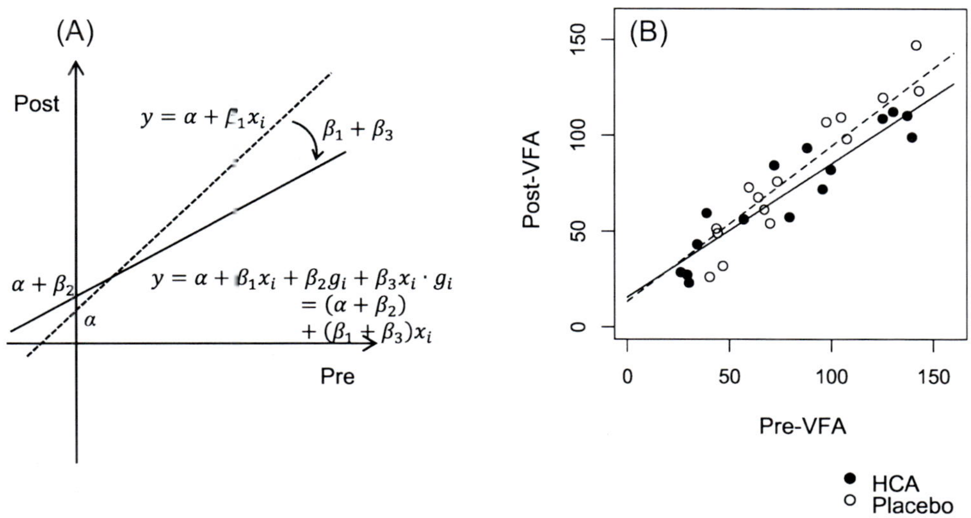

FIGURE 7.3 Schematic representation of common disease related to disturbance in the gut microbiota.

broad-spectrum antibiotics and a longer use of immune suppressive drugs have a great effect on human microbiota and are responsible for causing diseases due to disruptive changes in the gut environment.

7.5.1 *Clostridium difficile* infection

The spore-producing gram-positive bacteria *Clostridium difficile* is responsible for producing toxins A (TcdA) and B (TcdB), which causes cell death by damaging colonic membrane integrity (Genth et al., 2006). In the United States, more than 453,000 cases of *C. difficile* infections (CDI) were identified in 2011 as the major risk associated with antibiotic administration. Diarrhea, pseudomembranous colitis, and sepsis-related symptoms are associated with CDI (Bartlett, 1981). Two toxins are secreted by the bacteria; enterotoxin (toxin A) increases intestinal permeability and fluid secretion, and cytotoxin (toxin B) potentiates inflammation. Intracellular penetration of toxins occurred by binding with receptors through modifying Rho proteins, which are small glutamyl transpeptidase-binding proteins (Sun et al., 2010).

7.5.2 Inflammatory bowel disease

Inflammatory bowel diseases (IBD), which include Crohn's disease and ulcerative colitis (UC), lead to inflammation in the rectum and colon and, more specifically, in the terminal ileum and colon (Abraham and Cho, 2009). One of the causes of IBD is the hyperactivity of T lymphocytes against pathogens that are present in the gut. Antibody development against commensal microbial antigens and autoantigens causes the loss of different microbes in the gut that help to maintain the integrity of the gut mucus barrier (Landers et al., 2002). Gut microbiota and epithelial cell exposure lead to stimulation of the body's own immune system and produce inflammation. Firmicutes such as *Faecalibacterium prausnitzii* and *Roseburia* sp. are anti-inflammatory bacteria whose numbers decrease in the gut as a result of gut dysbiosis and are responsible for IBD formation. These bacteria decrease IL-12 and interferon gamma (IFN-γ) production and enhance antiinflammatory IL-10 production (Machiels et al., 2014). An in vivo mouse model study concludes that the presence of *Enterobacteriaceae* species and *B. fragilis*, both of which have high endotoxic lipopolysaccharides in their outer membranes, contribute to gut inflammation by suppressing T lymphocyte production through activating helper T cells by using the TLR4 signaling pathway (Gronbach et al., 2014). Secreted defensin, together with the mucus layer, works as a barrier against microbial attack for reducing inflammation and increasing the antimicrobial peptide defensin. Defensin is prone to developing mucolytic *Ruminococcus* sp. in the GIT, which induces inflammation through a deeper penetration and microbial infiltration (Salzman et al., 2010).

7.5.3 Colorectal cancer

Certain gut microbes interact with genetic and environmental factors that are capable of promoting a tumor-forming environment and are the fourth leading cause of colorectal cancer (CRC) worldwide (Arnold et al., 2017). Different studies have suggested that the densities of *Pseudomonas*, *Helicobacter*, *Acinetobacter*, and *Fusobacterium nucleatum* are increased and the densities of butyrate-releasing bacteria are deceased in the gut at the time of CRC (Flint et al., 2007). A study found that when the density of *F. nucleatum* was high, it significantly increased *Bacteroides massiliensis*, *B. ovatus*, *B. vulgatus*, and *Escherichia coli* in CRC from advanced adenoma to carcinoma (Feng et al., 2015). *F. nucleatum* developed cancer by increasing myeloid infiltration as well as expressing several inflammatory genes, including prostaglandin-endoperoxide synthase 2, known as PTGS2 (COX2), SCYB1 (IL8), IL6, tumor necrosis factor alpha, and MMP3 in a mouse model. FadA adhesin secreted from these bacteria enhance attachments that promote carcinoma through the upregulation of T cell factor, host NF-κB, oncogene c-Myc, and cyclin D1. Toxin fragilysin, which is produced by enterotoxigenic *B. fragilis* (ETBF), is capable of activating the Wnt/beta-catenin signaling pathway and NF-κB and causes the release of inflammatory mediators with the increasing of cellular differentiation. The ETBF enhances T helper 17 cell-dependent tumorigenesis by activating the degradation of zonula occludens (ZO)-1 protein and E-cadherin, causing damage to the intestinal wall (Kostic et al., 2013).

7.5.4 Rheumatoid arthritis

Gut microorganisms have an impact on the initialization of rheumatoid arthritis. The proposed mechanism includes exhilarating antigen-presenting cells via the effect on TLRs or NOD-like receptors, peptide citrullinization, changing intestinal permeability, and increasing T cell activation (Moya et al., 2015). *Streptococcus pyogenes*, *L. casei*, and *Eubacterium aerofaciens* were capable of producing polyarthritis when injected intraperitonially in murine arthritis models (Horta-Baas et al., 2017). Intradermal administration of *Mycobacterium bovis* and the murein layer of *Staphylococcus epidermidis* also induced arthritis in rats (Kohashi et al., 1979). Bacterial lipopolysaccharides stimulate TLR4, which enhances the adequacy of inflammatory cytokines in synovial tissue and causes the progression of autoimmune arthritis. It has been suggested that gram-negative bacteria help to cure adjuvant-induced arthritis, but gram-positive bacteria such as *Bifidobacterium*, *Propionibacterium acnes*, and *Lactobacillus* cause the progression of this disease, owing to the presence of their peptidoglycan membrane (Kohashi et al., 1985). Intestinal microbiota produce Th17 cells, which move to the peripheral lymphoid tissue and excrete signaling molecule IL-17 and increase systemic B cell and antibody production by acting directly on blood cells (Negi et al., 2017).

7.5.5 Obesity

Human intestinal microorganisms and host genomic variability work as important factors in the case of several metabolic disorders and obesity, with gut microbiota greatly influencing the metabolism of foods in humans (Ley et al., 2006). A higher number of Firmicutes and fewer Bacteroidetes were observed in obese individuals. In an energy balance study related to obesity, the Firmicutes percentage was increased to 20%, and the same percentage was decreased in the case of Bacteroidetes with an additional energy harvest of 150 kcal per day (Jumpertz et al., 2011). Antibiotics used against bacterial infection may change the balance of microorganisms into the GIT. Clindamycin administration for seven days was shown to irreversibly alter the *Bacteroides*, and individuals who were exposed to ciprofloxacin experienced disruptive changes in their gut microbial population (Jernberg et al., 2007). A study of Finnish preschool-age children revealed that childhood obesity is linked to antibacterial drug exposure (Korpela et al., 2016). Among intestinal microbiota endotoxins, *Enterobacter*-producing bacteria were predominant in obese patients (Fei and Zhao, 2013), and *Lactobacillus* was overabundant in the case of obese and overweight children. A study showed that an obese person has a higher amount of *Bacteroides* (P. Patil et al., 2012) and that the genus *Roseburia* is helpful for obese individuals, as it is increased in the stool at the time when obese individuals take in more indigestible polysaccharides (Duncan et al., 2007).

7.5.6 Esophageal cancer

Gut microbiota distribution varies from the oral cavity down to the colon ranging in numbers from 10^1 to 10^{12} (Korpela et al., 2016). The esophagus contains lower numbers of microorganisms in comparison to the oropharynx and stomach-ejected organisms (Gagliardi et al., 1998). About 95%−98% of them are *Streptococcus viridans* (phylum: Firmicutes); others include *Bacteroides*, *Actinobacteria*, *Proteobacteria*, *Fusobacteria*, and TM7. Currently, a 16S rDNA sequencing technology is being used to characterize the diversity of the esophagus microbiota in an individual's normal esophagus compared to individuals with reflux esophagitis and Barrett's esophagus (Yang et al., 2009) (Fig. 7.4). There are several human diseases associated with disturbances in the gut. One of these is the gastroesophageal reflux disease caused by *H. pylori*, which is capable of producing erosive esophagitis, leading to esophageal adenocarcinoma (Rubenstein and

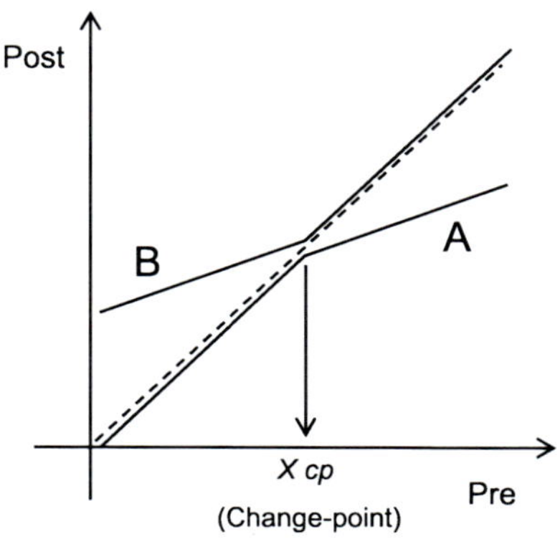

FIGURE 7.4 Microbes in the normal esophagus compared to reflux and Barrett's esophagus. Gram-positive bacteria are important for maintaining a normal environment in the esophagus, and gram-negative bacteria are associated with esophageal reflux and Barrett's esophagus.

Shaheen, 2015; Man, 2011). Zaidi et al. reported that, according to the results of a rat model, the carcinoma in the esophagus might be due to the presence of *E. coli* with esophagojejunal anastomosis. *Clostridiales* and *Erysipelotrichales* may produce gastric dysbiosis, leading to the development of abnormal cells in the esophagus that result in adenocarcinoma (Nasrollahzaden et al., 2015). Recently, it has been observed that the presence of *P. gingivalis* and *F. nucleatum* may lead to cancer in the esophagus with very few clinical symptoms (Yamamura et al., 2016).

7.6 Microorganism that can prevent human disease

Gram-positive microorganisms (species: *Lactobacillus* and *Bifidobacterium*) are being used aggressively to treat gastrointestinal disorders, but gram-negative bacteria are also used in certain circumstances (Marco et al., 2006). Recently, Mutaflor, a probiotic that containing *E. coli*, was used to treat constipation and chronic digestive disease. Currently, probiotics are prominently used to treat or prevent IBD, irritable bowel syndrome, viral or bacterial diseases, and antibiotic-related diarrhea (De Preter et al., 2011).

7.6.1 Colorectal cancer

Gut microbiota imbalance in the GIT may cause cancer. Microorganisms may prevent colon cancer through producing SCFAs, boosting the immune system and regulating blood cholesterol. After a large amount of fiber intake, different bacterial species, including *Lachnospiraceae* and *Ruminococcaceae* produce a high amount of butyrate bacterial metabolites that prevent colon cancer (Hester et al., 2015). When used alone, *Bifidobacteria* were shown to reduce tumor growth in mice. However, a synergistic effect was found when *Bifidobacteria* were used with programmed death-1 inhibitors. *Clostridium butyricum* and 1,2-two hydrazine hydrochloride reduced tumor growth by decreasing $CD4^+$ and $CD8^+$ T lymphocytes, acting over the cell cycle phase, decreasing inflammatory cytokine production, and increasing tumor apoptosis (Chen et al., 2015). By activating the immune system, probiotics play an important role in preventing common types of cancer by improving intestinal barrier function but also by producing antioxidants and SCFAs. It has been suggested that the use of probiotics with other drugs can significantly reduce cyclooxygenase-2 expression and prevent colon cancer. The treatments are different among healthy and CRC patients. Specifically, probiotics used in healthy people prevent the growth of colon carcinoma but, for colon cancer, influence intestinal ecology and promote effective treatment methods (Chen et al., 2015; Walia et al., 2015).

7.6.2 Gastric cancer

H. pylori is mainly responsible for producing gastric cancer, since it is able to disrupt the acid mucus barrier and colonize the gastric epithelium, which is where gastritis is initially produced, and then gradually develops into a peptic ulcer and ultimately gastric cancer (Salama et al., 2013). It has been observed in preclinical studies that *Lactobacillus johnsonii* was able to kill the free-floating *H. pylori* cells as well as the attached bacteria in epithelial cells in tissue culture (Michetti et al., 1999). In in vivo and in vitro studies, *L. acidophilus* was shown to reduce *H. pylori* through improving gastric ulcers, which might be a result of the production of autolysin (Lorca et al., 2001). It has been reported that *H. pylori*–infected volunteers were permitted to drink *Bifidobacterium lactis* Bb12 and *L. acidophilus* La5-containing yogurt and were then found to have moderately reduced *H. pylori* values in a urea breath test (Jungersen et al., 2014). In another study, school children who were infected with *H. pylori* were found to have reduced urea breath test values of nearly 12% after four weeks of treatment with *L. johnsonii* La1 (Cruchet et al., 2003).

7.6.3 Inflammatory bowel diseases

The effects of probiotics against chronic active pouchitis were evaluated in a one-year study of 36 patients. Among them, 20 patients received 6 grams of medical food VSL#3 (probiotic) once a day, and 16 patients received a placebo. The results demonstrated that the VSL#3-treated group significantly improved their condition from IBDs (Mimura et al., 2004). When the administered dose was decreased to 3 grams daily for one year, the results were shown to reduce pouchitis. The administration of antibiotics together with a fructooligosaccharide and *Lactobacillus GG* among patients for one month resulted in complete recovery from these symptoms. The fructooligosaccharide helps to increase the number of *Bifidobacteria* in the human colon (Gibson, 1999). Administration of 750 mg of *S. boulardii* for seven days reduced the intensity of Crohn's disease. Inflammation was decreased in the GIT when *Saccharomyces boulardii* was used in combination with the mesalazine for six months (Guslandi et al., 2000). In a study conducted on 19 patients for three months. *Lactobacillus salivarius* was able to reduce the intensity of intestinal inflammation. In this study, the Crohn's disease activity index mean values were decreased in treated individuals from 217 to 150. Probiotics are being used to treat UC. Treatment of patients with bowel diseases caused by nonpathogenic *E. coli* strains were significantly improved (Basso et al., 2019) as the number of pathogenic *E. coli* inversely decreased related to the number of non-pathogenic *E. coli*. This observation supports the use of nonpathogenic *E. coli* for the treatment of UC. The administration of 1500 mg of mesalazine or 200 mg of *E. coli* daily for 12 weeks reduced the UC in patients (Kruis et al., 1997). Another study was conducted on 1200 patients and showed a reduction in the intensity of UC. The probiotic Mutaflor, which contains *E. coli*, was administered along with mesalazine 2400 mg/day, for three months to one year. Both of the drugs were found to reduce UC (Rembacken et al., 1999).

7.6.4 Obesity

Changing gut microbes, such as a lower amount of *Bifidobacterium*, may cause the development of obesity after childhood. At the same time, obese mothers who have an insufficiency of *Bifidobacterium* may give birth to children who have inheritable obesity characteristic (Collado et al., 2009). The use of probiotics in obese individuals inhibits the expression of the adiposity-responsive genes of the host by enhancing the endocrinal L cell numbers of those that produce GPL-1 and GPL-2. GPL-2 reduces the permeability of the intestinal wall, and GPL-1 has been shown to reduce food intake and decrease body fat (Tilg et al., 2009).

7.6.5 Hepatic encephalopathy

Enteric bacterial flora plays an important role in producing ammonia, phenols, and mercaptans that might cause the development of hepatic encephalopathy by catalyzing systemic inflammatory responses. The combination of lactulose-containing probiotics with antibiotics is used to treat this disease. Lactulose produces butyric acid, lactic acid, acetic acid, and hydrogen gas by fermentation in the large intestine. The production of hydrogen gas causes the development of flatulence in the colon, resulting in greater removal of colon bacteria, in particular two types of bacteria: urease-positive bacteria and ammonia-producing bacteria (Bongaerts and Severijnen, 2001). Recent studies have shown that the use of *C. butyricum* and *Bifidobacterium infantis* can minimize the intensity of hepatic encephalopathy (Xia et al., 2018).

7.7 Dietary modulation of the gut microbiota

The structural and functional diversity of microbiota of the gut system can quickly respond to the change in macronutrient levels and daily food components. Dietary protein, fiber, fat, vegetables, fruits, and various food additives appear to be crucial determinants that control the microbiota environment and its composition (Chassaing et al., 2017) (Fig. 7.5 and Table 7.1). Dietary fibers are monosaccharides that determine the gut bacterial ecology, diversity, and function. Kovatcheva-Datchary et al. stated that densities of the gut microbiota *Prevotella* increased in participants with improved glucose metabolism following dietary fiber supplementation. It has also been indicated that high-fat diet–induced obesity has been prevented by dietary fiber intake by modifying the gut microbiota (Li et al., 2016).

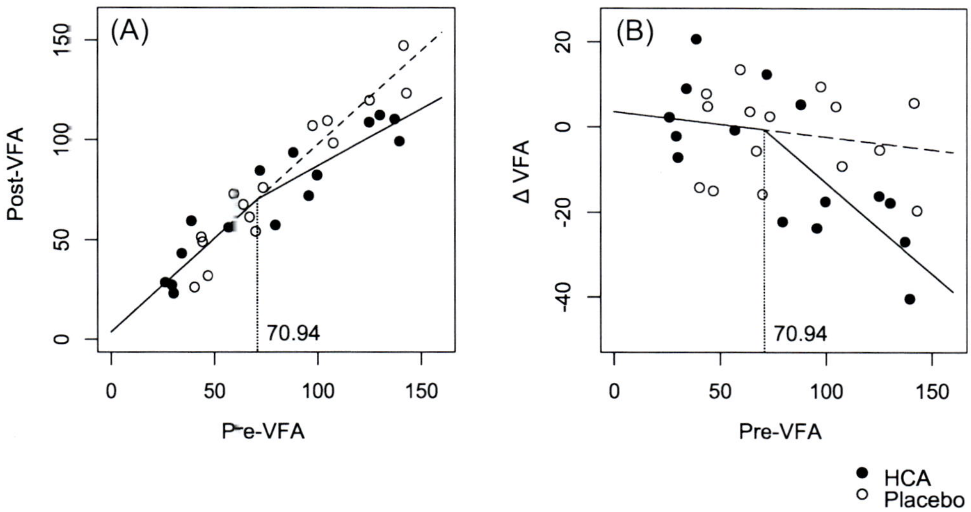

FIGURE 7.5 Schematic representation of the main dietary factors that determine the diversity of the gut microbiota.

TABLE 7.1 The effect of various dietary types on gut microbiomes and human health.

Dietary types	Effect on gut microbiomes	Effect on health mediated by microbiomes	Reference
Prebiotic	Increases the concentration of *Bifidobacterium*, *Lactobacillus*, and lactic acid bacteria	Production of beneficial metabolites	Reddy et al. (1992)
Vegetarian diet	Decreases the quantity and changes the diversity of *Clostridium* cluster IV and *Clostridium* ribosomal RNA clusters XIVa and XVIII	Maintains metabolic homeostasis	Liszt et al. (2009)
Starch diet	Dramatically Increases *Eubacterium rectale* and *Ruminococcus bromii*	Maintains metabolic homeostasis	Liszt et al. (2009)
High-fat diet	Enhances the abundance of Proteobacteria	Maintains metabolic homeostasis	Monira et al. (2011)
Cheese	Increases *Bifidobacteria* and decreases *Bacteroides* and *Clostridia*, some strains of which are associated with intestinal infections	Increases production of short-chain fatty acids, reduces production of trimethylamine N-oxide	Monira et al. (2011), Zheng et al. (2015)
Artificial sweeteners	Overgrowth of *Proteobacteria* and *Escherichia coli*. The growth of *Bacteroides, Clostridia*, and total aerobic bacteria was significantly lower, and fecal pH was significantly higher.	Induces glucose intolerance	Rodriguez-Palacios et al. (2018)

Dietary protein is the key source of nitrogen for the growth of microbes found in the colon and is crucial to the carbohydrate's assimilation and production of useful products. Hence a joint interaction between carbohydrates and protein in the large bowel can contribute to bowel health (Conlon and Bird, 2014). A high-fat diet also disturbs the structural as well as functional patterns of gut microbiota, which regulate fatty acid metabolism and intestinal absorption (Leone et al., 2015). Kaliannan et al. found that mice that were fed a high concentration of an omega-6 fatty acid–containing diet had systemic low-grade inflammation and a significantly higher degree of metabolic endotoxemia (Kaliannan et al., 2015).

Although regulatory agencies consider artificial sweeteners to safe for consumption as sugar alternatives, some laboratory animal studies have revealed that these sweeteners may have adverse effects on the gut microbiota. Sugar substitutes such as aspartame, saccharin, and sucralose have been shown to destabilize the diversity and stability of the gut microbiota. Rats that were fed sucralose for three months had a higher proportion a total aerobic bacteria count coupled with *Clostridia* spp. and *Bacteroides* in their guts and a significantly higher fecal pH than rats that were not fed sucralose (Perry et al., 2016). Mice that were fed sucralose for six months had altered fecal metabolites and a surge in the expression of gut bacterial proinflammatory genes (Beserra et al., 2015). Similarly, various food additives, such as emulsifiers, have been shown to alter gut microbes in animals. Mice that were fed relatively low concentrations of carboxymethylcellulose and polysorbate-80 displayed reduced microbial diversity compared to mice that were not fed with emulsifiers. *Verrucomicrobia* and *Bacteroidales* were reduced and inflammation-promoting *Proteobacteria* associated with mucus was enriched by feeding emulsifiers (Raman et al., 2013). Moreover, nonabsorbable apple procyanidins, vitamins, probiotics, synbiotics, and prebiotics are markedly shaping the diversity of the human gut microbiota and their associated functions.

Apart from diet, various disease-targeted therapeutic chemicals are principal modulators of the composition of the human gut microbiota. Antibiotics evidently have negative effects on the amount and diversity of normal gut flora, as low doses of antibiotics are usually given to livestock to enhance their weight and growth. Generally, inadequate clinical evidence exists to support comprehensive concluding remarks about the impact of antibiotics as well as dietary preferences on the proportion and/or the diversity of the gut microbiota.

7.8 Conclusion

Colonization of the human GIT with microorganisms starts immediately at birth and is characterized by a permanent adaptation to varying conditions. Those GIT microorganisms are crucial and participate in a wide range of activities, including digestion and metabolism of foods, modulation of the immune system, production of vitamins and SCFAs, and involvement in epithelial development. Several factors, such as age, sex, food habits, microbial interactions gut pH, bile acid, intestinal secretions, and application of drugs, directly influence the number and types of microbiotas that are found throughout the GIT. Disturbances in the composition of microorganisms that are found in the GIT may cause acute or chronic diseases in apparently healthy individuals. Recent studies have shown that the use of food to balance the microbial environment and to develop microbial-based therapies is becoming a more popular method of treating disease. Homeostasis can be achieved by reducing microbial consumption and by ensuring a modified and balanced diet for both the host and the microorganisms. Treatment of diseases such as inflammatory bowel disease, hepatic encephalopathy, obesity, peptic ulcer, colon cancer, and UC by using probiotics is gaining in popularity because of the beneficial effects on health. In most cases, probiotics contain *Lactobacillus* or *Bifidobacterium* spp. at higher concentrations, which plays an important role in the host defense system, production of antimicrobial defensins, maintaining gut permeability, and actively participating in vitamin and SCFA production. While excessive intake of prebiotics may cause flatulence, abdominal pain, and bloating, a prudent combination of probiotics and prebiotics ensures the synergistic amplification of therapeutic effects to decrease the pathogenic content in body.

References

Abraham, C., Cho, J.H., 2009. Inflammatory bowel disease. N. Engl. J. Med. 361 (21), 2066–2078.

Adlerberth, I., Wold, A.E., 2009. Establishment of the gut microbiota in Western infants. Acta Paediatr. 98 (2), 229–238.

Arnold, M., Sierra, M.S., Laversanne, M., Soerjomataram, I., Jemal, A., Bray, F., 2017. Global patterns and trends in colorectal cancer incidence and mortality. Gut 66 (4), 683–691.

Bartlett, J.G., 1981. Antimicrobial agents implicated in *Clostridium difficile* toxin-associated diarrhea or colitis. Johns. Hopkins Med. J. 149 (1), 5–9.

Basso, P.J., Câmara, N.O.S., Sales-Campos, H., 2019. Microbial-based therapies in the treatment of inflammatory bowel disease—an overview of human studies. Front. Pharmacol. 9, 1571. Jan 10.

Beserra, B.T.S., Fernandes, R., do Rosario, V.A., Mocellin, M.C., Kuntz, M.G.F., Trindade, E.B.S.M., 2015. A systematic review and *meta*-analysis of the prebiotics and synbiotics effects on glycaemia, insulin concentrations and lipid parameters in adult patients with overweight or obesity. Clin. Nutr. 34 (5), 845–858.

Bongaerts, G.P., Severijnen, R.S., 2001. The beneficial, antimicrobial effect of probiotics. Med. Hypotheses 56 (2), 174–177.

Braniste, V., Al-Asmakh, M., Kowal, C., Anuar, F., Abbaspour, A., Tóth, M., et al., 2014. The gut microbiota influences blood-brain barrier permeability in mice. Sci. Transl. Med. 6 (263), 263ra158.

Chassaing, B., Vijay-Kumar, M., Gewirtz, A.T., 2017. How diet can impact gut microbiota to promote or endanger health. Curr. Opin. Gastroenterol. 33 (6), 417–421.

Chen, T., Kim, C.Y., Kaur, A., Lamothe, L., Shaikh, M., Keshavarzian, A., et al., 2017. Dietary fibre-based SCFA mixtures promote both protection and repair of intestinal epithelial barrier function in a Caco-2 cell model. Food Funct. 8 (3), 1166–1173.

Chen, Z.F., Ai, L.Y., Wang, J.L., Ren, L.L., Yu, Y.N., Xu, J., et al., 2015. Probiotics *Clostridium butyricum* and *Bacillus subtilis* ameliorate intestinal tumorigenesis. Future Microbiol. 10 (9), 1433–1445.

Choi, S., Hwang, Y.J., Shin, M.J., Yi, H., 2017. Difference in the gut microbiome between ovariectomy-induced obesity and diet-induced obesity. J. Microbiol. Biotechno . 27 (12), 2228–2236.

Collado, M., Isolauri, E., Salminen, S., Sanz, Y., 2009. The impact of probiotic on gut health. Curr. Drug. Metab. 10 (1), 68–78.

Collins, S.M., Surette, M. Bercik, P., 2012. The interplay between the intestinal microbiota and the brain. Nat. Rev. Microbiol. 10 (11), 735–742.

Collins, J., Borojevic, R., Verdu, E.F., Huizinga, J.D., Ratcliffe, E.M., 2014. Intestinal microbiota influence the early postnatal development of the enteric nervous system. Neurogastroenterol. Motil. 26 (1), 98–107.

Conlon, M.A., Bird, A.R., 2014. The impact of diet and lifestyle on gut microbiota and human health. Nutrients 7 (1), 17–44.

Cruchet, S., Obregon, M.C., Salazar, G., Diaz, E., Gotteland, M., 2003. Effect of the ingestion of a dietary product containing *Lactobacillus johnsonii* La1 on *Helicobacter pylori* colonization in children. Nutrition 19 (9), 716–721.

Cuche, G., Malbert, C.H., 1999. Short-chain fatty acids present in the ileum inhibit fasting gastrointestinal motility in conscious pigs. Neurogastroenterol. Motil. 11 (3), 219–225.

De Preter, V., Hamer, H.M., Windey, K., Verbeke, K., 2011. The impact of pre- and/or probiotics on human colonic metabolism: does it affect human health? Mol. Nutr. Food Res. 55 (1), 46–57.

Degnan, P.H., Taga, M.E., Goodman, A.L., 2014. Vitamin B12 as a modulator of gut microbial ecology. Cell Metab. 20 (5), 769–778.

Dominguez-Bello, M.G., Costello, E.K., Contreras, M., Magris, M., Hidalgo, G., Fierer, N., et al., 2010. Delivery mode shapes the acquisition and structure of the initial microbiota across multiple body habitats in newborns. Proc. Natl Acad. Sci. USA 107 (26), 11971–11975.

Duncan, S.H., Belenguer, A., Holtrop, G., Johnstone, A.M., Flint, H.J., Lobley, G.E., 2007. Reduced dietary intake of carbohydrates by obese subjects results in decreased concentrations of butyrate and butyrate-producing bacteria in feces. Appl. Env. Microbiol. 73 (4), 1073–1078.

Elinav, E., Strowig, T., Kau, A.L., Henao-Mejia, J., Thaiss, C.A., Booth, C.J., et al., 2011. NLRP6 inflammasome regulates colonic microbial ecology and risk for colitis. Cell 145 (5), 745–757.

Fabbiano, S., Suárez-Zamorano, N., Chevalier, C., Lazarević, V., Kieser, S., Rigo, D., et al., 2018. Functional gut microbiota remodeling contributes to the caloric restriction-induced metabolic improvements. Cell Metab. 28 (6), 907–921. e7.

Fei, N. Zhao, L., 2013. An opportunistic pathogen isolated from the gut of an obese human causes obesity in germfree mice. ISME J. 7 (4), 880–884.

Feng, Q., Liang, S., Jia, H., Stadlmayr, A., Tang, L., Lan, Z., et al., 2015. Gut microbiome development along the colorectal adenoma-carcinoma sequence. Nat. Commun. 6 (1), 1–13.

Flint, H.J., Duncan, S.H., Scott, K.P., Louis, P., 2007. Interactions and competition within the microbial community of the human colon: links between diet and health. Env. Microbiol. 9 (5), 1101–1111.

Francino, M., 2014. Early development of the gut microbiota and immune health. Pathogens 3 (3), 769–790.

Frank St., D.N., Amand, A.L., Feldman, R.A., Boedeker, E.C., Harpaz, N., Pace, N.R., 2007. Molecular-phylogenetic characterization of microbial community imbalances in human inflammatory bowel diseases. Proc. Natl Acad. Sci. USA 104 (34), 13780–13785.

Gagliardi, D., Makihara, S., Corsi, P.R., Viana Ade, T., Wiczer, M.V., Nakakubo, S., et al., 1998. Microbial flora of the normal esophagus. Dis. Esophagus 11 (4), 248–250.

Geleijnse, J.M., Vermeer, C., Grobbee, D.E., Schurgers, L.J., Knapen, M.H., van der Meer, I.M., et al., 2004. Dietary intake of menaquinone is associated with a reduced risk of coronary heart disease: the Rotterdam Study. J. Nutr. 134 (11), 3100–3105.

Genth, H., Huelsenbeck, J. Hartmann, B., Hofmann, F., Just, I., Gerhard, R., 2006. Cellular stability of Rho-GTPases glucosylated by *Clostridium difficile* toxin B. FEBS Lett. 580 (14), 3565–3569.

Gibson, G.R., 1999. Dietary modulation of the human gut microflora using the prebiotics oligofructose and inulin. J. Nutr. 129 (7 Suppl), 1438S–1441SS.

Gill, S.R., Pop, M., DeBoy, R.T., Eckburg, P.B., Turnbaugh, P.J., Samuel, B.S., et al., 2006. Metagenomic analysis of the human distal gut microbiome. Science 312 (5778), 1355–1359.

Gronbach, K., Flade, I., Holst, O., Lindner, B., Ruscheweyh, H.J., Wittmann, A., et al., 2014. Endotoxicity of lipopolysaccharide as a determinant of T-cell-mediated colitis induction in mice. Gastroenterology 146 (3), 765–775.

Guslandi, M., Mezzi, G., Sorghi, M., Testoni, P.A., 2000. *Saccharomyces boulardii* in maintenance treatment of Crohn's disease. Dig. Dis. Sci. 45 (7), 1462–1464.

Hand, T.W., Dos Santos, L.M., Bouladoux, N., Molloy, M.J., Pagán, A.J., Pepper, M., et al., 2012. Acute gastrointestinal infection induces long-lived microbiota-specific T cell responses. Science 337 (6101), 1553–1556.

Hester, C.M., Jala, V.R., Langille, M.G.I., Umar, S., Greiner, K.A., Haribabu, B., 2015. Fecal microbes, short chain fatty acids, and colorectal cancer across racial/ethnic groups. World J. Gastroenterol. 21 (9), 2759–2769.

Hillman E.T., Lu, H., Yao, T., Nakatsu, C.H., 2017. Microbial ecology along the gastrointestinal tract. Microbes Env. 32 (4), 300–313.

Horta-Baas, G., Romero-Figueroa, M.D.S., Montiel-Jarquín, A.J., Pizano-Zárate, M.L., García-Mena, J., Ramírez-Durán, N., 2017. Intestinal dysbiosis and rheumatoid arthritis: a link between gut microbiota and the pathogenesis of rheumatoid arthritis. J. Immunol. Res. 2017, 4835189. Aug 30.

Ijssennagger, N., Belzer, C., Hooiveld, G.J., Dekker, J., van Mil, S.W., Müller, M., et al., 2015. Gut microbiota facilitates dietary heme-induced epithelial hyperproliferation by opening the mucus barrier in colon. Proc. Natl Acad. Sci. USA 112 (32), 10038–10043.

Jernberg, C., Löfmark, S., Edlund, C., Jansson, J.K., 2007. Long-term ecological impacts of antibiotic administration on the human intestinal microbiota. ISME J. 1 (1), 56–66.

Jones, B V., Begley, M., Hill, C., Gahan, C.G.M., Marchesi, J.R., 2008. Functional and comparative metagenomic analysis of bile salt hydrolase activity in the human gut microbiome. Proc. Natl. Acad. Sci. USA 105 (36), 13580–13585.

Jumpertz, R., Le, D.S., Turnbaugh, P.J., Trinidad, C., Bogardus, C., Gordon, J.I., et al., 2011. Energy-balance studies reveal associations between gut microbes, caloric load, and nutrient absorption in humans. Am. J. Clin. Nutr. 94 (1), 58–65.

Jungersen, M., Wind, A., Johansen, E., Christensen, J., Stuer-Lauridsen, B., Eskesen, D., 2014. The science behind the probiotic strain *Bifidobacterium animalis* subsp. lactis BB-12®. Microorganisms 2 (2), 92–110.

Kabouridis, P.S., Lasrado, R., McCallum, S., Chng, S.H., Snippert, H.J., Clevers, H., et al., 2015. Microbiota controls the homeostasis of glial cells in the gut lamina propria. Neuron 85 (2), 289–295.

Kaliannan, K., Wang, B., Li, X.Y., Kim, K.J., Kang, J.X., 2015. A host-microbiome interaction mediates the opposing effects of omega-6 and omega-3 fatty acids on metabolic endotoxemia. Sci. Rep. 5, 11276.

Koh, A., De Vadder, F., Kovatcheva-Datchary, P., Bäckhed, F., 2016. From dietary fiber to host physiology: short-chain fatty acids as key bacterial metabolites. Cell 165 (6), 1332–1345.

Kohashi, O., Kohashi, Y., Takahashi, T., Ozawa, A., Shigematsu, N., 1985. Reverse effect of gram-positive bacteria vs. gram-negative bacteria on adjuvant-induced arthritis in germfree rats. Microbiol. Immunol. 29 (6), 487–497.

Kohashi, O., Kuwata, J., Umehara, K., Uemura, F., Takahashi, T., Ozawa, A., 1979. Susceptibility to adjuvant-induced arthritis among germfree, specific-pathogen-free, and conventional rats. Infect. Immun. 26 (3), 791–794.

Korpela, K., Salonen, A., Virta, L.J., Kekkonen, R.A., Forslund, K., Bork, P., et al., 2016. Intestinal microbiome is related to lifetime antibiotic use in Finnish pre-school children. Nat. Commun. 7, 10410.

Kostic, A.D., Chun, E., Robertson, L., Glickman, J.N., Gallini, C.A., Michaud, M., et al., 2013. *Fusobacterium nucleatum* potentiates intestinal tumorigenesis and modulates the tumor-immune microenvironment. Cell Host Microbe 14 (2), 207–215.

Kruis, W., Schütz, E., Fric, P., Fixa, B., Judmaier, G., Stolte, M., 1997. Double-blind comparison of an oral *Escherichia coli* preparation and mesalazine in maintaining remission of ulcerative colitis. Aliment. Pharmacol. Ther. 11 (5), 853–858.

Landers, C.J., Cohavy, O., Misra, R., Yang, H., Lin, Y.C., Braun, J., et al., 2002. Selected loss of tolerance evidenced by crohn's disease-associated immune responses to auto- and microbial antigens. Gastroenterology 123 (3), 689–699.

Leone, V., Gibbons, S.M., Martinez, K., Hutchison, A.L., Huang, E.Y., Cham, C.M., et al., 2015. Effects of diurnal variation of gut microbes and high-fat feeding on host circadian clock function and metabolism. Cell Host Microbe 17 (5), 681–689.

Levenson, J.M., O'Riordan, K.J., Brown, K.D., Trinh, M.A., Molfese, D.L., Sweatt, J.D., 2004. Regulation of histone acetylation during memory formation in the hippocampus. J. Biol. Chem. 279 (39), 40545–40559.

Ley, R.E., Peterson, D.A., Gordon, J.I., 2006. Ecological and evolutionary forces shaping microbial diversity in the human intestine. Cell 124 (4), 837–848.

Ley, R.E., Turnbaugh, P.J., Klein, S., Gordon, J.I., 2006. Microbial ecology: human gut microbes associated with obesity. Nature 444 (7122), 1022–1023.

Li, J., Jia, H., Cai, X., Zhong, H., Feng, Q., Sunagawa, S., et al., 2014. An integrated catalog of reference genes in the human gut microbiome. Nat. Biotechnol. 32 (8), 834–841.

Li, X., Guo, J., Ji, K., Zhang, P., 2016. Bamboo shoot fiber prevents obesity in mice by modulating the gut microbiota. Sci. Rep. 6, 32953. Sep 7.

Ling, Z., Jin, C., Xie, T., Cheng, Y., Li, L., Wu, N., 2016. Alterations in the fecal microbiota of patients with HIV-1 infection: an observational study in a chinese population. Sci. Rep. 6 (1), 30673.

Liszt, K., Zwielehner, J., Handschur, M., Hippe, B., Thaler, R., Haslberger, A.G., 2009. Characterization of bacteria, clostridia and Bacteroides in faeces of vegetarians using qPCR and PCR-DGGE fingerprinting. Ann. Nutr. Metab. 54 (4), 253–257.

Lloyd-Price, J., Abu-Ali, G., Huttenhower, C., 2016. The healthy human microbiome. Genome Med. 8 (1), 51.

Lorca, G.L., Wadström, T., Font de Valdez, G., Ljungh, Å., 2001. *Lactobacillus acidophilus* autolysins inhibit *Helicobacter pylori* in vitro. Curr. Microbiol. 42 (1), 39–44.

Lynch, S.V., Pedersen, O., 2016. The human intestinal microbiome in health and disease. N. Engl. J. Med. 375 (24), 2369–2379.

Machiels, K., Joossens, M., Sabino, J., De Preter, V., Arijs, I., Eeckhaut, V., et al., 2014. A decrease of the butyrate-producing species roseburia hominis and Faecalibacterium prausnitzii defines dysbiosis in patients with ulcerative colitis. Gut. 63 (8), 1275–1283.

Mackie, R.I., Sghir, A., Gaskins, H.R., 1999. Developmental microbial ecology of the neonatal gastrointestinal tract. Am. J. Clin. Nutr. 69 (5), 1035S–1045S.

Man, S.M., 2011. The clinical importance of emerging Campylobacter species. Nat. Rev. Gastroenterol. Hepatol. 8 (12), 669–685.

Marco, M.L., Pavan, S., Kleerebezem, M., 2006. Towards understanding molecular modes of probiotic action. Curr. Opin. Biotechnol. 17 (2), 204–210.

Mazmanian, S.K., Round, J.L., Kasper, D.L., 2008. A microbial symbiosis factor prevents intestinal inflammatory disease. Nature 453 (7195), 620–625.

McVey Neufeld, K.A., Mao, Y.K., Bienenstock, J., Foster, J.A., Kunze, W.A., 2013. The microbiome is essential for normal gut intrinsic primary afferent neuron excitability in the mouse. Neurogastroenterol. Motil. 25 (2), 183–e88.

Michetti, P., Dorta, G., Wiesel, P.H., Brassart, D., Verdu, E., Herranz, M., et al., 1999. Effect of whey-based culture supernatant of *Lactobacillus acidophilus* (johnsonii) La1 on *Helicobacter pylori* infection in humans. Digestion 60 (3), 203–209.

Mimura, T., Rizzello, F., Helwig, U., Poggioli, G., Schreiber, S., Talbot, I.C., et al., 2004. Once daily high dose probiotic therapy (VSL#3) for maintaining remission in recurrent or refractory pouchitis. Gut 53 (1), 108–114.

Monira, S., Nakamura, S., Gotoh, K., Izutsu, K., Watanabe, H., Alam, N.H., et al., 2011. Gut microbiota of healthy and malnourished children in Bangladesh. Front. Microbiol. 2, 228. Nov 21.

Moya, P., Moya, P., Corominas, H., Salazar, J., Baiget, M., 2015. Consensus on the use of methotrexate beyond the clinical recommendation: adjusted dose and pharmacogenetics. Reumatología Clínica (Engl. Ed.) 11 (4), 261–262.

Nasrollahzadeh, D., Malekzadeh, R., Ploner, A., Shakeri, R., Sotoudeh, M., Fahimi, S., et al., 2015. Variations of gastric corpus microbiota are associated with early esophageal squamous cell carcinoma and squamous dysplasia. Sci. Rep. 5, 8820. Mar 6.

Negi, S., Singh, H., Mukhopadhyay, A., 2017. Gut bacterial peptides with autoimmunity potential as environmental trigger for late onset complex diseases: in–silico study. PLoS One 12 (7), e0180518.

Neuman, H., Debelius, J.W., Knight, R., Koren, O., 2015. Microbial endocrinology: the interplay between the microbiota and the endocrine system. FEMS Microbiol. Rev. 39 (4), 509–521.

Nie, Y.F., Hu, J., Yan, X.H., 2015. Cross-talk between bile acids and intestinal microbiota in host metabolism and health. J. Zhejiang Univ. Sci. B 16 (6) 436–446.

O'Hara, A.M., Shanahan, F., 2006. The gut flora as a forgotten organ. EMBO Rep. 7 (7), 688–693.

Paust, S., Lu, L., McCarty, N., Cantor, H., 2004. Engagement of B7 on effector T cells by regulatory T cells prevents autoimmune disease. Proc. Natl. Acad. Sci. USA 101 (28), 10398–10403.

Perry, R.J., Peng, L., Barry, N.A., Cline, G.W., Zhang, D., Cardone, R.L., et al., 2016. Acetate mediates a microbiome-brain-β-cell axis to promote metabolic syndrome. Nature 534 (7606), 213–217.

P. Patil, D., Pdhotre, D., Shchavan, S., Sultan, A., Jain, D.S., Lanjekar, V.B., et al., 2012. Molecular analysis of gut microbiota in obesity among Indian individuals. J. Biosci. 37 (4), 647–657.

Rajilić-Stojanović, M., de Vos, W.M., 2014. The first 1000 cultured species of the human gastrointestinal microbiota. FEMS Microbiol. Rev. 38 (5), 996–1047.

Raman, M., Ambalam, P., Kondepudi, K.K., Pithva, S., Kothari, C., Patel, A.T., et al., 2013. Potential of probiotics, prebiotics and synbiotics for management of colorectal cancer. Gut Microbes 4 (3), 181–192.

Reddy, B.S., Engle, A., Simi, B., Goldman, M., 1992. Effect of dietary fiber on colonic bacterial enzymes and bile acids in relation to colon cancer. Gastroenterology 102 (5), 1475–1482.

Rembacken, B.J., Snelling, A.M., Hawkey, P.M., Chalmers, D.M., Axon, A.T.R., 1999. Non-pathogenic *Escherichia coli* vs mesalazine for the treatment of ulcerative colitis: a randomised trial. Lancet 354 (9179), 635–639.

Rodriguez-Palacios, A., Harding, A., Menghini, P., Himmelman, C., Retuerto, M., Nickerson, K.P., et al., 2018. The artificial sweetener splenda promotes gut proteobacteria, dysbiosis, and myeloperoxidase reactivity in crohn's disease–like ileitis. Inflamm. Bowel Dis. 24 (5), 1005–1020.

Rubenstein, J.H., Shaheen, N.J., 2015. Epidemiology, diagnosis, and management of esophageal adenocarcinoma. Gastroenterology 149 (2), 302–317. e1.

Salama, N.R., Hartung, M.L., Müller, A., 2013. Life in the human stomach: persistence strategies of the bacterial pathogen *Helicobacter pylori*. Nat. Rev. Microbiol. 11 (6), 385–399.

Salzman, N.H., Hung, K., Haribhai, D., Chu, H., Karlsson-Sjöberg, J., Amir, E., et al., 2010. Enteric defensins are essential regulators of intestinal microbial ecology. Nat. Immunol. 11 (1), 76–83.

Schloss, P.D., Handelsman, J., 2004. Status of the microbial census. Microbiol. Mol. Biol. Rev. 68 (4), 686–691.

Sekirov, I., Russell, S.L., Antunes, L.C., Finlay, B.B., 2010. Gut microbiota in health and disease. Physiol. Rev. 90 (3), 859–904.

Sender, R., Fuchs, S., Milo, R., 2016. Revised estimates for the number of human and bacteria cells in the body. PLoS Biol. 14 (8), e1002533.

Sun, X., Savidge, T., Feng, H., 2010. The enterotoxicity of *Clostridium difficile* toxins. Toxins (Basel) 2 (7), 1848–1880.

Thaiss, C.A., Zmora, N., Levy, M., Elinav, E., 2016. The microbiome and innate immunity. Nature 535 (7610), 65–74.

Tilg, H., Moschen, A.R., Kaser, A., 2009. Obesity and the microbiota. Gastroenterology 136 (5), 1476–1483.

Tomasik, P.J., Tomasik, P., 2003. Probiotics and prebiotics. Cereal Chem. 80 (2), 113–117.

Topping, D.L., Clifton, P.M., 2001. Short-chain fatty acids and human colonic function: roles of resistant starch and nonstarch polysaccharides. Physiol. Rev. 81 (3), 1031–1064.

Walia, S., Kamal, R., Kanwar, S.S., Dhawan, D.K., 2015. Cyclooxygenase as a target in chemoprevention by probiotics during 1,2-dimethylhydrazine induced colon carcinogenesis in rats. Nutr. Cancer 67 (4), 603–611.

Xia, X., Chen, J., Xia, J., Wang, B., Liu, H., Yang, L., et al., 2018. Role of probiotics in the treatment of minimal hepatic encephalopathy in patients with HBV-induced liver cirrhosis. J. Int. Med. Res. 46 (9), 3596–3604.

Yamamura, K., Baba, Y., Nakagawa, S., Mima, K., Miyake, K., Nakamura, K., et al., 2016. Human microbiome *Fusobacterium nucleatum* in esophageal cancer tissue is associated with prognosis. Clin. Cancer Res. 22 (22), 5574–5581.

Yang, L., Lu, X., Nossa, C.W., Francois, F., Peek, R.M., Pei, Z., 2009. Inflammation and intestinal metaplasia of the distal esophagus are associated with alterations in the microbiome. Gastroenterology 137 (2), 588–597.

Yano, J.M., Yu, K., Donaldson, G.P., Shastri, G.G., Ann, P., Ma, L., et al., 2015. Indigenous bacteria from the gut microbiota regulate host serotonin biosynthesis. Cell 161 (2), 264–276.

Zheng, H., Yde, C.C., Clausen, M.R., Kristensen, M., Lorenzen, J., Astrup, A., et al., 2015. Metabolomics investigation to shed light on cheese as a possible piece in the French paradox puzzle. J. Agric. Food Chem. 63 (10), 2830–2839.

Section III

Microbiome and Metabolic Syndrome

Chapter 8

The beneficial role of healthy microbiome in metabolic syndrome and cardiovascular health

Smriti Arora[1], Nitika Bhambri[1], Samikshha Puri[1], Anamika[1], Nandini Dasgupta[2], Shradha Khater[2] and Gautam Das[2]

[1]*Department of Allied Health Sciences, University of Petroleum and Energy Studies (UPES), Dehradun, Uttarakhand, India,*
[2]*miBiome Therapeutics LLP, Mumbai, Maharashtra, India*

8.1 Introduction

The human gut microbiota is engaged in multiple interactions that affect our health in a multitude of ways. Microbial colonization occurs during delivery transit through the vaginal canal and immediately post parturition. The first colonizers of the gut are influenced by the mode of delivery (i.e., vaginal vs Cesarean) (Li et al., 2021) and specific compounds in human milk (Milani et al., 2017). The early colonization of the gut by microbes and their timely succession from childhood to adulthood indicate the coevolution of humans and their gut microbiota (Bokulich et al., 2016). The gut microbiota comprises members from three domains of life: Bacteria, Archaea, and Eukarya. Most of the investigations on microbiome so far have focused on bacteria. However, studies of other microorganisms, such as the collection of human gut viruses, called the virome, have gained impetus in recent years (Lim et al., 2015; Santiago-Rodriguez and Hollister, 2019). There are various in vivo and in vitro studies using mouse models that have highlighted the role of the gut microbiota in diseases such as obesity, diabetes, cardiovascular diseases (CVD), and inflammatory bowel disease (IBD). The gut microbiota aids in maintaining our health by protecting us from pathogen colonization, assisting in the breakdown of food, and increasing the availability of nutrients to human hosts, including production of short-chain fatty acids (SCFAs), aromatic amino acids, and synthesizing vitamins, steroid hormones, and neurotransmitters. The gut microbiota also promotes host cell differentiation, stimulation, and modulation of the immune system and metabolism of bile salts, detoxification of specific drugs, and elimination of exogenous toxins.

The microbiota of the gut is influenced by endogenous and exogenous factors (Zhernakova et al., 2016), such as use of xenobiotics or antibiotics (Amar et al., 2008; Maurice et al., 2013), lifestyle, hygiene, diet (David et al., 2014), household cohabitation (Lynch and Pedersen, 2016), host genetic factors (Goodrich et al., 2014), mode of delivery of the neonate (Amar et al., 2008), infections, diurnal rhythm, environmental microbial exposures (Fujimura et al., 2014), and relocation to a different country (Vangay et al., 2018).

Gut microbes that exist in peace with humans are called commensals, and a negative interaction can sometimes manifest, termed "pathogenic" or "parasitism." The gut harbors a small number of opportunistic pathogens too. The holobiont, consisting of the human and its microbiome together, is an ecosystem in which the microbiota is evolving and conflicts among the selected microbes are averted (Bordenstein and Theis, 2015). A disruption or imbalance in the gut microbiota is referred to as dysbiosis (Fig. 8.1). Dysbiosis has been associated with diseases ranging from localized gastroenterological disorders to metabolic diabetes, obesity, CVDs, hepatic pathologies, and IBD.

Until quite recently, human microbiology was synonymous with culturing the microbes and observing them by using microscopes. In recent years, advances in sequencing technology have allowed the study of microbial diversity, composition, and function in the human body without traditional culturing and microscopic procedures. These studies have revealed that studying microbes studied in isolation does not give an accurate account of their long-term influence on our health. It is well understood that humans are under the influence of multispecies microbial communities that act in

Microbiome, Immunity, Digestive Health and Nutrition. DOI: https://doi.org/10.1016/B978-0-12-822238-6.00003-0

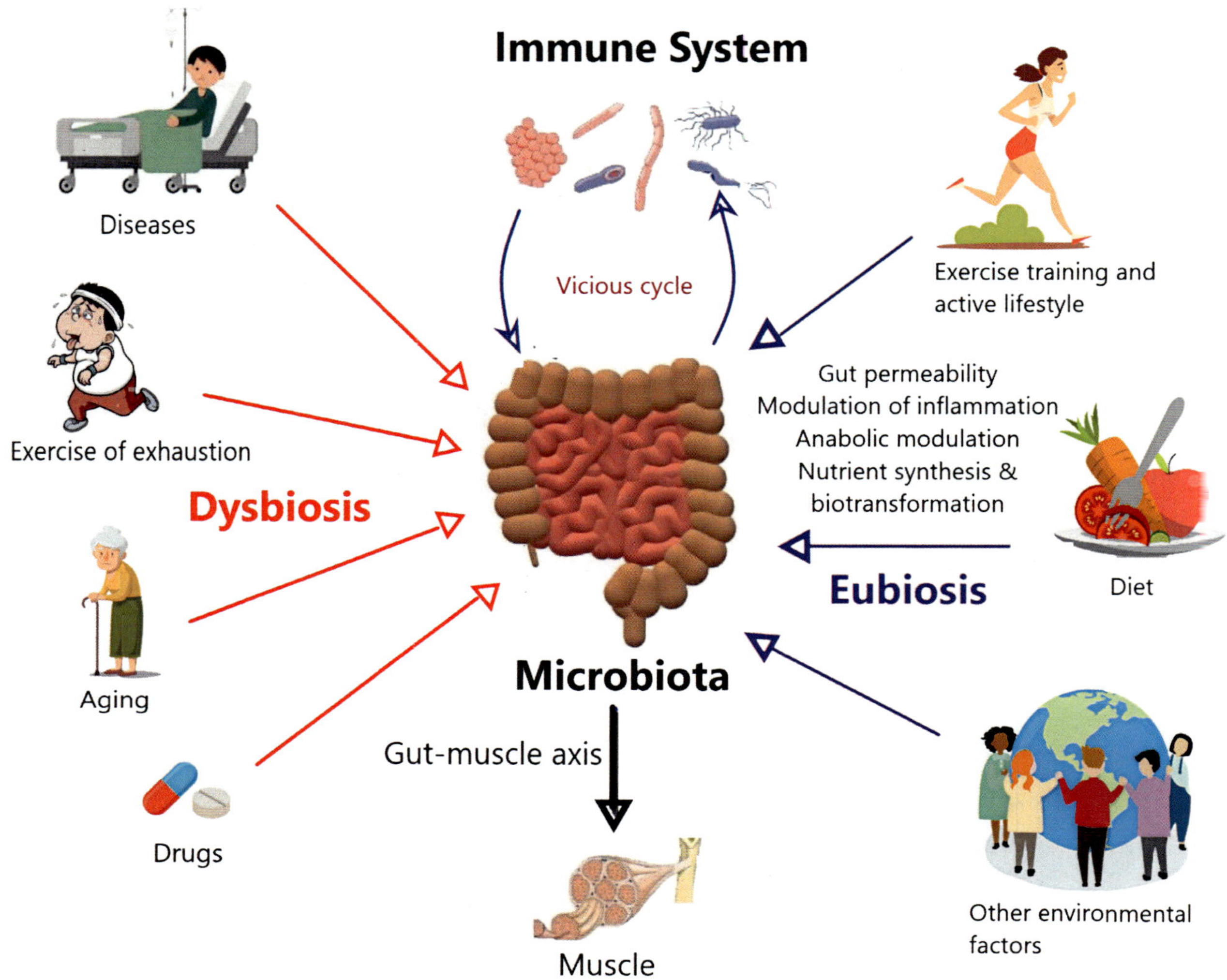

FIGURE 8.1 Dysbiosis is caused by several factors, such as diseases and an unhealthy diet, and can be corrected by a restorative diet, proper exercise, and other environmental factors.

coordination, such as microbes forming biofilms. Also, it is now quite evident that the chemical biology of the gut microbiome is as important as that of humans. 16S rRNA gene sequencing has been the technique of choice for understanding diversity, community composition, complexity, and enumeration of the microbiota (Fig. 8.2). However, it is swiftly being replaced by shot-gun sequencing (Fig. 8.2).

Although the investigations of the gut microbiota have provided a wealth of information about many diseases resulting from dysbiosis, this chapter focuses on the influence of the gut microbiota on health, immunity, and metabolic syndrome (MetS). The chapter is subdivided into sections that reveal key findings from various studies on the role of the gut microbiota and MetS, such as obesity, hyperglycemia and diabetes, dyslipidemia, hypertension, and CVD. The underlying mechanisms for these diseases and their correlation with dysbiosis are detailed along with possible interventions using diet, prebiotics, and probiotics.

8.2 Gut microbiota and metabolic syndrome

MetS is a cluster of five conditions in which the risks of heart diseases, stroke, and diabetes increase significantly. In a clinical setting, a patient is diagnosed with MetS when at least three of the following conditions hold true (Alberti et al., 2009):

1. Abdominal obesity: a waist circumference of more than 90 cm (35.43 inches) in men or more than 80 cm (31.4961 inches) in women

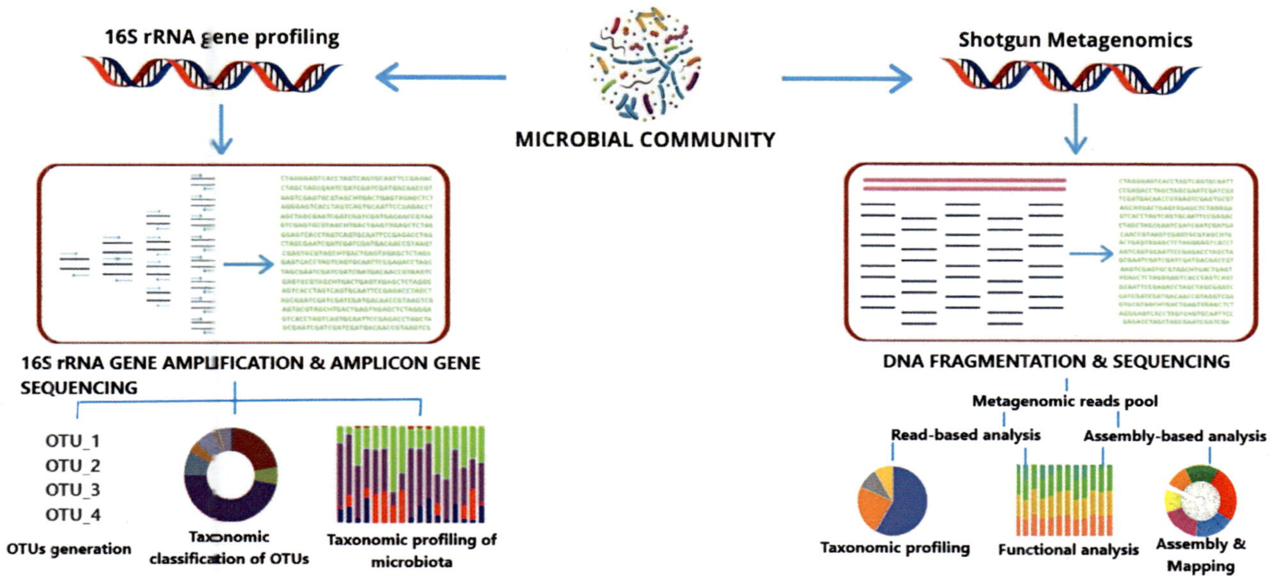

FIGURE 8.2 The 16S rRNA gene sequencing and shotgun sequencing approaches to studying the gut microbiome. The latter approach is rapidly taking over.

2. High blood pressure: more than 130/85 mmHg
3. Triglyceride level: more than 150 mg/dL
4. High-density lipoprotein cholesterol (HDL-C): less than 40 mg/dL in men or less than 50 mg/dL in women
5. High blood sugar: fasting glucose is more than 100 mg/dL

In addition to diet, lifestyle, genetic factors, and perturbation in the gut microbiota might play crucial roles in the development and progression of MetS.

8.2.1 Obesity

Overweight and obesity are disorders involving excessive body fat accumulation, which increases health risks. Since 1975 the global disease burden of obesity has tripled, reaching epidemic proportions (Bhupathiraju and Hu, 2016). Obesity has a multifactorial etiology. The factors that influence obesity are a combination of an imbalanced energy intake and energy expenditure, the nutritional value of the energy, genetics, and an individual's energy metabolism (Muscogiuri et al., 2019). The gut microbiota composition and diet also play crucial roles in influencing host metabolism, digestion, energy extraction, and hepatic fat storage.

Initial evidence of the development of obesity and its association with gut microbiota came from studies on mice. For example, 42% higher body fat in conventionally raised mice (CONV-R) than in germ-free (GF) mice was observed (Bäckhed et al., 2004), despite the fact that GF mice had significantly higher energy intake. A conventionalization or colonization of GF mice with CONV-R microbiota increased the body fat content by more than 50%. The researchers concluded that the gut microbiota is an important environmental factor in regulating fat storage. Other studies show an altered gut microbiota in obese mice compared to lean mice. A 50% decrease in the abundance of phylum Bacteroidetes and a 50% increase in the abundance of phylum Firmicutes were reported (Ley et al., 2005). However, a decreased Firmicutes to Bacteroidetes ratio in obese and overweight subjects was also reported (Schwiertz et al., 2010). A metaanalysis of seven studies did not show any difference in the abundance of Bacteroidetes between obese and lean individuals (Angelakis et al., 2012). These studies indicate that the difference at the phylum level might not be universal and that obesity-related differences at the genus or species level might play a role (Baothman et al., 2016). A few studies have identified taxonomic signatures at the species or genus level (Le Roy et al., 2018). A recent multicohort study involving a German population identified a decreased abundance of *Oscillibacter*, *Akkermansia*, *Faecalibacterium*, and *Alistipes* in obesity (Thingholm et al., 2019). At the species level, multiple studies show a decreased abundance of *Bacteroides thetaiotaomicron* and the anti-inflammatory *Faecalibacterium prausnitzii* in obese individuals (Castaner et al., 2018; Li et al., 2017). The identification of taxonomical signatures related to obesity can be the first step in developing approaches for early detection and management of the disease.

A role for SCFAs in the etiology of obesity has been proposed. SCFAs are hypothesized to reduce inflammation, regulate host energy balance, induce satiety, and inhibit fat accumulation (Kimura et al., 2014; Muscogiuri et al., 2019). Excessive body fat has been associated with increased levels of gut-derived lipopolysaccharides (LPS) as well. Higher levels of LPS might in turn activate proinflammatory cytokines, leading to low-grade inflammation and increasing the risk for type 2 diabetes (T2D) (Creely et al., 2007).

8.2.2 Gut microbiota and hyperglycemia and diabetes

Diabetes can occur either when the body does not produce enough insulin (type 1) or when the efficiency of insulin use decreases (type 2). Uncontrolled diabetes can lead to serious complications, such as CVD, stroke, kidney failure, retinopathy, and neuropathy (Lin et al., 2020).

8.2.2.1 Type 2 diabetes

Type 2 diabetes (T2D) occurs largely as a result of insulin resistance, excessive body fat, and/or a sedentary lifestyle. Increased central adiposity is one of the most important risk factors for T2D. Recent research has hinted at dysbiotic gut microbiota having been an overlooked but important driver of chronic inflammation, a prominent characteristics of obesity leading to T2D (Scheithauer et al., 2020). Multiple studies have indicated the role of anti-inflammatory bacteria from class Clostridia such as *Roseburia intestinalis* and *F. prausnitzii* in improving insulin sensitivity (Karlsson et al., 2012; Qin et al., 2012). The levels of the proinflammatory genus *Collinsella* decreased with weight loss in obese T2D individuals (Frost et al., 2014). These studies have provided a broader understanding of changes in the gut microbiota of patients with T2D.

To gain a deeper understanding of the mechanism underlying insulin resistance, Pedersen et al. studied metabolome signatures in addition to metagenomics (Pedersen et al., 2016). They found a significant contribution of serum branched-chain amino acids (BCAAs) and LPS to insulin resistance. Species such as *Prevotella copri* and *Bacteroides vulgatus*, having an increased potential for the biosynthesis of BCAA, were found to be associated with insulin resistance. Another small study revealed the role of perturbed gut microbiota in increased BCAA by transplanting microbiota from obese-lean twin pairs into mice (Arany and Neinast, 2018). Plasma BCAA was increased in mice when the fecal transplant came from the obese twin rather than from the lean counterpart. BCAA administration was found to increase insulin resistance in both mice and humans (Newgard et al., 2009), and a dietary restriction of BCAA improved insulin sensitivity (White et al., 2016). Combining the available evidence, causative role of BCAA in insulin resistance has been proposed.

LPS-induced inflammation has been observed at the outset of insulin resistance and diabetes (Festi et al., 2014). A high-fat diet was shown to increase plasma levels of LPS and LPS-containing species in the guts of mice. External infusion of LPS in mice, mimicking low-grade inflammation, induced insulin resistance, obesity, and diabetes (Cani and Delzenne, 2009). Similarly, human trials have shown increased circulating plasma LPS in individuals who consumed a high-fat diet (Amar et al., 2008). Circulating levels of LPS were increased in diabetic individuals, highlighting that LPS might be a mediator of inflammation in T2D (Creely et al., 2007).

8.2.3 Gut microbiota and dyslipidemia

Dyslipidemia is an umbrella term that refers to abnormal levels of plasma lipids or lipoproteins. Diet and lifestyle are the major causative factors for dyslipidemia; congenital factors are responsible for only 20% of cases (Moini et al., 2020). Although dyslipidemia does not cause any symptoms, plenty of evidence has identified dyslipidemia as a risk factor for vascular diseases, such as CVD, stroke, and peripheral arterial diseases (Hedayatnia et al., 2020).

Numerous studies have investigated the role of a dysbiotic gut in abnormal levels of serum lipids. High levels of serum triglycerides and low-density lipoprotein cholesterol (LDL-C) in obese patients with low microbial gene richness have been observed (Cotillard et al., 2013). A low-calorie, high-protein diet reduced the triglyceride levels with a concomitant increase in microbial gene richness. Similarly, individuals in a Danish population who had low microbial gene richness had higher insulin resistance and dyslipidemia (Le Chatelier et al., 2013). The order *Bacteroidales* and the Clostridiaceae family were associated with lower levels of triglyceride in plasma. Mice that were fed a high-fructose, high-glucose diet exhibited increased glucose intolerance, dyslipidemia, and lipid accumulation along with a loss in gut microbial diversity, a lower abundance of *Bacteroidetes*, and a significantly higher proportion of *Proteobacteria* (Do et al., 2018). A study involving 893 individuals found a positive correlation between the family Pasteurellaceae and genus *Coprococcus* and serum triglycerides levels and a negative correlation of species *Collinsella stercoris* (Fu et al., 2015).

Atherosclerosis (AS) is obstruction of arteries caused by the buildup of lipids. Species from the Erysipelotrichaceae and Lachnospiraceae families were found to be positively correlated with cholesterol and LDL-C levels, but individuals with AS did not have a taxonomic signature compared with controls (Koren et al., 2011). Conversely, a positive correlation of genus *Collinsella* and a negative correlation of *Roseburia* and *Eubacterium* were found in individuals with symptomatic AS (Karlsson et al., 2012). A study in non-alcoholic steatohepatitis (NASH) patients exhibited a strong positive correlation of *Collinsella* with triglycerides and total cholesterol and a negative correlation with HDL-C (Astbury et al., 2020). To conclude, the proinflammatory genus *Collinsella* was associated with dyslipidemia and obesity in multiple studies, suggesting that pathways in *Collinsella* might affect the lipid metabolism or inflammation of the host.

8.2.4 Gut microbiota and hypertension

Hypertension (high blood pressure) is a major risk factor for CVD. Genetic factors explain 5% of the hypertension occurrence (Verhaar et al., 2020); lifestyle and diet explain a larger proportion of hypertension incidence. Gut microbiota, an environmental factor, was linked to hypertension in mice and humans (Unger et al., 2020). In a mice model of hypertension the mice that were fed with a high-fiber diet showed decreased blood pressure (Marques et al., 2017). The consumption of a high-fiber diet improved gut dysbiosis, and there was an associated increase in the abundance of *Bacteroides acidifaciens*. The effect of fiber on blood pressure was hypothesized to be through bacterial SCFA acetate. Conversely, high SCFA levels were reported in individuals with hypertension compared to normotensive individuals (Huart et al., 2019). *Clostridium sensu* was the only species that was significantly associated with hypertensive patients. Species from the Ruminococcaceae and Clostridiales families were found to be negatively associated with hypertensive patients. A higher proportion of opportunistic *Klebsiella*, *Streptococcus*, and *Parabacteroides* species and a lower abundance of SCFA-producing *Roseburia* and *F. prausnitzii* were observed in 60 patients with hypertension. Gram-negative *Prevotella* and *Klebsiella* were frequently found in abundance in prehypertensive and hypertensive individuals (Li et al., 2017). These associations highlight the effect of the gut microbiome on hypertension via the microbial metabolites SCFA and LPS (Verhaar et al., 2020).

8.2.5 Gut microbiota and cardiovascular disease

CVD is a broad term that encompasses disorders affecting the heart or blood vessels. The disorders include arrhythmia or abnormal heart rhythms, coronary artery disease (CAD) or narrowing of arteries, heart attack, heart failure, rheumatic heart disease, stroke, and peripheral vascular disease. Although AS is the main pathological cause of CVD, each element of MetS, that is, obesity, insulin resistance, diabetes dyslipidemia, and hypertension, contributes individually in increasing the risk of CVD (Tune et al., 2017). CAD, a common type of heart disease, is a major cause of mortality worldwide. An imbalanced gut microbiota is being considered to be an important determinant contributing to CAD (Zhu et al., 2018). To characterize changes in the gut microbiota, individuals with CAD were compared with healthy controls and individuals with coronary risk factors such as hypertension, diabetes, or dyslipidemia (Emoto et al., 2016). The study revealed that the phylum Bacteroidetes was negatively associated with CAD and that the order Lactobacillales was positively associated with CAD. Another study showed a significant increase in the genera *Escherichia* and *Enterococcus*, while *Faecalibacterium*, *Subdoligranulum*, *Roseburia*, and *Eubacterium rectale* decreased in the CAD subjects. The depletion of *Roseburia* and *Eubacterium* was also reported in symptomatic AS patients. Interestingly, another study found that *Roseburia* and *Faecalibacterium* along with other members of the families Lachnospiraceae and Ruminococcaceae were depleted in CAD patients (Liu et al., 2019). Members of Ruminococcaceae and Lachnospiraceae, in particular *Roseburia*, complements each other in the production of butyrate and hence the regulation of inflammation (Morgan et al., 2012). The anti-inflammatory property could be the key reason that these bacteria are enhanced in healthy subjects compared to patients with MetS. The role of butyrate as an anti-inflammatory and an atheroprotective agent was studied in mice by giving them oral supplements of butyrate (Aguilar et al., 2014). Butyrate reduced the chances of aortic AS by 50%. Butyrate reduced the production of proinflammatory cytokines, decreasing macrophage adhesion and migration and increasing the stability of plaque. In addition to SCFA, trimethylamine-N-oxide (TMAO) was reported to be positively linked not just to CVD and T2D but to other MetS disorders as well (Lent-Schochet et al., 2018). Multiple studies have found increased TMAO concentration in subjects with coronary AS and other CVDs (Gao et al., 2020; Stubbs et al., 2016). TMAO was demonstrated to be a prognostic marker of adverse cardiovascular events (Matsuzawa et al., 2019). Trimethyllysine, a precursor of TMAO, was shown to be a strong predictor of cardiovascular events (Li et al., 2018, 2019).

Other CVDs, such as atrial fibrillation (AF) and heart failure, have been associated with dysbiotic gut microbiota. Tabata et al. found the genus *Enterobacter* to be decreased and the genera *Parabacteroides*, *Lachnoclostridium*, *Streptococcus*, and *Alistipes* to be increased in AF patients compared to controls (Tabata et al., 2021). Studies found a perturbed abundance of the bacterial species *F. prausnitzii*, *P. copri*, and *B. vulgatus* in AF subjects compared to non-AF subjects (Zuo et al., 2019; Zuo et al., 2020). Similarly, in patients with heart failure, the beneficial gut microbial species such as *E. rectale* and *Dorea longicatena* were found to be depleted (Kamo et al., 2017). Interestingly, *Faecalibacterium* was depleted in older heart failure patients too. Another study has found *Streptococcus* spp. and *Veillonella* spp. to be enriched in patients with heart failure (Katsimichas et al., 2018). Hence a strong correlation between imbalanced gut microbiota and CVD has been established. Whether these perturbations are causative or a consequence of the disease is still being debated. More human trials are required to better understand the effect of CVD on gut microbiota and vice versa.

8.2.6 Early gut microbiota and MetS

The human microbiome is dynamic. Factors such as age, geographic location, mode of birth, antibiotic use, and diet influence the microbiota. The most dramatic change in the gut microbiota occurs during infancy (Johnson and Versalovic, 2012). The gut microbiota stabilizes and starts attaining an adult-like microbiome at the age of 3 years. The microbiome at a very early age can drive health-related outcomes. Even during adolescence the gut microbiota was shown to dictate the magnitude of weight loss (Santacruz et al., 2009). The gut microbiota of overweight children was rich in pathogens such as *Staphylococcus aureus* and the Enterobacteriaceae family and depleted of microorganisms such as *Bifidobacteria*, *Desulfovibrio*, and *Akkermansia muciniphila* (Festi et al., 2014).

The microbiome at the age of 2 years could explain 50% of the variation in BMI at age 12 among 165 children (Stanislawski et al., 2018). Another study detected a significant change in the gut microbiota of children who gained weight within a four-year window, compared to children who remained normal weight (Rampelli et al., 2018). Although the research is still in its infancy, there is evidence suggesting that the early gut microbiota have a role as a marker of obesity development.

8.3 Mechanism at the molecular level

8.3.1 The gut barrier

The gastrointestinal (GI) tract occupies the largest area in the human body with the maximum number and diversity of microbiota. The human GI tract contains over 10^{14} microorganisms (Kavanagh et al., 2019). The parts of the alimentary canal—the stomach, small intestine, and colon—are referred to as the gut. Of all the components of the gut, the intestinal tract predominantly interacts with the microbiota (Fig. 8.3). The gut barrier is crucial for compartmentalization of the microbiota to the intestinal lumen. It comprises mechanical, immunological, neurological, chemical, and microbial components (Brescia and Rescigno, 2021). The epithelium plays the first line of defense along with associated cells, such as goblet cells, glial cells, and Paneth cells, along the tight and adherens junctions. The epithelium secretes antimicrobial peptides, which contribute to defense against luminal microbes.

The second line of defense is the immunological barriers in the lamina propia, where dendritic cells, lymphocytes, and macrophages reside and provide protection. The endothelium of the blood vessels serves as the third line of defense (Liu et al., 2020). The intestinal epithelium cells are renewed every three days in humans by apoptosis of enterocytes and their exfoliation. The underlying multipotent stem cells in crypts replace them. This turns over the damaged or infected cells and hence acts as a protective barrier (Bischoff et al., 2014). Functionally, the lower functional barrier consisting of immune cells is organized into a specialized network known as gut-associated lymphoid tissue, which comprises 70% of the body's immunocytes and is hence one of the largest lymphoid organs in the body (Scaldaferri et al., 2012). The mechanical components of the gut barrier also include the thick mucus lining of the colon, which may be overlaid by a thin mucus. The chemical component that fortifies the gut mucosa is the secretory immunoglobulin A (sIgA) along with the antimicrobial peptides that are synthesized by epithelium (Meyer-Hoffert et al., 2008) (see Fig. 8.3). Perturbations in the gut barrier increase intestinal permeability as measured by the rate of flux of molecules and consequently lead to several intestinal diseases such as IBD, irritable bowel syndrome, celiac disease, and development of the early stages of cancers of colon. The terms "gut barrier" and "intestinal permeability" are sometimes used

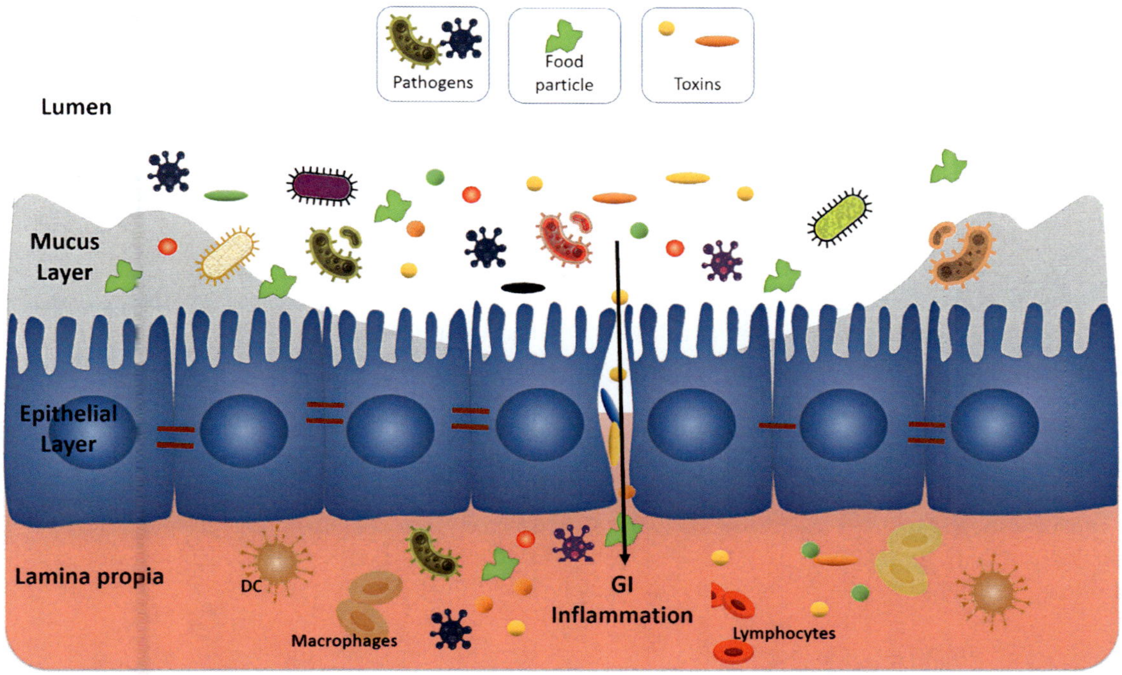

FIGURE 8.3 The gut barrier. The epithelium of intestine with projections, where microbiota in gut influence production of short-chain fatty acids, mucus, immunoglobulin A, and antimicrobial peptides.

interchangeably, although they describe different functions of the same gut mucosa. The perturbation of the gut barrier allows intestinal content into the lumen through the underlying tissues, referred to as "leaky gut." The microbes are restricted to the lumen, owing to antimicrobial peptides trapped in the mucosa. The number and diversity of microbiota increase from the small intestine to the colon (Heiss and Olofsson, 2018). The microbial count is 10^7-10^8 and $10^{10}-10^{11}$ per milliliter of content in the ileum and large intestine respectively (Hao and Lee, 2004; Ley et al., 2005; Sekirov et al., 2010). Bacteriodetes and Firmicutes are the two most predominant phyla of microbiota in humans that show intra and interpersonal flexibility at the genus level in the fecal sample (Wells et al., 2017). The other phyla occupying the human intestine include Actinobacteria, Proteobacteria Verrumicrobia, Fusobacteria, and Euryarchaeota (Bäckhed et al., 2015). The different composition of microbiota in small intestinal microbiota and that of the colon is also because of differences in oxygen tension (Donaldson et al., 2016). Bacteria such as *Bacteriodetes fragilis* are inhabitants as well as of crypts of the colon. Other mucosal-associated bacteria are localized to the mucosa and degrade mucin. The composition of mucus-associated bacteria and selection of specific species is strongly influenced by the mucin glycosylation profile (Paone and Cani, 2020). Mucin is the representation of a complex agglomerates of glycoproteins. Specified O-linked glycans are characteristic of these glycoproteins, which are produced by goblet cells (Antonini et al., 2019). The mucin degrading phyla include *B. thetaiotaomicron*, *Bifidobacterium bifidum*, *A. muciniphila*, *B. fragilis*, *Ruminococcus gnavus*, and *Ruminococcus torques* (Abildgaard et al., 2017). Gut bacteria contribute to human health by promoting the secretion of mucus and by production of SCFAs.

8.3.2 Energy metabolites

Gut microbiota regulates host processes, including energy balance, by secreting energy metabolites such as the SCFAs acetate, butyrate, and propionate.

8.3.2.1 Short-chain fatty acids

Digestion of food is initiated by host gut enzymes in the upper gut. However, complex carbohydrates, undigested proteins, and other food components escape digestion and move to the lower gut, leading to their digestion in the cecum and colon (Macfarlane and Macfarlane, 2012). SCFAs are fermentation products of nondigestible carbohydrates in the colon. Butyrate is used for upregulation of mucus production and cell proliferation of enterocytes (Burger-van Paassen et al., 2009; Martens et al., 2009). Acetate, propionate, and butyrate represent 90%–95% of the total SCFAs in the colon (Cummings et al., 1987). Microbes encounter either of two situations in the gut: a low concentration of complex carbohydrates or fermentable fibers or a high concentration of complex carbohydrates. While a diet that is rich in complex carbohydrate or fiber is the preferred substrate for the gut microbiota, when it is unavailable, microbes switch to energetically less favorable sources of growth, such as amino acids from endogenous proteins (Cummings and Macfarlane, 1991; Wall et al., 2009). In the latter case, SCFAs are minor products; the major products are the branched-chain fatty acids (BCFAs) 2-methylbutyrate, isobutyrate, and isovalerate, which originate from isoleucine, valine, and leucine (Smith and Macfarlane, 1997). Amino acids such as valine, leucine, and isoleucine contribute a small concentration (5%) to total SCFA production (Ríos-Covián et al., 2016). BCFAs are implicated in insulin resistance (Newgard et al., 2009). However, when consumed along with fiber, the same proteins or fats restore a good microbiota, reduce levels of toxic metabolites, and promote SCFA production (Sanchez et al., 2009). The concentration of SCFAs is higher in the cecum and proximal colon than in the distal colon (den Besten et al., 2013). This fermentation process of producing BCFAs occurs in the colon by *Lactobacilli* and *Bifidobacteria* under anaerobic conditions. Acetate is the most abundant SCFA in the colon and is also detected in feces. Bacterial cross-feeding fine-tunes fatty acid production and consumption. Cross-feeding occurs when one set of microbes degrade the complex carbohydrates, and the products are consumed by other species present in the colon (Belenguer et al., 2006; Flint et al., 2007). An in silico study suggested anoxic conditions promote more cross-feeding, which is prevalent in the large intestines (Heinken and Thiele, 2015). SCFA production can vary between individuals, owing to large differences in gut microbial composition and activity. In addition, SCFAs seem to beneficially affect glucose metabolism by normalizing plasma glucose levels, increasing glucose handling, and promoting cross-feeding (Bloemen et al., 2009).

8.3.2.1.1 Short-chain fatty acids and modulation of cell signaling and immune response

The SCFA concentration in the gut depends on the microbiota composition, intestinal transit time, the host-microbiota metabolic flux of SCFAs, and the fiber content of the host diet. SCFAs act as inhibitors of histone deacetylases (HDACs) (Schilderink et al., 2013; Wu et al., 2012). SCFAs through the G-protein coupled receptors (GPCRs) influence the expansion and function of hematopoietic and nonhematopoetic cell lineages (Rooks and Garrett, 2016). SCFA induced HDAC inhibition leads to regulation of NF-κB and proinflammatory response. In addition, the anti-inflammatory role of SCFA extends to macrophages (Chang et al., 2014) and dendritic cells (DCs) (Singh et al., 2010). T_{reg} cells frequency and function both are modulated by SCFA. SCFAs were measured in sputum of cystic fibrosis patients. It was found out that SCFA mediated recruitment and persistence of neutrophils, also led to an aggravation of inflammation and proliferation of *Pseudomonas aeruginosa*. Thus the role of SCFA in immunomodulation is context dependent (Ghorbani et al., 2015).

8.3.2.2 Bile salt hydrolase

Bile salts are products of cholesterol metabolism in the liver. Primary bile salts are stored and concentrated in gall bladder. Furthermore, in the distal ileum, bile salt hydrolases of gut microbes remove glycine or taurine from bile salt conjugated lipids. This prevents their reuptake by apical sodium-dependent bile acid transporter (Chiang, 2009; Wahlström et al., 2016). These get dehydroxylated by microbes to deoxycholic acid or lithocholic acid, which are known as secondary bile acids. These can be excreted in the feces to reduce cholesterol circulating in the blood, thereby reducing the risk of AS. Bile salts are known modulators of gut microbiota. Surprisingly, in patients with CVD, serum levels of primary bile salts were decreased and the ratio of secondary to primary bile salts increased (Mayerhofer et al., 2017). A dysregulation in bile acid metabolism leads to dysbiosis and gives rise to various disorder, such as obesity, IBD, colorectal cancer, hepatocellular carcinoma, and *Clostridium difficile* infection (Joyce and Gahan, 2016).

8.3.2.3 Trimethylamine oxide

TMAO is an important gut microbiota–derived metabolite that has been implicated in AS and other CVDs (Tang et al., 2013). TMAO production depends largely on three factors: dietary precursors, gut microbiota, and oxidation of trimethylamine (TMA) to TMAO by flavin monooxygenases (FMOs). The biosynthetic precursors for TMAO are choline, betaine, and L-carnitine in the human diet (Yang et al., 2019). In fact, high consumption of red meat has been linked to increased CVDs (Bernstein et al., 2010), and higher microbial production of TMA and TMAO was linked to L-carnitine from red meat (Wang et al., 2019). The dietary precursors are converted into TMA by the gut microbial enzymes (Koeth et al., 2013). TMA is further metabolized to TMAO in the liver by FMOs. Specific strains of gut microbiota were shown to convert dietary choline to TMA (Romano et al., 2015). Colonization of these TMA-producing bacteria in the intestines of germ-free mice was linked to increased accumulation of plasma TMAO.

A number of mechanisms have been proposed to explain the molecular basis of the link between TMAO and AS (Fig. 8.4). AS and CVD are characterized by low-grade inflammation. Studies show that plasma TMAO levels are associated with low-grade inflammation (Rohrmann et al., 2016). TMAO induces inflammation by activation of the NF-κB pathway; increased expression of proinflammatory genes such as TNF-α, IL-18, and IL-1β; and suppression of anti-inflammatory cytokine IL-10 (Chen et al., 2017a). TMAO also induced inflammation by stimulating mitochondrial reactive oxygen species (mtROS) (Yang et al., 2019). Another proposed mechanism through which TMAO might contribute to increased risk of AS and other cardiac events is through activation of platelets and thrombosis (Zhu et al., 2018). TMAO-dependent suppression of reverse cholesterol transport (RCT) was revealed. Suppression of RCT by TMAO increases the risks for AS and other coronary artery diseases.

We propose that differences in age, geographical location, ethnic origins, dietary habits, and, most important, study design might change the outcome of the experiments. Hence more clinical trials in human are required to understand the effect of TMAO.

8.4 Metabolic syndrome therapy by managing gut microbiota

8.4.1 Influence of diet

Diet drives the composition and metabolism of the gut microbiota. Food digestion by the microbiota can happen in two ways: saccharolytic and proteolytic. This depends on the concentration of substrate in the host colon and hence on the diet of the host (Sekirov et al., 2010). SCFAs are a result of carbohydrate fermentation by the gut microbiota. Proteolytic fermentation produces SCFA too, along with other cometabolites, such as ammonia, amines, phenols, thiols, and indoles. These are potentially toxic. They are cleared renally, and their accumulation is referred to as microbial uremic syndrome (Nallu et al., 2017). A diet that is rich in complex carbohydrates, such as that of Hadza hunter gatherers of Tanzania, influences the diversity of microbiota positively (Schnorr et al., 2014). By contrast, prolonged consumption of a diet that is rich in fats and sugars leads to reduced diversity or extinction of gut microbiota (Sonnenburg et al., 2016). In a particular study in which humans were fed a barley kernel–based diet, glucose tolerance improved. A corresponding increase in *P. copri* was observed with an increased fermentative capacity (Kovatcheva-Datchary et al., 2015). A Western diet, that is, a diet rich in saturated fats and sugars, is known to be associated with type 2 diabetes (T2D), chronic metabolic inflammation, and other noncommunicable diseases (Christ et al., 2019). A longitudinal and cross-sectional study of the multigenerational Asian American cohort gave insights into how immigration to the United States changes alpha diversity (Vangay et al., 2018; Hills et al., 2019). With time spent in the United States, the fiber-degrading beneficial *Prevotella* was replaced by *Bacteroides*. Furthermore, immigration into the United States led to a

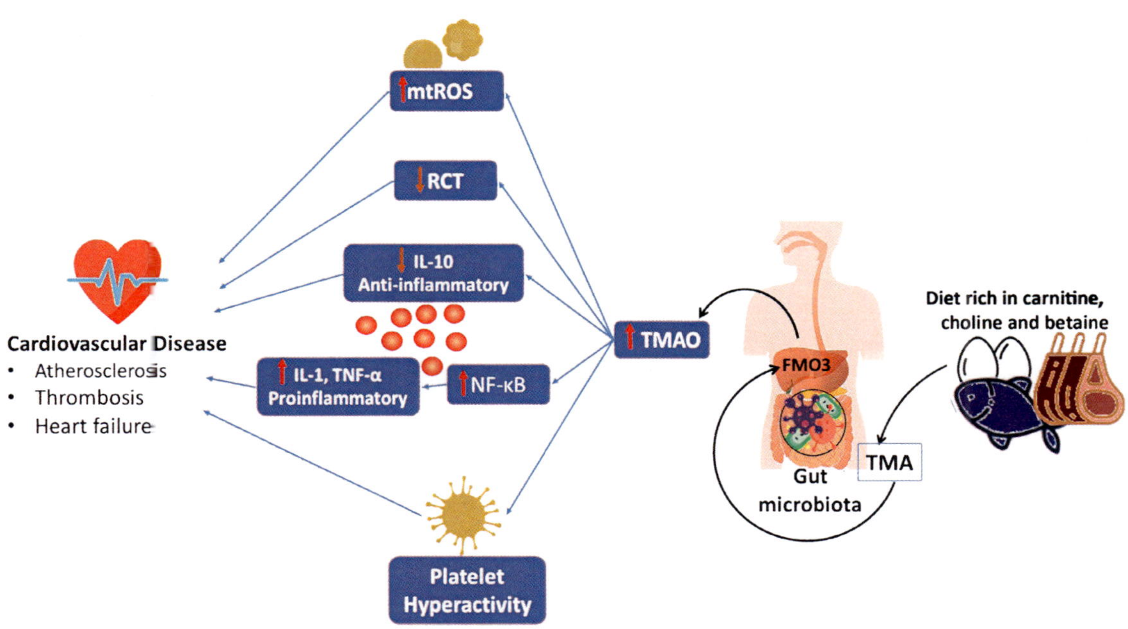

FIGURE 8.4 Putative molecular basis of link between trimethylamine-N-oxide and cardiovascular diseases.

fourfold increase in the risk for obesity as well as reduced gut diversity and function, compared to being in their home country. The underlying reason might be ingestion of a Western diet (Lauderdale and Rathouz, 2000). By contrast, a Mediterranean diet has been associated with an improved microbial profile and a low risk for obesity, diabetes, and reduced plasma cholesterol. Shotgun sequencing reflected gene richness and the presence of fiber-degrading *F. prausnitzii*, *Roseburia*, and butyrate producers. A reduced abundance of proinflammatory *R. gnavus* and *R. torques* was observed (Cani and Delzenne, 2009; Meslier et al., 2020). Another metagenomics-based study characterized and classified the gut microbiota into enterotypes. In the first enterotype, which had a high ratio of *Bacteroides* to *Prevotella*, the diet was rich in animal protein, nutrient choline, and saturated fat. In the second enterotype, in which the diet was rich in plant-based fiber, high amounts of *Prevotella* and low amounts of *Bacteroides* were observed (Rajilic-Stojanovic et al., 2015). A metagenomics analysis of twin data revealed that environmental factors such as diet and cohabitation play a more vital role in structuring the gut microbiota than heritable factors (Rothschild et al., 2018). In conclusion, a plant-based diet that is rich in fruit and fiber promotes alpha diversity, whereas a Western diet reduces it, promoting obesity (Menni et al., 2017; Simpson and Campbell, 2015). Furthermore, the type of fiber that is consumed in the diet also makes a difference in the composition of the gut microbiota. Fibers are classified into four types on the basis of whether they are readily fermented or form a viscous cross-linked gel (McRorie, 2015). Fermentation of fiber into SCFAs promoted microbial diversity and reduced weight gain and outweighed the heritable contribution to obesity (Fernandes et al., 2014). A diet that is rich in whey and pea protein has been associated with an increased abundance of *Bifidobacterium* and *Lactobacillus*, and whey has been shown to decrease the abundance of pathobionts *Bacteroides fragilis* and *Clostridium perfringens* (Singh et al., 2017). The inclusion of tea, coffee, and red wine is known to have bifidogenic effects on gut microbiota as a result of polyphenols and prebiotic effects (Zhernakova et al., 2016). A chemical nutritional study conducted on 381 nondiabetic individuals showed a correlation between noncalorie artificial sweeteners (NAS) and MetS. The study revealed that NAS can alter the gut microbiota (Suez et al., 2014). However, it is important to note that each individual has a unique gut microbiota, and therefore the response to a particular kind of diet may differ from individual to individual. This offers avenues for personalized microbiome therapies because one size does not fit all.

8.4.2 Influence of prebiotics

Prebiotics are nondigestible carbohydrates that are fermented by the gut microbiota, leading to increased abundance of *Bifidobacterium* and *Lactobacillus* spp. and the production of beneficial SCFAs in the colon (Cani and Delzenne, 2009; O'Connor et al., 2017). Another proposition is that prebiotics exert antiadhesive effects on pathogens. If the glycan is given as a prebiotic, it exerts an antiadhesive effect (Milani et al., 2017). The main prebiotics are inulin-type fructans, which include fructooligosaccharides (FOS), galactooligosaccharides (GOS), and lactulose (Valcheva and Dieleman, 2016). These prebiotics bind water and distend the stomach, thus increasing satiety (Slavin, 2013). The type and dose of prebiotic are important for treatment of MetS. The doses may vary between 5 and 20 g in humans. Human milk oligosaccharides exert antiadhesive effects on enteropathogenic *Escherichia coli* (Manthey et al., 2014). GOS and FOS exert prebiotic effects by increasing bifidobacterial abundance and decreasing the abundance of *E. coli* (Markowiak and Slizewska, 2017; Musilova et al., 2015). In a Canadian study, FOS and inulin supplementation decreased body fat, serum triglyceride levels, and obesity (Nicolucci et al., 2017). It is important to note that prebiotic therapies work on the basis of the individual's starting microbial profile. A study compared SCFA production in subjects in which high fiber-degrading microbiota (*Prevotella* and *Bacteroides*) dominated the microbiota. Equal levels of SCFAs were produced from any of the three prebiotics FOS, arabinoglycan, and sorghum. However, the levels of SCFAs depended on the type of prebiotic that was given (Chen et al., 2017b). With the right amounts of fiber in the diet, the consumption and circulation of omega-3 fatty acids positively correlated with alpha diversity. Prebiotics, when given in the correct dose with knowledge of prior gut microbiota, can aid MetS patients, owing to their alpha diversity−promoting effects. A diverse microbiota has high levels of SCFAs and thereby exerts health benefits.

8.4.3 Influence of probiotics

Probiotics are living microorganisms that, when administered in appropriate amounts, confer beneficial effects on the host. These include *Lactobacillus rhamnosus* GG, *L. reuteri*, certain strains of *L. casei*, and *Bifidobacteria* (Sartori et al., 2017). Daily supplementation of *A. muciniphila* is known to improve insulin sensitivity, reduce insulinemia, and decrease body weight over three months' time. *Akkermansia* could be the next-generation probiotic to resolve MetS syndrome (Dao et al., 2016). When a prebiotic is combined with a probiotic, the combination is called a synbiotic. High abundances of *Bifidobacterium* and *Akkermansia* are associated with leanness (Arora et al., 2019). Protective effects of flax seed fibers in mice with diet-induced obesity have been reported. Although the effects of probiotics have largely been shown to be favorable, one randomized clinical trial conducted on high-risk overweight and obese pregnant women who were administered probiotics *L. rhamnosus* and *Bifidobacterium animalis* subspecies lactis [BB-12] from the second trimester onwards did not show a higher efficacy in preventing gestational diabetes compared to the placebo group (Callaway et al., 2019). This study contradicted an earlier study on normal-weight pregnant women who were administered *L. rhamnosus* HN001 during early pregnancy; it succeeded in reducing prevalence of gestational diabetes in those women (Wickens et al., 2017). Clinical trials on probiotic treatment and glycemic control were conducted and led to the conclusion that administration of multistrain probiotics comprising *Lactobacillus acidophilus*, *Lactobacillus bulgaricus*, and/or *Bifidobacterium lactis* and *Streptococcus thermophiles* for 6–12 weeks controlled glycemic levels (Tiderencel et al., 2020). The data on the effects of probiotics in patients with CVD risk and a subgroup analysis revealed significantly favorable effects of probiotics on the treatment of cardiovascular risk factors (Dixon et al., 2020). Probiotic investigations promise treatment of dyslipidemia, hypertension, insulin resistance, obesity, and even heart failure and myocardial infarction (Ettinger et al., 2014; Sun and Buys, 2015). Kefir is a synbiotic that is effective in management of CVD and other metabolic disorders (Pimenta et al., 2018).

8.4.4 Influence of exercise

Exercise in moderation is known to positively modulate the intestinal microbiome composition. However, training and exhaustion have adverse effects. As is the case with other organs, the gut microbiota affects the pathophysiology of skeletal muscles as well (Ticinesi et al., 2019). A healthy microbiome represents limited numbers of highly represented taxa, namely, *Prevotella* and *Bacteroides*, with significant numbers of others minor players but with high metabolic activity (Greenhalgh et al., 2016). In older age, which also corresponds to lower amounts of exercise and physical activity, the diversity and richness of the gut microbiota declines, resilience to perturbations of the microbiota is reduced, and interindividual variability increases (Claesson et al., 2012). Dysbiosis can lead to MetS and CVDs. Preventing or treating MetS through exercise is possible. A recent study used high-throughput sequencing to delineate and differentiate between gut microbiota signatures of athletes evaluating body builders and healthy sedentary men (Jang et al., 2019). Other treatments, such as surgery (fecal microbiota transplantation/FMT), clinical trials, and drugs to modulate microbiota, are becoming commonplace. However, the results are not consistent in every case. For example, FMT is 90% efficacious in the case of *C. difficile*–induced recurrent infection. For other diseases such as chronic IBD, the clinical remission post FMT is modest.

8.5 Conclusion

MetS is a group of risk factors that can eventually lead to CVD, stroke, and other complications, such as nonalcoholic fatty liver disease. In addition to intrinsic factors, such as genetics and gut microbiome, extrinsic factors such as diet and lifestyle play crucial roles in the development of MetS. An imbalanced gut microbiota that induces low-grade inflammation is usually associated with MetS. Although some taxonomic signatures associated with obesity, insulin resistance, hypertension, and CVD are emerging, further research is required to identify a potential pathogenetic factor. Low-grade inflammation caused by a dysbiotic gut might be the link between MetS and CVD.

Probiotics and prebiotics are often used to revamp the gut microbiota and manage MetS. Few studies have shown positive effects of prebiotics and probiotics on the gut microbiota of MetS patients. Additional studies on a larger scale and randomized double-control trials are required to establish the clinical efficacy of probiotics and prebiotics in managing MetS.

References

Abildgaard, A., Elfving, B., Hokland, M., Wegener, G., Lund, S., 2017. Probiotic treatment reduces depressive-like behaviour in rats independently of diet. Psychoneuroendocrinology 79, 40–48.

Aguilar, E.C., Leonel, A.J., Teixeira, L.G., Silva, A.R., Silva, J.F., Pelaez, J.M.N., et al., 2014. Butyrate impairs atherogenesis by reducing plaque inflammation and vulnerability and decreasing NFκB activation. Nutr. Metab. Cardiovas. Dis 24 (6), 606–613.

Alberti, K.G., Eckel, R.H., Grundy, S.M., Zimmet, P.Z., Cleeman, J.I., Donato, K.A., et al., 2009. Harmonizing the metabolic syndrome: a joint interim statement of the international diabetes federation task force on epidemiology and prevention; National Heart, Lung, and Blood Institute; American Heart Association; World Heart Federation; International Atherosclerosis Society; and International Association for the Study of Obesity. Circulation 120 (16), 1640–1645.

Amar, J., Burcelin, R., Ruidavets, J.B., Cani, P.D., Fauvel, J., Alessi, M.C., et al., 2008. Energy intake is associated with endotoxemia in apparently healthy men. Am. J. Clin. Nutr. 87 (5), 1219–1223.

Angelakis, E., Armougom, F., Million, M., Raoult, D., 2012. The relationship between gut microbiota and weight gain in humans. Future Microbiol. 7 (1), 91–109.

Antonini, M., Lo Conte, M., Sorini, C., Falcone, M., 2019. How the interplay between the commensal microbiota, gut barrier integrity and mucosal immunity regulates brain autoimmunity. Front. Immunol.

Arany, Z., Neinast, M., 2018. Branched chain amino acids in metabolic disease. Curr. Diabetes Rep. 18 (10), 76.

Arora, T., Rudenko, O., Egerod, K.L., Husted, A.S., Kovatcheva-Datchary, P., Akrami, R., et al., 2019. Microbial fermentation of flaxseed fibers modulates the transcriptome of GPR41-expressing enteroendocrine cells and protects mice against diet-induced obesity. Am. J. Physiol. Endocrinol. Metab. 316 (3), E453-e63.

Astbury, S., Atallah, E., Vijay, A., Aithal, G.P., Grove, J.I., Valdes, A.M., 2020. Lower gut microbiome diversity and higher abundance of proinflammatory genus Collinsella are associated with biopsy-proven nonalcoholic steatohepatitis. Gut Microbes 11 (3), 569–580.

Bäckhed F, Ding H, Wang T, Hooper LV, Koh GY, Nagy A, et al., 2004. The gut microbiota as an environmental factor that regulates fat storage. 101 (44), 15718–15723.

Bäckhed, F., Roswall, J., Peng, Y., Feng, Q., Jia, H., Kovatcheva-Datchary, P., et al., 2015. Dynamics and stabilization of the human gut microbiome during the first year of life. Cell Host Microbe 17 (6), 852.

Baothman, O.A., Zamzami, M.A., Taher, I., Abubaker, J., Abu-Farha, M., 2016. The role of gut microbiota in the development of obesity and diabetes. Lipids Health Dis 15 (1), 108.

Belenguer, A., Duncan, S.H., Calder, A.G., Holtrop, G., Louis, P., Lobley, G.E., et al., 2006. Two routes of metabolic cross-feeding between Bifidobacterium adolescentis and butyrate-producing anaerobes from the human gut. Appl. Env. Microbiol 72 (5), 3593–3599.

Bernstein, A.M., Sun, Q., Hu, F.B., Stampfer, M.J., Manson, J.E., Willett, W.C., 2010. Major dietary protein sources and risk of coronary heart disease in women. Circulation 122 (9), 876–883.

Bhupathiraju, S.N., Hu, F.B., 2016. Epidemiology of obesity and diabetes and their cardiovascular complications. Circ. Res. 118 (11), 1723–1735.

Bischoff SC, Barbara G, Buurman W, Ockhuizen T, Schulzke J-D, Serino M, et al., 2014. Intestinal permeability—a new target for disease prevention and therapy. 14 (1), 1–25.

Bloemen, J.G., Venema, K., van de Poll, M.C., Olde Damink, S.W., Buurman, W.A., Dejong, C.H., 2009. Short chain fatty acids exchange across the gut and liver in humans measured at surgery. Clin. Nutr. 28 (6), 657–661.

Bokulich, N.A., Chung, J., Battaglia, T., Henderson, N., Jay, M., Li, H., et al., 2016. Antibiotics, birth mode, and diet shape microbiome maturation during early life. Sci. Transl. Med. 8 (343), 343ra82.

Bordenstein, S.R., Theis, K.R., 2015. Host biology in light of the microbiome: ten principles of holobionts and hologenomes. PLoS Biol. 13 (8), e1002226.

Brescia P, Rescigno MJTiMM, 2021. The gut vascular barrier: a new player in the gut–liver–brain axis.

Burger-van Paassen, N., Vincent, A., Puiman, P.J., van der Sluis, M., Bouma, J., Boehm, G., et al., 2009. The regulation of intestinal mucin MUC2 expression by short-chain fatty acids: implications for epithelial protection. Biochem. J. 420 (2), 211–219.

Callaway, L.K., McIntyre, H.D., Barrett, H.L., Foxcroft, K., Tremellen, A., Lingwood, B.E., et al., 2019. Probiotics for the prevention of gestational diabetes mellitus in overweight and obese women: findings from the SPRING double-blind randomized controlled trial. Diabetes Care 42 (3), 364–371.

Cani, P.D., Delzenne, N.M., 2009. The role of the gut microbiota in energy metabolism and metabolic disease. Curr. Pharm. Des. 15 (13), 1546–1558.

Castaner, O., Goday, A., Park, Y.M., Lee, S.H., Magkos, F., Shiow, S.T.E., et al., 2018. The gut microbiome profile in obesity: a systematic review. Int. J. Endocrinol. 2018, 4095789.

Chang, P.V., Hao, L., Offermanns, S., Medzhitov, R., 2014. The microbial metabolite butyrate regulates intestinal macrophage function via histone deacetylase inhibition. Proc. Natl. Acad. Sci. U S A 111 (6), 2247–2252.

Chen, M.L., Zhu, X.H., Ran, L., Lang, H.D., Yi, L., Mi, M.T., 2017a. Trimethylamine-N-oxide induces vascular inflammation by activating the NLRP3 inflammasome through the SIRT3-SOD2-mtROS signaling pathway. J. Am. Heart Assoc. 6 (9).

Chen, T., Long, W., Zhang, C., Liu, S., Zhao, L., Hamaker, B.R., 2017b. Fiber-utilizing capacity varies in Prevotella- vs Bacteroides-dominated gut microbiota. Sci. Rep. 7 (1), 2594.

Chiang, J.Y., 2009. Bile acids: regulation of synthesis. J. Lipid Res. 50 (10), 1955–1966.

Christ, A., Lauterbach, M., Latz, E., 2019. Western diet and the immune system: an inflammatory connection. Immunity 51 (5), 794–811.

Claesson, M.J., Jeffery, I.B., Conde, S., Power, S.E., O'Connor, E.M., Cusack, S., et al., 2012. Gut microbiota composition correlates with diet and health in the elderly. Nature 488 (7410), 178–184.

Cotillard, A., Kennedy, S.P., Kong, L.C., Prifti, E., Pons, N., Le Chatelier, E., et al., 2013. Dietary intervention impact on gut microbial gene richness. Nature 500 (7464), 585–588.

Creely, S.J., McTernan, P.G., Kusminski, C.M., Fisher, F.M., Da Silva, N.F., Khanolkar, M., et al., 2007. Lipopolysaccharide activates an innate immune system response in human adipose tissue in obesity and type 2 diabetes. Am. J. Physiol. Endocrinol. Metab. 292 (3), E740–E747.

Cummings, J.H., Macfarlane, G.T., 1991. The control and consequences of bacterial fermentation in the human colon. J. Appl. Bacteriol. 70 (6), 443–459.

Cummings, J.H., Pomare, E.W., Branch, W.J., Naylor, C.P., Macfarlane, G.T., 1987. Short chain fatty acids in human large intestine, portal, hepatic and venous blood. Gut 28 (10), 1221–1227.

Dao, M.C., Everard, A., Aron-Wisnewsky, J., Sokolovska, N., Prifti, E., Verger, E.O., et al., 2016. Akkermansia muciniphila and improved metabolic health during a dietary intervention in obesity: relationship with gut microbiome richness and ecology. Gut 65 (3), 426–436.

David, L.A., Maurice, C.F., Carmody, R.N., Gootenberg, D.B., Button, J.E., Wolfe, B.E., et al., 2014. Diet rapidly and reproducibly alters the human gut microbiome. Nature 505 (7484), 559–563.

den Besten, G., van Eunen, K., Groen, A.K., Venema, K., Reijngoud, D.J., Bakker, B.M., 2013. The role of short-chain fatty acids in the interplay between diet, gut microbiota, and host energy metabolism. J. Lipid Res. 54 (9), 2325–2340.

Dixon, A., Robertson, K., Yung, A., Que, M., Randall, H., Wellalagodage, D., et al., 2020. Efficacy of probiotics in patients of cardiovascular disease risk: a systematic review and *meta*-analysis. Curr. Hypertens. Rep. 22 (9), 74.

Do MH, Lee E, Oh M-J, Kim Y, Park H-Y., 2018. High-glucose or -fructose diet cause changes of the gut microbiota and metabolic disorders in mice without body weight change. 10 (6), 761.

Donaldson, G.P., Lee, S.M., Mazmanian, S.K., 2016. Gut biogeography of the bacterial microbiota. Nat. Rev. Microbiol. 14 (1), 20–32.

Emoto, T., Yamashita, T., Sasaki, N., Hirota, Y., Hayashi, T., So, A., et al., 2016. Analysis of gut microbiota in coronary artery disease patients: a possible link between gut microbiota and coronary artery disease. J. Atheroscler. Thromb. 23 (8), 908–921.

Ettinger, G., MacDonald, K., Reid, G., Burton, J.P., 2014. The influence of the human microbiome and probiotics on cardiovascular health. Gut Microbes 5 (6), 719–728.

Fernandes, J., Su, W., Rahat-Rozenbloom, S., Wolever, T.M., Comelli, E.M., 2014. Adiposity, gut microbiota and faecal short chain fatty acids are linked in adult humans. Nutr. Diabetes 4, e121.

Festi, D., Schiumerini, R., Eusebi, L.H., Marasco, G., Taddia, M., Colecchia, A., 2014. Gut microbiota and metabolic syndrome. World J. Gastroenterol. 20 (43), 16079–16094.

Flint, H.J., Duncan, S.H., Scott, K.P., Louis, P., 2007. Interactions and competition within the microbial community of the human colon: links between diet and health. Env. Microbiol. 9 (5), 1101–1111.

Frost, G., Sleeth, M.L., Sahuri-Arisoylu, M., Lizarbe, B., Cerdan, S., Brody, L., et al., 2014. The short-chain fatty acid acetate reduces appetite via a central homeostatic mechanism. Nat. Commun. 5, 3611.

Fu, J., Bonder, M.J., Cenit, M.C., Tigchelaar, E.F., Maatman, A., Dekens, J.A., et al., 2015. The gut microbiome contributes to a substantial proportion of the variation in blood lipids. Circ. Res. 117 (9), 817–824.

Fujimura, K.E., Demoor, T., Rauch, M., Faruqi, A.A., Jang, S., Johnson, C.C., et al., 2014. House dust exposure mediates gut microbiome Lactobacillus enrichment and airway immune defense against allergens and virus infection. Proc. Natl. Acad. Sci. U S A 111 (2), 805–810.

Gao, J., Yan, K.-T., Wang, J.-X., Dou, J., Wang, J., Ren, M., et al., 2020. Gut microbial taxa as potential predictive biomarkers for acute coronary syndrome and post-STEMI cardiovascular events. Sci. Rep. 10 (1), 2639.

Ghorbani, P., Santhakumar, P., Hu, Q., Djiadeu, P., Wolever, T.M., Palaniyar, N., et al., 2015. Short-chain fatty acids affect cystic fibrosis airway inflammation and bacterial growth. Eur. Respir. J. 46 (4), 1033–1045.

Goodrich, J.K., Waters, J.L., Poole, A.C., Sutter, J.L., Koren, O., Blekhman, R., et al., 2014. Human genetics shape the gut microbiome. Cell 159 (4), 789–799.

Greenhalgh, K., Meyer, K.M., Aagaard, K.M., Wilmes, P., 2016. The human gut microbiome in health: establishment and resilience of microbiota over a lifetime. Env. Microbiol. 18 (7), 2103–2116.

Hao, W.L., Lee, Y.K., 2004. Microflora of the gastrointestinal tract: a review. Methods Mol. Biol. 268, 491–502.

Hedayatnia, M., Asadi, Z., Zare-Feyzabadi, R., Yaghooti-Khorasani, M., Ghazizadeh, H., Ghaffarian-Zirak, R., et al., 2020. Dyslipidemia and cardiovascular disease risk among the MASHAD study population. Lipids Health Dis. 19 (1), 42.

Heinken, A., Thiele, I., 2015. Anoxic conditions promote species-specific mutualism between gut microbes in silico. Appl. Env. Microbiol. 81 (12), 4049–4061.

Heiss, C.N., Olofsson, L.E., 2018. Gut microbiota-dependent modulation of energy metabolism. J. Innate Immun. 10 (3), 163–171.

Hills Jr, R.D., Pontefract, B.A., Mishcon, H.R., Black, C.A., Sutton, S.C., Theberge, C.R., 2019. Gut microbiome: profound implications for diet and disease. Nutrients 11 (7).

Huart, J., Leenders, J., Taminiau, B., Descy, J., Saint-Remy, A., Daube, G., et al., 2019. Gut microbiota and fecal levels of short-chain fatty acids differ upon 24-hour blood pressure levels in men. Hypertension (Dallas, Tex: 1979) 74 (4), 1005–1013.

Jang, L.G., Choi, G., Kim, S.W., Kim, B.Y., Lee, S., Park, H., 2019. The combination of sport and sport-specific diet is associated with characteristics of gut microbiota: an observational study. J. Int. Soc. Sports Nutr. 16 (1), 21.

Johnson CL, Versalovic J., 2012. The human microbiome and its potential importance to pediatrics. 129 (5), 950–960.

Joyce, S.A., Gahan, C.G., 2016. Bile acid modifications at the microbe-host interface: potential for nutraceutical and pharmaceutical interventions in host health. Annu. Rev. Food Sci. Technol. 7, 313–333.

Kamo, T., Akazawa, H., Suda, W., Saga-Kamo, A., Shimizu, Y., Yagi, H., et al., 2017. Dysbiosis and compositional alterations with aging in the gut microbiota of patients with heart failure. PLoS One 12 (3), e0174099.

Karlsson, F.H., Fåk, F., Nookaew, I., Tremaroli, V., Fagerberg, B., Petranovic, D., et al., 2012. Symptomatic atherosclerosis is associated with an altered gut metagenome. Nat. Commun. 3 (1), 1245.

Katsimichas, T., Ohtani, T., Motooka, D., Tsukamoto, Y., Kioka, H., Nakamoto, K., et al., 2018. Non-ischemic heart failure with reduced ejection fraction is associated with altered intestinal microbiota. Circ. J. 82 (6), 1640–1650.

Kavanagh, K., Hsu, F.-C., Davis, A.T., Kritchevsky, S.B., Rejeski, W.J., Kim, S., 2019. Biomarkers of leaky gut are related to inflammation and reduced physical function in older adults with cardiometabolic disease and mobility limitations. Geroscience 41 (6), 923–933. Available from: https://doi.org/10.1007/s11357-019-00112-z.

Kimura I, Inoue D, Hirano K, Tsujimoto G. 2014. SCFA Receptor GPR43 Energy Metab. 5 (85).

Koeth, R.A., Wang, Z., Levison, B.S., Buffa, J.A., Org, E., Sheehy, B.T., et al., 2013. Intestinal microbiota metabolism of l-carnitine, a nutrient in red meat, promotes atherosclerosis. Nat. Med. 19 (5), 576–585.

Koren O, Spor A, Felin J, Fåk F, Stombaugh J, Tremaroli V, et al., 2011. Human oral, gut, and plaque microbiota in patients with atherosclerosis. 108 (Supplement 1), 4592–4598.

Kovatcheva-Datchary, P., Nilsson, A., Akrami, R., Lee, Y.S., De Vadder, F., Arora, T., et al., 2015. Dietary fiber-induced improvement in glucose metabolism is associated with increased abundance of prevotella. Cell Metab. 22 (6), 971–982.

Lauderdale, D.S., Rathouz, P.J., 2000. Body mass index in a US national sample of Asian Americans: effects of nativity, years since immigration and socioeconomic status. Int. J. Obes. Relat. Metab. Disord. 24 (9), 1188–1194.

Le Chatelier, E., Nielsen, T., Qin, J., Prifti, E., Hildebrand, F., Falony, G., et al., 2013. Richness of human gut microbiome correlates with metabolic markers. Nature 500 (7464), 541–546.

Le Roy, C.I., Beaumont, M., Jackson, M.A., Steves, C.J., Spector, T.D., Bell, J.T., 2018. Heritable components of the human fecal microbiome are associated with visceral fat. Gut Microbes 9 (1), 61–67.

Lent-Schochet, D., Silva, R., McLaughlin, M., Huet, B., Jialal, I., 2018. Changes to trimethylamine-N-oxide and its precursors in nascent metabolic syndrome. Horm. Mol. Biol. Clin. Investig. 35 (2).

Ley RE, Bäckhed F, Turnbaugh P, Lozupone CA, Knight RD, Gordon JI., 2005. Obesity alters gut microbial ecology. 102 (31), 11070–11075.

Li, J., Zhao, F., Wang, Y., Chen, J., Tao, J., Tian, G., et al., 2017. Gut microbiota dysbiosis contributes to the development of hypertension. Microbiome 5 (1), 14.

Li, X.S., Wang, Z., Cajka, T., Buffa, J.A., Nemet, I., Hurd, A.G., et al., 2018. Untargeted metabolomics identifies trimethyllysine, a TMAO-producing nutrient precursor, as a predictor of incident cardiovascular disease risk. JCI Insight 3 (6).

Li, X.S., Obeid, S., Wang, Z., Hazen, B.J., Li, L., Wu, Y., et al., 2019. Trimethyllysine, a trimethylamine N-oxide precursor, provides near- and long-term prognostic value in patients presenting with acute coronary syndromes. Eur. Heart J. 40 (32), 2700–2709.

Li, N., Liang, S., Chen, Q., Zhao, L., Li, B., Huo, G., 2021. Distinct gut microbiota and metabolite profiles induced by delivery mode in healthy Chinese infants. J. Proteom. 232, 104071.

Lim, E.S., Zhou, Y., Zhao, G., Bauer, I.K., Droit, L., Ndao, I.M., et al., 2015. Early life dynamics of the human gut virome and bacterial microbiome in infants. Nat. Med. 21 (10), 1228–1234.

Lin, X., Xu, Y., Pan, X., Xu, J., Ding, Y., Sun, X., et al., 2020. Global, regional, and national burden and trend of diabetes in 195 countries and territories: an analysis from 1990 to 2025. Sci. Rep. 10 (1), 14790.

Liu, H., Chen, X., Hu, X., Niu, H., Tian, R., Wang, H., et al., 2019. Alterations in the gut microbiome and metabolism with coronary artery disease severity. Microbiome 7 (1), 68.

Liu, S., Song, P., Sun, F., Ai, S., Hu, Q., Guan, W., et al., 2020. The concept revolution of gut barrier: from epithelium to endothelium. Int. Rev. Immunol. 1–8.

Lynch, S.V., Pedersen, O., 2016. The human intestinal microbiome in health and disease. N. Engl. J. Med. 375 (24), 2369–2379.

Macfarlane, G.T., Macfarlane, S., 2012. Bacteria, colonic fermentation, and gastrointestinal health. J. AOAC Int. 95 (1), 50–60.

Manthey, C.F., Autran, C.A., Eckmann, L., Bode, L., 2014. Human milk oligosaccharides protect against enteropathogenic Escherichia coli attachment in vitro and EPEC colonization in suckling mice. J. Pediatr. Gastroenterol. Nutr. 58 (2), 165–168.

Markowiak, P., Slizewska, K., 2017. Effects of probiotics, prebiotics, and synbiotics on human health. Nutrients. 9 (9).

Marques, F.Z., Nelson, E., Chu, P.Y., Horlock, D., Fiedler, A., Ziemann, M., et al., 2017. High-fiber diet and acetate supplementation change the gut microbiota and prevent the development of hypertension and heart failure in hypertensive mice. Circulation 135 (10), 964–977.

Martens, E.C., Roth, R., Heuser, J.E., Gordon, J.I., 2009. Coordinate regulation of glycan degradation and polysaccharide capsule biosynthesis by a prominent human gut symbiont. J. Biol. Chem. 284 (27), 18445–18457.

Matsuzawa, Y., Nakahashi, H., Konishi, M., Sato, R., Kawashima, C., Kikuchi, S., et al., 2019. Microbiota-derived trimethylamine N-oxide predicts cardiovascular risk after STEMI. Sci. Rep. 9 (1), 11647.

Maurice, C.F., Haiser, H.J., Turnbaugh, P.J., 2013. Xenobiotics shape the physiology and gene expression of the active human gut microbiome. Cell 152 (1–2), 39–50.

Mayerhofer, C.C.K., Ueland, T., Broch, K., Vincent, R.P., Cross, G.F., Dahl, C.P., et al., 2017. Increased secondary/primary bile acid ratio in chronic heart failure. J. Card. Fail. 23 (9), 666–671.

McRorie Jr., J.W., 2015. Evidence-based approach to fiber supplements and clinically meaningful health benefits, part 2: what to look for and how to recommend an effective fiber therapy. Nutr. Today 50 (2), 90–97.

Menni, C., Zierer, J., Pallister, T., Jackson, M.A., Long, T., Mohney, R.P., et al., 2017. Omega-3 fatty acids correlate with gut microbiome diversity and production of N-carbamylglutamate in middle aged and elderly women. Sci. Rep. 7 (1), 11079.

Meslier, V., Laiola, M., Roager, H.M., De Filippis, F., Roume, H., Quinquis, B., et al., 2020. Mediterranean diet intervention in overweight and obese subjects lowers plasma cholesterol and causes changes in the gut microbiome and metabolome independently of energy intake. Gut 69 (7), 1258–1268.

Meyer-Hoffert, U., Hornef, M.W., Henriques-Normark, B., Axelsson, L.G., Midtvedt, T., Putsep, K., et al., 2008. Secreted enteric antimicrobial activity localises to the mucus surface layer. Gut 57 (6), 764–771.

Milani, C., Duranti, S., Bottacini, F., Casey, E., Turroni, F., Mahony, J., et al., 2017. The first microbial colonizers of the human gut: composition, activities, and health implications of the infant gut microbiota. Microbiol. Mol. Biol. Rev. 81 (4).

Moini J, Ahangari R, Miller C, Samsam M. 2020. Chapter 18—Perspective on economics and obesity. In: Moini J, Ahangari R, Miller C, Samsam M, editors. Global Health Complications of Obesity: Elsevier. p. 411–423.

Morgan, X.C., Tickle, T.L., Sokol, H., Gevers, D., Devaney, K.L., Ward, D.V., et al., 2012. Dysfunction of the intestinal microbiome in inflammatory bowel disease and treatment. Genome Biol. 13 (9), R79.

Muscogiuri, G., Cantone, E., Cassarano, S., Tuccinardi, D., Barrea, L., Savastano, S., et al., 2019. Gut microbiota: a new path to treat obesity. Int. J. Obes. Suppl. 9 (1), 10–19.

Musilova, S., Rada, V., Marounek, M., Nevoral, J., Duskova, D., Bunesova, V., et al., 2015. Prebiotic effects of a novel combination of galactooligosaccharides and maltodextrins. J. Med. Food 18 (6), 685–689.

Nallu A, Sharma S, Ramezani A, Muralidharan J, Raj DJTR., 2017. Gut microbiome in chronic kidney disease: challenges and opportunities. 179, 24–37.

Newgard, C.B., An, J., Bain, J.R., Muehlbauer, M.J., Stevens, R.D., Lien, L.F., et al., 2009. A branched-chain amino acid-related metabolic signature that differentiates obese and lean humans and contributes to insulin resistance. Cell Metab. 9 (4), 311–326.

Nicolucci, A.C., Hume, M.P., Martinez, I., Mayengbam, S., Walter, J., Reimer, R.A., 2017. Prebiotics reduce body fat and alter intestinal microbiota in children who are overweight or with obesity. Gastroenterology 153 (3), 711–722.

O'Connor, S., Chouinard-Castonguay, S., Gagnon, C., Rudkowska, I., 2017. Prebiotics in the management of components of the metabolic syndrome. Maturitas 104, 11–18.

Paone, P., Cani, P.D., 2020. Mucus barrier, mucins and gut microbiota: the expected slimy partners? Gut 69 (12), 2232–2243.

Pedersen, H.K., Gudmundsdottir, V., Nielsen, H.B., Hyotylainen, T., Nielsen, T., Jensen, B.A.H., et al., 2016. Human gut microbes impact host serum metabolome and insulin sensitivity. Nature 535 (7612), 376–381.

Pimenta, F.S., Luaces-Regueira, M., Ton, A.M.M., Campagnaro, B.P., Campos-Toimil, M., Pereira, T.M.C., et al., 2018. Mechanisms of action of kefir in chronic cardiovascular and metabolic diseases. Cell. Physiol. Biochem. 48 (5), 1901–1914.

Qin, J., Li, Y., Cai, Z., Li, S., Zhu, J., Zhang, F., et al., 2012. A metagenome-wide association study of gut microbiota in type 2 diabetes. Nature 490 (7418), 55–60.

Rajilic-Stojanovic, M., Jonkers, D.M., Salonen, A., Hanevik, K., Raes, J., Jalanka, J., et al., 2015. Intestinal microbiota and diet in IBS: causes, consequences, or epiphenomena? Am. J. Gastroenterol. 110 (2), 278–287.

Rampelli, S., Guenther, K., Turroni, S., Wolters, M., Veidebaum, T., Kourides, Y., et al., 2018. Pre-obese children's dysbiotic gut microbiome and unhealthy diets may predict the development of obesity. Commun. Biol. 1 (1), 222.

Ríos-Covián, D., Ruas-Madiedo, P., Margolles, A., Gueimonde, M., delos Reyes-Gavilán, C.G., Salazar, N., 2016. Intestinal short chain fatty acids and their link with diet and humanhealth. Front. Microbiol. 7 (185).

Rohrmann, S., Linseisen, J., Allenspach, M., von Eckardstein, A., Müller, D., 2016. Plasma concentrations of trimethylamine-N-oxide are directly associated with dairy food consumption and low-grade inflammation in a German adult population. J. Nutr. 146 (2), 283–289.

Romano, K.A., Vivas, E.I., Amador-Noguez, D., Rey, F.E., 2015. Intestinal microbiota composition modulates choline bioavailability from diet and accumulation of the proatherogenic metabolite trimethylamine-N-oxide. mBio 6 (2), e02481.

Rooks, M.G., Garrett, W.S., 2016. Gut microbiota, metabolites and host immunity. Nat. Rev. Immunol. 16 (6), 341–352.

Rothschild, D., Weissbrod O., Barkan, E., Kurilshikov, A., Korem, T., Zeevi, D., et al., 2018. Environment dominates over host genetics in shaping human gut microbiota. Nature 555 (7695), 210–215.

Sanchez, J.I., Marzorati, M., Grootaert, C., Baran, M., Van Craeyveld, V., Courtin, C.M., et al., 2009. Arabinoxylan-oligosaccharides (AXOS) affect the protein/carbohydrate fermentation balance and microbial population dynamics of the simulator of human intestinal microbial ecosystem. Microb. Biotechnol. 2 (1), 101–113.

Santacruz, A., Marcos, A., Wärnberg, J., Martí, A., Martin-Matillas, M., Campoy, C., et al., 2009. Interplay between weight loss and gut microbiota composition in overweight adolescents. Obesity (Silver Spring) 17 (10), 1906–1915.

Santiago-Rodriguez, T.M., Hollister, E.B., 2019. Human virome and disease: high-throughput sequencing for virus discovery, identification of phage-bacteria dysbiosis and development of therapeutic approaches with emphasis on the human gut. Viruses 11 (7).

Scaldaferri, F., Pizzoferrato, M., Gerardi, V., Lopetuso, L., Gasbarrini, A., 2012. The gut barrier: new acquisitions and therapeutic approaches. J. Clin. Gastroenterol. 46 (Suppl), S12–S17.

Scheithauer, T.P.M., Rampanelli, E., Nieuwdorp, M., Vallance, B.A., Verchere, C.B., van Raalte, D.H., et al., 2020. Gut microbiota as a trigger for metabolic inflammation in obesity and type 2. Diabetes 11 (2546).

Schilderink, R., Verseijden, C., de Jonge, W.J., 2013. Dietary inhibitors of histone deacetylases in intestinal immunity and homeostasis. Front. Immunol. 4, 226.

Schnorr, S.L., Candela, M., Rampelli, S., Centanni, M., Consolandi, C., Basaglia, G., et al., 2014. Gut microbiome of the Hadza hunter-gatherers. Nat. Commun. 5, 3654.

Schwiertz, A., Taras, D., Schäfer, K., Beijer, S., Bos, N.A., Donus, C., et al., 2010. Microbiota and SCFA in lean and overweight healthy subjects. Obes. (Silver Spring) 18 (1), 190–195.

Sekirov, I., Russell, S.L., Antunes, L.C., Finlay, B.B., 2010. Gut microbiota in health and disease. Physiol. Rev. 90 (3), 859–904.

Simpson, H.L., Campbell, B.J., 2015. Review article: dietary fibre-microbiota interactions. Aliment. Pharmacol. Ther. 42 (2), 158–179.

Singh, N., Thangaraju, M., Prasad, P.D., Martin, P.M., Lambert, N.A., Boettger, T., et al., 2010. Blockade of dendritic cell development by bacterial fermentation products butyrate and propionate through a transporter (Slc5a8)-dependent inhibition of histone deacetylases. J. Biol. Chem. 285 (36) 27601–27608.

Singh, R.K., Chang, H.W., Yan, D., Lee, K.M., Ucmak, D., Wong, K., et al., 2017. Influence of diet on the gut microbiome and implications for human health. J. Transl. Med. 15 (1), 73.

Sirtori, C.R., Pavanello, C., Calabresi, L., Ruscica, M., 2017. Nutraceutical approaches to metabolic syndrome. Ann. Med. 49 (8), 678–697.

Slavin, J., 2013. Fiber and prebiotics: mechanisms and health benefits. Nutrients 5 (4), 1417–1435.

Smith, E.A., Macfarlane, G.T., 1997. Dissimilatory amino acid metabolism in human colonic bacteria. Anaerobe 3 (5), 327–337.

Sonnenburg, E.D., Smits, S.A., Tikhonov, M., Higginbottom, S.K., Wingreen, N.S., Sonnenburg, J.L., 2016. Diet-induced extinctions in the gut microbiota compound over generations. Nature 529 (7585), 212–215.

Stanislawski, M.A., Dabelea, D., Wagner, B.D., Iszatt, N., Dahl, C., Sontag, M.K., et al., 2018. Gut microbiota in the first 2 years of life and the association with body mass index at age 12 in a norwegian birth cohort. mBio 9 (5).

Stubbs, J.R., House, J.A., Ocque, A.J., Zhang, S., Johnson, C., Kimber, C., et al., 2016. Serum trimethylamine-N-oxide is elevated in CKD and correlates with coronary atherosclerosis burden. J. Am. Soc. Nephrol.: JASN 27 (1), 305–313.

Suez, J., Korem, T., Zeevi, D., Zilberman-Schapira, G., Thaiss, C.A., Maza, O., et al., 2014. Artificial sweeteners induce glucose intolerance by altering the gut microbiota. Nature 514 (7521), 181–186.

Sun, J., Buys, N., 2015. Effects of probiotics consumption on lowering lipids and CVD risk factors: a systematic review and *meta*-analysis of randomized controlled trials. Ann. Med. 47 (6), 430–440.

Tabata, T., Yamashita, T., Hosomi, K., Park, J., Hayashi, T., Yoshida, N., et al., 2021. Gut microbial composition in patients with atrial fibrillation: effects of diet and drugs. Heart Vessel 36 (1), 105–114.

Tang, W.H., Wang, Z., Levison, B.S., Koeth, R.A., Britt, E.B., Fu, X., et al., 2013. Intestinal microbial metabolism of phosphatidylcholine and cardiovascular risk. N. Engl. J. Med. 368 (17), 1575–1584.

Thingholm, L.B., Rühlemann, M.C., Koch, M., Fuqua, B., Laucke, G., Boehm, R., et al., 2019. Obese individuals with and without type 2 diabetes show different gut microbial functional capacity and composition. Cell Host Microbe 26 (2), 252–264. e10.

Ticinesi, A., Lauretani, F., Tana, C., Nouvenne, A., Ridolo, E., Meschi, T., 2019. Exercise and immune system as modulators of intestinal microbiome: implications for the gut-muscle axis hypothesis. Exerc. Immunol. Rev. 25, 84–95.

Tiderencel, K.A., Hutcheon, D.A., Ziegler, J., 2020. Probiotics for the treatment of type 2 diabetes: a review of randomized controlled trials. Diabetes/Metabolism Res. Rev. 36 (1), e3213.

Tune, J.D., Goodwill, A.G., Sassoon, D.J., Mather, K.J., 2017. Cardiovascular consequences of metabolic syndrome. Transl. Res. 183, 57–70.

Unger, T., Borghi, C., Charchar, F., Khan, N.A., Poulter, N.R., Prabhakaran, D., et al., 2020. International society of hypertension global hypertension practice guidelines. Hypertension (Dallas, Tex: 1979) 75 (6), 1334–1357. 2020.

Valcheva, R., Dieleman, L.A., 2016. Prebiotics: definition and protective mechanisms. Best. Pract. Res. Clin. Gastroenterol. 30 (1), 27–37.

Vangay, P., Johnson, A.J., Ward, T.L., Al-Ghalith, G.A., Shields-Cutler, R.R., Hillmann, B.M., et al., 2018. US immigration westernizes the human gut microbiome. Cell 175 (4), 962–972. e10.

Verhaar BJH, Prodan A, Nieuwdorp M, Muller M., 2020. Gut microbiota in hypertension and atherosclerosis: a review. 12 (10), 2982.

Wahlström, A., Sayin Sama, I., Marschall, H.-U., Bäckhed, F., 2016. Intestinal crosstalk between bile acids and microbiota and its impact on host metabolism. Cell Metab. 24 (1), 41–50.

Wall, R., Ross, R.P., Shanahan, F., O'Mahony, L., O'Mahony, C., Coakley, M., et al., 2009. Metabolic activity of the enteric microbiota influences the fatty acid composition of murine and porcine liver and adipose tissues. Am. J. Clin. Nutr. 89 (5), 1393–1401.

Wang, Y., Zhao, W., Shi, J., Wang, J., Hao, J., Pang, X., et al., 2019. Intestinal microbiota contributes to altered glucose metabolism in simulated microgravity mouse model. FASEB J. Off. Publ. Fed. Am. Soc. Exp. Biol. 33 (9), 10140–10151.

Wells, J.M., Brummer, R.J., Derrien, M., MacDonald, T.T., Troost, F., Cani, P.D., et al., 2017. Homeostasis of the gut barrier and potential biomarkers. Am. J. Physiol. Gastrointest. Liver Physiol 312 (3), G171–G193.

White, P.J., Lapworth, A.L., An, J., Wang, L., McGarrah, R.W., Stevens, R.D., et al., 2016. Branched-chain amino acid restriction in Zucker-fatty rats improves muscle insulin sensitivity by enhancing efficiency of fatty acid oxidation and acyl-glycine export. Mol. Metabol. 5 (7), 538–551.

Wickens, K.L., Barthow, C.A., Murphy, R., Abels, P.R., Maude, R.M., Stone, P.R., et al., 2017. Early pregnancy probiotic supplementation with Lactobacillus rhamnosus HN001 may reduce the prevalence of gestational diabetes mellitus: a randomised controlled trial. Br. J. Nutr. 117 (6), 804–813.

Wu, J., Zhou, Z., Hu, Y., Dong, S., 2012. Butyrate-induced GPR41 activation inhibits histone acetylation and cell growth. J. Genet. Genomics 39 (8), 375–384.

Yang, S., Li, X., Yang, F., Zhao, R., Pan, X., Liang, J., et al., 2019. Gut microbiota-dependent marker TMAO in promoting cardiovascular disease: inflammation mechanism, clinical prognostic, and potential as a therapeutic target. Front. Pharmacol. 10 (1360).

Zhernakova, A., Kurilshikov, A., Bonder, M.J., Tigchelaar, E.F., Schirmer, M., Vatanen, T., et al., 2016. Population-based metagenomics analysis reveals markers for gut microbiome composition and diversity. Science 352 (6285), 565–569.

Zhu, W., Buffa, J.A., Wang, Z., Warrier, M., Schugar, R., Shih, D.M., et al., 2018. Flavin monooxygenase 3, the host hepatic enzyme in the metaorganismal trimethylamine N-oxide-generating pathway, modulates platelet responsiveness and thrombosis risk. J. Thrombosis Haemostasis: JTH 16 (9), 1857–1872.

Zuo, K., Li, J., Wang, P., Liu, Y., Liu, Z., Yin, X., et al., 2019. Duration of persistent atrial fibrillation is associated with alterations in human gut microbiota and metabolic phenotypes. mSystems 4 (6).

Zuo, K., Yin, X., Li, K., Zhang, J., Wang, P., Jiao, J., et al., 2020. Different types of atrial fibrillation share patterns of gut microbiota dysbiosis. mSphere 5 (2).

Chapter 9

The role of the intestinal microbiota in weight loss in overweight and obese humans

Isabel Sospedra López, Gema Alfonso Sigüenza, Estela González Rodríguez, Mar Lozano Casanova and José Miguel Martínez-Sanz

Nursing Department, Faculty of Health Sciences, University of Alicante, Alicante, Spain

9.1 Causes of overweight and obesity

Obesity is a chronic disease characterized by an abnormal or excessive accumulation of fat caused by an imbalance between the energy ingested through food and the energy expended over a long period of time (Heinsen et al., 2016; Obesidad y sobrepeso, n.d.; Romieu et al., 2017; Xiao et al., 2014). Few chronic diseases have progressed as alarmingly in most countries in recent decades as obesity (Heinsen et al., 2016; Kong et al., 2013). The causes behind the obesity epidemic appear to be complex (Million et al., 2012b), since obesity is a multifactorial condition (Nehra et al., 2016; Rial et al., 2016; Xiao et al., 2014). At present, there are different interventions to combat this disease, such as physical exercise (Zheng et al., 2016), caloric intake restriction (Kong et al., 2013), drugs (Dong et al., 2017), and bariatric surgery (Furet et al., 2010). Among these, dietary intervention is the most widely used approach (Kong et al., 2013). While a wide variety of dietary approaches may efficiently induce weight reduction (Thaiss et al., 2016), interindividual variability in response to dietary changes and subsequent weight recovery are important challenges (Kong et al., 2013). In this context the identification of new factors involved in the regulation of the energy balance is essential in order to develop more efficient intervention strategies to control obesity and associated diseases (Dalmau et al., n.d.).

An emerging factor that affects human metabolic homeostasis and that, according to several studies, has a direct implication in obesity and its metabolic complications is the intestinal microbiota (Beaumont et al., 2016; Cardinelli et al., 2015; Dror et al., 2017; Rial et al., 2016; Thaiss et al., 2016).

9.2 The microbiota and obesity

The gut microbiota is a complex and dynamic ecosystem that hosts a wide variety of microorganisms (Remely et al., 2015)—mainly bacteria but also archaea, viruses, fungi, and protozoa (Heinsen et al., 2016)—which colonize the surfaces of the gastrointestinal tract (Nehra et al., 2016). The microbial density varies throughout the digestive tract, being lower in the stomach and duodenum, as a result of acidic, biliary, and pancreatic secretions, and higher in the colon (Zhang et al., 2015). It is known that under normal homeostatic conditions, the microbiota plays a symbiotic role by ensuring several metabolic functions essential to the host (Rial et al., 2016).

Studies such as the Human Microbiome Project and MetaHit have provided the most comprehensive view of the microbial diversity associated with humans to date (Barlow et al., 2015; Thursby and Juge, 2017). Of the intestinal bacterial species that have been identified, 98% are strictly anaerobic, belonging mostly to the phyla *Firmicutes* (64%), *Bacteroidetes* (23%), *Proteobacteria* (8%), and *Actinobacteria* (3%); *Verrucomycrobia*, *Fusobacteria*, and the phylum TM7 account for the remaining 2% (Angelakis and Lagier, 2017; Cardinelli et al., 2015). *Bacteroidetes* and *Firmicutes* are the most prevalent bacterial phyla (Barlow et al., 2015; Carlucci et al., 2016).

Microbiome, Immunity, Digestive Health and Nutrition. DOI: https://doi.org/10.1016/B978-0-12-822238-6.00016-9

Despite this general classification, the composition of the gut microbiota is subject to significant variations both between individuals and within the same individual, owing to the influence of various physiological and environmental factors, such as the consumption of antibiotics, lifestyle, and diet (Rial et al., 2016).

The potential influences of dietary habits on the gut microbiota have attracted interest (Liu et al., 2017; Nettleton et al., 2016). It has been observed that Western-type diets, which are low in fiber and high in saturated fat and sugars, modify the proportions of the main microbial species and reduce the microbial diversity (Nettleton et al., 2016). Alterations in the microbiota, or dysbiosis, have been associated with numerous metabolic diseases, including obesity (Liu et al., 2017; Nettleton et al., 2016). This relationship has been established in experiments on animals. Lean phenotype mice were found to gain weight without increasing energy consumption after receiving a transplant of the microbiota from obese phenotype mice (Nettleton et al., 2016). These results suggest that the profile of the microbiota is intricately involved in energy absorption. The intestinal microbiota associated with obesity may have a greater capacity to extract energy from the diet through the degradation of polysaccharides into short-chain fatty acids and monosaccharides that are then absorbed through the colonic mucosa. The change in the gut microbiota can also affect lipogenesis and fat storage and adiposity. In addition, this dysbiotic state usually causes a low-grade systemic inflammation, which has been identified as one of the possible causes of obesity (Heinsen et al., 2016; Rial et al., 2016).

However, on the basis of the current evidence, it is not clear which microbiota profile is associated with the presence of obesity, although a higher proportion of bacteria of the phylum *Firmicutes* and a lower proportion of bacteria of the phylum *Bacteroidetes* appear to coincide with obesity (Dror et al., 2017; Nettleton et al., 2016; Schwiertz et al., 2010). Observations of this type have stimulated a great deal of research aimed at understanding the differences between the microbiota of lean and obese phenotypes and how the microbiome communicates with the host to affect body weight and metabolic health (Nettleton et al., 2016). Recent studies suggest that targeting the microbiome in addition to restricting caloric intake could be a promising approach to achieve significant weight loss after a dietary intervention (Heinsen et al., 2016).

There have been several recent studies on the influence of the microbiota in overweight or obese subjects (BMI ≥ 25 kg/m^2) who are more than 18 years old (Chilloux and Dumas, 2017; Cotillard et al., 2013; Damms-Machado et al., 2015; de Souza et al., 2015; Dong et al., 2017; Heinsen et al., 2016; Jumpertz et al., 2011; Le Chatelier et al., 2013; Lee et al., 2014; Liu et al., 2017; Nehra et al., 2016; Salonen et al., 2014; Walker et al., 2011; Zhang et al., 2015).

Most of the published studies involved a dietary intervention; 50% of them also included a later maintenance phase. All had an initial evaluation of the intestinal microbiota, though using different techniques.

Quantitative polymerase chain reaction has been the most commonly used technique, followed by metagenomic sequencing, among the next-generation sequencing techniques and finally the analysis of metabolic profiles with nuclear magnetic resonance spectroscopy.

The relationship between the composition of the microbiota and the weight status has been analyzed in several studies. One of the works in which the intestinal microbiota of obese humans was compared with that of subjects of normal weight was carried out by Million et al. (2012b). Their results indicate that the gut microbiota at the species level is related to body weight and obesity, since reduced levels of *Methanobrevibacter smithii* and higher levels of *Lactobacillus*, mainly *L. reuteri*, were associated with obesity, while high levels of *Bifidobacterium animalis*, *L. paracasei*, and *L. plantarum* were associated with normal weight. These findings appear to be consistent with other research showing that the colonization of germ-free mice by *M. smithii* produced an increase in host adiposity (Samuel and Gordon, 2006). Similarly, other work has found a positive correlation of *L. reuteri* with weight gain (Million et al., 2012a) and of *L. plantarum* with weight loss (Karlsson et al., 2013) in humans (Drissi et al., 2017).

In their research, Jumpertz et al. (2011) found no significant differences in bacterial abundance between lean and obese phenotypes. This result may be because the study group was relatively small ($n = 21$) or because the analysis of the microbiota was not performed shortly after the dietary intervention but during the subsequent maintenance period. In lean subjects, the proportion of *Firmicutes* was decreased, and that of *Bacteroidetes* was increased, the opposite of what was found in obese individuals in the study of Million et al. (2012b). In addition, in lean individuals a 20% increase in the proportional representation of *Firmicutes* was associated with higher energy absorption (150 kcal), while a 20% increase in *Bacteroidetes* was associated with a decrease in absorption (150 kcal).

9.3 The bacterial composition of the microbiota according to weight status

Several of the studies cited investigated the relationship between the composition of the microbiota and weight status. Table 9.1 shows the results of three studies in which the initial sample was classified according to the BMI values to

TABLE 9.1 **Differences in the bacterial composition of the microbiota between lean and obese individuals.**

References	Bacteria studied	Composition of the microbiota	
		Lean participants	Obese participants
Le Chatelier et al. (2013)	Total microbiota	High levels of *Sporobacter termitidis* (*Firmicutes*)	High levels of *Clostridium symbiosum* and *Ruminococcus gnavus* (*Firmicutes*)
Million et al. (2012a, 2012b)	Firmicutes, Bacteroidetes, *Methanobrevibacter smithii, Lactococcus lactis, B. animalis,* and seven *Lactobacillus* species	High levels of *Bifidobacterium animalis (Actinobacteria), Lactobacillus paracasei,* and *Lactobacillus plantarum* (*Firmicutes*)	High levels of *Lactobacillus,* principally *L. reuteri* (*Firmicutes*) Lower abundance of *B. animalis* (*Actinobacteria*), the archaea *M. smithii,* and the phylum *Bacteroidetes.*
Jumpertz et al. (2011)	Total microbiota	No significant differences in bacterial abundance were observed between lean and obese individuals	

determine differences in the microbiota between the body phenotypes (Jumpertz et al., 2011; Le Chatelier et al., 2013; Million et al., 2012b).

This relationship between the microbiota and overweight and obesity has also been studied by using the reverse methodology (Cotillard et al., 2013; Le Chatelier et al., 2013), that is, by classifying the participants according to their high or low microbial genetic load and relating the results that are obtained to the subjects' weight status (Table 9.2).

The data that were provided by other studies in which the weight status was modified through dietary intervention are similar. The abundance of the phylum *Firmicutes* was shown to decrease after the weight loss phase (Lee et al., 2014), coinciding with a decline in the *Firmicutes/Bacteroidetes* ratio (F/B) (Damms-Machado et al., 2015; Lee et al., 2014; Remely et al., 2015). This relationship is considered a health marker and an imbalance in it is associated with different diseases. Moreover, it is usually higher in obese humans than in normal weight individuals, as reported in another study (Mariat et al., 2009).

9.4 Effect of dietary interventions on the microbiota

In addition to the relationship between the microbiota and the weight status, the effect of dietary interventions has also been studied widely. Some of these studies indicate that the most appropriate dietary interventions for weight loss are those that focus on diets that are low in calories (800–1200 kcal/day), are low in fat, and high in fiber and that are followed for at least 8 weeks, as reflected in Table 9.3 (Damms-Machado et al., 2015; Kong et al., 2013; Lee et al., 2014; Louis et al., 2016; Remely et al., 2015; Xiao et al., 2014).

Some authors observed that the abundances of *Lactobacillus*, *Leuconostoc*, and *Pediococcus* were higher in the group with lower weight loss, and that these bacteria correlated positively with weight recovery (Kong et al., 2013) However, the influence of *Lactobacillus* on weight gain could not be demonstrated, and it seems that the consumption of starch-rich products, and the nonconsumption of dairy products, influenced these correlations.

The same alterations in the intestinal microbiota were also observed in other studies where dietary intervention did not cause weight loss, as in that of Souza et al. (de Souza et al., 2015). The results they obtained after oral supplementation with glutamine were similar to those obtained in weight loss programs described elsewhere in the literature. The amino acid L-glutamine (GLN) plays a physiological role in the intestine and contributes to a nutritionally important part of the generation of intestinal energy (Reeds and Burrin, 2001; Wu, 2013). However, further studies are needed to verify the relationship between GLN and weight status.

Another important finding was that of Le Chatelier et al. (2013). After classifying the participants according to their microbial genetic load, he found that the low microbial gene count (LGC) group was composed of more obese participants. This group was associated with proinflammatory species, such as *Ruminococcus gnavus* and *Bacteroides*, whereas antiinflammatory species, such as *Faecalibacterium prausnitzii*, prevailed in the high microbial gene count

TABLE 9.2 Differences in the bacterial composition of the microbiota between individuals with a high (HGC) and low (LGC) microbial gene count.

Reference	Composition of the microbiota		Observations
	HGC	LGC	
Le Chatelier et al. (2013)	Greater abundance of *Verrucomicrobia, Actinobacteria* and *Euryarchaeota*. Greater frequency of antiinflammatory species such as *Faecalibacterium prausnitzii*	Greater abundance of *Proteobacteria* and *Bacteroidetes* Greater frequency of proinflammatory species such as *Bacteroides* and *Ruminococcus gnavus*	The LGC group was composed of more obese participants. In addition, this group was characterized by more marked global adiposity, insulin resistance, dyslipidemia, and a more pronounced inflammatory phenotype.
Cotillard et al. (2013)	This group housed an average of 561,499 genes.	This group housed an average of 379,436 genes.	The LGC group had higher insulin resistance and fasting serum levels of triglycerides as well as a tendency toward higher LDL cholesterol and inflammation than the HGC group.

(HGC) group. The LGC group was also characterized by being metabolically less healthy than the HGC group, as was also demonstrated in another study (Cotillard et al., 2013). Interestingly, the LGC subjects appeared to consume fewer fruits, vegetables, and fish products than the HGC subjects. This supports the idea that long-term dietary habits may modify the genetic richness and associated phenotypes. Although the study conducted by Cotillard et al. (2013) showed a transient effect, since the general trend was a return to values close to the baseline level at the end of the weight maintenance phase, it was observed that the abundance of several bacterial species could be influenced by nutritional intervention. This was also demonstrated in many of the other investigations that were included in this review (Dao et al., 2016; de Souza et al., 2015; Dror et al., 2017; Jumpertz et al., 2011; Louis et al., 2016; Salonen et al., 2014; Walker et al., 2011).

One of the dietary interventions that produced obvious changes in the intestinal microbiota was the diet rich in nondigestible carbohydrates. However, diets that were rich in nonstarchy polysaccharides (NPS) and resistant starch (RS) were found to have differing impacts on the ecology of the human gut microbial community. While the RS diet has been shown to cause an increase in *R. bromii*–related bacteria (*Firmicutes*) and a decrease in both the total abundance and the diversity of the microbiota (Salonen et al., 2014; Walker et al., 2011), as was seen in previous studies (Abell et al., 2008), the NSP-rich diet has been found to cause an increase in *Prevotella* (*Bacteroidetes*). It is also true that in the study of Salonen et al. (2014) the diet explained only about 10% of the total variance in the composition of the microbiota, a percentage that is substantially lower than the interindividual variance.

Other studies showed alterations in the composition and/or activity of the gut microbial community after the consumption of a diet that was rich in dairy products (Zheng et al., 2016) and after a treatment with nonivamide, a chemical compound that is present in hot peppers (Hochkogler et al., 2017), although more research is needed to confirm these results, as they were not entirely as expected.

Some studies agree that alteration of the intestinal microbiota through dietary intervention results in improvements in the metabolic state. Xiao et al. (2014) reported that the metabolic improvements that were produced in this way in the participants in their study were associated with a decrease in endotoxin-eliminating bacteria (*Enterobacteriaceae* and *Desulfovibrionaceae*) and an increase in bacteria that protect the intestinal barrier (*Bifidobacteriaceae*).

The results of the study conducted by Dao et al. (2016) show an association between *A. muciniphila* and a healthier insulin sensitivity profile. This study also indicated that subjects with higher *A. muciniphila* abundance and greater genetic richness were metabolically healthier before and after the dietary intervention, confirming in humans what has been observed in murine models (Everard et al., 2013; Shin et al., 2014). Louis et al. (2016) associated successful weight reduction in obese people with an increase in *Akkermansia* numbers in the stool.

TABLE 9.3 Relationship between diet and the gut microbiota and their effects on body weight status, body composition, and metabolism.

Reference	Nutritional intervention				Composition of the microbiota		Effects on body weight status and body composition	Effects on metabolism
	Dietary intervention		Maintenance phase					
	Duration	Diet	Duration	Diet	Before diet	After diet		
Hochkogler et al. (2017)	12 weeks	Usual diet plus the consumption of 0.15 mg of nonivamide in 450 mL of milkshake, distributed in three daily doses. This formulation supposes an additional energy intake of 375 kcal per day.	–	–	Domination by the phyla *Firmicutes*, *Bacteroidetes*, and, to a lesser extent, the phyla *Actinobacteria* and Proteobacteria. The dominant families included *Lachnospiraceae*, *Ruminococcaceae*, *Bacteroidaceae*, *Bifidobacteriaceae*, *Prevotellaceae*, and *Coriobacteriaceae*.	There were no changes in the composition of the intestinal microbiota.	No weight loss.	There was no effect on glucose metabolism. However, in the group that took nonivamide, the concentration of serotonin increased.
Dao et al. (2016)	6 weeks	Diet with caloric restriction, enriched with fiber and proteins.	6 weeks	Unspecified weight maintenance diet.	25 participants had a greater abundance of *A. muciniphila* in comparison with the other 24 participants.	The abundance of *A. muciniphila* declined in the Akk HI group but was consistently higher than in the Akk LO group.	No difference in weight loss between the Akk HI and Akk LO groups, although in the Akk HI group the circumference of the waist decreased; the Akk HI, HGC group had the lowest average values of android fat.	There was a decrease in total and LDL cholesterol in the Akk HI group and an improvement in the fasting levels of glucose and triglycerides in the Akk HI, HGC group.
Louis et al. (2016)	6 months	Hypocaloric diet (800 kcal) enriched with fiber (inulin). The first 3 months involved a formula-based diet, and the remaining 3 months involved food.	–	In the stabilization phase, energy consumption was slowly increased to a level that allowed weight maintenance.	In the PS group, *Alistipes*, *Pseudoflavonifractor*, *Ethanoligenens*, *Gordonibacter*, and *Symbiobacterium* were more abundant. At the species level, several *Bacteroides* species,such as *B. caccae* and *B. massiliensis*, were less abundant in the PS group, while *Clostridium leptum* was more abundant than in the NS group.	During the intervention the abundance of *Bacteroides* did not change; however, those of *B. vulgatus* and *B. dorei* decreased in the first 3 months, while those of *B. cellulosilyticus* and *B. intestinalis* increased. In the successful weight loss subgroup, *Prevotella* (*Bacteroidetes*) was less abundant, and *Akkermansia muciniphila* increased significantly from the start to the end of the program.	Strong weight reduction during the diet. The participants were classified into two categories with respect to long-term weight loss: those with persistent success and those without success.	Minor changes in cholesterol and more notable ones in triglycerides.
Zheng et al. (2016)	24 weeks	Diet rich in dairy products (4–5 per day) or low in dairy products (0–1 per day) with a deficit of 500 kcal according to the estimated energy requirement (52% carbohydrates, 30% fat, and 18% proteins).	–	–	Not reported.	The diet rich in dairy products was associated with alterations in the intestinal microbial activity.	No weight loss.	The changes induced by the dairy product–rich diet were reflected in the metabolome related to protein catabolism, endogenous energy, lipid metabolism, and gut microbes. This likely affected the maintenance of lean body mass, fat deposition, and intestinal health.

(*Continued*)

TABLE 9.3 (Continued)

Reference	Nutritional intervention				Composition of the microbiota		Effects on body weight status and body composition	Effects on metabolism
	Dietary intervention		Maintenance phase					
	Duration	Diet	Duration	Diet	Before diet	After diet		
Remely et al. (2015)	4 months	Diet low in fat and animal products, and rich in fruits and vegetables according to the food pyramid of the Austrian Ministry of Health.	–	–	The intestinal microbiota was less abundant.	During the study, the *Firmicutes/Bacteroidetes* ratio (F/B) declined, and the abundance of the lactobacilli *Faecalibacterium prausnitzii* and *A. muciniphila* rose.	The average BMI decreased, as did the body fat content. Also, there was an increase in lean mass.	No metabolic changes.
de Souza et al. (2015)	2 weeks	15 g of GLN or ALA diluted in 200 mL of water, twice a day.	–	–	Certain dominant groups of bacteria were detected, such as the phyla *Bacteroidetes* and *Firmicutes*, and the genera *Alistipes*, *Bacteroides*, *Megasphaera*, *Prevotella*, *Phascolarctobacterium*, and *Succinivibrio*.	ALA treatment: The phylum *Bacteriodetes* showed a tendency to decline in abundance, while the relative abundances of *Firmicutes* and *Actinobacteria* showed a tendency to increase. The F/B increased. GLN treatment: The proportion of *Bacteroidetes* showed a tendency to increase, while *Firmicutes* and *Actinobacteria* declined. *Prevotella* increased, and *Veillonella* decreased. *Proteobacteria* increased slightly in both groups.	There was no weight loss.	No metabolic changes.
Damms-Machado et al. (2015)	18 weeks	Very low calorie formula diet (800 kcal/day): 5 daily sachets of 160 kcal each (Optifast 800 formula, Nestlé Inc.) for 12 weeks. Followed by progressive adaptation to a low-fat, high-fiber, and energy-restricted diet (refeeding phase: 6 weeks).	7 weeks	The stabilization phase was not described.	Not reported.	A decline in *Bacteroidetes* and a rise in certain *Firmicutes* species, namely, butyrate-producing bacterial strains (*Butyrivibrio fibrisolvens*, *Clostridium saccharolyticum*, *Eubacterium limosum*, and *Blautia hydrogenotrophica*).	Mean weight loss of 30 kg.	An increase in the energy reabsorption potential of the intestinal microbiota as well as in the microbial capacity for the fermentation of butyrate. In addition, there were favorable changes in the blood lipid profile.
Salonen et al. (2014)	10 weeks	Standard weight maintenance diet followed by a diet high in resistant starch (RS), another high in nonstarch polysaccharides (NSP), and last, a weight loss diet with high levels of protein and moderate levels of carbohydrates (WL).	–	–	The microbiota of the nonresponder patients was notably more homogeneous from the point of view of composition and significantly higher in diversity compared to the responder group.	Increase of *Oscillospira guillermondii*, *R. bromii*, *Sporobacter termitis*, *Clostridium leptum*, Bacteroidetes (*Alistipes* spp., mainly), and *C. cellulosi* with the RS diet. Increase in various members of the family *Lachnospiraceae* and the genera *Eggerthella*, *Collinsella*, and *Corynebacterium* (members of the Actinobacteria), together with *Bacteroides vulgatus* and *Prevotella oralis*, in the NSP diet.	Not reported.	Not reported.

						Decline in *Bifidobacterias* due to the WL diet.		
Lee et al. (2014)	8 weeks	Diet of 20–25 kcal/kg/day + 3 g of BTS and probiotics (DUOLAC 7) or BTS and placebo, twice a day. The probiotic formulation contained seven species of lactic acid bacteria and Bifidobacteria	–	–	Not reported.	Increase in *B. breve*, *B. lactis* and *L. rhamnosus* in the probiotic group. Decrease in *Firmicutes* in the group that lost the most weight. An increase in *Bacteroidetes* and a decrease in the F/B in the groups that lost more and less weight.	There was one group that had the greatest weight loss (upper quartile) compared to the other group (lower quartile), regardless of the dietary intervention.	HDL cholesterol decreased in the placebo group but increased in the probiotic group.
Xiao et al. (2014)	9 weeks	Diet plan based on whole grains, traditional Asian medicinal foods, and prebiotics. Daily energy intake of 1000 kcal for women and 1600 kcal for men.	14 weeks	Diet with a high fiber content and a low content of foods of animal origin.	The microbiota was dominated by four phyla: *Firmicutes*, *Bacteroidetes*, *Proteobacteria*, and *Actinobacteria*.	There were no changes in the abundance of *Bacteroidetes* and *Firmicutes* or in the F/B. There were changes in the populations of *Actinobacteria* and *Proteobacteria*. At the genus level, *Bifidobacterium* increased, and the proportion of the microbiota containing lipopolysaccharides (*Escherichia/Shigella*, *Klebsiella*, and *Citrobacter*) decreased. The abundance of sulfate-reducing bacteria of the Desulfovibrionaceae family decreased.	1Eight participants had lost more than 10% of their initial body weight, 33 participants lost from 5% to 10%, and 30 participants lost from 0% to 5%. Eight participants gained an average of 0.84 ± 0.70 kg.	Reduction of metabolic syndrome. Improvements in systemic inflammatory tone and intestinal barrier function. Reduction in the antigen load for the host.
Kong et al. (2013)	6 weeks	The foods typically consumed with the addition of 4 daily dietary supplements (60–75 kcal). Low-calorie diet high in protein and rich in fiber with high-glycemic carbohydrates. Daily energy intake of 1200 kcal for women and 1500 kcal for men (40% carbohydrates, 25% fat, and 35% protein).	6 weeks	Weight maintenance diet prescribed individually by a dietician (20%–30% more kcal than the GER of each subject).	The group that lost the least weight during the dietary intervention and regained it quickly during the stabilization period (group C) had the highest number of bacteria from the *Lactobacillus*, *Leuconostoc*, and *Pediococcus* groups in the fecal samples.	Not reported.	The participants were classified according to weight loss. Groups A and B lost the most weight during the energy intake restriction. During the maintenance phase, group A continued to lose weight, while group B remained stable. Group C lost less weight and regained it quickly in the maintenance period. In addition, group A had a better reduction in waist and hip circumferences and in the diameter of the adipocytes.	The group that lost the most weight throughout the study (group A) exhibited a decrease in leptin and increased its insulin sensitivity more favorably than the other groups.
Cotillard et al. (2013)	6 weeks	Energy restricted, high protein diet with low glycemic carbohydrates and enriched with soluble fiber. Daily energy intake of 1200 kcal for women and 1500 kcal for men (40% carbohydrates, 25% fat and 35% protein).	6 weeks	Body weight stabilization period with a 20% increase in total energy intake, above the resting energy metabolic rate.	60% of the individuals in the sample had a high intestinal microbial gene richness (HGC) compared to the remaining 40%, who had a low microbial gene count (LGC).	The genetic richness increased significantly in the LGC group. *Eubacterium rectole* and *Bifidobacterium* spp. decreased during the caloric restriction phase.	Decreases in the values of adiposity indicators (hip circumference and total fat).	Improvement in the systemic metabolic state (decrease in circulating cholesterol and a tendency toward a decrease in inflammation (highly sensitive C-reactive protein)).
Jumpertz et al. (2011)	3 days	Random diet of 2400 or 3400 kcal per day. Both had 60% carbohydrates, 16% fat, and 24% protein, and the fiber contents were similar. These diets were preceded by a weight maintenance prediet for 3 days.	3 days	Wash diet after 3 days with the 2400-kcal diet or the 3400-kcal diet.	No significant differences in bacterial abundance between the lean and obese individuals.	In the lean individuals the abundance of *Firmicutes* fell and that of *Bacteroidetes* rose.	Not reported.	A 20% increase in Firmicutes was associated with an increase in nutrient absorption of 150 kcal, while a 20% increase in Bacteroidetes was associated with a decrease in absorption.

(Continued)

TABLE 9.3 (Continued)

Reference	Nutritional intervention				Composition of the microbiota		Effects on body weight status and body composition	Effects on metabolism
	Dietary intervention		Maintenance phase					
	Duration	Diet	Duration	Diet	Before diet	After diet		
Walker et al. (2011)	10 weeks	Control diet to maintain weight followed by a diet rich in NSP or RS, and finally a diet intended to produce weight loss, lower in carbohydrates and high in protein (WL).	–	–	Among the participants, the most abundant phylum was *Eubacterium rectale.*	Relatives of *Ruminococcus bromii* (*R-rhinococcci*) were more abundant with the RS diet than with the NSP diet. The Oscillibacter group was increased by the RS and WL diets. Relatives of *Eubacterium rectole* were increased by the RS diet but were decreased, along with *Collinsella aerofaciens*, by the WL diet.	Not reported.	Not reported.

Although in general all of the results that we have cited suggest that human obesity is associated with specific changes at the level of the phylum and with lower bacterial diversity, at the same time they are also associated with changes at the metabolic level, and contradictory findings have been reported. Damms-Machado et al. (2015) observed that after a very low calorie dietary intervention, body weight and the F/B ratio were reduced, while some butyrate-producing strains belonging to the phylum *Firmicutes* increased in number.

Despite the overall increase in this phylum, which is generally associated with increased energy absorption by the host, the specific growth of these species can result in positive metabolic effects. This increase can also be explained by the supply of soluble fiber. Something similar happened in the study of Louis et al. (2016), which showed an increase in *Prevotella* (belonging to the phylum *Firmicutes*) in the successful weight loss group, and in the study of Le Chatelier et al. (2013), in which high levels of *Sporobacter termitidis* (*Firmicutes*) were detected in lean individuals. Salonen et al. (2014) and Xiao et al. (2014) observed no changes in the proportions of these taxa after weight loss. Also, de Souza et al. (2015) and Remely et al. (2015) did not find metabolic changes after the microbial composition was altered by dietary intervention.

These results are consistent with other research, as there are conflicting data on the link between the gut microbiota and obesity in humans (Collado et al., 2008; Santacruz et al., 2009; Shin et al., 2014). Some studies have shown the opposite results (Turnbaugh and Gordon, 2009) or even claim that there is no evidence that the phyla *Bacteroidetes* and *Firmicutes* have a role in obesity (Collado et al., 2008). It appears that while the composition of the gut microbiota at the phylum level is related to weight gain in experimental animal studies (Nettleton et al., 2016), no clear evidence has been provided in human studies.

9.5 Conclusions

In the available literature there are consistent data on the relationship between diet and the intestinal microbiota, since diets that are designed to achieve weight loss in obese subjects can significantly alter the composition and microbial diversity. However, although most studies in this area have tended to confirm that there is a relationship between obesity and the gut microbiota at the level of the phylum, there are still controversies. More research is needed to understand and confirm this connection and its mechanisms, although targeting the microbiome could be a promising future approach to maintain significant weight loss, along with dietary intervention.

References

Abell, G.C.J., Cooke, C.M., Bennett, C.N., Conlon, M.A., McOrist, A.L., 2008. Phylotypes related to *Ruminococcus bromii* are abundant in the large bowel of humans and increase in response to a diet high in resistant starch. FEMS Microbiol. Ecol. 66, 505–515. Available from: https://doi.org/10.1111/j.1574-6941.2008.00527.x.

Angelakis, E., Lagier, J.-C., 2017. Samples and techniques highlighting the links between obesity and microbiota. Microb. Pathog. 106, 119–126. Available from: https://doi.org/10.1016/j.micpath.2016.01.024.

Barlow, G.M., Yu, A., Mathur, R., 2015. Role of the gut microbiome in obesity and diabetes mellitus. Nutr. Clin. Pract. Off. Publ. Am. Soc. Parenter. Enter. Nutr. 30, 787–797. Available from: https://doi.org/10.1177/0884533615609896.

Beaumont, M., Goodrich, J.K., Jackson, M.A., Yet, I., Davenport, E.R., Vieira-Silva, S., et al., 2016. Heritable components of the human fecal microbiome are associated with visceral fat. Genome Biol. 17, 189. Available from: https://doi.org/10.1186/s13059-016-1052-7.

Cardinelli, C.S., Sala, P.C. Alves, C.C., Torrinhas, R.S., Waitzberg, D.L., 2015. Influence of intestinal microbiota on body weight gain: a narrative review of the literature. Obes. Surg. 25, 346–353. Available from: https://doi.org/10.1007/s11695-014-1525-2.

Carlucci, C., Petrof, E.O., Allen-Vercoe, E., 2016. Fecal microbiota-based therapeutics for recurrent clostridium difficile infection, ulcerative colitis and obesity. EBioMedicine 13, 37–45. Available from: https://doi.org/10.1016/j.ebiom.2016.09.029.

Chilloux, J., Dumas, M.-E., 2017. Are gut microbes responsible for post-dieting weight rebound? Cell Metab. 25, 6–7. Available from: https://doi.org/10.1016/j.cmet.2016.12.016.

Collado, M.C., Isolauri, E. Laitinen, K., Salminen, S., 2008. Distinct composition of gut microbiota during pregnancy in overweight and normal-weight women. Am. J. Clin. Nutr. 88, 894–899. Available from: https://doi.org/10.1093/ajcn/88.4.894.

Cotillard, A., Kennedy, S.P., Kong, L.C., Prifti, E., Pons, N., Le Chatelier, E., et al., 2013. Dietary intervention impact on gut microbial gene richness. Nature 500, 585–588. Available from: https://doi.org/10.1038/nature12480.

Dalmau, J., Sanz, Y., Santacruz, A., n.d. Influencia de la microbiota intestinal en la obesidad y las alteraciones del metabolismo [WWW Document]. Acta Pediátrica Esp. Available from: <https://www.actapediatrica.com/index.php/secciones/nutricion-infantil/265-influencia-de-la-microbiota-intestinal-en-la-obesidad-y-las-alteraciones-del-metabolismo> (accessed 07.08.21).

Damms-Machado, A., Mitra, S., Schollenberger, A.E., Kramer, K.M., Meile, T., Königsrainer, A., et al., 2015. Effects of surgical and dietary weight loss therapy for obesity on gut microbiota composition and nutrient absorption. BioMed. Res. Int. 2015, 806248. Available from: https://doi.org/10.1155/2015/806248.

Dao, M.C., Everard, A., Aron-Wisnewsky, J., Sokolovska, N., Prifti, E., Verger, E.O., et al., 2016. Akkermansia muciniphila and improved metabolic health during a dietary intervention in obesity: relationship with gut microbiome richness and ecology. Gut 65, 426–436. Available from: https://doi.org/10.1136/gutjnl-2014-308778.

de Souza, A.Z.Z., Zambom, A.Z., Abboud, K.Y., Reis, S.K., Tannihão, F., Guadagnini, D., et al., 2015. Oral supplementation with L-glutamine alters gut microbiota of obese and overweight adults: a pilot study. Nutr. Burbank Los Angel. Cty. Calif. 31, 884–889. Available from: https://doi.org/10.1016/j.nut.2015.01.004.

Dong, Z., Xu, L., Liu, H., Lv, Y., Zheng, Q., Li, L., 2017. Comparative efficacy of five long-term weight loss drugs: quantitative information for medication guidelines. Obes. Rev. Off. J. Int. Assoc. Study Obes. 18, 1377–1385. Available from: https://doi.org/10.1111/obr.12606.

Drissi, F., Raoult, D., Merhej, V., 2017. Metabolic role of lactobacilli in weight modification in humans and animals. Microb. Pathog. 106, 182–194. Available from: https://doi.org/10.1016/j.micpath.2016.03.006.

Dror, T., Dickstein, Y., Dubourg, G., Paul, M., 2017. Microbiota manipulation for weight change. Microb. Pathog. 106, 146–161. Available from: https://doi.org/10.1016/j.micpath.2016.01.002.

Everard, A., Belzer, C., Geurts, L., Ouwerkerk, J.P., Druart, C., Bindels, L.B., et al., 2013. Cross-talk between Akkermansia muciniphila and intestinal epithelium controls diet-induced obesity. Proc. Natl. Acad. Sci. U. S. A. 110, 9066–9071. Available from: https://doi.org/10.1073/pnas.1219451110.

Furet, J.-P., Kong, L.-C., Tap, J., Poitou, C., Basdevant, A., Bouillot, J.-L., et al., 2010. Differential adaptation of human gut microbiota to bariatric surgery-induced weight loss: links with metabolic and low-grade inflammation markers. Diabetes 59, 3049–3057. Available from: https://doi.org/10.2337/db10-0253.

Heinsen, F.-A., Fangmann, D., Müller, N., Schulte, D.M., Rühlemann, M.C., Türk, K., et al., 2016. Beneficial effects of a dietary weight loss intervention on human gut microbiome diversity and metabolism are not sustained during weight maintenance. Obes. Facts 9, 379–391. Available from: https://doi.org/10.1159/000449506.

Hochkogler, C.M., Lieder, B., Rust, P., Berry, D., Meier, S.M., Pignitter, M., et al., 2017. A 12-week intervention with nonivamide, a TRPV1 agonist, prevents a dietary-induced body fat gain and increases peripheral serotonin in moderately overweight subjects. Mol. Nutr. Food Res. 61. Available from: https://doi.org/10.1002/mnfr.201600731.

Jumpertz, R., Le, D.S., Turnbaugh, P.J., Trinidad, C., Bogardus, C., Gordon, J.I., et al., 2011. Energy-balance studies reveal associations between gut microbes, caloric load, and nutrient absorption in humans. Am. J. Clin. Nutr. 94, 58–65. Available from: https://doi.org/10.3945/ajcn.110.010132.

Karlsson, F.H., Tremaroli, V., Nookaew, I., Bergström, G., Behre, C.J., Fagerberg, B., et al., 2013. Gut metagenome in European women with normal, impaired and diabetic glucose control. Nature 498, 99–103. Available from: https://doi.org/10.1038/nature12198.

Kong, L.C., Wuillemin, P.-H., Bastard, J.-P., Sokolovska, N., Gougis, S., Fellahi, S., et al., 2013. Insulin resistance and inflammation predict kinetic body weight changes in response to dietary weight loss and maintenance in overweight and obese subjects by using a Bayesian network approach. Am. J. Clin. Nutr. 98, 1385–1394. Available from: https://doi.org/10.3945/ajcn.113.058099.

Le Chatelier, E., Nielsen, T., Qin, J., Prifti, E., Hildebrand, F., Falony, G., et al., 2013. Richness of human gut microbiome correlates with metabolic markers. Nature 500, 541–546. Available from: https://doi.org/10.1038/nature12506.

Lee, S.J., Bose, S., Seo, J.-G., Chung, W.-S., Lim, C.-Y., Kim, H., 2014. The effects of co-administration of probiotics with herbal medicine on obesity, metabolic endotoxemia and dysbiosis: a randomized double-blind controlled clinical trial. Clin. Nutr. Edinb. Scotl. 33, 973–981. Available from: https://doi.org/10.1016/j.clnu.2013.12.006.

Liu, R., Hong, J., Xu, X., Feng, Q., Zhang, D., Gu, Y., et al., 2017. Gut microbiome and serum metabolome alterations in obesity and after weight-loss intervention. Nat. Med. 23, 859–868. Available from: https://doi.org/10.1038/nm.4358.

Louis, S., Tappu, R.-M., Damms-Machado, A., Huson, D.H., Bischoff, S.C., 2016. Characterization of the gut microbial community of obese patients following a weight-loss intervention using whole metagenome shotgun sequencing. PLoS One 11, e0149564. Available from: https://doi.org/10.1371/journal.pone.0149564.

Mariat, D., Firmesse, O., Levenez, F., Guimarães, V., Sokol, H., Doré, J., et al., 2009. The Firmicutes/Bacteroidetes ratio of the human microbiota changes with age. BMC Microbiol. 9, 123. Available from: https://doi.org/10.1186/1471-2180-9-123.

Million, M., Angelakis, E., Paul, M., Armougom, F., Leibovici, L., Raoult, D., 2012a. Comparative *meta*-analysis of the effect of Lactobacillus species on weight gain in humans and animals. Microb. Pathog. 53, 100–108. Available from: https://doi.org/10.1016/j.micpath.2012.05.007.

Million, M., Maraninchi, M., Henry, M., Armougom, F., Richet, H., Carrieri, P., et al., 2012b. Obesity-associated gut microbiota is enriched in Lactobacillus reuteri and depleted in Bifidobacterium animalis and Methanobrevibacter smithii. Int. J. Obes. 2005 (36), 817–825. Available from: https://doi.org/10.1038/ijo.2011.153.

Nehra, V., Allen, J.M., Mailing, L.J., Kashyap, P.C., Woods, J.A., 2016. Gut microbiota: modulation of host physiology in obesity. Physiol. Bethesda Md. 31, 327–335. Available from: https://doi.org/10.1152/physiol.00005.2016.

Nettleton, J.E., Reimer, R.A., Shearer, J., 2016. Reshaping the gut microbiota: impact of low calorie sweeteners and the link to insulin resistance? Physiol. Behav. 164, 488–493. Available from: https://doi.org/10.1016/j.physbeh.2016.04.029.

Obesidad y sobrepeso, n.d [WWW Document]. Available from: <https://www.who.int/es/news-room/fact-sheets/detail/obesity-and-overweight> (accessed 07.08.21).

Reeds, P.J., Burrin, D.G., 2001. Glutamine and the bowel. J. Nutr. 131, 2505S–2508SS. Available from: https://doi.org/10.1093/jn/131.9.2505S. discussion 2523S-4S.

Remely, M., Tesar, I., Hippe, B., Gnauer, S., Rust, P., Haslberger, A.G., 2015. Gut microbiota composition correlates with changes in body fat content due to weight loss. Benef. Microbes 6, 431–439. Available from: https://doi.org/10.3920/BM2014.0104.

Rial, S.A., Karelis, A.D., Bergeron, K.-F., Mounier, C., 2016. Gut microbiota and metabolic health: the potential beneficial effects of a medium chain triglyceride diet in obese individuals. Nutrients 8, E281. Available from: https://doi.org/10.3390/nu8050281.

Romieu, I., Dossus, L., Barquera, S., Blottière, H.M., Franks, P.W., Gunter, M., et al., 2017. Energy balance and obesity: what are the main drivers? Cancer Causes Control. CCC 28, 247–258. Available from: https://doi.org/10.1007/s10552-017-0869-z.

Salonen, A., Lahti, L., Salojärvi, J., Holtrop, G., Korpela, K., Duncan, S.H., et al., 2014. Impact of diet and individual variation on intestinal microbiota composition and fermentation products in obese men. ISME J. 8, 2218–2230. Available from: https://doi.org/10.1038/ismej.2014.63.

Samuel, B.S., Gordon, J.I., 2006. A humanized gnotobiotic mouse model of host-archaeal-bacterial mutualism. Proc. Natl. Acad. Sci. USA 103, 10011–10016. Available from: https://doi.org/10.1073/pnas.0602187103.

Santacruz, A., Marcos, A., Wärnberg, J., Martí, A., Martin-Matillas, M., Campoy, C., et al., 2009. Interplay between weight loss and gut microbiota composition in overweight adolescents. Obes. Silver Spring Md. 17, 1906–1915. Available from: https://doi.org/10.1038/oby.2009.112.

Schwiertz, A., Taras, D., Schäfer, K., Beijer, S., Bos, N.A., Donus, C., et al., 2010. Microbiota and SCFA in lean and overweight healthy subjects. Obes. Silver Spring Md. 18, 190–195. Available from: https://doi.org/10.1038/oby.2009.167.

Shin, N.-R., Lee, J.-C., Lee, H.-Y., Kim, M.-S., Whon, T.W., Lee, M.-S., et al., 2014. An increase in the Akkermansia spp. population induced by metformin treatment improves glucose homeostasis in diet-induced obese mice. Gut 63, 727–735. Available from: https://doi.org/10.1136/gutjnl-2012-303839.

Thaiss, C.A., Itav, S., Rothschild, D., Meijer, M.T., Levy, M., Moresi, C., et al., 2016. Persistent microbiome alterations modulate the rate of post-dieting weight regain. Nature 540, 544–551. Available from: https://doi.org/10.1038/nature20796.

Thursby, E., Juge, N., 2017. Introduction to the human gut microbiota. Biochem. J. 474, 1823–1836. Available from: https://doi.org/10.1042/BCJ20160510.

Turnbaugh, P.J., Gordon, J.I., 2009. The core gut microbiome, energy balance and obesity. J. Physiol. 587, 4153–4158. Available from: https://doi.org/10.1113/jphysiol.2009.174136.

Walker, A.W., Ince, J., Duncan, S.H., Webster, L.M., Holtrop, G., Ze, X., et al., 2011. Dominant and diet-responsive groups of bacteria within the human colonic microbiota. ISME J. 5, 220–230. Available from: https://doi.org/10.1038/ismej.2010.118.

Wu, G., 2013. Functional amino acids in nutrition and health. Amino Acids 45, 407–411. Available from: https://doi.org/10.1007/s00726-013-1500-6.

Xiao, S., Fei, N., Pang, X., Shen, J., Wang, L., Zhang, B., et al., 2014. A gut microbiota-targeted dietary intervention for amelioration of chronic inflammation underlying metabolic syndrome. FEMS Microbiol. Ecol. 87, 357–367. Available from: https://doi.org/10.1111/1574-6941.12228.

Zhang, C., Yin, A., Li, H., Wang, R., Wu, G., Shen, J., et al., 2015. Dietary modulation of gut microbiota contributes to alleviation of both genetic and simple obesity in children. EBioMedicine 2, 968–984. Available from: https://doi.org/10.1016/j.ebiom.2015.07.007.

Zheng, H., Lorenzen, J.K., Astrup, A., Larsen, L.H., Yde, C.C., Clausen, M.R., et al., 2016. Metabolic effects of a 24-week energy-restricted intervention combined with low or high dairy intake in overweight women: an NMR-based metabolomics investigation. Nutrients 8, 108. Available from: https://doi.org/10.3390/nu8030108.

Chapter 10

Role of microbial metabolites in cardiovascular and human health

Abhilasha Singh

Cardiovascular Research Centre, Massachusetts General Hospital, Harvard Medical School, Boston, MA, United States

Abbreviations

CVD cardiovascular disease
TMA trimethyl amine
TMAO trimethylamine N-oxide
SCFAs short-chain fatty acids
IS indoxyl sulfate
BA bile acids
DAT desaminotyrosine
EPS ethylphenylsulfate
LCFAs long-chain fatty acids

10.1 Introduction

The human gastrointestinal tract is home to a hundred trillion bacteria in different morphologies and metabolic preferences. These bacteria are collectively termed the gut microbiome, which is sometimes called "the forgotten organ" or pseudo-organ (Ma and Taylor, 2020). Most of the gut microbial community is composed of five phyla (*Bacteroidetes*, *Firmicutes*, *Actinobacteria*, *Proteobacteria*, and *Cerrucomicrobia*) (Shin et al., 2015). However, they make up an abundant and diverse population at the species level. Growing evidence suggests that they play roles in the pathogenesis of various intestinal and extraintestinal disorders, such as obesity, inflammatory bowel disease, nonalcoholic steatohepatitis, and CVD (Saltzman et al., 2018). The development of next-generation sequencing techniques and multiomics approaches has expanded the horizon of research directed toward microbial diversity and their metabolic potential for benefiting the host physiology (Fig. 10.1).

Metabolites of microbial origin stimulate the mucosal immune system or circulatory immune cells after translocation into it. Their absorption into the bloodstream initiates a series of biochemical reactions with several distal organs. In the host, such metabolites are capable of activating receptors on cells, modulate cellular excitability, and change its expression patterns by using epigenetic mechanisms.

Recent research has focused on recognition of the gut microbiome as a multifunctional organ with enormous potential in the pathophysiology of diseases and their therapeutics. cardiovascular diseases (CVDs) are no exception to this association. The intestinal microbiome is emerging as a novel risk factor for CVD (Yoshida et al., 2018). Recent experimental developments with germ-free mouse models have linked the gut microbiota and dysbiosis as causal factors for vascular dysfunction (Li et al., 2017). Human experimental data have revealed that primary hypertensive patients possess decreased microbial richness and diversity (Yang et al., 2015a,b). Furthermore, Li et al. (2017) have shown elevated blood pressure (BP) to be transferrable through microbiota, specifically when fecal samples from hypertensive human donors were transplanted into germ-free mice. Marques et al. (2019) have proposed the high-fiber diet concept which changes the murine gut microbiota, thus preventing the development of hypertension and heart failure. The authors of this study have also suggested that this might be due to short-chain fatty acids (SCFA) (e.g., acetate), which

Microbiome, Immunity, Digestive Health and Nutrition. DOI: https://doi.org/10.1016/B978-0-12-822238-6.00009-1

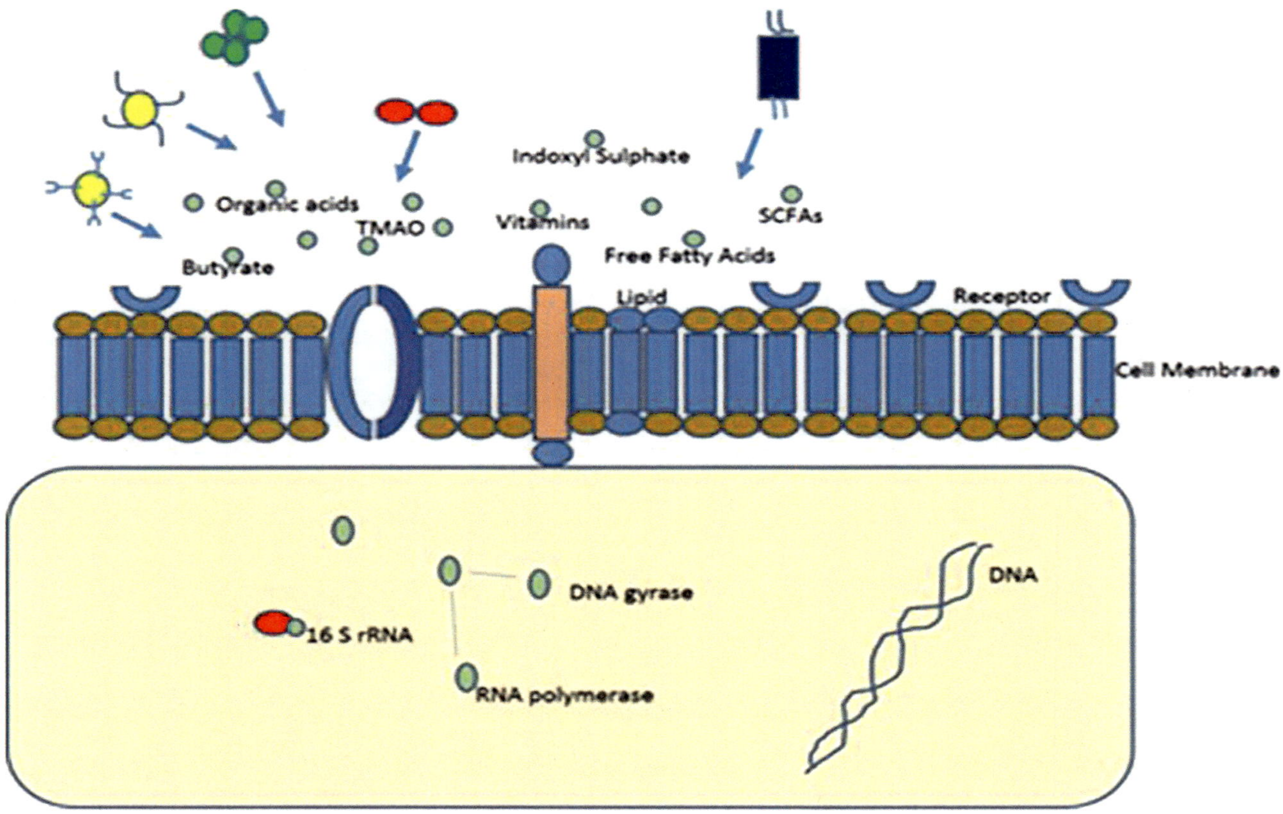

FIGURE 10.1 Bioactive metabolites in the gut microbiome and their implications for host pathology.

are increased by gut microbial populations associated with high-fiber diets. Furthermore, acetate supplementation downregulated the renin-angiotensin system. These findings support the functional role of microbial metabolites as one of the important causative factors of CVD. There is an endless number of metabolites formed from these microbial populations diverse in morphological and metabolic preferences. However, recent evidence has suggested the role of these important microbial metabolites SCFA (Brown and Hazen, 2018), indoxyl sulfate (IS) (Yang et al., 2015a,b), trimethylamine N-oxide (TMAO) (Warrier et al., 2015), betaine (Meyer and Shea, 2017), and other secondary metabolites of clinical significance to be some of the major determinants of CVD.

10.2 Gut microbial metabolites and cardiovascular disease

The metabolic health and its link with CVD risk are associated with the gut microbiome. The mechanistic insights into the microbial mediated production of metabolites have shown to have both beneficial and detrimental effects on metabolic health and CVD risk. Numerous studies have reported the concept of food and medicines, as the major contributor in influencing gut microbiota, including the composition and metabolism. For example, microbially mediated TMA production and its subsequent modification to hepatic TMAO have emerged as one of the prominent risk factors for CVD (Meyer and Shea, 2017). This literature suggests the growing possibility of modulating the gut microbiome via dietary, biological, or xenobiotic approaches, making this pseudo-organ an attractive target for a wide range of health outcomes, so categorization of the metabolites is important (Fig. 10.2). According to a classification method put forward by Postler and Ghosh (2017), these metabolites are classified into three different types: (1) metabolites from dietary sources or drugs that are transformed by gut microbiota, (2) metabolites of host metabolic origin that are secreted by the host and modified by microbiota in the gut, (3) metabolites that are de novo synthesized by the gut microbiota itself. An example of the first type of metabolites is compound K, which is transformed in the gut by microbes from ginsenosides Rb1, Rb2, and R. A typical example of the second type of metabolites is bile acids (BA), which are synthesized from cholesterol in the liver and transformed by gut microbes (Feng et al., 2018). The third type of metabolites includes SCFAs and polysaccharide A.

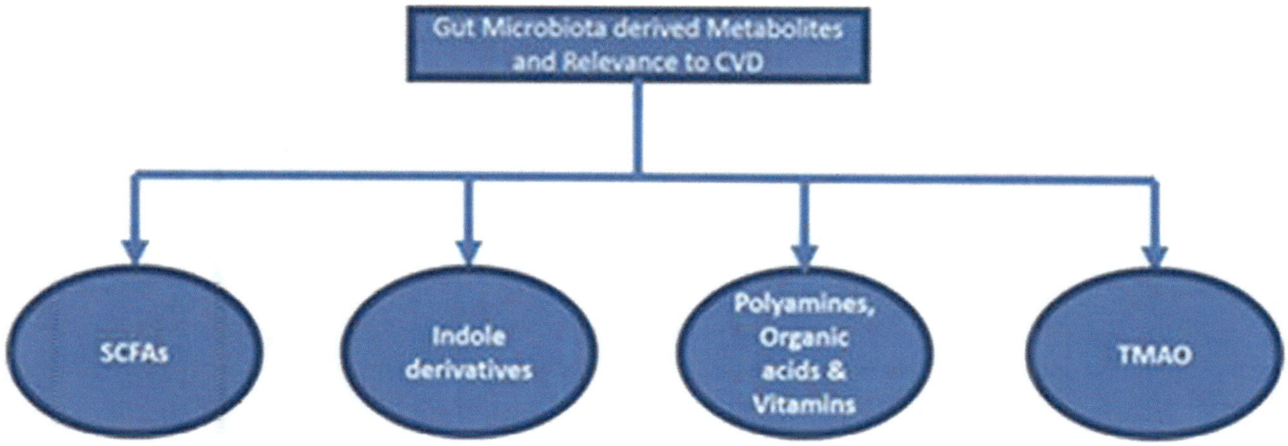

FIGURE 10.2 Categorizing metabolites on the basis of their metabolic preferences.

Overall, research in the past decade has focused primarily on identifying those metabolites of significance to human health and diseases. Of these arrays of metabolites, SCFAs, indole derivatives, polyamines, organic acids, vitamins, and TMAO are of clinical significance to CVD. One of the significant aspects of biology today is a focus on metabolic health and diseases via the role of dietary intake and the primary synthesized products of their breakdown by the intestinal gut microbiota into SCFAs.

10.2.1 Short-chain fatty acids

SCFA are products of the bacterial fermentation process in the colon of human and animal models mimicking human physiology (Feng et al., 2018). Both diet and the intestinal milieu interact in a complex process with the bacterial population in the gut and have a great impact on host physiology. The three main phyla, namely, Bacteroidetes (gram-negative), Firmicutes (gram-positive), and Actinobacteria (gram-positive), are the most abundant in the host intestine (Ríos-Covián et al., 2016). Among these, Firmicutes is the one that primarily produces butyrate as its metabolic result (Panda et al., 2014). Carbohydrates that are not utilized metabolically are fermented in the proximal colon by saccharolytic bacteria. This fermentation process within the intestine leads to SCFAs together with gaseous products, such as CO_2 and H_2 (Brahe et al., 2013). Dietary fibers that cause high amounts of SCFAs lower the pH within the colon, which, in turn, affects the composition of the colonic microbiota and thereby the SCFA production. Butyrate is one such product that plays a crucial role in maintaining the health of the host via regulation of the system and its downstream events (Nakamura et al., 2010); it also is involved in the maintenance of the epithelial barrier (Nakamura et al., 2010; Panda et al., 2014). In a study using a rat model, the addition of fiber to the diet resulted in increased cecal SCFA concentrations compared to pigs, which is considered a better model for the human gastrointestinal tract (Geirnaert et al., 2017). This increase in daily fiber intake is reflected in a smaller amount in the cecal SCFA in a pig model. However, in studies with pigs as a model, the rise in daily fiber intake per kilogram of body weight is far less when compared with the rodent model. Nevertheless, SCFAs are a necessary conclusion of bacterial fermentation, required to balance redox anaerobic environment of the gut.

10.2.1.1 Role of short-chain fatty acids in cardiovascular health

SCFA play a major role in maintaining metabolic health. The main arena of interest in relevance to CVD is their role in modulating obesity and metabolic responses to glucose and lipid homeostasis. The utmost intriguing role for SCFAs is in orchestrating an epithelial barrier function that preserves the gut integrity that prevents translocation of bacterial proinflammatory components across the gut cellular wall (Mu et al., 2017). Butyrate is another sort of SCFA that is produced within the colon of a host, and evidence from a plethora of animal studies supports a key regulatory effect of butyrate in maintaining epithelial integrity within the host and restoring the normal barrier function, primarily through orchestration of the tight junction proteins. These tight junction proteins are the proteins that govern paracellular permeability and solute transport through the channels between intestinal cells (Nakamura et al., 2010; Panda et al., 2014). Therefore an optimal gut barrier prevents the translocation of microbial cell wall components in the gut and prevents the host intestine from lipopolysaccharides (LPS), which are strongly proinflammatory, upholding homeostatic balance.

In addition to these pivotal roles, the butyrate induces mucin production, which creates a physical barrier between the epithelial cells and bacteria in the lumen of the host intestine (Nakamura et al., 2010). Interestingly, evidence from a recent human study showed that with increased gut permeability the levels of LPS were elevated in hypertensive individuals, indicating a strong correlation between gut permeability and SCFAs (Okumura and Takeda, 2018). A congruous study with a conjoined mouse model showed the extensive interplay of inulin, fermenting to SCFA and therefore improvising intestinal function in mice (Harbison et al., 2019). Butyrate is the preferred substrate for colonocytes. More recently, a study focused on identifying a role for SCFA in intercalating epithelial integrity through coordinated maintenance of tight junction proteins that regulate the intracellular molecular pathway in the lumen and hepatic portal system (Damms-Machado et al., 2015). Also, SCFAs are ligands for several cell surface receptors, such as free fatty acid receptor 2 and 3 (FFAR 2/3). These are found on a wide range of cell types, including enteroendocrine and immune cells, validating the role of SCFAs in human health and metabolism (Mañas et al., 2011). An array of human and animal research studies suggests that SCFAs act as a major regulator of cardiovascular pathology and its homeostasis by different mechanisms of action.

Evidence from the literature also suggests that deorphaned G protein—coupled receptors are natural ligands for SCFA (Mishra et al., 2020). Furthermore, free fatty acids receptors (FFAR) 2 and 3, which have a better affinity for acetate and propionate, are expressed within the intestine, adipose tissue, pancreas, and some immune cell subtypes, indicating their relevance as a protective mechanism for the host physiology, likely GPR109A, which as a higher affinity for butyrate and has been shown to play a role in inflammatory pathways in the intestine (Chambers et al., 2018). In mice, butyrate can even act on nucleotide-binding oligomerization domain-like receptors (NLRs), key modulators of inflammation in an FFAR2 dependent manner to manage key components of the tight junction complex (Sun et al., 2017). Validated research further explaining the concept was performed utilizing in vitro cell line models. The research revealed a possible role for p38 mitogen-activated protein kinase (MAPK) (Silva et al., 2018), receptor-mediated IL-10 expression (Fitsialos et al., 2007), and AMPK-mediated intracellular ATP formation (Gabryšová et al., 2014) via regulation of claudin proteins by butyrate as the sole substrate. Intriguingly, the involvement of FFAR2 during this pathway opens a place for other SCFA in regulating the barrier function (Sun et al., 2017). It is evident that the olfactory receptor 78 (Olfr78) receptors have a better affinity for acetate and propionate that are localized in autonomic nerves within the heart and gut, within the smooth muscle cells of arteries, and in renal juxtaglomerular cells to support vascular function (Fitsialos et al., 2007). SCFAs further function on histone deacetylase inhibitors via acetylation, which is assumed to be a key regulator of transcription factor activation and downstream organic phenomena through regulation of the chromatin structure (Kasubuchi et al., 2015). Similarly, studies with SCFA have shown its extensive role in the regulation of BP in animal models. Acetate and propionate appear to manage BP levels during a complex interplay involving the induction of renin production through Olfr78 and counter-regulation through FFAR3. Evidences from knock out models suggest that propionate induces the discharge of renal renin and an increase in BP through Olfr78 but can even induce a discount in BP that appears FFAR3 dependent (Gabryšová et al., 2014; Sun et al., 2017). In rats, butyrate has been appeared to lower BP by constricting angiotensin II-induced expression of renal prorenin receptors and renin (Brown and Hazen, 2018; Sun et al., 2017). By and large, intestine obstruction work on SCFA and butyrate which was long been perceived as critical substrates for keeping up a solid intestine. Hyperglycemia and increased gut permeability are related to translocation of bacteria and/or their semipermeable membrane components, which trigger an inflammatory cascade that has been related to obesity and insulin resistance (Gabryšová et al., 2014; Kasubuchi et al., 2015). As a high-fat diet—induced steatohepatitis mouse model showed, butyrate constricts steatohepatitis through advancement in high-fat diet—induced intestinal mucosa harm by upregulating zonulin and showed diminished endotoxin levels (Panda et al., 2014; Ververis et al., 2013). These improvements were related to the downregulation of endotoxin-associated genes (TLR4 and Myd88) and expression of proinflammatory genes (MCP-1, TNF-α, IL-1, IL-2, IL-6, and IFN-γ) in the liver (Kuryłowicz and Koźniewski, 2020; Singh et al., 2018). Therefore it is important to develop a basic part of a competent physical boundary inside the gut between the luminal microbes and then have a framework that is well designed enough for a host system to justify the role of SCFA in gut barrier function. Nevertheless, SCFA is a very important immune regulator of vascular events in human and animal physiology, and the literature has shown evidence of its formation from the microbiome of gut, signifying the emergence of gut microbial metabolites as a very important regulator of CVD.

10.2.1.2 *Indoxyl sulfate*

The gut microbiome and its metabolites are implicated in the pathogenesis of various chronic diseases. In humans with chronic renal disorder (CKD) the quantity of beneficial colonic bacteria that produce SCFAs are decreased with a concurrent increase in the bacterial population producing uremic toxins such as IS and p-cresol sulfate (pCS) (Summers et al., 2020). The gut microbial metabolite IS is another important microbial metabolite; it is produced by gut microbial tryptophanases that convert dietary tryptophan into indole, which is then converted to indoxyl and IS within the liver by the sequential actions of cytochrome P450 enzymes and sulfotransferase (Hayashi et al., 2018). IS has been shown to possess proinflammatory and prooxidant effects in cardiomyocytes and cardiac fibroblasts (Pieniazek et al., 2018). Furthermore, recent reports with rodents illustrated the effects of indole and IS on BP that were mediated by peripheral events and serotonin signaling (Sircana et al., 2019).

The large variety of microorganisms that reside in the intestinal tract (the gut microbiome) also plays a very important role in maintaining host health, and alteration of the gut microbiome has been related to many illnesses in humans, including CKD. Uremia related to CKD has been shown to negatively affect the gut microbiome in humans and rats, causing intestinal dysbiosis (Nallu et al., 2017). In humans and rats with CKD as pathology, these modulations shift the gut microbiota to a less diverse populace, that's dominated by specific bacterial colonies. Indoles are produced by the metabolism of dietary tryptophan by tryptophanase in intestinal bacteria like *Escherichia coli*, *Proteus vulgaris*, and *Bacteroides* spp. (Hirakawa et al., 2010) Indole and p-cresol are absorbed into the blood, metabolized to IS and pCS within the liver, and excreted within the urine. Emerging evidences suggest, IS to be a proinflammatory agent intercalating intestinal dysbiosis and host inflammation working locally and systemically. As for other immune cells, macrophage function is additionally inhibited by uremic toxins (Adesso et al., 2013). IS is additionally called a CKD-associated nephron-vascular toxin in uremia, a diminished immune defense contributes to the high prevalence of infections, whereas preactivation and priming of immune cells cause to inflammation and consequently lead to CVD (Yanai et al., 2019).

10.2.1.3 Indoxyl sulfate and its implications for cardiovascular disease pathology

Cardiovascular disease is the most typical secondary complication of CKD, and it is an independent risk factor for mortality in patients with CKD. It is also the leading explanation for death among patients who are on dialysis. In patients with CKD conditions, a decreased glomerular filtration rate leads to reduced IS levels and subsequently increases with the progression of CKD conditions (Drüeke and Massy, 2016). Moreover, IS is a protein-bound solute that cannot be fully eradicated by the method of routine dialysis; rather, a high concentration of IS within the circulation usually exerts its effect in patients with predialysis and postdialysis, which certainly justifies its implications for CVD pathology (Yu et al., 2011). It has also been shown that IS is involved in the development of atherosclerosis because it can induce vascular endothelial cell dysfunction by enhancing oxidative stress (Gelasco and Raymond, 2006). Additionally, an in vitro study revealed that IS has a hypertrophic effect on cultured cardiac myocytes through activation of the MAPK and NF-kB signaling pathways (Fujii et al., 2009). AST-120, an oral charcoal adsorbent that may reduce oxidative stress by lowering the serum IS level, was reported to stop the progression of cardiac hypertrophy, indicating that IS might play a very important role in the development of cardiac hypertrophy under uremic conditions. A further study indicated the role of Klotho, which is an antiaging gene that is predominantly expressed on renal, tubular, and epithelial cells. The proteins secreted by Klotho function as a humoral factor (Yang et al., 2015a,b). Numerous studies have demonstrated that the antiaging properties of Klotho are associated with the antioxidative potential of same. Klotho-deficient mice exhibit phenotypic features similar to those of uremia patients, having rapid CVD progression and premature death (Wakamatsu et al., 2018). Another typical characteristic of uremia is the formation of high oxidative stress mediated by Nox isoforms. Conversely, treatment with the Nox inhibitor DPI significantly constrained IS-induced reactive oxygen species (ROS) generation and activation of p38 and ERK1/2, thereby inhibiting cardiac hypertrophy in animal models (Adesso et al., 2013). Therefore IS has the property to promote cardiomyocyte hypertrophy, but further unmasking the underlying mechanism of the action of IS is necessary. Along with ROS production, IS is involved in calcium release, mostly from the mitochondrial compartment by enhancing the proinflammatory conditions (Pieniazek et al., 2018). Moreover, these effects are further increased in the presence of LPS. IS introduction also induces an increase in LPS-induced NF-kB activation (Yang et al., 2015a,b), NO release, iNOS and COX-2 expression (Wakamatsu et al., 2018), and proinflammatory cytokine (TNF-α and IL-6) (Pieniazek et al., 2018), the release of which is a very important marker of proinflammatory events in CVD establishing role of IS as an important metabolite in CVD pathology. All of these findings support the view that IS contributes to the deleterious effect associated with inflammation in CVD patients.

10.2.1.4 Gut microbiota and trimethylamine-N-oxide

Intestine microbial metabolites are embroiled as novel hazard variables for cardiovascular events and untimely death. TMAO is a gut microbiota–dependent metabolite that is produced from choline and phosphatidylcholine (Warrier et al., 2015). Several human and animal studies have identified the TMAO, metabolite (TMA) as a possible promoter of chronic diseases. TMA originates from microbial effects on dietary supplements, mainly choline, phosphatidylcholine, and L-carnitine, which are further ingested and changed into TMAO by hepatic flavin monooxygenases. Hence TMAO is transported into the tissues in the form of aggregated osmolytes or are commonly taken out by the kidney (Zeisel and Warrier, 2017). In humans, circulating TMAO and its precursors are related to the atherosclerotic burden and exerts extensive risk for CVD (Tang et al., 2019). Although most of the assembly of TMA occurs within the cecum and colon, oral flora (*Streptococcus sanguis*, a facultative anaerobe) can form TMA from choline (Jameson et al., 2018). A choline utilization gene cluster (Cut) that is responsible for anaerobic conversion of choline to TMA was identified in the sulfate-reducing bacterium *Desulfovibrio desulfuricans* (Zeisel and Warrier, 2017). Crucial genes within the cluster are CutC and CutD, encoding for choline TMA-lyase and its activating protein, respectively. Deletion of the CutC gene in *Desulfovibrio alaskensis* G20 and its heterologous expression from *D. alaskensis* G20 in *Escherichia* further confirmed that these genes are fundamental to the TMA conversion from choline (de Oliveira et al., 2020). The pathway explains a radical C–N bond cleavage of choline to form a replica of TMA and acetaldehyde (Cho et al., 2017). A microbial metabolic pathway generating TMA and malic semialdehyde from L-carnitine hydroxylation are found in human colon isolates belonging to the genera *Acinetobacter* and *Serratia* (Tang et al., 2019). Two specific genes, CntA and CntB, encode oxidoreductases (CntA encodes the carnitine oxidase, and CntB encodes a reductase) from L-carnitine, which later converts to TMA (Warrier et al., 2015). An alternative pathway converting carnitine to TMA involves the production of γ-butyrobetaine within the ileum with subsequent generation of γ-butyrobetaine, leading to TMA as its end product in the cecum and colon (Tang et al., 2019). Additional findings suggest the role of YeaW/YeaX genes, which encode for oxygenase and oxidoreductase to form a substrate that possesses promiscuity for γ-butyrobetaine, L-carnitine, choline, and betaine. Orthologs and homologs of CntA/CntB and YeaW/YeaX are located depending on the type of gut microbiota, such as Gammaproteobacteria (*E. coli*, *Citrobacter*, *Klebsiella pneumoniae*, *Providencia*, and *Shigella*), Firmicutes (*Sporosarcina*), Actinobacteria, and Betaproteobacteria (*Achromobacter*) (Zeisel and Warrier, 2017).

TMA is also obtained from bacterial metabolism of dietary betaine. Foods that include the fundamental betaine are of plant origin, such as spinach, beets, and grains that are grown in high osmotic environments (Janeiro et al., 2018). Plants use betaine in their roots as an osmolyte to help them continue intracellular water. The U.S. Department of Agriculture maintains a database on the betaine content in common foods, which can be used to calculate dietary betaine intake. For example, 100 g of spinach contains 577 mg of betaine, 100 g of cod or red meat contains 7 mg of betaine, and 100 g of eggs contains <1 mg of betaine. A recent study observed that men and women with greater Firmicutes to Bacteroidetes enrichment generated extra TMA after a mission meal (A mission meal is part of the experiment to achieve the goal.) (Cho et al., 2017). The elucidation of microbial TMA-generating pathways is ongoing, and a part of TMA-producing microbial metabolism is still to be determined. In foods such as meat, fish, and eggs, choline and carnitine are the principal natural sources of TMAO. Metabolized with the aid of bacteria within the intestine, choline and carnitine are shaped into TMA, which are then absorbed into the blood and transformed into TMAO by way of oxidizing flavin monooxygenase enzymes in the liver (Chou et al., 2019).

10.2.1.5 Trimethylamine-N-oxide and cardiovascular disease

A close relationship between the gut microbe–dependent production of TMAO, derived from specific dietary nutrients like choline and carnitine, and subsequent cardiovascular events is well known (Janeiro et al., 2018; Warrier et al., 2015). TMA, which is produced by the gut microbial enzymes TMA lyases, is a precursor of TMAO. TMAO levels can be measured by liquid chromatography–mass spectrometry. Elevated levels of TMAO in the blood are directly linked to CVDs, such as coronary artery disease (CAD) and acute or chronic heart disease (Warrier et al., 2015; Zeisel and Warrier, 2017). Tang et al. (2019) investigated the close association between fasting plasma levels and TMAO and subsequent adverse cardiovascular events (death, myocardial infarction, or stroke) for 3 years. In this follow-up study, 4007 patients were examined under elective cardiac catheterization, and it was found that the highest quartile for circulating TMAO levels had a 2.5-fold increased risk of major adverse cardiovascular events compared with the normal patients. Even after adjustment and treatment for traditional risk factors, an elevated TMAO level was predictive of an increased risk of major adverse CVDs and its pathophysiology (Warrier et al., 2015). Additionally, high concentrations of TMAO were observed in patients with stable heart disease when compared to healthy subjects. Such clinical evidence supports the notion that the gut microbiome may play a task in the development and progression of heart disease. These studies also showed elevated TMAO levels to be correlated to a 2.2-fold increase in the mortality rate of subjects even after an adjustment for traditional risk factors. Moreover, the blood TMAO levels were related to coronary plaque vulnerability, as assessed by optical coherence tomography, and therefore with long-term risks of cardiovascular events in patients with acute coronary syndrome (Wang and Zhao, 2018). The newest metagenome-wide association study demonstrated the microbial characterization of CAD patients and showed that the gut microbial enzymes that produce TMA were enriched within the patients with CAD compared to the healthy controls (Yang et al., 2019). TMAO regulates multiple aspects of cholesterol and sterol metabolism. Mice fed a diet supplemented with choline (or L-carnitine) or TMAO showed significant decreases ($\sim$30% and $\sim$16%, respectively) in reverse cholesterol transport (RCT), which is the movement of cholesterol from peripheral tissues back to the liver via the plasma (Chou et al., 2019). Suppression of intestinal microbiota (and therefore plasma TMAO concentrations) completely inhibited diet-dependent reductions in RCT. Studies also suggest that an increased TMAO level activates the NF-kappa B (NF-κB) expression pathway and downstream proinflammatory events by inducing inflammatory cytokines, adhesion molecules, and chemokines (Brufau et al., 2011). Studies performed by Sun et al. (2017) showed the relevance of oxidative stress and NLRP3 inflammasome activation for TMAO levels, as inflammatory cytokines like IL-18 and IL-1β were subsequently elevated. Additionally, TMAO contributed to platelet hyperreactivity and thrombosis (Brufau et al., 2011). In human clinical studies, elevated TMAO levels were related to an increased risk of CVD. Moreover, prospective cohort research findings have shown elevated plasma TMAO levels to increased the risk of major cardiovascular events such as myocardial infarction, stroke, and death (Ma et al., 2021). There is accumulating evidence that suggests a close association between TMAO and atherosclerotic potential. This postulates that TMAO causes atherosclerosis via three primary events:

1. CVD and stroke patients have high plasma TMAO levels and higher concentrations of TMAO. These conditions extensively suggest elevation in primary metabolites such as choline, betaine, and carnitine in plasma (Zeisel and Warrier, 2017; Janeiro et al., 2018).
2. Apolipoprotein E knockout mice, when treated on with TMAO precursors, develop chronic atherosclerosis (Haghikia et al., 2018).
3. Mouse peritoneal macrophages, when incubated with TMA, express cell surface receptors, scavenger receptors by cholesterol accumulation, and subsequent foam cell formation (Geng et al., 2018; Singh et al., 2019).

Choline and betaine are precursors for TMA formation by gut bacteria, and studies connect an increase in these two metabolites to loss of kidney function. Additionally, plasma concentrations of the TMAO precursor L-carnitine in subjects undergoing cardiac evaluation (n = 2,595) predicted increased risks for CVD and major adverse cardiac events (myocardial infarction, stroke, and death) (Zeisel and Warrier, 2017).

10.2.2 Other metabolites of clinical significance

Numerous metabolites are produced in the gut microbiome that are comparatively less recognized for their clinical implications in support of their proven role in CVD pathophysiology. However, basic science still promises their clinical significance to mark their extent in various forms of diseases directly or indirectly. Nevertheless, they have an increased effect on host physiology and metabolism via different signaling mediators.

10.2.2.1 Long-chain fatty acids

Long-chain fatty acids (LCFAs) are mainly produced from diet and are metabolized in the body to generate bioactive lipid mediators. These have been shown to have an effective role in reducing BP (via arachidonic acid) and modulating inflammatory events (Shibata et al., 2017). Accumulating evidence suggests that microbially generated LCFAs could be used for therapy as postbiotics.

10.2.2.2 Ethylphenylsulfate

Ethylphenylsulfate (4-EPS) is a bacterial fermentation product, also known as a uremic toxin. Specific pathogen-free and germ-free mice naturally possess very low concentrations of 4-EPS. The interest in this metabolite originated from a study that focused on maternal immune activation, the mouse model of autism spectrum disorder (ASD) (Vuillermot et al., 2017). This study showed elevated 4-EPS levels to be correlated with behavioral abnormalities. These altered pathologies were then observed to be ameliorated through treatment with *Bacteroides fragilis*, indicating the possibility of utilizing probiotic and postbiotic therapies as treatment methods for ASD (Descamps et al., 2019).

10.2.2.3 Imidazole propionate

Imidazole propionate is a histidine-derived metabolite. It has been identified to play a significant role in the pathological events of type 2 diabetes. Elevated concentrations of this metabolite tend to be a marker of impaired insulin signaling and thus acts by activating p38γ, p62, and mTORC1 in individuals with type 2 diabetes (Koh et al., 2020).

10.2.2.4 Desaminotyrosine

Desaminotyrosine (DAT) is a degradation product of tyrosine that is found to be naturally present in the urine of mammals. A recent research study showed the presence of *Clostridium orbiscendens* in an intact gut microbiome to be an essential requirement for detectable DAT levels in the serum and feces (Descamps et al., 2019). A reduction in DAT levels is associated with increased susceptibility to viral infection because of suboptimal amplification of interferon type 1 signaling.

10.2.2.5 Taurine

Taurine is an amino sulfonic acid. Commensal bacteria produce taurine by deconjugation of primary BA. Taurine was observed to induce NLRP6 inflammasome signaling and so has been utilized as a protective metabolite for colorectal cancer and diabetes (Wakamatsu et al., 2018; Yang and Cong, 2021). This was also observed as a risk factor for colorectal cancer, by mediating oxidative stress, primary and secondary bile acid pools, and an inflammation marker. In some recent clinical studies, taurine was administered to treat vascular complications, mainly diabetes. The protective efficacy of a ketogenic diet (reduced carbohydrates and a high-fat diet) are effective because of the presence of the gut microbiome, which plays an important role in the reduction of amino acid gamma-glutamylation and increased GABA/glutamate concentration in the brain (Kumar et al., 2020).

10.2.2.6 Polyamines

Polyamines such as spermine have been implicated in the regulation of inflammation pathways and induction of inflammatory bowel diseases, such as colitis (Kumar et al., 2020). These metabolites are generated either by the microbiota or by the host via diet and have pleiotropic effects on multiple tissues. Various members of the *Lactobacillus* family mediate polyamine synthesis via the biosynthetic spermine pathway that reduces colonic IL-18 expression and suppresses aggregation of NLRP6 inflammasome (Hocher and Adamski, 2017). These macrobiotic polyamines have also been found to be effective in regulating epithelial circadian rhythms.

10.2.2.7 Retinoic acid

Retinoic acid in the gut is maintained by *Clostridia* bacteria by suppression of retinol dehydrogenase 7 in intestinal epithelial cells A (Grizotte-Lake et al., 2018). Retinoic acid is a lipid metabolite that is derived from dietary vitamin A and has been implicated largely in the development and regulation of the immune system through IgA production. Also, retinoic acid maintains T-cell immunity through regulatory T cell differentiation and T helper 17 cell control. A deficiency in retinoic acid or its receptors has been associated with diseases such as colitis-associated colon cancer and inflammatory bowel disease (Neurath, 2019).

10.2.2.8 Bile acids

The primary BA, chenodeoxycholic acid and cholic acid, are synthesized in the liver from cholesterol by the hepatocytes (Johansson et al., 2021). In the small and large intestines the bacterial deconjugation, dehydrogenation, 7α-dehydroxylation, and epimerization of the primary BAs leads to the production of the secondary BAs. BAs play a key part in maintaining cholesterol homeostasis through the retention and emulsification of lipids. Microscopic organisms with bile salt hydrolases (*Lactobacillus* and *Clostridium*) encourage different aspects of BA digestion, including deconjugation, dehydrogenation, and dihydroxylation (Staley et al., 2017). Germ-free mouse models display higher by and large BA levels, for the most part by ursodeoxycholic acids and gallbladders by attenuation of the transmembrane G protein-coupled receptor 5 pathway (Jia et al., 2020). Presentation to cold temperatures in mice triggers the change of cholesterol to BAs within the liver, modifies intestine microbiome composition, fortifies lipoprotein handling in brown fat tissue, and generates thermal heat (Keipert et al., 2020). Recent studies have indicated that ursodeoxycholic acid improves insulin sensitivity and promotes weight loss in patients with type 2 diabetes. It is also known to be an antibacterial agent (Wang et al., 2019).

10.2.2.9 Flavonoids

Flavonoids are polyphenols obtained from plants as secondary metabolites. There are a large number of (over 50,000) naturally occurring flavonoids that have been characterized by various plants that are known to have a vast range of biological activities, including antioxidant, anticarcinogenic, and protective effects on different pathological diseases (Batiha et al., 2020). Members of colonic microbiota (*Bacteroides distasonis* and *Bacteroides uniformis*) are responsible for mediating fermentation and hydrolysis of ingested flavonoids (Descamps et al., 2019). Flavonoids being an antioxidant agent have disease-ameliorating property, several studies have shown the efficacy of flavonoids in altering lipid levels in hepatic steatosis, hyperlipidemia and insulin resistance in high-fat diet-fed mouse models (Wang et al., 2020). These flavonoids possess metabolizing activity in the gut microbiome, which becomes physiologically relevant in the postobesity situation Flavonoids such as luteolin, epigallocatechin gallate, hesperidin, and quercetin are powerful antioxidants that can inhibit inflammatory mediators (Nagula and Wairkar, 2019). Similar studies with human subjects reported decreased levels of IL-6 and TNF as well as improvements in social skills after treatment with flavonoids or flavonoid formulations (Descamps et al., 2019). Indeed, flavonoids seem to have beneficial effects across various types of disorders and diseases of the brain as well. For example, flavonoids have shown promising results in Alzheimer's disease by inhibiting amyloid-β oligomerization in Tg2576 mice and tau peptide aggregation (Prasanna and Upadhyay, 2021)

10.2.2.10 N-acyl amides

N-acyl amides are bioactive lipids that are known to have multiple roles in maintaining mammalian physiology (Cohen et al., 2017). N-acyl-3-hydroxy palmitoyl glycine, also known as commendamide, is one such metabolite from the human gut microbiome that was identified by functional metagenomic study for NF-kB activators (Lynch et al., 2017). It also resembles long-chain N-acyl-amides and has been shown to act as a G-protein-coupled receptor G2A/GPR132, modulating autoimmunity and atherosclerosis (Rodriguez et al., 2021). The microbial lipid N-acyl serinol is also similar to a human ligand for GPR119, which is a stimulator of glucagon-like peptide 1, which is produced by the intestinal L-cells and regulates metabolic hormone activity and glucose homeostasis. Similar results were observed in a mouse model when N-acyl serinol was administered to the diet of the mice (Cohen et al., 2017).

10.3 Conclusions

Overall, large-scale genome studies have revealed numerous genetic contributions to modern human diseases, such as metabolic, inflammatory, cardiovascular, and neurodegenerative disorders. As these genome studies reach their saturation curve concerning genetic associations, this modern era of biology is indicating lifestyle factors that play a major role in the pathogenesis of such life-threatening diseases. A significant component of lifestyle is its association with the external environment or environmental factors. One such factor is represented in the form of diet and its symbiotic association with microbes that reside in the gastrointestinal tract. The microbiome has been recognized as a major determinant in human disease, and the mechanistic approach by which these microorganisms contribute to host health has been extensively investigated. Commensal microorganisms have direct engagement to the host through physical interactions or through the expression of metabolites that gain access to the systemic circulation and thereby affect host physiology and disease. Recent technical advancements in the field of metabolomics coupled with mechanistic studies in human and animal models have enabled the characterization of human disease–associated gut microorganisms and their significant metabolites. Given the pervasiveness and pleiotropy of the effects of metabolites on human physiology, the translational potential of these molecules for postbiotic therapies offers enormous benefits to modern medicine. Together, these studies indicate that modulating the microbiome might be beneficial for a wide range of conditions.

References

Adesso, S., Popolo, A., Bianco, G., Sorrentino, R., Pinto, A., Autore, G., et al., 2013. The uremic toxin indoxyl sulphate enhances macrophage response to LPS. PLoS One 8 (9), e76778.

Batiha, G.E.S., Beshbishy, A.M., Ikram, M., Mulla, Z.S., El-Hack, M.E.A., Taha, A.E., et al., 2020. The pharmacological activity, biochemical properties, and pharmacokinetics of the major natural polyphenolic flavonoid: quercetin. Foods 9 (3), 374.

Brahe, L.K., Astrup, A., Larsen, L.H., 2013. Is butyrate the link between diet, intestinal microbiota and obesity-related metabolic diseases? Obes. Rev. 14 (12), 950–959.

Brown, J.M., Hazen, S.L., 2018. Microbial modulation of cardiovascular disease. Nat. Rev. Microbiol. 16 (3), 171.

Brufau, G., Groen, A.K., Kuipers, F., 2011. Reverse cholesterol transport revisited: contribution of biliary vs intestinal cholesterol excretion. Arterioscler. Thromb. Vasc. Biol. 31 (8), 1726–1733.

Chambers, E.S., Preston, T., Frost, G., Morrison, D.J., 2018. Role of gut microbiota-generated short-chain fatty acids in metabolic and cardiovascular health. Curr. Nutr. Rep. 7 (4), 198–206.

Cho, C.E., Taesuwan, S., Malysheva, O.V., Bender, E., Tulchinsky, N.F., Yan, J., et al., 2017. Trimethylamine-N-oxide (TMAO) response to animal source foods varies among healthy young men and is influenced by their gut microbiota composition: a randomized controlled trial. Mol. Nutr. Food Res. 61 (1), 1600324.

Chou, R.H., Chen, C.Y., Chen, I.C., Huang, H.L., Lu, Y.W., Kuo, C.S., et al., 2019. Trimethylamine N-oxide, circulating endothelial progenitor cells, and endothelial function in patients with stable angina. Sci. Rep. 9 (1), 1–10.

Cohen, L.J., Esterhazy, D., Kim, S.H., Lemetre, C., Aguilar, R.R., Gordon, E.A., et al., 2017. Commensal bacteria make GPCR ligands that mimic human signalling molecules. Nature 549 (7670), 48–53.

Damms-Machado, A., Mitra, S., Schollenberger, A.E., Kramer, K.M., Meile, T., Königsrainer, A., et al., 2015. Effects of surgical and dietary weight loss therapy for obesity on gut microbiota composition and nutrient absorption. BioMed. Res. Int. 2015.

de Oliveira, S.P.A., do Nascimento, H.M.A., Sampaio, K.B., de Souza, E.L., 2020. A review on bioactive compounds of beet (Beta vulgaris L. subsp. vulgaris) with special emphasis on their beneficial effects on gut microbiota and gastrointestinal health. Crit. Rev. Food Sci. Nutr. 1–12.

Descamps, H.C., Herrmann, B., Wiredu, D., Thaiss, C.A., 2019. The path toward using microbial metabolites as therapies. EBioMedicine 44, 747–754.

Drüeke, T.B., Massy, Z.A., 2016. Changing bone patterns with progression of chronic kidney disease. Kidney Int. 89 (2), 289–302.

Feng, W., Ao, H., Peng, C., 2018. Gut microbiota, short-chain fatty acids, and herbal medicines. Front. Pharmacol. 9, 1354.

Fitsialos, G., Chassot, A.A., Turchi, L., Dayem, M.A., LeBrigand, K., Moreilhon, C., et al., 2007. Transcriptional signature of epidermal keratinocytes subjected to in vitro scratch wounding reveals selective roles for ERK1/2, p38, and phosphatidylinositol 3-kinase signaling pathways. J. Biol. Chem. 282 (20), 15090–15102.

Fujii, H., Nishijima, F., Goto, S., Sugano, M., Yamato, H., Kitazawa, R., et al., 2009. Oral charcoal adsorbent (AST-120) prevents progression of cardiac damage in chronic kidney disease through suppression of oxidative stress. Nephrol. Dialysis Transplant. 24 (7), 2089–2095.

Gabryšová, L., Howes, A., Saraiva, M., O'Garra, A., 2014. The regulation of IL-10 expression. Interleukin-10 Health Dis. 157–190.

Geirnaert, A., Calatayud, M., Grootaert, C., Laukens, D., Devriese, S., Smagghe, G., et al., 2017. Butyrate-producing bacteria supplemented in vitro to Crohn's disease patient microbiota increased butyrate production and enhanced intestinal epithelial barrier integrity. Sci. Rep. 7 (1), 1–14.

Gelasco, A.K., Raymond, J.R., 2006. Indoxyl sulfate induces complex redox alterations in mesangial cells. Am. J. Physiol.-Renal Physiol. 290, F1551-8.

Geng, J., Yang, C., Wang, B., Zhang, X., Hu, T., Gu, Y., et al., 2018. Trimethylamine N-oxide promotes atherosclerosis via CD36-dependent MAPK/JNK pathway. Biomed. Pharmacother. 97, 941–947.

Grizotte-Lake, M., Zhong, G., Duncan, K., Kirkwood, J., Iyer, N., Smolenski, I., et al., 2018. Commensals suppress intestinal epithelial cell retinoic acid synthesis to regulate interleukin-22 activity and prevent microbial dysbiosis. Immunity 49 (6), 1103–1115.

Haghikia, A., Li, X.S., Liman, T.G., Bledau, N., Schmidt, D., Zimmermann, F., et al., 2018. Gut microbiota–dependent trimethylamine N-oxide predicts risk of cardiovascular events in patients with stroke and is related to proinflammatory monocytes. Arterioscler. Thromb. Vasc. Biol. 38 (9), 2225–2235.

Harbison, J.E., Roth-Schulze, A.J., Giles, L.C., Tran, C.D., Ngui, K.M., Penno, M.A., et al., 2019. Gut microbiome dysbiosis and increased intestinal permeability in children with islet autoimmunity and type 1 diabetes: a prospective cohort study. Pediatric Diabetes 20 (5), 574–583.

Hayashi, T., Yamashita, T., Watanabe, H., Kami, K., Yoshida, N., Tabata, T., et al., 2018. Gut microbiome and plasma microbiome-related metabolites in patients with decompensated and compensated heart failure. Circ. J. 83 (1), 182–192.

Hirakawa, H., Hayashi-Nishino, M., Yamaguchi, A., Nishino, K., 2010. Indole enhances acid resistance in Escherichia coli. Microb. Pathog. 49 (3), 90–94.

Hocher, B., Adamski, J., 2017. Metabolomics for clinical use and research in chronic kidney disease. Nat. Rev. Nephrol. 13 (5), 269–284.

Jameson, E., Quareshy, M., Chen, Y., 2018. Methodological considerations for the identification of choline and carnitine-degrading bacteria in the gut. Methods 149, 42–48.

Janeiro, M.H., Ramírez, M.J., Milagro, F.I., Martínez, J.A., Solas, M., 2018. Implication of trimethylamine N-oxide (TMAO) in disease: potential biomarker or new therapeutic target. Nutrients 10 (10), 1398.

Jia, W., Wei, M., Rajani, C., Zheng, X., 2020. Targeting the alternative bile acid synthetic pathway for metabolic diseases. Protein Cell 1–15.

Johansson, H., Søndergaard, J.N., Jorns, C., Kutter, C., Ellis, E., 2021. Chenodeoxycholic acid modulates bile acid synthesis independent of Fibroblast growth factor 19 in primary human hepatocytes. Front. Endocrinol. 11, 1106.

Kasubuchi, M., Hasegawa, S., Hiramatsu, T., Ichimura, A., Kimura, I., 2015. Dietary gut microbial metabolites, short-chain fatty acids, and host metabolic regulation. Nutrients 7 (4), 2839–2849.

Keipert, S., Lutter, D., Schroeder, B.O., Brandt, D., Ståhlman, M., Schwarzmayr, T., et al., 2020. Endogenous FGF21-signaling controls paradoxical obesity resistance of UCP1-deficient mice. Nat. Commun. 11 (1), 1–12.

Koh, A., Mannerås-Holm, L., Yunn, N.O., Nilsson, P.M., Ryu, S.H., Molinaro, A., et al., 2020. Microbial imidazole propionate affects responses to metformin through p38γ-dependent inhibitory AMPK phosphorylation. Cell Metab. 32 (4), 643–653.

Kumar, M., Singh, P., Murugesan, S., Vetizou, M., McCulloch, J., Badger, J.H., et al., 2020. Microbiome as an immunological modifier. Biomarkers for Immunotherapy of Cancer. Humana, New York, NY, pp. 595–638.

Kuryłowicz, A., Koźniewski, K., 2020. Anti-inflammatory strategies targeting metaflammation in type 2 diabetes. Molecules 25 (9), 2224.

Li, J., Zhao, F., Wang, Y., Chen, J., Tao, J., Tian, G., et al., 2017. Gut microbiota dysbiosis contributes to the development of hypertension. Microbiome 5 (1), 1–19.

Lynch, A., Crowley, E., Casey, E., Cano, R., Shanahan, R., McGlacken, G., et al., 2017. The Bacteroidales produce an N-acylated derivative of glycine with both cholesterol-solubilising and hemolytic activity. Sci. Rep. 7 (1), 1–10.

Ma, Z., Taylor, R.A., 2020. Human reproductive system microbiomes exhibited significantly different heterogeneity scaling with gut microbiome, but the intra-system scaling is invariant. Oikos 129 (6), 903–911.

Ma, R., Fu, W., Zhang, J. Hu, X., Yang, J., Jiang, H., 2021. TMAO: a potential mediator of clopidogrel resistance. Sci. Rep. 11 (1), 1–7.

Mañas, M., de Victoria, E.M., Gil, A., DYago, M., Mathers, J.C., 2011. The gastrointestinal tract. Susan A Lanham-New Ian A Macdonald, p. 205.

Marques, F.Z., Jama, H.A., Tsyganov, K., Gill, P.A., Rhys-Jones, D., Muralitharan, R.R., et al., 2019. Guidelines for transparency on gut microbiome studies in essential and experimental hypertension. Hypertension 74 (6), 1279–1293.

Meyer, K.A., Shea, J.W., 2017. Dietary choline and betaine and risk of CVD: a systematic review and meta-analysis of prospective studies. Nutrients 9 (7), 711.

Mishra S.P., Karunakar, P., Taraphder, S., Yadav, H., 2020. Free fatty acid receptors 2 and 3 as microbial metabolite sensors to shape host health: pharmacophysiological view. Biomedicines 8 (6), 154.

Mu, Q. Kirby, J., Reilly, C.M., Luo, X.M., 2017. Leaky gut as a danger signal for autoimmune diseases. Front. Immunol. 8, 598.

Nagula R.L., Wairkar, S., 2019. Recent advances in topical delivery of flavonoids: a review. J. Cont. Rel. 296, 190–201.

Nakamura, N., Lin, H.C., McSweeney, C.S., Mackie, R.I., Gaskins, H.R., 2010. Mechanisms of microbial hydrogen disposal in the human colon and implications for health and disease. Annu. Rev. Food Sci. Technol. 1, 363–395.

Nallu, A., Sharma, S., Ramezani, A., Muralidharan, J., Raj, D., 2017. Gut microbiome in chronic kidney disease: challenges and opportunities. Transl. Res. 179, 24–37.

Neurath, M.F., 2019. Targeting immune cell circuits and trafficking in inflammatory bowel disease. Nat. Immunol. 20 (8), 970–979.

Okumura, R., Takeda, K., 2018. Maintenance of intestinal homeostasis by mucosal barriers. Inflamm. Regen. 38 (1), 1–8.

Panda, S., Casellas, F., Vivancos, J.L., Cors, M.G., Santiago, A., Cuenca, S., et al., 2014. Short-term effect of antibiotics on human gut microbiota. PLoS One 9 (4), e95476.

Pieniazek, A., Gwozdzinski, L., Hikisz, P., Gwozdzinski, K., 2018. Indoxyl sulfate generates free radicals, decreases antioxidant defense, and leads to damage to mononuclear blood cells. Chem. Res. Toxicol. 31 (9), 869–875.

Postler, T.S., Ghosh, S., 2017. Understanding the holobiont: how microbial metabolites affect human health and shape the immune system. Cell Metab. 26 (1), 110–130.

Prasanna, P., Upadhyay, A., 2021. Flavonoid-based nanomedicines in Alzheimer's disease therapeutics: promises made, a long way to go. ACS Pharmacol. Transl. Sci. 4 (1), 74–95.

Ríos-Covián, D., Ruas-Madiedo, P., Margolles, A., Gueimonde, M., De Los Reyes-gavilán, C.G., Salazar, N., 2016. Intestinal short chain fatty acids and their link with diet and human health. Front. Microbiol. 7, 185.

Rodriguez, J., Olivares, M., Delzenne, N.M., 2021. Implication of the gut microbiota in metabolic inflammation associated with nutritional disorders and obesity. Mol. Nutr. Food Res. 65 (1), 1900481.

Saltzman, E.T., Palacios, T., Thomsen, M., Vitetta, L., 2018. Intestinal microbiome shifts, dysbiosis, inflammation, and non-alcoholic fatty liver disease. Front. Microbiol. 9, 61.

Shibata, N., Kunisawa, J., Kiyono, H., 2017. Dietary and microbial metabolites in the regulation of host immunity. Front. Microbiol. 8, 2171.

Shin, N.R., Whon, T.W., Bae, J.W., 2015. Proteobacteria: microbial signature of dysbiosis in gut microbiota. Trends Biotechnol. 33 (9), 496–503.

Silva, J.P., Navegantes-Lima, K.C., Oliveira, A.L., Rodrigues, D.V., Gaspar, S.L., Monteiro, V.V., et al., 2018. Protective mechanisms of butyrate on inflammatory bowel disease. Curr. Pharm. Des. 24 (35), 4154–4166.

Singh, A., Gowtham, S., Chakrapani, L.N., Ashokkumar, S., Kumar, S.K., Prema, V., et al., 2018. Aegeline vs Statin in the treatment of hypercholesterolemia: a comprehensive study in rat model of liver steatosis. Funct. Foods Health Dis. 8 (1), 1–16.

Singh, A., Srinivasan, A.K., Chakrapani, L.N., Kalaiselvi, P., 2019. LOX-1, the common therapeutic target in hypercholesterolemia: a new perspective of antiatherosclerotic action of aegeline. Oxid. Med. Cell. Longev. 2019.

Sircana, A., De Michieli, F., Parente, R., Framarin, L., Leone, N., Berrutti, M., et al., 2019. Gut microbiota, hypertension, and chronic kidney disease: recent advances. Pharmacol. Res. 144, 390–408.

Staley, C., Weingarden, A.R., Khoruts, A., Sadowsky, M.J., 2017. Interaction of gut microbiota with bile acid metabolism and its influence on disease states. Appl. Microbiol. Biotechnol. 101 (1), 47–64.

Summers, S., Quimby, J.M., Phillips, R.K., Stockman, J., Isaiah, A., Lidbury, J.A., et al., 2020. Preliminary evaluation of fecal fatty acid concentrations in cats with chronic kidney disease and correlation with indoxyl sulfate and p-cresol sulfate. J. Vet. Intern. Med. 34 (1), 206–215.

Sun, M., Wu, W., Liu, Z., Cong, Y., 2017. Microbiota metabolite short chain fatty acids, GPCR, and inflammatory bowel diseases. J. Gastroenterol. 52 (1), 1–8.

Tang, W.W., Bäckhed, F., Landmesser, U., Hazen, S.L., 2019. Intestinal microbiota in cardiovascular health and disease: JACC state-of-the-art review. J. Am. Coll. Cardiol. 73 (16), 2089–2105.

Ververis, K., Hiong, A., Karagiannis, T.C., Licciardi, P.V., 2013. Histone deacetylase inhibitors (HDACIs): multitargeted anticancer agents. Biol.: Targets Ther. 7, 47.

Vuillermot, S., Luan, W., Meyer, U., Eyles, D., 2017. Vitamin D treatment during pregnancy prevents autism-related phenotypes in a mouse model of maternal immune activation. Mol. Autism 8 (1), 1–13.

Wakamatsu, T., Yamamoto, S., Ito, T., Sato, Y., Matsuo, K., Takahashi, Y., et al., 2018. Indoxyl sulfate promotes macrophage IL-1β production by activating aryl hydrocarbon receptor/NF-κ/MAPK cascades, but the NLRP3 inflammasome was not activated. Toxins 10 (3), 124.

Wang, Z., Zhao, Y., 2018. Gut microbiota derived metabolites in cardiovascular health and disease. Protein Cell 9 (5), 416–431.

Wang, C., Zhu, C., Shao, L., Ye, J., Shen, Y., Ren, Y., 2019. Role of bile acids in dysbiosis and treatment of nonalcoholic fatty liver disease. Mediators Inflamm. 2019.

Wang, S.W., Sheng, H., Bai, Y.F., Weng, Y.Y., Fan, X.Y., Lou, L.J., et al., 2020. Neohesperidin enhances PGC-1α-mediated mitochondrial biogenesis and alleviates hepatic steatosis in high fat diet fed mice. Nutr. Diabetes 10 (1), 1–11.

Warrier, M., Shih, D.M., Burrows, A.C., Ferguson, D., Gromovsky, A.D., Brown, A.L., et al., 2015. The TMAO-generating enzyme flavin monooxygenase 3 is a central regulator of cholesterol balance. Cell Rep. 10 (3), 326–338.

Yanai, A., Uchiyama, K., Ishibashi, Y., 2019. Uremic encephalopathy in patients undergoing assisted peritoneal dialysis: a case series and literature review. CEN. Case Rep. 8 (4), 271–279.

Yang, W., Cong, Y., 2021. Gut microbiota-derived metabolites in the regulation of host immune responses and immune-related inflammatory diseases. Cell. Mol. Immunol. 1–12.

Yang, T., Santisteban, M.M., Rodriguez, V., Li, E., Ahmari, N., Carvajal, J.M., et al., 2015a. Gut dysbiosis is linked to hypertension. Hypertension 65 (6), 1331–1340.

Yang, K., Wang, C., Nie, L., Zhao, X., Gu, J., Guan, X., et al., 2015b. Klotho protects against indoxyl sulphate-induced myocardial hypertrophy. J. Am. Soc. Nephrol. 26 (10), 2434–2446.

Yang, Y., Wang, Y., Sun, J., Zhang, J., Guo, H., et al., 2019. Dietary methionine restriction reduces hepatic steatosis and oxidative stress in high-fat-fed mice by promoting H 2 S production. Food & Function 10 (1), 61–77.

Yoshida, N., Yamashita, T., Hirata, K.I., 2018. Gut microbiome and cardiovascular diseases. Diseases 6 (3), 56.

Yu, M., Kim, Y.J., Kang, D.H., 2011. Indoxyl sulfate–induced endothelial dysfunction in patients with chronic kidney disease via an induction of oxidative stress. Clin. J. Am. Soc. Nephrol. 6 (1), 30–39.

Zeisel, S.H., Warrier, M., 2017. Trimethylamine N-oxide, the microbiome, and heart and kidney disease. Annu. Rev. Nutr. 37, 157–181.

Chapter 11

Beneficial role of gut microbiome in metabolic syndrome, obesity, and cardiovascular diseases

Pradipta Banerjee[1,2], Sriya Choudhury[3], Komal Jalan[4], Amitava Das[2], Nandini Ghosh[2] and Debasis Bagchi[5,6,7]
[1]*Department of Biochemistry and Plant Physiology, Centurion University of Technology and Management, Parlakhemundi, Odisha, India,* [2]*Department of Surgery, Indiana University School of Medicine, Indianapolis, IN, United States,* [3]*Department of Biochemistry, Institute of Biomedical Sciences, The West Bengal University of Health Sciences, Kolkata, West Bengal, India,* [4]*Department of Agriculture, University of Calcutta, Kolkata, West Bengal, India,* [5]*College of Pharmacy and Health Sciences, Texas Southern University, Houston, TX, United States,* [6]*Department of Biology, Adelphi University, Garden City, NY, United States,* [7]*Department of R&D, Victory Nutrition International, Inc., Bonita Springs, FL, United States*

11.1 Introduction

11.1.1 Gut microbiome

The microbiome is a diverse community of bacteria, archaea, fungi, protozoa, viruses, and their collective genome that live on and within the body. In the last 15 years, scientists have been able to describe the structure and function of the intestinal microbiota as a result of the development of advanced genetic sequencing methods and bioinformatics platforms. The microbiome is dynamic and alters throughout the host's life in response to several factors, such as diet, climate, medical treatments, and associated diseases. Gut microorganisms are known to play a role in energy absorption, storage, and consumption following dietary intake.

11.1.2 Dietary intake for reducing the risk factor of metabolic syndrome, obesity, and cardiovascular disease

The use of bioactive dietary components such as prebiotics and probiotics to manipulate the gut microbiota can help to reduce the risk of metabolic syndrome (MetS), obesity, and cardiovascular disease (CVD). The risk factor of these diseases can be reduced by ingestion of certain foodstuffs that pass through the colon intact and maybe fermented directly in the hindgut. Such foodstuffs are termed prebiotics. Prebiotics and dietary fiber increase the relative abundance of beneficial anaerobic bacteria, increase butyrate fermentation, and have favorable metabolic effects. The use of probiotics as live microbial feed supplements has received the most attention in this regard.

Daily consumption of dietary fibers, vegetables, and legumes has also been shown to have an inverse association with metabolic dysregulation and its effects. Flavonols, catechins, phenolic acids, and flavones are dietary polyphenols with active antioxidant properties. Common foods with high polyphenol content include seeds, fruits, tea, cocoa products, and wine.

Disturbances in the intestinal microbiota can cause cellular dysfunction, metabolic changes, fat metabolism, adipose tissue formation, and inflammatory processes (gut dysbiosis) (Sun et al., 2020). By altering the host's immune responses and energy homeostasis, gut dysbiosis is one of the significant factors associated with the pathogenesis of GI disorders and MetS, which can result in the upstream control of inflammatory cascades, insulin resistance, and failure of the body's metabolism. Researchers also clarified the direct correlation between GI and metabolic disorders in addition to gut dysbiosis. Modification of the gut microbiota may be a target for preventing or reversing MetS in people and could offer a new way to treat a variety of chronic diseases. Individual foods have been shown to promote the growth of

Microbiome, Immunity, Digestive Health and Nutrition. DOI: https://doi.org/10.1016/B978-0-12-822238-6.00018-2

particular bacterial taxa, suggesting that intestinal bacteria could be guiding our food tastes, appetite, and feelings of fullness after eating (Fetissov, 2017). The purpose of this work was to highlight the close relationship between gut dysbiosis, GI disorders, and MetS as well as the role of bioactive dietary components in influencing a variety of pathways involved in the treatment of both disorders. Table 11.1 summarizes the benefits of dietary ingredients such as dietary fibers, prebiotics, probiotics, polyunsaturated fatty acids, and polyphenols.

11.2 Metabolic syndrome

MetS is a group of medical conditions that elevate the risk factor for CVDs and other health complications, such as stroke and type 2 diabetes. It is a condition of low-grade chronic inflammation, which includes serious diseases such as hyperglycemia, hyperlipidemia, hypertension, and obesity (de Toro-Martín et al., 2017)

MetS includes a congregation of at least three medical conditions among the following list of health issues: high blood pressure, high blood sugar, obesity, high level of serum triglycerides (TG), and low levels of serum high-density lipoproteins (HDL). MetS is a group of metabolic risk factors that come together in a single person. The World Health Organization (WHO) was the first organization to tie together the different metabolic risk factors associated with MetS and define this medical condition. There are some conflicts of interest among different organizations in defining MetS. The WHO definition makes it mandatory for insulin resistance to be present among the list of risk factors for the diagnosis of MetS in a patient (Huang, 2009). By contrast, the International Diabetes Foundation mandates obesity, but not necessarily insulin resistance, as a primary criterion for MetS.

The various conditions that are involved in MetS express themselves differently in different patients; hence the underlying pathophysiological state varies among patients. The mechanisms of pathways involved in MetS are quite complex and are still under investigation. The signs and symptoms do not manifest themselves in any particular order. Thus establishing a cause-effect relationship becomes difficult. The most important possible influential factors for MetS are genetics, maternal health, age, and ethnicity (Schnack and Romani, 2017). Cardiovascular complications (coronary heart disease, cardiac failure, and stroke), type 2 diabetes mellitus, and hepatic anomalies such as hepatic steatosis are the most common pathologies associated with an uncontrolled and recurrent condition of MetS. Most recently, the role of dietary factors, deficiency in key micronutrients and macronutrients, sedentary behavior, psychotropic medication use, and disrupted chronobiology have been asserted as high-risk factors for the onset and progression of MetS, which leads to hyperinsulinemia, hyperlipidemia, hyperglycemia, and hyperpertension, as shown in Fig. 11.1. Thus the factors that are responsible for MetS are a combination of complex intrinsic factors, such as genetics and the gut microbiome, and extrinsic factors, such as dietary habits and lifestyle. We can observe that while genetics are fixed for a particular individual, we have ample opportunity to alter the gut microbiome to bring about changes in the way the human body processes external cues, thereby making it possible to effectively reduce the risk for MetS (Dabke et al., 2019).

11.3 Composition of a healthy gut microbiome

Molecular biologist Joshua Lederberg defined the gut microbiome as "the totality of microorganisms, bacteria, viruses, protozoa, and fungi, and their collective genetic material present in the GI tract." These microbes have the intense capability to affect our physiology. In the past few years, tremendous research work has been done to evaluate the potential interactions between gut microbes and the host. These interactions can be the key factor in understanding the contribution of a healthy gut microbiome in metabolism, immunity, neuroendocrine responses, and various other direct and indirect physiological functions (Shreiner et al., 2015). The microbiota of our gut is quite diverse and heavily populated in comparison to the microbiota in other parts of the human body. Characterization of the gut microbiota in healthy individuals is an essential step in analyzing the beneficial influence of a healthy microbiome in different kinds of diseases. The gut microbiota has a large population, comprising approximately 1014 cells in the colon. The intestinal microbiome is an important genetic resource. It is usually considered to be our second genome, and it tremendously affects our physiology and metabolism (Salazar et al., 2017). The gut microbiome mainly includes microorganisms such as bacteria, archaea, viruses, and some unicellular eukaryotes. These microorganisms are believed to have coevolved with human beings in a commensal way. The microbial colonization of the human GI tract is unstable for a few months after birth. At first, the gut is colonized by facultative anaerobes, followed by subsequent colonization by strict anaerobes. The intestinal microbiota keeps on evolving, and at about 3 years of age, it reaches a stable, complex population. A normal healthy human adult harbors around 1000 species of bacteria, which belong mainly to the Bacteroidetes and Firmicutes phyla (Lozupone et al., 2012). Bacteriodetes and Firmicutes comprise 80%–90% of the total microbiota population. The relative abundance of *Lactobacillus* and *Bifidobacterium* decrease with age, while *Bacteroides* and

TABLE 11.1 Influence of dietary components in boosting healthy gut microbiome and reducing the risk of metabolic syndrome.

DIET	Gut microflora	Effects
Carbohydrates Digestible (sucrose, fructose, e.g., date fruits, lactose) Nondigestible dietary fibers Prebiotics and resistant starches (e.g., wheat, barley, soybeans, raw oats, nondigestible oligosaccharides such as fructans and fructooligosaccharides that occur naturally in a variety of plants, such as onion, asparagus, chicory, banana, and artichoke, polydextrose, galactooligosaccharides)	Date fruits (sucrose and fructose) ↑ Relative abundance of *Bifidobacteria*, ↓ *Bacteroides* (Eid et al., 2014). Lactose ↓ Pathogenic *Clostridia* Stimulates *Bifidobacteria*	Reduces the risk of irritable bowel syndrome. Lactose supplementation also increases the concentration of beneficial SCFAs (Jeffery and O'Toole, 2013) They travel directly to the large intestine and undergo fermentation by the microorganisms present there. Dietary fibers are the key source of microbiota accessible carbohydrates (MACs). Microbes can utilize these MACs to provide energy to the host and modify the intestinal environment (Sonnenburg and Sonnenburg, 2014). This property of dietary fibers validates them to be designated as prebiotics. A diet abundant in such substances enhances the total gut bacterial abundance and ensures microbiota gene richness (Halmos et al., 2015).
Proteins (beef-rich diet, beef-rich diet, animal proteins)	*Animal proteins* ↑ Concentration of bile tolerant anaerobes like, *Bilophila, Bacteroides and Alistipes* (Cotillard et al., 2013). *Red meat* *Whey and pea protein* ↑ *Bifidobacterium*, while additionally ↓ Pathogenic *Clostridium perfringens* *Beef-rich diet* ↑ *Bacteroides* and *Clostridia* ↓ *Bifidobacterium adolescentis*	• Positively correlated with overall microbial diversity • Increased levels of trimethylamine N-oxide, responsible for the increase in risk of CVD (De Filippis et al., 2016) • Pea protein is also capable of increasing short-chain fatty acid levels in the intestine, which in turn, have an antiinflammatory effect and play an important role in maintaining the mucosal barrier (Kim et al., 2014).
Unsaturated (healthy) fats: (polyunsaturated fat such as fish oil, monounsaturated and polyunsaturated fatty acids such as olive oil, peanut oil, corn oil, tuna, trout, sardines, and walnuts)	↑ *Bifidobacterium*, ↑ *Lactobacillus*, and ↑ *Streptococcus*. Found in fish oil fed mice, (Caesar et al., 2015).	• Reduces the risk of different metabolic disorders. • ↓ Fasting glucose • ↓ Total cholesterol levels • ↓ Plasma LDL-cholesterol levels (by intake of monounsaturated fats)
Polyphenols (flavonols, catechins, phenolic acids, and flavones present in foods such as seeds, fruits, tea, cocoa products, and wine)	↑ *Bifidobacterium* and ↑ *Lactobacillus* *Red wine polyphenols:* ↑ The relative concentration of *Bacteroides*. *Bifidobacterium*. *Fruit polyphenols:* ↓ Infections by enteropathogens by ↓ *Salmonella typhimurium and* ↓ *Staphylococcus aureus* *Fruit, wine, and tea-based polyphenols* ↓ *Clostridium* species in our microbiome (Cueva et al., 2013).	• Have active antioxidant properties. • *Red wine polyphenols* • Modulates the immune system, • Prevents cancer and • Manages metabolic disorders. • *Cocoa-derived polyphenols*: • ↑ Levels of plasma HDLs and • ↓ Levels of plasma triacylglycerol levels (Martin et al., 2012) • Reduces gut inflammation
Probiotics (e.g., yogurt)	↑ *Bifidobacteria* and *Lactobacilli*	• ↑ HDL cholesterol levels and insulin sensitivity and • ↑ Sensitivity C-reactive protein. • ↓ Levels of serum triglycerides, LDL cholesterol, and total cholesterol

(Continued)

TABLE 11.1 (Continued)

DIET	Gut microflora	Effects
		• Yogurt contains lactic acid bacteria, positively regulates the intestinal health, and can even treat disorders like inflammatory bowel disease. (Shen et al., 2014). • Induce antiinflammatory cytokines such as IL-10 • Shows rise in total bacterial load in our gut. • Boosts humoral immune response • Inhibits the adhesion of pathogen to intestinal mucosa and • Reduces ascites and abdominal distention in liver disease patients. • Prevention of obesity, reduction in body weight, and improvement in many metabolic parameters.
Selective diets: *Vegetarian diet* (plant based, rich in fibers) *Mediterranean diet* (polyphenol rich products such as vegetables, grains, extra-virgin olive oil, legumes, nuts, whole-grain cereals, high content of beneficial fatty acids, and low contents of refined sugar and processed meats) *Western diet* (high consumption of processed meats, sugar and salt)	↓ *Escherichia coli* ↑ higher ratio of *Bifidobacteria* to *E. Coli* ↓*Bifidobacterium* and *Eubacterium* species.	• Develops a diverse and stable gut microbial system. • Alters the microbial gene count and bring about reduction in body fat and improvement in insulin sensitivity. • Lowers the production of secondary bile acids and reduces the risk of inflammation, intestinal barrier dysfunction and liver cancer (Cao et al., 2017). • Reduces the occurrence of metabolic syndrome and obesity in patients with CVD (Martínez-González et al., 2019). • Decreases the mortality rate and morbidity in such patients. • The high contents of vegetables, legumes, and fruits in Mediterranean diet are major substrates for microbial production of SCFAs. • ↑ Risk of CVD and other metabolic disorders. • ↓ Microbial diversity • Western-style diet increases the risk of metabolic syndrome.

bacterial genes associated with the ability to metabolize complex carbon sources increase with age. Other bacterial species belonging to the Proteobacteria, Cyanobacteria, and Actinobacteria phyla are found at comparatively lower levels. Methanogenic archaea (mainly *Methanobrevibacter smithii*), eukarya (mainly yeasts), and viruses (primarily phage) are also present in the gut microbiome, as shown in Table 11.2. The concept of a core microbiome in healthy human adults has been brought into the limelight repeatedly by various culture-based studies. *Roseburia intestinalis*, *Bacteroides uniformis*, and *Faecalibacterium prausnitzii* are the key players in the core set of healthy gut microbiota, although their abundance can be as low as 0.5% in some people (Reyes et al., 2010). At the same time, it must be remembered that the concentration and relative proportions of these core set of species in a healthy gut microbiome varies to a great extent among individuals across the population; hence the concept of a uniform core microbiome composition in all human beings is not always plausible.

Table 11.2 shows the composition of gut microbiota, their phylum, genus, species, and relationship with the different disease. Apart from these, other gut microbiota includes Eukarya (yeast), viruses (primarily phage), and methanogenic archaea (mainly *M. smithii*) (Scotti et al., 2017)

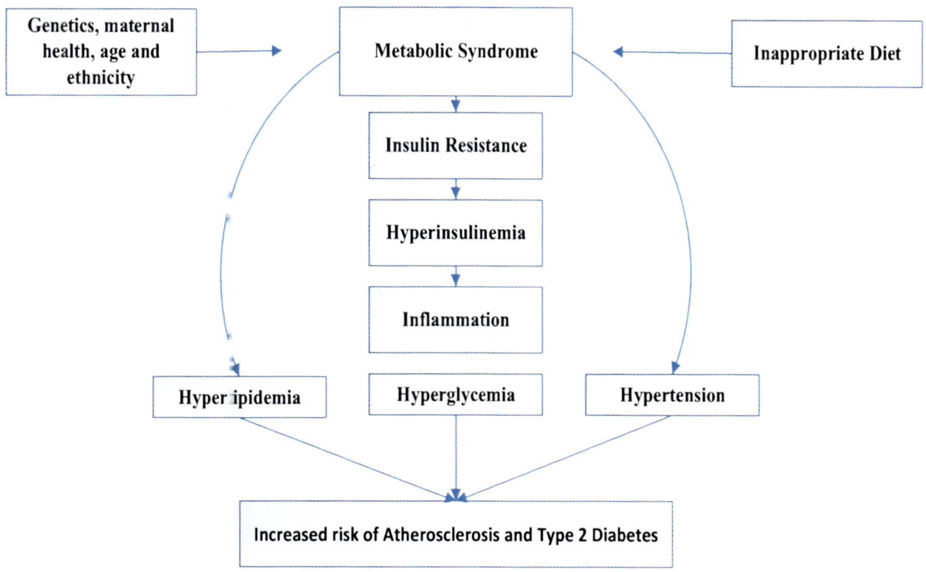

FIGURE 11.1 A flowchart depicting the factors that contribute to metabolic syndrome (MetS). Poor dietary habits and inadequate physical activity, combined with a genetic background that occurs throughout fetal development, contribute to insulin resistance, which is the significant factor in the development of hyperlipidemia, hyperglycemia, and hypertension. These effects eventually result in metabolic and inflammatory dysregulation, which manifests as MetS. Finally, MetS contributes to type 2 diabetes and cardiovascular disease, either indirectly or directly.

TABLE 11.2 Composition of taxonomic gut microbiota of human.

Phylum/class	Genus	Species	Relationship with disease
Actinobacteria (<10%)	*Bifidobacterium* (gram-positive obligate anaerobe)	*B. longum, B. bifidum*	1. Decrease in obesity 2. Decrease in intestinal LPS levels
Fermicutes (60%–80%)	*Roseburia Faecalibacterium* (gram-positive obligate anaerobe)	*R. intestinalis,* *F. prausnitzii,* *Lactobacillus reuteri*	*Faecalibacterium* shows 1. Decrease in obesity 2. Decrease in T2D
	Lactobacillus (gram-positive facultative anaerobe)		*Lactobacillus* shows 1. Decrease in metabolic disorder 2. Promotes antiinflammatory activity
	Eubacterium (gram-positive obligate anaerobe)		*Eubacterium* shows 1. Decrease in Atherosclerosis 2. Decrease in T2D
	Clostridium (gram-positive obligate anaerobe)		*Clostridium* shows 1. Positive correlation with obesity 2. Increase in T2D
Bacteroidetes (20%–30%)	*Bacteroides* (gram-negative obligate anaerobe)	*B. uniformis*	1. Increase in obesity 2. Increase in CVD
Proteobacteria (<1%)	*Desulfovibrio* *Bilophila* *Escherichia*	*D. intestinalis,* *B. wadsworthia,* *E. coli*	*E. coli* shows 1. Increase in T2D Increase in IBD
Verrucomicrobia	*Akkermansia* (gram-negative obligate anaerobe)	*A. muciniphila*	1. Decrease in obesity 2. Decrease in IBD 3. Decrease in T2D

11.4 Beneficial role of a healthy microbiome in preventing metabolic syndrome

11.4.1 How the microbiome is related to health and diseases

The microbiome plays a very important role in the physiological well-being of the host. The gut microbiota is capable of synthesizing different metabolic products that can influence human health in either a positive or negative way, upon interacting with the human body (Gomaa, 2020). The human microbiome constitutes a large number of prokaryotes, eukaryotes and viruses, which are closely related to the health and diseases of human beings. These microbial communities occupy various habitats inside the human body. We humans provide residential space for these microbes; in return, these microbes metabolize molecules that we are incapable of metabolizing ourselves. A symbiotic relationship is developed between the microbes and the host. We are dependent on these microbes to boost our immune system, fight pathogens, and aid in nutrition. Thus the human microbiome has a monumental impact on human health and diseases (Cong and Zhang, 2018). The intestinal microbiome has been considered to be an "internalized organ" by recent studies. Some of the most significant functions of the gut microbiota in maintaining proper human health are as follows:

1. The gut microbiota thrives and replicates on our intestinal surfaces, creates different antimicrobial substances, and forms a stable system that protects us from the invasion of pathogens (Mills et al., 2019);
2. It plays an important role in the metabolism and proper functioning of our digestive system (Rothschild et al., 2018);
3. It modifies insulin resistance and influences insulin secretion (Kelly et al., 2015);
4. It controls proliferation and differentiation of epithelial cells (Wiley et al., 2017);
5. It affects brain-gut communication and promotes proper neurological functioning in the host (Zheng et al., 2019).

Throughout the years a huge spectrum of research work has been carried out to establish the linkage between the intestinal microbiome and the risk, onset, and progression of different diseases. Our complex microbial ecosystem might be responsible for affecting (either inducing or aggravating) a variety of diseases.

11.4.2 Gut microbiome and metabolic syndrome

The intestinal microbiota is capable of synthesizing certain metabolites that have an influential impact on human health. A number of these microbiota-derived metabolites confer a positive effect on the host by displaying antioxidant and antiinflammatory activities. Some of them might act like vitamins, nutrients, and energy sources. One of the most important functions that these metabolites perform is the regulation of gut barrier activity. On the other hand, some metabolites, such as immunotoxins and cytotoxins, are capable of having a negative effect on the host (Moran-Ramos et al., 2017). Thus the balance of the gut microbiome is very important for maintaining proper health of human beings. The dysbiosis of the gut microbiome can lead to a variety of physiological changes and therefore amplify the risk of MetS. The interactions between the gut microbiome and host metabolism are mediated by numerous factors, such as the metabolism of short-chain fatty acids (SCFAs), the metabolism of bile acids, and inflammation due to gut barrier defects (Wang et al., 2020). Intake of a high-sugar or high-fat diet, stress, and a sedentary lifestyle are some of the most prominent factors that can cause gut dysbiosis. Dysbiosis of the gut microbiota leads to increased permeability, adiposity, systemic inflammation, and insulin resistance, which in turn result in an increase in risk factors for metabolic disorders such as type 2 diabetes mellitus, atherosclerosis, nonalcoholic fatty liver disease, peripheral artery disease, and CVDs (Pushpanathan et al., 2019). The fermentation of dietary polysaccharides that the human body cannot absorb occurs via the intestinal microbiota. Dietary fibers are insoluble and soluble fibers that make up the indigestible part of plant food carbohydrates. Soluble fibers are broken down into SCFAs by enzymes that originate in the gut microbiome. In the intestine, SCFAs (acetate, butyrate, and propionate) are consumed and used as energy by the host. SCFAs are physiologically important, as they regulate gut homeostasis and facilitate liver metabolism (Canfora et al., 2015). SCFAs maintain energy balance by controlling hormones such as trypsin, leptin, and glucagon-like peptide 1, thus inhibiting the development of metabolic disorders such as abnormal glucose metabolism, hypertension, and obesity.

Bile acid is an endocrine molecule that is responsible for regulating the balance between lipids, glucose, and energy (Molinaro et al., 2018). Bile acids play a crucial role in lipid and glucose metabolism by activating a nuclear receptor called farnesol X and a membrane receptor known as G protein–coupled membrane receptor. This activation can be either direct or indirect. Dynamic interactions take place between the gut microbial ecosystem and bile acids. This interaction can be either beneficial or harmful to the host's metabolism through dietary changes.

Thus MetS can also be influenced by the interactions among the gut microbiome, bile acids, and host metabolites, as changes in bile acid by microorganisms affects signals emerging from the bile acid receptor.

It has been found that the primary pathophysiological basis of MetS is chronic inflammation mainly dominated by insulin resistance. Bacteria and their components, like endotoxins, gain entry into the circulation and cause low levels of inflammation because of an imbalance created in the gut microbiome and gut barrier disruption (Schroeder and Bäckhed, 2016). If an imbalance occurs in the gut microbiota, it can lead to an increase in the expression of lipopolysaccharide, (LPS) degradation of the mucous layer, facilitation of endotoxemia, production of proarthenogenic trimethylamine (TMA) and promotion of several other pathways that cause metabolic disorders (Yoshida et al., 2018). These changes in the gut microbiota are responsible for translocating bacterial metabolites such as TMA, phenylacetic acid, imidazole propionate, and expressing pathogen-associated molecular patterns that are capable of inducing low-grade inflammation through inflammatory cytokines such as IL-1β. In individuals who have a healthy metabolism, the gut microbiota controls the integrity of the intestine by a variety of mechanisms (Cani, 2018). The dendritic cells are capable of extracting microbial antigens from the lumen of the gut and inducing the activation of different immune cells, which in turn promote the secretion of antimicrobial peptides, mucus, and immunoglobulin A (Brown et al., 2013). Adenosine monophosphate secretion by nucleotide-binding oligomerization domain, leucine-rich repeat domain, and domain-containing protein 3 inflammasome-sensing microbiota metabolites is promoted by the gut epithelium. Certain metabolites that are synthesized by the gut microbiota, such as secondary bile acids and aryl hydrocarbon receptor agonists, indirectly supervise the maintenance of the gut barrier function. The gut barrier integrity is also determined by host endogenous factors, such as insulinemia in blood, which can regulate the secretion of mucus by fatty acid synthase (Natividad et al., 2018). Overall, we can say that gut microbiota dysbiosis and metabolic inflammation are closely related and lead to MetS. The gut microbiome influences metabolic inflammation and diseases, which act as important markers of major metabolic disorders, such as obesity, type 2 diabetes mellitus, and atherosclerosis.

11.4.3 Gut microbiome: a new strategy to treat obesity

The gut microbiome greatly influences the protection of the intestinal mucosa by maintaining symbiotic relationships with the host. Such a close and intricate relationship enables the microbiome to contribute to the host's physiological homeostasis. Irregularities in the gut microbial composition lead to different diseases. Obesity is one such prominent metabolic disorder. The gut microbiome has an important role in regulating the fat storage mechanisms inside our body. In metabolic diseases such as obesity, our gut microbiota is altered so that there is an increase in the energy harvested from the food we consume (Patterson et al., 2016). The intestinal microbial community produces certain compounds that are absorbed into our systemic circulation and can enhance the risk of obesity-related complications, as they cause an increase in insulin resistance and aggravate damage due to tissue inflammation (Khan et al., 2016). Alterations of the gut microbiota in children have been seen to be responsible for the onset of obesity, as changes in gut microbiota lead to harmful effects on the mechanisms regulating appetite (Fetissov, 2017). Obesity is generally associated with a gut microbiome that has a reduction in diversity, which in turn reduces metabolic energy consumption as compared to the microbiome of lean people. The microbiota that facilitates proper fermentation of carbohydrates into SCFAs are responsible for the setting in of obesity by providing excess energy substrate to the host. Therefore low-efficiency bacterial breakdown of complex carbohydrates by our gut microbiota could be a strategy to avoid the development of obesity (Muscogiuri et al., 2019).

Clinical evidence suggests that there is a complex relationship between the gut microbiota and regulation of body weight. The microbiota of obese people is characterized by a lower concentration of Bacteroidetes and a higher concentration of Firmicutes as compared to people without obesity (Ley et al., 2005). A large number of microbes and their respective proportions in the gut microbiome are responsible for obesity-related differences among individuals. Some examples of such microbes are *Lactobacillus reuteri*, *Lactobacillus sakei*, *Clostridium innocuum*, *Catenibacterium mitsuokai*, and *Actinobacteria*, as well as some rare *Archaea* organisms such as *M. smithii*. If changes occur in the gut microbial population leading to a predominance of gram-negative bacteria, an alteration of the protein structure in the endothelial "tight junctions" is observed. This phenomenon can cause an increase in intestinal permeability and further promote the absorption of endotoxins such as LPS into the host's circulation, which in turn, induces an increase in LPS-signaling cascades. This leads to the formation of proinflammatory cytokines which are associated with the development of obesity. Hence it was observed that a person who consumes a healthy diet has high insulin sensitivity, high SCFA levels, lower serum cholesterol, lower inflammation, and a low risk of cardiovasular disease, as shown in Fig. 11.2.

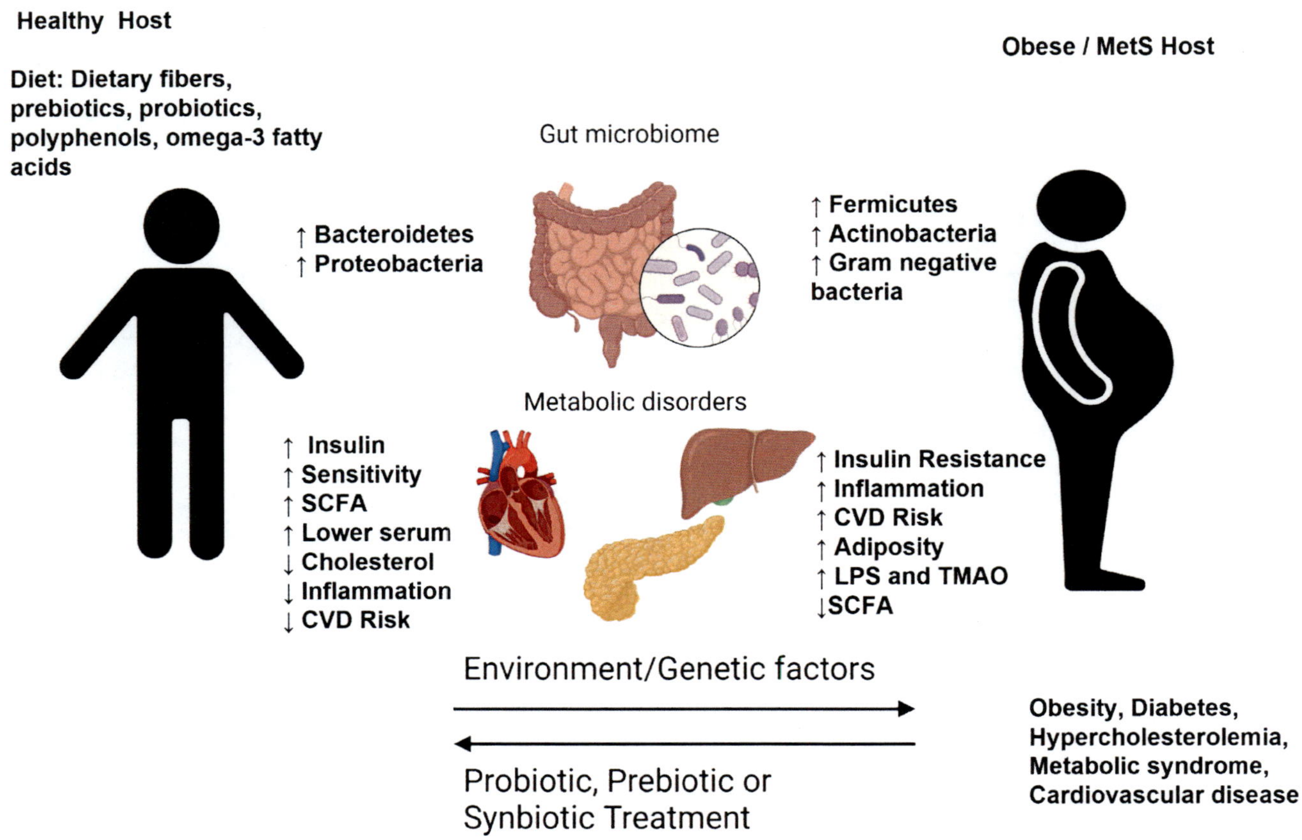

FIGURE 11.2 An overview of the role of the gut microbiome in metabolic syndrome and obesity. Up and down arrows indicate increase and decrease, respectively. A comparison of metabolic processes and disorders of healthy and obese human is shown.

The latest studies have shown that dysbiosis associated with obesity can be regulated by being on a low-energy diet for several months. It has been also reported that ingesting certain micronutrients, prebiotics, and probiotics can cause an alteration in the gut microbial composition and hence affect the regulation of gene expression at sites like muscle, liver and adipose tissue. Bariatric surgery is becoming a very crucial therapeutic option for the treatment of individuals suffering from morbid obesity. It is a reliable method for achieving huge reductions in body mass index, as shown in Fig. 11.3 (Rajjo et al., 2017). The reduction in body weight after bariatric surgery is related to gut microbiota modifications (Ejtahed et al., 2018). Thus prominent obesity treatment strategies, such as intake of calorie reduction diets and undergoing bariatric surgery, to modify the gut microbiome are based on the hypothesis that alteration in the composition of the intestinal microbiota can play a key role in developing an additional mechanism targeted to achieve stable weight loss.

11.4.4 Intestinal microbiota in cardiovascular diseases: the heart-gut axis

In recent times, there has been an increase in interest in the interactions between the intestinal microbiota and the host due to current discoveries about the influence of the gut microbiota on human health. Changes in the gut microbial composition are related to the pathogenesis of CVD, such as atherosclerosis, hypertension, and heart failure (Ahmadmehrabi and Tang, 2017). There is a significant contribution of molecules derived from the gut microbiota in the development and progression of heart diseases. Several studies have been undertaken to establish the linkage between our heart and gut, thereby leading to the formation of the concept of a heart-gut axis in organisms. A novel pathway linking the host gut microbiome and CVDs has been described recently. This pathway focuses primarily on the relationship between TMA-N-oxide (TMAO) and atherosclerosis (Wang et al., 2011). TMAO and choline were first identified by metabolomics as small-molecule metabolites in human plasma that are associated with the risk of CVD. Both of these molecules are generated by the metabolism of phosphatidylcholine. Studies involving mouse models established that the excess dietary intake of choline and TMAO leads to an increase in macrophage foam cell formation and the development of atherosclerosis. The atherosclerotic effects of the metabolism of phosphatidylcholine were found to be influenced by the composition of

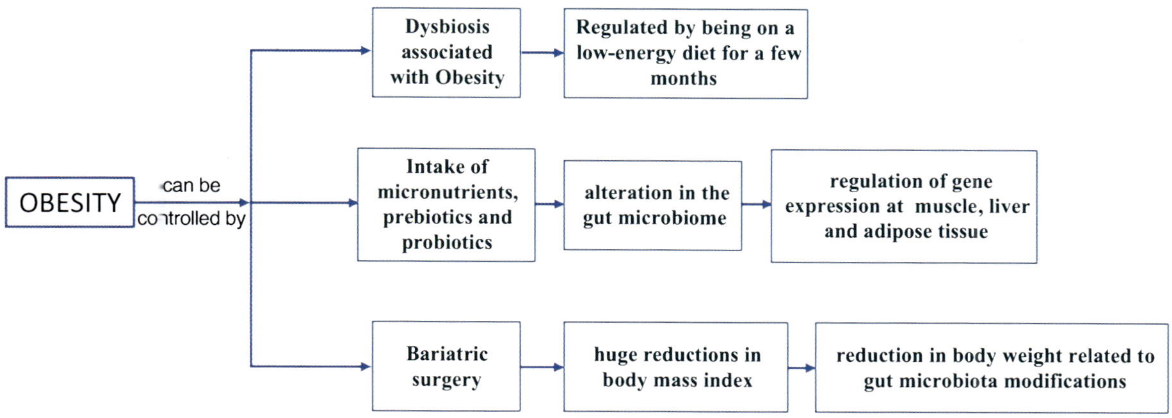

FIGURE 11.3 Obesity can be controlled by consuming a low-energy diet for several months; by consuming micronutrients, prebiotics, and probiotics; and by undergoing bariatric surgery.

the gut microbiota. Dysbiosis of the gut microbiota is a major factor that is responsible for CVD and other metabolic disorders that pose a risk for CVD. Imbalance in the microbiome can sometimes lead to a disastrous impact on host health. Dysbiosis, the decline in gut microbial diversity, and a decrease in the ratio of Bacteroidetes to Firmicutes have been observed in hypertensive animals. Patients suffering from symptomatic atherosclerosis have been found to have an increased concentration of *Collinsella* in their gut microbiome (Karlsson et al., 2012). In some instances, alterations in the gut microbiota can bring about a positive impact on the cardiovascular health of organisms. The establishment of the novel pathway linking our gut microbiota and the risk factors for CVD has led to the development of new opportunities for therapeutic interventions. Interventions that are targeted to change the composition of the gut microbiome represent a promising treatment approach. In diet-induced obese mice the abundance of *Akkermansia muciniphila* is inversely correlated to the metabolism of lipids and inflammation in their adipose tissue (Schneeberger et al., 2015), while Tenericutes and Christensenellaceae are associated with lower triglyceride (TG) and increased HDL levels in host organisms. Increased concentrations of Methanobrevibacter and Peptococcaceae are also linked to reduced triglyceride levels. A diet that is rich in fiber is capable of increasing the concentration of SCFA-producing microbes in the gut and hence lower the blood pressure in hypertensive patients (Marques et al., 2017). Intake of prebiotics can have a beneficial impact on the gut microbiota composition. Prebiotics are capable of increasing the ratio of Bacteroidetes to Firmicutes in the gut and reducing the efficacy of other characteristics linked to the diabetic phenotype.

Administration of probiotics such as *Lactobacilli* and *Bifidobacteria* in a mouse model has been seen to help lower their cholesterol, protecting them from obesity caused by a poor diet and increasing their insulin secretion (Tuohy et al., 2014).

Fecal microbiota transplantation (FMT) is a modern therapeutic strategy that has garnered much attention for treating cardiometabolic disorders. In FMT a patient's gut microbiota is replaced by that of a healthy donor, essentially displacing the pathogenic strains. It has been found that FMT is a quite effective treatment method for CVD. Allogenic fecal transplant has the potential to improve insulin sensitivity in patients suffering from MetS, irrespective of their weight variations. Another effective therapeutic strategy for the treatment and prevention of CVD involves the use of 3,3-dimethyl-1-butanol (DMB). DMB is a chemical analog of choline and is capable of inhibiting TMA production and hence reducing TMAO levels in the circulation in the case of mice on a choline-rich diet. This in turn caused a reduction in the formation of macrophage foam cells and ultimately decreased the risk of developing atherosclerosis in vivo, as shown in Fig. 11.4 (Wang et al., 2015).

11.5 Influence of dietary components in boosting a healthy microbiome and reducing the risk of metabolic syndrome

Currently, an enormous amount of research work is being carried out regarding the significance of the microbiome in modulating host health and controlling different pathological processes. Various studies have shown that it is possible to bring about certain alterations in the microbiota which can potentially rectify metabolic disorders resulting from dysbiosis. Diet can play an important role in remodeling the microbiome and can drastically alter the structure of the gut microbial community within just 24 hours (Green et al., 2020). The dietary patterns of human beings, particularly the

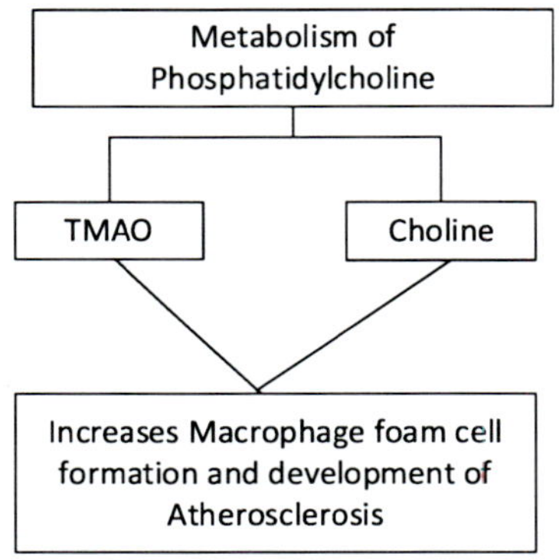

FIGURE 11.4 Trimethylamine-N-oxide and choline develops atherosclerosis.

composition of micronutrients and macronutrients and the nutritional effects of macronutrients, are the chief contributors in regulating the gut microbiome.

11.5.1 Carbohydrates

Carbohydrates are probably one of the most vigorously studied dietary components for their capability to make changes in the gut microbiome. Dietary fibers can play a role in the treatment of MetS through a variety of mechanisms involving their physicochemical characteristics and intestinal flora fermentability. There are two kinds of carbohydrates: digestible and nondigestible carbohydrates. Starch and sugars such as glucose, sucrose, fructose, and lactose as well as pectins, gums, mucilages, and hemicelluloses are digestible carbohydrates that are degraded in the intestine by enzymes, contain by-products such as SCFAs, and have high water-keeping ability. They release glucose directly into the bloodstream after degradation, triggering an insulin response (Saltiel, 2016). A human host who had been fed high levels of sucrose and fructose by eating date fruits had an increase in the relative abundance of *Bifidobacteria* and a decrease in the relative abundance of *Bacteroides*, according to a report (Eid et al., 2014). A separate study showed that the incorporation of lactose in the diet decreases the concentration of pathogenic *Clostridia* in our microbiome, thereby reducing the risk of irritable bowel syndrome. Nondigestible carbohydrates include fibers and resistant starches. They travel directly to the large intestine and undergo fermentation by the microorganisms present there. Dietary fibers are the key source of microbiota-accessible carbohydrates (MACs). Microbes can utilize these MACs to provide energy to the host and modify the intestinal environment (Sonnenburg and Sonnenburg, 2014). This property of dietary fibers validates them to be designated as prebiotics. Some sources of such prebiotics are wheat, barley, soybeans, raw oats, nondigestible oligosaccharides such as fructans, fructooligosaccharides that occur naturally in a variety of plants such as onion, asparagus, chicory, banana and artichoke, polydextrose, galactooligosaccharides and so on, which act as prebiotics by stimulating bifidobacteria. Studies have shown that a diet that is rich in such substances increases the total bacterial abundance in the gut and ensures microbiota gene richness (Halmos et al., 2015). In contrast, insoluble fibers are subjected to limited fermentation with minimal water-holding capacity. The effect of arabinoxylan oligosaccharides and polyunsaturated fatty acids on the gut microbiome have been studied (Kjølbæk et al., 2020). The gut microbial modulation by arabinoxylan oligosaccharides was positively correlated with blood concentration of insulin, markers of lipid metabolism (triglycerides, low-density lipoprotein, very low-density lipoprotein, total cholesterol, and apolipoprotein B), and hepatic function. It was found that dietary inulin can halt the progression of diabetes via suppression of the inflammatory cascade through gut microbial modulation (Li et al., 2019).

11.5.2 Proteins

The influence of dietary proteins on gut microbial composition was first explained in 1977. It was observed that the subjects who consumed a beef-rich diet had a lower concentration of *Bifidobacterium adolescentis* and a higher concentration of *Bacteroides* and *Clostridia* in their microbiome in comparison to the subjects who consumed a meatless diet. A maximum number of studies in this field noted that the consumption of proteins is positively correlated with overall microbial diversity (Clarke et al., 2014). The consumption of whey and pea protein has been observed to increase the number of gut-commensal *Bifidobacterium* while additionally decreasing the number of pathogenic *Clostridium perfringens*. Pea protein is also capable of increasing SCFA levels in the intestine, which in turn have an antiinflammatory effect and play an important role in maintaining the mucosal barrier, as shown in Fig. 11.5 (Kim et al., 2014). By contrast, it has been seen that an increase in consumption of animal protein leads to an increase in the concentration of bile tolerant anaerobes such as *Alistipes*, *Bilophila*, and *Bacteroides* (Cotillard et al., 2013). Many microbial genera that are associated with increased levels of TMAO are promoted by the intake of red meat (De Filippis et al., 2016). TMAO is responsible for the increase in the risk of CVD. It has been observed that plant-derived proteins are related to lower mortality as compared to animal-derived proteins.

11.5.3 Fats

Diets that are rich in saturated fatty acids and trans fatty acids increase the risk of CVD. Conversely, monounsaturated fatty acids and polyunsaturated fatty acids play a crucial role in alleviating the risk of chronic disorders. The Western diet contains high levels of trans and saturated fats and hence predisposes regular consumers to several health problems. Various human studies have shown that a high-fat diet increases the concentration of anaerobic microflora such as *Bacteroides* (Fava et al., 2013). It has been seen that consuming a low-fat diet leads to an increase in *Bifidobacterium* and a reduction in fasting glucose and total cholesterol levels. Intake of foods that are rich in monounsaturated fatty acids has been observed to reduce plasma low-density lipoprotein (LDL) cholesterol levels. Endotoxemia-induced inflammation can also be regulated by microbial changes, as shown in research on mice that were fed a high-fat diet, by inducing overgrowth of *Escherichia coli* and *Bilophila wadsworthia* (bile-tolerant gram-negative bacteria) as well as a decrease in Bifidobacteria (Marette and Jobin, 2015). Another mouse study, which compared the effects of consuming lard-derived lipids and fish oil—derived lipids, showed that the concentration of *Bacteroides* and *Bilophila* increased in lard-fed mice while the concentration of *Bifidobacterium*, *Lactobacillus*, and *Streptococcus* increased in mice that were fed fish oil (Caesar et al., 2015). These results further support the idea that the consumption of fats such as fish oil has a beneficial impact on human health by altering the gut microbiome in a good way as compared to the negative impact generated by the consumption of harmful fats such as lard.

11.5.4 Fatty acids

SCFA (butyrate, acetate, and propionate) are saturated fatty acids that have fewer than six carbon atoms and that have been shown to improve gut dysbiosis and the host's metabolism (Marette and Jobin, 2015) SCFAs have been shown to decrease inflammation and increase insulin sensitivity in the liver and muscles and to increase energy expenditure (Kimura et al., 2013).

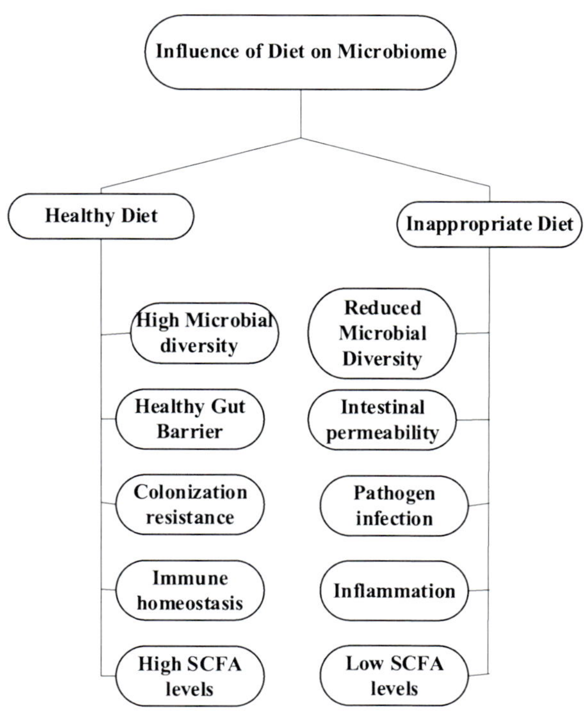

FIGURE 11.5 Effect of a healthy diet and a poor-quality diet on gut microbiota.

SCFAs are produced primarily by the microbial fermentation of indigestible carbohydrates or prebiotics, and supplementation with prebiotics has been shown to increase butyrate development and Bacteroidetes in Wistar rats. A high-fat diet, on the other hand, decreases butyrate formation while increasing liver fat and inflammation (Jakobsdottir et al., 2013). Prebiotics also improve the risk for MetS by restoring the excess of mucin-foraging bacteria (*A. muciniphila*), which was reduced in obese and diabetic mice (Everard et al., 2014). Butyrate has antiobesity properties, resulting in decreased lipoprotein lipase (LPL) expression and increased lipolysis (Korecka et al., 2013). Intake of an oleic acid–rich diet increased insulin sensitivity in young adults in a study. Huang et al. found that omega-6 fatty acids have a stronger effect than saturated fatty acids (Huang et al., 2013). Omega-3 fatty acids, on the other hand, modulate the gut microbiota, which, among other processes, has a greater ability to prevent metabolic dysregulations (Kaliannan et al., 2016). Omega-3 fatty acids reduced bacterial overgrowth and recruited regulatory T cells to the small intestine, reducing diet-induced inflammation (Ghosh et al., 2013). Thus foods that are abundant in monounsaturated and polyunsaturated fatty acids, such as olive oil, peanut oil, corn oil, salmon, trout, sardines, and walnuts, are capable of positively modulating the gut microbiota and lowering the risk of various metabolic disorders.

11.5.5 Polyphenols

Polyphenols are a group of bioactive compounds that are found in a wide range of fruits, vegetables, nuts, and beverages as well as in small amounts in dried legumes and cereal grains. They are secondary plant metabolites that provide a wide range of phenolics with several structural variants. Flavonoids (flavanones, isoflavones, flavones, and anthocyanidins) and nonflavonoids (curcuminoids, phenolic acids, coumarins, lignans, and xanthones) are the two class of flavonoids (Khan et al., 2020). Green tea, cranberries, wine, nuts, grape seeds, and cocoa, along with other polyphenol-rich foods and beverages, have a significant impact on various factors of metabolic disorders (Del Rio et al., 2013). Flavanones, isoflavones, anthocyanins, flavonols, and resveratrol are the most investigated types of polyphenols that have a beneficial effect on the prevention of MetS risk factors, most likely through enhanced insulin signaling, downregulation of oxidative damage, gut bacteria manipulation, improvement in dysfunction of endothelium, or influencing serum cholesterol and glucose levels. These food products are responsible for increasing the concentration of bacterial genera such as *Bifidobacterium* and *Lactobacillus* in the gut microbiome (Cuervo et al., 2014). Polyphenols are metabolized by gut microbial organisms into simpler compounds, helping to increase their bioavailability. Subjects who consumed red wine polyphenols were observed to have an increase in the relative concentration of *Bacteroides*. *Bifidobacterium* have many important health benefits like modulation of the immune system, prevention of cancer and management of metabolic disorders. Cocoa-derived polyphenols have been seen to be associated with increased levels of plasma HDLs and decreased levels of plasma triacylglycerols (Martin et al., 2012). Consumption of fruit polyphenols has been linked with a reduction in the risk of infections by enteropathogens like *Salmonella typhimurium* and *Staphylococcus aureus*. Several studies have shown that intake of fruit, wine, and tea-based polyphenols can decrease pathogenic *Clostridium* species in the human microbiome (Cueva et al., 2013). Resveratrol has been shown to reduce atherosclerosis by lowering TMAO synthesis and bile acid metabolism, which correlates with inhibiting microbial TMA production via gut microbiota remodeling (Chen et al., 2016). In mice, resveratrol treatment increased the levels of *Bifidobacterium* and *Lactobacillus*, which increased bile acid hydrolase activity. Researchers have found that quercetin along with resveratrol can reverse gut dysbiosis caused by excess fat intake in mice, preventing obesity (Zhao et al., 2017). The biochemical influence of polyphenol-rich cranberry extract on the microbiome of mice was studied. Cranberry extract was discovered to lower visceral obesity, body weight, liver weight, hepatic inflammation, and triglyceride deposition in the liver (Anhê et al., 2015). Cranberry extract improved insulin sensitivity and resistance in mice as well as hyperinsulinemia, intestinal TG, oxidative damage, and intestinal inflammation. It also increased the proportion of mucin-degrading *Akkermansia* species in the intestinal flora. Grape seed proanthocyanidin extract was found to reduce inflammation and adiposity in mice that were fed an excess-fat diet (Liu et al., 2017). It improved insulin sensitivity, thus lowering the production of inflammatory mediators in the blood.

11.5.6 Spices

Spices are extensively used in traditional medicine, as food preservatives, as natural coloring agents in cosmetics, and as dental formulations. Hence they play an influential role in the economy of the countries where they are manufactured (Li et al., 2020).

11.5.7 Prebiotics

Several dietary fibers are known as prebiotics, which has been shown to help with GI problems (Floch, 2018). These are compounds that contribute to the development of beneficial microbes that benefit their host's health (Kaliannan et al., 2016). Lactulose, oligosaccharides (fructooligosaccharide and galactooligosaccharide), psyllium, and inulin are all examples of prebiotics. Oligofructose intake increased glucose tolerance, decreased insulin resistance, and increased the activity of antiinflammatory cytokines in mice with modified gut microbiota and metabolic dysregulation (de Cossío et al., 2017). Prebiotic-rich foods include garlic, asparagus, onions, artichokes, leeks, tomatoes, oats, wheat, and soybeans (Davani-Davari et al., 2019). Prebiotics improve the host's metabolism by enhancing the development of *Bifidobacterium* and *Lactobacillus* species in the gut (Mäkeläinen et al., 2010). Prebiotics can help to stimulate the intestinal immune system, regulate pathogen and toxin development, increase SCFA output, reduce lactose intolerance, and relieve constipation (Al-Sheraji et al., 2013). The word "synbiotics" refers to a form of nutritional supplement that combines probiotics and prebiotics to provide synergistic health benefits (Eslamparast et al., 2014) Prebiotics can help with weight loss and improve insulin sensitivity.

11.5.8 Probiotics

Probiotics are live microorganisms that, when consumed, induce a beneficial effect in the host body by improving or restoring the gut microflora (Mishra et al., 2018). Fermented food products such as yogurt contain lactic acid bacteria, which can positively regulate intestinal health and can even treat disorders like inflammatory bowel disease (Shen et al., 2014). Probiotics can be ingested in the form of capsules that contain desired quantities of live bacterial cells. However, in the modern market situation, consumers also have access to different food products that are enriched with probiotic microbes that act as functional ingredients that are capable of improving human health. In addition to being incorporated into food products, probiotic bacterial strains can also be combined with prebiotic fibers to enhance their growth and activity. Such a combinatorial approach is termed "synbiotics." Probiotics can induce antiinflammatory cytokines such as IL-10. Regular consumption of fermented milk or yogurt has been reported to bring about a rise in the total bacterial load in the gut. Studies have shown that there have been significant increases in the concentration of beneficial gut microbes such as *Bifidobacteria* and *Lactobacilli* by the daily intake of different kinds of probiotics. Some studies conducted on obese adults have confirmed that probiotic supplementation is associated with improvement in HDL cholesterol levels and insulin sensitivity and reduction in the levels of serum TG, LDL cholesterol, total cholesterol, and high sensitivity C-reactive protein (Rajkumar et al., 2014). Some other significant health benefits of consuming probiotics are boosting the humoral immune response, inhibiting the adhesion of pathogens to the intestinal mucosa, and decreasing ascites and abdominal distention in patients with liver disease. Animal studies associated with the supplementation of probiotics have shown that ingestion of probiotics can lead to marked benefits such as prevention of obesity, reduction in body weight, and improvement in many metabolic parameters.

11.5.9 Selective diets

Apart from describing the overall influence of micronutrients and macronutrients on the composition and functioning of our microbiota, it is also important to analyze how some specific dietary patterns and eating habits can affect our microbiome and metabolic health. Some such selective dietary patterns include vegetarian diets, vegan diets, the Mediterranean diet, and the Western-style diet. A vegetarian diet is naturally rich in MACs and has a positive effect on the microbiome. It induces the development of a diversified and balanced gut microbial system. A calorie-restricted diet along with an increase in fiber intake has been seen to alter the microbial gene count and bring about a reduction in body fat and improvement in insulin sensitivity. It has also been noted that a plant-based diet leads to lower production of secondary bile acids and reduces the risk of inflammation, intestinal barrier dysfunction, and liver cancer (Cao et al., 2017). The traditional Mediterranean diet is characterized by polyphenol-rich products such as vegetables, grains, extra-virgin olive oil, legumes, nuts, whole-grain cereals, a high content of beneficial fatty acids, and low contents of refined sugar and processed meats. The Mediterranean diet is capable of reducing the incidence of obesity and MetS in patients with CVD (Martínez-González et al., 2019). It also decreases the mortality rate and morbidity in such patients. Individuals who consume a Mediterranean diet have lower counts of *E. coli* and a higher ratio of *Bifidobacteria* to *E. coli*. The high contents of vegetables, legumes, and fruits in the Mediterranean diet are major substrates for microbial production of SCFAs. By contrast, the Western-style diet involves high consumption of processed meats, sugar, and salt and hence is associated with an increase in risks for CVD and other metabolic disorders. It causes a decrease in microbial diversity, as shown in Fig. 11.5, and a significant reduction in the concentration of commensal *Bifidobacterium* and *Eubacterium* species.

Thus it can be concluded that the consumption of Mediterranean and plant-based diets (healthy diets) have a beneficial effect on the gut microbiome and help in preventing metabolic disorders, whereas a Western-style diet (unhealthy diet) increases the risk of MetS, as summarized in Fig. 11.5.

11.6 Conclusion

The human gut microflora is essential for ensuring the host's well-being, but it can contain both harmful and beneficial microflora. *Akkermansia*, *Bifidobacterium*, and *Lactobacillus*, species are useful to humans and are found in many probiotic formulations, but *Bacteroides* and *Ruminococcus* genera have been linked to negative health outcomes. Core microbial diversity and the Firmicutes to Bacteroidetes ratio are general markers of health that can vary according to age. The diet quality and environmental conditions play a significant role (Davenport et al., 2014). Dietary foodstuffs containing adequate levels of prebiotics and probiotics that promote the host's health can modulate this microflora. In implementing a new dietary plan, such as following a complete plant-based or animal-based diet, rapid changes in gut microbiota have been observed (David et al., 2014). Even though a variety of therapeutic options are available, none of them can provide a complete cure with a favorable mode of action. Because of the negative effects of medications, scientists are looking into alternative therapies, such as dietary supplements and functional foods. According to the literature that is available from clinical trials, the daily intake of bioactive dietary food in proper amounts could be seen to facilitate the growth and development of beneficial microflora, enhance lactose intolerance, reduce inflammatory response, regulate gut immunity, upregulate gastric juices, promote the digestive capability of the intestinal tract, and increase the digestive potential of the GI tract. However, based on the limited experimental evidence available from human studies, more extensive clinical trials on such bioactive food products in human populations are required to improve the efficiency and effectiveness of these therapies for healthy bioactive ingredients.

References

Ahmadmehrabi, S., Tang, W., 2017. Gut microbiome and its role in cardiovascular diseases. Curr. Opin. Cardiol. 32 (6), 761–766. Available from: https://doi.org/10.1097/HCO.0000000000000445.

Al-Sheraji, S.H., Ismail, A., Manap, M.Y., Mustafa, S., Yusof, R.M., Hassan, F.A., 2013. Prebiotics as functional foods: a review. J. Funct. Foods 5, 1542–1553. Available from: https://doi.org/10.1016/j.jff.2013.08.009.

Anhê, F.F., Roy, D., Pilon, G., Dudonné, S., Matamoros, S., Varin, T.V., et al., 2015. A polyphenol-rich cranberry extract protects from diet-induced obesity, insulin resistance and intestinal inflammation in association with increased Akkermansia spp. population in the gut microbiota of mice. Gut 64 (6), 872–883. Available from: https://doi.org/10.1136/gutjnl-2014-307142.

Brown, E.M., Sadarangani, M., Finlay, B.B., 2013. The role of the immune system in governing host-microbe interactions in the intestine. Nat. Immunol. 14 (7), 660–667. Available from: https://doi.org/10.1038/ni.2611.

Caesar, R., Tremaroli, V., Kovatcheva-Datchary, P., Cani, P.D., Bäckhed, F., 2015. Crosstalk between gut microbiota and dietary lipids aggravates WAT inflammation through TLR signaling. Cell Metab. 22 (4), 658–668. Available from: https://doi.org/10.1016/j.cmet.2015.07.026.

Canfora, E.E., Jocken, J.W., Blaak, E.E., 2015. Short-chain fatty acids in control of body weight and insulin sensitivity. Nat. Rev. Endocrinol. 11 (10), 577–591. Available from: https://doi.org/10.1038/nrendo.2015.128.

Cani, P.D., 2018. Human gut microbiome: hopes, threats and promises. Gut 67 (9), 1716–1725. Available from: https://doi.org/10.1136/gutjnl-2018-316723.

Cao, H., Xu, M., Dong, W., Deng, B., Wang, S., Zhang, Y., et al., 2017. Secondary bile acid-induced dysbiosis promotes intestinal carcinogenesis. Int. J. Cancer 140 (11), 2545–2556. Available from: https://doi.org/10.1002/ijc.30643.

Chen, M.L., Yi, L., Zhang, Y., Zhou, X., Ran, L., Yang, J., et al., 2016. Resveratrol attenuates trimethylamine-N-oxide (TMAO)-induced atherosclerosis by regulating TMAO synthesis and bile acid metabolism via remodeling of the gut microbiota. mBio 7 (2), e02210–e02215. Available from: https://doi.org/10.1128/mBio.02210-15.

Clarke, S.F., Murphy, E.F., O'Sullivan, O., Lucey, A.J., Humphreys, M., Hogan, A., et al., 2014. Exercise and associated dietary extremes impact on gut microbial diversity. Gut 63 (12), 1913–1920. Available from: https://doi.org/10.1136/gutjnl-2013-306541.

Cong, J., Zhang, X., 2018. How human microbiome talks to health and disease. Eur. J. Clin. Microbiol. Infect. Dis.: Off. Publ. Eur. Soc. Clin. Microbiol. 37 (9), 1595–1601. Available from: https://doi.org/10.1007/s10096-018-3263-1.

Cotillard, A., Kennedy, S.P., Kong, L.C., Prifti, E., Pons, N., Le Chatelier, E., et al., 2013. Dietary intervention impact on gut microbial gene richness. Nature 500 (7464), 585–588. Available from: https://doi.org/10.1038/nature12480.

Cuervo, A., Valdés, L., Salazar, N., de los Reyes-Gavilán, C.G., Ruas-Madiedo, P., Gueimonde, M., et al., 2014. Pilot study of diet and microbiota: interactive associations of fibers and polyphenols with human intestinal bacteria. J. Agric. Food Chem. 62 (23), 5330–5336. Available from: https://doi.org/10.1021/jf501546a.

Cueva, C., Sánchez-Patán, F., Monagas, M., Walton, G.E., Gibson, G.R., Martín-Álvarez, P.J., et al., 2013. In vitro fermentation of grape seed flavan-3-ol fractions by human faecal microbiota: changes in microbial groups and phenolic metabolites. FEMS Microbiol. Ecol. 83 (3), 792–805. Available from: https://doi.org/10.1111/1574-6941.12037.

Dabke, K., Hendrick, G., Devkota, S., 2019. The gut microbiome and metabolic syndrome. J. Clin. Invest. 129 (10), 4050–4057. Available from: https://doi.org/10.1172/JCI129194.

Davani-Davari, D., Negahdaripour, M., Karimzadeh, I., Seifan, M., Mohkam, M., Masoumi, S.J., et al., 2019. Prebiotics: definition, types, sources, mechanisms, and clinical applications. Foods (Basel, Switz.) 8 (3), 92. Available from: https://doi.org/10.3390/foods8030092.

Davenport, E.R., Mizrahi-Man, O., Michelini, K., Barreiro, L.B., Ober, C., Gilad, Y., 2014. Seasonal variation inhuman gut microbiome composition. PLoS One 9, e9073. Available from: https://doi.org/10.1371/journal.pone.0090731.

David, L., Maurice, C., Carmody, R., et al., 2014. Diet rapidly and reproducibly alters the human gut microbiome. Nature 505, 559–563. Available from: https://doi.org/10.1038/nature12820.

de Cossío, L.F., Fourrier, C., Sauvant, J., Everard, A., Capuron, L., Cani, P.D., et al., 2017. Impact of prebiotics on metabolic and behavioral alterations in a mouse model of metabolic syndrome. Brain Behav. Immun. 64, 33–49. Available from: https://doi.org/10.1016/j.bbi.2016.12.022.

De Filippis, F., Pellegrini, N., Vannini, L., Jeffery, I.B., La Storia, A., Laghi, L., et al., 2016. High-level adherence to a Mediterranean diet beneficially impacts the gut microbiota and associated metabolome. Gut 65 (11), 1812–1821. Available from: https://doi.org/10.1136/gutjnl-2015-309957.

de Toro-Martín, J., Arsenault, B.J., Després, J.P., Vohl, M.C., 2017. Precision nutrition: a review of personalized nutritional approaches for the prevention and management of metabolic syndrome. Nutrients 9 (8), 913. Available from: https://doi.org/10.3390/nu9080913.

Del Rio, D., Rodriguez-Mateos, A., Spencer, J.P., Tognolini, M., Borges, G., Crozier, A., 2013. Dietary (poly) phenolics in human health: structures, bioavailability, and evidence of protective effects against chronic diseases. Antioxid. Redox Signal. 18, 1818–1892. Available from: https://doi.org/10.1089/ars.2012.4581.

Eid, N., Enani, S., Walton, G., Corona, G., Costabile, A., Gibson, G., et al., 2014. The impact of date palm fruits and their component polyphenols, on gut microbial ecology, bacterial metabolites and colon cancer cell proliferation. J. Nutr. Sci. 3, e46. Available from: https://doi.org/10.1017/jns.2014.16.

Ejtahed, H.S., Angoorani, P., Hasani-Ranjbar, S., Siadat, S.D., Ghasemi, N., Larijani, B., et al., 2018. Adaptation of human gut microbiota to bariatric surgeries in morbidly obese patients: a systematic review. Microb. Pathog. 116, 13–21. Available from: https://doi.org/10.1016/j.micpath.2017.12.074.

Eslamparast, T., Zamani, F., Hekmatdoost, A., Sharafkhah, M., Eghtesad, S., Malekzadeh, R., et al., 2014. Effects of synbiotic supplementation on insulin resistance in subjects with the metabolic syndrome: a randomised, double-blind, placebo-controlled pilot study. Br. J. Nutr. 112 (3), 438–445. Available from: https://doi.org/10.1017/S0007114514000919.

Everard, A., Lazarevic, V., Gaïa, N., Johansson, M., Ståhlman, M., Backhed, F., et al., 2014. Microbiome of prebiotic-treated mice reveals novel targets involved in host response during obesity. ISME J. 8 (10), 2116–2130. Available from: https://doi.org/10.1038/ismej.2014.45.

Fava, F., Gitau, R., Griffin, B.A., Gibson, G.R., Tuohy, K.M., Lovegrove, J.A., 2013. The type and quantity of dietary fat and carbohydrate alter faecal microbiome and short-chain fatty acid excretion in a metabolic syndrome 'at-risk' population. Int. J. Obes. 37 (2), 216–223. Available from: https://doi.org/10.1038/ijo.2012.33.

Fetissov, S.O., 2017. Role of the gut microbiota in host appetite control: bacterial growth to animal feeding behaviour. Nat. Rev. Endocrinol. 13 (1), 11–25. Available from: https://doi.org/10.1038/nrendo.2016.150.

Floch, M.H., 2018. The role of prebiotics and probiotics in gastrointestinal disease. Gastroenterol. Clin. 47, 179–191. Available from: https://doi.org/10.1016/j.gtc.2017.09.011.

Ghosh, S., Molcan, E., DeCoffe, D., Dai, C., Gibson, D.L., 2013. Diets rich in n-6 PUFA induce intestinal microbial dysbiosis in aged mice. Br. J. Nutr. 110 (3), 515–523. Available from: https://doi.org/10.1017/S0007114512005326.

Gomaa, E.Z., 2020. Human gut microbiota/microbiome in health and diseases: a review. Antonie van. Leeuwenhoek 113 (12), 2019–2040. Available from: https://doi.org/10.1007/s10482-020-01474-7.

Green, M., Arora, K., Prakash, S., 2020. Microbial medicine: prebiotic and probiotic functional foods to target obesity and metabolic syndrome. Int. J. Mol. Sci. 21 (8), 2890. Available from: https://doi.org/10.3390/ijms21082890.

Halmos, E.P., Christophersen, C.T., Bird, A.R., Shepherd, S.J., Gibson, P.R., Muir, J.G., 2015. Diets that differ in their FODMAP content alter the colonic luminal microenvironment. Gut 64 (1), 93–100. Available from: https://doi.org/10.1136/gutjnl-2014-307264.

Huang, P.L., 2009. A comprehensive definition for metabolic syndrome. Dis. Model. Mech. 2 (5–6), 231–237. Available from: https://doi.org/10.1242/dmm.001180.

Huang, E.Y., Leone, V.A., Devkota, S., Wang, Y., Brady, M.J., Chang, E.B., 2013. Composition of dietary fat source shapes gut microbiota architecture and alters host inflammatory mediators in mouse adipose tissue. J. Parenter. Enteral. Nutr. 37, 746–754. Available from: https://doi.org/10.1177/0148607113486931. **2013**.

Jakobsdottir, G., Xu, J., Molin, G., Ahrne, S., Nyman, M., 2013. High-fat diet reduces the formation of butyrate, but increases succinate, inflammation, liver fat and cholesterol in rats, while dietary fibre counteracts these effects. PLoS One 8, e80476. Available from: https://doi.org/10.1371/journal.pone.0080476.

Jeffery, I.B., O'Toole, P.W., 2013. Diet-microbiota interactions and their implications for healthy living. Nutrients 5 (1), 234–252. Available from: https://doi.org/10.3390/nu5010234.

Kaliannan, K., Wang, B., Li, X.-Y., Bhan, A.K., Kang, J.X., 2016. Omega-3 fatty acids prevent early-life antibioticexposure-induced gut microbiota dysbiosis and later-life obesity. Int. J. Obes. 40, 1039–1042. Available from: https://doi.org/10.1038/ijo.2016.27.

Karlsson, F.H., Fåk, F., Nookaew, I., Tremaroli, V., Fagerberg, B., Petranovic, D., et al., 2012. Symptomatic atherosclerosis is associated with an altered gut metagenome. Nat. Commun. 3, 1245. Available from: https://doi.org/10.1038/ncomms2266.

Kelly, C., Zheng, L., Campbell, E., Saeedi, B., Scholz, C., Bayless, A., et al., 2015. Crosstalk between microbiota-derived short-chain fatty acids and intestinal epithelial HIF augments tissue barrier function. Cell Host Microbe 17 (5), 662–671. Available from: https://doi.org/10.1016/j.chom.2015.03.005.

Khan, M.J., Gerasimidis, K., Edwards, C.A., Shaikh, M.G., 2016. Role of gut microbiota in the aetiology of obesity: proposed mechanisms and review of the literature. J. Obes. 2016, 7353642. Available from: https://doi.org/10.1155/2016/7353642.

Khan, H., Ullah, H., Tundis, R., Belwal, T., Devkota, H.P., Daglia, M., et al., 2020. Dietary flavonoids in the management of huntington's disease: mechanism and clinical perspective. eFood 1, 38–52. Available from: https://www.atlantis-press.com/journals/efood/125934127.

Kim, C.H., Park, J., Kim, M., 2014. Gut microbiota-derived short-chain Fatty acids, T cells, and inflammation. Immune Netw. 14 (6), 277–288. Available from: https://doi.org/10.4110/in.2014.14.6.277.

Kimura, I., Ozawa, K., Inoue, D., Imamura, T., Kimura, K., Maeda, T., et al., 2013. The gut microbiota suppresses insulin-mediated fat accumulation via the short-chain fatty acid receptor GPR43. Nat. Commun. 4, 1–12. Available from: https://doi.org/10.1038/ncomms2852.

Kjølbæk, L., Benítez-Páez, A., Del Pulgar, E.M.G., Brahe, L.K., Liebisch, G., Matysik, S., et al., 2020. Arabinoxylan oligosaccharides and polyunsaturated fatty acid effects on gut microbiota and metabolic markers in overweight individuals with signs of metabolic syndrome: a randomized cross-over trial. Clin. Nutr. 39, 67–79. Available from: https://doi.org/10.1016/j.clnu.2019.01.012.

Korecka, A., de Wouters, T., Cultrone, A., Lapaque, N., Pettersson, S., Doré, J., et al., 2013. ANGPTL4 expression induced by butyrate and rosiglitazone in human intestinal epithelial cells utilizes independent pathways. Am. J. Physiol. Gastrointest. Liver Physiol. 304 (11), G1025–G1037. Available from: https://doi.org/10.1152/ajpgi.00293.2012.

Ley, R.E., Bäckhed, F., Turnbaugh, P., Lozupone, C.A., Knight, R.D., Gordon, J.I., 2005. Obesity alters gut microbial ecology. Proc. Natl Acad. Sci. U S A. 102 (31), 11070–11075. Available from: https://doi.org/10.1073/pnas.0504978102.

Li, K., Zhang, L., Xue, J., Yang, X., Dong, X., Sha, L., et al., 2019. Dietary inulin alleviates diverse stages of type 2 diabetes mellitus via anti-inflammation and modulating gut microbiota in db/db mice. Food Funct. 10 (4), 1915–1927. Available from: https://doi.org/10.1039/c8fo02265h.

Li, A.-L., Ni, W.-W., Zhang, Q.-M., Li, Y., Zhang, X., Wu, H.-Y., et al., 2020. Effect of cinnamon essential oil on gut microbiota in the mouse model of dextran sodium sulfate-induced colitis. Microbiol. Immunol. 64, 23–32. Available from: https://doi.org/10.1111/1348-0421.12749.

Liu, W., Zhao, S., Wang, J., Shi, J., Sun, Y., Wang, W., et al., 2017. Grape seed proanthocyanidin extract ameliorates inflammation and adiposity by modulating gut microbiota in high-fat diet mice. Mol. Nutr. Food Res. 61, 1601082. Available from: https://doi.org/10.1002/mnfr.201601082.

Lozupone, C.A., Stombaugh, J.I., Gordon, J.I., Jansson, J.K., Knight, R., 2012. Diversity, stability and resilience of the human gut microbiota. Nature 489 (7415), 220–230. Available from: https://doi.org/10.1038/nature11550.

Mäkeläinen, H., Saarinen, M., Stowell, J., Rautonen, N., Ouwehand, A.C., 2010. Xylo-oligosaccharides and lactitol promote the growth of Bifidobacterium lactis and Lactobacillus species in pure cultures. Benef. Microbes 1 (2), 139–148. Available from: https://doi.org/10.3920/BM2009.0029.

Marette, A., Jobin, C., 2015. SCFAs take a toll en route to metabolic syndrome. Cell Metab. 22, 954–956. Available from: https://doi.org/10.1016/j.cmet.2015.11.006.

Marques, F.Z., Nelson, E., Chu, P.Y., Horlock, D., Fiedler, A., Ziemann, M., et al., 2017. High-fiber diet and acetate supplementation change the gut microbiota and prevent the development of hypertension and heart failure in hypertensive mice. Circulation 135 (10), 964–977. Available from: https://doi.org/10.1161/CIRCULATIONAHA.116.024545.

Martin, F.P., Montoliu, I., Nagy, K., Moco, S., Collino, S., Guy, P., et al., 2012. Specific dietary preferences are linked to differing gut microbial metabolic activity in response to dark chocolate intake. J. Proteome Res. 11 (12), 6252–6263. Available from: https://doi.org/10.1021/pr300915z.

Martínez-González, M.A., Gea, A., Ruiz-Canela, M., 2019. The mediterranean diet and cardiovascular health. Circ. Res. 124 (5), 779–798. Available from: https://doi.org/10.1161/CIRCRESAHA.118.313348.

Mills, S., Stanton, C., Lane, J., Smith, G., Ross, R., 2019. Precision nutrition and the microbiome, part I: current state of the science. Nutrients 11 (4), 923–968. Available from: https://doi.org/10.3390/nu11040923.

Mishra, S.S., Behera, P.K., Kar, B., Ray, R.C., 2018. Advances in probiotics, prebiotics and nutraceuticals, Innovations in Technologies for Fermented Food and Beverage Industries, 2018. Springer, Cham, Switzerland, pp. 121–141. Available from: https://www.springer.com/gp/book/9783319748191.

Molinaro, A., Wahlström, A., Marschall, H.U., 2018. Role of bile acids in metabolic control. Trends Endocrinol. Metabol. TEM 29 (1), 31–41. Available from: https://doi.org/10.1016/j.tem.2017.11.002.

Moran-Ramos, S., López-Contreras, B.E., Canizales-Quinteros, S., 2017. Gut microbiota in obesity and metabolic abnormalities: a matter of composition or functionality? Arch. Med. Res. 48 (8), 735–753. Available from: https://doi.org/10.1016/j.arcmed.2017.11.003.

Muscogiuri, G., Cantone, E., Cassarano, S., Tuccinardi, D., Barrea, L., Savastano, S., et al., 2019. Gut microbiota: a new path to treat obesity. Int. J. Obes. Suppl. 9 (1), 10–19. Available from: https://doi.org/10.1038/s41367-019-0011-7.

Natividad, J.M., Agus, A., Planchais, J., Lamas, B., Jarry, A.C., Martin, R., et al., 2018. Impaired aryl hydrocarbon receptor ligand production by the gut microbiota is a key factor in metabolic syndrome. Cell Metab. 28 (5), 737–749. Available from: https://doi.org/10.1016/j.cmet.2018.07.001.

Patterson, E., Ryan, P.M., Cryan, J.F., Dinan, T.G., Ross, R.P., Fitzgerald, G.F., et al., 2016. Gut microbiota, obesity and diabetes. Postgrad. Med. J. 92 (1087), 286–300. Available from: https://doi.org/10.1136/postgradmedj-2015-133285.

Pushpanathan, P., Mathew, G.S., Selvarajan, S., Seshadri, K.G., Srikanth, P., 2019. Gut microbiota and its mysteries. Indian. J. Med. Microbiol. 37 (2), 268–277. Available from: https://doi.org/10.4103/ijmm.ijmm_19_373.

Rajjo, T., Mohammed, K., Alsawas, M., Ahmed, A.T., Farah, W., Asi, N., et al., 2017. Treatment of pediatric obesity: an umbrella systematic review. J. Clin. Endocrinol. Metab. 102 (3), 763–775. Available from: https://doi.org/10.1210/jc.2016-2574.

Rajkumar, H., Mahmood, N., Kumar, M., Varikuti, S.R., Challa, H.R., Myakala, S.P., 2014. Effect of probiotic (VSL#3) and omega-3 on lipid profile, insulin sensitivity, inflammatory markers, and gut colonization in overweight adults: a randomized, controlled trial. Mediat. Inflamm. 2014, 348959. Available from: https://doi.org/10.1155/2014/348959.

Reyes, A., Haynes, M., Hanson, N., Angly, F.E., Heath, A.C., Rohwer, F., et al., 2010. Viruses in the faecal microbiota of monozygotic twins and their mothers. Nature 466 (7304), 334–338. Available from: https://doi.org/10.1038/nature09199.

Rothschild, D., Weissbrod, O., Barkan, E., Kurilshikov, A., Korem, T., Zeevi, D., et al., 2018. Environment dominates over host genetics in shaping human gut microbiota. Nature 555 (7695), 210–215. Available from: https://doi.org/10.1038/nature25973.

Salazar, N., Valdés-Varela, L., González, S., Gueimonde, M., de Los Reyes-Gavilán, C.G., 2017. Nutrition and the gut microbiome in the elderly. Gut Microbes 8 (2), 82–97. Available from: https://doi.org/10.1080/19490976.2016.1256525.

Saltiel, A.R., 2016. Insulin signaling in the control of glucose and lipid homeostasis. Handb. Exp. Pharmacol. 233, 51–71. Available from: https://doi.org/10.1007/164_2015_14.

Schnack, L.L., Romani, A., 2017. The metabolic syndrome and the relevance of nutrients for its onset. Recent. Pat. Biotechnol. 11 (2), 101–119. Available from: https://doi.org/10.2174/1872208311666170227112013.

Schneeberger, M., Everard, A., Gómez-Valadés, A.G., Matamoros, S., Ramírez, S., Delzenne, N.M., et al., 2015. Akkermansia muciniphila inversely correlates with the onset of inflammation, altered adipose tissue metabolism and metabolic disorders during obesity in mice. Sci. Rep. 5, 16643. Available from: https://doi.org/10.1038/srep16643.

Schroeder, B.O., Bäckhed, F., 2016. Signals from the gut microbiota to distant organs in physiology and disease. Nat. Med. 22 (10), 1079–1089. Available from: https://doi.org/10.1038/nm.4185.

Scotti, E., Boue, S., Sasso, L.G., Flippo, Z., 2017. Exploring the microbiome in health and disease: implications for toxicology. Toxicol. Res. Appl. 2017. Available from: https://doi.org/10.1177/2397847317741884.

Shen, J., Zuo, Z.X., Mao, A.P., 2014. Effect of probiotics on inducing remission and maintaining therapy in ulcerative colitis, Crohn's disease, and pouchitis: meta-analysis of randomized controlled trials. Inflamm. Bowel Dis. 20 (1), 21–35. Available from: https://doi.org/10.1097/01.MIB.0000437495.30052.be.

Shreiner, A.B., Kao, J.Y., Young, V.B., 2015. The gut microbiome in health and in disease. Curr. Opin. Gastroenterol. 31 (1), 69–75. Available from: https://doi.org/10.1097/MOG.0000000000000139.

Sonnenburg, E.D., Sonnenburg, J.L., 2014. Starving our microbial self: the deleterious consequences of a diet deficient in microbiota-accessible carbohydrates. Cell Metab. 20 (5), 779–786. Available from: https://doi.org/10.1016/j.cmet.2014.07.003.

Sun, Z., Sun, X., Li, J., Li, Z., Hu, Q., Li, L., et al., 2020. Using probiotics for type 2 diabetes mellitus intervention: advances, questions, and potential. Crit. Rev. Food Sci. Nutr. 60 (4), 670–683. Available from: https://doi.org/10.1080/10408398.2018.1547268.

Tuohy, K.M., Fava, F., Viola, R., 2014. 'The way to a man's heart is through his gut microbiota'—dietary pro- and prebiotics for the management of cardiovascular risk. Proc. Nutr. Soc. 73 (2), 172–185. Available from: https://doi.org/10.1017/S0029665113003911.

Wang, Z., Klipfell, E., Bennett, B.J., Koeth, R., Levison, B.S., Dugar, B., et al., 2011. Gut flora metabolism of phosphatidylcholine promotes cardiovascular disease. Nature 472 (7341), 57–63. Available from: https://doi.org/10.1038/nature09922.

Wang, Z., Roberts, A.B., Buffa, J.A., Levison, B.S., Zhu, W., Org, E., et al., 2015. Non-lethal inhibition of gut microbial trimethylamine production for the treatment of atherosclerosis. Cell 163 (7), 1585–1595. Available from: https://doi.org/10.1016/j.cell.2015.11.055.

Wang, P.X., Deng, X.R., Zhang, C.H., Yuan, H.J., 2020. Gut microbiota and metabolic syndrome. Chin. Med. J. 133 (7), 808–816. Available from: https://doi.org/10.1097/CM9.0000000000000696.

Wiley, N.C., Dinan, T.G., Ross, R.P., Stanton, C., Clarke, G., Cryan, J.F., 2017. The microbiota-gut-brain axis as a key regulator of neural function and the stress response: implications for human and animal health. J. Anim. Sci. 95 (7), 3225–3246. Available from: https://doi.org/10.2527/jas2016.1256.

Yoshida, N., Emoto, T., Yamashita, T., Watanabe, H., Hayashi, T., Tabata, T., et al., 2018. Bacteroides vulgatus and Bacteroides dorei reduce gut microbial lipopolysaccharide production and inhibit atherosclerosis. Circulation 138 (22), 2486–2498. Available from: https://doi.org/10.1161/CIRCULATIONAHA.118.033714.

Zhao, L., Zhang, Q., Ma, W., Tian, F., Shen, H., Zhou, M., 2017. A combination of quercetin and resveratrol reduces obesity in high-fat diet-fed rats by modulation of gut microbiota. Food Funct. 8, 4644–4656. Available from: https://doi.org/10.1039/C7FO01383C.

Zheng, P., Zeng, B., Liu, M., Chen, J., Pan, J., Han, Y., et al., 2019. The gut microbiome from patients with schizophrenia modulates the glutamate-glutamine-GABA cycle and schizophrenia-relevant behaviors in mice. Sci. Adv. 5 (2), 8317–8326. Available from: https://doi.org/10.1126/sciadv.aau8317.

Further reading

Conlon, M.A., Bird, A.R., 2014. The impact of diet and lifestyle on gut microbiota and human health. Nutrients 7 (1), 17–44. Available from: https://doi.org/10.3390/nu7010017.

Gibson, G.R., Probert, H.M., Loo, J.V., Rastall, R.A., Roberfroid, M.B., 2004. Dietary modulation of the human colonic microbiota: updating the concept of prebiotics. Nutr. Res. Rev. 17 (2), 259–275. Available from: https://doi.org/10.1079/NRR200479.

Laugerette, F., Furet, J.-P., Debard, C., Daira, P., Loizon, E., Géloën, A., et al., 2012. Oil composition of high-fat diet affects metabolic inflammation differently in connection with endotoxin receptors in mice. Am. J. Physiol. Endocrinol. Metab. 302, E374–E386. Available from: https://doi.org/10.1152/ajpendo.00314.2011.

Lynch, S., 2016. The human intestinal microbiome in health and disease. Available from: <https://www.nejm.org/doi/10.1056/NEJMra1600266>.

Perry, R.J., Peng, L., Barry, N.A., Cline, G.W., Zhang, D., Cardone, R.L., et al., 2016. Acetate mediates a microbiome-brain-β-cell axis to promote metabolic syndrome. Nature 534 (7606), 213–217. Available from: https://doi.org/10.1038/nature18309.

Si, X., Shang, W., Zhou, Z., Strappe, P., Wang, B., Bird, A., et al., 2018. Gut microbiome-induced shift of acetate to butyrate positively manages dysbiosis in high fat diet. Mol. Nutr. Food Res. 62 (3). Available from: https://doi.org/10.1002/mnfr.201700670.

Singh, R.K., Chang, H.W., Yan, D., et al., 2017. Influence of diet on the gut microbiome and implications for human health. J. Transl. Med. 15, 73. Available from: https://doi.org/10.1186/s12967-017-1175-y.

Thorburn, A.N., Macia, L., Mackay, C.R., 2014. Diet, metabolites, and "western-lifestyle" inflammatory diseases. Immunity 40 (6), 833–842. Available from: https://doi.org/10.1016/j.immuni.2014.05.014.

Tirosh, A., Calay, E.S., Tuncman, G., Claiborn, K.C., Inouye, K.E., Eguchi, K., et al., 2019. The short-chain fatty acid propionate increases glucagon and FABP4 production, impairing insulin action in mice and humans. Sci. Transl. Med. 11 (489), eaav0120. Available from: https://doi.org/10.1126/scitranslmed.aav0120.

Chapter 12

Restoring gut biome balance for weight loss: clinical applications

Naras Bhat, Jyoti Bhat, Kusum Bhat, Imran Junaid, Anita Bhat and Joy Bhat
Jivahealth Clinics, Concord, CA, United States

12.1 Introduction

The gut microbiome comprises all the microbes in the intestines. The gut biome act as another organ that's crucial for your health. The gut microbiota represents a metabolic gateway and command center through interaction with the host nutritional environment, inflammation, the immune system, energy balance, and body weight. In short, a diverse gut microbiome is beneficial for human health. Gut dysbiosis means reduced diversity, reduced resilience, lower levels of beneficial bacteria, and more opportunistic pathogens (Aoun et al., 2020).

According to the World Health Organization, the worldwide prevalence of obesity nearly tripled between 1975 and 2016. Globally, 39% of adults aged 18 years and older were overweight or obese in 2016. In most high-income countries, around two-thirds of adults are overweight or obese (Hannah Ritchie and Max Roser, 2017). In the United States about 74% of adults are overweight or have obesity (US Department of Agriculture and US Department of Health and Human Services, 2020).

The so-called Western lifestyle is a "master" manipulator of the intestinal microbiota (i.e., epigenetic influence). The U.S.-style dietary pattern is also known as the standard American diet (SAD). We currently exist in a state of evolutionary discordance; our genome has failed to adapt quickly in response to the rapid change in its environment. A Western dietary pattern is characterized by overconsumption of ultraprocessed foods that are devoid of microbial accessible fermentable fiber and polyphenols, an abundance of pro-inflammatory omega-6 cooking oil and animal fat, a sedentary lifestyle, urbanization, and disrupted circadian rhythms in a computer-centered life. The process of Westernization is increasingly spreading across the world (Hold, 2014; Martínez Steele et al., 2016).

The prevalence of shift work is also increasing rapidly because of industrialization and modernization. The past decade has seen a considerable increase in the literature about shift work. Several studies have verified that shift work is associated with reduced glucose tolerance, increased insulin resistance, and an increased risk of obesity (Ding and Xiao, 2020).

12.2 Gut microbiome resiliency

Resilience is the ability of an ecosystem to maintain its state and recover from perturbations. A resilient microbiota could be used as a surrogate for a healthy microbiota. The gut microbiota as an organ faces challenges and insults, such as diet, drugs, physical activity (PA), and circadian misalignment. An impaired ability to bounce back to the prechallenge baseline underpins a lack of resilience and dysbiosis (Dogra et al., 2020).

Furthermore, resilience is determined by the capacity to reduce the impact on the microbiome (resistance) and to recover from the impact of disturbance (recovery). An example is when antibiotic-induced perturbation of the microbiota is counterbalanced by higher prechallenge microbiota diversity (Raymond et al., 2016).

This review focuses on the current evidence regarding gut microbial disruption by a Western lifestyle and pragmatic clinical interventions to restore and reclaim resilience. Because the microbiota is easily manipulable by prebiotics, probiotics, antibiotics, fecal transplants, and dietary changes, altering the microbiota offers a tractable approach to the otherwise intractable problems of obesity and unhealthy lifestyle.

Microbiome, Immunity, Digestive Health and Nutrition. DOI: https://doi.org/10.1016/B978-0-12-822238-6.00022-4

12.3 Obesogenic features of the gut microbiome

12.3.1 Obesogenic gut microbiome is transferrable

A groundbreaking study found that obesity-resistant germ-free mice become obese and increase their energy harvest and caloric uptake after receiving a microbiota transplant from obese mice (Turnbaugh et al., 2006).

When obese and lean subjects were fed an isocaloric diet, the obese subjects stored more calories as fat in comparison to their lean counterparts (Jumpertz et al., 2011). A fecal microbiota transfer from an obese donor to a lean recipient recapitulated the obese phenotype as well as associated metabolic dysfunctions (Napolitano and Covasa, 2020; Zhang et al., 2019).

12.3.2 Mother-to-child transmission of obesity

An offspring could be a product of the combination of the dietary patterns of both the offspring and mother. Prenatal exposure to an adverse (obesogenic) microbiota-metabolic environment may increase the offspring's susceptibility to obesity in postnatal life. The ability of prebiotics to encourage the growth and metabolism of distinct health-promoting gut microbiota was explored to mitigate the intergenerational transmission of obesity to offspring (Soderborg et al., 2016).

12.4 Dysbiosis as obesogenic signature

There are four parts to the obesogenic signature of gut biome: Decreased diversity, an increased Firmicutes/Bacteroidetes ratio, decreased keystone bacteria, and increased gut barrier permeability.

12.4.1 Decreased diversity

In overweight and obese humans, low fecal bacterial diversity is associated with more marked overall adiposity, impaired glucose homeostasis, and higher low-grade inflammation (Peters et al., 2018).

12.4.2 Firmicutes/bacteroidetes ratio

Across multiple studies, the ratio of Firmicutes/Bacteroidetes (two major bacteria phyla) has been correlated with obesity, with a higher ratio (more Bacteroidetes) associated with leanness and a lower ratio (more *Firmicutes*) associated with higher body fat (Kasai et al., 2015).

A shift in the gut microbiota in favor of Firmicutes, for example, with consumption of a high-fat, high-carbohydrate diet increases energy extraction from the diet with corresponding weight gain and obesity if left uncontrolled. In contrast, consumption of a diet that is low in fat and sugar increases Bacteroidetes dominance, which encourages weight loss by stimulating increased expression of fasting-induced adipocyte factor and subsequent increase in energy expenditure and reduced fat storage.

12.4.3 Keystone bacteria

The term "keystone" indicates the pivotal role of certain bacteria for a healthy gut (Castaner et al., 2018).

Akkermansia muciniphila, a mucin-degrading bacterium, resides in the mucus layer and is responsible for gut barrier thickness. Its presence inversely correlates with body weight in rodents and humans. Diets that are high saturated fat diet and alcohol consumption reduce it. Prebiotics and dietary polyphenols increase it.

Faecalibacterium prausnitzii has been consistently reported as one of the main butyrate producers found in the intestine (Lopez-Siles et al., 2017). Butyrate is the primary fuel for the cells of the gut lining and maintains its integrity. Butyrate has antiinflammatory properties and stimulates the body's production of glutathione, an antioxidant molecule. Notably, prebiotics such as inulin-type fructans and arabinoxylans increase the number of *F. prausnitzii*.

Bifidobacteria is also important to help the fermentation of prebiotics to produce short-chain fatty acids (SCFAs), such as acetate, butyrate, and propionate (O'Callaghan and van Sinderen, 2016). Certain lactic acid bacteria, in particular certain members of the genus *Lactobacillus*, and members of the *Bifidobacterium* genus make up most currently commercialized probiotic food products.

12.4.4 Gut barrier permeability

Dysbiosis and increased gut permeability are two vulnerable factors in producing diet-induced metabolic endotoxemia (ME) and diet-induced inflammation. The critical positioning of the intestinal barrier at the interface between environment and internal milieu have led to the popularity of the "leaky gut" concept. Bacterial translocation can occur through the paracellular route, which is regulated by tight junctions (TJs) or the transcellular route.

12.4.4.1 Gut biome–focused lab tests

Three clinically useful, obesity-focused gut health tests are: Gut barrier test, Stool microbiome test, and Omega-3 index.

12.4.4.2 Gut barrier tests

Gut barrier permeability tests: Lactulose/Mannose tests: An elevated lactulose to mannitol ratio is an indicator of intestinal barrier dysfunction.

Zonulin: Quantitative fecal zonulin is a more commonly used gut barrier test in the United States.

12.4.4.3 Stool microbiome test

The stool is by far the most accessible material for gut microbiome analysis. The stool microbiome test (PCR 16S rDNA testing and amplicon analysis) is generally a most informative analysis. Several commercial labs do stool microbiome tests in the United States.

12.4.4.4 Omega-3 index

The Omega-3 Index test measures the amount of eicosapentaenoic acid and docosahexaenoic omega-3 fatty acids in the blood. Studies have shown that 8% is the optimal level of omega-3 fatty acid. Requiring one drop of blood, the in-home test kit can give an unbiased view of dietary intake of omega-3 fatty acids and inflammation potential (Walker et al., 2019).

12.4.5 Concept of prebiotics, probiotics, and postbiotics

Gut biome–centric knowledge of prebiotics, probiotics, and postbiotics is a prerequisite for a clinical approach.

12.4.5.1 Prebiotics

Prebiotics are fermentable foods for probiotics. They include microbiota-accessible carbohydrates, fermentable dietary fiber, and polyphenols. They act like fertilizers that stimulate the growth of healthy bacteria in the gut. Humans do not normally digest prebiotics directly.

We use a clinically meaningful fiber classification based on solubility, fermentation, viscosity, and gel formation (McRorie, 2015):

1. Insoluble, poorly fermented (e.g., wheat bran): "plastic effect" as a roughage in stool
2. Soluble, nonviscous, readily fermented (e.g., wheat dextrin, oligosaccharides (inulin), resistant starches): prebiotic function
3. Soluble viscous/gel forming, readily fermented (e.g., β-glucan [oats, barley], raw guar gum): prebiotic function
4. Soluble viscous/gel forming, nonfermentable (e.g., psyllium): stool softener/bulking agents

Further relevant details about oligosaccharides, resistant starch (RS), and polyphenols are provided in the following sections.

12.4.5.2 Oligosaccharides

Sandwiched between simple sugars (monosaccharides) and starches (polysaccharides) are the oligosaccharides. Small amounts of oligosaccharides occur naturally in many plants, but chicory root and Jerusalem artichokes are especially rich in oligosaccharides. Inulin from these sources is a commonly used oligosaccharide as a prebiotic.

12.4.5.3 Resistant starch

RS is fermented only in the colon, where bacteria produce SCFAs. Resistant starches are classified according to structure or source, as follows (Sajilata et al., 2006):

1. RS-1 is physically inaccessible to digestive enzymes. Its sources include whole or partially milled grains, seeds, and legumes. Clinically, we use cooked beans and lentils to provide this group.

2. RS-2 resists digestion because of the granule's nature. Sources include raw potatoes, underripe or green bananas, some legumes, and high-amylose starches, such as high-amylose corn. Clinically, we use potato starch powder and green banana powder as prebiotic supplements from this group.
3. RS-3 is produced in the cooking-cooling process. Sources include bread, tortillas, cooked and cooled potatoes, rice, and pasta. Clinically, we recommend cooked and overnight refrigerated staple foods such as rice, wheat, and potato.
4. RS-4 is a chemically modified starch that is found in a wide range of products.

12.4.5.4 Polyphenols

Dietary polyphenols are compounds of natural origin that are present in food items such as vegetables, fruits, cereals, tea, coffee, dark chocolate, cocoa powder, and wine. More than 90% of ingested polyphenols remain unprocessed through the small intestine and reach the large intestine, where they act as prebiotics (Kumar Singh et al., 2019).

Recent studies suggest that polyphenols are candidates for prebiotics because consumption of polyphenols, especially catechins, anthocyanins, and proanthocyanins, increases the abundance of *Lactobacillus*, *Bifidobacterium*, *Akkermansia*, and *Faecalibacterium* spp.

12.4.5.5 Probiotics

Probiotics are live bacteria that are known to provide health benefits when present in the human gut. Like prebiotics, probiotics can be ingested through both food and supplements. The most common probiotic-containing foods are yogurt and fermented foods.

12.4.5.6 Postbiotics

Postbiotics are the bioactive compounds the probiotic bacteria produce when they consume prebiotics. Postbiotics are metabolites or fragments derived from microorganisms. Such postbiotics have pleiotropic effects, including host energy balance, immunomodulatory, antiinflammatory, antioxidant, anticancer, and autophagy properties (Żółkiewicz et al., 2020). Although postbiotics include SCFAs, lipopolysaccharide (LPS), enzymes, vitamins, and other cytokines, for simplicity's sake we will elaborate SCFAs and LPS, which have opposing biological effects.

SCFAs are products of fermentation of plant polysaccharides by intestinal microbiota. Well-known SCFAs include acetic, propionic, and butyric acids, which can form the corresponding fatty acid salts (i.e., acetate, propionate, and butyrate). SCFAs play a key role in regulation of appetite, nourishment of other types of bacteria in the microbiome, a healthy inflammatory balance, and regulation of the immune system and are the primary fuel for cells that make up the lining of the large intestine.

LPS, which are also known as endotoxins, are cell wall components of gram-negative bacteria. Binding of LPS-protein complexes to the toll-like receptor (TLR) 4 activates the cellular NF-κB signaling pathway, which in turn leads to the production of various proinflammatory cytokines and chemokines. Low-grade elevation in plasma LPS has been termed metabolic endotoxemia, and this state is associated with a heightened proinflammatory and oxidant environment that is often observed in obesity (Boutagy et al., 2016).

In short, SCFAs and LPS have opposing effects on gut health. The effects of SCFAs are favorable; the effects of LPS are unfavorable.

12.5 Obesogenic metabolic endotoxemia and insulin resistance

12.5.1 Diet-induced lipopolysaccharide endotoxemia

Metabolic endotoxemia is a condition that is estimated to affect approximately 33% of the Western population. The condition is characterized by increased LPS in the blood during the first 5 hours of the postprandial period following consumption of an ultraprocessed meal. Meals that are high in fat and dense in calories seem to affect the condition more than do low-fat and low-calorie meals. This increase in serum endotoxin level is followed by elevated inflammation that is marked by measurable increases in inflammation markers such as interleukin-6, interleukin-1-alpha, interferon-gamma, triglycerides, and postprandial insulin (Duan et al., 2018).

ME has three basic steps: dysbiosis, gut barrier permeability, and LPS translocation. Dysbiosis is characterized by decreased microbial diversity, decreased amounts of *A. muciniphila*, decreased Bacteroidetes, and increased Firmicutes. Gut barrier impairment occurs when a disrupted TLR signaling pathway, a decreased mucus layer, and a decreased tight junction leads to increased permeability. LPS escapes from gut lumen into the blood plasma. LPS translocation occurs when LPS, free fatty acids, and proinflammatory cytokines (IL-1B, IL-6, and TNF-α) enter the systemic circulation and activate proinflammatory M-1 macrophages. These signals are sent to multiple target organs along a different axis (Ghosh et al., 2020) (Figure 12.1).

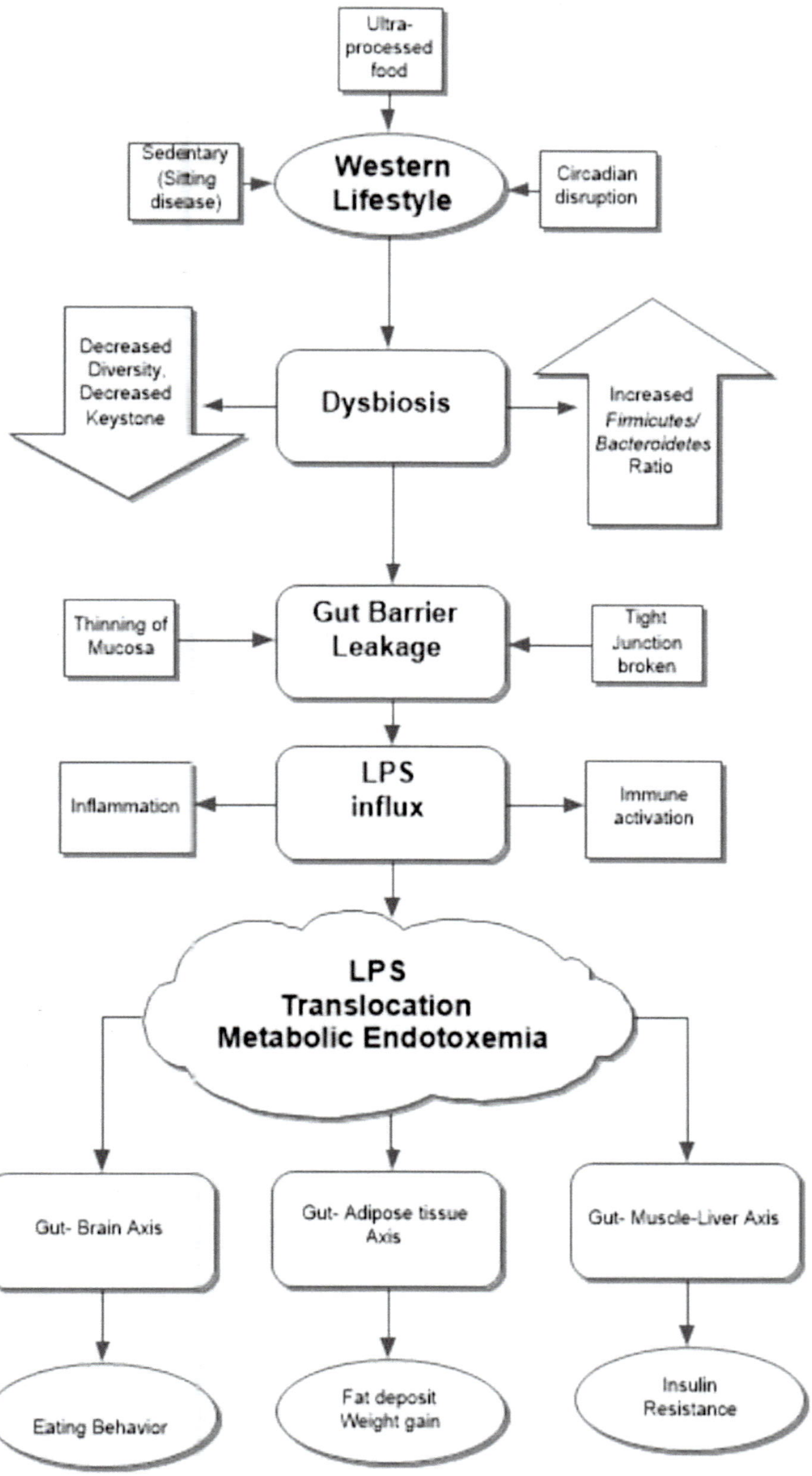

FIGURE 12.1 Delineates the process starting from Western Lifestyle, dysbiosis, gut barrier leakage, LPS influx, and ending with LPS translocation via different axis.

12.5.2 Insulin resistance and inflammation

Insulin helps to control the amount of glucose in the blood. When insulin resistance occurs, the body's cells do not respond normally to insulin, promoting a compensatory increase in insulin production and potentiating hyperinsulinemia. During insulin resistance, glucose cannot enter the cells as easily and builds up in the blood.

Insulin resistance and weight gain exacerbate one another. Weight gain—associated factors, such as excess body fat, visceral fat (fat around the organs), poor diet, low PA, and inflammation, contribute to insulin resistance and other comorbidities.

Insulin levels and action are regulated by both peripheral and central responses that involve the intestine and microbiome. Additionally, via the gut pancreatic islets axis, gut-derived inflammatory cytokines affect beta cell function.

Insulin resistance triggers hyperinsulinemia and fasting hyperglycemia and increases hepatic lipid synthesis, dyslipidemia, hypertension, and fat accumulation in adipose tissues (Schertzer and Lam, 2021; Cani et al., 2007).

12.5.3 Stress and metabolic endotoxemia

There two sides to stress and ME connection:

Stressful stimuli activate the sympathetic nervous system and hypothalamic-pituitary-adrenal (HPA) axis. Activation of both systems increases gut permeability to provide water, minerals, and energy-rich substances to meet the body's metabolic demands. Concurrently, LPS and ME increase as well (de Punder and Pruimboom, 2015).

Postbiotics can ameliorate stress-induced ME. The emerging concept of psychobiotics, that is, live microorganisms with potential mental health benefits, represents a novel approach for the management of stress-related conditions. Recent studies have identified the *Bifidobacterium longum* 1714 strain has a putative impact on stress-related behaviors, physiology, and cognitive performance (Allen et al., 2016). This psychobiotic exerts its effect by supporting NF-kB signaling and cytokine production, supporting antiinflammatory IL-10 levels, and targeting the HPA axis to support healthy cortisol levels.

12.5.4 Immune activation

The intestinal immune system is emerging as an important player in insulin resistance and obesity. A high-fat diet alters intestinal IgA and immune cells, and the IgA is a critical immune regulator of glucose homeostasis (Luck et al., 2019).

When a person eats a meal, the body needs to decide what to do with the energy in the meal. The immune cells calibrate that decision. The microbiota directly promotes local intestinal immunity through the effects on TLR expression and strengthening of gastrointestinal-associated lymphoid tissue (Creely et al., 2007).

12.6 Effect of dietary components on metabolic endotoxemia

Emerging human clinical studies have demonstrated that diet and dietary components are potent modifiers of circulating endotoxins and can be used to reduce the plasma levels of LPS significantly and improve metabolic health (Brown, 2017; Corrêa et al., 2019; McFarlin et al., 2017; Salazar et al., 2015; Sánchez-Tapia et al., 2020).

12.6.1 Effect of fasting on metabolic endotoxemia

Fasting and time-restricted feeding reduce LPS endotoxemia, gut dysbiosis, and inflammatory markers (Patterson and Sears, 2017).

12.6.2 Exercise and metabolic endotoxemia

Short-term exercise training modulates gut microbiota (e.g., reduction of the Firmicutes/Bacteroidetes ratio) profile and improves endotoxemia (Motiani et al., 2020). Before-meal PA reduces postprandial lipemic response and inflammatory markers (Teeman et al., 2016).

12.6.3 How long does it take to change the gut biome?

On average, it takes around 3 hours for a next generation of bacteria to exist in the gut (Myhrvold et al., 2015). In contrast, ME can last up to 5 hours after a saturated fatty meal. Thus before-meal, with-meal, or after-meal intervention with prebiotics or probiotics can alter ME favorably.

12.7 Microbial regulation of host energy balance

The gut microbiome has emerged as a key regulator of host metabolism. Collectively, there is a large body of evidence supporting the concept that gut microbiota–based therapies can be used to modulate the host metabolism (Cani et al., 2019; Amabebe et al., 2020).

12.7.1 Microbiota-associated energy harvest

Why do people gain weight differently on the exact same diet? The subjects' differing gut microbiota, differing amounts of SCFA, and LPS are key players. A shift in the gut microbiota in favor of Firmicutes, for example, with consumption of a high-fat, high-carbohydrate diet increases energy extraction from the diet with corresponding weight gain and obesity if left uncontrolled (Turnbaugh et al., 2006).

12.7.2 Cross-talk between the gut microbiota and the host metabolism

Cross-talk happens along multiple axes: the gut brain-axis, the gut-muscle axis, the gut–adipose tissue–liver axis, the gut–body clock axis, and the gut-autophagy axis.

12.7.3 Gut-brain axis of eating behavior

The gut microbiome is a "second brain" because it affects mood, motivation for PA, and eating behavior (Niccolai et al., 2019). The vagus nerve connects the gut to the brain. The gut microbiota controls the eating behavior by several mechanisms, including changes to receptors such as taste receptors, regulation of reward pathways, and production of toxins that alter mood.

Feeding, appetite, and energy expenditure are under the control of the central nervous system. SCFAs link receptors on enteroendocrine L cells, thus stimulating the release of glucagon-like peptiede-1 (GLP-1) and peptide tyrosine tyrosine, which is referred to as PYY. GLP-1 and PYY, in turn, modulate the activity of feeding centers in the brain. They can influence ghrelin, leptin, and other peptides that are involved in controlling food intake and energy expenditure.

Microbes are under selective pressure to manipulate the host's eating behavior to increase their own thriving, sometimes at the expense of host fitness. Microbes send messages to opioid, cannabinoid, dopamine, and serotonin receptors and influence the host's food idiosyncrasies (Alcock et al., 2014). In other words, food likes and dislikes, food cravings, and dependence are dictated by an individual's gut biome. Clinically, we alter the axiom "You are what you eat" to be "You are what your gut biome eats."

12.7.4 Gut-muscle axis and energy output

Several studies have suggested that bidirectional cross-talk exists between the skeletal muscles and the gut—the so-called muscle-gut or gut-muscle axis (Ortiz-Alvarez et al., 2020).

Sarcopenia could be the initial step in obesity or vice versa. The gut biome plays a role in sarcopenia. Sarcopenic obesity (SOB) is closely related to being elderly as a feature of aging, dysregulated nutrient sensing, and mitochondrial dysfunction. Along with insulin resistance and inflammation as the core pathogenesis of SOB, autophagy has recently gained attention as a significant mechanism of muscle aging in SOB (Ryu et al., 2020). Clinically, we focus on gut biome balance as an intervention to counter SOB.

PA (modest to moderate) enriches microflora diversity and improves the Bacteroidetes-Firmicutes ratio (Monda et al., 2017). Sedentary people have higher blood LPS levels than physically active people (Lira et al., 2010). Exercising in cold weather alters the gut biome and increases thermogenesis (Chevalier et al., 2015).

12.7.5 Gut-adipose-liver axis and energy storage

Adipose tissue is found in two different forms in mammals: white and brown. White fat (WAT) stores energy as triglycerides, whereas brown fat (BAT) burns extra calories to create heat. Additionally, there are beige, or brite, adipocytes, which are an inducible form of brown adipocytes. Accumulating evidence suggests that the gut microbiome affects the thermogenic capacity of brown fat and the formation of beige adipocytes (Li et al., 2019).

Butyrate has been shown to induce lipolysis, fatty acid oxidation, and thermogenesis, while acetate exerts an antilipolysis action and enhances beiging in WAT. In the liver, acetate limits fat accumulation and stimulates mitochondrial activity.

The gut microbiota and bile acids influence each other; the latter are bioactive lipids that are capable of modulating energy expenditure by inducing both BAT and WAT thermogenesis through TGR5 signaling. When the *A. muciniphila* population is increased, an enhanced beiging is often observed in WAT (Xiao and Kang, 2020).

12.7.6 Gut biome and circadian rhythm

The circadian clock system is composed of a central master clock in the suprachiasmatic nucleus of the hypothalamus and peripheral clocks, found in all tissues of the body, including gut, liver, pancreas, kidney, adipose tissue, and skeletal muscle. These clocks are also entrained by environmental and behavioral factors, such as light, sleep, PA, and feeding (Ding and Xiao, 2020).

Several characteristics of modern lifestyle, such as shift work, stress, jet lag, and unusual feeding patterns, have been shown to disrupt the biological clock. Blue light and artificial light exposure at night disrupt circadian rhythms by preventing melatonin release, which is required for regulating the activities of gut bacteria (Wei et al., 2020).

Time-restricted feeding (intermittent fasting) affects the gut biome. Cyclical changes in the gut microbiome from feeding/fasting rhythms contribute to the diversity of gut microflora and host metabolism (Zarrinpar et al., 2014).

Three key sleep-related neurotransmitters (melatonin, GABA, and dopamine) are made in the gut and modulated by gut biome. Five common sleep disorders, namely, insomnia, sleep apnea, shift work/social jet lag, night eating syndrome, and emotional eating, all pivot around these neurotransmitters.

12.7.7 Fasting and autophagy

Autophagy is a housecleaning tool that the body uses to eliminate damaged, dying, or dead cells that no longer function. By clearing away the "trash," autophagy enables the surviving cells to thrive. Autophagy is required to maintain antimicrobial defense, epithelial barrier integrity, and mucosal immune response. Disrupted autophagy worsens gut dysbiosis and vice versa.

A reduced-carbohydrate diet, intermittent fasting (IF), and increased PA are three main triggers of autophagy. The gut biome is the central modulator of autophagy.

The central upstream regulators of autophagy are that mammalian target of rapamycin complex 1 and adenosine monophosphate-activated protein kinase (AMPK). Angiopoietin-like 4, also known as fasting-inducible adipose factor, is a fasting-induced inhibitor of lipoprotein lipase (LPL) and a regulator of triglyceride metabolism and fat storage (Haq et al., 2019).

12.8 Western lifestyle as a menace to the gut biome

As cultures around the world have become more Westernized, the people have lost balanced gut bacteria (Cordain et al., 2005). Three key elements of a Western lifestyle are eating habits, sedentary habits, and circadian habits. In this section we will provide an overview of challenges for each element (Table 12.1).

Eating habits

Many of the dietary changes first started in the United States, where food industrialization was first developed, and then it became globalized (Martínez Steele et al., 2016; Singh et al., 2017). For that reason the ultraprocessed Western diet is also known as the standard American diet, for which an appropriate acronym is "SAD."

What is ultraprocessed food? The NOVA classification assigns a group to food products based on how much processing they have been through (Monteiro et al., 2018):

1. Group 1: Unprocessed or minimally processed foods: Unprocessed (or natural) foods are edible parts of plants (e.g., grains, seeds, fruits, leaves, stems, roots) or of animals (e.g., muscle, offal, eggs, milk). The goal is to consume foods mostly in their native forms.
2. Group 2: Processed culinary ingredients: Processed culinary ingredients, such as oils, butter, sugar, and salt, are substances that are derived from group 1 foods or from nature by processes that include pressing, refining, grinding, milling, and drying. The goal is to make freshly prepared drinks, dishes, and meals.
3. Group 3: Processed foods: Processed foods, such as bottled vegetables, canned fish, fruits in sirup, cheeses, and freshly made breads, are made essentially by adding salt, oil, sugar, or other substances from group 2 to group 1 foods. The goal is to increase the durability of group 1 foods or to modify their sensory qualities.

TABLE 12.1 Summarizing effect of each lifestyle factor disrupting or optimizing gut biome, and consequential clinical approach.

Lifestyle factor	Gut biome disrupters	Gut biome optimizers	Clinical approach	References
Ultraprocessed food	Removal of fermentable fiber, addition of antimicrobial additives	Minimally processed, whole food, RS, polyphenols	Fill half of the plate with RS-rich beans, lentils, whole grains; "feed and eat"	Martínez Steele et al. (2016); Monteiro et al. (2018); Sajilata et al. (2006); Sonnenburg and Sonnenburg (2014); What is a Whole Grain? The Whole Grains Council (2021)
Carbohydrates	Processed grains, refined sugar, and sweeteners	Whole grains, oligosaccharides, microbiota-accessible carbohydrates	Beans/lentils Cruciferous vegetables, cooked and cooled rice, wheat, root vegetables	Martínez Steele et al. (2016)
Fat	Saturated fats Omega-6 fats Trans fat	Omega-3 fats MCT oil	Change cooking oil from omega-6 to omega-3	Rial et al. (2016); Singh et al. (2017); Walker et al. (2019);
Protein	Animal protein	Plant proteins	Beans, lentils	David et al. (2014); Singh et al. (2017)
Micronutrient vitamins, minerals	Colon-focused micronutrient deficiency	Vitamin D, A, C, magnesium, zinc, selenium	Vitamin D and magnesium supplements, rest from food	Pham et al. (2021); Yang et al. (2020)
Phytonutrients	Polyphenol deficient	Rainbow-colored foods	5–6 servings of colored vegetables and fruits per day	Cheng et al. (2017); Kumar Singh et al. (2019); Lin et al. (2019); Yen et al. (2020)
Prebiotics	Processed, microbiota-accessible carbohydrates deficient foods	Fermentable fiber, oligosaccharides	Beans, lentils, RS powders, root vegetables	McRorie (2015)
Probiotics	Rare use of fermented foods	*Lactobacillus*, *Bifidobacterium*	Fermented foods, supplements	Amabebe et al. (2020)
Eating day and night, habitual snacking anytime	Constant metabolic endotoxemia	16/8 TRE fast 16 h, eat 8 h window	Eating window noon to 7:00 P.M. is commonly used	Cushing et al. (2017); Frank et al. (2021); Kesztyüs et al. (2021)
Sedentary sitting disease	Sitting > 6 h/day	NEAT, fidget, move	Stand and fidget: During computer, phone, TV time	Castellanos et al. (2020); Hamilton et al. (2004), (2007); Monda et al. (2017); Villablanca et al. (2015)
Sleep-wake circadian rhythm	Evening blue light, shift work, social jet lag, night eating, sleep loss	RS before dinner, NEAT increases sleep drive	Morning sun, evening blue light blockers, Circadian eating (sunrise to sunset)	Ding and Xiao (2020); Frank et al. (2021); Voigt et al. (2014); Wei et al. (2020); Zarrinpar et al. (2014)

4. Group 4: Ultraprocessed food and drink products: Ultraprocessed foods, such as soft drinks, sweet or savory packaged snacks, reconstituted meat products, and preprepared frozen dishes, are formulations that are made mostly or entirely from substances derived from foods and additives, with little if any intact Group 1 food. Ingredients of these formulations usually include those also used in processed foods, such as sugars, oils, fats, or salt and food additives and preservatives like hydrogenated oils, hydrolyzed proteins, soya protein isolate, maltodextrin, invert sugar and high-fructose corn sirup and other nonsugar sweeteners; and processing aids such as carbonating, firming, bulking and antibulking, defoaming, anticaking and glazing agents, emulsifiers, sequestrants and humectants. The goal of ultraprocessing is to create branded, convenient (durable, ready to consume), attractive (hyperpalatable) and highly profitable (low-cost ingredients) food products designed to displace all other food groups. Ultraprocessed food products are usually packaged attractively and marketed intensively. In short, ultraprocessed foods are industrial formulations with five or more cosmetic or sensory additives.

In a large cross-sectional study, ultraprocessed foods comprised 57.9% of energy intake and contributed 89.7% of the energy intake from added sugars in the United States (Martínez Steele et al., 2016).

Ultraprocessed foods have two opposing effects: removal and addition.

Removal of microbial accessible carbohydrates from whole grains and other plant foods starves our microbial self (Sonnenburg and Sonnenburg, 2014). All grains start life as whole seed (also called a "kernel") made up of three edible parts: the bran, the germ, and the endosperm. White flour and white rice are refined grains, for instance, because both have had their bran and germ removed, leaving only the endosperm (What is a Whole Grain? The Whole Grains Council, 2021). Most staple grains (e.g., rice, wheat, maize) are consumed in refined forms rather than as whole grains.

The addition of antimicrobial foods, such as refined sugars, artificial sweeteners, emulsifiers, antibiotics, and proinflammatory cooking oils (saturated fats, hydrogenated oils, omega-6 oils), contributes to dysbiosis. In the last 100 years, the amount of omega-3 polyunsaturated fats in the Western diet has declined, while the consumption of omega-6 fatty acids, especially in arachidonic acid–rich foods, has increased substantially. This is due to the development and use of vegetable and seed oils in our food supply. Almost all processed foods contain one or more vegetable oils.

The Western dietary style is deficient in antiinflammatory factors: omega-3 cooking oil, micronutrients, and phytonutrients (from vegetables and fruits).

Sedentary habits

Sitting has become the most common posture. A screen-centered sedentary lifestyle has led to sitting 6–15 hours of waking time, leading to sitting disease.

Circadian habits

The biological clock is disturbed by shift work, social jet lag, stress, artificial light at night, and eating against circadian rhythm.

12.9 Pragmatic mitigation of western lifestyle effects on the gut biome

A Western pattern lifestyle is inevitable. It is here to stay, and it is pervasive worldwide. The overall approach to mitigate this lifestyle is to restore and reclaim microbial balance, reduce ME inflammation, and optimize crosstalk between the gut microbiota and the host energy balance.

There are three low-hanging fruits in mitigating the gut biome dysbiosis and control damage caused by Western lifestyle:

1. Eating and fasting
2. Activity and sedentary
3. Wake and sleep

Nutritional management of the gut biome is a two-sided coin: what to eat (food contents) and when to eat (eating window and fasting window).

12.10 Eating and fasting

12.10.1 What to eat?

In determining what to eat, pragmatic factors include what foods are universally available, affordable, fast, and easy to prepare. It can help to use a dietary visual such as Myplate (US Department of Agriculture, 2021) but divide the plate equally with a microbiome focus. The feed and eat principle involves dividing the plate into two halves: the left half to feed the microbiome and the right half for the eater's indulgence. Apply this Myplate concept for each meal.

A good dietary philosophy is flexitarian, predominantly plant-based to foster the microbiome, with animal food allowed provided that half of the plate feeds the microbial self.

It is important to optimize dietary intake of the following five category foods:

1. Macronutrients: carbohydrates, protein, fat
2. Micronutrients: vitamins and minerals
3. Phytonutrients: Polyphenols
4. Prebiotic (food versus supplements)
5. Probiotics (food versus supplements)

Carbohydrates exist in two varieties: digestible and nondigestible. Digestible sugars and sweeteners are unfavorable for the microbiome Nondigestible carbohydrates contain fermentable fiber as microbiota-accessible carbohydrates, which are microbiome favorable

The following are sources of prebiotic whole foods (containing bran, germ, and endosperm):

Grains: wheat, rice, corn, oat, teff, buckwheat, millet
Beans and lentils black beans, garbanzo beans, soybeans, lentils (red, brown, green, etc.)
Root vegetables (tubers): potato, cassava, sweet potatoes, yam, taro

The simplest way to provide microbiota-accessible carbohydrates is as follows:

Eat 1 cup of cooked beans and/or lentils in each meal. In most pre-Westernized cultures, beans and lentils are widely used.
Cook and cool your staple food (rice, wheat, corn) and root vegetables overnight in the refrigerator. The next day, the grain has turned into RS.
RS powders: Potato starch, green banana, chicory (inulin).
Other sources: cruciferous vegetables (such as broccoli, Brussels sprouts, and cauliflower), collard greens, grains, nuts, and seeds

Protein

Microbial activity mirrors differences between herbivorous and carnivorous mammals, reflecting tradeoffs between carbohydrate and protein fermentation (David et al., 2014).

Individuals who consume pea protein exhibit increases in beneficial *Bifidobacterium* and *Lactobacillus* and consequently increases in intestinal SCFA levels. Plant-based protein fosters healthy bacteria compared to animal proteins (Singh et al., 2017).

The simplest way to provide plant protein is to eat 1 cup of beans and/or lentils with each meal.

Fats

Saturated fats, trans-fats, and omega-6 fats (especially arachidonic acid) increase dysbiosis, LPS endotoxemia, and inflammation (Singh et al., 2017). Omega-3 fats and medium-chain triglyceride (MCT) fats (Rial et al., 2016) counter such effects and increase gut biome resilience.

The simplest way to provide bioactive lipids is to do the following:

Change your cooking oil from omega-6-dominant oils (soy, corn, safflower, or sunflower oil) to omega-3-dominant oils (avocado or olive oil).
Eat an avocado as a part of your breakfast (Thompson et al., 2021).
Take fish oil supplements (to counter omega-6 excess in a Western-style diet).
Consume other plant-based omega-3 fatty acid foods, such as flaxseeds, chia seeds, walnuts, and hemp seeds.

Micronutrients

Micronutrient deficiency affects the gut biome, justifying the value of supplementing vitamins and minerals (Yang et al., 2020).

Vitamins

Colon-targeted vitamins (Pham et al., 2021) are vitamins A, B2, C and D, which modulate the gut biome. A decrease in the gut microbial diversity was reported with a dietary magnesium deficiency (Pyndt Jørgensen et al., 2015). Zinc and selenium are essential micronutrients for maintaining epithelial integrity, possibly by modulating the beneficial gut microbiota (Yang et al., 2020). Vitamin D, A, magnesium, and zinc are acquired from plant foods or supplements.

The simplest way to provide micronutrients is by consuming vitamin D and magnesium supplements as well as eating natural food sources.

Phytonutrients

Polyphenols (Kumar Singh et al., 2019) in plant foods increase the abundance of probiotic bacteria such as *Bifidobacterium* and *Lactobacillus*, increase SCFA, reduce gut permeability, and activate the AMPK metabolic master switch (Yen et al., 2020). Additionally, polyphenols reverse the gut biome disruption caused by saturated fatty meals (Cheng et al., 2017). Thus polyphenols are functional foods with potential therapeutic utility in manipulating intestinal microbiota, and they promote weight loss by restructuring the intestinal microecology (Lin et al., 2019).

The simplest way to provide phytonutrients is to eat rainbow foods, that is, red, yellow, blue, purple, and black colored vegetables, and fruits in the amount of 4–5 cups a day. An easy way to remember is to eat 1–2 cups with each meal.

Prebiotics

RS powders are used as "wild card" supplements to provide prebiotics. The simplest way to provide a prebiotic supplement is by ingesting 1–2 tablespoons of RS powders, such as inulin, potato starch, or green banana, twice a day.

Probiotics

The effect of *Lactobacillus*, *Bifidobacterium*, and *Saccharomyces* on weight loss and/or fat deposition in overweight adults has been reported (Amabebe et al., 2020).

The simplest way to provide probiotic supplement is by consuming fermented foods, such as yogurt, kefir, buttermilk, apple cider vinegar, aged cheese, sauerkraut, kimchi, miso, tempeh, pickles, kombucha, fermented vegetables, natto, and sourdough.

12.10.2 Clinical protocol for microbiome-centered eating

1. *Goal*: Feed your microbial self, and then indulge in your eating.
2. *Tool*: Microbiota-accessible carbohydrates and polyphenols: beans, lentils, and rainbow vegetables and fruits.
3. *Ritual*: Use the plate as a cue, and do it at every meal.

12.10.3 When to eat? Intermittent fasting

The ubiquity of highly palatable ultraprocessed food and relentless marketing have played crucial roles in shifting homeostatic eating toward hedonic eating through both central (disruptions in dopaminergic signaling) and gut microbiome (LPS, ME, SCFA) mechanisms (Frank et al., 2021). The balance between homeostatic and hedonic eating behaviors is additionally influenced by the timing and rhythmicity of food ingestion. IF and time-restricted eating (TRE) are effective interventions to counter both challenges.

IF has many protocols. They fall into three categories: time-restricted feeding (e.g., fasting/eating 16/8, 12/12) short IF up to 2 days (e.g., alternate day fasting, 5:2 fasting), and prolonged fasting (e.g., 4–5 days fasting).

The physiological goal of fasting can be immediate, intermediate, or prolonged. Immediately, fasting results in reduced postprandial ME. After 12–16 hours of fasting, nutritional ketosis and autophagy set in, modifying nutritional sensors such as insulin, IGF-1, mTOR, and stem cells. At this level, the fuel source shifts from stored glycogen in the liver and muscle to adipose tissues. We use the analogy of a hybrid car in which the fuel shifts from gasoline to batter power. Prolonged or recurrent fasting leads to keto-adaptation, in which metabolic flexibility of shifting from carbohydrate to fat fuel sustained (Cushing et al., 2017).

12.10.3.1 Fasting versus fasting mimicking

The biggest behavioral challenge with fasting is dropout due to hunger, fatigue, or headache. The fasting-mimicking diet (FMD) overcomes this challenge. The FMD tricks your body into thinking that it is fasting while allowing intake of low-carbohydrate, low-protein, plant protein–dominant food below the detection bar of nutrition sensors. Longevity researcher Valter Longo developed a 5-day FMD diet plan and tested it on yeast, mice, and humans (Brandhorst et al., 2015; Wei et al., 2017).

We have applied the FMD principle into a 16/8 TRE plan in our clinical practice. In principle, breakfast is replaced with a vegan source of fat, such as avocado, nuts, nut butters, and MCT oil.

TRE offers several advantages over other forms of dietary interventions to prevent or treat weight problems and associated disease or disease risks (Kesztyüs et al., 2021):

1. Low-threshold approach (meaning that the implementation does not necessarily have to be medically supervised)
2. No calorie counting
3. No dietary restrictions

4. Individually adaptable to the individual's daily rhythm
5. May improve health even without weight loss (Sutton et al., 2018)

The advantages of TRE make this approach particularly interesting for public health interventions, as the low barriers and ease of implementation can have a positive impact on both entry and adherence.

12.10.4 Clinical protocol for fasting-mimicking time-restricted eating

1. *Goal:* Switch fuel mode from carbohydrate burning to carbohydrate and fat burning for dual metabolic flexibility.
2. *Tool*: IF, 16/8 TRE.
3. *Ritual*: Skip breakfast and eat lunch 16 hours from the previous night's dinner. Or, you can skip dinner and eat breakfast 16 hours after your last meal.
4. Patient education: Remember that this is not a water-only fast, which makes you hungry and loopy. This is fasting mimicking to boost nutritional ketosis and keep you hungry and alert.
5. NO carbohydrates and NO protein with breakfast to turn off carbohydrate fuel.
6. Step 1. Eat your dinner at 8:00 p.m. or sooner.
7. Step 2. Turn on your metabolic switch at breakfast time by eating a plant-derived fat keto breakfast:
   a. Half of avocado
   b. One tablespoon MCT oil or powder
   c. Coffee or tea without sweetener
8. Step 3. Lunch at 12:00 or 1:00 P.M. (i.e., 16 hours from the previous night's dinner time)
9. Note: Avocado is best choice, but you can substitute other plant-based fats, such as five olives or 1 tablespoon of nut butter (macadamia, almond, peanut).

12.11 Physical activity versus physical inactivity

In terms of PA, there are two different clinically useful paradigm shifts: exercise versus nonexercise activity and sedentary versus sitting disease.

12.11.1 Exercise versus nonexercise

PA is characterized by any movement of the skeletal muscles that demands energy expenditure. Such PA can involve exercise or nonexercise activity thermogenesis (NEAT):

Exercise: Structured, planned, repetitive fitness-focused activities such as gym workouts, hiking, and tennis.

NEAT: In epidemiological studies, nonexercise behaviors were called *active living* until James Levine, an endocrinologist at the Mayo Clinic, did his research to introduce NEAT as a new paradigm. NEAT represents common daily activities, such as standing, housework, fidgeting, laughing, and singing, excluding planned exercise activities (Villablanca et al., 2015).

Physical inactivity (PI) is minimal movement of skeletal muscles that demands minimal energy expenditure. Such PI can be sedentary or sitting disease:

12.11.2 Sedentary versus sitting disease

Sedentary: Sedentary behavior has misleadingly been used as a synonym for not having the habit of exercising. Sedentary time should be defined as muscular inactivity rather than the absence of exercise.

Sitting disease: Marc T. Hamilton of the Pennington Biomedical Research Center in Baton Rouge, Louisiana, United States, is a pioneer in inactivity physiology research. Inactivity physiology is a paradigm shift about a sedentary lifestyle, or sitting too much, also known as sitting disease. This paradigm argues that "sitting too much is not the same as exercising too little" and that there are potent pathophysiological effects caused by too much sitting that cannot be cured by typical exercise training (Hamilton et al., 2007).

The most common posture during waking hours is sitting. We sit to do desk work or use a computer or telephone, to watch TV, to eat, to travel, to socialize, to read, and so on, totaling 12–15 hours a day. Sitting is labeled as a disease because of its adverse metabolic consequences. Sedentary behavior and sitting disease affect the gut biome. Research has shown that the enzyme LPL begins to "turn off" after 20 minutes of sitting. LPL breaks down triglycerides in the blood. Less LPL increases weight gain. "Sitting is the new smoking" is a popular expression showing just how bad sitting is for the body (Hamilton et al., 2004).

The gut microbiome–muscle axis is affected by PA or PI. PA enriches the microflora diversity and improves the Bacteroidetes/Firmicutes ratio (Monda et al., 2017). Sedentary people have higher blood LPS levels than do people who do more exercise (Lira et al., 2010). The diversity network complexity of the gut microbiota was significantly lower in sedentary compared to active individuals (Castellanos et al., 2020; Monda et al., 2017). Exercising in cold weather alters the gut biome and increases thermogenesis (Chevalier et al., 2015).

AMPK is a gut biome–linked cellular energy sensor that is activated mostly in low-energy states. AMPK activation closely resembles those of calorie restriction and exercise. A 20-minute walk before a meal can activate the AMPK system and optimize the effects on the gut biome of a forthcoming meal.

Clinically, we emphasize the health benefits to support purposeful aerobic exercise. NEAT must be an add-on, based on the following arguments:

1. A common statement in weight loss clinics is "I regularly exercise 1 hour a day yet do not lose weight." The problem lies in the fact that 1 hour of exercise is only 4% of the 24-hour day.
2. What matters is how physically active you are during the remaining 96% of your 24-hour day. If you deduct 8 hours of sleep time, you still have 9 hours, or 63% of your waking time, to be used in activity rather than inactivity.
3. The flip side is how much time you sit during that 96% of time (sitting disease).
4. A unique goal of inactivity physiology is to improve health outcomes by greatly increasing levels of PA in all people, even those who cannot or will not engage in the classical recommendations of 150 minutes a week of moderate to vigorous PA.
5. Nonsitting is extremely challenging, since sitting time has become a monumental part of life in the modern world. The holy grail of healthy active living would come from a breakthrough discovery for how to safely replace most of the sedentary and sitting time with PA in everyone, regardless of age and current health status (Hamilton et al., 2007).

12.11.3 Clinical protocol for physical activity

Goal: Reduce your sitting time by increasing NEAT and continue planned exercise and sports activities.

Tools: NEAT such as like standing, fidgeting, and moving.

1. Stand and move during these top three key situations: computer or desk work, TV, and phone time.
2. Reduce your sitting time to at least 4–6 hours in 24 hours.
3. During desk work, computer work, and the like, stand up for 3 minutes after 20 minutes of sitting (to activate LPL).
4. Walk for 20 minutes before meals to activate your gut-nutrition sensor system (AMPK).
5. Walk in the morning sun (to align the circadian clock), ideally in nature (forest bathing).

12.12 Wake-sleep circadian rhythm

Shift work, artificial light, social jet lag, and the like, disrupt the gut biome. Feeding your gut biome with RS increases your resilience against circadian disruption. Furthermore, the gut microbiome of animals fed a high-fat or high-sugar diet is more prone to circadian rhythm disruption (Voigt et al., 2014).

The simplest way to balance circadian rhythm is to take a morning walk in the sun and use blue light–blocking goggles in the evening and RS at night.

12.12.1 Clinical protocol for circadian rhythm balance

Goal: Align the gut clock to the central clock in the brain.

Tools: Eating, fasting, light exposure, NEAT.

Ritual:

1. Adjust eating time frames. Try to align the eating window between sunrise and sunset. The easiest way is to start 16/8 fasting as an extension of nighttime and make the eating window after 12:00 noon but before sunset.
2. Avoid high-fat foods at dinner to avoid dysbiosis and LPS endotoxemia, which disrupts sleep. Instead, increase prebiotics at dinner (1 cup of beans and/or lentils), which increase gut resilience and enable better sleep. Limit alcohol

intake closer to bedtime; if you do drink, combine the alcohol with a prebiotic to offset alcohol effects on the gut biome.

3. Avoid exposure to blue light after sunset (using blue light–blocking goggles or computer and smartphone light blockers). Get plenty of exposure to sunlight in the morning.
4. Shift work and social jet lag change the gut clock and disrupt the metabolism, increasing weight gain. That can be solved by working on the gut biome.
5. Replace sitting time with NEAT activities to increase the sleep drive.

12.13 Conclusion

The gut biome is the central hub and command center of metabolism: energy intake and expenditure. The ultraprocessed food habits, sedentary habits, and circadian habits of the Western lifestyle are in discordance with the evolutionary habits of the gut biome. Obesogenic dysbiosis leads to ME, inflammation, immune activation, increased energy harvest, and weight gain. If you focus on feeding a trillion members of your microbial self (philanthropically!) on the first half of your plate, then eating your meal is an easy-to-follow principle. Beans, lentils, RS powders, and rainbow-colored vegetables and fruits are the mainstay of microbially accessible food. Shifting from proinflammatory omega-6 to antiinflammatory omega-3 cooking oils is a simple bioactive lipid modulation. Replacing sitting time with NEAT is pragmatic PA regardless of physical or psychological limitations. TRE is the IF of choice. A morning walk in the sun, evening blocking of blue light exposure, and a circadian-aligned eating pattern are simple tools to reclaim microbial balance and resilience.

References

Alcock J., Maley, C.C., Aktipis, C.A., 2014. Is eating behavior manipulated by the gastrointestinal microbiota? Evolutionary pressures and potential mechanisms. BioEssays: N. Rev. Mol. Cell. Dev. Biol. 36 (10), 940–949. Available from: https://doi.org/10.1002/bies.201400071.

Allen, A., Hutch, W., Borre, Kennedy, P.J., Temko, A., Boylan, G., et al., 2016. *Bifidobacterium longum* 1714 as a translational psychobiotic: modulation of stress, electrophysiology and neurocognition in healthy volunteers. Transl. Psychiatry 6, e939. Available from: https://doi.org/10.1038/tp.2016.191.

Amabebe, E., Robert, F.C., Agbalalah, T., Orubu, E., 2020. Microbial dysbiosis-induced obesity: role of gut microbiota in homoeostasis of energy metabolism. Br. J. Nutr. 123 (10), 1127–1137. Available from: https://doi.org/10.1017/S0007114520000380.

Aoun, A., Darwish, F., Hamod, N., 2020. The influence of the gut microbiome on obesity in adults and the role of probiotics, prebiotics, and synbiotics for weight loss. Prevent. Nutr. Food Sci. 25 (2), 113–123. Available from: https://doi.org/10.3746/pnf.2020.25.2.113.

Boutagy, N.E., McMillan, R.P., Frisard, M.I., Hulver, M.W., 2016. Metabolic endotoxemia with obesity: is it real and is it relevant? Biochimie 124, 11–20. Available from: https://doi.org/10.1016/j.biochi.2015.06.020.

Brandhorst, S., Choi, I.Y., Wei, M., Cheng, C.W., Sedrakyan, S., Navarrete, G., et al., 2015. A periodic diet that mimics fasting promotes multi-system regeneration, enhanced cognitive performance, and healthspan. Cell Metab. 22 (1), 86–99. Available from: https://doi.org/10.1016/j.cmet.2015.05.012.

Brown, B.I., 2017. Nutritional management of metabolic endotoxemia: a clinical review. Altern. Ther. Health Med. 23 (4), 42–54.

Cani, P.D., Amar, J., Iglesias, M.A., Poggi, M., Knauf, C., Bastelica, D., et al., 2007. Metabolic endotoxemia initiates obesity and insulin resistance. Diabetes 56 (7), 1761–1772. Available from: https://doi.org/10.2337/db06-1491.

Cani, P.D., Van Hul, M., Lefort, C., Depommier, C., Rastelli, M., Everard, A., 2019. Microbial regulation of organismal energy homeostasis. Nat. Metab. 1 (1), 34–46. Available from: https://doi.org/10.1038/s42255-018-0017-4.

Castaner, O., Goday, A., Park, Y.M., Lee, S.H., Magkos, F., Shiow, S., et al., 2018. The gut microbiome profile in obesity: a systematic review. Int. J. Endocrinol. 2018, 4095789. Available from: https://doi.org/10.1155/2018/4095789.

Castellanos, N., Diez, G.G., Antúnez-Almagro, C., Bailén, M., Bressa, C., González Soltero, R., et al., 2020. A critical mutualism - competition interplay underlies the loss of microbial diversity in sedentary lifestyle. Front. Microbiol. 10, 3142. Available from: https://doi.org/10.3389/fmicb.2019.03142.

Cheng, M., Zhang, X., Miao, Y., Cao, J., Wu, Z., Weng, P., 2017. The modulatory effect of (-)-epigallocatechin 3-O-(3-O-methyl) gallate (EGCG3″Me) on intestinal microbiota of high fat diet-induced obesity mice model. Food Res. Int. (Ottawa, Ont.) 92, 9–16. Available from: https://doi.org/10.1016/j.foodres.2016.12.008.

Chevalier, C., Stojanović, O., Colin, D.J., Suarez-Zamorano, N., Tarallo, V., Veyrat-Durebex, C., et al., 2015. Gut microbiota orchestrates energy homeostasis during cold. Cell 163 (6), 1360–1374. Available from: https://doi.org/10.1016/j.cell.2015.11.004.

Cordain, L., Eaton, S.B., Sebastian, A., Mann, N., Lindeberg, S., Watkins, B.A., et al., 2005. Origins and evolution of the Western diet: health implications for the 21st century. Am. J. Clin. Nutr. 81 (2), 341–354. Available from: https://doi.org/10.1093/ajcn.81.2.341.

Corrêa, T., Rogero, M.M., Hassimotto, N., Lajolo, F.M., 2019. The two-way polyphenols-microbiota interactions and their effects on obesity and related metabolic diseases. Front. Nutr. 6, 188. Available from: https://doi.org/10.3389/fnut.2019.00188.

Creely, S.J., McTernan, P.G., Kusminski, C.M., Fisher, f M., DaSilva, N.F., Khanolkar, M., et al., 2007. Lipopolysaccharide activates an innate immune system response in human adipose tissue in obesity and type 2 diabetes. Am. J. Physiol. Endocrinol. Metab. 292 (3), E740–E747. Available from: https://doi.org/10.1152/ajpendo.00302.2006.

Cushing, E.M., Chi, X., Sylvers, K.L., Shetty, S.K., Potthoff, M.J., Davies, B., 2017. Angiopoietin-like 4 directs uptake of dietary fat away from adipose during fasting. Mol. Metab. 6 (8), 809–818. Available from: https://doi.org/10.1016/j.molmet.2017.06.007.

David, L.A., Maurice, C.F., Carmody, R.N., Gootenberg, D.B., Button, J.E., Wolfe, B.E., et al., 2014. Diet rapidly and reproducibly alters the human gut microbiome. Nature 505 (7484), 559–563. Available from: https://doi.org/10.1038/nature12820.

de Punder, K., Pruimboom, L., 2015. Stress induces endotoxemia and low-grade inflammation by increasing barrier permeability. Front. Immunol. 6, 223. Available from: https://doi.org/10.3389/fimmu.2015.00223.

Ding, L., Xiao, X.H., 2020. Gut microbiota: closely tied to the regulation of circadian clock in the development of type 2 diabetes mellitus. Chin. Med. J. 133 (7), 817–825. Available from: https://doi.org/10.1097/CM9.0000000000000702.

Dogra, S.K., Doré, J., Damak, S., 2020. Gut microbiota resilience: definition, link to health and strategies for intervention. Front. Microbiol. 11, 572921. Available from: https://doi.org/10.3389/fmicb.2020.572921.

Duan, Y., Zeng, L., Zheng, C., Song, B., Li, F., Kong, X., et al., 2018. Inflammatory links between high fat diets and diseases. Front.Immunol. 9, 2649. Available from: https://doi.org/10.3389/fimmu.2018.02649.

Frank, J., Gupta, A., Osadchiy, V., Mayer, E.A., 2021. Brain-gut-microbiome interactions and intermittent fasting in obesity. Nutrients 13 (2), 584. Available from: https://doi.org/10.3390/nu13020584.

Ghosh, S.S., Wang, J., Yannie, P.J., Ghosh, S., 2020. Intestinal barrier dysfunction, LPS translocation, and disease development. J. Endocr. Soc. 4 (2), bvz039. Available from: https://doi.org/10.1210/jendso/bvz039.

Hamilton, M.T., Hamilton, D.G., Zderic, T.W., 2007. Role of low energy expenditure and sitting in obesity, metabolic syndrome, type 2 diabetes, and cardiovascular disease. Diabetes 56 (11), 2655–2667. Available from: https://doi.org/10.2337/db07-0882.

Hamilton, M.T., Hamilton, D.G., Zderic, T.W., 2004. Exercise physiology vs inactivity physiology: an essential concept for understanding lipoprotein lipase regulation. Exerc. sport. Sci. Rev. 32 (4), 161–166. Available from: https://doi.org/10.1097/00003677-200410000-00007.

Hannah Ritchie and Max Roser, 2017. Obesity. Published online at OurWorldInData.org. Available from <https://ourworldindata.org/obesity>.

Haq, S., Grondin, J., Banskota, S., Khan, W.I., 2019. Autophagy: roles in intestinal mucosal homeostasis and inflammation. J. Biomed. Sci. 26 (1), 19. Available from: https://doi.org/10.1186/s12929-019-0512-2.

Hold, G.L., 2014. Western lifestyle: a 'master' manipulator of the intestinal microbiota? Gut 63, 5–6. Available from: https://doi.org/10.1136/gutjnl-2013-304969.

Jumpertz, R., Le, D.S., Turnbaugh, P.J., Trinidad, C., Bogardus, C., Gordon, J.I., et al., 2011. Energy-balance studies reveal associations between gut microbes, caloric load, and nutrient absorption in humans. Am. J. Clin. Nutr. 94 (1), 58–65. Available from: https://doi.org/10.3945/ajcn.110.010132.

Kasai, C., Sugimoto, K., Moritani, I., Tanaka, J., Oya, Y., Inoue, H., et al., 2015. Comparison of the gut microbiota composition between obese and non-obese individuals in a Japanese population, as analyzed by terminal restriction fragment length polymorphism and next-generation sequencing. BMC Gastroenterol. 15, 100. Available from: https://doi.org/10.1186/s12876-015-0330-2.

Kesztyüs, D., Vorwieger, E., Schönsteiner, D., Gulich, M., Kesztyüs, T., 2021. Applicability of time-restricted eating for the prevention of lifestyle-dependent diseases in a working population: results of a pilot study in a pre-post design. German Medical Science: GMS e-Journal 19, Doc04. Available from: https://doi.org/10.3205/000291.

Kumar Singh, A., Cabral, C., Kumar, R., Ganguly, R., Kumar Rana, H., Gupta, A., et al., 2019. Beneficial effects of dietary polyphenols on gut microbiota and strategies to improve delivery efficiency. Nutrients 11 (9), 2216. Available from: https://doi.org/10.3390/nu11092216.

Lin, S., Wang, Z., Lam, K.L., Zeng, S., Tan, B.K., Hu, J., 2019. Role of intestinal microecology in the regulation of energy metabolism by dietary polyphenols and their metabolites. Food Nutr. Res. 63. Available from: https://doi.org/10.29219/fnr.v630.151810.29219/fnr.v63.1518.

Lira, F.S., Rosa, J.C., Pimentel, G.D., Souza, H.A., Caperuto, E.C., Carnevali Jr, L.C., et al., 2010. Endotoxin levels correlate positively with a sedentary lifestyle and negatively with highly trained subjects. Lipids Health Dis. 9, 82. Available from: https://doi.org/10.1186/1476-511X-9-82.

Li, B., Li, L., Li, M., Lam, S.M., Wang, G., Wu, Y., et al., 2019. Microbiota depletion impairs thermogenesis of brown adipose tissue and browning of white adipose tissue. Cell Rep. 26 (10), 2720–2737. Available from: https://doi.org/10.1016/j.celrep.2019.02.015. e5.

Lopez-Siles, M., Duncan, S.H., Garcia-Gil, L.J., Martinez-Medina, M., 2017. Faecalibacterium prausnitzii: from microbiology to diagnostics and prognostics. ISME J. 11 (4), 841–852. Available from: https://doi.org/10.1038/ismej.2016.176.

Luck, H., Khan, S., Kim, J.H., Copeland, J.K., Revelo, X.S., Tsai, S., et al., 2019. Gut-associated IgA^+ immune cells regulate obesity-related insulin resistance. Nat. Commun. 10 (1), 3650. Available from: https://doi.org/10.1038/s41467-019-11370-y.

Martínez Steele, E., Baraldi, L.G., Louzada, M.L., Moubarac, J.C., Mozaffarian, D., Monteiro, C.A., 2016. Ultra-processed foods and added sugars in the US diet: evidence from a nationally representative cross-sectional study. BMJ Open 6 (3), e009892. Available from: https://doi.org/10.1136/bmjopen-2015-009892.

McFarlin, B.K., Henning, A.L., Bowman, E.M., Gary, M.A., Carbajal, K.M., 2017. Oral spore-based probiotic supplementation was associated with reduced incidence of post-prandial dietary endotoxin, triglycerides, and disease risk biomarkers. World J. Gastrointest.Pathophysiol. 8 (3), 117–126. Available from: https://doi.org/10.4291/wjgp.v8.i3.117.

McRorie Jr, J.W., 2015. Evidence-based approach to fiber supplements and clinically meaningful health benefits, part 1: what to look for and how to recommend an effective fiber therapy. Nutr. Today 50 (2), 82–89. Available from: https://doi.org/10.1097/NT0.0000000000000082.

Monda, V., Villano, I., Messina, A., Valenzano, A., Esposito, T., Moscatelli, F., et al., 2017. Exercise modifies the gut microbiota with positive health effects. Oxid. Med. Cell. Longev. 2017, 3831972. Available from: https://doi.org/10.1155/2017/3831972.

Monteiro, C.A., Cannon, G., Moubarac, J.C., Levy, R.B., Louzada, M., Jaime, P.C., 2018. The UN decade of nutrition, the NOVA food classification and the trouble with ultra-processing. Public. Health Nutr. 21 (1), 5−17. Available from: https://doi.org/10.1017/S1368980017000234.

Motiani, K.K., Collado, M.C., Eskelinen, J.J., Virtanen, K.A., Löyttyniemi, E., Salminen, S., et al., 2020. Exercise training modulates gut microbiota profile and improves endotoxemia. Med. Sci.Sports Exerc. 52 (1), 94−104. Available from: https://doi.org/10.1249/MSS.0000000000002112.

Myhrvold, C., Kotula, J.W., Hicks, W.M., Conway, N.J., Silver, P.A., 2015. A distributed cell division counter reveals growth dynamics in the gut microbiota. Nat. Commun. 6, 10039. Available from: https://doi.org/10.1038/ncomms10039.

Napolitano, M., Covasa, M., 2020. Microbiota transplant in the treatment of obesity and diabetes: current and future perspectives. Front. Microbiol. 11, 590370. Available from: https://doi.org/10.3389/fmicb.2020.590370.

Niccolai, E., Boem, F., Russo, E., Amedei, A., 2019. The gut−brain axis in the neuropsychological disease model of obesity: a classical movie revised by the emerging director "microbiome.". Nutrients 11 (1), 156. Available from: https://doi.org/10.3390/nu11010156.

Ortiz-Alvarez, L., Xu, H. Martinez-Tellez, B., 2020. Influence of exercise on the human gut microbiota of healthy adults: a systematic review. Clin. Transl. Gastroenterol. 11 (2), e00126. Available from: https://doi.org/10.14309/ctg.0000000000000126.

O'Callaghan, A., van Sinderen, D., 2016. Bifidobacteria and their role as members of the human gut microbiota. Front. Microbiol. 7, 925. Available from: https://doi.org/10.3389/fmicb.2016.00925.

Patterson, R.E., Sears, D.D., 2017. Metabolic effects of intermittent fasting. Annu. Rev. Nutr. 37, 371−393. Available from: https://doi.org/10.1146/annurev-nutr-071816-064634.

Peters, B.A., Shapiro, J.A., Church, T.R., Miller, G., Trinh-Shevrin, C., Yuen, E., et al., 2018. A taxonomic signature of obesity in a large study of American adults. Sci. Rep. 8, 9749. Available from: https://doi.org/10.1038/s41598-018-28126-1.

Pham, V.T., Fehlbaum, S. Seifert, N., Richard, N., Bruins, M.J., Sybesma, W., et al., 2021. Effects of colon-targeted vitamins on the composition and metabolic activity of the human gut microbiome- a pilot study. Gut Microbes 13 (1), 1−20. Available from: https://doi.org/10.1080/19490976.2021.1875774.

Pyndt Jørgensen, B., Winther, G., Kihl, P., Nielsen, D.S., Wegener, G., Hansen, A.K., et al., 2015. Dietary magnesium deficiency affects gut microbiota and anxiety-like behaviour in C57BL/6N mice. Acta Neuropsychiatrica 27 (5), 307−311. Available from: https://doi.org/10.1017/neu.2015.10.

Raymond, F., Ouameur, A.A., Déraspe, M., Iqbal, N., Gingras, H., Dridi, B., et al., 2016. The initial state of the human gut microbiome determines its reshaping by antibiotics. ISME J. 10 (3), 707−720. Available from: https://doi.org/10.1038/ismej.2015.148.

Rial, S.A., Karelis, A.D., Bergeron, K.F., Mounier, C., 2016. Gut microbiota and metabolic health: the potential beneficial effects of a medium chain triglyceride diet in obese individuals. Nutrients 8 (5), 281. Available from: https://doi.org/10.3390/nu8050281.

Ryu, J.Y., Choi, H.M., Yang, H.I., Kim, K.S., 2020. Dysregulated autophagy mediates sarcopenic obesity and its complications via AMPK and PGC1α signaling pathways: potential involvement of gut dysbiosis as a pathological link. Int. J. Mol. Sci. 21 (18), 6887. Available from: https://doi.org/10.3390/ijms21186887.

Sajilata, M.G., Singhal, R.S., Kulkarni, P.R., 2006. Resistant starch-a review. Compr. Rev. Food Sci. Food Saf. 5 (1), 1−17. Available from: https://doi.org/10.1111/j.1541-4337.2006.tb00076.x.

Salazar, N., Dewulf, E.M., Neyrinck, A.M., Bindels, L.B., Cani, P.D., Mahillon, J., et al., 2015. Inulin-type fructans modulate intestinal Bifidobacterium species populations and decrease fecal short-chain fatty acids in obese women. Clin. Nutr. (Edinburgh, Scotl.) 34 (3), 501−507. Available from: https://doi.org/10.1016/j.clnu.2014.06.001.

Schertzer, J.D., Lam, T., 2021. Peripheral and central regulation of insulin by the intestine and microbiome. Am. J. Physiol. Endocrinol. Metab. 320 (2), E234−E239. Available from: https://doi.org/10.1152/ajpendo.00547.2020.

Singh, R.K., Chang, H.W., Yan, D., Lee, K.M., Ucmak, D., Wong, K., et al., 2017. Influence of diet on the gut microbiome and implications for human health. J. Transl. Med. 15 (1), 73. Available from: https://doi.org/10.1186/s12967-017-1175-y.

Soderborg, T.K., Borengasser, S.J., Barbour, L.A., Friedman, J.E., 2016. Microbial transmission from mothers with obesity or diabetes to infants: an innovative opportunity to interrupt a vicious cycle. Diabetologia 59 (5), 895−906. Available from: https://doi.org/10.1007/s00125-016-3880-0.

Sonnenburg, E.D., Sonnenburg, J.L., 2014. Starving our microbial self: the deleterious consequences of a diet deficient in microbiota-accessible carbohydrates. Cell Metab. 20 (5), 779−786. Available from: https://doi.org/10.1016/j.cmet.2014.07.003.

Sutton, E.F., Beyl, R., Early, K.S., Cefalu, W.T., Ravussin, E., Peterson, C.M., 2018. Early time-restricted feeding improves insulin sensitivity, blood pressure, and oxidative stress even without weight loss in men with prediabetes. Cell Metab. 27 (6), 1212−1221. Available from: https://doi.org/10.1016/j.cmet.2018.04.010. e3.

Sánchez-Tapia, M., Miller, A.W., Granados-Portillo, O., Tovar, A.R., Torres, N., 2020. The development of metabolic endotoxemia is dependent on the type of sweetener and the presence of saturated fat in the diet. Gut Microbes 12 (1), 1801301. Available from: https://doi.org/10.1080/19490976.2020.1801301.

Teeman, C.S., Kurti, S.P., Cull, B.J., Emerson, S.R., Haub, M.D., Rosenkranz, S.K., 2016. Postprandial lipemic and inflammatory responses to high-fat meals: a review of the roles of acute and chronic exercise. Nutr. Metab. 13, 80. Available from: https://doi.org/10.1186/s12986-016-0142-6.

Thompson, S.V., Bailey, M.A., Taylor, A.M., Kaczmarek, J.L., Mysonhimer, A.R., Edwards, C.G., et al., 2021. Avocado consumption alters gastrointestinal bacteria abundance and microbial metabolite concentrations among adults with overweight or obesity: a randomized controlled trial. J. Nutr. 151 (4), 753−762. Available from: https://doi.org/10.1093/jn/nxaa219.

Turnbaugh, P.J., Ley, R.E. Mahowald, M.A., Magrini, V., Mardis, E.R., Gordon, J.I., 2006. An obesity-associated gut microbiome with increased capacity for energy harvest. Nature 444 (7122), 1027−1031. Available from: https://doi.org/10.1038/nature05414.

U.S. Department of Agriculture and U.S. Department of Health and Human Services, 2020, December. Dietary Guidelines for Americans, 2020-2025. 9th Edition. Available from <http://dietaryguidelines.gov>.

U.S. Department of Agriculture. 2021, September 5. ChooseMyPlate.gov Website. Washington, DC. What is MyPlate? Accessed from: <https://www.myplate.gov/eat-healthy/what-is-myplate>.

Villablanca, P.A., Alegria, J.R., Mookadam, F., Holmes Jr, D.R., Wright, R.S., Levine, J.A., 2015. Nonexercise activity thermogenesis in obesity management. Mayo Clin. Proc. 90 (4), 509–519. Available from: https://doi.org/10.1016/j.mayocp.2015.02.001.

Voigt, R.M., Forsyth, C.B., Green, S.J., Mutlu, E., Engen, P., Vitaterna, M.H., et al., 2014. Circadian disorganization alters intestinal microbiota. PLoS One 9 (5), e97500. Available from: https://doi.org/10.1371/journal.pone.0097500.

Walker, R.E., Jackson, K.H., Tintle, N.L., Shearer, G.C., Bernasconi, A., Masson, S., et al., 2019. Predicting the effects of supplemental EPA and DHA on the omega-3 index. Am. J. Clin. Nutr. 110 (4), 1034–1040. Available from: https://doi.org/10.1093/ajcn/nqz161.

Wei, M., Brandhorst, S., Shelehchi, M., Mirzaei, H., Cheng, C.W., Budniak, J., et al., 2017. Fasting-mimicking diet and markers/risk factors for aging, diabetes, cancer, and cardiovascular disease. Sci. Transl. Med. 9 (377), eaai8700. Available from: https://doi.org/10.1126/scitranslmed.aai8700.

Wei, L., Yue, F., Xing, L., Wu, S., Shi, Y., Li, J., et al., 2020. Constant light exposure alters gut microbiota and promotes the progression of steatohepatitis in high fat diet rats. Front. Microbiol. 11, 1975. Available from: https://doi.org/10.3389/fmicb.2020.01975.

What is a Whole Grain? The Whole Grains Council, 2021, September 5. Available from <https://wholegrainscouncil.org/what-whole-grain>.

Xiao, H., Kang, S., 2020. The role of the gut microbiome in energy balance with a focus on the gut-adipose tissue axis. Front. Genet. 11, 297. Available from: https://doi.org/10.3389/fgene.2020.00297.

Yang, Q., Liang, Q., Balakrishnan, B., Belobrajdic, D.P., Feng, Q.J., Zhang, W., 2020. Role of dietary nutrients in the modulation of gut microbiota: a narrative review. Nutrients 12 (2), 381. Available from: https://doi.org/10.3390/nu12020381.

Yen, G.-C., Cheng, H.-L., Lin, L.-Y., Hsu, C.L., 2020. The potential role of phenolic compounds on modulating gut microbiota in obesity. J. Food Drug. Anal. 28 (2). Available from: https://doi.org/10.38212/2224-66140.1054Article 1.

Zarrinpar, A., Chaix, A., Yooseph, S., Panda, S., 2014. Diet and feeding pattern affect the diurnal dynamics of the gut microbiome. Cell Metab. 20 (6), 1006–1017. Available from: https://doi.org/10.1016/j.cmet.2014.11.008.

Zhang, Z., Mocanu, V., Cai, C., Dang, J., Slater, L., Deehan, E.C., et al., 2019. Impact of fecal microbiota transplantation on obesity and metabolic syndrome-a systematic review. Nutrients 11 (10), 2291. Available from: https://doi.org/10.3390/nu11102291.

Żółkiewicz, J., Marzec, A., Ruszczyński, M., Feleszko, W., 2020. Postbiotics-a step beyond pre- and probiotics. Nutrients 12 (8), 2189. Available from: https://doi.org/10.3390/nu12082189.

Further reading

Database on Polyphenol Content in Foods, 2021, September 5. Phenol- Explorer. Available from: <http://phenol-explorer.eu/>.

de La Serre, C.B., de Lartigue, G., Raybould, H.E., 2015. Chronic exposure to low dose bacterial lipopolysaccharide inhibits leptin signaling in vagal afferent neurons. Physiol. Behav. 139, 188–194. Available from: https://doi.org/10.1016/j.physbeh.2014.10.032.

Hamilton, M.T., 2018. The role of skeletal muscle contractile duration throughout the whole day: reducing sedentary time and promoting universal physical activity in all people. J. Physiol. 596 (8), 1331–1340. Available from: https://doi.org/10.1113/JP273284.

Smith, R.P., Easson, C., Lyle, S.M., Kapoor, R., Donnelly, C.P., Davidson, E.J., et al., 2019. Gut microbiome diversity is associated with sleep physiology in humans. PLoS One 14 (10), e0222394. Available from: https://doi.org/10.1371/journal.pone.0222394.

Chapter 13

Harnessing gut friendly microbiomes to combat metabolic syndrome

Chitra Selvan[1] and Satinath Mukhopadhyay[2]

[1]*Department of Endocrinology, Ramaiah Medical College, Bengaluru, Karnataka, India,* [2]*Department of Endocrinology, Institute of Postgraduate Medical Education & Research, Kolkata, West Bengal, India*

13.1 Introduction

The human body is colonized by 10–200 trillion microbes in the intestine, skin, and oral and genital cavities; the majority of these microbes are anaerobes. The human intestine is colonized mostly by bacteria belonging to the two phyla Bacteroidetes and Firmicutes (Gomes et al., 2018). The gram-negative Bacteroidetes phylum includes the genera *Bacteroides*, *Prevotella*, *Parabacteroides*, and *Alistipes*, while the gram-positive Firmicutes includes species such as *Faecalibacterium prausnitzii*, *Eubacterium rectale*, and *Eubacterium hallii* (Louis et al., 2010). The term "microbiome" refers to the gene pool of these bacteria and microbes that colonize our bodies. This gene pool has 150-fold more genes than the host genomes. There has been a particular interest in the gut microbiota, as these genes could potentially alter nutrient absorption in the intestine, induce a state of low-grade chronic inflammation, affect satiety signaling, and modulate the risk of metabolic disorders.

Metabolic syndrome is a conglomeration of risk factors to identify individuals with increased risk of cardiovascular events, type 2 diabetes, and nonalcoholic steatohepatitis (Eckel et al., 2005). With the increasing prevalence of obesity and metabolic syndrome all over the world, there is a compelling need to identify factors that could potentially be targeted to reduce the risk of metabolic syndrome.

In this chapter we attempt to review the potential mechanisms of interaction of the gut microbiome and metabolic syndrome and review the available evidence suggesting the possibilities of harnessing the gut-friendly microbiome to combat metabolic syndrome.

13.2 Gut microbiome

The intestine is usually sterile at birth and gets colonized by microbiota in phases. The initial acquisition is usually from the mother's birth canal if baby was born via a vaginal delivery or from the mother's skin if is the baby was born via caesarean section and from the immediate environment in the hospital. The next phase of colonization occurs on initiation of feeding (breast feeding vs bottle feeding), and the final phase of colonization occurs on initiation of solid food, when the microbiota begins to resemble the adult gut flora. The most common colonizing microbiota in the intestine are Bacteroidetes (e.g., *Bacteroides* and *Prevotella*), Firmicutes (e.g., *Clostridium*, *Enterococcus*, *Lactobacillus*), Actinobacteria (e.g., *Bifidobacterium*), and Proteobacteria (e.g., *Helicobacter* and *Escherichia*) (Ley et al., 2005). The microbiota that colonizes human beings is very similar to the microbiota that colonizes mice (superkingdom). Much of the initial learning about the effect of the gut microbiota has come from experiments carried out on laboratory-bred mice.

Germ-free (GF) mice that received fecal transplantation from obese mice consumed more calories and gained a greater fat percentage in comparison to the GF mice that received fecal transplantation from normal-weight mice. Bäckhed et al. posited for the first time that factors beyond the realm of energy expenditure could influence the way in which energy is assimilated into the body for maintenance of energy homeostasis (Bäckhed et al., 2004). These researchers conducted experiments that involved GF mice to make sure that their intestines were not colonized by any

Microbiome, Immunity, Digestive Health and Nutrition. DOI: https://doi.org/10.1016/B978-0-12-822238-6.00033-9

bacteria at all. The food intake and weight gain were monitored at baseline. GF mice were maintained in gnotobiotic isolators under a strict 12-hour light cycle (lights on at 0600 hours) and fed an autoclaved chow diet ad libitum. The researchers then conventionalized adult GF mice with a normal microbiota harvested from the cecum of conventionally raised animals and noticed that this produced a 60% increase in body fat content and insulin resistance within 14 days despite reduced food intake. This experiment highlighted the effects of the microbiota on obesity and insulin resistance (Bäckhed et al., 2004).

The team then continued to explore whether GF mice would be resistant to diet-induced obesity (DIO). Adult GF mice were conventionalized with an unfractionated cecal microbiota from a donor that had also been fed low-fat chow. Three weeks later, half of the animals were switched to a Western diet, in which 41% of the calories came from fat, 41% from readily digested carbohydrates including simple sugars (sucrose); and 18% from protein (caloric density: 4.8 kcal/g). Chow consumption and weight gain were recorded weekly. After 8 weeks, conventionalized animals on the Western diet had gained significantly more weight than their GF counterparts (5.3 ± 0.8 vs 2.1 ± 0.5 g; $p < 0.05$), although both GF and conventionalized mice consumed similar amounts of the Western diet (2.61 ± 0.05 vs 2.48 ± 0.07 g per mouse per day; $p = 0.24$). Epididymal fat-pad weights were also significantly greater in conventionalized mice that were fed the Western diet (37 ± 5 vs 22 ± 1 mg/g of body weight; $p < 0.05$). Weight gain in the GF mice that were fed the Western diet was not significantly different from the weight gain observed in GF mice that had been maintained on the standard low-fat polysaccharide-rich diet. Bäckhed and team thus demonstrated that in contrast to mice with a gut microbiota, GF animals are protected against obesity that develops after consuming a Western-style high-fat, sugar-rich diet. Their results highlight the role of the gut microbiota in the obesity pandemic (Cani and Delzenne, 2009).

Further, Ley et al. studied the cecal microbiota of ob/ob mice, lean-ob and wild-type siblings, and their ob mothers, all of which fed the same polysaccharide-rich diet. First, the researchers demonstrated that the mouse and human microbiota are similar at the division (superkingdom) level, with Firmicutes and Bacteroidetes dominating. Second, compared with lean mice, ob/ob animals were shown to have a 50% reduction in the abundance of Bacteroidetes and a proportional increase in Firmicutes, suggesting that there were differences in the gut microbiota between lean and obese mice (Ley et al., 2005).

Turnbaugh et al. (2009) went on to study the effect of the human gut microbiota on GF mice by transplanting adult human fecal microbiota into GF mice. A stable human gut community was established in the recipient mice that were maintained on a standard low-fat, plant polysaccharide-rich (LF/PP) diet. Fecal samples were collected from these animals 1 day, 1 week, and 1 month after colonization; half of the mice were then switched to a high-fat, high-sugar Western diet. Mice from the LF/PP diet group and the Western diet group were kept on their respective diets for 2 additional months, with weekly fecal sampling (Turnbaugh et al., 2009).

Administration of a Western diet to humanized mice produced a significant increase in adiposity relative to age- and gender-matched humanized mice that consumed the LF/PP diet (epididymal fat-pad weight, $2.29 \pm 0.15\%$ vs $1.42 \pm 0.05\%$ of body weight; two independent experiments; $p < 0.05$). That was not a surprise. Consumption of the Western diet was also associated with a significant increase in the relative abundance of the Firmicutes phylum, especially the Erysipelotrichia class of bacteria and *Enterococcus*, and a significant decrease in the proportional representation of members of the Bacteroidetes ($p < 0.01$) (Cani and Delzenne, 2009).

It would be interesting to transplant the microbiota isolated from the mice that gained weight on a Westernized diet or DIO into GF mice that were consuming conventional meals to determine whether we could "transmit" obesity.

Turnbaugh et al. (2009) transplanted cecal microbiota harvested from obese, conventionally raised wild-type donors that had been on the Western diet for 8 weeks or more since weaning or the cecal microbiota from lean CHO-fed controls into 8- to 9-week-old GF carbohydrate-fed recipients, including a carbohydrate-fed control group. All recipients (after transplantation) were maintained on a 61% carbohydrate diet (of which 2% were from fructose, glucose, lactose, maltose, and sucrose combined) and then euthanized 14 days after receiving the microbiota transplant. Mice that were colonized with a DIO-associated microbiota exhibited a significantly greater percentage increase in body fat, as defined by dual-energy X-ray absorptiometry, than mice receiving microbiota from carbohydrate-fed donors (43.0 ± 7.1 vs $24.8 \pm 4.9\%$ increase; $p < 0.05$). Importantly, there were no statistically significant differences in chow consumption (14.5 ± 0.3 vs 14.7 ± 0.8 kcal/day) or initial weight (22.9 ± 0.3 vs 23.8 ± 0.7 g) between recipients of the obese and lean cecal microbiota.

Turnbaugh et al. (2009) were able to demonstrate similar results in humanized microbiota transplantation studies. Mice that were colonized with a microbiota from obese Western diet–fed humanized donors gained significantly more adiposity during the 2 weeks after transplantation than did mice that were colonized with the humanized microbiota from LF/PP-fed donors ($68.5\% \pm 8.2\%$ vs $34.55 \pm 3.7\%$ increase in total body fat; $p < 0.05$). Thus the ability to gain adiposity could be horizontally transmitted by intestinal transplantation, revolutionizing the way in which the etiological basis of obesity is understood (Turnbaugh et al., 2009).

At this point, it is fair to wonder about the evidence from human studies. Ley et al. (2005) studied 12 obese people who were randomly assigned to either a fat-restricted (FAT-R) or a carbohydrate-restricted (CARB-R) low-calorie diet. The composition of their gut microbiota was monitored over the course of 1 year by sequencing 16S ribosomal RNA genes from stool samples. The resulting dataset revealed that most (70%) of the 4074 identified species-level phylogenetic types (phylotypes) were unique to each person. Bacteroidetes and Firmicutes divisions dominated the microbiota (92.6%). At baseline, obese people had fewer Bacteroidetes ($p < 0.001$) and more Firmicutes ($p = 0.002$) than did lean controls. Over time, the relative abundance of Bacteroidetes increased ($p < 0.001$), and the abundance of Firmicutes decreased ($p = 0.002$), irrespective of diet type. Increased abundance of Bacteroidetes correlated with percentage loss of body weight (R2 was 0.8 for the CARB-R diet and 0.5 for the FAT-R diet, $p < 0.05$) and not with changes in dietary calorie content over time (R2 was 0.06 for the CARB-R diet and 0.09 for the FAT-R diet). This correlation held only after the person had lost at least 6% of his or her body weight on the FAT-R diet and at least 2% on the CARB-R diet. This experiment opened up the possibility of potentially altering the composition of the gut microbiota to alter metabolic health (Bäckhed et al., 2004).

13.3 Gut microbiota and metabolic disorders: mechanisms

13.3.1 Gut microbiome and energy homeostasis

Since gut bacteria play a crucial role in fermentation and degradation of food, increased energy harvest is an attractive hypothesis to explain the mechanistic link between the microbiota and obesity. The hypothesis is that the gut microbiota promotes the absorption of monosaccharides and short-chain fatty acids, which are the end products of fermentation (Le et al., 2003). Short-chain fatty acids are also important signaling molecules at two G protein–coupled receptors in the intestinal wall, that is, GPR 41 and GPR 43. There is some evidence to suggest that short-chain fatty acids can act on these receptors and modulate the secretion of peptide YY, which could be involved in signaling satiety and intestinal permeability (Karra et al., 2009).

Butyrate, a short-chain fatty acid that is produced by the gut microbiome, has long been known to promote growth in farmed animals and more recently has been reported to improve body weight and composition, lipid profile, insulin sensitivity, and glycemia in animal models of metabolic syndrome.

Bäckhed and team also added to our understanding of the relationship between the microbiota and obesity. They noted that wild-type mice had decreased expression of fasting-induced adipocyte factor, also referred to as angiopoetin-like factor 4 (Angptl4), a molecule that is produced in the intestine whose primary action is to inhibit lipoprotein lipase activity in the body. When GF mice were colonized with gut microbiota, the expression of fasting-induced adipocyte factor came down with resultant increased expression of lipoprotein lipase activity that correlated with increased fat deposition in the body (Ley et al., 2005).

In addition to Angptl4, AMP-activated protein kinase (AMPK), one of the main metabolic fuel sensors in liver and skeletal muscle, is impacted by the gut microbiota. Activated AMPK indicates increased levels of fatty acid oxidation. AMPK activity is increased in both skeletal muscle and the liver of GF mice in comparison with conventionalized mice, suggesting that fat utilization is high in GF mice, protecting them against DIO (Turnbaugh et al., 2009).

13.4 Dysbiosis

The altered composition of the gut microbiota secondary to changes in the diet is called dysbiosis (Cani and Delzenne, 2009). Research showed that ob/ob mice exhibited a 50% reduction in the abundance of Bacteroidetes and a proportional increase in Firmicutes; similar changes in the gut microbiota were seen in obese people (Bäckhed et al., 2004).

In a recent metaanalysis of 10 independent studies examining the gut microbiota in obese human subjects, a weak association was seen between the human fecal microbial communities and obesity status. This association was confounded by large interpersonal variation and insufficient sample size. Therefore the use of the Bacteroides to Firmicutes ratio to identify the link between gut dysbiosis and obesity might be an oversimplification of a complex problem (Sze and Schloss, 2016).

There has also been interest in a single strain of bacteria that could confer benefit or be detrimental to metabolic health. One such strain is *Bifidobacteria*, with reports showing an inverse relationship between its numbers and the severity of diet-induced or genetically determined obesity. *Bifidobacteria* have also been shown to reduce intestinal lipopolysaccharide (LPS) levels in mice and to improve the mucosal barrier function (Martinez et al., 2017). Other single bacterial strains that have been documented to increase obesity in GF mice are *Enterobacter cloacae*, *Clostridium ramosum*, and *Lachnospiraceae strain*. In contrast, the introduction of *Akkermansia muciniphila*, phylum Verrucomicrobia, into a complex community of gut microbiota has been directly implicated in improving metabolic syndrome and atherosclerosis (Everard et al., 2013).

13.5 Gut microbiome and inflammation

Obesity is a state of chronic low-grade inflammation, and the intestinal gut microflora, which are different between lean and obese individuals, appear to account for the difference in the degree of inflammation between the two groups. By administering antibiotics to high-fat-fed and ob/ob mice, Cani et al. demonstrated less metabolic endotoxemia and decrease in the cecal content of LPS, a component of the outer membrane of gram-negative bacteria. Furthermore, this correlated with reduced glucose intolerance, body weight gain, fat mass development, inflammation, oxidative stress, and macrophage infiltration marker mRNA expression in visceral adipose tissue. High-fat feeding strongly increased intestinal permeability and reduced the expression of genes coding for proteins of the tight junctions. The absence of CD14 in ob/ob CD14 mutant mice mimicked the metabolic and inflammatory effects of antibiotics (Cani et al., 2008).

Mice that were fed a lard diet that was high in saturated fat exhibited increased expression of toll-like receptor 4 (TLR4) in white adipose tissue as well as increased circulating levels of LPS when compared with fish oil–fed mice. We have shown that fetuin-A acts as an adaptor protein between free fatty acids and TLR4, leading to adipose tissue inflammation (Pal et al., 2012). These changes appear to be mediated by induction of monocyte chemoattractant protein, CCL2 in WAT by gut microbes, promoted specifically by a lard-based but not by a fish oil–based diet (Caesar et al., 2015).

13.6 Future: harnessing the gut microbiota to treat metabolic disorders

There are numerous ways in which we could potentially alter the gut microbiota to our benefit to address metabolic diseases. The most dramatic of them is to consider the effects of a fecal transplantation of gut microbiota to study the metabolic effects.

13.7 Intestinal transplantation

Treatment-naive male subjects with metabolic syndrome underwent small intestine biopsies and subsequent bowel lavage through a duodenal tube, followed by random assignment to either an allogenic (from lean male donors with a body mass index less than 23 kg/m^2) or autologous (reinfusion of their own collected feces) gut microbiota infusion. Insulin sensitivity was measured before and 6 weeks after intervention by means of a hyperinsulinemic euglycemic clamp using [6,6 2H2]-glucose to measure endogenous glucose production (EGP) and hepatic and peripheral insulin sensitivity (rate of disappearance [Rd]). Large (fecal samples) and small (duodenal biopsies) intestinal gut microbiota composition and fecal short-chain fatty acids were measured at baseline and 6 weeks after infusion. The authors found an improvement in peripheral insulin sensitivity 6 weeks after allogenic gut microbiota infusion (median rate of glucose disappearance, Rd: 26.2–45.3 mol/kg/min; $p < 0.05$). A trend toward improvement in hepatic insulin sensitivity, expressed as EGP suppression, was observed in the allogenic group (median EGP suppression: 51.5%–61.6%; $p < 0.08$). No differences in diet composition, resting energy expenditure, or counterregulatory hormones were found between the groups. Gut microbial diversity increased significantly after allogenic gut microbiota transfer but was unchanged in the autologous group. Thus intestinal transplantation from lean individuals resulted in improved insulin sensitivity in individuals with metabolic syndrome. The long-term sustainability of this effect and the safety of this procedure are yet to be studied in detail, but this procedure does present an interesting treatment option for metabolic syndrome (Vrieze et al., 2012).

In a randomized, double-blind, placebo-controlled clinical trial of 87 adolescents with obesity, fecal microbiota transfer (FMT) from four healthy lean donors of the same gender in oral encapsulated form or placebo was administered, and metabolic parameters and gut microbiota were monitored up to 26 weeks. There was no effect of FMT on body mass index (BMI) standard deviation score at 6 weeks (adjusted mean difference [aMD] −0.026; 95% CI: −0.074, 0.022). FMT alone was associated with a reduction in android-to-gynoid-fat ratio that was sustained for at least 26 weeks, particularly in female adolescents, and was associated with favorable changes in the overall gut microbiome composition. In post hoc exploratory analyses among participants with metabolic syndrome at baseline, FMT led to greater resolution of this condition (18 to 4) compared with placebo (13 to 10) by 26 weeks (adjusted odds ratio, 0.06; 95% CI: 0.01–0.45; $p = 0.007$) (Leong et al., 2020).

FMT-TRIM was another recently published 12-week randomized, double-blind, placebo-controlled pilot trial of oral FMT capsules with 24 adults with obesity and mild-to-moderate insulin resistance (homeostatic model assessment of insulin resistance [HOMA-IR] between 2.0 and 8.0) to weekly healthy lean donor FMT vs placebo capsules for 6 weeks. There were no statistically significant improvements in insulin sensitivity or any other metabolic outcomes in the FMT group compared to the placebo group (Yu et al., 2020).

One of the reasons for the divergent results on the FMT studies could be different routes of administering donor feces (frozen capsules instead of duodenal infusion of fresh feces potentially losing beneficial strains). Larger studies with robust evidence are needed before FMT can be recommended as a mainstream treatment to address metabolic syndrome.

13.8 Oral butyrate

Butyrate's effects on cells have been attributed primarily to its epigenetic effects as a histone deacetylase inhibitor as well as its role as an agonist of free fatty acid receptors, but clear mechanistic evidence is lacking. Small-scale human clinical trials have failed to show significant benefits in patients with metabolic syndrome (Gao et al., 2009).

The failure of these trials is attributed to the rapid clearance and poor bioavailability of the oral butyrate. Better-planned studies are needed to understand the effects of oral butyrate on metabolic syndrome. It is also interesting to note that the intestinal transplant experiments described earlier did document a 2.5-fold increase in the butyrate-producing strains, namely, *Roseburia intestinalis* and *E. hallii*, in the group that received allogenic lean intestinal transplantation (Bridgeman et al., 2020).

13.8.1 Probiotics

Probiotics are live microorganisms that are delivered individually or in combinations such as VSL#3 that positively affect health outcomes in the host. They have been used to manage antibiotic-induced gastrointestinal symptoms. Studies looking at the effects of probiotics on obesity and overweight are made difficult by the fact that there is no known bacterial strain that is of universal benefit. Numerous metaanalyses have been published recently that look at the use of probiotics and their effects on obesity. Two important metaanalyses looking at adults showed a benefit of probiotics on weight loss; one study looking at obese adolescents also documented benefits.

In a metaanalysis by Koutnikova et al. (2019), 105 articles met the inclusion criteria, representing 6826 subjects. In overweight but not obese subjects, probiotics induced improvements in body weight ($k = 25$ trials, $d = -0.94$ kg mean difference, 95% CI: -1.17 to -0.70, $I^2 = 0.0\%$), body mass index ($k = 32$, $d = -0.55$ kg/m^2, 95% CI: -0.86 to -0.23, $I^2 = 91.9\%$), waist circumference ($k = 13$, $d = -1.31$ cm, 95% CI: -1.79 to -0.83, $I^2 = 14.5\%$), body fat mass ($k = 11$, $d = -0.96$ kg, 95% CI: -1.21 to -0.71, $I^2 = 0.0\%$), and visceral adipose tissue mass ($k = 5$, $d = -6.30$ cm^2, 95% CI: -9.05 to -3.56, $I^2 = 0.0\%$). In the same metaanalysis, in type 2 diabetes, probiotics reduced fasting glucose ($k = 19$, $d = -0.66$ mmol/L, 95% CI: -1.00 to -0.31, $I^2 = 27.7\%$), glycated hemoglobin ($k = 13$, $d = -0.28$ pp, 95% CI: -0.46 to -0.11 $I^2 = 54.1\%$), and HOMA-IR ($k = 10$, $d = -1.05$ pp, 95% CI: -1.48 to -0.61, $I^2 = 18.2\%$). In subjects with NAFLD, probiotics reduced alanine ($k = 12$, $d = -10.2$ U/L, 95% CI: -14.3 to -6.0, $I^2 = 93.50\%$) and aspartate aminotransferases ($k = 10$, $d = -9.9$ U/L, 95% CI: -14.1 to -5.8, $I^2 = 96.1\%$). These improvements were observed mostly with bifidobacteria (*Bifidobacterium breve*, *B. longum*), *Streptococcus salivarius* subsp. *thermophilus* and lactobacilli (*Lactobacillus acidophilus*, *L. casei*, and *L. delbrueckii*) containing mixtures and influenced by trials conducted in one country. In this metaanalysis the intake of probiotics resulted in minor but consistent improvements in several metabolic risk factors in subjects with metabolic diseases (Koutnikova et al., 2019). Similar results were noted in another metaanalysis by Wang et al. (2019).

A recently published encouraging result from a single-strain probiotic utilized *A. muciniphila* bacteria. In a randomized, double-blind, placebo-controlled pilot study in overweight or obese insulin-resistant volunteers, 40 subjects were enrolled with daily oral supplementation of *A. muciniphila* bacteria or placebo. Compared to placebo, pasteurized *A. muciniphila* improved insulin sensitivity ($+28.62\% \pm 7.02\%$, $p = 0.002$) and reduced hyperinsulinemia ($-34.08\% \pm 7.12\%$, $p = 0.006$) and plasma total cholesterol ($-8.68\% \pm 2.38\%$, $p = 0.02$). Pasteurized *A. muciniphila* supplementation also decreased body weight (-2.27 ± 0.92 kg, $p = 0.091$), fat mass (-1.37 ± 0.82 kg, $p = 0.092$), and hip circumference (-2.63 ± 1.14 cm, $p = 0.091$). After 3 months of supplementation, *A. muciniphila* reduced the levels of the relevant blood markers for liver dysfunction and inflammation (Depommier et al., 2019).

13.9 Heat-killed probiotics

A concern associated with the use of probiotics is that of systemic infections caused by translocation, especially in immunodeficient individuals and children, or the acquisition of antibiotic resistance genes. To avoid this risk, there is a growing interest in nonviable thermally killed microorganisms. Pedret et al. (2018) have determined the effect of *Bifidobacterium animalis* subsp. *lactis* CECT 8145 (Ba8145) (1010 CFU per day for 12 weeks) in the form of live cells subjected to high temperature on body weight. In obese individuals the ingestion of Ba8145, both the live and modified forms, improves anthropometric parameters. Similar effects in the form of weight reduction, waist circumference, and BMI were obtained by using *Pedicoccus pentosaceus* (1011 CFU per day for 12 weeks) (Pedret et al., 2018).

13.10 Prebiotics

Prebiotics are nonliving digestive microbial stimulants that are selectively fermented by the resident microbiota. Hitherto, the term was used in essence for nondigestible fibers. However, the current definition also extends to bioactive compounds of different origin, such as polyunsaturated fatty acids and polyphenols.

Dehghan et al. demonstrated the hypoglycemic and antiinflammatory effects of oligofructose-enriched inulin in patients with type 2 diabetes. In their study, subjects showed marked reductions in fasting glucose, body weight, and inflammatory markers of endotoxemia (Dehghan et al., 2014).

13.11 Synbiotics

Interesting observations were noted in a randomized, double-blind, placebo-controlled clinical trial. Supplementation of a synbiotic containing *L. case*i (3.5×10^9 CFU), *L. rhamnosus* (7.5 × 108 CFU), *L. bulgaricus* (108 CFU), *L. acidophilus* (109 CFU), *B. breve* (1010 CFU), *B. longum* (3.5×10^9 CFU), *S. thermophilus* (108 CFU), and FOS for 16 weeks (from 4 weeks before surgery to 12 weeks after surgery) in patients after gastric bariatric surgery compared to placebo significantly improved anthropometric parameters (Wiciński et al., 2020).

13.12 Postbiotics

Postbiotics are defined as functional bioactive compounds obtained from food-grade microorganisms that confer health benefits when administered in adequate amounts. They include cell structures, secreted molecules or metabolic byproducts, and inanimate microorganisms. This heterogeneous group of molecules presents a broad range of mechanisms and may exhibit some advantages over traditional "biotics" such as probiotics and prebiotics.

The advantages of postbiotics are that they are safe, are well tolerated, and are associated with reduced risk for adverse effects in vulnerable individuals (pregnant women, premature children, old adults, and immunocompromised subjects) with no risk for transferring antibiotic-resistant genes to pathogenic or commensal bacteria. They also have longer stability with respect to their biopreservative effects (Salminen et al., 2021). There has been interest in the use of postbiotics to improve metabolic outcomes in type 2 diabetes mellitus with or without obesity (Cabello-Olmo et al., 2021).

13.13 Summary and concluding remarks

The human intestine is home to millions of microbes with distinct differences between the composition of the microbiota in various metabolic conditions such as obesity and type 2 diabetes. Encouraging results from animal studies have led to an exciting flurry of activities in research attempting to find ways to modulate the gut microbiota to tackle metabolic disorders. Use of prebiotics and postbiotics has shown reasonable benefits for metabolic disorders. However, we must wait to learn more about gut dysbiosis and the contribution of specific strains of microbiota to metabolic disorders. Intestine transplantation or fecal microbiota transplantation are attractive treatment options. While studies using duodenal infusion have shown encouraging results, those using oral capsulated forms of the same infusate have not. As the obesity pandemic continues to spread across the globe, research on the gut microbiota will be eagerly watched for breakthroughs in the management of obesity and metabolic disorders such as type 2 diabetes.

References

Bridgeman, S.C., Northrop, W., Melton, P.E., Ellison, G.C., Newsholme, P., Mamotte, C.D.S., 2020. Butyrate generated by gut microbiota and its therapeutic role in metabolic syndrome. Pharmacol. Res. 160, 105174.

Bäckhed, F., Ding, H., Wang, T., Hooper, L.V., Koh, G.Y., Nagy, A., et al., 2004. The gut microbiota as an environmental factor that regulates fat storage. Proc. Natl. Acad. Sci. USA 101 (44), 15718–15723.

Cabello-Olmo, M., Araña, M., Urtasun, R., Encio, I.J., Barajas, M., 2021. Role of postbiotics in diabetes mellitus: current knowledge and future perspectives. Foods 10 (7), 1590.

Caesar, R., Tremaroli, V., Kovatcheva-Datchary, P., Cani, P.D., Bäckhed, F., 2015. Crosstalk between gut microbiota and dietary lipids aggravates WAT inflammation through TLR signaling. Cell Metab. [Internet] [cited 2016 Sep 26]; 22:658-68.

Cani, P.D., Delzenne, N.M., 2009. The role of the gut microbiota in energy metabolism and metabolic disease. Curr. Pharm. Des. 15, 1546–1558.

Cani, P.D., Delzenne, N.M., 2009. Interplay between obesity and associated metabolic disorders: new insights into the gut microbiota. Curr. Opin. Pharmacol. 9 (6), 737–743.

Cani, P.D., Bibiloni, R., Knauf, C., Waget, A., Neyrinck, A.M., Delzenne, N.M., et al., 2008. Changes in gut microbiota control metabolic endotoxemia-induced inflammation in high-fat diet-induced obesity and diabetes in mice. Diabetes 57, 1470–1481.

Dehghan, P., Pourghassem Gargari, B., Asghari Jafar-abadi, M., 2014. Oligofructose-enriched inulin improves some inflammatory markers and metabolic endotoxemia in women with type 2 diabetes mellitus: a randomized controlled clinical trial. Nutrition 30, 418–423.

Depommier, C., Everard, A., Druart, C., et al., 2019. Supplementation with *Akkermansia muciniphila* in overweight and obese human volunteers: a proof-of-concept exploratory study. Nat. Med. 25, 1096–1103.

Eckel, R.H., Grundy, S.M., Zimmet, P.Z., 2005. The metabolic syndrome. Lancet. 365 (9468), 1415–1428.

Everard, A., Belzer, C., Geurts, L., Ouwerkerk, J.P., Druart, C., Bindels, L.B., et al., 2013. Cross-talk between *Akkermansia muciniphila* and intestinal epithelium controls diet-induced obesity. Proc. Natl. Acad. Sci. U S A [Internet] [cited 2016 Sep 26]; 110:9066-71.

Gao, Z., Yin, J., Zhang, J., Ward, R.E., Martin, R.J., Lefevre, M., et al., 2009. Butyrate improves insulin sensitivity and increases energy expenditure in mice. Diabetes. 58 (7), 1509–1517.

Gomes, A.C., Hoffmann, C., Mota, J.F., 2018. The human gut microbiota: metabolism and perspective in obesity. Gut Microbes 9 (4), 308–325.

Karra, E., Chandarana, K. Batterham, R.L., 2009. The role of peptide YY in appetite regulation and obesity. J. Physiol. 587, 19–25.

Koutnikova, H., Genser, B., Monteiro-Sepulveda, M., et al., 2019. Impact of bacterial probiotics on obesity, diabetes and non-alcoholic fatty liver disease related variables: a systematic review and *meta*-analysis of randomised controlled trials. BMJ Open. 9, e017995.

Le, P.E., Loison, C., Struyf, S., Springael, J.Y., Lannoy, V., Decobecq, M.E., et al., 2003. Functional characterization of human receptors for short chain fatty acids and their role in polymorphonuclear cell activation. J. Biol. Chem. 278, 25481–25489.

Leong, K.S.W., Jayasinghe, T.N., Wilson, B.C.;, Derraik, J.G.B., Albert, B.B., Chiavaroli, V., et al., 2020. Effects of fecal microbiome transfer in adolescents with obesity: the gut bugs randomized controlled trial. JAMA Netw. Open. 3, e2030415.

Ley, R.E., Bäckhed, F., Turnbaugh, P., Lozupone, C.A., Knight, R.D., Gordon, J.I., 2005. Obesity alters gut microbial ecology. Proc. Natl. Acad. Sci. U S A 102 (31), 11070–11075.

Louis, P., Young, P., Holtrop, G., Flint, H.J., 2010. Diversity of human colonic butyrate-producing bacteria revealed by analysis of the butyryl-CoA: acetate CoA-transferase gene. Env. Microbiol. 12, 304–314.

Martinez, K.B., Leone, V., Chang, E.B., 2017. Western diets, gut dysbiosis, and metabolic diseases: are they linked? Gut Microbes 8 (2), 130–142.

Pal, D., Dasgupta, S., Kundu, R., et al., 2012. Fetuin-A acts as an endogenous ligand of TLR4 to promote lipid-induced insulin resistance. Nat. Med. 18, 1279–1285.

Pedret, A., Valls, R.M., Calderón-Pérez, L., Llauradó, E., Companys, J., Pla-Pagà, L., et al., 2018. Effects of daily consumption of the probiotic *Bifidobacterium animalis* subsp. lactis CECT 8145 on anthropometric adiposity biomarkers in abdominally obese subjects: a randomized controlled trial. Int. J. Obes. 43, 1863–1868.

Salminen, S., Collado, M.C., Endo, A., Hill, C., Lebeer, S., Quigley, E.M.M., et al., 2021. The International scientific association of probiotics and prebiotics (ISAPP) consensus statement on the definition and scope of postbiotics. Nat. Rev. Endocrinol.

Sze, M.A., Schloss, P.D., 2016. Looking for a signal in the noise: revisiting obesity and the microbiome. mBio. 7 (4), e01018. -16.

Turnbaugh, P.J., Ridaura, V.K., Faith, J.J., Rey, F.E., Knight, R., Gordon, J.I., 2009. The effect of diet on the human gut microbiome: a metagenomic analysis in humanized gnotobiotic mice. Sci. Transl. Med. 1 (6), 6ra14.

Vrieze, A., Van Nood, E., Holleman, F., Salojärvi, J., Kootte, R.S., Bartelsman, J.F., et al., 2012. Transfer of intestinal microbiota from lean donors increases insulin sensitivity in individuals with metabolic syndrome. Gastroenterology. 143 (4), 913. -6.e7.

Yu, E.W., Gao, L., Stastka, P., Cheney, M.C., Mahabamunuge, J., Torres Soto, M., et al., 2020. Fecal microbiota transplantation for the improvement of metabolism in obesity: the FMT-TRIM double-blind placebo-controlled pilot trial. PLoS Med. 17, e1003051.

Wang, Z.-B., Xin, S.-S., Ding, L.-N., Ding, W.-Y., Hou, Y.-L., Liu, C.-Q., et al., 2019. The potential role of probiotics in controlling overweight/obesity and associated metabolic parameters in adults: a systematic review and *meta*-analysis. Evid. Based Complement. Alternat. Med. Article ID 3862971, 14 pages, 2019.

Wiciński, M., Gębalski, J., Gołębiewski, J., Malinowski, B., 2020. Probiotics for the treatment of overweight and obesity in humans—a review of clinical trials. Microorganisms. 8 (8), 1148.

Chapter 14

The effect of resveratrol-mediated gut microbiota remodeling on metabolic disorders

Marzena Wojcik and Lucyna Wozniak
Department of Structural Biology, Medical University of Lodz, Lodz, Poland

14.1 Introduction

The adult human gastrointestinal tract is a complex ecosystem of microorganisms composed of the bacteria, archaea, and eukarya that are collectively termed the gut microbiota (GM). Bacterial species are the most dominant, especially in the large intestine. Currently, the human GM is considered to be a "second genome," since its genetic content, known as the gut microbiome, has been estimated to exceed more than 100-fold that of the human genome (Bäckhed et al., 2005). Although healthy individuals have unique microbiota compositions that can be affected by a wide variety of host, microbiological, environmental, and dietary factors, two bacterial phyla, namely, Firmicutes (gram-positive bacteria containing over 200 genera, including *Lactobacillus*, *Mycoplasma*, *Streptococcus*, and *Clostridium*) and Bacteroidetes (gram-negative bacteria containing approximately 20 genera and species), have been found to dominate the GM (87% of the microbial community), followed by other phyla that are represented at subdominant levels, including *Proteobacteria* (8%) and *Actinobacteria* (3%) as well as *Verrucomicrobia*, *Fusobacteria*, and the TM7 phylum, which constitute 2% of bacterial populations (Frank et al., 2007). Among the archaea, *Methanobrevibacter smithii*, belonging to Methanobacteriales, is the most prevalent and abundant methanogen in the human gut (Eckburg et al., 2005).

It is now widely accepted that a healthy GM plays several important structural, protective, and metabolic functions in vital processes of the host (Prakash et al., 2011). Among the metabolic functions is nutrient metabolism, including the fermentation of indigestible polysaccharides by various gut microbes, producing short-chain free fatty acids (SCFAs), mainly acetate, propionate, and butyrate, as the end products. These three main SCFAs serve as an energy source for the colonic epithelia (butyrate), liver (propionate), and peripheral tissues (acetate); thus they are responsible for providing up to 10% of the daily caloric requirements in humans (Riley et al., 2013). SCFAs signal via the central nervous system and several G protein–coupled receptors to modulate a range of physiological processes, including energy homeostasis, lipid and carbohydrate metabolism, and the inflammatory response, as well as the integrity of the epithelial barrier through coordinated regulation of tight junction proteins (TJPs). Additionally, acting as signaling molecules, SCFAs promote secretion of leptin by adipocytes as well as glucagon-like peptide (GLP-1) and peptide YY by enteroendocrine cells to regulate host satiety (Amabebe et al., 2020; Morrison and Preston, 2016).

Various taxa of the GM can also metabolize dietary components such as choline and carnitine to trimethylamine (TMA), which is subsequently oxidized to trimethylamine-*N*-oxide (TMAO) by hepatic enzymes.

Apart from the production of SCFAs and TMA/TMAO, the GM is implicated in the synthesis of vitamin K and some B vitamins (i.e., folate, B_{12}, and biotin); protection against luminal colonization by pathogenic bacteria such as *Escherichia coli*, *Clostridia*, *Salmonella*, and *Shigella* species; the conversion and metabolism of bile acids; the biotransformation and elimination of xenobiotics and drugs; and the development and modulation of the immune system (Collins and Patterson, 2020; Prakash et al., 2011).

Disruption or an imbalance of the composition of intestinal microbiota, termed "dysbiosis," is an important pathological process that has been implicated in the development of obesity and type 2 diabetes mellitus (T2DM), two serious metabolic disorders that have become alarming public health problems worldwide. As diet is thought to be a major

Microbiome, Immunity, Digestive Health and Nutrition. DOI: https://doi.org/10.1016/B978-0-12-822238-6.00023-6

determinant of the colonic microbiome linked to energetic and metabolic homeostasis of the host organism, modulation of the gut microbiome through dietary intervention, particularly eating a diet that is rich in plant polyphenols, is currently one of the most promising options in treating the metabolic disorders (Catalkaya et al., 2020).

Polyphenols are a broad and heterogeneous group of compounds that have numerous health-promoting effects and are found in both animal and human studies (Catalkaya et al., 2020). Resveratrol (*trans*-3,5,4′-trihydroxystilbene, RSV), a phytoalexin that is highly abundant in the skins of red grapes and moderately abundant in peanuts and various berries, has been one of the most frequently studied polyphenols in recent years because of its pleiotropic activity (Fig. 14.1). RSV has been shown to exhibit numerous health beneficial effects, including cardioprotective, neuroprotective, antioxidant, antiinflammatory, and antitumor activities (Colica et al., 2018). Additionally, this stilbene may mimic the beneficial effects of dietary restriction, extending the life span of mice and other species (Bass et al., 2007; Baur et al., 2006). RSV has also attracted increasing attention because of its antiobesity and antidiabetic properties. Numerous studies ranging from cell cultures to animal studies as well as some clinical trials have demonstrated that RSV exhibits multitargeted antiobesity and antidiabetic effects, mainly resulting from its ability to modulate various cellular pathways linked to metabolic and energy homeostasis as well as inflammatory response. Although the precise molecular mechanisms underlying these effects are still not completely understood, the two key metabolic sensors, NAD + -dependent deacetylase sirtuin 1 (SIRT1) and adenosine monophosphate-(AMP)-activated protein kinase (AMPK), that regulate gene expression according to the metabolic state of the cell have been recognized as an important molecular targets of RSV action (Aguirre et al., 2014; Szkudelska and Szkudelski, 2010). In recent years, evidence has emerged indicating that a significant part of the antiobesity and antidiabetic actions of RSV is mediated by changes in the GM composition that favorably affect the host's physiology, improving its metabolic and energy parameters. Additionally, the human intestinal bacteria participate in RSV metabolism, converting this stilbene to the three main metabolites dihydroresveratrol, 3,4′-dihydroxy-*trans*-stilbene, and 3,4′-dihydroxybibenzyl (lunularin), which can be absorbed in the gut and are excreted in urine, the extent of which varies interindividually (Fig. 14.1) (Bode et al., 2013).

In this chapter we provide an up-to-date review on the antiobesity and antidiabetic effects of RSV that are mediated through the GM, especially focusing on the molecular and cellular mechanisms by which an interaction between RSV and the GM may affect metabolic relevant outcomes.

14.2 Gut microbiota and obesity

Obesity, defined as a body mass index (BMI) greater than or equals to 30, is a result of an energy imbalance between calories consumed and calories expended and is now reaching pandemic proportions across much of the world.

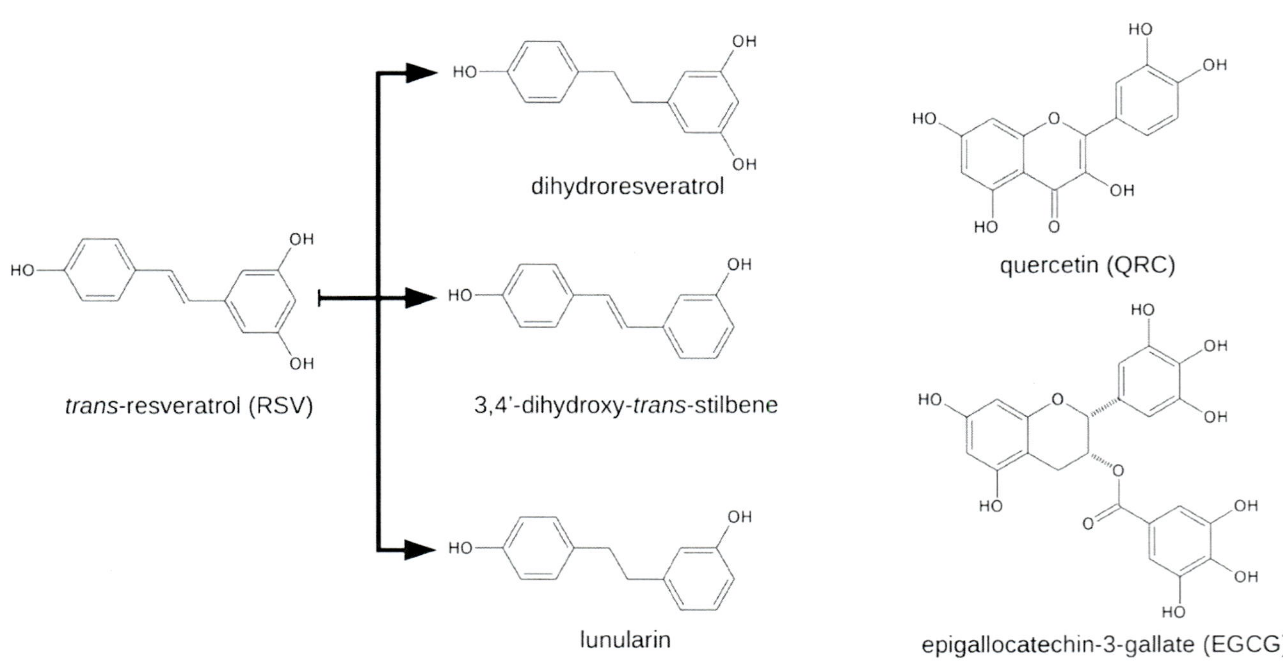

FIGURE 14.1 Chemical structures of QRC, EGCG, and resveratrol (RSV) and metabolites formed from RSV by the human gut microbiota.

According to the latest data from the World Health Organization (WHO), the worldwide prevalence of obesity nearly tripled between 1975 and 2016, and more than 1.9 billion adults aged 18 years and older are overweight, and of those, over 650 million adults are obese. Furthermore, most of the world's population lives in countries where overweight and obesity pose a greater risk of death than underweight (WHO, 2021; https://www.who.int/en/news-room/fact-sheets/detail/obesity-and-overweight).

It is now well established that obesity, particularly central obesity, is the prominent risk factor for insulin resistance and leads to T2DM and other features of metabolic syndrome, such as dyslipidemia and hypertension, with serious socioeconomic implications.

The global prevalence of obesity among children has also increased dramatically in recent decades, with evidence indicating that more than 330 million children aged 5–19 years were overweight or obese in 2016. Obesity in childhood is associated with greater risk factors and increased prevalence of cardiovascular diseases, coronary heart disease, hypertension, and diabetes in adulthood, suggesting the need for implementing health-promoting interventions and programs for the treatment and prevention of childhood obesity (Llewellyn et al., 2016).

Obesity is a multifactorial disease with genetic, metabolic, behavioral, and environmental influences. Among the environmental factors, a sedentary lifestyle and increased intake of a Western diet are thought to be major causative factors for obesity (Fock and Khoo, 2013). There is an increasing body of evidence that the GM is also an important environmental factor in obesity through regulating energy and metabolic homeostasis. An early animal study showed that adult germ-free (GF) C57BL/6J mice had significantly lower weight and body fat percent, despite greater food consumption than mice colonized with conventional GM (so-called conventionalized (CONV) mice) (Bäckhed et al., 2004). Moreover, colonization of GF mice with conventional GM led to the increased body fat content that was accompanied by increased levels of the adipocyte-derived hormone leptin, despite reduced food intake by these animals (Bäckhed et al., 2004). Importantly, in the CONV mice, the microbiota promoted absorption of monosaccharides from the gut and stimulated hepatic lipogenesis de novo as well as increased triglyceride (TG) storage in adipocytes, The latter issue was a consequence of enhanced catalytic activity of lipoprotein lipase (LPL), that is, a key enzyme engaged in lipoprotein metabolism that hydrolyzes TG-rich lipoproteins such as chylomicrons and very low-density lipoproteins to free fatty acids and glycerol for tissue use. Fasting-induced adipocyte factor (Fiaf) from the angiopoietin-like family of proteins was also implicated in this process, since its intestinal expression was decreased and it associated with increased LPL activity in adipocytes, promoting energy storage as fat in conventionalized GF mice. Another factor that modulates microbiota-mediated obesity is AMPK. By comparing GF and CONV mice that were fed low-fat versus high-fat diets, Bäckhed and colleagues (Bäckhed et al., 2007) revealed that GF mice were resistant to diet-induced obesity due to increased skeletal muscle and liver levels of phosphorylated AMPK and its downstream targets involved in fatty acid oxidation (i.e., acetyl CoA carboxylase and carnitine palmitoyltransferase-1). Hence two independent but complementary mechanisms have been proposed in the protection against obesity: high Fiaf levels and increased AMPK activity (Bäckhed et al., 2007).

Investigations done on genetically obese leptin-deficient (*ob/ob*) mice and lean mice showed that obesity is associated with altering the relative abundance of Bacteroidetes and Firmicutes in the gut, as evidenced by a 50% reduction in the abundance of Bacteroidetes and a proportional increase in Firmicutes in *ob/ob* mice versus their lean littermates (Ley et al., 2005). Furthermore, using the same mouse model and metagenomic approach, Turnbaugh et al. (2006) revealed the existence of a positive relationship between altered obese microbiota (a high Firmicutes-to-Bacteroidetes (F/B) ratio) and increased SCFA production, including butyrate and acetate, owing to the enrichment of the *ob/ob* microbiome in genes that encode enzymes that are capable of breaking down indigestible dietary polysaccharides. Hence compared to lean mice, obese mice displayed more capacity for harvesting energy. Moreover, they had far less energy remaining in their feces, a result suggesting that the enhanced extraction of energy from food may partly contribute to the excessive weight gain of these animals.

Many but not all (Duncan et al., 2008; Finucane et al., 2014; Schwiertz et al., 2010) human studies have confirmed the association of the increased F/B ratio with obesity (Crovesy et al., 2020). Several possible reasons for these conflicting results have been proposed in the literature. One possibility is that sample-processing and data analysis methodologies (i.e., the DNA isolation method, the selection of the amplified 16S rRNA region, the sequencing method, and bioinformatics analysis, including taxonomy database and taxonomy assignment algorithm used) may have differential effects across studies. Another possibility is that the insufficient size of the study groups can negatively affect the statistical power of a study to detect small variations. Finally, not taking into account important lifestyle-associated factors in the studied populations (e.g., diet, physical activity, antibiotic consumption) that contribute to shaping the GM composition might generate bias and lead to result misinterpretations (Magne et al., 2020). These drawbacks must be taken into account in future human studies on the relationship between the F/B ratio and obesity status.

It is notable that the F/B ratio can be affected not only by BMI, but also by gender. By comparing the microbiota of lean and obese men and women with similar nutritional backgrounds, matched by age and stratified into three groups according to BMI (BMI < 30, 30 ≤ BMI ≤ 33, and BMI > 33), Haro et al. (2016) detected a higher F/B ratio in men than women under a BMI of 33, which decreased surprisingly in men when BMI was greater than 33. Differences in the GM composition between men and women were also observed at the genus level (*Bacteroides*, *Bilophila*, *Veillonella*, and *Methanobrevibacter*) and at the species level (*Bacteroides plebeius*, *B. caccae*, and *Conus catus*). This highlights the importance of gender-associated differences in the bacterial community pattern linked to obesity, which might be partially explained by differential influences of male and female sex hormones on the gut microbiome (Shin et al., 2019).

Emerging human data have linked low microbial richness to obesity (Le Chatelier et al., 2013), and dietary intervention including increased consumption of high-fiber foods has been proposed to restore the complexity of bacterial communities and improve some clinical phenotypes associated with obesity (Cotillard et al., 2013). It is noteworthy that from overweight to morbid obesity, the loss of GM richness was associated with adverse body composition, adipocyte hypertrophy, and overt metabolic complications (Aron-Wisnewsky et al., 2019).

In recent years there has been a growing appreciation for microbiota-mediated inflammation in obesity. On the basis of animal studies, it is now widely accepted that the intestinal gram-negative-derived lipopolysaccharide (LPS, also known as endotoxin) is one of the most potent activators of innate immune signaling and an important mediator of the microbiome's influence on host physiology (Fuke et al., 2019). LPS is taken up by LPS-binding protein (LBP) in the blood and transferred to CD14, which in turn interacts with toll-like receptor (TLR)-4 to trigger a downstream signaling cascade, leading to activation of the nuclear factor-κB (NF-κB) pathway and subsequently an increased expression of inflammatory proteins such as tumor necrosis factor-α (TNF-α), monocyte chemotactic protein-1, and interleukin (IL)-6 (Zweigner et al., 2006). A high-fat diet (HFD) has been shown to increase the proportion of an LPS-containing microbiota in the gut as well as LPS translocation and circulation concentration up to twofold to threefold, a level that is defined as "metabolic endotoxemia" (Cani et al., 2007). The elevated circulating LPS was accompanied by an increase in gut permeability and corresponding reductions in the mRNA expression of tight junction-related factors such as zonula occludens-1 (ZO-1) and occludin (Cani and Delzenne, 2011). Furthermore, the plasma LPS level positively correlated with several markers of inflammation and oxidative stress, supporting the existence of a strong relationship between the GM, endotoxemia, inflammation, and oxidative stress under HFD conditions (Cani et al., 2008).

14.2.1 Resveratrol, the gut microbiota, and obesity

To date, great efforts have been made to explore in vivo, especially in animal models with diet-induced obesity, the mechanisms by which RSV may affect GM-mediated host fat storage and metabolism. A summary of the main studies evaluating different modes of action of RSV alone or used in a combination with other polyphenolic compounds on the GM composition and associated mechanisms is presented in Table 14.1.

Qiao et al. (2014) demonstrated that treatment with RSV (200 mg/kg/day) leads to a significant reduction in body and visceral adipose weights as well as blood glucose and lipid levels in HFD-fed mice compared with normal chow diet-fed mice. The reason for this finding was the altered colonic microbiota composition by dietary RSV, with the increased B/F ratio that correlated negatively with the body weight. Additionally, an increase in the *Lactobacillus* and *Bifidobacterium* frequencies and a reduction in *Enterococcus faecalis* (i.e., an important, multidrug-resistant nosocomial pathogen) were also detected. At the molecular level, the GM modulation by RSV in HFD mice led to increased Fiaf expression in their intestines accompanied by downregulation of the expression of *Lpl* and the stearoyl-CoA desaturase 1 gene (*Scd1*) in the liver as well as suppressing expression of adipogenesis or lipogenesis genes in visceral adipose tissues, including those encoding peroxisome proliferator-activated receptor gamma (PPARγ), acetyl-CoA carboxylase 1 (Acc-1), and fatty acid synthase (Fas). Hence the authors suggested that RSV may exert antiobesity effects that are partly mediated by the mechanisms of regulating the composition of GM, and this in turn may affect their functional relationships with host fat storage and metabolism.

Another interesting study with RSV supplementation and using a HFD-fed mouse model demonstrated that animals treated with RSV (200 mg/kg/day) for 8 weeks displayed reduced body weight gain, fat deposition, and the growth of obesity-associated GM such as *Lactococcus*, *Clostridium XI*, *Oscillibacter*, and *Hydrogenoanaerobacterium* as compared with mice receiving a HFD alone. These effects seem to be attributed to changes in of the mechanistic target of rapamycin (mTOR) signaling pathway, a well-known key regulator of energy storage and consumption. The authors observed that RSV inhibited mTORC1 and activated mTORC2 in HFD-fed mice (Jung et al., 2016).

TABLE 14.1 Main preclinical and human data reporting the antiobesity effects of resveratrol mediated by gut microbiota remodeling, along with associated mechanisms.

Animal or human study	Dose and duration	Impact on gut microbiota	Impact on host metabolism and immune status	References
Kunming mice	RSV: 200 mg/kg/day for 12 weeks	↑ Bacteroidetes, *Lactobacillus* and *Bifidobacterium* ↓ Firmicutes and *Enterococcus faecalis*	↓ Body and visceral adipose weight, serum glucose and lipid levels, hepatic *Lpl* and *Scd1* expression, and adipocyte *Ppar-g*, *Acc1*, and *Fas* expression ↑ Intestine *Fiaf* expression	Qiao et al. (2014)
HFD-fed C57Bl/6J mice	RSV: 200 mg/kg/day for 8 weeks	↓ *Lactococcus*, *Clostridium XI*, *Oscillibacter*, and *Hydrogenoana erobacterium*	↓Fat deposition and body weight gain −|mTORC1 →mTORC2	Jung et al. (2016)
HFD-fed C57BL/6J mice	RSV: 400 mg/kg/day for 4 weeks	↓ F/B ratio and *Proteobacteria* ↑*Lactobacillus* and *Bifidobacterium*	↓ Body weight gain, pgVAT and ingSAT, the size of adipocytes in WAT ↑ Insulin sensitivity, the expression of UCP1 and SIRT1 in ingSAT and pgVAT depots	Liao et al. (2018)
HFS-fed Wistar rats	RSV (15 mg/kg BW/day) or QRC (30 mg/kg BW/day) or RSV(15 mg/kg BW/day) + QRC (30 mg/kg BW/day) for 6 weeks	RSV + QRC ↑ *Clostridium clariflavum*, *C. methylpentosum*, and *Blautia stercoris* species	RSV + QRC ↓ body weight gain and serum insulin ↑ Colonic TJP-2 and Ocln expression	Etxeberria et al. (2015)
HFD-fedWistar rats	RSV (15 mg/kg BW/day) + QRC (30 mg/kg BW/day) for 10 weeks	↓ Firmicutes and F/B ratio ↓ At the family level: *Lachnospiraceae*, *Desulfovibrionaceae*, *Acidaminococcaceae*, *Coriobacteriaceae*, *Bilophila*, and *Lachnospiraceae* ↑At the family level: *Ruminococcaceae*, *Bacteroidales_S24−7_* group, *Akkermansia* and *Christensenellaceae*	↓Body weight gain and adipose tissue weight ↓ Serum TC, TGs, LDL-C, insulin, TNF-α, IL-6, and MCP-1, ↑ Serum HDL-C, adiponectin	Zhao et al. (2017)
healthy adults with overweight and obesity (18 men and 19 women)	EGCG (282 mg/day) + RES (80 mg/day) for 12 weeks	↓ Bacteroidetes in men but not in women; no significant changes in the abundance of *Firmicutes*,	Baseline Bacteroidetes abundance correlated with the EGCG + RES-induced increase in fat	Most et al. (2017)

(Continued)

TABLE 14.1 (Continued)

Animal or human study	Dose and duration	Impact on gut microbiota	Impact on host metabolism and immune status	References
		Actinobacteria, *γ-Proteobacteria*, and *Akkermansia muciniphila* (phylum *Verrucomicrobiae*), sulfate-reducing bacteria, acetogenic bacteria, and the archeon *Methanobrevibacter smithii* in both men and women	oxidation in men but not in women	

Acc1, acetyl-CoA carboxylase 1; *BMI*, body mass index; *BW*, body weight; *EGCG*, epigallocatechin-3-gallate; *Fas*, fatty acid synthase; *F/B ratio*, Firmicutes-to-Bacteroidetes ratio; *Fiaf*, fasting-induced adipocyte factor; *HDL*, high-density lipoprotein cholesterol; *HFD*, high-fat diet; *HFS*, high-fat sucrose diet; *IL-6*, interleukin 6; *ingSAT*, inguinal adipose tissue; *LDL-C*, low-density lipoprotein cholesterol; *Lpl*, lipoprotein lipase; *MCP-1*, monocyte chemotactic protein-1; *mTOR*, mechanistic target of rapamycin; *Ocln*, occludin; *pgVAT*, perigonadal visceral adipose tissue; *Ppar-g*, peroxisome proliferator-activated receptor gamma; *QRC*, quercetin; *Scd1*, stearoyl-CoA desaturase 1; *SIRT1*, sirtuin 1; *TC*, total cholesterol; *TGs*, TJP-2, tight-junction protein 2; triglyceride; *TNF-α*, tumor necrosis factor α; *UCP-1*, uncoupling protein 1; *WAT*, white adipose tissue → induction; –| inhibition; ↑ increase; ↓ decrease.

Over the last few years, the GM has emerged as an important factor linking RSV consumption to enhancing energy expenditure capacity by inducing the browning of white adipose tissue (WAT), thereby resulting in slimming. More specifically, the supplementation with RSV for 4 weeks to HFD-fed mice reversed the gut dysbiosis observed in HFD-fed mice by modulating the GM composition and significantly reduced the body weight gain, perigonadal visceral adipose tissue (pgVAT) and inguinal adipose tissue (ingSAT) accumulation, the size of adipocytes in WAT, and insulin resistance. Transplanting feces from RSV-treated mice to HFD-fed mice reproduced the beneficial effects of RSV, accompanied by a marked increase in the expression of uncoupling protein 1 (UCP1) and SIRT1, which are involved in mediating the effect of thermogenesis and promoting WAT browning (Liao et al., 2018). Altogether, these data reveal that RSV-induced browning of white fat is partly mediated by GM remodeling.

Like RSV, quercetin (3,3′,4′,5,7-pentahydroxyflavone; QRC), being one of the most common flavonoids present in a wide variety of foods, has attracted attention for its antiobesity and antiinflammatory effects. Hence both polyphenols are currently considered as potential candidates for the amelioration of metabolic disturbances in obesity, acting through a GM-dependent and GM-independent mechanisms (Peredo-Escárcega et al., 2015; Zhao et al., 2017). In the context of the antiobesity effects of RSV and QRC mediated by regulating the composition of gut microbes, Zhao et al. (2017) demonstrated that dietary administration of RSV (15 mg/kg body weight) plus QRC (30 mg/kg body weight) in HFD-induced obesity rats reduced *Firmicutes* abundance, the F/B ratio and the abundance of obesity-associated bacteria, such as *Lachnospiraceae*, *Desulfovibrionaceae*, *Acidaminococcaceae*, *Coriobacteriaceae*, *Bilophila*, and *Lachnospiraceae*. In addition, there was also increasing the abundance of *Bacteroidales S24−7* group, *Christensenellaceae*, *Akkermansia*, and *Ruminococcaceae*, which have all been related to reducing HFD-induced obesity. These changes were accompanied by decreasing the body weight gain, adipose tissue weight, and serum inflammatory markers as well as by improving the serum lipid profile and adiponectin level in HFD-fed rats. The authors concluded that supplementation of RSV plus QRC may have beneficial effects on ameliorating HFD-induced obesity and reducing HFD-induced GM dysbiosis. In the previous study, in which those same doses of both polyphenols were employed to treat rats on a high-fat sucrose diet, Etxeberria et al. (2015) revealed that the combination of RSV with QRC scarcely modified the profile of gut bacteria (it enhanced the abundance of two *Clostridia* and *Blautia stercoris* species) but acted at the intestinal level, altering the mRNA expression of TJP. Furthermore, the ingestion of both polyphenols together prevented the body weight gain and reduced serum insulin levels. Interestingly, supplementation with QRC was more effective than that with RSV alone or in the combination with QRC in remodeling the gut microbial community structure, related to its antiinflammation and metabolic modulation function.

As far as we know, there has been only one human study primarily investigating the potential implication of the combination of RSV (80 mg/day) and epigallocatechin-3-gallate (EGCG; 282 mg/day), that is, the main catechin of green tea that has been widely investigated in terms of its antiobesity effects among other things (Fig. 14.1) (Li et al., 2018) on the GM composition and, subsequently, the metabolic outcome in healthy untrained men and women with overweight and obesity (Most et al., 2017). After a 12-week polyphenol supplementation, there was a significant reduction in Bacteroidetes abundance in men but not in women, whereas the abundance of other bacterial phyla and species (i.e., Firmicutes, *Actinobacteria*, *γ-Proteobacteria*, *Akkermansia muciniphila* (phylum Verrucomicrobiae), sulfate-reducing bacteria, acetogenic bacteria, and the archeon *Methanobrevibacter smithii*) was not affected by the intervention in men or in women. Interestingly, Bacteroidetes abundance at baseline was linked to the RSV + EGCG-induced increase in fat oxidation in men but not in women (Most et al., 2017). However, the assessment of the so-called pure effect of RSV in this study is not possible, since RSV and EGCG were tested together. Therefore further clinical studies on a larger scale and with longer periods of supplementation with RSV and EGCG, taken alone and in the combination, and tighter controls are needed to confirm gender-related differences in the beneficial effects of these two polyphenols in subjects with overweight and obesity.

14.3 Gut microbiota and type 2 diabetes mellitus

T2DM is a chronic and multifactorial metabolic disorder characterized by hyperglycemia resulting from progressive beta cell dysfunction in the presence of insulin resistance. Currently, T2DM is thought to be one of the most abundant metabolic diseases worldwide. It has reached alarming levels, especially in low- and middle-income countries, with obesity and aging seen as the main driving forces (Galicia-Garcia et al., 2020). It has been reported that weight loss as a result of lifestyle, pharmaceutical, and obesity surgical interventions reduces the risk of developing T2DM in the long term by approximately 32%, 10%–74% and 63%, respectively (Aucott, 2008). T2DM is also a well-known risk factor for cardiovascular diseases and other secondary complications, such as nephropathy, neuropathy, and retinopathy, as well as developing certain types of cancers, including liver, pancreas, colorectal, bladder, and breast cancers. Although the pathophysiological mechanisms related to the course of T2DM are multifactorial, the key contributors appear to be adipokine dysregulation, inflammation, and oxidative stress, stemming from overproduction of reactive oxygen species and/or decreased efficiency of the antioxidant defense systems (Galicia-Garcia et al., 2020).

Currently, there is substantial evidence supporting the hypothesis that disturbances in the GM are associated with T2DM development. As early as in 2010, Larsen et al. (2010) found a significant difference between the GM composition of T2DM patients and healthy individuals, mainly at the phylum and class levels. The ratios of B/F and the *Bacteroides-Prevotella* group to the *Clostridium coccoides–Eubacterium rectale* group were positively correlated with blood glucose concentration; furthermore, the class of *Betaproteobacteria* was highly enriched in T2DM patients and positively associated with plasma glucose. Subsequently, the first metagenome-wide association study by Qin et al. (2012) revealed moderate intestinal dysbiosis in T2DM subjects, which was characterized by a reduction in butyrate-producing bacteria (*Roseburia intestinalis* and *Faecalibacterium prausnitzii*) and an increase in various opportunistic pathogen, as well as enrichment of other microbial functions conferring sulfate reduction and oxidative stress resistance. This clearly points out a linkage between reduced abundance of beneficial butyrate-producing bacteria and T2DM. Consistent with this, Forslund et al. (2015) observed that metformin-untreated T2DM was associated with a decrease in genera containing known butyrate producers, such as *Roseburia* spp., *Subdoligranulum* spp., and a cluster of butyrate-producing *Clostridiales* spp. Decreasing the abundance of butyrate-producing bacteria in individuals with T2DM might have deleterious effects on the intestinal immune system, as butyrate is the most important SCFA, owing to its role in promoting gut epithelial integrity and antiinflammatory response, particularly in intestinal cells (Xu et al., 2018).

While there is no doubt that T2DM patients display intestinal dysbiosis, the associations between the taxonomic groups and T2DM remain unclear because of substantial differences in the findings in the literature. By reviewing the results concerning microbial associations with T2DM from 42 human studies, Gurung et al. (2020) established that the genera *Bifidobacterium*, *Bacteroides*, *Faecalibacterium*, *Akkermansia*, and *Roseburia* are negatively associated with T2DM, whereas the genera *Ruminococcus*, *Fusobacterium*, and *Blautia* are positively associated with T2DM. Additionally, the authors extensively discussed different potential mechanisms by which the aforementioned taxa can affect T2DM.

Growing evidence supports the association between metabolic endotoxemia and T2DM. In this regard, most of the studies found higher fasting and postprandial LPS concentrations in diabetic subjects than in healthy controls as a consequence of enhanced intestinal permeability and increased LPS absorption (Gomes et al., 2017). The elevated circulating level of endotoxin may then elicit an inflammatory response, leading to the state of subclinical inflammation that is often present in T2DM subjects, which promotes metabolic dysregulation (Galicia-Garcia et al., 2020). It is worth noting that among the most cited variables that correlated with serum LPS or LBP concentrations were TGs, fasting glycemia, insulinemia, C-reactive protein, glycosylated hemoglobin, and total cholesterol (Gomes et al., 2017). In general, antidiabetic medications have been shown to decrease fasting LPS concentrations in T2DM patients, the most favorable effect being seen for rosiglitazone (Gomes et al., 2017).

14.3.1 Resveratrol, the gut microbiota, and type 2 diabetes mellitus

The relationship between RSV, the GM and improved glucose homeostasis has been confirmed in two fecal microbiota transplant (FMT) studies carried out by the same research group (Kim et al., 2018; Sung et al., 2017). The transplantation of fecal matter from RSV-fed donor mice to conventional mice that were fed an obesogenic high-fat, high-sugar diet caused GM remodeling with an increase in the abundance of *Parabacteroides* and a decrease in the abundance of *Moryella* and *Akkermansia* as well as improvement in glucose homeostasis. The improvement in glucose homeostasis occurred in the absence of a reduction in body weight of animals; therefore the metabolic improvements were not secondary to weight loss. Additionally, obese mice that received RSV-FMT had reduced inflammatory cytokine levels in the colon, suggesting the antiinflammatory effect of RSV, at least at the level of the colon. Moreover, the authors highlighted the fact that FMTs from RSV-fed mice to obese mice was more efficient than oral supplementation for the regulation of glucose homeostasis, showing the potential for such approach in the treatment of metabolic disturbances.

GLP-1 is a peptide hormone from the intestinal tract that plays an essential role in regulating glucose homeostasis, gut motility, appetite, and islet function. Impairment of its action is thought to be linked to the development of T2DM. Dao et al. (2011) have looked at the glucoregulatory effect of RSV in a HFD-fed mouse model of diabetes. According to the results, supplementation with RSV (60 mg/kg/day) for 5 weeks significantly improved oral glucose tolerance and insulin secretion and was related to increased GLP-1 concentration in the portal vein in HFD-induced diabetic mice. These alterations were accompanied by changes in the GM composition and reduced the inflammatory status, as evidenced by markedly increased IL-10 expression and decreased TNF-α expression in the colon, liver, and muscle. The authors concluded that RSV may affect the enteroendocrine axis in mice. However, a double-blind, randomized, crossover study of patients with T2DM failed to confirm these findings. When RSV was administered at a dose of 500 mg twice daily for 5 weeks, there was no effect on GLP-1 secretion, glycemic control, gastric emptying, or body weight (Thazhath et al., 2016). Thus clear inconsistencies in how RSV influences GLP-1 secretion exist between animal and human studies, and further studies in this area are sorely needed.

To sum up, owing to the present and predicted future health and economic burden, there is an increasing focus on the prevention and treatment of lifestyle-related diseases, such as obesity and T2DM. Evidence links HFD intake to the GM dysbiosis, which contributes to disturbed energy homeostasis and low-grade inflammation and, ultimately, obesity and T2DM. The modulation of the GM by RSV emerges as a promising strategy in the treatment of obesity and T2DM. Although not conclusive, research strongly suggests that RSV consumption may beneficially alter the composition of the GM and positively affect energy metabolism and the inflammatory status that is associated with Western lifestyles. What has been lacking is clinical evidence in support of this idea. Therefore well-defined dietary intervention studies are needed to better understand the link between RSV consumption, the GM, and metabolic disorders, especially obesity and T2DM, which may lead to advances in treatment approaches.

Acknowledgments

The authors would like to extend their sincere thanks to Andrzej Zieleniak for technical preparation of the figure.

References

Aguirre, L., Fernández-Quintela, A., Arias, N., Portillo, M.P., 2014. Resveratrol: anti-obesity mechanisms of action. Molecules 19, 18632–18655.

Amabebe, E., Robert, F.O., Agbalalah, T., Orubu, E., 2020. Microbial dysbiosis-induced obesity: role of gut microbiota in homoeostasis of energy metabolism. Br. J. Nutr. 123, 1127–1137.

Aron-Wisnewsky, J., Prifti, E., Belda, E., Ichou, F., Kayser, B.D., Dao, M.C., et al., 2019. Major microbiota dysbiosis in severe obesity: fate after bariatric surgery. Gut 68, 70–82.

Aucott, L.S., 2008. Influences of weight loss on long-term diabetes outcomes. Proc. Nutr. Soc. 67, 54–59.

Bass, T.M., Weinkove, D., Houthoofd, K., Gems, D., Partridge, L., 2007. Effects of resveratrol on lifespan in *Drosophila melanogaster* and *Caenorhabditis elegans*. Mech. Ageing Dev. 128, 546–552.

Baur, J.A., Pearson, K.J., Price, N.L., Jamieson, H.A., Lerin, C., Kalra, A., et al., 2006. Resveratrol improves health and survival of mice on a high-calorie diet. Nature 444, 337–342.

Bode, L.M., Bunzel, D., Huch, M., Cho, G.S., Ruhland, D., Bunzel, M., et al., 2013. *In vivo* and *in vitro* metabolism of *trans*-resveratrol by human gut microbiota. Am. J. Clin. Nutr. 97, 295–309.

Bäckhed, F., Ding, H., Wang, T., Hooper, L.V., Koh, G.Y., Nagy, A., et al., 2004. The gut microbiota as an environmental factor that regulates fat storage. Proc. Natl. Acad. Sci. USA. 101, 15718–15723.

Bäckhed, F., Ley, R.E., Sonnenburg, J.L., Peterson, D.A., Gordon, J.I., 2005. Host-bacterial mutualism in the human intestine. Science 307, 1915–1920.

Bäckhed, F., Manchester, J.K., Semenkovich, C.F., Gordon, J.I., 2007. Mechanisms underlying the resistance to diet-induced obesity in germ-free mice. Proc. Natl. Acad. Sci. USA. 104, 979–984.

Cani, P.D., Amar, J., Iglesias, M.A., Poggi, M., Knauf, C., Bastelica, D., et al., 2007. Metabolic endotoxemia initiates obesity and insulin resistance. Diabetes 56, 1761–1772.

Cani, P.D., Bibiloni, R., Knauf, C., Waget, A., Neyrinck, A.M., Delzenne, N.M., et al., 2008. Changes in gut microbiota control metabolic endotoxemia-induced inflammation in high-fat diet-induced obesity and diabetes in mice. Diabetes 57, 1470–1481.

Cani, P.D., Delzenne, N.M., 2011. The gut microbiome as therapeutic target. Pharmacol. Ther. 130, 202–212.

Catalkaya, G., Venema, K., Lucini, L., Rocchetti, G., Delmas, D., Daglia, M., et al., 2020. Interaction of dietary polyphenols and gut microbiota: microbial metabolism of polyphenols, influence on the gut microbiota and implications on host health. Food Front. 1, 109–133.

Le Chatelier, E., Nielsen, T., Qin, J., Prifti, E., Hildebrand, F., Falony, G., et al.,MetaHIT consortium 2013. Richness of human gut microbiome correlates with metabolic markers. Nature 500, 541–546.

Colica, C., Milanović, M., Milić, N., Aiello, V., De Lorenzo, A., Abenavoli, L., 2018. A systematic review on natural antioxidant properties of resveratrol. Nat. Product. Commun. 3, 1195–1203.

Collins, S.L., Patterson, A.D., 2020. The gut microbiome: an orchestrator of xenobiotic metabolism. Acta Pharm. Sin. B 10, 19–32.

Cotillard, A., Kennedy, S.P., Kong, L.C., Prifti, E., Pons, N., LeChatelier, E., et al.,ANR MicroObes consortium 2013. Dietary intervention impact on gut microbial gene richness. Nature 500, 585–588.

Crovesy, L., Masterson, D., Rosado, E.L., 2020. Profile of the gut microbiota of adults with obesity: a systematic review. Eur. J. Clin. Nutr. 74, 1251–1262.

Dao, T.M., Waget, A., Klopp, P., Serino, M., Vachoux, C., Pechere, L., et al., 2011. Resveratrol increases glucose induced GLP-1 secretion in mice: a mechanism which contributes to the glycemic control. PLoS One 6 (6), e20700.

Duncan, S.H., Lobley, G.E., Holtrop, G., Ince, J., Johnstone, A.M., Louis, P., et al., 2008. Human colonic microbiota associated with diet, obesity and weight loss. Int. J. Obes. 32, 1720–1724.

Eckburg, P.B., Bik, E.M., Bernstein, C.N., Purdom, E., Dethlefsen, L., Sargent, M., et al., 2005. Diversity of the human intestinal microbial flora. Science 308, 1635–1638.

Etxeberria, U., Arias, N., Boqué, N., Macarulla, M.T., Portillo, M.P., Martínez, J.A., et al., 2015. Reshaping faecal gut microbiota composition by the intake of trans-resveratrol and quercetin in high-fat sucrose diet-fed rats. J. Nutr. Biochem. 26, 651–660.

Finucane, M.M., Sharpton, T.J., Laurent, T.J., Pollard, K.S., 2014. A taxonomic signature of obesity in the microbiome? Getting to the guts of the matter. PLoS One 9 (1), e84689.

Fock, K.M., Khoo, J., 2013. Diet and exercise in management of obesity and overweight. J. Gastroenterol. Hepatol. 28 (Suppl 4), 59–63.

Forslund, K., Hildebrand, F., Nielsen, T., Falony, G., LeChatelier, E., Sunagawa, S., et al.,MetaHIT consortium 2015. Disentangling type 2 diabetes and metformin treatment signatures in the human gut microbiota. Nature 528, 262–266.

Frank, D.N., St Amand, A.L., Feldman, R.A., Boedeker, E.C., Harpaz, N., Pace, N.R., 2007. Molecular-phylogenetic characterization of microbial community imbalances in human inflammatory bowel diseases. Proc. Natl. Acad. Sci. USA. 104, 13780–13785.

Fuke, N., Nagata, N., Suganuma, H., Ota, T., 2019. Regulation of gut microbiota and metabolic endotoxemia with dietary factors. Nutrients 11 (10), 2277.

Galicia-Garcia, U., Benito-Vicente, A., Jebari, S., Larrea-Sebal, A., Siddiqi, H., Uribe, K.B., et al., 2020. Pathophysiology of type 2 Diabetes Mellitus. Int. J. Mol. Sci. 21 (17), 6275.

Gomes, J., Costa, J.A., Alfenas, R., 2017. Metabolic endotoxemia and diabetes mellitus: a systematic review. Metabolism 68, 133–144.

Gurung, M., Li, Z., You, H., Rodrigues, R., Jump, D.B., Morgun, A., et al., 2020. Role of gut microbiota in type 2 diabetes pathophysiology. EBioMedicine 51, 102590.

Haro, C., Rangel-Zúñiga, O.A., Alcalá-Díaz, J.F., Gómez-Delgado, F., Pérez-Martínez, P., Delgado-Lista, J., et al., 2016. Intestinal microbiota is influenced by gender and body mass index. PLoS One 11 (5), e0154090.

Jung, M.J., Lee, J., Shin, N.R., Kim, M.S., Hyun, D.W., Yun, J.H., et al., 2016. Chronic repression of mTOR complex 2 induces changes in the gut microbiota of diet-induced obese mice. Sci. Rep. 6, 30887.

Kim, T.T., Parajuli, N., Sung, M.M., Bairwa, S.C., Levasseur, J., Soltys, C.M., et al., 2018. Fecal transplant from resveratrol-fed donors improves glycaemia and cardiovascular features of the metabolic syndrome in mice. Am. J. Physiol. Endocrinol. Metab. 315, E511–E519.

Larsen, N., Vogensen, F.K., van den Berg, F.W., Nielsen, D.S., Andreasen, A.S., Pedersen, B.K., et al., 2010. Gut microbiota in human adults with type 2 diabetes differs from non-diabetic adults. PLoS One 5 (2), e9085.

Ley, R.E., Bäckhed, F., Turnbaugh, P., Lozupone, C.A., Knight, R.D., Gordon, J.I., 2005. Obesity alters gut microbial ecology. Proc. Natl. Acad. Sci. USA. 102, 11070–11075.

Liao, W., Yin, X., Li, Q., Zhang, H., Liu, Z., Zheng, X., et al., 2018. Resveratrol-induced white adipose tissue browning in obese mice by remodeling fecal microbiota. Molecules 23 (12), 3356.

Li, F., Gao, C., Yan, P., Zhang, M., Wang, Y., Hu, Y., et al., 2018. EGCG reduces obesity and white adipose tissue gain partly through AMPK activation in mice. Front. Pharmacol. 9, 1366.

Llewellyn, A., Simmonds, M., Owen, C.G., Woolacott, N., 2016. Childhood obesity as a predictor of morbidity in adulthood: a systematic review and *meta*-analysis. Obes. Rev. 17, 56–67.

Magne, F., Gotteland, M., Gauthier, L., Zazueta, A., Pesoa, S., Navarrete, P., et al., 2020. The firmicutes/bacteroidetes ratio: a relevant marker of gut dysbiosis in obese patients? Nutrients 12, 1474.

Morrison, D.J., Preston, T., 2016. Formation of short chain fatty acids by the gut microbiota and their impact on human metabolism. Gut Microbes 7, 189–200.

Most, J., Penders, J., Lucchesi, M., Goossens, G.H., Blaak, E.E., 2017. Gut microbiota composition in relation to the metabolic response to 12-week combined polyphenol supplementation in overweight men and women. Eur. J. Clin. Nutr. 71, 1040–1045.

Peredo-Escárcega, A.E., Guarner-Lans, V., Pérez-Torres, I., Ortega-Ocampo, S., Carreón-Torres, E., Castrejón-Tellez, V., et al., 2015. The combination of resveratrol and quercetin attenuates metabolic syndrome in rats by modifying the serum fatty acid composition and by upregulating SIRT 1 and SIRT 2 expression in white adipose tissue. Evid. Based Complement. Altern. Med. 2015, 474032.

Prakash, S., Rodes, L., Coussa-Charley, M., Tomaro-Duchesneau, C., 2011. Gut microbiota: next frontier in understanding human health and development of biotherapeutics. Biologics 5, 71–86.

Qiao, Y., Sun, J., Xia, S., Tang, X., Shi, Y., Le, G., 2014. Effects of resveratrol on gut microbiota and fat storage in a mouse model with high-fat-induced obesity. Food Funct. 5, 1241–1249.

Qin, J., Li, Y., Cai, Z., Li, S., Zhu, J., Zhang, F., et al., 2012. A metagenome-wide association study of gut microbiota in type 2 diabetes. Nature 490, 55–60.

Riley, L.W., Raphael, E., Faerstein, E., 2013. Obesity in the United States - dysbiosis from exposure to low-dose antibiotics?. Front. Public. Health 1, 69.

Schwiertz, A., Taras, D., Schäfer, K., Beijer, S., Bos, N.A., Donus, C., et al., 2010. Microbiota and SCFA in lean and overweight healthy subjects. Obesity 18, 190–195.

Shin, J.H., Park, Y.H., Sim, M., Kim, S.A., Joung, H., Shin, D.M., 2019. Serum level of sex steroid hormone is associated with diversity and profiles of human gut microbiome. Res. Microbiol. 170, 192–201.

Sung, M.M., Kim, T.T., Denou, E., Soltys, C.M., Hamza, S.M., Byrne, N.J., et al., 2017. Improved glucose homeostasis in obese mice treated with resveratrol is associated with alterations in the gut microbiome. Diabetes 66, 418–425.

Szkudelska, K., Szkudelski, T., 2010. Resveratrol, obesity and diabetes. Eur. J. Pharmacol. 635, 1–8.

Thazhath, S.S., Wu, T., Bound, M.J., Checklin, H.L., Standfield, S., Jones, K.L., et al., 2016. Administration of resveratrol for 5 wk has no effect on glucagon-like peptide 1 secretion, gastric emptying, or glycemic control in type 2 diabetes: a randomized controlled trial. Am. J. Clin. Nutr. 103, 66–70.

Turnbaugh, P.J., Ley, R.E., Mahowald, M.A., Magrini, V., Mardis, E.R., Gordon, J.I., 2006. An obesity-associated gut microbiome with increased capacity for energy harvest. Nature 444, 1027–1031.

World Health Organization, 2021. Obesity and overweight. Fact sheet 311. Available from: https://www.who.int/en/news-room/fact-sheets/detail/obesity-and-overweight.

Xu, Y.H., Gao, C.L., Guo, H.L., Zhang, W.Q., Huang, W., Tang, S.S., et al., 2018. Sodium butyrate supplementation ameliorates diabetic inflammation in db/db mice. J. Endocrinol. 238, 231–244.

Zhao, L., Zhang, Q., Ma, W., Tian, F., Shen, H., Zhou, M., 2017. A combination of quercetin and resveratrol reduces obesity in high-fat diet-fed rats by modulation of gut microbiota. Food Funct. 8, 4644–4656.

Zweigner, J., Schumann, R.R., Weber, J.R., 2006. The role of lipopolysaccharide-binding protein in modulating the innate immune response. Microbes Infect. 8, 946–952.

Section IV

Microbiome and Immune Health

Chapter 15

The microbiome, immunity, anaerobism, and inflammatory conditions: a multifaceted systems biology intervention

Bernard William Downs[1], Samudra Prosad Banik[2], Manashi Bagchi[3], Rituparna Banik Ghosh[4], Steve Kushner[5] and Debasis Bagchi[6,7,8]

[1]*Division of Precision Nutrition, Victory Nutrition International, Inc., Bonita Springs, FL, United States,* [2]*Department of Microbiology, Maulana Azad College, Kolkata, West Bengal, India,* [3]*Department of R&D, Dr. Herbs LLC, Concord, CA, United States,* [4]*Department of Physiology, Bhairab Ganguly College, Kolkata, West Bengal, India,* [5]*Department of R&D, ALM R&D, Oldsmar, FL, United States,* [6]*College of Pharmacy and Health Sciences, Texas Southern University, Houston, TX, United States,* [7]*Department of Biology, Adelphi University, Garden City, NY, United States,* [8]*Department of R&D, Victory Nutrition International, Inc., Bonita Springs, FL, United States*

15.1 Introduction

In industrialized countries the incidence of chronic degenerative diseases (CDD) is now an epidemic. According to the Centers for Disease Control and Prevention, of the ten most prevalent health maladies, collectively, six are due to digestive disorders, making them the most prevalent health problem in our society [Centers for Disease Control and Prevention (CDC), 2021; Digestive Diseases Statistics for the United States. NIDDK, 2021]. Digestive maladies rank as the second most frequent reason for emergency room visits after accidents, the number one reason for the most purchased OTC remedies, the number one reason for doctor visits, and the number one reason for the most prescribed medications for a legitimate health malady. Our medical and healthcare institutions are overburdened in their attempts to arrest suffering from CDD [Centers for Disease Control and Prevention (CDC), 2021; Digestive Diseases Statistics for the United States. NIDDK, 2021]. The natural products and functional foods industries offer a plethora of products to promote a healthier quality of life, including a vast supply of prebiotics, probiotics, synbiotics, psychobiotics, and various types of enzyme products. The dairy section of most grocery stores offers an abundant array of dairy and nondairy forms of probiotic-rich yogurts, kefir products, drinks, and the like (Cencic and Chingwaru, 2010; El Sohaimy SA, 2012). These types of products target condition-specific or system-specific benefits, such as improving gut and immune functions. However, taken together, the pharmacological and natural interventions have not been able to impede the increase in gastrointestinal (GI) tract disorders (Cencic and Chingwaru, 2010). It is incumbent on researchers and health professionals to find more effective solutions.

15.2 Restoring the ideal biological environment and optimal homeostatic equilibrium

To ensure the greatest beneficial contribution to overall health, the functioning of multiple systems must be simultaneously addressed, optimized, and integrated into a therapeutic intervention. All CDD are anaerobic pathologies (Di Renzo et al., 2021; Downs et al., 2020). Therefore the primary objectives are the following:

1. Promote digestive competence (i.e., disintegration and dissolution of foodstuffs, absorption of nutrients along with important dietary cofactors)
2. Restore aerobic metabolism (i.e., cellular oxidative respiration via aerobic glycolysis)

Microbiome, Immunity, Digestive Health and Nutrition. DOI: https://doi.org/10.1016/B978-0-12-822238-6.00024-8

3. Promote and restore a healthy microbiome to curtail facultative anaerobic microbial overgrowth and restore the optimal ratio of anaerobic to aerobic microbes in the GI tract, curtailing the need for an increase in dendritic cell and cytokine activity
4. Reduce the potential biological incubation of opportunistic pathogens and anaerobes (e.g., yeasts, worms, molds, parasites, dangerous bacteria, viral species)
5. Ensure competent immune function and therefore reduce inflammatory sequelae.

The systems biology paradigm makes these five factors intricately interconnected, interactive, and interconsequential; that is, whatever affects (or detracts from) one has the potential to affect them all (Fig. 15.1). For example, trying to boost immune competence (i.e., objective 4) without accounting for any impairments in the other four factors can result only in limited short-term success at best. If one system or factor is impaired, it is most probable that other systems or factors are also impaired, though only with subclinical manifestations. Clearly, a reductionist intervention, while possibly achieving some level of phase 1 short-term success for a particular factor, would ultimately be doomed to failure in phase 2 along with the failure of subsequent codependent or collateral events. This is not to suggest that improvements in symptoms could not be achieved by targeting an intervention of any one of the five factors. However, the underlying etiology of the problem remains either undetected and/or unresolved. The symptoms for which the person is being treated are simply the consequence of a dysregulation or metabolic disruption; they are not necessarily the cause of the health malady. Multiple systems need to be addressed simultaneously in a competent, comprehensive, and sufficient manner to enable a multisystems resolution (Cencic and Chingwaru, 2010; Di Renzo et al., 2021; Downs et al., 2020).

15.3 Digestion: the first step in metabolism

Aside from breathing and drinking water, the first step in human metabolism is digesting consumed foods and beverages. This involves the disintegration, dissolution, and absorption of consumed foodstuffs. An impairment in any of these actions has system-wide biological consequences. This process is encumbered by alterations in the natural state of foodstuffs caused by genetic modifications; agribusiness chemicals (chemical fertilizers, pesticides, bactericides, fungicides, growth enhancers, etc.); irradiation, gassing, and coloring; food processing (heating/blanching, flavor enhancement/additives, preservatives, etc.); and fast foods (Cencic and Chingwaru, 2010; Corbier et al., 2019; Di Renzo et al., 2021; Downs et al., 2020). The objective of these methods of food growing, processing, handling, and preparation is to make food bigger, look better, last longer (all for commercial viability), and taste better. The nutritional value of food is an incidental matter.

15.4 Antioxidant therapy: firing at the wrong target

Copious amounts of various antioxidants are routinely recommended to mitigate inflammation, which may promote at least a modicum of improvement. However, as has already been mentioned, all chronic degenerative disorders

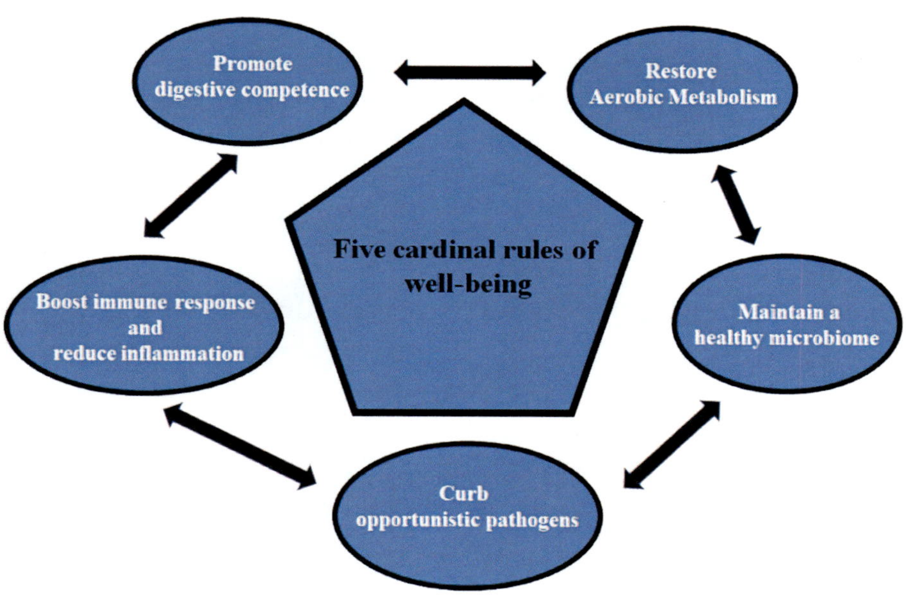

FIGURE 15.1 Five cardinal rules to restore homeostasis. Establishment of an aerobic regimen within the body is fundamental to maintenance of a healthy lifestyle, since it boosts five intricately associated factors to ensure the body's well-being. The human microbiome probably constitutes the most significant checkpoint in keeping us healthy and happy.

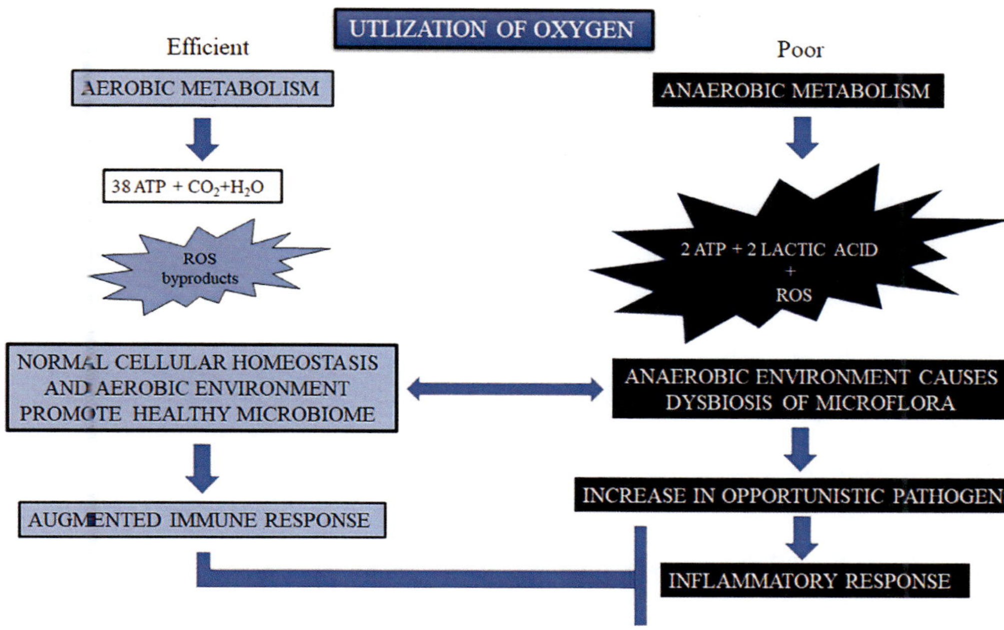

FIGURE 15.2 Alternative fates of oxygen utilization. Judicious use of oxygen is central to the establishment of a healthy state. Aerobic metabolism is energy efficient and suppresses the inflammatory response by keeping the gut microbes healthy. However, if there is a lack of proper distribution of oxygen, anaerobic metabolism ensues, and this leads to dysbiosis of the intestinal microflora followed by acute or chronic inflammatory response. *ROS* reactive oxygen species.

(including chronic inflammation) are caused by an inability to effectively utilize available oxygen; they are anaerobic pathological events (Downs et al., 2020; Corbier et al., 2019). Now the question is what happens to inhaled oxygen that cannot be effectively used? Since an important function of oxygen is for oxidation reactions and the cellular oxygen utilization mechanisms are impaired in anaerobic metabolism, the available oxygen oxidizes cellular components, resulting in oxidative damage, such as damage to lipids, cell membranes, neurons, and cross-linking proteins, as well as damaging DNA, and inducing inflammatory events. These are all sequelae of anaerobic-/hypoxic-induced pathological events. The anaerobic environment generates a much greater number of reactive oxygen species (ROS) than a healthy cellular aerobic environment. Aerobic metabolism produces 38 ATP molecules per glucose molecule and generates some ROS byproducts, such as singlet oxygen, as well as CO_2 and water (Corbier et al., 2019; Downs et al., 2020). In the process of making 38 ATP molecules per glucose molecule, ROS byproducts of oxidative respiration (aka aerobic glycolysis) are generated, but at a much lower rate than anaerobic metabolism. Anaerobic metabolism produces only two ATP molecules per glucose molecule; but also produces two lactic acid molecules in the process (Downs et al, 2020.; Corbier et al., 2019), perpetuating a pathogenic anaerobic/hypoxic state (Fig. 15.2). The unusable oxygen that is present in the anaerobic environment inflicts a significantly greater amount of oxidative damage. Therefore antioxidants, while neutralizing some ROS, are not the best or most effective solution for correcting an anaerobic environment, which generates the greatest amount of ROS (Bullard et al., 2019; Corbier et al., 2019; Downs et al., 2020). Restoring aerobic cellular metabolism is crucial to reduce ROS generation, reduce inflammation, and achieve optimal gene expression and cellular functionality for a healthy state.

15.5 Role of the microbiome

The human microbiome comprises the entire habitat of all forms of microbial cells, including bacteria, archaea, and lower and higher eukaryotes that inhabit the human body, including the microniches composed of cells and tissue fluids (Anderson and Durstine, 2019). It is a cumulative representation of the genomes of all microbial communities encompassing the three domains of microbial life (i.e., Eukaryota, Archaea, Bacteria) living on or in any part of the human body as allochthonous or autochthonous communities. This includes microbial fungi, bacteria, viruses, bacteriophages, archaea, and protozoa. Combined with the human genome, they make up the human-microbial superorganism (Dietert and Silbergeld, 2015), sometimes referred to as the holobiont. This redefined perspective expands our thinking about how we perceive human health and diseases. This new perspective is evidenced by (1) an elevated risk of specific diseases resulting from microbial dysbiosis and (2) increasing evidence that components of the microbiome, such as gut microbiota, have a profound influence over behavior, immunity, metabolism, and food preferences and intolerances (Anderson and Durstine, 2019).

According to the most recent detailed published calculations, the number of somatic cells in the human body is estimated to be approximately 30 trillion. Additionally, the number of microbial cells (identified as bacterial cells) is estimated to be approximately 38 trillion. This more recent data indicates a ratio of human cells to bacterial cells of 1:1.3, a significantly different number than previous comparatively inaccurate "back of envelope" estimates (Sender et al., 2016). Based on the greater number of microbial cells over somatic cells, one interesting perspective suggests that a human would be more accurately defined as a giant microbe.

As has already been indicated, microbial species are present on all body surfaces, including the GI tract, respiratory tract, skin, and urogenital tract (Costello et al., 2009). Other sites that are populated by microbial species include the mammary glands, ovarian follicles, placenta, seminal fluid, saliva, oral mucosa, nasal passages, conjunctiva, and biliary tract (Marchesi and Ravel, 2015; NIH Human Microbiome Project defines normal bacterial makeup of the body; Sherwood et al., 2013). The addition of viral species and their genomes includes the habitat of other tissues, such as the nervous and lymphatic systems, substantially increasing the number of microbial cells. This does not include the vast array of viral genes, which, when combined with other microbes, amount to about 100 times the number of genes of the human genome (Sekirov et al., 2010).

An interesting conundrum is that aerobic metabolism, for example, in muscles, drives normal cellular function and optimal health, whereby, as has already been mentioned, oxygen oxidizes one glucose molecule (also called oxidative respiration or aerobic glycolysis) to produce 38 ATP molecules through the glycolytic cascade, the citric acid cycle, and the electron transport chain. In contrast, anaerobic glycolysis is essentially forced to occur when cells have exhausted the energetic fuels, intracellular machinery, and/or alkalinizing resources (i.e., "buffers") that are required for effective oxidative respiration (Corbier et al., 2019; Downs et al., 2020). The conundrum is that the GI microbiome is composed mostly of anaerobic organisms that are essential to optimal health (Donelli et al., 2012; Yao et al., 2021). Facultative anaerobes are unique in that they can function in both anaerobic and aerobic tissue environments. However, habitual diet and lifestyle abuses, negative epigenetic influences, and excessive stress can induce dysbiosis, in which facultative anaerobic bacteria can overgrow and transition from normal homeostatic contributors to toxic anaerobic pathogens. This pathogenic transition can lead to damage to the intestinal mucosal border, compromised immunity, formation of antibiotic-resistant adherent biofilms, impaired synthesis of various fibers and short-chain fatty acids (SCFAs), impaired neurotransmitter receptivity/functionality, increased inflammatory events (e.g., inflammatory bowel disease), and even lethal sepsis (Carpentier and Cerf, 1993; Yao et al., 2021). It is apparent then that the functional competence of anaerobic microbes throughout the microbiome is a symbiotic requirement to ensure the functional competence of healthy cellular aerobic metabolic events. Conversely, the dysfunctional incompetence of anaerobic microbes leads to a dysfunction in cellular aerobic metabolic events and a transition to an increase in anaerobic pathologies culminating in the subclinical and clinical development of CDD (Corbier et al., 2019; Downs et al., 2020).

Improving habitual diet and lifestyle choices (healthy food choices, moderate exercise, sufficient restful sleep, etc.), promoting positive epigenetic influences, and improving stress tolerance, among other factors, are prudent strategies to promote and protect the many functions of a healthy microbiome. The importance of the microbiome to human health includes its benefits for healthy physiological function and detriments to it that cause significant consequences to physiological health. The mechanisms through which microbial organisms exert their beneficial or detrimental effects are yet to be fully elucidated. A more detailed understanding of the role of signaling molecules, the recognition of bacterial epitopes by both intestinal epithelial and mucosal immune cells, and the influence of changes in the contextual bioenvironment in which microbial cells live is still needed (Sekirov et al., 2010). Epigenetic influences are important factors that need to be considered in assessing the benefits or causes of impairments of the microbiome and their beneficial or detrimental effects, respectively, on physiological health.

While reports vary, it is generally accepted that the human gut contains between 500 and 1000 species of microbes (Xu and Gordon, 2003; Turnbaugh et al., 2007). However, other early estimates suggested that there are over 35,000 species in the human body (Frank et al., 2007), with over 70% of those residing in the colon (Ley et al., 2006; Whitman et al., 1998). This is additional confirmation of the vastness, diversity, and functional complexity of the microbiome. An important consideration in assessing the effects on human health of any one of the organisms is the synergistic and mutualistic effects they exert in combination with other nutritional, environmental, and biofactors (i.e., epigenetics) influencing tissue functionality. We have also advanced in our understanding that abnormal changes in the microbiome, the microbial metabolome, and their vast array of interactions with the immune, endocrine, digestive, and neurological systems are closely correlated with an expansive range of chronic illnesses, including inflammatory bowel disease (Frank et al., 2007; Gevers et al., 2014; Ni et al., 2017), cancer (Kostic et al., 2013), and, owing to gut-brain interconnectivity, a constellation of conditions classified under the rubric of reward deficiency syndrome, including major depressive disorders, ADHD, OCD, and ODD, among others (Blum et al., 2018; Jiang et al., 2015; Zheng et al., 2016).

15.6 Epigenetic influences

The growth rate and survival of each of the members of the living microbiome ecosystem experience fluctuations due to the dynamics of epigenetic influences, such as dietary changes, stress, and vigorous personal hygiene habits or the lack thereof (Gilbert et al., 2018) For example, the same dietary ingredient can have different effects on blood glucose measurements of different people; an effect that is strongly influenced by the microbiome and epigenetic factors. Chronic lifestyle habits exert a strong influence on microbiome composition and overall health. For example, sleep deprivation, with a consequential reduction in stress tolerance, has been shown to adversely affect the gut microbiome (Benedict et al., 2016), with a greater ratio of Firmicutes to Bacteroidia (formerly known as Bacteroidetes, a name that is still routinely used). Overgrowth of Firmicutes has been associated with the development of diabetes and obesity (Komaroff, 2017; Ley et al., 2005, 2006). Maintaining the optimal homeostatic environment is a delicate biological balancing act. A reduced quantity of Bacteroidia in some cases is associated with obesity, while an overabundance is associated with various lower GI tract maladies, such as irritable bowel syndrome and type 1 and type 2 diabetes (Rajilić-Stojanović and de Vos, 2014). The greater amount of Firmicutes also correlates with an elevated amount of Coriobacteriaceae and Erysipelotrichaceae, which are associated with loss of sleep (Dietert, 2014). Stress increases intestinal permeability and is correlated with changes in Bacteroidia and Actinobacteria, with corresponding shifts in metabolite concentrations and inflammatory markers (Zeng et al., 2017).

15.7 Microbiome functional competence

Various factors that affect the functional competence of the microbiome include the type of microbial species constituting the microbiome; the ratio of anaerobic to aerobic microorganisms populating the microbiome; the amount or concentration of the various strains of microbial species; the health and functional competence of the GI mucosal border and its microbial attachment sites; the nature and quality of the environment housing the microbiome; the nature (positive and/or negative) of epigenetic influences on the microbiome; stress; positive and negative neurochemical factors, that is, neurotransmitter receptor types and density (and the role of psychobiotics); and other factors (Dietert, 2014; Ley et al., 2005, 2006; Rajilić-Stojanović and de Vos, 2014; Zeng et al., 2017).

Essentially, the digestive system, from the oral cavity to the anus, is an external organ on the inside of the body, since it is open at both ends. As such, the GI tract holds approximately 70%−80% of immune system lymphatic tissues in the body and houses the vast majority of the body's microbiome. Site-specific microbiomes located in barrier tissues are part of the front line of exposure that encounters xenobiotics prior to absorption across the gut, skin, or respiratory system (Dietert and Silbergeld, 2015). These microbiomes interact with and within the immediate environment in which they reside to modify responses to and from the host organism. The microbial communities have also been shown to participate in molecular biotransformations, such as turning fibers into fatty acids. A compromised microbiome translates to compromised immunity and digestive incompetence.

A competent microbiome in infancy and beyond is important for the functional completion of the "self." Missing microbes and a consequential lack of intended self-completion in the infant is an etiological factor that subsequently results in a compromised health status of both children and adults according to studies reported (Dietert, 2014). Moreover, epigenetic (environmental, lifestyle, and genetic) influences are important factors that have a crucial role in determining the diversity and functional competence of the gut microbiota (Arrieta et al., 2014). Biomarkers have been identified in the microbiome that have been shown to be accurate indicators of disorders such as autoimmune disease, asthma, metabolic syndrome, and obesity (Finucane et al., 2014).

The gut microbiome plays an important role in the metabolism of food. Microbial metabolism of nutrients and other environmental factors are involved in regulating such biological achievements as fundamental brain development and function (Selkrig et al., 2014), behaviors (Cryan and Dinan, 2012), immune-mediated intestinal homeostasis, mucosal immunity (McDermott and Huffnagle, 2014), the hypothalamic-pituitary-adrenal axis (Clarke et al., 2014), and hepatic and cardiovascular status (Dietert and Silbergeld, 2015).

15.8 Inflammatory events

Derangement, deterioration, and/or erosion of the anaerobic/aerobic microbial equilibrium in the microbiome predisposes the microbiome to an overgrowth of facultative anaerobic microbes, a further deterioration in the beneficial microbial bioenvironment, disruption of the gut barrier, and a reduction in the competence of many bodily functions (impaired gut/brain interconnectivity, inducing mental and behavioral disorders, etc.), increasing cellular anaerobic events and consequential inflammatory sequela, and an increase in the incidence of chronic disease pathologies, the vast majority (potentially all) of which are anaerobic events. Inflammation is induced by an increasingly anaerobic bioenvironment (impairing aerobic metabolism) (Corbier et al., 2019), a continuing overgrowth of opportunistic pathogens and anaerobes (yeasts, worms, parasites, viral species, etc.), an alteration in gene expression, and a consequent increase in cellular anaerobic pathogenic inflammatory disorders, such as inflammatory bowel diseases, irritable bowel syndrome characterized by impaired gut/brain interconnectivity (Panza et al., 2019), increased stress intolerance, autoimmune disorders (Brandl et al., 2021; Fattizzo et al., 2021; Moles et al., 2021), activation and increased virulence of herpes viral species (Mostyka et al., 2021; Panza et al., 2019), cardiac disturbances (atrial fibrillation including palpitations, arrhythmias, etc. (Battaglini et al., 2020; Zhang et al., 2021; Zuo et al., 2020), neuromuscular-endocrine dysfunction, and other conditions that were previously reported as idiopathic disorders.

Pathogens such as the herpesvirus, a DNA virus, can cross a weakened blood-brain barrier into the central nervous system, causing neurological damage by catalyzing cytokine production and inducing neuroinflammation (Downs et al., 2021). In most cases, treatments are targeted at the inflammatory condition while the herpetic etiology remains undetected. The resulting disorders can include a variety of escalating degenerative and debilitating inflammatory chronic pathologies, which are classified as idiopathic but are of viral origins, including dementia, Alzheimer's disease (AD), Parkinson's disease, amyotrophic lateral sclerosis, multiple sclerosis, sciatic neural inflammation, Bell's palsy, fibromyalgia, joint deterioration, and baffling neuromuscular disorders (Chronic Inflammation et al., 2012; Immunity and Inflammation in Health and Disease et al., 2017; Inflammation, Advancing Age and Nutrition et al., 2013). Moreover, pathogens can also cross a weakened intestinal barrier, penetrating vascular circulation and crossing the blood-brain barrier. They can also initially cause low-grade chronic inflammation that, without intervention, ultimately escalates to more serious chronic inflammatory pathologies and neuroinflammation. Increased permeability of the gut and blood-brain barrier induced by herpetic infectivity and microbial dysbiosis can impact AD pathogenesis. Inflammatory microorganisms, including viral species, in the gut microbiome can induce peripheral inflammation and cause brain amyloid-β deposition in subjects with cognitive impairment (Panza et al., 2019). Interestingly, certain bacteria in the gut microbiota are able to secrete copious amounts of amyloids and lipopolysaccharides that can induce the production of proinflammatory cytokines associated with the pathogenesis of AD (Jiang et al., 2017). Moreover, research suggests that the microbial composition of the gut in patients with AD, the most prevalent neurodegenerative disease, is different from subjects without AD, especially in regard to having lower amounts of *Eubacterium rectale* and *Bacteroides fragilis*, both of which have antiinflammatory activity, leading to a greater incidence and severity of inflammatory pathologies (Zhang et al., 2020).

15.9 Multifaceted systems biology intervention: conceptual overview

Until now, we have established the value and effects (both good and bad) of the microbiome in immune and neurological functions; the consequences of anaerobic metabolism and dysbiosis; and the influence of opportunistic anaerobes in inflammation, digestive disorders, and CDD, including neuroinflammatory diseases. For any corrective intervention to be effective against multiple pathological dysfunctions, multiple system targets must be simultaneously addressed, that is, a systems biology paradigm. This conceptual perspective significantly eclipses, with minor variations, a reductionist paradigm (i.e., reducing the target to a single molecular locus; reducing the intervening substance to a single active ingredient; reducing the mechanism of action to a single (primary) mechanistic action; and reducing the outcome to a single beneficial effect). A systems biology paradigm depends on providing nutritional resources that enable the body to create the ideal biological environment to optimize gene expression; molecular arrangements; and cellular, tissue, organ, and system functions to avail overall health and optimal systemic functionality (Blum et al., 2020). Homeostatic regulation is critically dependent on mechanistic feedback. Systems biology provides the crucial information that is necessary to ensure optimal biological functioning via the effective monitoring and interactive feedback of biophysiochemical signaling and functions of genes, proteins, and their metabolites that assess and regulate metabolic and signaling pathways, guiding biological behaviors (Chuang et al., 2010). The definition of "optimal health" given above must be viewed in the context of a systems biology perspective and with an understanding that biologically competent and accurate feedback requires the ideal biological environment. The microbiome is a crucially important participant in the creation, maintenance, and functional competence of the ideal biological environment.

15.10 Microbial dysbiosis

The overgrowth of facultative anaerobes in the GI tract leads to dysbiosis and predisposes excessive anaerobic pathogenic distress on other tissues that can erode GI mucosal barrier function; impair appropriate immune "recognition and response" activities in lymphoid tissue, such as Peyer's patches; allow the passage of proteins, such as lectins, through the barrier into the bloodstream, inducing defensive immunoreactivities; and impair the homeostatic regulation of gut-brain interconnectivity, which can contribute to depression, mood swings, cognitive deficits, bowel irregularities, and headaches, among other detriments. Popular species and strains include but are not limited to those represented in Table 15.1.

TABLE 15.1 Major microbial strains.

Lactic acid bacteria (firmicutes)	*Bifidobacteria* (actinobacteria)	Bacteriodes (bacteriodia)	Other firmicutes
Lactobacillus sp. JNU 3829	*B. adolescentis* ATCC15703	*B. acidifaciens*	Bacillus
Lactobacillus acidophilus KU41	*B. adolescentis* 22L	*B. caccae*	Clostridium
Lactobacillus acidophilus M23	*B. adolescentis* BBMN23	*B. coprocola*	Enterococcus
Lactobacillus brevis CH7	*B. bifidum* PRL2010	*B. coprosuis*	Ruminicoccus
Lactobacillus casei MB3	*B. bifidum* S17	*B. eggerthii*	Erysipelotrichia
Lactobacillus fermentum NS2	*B. bifidum* BGN4	*B. finegoldii*	Negativicutes
Lactobacillus plantarum M13	*B. bifidum* ATCC29521	*B. fragilis*	Limnochordia
Lactobacillus plantarum NS3	*B. bifidum* BF3	*B. helcogenes*	Tissierellia
Lactobacillus sakei CH3	*B. breve* UCC2003	*B. intestinalis*	Thermolithobacteria
Lactobacillus sakei MA9	*B. breve* ACS-071-V-Sch8b	*B. massiliensis*	Mollicutes
	B. breve 12L	*B. nordii*	Bacillus
	B. breve JCM7017	*B. ovatus*	Clostridium
	B. breve JCM701	*B. thetaiotaomicron*	Enterococcus
	B. breve NCFB2258	*B. vulgatus*	Ruminicoccus
	B. breve 689b	*B. plebeius*	Erysipelotrichia
	B. breve S27	*B. uniformis*	Negativicutes
	B. breve DSM20213	*B. salyersai*	Limnochordia
	B. breve BR3	*B. pyogene*	
	B. longum NCC2705	*B. goldsteinii*	
	B. longum DJO10A	*B. dorei*	
	B. longum subsp.infantisATCC15697	*B. johnsonii*	
	B. longum subsp. *longum* JDM301		
	B. longum subsp. *longum* BBMN68		
	B. longum subsp. *longum* JCM1217		
	B. longum subsp.infantis157F		
	B. longum subsp. *longum* KACC91563		

(Continued)

TABLE 15.1 (Continued)

Lactic acid bacteria (firmicutes)	*Bifidobacteria* (actinobacteria)	Bacteriodes (bacteriodia)	Other firmicutes
	B. longum BXY01		
	B. longum subsp. *longum* GT15		
	B. longum 105-A		
	B. longum subsp. *infantis* BT1		
	B. longum BG7		
	B. longum subsp. *longum* NCIMB8809		
	B. longum subsp. *longum* CCUG30698		
Upper GI tract	Mid-lower GI tract	[a]Lower GI tract	[a]Diverse sites

[a]*Site of primary activity.*

A critical review of published research has demonstrated that the intestinal flora profoundly exerts a significant beneficial effect on the healthy structure and function of the blood-brain barrier (Ley et al., 2006; Zhang et al., 2021).

15.11 Functions of microbial strains

Lactic acid bacteria (LAB) and *Bifidobacteria* are the most commercially popular strains. The genus Lactobacillus belongs to the phylum Firmicutes, class Bacilli, order II Lactobacillales, and family Lactobacillaceae. *Lactobacilli* are gram-positive, catalase-negative, nonspore-forming, rod-shaped bacteria that produce lactic acid as the major end product of fermentation (De Angelis and Gobbetti, 2011).

Lactic acid bacteria (LAB) have demonstrated inhibition of pathogenic microorganisms, protection against GI diseases, antimutagenic and anticarcinogenic activities, enhancement of the host immune response (du Toit et al., 1998), and reduction of cholesterol. In addition to cholesterol-lowering effects, LAB were shown to be acid resistant, to improve bile tolerance, and to achieve greater adherence to HT-29 cells, a human colon adenocarcinoma cell line (Song et al., 2015). Select species of LAB are ideal candidates for probiotic formulations (see Table 13.1).

Bifidobacteria are gram-negative anaerobic bacteria that generally thrive in the lower oxygen-containing habitats of the GI tract, that is, in anaerobic and microaerophilic (low-oxygen) environments (Brooks et al., 2007).

Carbohydrate fermentation by Bacteriodes results in the production of a pool of SCFAs that are reabsorbed through the large intestine and utilized by the host as an energy source, providing a significant proportion of the host's daily energy requirement (Hooper et al., 2002).

15.12 Microbial restoration

Microbial restoration exhibits the regulatory aspects of utilizing microbial capability in restoring degraded ecosystems, and an appropriate diet can uplift microbiome profiles. Research demonstrates that abundance of probiotic, prebiotic, synbiotic, and fiber-rich ingredients can restore microbial homeostasis in the GI tract (Georgieva et al., 2015). Most of these strains are prevalent in the small intestines, such as *Lactobacillus* strains, like *acidophilus, rhamnosus, casei, thermophilus*, (and other lactobacilli strains); *Bacillus coagulans* (formerly *Lactobacillus sporogenes*); *Bifidobacteria* strains (i.e., *bifidum, longum, adolescentis, infantis*, etc.); *Bulgaricus* strains; and some others (Georgieva et al., 2015).

Georgieva et al. Georgieva et al. (2015) assessed the antimicrobial activity of numerous strains against two pathogens in both broth and milk and reported that *L. acidophilus* L-1; *L. bulgaricus* 6; *L. helveticus* N11; *L. helveticus* 3; *L. plantarum* 24–2L; *L. plantarum* 24%–4%; *L. plantarum* 24–5D; *L. fermentum* 1; *L. brevis* 1 and *B. animalis subsp. lactis* L-3, exhibited the most potent efficacy. These researchers indicated that these cultures may play a potential therapeutic role in clinical practice as well as utilizing these strains as cocultures, starter cultures, or bioprotective cultures in the food industries to enhance food safety and improve food quality (Georgieva et al., 2015).

15.13 Butyric acid

Butyric acid constitutes one of the most significant SCFAs produced and secreted by bacteria as a major secondary metabolite. It is implicated in augmenting the immune response. Butyrate along with propionate, another major SCFA, mediates the differentiation of regulatory T cells (Tregs) by inhibiting histone deacetylase. Butyrate also prevents expression of ICAM-1 and other lymphocyte associated antigens and consequently suppresses proinflammatory immune response The biomarkers of butyric acid are used as clinical treatment tools. Bacterial butyryl-CoA/acetate-CoA transferase activity regulates butyrate production in the large intestine and the ruminant forestomach, while butyrate production is maximized when fermentable fiber and nonstructural carbohydrates are appropriately balanced. Butyric acid and microbial butyrate have vital roles in the treatment of ulcerative colitis and inflammatory bowel disease by decreasing TNF-α and IL-13 and modulating the expression of claudin-2, occludin, cingulin, and zonula occludens proteins ZO-1 and ZO-2 (Plöger et al., 2012).

Hiippala et al. (2018) have exhibited that butyric acid enhances the integrity of the colon epithelium by boosting mucin biosynthesis, the growth and differentiation of intestinal cells, and the tightening of cell junctions. Kong et al. (2019) have demonstrated that microbial diversity and affluence of bacteria producing butyric acid, as well as friendly bacteria such as *Lactobacillus*, *Clostridium sensu stricto*, *Prevotella*, and *Alloprevotella*, decrease with both high-fat diets (HFD) and high-sucrose diets (HSD). Under the same conditions the abundance of diverse pathogenic bacteria, such as *Bacteroides*, *Alistipes*, and *Anaerotruncus* increases. These researchers also reported that abundance of *Oscillibacter*, *Escherichia/Shigella*, *Acinetobacter*, and *Blautia* significantly increased in the group that were fed a high-sucrose diet, while significant increase in *Allobaculum*, *Olsenella*, and *Ruminococcus* bacteria were reported in the HFD group. Probiotics supplementation restored and revitalized bacterial colonization, which were affected by HFD and HSD, as well as reduced weight gain in both the HFD and HSD groups (Kong et al., 2019).

Oh et al. (2019) reported that a mixed biomass of macroalgae/lignocellulose, derived from brown algae (*Saccharina japonica*) and rice straw, can enhance the production of butyric acid compared to a single biomass. In another independent investigation, Jacob et al. (2020) showed that early administration of butyrate restores immunological tolerance in patients with type 1 diabetes through modulation of Tregs with migratory capabilities. Butyrate administration boosts Tregs in the colon, mesenteric lymph nodes, and Peyer's patches and its protective ability following depletion of Tregs.

15.14 Conclusion

In spite of sustained initiatives taken by the research and medical fraternity as well as the healthcare industry, chronic digestive disorders are still six of the ten most prevalent health maladies. The NIDDK reports that approximately 60–70 million people are affected by chronic diverse digestive disorders and problems, including irritable bowel syndrome, bloating and gas, constipation, diarrhea, lactose intolerance, fecal incontinence, acid reflux, Crohn's disease, celiac disease, diverticulosis, and diverticulitis. In industrialized countries the incidence of CDD is now an epidemic.

The natural products and functional foods industries offer a plethora of products to promote a healthier quality of life, including a vast array of prebiotics, probiotics, synbiotics, psychobiotics, and various types of fibers and enzyme products. The dairy section of most grocery stores offers an abundant array of dairy and nondairy forms of probiotic-rich yogurts, kefir products, drinks, and the like. However, combined, pharmacological and and natural interventions have not led to a decrease in GI tract disorders.

Recent research confirms the invaluable contribution that human microbiomes make to achieve optimal gene expression and competent neurological, immunological, endocrinological, dermal, oral, and GI systems interconnectivity, functionality, and overall optimal health. Emerging research continues to provide a greater understanding of the role of therapeutic products and treatment options that can restore healthy body homeostasis through contributions made by bolstering a healthy systemic microbiome. The process of microbial restoration starts with consuming a healthy diet that is rich in plant-based foods. These types of foods contain a range of digestible and indigestible plant fibers that serve as prebiotics, owing to their saccharide and indigestible fiber content. Moreover, eating a good portion of the appropriate plant-based foods in raw form ensures a reasonable supply of live food enzymes to facilitate disintegration and dissolution of consumed foodstuffs. These important first steps concomitantly reduce digestive demands and burden on the GI tract and its neuroimmune system constituents while boosting GI tract homeostasis, functional competence, and the microbiome's contribution to overall health.

Conventional institutionalized agribusiness practices, including chemical fertilizers, pesticides, herbicides, fungicides, bactericides, growth enhancers, GMO technologies, coloring, gassing, irradiation; food processing technologies, including cooking/heating/blanching, functional food additives (texturizers, mouthfeel enhancers, etc.), preservatives, flavor enhancers and sodium-based seasonings, and coloring agents; and food distribution treatments, including chopping, cutting, preservatives, and vacuum packing, are collectively designed to make foods bigger, look better, last longer (for commercial viability), and taste better. The nutritional value is just an innate and incidental feature. These food enhancement and treatment technologies do not include the methods that are used to increase and enhance livestock production parameters (e.g., antibiotics, growth hormones). These food technologies can and do diminish the foods' nutritional value to human and animal health while increasing the potential for greater duress and adverse events. Given the documented state of the food supply and its impact on the GI tract and its functional constituents (i.e., an epidemic of GI tract disorders), it is prudent to use evidence-based health-promoting microbial supplements, including prebiotics, probiotics, synbiotics, psychobiotics, and enzymes. Dietary microbial supplements should be accompanied by fibers and plant-based food supplements to augment food resources and synergistically promote microbiome reconstitution and optimal homeostasis.

References

Anderson, E., Durstine, J.L., 2019. Physical activity, exercise, and chronic diseases: a brief review. Sports Med. Health Sci. 1 (1), 3–10. Available from: https://doi.org/10.1016/j.smhs.2019.08.006.

Arrieta, M.C., Stiemsma, L.T., Amenyogbe, N., Brown, E.M., Finlay, B., 2014. The intestinal microbiome in early life: health and disease. Front. Immunol. Front Immunol. 5, 427. Available from: https://doi.org/10.3389/fimmu.2014.00427.

Battaglini, D., Robba, C., Lopes da Silva, A., Dos Santos Samary, C., Leme Silva, P., Dal Pizzol, F., et al., 2020. Brain-heart interaction after acute ischemic stroke. Crit. Care 24 (1), 163. Available from: https://doi.org/10.1186/s13054-020-02885-8.

Benedict, C., Vogel, H., Jonas, W., Woting, A., Blaut, M., Schürmann, A., et al., 2016. Gut microbiota and glucometabolic alterations in response to recurrent partial sleep deprivation in normal-weight young individuals. Mol. Metab. 5 (12), 1175–1186. Available from: https://doi.org/10.1016/j.molmet.2016.10.003.

Blum, K., Downs, B.W., Bagchi, D., Thanos, P.K., Baron, D., Bowirrat, A., et al., 2020. A systems biological approach coupling reward gene polymorphisms and a nutraceutical (KB220Z) pro-dopamine regulator (PDR) as putative "standard of care" to treat reward deficiency syndrome (RDS): emerging evidence - based scientific induction of "dopamine homeostasis.". Psychol. Psyc. 9 (10), 73–80. Available from: https://www.ecronicon.com/ecpp/pdf/ECPP-09-00780.pdf.

Blum, K., Modestino, E.J., Gondre-Lewis, M.C., Baron, D., Steinberg, B., Panayotis, T.K., et al., 2018. Pro-dopamine regulator (KB220) a fifty year sojourn to combat reward deficiency syndrome (RDS): evidence based bibliography (annotated). CPQ Neurol. Psychol. 1 (2), 1–26. PMC: 2019 PMID: 30957097 PMCID: PMC6448775 NIHMSID: NIHMS1000396.

Brandl, C., Bucci, L., Schett, G., Zaiss, M.M., 2021. Crossing the barriers: revisiting the gut feeling in rheumatoid arthritis. Eur. J. Immunol. 51 (4), 798–810. Available from: https://doi.org/10.1002/eji.202048876.

Brooks GF, Carroll KC, Butel JS, Morse SA., 2007. Jawetz, Melnick & Adelberg's Medical Microbiology (24th ed.). McGraw Hill. Sultan Qaboos Univ Med J. 7 (3), 273–275. ISBN-13: PMCID: PMC3074881 ISBN 978-0-07-128735-3.

Bullard, T., Ji, M., An, R., Trinh, L., Mackenzie, M., Mullen, S.P., 2019. A systematic review and *meta*-analysis of adherence to physical activity interventions among three chronic conditions: cancer, cardiovascular disease, and diabetes. BMC Public Health 19, 636. Available from: https://doi.org/10.1186/s12889-019-6877-z.

Carpentier, B., Cerf, O., 1993. Biofilms and their consequences, with particular reference to hygiene in the food industry. J. Appl. Bacteriol. 75 (6), 499–511. Available from: https://doi.org/10.1111/j.1365-2672.1993.tb01587.x.

Cencic, A., Chingwaru, W., 2010. The role of functional foods, nutraceuticals, and food supplements in intestinal health. Nutrients 2 (6), 611–625. Available from: https://doi.org/10.3390/nu2060611.

Centers for Disease Control and Prevention (CDC). 2021. National Center for Health Statistics: Digestive Diseases. https://www.cdc.gov/nchs/fastats/digestive-diseases.htm (accessed 18.08.21).

Chronic Inflammation, 2012. In: Roy, S., Bagchi, D., Raychaudhuri, S.P. (Eds.), Molecular Pathophysiology, Nutritional and Therapeutic Interventions. CRC Press LLC/Taylor & Francis, Boca Raton, FL, pp. 1–433.

Chuang, H.Y., Hofree, M., Idekar, T., 2010. A decade of systems biology. Annu. Rev. Cell Dev. Biol. 26, 721–744. Available from: https://doi.org/10.1146/annurev-cellbio-100109-104122.

Clarke, G., Stilling, R.M., Kennedy, P.J., Stanton, C., Cryan, J.F., Dinan, R.G., 2014. Minireview: gut microbiota: the neglected endocrine organ. Mol. Endocrinol. 28 (8), 1221–1238. Available from: https://doi.org/10.1210/me.2014-1108.

Corbier, J.R., Downs, B.W., Kushner, S., Aloisio, T., Bagchi, D., Bagchi, M., 2019. VMP35 MNC, a novel iron-free supplement, enhances cytoprotection against anemia in human subjects: a novel hypothesis. Food Nutr. Res. 63, 3410. Available from: https://doi.org/10.29219/fnr.v63.3410.

Costello, E.K., Lauber, C.L., Hamady, M., Fierer, N., Gordon, J.I., Knight, R., 2009. Bacterial community variation in human body habitats across space and time. Science 326 (5960), 1694–1697. Available from: https://doi.org/10.1126/science.1177486.

Cryan, J.F., Dinan, T.G., 2012. Mind-altering microorganisms: the impact of the gut microbiota on brain and behavior. Nat. Rev. Neurosci. 13, 701–712. Available from: https://doi.org/10.1038/nrn3346.

De Angelis, M., Gobbetti, M., 2011. Lactic acid bacteria/ Lactobacillus spp.: general characteristics. In: Encyclopedia of Dairy Sciences (Second ed.), Available from: https://doi.org/10.1016/B978-0-12-374407-4.00259-4.

Dietert, R.R., Silbergeld, E.K., 2015. Biomarkers for the 21st century: listening to the microbiome. Toxicol. Sci. 144 (2), 208–216. Available from: https://doi.org/10.1093/toxsci/kfv013.

Dietert, R.R., 2014. The microbiome in early life: self-completion and microbiota protection as health priorities. Birth Defects Res. B Dev. Reprod. Toxicol. 101 (4), 333–340. Available from: https://doi.org/10.1002/bdrb.21116.

Digestive Diseases Statistics for the United States. NIDDK. 2021. https://www.niddk.nih.gov/health-information/health-statistics/digestive-diseases (accessed 18.08. 21).

Di Renzo, L., Gualtieri, P., De Lorenzo, A., 2021. Diet, nutrition and chronic degenerative diseases. Nutrients 13 (4), 1372. Available from: https://doi.org/10.3390/nu13041372.

Donelli, G., Vuotto, C., Cardines, R., Mastrantonio, P., 2012. Biofilm-growing intestinal anaerobic bacteria. FEMS Immunol. Med. Microbiol. 65, 318–325. Available from: https://doi.org/10.1111/j.1574-695X.2012.00962.x.

Downs, B.W., Bagchi, M., Morrison, B.S., Galvin, J., Kushner, S., Bagchi, D., et al., 2021. A treatise on the role of Herpesvirus in neurodegeneration. In: Kumar, A., Bagchi, D. (Eds.), Antioxidants and Functional Foods for Neurodegenerative Disorders: Uses in Prevention and Therapy. CRC Press/Taylor & Francis, Boca Raton, FL, pp. 85–99.

Downs, B.W., Corbier, J.-R., Speight, N., Kushner, S., Aloisio, T., Bagchi, M., et al., 2020. Anemia: influence of dietary fat, sugar, salt on hemoglobin and blood health. In: Preuss, H.G., Bagchi, D. (Eds.). Dietary Sugar, Salt, and Fat in Human Health. Elsevier/Academic Press, San Diego, California, USA, pp. 103–128.

du Toit, M., Franz, C.M., Dicks, L.M., Schillinger, U., Haberer, P., Warlies, B., et al., 1998. Characterisation and selection of probiotic lactobacilli for a preliminary minipig feeding trial and their effect on serum cholesterol levels, faeces pH and faeces moisture content. Int. J. Food Microbiol. 40 (1–2), 93–104. Available from: https://doi.org/10.1016/s0168-1605(98)00024-5.

El Sohaimy, S.A., 2012. World Appl. Sci. J. 20(5): 691-708. Available from: https://doi.org/10.5829/idosi.wasj.2012.20.05.66119.

Fattizzo, B., Cavallaro, F., Folino, F., Barcellini, W., 2021. Recent insights into the role of the microbiome in malignant and benign hematologic diseases. Crit. Rev. Oncol. Hematol. 160, 103289. Available from: https://doi.org/10.1016/j.critrevonc.2021.103289.

Finucane, M.M., Sharpton, T.J., Laurent, T.J., Pollard, K.S., 2014. A taxonomic signature of obesity in the microbiome? Getting to the guts of the matter. PLoS One 9, e84689. Available from: https://doi.org/10.1371/journal.pone.0084689.

Frank, D.N., St Amand, A.L., Feldman, R.A., Boedeker, E.C., Harpaz, N., Pace, N.R., 2007. Molecular-phylogenetic characterization of microbial community imbalances in human inflammatory bowel diseases. Proc. Natl. Acad. Sci. USA 104 (34), 13780–13785. Available from: https://doi.org/10.1073/pnas.0706625104.

Georgieva, R., Yocheva, L., Tserovska, L., Zhelezova, G., Stefanova, N., Atanasova, A., et al., 2015. Antimicrobial activity and antibiotic susceptibility of Lactobacillus and Bifidobacterium spp. intended for use as starter and probiotic cultures. Biotechnol. Biotechnol. Equip. 29 (1), 84–91. Available from: https://doi.org/10.1080/13102818.2014.987450.

Gevers, D., Kugathasan, S., Denson, L.A., Vázquez-Baeza, Y., Van Treuren, W., Ren, B., et al., 2014. The treatment-naive microbiome in new-onset Crohn's disease. Cell Host Microbe 15 (3), 382–392. Available from: https://doi.org/10.1016/j.chom.2014.020.005.

Gilbert, J., Blaser, M.J., Caporaso, J.G., Jansson, J., Lynch, S.V., Knight, R., 2018. Current understanding of the human microbiome. Nat. Med. 24 (4), 392–400. Available from: https://doi.org/10.1038/nm.4517.

Hiippala, K., Jouhten, H., Ronkainen, A., Hartikainen, A., Kainulainen, V., Jalanka, J., et al., 2018. The potential of gut commensals in reinforcing intestinal barrier function and alleviating inflammation. Nutrients 10 (8), 988. Available from: https://doi.org/10.3390/nu10080988.

Hooper, L.V., Midtvedt, T., Gordon, J.I., 2002. How host-microbial interactions shape the nutrient environment of the mammalian intestine. Annu. Rev. Nutr. 22, 283–307. Available from: https://doi.org/10.1146/annurev.nutr.22.011602.092259.

Immunity and Inflammation in Health and Disease, 2017. Emerging roles of nutraceuticals and functional foodsISBN 978-0-1280–5417-8 In: Chatterjee, S., Jungraithmayr, W., Bagchi, D. (Eds.), Emerging Support. Elsevier/Academic Press, Amsterdam, Holland, pp. 1–459.

Inflammation, Advancing Age and Nutrition, 2013. In: Rahman, I., Bagchi, D. (Eds.), Research and Clinical Interventions. Elsevier/Academic Press, Amsterdam, Holland, pp. 1–364.

Jacob, N., Jaiswal, S., Maheshwari, D., Nallabelli, N., Khatri, N., Bhatia, A., et al., 2020. Butyrate induced Tregs are capable of migration from the GALT to the pancreas to restore immunological tolerance during type-1 diabetes. Sci. Rep. 10 (1), 19120. Available from: https://doi.org/10.1038/s41598-020-76109-y.

Jiang, H., Ling, Z., Zhang, Y., Mao, H., Ma, Z., Yin, Y., et al., 2015. Altered fecal microbiota composition in patients with major depressive disorder. Brain Behav. Immun. 48, 186–194. Available from: https://doi.org/10.1016/j.bbi.2015.03.016.

Jiang, C., Li, G., Huang, P., Liu, Z., Zhao, B., 2017. The gut microbiota and Alzheimer's disease—a review. J. Alzheimers Dis. 58 (1), 1–15. Available from: https://doi.org/10.3233/JAD-161141.

Komaroff, A.L., 2017. The microbiome and risk for obesity and diabetes. JAMA 317 (4), 355–356. Available from: https://doi.org/10.1001/jama.2016.20099.

Kong, C., Gao, R., Yan, X., Huang, L., Qin, H., 2019. Probiotics improve gut microbiota dysbiosis in obese mice fed a high-fat or high-sucrose diet. Nutrition 60, 175–184 Available from: https://doi.org/10.1016/j.nut.2018.100.002.

Kostic, A.D., Chun, E., Robertson, L., Glickman, J.N., Galliani, C.A., Michaud, M., et al., 2013. Fusobacterium nucleatum potentiates intestinal tumorigenesis and modulates the tumor-immune microenvironment. Cell Host Microbe 14 (2), 207–215. Available from: https://doi.org/10.1016/j.chom.2013.070.007.

Ley, R.E., Bäckhed, F., Turnbaugh, P., Lozupone, C.A., Knight, R.D., Gordon, J.I., 2005. Obesity alters gut microbial ecology. Proc. Natl. Acad. Sci. USA 102 (31), 11070–11075. Available from: https://doi.org/10.1073/pnas.0504978102.

Ley, R.E., Peterson, D.A., Gordon, J.I., 2006. Ecological and evolutionary forces shaping microbial diversity in the human intestine. Cell 124 (4), 837–848. Available from: https://doi.org/10.1016/j.cell.2006.020.017.

Ley, R.E., Turnbaugh, P.J., Klein, S., Gordon, J.I., 2006. Microbial ecology: human gut microbes associated with obesity. Nature 444 (7122), 1022–1023. Available from: https://doi.org/10.1038/4441022a.

Marchesi, J.R., Ravel, J., 2015. The vocabulary of microbiome research: a proposal. Microbiome 3, 31. Available from: https://doi.org/10.1186/s40168-015-0094-5.

McDermott, A.J., Huffnagle, G.B., 2014. Immunology 142 (1), 24–31. Available from: https://doi.org/10.1111/imm.12231.

Moles, L., Egimendia, A., Osorio-Querejeta, I., et al., 2021. Gut microbiota changes in experimental autoimmune encephalomyelitis and cuprizone mice models. ACS Chem. Neurosci. 12 (5), 893–905. Available from: https://doi.org/10.1021/acschemneuro.0c00695.

Mostyka, M., Shia, J., Neumann, W.L., Whitney-Miller, C.L., Feely, M., Yantiss, R.K., 2021. Clinicopathologic features of varicella zoster virus infection of the upper gastrointestinal tract. Am. J. Surg. Pathol. 45 (2), 209–214. Available from: https://doi.org/10.1097/PAS.0000000000001576.

NIH Human Microbiome Project defines normal bacterial makeup of the body. Genome sequencing creates first reference data for microbes living with healthy adults. NIH News June 13, 2012. https://www.nih.gov/news-events/news-releases/nih-human-microbiome-project-defines-normal-bacterial-makeup-body (accessed 17.08.21).

Ni, J., Shen, T.D., Chen, E.Z., Bittinger, K., Bailey, A., Roggiani, M., et al., 2017. A role bacterial urease gut dysbiosis Crohn's disease. Sci. Transl. Med. 9 (416), eaah6888. Available from: https://doi.org/10.1126/scitranslmed.aah6888.

Oh, H.J., Kim, K.Y., Lee, K.M., Lee, S.M., Gong, G., Oh, M.K., et al., 2019. Enhanced butyric acid production using mixed biomass of brown algae and rice straw by *Clostridium tyrobutyricum* ATCC25755. Bioresour. Technol. 273, 446–453. Available from: https://doi.org/10.1016/j.biortech.2018.110.037.

Panza, F., Lozupone, M., Solfrizzi, V., Watling, M., Imbimbo, B.P., 2019. Time to test antibacterial therapy in Alzheimer's disease: a review. Brain 142 (10), 2905–2929. Available from: https://doi.org/10.1093/brain/awz244.

Plöger, S., Stumpff, F., Penner, G.B., Schulzke, J.D., Gäbel, G., Martens, H., et al., 2012. Microbial butyrate and its role for barrier function in the gastrointestinal tract. Ann. N. Y. Acad. Sci. 1258, 52–59. Available from: https://doi.org/10.1111/j.1749-6632.2012.06553.x.

Rajilić-Stojanović, M., de Vos, W.M., 2014. The first 1000 cultured species of the human gastrointestinal microbiota. FEMS Microbio Rev. 38 (5), 996–1047. Available from: https://doi.org/10.1111/1574-6976.12075.

Sekirov, I., Russell, S.L., Antunes, L.C., Finlay, B.B., 2010. Gut microbiota in health and disease. Physiol. Rev. 90 (3), 859–904. Available from: https://doi.org/10.1152/physrev.00045.2009.

Selkrig, J., Wong, P., Zhang, X., Pettersson, S., 2014. Metabolic tinkering by the gut microbiome: implications for brain development and function. Gut Microbes 5 (3), 369–380. Available from: https://doi.org/10.4161/gmic.28681.

Sender, R., Fuchs, S., Milo, R., 2016. Revised estimates for the number of human and bacteria cells in the body. PLoS Biol. 19 (14(8)), e1002533. Available from: https://doi.org/10.1371/journal.pbio.1002533.

Sherwood, L., Willey, J., Woolverton, C., 2013. Prescott's Microbiology, ninth ed. McGraw Hill, New York, pp. 713–721, ISBN 9780073402406. OCLC 886600661.

Song, M., Yun, B., Moon, J.H., Park, D.J., Lim, K., Oh, S., 2015. Characterization of selected lactobacillus strains for use as probiotics. Korean J. Food Sci. Anim. Resour. 35 (4), 551–556. Available from: https://doi.org/10.5851/kosfa.2015.35.4.551.

Turnbaugh, P.J., Ley, R.E., Hamady, M., Fraser-Liggett, C.M., Knight, R., Gordon, J.I., 2007. The human microbiome project. Nature 449, 804–810. Available from: https://doi.org/10.1038/nature06244.

Whitman, W.B., Coleman, D.C., Wiebe, W.J., 1998. Prokaryotes: the unseen majority. Proc. Natl Acad. Sci. U. S. A 95 (12), 6578–6583. Available from: https://doi.org/10.1073/pnas.95.12.6578.

Xu, J., Gordon, J.I., 2003. Inaugural article: honor thy symbionts. Proc. Natl. Acad. Sci. USA 100 (18), 10452–10459. Available from: https://doi.org/10.1073/pnas.1734063100.

Yao, Q., Li, H., Fan, L., Zhang, Y., Zhao, S., Zheng, N., et al., 2021. Dietary regulation of the crosstalk between gut microbiome and immune response in inflammatory bowel disease. Foods 10 (2), 368. Available from: https://doi.org/10.3390/foods10020368.

Zeng, M., Inohara, N., Nuñez, G., 2017. Mechanisms of inflammation-driven bacterial dysbiosis in the gut. Mucosal Immunol. 10, 18–26. Available from: https://doi.org/10.1038/mi.2016.75.

Zhang, L., Wang, L., Huanga, L., Zhao, Y., Ding, H., Li, B., et al., 2021. Antidepressant effects of total iridoids of Valeriana jatamansi via the intestinal flora-blood–brain barrier pathway. Pharm. Biol. 59 (1), 912–921. Available from: https://doi.org/10.1080/13880209.2021.1944222.

Zhang, Y., Zhang, S., Li, B., Luo, Y., Gong, Y., Jin, X., et al., 2021. Gut microbiota dysbiosis promotes age-related atrial fibrillation by lipopolysaccharide and glucose-induced activation of NLRP3-inflammasome. Cardiovasc. Res. 118, 785–797. Available from: https://doi.org/10.1093/cvr/cvab114.

Zhang, M., Zhao, D., Zhou, G., Li, C., 2020. Dietary pattern, gut microbiota, and Alzheimer's disease. J. Agric. Food Chem. 68 (46), 12800–12809. Available from: https://doi.org/10.1021/acs.jafc.9b08309.

Zheng, P., Zeng, B., Zhou, C., Liu, M., Fang, Z., Xu, X., et al., 2016. Gut microbiome remodeling induces depressive-like behaviors through a pathway mediated by the host's metabolism. Mol. Psyc. 21, 786–796. Available from: https://doi.org/10.1038/mp.2016.44.

Zuo, K., Yin, X., Li, K., Zhang, J., Wang, P., Jiao, J., et al., 2020. Different types of atrial fibrillation share patterns of gut microbiota dysbiosis. mSphere 5 (2), e00071–20. Available from: https://doi.org/10.1128/mSphere.00071-20.

Further reading

Human Microbiome Project Consortium, 2012. A framework for human microbiome research. Nature 486 (7402), 215–221. Available from: https://doi.org/10.1038/nature11209.

Chapter 16

Preclinical data support the strong specificity and advertising ability to control the immune reactions at mucosal sites

Tamara Meleshko[1,2], Oleksandra Pallah[1,2] and Nadiya Boyko[1,2,3]

[1]*Department of Clinical Laboratory Diagnostics and Pharmacology, Faculty of Dentistry, Uzhhorod National University, Uzhhorod, Ukraine,* [2]*Research Development and Educational Centre of Molecular Microbiology and Mucosal Immunology, Uzhhorod National University, Uzhhorod, Ukraine,* [3]*Ediens LLC, Velyki Lasy, Ukraine*

16.1 Introduction

The mechanisms and/or the pathways by which human (host) microbes are involved (participate in or regulate) in cross-talk communication with the host cells have mainly been defined (Boyko et al., 2004; Cebra et al., 2005a, 2005b; Jiang et al., 2004; Keilbaugh et al., 2005). On the basis of the "classical" point of view, one may conclude that there are numerous data that provide the evidence or background for a strong direct connection between microorganisms and mucosal sites, and these microorganisms are responsible for the regulation of normal physiological inflammation versus pathological inflammation. They play a regulatory role via maintaining this balance and are the most elegant model for such interplaying and interrelation between commensal microbes and human health is inflammatory bowel diseases that can be classified as immune disorders caused by the specific changes to the gut microbial composition (Martín et al., 2013; Peloquin and Nguyen, 2013; Talham et al., 1999). In line with such direct regulation of the immune response at mucosal sites, commensal microbes act by producing different substances, including antibiotics, bacteriocides, and other metabolites, that affect the host's health (Dobson et al., 2012).

On the basis of these data the criteria for the classification and the definition of the human-relevant microbes had been divided into beneficial or pathogenic or potentially pathogenic agents. An interesting paper had been published about the inducing of different cytokine production by the same strain of genetically modified lactobacilli that caused significant and opposite changes of their immune effects from beneficial to detrimental as detected by producing of proinflammatory and antiinflammatory cytokines (Behnsen et al., 2013).

Recognition molecules (signals and structures) that are responsible for microbial recognition, such as toll-like receptors (TLRs) have been identified (Medzhitov, 2007; Pasare and Medzhitov, 2004). Also, all the other pathways of cross-talk between the host and the probiotics have been determined and explained at the genetic level (Lebeer et al., 2008).

With new microbiome data from the Human Microbiome Project (HMP) the majority of the principles of the microbial communication with the host immune and nonimmune cells have been revised or updated (Turnbaugh et al., 2007). The most new data have been provided about of the importance of microbial composition and their interaction based on simple physiological microbial characteristics: growing speed, synergy, and biofilm formation.

Alterations of relative microbial abundance and diversity are now well defined for the majority of diseases and their comorbidities. The disruptions of the microbiota biofilm structure and function, the dispersion of commensal bacteria, and the mechanisms whereby these dispersed commensals may become pathobionts have also been reported (Buret et al., 2019).

Microbiome, Immunity, Digestive Health and Nutrition. DOI: https://doi.org/10.1016/B978-0-12-822238-6.00002-9

A new vision and an updated description of the balanced ratio of microbes and its connection to human health have been proposed (Ding et al., 2019). The mechanism by which the changed microbial ratio can affect the microbiome-connected diseases is low-grade inflammation.

A very important issue in this process is that nutrition and energy mechanisms are controlling factors and crucial bases for the results of microorganism and macroorganism communication. Thus the new data could be used to supporting the idea that the proinflammatory or antiinflammatory efficacy can be directly modulated, as was shown recently for the reprogramming of dendritic cells (DCs) by modulation of the environmental stimuli such as all-trans retinoic acid (ATRA) (Bene et al., 2017).

Nevertheless, the ability to penetrate the beneficial commensal microbes through the epithelial barriers and the participation of their external structures in this process is still poorly characterized. There are data about the secretory-specific IgA level at mucosal surfaces and its regulatory role in this process (Boyko et al., 2006).

Translocation and colonization are mainly dependent on microbial architecture, direct cell-to-cells communication, and various environmental stimuli. These topics are important subjects for further experiments and testing.

16.2 Microbiota detection tools

Microbiota of the mucosal sites were plated on growth media, namely, anaerobic blood agar, bile esculin agar, *Streptococcus*-selective agar, Bacteroides bile esculin agar, MacConkey agar, blood agar, nutrient agar, mannitol salt agar, Wilson-Blair agar, Sabouraud dextrose agar, *Lactobacillus* MRS agar, and *Bifidobacterium* agar (all produced by HiMedia Laboratories, India) as well as on chromogenic chromatic detection (Liofilchem, Italy). Identification of isolated microorganisms was performed by using biochemical test systems: ANAERO-23, ENTERO-24, NEFERM-test, Candida-23, STAPHYtest 16, and STREPTOtest 24 (Erba Lachema s.r.o., Czech Republic).

Local immune responses to components of ethnic foods, such as beneficial microorganisms and edible plants extracts that are rich in biologically active compounds, that were administered orally have been detected in different mice models using fragment culture technique; their effect on systemic response was investigated on human DCs derived from peripheral blood monocytes. Secreted cytokines (IL-2, IL-4, IL-8, IL-10, IL-12, IL-17, IL-1b, TNF-α) were determined by enzyme-linked immunosorbent assay (ELISA). Levels of expression of NALP3, NAD1, NAD2, caspase 1 and 5 by DCs after their exposure to different microbial and food antigens were estimated by quantitative reverse transcription polymerase chain reaction (qRT-PCR).

16.2.1 Technological approaches

Ex vivo models (fragment culture method) are used for detection of strongly correlated biomarkers of low-grade initiation (Bomba et al., 2016). The production of proinflammatory and antiinflammatory cytokines in the culture of small bowel fragments and intestinal-associated lymphoid tissues was measured, including Peyer's patches (PP), the peritoneal cavity, and the mesenteric lymph nodes.

16.2.2 Mice models

Germ-free (GF) BALB/c and C.B17 severe combined immune deficient (SCID) mice were removed and placed under sterile conditions in isolators. Colonization of mice (mean age: 8 weeks) with Schaedler's *Escherichia coli* (kindly provided by Russell Schaedler) was achieved by intragastric administration of $10^7 - 10^8$ colony-forming units (CFU) of bacteria. *E. coli* colonization levels were quantified by serial dilutions of stool plated on Schaedler's agar. Total CFUs per gram of fecal contents $\times 10^5$ were then calculated. The mice were inoculated with segmented filamentous bacteria and altered Schaedler flora (ASF) at 8 weeks of age.

Data for commensal microbe translocation, dissemination, and colonization level of commensal bacterial strains that were tested after their oral administration into immunocompetent and IgA knockout (formerly GF) mice include the following.

The complicated mechanisms of the bacterial-host interaction were evaluated, including immune response to the bacterial colonization and persistence and ability of commensal microbes to translocate to different gut and immune cells and tissues.

The list of bacterial enteric pathogens that can cause or initiate wide-spectrum gut disorders or other infectious diseases is well known. Enterohemorrhagic *E. coli* (EHEC, serotypes O157:H7, O157:H-, and O26:H11) (Taguchi et al., 2002), *Salmonella typhimurium*, *Shigella flexneri*, and *Clostridium difficile* are causative agents of colitis and diarrhea in dysentery. *Helicobacter muridarum/hepaticus* provokes inflammatory bowel disease (IBD), but of the listed pathogens, only *Salmonella typhimurium* (as well as other salmonella species, such as *S. enterica*) leads to septicemia and translocates through the bloodstream into the liver, kidney, bone, brain and via secondary invasion to the gut epithelium cells. In comparison to *Shigella*, which can penetrate by themselves into epithelial cells, *Salmonella* cannot; they enter it by endocytosis. Their adhesion properties depend on mannose-sensitive receptors.

Listeria monocytogenes, another type of intracellular pathogenic bacteria, is an opportunistic pathogen that is able to disseminate into the blood and lymphoid streams and can also penetrate into different organs and tissues, including the central nervous system.

Bacteria (usually commensal gut normal flora representatives) that are chosen as probiotic remedies seem unable to translocate. The main demands for candidates to be a probiotic are good adhesive properties and a high level of colonization ability. GF mice were colonized with *Lactobacillus johnsonii* or *L. paracasei* that showed in vitro a level of adhesive properties similar to that of differentiated Caco-2 cells. In vivo experiments clearly indicated that the strains significantly differed in their colonization ability, but both of them were able to activate mucosal B cell responses (similar GCR and IgA-bearing lymphocytes formation in mucosa). However, a strong correlation was found between the colonization level and production of each lactobacillus-specific IgA in PPs and intestinal contents (Ibnou-Zekri et al., 2003). Viable cell counts of *L. johnsonii* were much higher than those of *L. paracasei* in the small intestine and colon of monoxenic mice, and correspondingly, *L. johnsonii* promoted substantial production of specific secretory IgA, whereas *L. paracasei* induced only marginal secretory IgA at levels that were in the same range as the cross-reactive IgA found in choroidal neovascularization mice. Translocation of these bacteria was also detected but only into mucosal lymphoid organs (the small intestine, colon, PP, and mesenteric lymph nodes).

The mechanism of the bacterial translocation phenomena is still unclear. Early colonization of the neonatal gut by commensal bacteria engenders a transient and controlled immune response at weaning, characterized by IgA secretion, that could be implicated in controlling bacterial adherence to the mucosal epithelium or cause clumping of bacteria near the mucosa to enhance washout by peristalsis. But little is known about whether secretion of IgA is also important for preventing translocation of resident or transient bacteria into mucosal tissues, blood, and internal organs (Macpherson et al., 2001). In vitro experiments showed that secretory IgA prevented transepithelial passage of *E. coli* across the epithelial layer (Albanese et al., 1994). In contrast, earlier work (Berg and Staaf, 1980) demonstrated that oral immunization of GF mice with heat-killed *E. coli* vaccines before colonization with the same strains did not reduce the incidence of translocation despite markedly increased levels of intestinal IgA.

Short-term translocation into the blood immediately (2 minutes) after oral administration of some bacteria (e.g., *Bacillus subtilis*) and their persistence usually 12 hours long with next fast elimination was described and termed "no symptomatic bacteremia." Such observation is usually used for one possible explanation of how probiotics work (Boyko et al., 2000).

Closely related to this subject are investigations in a few studies that proposed different ways and general principles for intestine immune homeostasis regulation. The regulatory role of T cells has been very well demonstrated in an experimental system of T cell–mediated inflammation in IBD as a consequence of a dysregulated immune response (Annacker and Powrie, 2002; Jiang et al., 2004).

On the other hand, some work declared that IgA responses to commensal bacteria are supported by T cell-independent (primitive) mechanisms (Macpherson et al., 2001). Well-founded argumentation with a critical point of view on this hypothesis can be found in a comment (Bos et al., 2001).

It can be proved by using mice that are deficient in Myd88, an adaptor molecule that is essential for TLR-mediated induction of inflammatory cytokines, as well as mice that are deficient in TLR-2 and TLR-4 that recognition of commensal microbiota by TLRs is also required for intestinal homeostasis (Rakoff-Nahoum et al., 2004). We detected that oral inoculation with two strains of lactobacilli (*Lactobacillus acidophilus* ASF#360 and *L. salivarius* ASF#361 in IgA knockout mice and B6, respectively) resulted in their translocation at early time points into some internal organs of IgA-deficient mice but not their wild-type (WT) counterparts (Fig. 16.1A). Later (on day 14) and especially after additional colonization with other ASF bacteria (*Spirochete*, *Bacteroides*, and four extremely oxygen-sensitive fusiforms, which are mostly closely related to *Clostridium* genera), translocation was not found.

With respect to understanding how the translocation phenomenon may be affected by IgA production in the gut and whether the ability of tested bacteria to translocate is dependent on their structural characteristics, such as surface antigenic structure (presence or absence of lipopolysaccharide (LPS), S-protein, etc.), and adhesive properties, we carried out a few additional studies. We checked out the translocation ability of Schaedler's *E. coli*, *Morganella morganii*, and *L. salivarius* in monoassociated IgA knockout mice compared with their WT B6 mice. In a case with Schaedler's *E. coli* oral administration such formerly GF mice we didn't find translocation (Fig. 16.1B).

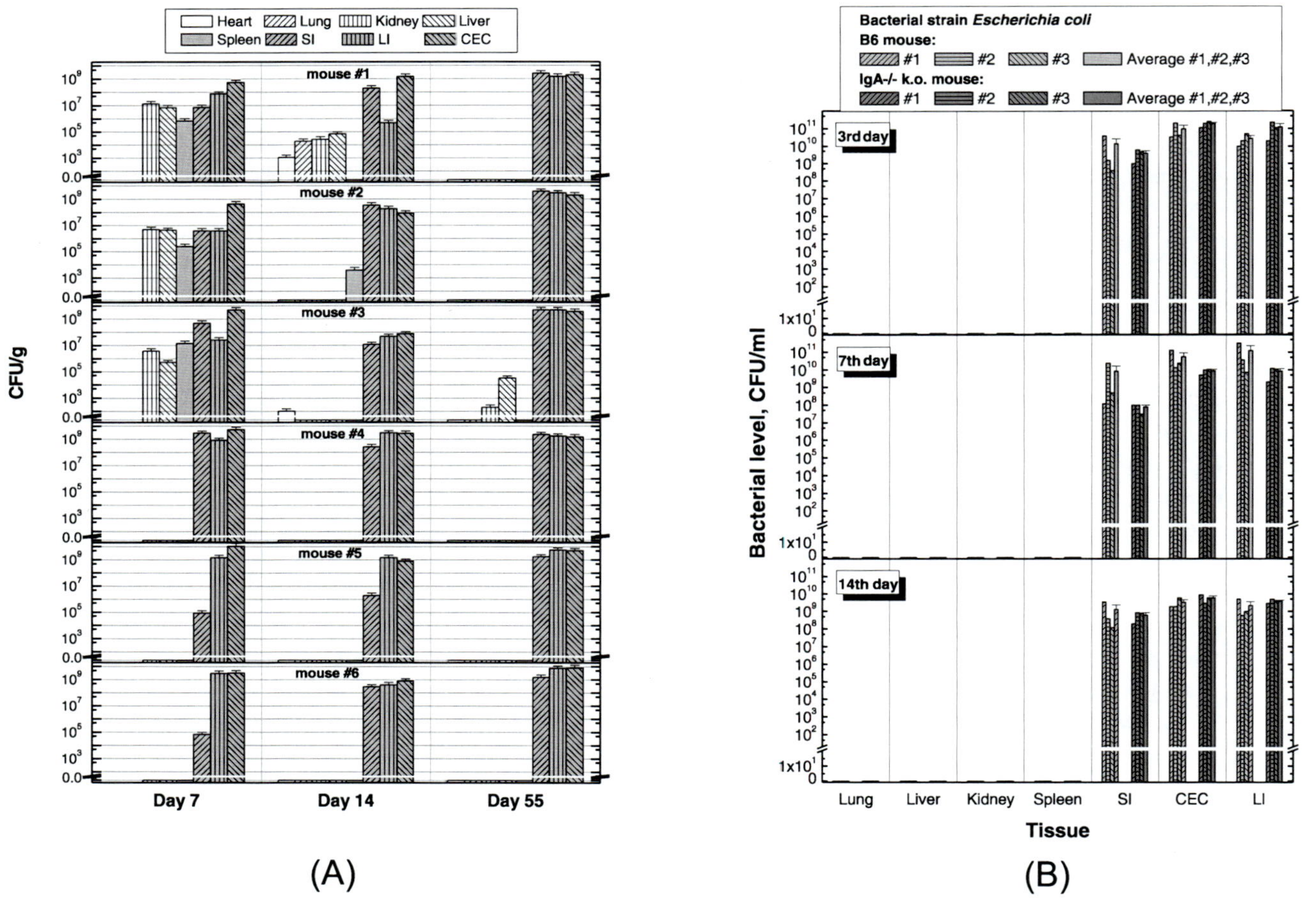

FIGURE 16.1 (A) Translocation of two lactobacilli strain on day 7 after oral inoculation of IgA knockout mice into kidney, liver and spleen and their colonization ability. (B) Colonization of formerly germ-free IgA knockout and B6 mice with Schaedler's *E. coli*. Translocation not detected.

In another experiment we were able to demonstrate that after previous oral inoculation of IgA knockout and B6 with Schaedler's *E. coli* and cocolonization of the same mice on day 3 with *L. salivarius*, the translocation phenomenon does not occur (Fig. 16.2A and B).

To detect how quickly and effectively *E. coli* could protect translocation of *L. salivarius* in such mice, we repeated this experiment in same conditions except that the *E. coli* was administered 24 hours before the next challenge with only one strain of lactobacilli. We got almost the same result: no translocation in IgA knockout mice that had previously been inoculated with *E. coli*, compared to the control group—IgA knockout mice, colonized with *L. salivarius*, but the level of translocation of lactobacilli into parenchymal organs in this experiment was lower—up to 10^4 CFU/g homogenate tissues from the spleen, liver, and kidneys—than in the case with two lactobacilli strains. These genetically modified mice could have age-dependent mechanisms of some intestinal disorders without regulative IgA secretory function, which can also influence the appearance of the translocation phenomenon.

The same results, that is, no translocation, were obtained when we used other enteric bacteria *M. morganii* for previous oral inoculation of GF IgA knockout and B6 mice and then (after 24 hours) orally inoculated them with one strain of lactobacilli (*L. salivarius*) (Table. 16.1).

We hypothesized that there might be levels of "natural" and specific IgA reactive with each bacteria in the gut compartment of B6 mice and IgM (correspondingly in both IgA knockout and B6 mice) that are responsible for regulation of the translocation that was detected in IgA knockout mice but was not found in B6 mice monoassociated with *L. salivarius*. To answer to this question, the levels of total and specific secretory IgA (B6) and IgM (both groups) were determined at mucosal sites.

B6 mice produced locally significant amounts of IgA but not IgM on day 14 after inoculation with two lactobacilli strains (Fig. 16.3). Compare these data with results obtained at the same time points from mucosal sites of mice IgA knockout mice, in which increasing IgM production in the gut was detected (Fig. 16.4).

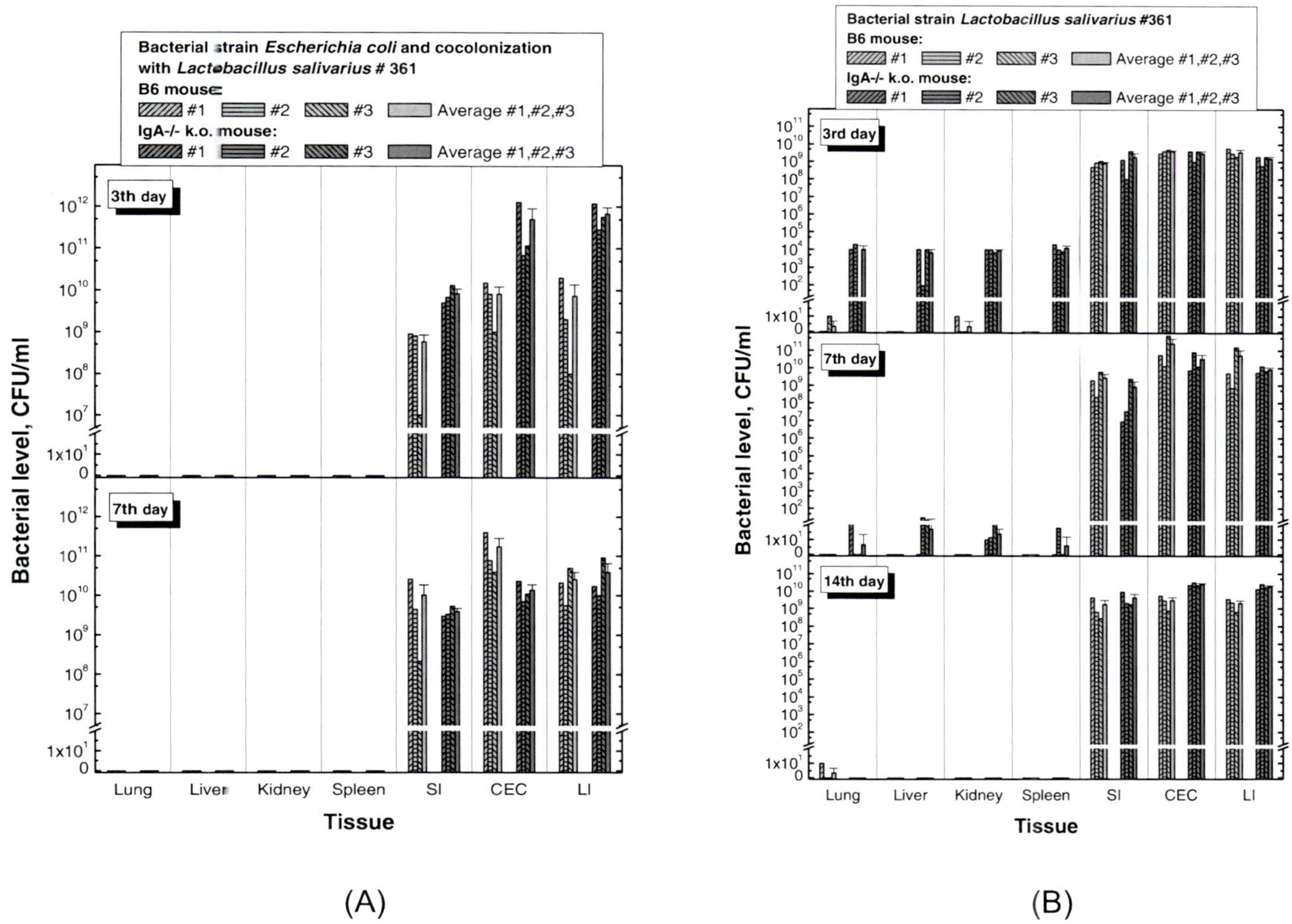

FIGURE 16.2 (A) Protection of translocation of lactobacilli in IgA knockout mice by the previous oral inoculation with Schaedler's *E. coli*. (B) Control mice, inoculated orally only with *L. salivarius*.

In another study, formerly germ-free B6 and IgA knockout mice were colonized with Schaedler's *E. coli*, and the total IgA and IgM response in the gut was measured (Fig. 16.5).

As can clearly be seen from the data presented in the figures, the level of IgM production in IgA knockout mice was profoundly higher even for total IgA and IgM produced in the gut compartments of WT mice after inoculation with Schaedler's *E. coli*.

16.2.3 Schaedler's *E. coli* induces diminished translocation of *Lactobacillus salivarius* into the lung: the role of surfactant protein D

The innate immune molecule surfactant protein D (SP-D) may play an important protective role during dissemination of bacteria to the lung. GF IgA knockout (IgA−/−) and WT adult mice were colonized orally with the commensal Schaedler's *E. coli* (O81:21) or the pathogenic *Lactobacillus salivarius* ASF #361. Translocation of these bacteria into the lung was assessed by organ homogenate plating. SP-D protein levels in the bronchoalveolar lavage fluid were assessed by Western blotting. Expression of the autocrine epithelial growth factor Reg III was studied by real-time PCR. Levels of SP-D were higher in WT GF mice than in IgA−/− mice, suggesting impairment of the pulmonary innate immune system of these animals. On days 3−7 after inoculation, translocation of *L. salivarius* to the lung was observed in the lactobacilli-colonized IgA−/− mice but not in their WT counterparts. *E. coli* showed colonization levels similar to those of lactobacilli but no translocation in either IgA−/− or WT mice. Colonization of mice with *E. coli* prior to *L. salivarius* resulted in diminished pulmonary translocation of the latter in IgA−/− mice. There was an increase in Reg IIIγ/Reg IIIβ expression mostly in the cecum of colonized IgA−/− mice compared to GF mice or mice that were monoassociated with *E. coli*. Prophylactic use of probiotics (Schaedler's *E. coli*) has a significant protective

TABLE 16.1 Results of microbiological investigation of germ-free IgA knockout and B6 mice.

Bacteria/day after inoculation	Content, organ	CFU/g tissue					
		Mice					
		IgA knockout (1–3)			Wild-type (WT) (4–6)		
Morganella morganii, 24 hours, 1st day (isolator#57)	Small intestine (SI)	3×10^8	5×10^8	1×10^7	1×10^7	2×10^8	4.7×10^9
	Large intestine (LI)	$1.88–2.0 \times 10^{10}$	$1.0–2.31 \times 10^{10}$	$9.84 \times 10^9–1.1 \times 10^{10}$	2.2×10^{11}	$1.03–2.0 \times 10^{10}$	2.0×10^{10}
	Cecum	$8.9 \times 10^9–1.0 \times 10^{10}$	$1.27–2.0 \times 10^{10}$	$2.0–2.56 \times 10^{10}$	$2.0–2.4 \times 10^{10}$	1.64×10^{10}	$2.28–5.0 \times 10^{10}$
	Liver	0	0	0	0	0	0
	Kidney	0	0	0	0	0	0
	Spleen	0	0	0	0	0	0
Lactobacillus salivarius third day (isolator#70)	SI	2.4×10^8	4×10^8	1.8×10^9	No WT control		
	LI	$4.16–5.0 \times 10^9$	5×10^9	$2.56–5.0 \times 10^9$			
	Cecum	$3.8–5.0 \times 10^9$	$3.0–4.8 \times 10^9$ 1×10^{10}	$2.16–3.0 \times 10^9$			
	Liver	1×10^2	1×10^2	1×10^2			
	Kidney	1×10^4	1×10^2	7×10^2			
	Spleen	1×10^2	1×10^2	1×10^2			
M. morganii + *L. salivarius* 24 h after second inoculation, third day after second inoculation, fourth day of experiment (isolator#57)		*M. morganii* titer					
	SI	5×10^8	1.2×10^9	8×10^6	$3.5–4.0 \times 10^7$	3.5×10^9	4×10^9
	LI	5×10^9	4×10^9	$2 \times 10^9/2 \times 10^{10}$	3×10^9	5×10^9	$3 \times 10^9–2 \times 10^{10}$
	Cecum	$4 \times 10^9–1 \times 10^{10}$	2×10^9	$3 \times 10^9–1 \times 10^{10}$	1×10^9	1×10^{10}	3×10^9
	Liver	0	0	0	0	0	0
	Kidney	0	0	0	0	0	0
	Spleen	0	0	0	0	0	0
		L. salivarius titer					
	SI	6×10^8	8×10^8	6.4×10^7	1.31×10^8	3×10^8	1×10^8
	LI	6×10^9	4×10^9	$9 \times 10^9–1 \times 10^{10}$	1×10^{10}	4×10^9	3×10^9
	Cecum	1.4×10^{10}	2×10^9	4×10^9	$3 \times 10^9–1 \times 10^{10}$	1.5×10^{10}	5×0^9
	Liver	0	0	0	0	0	0
	Kidney	0	0	0	0	0	0
	Spleen	0	0	0	0	0	0

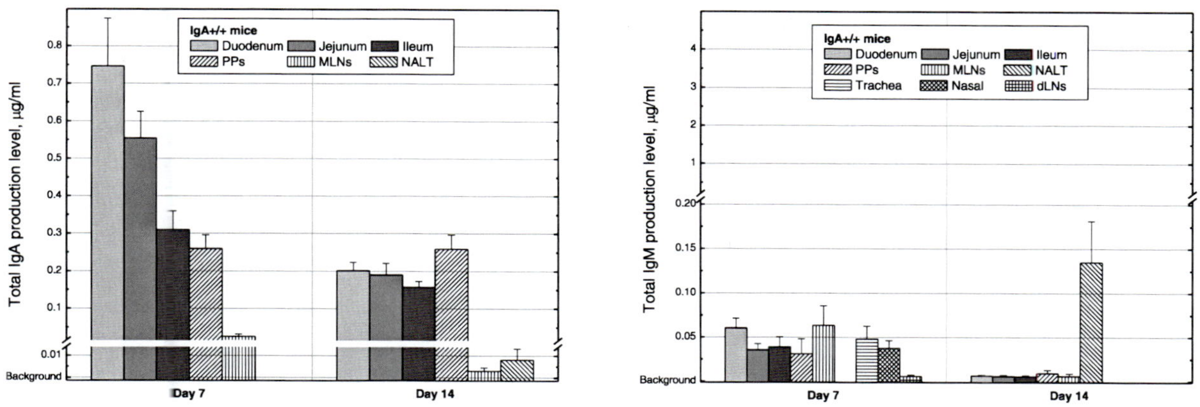

(A) (B)

FIGURE 16.3 Total level of (A) IgA and (B) IgM production in the gut compartments from wild-type mice inoculated with two lactobacilli strains.

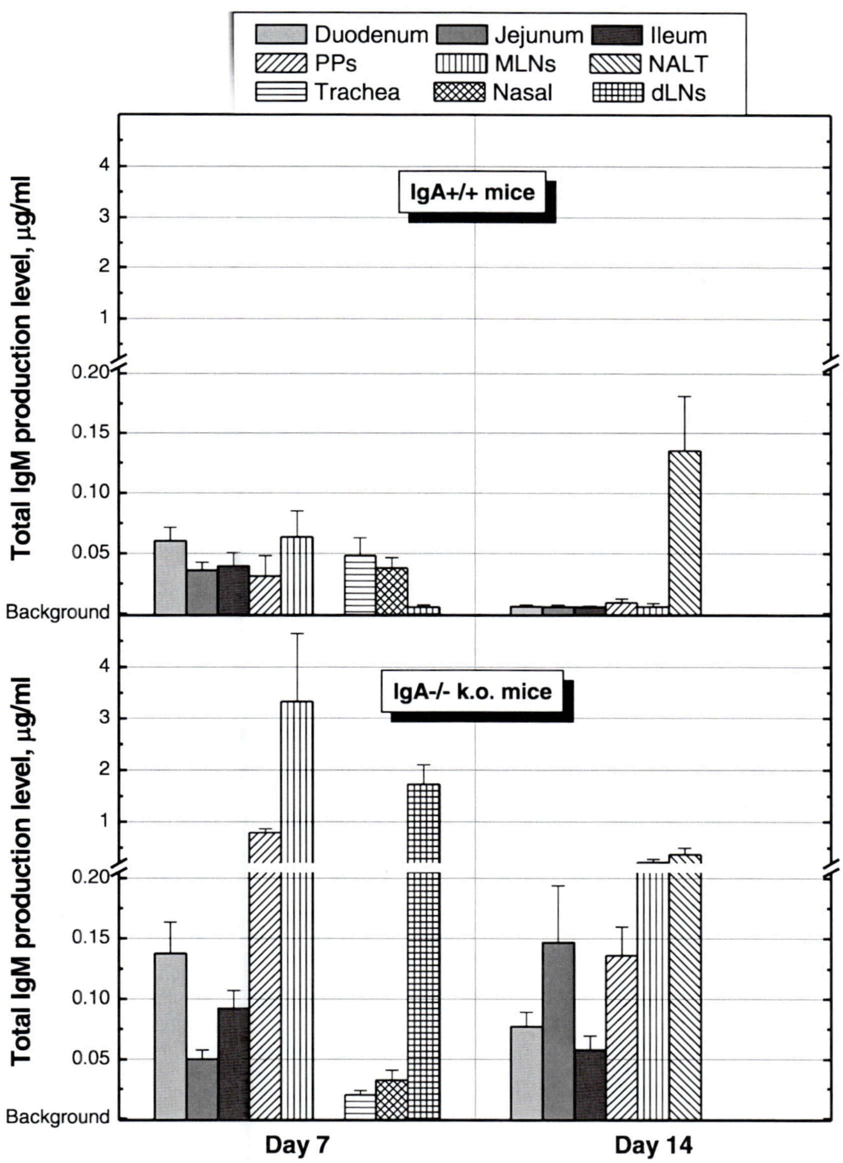

FIGURE 16.4 Comparative level of total IgM produced by gut mucosal sites of wild-type and IgA knockout mice inoculated with two lactobacilli strains in dynamic.

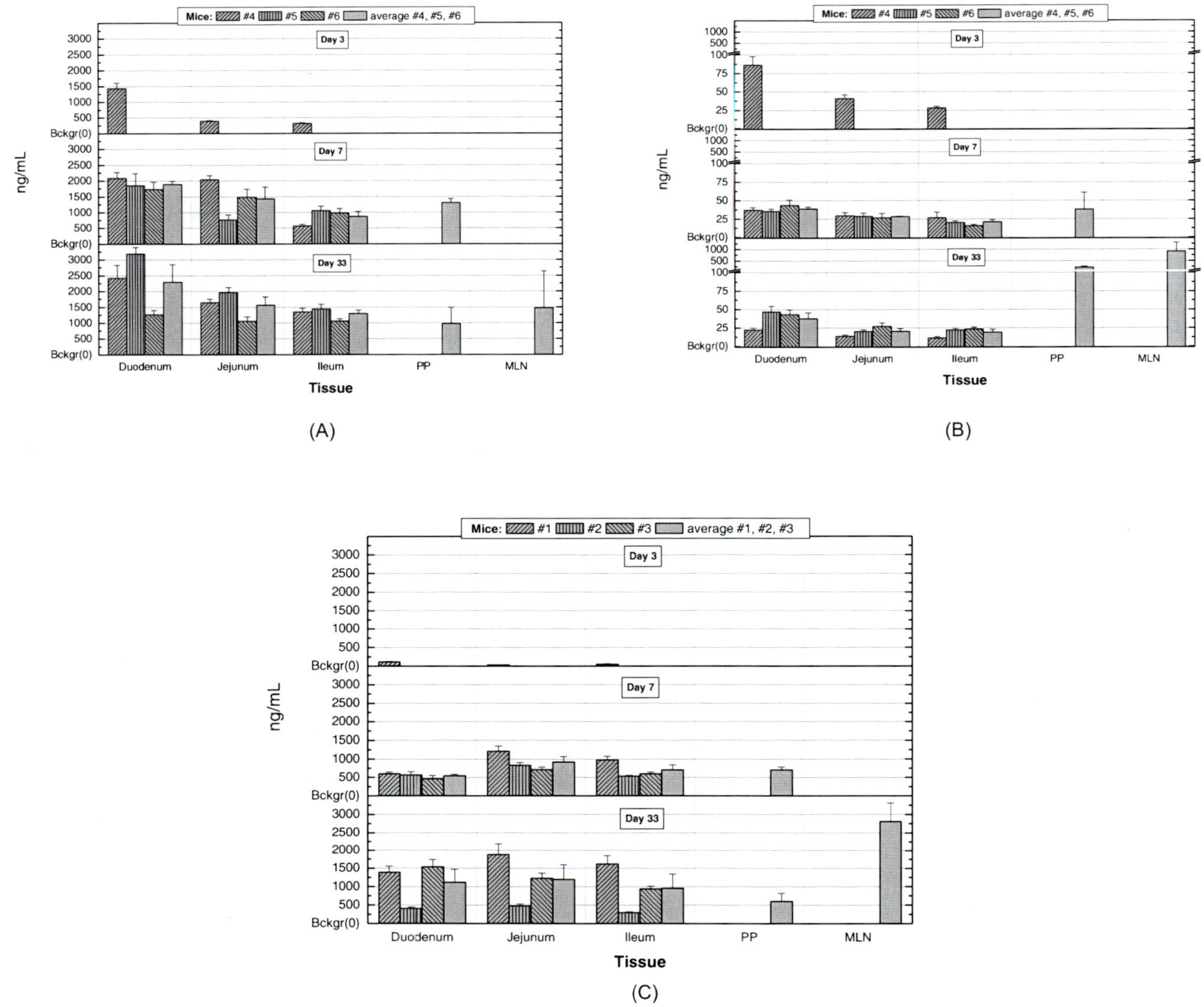

FIGURE 16.5 (A) Total IgA and (B) IgM production in the gut compartments of B6 (wild-type) mice inoculated by Schaedler's *E. coli* in dynamic. (C) Total IgM production in the gut compartments of IgA knockout mice inoculated by Schaedler's *E. coli* in dynamic.

effect against dissemination of bacteria. Reduced levels of SP-D in IgA−/− mice may contribute to greater susceptibility of these mice to bacterial translocation to the lung.

Several phenotypically and functionally distinguishable B cell subsets have been identified in some mammalian species, most particularly in mice. B1 cells frequently express minimally mutated germline VH and VL genes, while at least some B2 cells express V genes that are extremely mutated in a pattern that suggests antigen (Ag) selection and affinity maturation. Although such B2 cells are clearly selected for development by specific Ag interacting with their B cell receptors, demonstration of similar Ag-selection and stimulation by B1 cells has remained equivocal. We transferred FACS purified B1 and/or B2 cells, distinguishable by Igh-allotype markers, into formerly germ-free SCID mice that had been monoassociated with *Haemophilus influenzae*. *H. influenzae* expressing LPS, with or without phosphoryl choline (PC) prosthetic groups, was used to chronically colonize the upper respiratory tract, although the gastrointestinal tract remained free of viable bacteria. Using tissue fragment cultures taken from nasal and respiratory-associated lymphoid tissue versus gastrointestinal tissue, we have found evidence for specific Ag selection and local stimulation of IgA anti-PC-producing cells of the B1 lineage by a type 1 thymus-independent Ag. This IgA Ab expression requires "bystander" T cell "help," which is likely stimulated by other microbial antigens (Boyko et al., 2004).

Research data about specific responses in such experiments will be helpful for understanding the mechanisms of translocation phenomena. Reduced levels of SP-D in IgA−/− mice may contribute to greater susceptibility of these mice to bacterial translocation to the lung (Krytska et al., 2007). The cellular source for IFN-γ on monoassociation of SCID mice with Schaedler's *E. coli* was localized to a subset of intraepithelial natural killer cells (Keilbaugh et al., 2005), Moreover, physiological concentrations of ATRA are able to reprogram the differentiation of monocyte-derived dendritic cells (moDCs), resulting in altered gene expression profiles of the master transcription factors RARα and interferon regulatory factor 4, and concomitantly regulate the cell surface expression levels of CD1 proteins and also the mucosa-associated CD103 integrin to different directions. It was also demonstrated that the ATRA-conditioned moDCs exhibited enhanced proinflammatory cytokine secretion while their costimulatory and antigen-presenting capacity was reduced, thus reducing Th1 and presenting undetectable Th17 type responses against the tested microbiota strains. Importantly, these regulatory circuits could be prevented by the selective inhibition of RARα functionality. These results together demonstrate that selected commensal bacterial strains are able to drive strong effector immune responses by moDCs, while in the presence of ATRA, they support the development of both tolerogenic and inflammatory moDC in a RARα-dependent manner (Bene et al., 2017).

It will be interesting to detect whether previous inoculation of GF IgA knockout mice with Schaedler's *E. coli* or *M. morganii* will be effective against translocation into the internal organs by other gram-positive (*Listeria monocytogenes*) and Gram-negative (*Salmonella typhimurium*) virulent bacterial agents following their subsequent challenge.

We have also conducted studies on Balb/c line mice that were 18–20 g in weight using lyophilized strains of lactic acid bacteria (LAB) (*Lactobacillus acidophilus* IMV B-7279, *L. casei* IMV B-7280, *L. delbrueckii* subsp. *bulgaricus* IMV B-7281) and bifidobacteria (*Bifidobacterium animalis* VKL and *B. animalis* VKB). We cultivated the macrophages that were obtained from the peritoneal cavity of mice individually with the strains of LAB and bifidobacteria and evaluated their effect on macrophages, oxygen-dependent bactericidal activity, nitric oxide production, and immunoregulatory cytokines. We used atomic force microscope (AFM) scanning to estimate bacterial cell wall elasticity (Mokrozub and Nemtinov, 2015).

All strains had a stimulating effect on the functional activity of macrophages and ability to produce NO/NO_2 in vitro. Lactobacilli strains increased the production of IL-12 and IFN-γ in vitro. The AFM demonstrated different cell wall elasticity levels in various strains of LAB and bifidobacteria. The rigidity of the cell walls among lactobacilli was distributed as follows: *Lactobacillus acidophilus* IMV B-7279 > *Lactobacillus casei* IMV B-7280 > *Lactobacillus delbrueckii* subsp. *bulgaricus* IMV B-7281; among the strains of bifidobacteria, *B. animalis* VKB > *B. animalis* VKL. Probiotic strain survival in the macrophages depended on the bacterial cell wall elasticity and on the time of their joint cultivation. LAB and bifidobacteria strains stimulate immunomodulatory cytokines and active oxygen and nitrogen oxide compound production in macrophages. Strains with a more elastic cell wall according to AFM data demonstrated higher resistance to intracellular digestion in macrophages and a higher level of their activation.

AFM might be considered as a fast and accurate method to assess parameters of probiotic strain cell wall to predict their immunomodulatory properties. Further studies including sufficient cohorts of volunteers are necessary to translate model data of probiotic efficacy to human populations as well as conduction of epidemiological studies for obtaining evidence regarding adverse effect of particular food additives. In addition, research is needed on microbiota in relevant models related to specific processes (autoimmunity, asthma, diabetes, gout, hepatitis, cancer, NDD, CVD, thyroid, rheumatic diseases, etc.). The results of this research may indicate the effectiveness of probiotic therapy for prevention in humans based on personalized microbiota additives. To achieve the personalized approach, the development and application of products manufactured from an individual's own strains of organisms are considered promising. The potential alternative for probiotics might be the suggested lysates of probiotic strains that also exhibit immunomodulatory activity.

16.3 Conclusion

We suggest the further studies on dietary additives and immune and inflammatory responses as well as development and testing of promising nanomaterials for the prospect of a complex multiparametric impact on gastrointestinal microbiota and the gut-brain axis.

After approval, development of safe and effective person-related medications and dietary additives with future clinical testing should be initiated to implement results for routine practice.

References

Albanese, C.T., Smith, S.D., Watkins, S., Kurkchubasche, A., Simmons, R.L., Rowe, M.I., 1994. Effect of secretory IgA on transepithelial passage of bacteria across the intact ileum in vitro. J. Am. Coll. Surg. 179 (6), 679–688.

Annacker, O., Powrie, F., 2002. Homeostasis of intestinal immune regulation. Microbes Infection 4, 567–574.

Behnsen J., Deriu, E., Sassone-Corsi, M., Raffatellu, M., 2013. Probiotics: properties, examples, and specific applications. Cold Spring Harbor Perspect. Med. 3 (3), a010074.

Bene, K., Varga, Z., Petrov, V.O., Boyko, N., Rajnavolgyi, E., 2017. Gut microbiota species can provoke both inflammatory and tolerogenic immune responses in human dendritic cells mediated by retinoic acid receptor alpha ligation. Front. Immunol. 8, 427.

Berg, B., Staaf, H., 1980. Decomposition rate and chemical changes of Scots pine needle litter. II. Influence of chemical composition. Ecol. Bull. 373–390.

Bomba, A., Petrov, V.O., Drobnych, V.G., Bubnov, R.V., Bykova, O.M., Boyko, N., 2016. Cells, animal, SHIME and In silico models for detection and verification of specific biomarkers of non-communicative chronic diseases. EPMA J. 7 (1), A8, 1.

Bos, N.A., Jiang, H.-Q., Cebra, J.J., 2001. T cell control of the gut IgA response against commensal bacteria. Gut 48, 762–764.

Boyko, N.V., Kobuley, D., Weiser, J.N., Cebra, J.J., 2004. Are B1 cells selectively and specifically stimulable by antigen based on their BCR? Clin. Inv. Med. (Suppl.) 27 (4).

Boyko, N., Krytska, K., Kushnir, N., Kobuley, D., Stoel, M., Bos, N., et al., 2006. Is bacterial translocation depending of IgA production in a gut? J. Immunol. 176, 222.

Boyko, N.V., Lysetska, M.V., Arkadiev, V.G., Maksimov, Y.M., Danylenko, V.P., 2000. Study of immunotropic activities of bacillus subtilis as the basis of anti-scleroma probiotics. Microbiol. J. 62 (6), 22–25.

Buret, A.G., Motta, J.P., Allain, T., Ferraz, J., Wallace, J.L., 2019. Pathobiont release from dysbiotic gut microbiota biofilms in intestinal inflammatory diseases: a role for iron? J. Biomed. Sci. 26 (1), 1–14.

Cebra, J.J., Bos, N.A., Boyko, N.V., Kushnir, N., Stoel, M., Surh, C., et al., 2005a. Mutualism between the mucosal immune system and commensal intestinal bacteria. I. OHUS Monograph 9, 23–43.

Cebra, J.J., Jiang, H.-Q., Boyko, N., Tlaskalova-Hogenova, H., 2005b. The role of mucosal microbiota in the development, maintenance, and pathologies of the mucosal immune system, Mucosal Immunology, third ed. Elsevier Press, pp. 335–368.

Ding, R.-X., Goh, W.-R., Wu, R.-N., Yue, X.-Q., Luo, X., Khine, W.W.T., et al., 2019. Revisit gut microbiota and its impact on human health and disease. J. Food Drug Anal. 27 (3), 623–631. ISSN 1021–9498.

Dobson, A., Cotter, P.D., Ross, R.P., Hill, C., 2012. Bacteriocin production: a probiotic trait? Appl. Environ. Microbiol. 78 (1), 1–6.

Ibnou-Zekri, N., Blum, S., Schiffrin, E.J., von der Weid, T., 2003. Divergent patterns of colonization and immune response elicited from two intestinal *Lactobacillus strains* that display similar properties in vitro. Infect. Immun. 71 (1), 428–436.

Jiang, H.,Q., Thurnheer, M.C., Zuercher, A.V., Boyko, N.V., Bos, N.A., Cebra, J.J., 2004. Interactions of commensal gut microbes with subsets of B- and T-cells in the murine host. Vaccine 22 (7), 805–811.

Keilbaugh, S.A., Shin, M.E., Banchereau, R.F., McVay, L.D., Boyko, N., Artis, D., et al., 2005. Activation of RegIIIβ/γ and interferon γ expression in the intestinal tract of SCID mice: an innate response to bacterial colonisation of the gut. Gut 54 (5), 623–629.

Krytska, K., Boyko, N.V., Kushnir, N., Kobuley, D., Keilbaugh, S.A., Wu, G.D., et al., 2007. The role of surfactant protein D (SP-D) and IgA in Schaedler's *E. coli* induced diminished translocation of *Lactobacillus salivarius* into the lung: crosstalk between innate and adaptive immunity. J. Allergy Clin. Immunol. 119 (1), 177.

Lebeer, S., Vanderleyden, J., De Keersmaecker, S.C., 2008. Genes and molecules of lactobacilli supporting probiotic action. Microbiol. Mol. Biol. Rev. 72 (4), 728–764.

Macpherson, A.J., Hunziker, L., McCoy, K., Lamarre, A., 2001. IgA responses in the intestinal mucosa against pathogenic and non-pathogenic microorganisms. Microbes Infect. 3, 1021–1035.

Martín, R., Miquel, S., Ulmer, J., Kechaou, N., Langella, P., Bermúdez-Humarán, L.G., 2013. Role of commensal and probiotic bacteria in human health: a focus on inflammatory bowel disease. Microb. Cell Fact. 12, 71.

Medzhitov, R., 2007. Recognition of microorganisms and activation of the immune response. Nature 449, 819–826.

Mokrozub, V.G., Nemtinov, V.A., 2015. An approach to smart information support of decision-making in the design of chemical equipment. Chem. Pet. Eng. 51 (7), 487–492.

Pasare, C., Medzhitov, R., 2004. Toll-like receptors: linking innate and adaptive immunity. Microbes Infect. 6 (15), 1382–1387.

Peloquin, J.M., Nguyen, D.D., 2013. The microbiota and inflammatory bowel disease: insights from animal models. Anaerobe 24, 102–106.

Rakoff-Nahoum, S., Paglino, J., Eslami-Varzaneh, F., Edberg, S., Medzhitov, R., 2004. Recognition of commensal microflora by Toll-like receptors is required for intestinal homeostasis. Cell. 118, 229–241.

Taguchi, H., Takahashi, M., Yamaguchi, H., Osaki, T., Komatsu, A., Fujioka, Y., et al., 2002. Experimental infection of germ-free mice with hypertoxigenic enterohaemorrhagic *Escherichia coli* O157: H7, strain 6. J. Med. Microbiol. 51 (4), 336–343.

Talham, G.L., Jiang, H.Q., Bos, N.A., Cebra, J.J., 1999. Segmented filamentous bacteria are potent stimuli of a physiologically normal state of the murine gut mucosal immune system. Infect. Immun. 67, 1992–2000.

Turnbaugh, P.J., Ley, R.E., Hamady, M., Fraser-Liggett, C.M., Knight, R., Gordon, J.I., 2007. The human microbiome project. Nature 449 (7164), 804–810.

Section V

Microbiome and Cognitive Health

Chapter 17

Achieving dopamine homeostasis to combat brain-gut functional impairment: behavioral and neurogenetic correlates of reward deficiency syndrome

Kenneth Blum[1], Bernard William Downs[2], Debasis Bagchi[3,4,5], Gene-Jack Wang[6], Panyotis K. Thanos[7], Igor Elman[8], Rehan Jalali[9], Abdalla Bowirrat[10], David Baron[1], Diwanshu Soni[1], Rajendra D. Badgaiyan[11,12] and Mark S. Gold[13]

[1]*Department of Psychiatry, College of Osteopathic Medicine of the Pacific, Western University of Health Sciences, Pomona, CA, United States,* [2]*Division of Precision Nutrition, Victory Nutrition International, Inc., Bonita Springs, FL, United States,* [3]*College of Pharmacy and Health Sciences, Texas Southern University, Houston, TX, United States,* [4]*Department of Biology, Adelphi University, Garden City, NY, United States,* [5]*Department of R&D, Victory Nutrition International, Inc., Bonita Springs, FL, United States,* [6]*Laboratory of Neuroimaging, National Institute of Alcohol Abuse and Alcoholism, National Institutes of Health, Bethesda, MD, United States,* [7]*Department of Psychology and Behavioral Neuropharmacology and Neuroimaging Laboratory on Addictions (BNNLA), Research Institute on Addictions, University at Buffalo, Buffalo, NY, United States,* [8]*Behavioral Neuropharmacology and Neuroimaging Laboratory on Addictions, Clinical Research Institute on Addictions, Department of Pharmacology and Toxicology, Jacobs School of Medicine and Biosciences, State University of New York at Buffalo, Buffalo, NY, United States,* [9]*Division of Nutrigenomics, The Kenneth Blum Behavioral Neurogenetic Institute, Austin, TX, United States,* [10]*Department of Molecular Biology, Adelson School of Medicine, Ariel University, Ariel, Israel,* [11]*Department of Psychiatry, South Texas Veteran Health Care System, Audie L. Murphy Memorial VA Hospital, and Long School of Medicine, University of Texas Health Science Center, San Antonio, TX, United States,* [12]*Department of Psychiatry, MT. Sinai School of Medicine, New York, NY, United States,* [13]*Department of Psychiatry, Washington University School of Medicine, St. Louis, MO, United States*

17.1 Introduction

It is not uncommon for individuals with mood disorders or with addictions, impulsivity, obsessive-compulsive behavior, and some personality disorders—collectively referred to as reward deficiency syndrome (RDS)—also to experience digestive disorders. A significant reason for this is that the enteric nervous system is the largest and most complex subdivision of the peripheral nervous system (Gershon, 1998; Nezami and Srinivasan, 2010). The enteric nervous system and the central nervous system secrete many identical neurotransmitters, such as acetylcholine (ACh), dopamine, and serotonin, and these neurotransmitters can have impaired functional competence, owing to digestive dysfunction. Individuals with RDS can experience a similar dysfunction in how the brain perceives reward. RDS, which is generally induced by dopamine resistance or hypodopaminergia, can impair the natural processing of endorphins or diminish the response to exogenous dopamine stimulants. These disorders assemble underlying reward deficiency mechanisms in multiple brain centers. Since the widespread and remarkable array of associated and overlapping behavioral manifestations with a common root of hypodopaminergia, the basic endophenotype that is recognized as RDS can be likened to a behavioral octopus (Comings and Blum, 2000; Gondré-Lewis et al., 2020) (see Fig. 17.1).

In previous publications (Gondré-Lewis et al., 2020) our group evaluated well-characterized animal models for their construct validity and as potential models for RDS. In this chapter we discuss animal models that were used to study substance use disorder, major depressive disorder, early life stress, immune dysregulation, attention deficit hyperactivity disorder (ADHD), posttraumatic stress disorder (PTSD), compulsive gambling, and compulsive eating disorders. These disorders assemble underlying reward deficiency mechanisms in multiple brain centers.

Several neurotransmitters are involved in the standard processing of reward and punishment. They include pathways involving at least seven major neurotransmitters and many secondary messengers. These neurotransmitter networks

Microbiome, Immunity, Digestive Health and Nutrition. DOI: https://doi.org/10.1016/B978-0-12-822238-6.00001-7

FIGURE 17.1 Reward deficiency syndrome as a behavioral octopus.

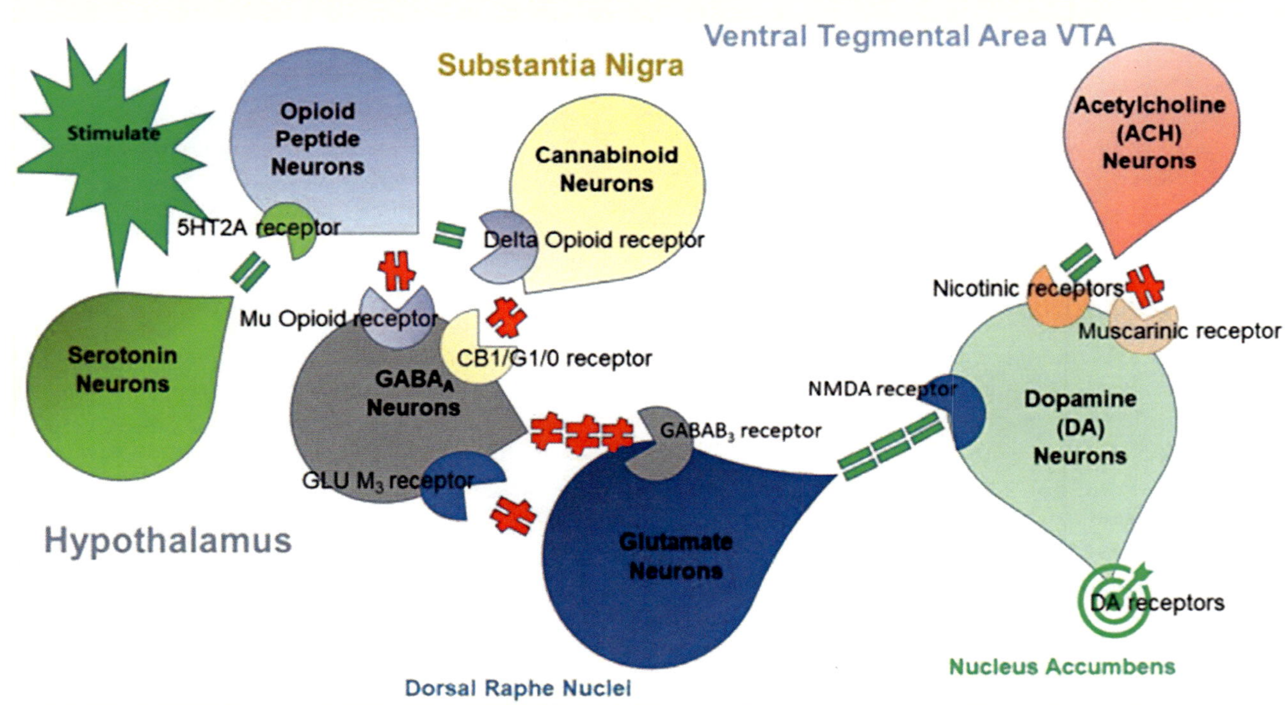

FIGURE 17.2 The brain reward cascade.

function within the mesolimbic and prefrontal cortex, where they regulate the final net pathway of "wanting," causing neuronal dopamine release (Berridge and Robinson, 2016; Comings and Blum, 2000; Gondré-Lewis et al., 2020).

Fig. 17.2 depicts the well-established brain reward cascade (BRC) and the correlates and neurotransmitters that are associated with addictions. The figure illustrates the interaction of at least seven major neurotransmitter pathways that are involved in the BRC. In the hypothalamus, environmental stimulation results in the release of serotonin, which in turn via, for example, 5HT-2a receptors activate (green equal sign in Fig. 17.2) the subsequent release of opioid peptides from opioid peptide neurons, also in the hypothalamus. Then, in turn, the opioid peptides have two distinct effects, possibly via two different opioid receptors. One that inhibits (red hash sign in Fig. 17.2) through the mu-opioid receptor (possibly via enkephalin) and projects to the substantia nigra to GABAA neurons. Another stimulates (green equal sign in Fig. 17.2) cannabinoid neurons (e.g., the anandamide and 2-arachidonoylglycerol) through beta-endorphin-linked delta receptors, which in turn inhibit GABAA neurons at the substantia nigra. Also, when activated, cannabinoids, primarily 2-arachidonoylglycerol, can indirectly disinhibit (red hash sign in Fig. 17.2) GABAA neurons through activation of G1/0 coupled to CB1 receptors in the substantia nigra. In the dorsal raphe nuclei, glutamate neurons can then indirectly disinhibit GABAA neurons in the substantia nigra through activation of GLU M3 receptors (red hash sign in Fig. 17.2). GABAA neurons, when stimulated, will, in turn, powerfully inhibit (red hash signs in Fig. 17.2) ventral tegmental area (VTA) glutaminergic drive via GABAB 3 neurons. It is also possible that stimulation of ACh neurons that at the nucleus accumbens (NAc) ACh, can stimulate both muscarinic (red hash) or nicotinic receptors (green hash in Fig. 17.2). It is indeed nicotinic receptors that help regulate dopamine release at the NAc. Finally, glutamate neurons in the VTA will project to dopamine neurons through N-methyl-D-aspartate receptors (green equal sign in Fig. 17.2) to preferentially release dopamine at the NAc), shown in Fig. 17.2 as a bulls-eye indicating a euphoria, or "wanting" response. The result is that when dopamine release is low (endorphin deficiency), unhappiness is felt, while general (healthy) happiness depends on the dopamine homeostatic tonic set point (see Fig. 17.2).

The cascading interaction of these neurotransmitters and secondary messengers results in the correct release of dopamine within the NAc and across many brain regions. These regions are involved in motivation, cognition (memory), pleasure, stress reduction, drug reinstatement, decision making, recall, well-being, and especially cravings (Myers and Cicero, 1969). The result is to provide *Homo sapiens* with a usual happiness set point (Fig. 17.3) identified as resting-state functional connectivity.

Over the last several decades starting in 1968, there has been a remarkable change in understanding the science behind how psychoactive drugs influence behavior through extraordinarily complex actions on neuronal pathways, especially in the mesolimbic system and the prefrontal cortex—cingulate gyrus of the brain. Ideas about recovery, incorporating the 12-step program and fellowship, and introducing the new definition of the term "addiction" from the American Society of Addiction Medicine has had a tremendous impact on our younger generation. The general acceptance of the now well-established phenomena that elaborate on the undeniable fact that addiction is indeed a brain disorder is changing the face of addiction medicine (Smith, 2012).

Early work from several scientists centered on serotonin and brain reward circuitry, revealing that depletion of brain serotonin augments stresslike reactions in rodents. They used *p*-chlorophenyl alanine (PCA), a tryptophan hydroxylase inhibitor that selectively depletes brain serotonin content (Geller and Blum, 1970). This experiment suggested that PCA promotes anxiety and that serotonin is an antianxiety brain substance. Along similar lines following the work of Myers and Cicero (1969), knowing at the time that during the dark phase, levels of serotonin in the pineal gland are low led Geller's group found strong evidence in rodents of enhanced darkness-induced ethanol drinking. The first experiment showed that rats that were placed in a dark closet drank more alcohol than those that were housed in the light (Reiter et al., 1973; Geller, 1971). The hypothesis was that an increase in pineal melatonin caused increased drinking. Subsequently, injections of melatonin in rats that were exposed to "normal" photoperiods (9 hours of darkness during a 24-hour day) revealed this to be accurate, since they also displayed augmented ethanol intake; several experiments

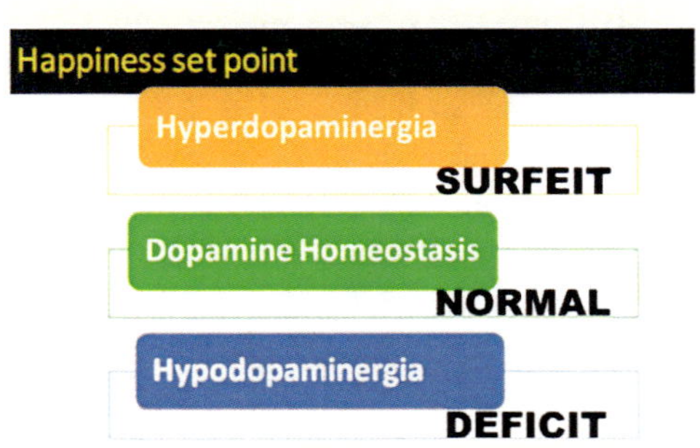

FIGURE 17.3 A simple schematic of tonic happiness set point based on dopamine release as explained in Fig. 17.2.

validated these results (Blum et al., 1973; Reiter et al., 1973, 1974). With experiments revealing that conversion to melatonin via adenylate cyclase stimulation of N-acetyltransferase increased threefold at night, our understanding of these issues improved (Coon et al., 2012; Namboodiri et al., 1985; Raiewski et al., 2012). Another theory concerning the neurochemical mechanism(s) of ethanol consumption suggested a role for dopamine condensation products, known as isoquinolines.

In 1970, Davis and Walsh (1970) proposed that a product of alcohol, tetrahydropapaverline, a benzyltetrahydroisoquinoline alkaloid derivative of the biogenic amine, dopamine, and acetaldehyde, condenses and can induce ethanol intake in rodents. Simultaneously, others reported a role in alcoholism for biogenic amines (for example, norepinephrine) and later indoleamine (serotonin) aldehyde condensation products salsolinol and carbochol, respectively (Cohen and Collins, 1970; Myers, 1989). This work suggested that alcohol and opiate addiction might have common neurochemical mechanisms. Indeed, at the time, controversy stimulated research by neuroscientists that provided evidence that: ethanol intake augmented the salsolinol metabolite in the rat brain (Blum et al., 1977; Hamilton et al., 1978); salsolinol induced an increase in ethanol intake (Blum et al., 1977); salsolinol acts like an opiate agonist (Blum et al., 1977; Hamilton et al., 1978; Marshall et al., 1977); and salsolinol-induced alcohol withdrawal tremors blocked by narcotic antagonism (Blum, 1988; Blum et al., 1982; Hamilton et al., 1979).

Following these early findings, the 1970s promised the world a new era of "sex, drugs, and rock-and-roll" (Blum et al., 2012a). Then in the 1980s, following comments about the recreational use of cocaine, Gold's group was the first to propose the dopamine depletion hypothesis and suggested the use of bromocriptine to reduce cocaine cravings (Dackis and Gold, 1985). Finally, in the 1990s, researchers began to see attempts to understand the neurogenetic and neurobiological mechanisms related to substance abuse through neuroimaging techniques (Boundy et al., 1998; Wang et al., 1993). While we contend that RDS is indeed a nosology to reflect all addictive behaviors, there are of course specific drug or no-drug behaviors that require addressing as well.

17.2 Understanding endorphin deficiency syndrome and opioid deficiency syndrome

Induction of an opioid peptide deficiency syndrome may involve both genetic and epigenetic insults to the brain reward circuitry specific to opioid peptides (i.e., endorphins, enkephalins, and dynorphins). Genetic deficits may include polymorphisms in the membrane metal- endopeptidase (MME), a carboxypeptidase enzyme, responsible for the inactivation of endogenous opioid peptides. Comings et al. (2000) identified a dinucleotide polymorphism in the 5′ region of the MME gene that provides a high activity of this enzyme and subsequently reduced endogenous opioid peptides by substantial inactivation. The resultant general hypoopioidergia or a specific hypoendorphinergia function at both the delta or mu receptors and a hypodynophinergia at kappa receptors (not shown in Fig. 17.4). Also, at birth, an individual may carry the rs260997 C allele of the proenkephalin (PENK) gene, which causes significantly reduced enkephalin synthesis across the BRC (Comings et al., 1999). Moreover, Niikura et al. (2010) reported that there are epigenetic insults that reduce mRNA expression involving downregulation of mu receptor numbers or availability by chronic use/misuse of opiate analgesics like heroin or potent synthetic opioids like methadone, buprenorphine, etc. Furthermore, Nylander et al. (1995) showed reduced leu-enkephalin in the VTA of Sprague-Dawley rats with chronic morphine treatment.

Induction of an opioid peptide deficiency syndrome may involve both genetic and epigenetic insults to the brain reward circuitry specific to opioid peptides (i.e., endorphins, enkephalins, and dynorphins). Genetic deficits may include polymorphisms in the membrane MME, a carboxypeptidase enzyme. Also, at birth, an individual may carry the rs260997 C allele of the proenkephalin (PENK) gene, which causes significantly reduced enkephalin synthesis across the BRC (red hash sign in Fig. 17.4). Nylander et al. showed reduced leu-enkephalin in the VTA of Sprague-Dawley rats with chronic morphine treatment (red hash sign in Fig. 17.4) The dark green arrows represent the normal flow.

Various human activities trigger the production of endorphins or enkephalins. Laughter is known to increased resistance to pain, stimulate endorphin production (Manninen et al., 2017), and vigorous aerobic exercise releases beta-endorphin (Hicks et al., 2019; Kami et al., 2017). Endorphins are released by pleasurable behaviors such as eating chocolate, which contains TIQs, sex, orgasm, eating food, yoga, meditation, and listening to music (Blum et al., 2017; Tuenter et al., 2018).

Enkephalins are in high concentrations in the brain, especially in the hippocampus and prefrontal cortex. Stressors affect neuropeptides and their action in metabotropic enkephalin G-protein-coupled receptors, including delta-opioid and mu receptors (Henry et al., 2017; Mao and Wang, 2001). Other endogenous opioid ligands, such as dynorphins bind to kappa receptors, while endorphins per se bind to mu receptors (Opioid peptides, 1986; Wagner et al., 1991).

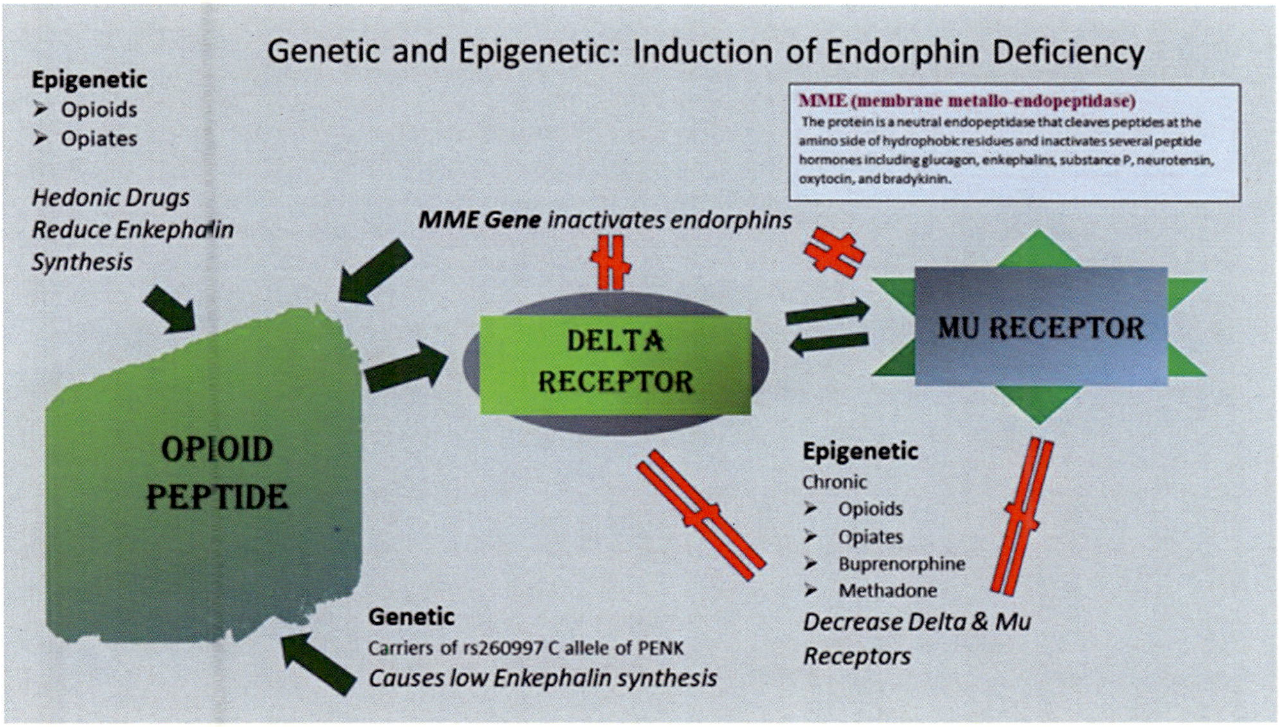

FIGURE 17.4 Schematic of genetic and epigenetic induction of opioid peptide deficiency syndrome. *Need permission from Neurological sciences.*

Within the mesolimbic reward process, enkephalins inhibit the release of the neurotransmitter GABA when they bind to the mu receptor to increase the production and release of dopamine (Watanabe et al., 2018). Endorphin or opioid deficiency will result in a lowered inhibitory control over GABAergic inhibition and thus through the raphe nucleus interaction at the substantia nigra–VTA junction, reduces dopamine release at the NAc. The resultant hypodopaminergia will set people up for addiction liability and other RDS behaviors.

It is now well known that the interaction of at least seven neurotransmitter systems (serotonin, opioid peptides, cannabinoids, GABA, glutamate, ACh, and dopamine) and associated secondary messengers play a role in reward processing in the brain of *H. sapiens*. The BRC, which provides dopamine release, illustrated in Fig. 17.1, is the basis of pleasures from everyday natural rewards. The suggestion is that balanced neurotransmission will animate feelings of well-being and reduced stress, while impairments in the BRC can result in hypodopaminergia and or hyperdopaminergia. For example, an individual that carries a polymorphism in the OPM1 (rs1799971-A118G), should display a reduced function of this mu-opioid receptor. This functional reduction will result in an overexpression of GABAergic signaling. In turn, this excess GABA signaling will significantly reduce the net preferential neuronal release of dopamine at the NAc. Also, there could be epigenetically induced methylation that can cause reduced mRNA expression for at least two generations (Szutorisz and Hurd, 2016). Having an antecedent opioid/endorphin deficiency configures the individual with the potential to have uncontrolled dependence once exposed to a powerful opioid such as Oxycontin. This risk reflected the challenges the entire scientific community has in the face of the worst opioid epidemic ever seen in the United States (Dolbin-MacNab and O'Connell, 2021).

17.3 Understanding obesity as a hypodopaminergia

Another example of a specific RDS behavior is obesity. Obesity is affecting the lives of more than 300 million individuals worldwide (Obesity and Overweight, 2017). Maintaining a healthy body composition using popular weight loss tactics has proven to be extremely challenging with an extraordinarily high rate of rebound weight gain contributing to the obesity epidemic. The vast number of weight loss tactics and programs are based primarily on the premise that an individual eats too much and does not exercise enough. If this were true, then everyone who ate less and exercised more would be thinner than the people who did not do those things. This is not the case.

A more comprehensive systems biology approach is necessary to simultaneously address multiple nutrigenomic (nutriepigenetics), hormonal, metabolic, and immune pathways. Managing obesity seems within reach as a better understanding of the function of our genome in drug and nutrient responses develops. Strategies indicated by this understanding of nutriepigenomics and neurogenetics in the treatment and prevention of metabolic syndrome and obesity include regulation of mRNA expression by promoting histone deacetylation and DNA methylation. Personalized to an individual's genetic makeup, dietary supplementations such as phytochemicals, vitamins, certain minerals, and importantly, functional amino acids can be utilized to epigenetically target deficient metabolic pathways. Furthermore, the chromatin structure of imprinted genes that control nutrients during fetal development can be altered (via "nutriepigenetics" or "nutrigenomics"). Pathways that affect dopamine signaling, molecular transport, and nervous system development are implicated in these strategies and affect stress, survival, and immune responses. Obesity is a subtype of RDS, and the discussed strategies in the treatment and prevention of obesity target the enhanced function of dopamine. It is a matter not only of gastrointestinal signaling linked to hypothalamic peptides, but also of finding novel ways to improve VTA dopaminergic function and homeostasis.

The literature is rife with research and reviews related to the role of dopamine and other neurotransmitters and their neurological interactions within the reward circuitry (Athanasiu et al., 2015; Blum, 1969; Blum et al., 1976a,b, 1990; Melchior and Myers, 1976; Myers et al., 1980; Tolosa et al., 1998; Yoshimura, 1973). Indeed, following their first genetic study that found severe alcoholism to be associated with the Taq A1 allele of the dopamine D2 receptor gene (D2R) (Blum et al., 1996a). While Volkow et al. (2013) helped delineate RDS, it is generaly agreed that RDS is the result of a malfunction (or an inability to fulfill genetic instructions) to achieve optimal BRC neurotransmission (Blum et al., 1990). The dysfunction is specifically characterized by abnormal craving behavior and reduced executive function (decision making) and involves several reward genes and secondary messenger deficits across the prefrontal cortex and mesolimbic systems. In contrast, the functional collaboration of dopamine with other neurotransmitters, such as serotonin and opioids (neuropeptides), is responsible for producing a sense of well-being, even in cocaine abuse (Sadakierska-Chudy et al., 2017).

In our opinion there are several important strategies to overcome obesity. But first, we must identify some factors that contribute to obesity, which include the following:

1. Poor dietary habits, including overfed and undernourished (i.e., nutrient-deficient high calorie intake)
2. Insufficient or inadequate physical activity
3. Excess exposure and/or intake of environmental toxins
4. Overburdening imposition on anaerobic metabolism (cannot effectively use O_2 and H_2O)
5. An increase in inflammatory conditions
6. Increase in reactive oxygen species and excessive oxidative damage to tissues.
7. Immune system distress and exhaustion
8. Stress (inducing stress hormones and "survival panic" bioresponses)
9. Digestive malfunction and dysbiosis.

By default, digestive malfunction is a collateral consequence of all the other factors. Collectively, most of these factors are chronically persistent, insulting, and an erosive strain on immune competence, neurochemistry, and the interconnectivity of the gut-brain axis. Moreover, the specificity and magnitude of effects of these insulting and functionally degrading factors merit further rigorous evaluation of (1) hypothalamic energy regulation and the impact of metabolic hormones on the VTA dopamine neurons; (2) neurobiological mechanisms shared between food and drug addiction identified; (3) neuroimaging data regarding the function of dopamine in food addiction, brain metabolism, and addiction transfer, following bariatric surgery, and the role of noncoding miRNAs in drug addiction and the epigenetics of metabolic syndrome linked to food addiction, requires focused investigation; (4) chromatin structural modifications of imprinted genes that control nutrients during fetal development need to be identified early in life; and (5) treatment supported by this research that targets deficient metabolic pathways epigenetically through the provision of dietary supplementation that includes phytochemicals, vitamins, and, functional amino acids.

However, to appreciate the role of the neural reward circuitry and dopamine regulation in the relationship of digestive competence to obesity, we must revisit the past regarding gut-hypothalamic mechanisms.

17.4 Gut-hypothalamic homeostatic energy regulation

The first reported connection between the ventromedial hypothalamus (VMH) and obesity came from Hetherington and Ranson, who in 1940 found that lesions of the VHM resulted in overeating and obesity in rats (Dube et al., 1999).

The lesion damage was extensive and included the ventromedial neurons and arcuate (ARC) nuclei. The importance of the ARC was the subject of intense research. The regulation of energy (glucose homeostasis) was attributed to the Proopiomelanocortin (POMC) and neuropeptide Y (NPY)/agouti-related peptide (AgRP) neurons located in the ARC (Scallet and Olney, 1986). Profound anorexia and weight loss were produced when Anand and Brobeck experimented with extensive lesions of the lateral hypothalamic area (LHA). The hypothalamic dual center hypothesis with the "satiety center" located in the VMH and the "feeding center" at the LHA site suggested by Stellar was dominant (Stellar, 1954). However, a newer view seems more likely. Grill and Kaplan (2002) began to realize that control of energy is distributed within a network of sites. Lesions in one part of the brain can alter body weight and adiposity (Keesey et al., 1979).

17.4.1 The set point dilemma

Keesey and others suggested that VMH and LHA lesions provided evidence that a weight set point might depend on external and internal environments (neural substrates) (Keesey et al., 1979) and that a feedback mechanism to monitor peripheral metabolic status for the control of energy homeostasis seems more likely. Kennedy (1953) proposed a lipostatic hypothesis suggesting that body fat storage might provide feedback. Kennedy suggested that negative feedback system-based signaling was transmitted in proportion to adipose tissue mass to regulate intake or expenditure and maintain body fat within a predetermined set point (see Fig. 17.2) (Kennedy, 1953).

The hormone leptin, which is produced in proportion to adipose tissue mass, was eventually found (Zhang et al., 1994). The idea of a set point (preserved body weight) is controversial, with articles for (Keesey and Powley, 2008) and against (Payne and Dugdale, 1977; Speakman et al., 2011). Interestingly, researchers have found that in certain rodent strains and most humans who become obese the "preserved" body weight can be easily stimulated upward, while efforts to move beneath a higher set point fail in approximately 90% of people (Bouret et al., 2015; Levin and Keesey, 1998; Thanos et al., 2008b). This set point dilemma remains the focus of weight loss researchers and may have some relationship to a similar set point that has been observed in the drug abuse literature (Zou et al., 2017).

17.4.2 Homeostatic systems

The hypothalamus integrates the hormonal, autonomic, and somatomotor regulation of nutrient levels in the body by stimulating neuroendocrine homeostatic reactions (Bouret et al., 2015). The networks that control central system energy homeostasis are distributed and interconnected neural systems (Andrews et al., 2008; Grill and Hayes, 2012; Levin et al., 2006; Sohn et al., 2013). Two neuronal types that control energy balance are the hypothalamic anorexigenic POMC neurons, and the orexigenic NPY/AgRP neurons (Broberger et al., 1998; Cowley et al., 2003). Transsynaptic GABA neurons can modulate POMC neuronal activity indirectly via inputs arising from NPY neurons (Beck et al., 1990; Broberger et al., 1998). Consumption is controlled by arcuate POMC and NPY neurons that project into several hypothalamic and extrahypothalamic sites (Beck et al., 1990). Projections from the hypothalamic paraventricular nucleus to the endocrine, autonomic, and somatomotor systems (Sawchenko, 1998; Watts, 2000), the preautonomic brain stem, and spinal cord (Abizaid et al., 2006) are involved in feeding and energy balance regulation.

VTA dopamine neurons are modified by metabolic hormones, such as leptin and ghrelin. Leptin has an inhibitory impact on VTA dopamine neurons, and hyperphagia of leptin-deficient mice is diminished without dopamine (Brunetti et al., 1999; Muelbl et al., 2016; Pfaffly et al., 2010). Conversely, the effect of the direct introduction of ghrelin into the VTA dopaminergic neurons encourages food consumption (Furness et al., 2020; Lookingland and Moore, 1984).

17.4.3 Feeding behavior regulation by dopaminergic systems

With the enormous complexity of neuronal processing, especially for dopamine signaling across different brain regions as described in this review, we would be remiss if we did not address the differences in sensitivity and regulation of dopaminergic function. The tuberoinfundibular dopaminergic neurons connect the hypothalamus to the pituitary gland (hormonal regulation, nurturing, pregnancy and sensory processes), and the nigrostriatal systems connect the substantia nigra to the striatum responsible for motor control, and indirectly the mesolimbic and mesocortical (reward) areas in the VTA.

As early as 1984, Moore's group (Moore, 1992) reported differences in dopamine-mediated receptors in the nigrostriatal and mesolimbic systems compared to the hypothalamic neurons, which comprise the tuberoinfundibular dopamine system. Different functionally appropriate mechanisms regulate these distinct dopaminergic neuronal systems. Annunziato (1979) revealed that there are different dopamine neurons between these systems, and, for example, dopamine receptors in the anterior pituitary and striatum vary in that the receptors in the anterior pituitary are more sensitive to the actions of antagonists and agonists than the ones in the striatum. Moreover, Demarest and Moore (1979) pointed out that the regulation of dopamine synthesis in terminals of nigrostriatal, mesolimbic, and tuberhypophyseal nerves differ from that in tuberoinfundibular nerves that do not have an autoreceptor regulatory mechanism. This finding provides information that dopamine neurons and their location may have different responses rather than having a similarity. These differences must be considered in any explanation of the relationship between food and drugs.

Most animals and humans will ignore the satiety effect of homeostatic systems when confronted with highly palatable foods, that is, foods that are high in sugar, salt, and fat (Schulte et al., 2015). The reward system comes into play, whereby rewarding (hedonic) stimuli encode neural circuits that are involved in "liking" (pleasure) and "wanting" (motivation) (Blum et al., 2012b). The VTA dopamine neurons have an important role in obesity (Blum et al., 2012b). The mesolimbic dopaminergic pathway composed of dopamine neurons connects the hippocampus and VTA with the limbic system medial prefrontal cortex, nucleus accumbens, and amygdala.

17.4.4 Food and drug addiction identification of common neurobiological mechanisms

It is essential for people who are in treatment and recovery to recognize that permitting candy at treatment centers may very well initiate unwanted addiction transfer from one substance, such as alcohol, opiates, or cocaine, to another, such as glucose. We now understand that an estimated 20% of individuals who undergo lap-band surgery will acquire full-blown drug addiction after surgery. This phenomenon of substituting food for drugs has been studied in both human and animal clinical trials. It is not shocking to contemplate those variations of genes involved in the reward circuitry functioning can be responsible for decreases in dopamine function, which augment craving for both glucose and alcohol. There is, however, also the possibility that during protracted (greater than three weeks) abstinence from alcohol and other substances, those affected may show a "hyperdopaminergic" epigenetically impacted state discerned by higher extra cellular DA levels (Hirth et al., 2016).

Regarding RDS, it is also noteworthy that addictive behaviors involving both substances and nonsubstances, such as food (glucose), opiates, alcohol, nicotine, cannabis and others. Internet gaming, gambling, music, and sex, all release dopamine at the nucleus accumbens. This acute release of dopamine can lead to feelings of well-being in the very short term, especially in individuals who, through either gene polymorphisms (variants) or the environment or both, have compromised dopaminergic function. The environmental compromise of dopaminergic function is provoked epigenetically by expression of many genes, by methylation (reduced expression), or by inhibition of deacetylation (increase expression) on histones in the chromatin material (Hirth et al., 2016). The core concept is represented by the formula $P = G + E$ (phenotype equals genetics plus environment). This is a complex interaction involving many polymorphic genes and their subsequent interaction with the environment. Thus if the hypodopaminergic effect of specific genetic reward polymorphisms is compounded by an unsatisfactory environment, further decreasing dopamine function, the affected person may try to boost their brain dopamine. They may self-medicate to obtain their required dopamine fix through behaviors, such as food, drugs, or sex, that cause the acute release of dopamine (Hirth et al., 2016; Salimpoor et al., 2011; Xu et al., 2016).

Within the framework of the brain's reward system with regard to food consumption, the "thrifty gene hypothesis" (a survival gene that is linked to famine and fat metabolism) and the commonality between food and drug addictions (Blum et al., 2017; Xu et al., 2016) are important. Food and drug consumption trigger reinforcers, both homeostatic and rewarding, that stimulate specific connections in many similar neural systems (Prentice et al., 2008; Solanki et al., 2020; Speakman, 2006). Several regions of the brain are involved in the reinforcement of both drug and food intake (Hernandez and Hoebel, 1990; Kelley et al., 2002), and several neurotransmitters, as well as hormones, have been studied in these and related brain areas (Kalivas and Volkow, 2005; Koob and Le Moal, 2005; Simpson et al., 2012; Spangler et al., 2004; Volkow and Wise, 2005). There have been numerous studies in the scientific literature that explicitly maintain these commonality concepts as offered by Avena's group (Wiss et al., 2020; Hill et al., 2014) and by Blum et al. (2016).

In clinical situations, reciprocal comorbidity of food and drug addiction typically appear in the literature. This common comorbidity deserves intensive investigation (Spangler et al., 2004).

Most recently, changes in the nomenclature of addictions indicate a significant shift in the understanding of addictions, where nonsubstance-related behaviors can also be distinguished as addictions. A plethora of data provides empirical evidence that there are overlaps of different types of addictive behaviors in etiology, phenomenology, and in the underlying psychological and biological mechanisms. One study by (Kotyuk et al., 2020) presents an epidemiological analysis carried out as part of the Psychological and Genetic Factors of Addictive Behaviors Study, in which data were collected from 3003 adolescents and young adults (42.6% males; mean age: 21 years). Addictions to psychoactive substances and behaviors were rigorously assessed. Data is provided on lifetime occurrences of the assessed substance uses, their cooccurrences, the prevalence estimates of specific behavioral addictions, and cooccurrences of different substance use and potentially addictive behaviors. Associations were found between (1) smoking and problematic Internet use, exercising, eating disorders, and gambling (2) alcohol consumption and problematic Internet use, problematic online gaming, gambling, and eating disorders, and (3) cannabis use and problematic online gaming and gambling. The results suggest a large overlap between the occurrence of these addictions and behaviors and underlies the importance of investigating the possible common psychological, genetic, and neural pathways. These data further support concepts such as the RDS and the component model of addictions that propose a common phenomenological and etiological background of different addictive and related behaviors.

Although overeating may have significant neurochemical associations with drug abuse, less is known about other eating disorders, such as bulimia (Ram et al., 2008) and anorexia (Mann et al., 2014). Nonetheless, there is growing support that equivalent gene polymorphisms influence food in similarly to how they affect drug abuse. Common risk alleles such as the A1 form of the dopamine type 2 receptor (DRD2) gene, which has been found to associate with eating disorders, such as binge eating disorder, bulimia, and anorexia, are similar to drug abuse as observed in genetic studies. Utilizing sophisticated neuroimaging, especially positron emission tomography (PET), Gene-Jack Wang, Nora Volkow, Peter K. Thanos, and others have contributed to the field with valuable insights regarding the molecular role of dopamine function in obesity (Thanos et al., 2008a,b, 2010; Tomasi et al., 2009; Volkow et al., 2008; Wang et al., 2009a,b). This enormous contribution started with a study, which used PET, that revealed an inverse relationship between body mass index (BMI) and dopamine in obese humans. The same group found that the availability of dopamine D2 receptors was reduced in obese individuals in inverse proportion to the severity of their BMI. Dopamine moderates reward and incentive circuits; thus dopamine deficits in obese individuals may prolong disordered eating to balance the reduced stimulation of these circuits. In an investigation, Thanos et al. (2015) provided strong evidence from animal studies that dopamine D2 gene expression can affect behaviors throughout an entire life span. They found that both locomotor activity and body weight were moderated by an enriched environment and that mice from an enriched environment showed greater behavioral variability between genotypes compared to mice from a deprived environment.

17.5 Summary

Genetic researchers wonder whether some people might have genetically fragile opioid systems that may implode when exposed to exogenous opioids. Results of the genetic addiction risk severity (GARS) test is a predictor of vulnerability to opioid dependence (Tomasi et al., 2015). These theorems argue that endorphinergic deficiency syndrome, a subset of RDS (Tomasi et al., 2015) rather than generalized opioid deficiency syndrome, is the process underlying the opioid epidemic. Gold's work on concurrent inhibition of the nucleus locus coeruleus by the alpha-2 adrenergic system provides for novel pharmacological therapies such as clonidine and lofexidine (Zhang et al., 2016), rapid and slow nonopioid detoxification and possible new treatments for craving and anxiety. The current emphasis on opioid replacement in opioid use disorders (OUD) is a treatment based on a disease model that has not been proven. Oddly, the most efficacious treatments for OUDs are the treatments that stimulate compliance (Thanos et al., 2010), which are so closely related to opioid drugs of abuse that the patient becomes addicted to the treatment, so discontinuation becomes unlikely (Tomasi et al., 2009). The brain has not carved out an opiate (e.g., codeine) or opioids (e.g., buprenorphine) deficiency as a phenotypic target per se.

However, methadone and buprenorphine treatments are the first options for most, if not all, patients with OUDs (Wang et al., 2009b). As in the early days of methadone maintenance, opioid replacement is seen in the same perspective in which a diabetic might see insulin but provided in dose equivalents that are not related to endogenous opioids. Rather, opioid replacement is used for saturation or blocking exogenous opioid reinforcement doses (Carpenter et al., 2013). Dose escalation of the exogenous opioids that are provided as treatments for OUDs may further compromise the patient's ability to ever recover from an underlying Endorphinergic Deficiency Syndrome. Relatively opioid-naive OUD patients may acquire an iatrogenic endorphin deficiency, which will downregulate dopamine synthesis. Exogenous opioids produce dopamine deficiency syndrome, also a subset of RDS, that includes depression, anhedonia, and may also be a cause of exogenous opioid seeking, substance abuse, and overeating (Badgaiyan et al., 2015; Wright et al., 2015).

Most guidelines around the world recommend psychosocially assisted agonist maintenance therapy, and not medication alone as a treatment for OUD. This assistance has been difficult or impossible for many patients during the viral crisis. Many have dropped out of treatment, and others have tried telemedicine and technologically assisting behavioral therapies while continuing their medically assisted treatments.

The finding that dopamine neurons in their locations may have dissimilar responses to the same substance must be considered in any explanation of the relationship between food and drugs. One pharmacological example,

(Syvälahti et al., 2000) revealed that the antipsychotic activity of neuroleptic drugs has different effects. Both nigrostriatal mechanisms (which induce extrapyramidal symptoms) and tuberoinfundibular mechanisms (which increase prolactin levels) may cooccur when the mesocortical effects (reduced symptoms of schizophrenia) of antipsychotic drugs are sought. Therefore we must be mindful of these differences, especially when we are thinking of therapeutic targets and common mechanisms between food and drugs.

While the hypothalamic-gut axis plays a very significant role in nutrient selection, therapeutic targets to prevent food addiction and neuroendocrine-mediated obesity stem from prodopamine regulation (see Fig. 17.1). Support from fMRI research in humans and animals encourages support for the premise that distributed neurobiological circuits are disturbed in addiction, metabolic syndrome, and obesity. Altered dopamine reward circuits are an antecedent to genetically induced survival panic, stress intolerance, microbiome disruption, pathologic food consumption, increased fat synthesis, and storage. For example, Chen et al. (2012) reported that compared to dopamine D2 Taq A2 allele, there is a high correlation of high percentage of body fat and the DRD2 Taq A1 allele (see Fig. 17.5 and many associations of DRD2 A1 allele). Based on the commonality of neuromechanisms between drug and glucose addiction and evidence from neuroimaging epigenetic studies, dopamine agonist treatment strategies that epigenetically target disrupted dopamine pathways as opposed to the present methods that utilize antagonistic dopamine therapy may succeed and make a significant contribution toward remitting digestive distress and dysbiosis.

FIGURE 17.5 Schematic of the commonality of a significant array of manifestations due to DRD2 A1 allele.

Pathways that affect dopamine signaling, molecular transport, and nervous system development are implicated in these strategies. Implementation of new dopamine agonist strategies, presented in this chapter, for the prevention and treatment of obesity, a subtype of RDS, target improved dopamine function. Neurogenetic evidence that supports prodopamine regulation is validated by the published clinical evidence and appears to be sensible (Blum et al., 1996b).

Balanced dopaminergic activity facilitates resting-state functional connectivity in the mesolimbic system and can facilitate healthy choices and reduced craving (Blum et al., 1996b; Febo et al., 2017; Fried et al., 2020). Moreover, novel approaches aimed at balancing dopamine activity can induce feelings of well-being and are called for, and even encouraged, to prevent and treat obesity and enable better microbiome health. Implementing this therapeutic message could help to change addiction to food and drugs as well as other behavioral addictions. It is not just a matter of gastrointestinal signaling linked to hypothalamic peptides but, alternatively, finding novel ways to improve VTA dopaminergic function and homeostasis, as illustrated in Fig. 17.1.

On the basis of this seminal research, and unlike other existing genetic tests, U.S. and foreign patented GARS test is, to our knowledge, the first proven and validated genetic addiction risk severity test that will have important clinical benefits. These benefits include the advancement of precision in personalized medicine and an assessment of the severity of RDS risk, including alcoholism, as was first suggested by our group. The test will also include the cytochrome P450 system of genetic variants that influence how individuals metabolize opioids. Our laboratory continues to develop other specific genetic tests for obesity, ADHD, and PTSD. As has been discussed, early identification of genetic antecedents will provide important information to help target systems therapeutics (Blum et al., 2020; Febo et al., 2017; Fried et al., 2020).

17.6 Conclusion

Clinically, the future is here. The microbiome treatment of chronic addiction depends on scientifically sound appropriate evidence-based early genetic risk determinations, leading to high-precision personalized care of the patient that optimize the sequela of genetically mediated neurotransmitter transactions and rebalance neurochemical cross-talk. These are not fictional, hypothetical, or theoretical practices.

References

Abizaid, A., Liu, Z.W., Andrews, Z.B., et al., 2006. Ghrelin modulates the activity and synaptic input organization of midbrain dopamine neurons while promoting appetite. J. Clin. Invest. 116, 3229–3239.

Andrews, Z.B., Liu, Z.W., Walllingford, N., et al., 2008. UCP2 mediates ghrelin's action on NPY/AgRP neurons by lowering free radicals. Nature 454, 846–851.

Annunziato, L., 1979. Regulation of the tuberoinfundibular and nigrostriatal systems. Evidence for different kinds of dopaminergic neurons in the brain. Neuroendocrinology 29 (1), 66–76.

Athanasiu, L., Smorr, L.L. Tesli, M., et al., 2015. Genome-wide association study identifies common variants associated with pharmacokinetics ofpsychotropic drugs. J. Psychopharmacol. 29, 884–891.

Badgaiyan, R.D., Sinha, S., Sajjad, M., Wack, D.S., 2015. Attenuated tonic and enhanced phasic release of dopamine in attention deficit hyperactivity disorder. See comment in *PubMed commons* below. PLoS One 10, e0137326.

Beck, B., Burlet, A., Nicolas, J.P., Burlet, C., 1990. Hyperphagia in obesity is associated with a central peptidergic dysregulation in rats. J. Nutr. 120, 806–811.

Berridge, K.C., Robinson, T.E., 2016. Liking, wanting, and the incentive-sensitization theory of addiction. Am. Psychol. 71 (8), 670–679.

Blum, K., 1969. The effect of dopamine and other catecholamines on neuromuscular transmission. Arch. Int. Pharmacodyn. Ther. 181, 297–306.

Blum, K., 1988. Narcotic antagonism of seizures induced by a dopamine-derived tetrahydroisoquinoline alkaloid. Experientia 44 (9), 751–753.

Blum, K., Merritt, J.H., Reiter, R.J., Wallace, J.E., 1973. A possible relationship between the pineal gland and ethanol preference in the rat. Curr. Ther. Res. Clin. Exp. 15 (1), 25–30.

Blum, K., Eubanks, J.D., Wallace, J.E., Schwertner, H.A., 1976a. Suppression of ethanol withdrawal by dopamine. Experientia 32, 493–495.

Blum, K., Wallace, J.E., Schwerter, H.A., Eubanks, J.D., 1976b. Morphine suppression of ethanol withdrawal in mice. Experientia 32, 79–82.

Blum, K., Hamilton, M.G. Meyer, E.K., Hirst, M., Marshall, A., 1977. Isoquinoline alkaloids as possible regulators of alcohol addiction. Lancet 1 (8015), 799–800.

Blum, K., DeLallo, L., Briggs, A.H., Hamilton. M.G., 1982. Opioid responses of isoquinoline alkaloids (TIQs). Prog. Clin. Biol. Res. 90, 387–398.

Blum, K., Noble, E.P., Sheridan, P.J., et al., 1990. Allelic association of human dopamine D2 receptor gene in alcoholism. JAMA 263, 2055–2060.

Blum, K., Sheridan, P.J., Wood, R.C., et al., 1996a. The D2 dopamine receptor gene as a determinant of reward deficiency syndrome. J. R. Soc. Med. 89, 396–400.

Blum, K., Braverman, E.R. Wood, R.C., Gill, J., Li, C., Chen, T.J., et al., 1996b. Increased prevalence of the Taq I A1 allele of the dopamine receptor gene (DRD2) in obesity with comorbid substance use disorder: a preliminary report. Pharmacogenetics 6 (4), 297–305. Available from: https://doi.org/10.1097/00008571-199608000-00003.

Blum, K., Werner, T., Carnes, S., Carnes, P., Bowirrat, A., Giordano, J., 2012a. Sex, drugs, and rock 'n' roll: hypothesizing common mesolimbic activation as a function of reward gene polymorphisms. J. Psychoact. Drugs 44 (1), 38–55.

Blum, K., Gardner, E., Oscar-Berman, M., Gold, M., 2012b. "Liking" and "wanting" linked to reward deficiency syndrome (RDS): hypothesizing differential responsivity in brain reward circuitry. Curr. Pharm. Des. 18, 113–118.

Blum K, Febo M, Badgaiyan RD, et al. 2016. Common neurogenetic diagnosis and meso-limbic manipulation of hypodopaminergic function in reward deficiency syndrome (RDS): changing the recovery landscape. Curr. Neuropharmacol.

Blum, K., Simpatico, T., Febo, M., Rodriquez, C., Dushaj, K., Li, M., et al., 2017. Hypothesizing music intervention enhances brain functional connectivity involving dopaminergic recruitment: common neuro-correlates to abusable drugs. Mol. Neurobiol. 54 (5), 3753–3758.

Blum, K., Bowirrat, A., Baron, D., Lott, L., Ponce, J.V., Brewer, R., et al., 2020. Biotechnical development of genetic addiction risk score (GARS) and selective evidence for inclusion of polymorphic allelic risk in substance use disorder (SUD). J. Syst. Integr. Neurosci. 6 (2). Available from: https://doi.org/10.15761/JSIN.1000221.

Boundy, V.A., Gold, S.J., Messer, C.J., Chen, J., Son, J.H., Joh, T.H., 1998. Regulation of tyrosine hydroxylase promoter activity by chronic morphine in TH9.0-LacZ transgenic mice. J. Neurosci. 18 (23), 9989–9995.

Bouret, S., Levin, B.E., Ozanne, S.E., 2015. Gene-environment interactions controlling energy and glucose homeostasis and the developmental origins of obesity. Physiol. Rev. 95 (1), 47–82.

Broberger, C., Johansen, J., Johansson, C., Schalling, M., Hokfelt, T., 1998. The neuropeptide Y/agouti gene-related protein (AGRP) brain circuitry in normal, anorectic, and monosodium glutamate-treated mice. Proc. Natl. Acad. Sci. USA 95, 15043–15048.

Brunetti, L., Michelotto, B., Orlando, G., Vacca, M., 1999. Leptin inhibits norepinephrine and dopamine release from rat hypothalamic neuronal endings. Eur. J. Pharmacol. 372, 237–240.

Carpenter, C.L., Wong, A.M., Li, Z., Noble, E.P., Heber, D., 2013. Association of dopamine D2 receptor and leptin receptor genes with clinically severe obesity. Obesity (Silver Spring) 21, E467–E473.

Chen, A.L.C., Blum, K., Chen, T.J.H., Giordano, J., Downs, B.W., Han, D., et al., 2012. Correlation of the Taq1 dopamine D2 receptor gene and percent body fat in obese and screened control subjects: a preliminary report. Food Funct. 3 (1), 40–48. Available from: https://doi.org/10.1039/c1fo10089k.

Cohen, G., Collins, M., 1970. Alkaloids from catecholamines in adrenal tissue: possible role in alcoholism. Science 167 (3926), 1749–1751.

Comings, D.E., Blum, K., 2000. Reward deficiency syndrome: genetic aspects of behavioral disorders. Prog. Brain Res. 126, 325–341.

Comings, D.E., Blake, H., Dietz, G., Gade-Andavolu, R., Legro, R.S., Saucier, G., 1999. The proenkephalin gene (PENK) and opioid dependence. Neuroreport 10 (5), 1133–1135.

Comings, D.E., Dietz, G., Gade-Andavolu, R., Blake, H., Muhleman, D., Huss, M., 2000. Association of the neutral endopeptidase (MME) gene with anxiety. Psychiatr. Genet. 10 (2), 91–94.

Coon, S.L., Munson, P.J., Cherukuri, P.F., Sugden, D., Rath, M.F., Moller, M., 2012. Circadian changes in long noncoding RNAs in the pineal gland. Proc. Natl. Acad. Sci. U. S. A 109 (33), 13319–13324.

Cowley, M.A., Smith, R.G., Diano, S., et al., 2003. The distribution and mechanism of action of ghrelin in the CNS demonstrates a novel hypothalamic circuit regulating energy homeostasis. Neuron 37, 649–661.

Dackis, C.A., Gold, M.S., 1985. Pharmacological approaches to cocaine addiction. J. Subst. Abus. Treat. 2 (3), 139–145.

Davis, V.E., Walsh, M.J., 1970. Alcohol, amines, and alkaloids: a possible biochemical basis for alcohol addiction. Science 167 (3920), 1005–1007.

Demarest, K.T., Moore, K.E., 1979. Comparison of dopamine synthesis regulation in the terminals of nigrostriatal, mesolimbic, tuberoinfundibular and tuberohypophyseal neurons. J. Neural Transmiss 46 (4), 263–277.

Dolbin-MacNab, M.L., O'Connell, L.M., 2021. Grandfamilies and the opioid epidemic: a systemic perspective and future priorities. Clin. Child. Fam. Psychol. Rev. 1–17. Available from: https://doi.org/10.1007/s10567-021-00343-7.

Dube, M.G., Xu, B., Kalra, P.S., Sninsky, C.A., Kalra, S.P., 1999. Disruption in neuropeptide Y and leptin signaling in obese ventromedial hypothalamic-lesioned rats. Brain Res. 816, 38–46.

Febo, M., Blum, K., Badgaiyan, R.D., Perez, P.D., Colon-Perez, L.M., Thanos, P.K., et al., 2017. Enhanced functional connectivity and volume between cognitive and reward centers of naïve rodent brain produced by pro-dopaminergic agent KB220Z. PLoS One 12 (4), e0174774. Available from: https://doi.org/10.1371/journal.pone.01747.

Fried, L., Modestino, E.J., Siwicki, D., Lott, L., Thanos, P.K., Baron, D., et al., 2020. Hypodopaminergia and "precision behavioral management" (PBM): it is a generational family affair. Curr. Pharm. Biotechnol. 21 (6), 528–541. Available from: https://doi.org/10.2174/1389201021666191210112108.

Furness, J.B., Pustovit, R.V., Syder, A.J., Ringuet, M.T., Yoo, E.J., Fanjul, A., et al., 2020. Dopamine and ghrelin receptor co-expression and interaction in the spinal defecation centers. Neurogastroenterol. Motil. e14051. Available from: https://doi.org/10.1111/nmo.14051. Dec 2.

Geller, I., 1971. Ethanol preference in the rat as a function of photoperiod. Science 173 (3995), 456–459.

Geller, I., Blum, K., 1970. The effects of 5-HTP on Para-Chlorophenylalanine (p-CPA) attenuation of "conflict" behavior. Eur. J. Pharmacol. 9 (3), 319–324.

Gershon, M.D., 1998. The second brain: a groundbreaking new understanding of nervous disorders of the stomach and intestine. Quill: An Imprint of Harper Collins Publishers. HarperCollins, New York, NY.

Gondré-Lewis, M.C., Bassey, R., Blum, K., 2020. Pre-clinical models of reward deficiency syndrome: a behavioral octopus. Neurosci. Biobehav. Rev. 115, 164–188.

Grill, H.J., Hayes, M.R., 2012. Hindbrain neurons as an essential hub in the neuroanatomically distributed control of energy balance. Cell Metab. 16, 296–309.

Grill, H.J., Kaplan, J.M., 2002. The neuroanatomical axis for control of energy balance. Front. Neuroendocrinol. 23, 2–40.

Hamilton, M.G., Blum, K., Hirst, M., 1978. Identification of an isoquinoline alkaloid after chronic exposure to ethanol. Alcohol. Clin. Exp. Res. 2 (2), 133–137.

Hamilton, M.G., Hirst, M., Blum, K., 1979. Opiate-like activity of salsolinol on the electrically stimulated Guinea pig ileum. Life Sci. 25 (26), 2205–2210.

Henry, M.S., Gendron, L., Tremblay, M.E., Drolet, G., 2017. Enkephalins: endogenous analgesics with an emerging role in stress resilience. Neural. Plast. 2017, 1546125.

Hernandez, L., Hoebel, B.G., 1990. Feeding can enhance dopamine turnover in the prefrontal cortex. Brain Res. Bull. 25, 975–979.

Hicks, S.D., Jacob, P., Perez, O., Baffuto, M., Gagnon, Z., Middleton, F.A., 2019. The transcriptional signature of a runner's high. Med. Sci. Sports Exerc. 51 (5), 970–978.

Hill, J.O., Berridge, K., Avena, N.M., et al., 2014. Neurocognition: the food–brain connection. Adv. Nutr. 5, 544–546.

Hirth, N., Meinhardt, M.W., Noori, H.R., et al., 2016. Convergent evidence from alcohol-dependent humans and rats for a hyperdopaminergic state in protracted abstinence. Proc. Natl. Acad. Sci. USA 113, 3024–3029.

Kalivas, P.W., Volkow, N.D., 2005. The neural basis of addiction: a pathology of motivation and choice. Am. J. Psychiatry 162, 1403–1413.

Kami, K., Tajima, F., Senba, E., 2017. Exercise-induced hypoalgesia: potential mechanisms in animal models of neuropathic pain. Anat. Sci. Int. 92 (1), 79–90.

Keesey, R.E., Powley, T.L., 2008. Body energy homeostasis. Appetite 51, 442–445.

Keesey, R.E., Mitchell, J.S., Kemnitz, J.W., 1979. Body weight and body composition of male rats following hypothalamic lesions. Am. J. Physiol. Regul. Integr. Comp. Physiol. 237, R68–R73.

Kelley, A.E., Bakshi, V.P., Haber, S.N., Steininger, T.L., Will, M.J., Zhang, M., 2002. Opioid modulation of taste hedonics within the ventral striatum. Physiol. Behav. 76, 365–377.

Kennedy, G.C., 1953. The role of depot fat in the hypothalamic control of food intake in the rat. Proc. R. Soc. Lond. B Biol. Sci. 140, 578–592.

Koob, G.F., Le Moal, M., 2005. Plasticity of reward neurocircuitry and the 'dark side' of drug addiction. Nat. Neurosci. 8, 1442–1444.

Kotyuk, E., Magi, A., Eisinger, A., Király, O., Vereczkei, A., Barta, C., et al., 2020. Co-occurrences of substance use and other potentially addictive behaviors: epidemiological results from the psychological and genetic factors of the addictive behaviors (PGA) study. J. Behav. Addict. 9 (2), 272–288. Available from: https://doi.org/10.1556/2006.2020.00033.

Levin, B.E., Keesey, R.E., 1998. Defense of differing body weight set-points in diet-induced obese and resistant rats. Am. J. Physiol. Regul. Integr. Comp. Physiol. 274, R412–R419.

Levin, B.E., Kang, L., Sanders, N.M., Dunn-Meynell, A.A., 2006. Role of neuronal glucosensing in the regulation of energy homeostasis. Diabetes 55, S122–S130.

Lookingland, K.J., Moore, K.E., 1984. Dopamine receptor-mediated regulation of incertohypothalamic dopaminergic neurons in the male rat. Brain Res. 304 (2), 329–338.

Mann, A.P., Accurso, E.C., Stiles-Shields, C., et al., 2014. Factors associated with substance use in adolescents with eating disorders. J. Adolesc. Health 55, 182–187.

Manninen, S., Tuominen, L., Dunbar, R.I., Karjalainen, T., Hirvonen, J., Arponen, E., 2017. Social laughter triggers endogenous opioid release in humans. J. Neurosci. 37 (25), 6125–6131.

Mao, L., Wang, J.Q., 2001. Selective activation of group I metabotropic glutamate receptors upregulates preprodynorphin, substance P, and preproenkephalin mRNA expression in rat dorsal striatum. Synapse (New York, NY) 39 (1), 82–94.

Marshall, A., Hirst, M., Blum, K., 1977. Analgesic effects of 3-carboxysalsolinol alone and in combination with morphine. Experientia 33 (6), 754–755.

Melchior, C.L., Myers, R.D., 1976. Genetic differences in ethanol drinking of the rat following injection of 6-OHDA, 5,6-DHT or 5,7-DHT into the cerebral ventricles. Pharmacol. Biochem. Behav. 5, 63–67.

Moore, K.E., 1992. Differential regulation of dopaminergic neurons in the mammalian brain: a brief review. Chin. J. Physiol. 35 (1), 67–76.

Muelbl, M.J., Nawarawong, N.N., Clancy, P.T., Nettesheim, C.E., Lim, Y.W., Olsen, C.M., 2016. Responses to drugs of abuse and non-drug rewards in leptin deficient ob/ob mice. Psychopharmacology. (Berl) 233, 2799–2811.

Myers, R.D., 1989. Isoquinolines, beta-carbolines and alcohol drinking: involvement of opioid and dopaminergic mechanisms. Experientia 45 (5), 436–443.

Myers, R.D., Cicero, T.J., 1969. Effects of serotonin depletion on the volitional alcohol intake of rats during a condition of psychological stress. Psychopharmacologia 15 (5), 373–381.

Myers, R.D., Melchior, C., Swartzwelder, H.S., 1980. Amine-aldehyde metabolites and alcoholism: fact, myth or uncertainty. Subst. Alcohol. Actions Misuse 1, 223–238.

Namboodiri, M.A., Sugden, D., Klein, D.C., Tamarkin, L., Mefford, I.N., 1985. Serum melatonin and pineal indoleamine metabolism in a species with a small day/night N-acetyltransferase rhythm. Comp. Biochem. Physiol. B. 80 (4), 731–736.

Nezami, B.G., Srinivasan, S., 2010. Enteric nervous system in the small intestine: pathophysiology and clinical implications. Curr. Gastroenterol. Rep. 12 (5), 358–365.

Niikura, K., Narita, M., Butelman, E.R., Kreek, M.J., Suzuki, T., 2010. Neuropathic and chronic pain stimuli downregulate central mu-opioid and dopaminergic transmission. Trends Pharmacol. Sci. 31 (7), 299–305.

Nylander, I., Vlaskovska, M., Terenius, L., 1995. The effects of morphine treatment and morphine withdrawal on the dynorphin and enkephalin systems in Sprague-Dawley rats. Psychopharmacology 118 (4), 391–400.

Obesity and Overweight [PDF on the internet] World Health Organization [cited 2017 May 20th]. Available from: <http://www.who.int/dietphysicalactivity/media/en/gsfs_obesity.pdf>

Opioid peptides: molecular pharmacology, biosynthesis, and analysis. *NIDA Res. Monogr.* 1986;70:1–414.

Payne, P.R., Dugdale, A.A., 1977. Mechanisms for the control of body weight. Lancet 8011, 583–68.

Pfaffly, J., Michaelides, M., Wang, G.J., Pessin, J.E., Volkow, N.D., Thanos, P.K., 2010. Leptin increases striatal dopamine D2 receptor binding in leptin-deficient obese (ob/ob) mice. Synapse 64, 503–510.

Prentice, A.M., Hennig, B.J., Fulford, A.J., 2008. Evolutionary origins of the obesity epidemic: natural selection of thrifty genes or genetic drift following predation release? Int. J. Obes. (Lond.) 32, 1607–1610.

Raiewski, E.E., Elliott, J.A., Evans, J.A., Glickman, G.L., Gorman, M.R., 2012. Twice daily melatonin peaks in Siberian but not Syrian hamsters under 24 h light:dark:light:dark cycles. Chronobiol. Int. 29 (9), 1206–1215.

Ram, A., Stein, D., Sofer, S., Kreitler, S., 2008. Bulimia nervosa and substance use disorder: similarities and differences. Eat. Disord. 16, 224–240.

Reiter, R.J., Blum, K., Wallace, J.E., Merritt, J.H., 1973. Effect of the pineal gland on alcohol consumption by congenitally blind male rats. Q. J. Stud. Alcohol. 34 (3), 937–939.

Reiter, R.J., Blum, K., Wallace, J.E., Merritt, J.H., 1974. Pineal gland: evidence for an influence on ethanol preference in male Syrian hamsters. Comp. Biochem. Physiol. A Comp. Physiol. 47 (1), 11–16.

Sadakierska-Chudy, A., Frankowska, M., Miszkiel, J., Wydra, K., Jastrzębska, J., Filip, M., 2017. Prolonged induction of miR-212/132 and REST expression in rat striatum following cocaine self-administration. Mol. Neurobiol. 54 (3), 2241–2254.

Salimpoor, V.N., Benovoy, M., Larcher, K., Dagher, A., Zatorre, R.J., 2011. Anatomically distinct dopamine release during anticipation and experience of peak emotion to music. Nat. Neurosci. 14 (2), 257–262.

Sawchenko, P.E., 1998. Toward a new neurobiology of energy balance, appetite, and obesity: the anatomists weigh in. J. Comp. Neurol. 402, 435–441.

Scallet, A.C., Olney, J.W., 1986. Components of hypothalamic obesity: bipiperidyl-mustard lesions add hyperphagia to monosodium glutamate-induced hyperinsulinemia. Brain Res. 374, 380–384.

Schulte, E.M., Avena, N.M., Gearhardt, A.N., 2015. Which foods may be addictive? The roles of processing, fat content, and glycemic load. PLoS One 10, e0117959.

Simpson, N., Maffei, A., Freeby, M., et al., 2012. Dopamine-mediated autocrine inhibitory circuit regulating human insulin secretion in vitro. Mol. Endocrinol. 26, 1757–1772.

Smith, D.E., 2012. The process addictions and the new ASAM definition of addiction. J. Psychoact. Drugs 44 (1), 1–4.

Sohn, J.W., Elmquist, J.K., Williams, K.W., 2013. Neuronal circuits that regulate feeding behavior and metabolism. Trends Neurosci. 36, 504–512.

Solanki, N., Abijo, T., Galvao, C., Darius, P., Blum, K., Gondré-Lewis, M.C., 2020. Administration of a putative pro-dopamine regulator, a neuronutrient, mitigates alcohol intake in alcohol-preferring rats. Behav. Brain Res. 385, 112563. Available from: https://doi.org/10.1016/j.bbr.2020.112563.

Spangler, R., Wittkowski, K.M., Goddard, N.L., Avena, N.M., Hoebel, B.G., Leibowitz, S.F., 2004. Opiate-like effects of sugar on gene expression in reward areas of the rat brain. Brain Res. Mol. Brain Res. 124, 134–142.

Speakman, J.R., 2006. Thrifty genes for obesity and the metabolic syndrome—time to call off the search? Diab Vasc. Dis. Res. 3, 7–11.

Speakman, J.R., Levitsky, D.A., Allison, D.B., et al., 2011. Set points, settling points and some alternative models: theoretical options to understand how genes and environments combine to regulate body adiposity. Dis. Model. Mech. 4, 733–745.

Stellar, E., 1954. The physiology of motivation. Psychol. Rev. 5, 5–22.

Syvälahti, E.K., Räkköläinen, V., Aaltonen, J., Lehtinen, V., Hietala, J., Striatal, D., 2000. Dopamine receptor density and psychotic symptoms in schizophrenia: a longitudinal study. Schizophr. Res. 43 (2–3), 159–161. Available from: https://doi.org/10.1016/s0920-9964(99)00027-4.

Szutorisz, H., Hurd, Y.L., 2016. Epigenetic effects of cannabis exposure. Biol. Psychiatry 79 (7), 586–594.

Thanos, P.K., Michaelides, M., Ho, C.W., et al., 2008a. The effects of two highly selective dopamine D3 receptor antagonists (SB-277011A and NGB-2904) on food self-administration in a rodent model of obesity. Pharmacol. Biochem. Behav. 89, 499–507.

Thanos, P.K., Michaelides, M., Piyis, Y.K., Wang, G.J., Volkow, N.D., 2008b. Food restriction markedly increases dopamine D2 receptor (D2R) in a rat model of obesity as assessed with in-vivo muPET imaging ([11C] raclopride) and in-vitro ([3H] spiperone) autoradiography. Synapse 62, 50–61.

Thanos, P.K., Kim, R., Cho, J., et al., 2010. Obesity-resistant S5B rats showed greater cocaine conditioned place preference than the obesity-prone OM rats. Physiol. Behav. 101, 713–718.

Thanos, P.K., Michaelides, M., Subrize, M., et al., 2015. Roux-en-Y gastric bypass alters brain activity in regions that underlie reward and taste perception. PLoS One 10, e0125570.

Tolosa, E., Martí, M.J., Valldeoriola, F., Molinuevo, J.L., 1998. History of levodopa and dopamine agonists in Parkinson's disease treatment. Neurology 50 (S2–10), S44–S48.

Tomasi, D., Wang, G.J., Wang, R., et al., 2009. Association of body mass and brain activation during gastric distention: implications for obesity. PLoS One 4, e6847.

Tomasi, D., Wang, G.J., Wang, R., Caparelli, E.C., Logan, J., Volkow, N.D., 2015. Overlapping patterns of brain activation to food and cocaine cues in cocaine abusers: association to striatal D2/D3 receptors. Hum. Brain Mapp. 36, 120–136.

Tuenter, E., Foubert, K., Pieters, L., 2018. Mood components in cocoa and chocolate: the mood pyramid. Planta Med. 84 (12 – 13), 839–844.

Volkow, N.D., Wise, R.A., 2005. How can drug addiction help us understand obesity? Nat. Neurosci. 8, 555–560.

Volkow, N.D., Wang, G.J., Fowler, J.S., Telang, F., 2008. Overlapping neuronal circuits in addiction and obesity: evidence of systems pathology. Philos. Trans. R. Soc. Lond. B Biol. Sci. 363, 3191–3200.

Volkow, N.D., Tomasi, D., Wang, G.J., et al., 2013. Predominance of D2 receptors in mediating dopamine's effects in brain metabolism: effects of alcoholism. J. Neurosci. 33, 4527–4535.

Wagner, J.J., Evans, C.J., Chavkin, C., 1991. Focal stimulation of the mossy fibers releases endogenous dynorphins that bind kappa 1-opioid receptors in Guinea pig hippocampus. J. Neurochem. 57 (1), 333–343.

Wang, G.J., Volkow, N.D., Fowler, J.S., Wolf, A.P., MacGregor, R.R., Shea, C.E., 1993. Comparison of two PET radioligands for imaging extrastriatal dopamine receptors in the human brain. Synapse. (New York, NY.) 15 (3), 246–249.

Wang, G.J., Volkow, N.D., Telang, F., et al., 2009a. Evidence of gender differences in the ability to inhibit brain activation elicited by food stimulation. Proc. Natl. Acad. Sci. USA 106, 1249–1254.

Wang, G.J., Volkow, N.D., Thanos, P.K., Fowler, J.S., 2009b. Imaging of brain dopamine pathways: implications for understanding obesity. J. Addict. Med. 3, 8–18.

Watanabe, Y., Aono, Y., Komiya, M., Waddington, J.L., Saigusa, T., 2018. Stimulation of accumbal GABA(B) receptors inhibits delta1- and delta2-opioid receptor-mediated dopamine efflux in the nucleus accumbens of freely moving rats. Eur. J. Pharmacol. 837, 88–95.

Watts, A.G., 2000. Understanding the neural control of ingestive behaviors: helping to separate cause from effect with dehydration-associated anorexia. Horm. Behav. 37, 261–283.

Wiss, D.A., Avena, N., Gold, M., 2020. Food addiction and psychosocial adversity: biological embedding, contextual factors, and public health implications. Nutrients 12 (11), 3521. Available from: https://doi.org/10.3390/nu12113521.

Wright, K.N., Hollis, F., Duclot, F., et al., 2015. Methyl supplementation attenuates cocaine-seeking behaviors and cocaine-induced c-Fos activation in a DNA methylation-dependent manner. J. Neurosci. 35, 8948–8958.

Xu, Q., Zhu, C., Fan, Y., Song, Z., Xing, S., Liu, W., et al., 2016. Population transcriptomics uncovers the regulation of gene expression variation in adaptation to changing environment. Sci. Rep. 6, 25536.

Yoshimura, K., 1973. Activation of Na-K activated ATPase in rat brain by catecholamine. J. Biochem. 74 (2), 389–391.

Zhang, Y., Proenca, R., Maffei, M., Barone, M., Leopold, L., Friedman, J.M., 1994. Positional cloning of the mouse obese gene and its human homologue. Nature 372, 425–432.

Zhang, Y., Ji, G., Xu, M., et al., 2016. Recovery of brain structural abnormalities in morbidly obese patients after bariatric surgery. Int. J. Obes. (Lond.) 40 (10), 1558–1565.

Zou, Z., Wang, H., d'Oleire Uquillas, F., Wang, X., Ding, J., Chen, H., 2017. Definition of substance and non-substance addiction. Adv. Exp. Med. Biol. 1010, 21–41.

Chapter 18

Influence of gut microbial flora in body's serotonin turnover and associated diseases

Sabyasachi Choudhuri[1], Jyotirmoy Panda[1] and Sayantan Maitra[2]
[1]Department of Pharmaceutical Science and Technology, Maulana Abul Kalam Azad University of Technology, Haringhata, West Bengal, India,
[2]Institute of Pharmacy, Jalpaiguri, Government of West Bengal, Department of Health and Family Welfare, West Bengal, India

List of abbreviation

GIT	gastrointestinal tract
ANS	autonomic nervous system
BDNF	brain-derived neurotrophic factor
MDD	major depressive disorder
SCFA	short-chain fatty acids
NF-κB	nuclear factor kappa-light-chain-enhancer of activated B cells
hdcA	histidine decarboxylase A gene
PABA	para-aminobenzoic acid
CNS	central nervous system
ADHD	attention deficit hyperactivity disorder
HDC	histidine decarboxylase
GABA	gamma aminobutyric acid
ENS	enteric nervous system
IBD	inflammatory bowel disease

18.1 Introduction

In recent years there has been a significant upsurge in interest in establishing a close association between the colonization and behavioral pattern of the gut microbiota and their implications for the body's total serotonin turnover. In the neonatal phase of life, the human gut gets colonized by various microbial infestations and which ultimately leads to the formation of gut microbial flora. Commensal bacteria offer vital nutrients to the host. They metabolize indigestible chemicals, protect against opportunistic pathogen colonization, and contribute to the formation of intestinal architecture as well as immune system activation, among other things. The host, on the other hand, offers nutrition and a stable habitat for the bacteria. In a unique example of microevolution, the host and the indigenous microbes have adapted to each other in order to retain the reciprocal benefits that this mutualism provides. Because of these beneficial roles in human physiology the gut microbiota serves a great role in the maintenance of the body's hemostasis. Recent research has demonstrated that the influence of commensal gut microbiota can be traced from the body's innate immune system to metabolic function. However, some emerging research has demonstrated the potential of the gut microbiota to influence human behavior, and evidence suggests that noninvasive contact between bacteria and the host neurophysiological system is quickly expanding (Martin et al., 2018). In this connection, Berdoyet al. showed that mice infected with *Toxoplasma gondii* demonstrated manipulative behavior such as a drastic reduction in the antianxiety effect that makes the rodents courageous toward predators (Berdoy et al., 2000).

Microbiome, Immunity, Digestive Health and Nutrition. DOI: https://doi.org/10.1016/B978-0-12-822238-6.00030-3

The capacity of bacteria to generate neurohumoral compounds indicates that the microbiome's relationship with the host can extend well beyond the function of certain bacterial and host neurophysiological interactions in infection. The gut microbiota offers a plethora of chemical substances such as histamines and catecholamines, such as compounds that can activate the host neurophysiology either directly by interacting with the receptors inside the GIT or by diffusion through the gastrointestinal wall to portal circulation (Berdoy et al., 2000).

This chapter focuses on neurotransmitter-modulating activities of the gut microbiota with the help of evidence presented by scientific studies, with special emphasis on the serotonergic system. In terms of relevance to the action and level of perception, serotonin (5-hydroxytryptamine, 5-HT) is perhaps most ambivalent among the neuromodulators in the brain. Serotonergic medications are used extensively in mental diseases and are considered among the most significant etiological influences in many neurological conditions and are also used as recreational drugs (Fischer and Ullsperger, 2017). Serotonin is a foundation for the perception of various neurophysiological behavior and a key factor in numerous neurodegenerative disorders. In almost all forms of behavior, such as appetitive, motor, emotional, and cognitive, and in the autonomic nervous system (ANS), serotonin has been involved. In social activity involving aggression and anxiety and in depression the role of serotonin in behavioral repression is found to be important.

The gastrointestinal tract (GIT) has the largest pool of serotonin inside the body, and the majority of serotonin, that is, 90%–95% of the total body's serotonin, is stored and located in the enterochromaffin (EC) cells of the gut and is synthesized there with the help of tryptophan innervated through enteric nervous system (ENS). Thus the gut microbial flora is crucial in serotonin turnover in the body. The microbiota involvement results in serotonergic system regulation, and studies have found that alteration in serotonin turnover results in changes in the limbic system of germ-free animals (Fischer and Ullsperger, 2017).

Neurobehavioral references studies have found a correlation between gut microbiota and alteration of brain-derived neurotrophic factor (BDNF). In anxiety and depression BDNF acts as a key modulator. Therefore if we correlate all these, it is a fair finding that differentiation in neurotransmitter receptor expression and alteration in serotonin turnover have a deep relationship with the gastrointestinal microbial flora.

This microbiota-gut-brain axis provides new trends for prevention and treatment of many mental and physical disorders, such as depression, mania, mood disorders, anxiety disorders, schizophrenia, attention deficit hyperactivity disorder (ADHD), inflammatory bowel disease (IBD), major depressive disorder (MDD), and autism (Lin et al., 2014).

In this chapter we will evaluate and discuss the evidence that supports the role of the gut microbiota in serotonin turnover and will establish a correlation between these two and will also discuss about the implication of gut microbiota in occurrence of serotonergic system associated disorder and will try to find a rational way of solution which justifies further scientific studies.

18.2 The gut microbiota

The human GIT is one of the most fascinating parts of the body. The GIT provides one of the greatest connections between the host, environmental factors, and the body's antigens. The initial understanding of the GIT related only to the digestive system. However, with the advanced understanding of medical sciences and human physiology along with advances in scientific tools, the scientific community has gathered a plethora of information and evidence about the role of the GIT and its components in various physiological systems. The fascinating role of the gastrointestinal microbial flora or gut microbiota is a focus of research. The human GIT is home to a diverse and active group of microorganisms known as the gut microbiota. The gut microbiota is considered the "chamber of secrets" of the body, owing to its various complex roles in most of the body's physiological system homeostasis. The distribution of human gut microbes is shown in Table 18.1. In an average lifetime, approximately 60 tons of food, along with a wealth of microorganisms from the atmosphere, move through the human GIT and threaten intestinal integrity (Bengmark, 1998).

The constituents of the human gut microbiota are related to overall human health, and knowledge of different microbial species is essential for better understanding of the relationships between them. It has been calculated that the number of microorganisms living in the GIT exceeds 10^{14}, which amounts to 10 times more microbes than human cells and 100 times more genomes (microbiome) than the human genome. The gut microbiota exists as a diverse ecosystem of around 300–500 bacterial organisms, with almost two million genes (Fischer and Ullsperger, 2017). Recent scientific studies of the human gut microbiota have found that the gut microbial flora include over 35,000 species of bacteria (Bengmark, 1998). The Human Microbiome Project and the Human Intestinal Tract Metagenome research also proposes that there may be around 10 million genes in the gut microbiota when defined from a viewpoint that includes total bacteria count. Data compiled from the studies have been used to identify 2172 human-isolated species, of which 93.5% were from Proteobacteria, Firmicute, Actinobacteria, and Bacteroidetes listed as 12 different phyla. Three of the 12 phyla included only one human-isolated species, such as the intestinal species of *Akkermansia muciniphila*, the only recognized representative of the phylum Verrucomicrobia. Of the identified

TABLE 18.1 Human gut microbiota and their distribution.

Site of gastrointestinal tract harboring microbes	Nature of gut microbes			
	Gram-positive bacilli	Gram-negative bacilli	Gram-positive cocci	Gram-negative cocci
Esophagus	*Rothia* spp.	*Bacteroides, Prevotella, Pseudomonas,*	*Streptococcus, Gemella*	*Veillonella*
Stomach	*Lactobacillus*	*Helicobacter pylori, Prevotella*	*Streptococcus, Enterococcus*	
Colon	*Clostridium, Lactobacillus*	*Bacteroides, Prevotella, Porphyromonas, Fusobacteria*	*Eubacterium, Enterococcus, Ruminococcus*	
Small intestine	*Clostridium, Lactobacillus*	*Bacteroides*	*Streptococcus, Enterococcus*	*γ-Proteobacteria*
Cecum	*Roseburia*	*Fusobacteria*	*Faecalibacterium, Ruminococcus*	

species 386 species are strictly anaerobic in humans and are often mostly located in mucosal areas such as the oral mucosa or GIT (Hugon et al., 2015). The existence of country-specific microbial signatures was discovered, implying that gut microbiota composition is influenced by environmental factors such as diet as well as likely host genetics. It should be remembered, however, that microbiotas with different compositions can share some functional redundancy, resulting in similar protein or metabolite profiles.

The composition, structure, and density of the gut microbiome are influenced by chemical, dietary, and immunological gradients. Antimicrobials, oxygen, and stomach acids are also known to play a major role and can limit the bacterial development, which enables characteristics anaerobes to grow rapidly in the gut microbiota, owing to the capacity to bind with the epithelial and mucosal layers of the GIT. In rodents, Lactobacillaceae are the main dominating microbes in the small intestine microbial group, and Prevotellaceae, Lachnospiraceae, and Rikenellaceae have shown domination in the colon (Gu et al , 2013). The phyla Firmicutes and Bacteroidetes consist largely of the stable gut microbiota. The phyla Actinobacteria and Verrucomicrobia are preceded by this. While this overall profile remains consistent, the intestinal microbiota shows both temporal and spatial disparities in genus-level distribution (Gu et al., 2013). Knowledge of the bacterial distribution in gut microbiota is critical for the development of therapeutic techniques to change and shape the microbial environment in disease (Fig. 18.1).

Since the taxonomic heterogeneity within the GIT is dependent on many variables, including genetic, physiological, psychological, and environmental determinants, the gut microbiome is highly variable in healthy individuals. About the fact that each person's microbiota is distinct, scientists believe that humans share a core microbiome and that the microbiota colonizes the GIT in a similar way across the life span. The entire intestinal tract is sterile at birth, but bacteria invade the system with the first feed. The initial colonizing bacteria differ depending on the infant's food intake. Bifidobacteria make up more than 90% of the overall intestinal bacteria of breast-fed babies. *Bacteroides*, *Staphylococci*, *Lactobacilli*, and *Clostridia* are almost absent, while Enterobacteriaceae and *Enterococci* are usually present but in low quantities. *Bifidobacteria* are not common in bottle-fed babies. Bifidobacteria are supplanted by enterics, *Bacteroides*, *Enterococci*, *Lactobacilli*, and *Clostridia* as breast-fed babies are shifted to a diet of cow's milk or solid food. Also, human milk contains a growth factor that promotes the growth of bifidobacteria, which play a significant role in preventing nonindigenous or pathogenic microbes from colonizing the infant intestinal tract (Todar, 2015).

The developing embryos are exposed to bacteria in utero. Infants who are born vaginally obtain a seed of their microbiota during passage through the birth canal from maternal vaginal and possibly fecal microbes, while infants who are born via cesarean delivery get their first significant introduction of bacteria through their mother's skin and the hospital environment (Mohajeri et al., 2018). The most common species in the GIT of a stable, vaginally born infant

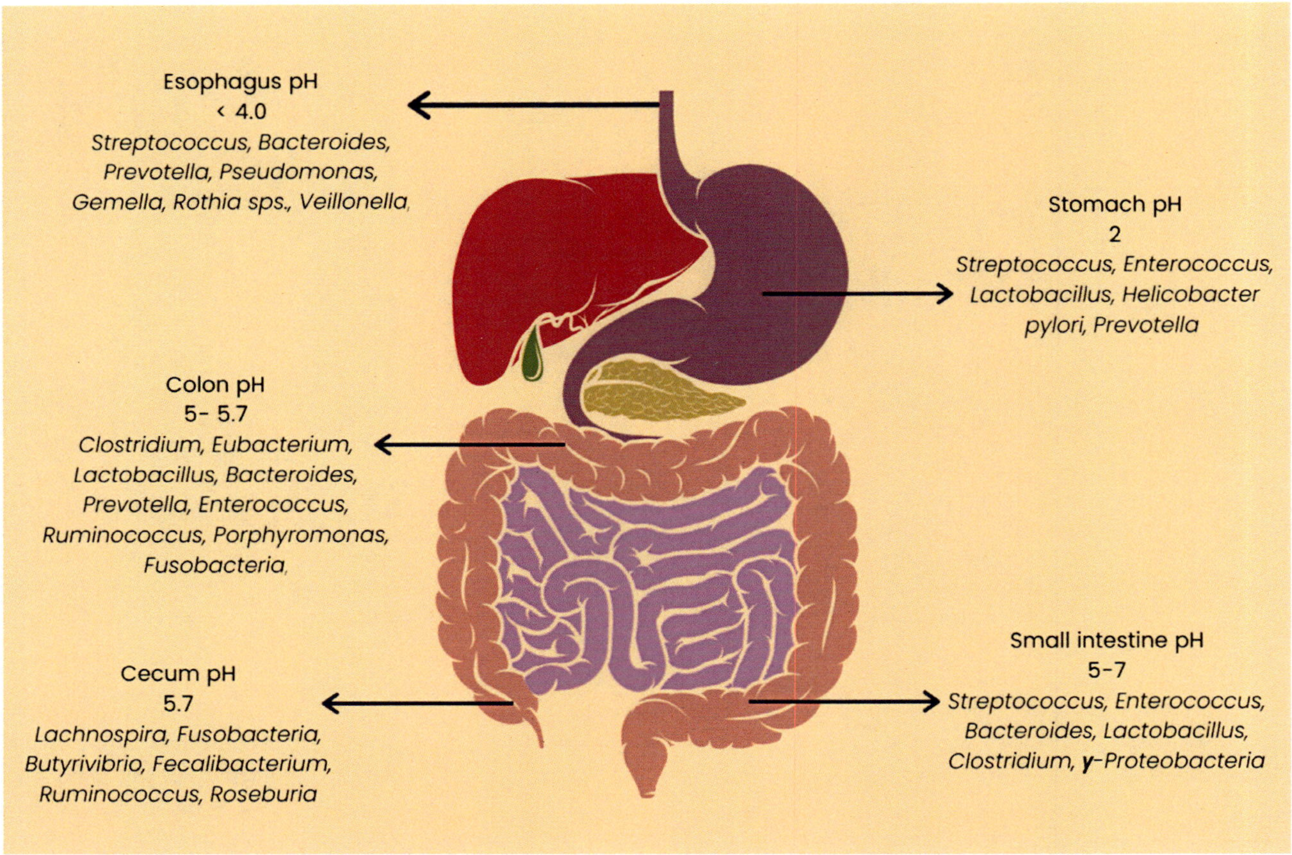

FIGURE 18.1 Distribution of human gut microbiota.

are *Bifidobacterium*, *Lactobacillus*, *Enterobacteriaceae*, and *Staphylococcus*, followed by Veillonellaceae and Lachnospiraceae. The composition of an infant's intestinal microbiota is unstable until the child is 2 years old or until the child starts eating solid food. Breast-fed babies have a different microbiome than formula-fed infants, and most infants' microbiota stabilize and evolve toward the adult microbial structure by the age of 3 (Sonnenberg and Artis, 2012). The intestinal microbiota establishes a symbiotic connection with the intestinal mucosa and provides the healthy person with considerable metabolic, immunological, and intestinal defensive functions. In the formation and maintenance of a functional immune system, an infant's gut microbiota serves an important role with the help of antigens that are present in cells, neutrophils cells, and other innate cells such as natural killer cells. The immunomodulatory contribution of the gut microbiota is seen in both the adaptive innate immune system and in key components such as gut-associated lymphoid tissues, regulatory T cells, group 3 innate T cells, insulin-like growth factor–producing beta cells, resident macrophages, and the dendritic cells that are found in the lamina propria (Sonnenberg and Artis, 2012) (Fig. 18.2).

The role of intestinal microbiota in lipid metabolism by preventing adipocytes from inhibiting lipoprotein lipase activity has been reported. It is also evidently found that *Bacteroides thetaiotaomicron* increased the efficiency of lipid hydrolysis by upregulating the expression of the colipase that is needed for lipid digestion by pancreatic lipase (Sonnenberg and Artis, 2012). *Bacteroides*, the primary species involved in carbohydrate metabolism, by exhibiting enzymes such as glycoside hydrolases, glycosyl transferases, and polysaccharide lyases, etc. The ability to ferment non-digestible fibers is an important part of the intestinal microbiota. These fibers are nutritious for the group of short-chain fatty acids (SCFAs). Propionate, acetate, and butyrate are the most essential SCFAs, and the most prevalent is acetate. Butyrate is an essential source of energy to those cells that form the internal lining of the human colon (colonocytes). In many microflorae caused issues such as the leaky gut and local inflammation, SCFA is thought to play a vital role (Støy, 2021). Butyrate has a variety of essential roles, including the capacity to cause apoptosis of carcinoma cells and modulation of the energy balance and diabetes-related gastrointestinal gluconeogenesis. Propionate is also essential for

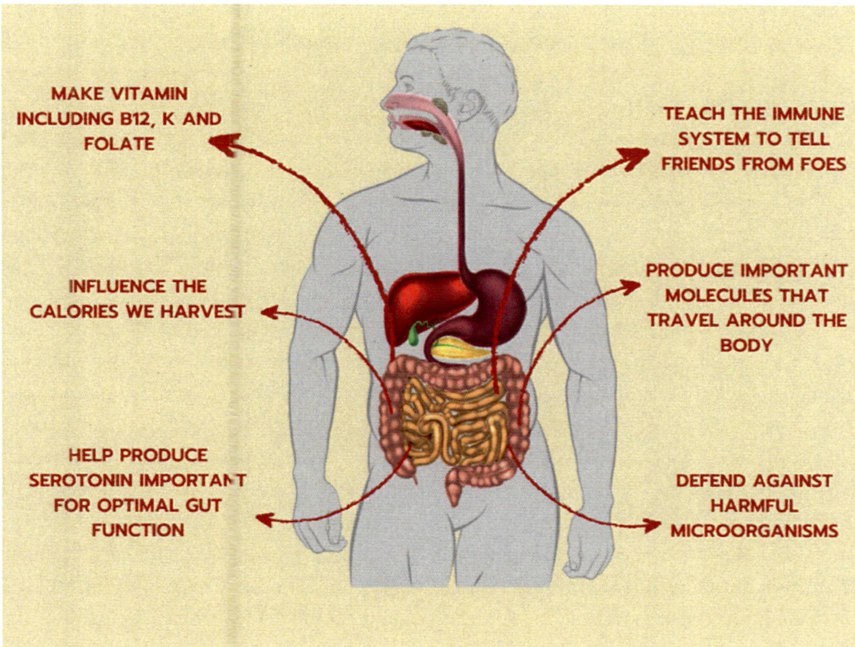

FIGURE 18.2 Function of the gut microbiota.

homeostasis of glucose because it controls liver gluconeogenesis and is involved in satiety signals. Finally, the metabolism regulation of extra gut tissues, including the metabolism of cholesterol and lipogenesis, depends on acetate (Støy, 2021).

The gut microbiota is abundant in a protein metabolizing machinery that works in parallel with human proteinases, using microbial proteinases and peptidases. Many amino acid transporters on the bacterial cell wall enable amino acids to enter the bacteria from the intestinal lumen, where they are converted into small signaling molecules and antimicrobial peptides (bacteriocins) by several gene products. For example, the bacterial enzyme histamine decarboxylase, which is encoded by the bacterial hdcA genes, converts L-histidine to histamine (Thomas et al., 2012).

It has been recently discovered that bacteria are found in newborns' amniotic fluid, placenta, and meconium; *Streptococcus mitis* and *Lactobacillus plantarum* were found in the meconium of fetuses, as well as *Staphylococci*, *Enterococci*, *Escherichia coli*, *Klebsiella pneumoniae*, and *Serratia marcescens* in the first stool of newborns, according to subsequent studies (Hu et al., 2013), which may help to understand why their microbiomes are so close after a time of adaptation.

Another significant metabolic role of the gut microbiota is the synthesis of vitamin K and other components of vitamin B. *Bacteroides* members have been shown to produce conjugated linoleic acid, which is antidiabetic, antiatherogenic, antiobesogenic, and hypolipidemic and has immunomodulatory properties.

The human gut microbiota is also engaged in the breakdown of various polyphenols (phenolic compounds) that are consumed in the diet. Secondary metabolites of polyphenolic origin can be found in a wide range of herbs, fruits, and plant-derived products (e.g., tea, cocoa, wine). Some of these metabolites are flavanones, flavan-3-ols, isoflavones, flavones, flavanols, tannins, lignans, anthocyanidins, and chlorogenic acids (Mohajeri et al., 2018). Among them, flavonoids and flavonoid subfamilies are the ones that the intestine absorbs the most. Polyphenols are sugar-bound glycosylated derivatives of glucose, galactose, ribulose, rhamnose, arabinofuranose, and arabinopyranose (Mohajeri et al., 2018).

18.3 Role of the gut microbiota in maintaining the body's normal physiological homeostasis

Homeostasis is a term derived from the Greek words for "same" and "steady," and it refers to any mechanism by which living beings are able to sustain the relatively stable conditions required for life. Walter Cannon, a dentist, coined the word in 1930. In his novel *The Wisdom of the Body* he explains how the human body retains constant temperatures and other essential factors, including the body's water, salt, sugar, protein, fat, calcium, and oxygen levels.

Diverse roles are played by the human microbiome in the maintenance and development of the body's normal and physiological homeostasis. The contribution of microorganisms of the gut microbiome is found in the formation and preservation of the immune system in the fulfillment of the body's nutritional requirement, xenobiotics metabolism in the regulation of inflammatory responses in the synthesis of neurotransmitter and in the maintenance of body's organ homeostasis (Hu et al., 2013). Recent research on the gut microbiota clearly demonstrates that this organism is also involved in the breakdown of exogenous pollutants, such as polycyclic aromatic hydrocarbons, heavy metals, plastic monomers, pesticides, ochratoxins, and organic compounds, which leads to elimination of toxin from the bloodstream through renal filtration (Hu et al., 2013).

Dietary nutrition has a major role in gut microbiota interactions with the host, microbial metabolites, and energy balance. In the gut microbiota a combination of microbial composition, metabolites, and energy supports commensal homeostasis. The host genotype, diet, and metabolites mediating this relationship all play a role in the diversity and specificity of the human gut microbiota (Thomas et al., 2017). By influencing the gut microbiota, the host genotype may affect the overall condition of the body. The genotype and richness of the gut microbiota of lean and obese people were found to vary significantly. Researchers were able to use a few bacterial markers to determine the extent of individual microbial species and genotype richness (Thomas et al., 2012). Meanwhile, early life antibiotic therapy altered the gut microbial composition and activity of genes involved in carbohydrate and lipid metabolism, as well as the levels of specific hormones, resulting in metabolic disorders (Hu et al., 2013). The gut microbiota contributes greatly in various body homeostatic pathways among which its role in our body's nutritional intake and metabolism. One of the main biochemical mechanisms used by the enteric microbiota is fermentation, which creates SCFAs such as propionate, acetate, and butyrate (Nicholson et al., 2012). SCFA promotes pathogen-competitive exclusion by stimulating protective mucus and IgA formation, promoting tolerance through regulatory T cell (Treg) induction, inhibiting the inflammatory mediator nuclear factor kappa-light-chain-enhancer of activated B cells (NF-κB), enhancing epithelial barrier integrity and repair, and inhibiting the inflammatory mediator NF-κB. However, in various models of colorectal cancer, butyrate has been shown to be both beneficial and deleterious, demonstrating the complexities of SCFA biology with a disease-specific impact. As a result, even mostly beneficial molecules or microbes may have harmful consequences in diverse ecosystems, such as the GIT (Krajmalnik-Brown et al., 2012). The production of SCFAs raises the luminal pH, which reduces mineral absorption, especially absorption of Ca^{2+} and Mg^{2+}. Furthermore, metabolites such as ammonia, phenols, indoles, N-nitroso compounds, and sulfides produced during putrefaction can increase the risk of colon cancer. Various preclinical studies have shown that bacteria in the intestines can regulate lipid, glucose, and energy homeostasis by several pathways, including improvements in intestinal obstacles and metabolic inflammation. Therefore an attractive way to deal with metabolism might be by modifying the gut microbiota composition and its metabolic function. Studies showed in a recent analysis that interactions and causal links between the variations in microbiota and the metabolic movement of the host need to be closer examined in human subject matter (Cani, 2016).

Vitamins are not generated by the human body, so they must be obtained from outside sources. In the body's vitamin requirements, the microorganism of the gut microbiome serves an important role. It not only synthesizes but also maintains the levels of various vitamins that are essential for physiological homeostasis. Vitamin K and most water-soluble B vitamins, such as biotin, nicotinic acid, pantothenic acid, cobalamin, folates, pyridoxine, riboflavin, and thiamine, are synthesized by some gut microbiota (Ichihashi et al., 1992). The colon produces the most of these vitamins.

Strains with the genes for de novo para-aminobenzoic acid (PABA) synthesis should be able to synthesize folate. Certain strains of *Bifidobacterium bifidum* and *B. longum* subsp. *infantis* contain higher amounts of folate than strains of *B. breve*, *B. longum* subsp. *longum*, and *B. adolescentis*. Except in the presence of PABA, *B. animalis* subsp. *lactis* is an auxotroph for 6-hydroxymethyl-7,8-dihydropterin pyrophosphate and does not contain folate (Bäckhed et al., 2015). In newborns a diverse variety of microorganisms have the capacity to synthesize folate. Biosynthesis of pyridoxal (vitamin B6) and biotin (vitamin B7) has shown similar findings. Microorganisms that contain thiamine (vitamin B1), pantothenate (vitamin B5), and cobalamin (vitamin B12) were found in lower numbers in younger children, but their cell numbers increased as they grew older. Newborns have a healthy microbiota and strong vitamin B12 transport mechanism, but this declines with age (Bäckhed et al., 2015). Therefore it is evident that vaginally delivered babies have a tangible benefit of vitamin production. Given the importance of vitamin K2 in bone and heart development, it is reasonable to assume that vaginally delivered babies are healthier than newborns delivered by cesarean birth. The structure of the microbiota in the gut determines the growth of the adaptive immune system; hence the mammalian immune system, which is in charge of regulating microorganisms, is influenced by the human microbiota (Rojo et al., 2017). The proinflammatory cells and Tregs in the intestine are activated by the gut microbiota. Furthermore, the human microbiota has a major impact on innate immunity.

One example is neutrophil aging, which reduces the body's proinflammatory properties. Toll-like receptors (TLRs) and MYD88-mediated signaling mechanisms are used by these cells to promote neutrophil ageing. In models of sickle cell disease or endotoxin-induced septic shock, microbial changes cause older neutrophils to circulate less, resulting in inflammation-related organ damage. As a result, these species aggressively regulate disease-promoting neutrophils, which are needed in the development of inflammatory diseases (Rojo et al., 2017). In addition, the microbiota in the intestine aids in the defense against pathogenic species. It encourages the development of antimicrobial compounds to combat invading pathogens and resist colonization. For instance, healthy gut microbiome antibodies (CD8-T and cD4) that respond to influenza virus infection in the respiratory tract may be responsible for controlling antibodies. Thus not only does the gut microbiota take part in the formation of body immunity, but it also contributes hugely to its maintenance.

During a disease state, the gut microbiota may alter, and this generally leads to characteristic changes in the body's normal physiology and affects homeostasis of the gut. This kind of disturbance is often seen during antibiotic therapy, and it is also associated with antibiotic resistance, which is a big problem. With so many organisms found in the GIT soon after birth, it is not shocking to see a number of genes coding for antibiotic resistance in a newborn's microbiome. One cause is the large number of Proteobacteria, which are believed to be naturally antibiotic-resistant. Early-colonizing bacteria have genes encoding susceptibility to tetracycline, bacitracin, and macrolides. Antibiotic resistance is also present in the microbiota in the adult GIT, implying that antibiotic resistance develops early in existence. Antibiotic resistance has been shown to be higher in the microbiome of babies who were delivered by cesarean birth (Ichihashi et al., 1992).

Diet, age, diet, stress, and diseases all influence the relative abundance and diversity of bacterial species in the GIT and other parts of the body. Dysbiosis, or a chronic disorder in the gut's microbial population, has been discovered in animal models and humans. Dysbiosis is caused by a shift in the population of microbes, with harmful bacteria outnumbering others that are more beneficial. For mutual advantage, host and microbial populations have developed strategies that foster a steady state. Dysbiosis is a microbial species abundance deficit that is attributed to compromised gut barrier activity and inflammatory cell activation (Chang and Lin, 2016).

The commensal microbiota can contribute to inflammatory responses unless homeostatic conditions in the GIT are disrupted. The secretion of proinflammatory cytokines will cause further damage to the epithelial barrier and host tissues (Turroni et al., 2014). The commensal microbial culture must be preserved by oral probiotic strains to restore intestinal homeostasis. Extracellular probiotic molecules (i.e., microbe-associated molecular patterns) secreted by probiotic cells interact with intestinal epithelial cells and are recognized by pattern recognition receptors. Peptidoglycan, lipoteichoic acids, bacteriocins, and fimbriae/pili complexes are examples of extracellular probiotic molecules (Rojo et al., 2017).

Lactobacillus sp. can recover gut homeostasis and reverse microbial dysbiosis. The secretion of IL-12 and IL-23 by dendritic cells involved in IBD was reduced by peptidoglycan from *L. casei* (Littman and Pamer, 2011).

By fostering food absorption, xenobiotic metabolism, and the regulation of innate and adaptive immune responses, gut microbiomes play a permanent role. Many cell-signaling pathways are triggered by proteins, peptides, and metabolites released locally and at distant sites. The host-microbial homeostasis is maintained by this intense cross-talk.

18.4 Influence of neurotransmitters on the gut microbiome population through the gut-brain axis

The gut-brain axis comprises the central nervous system (CNS), the neuroendocrine and neuroimmune systems, including the hypothalamic-pituitary-adrenal (HPA) axis, sympathetic and parasympathetic arms of the ANS, including the ENS, the vagus nerve, and the gut microbiota (Wang and Kasper, 2014). In the early 21st century, considerable evidence was accumulated to show the connection between the gut microbiome and the human brain. Research shows that the gut microbiota and various gastrointestinal disorders are the root cause of alteration in the stress level of various animal species. The importance of the gut microbiota in the maintenance of neuronal homeostasis can be referred to the studies performed by Sudo et al., (2004) that germ-free mice have higher stress hormone responses and lower hippocampal levels of BDNF, a protein that is essential for neurogenesis (Wang and Kasper, 2014). Currently, the most widely researched host patterns in microbiota-gut-brain studies have been anxietal improvements in explorational motivation and risk avoidance. Through early research into microbial regulation of anxiety-related behavior showed improvisation in stress hormones and the BDNF level of the brain, it appeared that the modulation of neurotransmitters may cause microbial alteration in host behavior. When contemplating how the microbiota could communicate with the

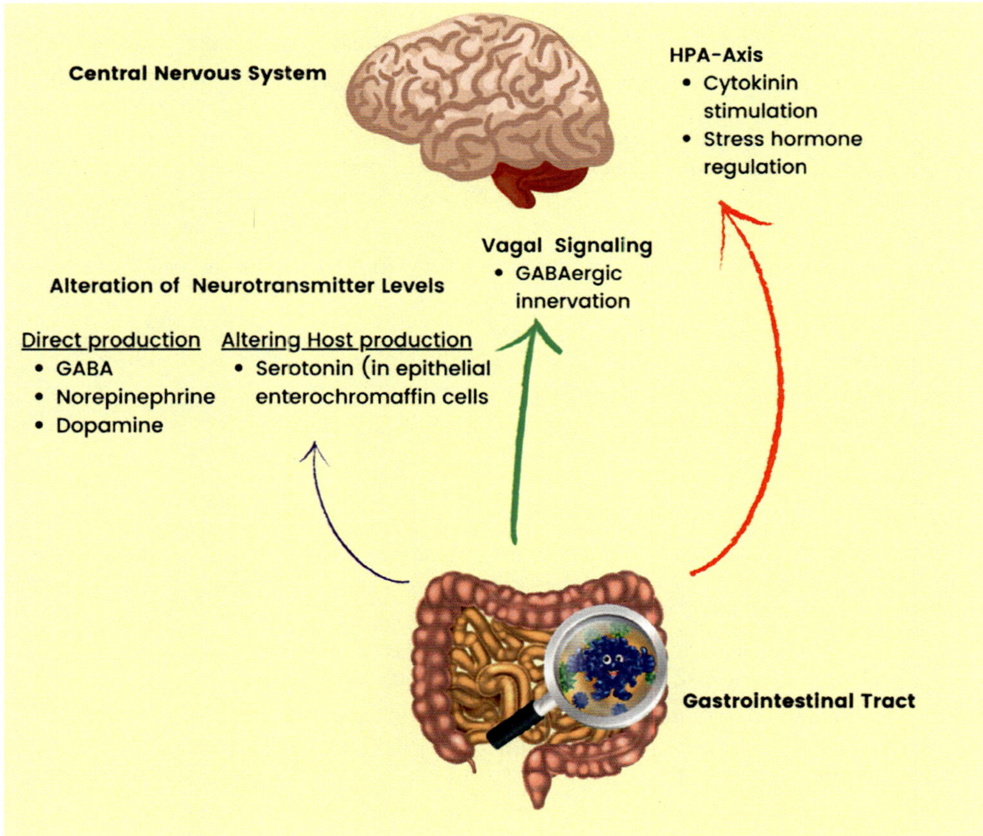

FIGURE 18.3 Communication route of the gut-brain axis.

nervous system, regulation of host neurotransmitters and/or associated pathways is probably the most obvious scenario. Bacteria have been discovered to be capable of producing a variety of major neurotransmitters. The gut-brain axis is a bidirectional neurohumoral communication mechanism that regulates homeostasis through the CNS and the ENS as well as neurological, endocrine, immune, and metabolic pathways, including the HPA (Littman and Pamer, 2011).

There are various routes of communication between the gut microbiota and the brain (Fig. 18.3). These include neurotransmitter synthesis or regulation of host neurotransmitter catabolism, vagal innervation, and HPA axis activation (Strandwitz, 2018). Notably, although Fig. 18.3 shows how the gut microbiota can interact with the brain, similar processes tend to be used to communicate with the ENS.

18.4.1 Role of biogenic amines in the gut microbiota

Changes in the gut microbiota can influence dopamine (DA), norepinephrine, and serotonin to affect the brain structures in different rodent stress models to exhibit anxiety and depression-like behaviors. The vagus nerve plays an important role in establishing gut-brain signaling. Commensal bacteria cell wall components that translocate to the brain, bacterial metabolites such as SCFAs (Strandwitz, 2018; Wang and Kasper, 2014) that have been linked to microglial activation, and neurotransmitters formed by gut microbes such as acetylcholine (Kawashima and Fujii, 2008), histamine, gamma aminobutyric acid (GABA), serotonin, noradrenaline, and DA are all involved. These neurotransmitters seem to play a key role in microbial ecology, complicating our understanding of how intestinal microbes interact with the CNS. However, there is good evidence that gut symbionts are essential for CNS function and that changes in their structure, diversity, and richness can play a role in the pathophysiology of neuropsychiatric disorders (Kawashima and Fujii, 2008).

Biogenic amines are low-molecular-weight organic nitrogen compounds. They are formed by the decarboxylation of amino acids or by the amination and transamination of aldehydes and ketones during normal metabolic processes in living cells and therefore are ubiquitous in animals, plants, microorganisms, and humans. Epinephrine, norepinephrine,

DA, serotonin, and histamine are five members of this class of amine that have been shown to act as neurotransmitters. Generally, these amines are used in the peripheral nervous system and the CNS to regulate mood, cognitive functions, gut mortality, and other physiological functions. Additions to that accumulated evidence suggest that these amines could also serve as essential signaling molecules between the commensal microbiota and the host intestine (Hughes and Sperandio, 2008).

18.4.1.1 Role of catecholamines in the gut microbiota

Dopamine is a precursor of other catecholamines, including norepinephrine and epinephrine, and is one of the most important neurotransmitters in reward-motivated actions. Norepinephrine has long been associated with arousal and alertness in the waking state as well as sensory signal recognition, but new research has discovered that it also plays a role in actions and processing, including memory, attention, motivation, reward, locomotor activity, neuroendocrine regulation, ingestion of water, learning, and focus (Borodovitsyna et al., 2017). Although it has not been shown that the microbiota modulates norepinephrine or DA in vivo, there is growing evidence that it can, or at the very least may play a role in host biosynthesis or catabolism. Despite demonstrating that the DA is a modulation factor in many gastrointestinal functional activities, its origin has not been identified. Studies reveal that the mesenteric organs (spleen, pancreas, GIT) are responsible for extensive DA production. Tyrosine hydroxylase, a novel nonneuronal dopaminergic pathway in the GIT, suggests the existence of this development and cellular distribution of the rate limiting enzyme for DA synthesis (Kawashima and Fujii, 2008). Dopaminergic signaling changes have been linked to extreme CNS symptoms including anxiety, ADHD, Parkinson's disease, and eating disorders. The HPA axis, the immune system, and the vagus nerve are all involved in the contact between intestinal microbes and the CNS. Studies demonstrate that stimulation of vagal afferent fibers from the upper GIT induces DA release throughout the brain (Kawashima and Fujii, 2008). DA is released in the brain as vagal afferent fibers from the upper intestinal tract are stimulated. Furthermore, activation of the right nodose ganglia rather than the left means that the right vagus nerve is implicated in nutrient sensitivity as well as gut proprioception. Further research suggests that food satiation and reward are linked by central DA release, which is mediated by vagal stimulation (Kawashima and Fujii, 2008). Finally, the propensity for DA metabolite microbial synthesis is positively linked to mental health quality, likely connecting the gut microbiome's neuroactive metabolic capability with mental well-being.

Norepinephrine or noradrenaline is an organic chemical compound that belongs to the family of catecholamine functions as a neurotransmitter and hormone. The release of norepinephrine is lowest during sleep and increases during waking. It achieves a much higher level in the so-called fight-or-flight response in situations of stress and risk. In the brain, norepinephrine increases excitement, encourages alertness, stimulates memory formation and restoration, and increases concentration and restlessness. In the remaining parts of the body, norepinephrine raises the heart rate and blood pressure, causes glucose release, increases the blood supply to the skeletal muscles, decreases blood flow to the digestive tract, and inhibits bladder voiding and gastrointestinal motility. In the pathophysiology of depression and anxiety the role of norepinephrine has been known for a long time, and satiation has also been found to be controlled by norepinephrine (Asarian and Bächler, 2014). There are several bacteria that have been shown to produce DA and norepinephrine (Table 18.2). *Proteus vulgaris*, *E. coli*, *Bacillus subtilis*, *S. marcescens*, and *Bacillus mycoides* all have fairly high levels of norepinephrine (0.45–2.13 mM) in their biomass in vitro (Asarian and Bächler, 2014; Asano et al., 2012). Norepinephrine tends to be formed as a quorum-sensing molecule in bacteria, but the synthesis of DA is still unknown. In the case of norepinephrine a recent study using germ-free mice determined that mice without bacteria had significantly lower levels of norepinephrine in the cecal lumen (35 ± 5 ng/g vs 3.8 ± 1.3 ng/g) and tissue (115 ± 14 ng/g vs 5.0 ± 0.5 ng/g) and that cecal levels of norepinephrine could be restored through colonization by a microbiota (Asano et al., 2012). This study showed that there is an influence of gut microbiota on the norepinephrine level in the GIT lumen of mice. But it is still not clear that bacteria produces norepinephrine to alter species behavior. The evidence that is available on norepinephrine in depression and anxiety-like behavior is very limited, but it suggests that Norepinephrine can be influenced by gut microbiota.

18.4.1.2 Role of 5-hydroxytryptamine or serotonin in the gut microbiota

Serotonin is a hormone and excitatory neurotransmitter that is released in massive amounts in the GIT, although it is best known for its central role in anxiety and depression. Gut bacteria produce almost 95% of the body's serotonin, which affects mood as well as gastrointestinal activity. Serotonin regulates gastrointestinal secretion and peristalsis, respiration, vasoconstriction, behavior, and neurological activity (Berger et al., 2009), among other physiological processes. Although serotonin is found in the body, it is mainly found throughout epithelial ECs in the GIT, where it

TABLE 18.2 Neurotransmitter-producing bacteria.

Dopamine	*Escherichia coli* (Hughes and Sperandio, 2008)
	Proteus vulgaris (Hughes and Sperandio, 2008)
	Bacillus mycoides (Hughes and Sperandio, 2008)
	Staphylococcus aureus (Hughes and Sperandio, 2008)
	E. coli (K-12) (Strandwitz, 2018)
Norepinephrine	*B. mycoides* (Hughes and Sperandio, 2008)
	P. vulgaris (Hughes and Sperandio, 2008)
	E. coli (K-12) (Strandwitz, 2018)
	Serratia marcescens (Asarian and Bächler, 2014)
	Bacillus subtilis (Asarian and Bächler, 2014)
Serotonin	*E. coli* (K-12) (Strandwitz, 2018)
	Klebsiella pneumoniae (NCIMB, 673) (Strandwitz, 2018)
	Morganella morganii (NCIMB, 10466) (Strandwitz, 2018)
	Hafnia alvei (NCIMB, 11999) (Strandwitz, 2018)
	Streptococcus thermophilus (NCFB2392) (Strandwitz, 2018)
Acetylcholine	*Lactobacillus plantarum* (Stanaszek et al., 1977)
Histamine	*M. morganii* (Stanaszek et al., 1977)
	E. coli (Stanaszek et al., 1977)
	P. vulgaris (Stanaszek et al., 1977)
	Raoultella planticola (Stanaszek et al., 1977)
	Citrobacter freundii (Stanaszek et al., 1977)
GABA	*Bifidobacterium adolescentis* (DPC6044) (Azbill et al., 2000)
	Bifidobacterium dentium (DPC6333) (Azbill et al., 2000)
	Lactobacillus brevis (DPC6108) (Azbill et al., 2000)
	Lactobacillus rhamnosus (YS9) (Azbill et al., 2000)
	Monascus purpureus (CCRC 31615) (Azbill et al., 2000)
Glutamate	*B. subtilis* (Stanaszek et al., 1977)

accounts for 90%−95% of serotonin (Asano et al., 2012; Berger et al., 2009). It has not been investigated whether its microbial effects on the peripheral serotonergic system were comparable to those found in the central serotonergic system.

The majority of the serotonin that is present in the gut is synthesized from tryptophan that is present in the ECs of the GIT. They are also present in the ENS. The synthetic cascade is the same regardless of where it is in the gut-brain axis. The rate-limiting enzyme, tryptophan hydroxylase, converts tryptophan to 5-hydroxytryptophan (5-HTP), which is not saturated at usual tryptophan concentrations (Fig. 18.4). Increased tryptophan concentrations can lead to increased metabolic activity. Following aromatic amino acid decarboxylase is involved in the conversion of the short-lived 5-HTP intermediate product to serotonin (Berger et al., 2009). The brain and the gut have a bidirectional coordination network, and the serotonin mechanism plays an important part in the processes and behavior of these two main organs. Disorders of the brain-gut axis, which manifest with dysfunctional connectivity between the brain and the gut and similar signs, are associated with changes in the serotonin system (Martin et al., 2018). Microbiota-deficient germ-free animals are raised in a sterile environment. These species can provide unique insights into the gut-brain axis components

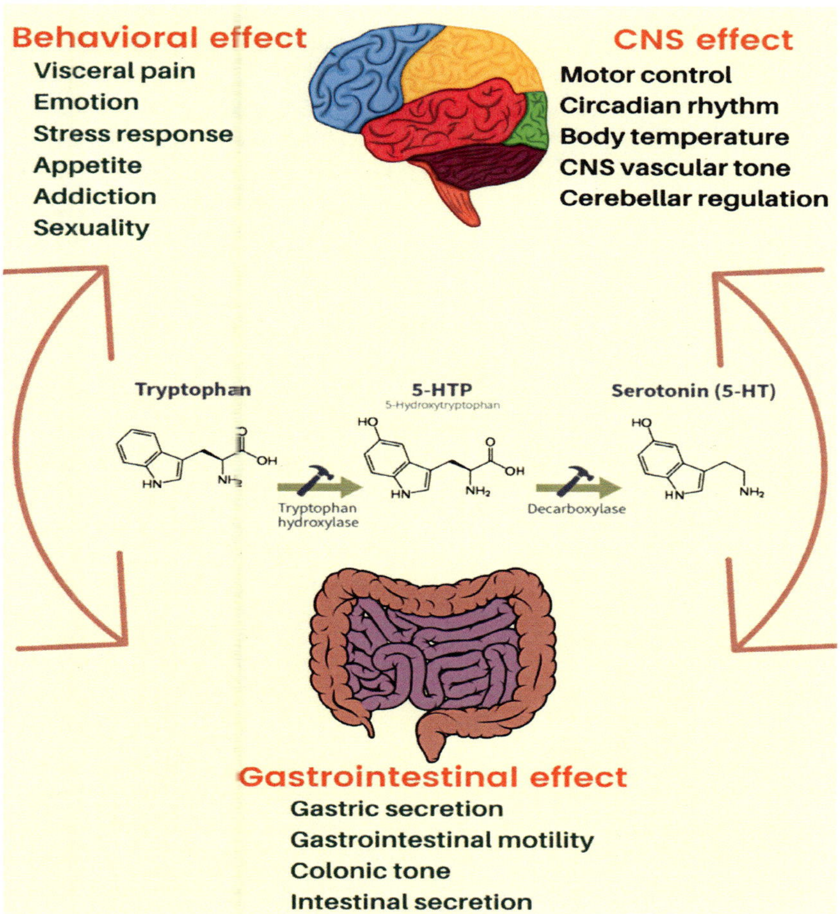

FIGURE 18.4 The brain-gut axis and serotonergic metabolism.

that are influenced by the microbiota. They have recently been used to show that the gut microbiota is essential for normal brain growth and behavior and that it can affect tryptophan and serotonin levels.

The proper functioning of the brain-gut-microbiome axis, which determines an individual's quality of life, is dependent on adequate serotonin signaling during one's life to nurture the development, operation, and maintenance of the axis. In addition, a series of preclinical laboratory techniques have shown that the structure of the gut microbiota has a significant impact on both tryptophan abundance and serotonin signaling. This raises a successful treatment approach to the interesting prospect of clinical targeting of the gut microbiota for serotonin-related brain-gut axis.

18.4.1.3 Role of histidine derivative on gut microbiota

Histamine is a nitrogenous organic compound that is implicated in local immune responses and regulates the physiological activity of the gut and acts as a neurotransmitter in the nervous system (Berger et al., 2009). Histamine is formed in cells not only by the enzyme histidine decarboxylase (HDC), but also by microbial decarboxylation of amino acids (Stratton et al., 1991). A variety of bacterial stresses have been identified in addition to mammalian stresses as causing secretion of histamine after decarboxylated of histidine by the action of HDC. This kind of production is clinically significant, since microorganism-produced histamine also causes scombroid poisoning, which happens after consumption of amine-contaminated food. The bacterial species that have HDC activity and that have been implicated in histamine fish poisoning are listed in Table 18.2. Under physiological conditions, commensal microorganisms in the gut can produce histamine and related compounds, which indicate an important role in luminal histamine in the immunoregulation of the gut. Recent studies have shown that histamine can also exert an antiinflammatory effect on the host by suppressing IL-18 production in the gut (Levy et al., 2015). While the existence of histamine-secreting bacteria in foods has been well established, their presence in the human gut microbiota has not been thoroughly examined. Recent studies have used primers that do not amplify human HDC to analyze fecal samples from 161 volunteers to measure the prevalence of bacterial HDC. Microbial-specific HDC was found in all feces samples, but there was a wide interindividual variance. Those feces have the bacteria species *E. coli*, *Morganella morganii*, and *Limosilactobacillus vaginalis* secretes the highest amount of histamine (Wessler et al., 2001). Given histamine signaling's powerful immunological impacts, it's not shocking that histamine from the microbiota may affect immune responses as well. In vitro, cultures that were supernatant to histamine-producing bacteria could promote the use of H1R reporters in CHO cells. Bacterial derived histamine specific H1R activation has been reported as action were blocked by H1R antagonist diphenhydramine (Wessler et al., 2001). Furthermore, it is found that *Lactobacillus reuteri* produce histamine has inhibitory activity in TLRs induced TNF-α production using H2R signaling which reduces PKA and subsequent cAMP activity (Levy et al., 2015; Wessler et al., 2001). TLRs from the microorganism-sensing PRR family are vital in maintaining the gut microbiota-host symbiotic relationship and intestinal homeostasis. Administering the *Lactobacillus* strain 30A secretive of histamine to mice resulted in quick loss of weight and decreased cytokine secretion of Peyer's patches. These figures indicate that histamine-driven intestinal microbes can have immunological effects in the mucosa, but the existence of such effects is based on the stimulated histamine receptor. In a separate research analysis a bacterium and its H2R activation were shown to rely on the elimination of the inflammation of the body by *L. reuteri* strains (Gominak, 2016).

Histamine is clearly present in the gastrointestinal mucosa, and elevated amounts have been linked to a variety of mucosal inflammatory disorders. The microbiota's role in the gut's histamine content and histamine receptor activation is currently unexplained. Although histamine was discovered more than a century ago, there are still significant gaps in our understanding of its immunoregulatory function, especially in the GIT.

18.4.2 Role of esters in the gut microbiota

18.4.2.1 Role of quaternary ammonium esters in the gut microbiota

Acetylcholine is an ester of acetyl and choline and it is the most abundant neurotransmitter in the body which is found both in the CNS and the Peripheral nervous system. Acetylcholine and the enzyme choline acetyltransferase have been found in epithelial, mesothelial, endothelial, muscle, and immune cells, mostly in granulocytes, lymphocytes, macrophages, and mast cells (Barcik et al., 2016). Acetylcholine is a neurotransmitter that influences our alertness throughout the day and the natural completion of rapid eye movement sleep at night. Acetylcholine is also referred to as a microbial neurometabolite because bacteria can produce a wide range of neurotransmitters and neuromodulators and acetylcholine is produced by *Lactobacillus* sp., as is shown in Table 18.2. The gut microbiota also provides the host with essential B vitamins (Rakoff-Nahoum et al., 2004). Vitamins B5 and B12, which are synthesized only by the intestinal microbiota, serve as coenzymes in a wide variety of host biochemical processes, including the synthesis of acetylcholine and cortisol, all of which are essential for normal nervous system operation. Vitamin B5 and B12 deficiency has been attributed to gastrointestinal pain, insomnia, and neuropsychological and hematological conditions (Rakoff-Nahoum et al., 2004). However, no correlation has been established between the lack of vitamin-producing gut microbiota and the onset of disease. Acetylcholine plays a major role in production of these vitamins, so these are indirectly linked to the level of acetylcholine in the body. Thus changes in acetylcholine in the human body physiology may lead to nerve disorders.

18.4.2.2 Role of glutamic acid ester in the gut microbiota

L-glutamate is the most abundant excitatory neurotransmitter in the CNS (Conn, 2003). Ionotropic receptors, which open ion channels to mediate rapid synaptic responses, and metabotropic (mGlu) receptors, which are coupled to G proteins and modulate signal transduction cascades, are two types of glutamate receptors. mGlu receptors are mediated by neurons and glia in the CNS, where they are found near the synaptic cleft. Consequently, mGlu receptors will control the release of glutamate and other neurotransmitters in presynaptic and heterosynaptic localizations as well as the impact of glutamate in postsynaptic neurons and accessory cells (Conn, 2003). The neurotransmitter glutamate is one of the most important chemical signals sent between the brain and the gut as well as through the brain-gut axis (Conn, 2003). Since glutamate does not cross the blood-brain barrier, it must be synthesized locally for use in the CNS. The synthesis, release, and reuptake of glutamate into neurons and glial cells influence synaptic glutamate levels. The enzyme glutamate synthase (glutamate-oxoglutarate amidotransferase) synthesizes glutamate in the gram-positive soil bacterium *B. subtilis* (see Table 18.2) by reductive amination of alpha-ketoglutarate (Mckenna, 2007). The vesicular glutamate transporter VGlut then transports it into synaptic vesicles. As an action potential reaches the synaptic membrane, the synaptic vesicles fuse, and glutamate is released into the synaptic cleft. Sodium-dependent transporters (excitatory amino acid transporters) clear most glutamate from the synaptic cleft (Azbill et al., 2000). This has led to an increased drug discovery effort in developing selective mGlu receptor ligands for a variety of neurological and psychiatric disorders. Group III mGlu receptors, which are often found presynaptically, regulate glutamate release and hence neuronal excitability. As a result, these receptors can play a role in the control of anxiety levels.

18.4.3 Role of miscellaneous neurotransmitters in the gut microbiota

18.4.3.1 Role of gamma-aminobutyric acid derivatives in the gut microbiota

GABA is a nonprotein amino acid and a primary regulator of the CNS that is found in plants, mammals, and microorganisms. A large body of evidence links altered GABAergic neurotransmission to a variety of CNS conditions, such as behavioral disorders, pain, and sleep disturbances, as well as the dysfunction of essential ENS functions like intestinal motility, gastric emptying, nociception, and acid secretion (Azbill et al., 2000). GABA is produced by irreversible α-decarboxylation of L-glutamate and the consumption of one cytoplasmic proton by a pyridoxal-5′-phosphate–dependent glutamate decarboxylase enzyme. From various studies, it is mostly found that it is present in the hippocampus, where it acts as a major inhibitory neurotransmitter in the mammalian CNS (Barrett et al., 2012; Conn, 2003).

Recent research has shown that the capacity of the intestinal microbiota or consumed probiotics, such as bifidobacteria and lactobacilli, to metabolize dietary monosodium glutamate is responsible for the elevated amount of GABA throughout the human gut (Barrett et al., 2012). In vitro experiments revealed that mostly three *Bifidobacteria* species, namely, *Bifidobacterium dentium*, *B. longum* subsp. *infantis*, and *B. adolescentis*, and many more shown in Table 18.2 synthesize GABA. The complex relationship of GABA and gut microbiota are involved in human health and diseases still remain unknown but GABA levels can be affected by manipulating the human microbiota.

18.5 Association between the microbiome-gut-brain axis and serotonin turnover

An understanding of the brain-gut coordination has shown a nuanced communication mechanism that not only ensures that gastrointestinal homeostasis is kept properly maintained but also serves various neurological functions like motor coordination, motivation, reward, encouragement and other higher cognitive functions. This complex interaction is known as the gut-brain axis. The gut-brain axis consists of the CNS, neuroendocrine and neuroimmune systems, which includes the HPA axis, sympathetic and parasympathetic arm of ANS which includes the ENS, vagus nerve and gut microbiota (Wang and Kasper, 2014). Gut microbiomes communicate with the CNS through at least three parallel and interconnected channels involving nervous, endocrine, and immune signaling. This center is not only tasked with monitoring an integration of gastrointestinal functions also to connect emotional and cognitive brain centers along with peripheral intestinal functions and mechanisms including immune homeostasis, intestinal reflex and intestinal endocrine signaling. Neuro-immuno-endocrine mediators are part of the pathways that drive gut-brain axis communication (Rhee et al., 2009), which means that both hormonal and neuronal communications allow the brain to influence various activities of the functional gastrointestinal effector cells such as epithelial cells, smooth muscle cells, immune cells, enteric neurons, gut cells of EC, and kajal cells, and these cells also greatly influence the gut microbiota, the function of which has been recently evaluated as mutual brain-gut communication.

The generation of neurons in the CNS is influenced by different environmental factors during development, and the host microbiota also exhibits complex variation in composition during brain maturation (Rhee et al., 2009). Previous research indicates that the permeation of the maternal-fetal system causes gut bacteria regulators to activate TLR2, which facilitates fetal neuronal growth and may have an influence on cognitive function in adulthood (Rhee et al., 2009; Stanaszek et al., 1977). Previous research has also shown that gut microorganisms play a part in modulating and guiding the developmental progress of neurogenesis in the CNS, with this complex activity taking place primarily in the hippocampus (Dinan and Cryan, 2017).

The ENS is composed of motor neurons, sensory neurons, and interneurons placed along the walls of the gastrointestinal system, stretching all the way from the lower third of the esophagus to the rectum. The ENS connects the brain's cognitive and emotional centers with peripheral intestinal functions through direct and indirect pathways. The gut microbiome greatly affects the way in which the ENS to connect with the gut-brain axis. Various animal model studies have shown that rodents with dysbiosis and/or those treated with antibiotics show drastic results in terms of neuronal signaling, which can be modulated by using probiotics or cultural colonization (Rhee et al., 2009). Approaching the microbiome from the perspective of microbial endocrinology can contribute to a better understanding of the specific mechanisms through which microorganisms influence behavior and hence to new treatments for specific mental illnesses based on regulation of the microbiome-gut-brain axis (Lyte, 2013).

In exploring how the microbes can communicate with the nervous system, the most apparent scenario will probably be manipulation of host neurotransmitters and/or similar forms. In general, a number of important neurotransmitters are found to be produced by bacteria.

Bacterial colonization studies of germ-free animal gut have shown that the colonization of microbes is critical for the growth and maturation of both the CNS and the ENS (Lyte, 2013). Owing to the lack of microbial colonization, the expression and turnover of neurotransmitters has been impaired, as well as changes in gut sensory motor function. But the GIT also contains trillions of microbes, which have been discovered to be an important component of the body's serotonin turnover. The immune system and the vagus nerve are two of the several signaling pathways that are found along the gut-brain axis (Strandwitz, 2018).

The brain-gut axis is a two-way coordination mechanism that connects the brain and the GIT. At both ends of the terminal network, serotonin serves as a primary neurotransmitter. As a result, it has been linked to a variety of neurodegenerative disorders, including Alzheimer's disease, anxiety, depression, and psychosis (Kim and Camilleri, 2000). The GIT contains about 95% of the serotonin in the body, with a gross serotonin content of around 10 mg; 90% is present in EC cells and 10% in enteric neurons. The hippocampus contains the remaining 5% of serotonin. The GIT produces almost half of the serotonin in the blood (Bertaccini, 1960). Enteric nerves include serotonin, which is synthesized in the ENS (Bertaccini, 1960). Serotonergic neurons account for about 2% of all myenteric neurons. When enteric nerves are activated, serotonin is released from the gut (Bertaccini, 1960).

The enzyme tryptophan hydroxylase 2 (TpH2) produces the neurotransmitter serotonin in the brain, where it plays an important role. In the intestine, specifically ECs, a specialized subtype of intestinal epithelial cell, produce the majority of the body's serotonin. The enzyme tryptophan hydroxylase 1 (TpH1) is involved in this process, as it creates 5-HTP, which is then metabolized into serotonin (Agus et al., 2018).

Intestinal serotonin activity is heavily influenced by the gut microbiota, as has been shown in germ-free mice with impaired serotonin activity in the colon (but not in the small intestine) and low serotonin concentrations in the blood (Agus et al., 2018; Bertaccini, 1960). The mechanisms by which the gut microbiota regulates serotonin activity remain unknown, but SCFAs (i.e., butyrate and acetate), directly functioning on ECs, have been implicated in the stimulation of TpH1 expression (Fukumoto et al., 2003). SCFAs have been shown to influence the transcription of the Chga gene, which codes for the neuroendocrine secretory protein chromogranin A, which is released in combination with serotonin (Bertaccini, 1960). Furthermore, some secondary bile acids, such as deoxycholate, which is formed by microbial cholate biotransformation, can induce serotonin biosynthesis. In the body's serotonin turnover, the gut microbiota shows the greatest crucial role, alteration of which can lead to the increase or decrease in serotonin level, which subsequently results in the stimulation or inhibition of the serotonergic system. The first direct evidence about blood serotonin origination from the intestine was established by Toh. Studies performed on dogs showed that the level of serotonin is almost three times higher in portal circulation than in arterial blood (Agus et al., 2018). The relationship between serotonin and the gut microbiome in the context of gut motility in germ-free mice was first shown by Abrams and Bishop in 1967 (Fukumoto et al., 2003). The experiment showed that the animals that were excluded from having live microbes in the body (germ free) had comparatively reduced gut motility in comparison with mice that had normal gut microbiota. These results were later explained by Wikoff et al., who established that the absence of microbes in the gut microbiota results in reduction in the serotonin level of the body (Toh, 1954).

In 1959, Erspamer and Testini (Erspamer and Testini, 1959) performed an experiment on three groups of rabbit, dogs, and humans. These experiments strengthened the hypothesis about the gut origination of the body's serotonin production. This study helps greatly in understanding the serotonin turnover rate in the gastrointestinal mucosa. To understand the study, a set of data from their experiment can be borrowed for the discussion. In that experiment they anesthetized two dogs (DOG 1 and DOG 2) with pentobarbitone for the removal of their entire GIT, spleen, and pancreas. Before the procedure, arterial blood was sampled for serum 5-HT estimation, and estimation of 5-HIAA was done by collecting urine for a 24-hours period. After the operation, they gave sufficient means to stabilize the dog and started the postoperative sample collection. DOG 1 died after 48 hours, and DOG 2 died after 25 hours. Blood samples were collected during the survival period of 20 hours and 40 hours, and their results are shown in Table 18.3.

Interpretation of the results shows a marked decrease in serum 5-HT and urinary 5-HIAA after the operation in both dogs, which supports the production of 5-HT by ECs of the GIT (Abrams and Bishop, 1967). A similar observation is found in the experiment of Hata et al., who compared the levels of conjugated and free serotonin in the gut lumen of ex-germ-free and germ-free mice and showed an increase in total 5-HT, that is, 563 ± 259 ng/g in ex-germ-free mice compared to a total serotonin level of 252 ± 89.0 ng/g in germ-free mice (Wikoff et al., 2009). These two studies show a direct relationship between the EC cells of the gut microbiota of the intestine and serotonin turnover. It is also worth

TABLE 18.3 Alteration in serum 5-HT levels and urinary 5-HIAA excretion in dogs after removal of the gastrointestinal tract (Abrams and Bishop, 1967).

	Serum 5-HT (μg/mL)	Urine volume (mL/24 h)	Urinary 5-HIAA (μg/24 h)
DOG 1 (12 kg)	0.42	736	674
Preoperation	0.47	230	170
0 h–24 h after operation	0.30	400	<40
DOG 2 (14.5 kg)			
Preoperation	0.20	1013	860
0 h–25 h after operation	0.23	200	180

mentioning that evidence has been compiled from various studies regarding the factors that affect this serotonin turnover.

The study of serotonin turnover can lead to the understanding of different clinical scenarios. Unmedicated patients with MDD have higher brain serotonin turnover, which is affected by the 5-HTT genotype. High brain serotonin turnover can be a biochemical substrate of MDD, as shown by the marked decrease in serotonin turnover following therapy with selective serotonin reuptake inhibitors and the associated change in symptoms (Fukumoto et al., 2003; Toh, 1954).

It is also evident that some serotonin antagonists can be also used in chemotherapy-induced emesis and some serotonin agonists can be used as prokinetic drugs. Prokinetic receptors ultimately degranulate enterochromaffil granulation. They release and ultimately sensitize the serotonergic nerve, and the sensation moves to the chemoreceptor trigger zone. Serotonin will work as a nonspecific bioamine. This will stimulate DA receptors and also stimulate 5HT3 receptors and bind with it. It will also stimulate dopaminergic receptors. In this process, practically emesis occurs; stimulation will lead to serotonin release and concurrent release of neuropeptide Y, ghrelin, and leptin/leptin. These releases further control absorption of saccharin and peristaltic movement. Gut microbial flora or gut microbes or *Enterobacter* influence the amount of release of ghrelin or leptin, which indirectly influence this pathway (Abrams and Bishop, 1967; Wikoff et al., 2009) (Fig. 18.5).

18.6 Dysregulation in the occurrence of gut microbiota and its implications in diverse pathological conditions

In recent decades, because of the rapid advancement of sequence-based screening and the human gnotobiotic model in commensal bacteria, interest in the human gut microbiome has increased rapidly. Although this field remains less understood, research has provided insight into possibilities for transforming diagnostic and treatment procedures. Researchers are elucidating the role of the gut microbiota and its specific metabolites in providing the host with defense against invading pathogens as well as control of diverse host physiological functions, such as metabolism, development of immunity, and nervous system homeostasis. Dysbiosis of the gut microbiota may cause the host machinery to malfunction, leading to pathogenesis and/or development of a wide range of diseases, including *Clostridium difficile* infection (an infectious disease), celiac disease (a multisystemic autoimmune disorder), IBD (an intestinal immune-mediated disease), obesity (a metabolic disease), and autism spectrum disorder (a neuropsychiatric disorder) (Erspamer and Testini, 1959). Many novel therapies originating from microbiome experiments to target related illnesses, such as fecal microbiota transplantation and probiotics and prebiotics for targeting associated disease, have been reviewed to incorporate the concept of how certain disease effects can be alleviated by dysbiosis reversal, revealing a new scientific method of disease management.

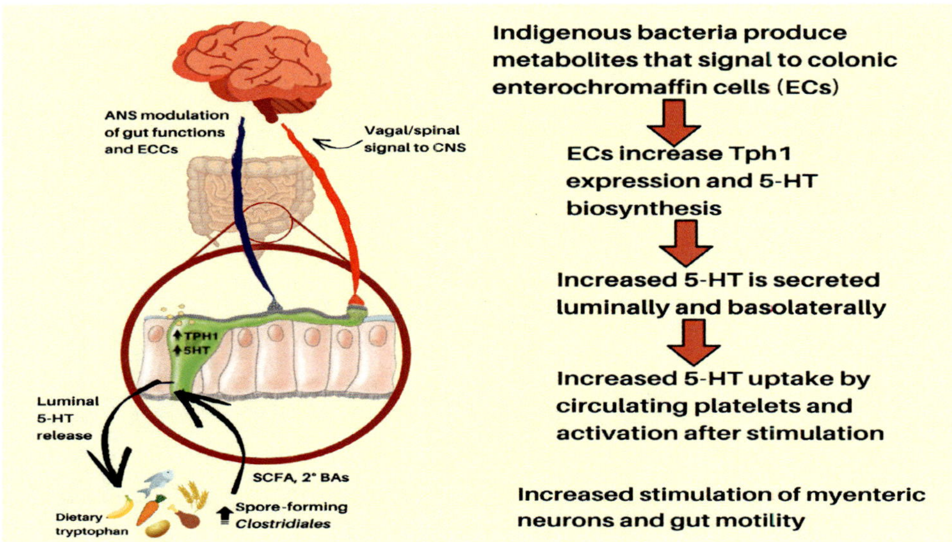

FIGURE 18.5 Brain-gut microbiome interaction and its influence on serotonin turnover.

The brain is not only the organ that is responsible for Parkinson's disease. Studies have shown that comorbid gastrointestinal dysfunction is common in Parkinson's patients and that about 80% of them suffer from constipation (Erspamer and Testini, 1959). Furthermore, intestinal microbes such as *Bacillus* spp. are known to produce DA, and the gut microbiota produces almost half of the body's DA. The level of DA in the brain is found to be twice as high in germ-free mice as in control mice (Hata et al., 2017). The germ-free mice showed that the locomotor activities were increased in comparison to control mice (Kho and Lal, 2018). Similarly, germ-free rats had a reduced amount of DA turnover in the striatum, hippocampus, and frontal cortex in comparison to control rats (Kho and Lal, 2018). Interestingly, the gut-microbiota-brain axis regulates the synthesis and inhibition of enzymes that convert tyrosine to DA, such as tyrosine hydroxylase and DOPA decarboxylase (Kho and Lal, 2018; Matsumoto et al., 2013). As a result, it is possible that commensal microbes regulate brain DA levels, which are reduced in Parkinson's disease. Furthermore, the gut microbiota regulates brain tyrosine levels; germ-free mice had lower tyrosine levels than control mice (Kho and Lal, 2018). The fact that DA synthesis in the brain is caused by DA-generating enzymes, whose synthesis and inhibition are mediated by the tut microbiota through the gut-microbiota-brain axis, suggests the involvement of the gut microbiota in Parkinson's disease.

Reid and Rand, (1951) were among the first to recognize the function of serotonin in the central control of breathing. They discovered that injecting serotonin intravenously induces apnea in cats. According to a follow-up report, low levels of serotonin, when injected into the carotid sinus area, caused brief apnea, accompanied by respiratory relaxation and a decrease in blood pressure. Serotonin is also essential in other breathing-related autonomic circuits. The serotonin regulation of CO_2-pH homeostasis is particularly intriguing. Serotonin is also essential for carotid body chemoreceptors' arterial oxygen sensing (Diaz Heijtz et al., 2011). Excess serotonin reduces ventilatory responses to hypoxia in MAOA-KO mice, implying that excessive serotonin inhibits the central respiratory network's responsiveness to hypoxia (Diaz Heijtz et al., 2011).

Serotonin interacts with the innate and adaptive immune system. During acute inflammation, serotonin recruits innate immune cells (immature dendritic cells, monocytes, mast cells, eosinophils, etc.) (Reid and Rand, 1951). In mice with colitis induced by sodium dextran sulfate, administration of serotonin increases the release of proinflammatory cytokines from macrophages (Prabhakar et al., 2006). In mice with collagen-induced arthritis, a lack of serotonin can cause an imbalance of T helper (Th) 17 and Tregs and can exacerbate the development of the disease (Chabbi-Achengli et al., 2016). The 5-HT1A receptor is more strongly expressed in CD4 + T cells in patients with multiple sclerosis; these cells produce more serotonin than the cells of healthy individuals (Montero et al., 2004). Therefore serotonin is closely related to immune cells in autoimmune diseases. As serotonin has been linked to the gut microbiota, it is evident that the gut microbiome has an impact in autoimmune diseases.

Serotonin is a naturally occurring vasoactive agent with a wide range of cardiovascular and physiologic effects. Serotonin's vasoconstrictive effects are mediated by 5-HT2 serotonergic receptors, and serotonin also enhances the release and actions of other vasoconstrictors like angiotensin and norepinephrine. Selective 5-HT2 serotonergic receptor blockers have been developed, and they tend to be effective vasodilators with therapeutic potential in a variety of cardiovascular diseases (Chabbi-Achengli et al., 2016; Montero et al., 2004).

Alzheimer's disease, the most prevalent cause of dementia, is a neurodegenerative condition that is characterized by diminished memory and amyloid-β peptide (Aβ) formation in the brain. After intranasal infection with *Chlamydia pneumoniae* that had been isolated from the brain of a deceased person with Alzheimer's disease, mice developed amyloid accumulation in their brains, suggesting that infection may play a role in the early pathogenesis of the disease. Research has shown that many bacterial species, including *B. subtilis*, *E. coli*, *Salmonella enterica*, *Salmonella typhimurium*, *Staphylococcus aureus*, and *Mycobacterium tuberculosis*, may produce large amounts of functional amyloid, which may lead to the pathology of Alzheimer's disease by accumulating proteinaceous misfolded Aβ oligomers and fibrils (Prabhakar et al., 2006). From all these scientific studies, it is evident that the gut microbiota plays an important role in the pathogenesis of Alzheimer's disease.

Dysbiosis of the gastrointestinal microbiota is linked to the pathogenesis of both intestinal and extraintestinal disorders. Irritable bowel syndrome (IBS), celiac disease, and IBD are examples of intestinal conditions; allergies, asthma, metabolic syndrome, cardiovascular disease, and obesity are examples of extraintestinal disorders.

The coexistence of the gut microbiota and the host shows the gut microbial flora's critical position in host wellbeing, and maintaining the gut microbiota's balance is essential for the host gut and overall systemic physiology. Dysbiosis is described as a shift in the steady-state structural composition of the gut microbiota that can disrupt microbial equilibrium and is associated with a variety of gut pathologies and intestinal inflammation (Buttó and Haller, 2016). Dysbiosis, or an alteration in the composition and/or function of the gut microbiota, is thought to be a cause of the breakdown of the microbe-host homeostatic relationship. Crohn's disease and ulcerative colitis are two forms of IBD, all of which are correlated with severe morbidity. Various mouse models of intestinal inflammation have presented strong support for the gut microbiota playing a central causal function in the formation of IBD. Antiinflammatory and Treg cell responses are induced by bacteria from the *Clostridia* and *Bacteroides* genera (Montero et al., 2004). The balance of proinflammatory and antiinflammatory cytokines generated by Th17 and Treg cells is essential for the host's intestinal homeostasis and is directly influenced by the normal gut microbiota. Evidence from segmented filamentous bacteria in mice has shown a high level of T helper 1 (Th1) and T helper 17 (Th17) (Buttó and Haller, 2016) proinflammatory cytokine aggregation. The discovery of microbial-derived SCFAs that can induce the expansion of Treg cells in mice (Buttó and Haller, 2016) and the isolation of 17 human clostridia organisms opens new therapeutic opportunities for the treatment of IBD.

Extraintestinal diseases have been linked to intestinal microbial dysbiosis, especially diseases that affect the gut-brain axis, affecting the CNS, behavior, and cognitive function. Several studies have focused on the hypothesis that the intestinal microbiota influences cognitive function and behavior by direct reprogramming of the HPA axis, which is triggered in response to infection and disrupted by psychological stressors. Enteric infections are known to induce fear, depression, and cognitive dysfunction; in contrast to conventional mice, germ-free mice with no intestinal microbiota show changes in stress responsiveness, central neurochemistry, and behavior, indicating a decrease in anxiety (Ivanov et al., 2009). Changes in the synthesis of neurotrophic factors and hormones, as well as the expression of their receptors, have been linked to increased anxiety-like behavior in germ-free mice (Buttó and Haller, 2016; Ivanov et al., 2009). *Campylobacter jejuni* (a bacterium that causes gastroenteritis) can cause anxiety-like behavior and brain stem stimulation (the nucleus tractus solitarius and lateral parabrachial nucleus) in pathogen-infected mice (Olivares et al., 2015). GABA, which may specifically affect immune and neural receptors in the ENS and CNS, can be influenced by commensal bacteria (Gaykema et al., 2004).

A maternal immune activation (MIA) mouse model in which pregnant animals receives the viral mimetic poly(I:C) show increased intestinal permeability and develop stereotypical anomalies in behavior, social ability, and communication that mimic autism spectrum disorder (ASD). These results suggest that the intestinal microbiota may play a causal role in the development of ASD (Hsiao et al., 2013). In comparison to control infants, MIA infants have intestinal dysbiosis and a changed serum metabolomic profile, which is characterized by elevated amounts of microbiota-derived 4-ethylphenyl sulfate (4EPS). The intestinal barrier function was restored and ASD-like symptoms were alleviated after administration of probiotic bacteria. Exogenously administered 4EPS, which is structurally like the toxic sulfated form of p-cresol, caused anxiety-like behavior in naive mice, implying that the GIT plays a role in autism and perhaps in other behavioral disorders by influencing the immune, metabolic, and nervous systems (Hsiao et al., 2013).

It is perhaps no surprise that GIT conditions such as IBD and IBS are frequent comorbidities in debilitating stress-related disorders like depression and anxiety, given the evolving preclinical evidence and indications in developmental disorders. Recent research suggests that intestinal permeability and bacterial translocation can play a role in the pathophysiology of depression by driving immune inflammatory and oxidative and nitrosative stress pathways. Chronic depression in humans has been linked to an improvement in the immune response (serum IgM and IgA responses) to lipopolysaccharide products of gram-negative bacteria gut enterobacteria, such as *Hafnia alvei*, *Pseudomonas aeruginosa*, *M. morganii*, *Pseudomonas putida*, *Citrobacter koseri*, and *K. pneumoniae* (Maes et al., 2012). In germ-free and pathogen-infected mice, attempts have been made to investigate the possible CNS and behavioral effects of bacteriotherapy. When compared to specific-pathogen free mice, germ-free mice show hyperresponsive HPA axis function in response to stress, and this hyperresponse of the HPA axis was reversed by *Bifidobacterium infantis* (Hsiao et al., 2013). *B. infantis* raised plasma tryptophan levels while decreasing serotonin metabolite concentrations in the frontal cortex and DA metabolite concentrations in the amygdaloid cortex, the two factors linked to depression. The effectiveness of probiotics for mood control in humans was indicated in a study of *Lactobacillus casei*, which found that participants with the lowest scores in the depressed/elated axis improved significantly in mood after taking the probiotic relative to the placebo sample (Benton et al., 2007). In healthy human volunteers, a mixture of *L. helveticus* and *B. longum* lowered anxiety and had positive psychological effects as well as lowering serum cortisol levels.

18.7 Conclusion

In this review, we have discussed the intricate and bidirectional interactions between the gut microbiota and the brain through several visceral organs and possible contributions by the dysregulated gut microbiota in the occurrence of serotonergic system–associated pathological conditions.

The mammalian brain cells elicit significant functional differences in individuals with normal and altered or absent gut microbiota. These include diminution in the difference in voltage between the resting and threshold potential in the cell of the main brain stem, focal changes in the hemisphere, differences in the expression of various ligand receptors for the interaction with peripheral and central neurotransmitters, and alterations in the body's total turnover of certain biogenic amines, including serotonin. Experimental results in the recent past have revealed a paradigm shift in our understanding of the brain-gut axis with special attention on serotonin, but exactly how the spore-forming gut microbiota interacts with EC granules of the ENS is not clearly known. Moreover, it is not clear how gender variation affects the influence of the gut microbiota on hippocampal and peripheral levels of serotonin. Data from preclinical and clinical studies have paved the way for research in the gray areas of the gut-microbiota-brain axis and offer remarkable potential for novel treatment targets not only in functional gastrointestinal disorders but in a wide range of neuropsychiatric disorders.

References

Abrams, G.D., Bishop, J.E., 1967. Effect of the normal microbial flora on gastrointestinal motility. Exp. Biol. Med. 126 (1), 301–304.

Agus, A., Planchais, J., Sokol, H., 2018. Gut microbiota regulation of tryptophan metabolism in health and disease. Cell Host Microbe 23 (6), 716–724.

Asano, Y., Hiramoto, T., Nishino, R., Aiba, Y., Kimura, T., Yoshihara, K., et al., 2012. Critical role of gut microbiota in the production of biologically active, free catecholamines in the gut lumen of mice. Am. J. PhysiolGastrointest. Liver Physiol. 303 (11), G1288–G1295.

Asarian, L., Bächler, T., 2014. Neuroendocrine control of satiation. Hormone Mol. Biol. Clin. Invest. 19 (3), 163–192.

Azbill, R., Mu, X., Springer, J., 2000. Riluzole increases high-affinity glutamate uptake in rat spinal cord synaptosomes. Brain Res. 871 (2), 175–180.

Bäckhed, F., Roswall, J., Peng, Y., Feng, Q., Jia, H., Kovatcheva-Datchary, P., et al., 2015. Dynamics and Stabilization of the human gut microbiome during the first year of life. Cell Host Microbe 17 (5), 690–703.

Barcik, W., Pugin, B., Westermann, P., Perez, N.R., Ferstl, R., Wawrzyniak, M., et al., 2016. Histamine-secreting microbes are increased in the gut of adult asthma patients. J. Allergy Clin. Immunol. 138 (5), 1491–1494.e7.

Barrett, E., Ross, R., O'toole, P., Fitzgerald, G., Stanton, C., 2012. γ-Aminobutyric acid production by culturable bacteria from the human intestine. J. ApplMicrobiol. 113 (2), 411–417.

Bengmark, S., 1998. Ecological control of the gastrointestinal tract. Role Probiotic Flora Gut. 42 (1), 2–7.

Benton, D., Williams, C., Brown, A., 2007. Impact of consuming a milk drink containing a probiotic on mood and cognition. Eur. J. Clin. Nutr. 61 (3), 355–361.

Berdoy, M., Webster, J.P., Macdonald, D.W., 2000. Fatal attraction in rats infected with Toxoplasma gondii. Proc. R. Soc. Lond. B 267 (1452), 1591–1594.

Berger, M., Gray, J.A., Roth, B.L., 2009. The expanded biology of serotonin. Annu. Rev. Med. 60, 355–366.

Bertaccini, G., 1960. Tissue 5-hydroxytryptamine and urinary 5-hydroxyindoleacetic acid after partial or total removal of the gastrointestinal tract in the rat. J. Physiol. (Lond.) 153, 239–249.

Borodovitsyna, O., Flamini, M., Chandler, D., 2017. Noradrenergic modulation of cognition in health and disease. Neural Plast. 2017, 6031478.

Buttó, L.F., Haller, D., 2016. Dysbiosis in intestinal inflammation: cause or consequence. Int. J. Med. Microbiol. 306 (5), 302–309.

Cani, P.D., 2016. Changes in gut microbes and host metabolism: squaring the circle. Nat. Rev. Gastroenterol. Hepatol. 13 (10), 563–564.

Chabbi-Achengli, Y., Coman, T., Collet, C., Callebert, J., Corcelli, M., Lin, H., et al., 2016. Serotonin Is involved in autoimmune arthritis through Th17 immunity and bone resorption. Am. J. Pathol. 186 (4), 927–937.

Chang, C., Lin, H., 2016. Dysbiosis in gastrointestinal disorders. Best. Pract. Res. Clin. Gastroenterol. 30, 3–15.

Conn, P.J., 2003. Physiological roles and therapeutic potential of metabotropic glutamate receptors. Ann. N. Y. Acad. Sci. 1003 (1), 12–21.

Diaz Heijtz, R., Wang, S., Anuar, F., Qian, Y., Bjorkholm, B., Samuelsson, A., et al., 2011. Normal gut microbiota modulates brain development and behavior. Proc. Natl. Acad. Sci. USA 108 (7), 3047–3052.

Dinan, T.G., Cryan, J.F., 2017. Gut instincts: microbiota as a key regulator of brain development, ageing and neurodegeneration. J. Physiol. 595 (2), 489–503.

Erspamer, V., Testini, A., 1959. Observations on the release and turnover rate of 5-hydroxytryptamine in the gastrointestinal tract. J. Pharm. Pharmacol. 11 (1), 618–623.

Fischer, A.G., Ullsperger, M., 2017. An update on the role of serotonin and its interplay with dopamine for reward. Front. Hum. Neurosci. 11, 1–6.

Fukumoto, S., Tatewaki, M., Yamada, T., Fujimiya, M., Mantyh, C., Voss, M., et al., 2003. Short-chain fatty acids stimulate colonic transit via intraluminal 5-HT release in rats. Am. J. Physiol.-Regul. Integr. Comp. Physiol. 284 (5), R1269–R1276.

Gaykema, R., Goehler, L.E., Lyte, M., 2004. Brain response to cecal infection with Campylobacter jejuni: analysis with Fos immunohistochemistry. Brain Behav. Immun. 18, 238–245.

Gominak, S., 2016. Vitamin D deficiency changes the intestinal microbiome reducing B vitamin production in the gut. The resulting lack of pantothenic acid adversely affects the immune system, producing a "pro-inflammatory" state associated with atherosclerosis and autoimmunity. Med. Hypotheses 94, 103–107.

Gu, S., Chen, D., Zhang, J., Lv, X., Wang, K., Duan, L., et al., 2013. Bacterial community mapping of the mouse gastrointestinal tract. PLoS One 8 (10), e74957.

Hata, T., Asano, Y., Yoshihara, K., Kimura-Todani, T., Miyata, N., Zhang, X., et al., 2017. Regulation of gut luminal serotonin by commensal microbiota in mice. PLoS One 12 (7), e0180745.

Hsiao, E.Y., Mcbride, S.W., Hsien, S., Sharon, G., Hyde, E.R., Mccue, T., et al., 2013. Microbiota modulate behavioral and physiological abnormalities associated with neurodevelopmental disorders. Cell 155 (7), 1451–1463.

Hu, J., Nomura, Y., Bashir, A., Fernandez-Hernandez, H., Itzkowitz, S., Pei, Z., et al., 2013. Diversified microbiota of meconium is affected by maternal diabetes status. PLoS One 8 (11), e78257.

Hughes, D.T., Sperandio, V., 2008. Inter-kingdom signalling: communication between bacteria and their hosts. Nat. Rev. Microbiol. 6 (2), 111–120.

Hugon, P., Dufour, J., Colson, P., Fournier, P., Sallah, K., Raoult, D., 2015. A comprehensive repertoire of prokaryotic species identified in human beings. Lancet Infect. Dis. 15 (10), 1211–1219.

Ichihashi, T., Takagishi, Y., Uchida, K., Yamada, H., 1992. Colonic absorption of menaquinone-4 and menaquinone-9 in rats. J. Nutr. 122 (3), 506–512.

Ivanov, I.I., Atarashi, K., Manel, N., Brodie, E.L., Shima, T., Karaoz, U., et al., 2009. Induction of intestinal Th17 cells by segmented filamentous bacteria. Cell 139 (3), 485–498.

Kawashima, K., Fujii, T., 2008. Basic and clinical aspects of non-neuronal acetylcholine: overview of non-neuronal cholinergic systems and their biological significance. J. Pharmacol. Sci. 106 (2), 167–173.

Kho, Z.Y., Lal, S.K., 2018. The human gut microbiome—a potential controller of wellness and disease. Front. Microbiol. 9.

Kim, D., Camilleri, M., 2000. Serotonin: a mediator of the brain–gut connection. Am. J. Gastroenterol. 95 (10), 2698–2709.

Krajmalnik-Brown, R., Ilhan, Z., Kang, D., Dibaise, J.K., 2012. Effects of gut microbes on nutrient absorption and energy regulation. Nutr. Clin. Pract. 27 (2), 201–214.

Levy, M., Thaiss, C.A., Zeevi, D., Dohnalová, L., Zilberman-Schapira, G., Mahdi, J.A., et al., 2015. Microbiota-modulated metabolites shape the intestinal microenvironment by regulating NLRP6 inflammasome signaling. Cell 163 (6), 1428–1443.

Lin, S., Lee, L., Yang, Y.K., 2014. Serotonin and mental disorders: a concise review on molecular neuroimaging evidence. Clin. Psychopharmacol. Neurosci. 12 (3), 196–202.

Littman, D.R., Pamer, E.G., 2011. Role of the commensal microbiota in normal and pathogenic host immune responses. Cell Host Microbe 10 (4), 311–323.

Lyte, M., 2013. Microbial endocrinology in the microbiome-gut-brain axis: how bacterial production and utilization of neurochemicals influence behavior. PLoS Pathog. 9 (11), e1003726.

Maes, M., Kubera, M., Leunis, J.C., Berk, M., 2012. Increased IgA and IgM responses against gut commensals in chronic depression: further evidence for increased bacterial translocation or leaky gut. J. Affect. Disord. 141 (1), 55–62.

Martin, C.R., Osadchiy, V., Kalani, A., Mayer, E.A., 2018. The brain-gut-microbiome axis. Cell. Mol. Gastroenterol. Hepatol. 6 (2), 133–148.

Matsumoto, M., Kibe, R., Ooga, T., Aiba, Y., Sawaki, E., Koga, Y., et al., 2013. Cerebral low-molecular metabolites influenced by intestinal microbiota: a pilot study. Front. Syst. Neurosci. 7, 9.

Mckenna, M.C., 2007. The glutamate-glutamine cycle is not stoichiometric: Fates of glutamate in brain. J. Neurosci. Res. 85 (15), 3347–3358.

Mohajeri, M.H., La Fata, G., Steinert, R.E., Weber, P., 2018. Relationship between the gut microbiome and brain function. Nutr. Rev. 76 (7), 481–496.

Montero, E., Nussbaum, G., Kaye, J.F., Perez, R., Lage, A., Ben-Nun, A., et al., 2004. Regulation of experimental autoimmune encephalomyelitis by CD4 + , CD25 + and CD8 + T cells: analysis using depleting antibodies. J. Autoimmun. 23 (1), 1–7.

Nicholson, J.K., Holmes, E., Kinross, J., Burcelin, R., Gibson, G., Jia, W., et al., 2012. Host-gut microbiota metabolic interactions. Science 336 (6086), 1262–1267.

Olivares, M., Neef, A., Castillejo, G., Palma, G.D., Varea, V., Capilla, A., et al., 2015. The HLA-DQ2 genotype selects for early intestinal microbiota composition in infants at high risk of developing coeliac disease. Gut 64 (3), 406–417.

Prabhakar, N.R., Peng, Y.J., Yuan, G., Kumar, G.K., 2006. Reactive oxygen species facilitate oxygen sensing. Novartis Found. Symp. 272, 95–99. discussion 100-5, 131-40.

Rakoff-Nahoum, S., Paglino, J., Eslami-Varzaneh, F., Edberg, S., Medzhitov, R., 2004. Recognition of commensal microflora by toll-like receptors is required for intestinal homeostasis. Cell 118 (2), 229–241.

Reid, G., Rand, M., 1951. Physiological actions of the partially purified serum vasoconstrictor (serotonin). Aust. J. Exp. Biol. Med. Sci. 29 (6), 401–415.

Rhee, S.H., Pothoulakis, C., Mayer, E.A., 2009. Principles and clinical implications of the brain–gut–enteric microbiota axis. Nat. Rev. Gastroenterol. Hepatol. 6 (5), 306–314.

Rojo, D., Méndez-García, C., Raczkowska, B.A., Bargiela, R., Moya, A., Ferrer, M., et al., 2017. Exploring the human microbiome from multiple perspectives: factors altering its composition and function. FEMS Microbiol. Rev. 41 (4), 453–478.

Sonnenberg, G.F., Artis, D., 2012. Innate lymphoid cell interactions with microbiota: implications for intestinal health and disease. Immunity 37 (4), 601–610.

Stanaszek, P.M., Snell, J.F., O'neill, J.J., 1977. Isolation, extraction, and measurement of acetylcholine from Lactobacillus plantarum. Appl. Env. Microbiol. 34 (2), 237–239.

Støy S. 2021. Role of gut microbiome: composition and factors: read here [Internet]. Biomcare. [cited 2021 April 10]. Available from: <https://biomcare.com/life-science/gut-microbiome/>

Strandwitz, P., 2018. Neurotransmitter modulation by the gut microbiota. Brain Res. 1693. 128–133.

Stratton, J.E., Hutkins, R.W., Taylor, S.L., 1991. Biogenic amines in cheese and other fermented foods: a review. J. Food Prot. 54 (6), 460–470.

Sudo, N., Chida, Y., Aiba, Y., Sonoda, J., Oyama, N., Yu, X.-N., Kubo, C., Koga, Y., 2004. Postnatal microbial colonization programs the hypothalamic-pituitary-adrenal system for stress response in mice. J Physiol 558 (Pt. 1), 263–275. Available from: https://doi.org/10.1113/jphysiol.2004.063388. 15133062.

Thomas, C.M., Hong, T., Van Pijkeren, J.P., Hemarajata, P., Trinh, D.V., Hu, W., et al., 2012. Histamine derived from probiotic lactobacillus reuteri suppresses TNF via modulation of PKA and ERK signaling. PLoS One 7 (2), e31951.

Thomas, S., Izard, J., Walsh, E., Batich, K., Chongsathidkiet, P., Clarke, G., et al., 2017. The host microbiome regulates and maintains human health: a primer and perspective for non-microbiologists. Cancer Res. 77 (8), 1783–1812.

Todar K., 2015. [Internet]. The Normal Bacterial Flora of Humans. University of Wisconsin; [cited 2021 April 10]. Available from: <http://www.textbookofbacteriology.net/normalflora_3.html>.

Toh, C.C., 1954. Release of 5-hydroxytryptamine (serotonin) from the dog's gastro-intestinal tract. J. Physiol. 126 (2), 248–254.

Turroni, F., Ventura, M., Buttó, L.F., Duranti, S., O'toole, P.W., Motherway, M.O., et al., 2014. Molecular dialogue between the human gut microbiota and the host: a Lactobacillus and Bifidobacterium perspective. Cell Mol. Life Sci. 71 (2), 183–203.

Wang, Y., Kasper, L.H., 2014. The role of microbiome in central nervous system disorders. Brain Behav. Immun. 38, 1–12.

Wessler, I., Kilbinger, H., Bittinger, F., Kirkpatrick, C.J., 2001. The non-neuronal cholinergic system. Jpn. J. Pharmacol. 85 (1), 2–10.

Wikoff, W.R., Anfora, A.T., Liu, J., Schultz, P.G., Lesley, S.A., Peters, E.C., et al., 2009. Metabolomics analysis reveals large effects of gut microflora on mammalian blood metabolites. Proc. Natl Acad. Sci. 106 (10), 3698–3703.

Chapter 19

The connection between diet, gut microbes, and cognitive decline

Abhai Kumar[1], Smita Singh[2] and Rameshwar Nath Chaurasia[1]

[1]*Department of Neurology, Institute of Medical Sciences, Banaras Hindu University, Varanasi, India,* [2]*Department of Geriatric Medicine, Institute of Medical Sciences, Banaras Hindu University, Varanasi, India*

19.1 Introduction

Diet and cognition are strongly associated, as it is well documented that a diet that is high in fat and sugar has a detrimental effect on cognition. The disturbance in gut-brain axis physiology causes abnormalities in cognition. Depression, memory impairment, decreased planning and problem-solving abilities, and decreased mental flexibility resulted from intake of a poor diet. Research studies using a mouse model have reported that such disturbances are the result of changes in the hippocampus and frontal lobe (Castanon et al., 2014). Neurite growth in the brain is regulated by brain-derived neurotropic factors (BDNF), synapsin, and CREB which was found downregulated after long-term consumption of a high-fat diet (Molteni et al., 2002; Wu et al., 2004). The low expression of these proteins influences the synaptic plasticity of neurons, thereby affecting learning and memory (Wu et al., 2004). Reactive oxygen species were found to be elevated after consumption of a diet that was rich in fat (Molteni et al., 2004; Pipatpiboon et al., 2012); these lead to a reduction in neurotropic factors. Reactive oxygen species in the brain result in a decrease in spatial memory measured by the Morris water maze test, reduced dendritic spine density (mainly in the hippocampus, perirhinal cortex, and prefrontal cortex), and reduced long-term potentiation in CA1 synapses (Bocarsly et al., 2015; Pintana et al., 2013; Sripetchwandee et al., 2014; Stranahan et al., 2008). The high-fat diet also results in lipid peroxidation of neuronal membrane and malondialdehyde release, causing neurotoxicity and preventing neurogenesis (Park et al., 2010). Blood-brain permeability has been found to be increased as a result of the consumption of a high-fat diet, which influences the transport of the hormones ghrelin and leptin across the blood-brain barrier; leptin resistance has been found in obese individuals (Klok et al., 2007). In rats that were fed a high-fat diet for 8 weeks, changes in microglia morphology were seen in areas of synaptic loss in the prefrontal cortex. The loss of microglia caused synaptic loss due to pruning of synapses. Furthermore, the decreased level of corticosterone in the obese rats has been linked to reduced dendritic spine density in the prefrontal cortex (Bocarsly et al., 2015). Studies that have been done on individuals with obesity have shown a detrimental effect on cognition throughout childhood development (Yates et al., 2012). A study was done on P300 auditory event-related potentials in obese prepubescent children and the response of brain waves to an auditory stimulus as well as level of cognition revealed that there was a decrease in auditory event-related potentials and that those potentials reflected the amplitude of the response of brain waves to an auditory stimulus as well as the level of cognition (Medvidovic et al., 2013). Furthermore, they found that the level of cognition is obese children was significantly low in comparison to that of nonobese children. The insulin-resistant condition and body mass index are negatively correlated with prefrontal inhibitory control, general mental ability, and visuospatial organization in children (Tascilar et al., 2011). Chronic systemic inflammation and production of cytokine production are directly correlated with dietary habits and the gut microbiome. The number of microbes in the gut is 10 times more than the total number of cells present in human body, and any changes in the physiology of gut microbes causes dysbiosis. The effect of cytokine and chronic inflammation on the central nervous system is well studied, but the composition of the gut microbiota is still poorly understood. This chapter provides a detailed overview of diet, the gut microbiome and altered cognitive function.

19.2 Microbiome composition and evolution

A variety of microorganisms live in stomach and intestine in commensal association with the host throughout the life of the living organism. The microorganisms include bacteria, viruses, and some other unicellular eukaryotes that have

Microbiome, Immunity, Digestive Health and Nutrition. DOI: https://doi.org/10.1016/B978-0-12-822238-6.00006-6

coevolved with humans. The colon microbiota is huge, numbering 10^{14} cells, and its genetic catalog is called the intestinal microbiome. About 10 million different genes have been found in the human microbiome (Roager et al., 2014). There are 1000–1500 bacterial species in the human gut (Qin et al., 2010); therefore the total genes outnumber the human genome. Because of the diversity in species and huge genome of bacterial species that are present in the gut, the microbiome is considered a second genome of humans, and the gut microbiome plays a significant role in human physiology and metabolism (Qin et al., 2010). The development of high-throughput analytical tools and meta-omics technologies has made possible the deciphering of the gut microbiome. The differences between a healthy gut and an unhealthy gut can be revealed only when the composition and functionality of gut microbiome has been elucidated. The colonization of microbes starts at birth, and this determines the development of disease in early and later life (Arboleya et al., 2012; Rodrıguez et al., 2015). The colonization of the human gut starts gradually with the succession of facultative anaerobes, which create a reducing environment that provides a strict environment for strict anaerobes (Fanaro et al., 2003). The gut microbiome is unstable at first, but the complexity and diversity of the gut microbiota gradually increases until it acquires a stable population at about 3 years of age (Salazar et al., 2014; Yatsunenko et al., 2012). The healthy human gut is composed of Firmucutes and Bacteroidetes, which constitute 80%–90% of the total microbiota, with representation of Proteobacteria, Fusobacteria, Cyanobacteria, Verrucomicrobia, and Actinobacteria phyla, while other species are found considerably lower levels (Yatsunenko et al., 2012). The three bacterial genera of bacteria (*Bacteroides*, *Prevotella*, and *Ruminococcus*) are abundantly colonized in the gut and are independent of nationality, sex, age, or body mass index (Arumugam et al., 2011). The enterotypes are influenced by long-term dietary pattern and short-term diet (David et al., 2014; Roager et al., 2014). However, there are controversies in the classification of the enterotypes that are present in human gut due to difference in the methodological techniques that have been employed in their identification (Huse et al., 2012). The intestinal microbiota of adults is stable, but stress, use of antibiotics, diet, and lifestyle changes might affect the composition and colonization (Zhang et al., 2015). Changes in the gut microbiota with advancing age occur as a result of changes in the physiology of the gastrointestinal tract. A change in dietary pattern results in disruption of the normal functions of the immune system that may contribute to increased risk of infection and frailty (Claesson et al., 2012; Odamaki et al., 2016). The gut microbiota of adults is composed of reduced microbial diversity, a shift in the dominant species, and a decline in beneficial microorganism increase the facultative anaerobic bacteria and decrease the availability of total short-chain fatty acids (SCFAs) (Salazar et al., 2013; Hayashi et al., 2003). The species that are abundantly found in the gut of older adults are lower levels of Firmicutes, mainly *Clostridium* cluster XIVa and *Faecalibacterium prausnitzii*, and *Actinobacteria* (mainly *Bifidobacteria*) and increased populations of *Proteobacteria* (Tiihonen et al., 2010; Woodmansey, 2007). Different studies have found varying numbers of *Lactobacilli* in the elderly population. Some studies have suggested that the population of *Lactobacilli* increases with advancing age, while other studies have shown that the population of *Lactobacilli* decreases with aging (Bartosch et al., 2004; Hayashi et al., 2003; Tiihonen et al., 2010; Woodmansey, 2007). The SCFAs, the main metabolite of microbes in the gut, decrease in adults as compared with young individuals (Claesson et al., 2012; Salazar et al., 2013). These changes mainly correspond with changes in bacterial colonization and differences in their metabolites, which shift from the predominantly saccharolytic metabolism that is normally observed in adults toward a predominantly putrefactive metabolism (Woodmansey et al., 2004). The change in gut microbiota does not necessarily mean a detrimental effect on health. Functional redundancy exists in the composition of the gut microbiota; therefore any change in microbes does not equate to functional deficiency. The changes in the gut microbiota with advancing age are correlated with fragility in older adults (Jackson et al., 2016). The composition of the gut microbiota is being elucidated gradually by different researchers, but owing to limitations in studies that could define the complete composition of the gut microbiota in the elderly population, a cause-and-effect relationship has not been established. Changes in dietary habits in older adults increases the risk of malnutrition as well as change in the immune system (Candore et al., 2010; Vallejo, 2011). The decline in the immune system and imbalance in dietary patterns increase the susceptibility of the elderly population to disease. Difference among elderly people from different geographical locations are also evident (Mueller et al., 2006). The differences in the elderly microbiota might be due to changes in dietary habits and subsequent increases in the risk of malnutrition as well as changes in the immune system (Candore et al., 2010). The changes in the gut microbiota make the elderly population more susceptible to disease (Vallejo, 2011).

19.3 Microbiota and brain functioning

Various studies have proven the association between the gut microbiota and brain functioning. The gastrointestinal tract is sterile in utero but rapidly colonized by bacteria during birth through maternal contact. The number of bacteria that are present in the colon of an adult is estimated to be about 10^{12} per gram of colonic content (Dominguez-Bello et al., 2011).

The gastrointestinal microbiome plays a critical role in regulating the mucosal immune system, thereby maintaining normal gut physiology. Various diseases, such as diabetes, nonalcoholic fatty liver disease, autism, multiple sclerosis, and cardiovascular disease, are associated with dysbiosis in the gastrointestinal microbiome (de Vos and de Vos, 2012). That gut microbiome regulates brain function has been proved by research in which the administration of antibiotics improved brain functions in patients suffering from hepatic encephalopathy, and it was recently found that cognitive decline in hepatic encephalopathy was due to growth of a specific type of bacteria (Bajaj et al., 2012; Riordan and Williams, 1997). The first study of stress and the gut microbiome found that an exaggerated hypothalamic pituitary response normalized after colonization of specific pathogen-free bacteria (SPF) (Sudo et al., 2004). Further, many studies concluded that brain chemistry and phenotypical behavior changes in germ-free mice. The germ-free mouse performs anxiolytic behavior and is corrected by early colonization of SPF (Diaz Heijtz et al., 2011; Neufeld et al., 2011). The neurotrophic factor BDNF is upregulated and decreases expression of serotonin receptor 1A (5HT1A) in the hippocampus and N methyl-D-aspartate (NMDA) receptor subunit NR2BmRNA expression in the amygdala (Neufeld et al., 2011). The connection between the gut microbiome and brain plasticity in later stage of life was demonstrated by another study which depicted reduced expression of PSD-95 and synaptophysin in the striatum of germ-free mice (Diaz Heijtz et al., 2011). The postnatal colonization and development of the gut microbiome remain stable in normal conditions, and the only variations occur as a result of diet and antimicrobial drugs (Dominguez-Bello et al., 2011). Studies suggested that changes in dietary pattern and use of antimicrobial drugs are sufficient to bring about changes in brain chemistry and functions (Bercik et al., 2011; Li et al., 2009). The effect of orally ingested antimicrobials changes BDNF in the hippocampus, but intraperitoneal treatment did not show any effect; these finding are conclusive evidence that the gut microbiome affects the brain (Bercik et al., 2011). The effect of the gut microbiome was also strongly correlated as shown by studies showing that mice that were fed a diet consisting of 50% lean ground beef–enriched chow had a significantly more diverse microbiome than mice that were fed with standard rodent chow and had better performance on a hole-board open field test (Li et al., 2009). The diet has a direct influence on the gut microbiome, which in turn improves learning and memory in mice, although the effect of host protein metabolism on brain functions could not be excluded. A study showed that mice deficient in interleukin-10 that were placed on a Western-style diet exhibited anxiety-like behavior and impaired memory, and these changes were prevented by treatment with the probiotics bacterium *Lactobacillus helveticus* (Ohland et al., 2013). That probiotic bacteria improves spatial memory has been shown in diabetic rats (Davari et al., 2013). The probiotics have a significant role in improving the gut microbiome, and the metabolite secreted from gut microbes directly or indirectly affect the brain-gut axis, immune activation, and endocrine and neural pathways, including vagus afferent neurons (Collins et al., 2012). Studies of humans suffering from hepatic encephalopathy have shown that they had an altered intestinal microbiome and disturbed cognitive functions as compared with people who had a stable gut microbiome. The bacterial families Alcaligeneceae and Porphyromonadaceae were positively correlated with cognitive impairment (Bajaj et al., 2012). The cognitive function and emotive intelligence were augmented in women who ate a 4-week diet of a fermented milk beverage containing several probiotic bacterial strains (Tillisch et al., 2013). The gut microbiome changes with age, and a lot of research is going on to determine how the composition of the gut microbiome changes with age, although this is not a linear process, and the longevity is associated with different microbiomes (Claesson et al., 2011, 2012). The microbial composition becomes less diverse with age but remains stable. The two dominant phyla in younger age groups, Bacteroides and Firmicutes, remain prominent in older age, but possible changes in the ratio of these phyla are still under debate. The reduction in *Bifidobacteria* as compared with *Proteobacteria* signifies that with aging, the number of pathogenic bacteria (pathobionts) is significantly more that of beneficial bacteria (symbionts). The SCFAs, in particular butyrate, are crucial for the maintenance of colonial epithelial integrity and inflammation. The microbial composition changes and thereby changes the metabolism of the bacteria that are colonized in the gut; this process is also called inflamm-aging (Franceschi et al., 2000). In a study done on older adults living in the community as compared with institutionalized subjects, the reduction in the expression of the gene for SCFAs was reduced, and the proinflammatory cytokines tumor necrosis factor alpha, interleukin-6, interleukin-8, and C-reactive protein increased (Franceschi et al., 2000). A strong correlation was found in the gut microbiome of institutionalized individuals with poor health and indices of frailty. Changes in dietary pattern are responsible for changes in diversity of the gut microbiome (Franceschi et al., 2000).

19.4 Microbiome-gut-brain axis

The gut not only receive signals from brain but also sends signal to the central nervous system. These encompass efferent and afferent neural connections and immune and metabolic signals. Based on the published research of the concept microbiome-gut-brain axis (Bercik et al., 2011; Bruce-Keller et al., 2015; Davari et al., 2013; Hsiao et al., 2013), the intestinal microbiome can affect both the peripheral nervous system and the central nervous system. The two-way communication system maintains normal organism homeostasis. A disturbance in gut microbial stability could lead to a disturbance of the central or autonomic nervous system or immune system and might be one cause of various diseases (Bauer et al., 2016; Dinan and Cryan, 2017; Fasano, 2012; Hsiao et al., 2013; Schroeder and Bäckhed, 2016; Turnbaugh et al., 2009). The brain and intestines are connected by various physiological and metabolic pathways, including neuronal, endocrine, immune, and metabolic pathways.

19.5 Microbiome and cognitive decline

The research that has been done on the gut microbiome has clearly established the association between the intestinal microbiome, dysbiosis, and neurological and psychiatric diseases. The neurological, neurological autoimmune, neurodegenerative, and psychiatric diseases include neuromyelitis, Parkinson's disease, Alzheimer's disease, amyotrophic lateral sclerosis, anxiety, schizophrenia, and autism. Cognitive impairment occurs as a result of these diseases. Research on the effects of the microbiome on cognition is ongoing, and extensive research is bringing about newer theories (Bajaj et al., 2012; Dinan et al., 2015). Behavioral disorders and cognitive dysfunction were observed in germ-free mice, but when these mice were fed with probiotics, the behavioral function and cognition were intact (Gareau et al., 2011; Liang et al., 2015). Studies shown have that infection with *Citrobacter rodentium* in combination with acute stress had resulted in memory failure in mice (Liang et al., 2015). The proper functioning of memory and learning is intact when HPA axis functioning is proper. It has been reported that *Bifidobacterium* can modulate the axis during stress, although the level of corticosteroids and cortex hormone were much higher as compared with conventional mice (Liang et al., 2015). The other factors that are important for learning and memory, such as BDNF, NMDA receptors (N-methyl-D-aspartate) and c-fos, are reduced in sterile mice (Intlekofer et al., 2013; Stefanko et al., 2009; Wu et al., 2008). The effect of serotonin on cognition and socializing has already been reported. The metabolites released from kynurenine pathway related to serotonin release and mental health are evident from various research studies (Bajaj et al., 2012). Cross-sectional studies that were conducted in Japan in groups of demented and nondemented patients has demonstrated that the number of Bacteroides (enterotype I) was lower and the number of other bacteria (enterotype III) was higher in demented patients than in nondemented patients. Further analysis revealed that a lower prevalence of Bacteroides and a high prevalence of other bacteria were associated with higher odds ratios than the traditional dementia biomarkers APOE4, SLI, and high VSRAD score (Saji et al., 2019). The level of serotonin and cognition were found to be enhanced in individuals on a regular diet of probiotics as compared with a group without probiotics (Saji et al., 2019).

19.6 Diet and cognition

The accumulating evidence suggests that not only single nutrients but also the combination of different vital compounds taken by individuals can have beneficial effect on the brain and overall health. A trial done on improving cognition by supplementation of vitamins and antioxidants was more successful (Kesse-Guyot et al., 2013; Summers et al., 2010). The effect of eating fruits and vegetable on cognition was determined by four longitudinal studies that investigated the development of mild cognitive impairment and dementia in connection with consumption of fruits and vegetables (Engelhart et al., 2002; Kang et al., 2005; Morris et al., 2006a; Sachdev et al., 2013). The studies concluded that vegetables that are rich in polyphenols along with fruit had better effects on cognition as compared with intake of vegetable only (Loef and Walach, 2013). Vegetables may also have a neuroregulatory role by inducing satiety and ultimately reducing the risk of obesity (Panickar, 2013). Vegetables can also influence the gut microbiota, which influences insulin secretion and obesity (Esteve et al., 2011). Vegetables also have neuroregulatory role by inducing satiety and reducing obesity (Panickar, 2013). The Mediterranean diet, which is composed of fruit and vegetables (rich in antioxidants and polyphenols), olive oil (rich in unsaturated fatty acid, vitamin E and polyphenols), and fish (rich in fatty acids, vitamin B12, and selenium), improves cognitive function and reduces the risk of Alzheimer's disease. A randomized trial done on Mediterranean diet and cognitive functions in individuals suffering from cardiovascular disease showed improved brain functions (Panickar, 2013). New dietary strategies addressing the specific needs of the elderly population for healthy aging are a focus of the European Project on Nutrition in Elderly People (NU-AGE), a European multicenter study that is focused on older adults and includes in-depth cognitive assessments (Santoro et al., 2014).

19.7 Conclusion

The composition of the gut microbiota has been elucidated with the help of high-throughput technologies, and various studies have found an association between the gut microbiota and brain function. The change in microbial composition with advancing age and comorbidities needs more comprehensive insight into the role of microbial communities and their metabolites that regulate the gut-brain axis and mucosal immune and endocrine responses. Dietary patterns differ from one geographical region to another, and diet is highly correlated with mental health if the diet encompass nutrition rich in antioxidants and polyphenols. The Mediterranean diet has been found to be very effective in maintaining balance between gut pathobionts and eubionts, which determine the secretion of SCFAs, bacterial metabolites that are beneficial for the growth of bacteria that promote good health. The importance of probiotics and prebiotics in the treatment of cognitive disorders is significant; the design of specific dietary interventions that take into account the roles of specific nutrients and their interaction with the microbiota will provide new methods of personalized treatment of specific diseases that develops as a result of to bacterial dysbiosis. The world population is aging rapidly, and changes in the gut microbiome with age have been well studied and documented through research. The development of specific food products for older adults will help in providing cost-effective personalized treatment and products.

References

Arboleya, S., Binetti, A., Salazar, N., Fernandez, N., Solıs, G., Hern_andez-Barranco, A., et al., 2012. Establishment and development of intestinalmicrobiota in preterm neonates. FEMSMicrobiol. Ecol. 79, 763–772.

Arumugam, M., Raes, J., Pelletier, E., Le Paslier, D., Yamada, T., Mende, D.R., et al., 2011. Enterotypes of the human gut microbiome. Nature 473, 174–180.

Bajaj, J.S., Ridlon, J.M., Hylemon, P.B., Thacker, L.R., Heuman, D.M., Smith, S., et al., 2012. Linkage of gut microbiome with cognition in hepatic encephalopathy. Am. J. Physiol. Gastrointest. Liver Physiol. 302, G168–G175.

Bartosch, S., Fite, A., Macfarlane, G.T., McMurdo, M.E., 2004. Characterization of bacterial communities in feces from healthy elderly volunteers and hospitalized elderly patients by using real-time PCR and effects of antibiotic treatment on the fecal microbiota. Appl. Env. Microbiol. 70, 3575–3581.

Bauer, K.C., Huus, K.E. Finlay, B.B., 2016. Microbes and the mind:emerging hallmarks of the gut microbiota-brain axis. Cell. Microbiol. 18, 632–644.

Bercik, P., Denou, E., Collins, J., Jackson, W., Lu, J., Jury, J., et al., 2011. The intestinal microbiota affect central levels of brain-derived neurotropic factor and behavior in mice. Gastroenterology 141, 599–609. e1–3.

Bocarsly, M.E., et al., 2015. Obesity diminishes synaptic markers, alters microglial morphology, and impairs cognitive function. Proc. Natl. Acad. Sci. USA 112, 15731–15736.

Bruce-Keller, A.J., Salbaum, J.M., Luo, M., Blanchard, E., Taylor, C.M., Welsh, D.A., et al., 2015. Obese-type gut microbiota induce neurobehavioral changes in the absence of obesity. Biol. Psychiatr. 77, 607–615.

Candore, G., Caruso, C., Jirillo, E., Magrone, T., Vasto, S., 2010. Low grade inflammation as a common pathogenetic denominator in age-related diseases: novel drug targets for anti-ageing strategies and successful ageing achievement. Curr. Pharm. Des. 16, 584–596.

Castanon, N., Lasselin, J. Capuron, L., 2014. Neuropsychiatric comorbidity in obesity: role of inflammatory processes. Front. Endocrinol. 5–74.

Claesson, M.J., Cusack, S., O'Sullivan, O., Greene-Diniz, R., de Weerd, H., Flannery, E., et al., 2011. Composition, variability, and temporal stability of the intestinal microbiota of the elderly. Proc. Natl. Acad. Sci. U. S. A. 108, 4586–4591.

Claesson, M.J., Jeffery, I.B., Conde, S., Power, S.E., O'Connor, E.M., Cusack, S., et al., 2012. Gut microbiota composition correlates with diet and health in the elderly. Nature 488, 178–184.

Collins, S.M., Surette, M., Bercik, P., 2012. The interplay between the intestinal microbiota and the brain. Nat. Rev. Microbiol. 10, 735–742.

Davari, S., Talaei, S.A., Alaei, H., Salami, M., 2013. Probiotics treatment improves diabetes-induced impairment of synaptic activity and cognitive function: behavioral and electrophysiological proofs for microbiome-gut-brain axis. Neuroscience 240, 287–296.

David, L.A., Maurice, C.F., Carmody, R.N., Gootenberg, D.B., Button, J.E., Wolfe, B.E., et al., 2014. Diet rapidly and reproducibly alters the human gut microbiome. Nature 505, 559–563.

de Vos, W.M., de Vos, E.A., 2012. Role of the intestinal microbiome in health and disease: from correlation to causation. Nutr. Rev. 70 (Suppl. 1), S45–S56.

Diaz Heijtz, R., Wang, S., Anuar, F., Qian, Y., Bjorkholm, B., Samuelsson, A., et al., 2011. Normal gut microbiota modulates brain development and behavior. Proc. Natl. Acad. Sci. USA 108, 3047–3052.

Dinan, T.G., Cryan, J.F., 2017. The microbiome-gut-brain axis in health and disease. Gastroenterol. Clin. North. Am. 46 (1), 77–89.

Dinan, T.G., Stilling, R.M., Stanton, C., Cryan, J.F., 2015. Collective unconscious: how gut microbes shape human behavior. J. Psychiatr. Res. 63, 1–9.

Dominguez-Bello, M.G., Blaser, M.J., Ley, R.E., Knight, R., 2011. Development of the human gastrointestinal microbiota and insights from high-throughput sequencing. Gastroenterology 140, 1713–1719.

Engelhart, M.J., Geerlings, M.I., Ruitenberg, A., Van Swieten, J.C., Hofman, A., Witteman, J.C., et al., 2002. Diet and risk of dementia: does fat matter? The Rotterdam study. Neurology 59, 1915–1921.

Esteve, E., Ricart, W., Fernandez-Real, J.M., 2011. Gut microbiota interactions with obesity, insulin resistance and type 2 diabetes: did gut microbiote co-evolve with insulin resistance? Curr. Opin. Clin. Nutr. Metab. Care 14, 483–490.

Fanaro, S., Chierici, R., Guerrini, P., Vigi, V., 2003. Intestinal microflora in early infancy: composition and development. Acta Paediatr. Suppl. 91, 48–55.

Fasano, A., 2012. Leaky gut and autoimmune diseases. Clin. Rev. Allergy Immunol. 42, 71–78.

Franceschi, C., Bonafè, M., Valensin, S., Olivieri, F., De, Luca, M., Ottaviani, E., et al., 2000. Inflamm-aging. An evolutionary perspective on immunosenescence. Ann. N. Y. Acad. Sci. 908, 244–254.

Gareau, M.G., Wine, E., Rodrigues, D.M., Cho, J.H., Whary, M.T., Philpott, D.J., et al., 2011. Bacterial infection causes stress-induced memory dysfunction in mice. Gut 60, 307–317.

Hayashi, H., Sakamoto, M., Kitahara, M., Benno, Y., 2003. Molecular analysis of fecal microbiota in elderly individuals using 16S rDNA library and T-RFLP. Microbiol. Immunol. 47, 557–570.

Hsiao, E.Y., McBride, S.W., Hsien, S., Sharon, G., Hyde, E.R., McCue, T., et al., 2013. Microbiota modulate behavioral and physiological abnormalities associated with neurodevelopmental disorders. Cell 155, 1451–1463.

Huse, S.M., Ye, Y., Zhou, Y., Fodor, A.A., 2012. A core human microbiome as viewed through 16S rRNA sequence clusters. PLoS One 7, e34242.

Intlekofer, K.A., Berchtold, N.C., Malvaez, M., Carlos, A.J., McQuown, S.C., Cunningham, M.J., et al., 2013. Exercise and sodium butyrate transform a subthreshold learning event into long-term memory via a brain-derived neurotrophic factor-dependent mechanism. Neuropsychopharmacology 38, 2027–2034.

Jackson, M.A., Jeffery, I.B., Beaumont, M., Bell, J.T., Clark, A.G., Vley, R.E., et al., 2016. Signatures of early frailty in the gut microbiota. Genome Med. 8, 8.

Kang, J.H., Ascherio, A., Grodstein, F., 2005. Fruit and vegetable consumption and cognitive decline in aging women. Ann. Neurol. 57, 713–720.

Kesse-Guyot, E., Andreeva, V.A., Lassale, C., Ferry, M., Jeandel, C., Hercberg, S., et al., 2013. Mediterranean diet and cognitive function: a French study. Am. J. Clin. Nutr. 97, 369–376.

Klok, M.D., Jakobsdottir, S., Drent, M.L., 2007. The role of leptin and ghrelin in the regulation of food intake and body weight in humans:a review. Obes. Rev. 8, 21–34.

Li, W., Dowd, S.E., Scurlock, B., Acosta-Martinez, V., Lyte, M., 2009. Memory and learning behavior in mice is temporally associated with diet-induced alterations in gut bacteria. Physiol. Behav. 96, 557–567.

Liang, S., Wang, T., Hu, X., Luo, J., Li, W., Wu, X., et al., 2015. Administration of Lactobacillus helveticus NS8 improves behavioral, cognitive, and biochemical aberrations caused by chronic restraint stress. Neuroscience 310, 561–577.

Loef, M., Walach, H., 2013. Midlife obesity and dementia: meta-analysis and adjusted forecast of dementia prevalence in the United States and China. Obesity (Silver Spring) 21, E51–E55.

Medvidovic, S., Titlic, M., Maras-Simunic, M., 2013. P300 evoked potential in patients with mild cognitive impairment. Acta Inform. Med. 21, 89–92.

Molteni, R., Barnard, R.J., Ying, Z., Roberts, C.K., Gómez-Pinilla, F., 2002. A high-fat, refined sugar diet reduces hippocampal brain-derived neurotrophic factor, neuronal plasticity, and learning. Neuroscience 112, 803–814.

Molteni, R., Wu, A., Vaynman, S., Ying, Z., Barnard, R.J., Gómez-Pinilla, F., 2004. Exercise reverses the harmful effects of consumption of a high-fat diet on synaptic and behavioral plasticity associated to the action of brain-derived neurotrophic factor. Neuroscience 123, 429–440.

Morris, M.C., Evans, D.A., Tangney, C.C., Bienias, J.L., Wilson, R.S., 2006. Associations of vegetable and fruit consumption with age-related cognitive change. Neurology 67, 1370–1376.

Mueller, S., Saunier, K., Hanisch, C., Norin, E., Alm, L., Midtvedt, T., et al., 2006. Differences in fecal microbiota in different European study populations in relation to age, gender, and country: a cross-sectional study. Appl. Env. Microbiol. 72, 1027–1033.

Neufeld, K.M., Kang, N., Bienenstock, J., Foster, J.A., 2011. Reduced anxiety-like behavior and central neurochemical change in germ-free mice, Neurogastroenterol. Motil. Off. J. Eur. Gastrointest. Mot. Soc., 23. , pp. 255–264. e119Diaz.

Odamaki, T., Kato, K., Sugahara, H., Hashikura, N., Takahashi, S., Xiao, J.Z., et al., 2016. Age-related changes in gut microbiota composition from newborn to centenarian:a cross-sectional study. BMC Microbiol. 16, 90.

Ohland, C.L., Kish, L., Bell, H., Thiesen, A., Hotte, N., Pankiv, E., et al., 2013. Effects of Lactobacillus helveticus on murine behavior are dependent on diet and genotype and correlate with alterations in the gut microbiome. Psychoneuroendocrinology 38 (9), 1738–1747.

Panickar, K.S., 2013. Effects of dietary polyphenols on neuroregulatory factors and pathways that mediate food intake and energy regulation in obesity. MolecularNutrition Food Res. 57, 34–47.

Park, H.R., Park, M., Choi, J., Park, K.Y., Chung, H.Y., Lee, J., 2010. A high-fat diet impairs neurogenesis: involvement of lipid peroxidation and brain-derived neurotrophic factor. Neurosci. Lett. 482, 235–239.

Pintana, H., Apaijai, N., Chattipakorn, N., Chattipakorn, S.C., 2013. DPP-4 inhibitors improve cognition and brain mitochondrial function of insulin-resistant rats. J. Endocrinol. 218, 1–11.

Pipatpiboon, N., Pratchayasakul, W., Chattipakorn, N., Chattipakorn, S.C., 2012. PPARgamma agonist improves neuronal insulin receptor function in hippocampus and brain mitochondria function in rats with insulin resistance induced by long term high-fat diets. Endocrinology 153, 329–338.

Qin, J., Li, R., Raes, J., Arumugam, M., Burgdorf, K.S., Manichanh, C., et al., 2010. A human gut microbial gene catalogue established by metagenomic sequencing. Nature 464, 59–65.

Riordan, S.M., Williams, R., 1997. Treatment of hepatic encephalopathy. N. Engl. J. Med. 337, 473–479.

Roager, H.M., Licht, T.R., Poulsen, S.K., Larsen, T.M., Bahl, M.I., 2014. Microbial enterotypes, inferred by the prevotella-to-bacteroides ratio, remained stable during a 6-month randomized controlled diet intervention with the new nordic diet. Appl. EnvironMicrobiol. 80, 1142–1149.

Rodrıguez, J.M., Murphy, K., Stanton, C., Ross, R.P., Kober, O.I., Juge, N., et al., 2015. The composition of the gut microbiota throughout life, with an emphasis on early life. Microb. Ecol. Health Dis. 26, 26050.

Sachdev, P.S., Lipnicki, D.M., Kochan, N.A., Crawford, J.D., Rockwood, K., Xiao, S., et al., 2013. COSMIC (Cohort Studies of Memory in an International Consortium): an international consortium toidentify risk and protective factors and biomarkers of cognitive ageing anddementia in diverse ethnic and sociocultural groups. BMC Neurol. 13, 165.

Saji, N., Niida, S., Murotani, K., Hisada, T., Tsuduki, T., Sugimoto, T., et al., 2019. Analysis of the relationship between the gut microbiome and dementia: a crosssectional study conducted in Japan. Sci. Rep. 9 (1008), 38218-7.

Salazar, N., Arboleya, S., Valdes, L., Stanton, C., Ross, P., Ruiz, L., et al., 2014. The human intestinal microbiome at extreme ages of life. Dietary intervention as a way to counteract alterations. Front. Genet. 406.

Salazar, N.L., Opez, P., Valdes, L., Margolles, A., Suarez, A., Patterson, A.M., et al., 2013. Microbial targets for the development of functional foods accordingly with nutritional and immune parameters altered in the elderly. JAm. Coll. Nutr. 32, 399–406.

Santoro, A., Pini, E., Scurti, M., Palmas, G., Berendsen, A., Brzozowska, A., et al.,The NU-AGE Consortium 2014. Combating inflammaging through a Mediterranean whole diet approach: the NU-AGE project's conceptual framework and design. Mech. Ageing Dev. 136–137, 3–13.

Schroeder, B.O., Bäckhed, F., 2016. Signals from the gut microbiota to distant organs in physiology and disease. Nat. Med. 22, 1079–1089.

Sripetchwandee, J., Pipatpiboon, N., Pratchayasakul, W., Chattipakorn, N., Chattipakorn, S.C., 2014. DPP-4 inhibitor and PPARγ agonist restore the loss of CA1 dendritic spines in obese insulin-resistant rats. Arch. Med. Res. 45, 547–552.

Stefanko, D.P., Barrett, R.M., Ly, A.R., Reolon, G.K., Wood, M.A., 2009. Modulation of long-term memory for object recognition via HDAC inhibition. Proc. Natl. Acad. Sci. USA 106, 9447–9452.

Stranahan, A.M., Norman, E.D., Lee, K., Cutler, R.G., Telljohann, R.S., Egan, J.M., et al., 2008. Diet-induced insulin resistance impairs hippocampal synaptic plasticity and cognition in middle-aged rats. Hippocampus 18, 1085–1088.

Sudo, N., Chida, Y., Aiba, Y., Sonoda, J., Oyama, N., Yu, X.N., et al., 2004. Postnatal microbial colonization programs the hypothalamic-pituitary-adrenal system for stress response in mice. J. Physiol. 558 (Pt. 1), 263–275.

Summers, W.K., Martin, R.L., Cunningham, M., DeBoynton, V.L., Marsh, G.M., 2010. Complex antioxidant blend improves memory in community-dwelling seniors. J. Alzheimer's Dis. 19, 429–439.

Tascilar, M.E., et al., 201 . P300 auditory event-related potentials in children with obesity: is childhood obesity related to impairment in cognitive functions? Pediatr. Diabetes 12, 589–595.

Tiihonen, K., Ouwehand, A.C., Rautonen, N., 2010. Human intestinal microbiota and healthy ageing. Ageing Res. Rev. 9, 107–116.

Tillisch, K., Labus, J., Kilpatrick, L., Jiang, Z., Stains, J., Ebrat, B., et al., 2013. Consumption of fermented milk product with probiotic modulates brain activity. Gastroenterology 144, 1394–1401. e4.

Turnbaugh, P.J., Hamady, M., Yatsunenko, T., Cantarel, B.L., Duncan, A., Ley, R.E., et al., 2009. A core gut microbiome in obese and lean twins. Nature 457, 480–484.

Vallejo, A.N., 2011. Immunological hurdles of ageing: indispensable research of the human model. Ageing Res. Rev. 10, 315–318.

Woodmansey, E.J., 2007. Intestinal bacteria and ageing. J. Appl. Microbiol. 102, 1178–1186.

Woodmansey, E.J., McMurdo, M.E., Macfarlane, G.T., Macfarlane, S., 2004. Comparison of compositions and metabolic activities of fecal microbiotas in young adults and in antibiotictreated and non-antibiotic-treated elderly subjects. Appl. Env. Microbiol. 70, 6113–6122.

Wu, A., Ying, Z., Gomez-Pinilla, F., 2004. The interplay between oxidative stress and brain-derived neurotrophic factor modulates the outcome of a saturated fat diet on synaptic plasticity and cognition. Eur. J. Neurosci. 19, 1699–1707.

Wu, X., Chen, P.S., Dallas, S., Wilson, B., Block, M.L., Wang, C.C., et al., 2008. Histone deacetylase inhibitors up-regulate astrocyte GDNF and BDNF gene transcription and protect dopaminergic neurons. Int. J. Neuropsychopharmacol. 11, 1123–1134.

Yates, K.F., Sweat, V., Yau, P.L., Turchiano, M.M., Convit, A., 2012. Impact of metabolic syndrome on cognition and brain: a selected review of the literature. Arterioscler Thromb. Vasc. Biol. 32, 2060–2067.

Yatsunenko, T., Rey, F.E. Manary, M.J., Trehan, I., Dominguez-Bello, M.G., Contreras, M., et al., 2012. Human gut microbiome viewed across age and geography. Nature 486, 222–227.

Zhang, Y.J., Li, S., Gan, R.Y., Zhou, T., Xu, D.P., Li, H.B., 2015. Impacts of gut bacteria on human health and diseases. Int. J. Mol. Sci. 16, 7493–7519.

Chapter 20

Role of the gut microbiome in Rett syndrome

Carly Albright[1], Amanda Friend[1], Benjamin Peters[1], Miranda Reed[1,2], Vishnu Suppiramaniam[1,2] and Subhrajit Bhattacharya[3]

[1]Department of Drug Discovery and Development, Auburn University, Auburn, AL, United States, [2]Center for Neuroscience Initiative, Auburn University, Auburn, AL, United States, [3]School of Pharmacy and Health Sciences, Keck Graduate Institute, Claremont, CA, United States

20.1 Introduction to the gut microbiome

Trillions of microorganisms reside in the human gastrointestinal tract (GIT); this population of microorganisms is known as the gut microbiome (Barko et al., 2018). Microorganisms can be found almost anywhere in or outside of the body, but the largest collection is found within the GIT, allowing extensive research into its origin and its involvement in diseases. The vast number of microbes living in the GIT consist of a collection of bacteria, archaea, and eukaryotes. The bacterial heterogeneity of the gut is extensive; however, two dominant phyla include Firmicutes and Bacteroidetes, which account for 90% of the microbes (Barko et al., 2018; Koh et al., 2016). It is important to note that while there are some similarities, gut microbiome composition varies across individuals, differences arising as early as at birth (e.g., according to mode of delivery at birth and whether the baby is breastfed or not) (Barko et al., 2018). Diet, medications, diseases, and various environmental factors also play crucial roles in the observed diversity. This section will give an introductory overview of the physiological roles of the gut microbiome and its influence on several diseases.

20.1.1 Digestion and the gut microbiome

The gut microbiome plays a critical role in various physiological processes that contribute to the health and well-being of the host. These processes expand beyond their classical role in digestion. Some of the additional processes include the role of the gut microbiome in drug absorption, distribution, metabolism and excretion (ADME); protection from pathogens; strengthening the intestinal barrier; and influence on various diseases (Barko et al., 2018). Digestion is intrinsically related to the gut microbiome because of its influence on various metabolic processes. The gut microbiome lines the entire digestive tract, but most of the bacteria dwell in the large intestine (Koh et al., 2016). Bacteria within the GIT are vital for digestion. Gut bacteria can help to break down indigestible materials (dietary fiber) (Koh et al., 2016). Dietary fiber is a key substrate for microbial growth (Koh et al., 2016). The ingested dietary fiber gets fermented by the microbiota in the colon and yields short-chain fatty acids (SCFAs); the three main SCFAs that are formed are acetate, propionate, and butyrate (C2, C3, and C4, respectively) (Barko et al., 2018; Koh et al., 2016; Yang et al., 2015). These SCFAs play a vital role in positive metabolic effects such as regulation of blood pressure, glucose, and lipid levels (Koh et al., 2016; Yang et al., 2015). The SCFAs either enter the circulation or remain in the colonocytes, as in the case of butyrate (Koh et al., 2016; Sánchez-Alcoholado et al., 2020; Yang et al., 2015). The other SCFAs are transported into the portal vein, acetate being the richest SCFA in circulation (Koh et al., 2016). The metabolic actions in which the SCFA are involved can be related back to interaction with G protein–coupled receptors and inhibition of histone deacetylase (HDAC); these signaling processes induce enhanced integrity of the gut barrier and prompt anti-inflammatory effects (Koh et al., 2016). The enhanced intestinal barrier is linked to butyrate having the ability to increase the assembly of tight junctions of the epithelial cells (Barko et al., 2018; Sánchez-Alcoholado et al., 2020). If the gut barrier is impaired, invasion of pathogens give rise to immune responses and inflammation (Koh et al., 2016; Sánchez-Alcoholado et al., 2020). For example, an animal study showed that hypertensive rats have fewer

Microbiome, Immunity, Digestive Health and Nutrition. DOI: https://doi.org/10.1016/B978-0-12-822238-6.00012-1

SCFA-producing microbes (Yang et al., 2015). Although additional research is needed before this finding can be applied to humans, it has been tied to the idea that patients with hypertension have increased gut permeability and inflammation due to an altered number of SCFAs (Koh et al., 2016; Yang et al., 2015). Additionally, the anaerobic bacteria in the colon are able to synthesize secondary bile acids (Barko et al., 2018; Sánchez-Alcoholado et al., 2020). Bile acids can assist in the absorption of dietary lipids and fat-soluble vitamins (Barko et al., 2018). Specifically, secondary bile acids have been associated with carcinogenic risk, such as in the case of colorectal cancer (CRC). SCFAs are able to suppress the oversynthesis of secondary bile acids, indicating an additional benefit of these bacterial metabolites (Sánchez-Alcoholado et al., 2020) (Fig. 20.1).

20.1.2 Absorption, distribution, metabolism and excretion

The gut microbiome also has an important role in pharmacology; the idea of ADME is vital to understanding the intended and unintended effects of medications. The small intestine is the site of absorption of most orally ingested drugs. The amount of drug that reaches the target organ is dependent on first-pass metabolism. First-pass metabolism by the liver reduces the concentration of the original drug. However, the gut microbiome can also be a site of the first-pass effect (Spanogiannopoulos et al., 2016). The gut microbiome can induce modifications on the drug molecule through the expression of enzymes. The enzymes that are involved in drug metabolism have been seen to be indirectly regulated by the gut microbiome, owing to metabolism of uremic solutes, steroids, and bile acids (Barko et al., 2018; Sánchez-Alcoholado et al., 2020; Spanogiannopoulos et al., 2016). Overall, it has been seen that the microbiome can influence drug bioavailability before the drug molecule is even absorbed, ultimately leading to the possibility of decreased efficacy of the drug (Spanogiannopoulos et al., 2016) (Fig. 20.2).

20.1.3 Gut microbiome altering medication efficacy

The gut microbiota is known to modify the action of certain medications, a concept known as pharmacodynamics. A well-known microbiome-drug interaction involves digoxin, which is used in the treatment of heart failure

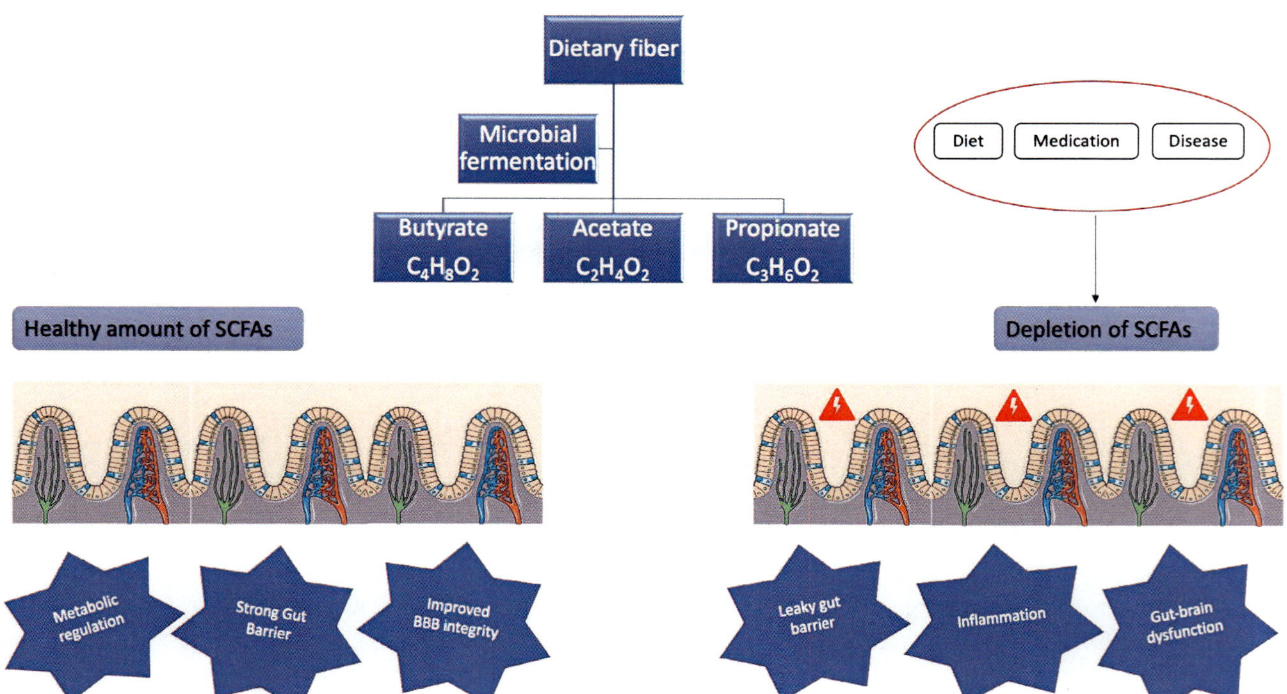

FIGURE 20.1 Importance of short-chain fatty acids (SCFAs). SCFAs contribute to an assortment of functions that are crucial in keeping the human host healthy. A variety of factors can alter the levels of SCFAs, such as diet, medications, and disease. An imbalance in SCFAs can compromise the positive roles they have in human health. In gut dysbiosis, SCFAs may be excreted from the body more frequently, owing to increased intestinal permeability, depleting the human host of these metabolites.

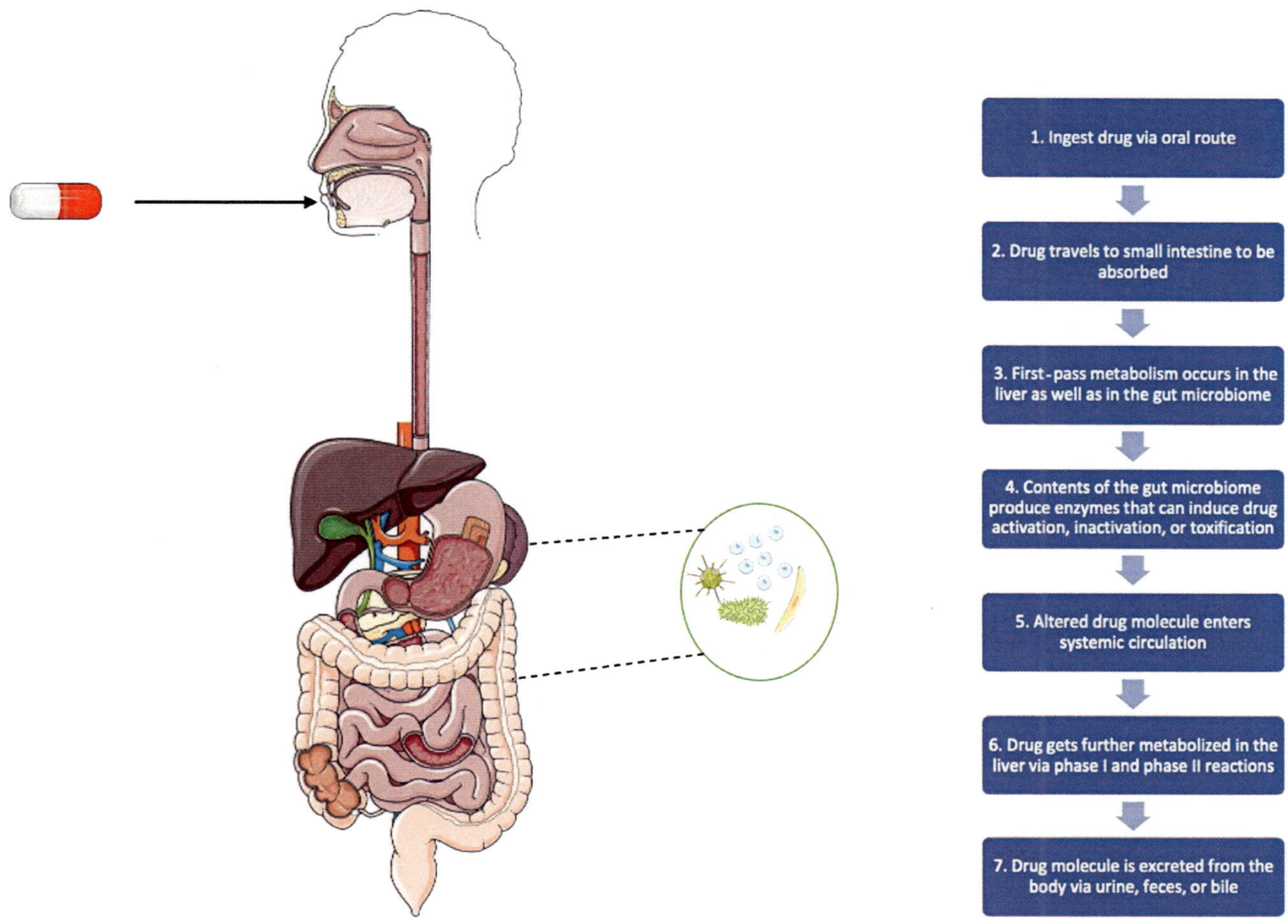

FIGURE 20.2 Representation of ADME. Along with the liver, the gut microbiome can also metabolize xenobiotics. In particular, it can be involved in first-pass metabolism. First-pass metabolism is the first occurrence of metabolism after the drug has been administered and before it is completely absorbed. The metabolism induced on the drug can result in an altered molecule, which can in turn have altered medication efficacy and/or safety, creating a changed pharmacokinetic profile of the drug. When a drug is taken orally, it is susceptible to first-pass metabolism, depicted here. Other routes of administration are not subject to this effect.

(Spanogiannopoulos et al., 2016). While it does not occur in every patient who receives digoxin, some will reduce digoxin to an inactive microbial metabolite during the absorption phase (Koppel et al., 2018; Spanogiannopoulos et al., 2016). The inactivated metabolite that is formed is dihydrodigoxin, which has decreased efficacy (Koppel et al., 2018; Spanogiannopoulos et al., 2016). The bacterium that is responsible for this process is *Eggerthella lenta*, an Actinobacterium (Koppel et al., 2018; Spanogiannopoulos et al., 2016). Another example includes the use of levodopa (L-DOPA), which is used in treating Parkinson's disease. L-DOPA is converted to active dopamine by aromatic amino acid decarboxylase enzymes, but it must cross the blood-brain-barrier (BBB) to have its therapeutic effects for Parkinson's patients. Converting L-DOPA to dopamine too soon in the periphery can cause unwanted side effects and decreased delivery to the central nervous system (CNS), which is why carbidopa is generally added to the L-DOPA; carbidopa can block decarboxylase enzymes and hence block the peripheral conversion of L-DOPA so that it can be transported across the BBB more efficiently. Despite this, carbidopa does not block bacterial decarboxylase enzymes, such as tyrosine decarboxylase (TDC); TDC genes have particularly been associated with *Lactobacillus* and *Enterococcus* bacteria (van Kessel et al., 2019). TDC usually decarboxylates tyrosine into tyramine, but it also can decarboxylate L-DOPA to dopamine in the small intestine, which can lead to premature activation and decreased delivery to the CNS (van Kessel et al., 2019). This alteration can lead to decreased efficacy of the medication, increased side effects, and increased doses required by patients (van Kessel et al., 2019).

20.1.4 Medication-induced gut microbiome dysbiosis

The gut microbiome itself can be altered in a way that has been associated with various disease states. Any alterations that disrupt the equilibrium of the microbiome, leading to an imbalance of microbes, is known as dysbiosis (Barko et al., 2018; Yang et al., 2015). Dysbiosis can be to the result of many factors; here, we will explore the role of medications. Antibiotics are widely recognized as effective agents against infectious diseases (Zhang et al., 2019). Antibiotics exert their action either by destroying bacteria completely or by slowing their rate of reproduction; however, antibiotics also are known for their damage to the gut microbiota (Dethlefsen et al., 2008; Spanogiannopoulos et al., 2016; Zhang et al., 2019). The extensive activity of these agents has been correlated with significant decreases in the diversity of the bacteria that dwell in the microbiome (Barko et al., 2018; Dethlefsen et al., 2008). In a study conducted by Dethlefsen et al., the effects of a short course of oral ciprofloxacin on three healthy adults were observed. Stool samples from the pretreatment, treatment, and posttreatment periods were analyzed by using rRNA sequencing (Dethlefsen et al., 2008). Results indicated that ciprofloxacin affected the taxa of the gut by around 30% in terms of decreased richness and evenness (Dethlefsen et al., 2008). These agents not only kill the infection-causing pathogen, but also can disrupt other bacteria, increasing the risk of antimicrobial resistance (Dethlefsen et al., 2008; Spanogiannopoulos et al., 2016). The mechanism of antimicrobial-induced resistance is elusive, but a key concept involves the ability of pathogens in the gut microbiome to transfer these antibiotic resistance genes in a horizontal manner (Dethlefsen et al., 2008; Zhang et al., 2019). Despite this knowledge, antibiotic resistance continues to be a growing public health issue requiring further research. Another example of drug-induced dysbiosis involves the use of proton pump inhibitors (PPIs). PPIs are commonly used to treat acid-related disorders and have been associated with an increase in the bacterium *Clostridium difficile* (Freedberg et al., 2015). *C. difficile* is notorious for causing an enteric infection that causes diarrhea and colonic inflammation; it is also the main culprit behind antibiotic-induced diarrhea (Freedberg et al., 2015; Zhang et al., 2019). The exact mechanism behind this remains unclear, but one explanation could be that it is due to increased vulnerability of the gut microbiome and pathogen colonization. Specifically, increased Enterococcaceae and Streptococcaceae growth has been associated with patients taking PPIs (Freedberg et al., 2015). These two bacterial taxa are particularly associated with *C. difficile* infections, increased bacterial overgrowth, and hence a less healthy gut microbiome composition (Freedberg et al., 2015; Spanogiannopoulos et al., 2016).

20.2 Diseases of the gut

20.2.1 Diseases central to the gut

Since the gut microbiome resides in the GIT, several disease states related to the digestive system are linked to the microbiome. Specifically, a common finding in the relationship between the gut microbiome and disease includes the idea of gut dysbiosis as well an altered level of intestinal permeability. Crohn's disease (CD) is a type of inflammatory bowel disease (IBD) that is characterized by prolonged inflammation and irritation of the GIT tract; the inflammation can affect any part of the GIT, but most commonly the ileum or colon are the prime sites (Pascal et al., 2017). The cause of CD is considered to be idiopathic, but factors involved in the development include a combination of environmental, microbial, and genetic factors (Pascal et al., 2017). In concern to the microbial aspect of CD, patients have been reported to have a decrease in *Faecalibacterium prausnitzii* and increased levels of *Escherichia coli* (Pascal et al., 2017). *F. prausnitzii* is associated with potent antiinflammatory properties and is a marker of a healthy gut, while *E. coli* is the most common cause of intestinal and extraintestinal disease (Pascal et al., 2017). Alongside this, increased intestinal barrier permeability contributes to enhanced activity of proinflammatory cytokines (Raskov et al., 2014). Owing to the idiopathic nature of this form of IBD, there is still a question as to whether these findings are the cause or consequence of CD (Raskov et al., 2014).

Additionally, patients with CRC are seen to present with dysbiosis of the gut microbiome. It is important to note that IBD is a risk factor for the development of CRC (Raskov et al., 2014; Sánchez-Alcoholado et al., 2020). CRC is linked to a decrease in SCFA-producing bacteria and an increased amount of proinflammatory pathogens; this link of decreased SCFAs ties back in the role of bile acids (Sánchez-Alcoholado et al., 2020). Similar to CD, the inflammation with CRC is related to increased intestinal permeability, which allows a cascade of inflammation to unfold (Raskov et al., 2014; Sánchez-Alcoholado et al., 2020). Despite the lack of a definitive cure for CRC, preventive measures include the consumption of a diet that is rich in fiber, leading to increased production of SCFAs and improved gastrointestinal function (Raskov et al., 2014; Sánchez-Alcoholado et al., 2020). Overall, in diseases that are close to the digestive system, two key features are present: increased inflammation and a gut microbiome of decreased diversity. Whether these findings are consequences or causes of the diseases remains an important area of research.

20.2.2 Diseases distant from the gut: psychiatric and neurological disorders

Besides intestinal disorders, extraintestinal disorders have also been linked to gut dysbiosis. Both psychiatric and neurological disorders likely have a relationship with the gut microbiome. For the purposes of this chapter, psychiatric disorders are defined as disorders of thought or behavior, while neurological disorders involve the CNS. It is important to note that the gut itself has a nervous system, termed the enteric nervous system (ENS) (Liang et al., 2018a). The gut and brain can communicate with one another via the gut-brain axis; notably, the ENS can work even without instruction from the brain. This communication network is very complex, but the overarching idea is that the gut and brain communicate via a bidirectional pathway. This network can influence the processes of the brain, behavior, and even the development of the CNS itself (Liang et al., 2018a). The gut-brain axis is a key component of the idea that the gut microbiome does not solely affect the GIT. Besides the importance of SCFAs in relation to the gut intestinal barrier, SCFAs have also been seen to have a role in this gut-brain communication system. SCFAs can both cross the BBB and improve its structural integrity (Braniste et al., 2014; Koh et al., 2016; Liang et al., 2018a). Studies of gut-brain communication have been widely completed in germ-free mice. In a study by Braniste et al., the lack of a gut microbiome in germ-free mice demonstrated reduced tight junction proteins in the BBB, leading to increased permeability (Braniste et al., 2014; Liang et al., 2018a). This finding was further supported by colonization of the germ-free mice with microbes that produced SCFAs, which was shown to increase the integrity of the BBB (Braniste et al., 2014; Quigley, 2017). In short, it is believed that increased gut permeability can lead to systemic inflammation, which can then affect BBB integrity and result in neuroinflammation (Quigley, 2017), which is critical in making the connection between the gut microbiome and disorders of the CNS.

20.2.2.1 Psychiatric disorders

The study of psychiatric disorders in relation to the gut microbiome is extensive. The psychiatric-related disorders that will be discussed in this chapter are major depressive disorder (MDD) and autism spectrum disorder (ASD). The role of the gut in these two disorders warrants more conclusive investigation, but research advances in this area are important to point out. MDD can be classified on the basis of an individual experiencing specific criteria in the *Diagnostic and Statistical Manual of Mental Disorders* (DSM-5). The most common hypothesis for the pathophysiology of MDD involves a depletion of monoamines (serotonin and norepinephrine). Individuals with MDD exhibit gastrointestinal symptoms; likewise, a high percent of people with IBD also have comorbid depression (Bear et al., 2020). Studies with MDD patients show inconsistent findings of altered gut microbiomes, but research points toward increases in fecal *Bacteroidetes* and *Proteobacteria* as well as decreases in levels of Firmicutes (Bear et al., 2020; Dinan and Cryan, 2012; Liang et al., 2018b). MDD-associated depression is also strongly linked to dysfunction of the hypothalamic-pituitary-adrenal (HPA) axis, which is an additional aspect of the gut-brain axis (Bear et al., 2020; Dinan and Cryan, 2012; Liang et al., 2018b; Sudo et al., 2004). The HPA axis is critical in the formation of a stress response, while the microbial content of the gut has been seen to be important in coordinating stress as well (Dinan and Cryan, 2012). Chronic stress not only can overactivate the HPA axis and disrupt its regulation but also can increase gut permeability, which can prompt the entry of excessive amounts of proinflammatory cytokines (Dinan and Cryan, 2012; Sudo et al., 2004). A study by Sudo et al. involving germ-free mice showed higher levels of stress while being restrained (Sudo et al., 2004). HPA dysfunction is an important mechanism behind depression, and growing amounts of data are pointing toward increased intestinal permeability as being the cause of HPA dysfunction (Dinan and Cryan, 2012) (Fig. 20.3).

ASD is a neurodevelopmental disorder. ASD is typically characterized by deficits in both social interaction and communication (Li et al., 2017). While environmental and genetic factors can potentially influence the development of ASD, there is no single cause (Li et al., 2017). Although researchers have been unable to pinpoint an exact etiology of ASD, the gut microbiome may have some influence. As in MDD, individuals with ASD frequently have GI symptoms (Kandeel et al., 2020; Li et al., 2017). ASD has been linked to the presence of a weak intestinal barrier, allowing the entry of toxins and bacteria into the blood that can signal an immune response and induce inflammation (Li et al., 2017). One study found that first-degree relatives of patients with autism also presented with increased gut permeability, which could be a cause of ASD rather than a consequence (Kandeel et al., 2020; Parracho et al., 2005). As for the microbial population in an individual with ASD, conflicting data exist, but several studies have found a higher abundance of the *Clostridium* species present in the stool of ASD children (Kandeel et al., 2020; Parracho et al., 2005). Specifically, researchers found an abundance of *C. histolyticum*, a known toxin producer, sparking the idea that toxins produced by *Clostridia* may travel along the vagal nerve into the CNS (Kandeel et al., 2020; Parracho et al., 2005). Additionally, maternal factors can shape the neonatal microbiome and influence the development of ASD. One case-control study concluded that breastfeeding for over 6 months is seen to lower the risk of development of ASD when

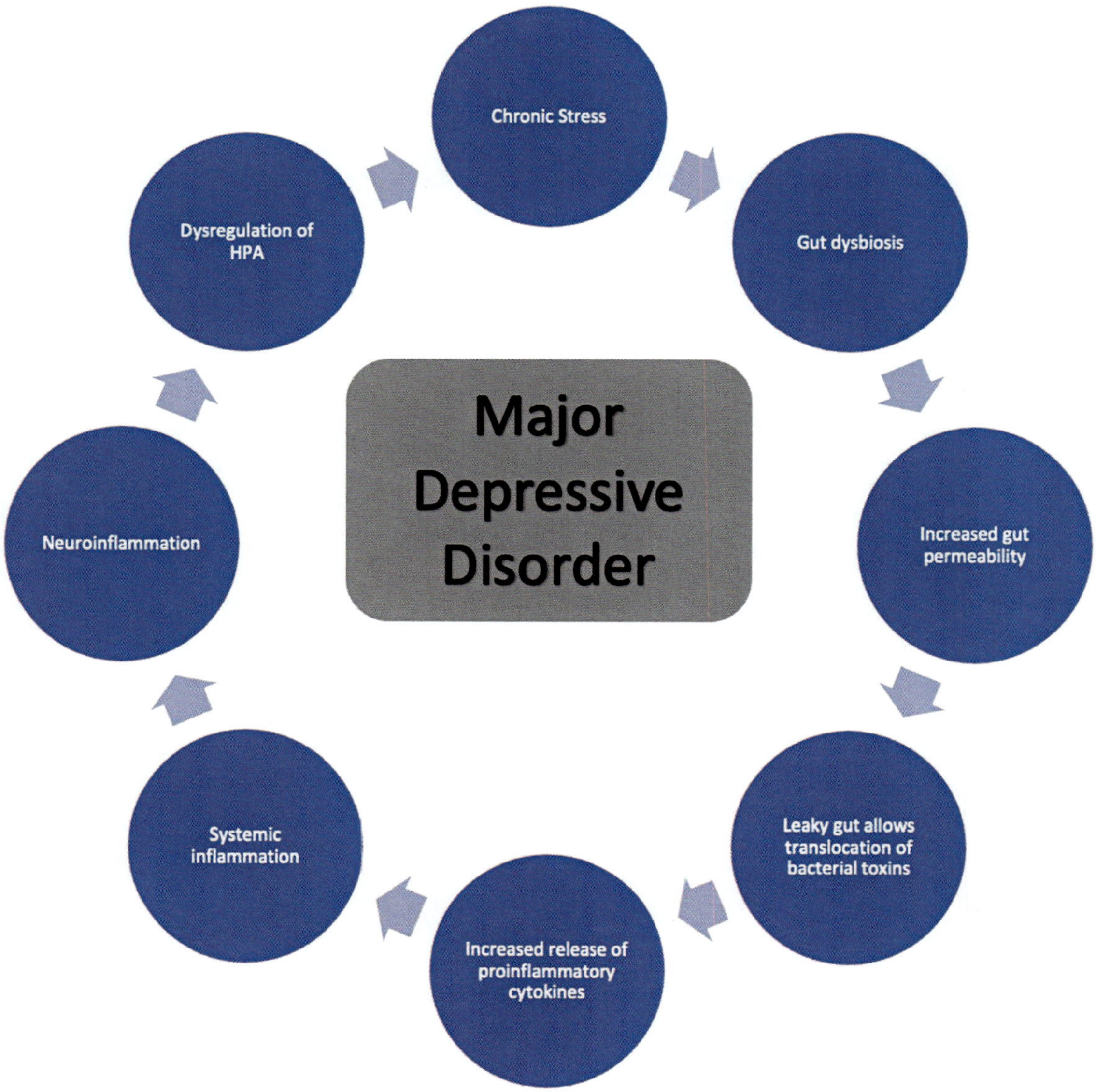

FIGURE 20.3 An alternative theory of major depressive disorder (MDD) involving the gut microbiome. This cycle is a suggested path for MDD that involves the gut microbiome. Chronic stress can lead to dysbiosis of the gut microbiome, unleashing a cascade of events.

compared to an absence of breastfeeding (Schultz et al., 2006). A possible explanation for this includes enhanced immune protection provided by the breast milk (Schultz et al., 2006). As for method of delivery and ASD, birth by caesarian delivery is associated with a significant increase in the risk of ASD (Dominguez-Bello et al., 2016; Li et al., 2017). Individuals who were born via caesarian delivery appear to have a less diverse microbial environment; they lack the exposure to vaginal bacteria that is key in shaping the early microbiome (Dominguez-Bello et al., 2016). This lack of diversity is a common finding in individuals who present with ASD.

20.2.2.2 Neurological disorders

Neurological disorders have mechanisms in relation to the role of the gut microbiome that are very similar to those in psychiatric disorders. The common theme is the idea of gut dysbiosis and disturbances of the gut-brain axis. The neurological disorders that are discussed in this section are Parkinson's disease (PD) and Alzheimer's disease (AD). PD is a progressive neurodegenerative disorder and is considered to be a movement disorder (Yang et al., 2019). Hallmark features of PD include loss of dopaminergic neurons in the substantia nigra and the presence of Lewy bodies in the subcortical region of the brain (Fasano et al., 2015; Yang et al., 2019). While the exact cause of PD is unknown, genetics and environmental factors are probable triggers. Recently, the gut microbiome has become an area of interest in PD. Constipation is a large issue in individuals with PD; it has been reported to occur in 80%–90% of patients, suggesting that alterations to the microbial composition may be responsible (Fasano et al., 2015; Keshavarzian et al., 2015). Unsurprisingly, the gut microbiome in PD is postulated to exhibit an increase in intestinal permeability as well as a cascade of persistent inflammation; however, a unique aspect to PD includes the relationship of the ENS and alpha-synuclein. While Lewy bodies are found in the brain, a study by Keshavarzian et al. showed that patients with PD also had alpha-synuclein deposits in their colon, which were found 2–5 years before the onset of PD symptoms (Keshavarzian et al., 2015; Yang et al., 2019). This study hypothesized that alpha-synuclein deposits begin in the intestines, pointing to the idea that PD begins in the gut. The aggregated proteins can then travel from the gut to the brain via the vagal nerve (Keshavarzian et al., 2015; Yang et al., 2019). Keshavarzian et al. also found that butyrate-forming bacteria such as *Blautia*, *Coprococcus*, and *Roseburia* were decreased in the fecal matter of PD patients, suggesting the importance of this SCFA in relation to gut wall integrity and disease (Keshavarzian et al., 2015) (Fig. 20.4).

AD is the most common form of dementia. It is also important to note that patients with PD can eventually develop a form of dementia, known as PD dementia. Like PD, AD is a progressive, neurodegenerative disease that typically begins with cognitive decline, while motor decline is experienced in later stages. The most prominent characteristic of AD is loss of memory. A few risk factors that have been associated with AD are increasing age, head injury, depression, cardiovascular disease, family history of dementia, and the presence of apolipoprotein E4 (Kim et al., 2009). The apolipoprotein E4 allele appears to be the strongest genetic component in the development of AD (Kim et al., 2009). The pathophysiology of AD includes the presence of extracellular plaques made of beta-amyloid in the brain, intracellular neurofibrillary tangles, and neuronal loss (Kilgore et al., 2010; Kim et al., 2009). Despite the absence of a definitive cause, dysregulation of the gut-brain axis may play a part in AD. In relation to SCFAs, mice serving as models for AD were treated with sodium butyrate and other HDAC inhibitors; the treatment resulted in stabilization of contextual memory and halted cognitive deficits of the mice (Kilgore et al., 2010). An additional study observed an increase of the phylum Bacteroidetes, a gram-negative bacterium in the gut (Kilgore et al., 2010; Vogt et al., 2017). The main component that makes up the outer membrane of gram-negative bacteria is lipopolysaccharide (LPS); LPS itself is an endotoxin and can induce the release of proinflammatory cytokines, which in turn can cause systemic inflammation (Vogt et al., 2017). The findings from this study can be compared to additional research that injected LPS into model AD mice and found that amyloid-beta deposition increased, as did the accumulation of tau, which is the component of the

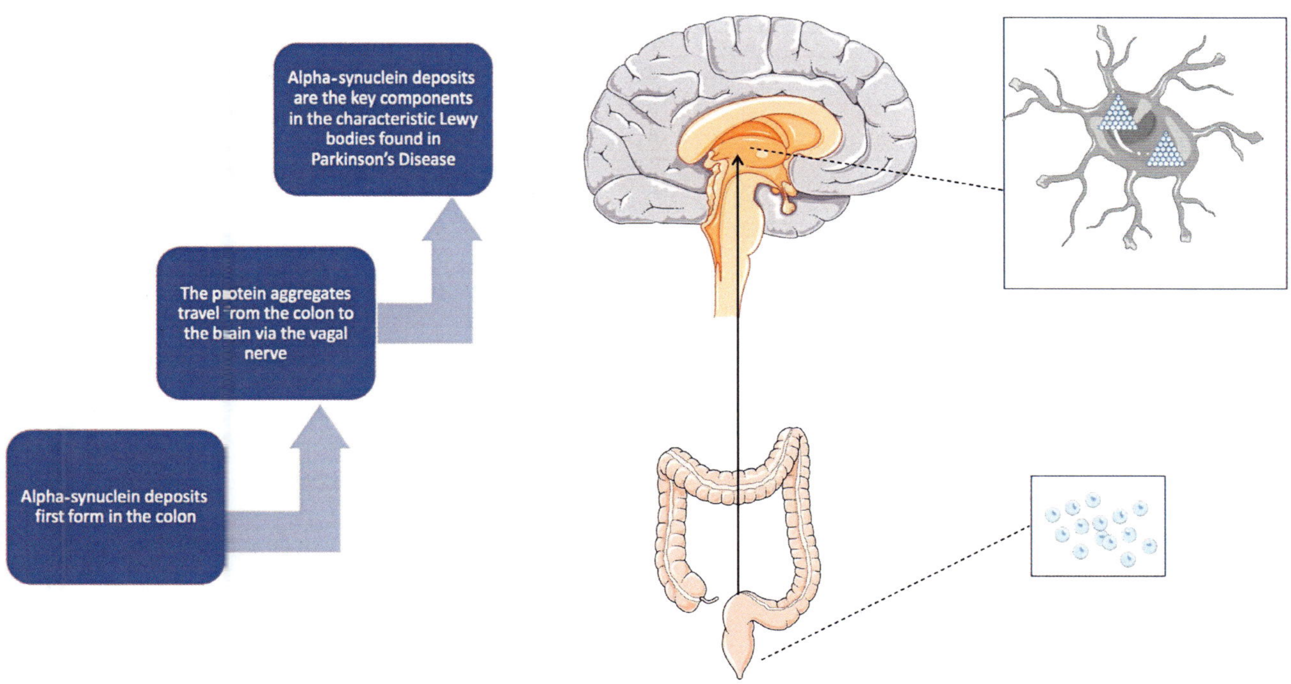

FIGURE 20.4 Parkinson s disease and the gut. Alpha-synuclein deposits have been discussed as first forming in the colon, before any obvious signs of the disease. This indicates that PD may originate in the gut. The aggregate of proteins can travel from the colon to the central nervous system, leading to the characteristic Lewy bodies in Parkinson's disease.

neurofibrillary tangles that are characteristic of AD (Sheng et al., 2003; Vogt et al., 2017). These studies demonstrate the powerful role that chronic inflammation induced by LPS may play in the development of AD. While PD and AD differ in pathophysiology, the mechanisms by which the gut-brain axis and the gut microbiome are involved are similar. Gut dysbiosis, increased intestinal and BBB permeability, and inflammation tend to be the central processes behind these disorders (Fig. 20.5).

20.3 The gut microbiome and Rett syndrome

Now that the gut microbiome has been defined and extensively examined in therapeutics and disease states, let us elucidate the role of the gut microbiome in Rett syndrome (RTT), a rare genetic neurodevelopmental disorder (Smeets et al., 2012). RTT almost exclusively affects females (Amir et al., 1999; Kyle et al., 2018; Smeets et al., 2012). The genetic aspect of RTT comes from researchers' conclusion that RTT is an X-linked dominant disorder due to mutations in the methyl-CpG binding protein 2 (MeCP2) gene (Amir et al., 1999). Owing to its X-linked nature, males who present with the MeCP2 gene mutation that is seen in classic RTT often have a shortened life expectancy (Amir et al., 1999; Kyle et al., 2018). Unlike PD and AD, RTT is not considered to be a neurodegenerative disorder; patients with RTT can live well into adulthood, and while there is early neurological regression, there is no evidence of neuronal death (Kyle et al., 2018). RTT was previously classified as an ASD according to the criteria of the *Diagnostic and Statistical Manual of Mental Disorders* (DSM) but was removed from that classification in the current DSM-5 edition (Tsai, 2015). This removal eliminates the universal classification of individuals who have RTT as also having ASD; this change also served to emphasize the genetic component of RTT (Neul et al., 2010; Tsai, 2015).

20.3.1 Background and history of Rett syndrome

In 1999, RTT was found to be caused by mutations in the X-linked gene that encodes MeCP2, which can activate and repress transcription (Amir et al., 1999; Chahrour and Zoghbi, 2007). RTT occurs almost exclusively in females, affecting 1 in 10,000 female births, leaving patients with severe impairment, including loss of speech, uncontrolled movements, and other abnormalities (Amir et al., 1999; Chahrour and Zoghbi, 2007). Following Down syndrome, RTT is the second most common cause of significant intellectual disability in females (Gold et al., 2018). RTT is generally characterized by normal growth and development early in life, which is followed by slow development. Once this occurs, a wide variety of clinical issues can exist: distinctive hand gestures, loss of purposeful hand movements, gait issues, slowed brain growth, seizure disorders, breathing issues, scoliosis, gastrointestinal issues, and intellectual disability (Smeets et al., 2012; What is Rett Syndrome?, n.d.). More specifically, patients with this disorder will grow and develop normally until the age of 6–18 months, when development is slowed and previous skills that they have achieved will

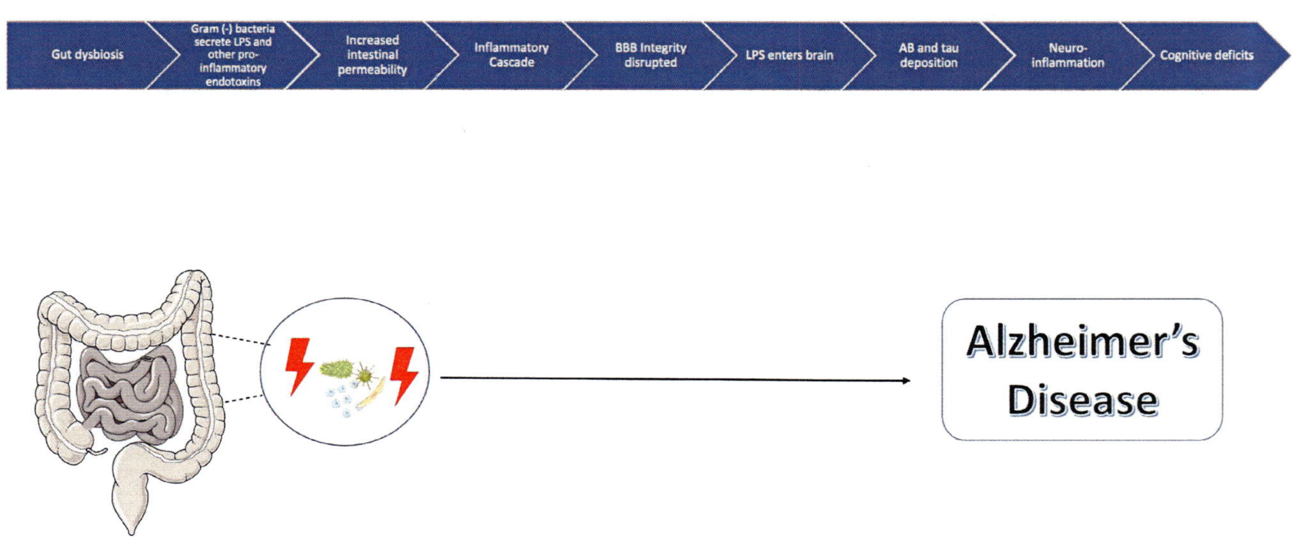

FIGURE 20.5 Gut toxins and Alzheimer's disease. Gram-negative bacteria are capable of releasing lipopolysaccharide (LPS) and other endotoxins upon any kind of disruption, such as gut dysbiosis. The release of LPS can lead to an inflammatory cascade throughout the body. It has been hypothesized that LPS can travel toward the brain, alter its permeability, and induce the formation of amyloid beta and tau.

begin to regress. The condition has defined stages of progression but eventually reaches stagnation. Patients with RTT can live until their seventies but become physically disabled (Chahrour and Zoghbi, 2007). This disorder was first observed in the 1950s by Austrian pediatrician Dr. Andreas Rett. In 1966, Dr. Rett published the first description of RTT. However, RTT did not become widely recognized until 1983, when Swedish neurologist Dr. Bengt Hagberg published a report of RTT (Chahrour and Zoghbi, 2007). The next advancement in the understanding of RTT occurred in 1999, when research fellow Ruthie Amir discovered that the disorder was caused by a mutation in the MeCP2 gene. This led Dr. Huda Y. Zoghbi to discover the location of the gene on the Xq28 site on the X chromosome, proving the disorder to be X-linked (What is Rett Syndrome?, n.d.).

20.3.2 Pathology, symptoms, and progression

The most common method to confirm a diagnosis of RTT is by using a blood test to identify a MeCP2 mutation (Chahrour and Zoghbi, 2007; Neul et al., 2010). RTT can also be diagnosed by way of a fulfillment of the specific diagnostic criteria. The four main criteria for diagnosis are partial or complete loss of acquired purposeful hand movements, partial or complete loss of acquired spoken language, impaired or lost ability to walk, and stereotypic hand movements (Neul et al., 2010). Patients with a mutation in the MeCP2 gene who do not meet the clinical criteria should not be diagnosed with RTT but instead should be diagnosed with a MeCP2-related disorder. Some patients are diagnosed with atypical RTT if they present with symptoms of classic RTT but do not meet all the diagnostic criteria (What is Rett Syndrome?, n.d.).

The symptoms of RTT usually present in girls during early childhood. There are four stages of RTT; stage 1 begins around 6–18 months of age, followed by a rapid regression stage, a plateau phase, and a late motor deterioration stage (Sandweiss et al., 2020; Smeets et al., 2012). As was mentioned previously, patients will initially meet developmental milestones and can begin to walk and speak before the first sign of illness (Chahrour and Zoghbi, 2007). The stalling of development is considered the first stage of RTT and is manifested as a deceleration of head growth known as microcephaly, slowed growth, weight loss, and poor posture (Chahrour and Zoghbi, 2007). As the disease progresses, regression can be seen as patients start to lose communication and socialization skills as well as motor skills (Gold et al., 2018). This marks the second stage of RTT, which most commonly occurs between the ages of 1 and 4 years (Sandweiss et al., 2020). A characteristic symptoms of RTT is loss of purposeful movements of the hands and the development of stereotypic hand movements (Sandweiss et al., 2020). Autistic features can also be seen in RTT, including dysphasia and social withdrawal. An unusual symptom associated with RTT is breathing anomalies in which some patients hold their breath, forcibly expulse air and saliva, and swallow large amounts of air. At this stage, patients will also start to experience a loss of motor coordination, starting with an abnormal gait (Chahrour and Zoghbi, 2007). The third stage, the postregression stage, usually occurs between the ages of 2 and 10 years. In this stage, the phenotype stabilizes, and the patient may even begin to regain some of the skills that were lost (Gold et al., 2018). Seizures of varying severity can be seen in this stage, their occurrence declining as patients reach adulthood. The fourth and final stage of RTT usually begins around the age of 10 years. Patients experience motor deterioration, which can last for decades (Gold et al., 2018). Specifically, patients at this stage can experience motor symptoms of Parkinsonism (Sandweiss et al., 2020). Although these symptoms can be debilitating and leave the patient physically and mentally disabled, RTT is not considered to be a neurodegenerative disorder (Chahrour and Zoghbi, 2007). The most common causes of death in patients with RTT are cardiorespiratory problems (Sandweiss et al., 2020).

20.3.3 Genetic links

The most prevalent genetic origin of RTT includes sporadic loss-of-function mutations in the X-linked gene that encodes for MeCP2 protein (Amir et al., 1999). The gene has been seen to be affected by a large array of different mutations, but eight missense and nonsense mutations make up around 70% of the mutations in RTT. MeCP2 is an epigenetic modulator expressed in the brain that helps to regulate gene expression by binding to methylated DNA (Gold et al., 2018). MeCP2 is thought to affect different phases of neurodevelopment; hence RTT patients have normal early development (Sandweiss et al., 2020). MeCP2 mutations are said to be mostly responsible for RTT; however, this mutation alone does not warrant a diagnosis of classic RTT. Patients can still have RTT if this mutation does not exist and may even have atypical RTT with the presence of a MeCP2 mutation (Gold et al., 2018; Sandweiss et al., 2020). The type and region of the MeCP2 mutation can determine the severity of the RTT phenotype (Amir et al., 1999; Chahrour and Zoghbi, 2007; Sandweiss et al., 2020). Several subtypes of atypical RTT exist, ranging widely in degree of severity (Rett Syndrome Fact Sheet, n.d.).

20.3.4 DNA methylation in Rett syndrome

Around the same time that RTT was being recognized worldwide, advancements in the understanding of DNA methylation were occurring, providing a potential link between the two (Chahrour and Zoghbi, 2007). DNA methylation is an epigenetic mechanism that aids in many crucial cellular processes of the body (Robertson, 2005). Abnormal DNA methylation has been associated with several human diseases, including RTT (Robertson, 2005). Most of the DNA methylation in mammals occurs on CpG dinucleotides (Robertson, 2005). This process of CpG methylation is responsible for the regulation of gene expression via two main mechanisms: the direct inhibition of transcription factors or the recruitment of proteins involved in gene repression (Chahrour and Zoghbi, 2007; Robertson, 2005). While the relationship between MeCP2 and RTT was revealed in 1999, earlier in 1992 British geneticist Dr. Adrian Bird and his colleagues initially discovered MeCP2, a protein that preferentially binds to the methylated CpG (Chahrour and Zoghbi, 2007; Bird, n.d.). It found not only that the gene that encodes for MeCP2 is present on the X chromosome, but also that it represses transcription. Despite this initial finding, further research has revealed MeCP2 to be a multifunctional protein, given its additional ability to activate genes and bind nonspecifically to unmethylated DNA (Hansen et al., 2010). For example, MeCP2 is highly expressed in the CNS, which can explain some of the neurological symptoms that are associated with RTT (Hansen et al., 2010). Generally, symptoms of RTT do not manifest until the age of 6 months, leading to the idea that MeCP2 is more involved with neuronal maintenance and synapse function than with neurogenesis (Ehrhart et al., 2016). Since proper MeCP2 function is lost in RTT, the importance of appropriate DNA methylation and gene expression is suggested in the development of normal brain function.

20.3.5 Biogenic amine systems in Rett syndrome

With increased understanding of the pathophysiology of RTT, a link between MeCP2 and biogenic amine metabolism started to be investigated. The biogenic amine neurotransmitters include dopamine, epinephrine, norepinephrine, histamine, and serotonin (Purves et al., n.d.). Deficits or overexpression of these neurotransmitters are implicated in several disease states. Patients with RTT have been shown to have a reduction in the levels of certain biogenic amines and their respective metabolites (Roux and Villard, 2010; Samaco et al., 2009; Zoghbi et al., 1989). Various research results point toward MeCP2 mutations being able to alter the metabolism of these neurotransmitters and their metabolites. Furthering this notion, one study revealed that the synthetic enzymes that are responsible for the production of dopamine, norepinephrine, and serotonin were reduced in MeCP2-null mice (Samaco et al., 2009). Additionally, an analysis of the cerebrospinal fluid of patients with RTT revealed reduced levels of the metabolites of dopamine, serotonin, and norepinephrine, as compared to control patients (Roux and Villard, 2010; Zoghbi et al., 1989). This study found reduced levels of homovanillic acid, 5-hydroxyindole acetic acid, and 3-methoxy-4-hydroxyphenylglycol; these are the metabolites of dopamine, serotonin, and norepinephrine, respectively. It has been proposed that the characteristic motor movements that are seen in patients could be due to issues with dopaminergic metabolism, with evidence of MeCP2 dysfunction inducing synaptic deficits in the substantia nigra area of the brain (Roux and Villard, 2010; Samaco et al., 2009; Viemari et al., 2005). Research also points toward decreases in norepinephrine being responsible for autonomic nervous system dysfunctions in RTT, such as the breathing abnormalities (Roux and Villard, 2010). One experiment studied the relationship between the respiratory system and norepinephrine in MeCP2-null mice (Viemari et al., 2005). The results showed that MeCP2 deficiency in mice resulted in slow, irregular breathing patterns as well as a reduction in medullary norepinephrine and serotonin levels. The reduction in serotonin is assumed to be due to norepinephrine modulating serotonin metabolism. The researchers in this study also suggested that the reduction in these biogenic amines occurred postnatally (Viemari et al., 2005).

20.3.6 Rett syndrome in males

The X-linked dominant nature of RTT led to the initial thought that the disorder could affect only females. Early research even described RTT as a lethal condition in males; however, this idea was disproved with the discovery of MeCP2 mutations and classic RTT symptoms being present in males. In the case of females with RTT, only one of their two X chromosomes will have the mutated MeCP2 gene, meaning that around half of the female patient's cells will have an improperly functioning MeCP2 gene. Only one of the female X chromosomes will remain active in each cell, while the other is silenced; this phenomenon is called X-chromosome inactivation and can play a large role in the severity of the expressed RTT phenotype [34, 51]. Males, however, are hemizygous for X-linked genes; they only have one X chromosome, causing all cells to be affected by the dysfunctional MeCP2 gene (Trappe et al., 2001; Villard, 2007). This explains the typically more severe and debilitating presentation of RTT in males; since in most cases there is no additional X chromosome in males as there is in females, there is no functioning MeCP2 gene to compensate for the mutation (Sandweiss et al., 2020; What is Rett Syndrome?, n.d.). When this gene mutation is present in males, it can result in a variety of clinical presentations (Chahrour and Zoghbi, 2007; Villard, 2007). Males who present with the full MeCP2 mutation that is seen in classic RTT typically suffer from severe neonatal encephalopathy, which can lead to death in the first year of life. However, this same mutation can also arise from somatic mosaicism or in the instance of an extra X-chromosome (Chahrour and Zoghbi, 2007; Trappe et al., 2001; Villard, 2007). For other MeCP2 mutations the clinical presentation can greatly vary; criteria for RTT are typically not met in these additional circumstances. Overall, the misconception that RTT could be present only in females came from the idea, which was later proven false, that mutations in the MeCP2 gene would be lethal for males, explaining the lack of male RTT diagnoses in the early research of RTT. Studies have also discovered that in the majority of RTT cases, MeCP2 mutations arise from the male germ cell (What is Rett Syndrome?, n.d.). In a study by Trappe et al., it was discovered that the mutations in sporadic RTT are derived from the paternal X chromosome, which is inherited by daughter offspring only (Clarke and Abdala Sheikh, 2018; Trappe et al., 2001). This finding serves as one explanation of the high number of female RTT patients in comparison to males.

20.3.7 Role of different excitatory and inhibitory receptor systems: GABA, NMDAR, and AMPAR

Expanding from the discussion of deficits in the biogenic neurotransmitters in RTT patients, an additional neurotransmitter system has also been linked in this disorder. Gamma-aminobutyric acid (GABA) and glutamate are both neurotransmitters; however, unlike the biogenic amines, GABA and glutamate are amino acid neurotransmitters (El-Khoury et al., 2014). Glutamate is the major excitatory neurotransmitter, while GABA is the major inhibitory neurotransmitter of the CNS (El-Khoury et al., 2014; Johnston et al., 2015). The receptors for glutamate include N-methyl-D-aspartate (NMDA) and alpha-amino-3-hydroxy-5-methyl-4-isoxazole propionic acid (AMPA) receptors; GABA has two subtypes: $GABA_A$ and $GABA_B$ (El-Khoury et al., 2014; Blue et al., 1999). Imbalances between these excitatory and inhibitory systems have been implicated in RTT. Studies have shown various abnormal anatomical findings that point toward this idea of excitatory and imbalance in RTT patients. Postmortem brain tissue of RTT patients was studied to identify the levels of GABA and glutamate receptors (Blue et al., 1999). The researchers related their findings to the ages of the patients to potentially explain the progressive changes in RTT patients. In older RTT patients, the NMDA and AMPA receptor densities were seen to be markedly decreased in the putamen region of the basal ganglia in comparison to that of younger individuals with RTT. The density of GABA receptors was found to be significantly increased in the caudate of the younger RTT patients. The researchers in this study suggested that the overall decreases in receptor density that were found in the older RTT group could be indicative of the plateau phase of the disorder (Blue et al., 1999; Johnston et al., 2015). Additionally, one of the defining features of RTT patients includes the presentation of microencephaly (Blue et al., 1999; Kyle et al., 2018).

These research studies led to potential therapeutic agents. Dextromethorphan, an NMDA receptor antagonist, has been studied for use in treating seizures and cognitive dysfunction in girls with RTT. While a blinded trial is in progress, the unblinded trial found that the dextromethorphan improved some key features of RTT, such as receptive language, behavioral hyperactivity, and the occurrence of seizures (Smith-Hicks et al., 2017). However, it should be noted that this study lacked a placebo arm. Other NMDA receptor antagonists, including memantine and ketamine, are also in discussion as potential therapeutics (Johnston et al., 2015). A separate study discovered that treatment with a GABA reuptake inhibitor increased the lifespan of MeCP2-deficient mice (El-Khoury et al., 2014). When a serotonin 1a reuptake blocker was added to the GABA reuptake blocker, improvements in breathing were also noted (El-Khoury et al., 2014). These results further indicate that GABA and glutamate dysfunction could play a role in RTT (El-Khoury et al., 2014).

20.3.8 Treatment paradigm

Currently, there is no cure for RTT. Treatment for RTT generally focuses on symptom management, which can widely vary between patients (Rett Syndrome Fact Sheet, n.d.). Various curative approaches for RTT have been studied, but they are not without challenges. One approach includes the idea of restoring gene function, which could reverse the neurological deficits from the loss of MeCP2. This therapy could potentially restore neuronal function in patients and reduce or halt symptoms of cognitive impairment and social behavior (Chahrour and Zoghbi, 2007). However, this potentially promising therapy brings about the issue of overcorrecting MeCP2 gene function; the correct dose is critical. Providing too much MeCP2 to cells can be just as harmful as having too little MeCP2 protein (Clarke and Abdala Sheikh, 2018; Sandweiss et al., 2020). This issue is especially relevant in females, as they are not completely devoid of

MeCP2; half of their cells are normally functioning, so delivery of MeCP2 would need to avoid the healthy cells (Sandweiss et al., 2020). An additional treatment option is based on the principle of X-chromosome inactivation discussed previously. The goal of this therapy would be to reactivate the inactive X chromosome, essentially reactivating the silent MeCP2 in every cell (Clarke and Abdala Sheikh, 2018). However, an area of concern for this method includes the risk of activating other X chromosome genes besides the MeCP2; hence this approach would require extreme precision so that only the MeCP2 gene is activated (Clarke and Abdala Sheikh, 2018).

The treatment goals of RTT include providing the most effective pharmacological and nonpharmacological options to improve the patient's quality of life (Brahm et al., 2017). Selecting the right treatment can be accomplished by identifying the patient's developmental changes at each stage of the disorder; as clinical presentations can vary between individuals, there is no treatment algorithm that can effectively encompass every patient. Although these clinical differences exist, there are several common interventions for RTT patients that are important to discuss. For example, RTT patients may undergo therapy for respiratory complications, such as obstructive sleep apnea or other breathing anomalies. Surgery may even be indicated for cases of severe scoliosis (Brahm et al., 2017). Occupational therapy is often used to help patients improve their skills for performing daily activities, while physical therapy is utilized to help prolong mobility and assist with gait issues (Gold et al., 2018; Sandweiss et al., 2020). Often patients need special equipment such as bodily braces or mobility aids such as wheelchairs or walkers. Chronic constipation is extremely common in RTT patients, often prompting intervention from gastroenterologists (Gold et al., 2018). Nutritional programs, social and support services, and special academic care may also be warranted (Rett Syndrome Fact Sheet, n.d.). Up to 60% of patients with RTT experience seizures, making use of antiepileptic medications common in this disorder (Sandweiss et al., 2020; Brahm et al., 2017). Commonly used antiepileptics for RTT are valproate, carbamazepine, lamotrigine, levetiracetam, and topiramate (Brahm et al., 2017; Smith-Hicks et al., 2017). Overall, treatment for individuals with RTT requires a multidisciplinary effort of healthcare providers to improve the patient's quality of life. While no medical cure currently exists, research is advancing, and treatment may begin to extend from symptom management alone.

20.4 Relationship between gut microbiome and Rett syndrome

This section will tie together the seemingly separate concepts discussed above: the gut microbiome and RTT and try to establish a common link between them. The relationship between the two may be surprising initially, given that RTT is a progressive neurodevelopmental disorder, while the gut microbiome would seem to be an altogether different system (Borghi et al., 2017; Strati et al., 2016). To understand the relationship between the gut microbiome and RTT, it is prudent to recall how the CNS and the gut microbiome are intertwined. Again, the gut-brain axis communication network is considered to be bidirectional. This bidirectionality essentially means that the gut microbiome can influence CNS function and, in turn, the CNS can influence the microbial gut composition (Borghi et al., 2017; Liang et al., 2018a). To elaborate on this idea further, the CNS controls gut microbiome composition through satiation signaling peptides that affect nutrient availability, endocrines that affect gut function, and neural pathways. In the other direction, the gut microbiome controls CNS activities through neural (direct activation of neurons by microbiome), endocrine (ECC release of serotonin), metabolic (microbiota synthesis of neuroactive molecules), and immune (CNS-infiltrating immune cells and systemic inflammation) pathways (Borghi et al., 2017; Liang et al., 2018b; Strati et al., 2016). This linkage between the gut microbiome and the CNS is the first step in expanding upon the relationship of the gut microbiome and RTT.

As was discussed previously, RTT is primarily and directly caused by loss of function mutations on the MeCP2 gene, and mutations on this gene are commonly associated with the neurodevelopmental issues that are frequently observed in RTT (Borghi et al., 2017; Strati et al., 2016). The multifunctionality of MeCP2 has prompted research to uncover its various functions, extending beyond its widely known role of suppressing transcription. For example, it has been discovered that MeCP2 plays a role in regulating synaptic communication in the CNS. Specifically, it can facilitate spontaneous neurotransmission and short-term synaptic plasticity, pointing toward a possible explanation of a MeCP2 mutation influencing the slowing of development that is seen in RTT patients (Borghi et al., 2017; Johnston et al., 2015). Aside from neurological symptoms in RTT, it is known that patients frequently have an array of GIT issues. The GIT symptoms of RTT have mainly been attributed to the presence of gut dysbiosis. We know that a commonly described idea includes the presence of a dysbiotic gut microbiome inducing increased permeability of the intestinal lining, which can lead to a systemic spread of inflammation, including neuroinflammation. However, as in other disorders that point toward gut dysbiosis, the exact etiology of this is not completely understood, but it is a potential step in understanding the widespread of presentations seen in RTT patients.

More specific to RTT, some researchers point toward the idea that the MeCP2 mutation influences gut dysbiosis via ENS impairment. MeCP2 is widely expressed in ENS neurons, suggesting a correlation between its mutative state and the presence of GIT abnormalities and a faulty communication pathway between the ENS and CNS, but this has yet to be definitively confirmed (Borghi et al., 2017; Strati et al., 2016). An important point to remember is that there are variations of the MeCP2 mutation that can lead to varying degrees of severity of the RTT phenotype, and there are also mutations that do not involve MeCP2 at all (Sandweiss et al., 2020; Strati et al., 2016). This is significant to note because patients with higher clinical severity are generally shown to have a more reduced microbial diversity (Borghi et al., 2017; Borghi and Vignoli, 2019). Once more definitive research of the role of the gut microbiome role in RTT becomes available, increased therapeutic options may be able to be discovered.

20.4.1 Effect on gut microbiome diversity and dysbiosis

Current research is targeted at determining the role that diet, genetics, age, and physical factors play in the gut microbiome dysbiosis that is observed in RTT. By understanding the significance of the previously listed factors, one can start to understand potential treatment options for RTT. The reduced microbial diversity that is associated with RTT is a feature similar to that of psychiatric and neurological disease states that were examined previously. The altered microbial content of the gut may influence the common GIT side effects as well as the severity of the RTT phenotype (Borghi and Vignoli, 2019). Specific to RTT, constipation is a well-documented concern; a recent survey of females with RTT found that approximately 80% of the cohort experienced constipation, as well as 92% of the group having GIT problems in general (Motil et al., 2012). GIT issues in RTT patients have led to the study of gut dysbiosis as potentially being the causative factor. Specifically, RTT patients have been found to have less alpha diversity in their gut microbiome (Borghi et al., 2017; Borghi and Vignoli, 2019). Alpha diversity metrics allow for the study of microbial diversity, such as in terms of richness and evenness (Borghi and Vignoli, 2019; Strati et al., 2016;). Decreased alpha diversity was found in separate studies by Strati et al. and Borghi et al.; both researchers examined alterations in the gut microbiome of RTT patients (Borghi et al., 2017; Strati et al., 2016). Strati used three different alpha diversity estimators; all three estimators displayed a significantly less diverse bacterial gut microbiota compared to a healthy cohort (Strati et al., 2016). The results from Borghi had similar alpha diversity tests, but Borghi did not observe a statistically significant change in alpha diversity (Borghi et al., 2017). His team concluded that this was because they did not have enough subjects enrolled in the study to reach statistical significance (Borghi et al., 2017). Beta diversity analyses were also incorporated into both of these studies. Beta diversity involves the differences present between microbial communities (Borghi and Vignoli, 2019; Strati et al., 2016). The Strati group used weighted and unweighted UniFrac distances to determine beta diversity between a healthy cohort, an RTT with constipation cohort (RTT-C), and an RTT without constipation cohort (RTT-NC) (Strati et al., 2016). The beta diversity test found a significant difference in gut microbial diversity between the healthy cohort and both RTT cohorts (Strati et al., 2016). However, the beta diversity tests did not find a significant difference between the RTT-NC and RTT-C cohorts (Strati et al., 2016). This finding is important because it indicates that the changes in the gut microbiome in individuals with RTT seem to be independent of the presence of constipation (Strati et al., 2016). Borghi et al.'s study included similar beta diversity tests; it also used unweighted and weighted UniFrac distances to determine beta diversity. The researchers observed a separation between the centroids of the RTT and control group based on the unweighted UniFrac distance, but it was not statistically significant; the weighted UniFrac results were also not significant (Borghi et al., 2017).

To compare the gut microbial contents found in the two studies, Strati et al.'s findings include increases seen in *Bifidobacterium*, Clostridia, and *Candida* as being the main bacterial components responsible for the RTT dysbiotic gut (Strati et al., 2016). As for Borghi's findings, his study highlighted an abundance of *Bacteroidaceae*, *Clostridium* spp., and *Sutterella* spp. as being influencers of gut dysbiosis (Borghi et al., 2017). Borghi et al. also observed some bacterial taxa that increased with the patient's degree of clinical severity, including *Bacteroidaceae*, *Enterobacteriaceae*, and *Erysipelotrichaceae* (Strati et al., 2016). This is an important point of overlap between the two studies. While Strati et al. do not display these particular results at the family level, they observed that a few genera, *Escherichia*, *Shigella*, and *Erysipelotrichaceae incertae sedis*, were significantly elevated in RTT; these bacterial genera are classified in the bacterial families Enterobacteriaceae and Erysipelotrichaceae mentioned by Borghi et al. (Borghi et al., 2017; Strati et al., 2016). A key point to highlight is the presence of an abundance of *Clostridium* in both studies. *Clostridium* is also found in abundance in patients with ASD, as previously examined. Despite RTT no longer being considered an ASD, the two disorders share GIT side effects, suggesting the importance of an altered level of *Clostridium* in this aspect of symptomatology.

Another key observation found in the Strati and Borghi studies of patients with RTT was about increased levels of fecal SCFAs. The main SCFAs that were found to have elevated fecal concentrations in the Borghi study were butyrate and propionate, despite their increase not being significant; however, significant increases were seen in isobutyrate and isovalerate (Borghi et al., 2017). Isobutyrate and isovalerate are considered to be branched SCFAs, commonly referred to as branched-chain fatty acids (BCFAs) (Borghi et al., 2017; Koh et al., 2016). The BCFAs are products of protein fermentation. Strati et al.'s study had similar SCFA findings. They observed a significant increase in fecal propionate, isovalerate/2-methylbutyrate and isobutyrate (Strati et al., 2016). These findings are comparable to the findings from Borghi et al. It is clear that SCFAs are important for colonic health and various metabolic processes (Barko et al., 2018; Borghi and Vignoli, 2019; Koh et al., 2016; Strati et al., 2016). The increased fecal amounts of SCFAs in relation to the GIT effects in RTT, such as constipation, have yet to be fully elucidated. In general, studies of fecal SCFA concentrations vary in their findings and rationale. Despite this, some researchers have observed increased fecal SCFA concentrations as being linked with gut dysbiosis (Barko et al., 2018; Koh et al., 2016; Motil et al., 2012). One explanation could be due to the intestinal inflammation seen in RTT causing reduced SCFA absorption by the colonocytes as a result of an impaired ability to utilize the SCFAs (Koh et al., 2016; Yang et al., 2015). Additionally, the fermentation process to produce SCFAs and BCFAs are completed by some *Bifidobacterium*, *Bacteroides*, and *Clostridium* species (Magne et al., 2020; Motil et al., 2012; Strati et al., 2016). Owing to the overall increase in these species of bacteria throughout the two studies, it could allow for an enhanced rate of fermentation that allows high levels of SCFAs to be delivered to the colon and hence the higher fecal amounts.

Overall, there are common findings from the Strati and Borghi studies in terms of gut microbiome dysbiosis in RTT patients. Both studies found a decrease in alpha diversity as well as an increase in fecal SCFA/BCFA concentrations. These findings help to further the theorized idea of MeCP2 mutations inducing gut dysbiosis, leading to inflammation and GIT dysfunction. The two studies displayed important key similarities of bacterial abundance in RTT patients, but differences were also present. Potential explanations will be discussed further, but it should be noted that the relationship between the gut microbiome and RTT is still a relatively new concept, warranting additional investigation to address the disparities between the two studies.

20.4.2 Recent and past studies

At this point in time, few studies have been completed that examine which bacterial phyla contribute to the gut microbiome dysbiosis in RTT; the research by Strati et al. and Borghi et al. are two of the only studies available on this topic. This section will help to explain in greater detail the significance of their findings in terms of clinical aspects of RTT. In reviewing the phylum-level analyses from Borghi et al. and Strati et al., a significant discrepancy is seen immediately. Strati et al. observed that Actinobacteria was the most abundant phyla within their RTT cohort (Strati et al., 2016). Borghi's team observed that the predominant bacterial phyla were Bacteroidetes and Firmicutes (Borghi et al., 2017). This observation made by Borghi et al. is not unusual as the phyla of Bacteroidetes and Firmicute typically make up 90% of the gut's bacterial community in healthy subjects (Koh et al., 2016; Magne et al., 2020; Mariat et al., 2009). While these findings differ, this is not entirely unusual in discussing a novel disease. Studies such as these tend to have difficulty accounting for an array of confounders. In regard to bacterial, fungal, and SCFA concentrations, factors such as age, body mass index (BMI), diet, and medications can all affect a host's gut microbiota concentration (Magne et al., 2020; Mariat et al., 2009; Riva et al., 2017; Wong et al., 2006). SCFA concentrations can also be affected by the concentrations of some bacterial species (Wong et al., 2006).

The Firmicutes/Bacteroidetes ratio can be said to be an estimator of intestinal dysbiosis (Magne et al., 2020; Mariat et al., 2009). This makes sense from a statistical point of view; if 90% of the bacterial gut flora come from two phyla and the ratio changes considerably, this can mean that the concentrations of the gut bacterial flora must also be changing. Strati et al. discussed observing a change in the ratio caused by a significant increase in the Actinobacteria phylum (Strati et al., 2016). A change in the Firmicutes/Bacteroidetes ratio was not determined to be significant by Borghi et al. (2017). Once again, disparate results are not entirely uncommon, and in this case, there may be a few explanations. One of the possible confounders that could explain the discrepancies in the gut bacterial flora is the age of the patient population. The average age of subjects in the Strati et al. study was 12 years, compared to the average age of 23 years in the Borghi et al. study (Borghi et al., 2017; Borghi and Vignoli, 2019; Strati et al., 2016). An additional study done by Mariat et al. characterized how changes in the gut microbiome occur with age; RTT was not a factor in this study (Mariat et al., 2009). This study found that *Bifidobacterium* species were the most abundant group measured in the fecal content of infants. It is important to note that this study observed that the abundance of *Bifidobacterium* decreases significantly between infancy and adulthood (Mariat et al., 2009). Because the Strati et al. study predominantly used adolescent RTT patients, it is worth examining how the youth of their subjects influenced the presence of a large abundance of *Bifidobacterium* (Strati et al., 2016).

The study done by Mariat et al. additionally found that the Firmicutes/Bacteroidetes ratio changes with age (Mariat et al., 2009). They found that the Firmicutes/Bacteroidetes ratio for infants, adults, and the elderly was 0.4, 10.9, and 0.6, respectively (Strati et al., 2016). In relation to age, the Bacteroidetes concentration seems to be roughly similar between infants and adults but is increased in the elderly (Strati et al., 2016). The concentration of Firmicutes appears to decrease progressively with age. Because the concentrations of Bacteroides are relatively similar between infants and adults, the observed decrease in the Strati et al. study appears to not be attributable to age. Borghi et al., in an attempt to account for confounders, measured changes in flora based on their patients' BMI or diet (Borghi et al., 2017). Borghi et al. decided to include BMI because a previous study by Riva et al. focused on how changes in BMI are correlated with pediatric gut microbiome changes (Riva et al., 2017). Riva et al. observed that obese subjects were enriched in bacterial species of the Firmicutes phylum and depleted in species belonging to the Bacteroidetes phylum (Riva et al., 2017). A loss of Bacteroidetes and an increase in Firmicutes caused an increase in the Firmicutes/Bacteroidetes ratio (Riva et al., 2017). Because Strati et al. did not account for the BMI of their subjects, it is difficult to say with any certainty whether patient BMI had a role in the gut microbiota dysbiosis.

With diet, a factor that can affect bacterial and SCFA concentrations, Borghi et al. and Strati et al. took different approaches to account for the impact that diet could have on results. Strati et al. ensured that each patient was following a Mediterranean diet (Strati et al., 2016). While this does not mean that each patient was eating the exact same meals, it limited the variety of food to a narrower spectrum. Borghi et al., by contrast, recorded the patients' meals and reviewed the results afterwards (Borghi et al., 2017). By not keeping subjects to a specific diet, Borghi et al. were able to identify specific trends (Borghi et al., 2017). They observed that the RTT patients had a decrease in consumption of carbohydrates and fiber but a greater consumption of protein, primarily from animal protein, compared to their healthy counterparts (Borghi et al., 2017; Borghi and Vignoli, 2019). The difference in diet between the two studies indicate the possibility of diet being contributory to the variation in findings in regard to bacterial and SCFA concentrations. The method by which Borghi et al. observed the influence of diet allows for an interesting clinical aspect. For example, a study by Wong et al. stated that SCFAs were created primarily by carbohydrate fermentation, while BCFAs were produced primarily from protein fermentation (Wong et al., 2006). This principle, combined with the findings from Borghi et al., could help to explain why RTT subjects were found to have increased BCFAs (Borghi et al., 2017; Strati et al., 2016; Wong et al., 2006).

The information provided in the preceding discussion can assist in illustrating how multiple factors can contribute to the gut microbiome dysbiosis that is seen in RTT. A few of these changes may provide some clinical significance; by observing how diet can change microbial concentrations, clinicians may be able to target diets that could result in desirable microbial changes. Essentially, these clinical aspects are centered upon improving a patient's disease status by changing the composition of the gut microbiome.

20.4.3 Rett syndrome and other neurological and psychiatric disorders that share gut microbiome alterations

While changes to the gut microbiome have only recently been studied in RTT, other disorders that have been more extensively studied seem to share a common pattern of gut microbiome dysbiosis. These disorders include ASD, PD, and epilepsy. ASD and PD were examined earlier in the chapter; here, findings discussed earlier will be revisited to serve as a comparison to RTT. It is necessary to recall the clinical definitions of these disorders; ASD is a neurodevelopmental disorder, like RTT, but it is also classified as a psychiatric disorder, while PD is a neurodegenerative disorder. Seizures are common in RTT, but epilepsy has yet to be examined and will be introduced here. Epilepsy is defined as neurological disorder characterized by recurrent seizures that are seemingly unprovoked (Stafstrom and Carmant, 2015). These definitions are important in understanding the pathophysiology of each disease, which aids in determining the role of the gut microbiome in these conditions. The role of the microbiome-gut-brain axis has been implicated in all of these disorders as well as other neurological and psychiatric conditions; specifically, disturbances to this communication pathway resulting in widespread inflammation seem to be the main theory (Quigley, 2017).

To briefly revisit gut microbiome alterations in ASD and discuss its similarities with RTT, ASD patients, like individuals with RTT, commonly present with GIT issues, such as abdominal pain, constipation, diarrhea, and bloating (Bird, n.d.; Brahm et al., 2017; Johnston et al., 2015; Villard, 2007). The increased severity of GIT dysfunction in RTT has been linked to a more severe disease presentation; this is also commonly seen in ASD (Borghi and Vignoli, 2019).

A study completed in Slovakia observed a strong, statistically significant, positive correlation between GIT symptoms and severity of ASD based on scores from the Autistic Diagnostic Interview scale (Tomova et al., 2015). Furthermore, on the basis of the study by Parracho et al. that was mentioned previously, it was determined that high levels of *C. histolyticum* were found in fecal samples of ASD patients, suggesting that this overabundance may have some impact on the GIT dysfunction in ASD (Parracho et al., 2005; Tomova et al., 2015) The Slovakian study also determined that *Clostridium* was more common in patients with ASD than in a healthy cohort (Tomova et al., 2015). Additionally, it is important to keep in mind that *C. histolyticum* is a known toxin producer. We previously examined the possibility of its metabolites spreading to the brain and causing systemic effects (Parracho et al., 2005). Along with the GIT effects, this bacterial abundance can perhaps explain the behavioral and other neurological effects in RTT, as high fecal levels of *Clostridium* species were also found in the studies of both Strati et al. and Borghi et al. Despite the differences between RTT and ASD in terms of etiology and certain aspects of clinical presentation, the role of the gut microbiome in the two conditions does appear to be rather similar.

Recall that PD patients often experience severe constipation, leading to the research into how the gut microbiome may be involved. Constipation is often the most reported GIT effect in patients with RTT; it has been said that 80%–90% of patients with PD experience constipation (Fasano et al., 2015; Keshavarzian et al., 2015; Wong et al., 2006). This GIT issue largely follows the hypothesis that was previously mentioned, including increases and intestinal permeability and inflammation. In the studies by Strati et al. and Borghi et al., both groups saw increases in certain SCFAs in the fecal matter of patients with RTT. For PD we mentioned earlier that Keshavarzian et al. found a decrease in bacteria that formed the SCFA butyrate (Keshavarzian et al., 2015). This study hypothesized this finding as a reason for the GIT issues that are seen in PD, as butyrate is important in GIT function. The findings from the RTT studies by Strati et al. and Borghi et al. do not follow this finding of Keshavarzian et al. exactly, particularly those of Borghi et al.; Strati et al. do not mention butyrate as a finding, while Borghi et al. found a nonsignificant increase in butyrate. Despite the dissimilarities in this aspect of SFCAs, it is clear that SCFAs play a role in GIT health, and various factors can affect the amounts of SCFAs that are present in fecal matter. Both PD and RTT have movement issues, such as gait difficulties. Some RTT patients can even be described as having Parkinson-like symptoms. Scheperjans et al. investigated the gut microbiome in PD and found that "the relative abundance of Enterobacteriaceae was positively associated with the severity of postural instability and gait difficulty" (Scheperjans et al., 2015). This can be applied to Strati et al. and Borghi et al., who both found increases in Enterobacteriaceae as well.

Seizures are common in RTT patients, and some patients with RTT also have epilepsy; however, it is common for a misdiagnosis of epilepsy to occur in this patient group. Many of the stereotypical movements of RTT can be mistaken for epileptic seizures (Operto et al., 2019). Epilepsy can be accompanied by abdominal issues or complaints of cramping, nausea, and vomiting (Kocamaz and Işıkay, 2019; Operto et al., 2019). A specific type of epilepsy, termed abdominal epilepsy, is generally defined by these symptoms plus the typical CNS disturbances (Kocamaz and Işıkay, 2019). The cause of abdominal epilepsy is unknown, but dysfunction of the gut-brain axis may be implicated, owing to the presence of both autonomic nervous system (which encompasses the ENS) and CNS malfunctions. Any alteration to the gut microbiome can increase the person's susceptibility to having seizures, such as the use of antibiotics, infection, and stress (Lum et al., 2020). These factors have all been seen to alter the gut microbiome, driving it toward a state of dysbiosis that can potentially lead to a cascade of inflammation; the influx of inflammatory cytokines is key in epilepsy (Gong et al., 2020; Lum et al., 2020). In the gut microbiome of patients with epilepsy, Gong et al. observed a decreased alpha diversity in comparison to their healthy cohort, which again ties in similar findings from Strati et al. and Borghi et al. Gong et al. (2020) specifically found that patients with epilepsy had increases in fecal *Actinobacteria* and *Verrucomicrobia* and decreases in *Proteobacteria* at the phylum level; at the genus level, there were increases in *Prevotella_9*, *Blautia*, *Bifidobacterium* (Gong et al., 2020). When these results are compared to those of Strati et al. and Borghi et al., similarities are found in terms of increases in *Actinobacteria*, *Verrucomicrobia*, *Blautia*, and *Bifidobacterium*. However, unlike Gong et al., Strati et al. and Borghi et al. did not find decreases in *Proteobacteria* or an increase in *Prevotella* (Borghi et al., 2017; Gong et al., 2020; Strati et al., 2016). Of importance, *Prevotella* is known to promote inflammation, possibly as a result of overproduction of the proinflammatory cytokine IL-6 (Gong et al., 2020). The absence of this finding in the work of Strati et al. and Borghi et al. should not be deemed unusual. While seizure disorders can occur in RTT, it is not a defining criterion of the disorder. However, a comparison of these studies can still serve as a model of gut microbiome dysbiosis between epilepsy and RTT. While studies of each of the above disease states may contain differing results, they all echo the same theme: Each of these disease states, while uniquely different, can be possibly characterized by a dysbiosis of their respective gut microbiomes. RTT, as a neurodevelopmental disorder, seems to follow a pattern of gut microbiome dysbiosis similar to those of these neurological and psychiatric disorders.

20.4.4 Intestinal inflammation in Rett syndrome

So far, the idea of inflammation has been heavily connected to gut microbiome dysbiosis in various disease states, ranging from the gut itself to the CNS. In RTT the presence of intestinal inflammation has been seen to influence patient presentation and even the pathophysiology of the disease (Pecorelli et al., 2020). As we discussed earlier in the chapter, when the gut microbiome's equilibrium is disturbed, a cascade of events can unfold; the dysbiotic gut can lead to increases in the permeability of the gut lining, which can then allow the entry of harmful substances that can result in systemic inflammation (Braniste et al., 2014; Li et al., 2017). In a review of the inflammatory statuses of RTT patients by Pecorelli et al., findings show that abnormalities of humoral and cell-mediated immunity together with chronic low-grade inflammation in multiple organs not only represent clinical manifestations of RTT, but also can contribute to its development and deteriorating course (Pecorelli et al., 2020). From the work of Strati et al., they determined that patients with RTT suffered from mild intestinal inflammation because of the finding of elevated erythrocyte sedimentation rates, fecal calprotectin, and IgA concentrations among their RTT subjects (Strati et al., 2016). All of these can be considered to be biomarkers of inflammation; in particular, IgA abundance can be due to increased protein uptake from a damaged intestinal lining (Pecorelli et al., 2020). Borghi et al. do not mention inflammatory biomarkers but rather noted that Erysipelotrichaceae, a bacterial family that is linked to inflammation, had a severity-related increase in abundance (Borghi et al., 2017). The *Erysipelotrichaceae* species have been noted to alter GIT permeability through their proinflammatory effects (Borghi and Vignoli, 2019).

In studies of MeCP2-null mice, dysregulated inflammatory responses have been noted, with increase of TNF-alpha release (Pecorelli et al., 2020). These results work show that the MeCP2 gene mutation in RTT can have a direct effect on increased inflammation. Additionally, to study immune dysregulation in RTT, Byiers et al. observed an increase in the salivary cytokines of IL-1β, IL-6, IL-8, IL-10, GM-CSF, TNF-α, and VEGF in RTT patients (Byiers et al., 2020). These results showed that increases in salivary cytokines had a significant positive correlation with RTT clinical severity, indicating further research on inflammatory responses in RTT (Byiers et al., 2020).

In research by Shin et al., gut inflammation is said to be caused primarily by a divergence from proper interactions between the epithelium, gut microbiome, and immune components within the gut (Shin and Kim, 2018). This study viewed the effects of epithelial barrier dysfunction on inflammatory responses in a human gut on a chip to demonstrate the human intestine and its cross-talk with the microbiome; it was not a study specific to RTT (Shin and Kim, 2018). Their research found that proinflammatory cytokines such as IL-1β, IL-6, and TNF-α were not produced when the villus epithelium layer was challenged with dextran sulfate sodium (DSS), which is commonly used to cause colitis in animal models, or peripheral blood mononuclear cells (commonly lymphocytes or monocytes) (Shin and Kim, 2018). These additions induced oxidative stress but not the production of the above cytokines (Shin and Kim, 2018). When luminal components, *E. coli* or LPS, were simultaneously added to the DSS-epithelium, there was a large production of the proinflammatory cytokines (Shin and Kim, 2018). In simpler terms, the inflammatory response from an impaired gut barrier occurred in the presence of gut microbiome components: *E. coli* and LPS (Shin and Kim, 2018). In this study, intact epithelial barriers, those not treated with DSS, were seen to avoid the inflammatory cascade when treated with *E. coli* or LPS. This work done by Shin et al. can potentially be applied to the work done by Strati et al. Strati et al. described RTT patients as having a reduced mucin layer caused by the inhibition of goblet cells as a result of abnormally high SCFA levels (Strati et al., 2016). DSS-treated gut epithelium resulted in goblet cell loss, and the loss of goblet cells results in a weak intestinal barrier that is vulnerable to microbial attacks and hence inflammation (Strati et al., 2016; Shin and Kim, 2018). The pathogenesis of the inflammation that is seen in RTT is not fully understood, but from all of these studies, it is clear that the gut microbiome can play a role in intestinal inflammation, whether induced by the MeCP2 gene itself or through issues stemming from gut dysbiosis. Research into why RTT patients have intestinal inflammation is scarce; however, as we have discussed, research from other disease states and the gut microbiome can be applied in considering RTT.

20.4.5 Rett syndrome therapeutics and the gut microbiome

As RTT currently does not have a cure or a proper treatment regimen to fully restores cognitive and physical functioning, patients are left with few options. In terms of therapy targeted toward the gut microbiome, few options have been studied. Restoring the equilibrium to the gut microbiome may be able to decrease the severity of the RTT presentation, given that a more dysbiotic gut microbiome is related to a more severe disease (Borghi et al., 2017; Borghi and Vignoli, 2019). Normalizing the gut microbiome could lead to enhanced gut-brain communication, providing improvements in GIT disturbances as well as cognitive function. Given that RTT is caused by MeCP2 mutations, these have remained the target of curative treatment. To briefly revisit the MeCP2 treatment approaches, the two main ideas have been through direct MeCP2 restoration or by reactivation of the inactive X chromosome (Clarke and Abdala Sheikh, 2018; Vashi and Justice, 2019). Despite restoration of MeCP2 being the desired drug target, these two treatment approaches have proven to be difficult to implement both safely and effectively. The first approach comes with the risk of essentially overdosing on MeCP2, and the second presents concerns involving other X-chromosome loci being activated along with the MeCP2 gene (Vashi and Justice, 2019). Both treatment approaches require careful precision. Treatment options involving downstream pathways of MeCP2 have also been brought to attention. For the most part, the downstream treatment of RTT is targeted at improving the patient's symptoms. Ketamine acts on one of the downstream targets. It is an NMDA receptor agonist and can affect neurotransmitter signaling, which has been shown in mice studies to improve cortical processing and connectivity in RTT (Getachew et al., 2018; Vashi and Justice, 2019). Unfortunately, these studies did not discuss changes to the gut microbiome as a result of therapy. Because these studies were completed in MeCP2-null mice, the effectiveness of ketamine in RTT human patients is not fully known.

Despite the lack of data on how treatment of RTT with ketamine might change the gut microbiome, studies of ketamine as an antidepressant have discussed its interaction with the gut microbiome in rat models (Getachew et al., 2018). To expand upon this, Getachew et al. observed that the administration of ketamine in rats caused increases in *Lactobacillus*, *Turicibacter*, and *Sarcina* [153]. The researchers also determined that the concentrations of the opportunistic pathogens *Mucispirillum* and *Ruminococcus* were significantly decreased (Getachew et al., 2018). Low levels of *Lactobacillus* and *Turicibacter* are commonly found in depression (Getachew et al., 2018). High levels of *Ruminococcus* are associated with inflammation and severe IBS; some *Mucispirillum* species are associated with intestinal inflammation (Getachew et al., 2018). While these findings are not specific to RTT, the findings may still be applicable. Patients with RTT are estimated to have comorbid depression at a rate of 14%–16%; patients with MDD frequently have GIT disorders, possibly resulting from intestinal inflammation (Borghi et al., 2017; Shin and Kim, 2018; Strati et al., 2016). Given that depression and intestinal inflammation are associated with RTT, ketamine may prove to be an effective option for patient symptomatology. Despite this, in a comparison of the results of Getachew et al. to those of Strati et al. and Borghi et al., important differences are present that can undermine the use of ketamine as an RTT therapeutic. From the work by Strati et al., *Lactobacillus* and *Turicibacter* were found to be elevated in RTT, the opposite of what is found in depression. Also, *Ruminococcus* was slightly depleted in RTT patients in comparison to controls, which somewhat weakens the argument of ketamine being helpful for intestinal inflammation in RTT (Strati et al., 2016). Borghi et al. also noted a decrease in *Ruminococcus* (Borghi et al., 2017). Overall, the changes that are seen in the gut microbiome from ketamine treatment in RTT patients requires investigation, as there are large disparities in comparing posttreatment gut microbiomes of rat models to that of untreated RTT patients. Some patients with other disorders report symptom relief from probiotic supplementation. A review article looking at probiotic supplementation use for IBS determined that when using multistrand probiotics, most patients found a significant improvement in symptoms and overall quality of life (Fjeldheim Dale et al., 2019). While patients with IBS do not necessarily suffer from a neurological disorder such as RTT, they do suffer from intestinal inflammation and a dysbiotic gut microbiome (Fjeldheim Dale et al., 2019; Sivamaruthi et al., n.d.). A literature search for the use of probiotics in RTT yielded scarce results, indicating that there is little to no published evidence at this time for their use. However, research has been completed to observe the effects of probiotics in ASD. Given that RTT was once considered to be an ASD, certain aspects of ASD studies may be applicable for RTT. The results from probiotic use in ASD showed patient improvement in anxiety, antisocial behavior, communication issues, and some gastrointestinal discomfort (Fjeldheim Dale et al., 2019; Sivamaruthi et al., n.d.). The relationship of the some of the signs and symptoms that the two disorders share may provide insight into beneficial symptom relief options for RTT.

20.5 Summary

To summarize, treatment for RTT remains limited. While therapeutics targeted to improve common symptoms are available, they are not cures. Genetic modification is at the center of a potential RTT cure, but successful treatment in animal models must occur first. RTT is a fairly rare disorder, and as result, the available research is unfortunately less abundant than is the case for other neurological disorders. Further RTT-specific research into the role of the gut microbiome and its communication network is needed. The two RTT-specific studies discussed, those by Strati et al. and Borghi et al., make up some of the only data available on the interplay between RTT and alterations of the gut microbiome. Many research projects and clinical trials working on therapeutic options for RTT appear to be coming in the near future. We greatly hope that from their work we can obtain a better understanding of how the gut microbiome and RTT are connected and thus be able to provide treatment options that expand from treating symptoms alone.

References

Amir, R.E., Van Den Veyver, I.B., Wan, M., Tran, C.Q., Francke, U., Zoghbi, H.Y., 1999. Rett syndrome is caused by mutations in X-linked MECP2, encoding methyl- CpG-binding protein 2. Nat. Genet. 23. Available from: https://doi.org/10.1038/13810.

Barko, P.C., McMichael, M.A., Swanson, K.S., Williams, D.A., 2018. The gastrointestinal microbiome: a review. J. Vet. Intern. Med. 32, 9–25. Available from: https://doi.org/10.1111/jvim.14875.

Bear, T.L.K., Dalziel, J.E. Coad, J., Roy, N.C., Butts, C.A., Gopal, P.K., 2020. The role of the gut microbiota in dietary interventions for depression and anxiety. Adv. Nutr. 11, 890–907. Available from: https://doi.org/10.1093/advances/nmaa016.

Bird, A., n.d. Wellcome Centre for Cell Biology. Available from: <https://www.wcb.ed.ac.uk/research/bird> (accessed 09.01.21).

Blue, M.E., Naidu, S., Johnston, M.V., 1999. Altered development of glutamate and GABA receptors in the basal ganglia of girls with Rett syndrome. Exp. Neurol. 156, 345–352. Available from: https://doi.org/10.1006/exnr.1999.7030.

Borghi, E., Vignoli, A., 2019. Molecular sciences Rett syndrome and other neurodevelopmental disorders share common changes in gut microbial community: a descriptive review. Available from: <https://doi.org/10.3390/ijms20174160>.

Borghi, E., Borgo, F., Severgnini, M., Savini, M., Casiraghi, M., Vignoli, A., 2017. Rett syndrome: a focus on gut microbiota. Int. J. Mol. Sci. 18, 344. Available from: https://doi.org/10.3390/ijms18020344.

Brahm, N.C., Erickson, S.R., Stewart, D.W., 2017. Disorders associated with intellectual disabilities. In: DiPiro, J.T., Talbert, R.L., Yee, G.C., Matzke, G.R., Wells, B.G., Posey, L. (Eds.), Pharmacotherapy: A Pathophysiologic Approach, 10e. McGraw-Hill; Accessed November 30, 2020. Available from: <https://accesspharmacy.mhmedical.com/content.aspx?bookid = 1861&sectionid = 146065733>.

Braniste, V., Al-Asmakh, M., Kowal, C., Anuar, F., Abbaspour, A., Tóth, M., et al., 2014. The gut microbiota influences blood-brain barrier permeability in mice. Sci. Transl. Med. 6, 263ra158. Available from: https://doi.org/10.1126/scitranslmed.3009759.

Byiers, B.J., Merbler, A.M., Barney, C.C., Frenn, K.A., Panoskaltsis-Mortari, A., Ehrhardt, M.J., et al., 2020. Evidence of altered salivary cytokine concentrations in Rett syndrome and associations with clinical severity. Brain Behav. Immun. Heal. 1, 100008. Available from: https://doi.org/10.1016/j.bbih.2019.100008.

Chahrour, M., Zoghbi, H.Y., 2007. The story of rett syndrome: from clinic to neurobiology. Neuron 56, 422–437. Available from: https://doi.org/10.1016/j.neuron.2007.10.001.

Clarke, A.J., Abdala Sheikh, A.P., 2018. A perspective on "cure" for Rett syndrome. Orphanet J. Rare Dis. 13, 44. Available from: https://doi.org/10.1186/s13023-018-0786-6.

Dethlefsen, L., Huse, S., Sogin, M.L., Relman, D.A., 2008. The pervasive effects of an antibiotic on the human gut microbiota, as revealed by deep 16s rRNA sequencing. PLoS Biol. 6, 2383–2400. Available from: https://doi.org/10.1371/journal.pbio.0060280.

Dinan, T.G., Cryan, J.F., 2012. Regulation of the stress response by the gut microbiota: implications for psychoneuroendocrinology. Psychoneuroendocrinology 37, 1369–1378. Available from: https://doi.org/10.1016/j.psyneuen.2012.03.007.

Dominguez-Bello, M.G., De Jesus-Laboy, K.M., Shen, N., Cox, L.M., Amir, A., Gonzalez, A., et al., 2016. Partial restoration of the microbiota of cesarean-born infants via vaginal microbial transfer. Nat. Med. 22, 250–253. Available from: https://doi.org/10.1038/nm.4039.

Ehrhart, F., Coort, S.L.M. Cirillo, E., Smeets, E., Evelo, C.T., Curfs, L.M.G., 2016. Rett syndrome - Biological pathways leading from MECP2 to disorder phenotypes. Orphanet J. Rare Dis. 11, 1–13. Available from: https://doi.org/10.1186/s13023-016-0545-5.

El-Khoury, R., Panayotis, N., Matagne, V., Ghata, A., Villard, L., Roux, J.-C., 2014. GABA and glutamate pathways are spatially and developmentally affected in the brain of Mecp2-deficient mice. PLoS One 9, e92169. Available from: https://doi.org/10.1371/journal.pone.0092169.

Fasano, A., Visanji, N.P., Liu, L.W.C., Lang, A.E., Pfeiffer, R.F., 2015. Gastrointestinal dysfunction in Parkinson's disease. Lancet Neurol. 14, 625–639. Available from: https://doi.org/10.1016/S1474-4422(15)00007-1.

Fjeldheim Dale, H., Rasmussen, S.H., Asiller, Ö.Ö., Lied, G.A., 2019. Nutrients communication probiotics in irritable bowel syndrome: an up-to-date systematic review. Available from: <https://doi.org/10.3390/nu11092048>.

Freedberg, D.E., Toussaint, N.C., Chen, S.P., Ratner, A.J., Whittier, S., Wang, T.C., et al., 2015. Proton pump inhibitors alter specific taxa in the human gastrointestinal microbiome: a crossover trial. Gastroenterology 149, 883–885. Available from: https://doi.org/10.1053/j.gastro.2015.06.043. e9.

Getachew, B., Aubee, J.I., Schottenfeld, R.S., Csoka, A.B., Thompson, K.M., Tizabi, Y., 2018. Ketamine interactions with gut-microbiota in rats: relevance to its antidepressant and anti-inflammatory properties. BMC Microbiol. 18, 222. Available from: https://doi.org/10.1186/s12866-018-1373-7.

Gold, W.A., Krishnarajy, R., Ellaway, C., Christodoulou, J., 2018. Rett syndrome: a genetic update and clinical review focusing on comorbidities. ACS Chem. Neurosci. 9, 167–176. Available from: https://doi.org/10.1021/acschemneuro.7b00346.

Gong, X., Liu, X., Chen, C., Lin, J., Li, A., Guo, K., et al., 2020. Alteration of gut microbiota in patients with epilepsy and the potential index as a biomarker. Front. Microbiol. 11, 2306. Available from: https://doi.org/10.3389/fmicb.2020.517797.

Hansen, J.C., Ghosh, R.P., Woodcock, C.L., 2010. Binding of the Rett syndrome protein, MeCP2, to methylated and unmethylated DNA and chromatin. IUBMB Life 62, 732–738. Available from: https://doi.org/10.1002/iub.386.

Johnston, M., Blue, M.E., Naidu, S., 2015. Recent advances in understanding synaptic abnormalities in Rett syndrome. F1000Research 4, 1490. Available from: https://doi.org/10.12688/f1000research.6987.1.

Kandeel, W.A., Meguid, N.A., Bjørklund, G., Eid, E.M., Farid, M., Mohamed, S.K., et al., 2020. Impact of clostridium bacteria in children with autism spectrum disorder and their anthropometric measurements. J. Mol. Neurosci. 70, 897–907. Available from: https://doi.org/10.1007/s12031-020-01482-2.

Keshavarzian, A., Green, S.J., Engen, P.A., Voigt, R.M., Naqib, A., Forsyth, C.B., et al., 2015. Colonic bacterial composition in Parkinson's disease. Mov. Disord. 30, 1351–1360. Available from: https://doi.org/10.1002/mds.26307.

Kilgore, M., Miller, C.A., Fass, D.M., Hennig, K.M., Haggarty, S.J., Sweatt, J.D., et al., 2010. Inhibitors of class 1 histone deacetylases reverse contextual memory deficits in a mouse model of alzheimer's disease. Neuropsychopharmacology 35, 870–880. Available from: https://doi.org/10.1038/npp.2009.197.

Kim, J., Basak, J.M., Holtzman, D.M., 2009. The role of apolipoprotein E in Alzheimer's disease. Neuron 63, 287–303. Available from: https://doi.org/10.1016/j.neuron.2009.06.026.

Kocamaz, H., Işıkay, S. 2019. Epilepsy and GI disorders. Epilepsy - Advances in Diagnosis and Therapy, IntechOpen. Available from: https://doi.org/10.5772/intechopen.85813.

Koh, A., De Vadder, F., Kovatcheva-Datchary, P., Bäckhed, F., 2016. From dietary fiber to host physiology: short-chain fatty acids as key bacterial metabolites. Cell 165, 1332–1345. Available from: https://doi.org/10.1016/j.cell.2016.05.041.

Koppel, N., Bisanz, J.E., Pandelia, M.E., Turnbaugh, P.J., Balskus, E.P., 2018. Discovery and characterization of a prevalent human gut bacterial enzyme sufficient for the inactivation of a family of plant toxins. Elife 7. Available from: https://doi.org/10.7554/eLife.33953.

Kyle, S.M., Vashi, N., Justice, M.J., 2018. Rett syndrome: a neurological disorder with metabolic components. Open. Biol. 8. Available from: https://doi.org/10.1098/rsob.170216.

Li, Q., Han, Y., Dy, A.B.C., Hagerman, R.J., 2017. The gut microbiota and autism spectrum disorders. Front. Cell Neurosci. 11, 120. Available from: https://doi.org/10.3389/fncel.2017.00120.

Liang, S., Wu, X., Jin, F., 2018a. Gut-brain psychology: rethinking psychology from the microbiota–gut–brain axis. Front. Integr. Neurosci. 12, 33. Available from: https://doi.org/10.3389/fnint.2018.00033.

Liang, S., Wu, X., Hu, X., Wang, T., Jin, F., 2018b. Recognizing depression from the microbiota–gut–brain axis. Int. J. Mol. Sci. 19. Available from: https://doi.org/10.3390/ijms19061592.

Lum, G.R., Olson, C.A., Hsiao, E.Y., 2020. Emerging roles for the intestinal microbiome in epilepsy. Neurobiol. Dis. 135, 104576. Available from: https://doi.org/10.1016/j.nbd.2019.104576.

Magne, F., Gotteland, M., Gauthier, L., Zazueta, A., Pesoa, S., Navarrete, P., et al., 2020. The firmicutes/bacteroidetes ratio: a relevant marker of gut dysbiosis in obese patients? Nutrients 12, 1474. Available from: https://doi.org/10.3390/nu12051474.

Mariat, D., Firmesse, O., Levenez, F., Guimarães, V.D., Sokol, H., Doré, J., et al., 2009. The firmicutes/bacteroidetes ratio of the human microbiota changes with age. BMC Microbiol. 9, 123. Available from: https://doi.org/10.1186/1471-2180-9-123.

Motil, K.J., Caeg, E., Barrish, J.O., Geerts, S., Lane, J.B., Percy, A.K., et al., 2012. Gastrointestinal and nutritional problems occur frequently throughout life in girls and women with rett syndrome. J. Pediatric Gastroenterol. Nutr. Available from: https://doi.org/10.1097/MPG.0b013e31824b6159.

Neul, J.L., Kaufmann, W.E., Glaze, D.G., Christodoulou, J., Clarke, A.J., Bahi-Buisson, N., et al., 2010. Rett syndrome: revised diagnostic criteria and nomenclature. Ann. Neurol. 68, 944–950. Available from: https://doi.org/10.1002/ana.22124.

Operto, F.F., Mazza, R., Pastorino, G.M.G., Verrotti, A., Coppola, G., 2019. Epilepsy and genetic in Rett syndrome: a review. Brain Behav. 9. Available from: https://doi.org/10.1002/brb3.1250.

Parracho, H.M.R.T., Bingham, M.O., Gibson, G.R., McCartney, A.L., 2005. Differences between the gut microflora of children with autistic spectrum disorders and that of healthy children. J. Med. Microbiol. 54, 987–991. Available from: https://doi.org/10.1099/jmm.0.46101-0.

Pascal, V., Pozuelo, M., Borruel, N., Casellas, F., Campos, D., Santiago, A., et al., 2017. A microbial signature for Crohn's disease. Gut 66, 813–822. Available from: https://doi.org/10.1136/gutjnl-2016-313235.

Pecorelli, A., Cervellati, C., Cordone, V., Hayek, J., Valacchi, G., 2020. Compromised immune/inflammatory responses in Rett syndrome. Free. Radic. Biol. Med. 152, 100–106. Available from: https://doi.org/10.1016/j.freeradbiomed.2020.02.023.

Purves, D., Augustine, G.J., Fitzpatrick, D., Katz, L.C., LaMantia, A.-S., McNamara, J.O., n.d. The Biogenic Amines—Neuroscience—NCBI Bookshelf <https://www.ncbi.nlm.nih.gov/books/NBK11035/> (accessed 09.01.21).

Quigley, E.M.M., 2017. Microbiota-brain-gut axis and neurodegenerative diseases. Curr. Neurol. Neurosci. Rep. 17. Available from: https://doi.org/10.1007/s11910-017-0802-6.

Raskov, H., Pommergaard, H.C., Burcharth, J., Rosenberg, J., 2014. Colorectal carcinogenesis-update and perspectives. World J. Gastroenterol. 20, 18151–18164. Available from: https://doi.org/10.3748/wjg.v20.i48.18151.

Rett Syndrome Fact Sheet, n.d. National institute of neurological disorders and stroke. Available from: <https://www.ninds.nih.gov/disorders/patient-caregiver-education/fact-sheets/rett-syndrome-fact-sheet> (accessed 19.01. 21).

Riva, A., Borgo, F., Lassandro, C., Verduci, E., Morace, G., Borghi, E., et al., 2017. Pediatric obesity is associated with an altered gut microbiota and discordant shifts in Firmicutes populations. Env. Microbiol. 19, 95–105. Available from: https://doi.org/10.1111/1462-2920.13463.

Robertson, K.D., 2005. DNA methylation and human disease. Nat. Rev. Genet. 6, 597–610. Available from: https://doi.org/10.1038/nrg1655.

Roux, J.C., Villard, L., 2010. Biogenic amines in rett syndrome: the usual suspects. Behav. Genet. 40, 59–75. Available from: https://doi.org/10.1007/s10519-009-9303-y.

Samaco, R.C., Mandel-Brehm, C., Chao, H.T., Ward, C.S., Fyffe-Maricich, S.L., Ren, J., et al., 2009. Loss of MeCP2 in aminergic neurons causes cell-autonomous defects in neurotransmitter synthesis and specific behavioral abnormalities. Proc. Natl. Acad. Sci. U. S. A. 106, 21966–21971.

Sánchez-Alcoholado, L., Ramos-Molina, B., Otero, A., Laborda-Illanes, A., Ordóñez, R., Medina, J.A., et al., 2020. The role of the gut microbiome in colorectal cancer development and therapy response. Cancers (Basel) 12. Available from: https://doi.org/10.3390/cancers12061406.

Sandweiss, A.J., Brandt, V.L., Zoghbi, H.Y., 2020. Advances in understanding of Rett syndrome and MECP2 duplication syndrome: prospects for future therapies. Lancet Neurol. 19, 689–698. Available from: https://doi.org/10.1016/S1474-4422(20)30217-9.

Scheperjans, F., Aho, V., Pereira, P.A.B., Koskinen, K., Paulin, L., Pekkonen, E., et al., 2015. Gut microbiota are related to Parkinson's disease and clinical phenotype. Mov. Disord. 30, 350–358. Available from: https://doi.org/10.1002/mds.26069.

Schultz, S.T., Klonoff-Cohen, H.S., Wingard, D.L., Akshoomoff, N.A., Macera, C.A., Ji, M., et al., 2006. Breastfeeding, infant formula supplementation, and autistic disorder: the results of a parent survey. Int. Breastfeed. J. 1, 16. Available from: https://doi.org/10.1186/1746-4358-1-16.

Sheng, J.G., Bora, S.H., Xu, G., Borchelt, D.R., Price, D.L., Koliatsos, V.E., 2003. Lipopolysaccharide-induced-neuroinflammation increases intracellular accumulation of amyloid precursor protein and amyloid β peptide in APPswe transgenic mice. Neurobiol. Dis. 14, 133–145. Available from: https://doi.org/10.1016/S0969-9961(03)00069-X.

Shin, W., Kim, H.J., 2018. Intestinal barrier dysfunction orchestrates the onset of inflammatory host-microbiome cross-talk in a human gut inflammation-on-a-chip. Proc. Natl. Acad. Sci. U. S. A. 115, E10539–E10547. Available from: https://doi.org/10.1073/pnas.1810819115.

Sivamaruthi, B.S., Suganthy, N., Kesika, P., Chaiyasut, C., n.d. The role of microbiome, dietary supplements, and probiotics in autism spectrum disorder. Available from: <https://doi.org/10.3390/ijerph17082647>.

Smeets, E.E.J., Pelc, K., Dan, B., 2012. Rett syndrome. Mol. Syndromol. 2, 113–127. Available from: https://doi.org/10.1159/000337637.

Smith-Hicks, C.L., Gupta, S., Ewen, J.B., Hong, M., Kratz, L., Kelley, R., et al., 2017. Randomized open-label trial of dextromethorphan in Rett syndrome. Neurology 89, 1684–1690. Available from: https://doi.org/10.1212/WNL.0000000000004515.

Spanogiannopoulos, P., Bess, E.N., Carmody, R.N., Turnbaugh, P.J., 2016. The microbial pharmacists within us: a metagenomic view of xenobiotic metabolism. Nat. Rev. Microbiol. 14, 273–287. Available from: https://doi.org/10.1038/nrmicro.2016.17.

Stafstrom, C.E., Carmant, L., 2015. Seizures and epilepsy: an overview for neuroscientists. Cold Spring Harb. Perspect. Biol. 7, 1–19. Available from: https://doi.org/10.1101/cshperspect.a022426.

Strati, F., Cavalieri, D., Albanese, D., De Felice, C., Donati, C., Hayek, J., et al., 2016. Altered gut microbiota in Rett syndrome. Microbiome 4, 41. Available from: https://doi.org/10.1186/s40168-016-0185-y.

Sudo, N., Chida, Y., Aiba, Y., Sonoda, J., Oyama, N., Yu, X.N., et al., 2004. Postnatal microbial colonization programs the hypothalamic-pituitary-adrenal system for stress response in mice. J. Physiol. 558, 263–275. Available from: https://doi.org/10.1113/jphysiol.2004.063388.

Tomova, A., Husarova, V., Lakatosova, S., Bakos, J., Vlkova, B., Babinska, K., et al., 2015. Gastrointestinal microbiota in children with autism in Slovakia. Physiol. Behav. 138, 179–187. Available from: https://doi.org/10.1016/j.physbeh.2014.10.033.

Trappe, R., Laccone, F., Cobilanschi, J., Meins, M., Huppke, P., Hanefeld, F., et al., 2001. MECP2 mutations in sporadic cases of Rett syndrome are almost exclusively of paternal origin. Am. J. Hum. Genet. 68, 1093–1101. Available from: https://doi.org/10.1086/320109.

Tsai, L.Y., 2015. The impact of DSM-5 on autism spectrum disorder. Psychopathol. Rev. PR. 2, 3–16. Available from: https://doi.org/10.5127/pr.035414.

van Kessel, S.P., Frye, A.K., El-Gendy, A.O., Castejon, M., Keshavarzian, A., van Dijk, G., et al., 2019. Gut bacterial tyrosine decarboxylases restrict levels of levodopa in the treatment of Parkinson's disease. Nat. Commun. 10. Available from: https://doi.org/10.1038/s41467-019-08294-y.

Vashi, N., Justice, M.J., 2019. Treating Rett syndrome: from mouse models to human therapies. Mamm. Genome 30, 90–110. Available from: https://doi.org/10.1007/s00335-019-09793-5.

Viemari, J.C., Roux, J.C., Tryba, A.K., Saywell, V., Burnet, H., Peña, F., et al., 2005. Mecp2 deficiency disrupts norepinephrine and respiratory systems in mice. J. Neurosci. 25, 11521–11530. Available from: https://doi.org/10.1523/JNEUROSCI.4373-05.2005.

Villard, L., 2007. MECP2 mutations in males. J. Med. Genet. 44, 417–423. Available from: https://doi.org/10.1136/jmg.2007.049452.

Vogt, N.M., Kerby, R.L., Dill-McFarland, K.A., Harding, S.J., Merluzzi, A.P., Johnson, S.C., et al., 2017. Gut microbiome alterations in Alzheimer's disease. Sci. Rep. 7. Available from: https://doi.org/10.1038/s41598-017-13601-y.

What is Rett Syndrome?, n.d. Rettsyndrome.org. Available from: <https://www.rettsyndrome.org/about-rett-syndrome/what-is-rett-syndrome/> (accessed 19.01.21).

Wong, J.M.W., De Souza, R., Kendall, C.W.C., Emam, A., Jenkins, D.J.A., 2006. Colonic health: fermentation and short chain fatty acids. J. Clin. Gastroenterol. 40, 235–243. Available from: https://doi.org/10.1097/00004836-200603000-00015.

Yang, T., Santisteban, M.M., Rodriguez, V., Li, E., Ahmari, N., Carvajal, J.M., et al., 2015. Gut dysbiosis is linked to hypertension. Hypertension 65, 1331–1340. Available from: https://doi.org/10.1161/HYPERTENSIONAHA.115.05315.

Yang, D., Zhao, D., Ali Shah, S.Z., Wu, W., Lai, M., Zhang, X., et al., 2019. The role of the gut microbiota in the pathogenesis of Parkinson's disease. Front. Neurol. 10, 1155. Available from: https://doi.org/10.3389/fneur.2019.01155.

Zhang, S., Chen, D.C., Chen, L.M., 2019. Facing a new challenge: the adverse effects of antibiotics on gut microbiota and host immunity. Chin. Med. J. (Engl.) 132, 1135–1138. Available from: https://doi.org/10.1097/CM9.0000000000000245.

Zoghbi, H.Y., Milstien, S., Butler, I.J., Smith, E.O., Kaufman, S., Glaze, D.G., et al., 1989. Cerebrospinal fluid biogenic amines and biopterin in rett syndrome. Ann. Neurol. 25, 56–60. Available from: https://doi.org/10.1002/ana.410250109.

Section VI

Microbiome, Dermal Health and Wound Healing

Chapter 21

Skin microbiota and its role in health and disease with an emphasis on wound healing and chronic wound development

J.H. Kim, A. Ahamed, K. Chen, E.G. Lebig, B. Petros, S. Saeed and M. Martins-Green
Department of Molecular, Cell and Systems Biology, University of California, Riverside, CA, United States

21.1 Introduction

The skin is the largest organ in the human body. It is a barrier that protects the organs from the environment, functioning as an interface between the outside and the inside of the body. The cellular structure and composition of the skin evolved to act as a physical, chemical, and biological barrier with support from both the innate and adaptive immune systems (Grice and Segre, 2011; Huang et al., 2005; Knackstedt et al., 2020; Kong and Segre, 2012; Schmidt, 2020). A myriad of chemical and physical pollutants, toxins, and biological agents, such as pathogens, are barred by the skin from entering the body and causing damage. If the integrity of the skin is compromised, the body relies on the highly complex process of cutaneous wound healing to quickly close and strengthen the damaged area to restore the protective functions of the skin

Residing on and in the body is a complex and diverse community of microbes called the microbiota (Ashrafi et al., 2020; Johnson et al., 2018; Kong, 2011; Nakatsuji et al., 2013; Yu et al., 2020). Trillions of microbes across bacterial and fungal families use a wide range of aerobic and anaerobic metabolic pathways to survive. Viruses also inhabit various locations of the body; some viruses are pathogenic to humans whereas others are pathogenic to bacteria (also called bacteriophages) (Johnson et al., 2018; Kali, 2015; Kong, 2011; Kong and Segre, 2012; Schommer and Gallo, 2013). Single-celled and multicellular parasites are also present in the body and contribute to the microbiota. The collective genome of the microbiota (termed the microbiome) results in unique metabolic and proteomic profiles specific to each location of the body. The contributions and impact of the microbiome on health and disease are still under investigation (Grice and Segre, 2011; Grice et al., 2009; Hannigan and Grice, 2013).

Considerable effort is being put into studying the role of the microbiota in skin-related infections, disorders, and disease (Krishna and Miller, 2012; Tett et al., 2017; Williams et al., 2017). In particular, how host-microbe interactions are specifically relevant to wound healing is an urgent topic of research because specific microbes may play an important role in wound healing and the overall health of the skin. If the homeostasis between human skin and the microbiota is disturbed, dysbiosis may result in the development of a number of human skin diseases and ailments, including the development of chronic wounds.

Historically, many important wound microbes were identified by relying on isolation and culturing protocols under artificial laboratory conditions, emphasizing only a small fraction of the entire skin microbiota. The limited growth conditions that can be replicated in culture cannot capture and sustain the growth of the majority of microbes, greatly limiting the ability to identify and describe the whole microbiota (Batista et al., 2002; Grice et al., 2009; Misic et al., 2014; Verbanic et al., 2020). In recent years, advances in high-throughput DNA sequencing platforms have led to the identification of many microbes in normal healthy human skin and have characterized the microbiome of wounded skin (Kalan et al., 2016; Mcmurdie et al., 2015; Pang et al., 2020; Tsilimigras and Fodor, 2016). This chapter discusses how the relationship between the skin and the residing microbiota plays a role in skin disease, normal wound healing, and chronic wound development. The references that are cited were chosen as examples of several possible references for the same subject.

Microbiome, Immunity, Digestive Health and Nutrition. DOI: https://doi.org/10.1016/B978-0-12-822238-6.00027-3

21.2 Skin structure and function

The skin is important in carrying out very important functions, such as protection, preservation of water and electrolytes, and regulation of temperature, water, and fat storage, in addition to playing a major role in the endocrine and immunological systems (Kolarsick et al., 2011; Salvo et al., 2015). The skin is the largest organ in humans and functions as an interface between the connective tissue and organs and the environment, so it must be waterproof and protect the body against environmental stresses. It is a continuously developing organ composed of three main structures, namely, the epidermis, the dermis, and the hypodermis. It also includes other anatomical structures such as hair follicles, nails, sebaceous glands, and sweat glands (Gilaberte et al., 2016; Khavkin and Ellis, 2011; Menon, 2015). The skin is mostly acidic and dry, but different areas of the body have different topography.

The epidermis is the outer layer of the skin, and it interact directly with the environment. It is a physical barrier, resisting penetration by microorganisms and potential toxins while retaining moisture and nutrients inside the body. The thickness of this layer depends on the location on the body; for example, the skin is thin on the eyelids and thick on the palms of the hands and soles of the feet (Kolarsick et al., 2011). The epidermis is composed predominantly of keratinocytes and melanocytes but includes other types of cells such as Langerhans and Merkel cells (Gilaberte et al., 2016). Because this layer of the skin interacts with the environment, the epidermis is continuously renewing itself as these cells undergo proliferation and terminal differentiation, followed by cell death. The epidermis is composed of sublayers called the stratum corneum, stratum granulosum, stratum spinosum, and basal layer based on the morphology and location of the keratinocytes (Gilaberte et al., 2016; Khavkin and Ellis, 2011). The most superficial layer of the skin, the stratum corneum, functions as the major constituent of the skin barrier. This layer is made up of corneocytes, which are flat, enucleated dead keratinocytes that have undergone terminal differentiation. These cells are connected together by desmosomes, which help to create the cornified envelop. In combination with proteins and lipids, these cells create an insoluble barrier that has a critical role in the skin barrier (Menon, 2015). Underneath the stratum corneum is the stratum granulosum, which is a cell layer that contains live keratinocytes that arise as differentiated cells from the layer below, the stratum spinosum, and ascend outward from the basal layer, a layer of rapidly proliferating columnar-shaped keratinocytes (Gilaberte et al., 2016; Khavkin and Ellis, 2011; Menon, 2015). The basal keratinocytes are attached to the basement lamina, which separates and connects the epidermis to the dermis. The basement lamina is a thin extracellular matrix sheet that contains three structural components: the lamina lucida, the lamina densa, and the sublamina densa (Gilaberte et al., 2016; Khavkin and Ellis, 2011; Menon, 2015; Yurchenco, 2011). These layers are composed of collagen type IV, laminin, entactin, and fibronectin that are anchored to the underlying dermis by filaments composed of collagens types VI, VII, XV, and XVIII and chondroitin sulfate proteoglycans (Yurchenco, 2011). The basal lamina plus the molecules connecting it to the underlying dermis constitute the basement membrane (Gilaberte et al., 2016; Khavkin and Ellis, 2011; Lai-Cheong and McGrath, 2017; Martins-Green, 1997; Yurchenco, 2011).

Beneath the basement membrane is the dermis, which is composed of the papillary dermis and the reticular dermis. The papillary dermis is a thin layer of loosely arranged collagen fibers beneath the basement membrane; the reticular dermis is a thick layer of collagen bundles that align parallel to the surface of the skin. Seventy percent of the dermis is made up of collagen fibers, which give the skin its strength and toughness. Elastin fibers, by contrast, maintain the elasticity and flexibility of the skin (Gilaberte et al., 2016; Khavkin and Ellis, 2011; Menon, 2015; Morton and Phillips, 2016). Throughout the collagen and elastin fibers, proteoglycans provide viscosity and hydration to the extracellular matrix. The dermis primarily contains fibroblasts, myeloid cells, and lymphoid immune cells that reside in and maintain the dermal layer (Kolarsick et al., 2011). One immune cell type that is present in the dermis is the mast cell. Mast cells are important in both innate and adaptive immune responses. They have an important role in chemoattracting other immune cells by secreting chemokines and cytokines. These mast cells also secrete histamine during allergic responses (Gilaberte et al., 2016). Lymph and blood vessels are also present in the dermis. These vessels have important homeostatic functions because they provide nutrients for the epidermis and dermis and support immune processes. In addition, the dermis contains a network of sensory nerves that has an important role in maintaining cutaneous homeostasis and interfacing with the environment. The skin is also a critical sensory organ that allows for the perception of touch, pain, and temperature (Gilaberte et al., 2016; Khavkin and Ellis, 2011). Moreover, sebaceous glands, located in the dermis, are connected to hair follicles, forming the pilosebaceous unit, and secrete lipid-rich sebum. Sebum is a hydrophobic coating that protects and lubricates the skin and hair and provides antibacterial shielding.

The skin also contains sweat glands (Kolarsick et al., 2011). There are two types of sweat gland, which are distinguished by their development, morphology, and function. Eccrine sweat glands constitute the majority of these glands and are distributed throughout nearly the entire body, whereas apocrine sweat glands are present in a just few areas on the body (Gilaberte et al., 2016; Khavkin and Ellis, 2011; Kolarsick et al., 2011). Eccrine sweat glands have an important thermoregulatory function and can respond to emotional stimuli. Under stress, such as heat, eccrine sweat glands are capable of producing up to 10 liters of sweat per day. Sweat is a clear, odorless solution that contains salts such as sodium, calcium, chloride, and potassium. Inorganic and organic chemicals, such as urea, lactate, amino acids, and bicarbonate, are also present in smaller concentrations. Larger proteins structures, such as immunoglobulins, make up less than 1% of sweat by weight (Baker, 2019). Apocrine sweat glands are present in limited areas: the axilla, perianal region, areolae, periumbilical skin, prepuce, scrotum, mons pubis, and labia minora (Groscurth, 2002). They produce a sterile, thick, milky, and odorless fluid that contains many proteins, fatty acids, carbohydrates, ammonia, lipids, and ferric ions. This secretion can undergo bacterial decomposition to produce potent odorous compounds.

Beneath the dermis is the hypodermis or subcutaneous tissue, which contains fibroblasts, adipose cells, and macrophages (Gilaberte et al., 2016; Khavkin and Ellis, 2011; Menon, 2015). A fibrous network of connective tissue provides structural support for additional nerve cells, blood vessels, and lymphatic vessels located within the hypodermis. The subcutaneous adipocytes give the hypodermis some characteristics of an endocrine tissue because these adipocytes can produce hormones such as adiponectin, which regulates glucose levels and leptin which regulates hunger.

21.3 Skin microbiome diversity in different parts of the body

Using a variety the techniques, scientists have found that the skin microbiome consists of both benign commensal microbes that contribute to the health of the skin and pathogenic microbes that can cause infection and disease if not controlled (Fyhrquist et al., 2016; Grice and Segre, 2011; Kali, 2015; Kong and Segre, 2012; Schommer and Gallo, 2013). The microbes that are present in skin include bacteria, fungi, and viruses. Commensal bacteria have recently been shown to stimulate the adaptive and innate immune system and directly or indirectly prevent pathogens from causing infections (Chen and Tsao, 2013; Coates et al., 2019; Grice and Segre, 2011; MacLeod and Mansbridge, 2016; Meisel et al., 2018). Therefore, understanding how the skin microbiota colonizes different sites in the body may provide insight into the pathophysiology of cutaneous infections and could explain how the balance between skin health and disease can be modulated by the microbiota.

Specific dermatological disorders manifest themselves at different skin sites. As was discussed earlier in the chapter, the skin provides many niches in which large populations of microbes are subjected to a variety of ecological pressures, including humidity, temperature, and pH (Fyhrquist et al., 2016; Grice and Segre, 2011; Nakatsuji et al., 2013; Schommer and Gallo, 2013). High temperature and high humidity are associated with increased quantities of bacteria and increased diversity (Knackstedt et al., 2020). These ecological factors contribute to the uniqueness of the various skin structures, such as the hair follicles and sebaceous, eccrine, and apocrine glands, which produce a variety of antimicrobial peptides and lipids that can chemically control and regulate the population of microbes on the skin. Competition between microbial species is important for the development and maintenance of a healthy microbiome (Gilbert et al., 2018; Kolarsick et al., 2011; Lunjani et al., 2019; Schommer and Gallo, 2013).

The bacterial species that populate the different regions of the skin are directly related to the microenvironment (Knackstedt et al., 2020). Owing to the complexity of the microenvironment, the normal microbiota of the skin is very complex and likely developed through the cooperation of commensal bacteria and competitive exclusion of pathogens (Ouwehand et al., 2003). Major bacterial families that are present on the skin include Actinobacteria (51.8%), Firmicutes (24.4%) Proteobacteria (16.5%), and Bacteroidetes (6.3%) with Actinobacteria most abundant on the skin (Grice et al., 2009). Major bacterial contributors at the genera level to the skin microbiota are *Corynebacterium*, *Propionibacterium*, and *Staphylococcus* (Grice et al., 2009).

Areas of the skin with abundant sebaceous glands, such as the skin on the face, are predominantly populated by *Cutibacterium acnes* (formally known as *Propionibacterium acnes*) and *Staphylococcus* species. *C. acnes* is a common skin commensal bacterium and has encoded in the genome multiple lipases that degrade lipids in sebum (Grice and Segre, 2011). In areas of the skin that are considered moist, such as the axilla, the predominate members of the microbiota are *Propionibacterium, Corynebacterium*, and *Staphylococcus* species. On the palm of the hand, *Propionibacterium* (31.6%), *Streptococcus* (17.2%), *Staphylococcus* (8.3%), *Corynebacterium* (4.3%), and *Lactobacillus* (3.1%) are present (Egert and Simmering, 2016; Egert et al., 2017). The dominant bacteria in the pilosebaceous units, or pores, on the nose is *C. acnes*, which accounts for 87% of the clones. Other bacteria that have been isolated from the pores include *Staphylococcus epidermidis*, *P. humerusii*, and *P. granulosum* (Fitz-Gibbon et al., 2013).

Areas of the skin that are not characterized by high moisture content or sebum production have a mix of bacterial populations from the Proteobacteria and Flavobacteriales (Grice and Segre, 2011). On the forearm, *Propionibacterium* (22.0%), *Corynebacterium* (19.0%), *Staphylococcus* (11.1%), *Streptococcus* (5.8%), *Acinetobacter* (3.7%), and

Finegoldia (1.3%) can be found (Gao et al., 2007). On the skin surrounding the ankles, common bacteria that have been identified include *Pseudomonas*, *Corynebacterium*, *Staphylococcus*, *Chryseobacterium*, *Acinetobacter*, *Methylophilus*, *Acidithiobacillus*, *Segetibacter*, *Wautersiella*, and *Psychrobacter* (Grice and Segre, 2011). On the skin of the feet, the microbiota is dominated by bacteria from the genera *Staphylococcus*, *Acinetobacter*, *Kocuria*, *Corynebacterium*, and *Micrococcus* (Grice and Segre, 2011). The predominant fungi species that have been detected by using phylogenetic marker 18S rRNA belong to the *Malassezia* genus, including the most frequent isolates *Malassezia globosa*, *Malassezia restricta*, and *Malassezia sympodialis* (Schommer and Gallo, 2013). These are lipophilic microbes that are associated with areas of the skin that produce significant levels of sebum (Byrd et al., 2018). Little is known about the role of viruses as a part of the human skin, which is still under investigation (Hannigan and Grice, 2013).

21.4 Microbial dysbiosis in skin-related diseases

An altered cutaneous microbiome has been shown to contribute to the high susceptibility of the skin to infections. Bacteria on the skin have important functions for skin health and healing (Kong and Segre, 2012; Loesche et al., 2017; Sanmiguel and Grice, 2015). However, the impact of these microbes on the health and disease of the skin is still under investigation (Grice and Segre, 2011; Grice et al., 2009; Hannigan and Grice, 2013). When disturbed, the skin microbiota can play a role in disease progression of many skin ailments, including atopic dermatitis, acnes vulgaris, rosacea, psoriasis, and some skin cancers (Table 21.1, Fig. 21.1).

21.4.1 Atopic dermatitis

Atopic dermatitis is a chronic skin disorder that affects approximately 15% of U.S. children and also affects adults (Williams and Gallo, 2015). This disorder is commonly associated with irritated skin that is colonized and/or infected by *Staphylococcus aureus*. The irritated skin can be found on almost any part of the body but is commonly localized on the antecubital fossa and the popliteal fossa. This skin disease can be effectively treated with topical or systemic antibiotics and steroids. Medicated baths are also used to decrease the bacterial burden on the affected skin.

21.4.2 Acne vulgaris

Acne vulgaris is the most common skin disease in the United States and affects 80% of the population (Roudsari et al., 2015). Acne develops frequently on the face and upper torso and most often affects adolescents. It occurs when hair follicles and sebaceous glands are clogged and inflamed, resulting in the formation of comedones, papules, pustules, nodules, and/or cysts. Lesions at different stages frequently coexist, often in the same location. Nodules and cysts can

TABLE 21.1 Common bacteria in skin-related diseases.

Skin-related disease	Common bacteria genera
Atopic dermatitis	*Staphylococcus aureus*
Acne vulgaris	*Propionibacterium acnes* *Corynebacterium* *Propionibacterium* *Staphylococci*
Rosacea	*Bacillus oleronius* *Staphylococcus epidermidis*
Psoriasis	*Staphylococcus* *Micrococcus* *Planococcaceae*
Skin cancers	*Propionibacterium* *Staphylococcus* *Corynebacterium* *Halomonas* *Salinicola* *Shewanella*

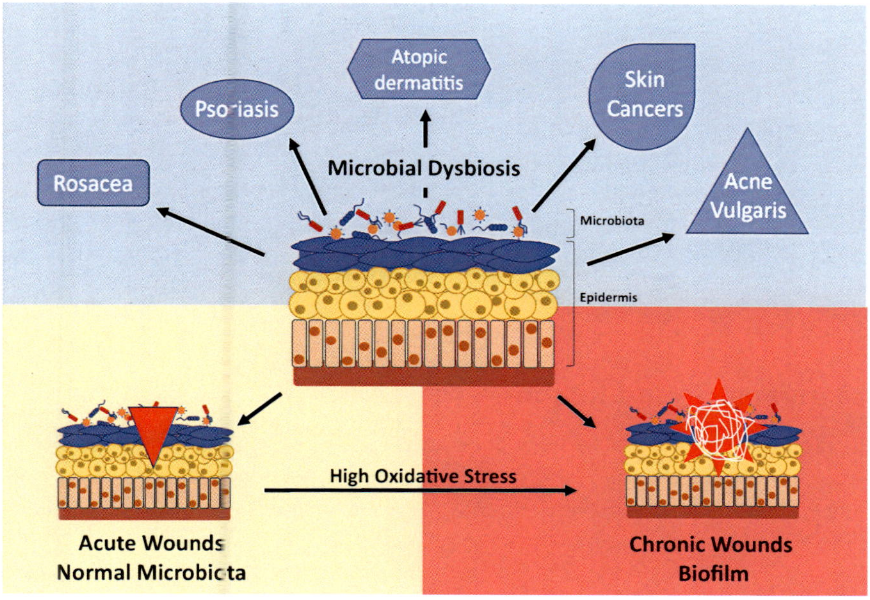

FIGURE 21.1 (*Top*) The skin microbiota plays a key role in maintaining skin integrity and health. However, microbial dysbiosis is commonly seen in many skin-related diseases, including rosacea, psoriasis, atopic dermatitis, acne vulgaris, and skin cancers. (*Bottom left*) Normal microbial presence in the wound tissue is critical for proper healing, as bacteria initiate key inflammatory responses such as those performed by neutrophils and macrophages. (*Bottom right*) Under high oxidative stress conditions, chronic inflammation sets in, causing extensive damage to the wound tissue. This, in addition to the presence of pathogen opportunistic bacteria, hinders the tissue's ability to fight infection and results in the formation of biofilm (*white squiggles*) on a wound that fails to heal.

be quite painful and may require medical attention. Scarring as a result of these lesion can be a source of significant emotional distress. Treatment options can involve a combination of topical and systemic medications directed at treating or preventing the formation of acne. These treatment options are aimed at reducing sebum production, comedone formation, inflammation, and bacterial counts and at normalizing keratinization. *P. acnes* is a dominant skin bacterial species in this disease. Other bacteria present in acne are *Corynebacteria*, *Propionibacteria*, and *Staphylococci* (Dréno et al., 2018).

21.4.3 Rosacea

Approximately 3% of the world population suffers from rosacea. Rosacea is commonly associated with flushing but also presents itself as inflammatory nodules, nontransient erythema, papules, pustules, and telangiectasia (Culp and Scheinfeld, 2009; Daou et al., 2021). There are four clinically based subtypes of rosacea: erythematotelangiectatic, papulopustular, phymatous, and ocular. Patients experience the onset of rosacea usually after age 30 with increased prevalence in women and individuals with fair skin. In addition to genetic elements, other risk factors that contribute to rosacea include increased alcohol consumption and excessive ultraviolet (UV) exposure. While there is a genetic predisposition for rosacea, dysbiosis of the affected skin microbiota may be contributors. Rosacea occurs when both the innate and adaptive immune systems are dysregulated (associated with chronic inflammation), which leads to the dysbiosis of microbes on the skin. Specifically, disturbances to the skin include altered TLR2 expression, high levels of the serine protease kallikrein 5, and abnormal expression of the cathelicidin antimicrobial peptide LL-37 (Margalit et al., 2016). Bacterial species that are found on the skin of rosacea patients include *Bacillus oleronius* and *S. epidermidis* (Daou et al., 2021).

21.4.4 Psoriasis

Psoriasis affects 2% of the world population and results from a combination of genetic and environmental factors. The pathology of this disorder includes hyperkeratosis, inflammation, hyperproliferation of keratinocytes, and increased angiogenesis. Although little is known about the role of the skin microbiome in psoriasis, the overall diversity of the microbiota overlaying psoriasis lesions was found to be greater than in healthy skin (Dréno et al., 2018). Firmicutes are predominant, whereas Actinobacteria are underrepresented in psoriatic lesions in comparison to healthy skin. It is not known whether the alterations in the bacterial population in psoriasis are a consequence of the disease or contribute to its pathogenesis. *Staphylococcus*, *Micrococcus*, and Planococcaceae are the most abundant bacteria that have been identified in patients with psoriasis (Stehlikova et al., 2019).

21.4.5 Skin cancers

The association between microbiota and skin cancer is a recent proposition (Squarzanti et al., 2020; Yu et al., 2015). Although no pathogen has been found to cause or initiate the development of skin cancers, patients with certain chronic skin diseases may have an altered skin microbiota with colonization of specific pathogenic bacteria that increase the risks for developing cutaneous malignancies. For example, the microbiome of cutaneous melanomas and benign melanocytic nevi has been characterized by the presence of *Propionibacterium* as the most common bacterial genus in these lesions (Salava et al., 2016). Common skin colonizers such as *Staphylococcus* and *Corynebacterium* are found in skin cancer, but various bacterial taxa that are not typically found on human skin were also identified. These bacterial genera include *Halomonas*, *Salinicola*, and *Shewanella*.

21.5 The microbiome and cutaneous wound healing

Wound healing is a dynamic process that involves cellular and molecular processes that must be precisely regulated both temporally and spatially. Healing of the skin is organized into several overlapping stages that require the concerted activation, proliferation, differentiation, and migration of many different cell types that reside in the skin tissue. The major stages of cutaneous wound healing include hemostasis, an inflammatory phase, a proliferative phase, and a remodeling phase.

Hemostasis requires the release of factors that attract and activate circulating platelets to injured endothelial cells, forming the initial plug to stop the loss of blood. Clotting is achieved with additional molecular cascades to form fibrin in order to fortify the platelet plug with platelets trapped in the clot. The platelets release growth factors and chemokines into the immediate wound environment. Hemostasis is achieved once blood is no longer lost from the damaged blood vessels in the wound. At the same time, blood vessels begin to dilate, allowing oxygen, nutrients, and chemotactic cytokines and leukocytes such as neutrophils, monocytes, and macrophages into the wound tissue. Activated platelets also secrete growth factors that stimulate the keratinocytes at the wound edge to proliferate and migrate to reepithelialize the wound.

The first inflammatory cells that are recruited to the wound bed are neutrophils (Eming et al., 2007). Neutrophils clear the wound bed of cellular debris and microorganisms by secreting proteolytic enzymes such as matrix metalloproteases, elastase, reactive oxygen species, and cytokines (Egozi et al., 2003; Wlaschek et al., 1997). Neutrophils also produce antimicrobial peptides which are multifunctional molecules that are capable of directly killing pathogens and stimulating cytokine production (Büchau et al., 2007; Gallo et al., 1994; Lai and Gallo, 2009; Lai et al., 2010; Sumikawa et al., 2006). Some of these cytokines attract monocytes, which are leukocytes that circulate in the blood but, when in the tissue, differentiate into macrophages that play a critical role in the initiation, maintenance, and resolution of inflammation. Inflammatory macrophages (M1) play an important role in antigen presentation and have phagocytic abilities to clear dead neutrophils, bacteria, and the debris that neutrophils leave behind in the process of killing pathogens (Schäfer and Werner, 2008). Important genes that are involved in macrophage function include toll-like receptors, complement receptors, and Fc receptors, allowing macrophages to sense and respond to their environment (Büchau et al., 2007; Gordon, 2003; Karin et al., 2006). They are also considered immunomodulators because they produce various cytokines and growth factors such as TNF-α, IL-1, IL-6, IL-8, and IL-12. In conjunction with leukotrienes, prostaglandins increase vascular permeability and additional recruitment of inflammatory and immune cells.

Successful wound repair requires resolution of the inflammatory phase (Wlaschek and Scharffetter-Kochanek, 2005). In the process, M1 macrophages secrete factors that result in either attracting a new wave of monocytes that in the tissue differentiate in M2 macrophages or stimulating the M1 macrophages to become M2 macrophages. Prohealing M2 macrophages are cells that resolve inflammation and prepare the tissue for the next phase of healing. This can be achieved by downregulation of inflammatory molecules by antiinflammatory cytokines such as IL-10 and TGF-β1 or upregulation of antiinflammatory molecules such as IL-1 receptor antagonist. The mechanism of controlling the termination of the inflammatory phase of wound healing is still under investigation (Schäfer and Werner, 2008).

The inflammatory phase of wound repair overlaps with the proliferative phase, which involves reepithelialization, formation of the healing tissue, and angiogenesis. During this phase, new tissue forms that is composed of keratinocyte, fibroblasts, macrophages, lymphocytes, and endothelial cells, which work in concert to rebuild the new tissue (Barrientos et al., 2008; Kolluru et al., 2012). To begin the repair of either damaged or missing dermal tissue, fibroblasts and endothelial cells migrate and proliferate beneath the newly formed epidermis to generate new dermal tissue called granulation tissue (Nouvong et al., 2016; Schäfer and Werner, 2008; Stadelmann et al., 1998).

Fibroblasts can proliferate and differentiate into myofibroblasts, which assist in wound healing by contracting the healing tissue. They also secrete extracellular matrix molecules, such as collagen, that contribute to strengthen the tissue and begin the formation of the scar. Angiogenesis, a process by which new blood vessels develop by proliferation and migration of endothelial cells, bring nutrients and oxygen to the wound site to support the reconstruction of the healing tissue (Nouvong et al., 2016; Schäfer and Werner, 2008; Stadelmann et al., 1998).

Remodeling is the final stage of wound healing and follows the completion of reepithelization and granulation tissue formation. The excess cells and ECM that are present in the healing tissue must now be removed for complete formation of the scar. This primarily occurs by cell apoptosis (Schäfer and Werner, 2008). Apoptosis, a form of cell death that is tightly regulated and controlled, allows for the removal of the remaining inflammatory cells, excess fibroblasts, and endothelial cells as well as ECM without causing additional damage to the wound tissue. The tensile strength of the tissue is enhanced by myofibroblasts remodeling various components of the extracellular matrix so that the healing tissue becomes stronger and the skin barrier is reestablished to protect the newly formed tissue from the environment (Loots et al., 1998; Stadelmann et al., 1998).

In contrast to what is currently understood by those outside the wound-healing field, the presence of microbes in the wound is critical for proper wound healing. A sterile wound does not lead to a better healing outcome. This is because microbes induce neutrophil and macrophage infiltration, which, when properly regulated, clear the wound of pathogens and activate proper healing.

Studies of host-microbe interactions in the skin further stress the complex relationship between wound-healing processes and skin-residing microbes (Krishna and Miller, 2012; Williams et al., 2017). For example, patients with diabetes are at risk of developing wounds that become infected and chronic. This increased risk may stem from a dysbiosis of the skin microbiota resulting from a different activated innate immune response. Studies surveying the skin over the toe web space in 41 patients showed that the microbial composition and bacterial growth were different in patients with diabetes (Thimmappaiah Jagadeesh et al., 2017). Of the 220 microbial colonies that were isolated from all samples, 95% of all colonies isolated were gram-positive bacteria. Gram-negative bacteria and fungal colonies accounted for only 2.7% and 0.9%, respectively. *S. epidermidis* was highly prevalent among diabetic patients. 77.5% of diabetic patients had *S. epidermidis* compared to 53.7% of patient without diabetes. On the other hand, *Bacillus* species were isolated from 34% of patients without diabetes; only 15% of diabetic patients was found to have *Bacillus* species.

The importance of the microbiome in the skin for proper healing has become more and more apparent. Microorganisms on the skin, whether commensal or pathogenic, have a close relationship with wound healing because they interact with the immune system and, they can play a major role in determining the strength of the inflammatory response. However, if the inflammatory response of wound healing is not properly regulated, chronic wounds can develop.

One study surveying the bacterial microbiome in acute wounds shows that as wounds heal, the relative abundances of *Bevibacterium*, *Microbacterium*, *Mycobacterium*, and *Paracoccus* species significantly change over time (Ashrafi et al., 2020). A reciprocal relationship between the relative abundance of *Staphylococcus* and *Propionibacterium* was found to be significant 21 days post injury. The relative abundance of *Staphylococcus* decreased as the abundance of *Propionibacterium* increased. The increase in *Propionibacterium* was positively correlated with collagen levels in the wound as a result of stimulating the immune system. Ultimately, the predominant genera in wounds that heal are similar to those of healthy skin. How the process of wound healing leads to this outcome is still under investigation.

Several animal models have been used to elucidate the fundamentals of host-microbe interactions during wound healing. Using a diabetic mice model for cutaneous wound healing, wounds swabbed during the course of healing show that the wound microbiome is very complex and dynamic as the wound begins to heal and ultimately close (Kim et al., 2019). The diverse microbiome of the healing wounds includes *Cutibacterium acnes*, *Achromobacter*, *Delftia*, and *Escherichia coli*. Results from these studies have only scratched the surface of the complexity between the microbiota and the host during cutaneous wound healing.

21.5.1 The microbiome and cutaneous chronic wound development

Chronic wounds develop as a result of defective wound-healing regulation, that is, deviation from the highly coordinated processes describe above that are involved in wound healing. Chronic wounds are characterized by having increased levels of oxidative stress, chronic inflammation, damaged microvasculature, and abnormal collagen matrix deposition (Demidova-Rice et al., 2012; Kim et al., 2019). Most chronic wounds occur in patients with underlying pathologies or systemic diseases, leading to an aberrant and/or impaired immune response. Tissues that are ischemic, necrotic, and hypoxic have high levels of oxidative stress that contributes to a microenvironment that is conducive to colonization and proliferation of potentially pathological microbial populations.

Oxidative stress, both endogenously and exogenously produced in the wound, can delay wound closure, providing time for infection to occur and further preventing healing in a timely manner. During wound healing, if the inflammatory phase becomes chronic, the host tissue undergoes extensive damage due to a continuous influx of inflammatory cells that release cytotoxic enzymes and an increase in free oxygen radicals and inflammatory mediators, resulting in cell death. The physiological balance between tissue destruction and tissue reassembly is significantly disturbed, stemming from prolonged inflammation and overall catabolic proteolytic enzyme activity (Schneider et al., 2007). Such disturbances are observed in humans with chronic metabolic diseases such as diabetes, which have many complications related closely to elevated levels of oxidative stress.

The most common chronic wounds found in patients are venous and arterial ulcers, pressure ulcers, and diabetic foot ulcers. In these wounds, the healing process is complicated by pathogenic, opportunistic bacteria and fungi that take advantage of host nutrients that are leeched in the destructive inflammatory microenvironment. These microbes contribute to the damaging of the host tissue when they form biofilm (Gjødsbøl et al., 2006; James et al., 2008; Loesche et al., 2017; Misic et al., 2014; Schäfer and Werner, 2008; Wolcott et al., 2016; Zhao et al., 2013). These biofilms are recalcitrant to conventional antibiotic therapies because the structure of the biofilm decreases the efficacy of antibiotic therapy by significantly decreasing their diffusion rate. The biofilm extracellular polymeric substance (EPS) also helps the bacteria in the biofilm to evade the host innate and adaptive immune system (Dhall et al., 2014; Gontcharova, 2010; Raghav et al., 2018).

In the destructive microenvironment of chronic wounds, biofilm-forming bacteria take advantage of host nutrients and contribute to the damage of the host tissue. These biofilms are difficult to control and remove because the EPS, which is composed of proteins, DNA, RNA, lipids, and polysaccharides, allows bacteria harbored within it to be tolerant to conventional antibiotic therapies and evade the host's innate and adaptive immune responses. A multitude of aerobic and anaerobic bacterial species across the phylogenetic tree aggregate into sessile microcolonies that are harbored within these biofilms. Each species in the biofilm differs significantly from its planktonic counterparts in morphology, mode of communication, and metabolism with very different nutritional demands and roles. At the same time, stochastic processes and nonuniform gene expression can lead to the appearance of subpopulations of bacteria with different phenotypes and environmental responses. The biofilm provides a unique environment for cell signaling by the production and detection of quorum-sensing molecules, which promote collective behavior, such as optimizing nutrient acquisition and regulation of virulence, leading to sustained pathogenicity in the wound (Stewart and Franklin, 2008; Zhao et al., 2013). They can also evade the host's innate immune response (Costerton et al., 2007; Fux et al., 2005).

In addition, many bacterial species rely on quorum sensing to regulate much of their physiological processes; it is also a central mechanism in regulating the social activities of a population (Li and Tian, 2012; Yan and Wu, 2019). This mechanism relies on small, diffusible signaling molecules that are produced by bacterial cells to detect and respond to other bacterial cells in the population (Li and Tian, 2012). Quorum sensing allows bacteria to organize the individual activities of a single bacterium in favorable microbial interactions to achieve population-level virulence in bacterial infections (Li and Tian, 2012). Their use of autoinducer signal molecules significantly increases virulence by mediating subsequent transcription of hundreds of virulence genes. Studies have shown that quorum sensing is necessary for pathogenic bacteria to reach a critical cell density that overwhelms the immune system or defense mechanisms through the expression of virulence factors, initiating the symptoms of infection and resulting in the development of the infectious disease. Quorum sensing has a close relationship with bacterial biofilms because the development of biofilm is induced and regulated by many quorum sensing genes in addition to environmental factors in the host (Li and Tian, 2012). *Pseudomonas aeruginosa* is a common opportunistic pathogen that is highly associated with various chronic infections, including various etiologies of chronic wounds. A prominent feature of *P. aeruginosa* is the use of multiple quorum-sensing systems to increase virulence and establish infection quickly by evading the host innate and adaptive immune systems.

21.5.2 Specific chronic wounds and their wound microbiome

The microenvironment of chronic wounds provides optimal conditions that can support different bacterial species with multiple abiotic and nutritional needs. Sequencing for bacterial 16S rRNA in several chronic wound studies show that no single microorganism has been identified to be the causal agent of infection or biofilm in chronic wounds, adding to the complexity of chronic wounds. Multiple studies surveying cutaneous chronic wounds in humans show that chronic wounds are colonized by many bacterial species (Table 21.2, Fig. 21.1). Various types of chronic wounds can be colonized by the same bacterial genus and species. Common bacteria that are found in human chronic wounds include but are not limited to Corynebacteria, Bacteroides, *Peptoniphilus* sp., *Fingoldia* sp., *Anaerococcus* sp., *Streptococcus* sp.,

TABLE 21.2 Common bacteria in chronic ulcers.

Chronic ulcers	Common bacteria genera	
Venous ulcers	*Staphylococcus aureus* *Corynebacterium* *Pseudomonas aeruginosa* *Serratia* *Streptococci*	
Venous ulcers	*Streptococcus* *Peptoniphilus* *Anaeroccocus* *Gemella* *Staphylococcus* *Corynebacterium* *Enterobacteraceae* *Anaerococcus* *Enterococcus*	*Acinetobacter* *Finegoldia* *Micrococcs* *Enhydrobacter* *Prevotella* *Lachnospiraceae* *Blautia* *Lactobacillus*
Diabetic foot ulcers	*Acinetobacter* *Bacillus* *Enterobacter* *Escherichia*	*Propionibacterium* *Enterococcus* *Pseudomonas* *Staphylococcus*

Serratia sp., Clostridiales, and Streptococcaceae. Many of the bacteria that have been identified in human wounds have also been identified in a diabetic chronic wound murine model. Nonhealing chronic wounds in 5- to 6-month-old diabetic mice had wounds colonized by biofilm-forming bacterial species such as *P. aeruginosa*, *Enterobacter cloacae*, *Corynebacterium frankenforstense*, and *Acinetobacter* sp. (Kim et al., 2020).

21.5.2.1 *Venous ulcers*

Venous ulcers form as a result of poor blood circulation and chronic venous insufficiency (Kirsner and Vivas, 2015) and develop due to poor venous pressure or sustained venous hypertension because valves in the calf are unable to maintain proper blood pressure (Kirsner and Vivas, 2015; Grey et al., 2006). Up to 50% of venous ulcers form because of superficial venous insufficiency and/or incompetent perforating veins (Grey et al., 2006). They usually form on the calf and ankle, with 95% of venous ulceration in the gaiter area of the leg. The wound bed of the ulcer usually has a fibrinous layer mixed with granulation tissue, and in general with an irregular wound edge (Grey et al., 2006). Patients with history of smoking or diabetes are at risk of developing arterial ulcers. Other important risk factors include hypertension and obesity. Venous leg ulcers are often infected; the predominant microbes that are found in these ulcers include *S. aureus*, *Corynebacterium*, *P. aeruginosa*, *Serratia*, and *Streptococci* bacteria.

21.5.2.2 *Pressure ulcers*

Pressure ulcers are injuries of the skin or underlying tissue that occur as a result of pressure alone or pressure in combination with shearing forces over bony prominences (Kirsner and Vivas, 2015). Age is an important risk factor for pressure ulcers in elderly or disabled patients. Patients who constantly reside in a bed or wheelchair can present with pressure ulcers. These ulcers start developing when soft tissue, most commonly between a surface and a bony prominence, is compressed for 2–6 hours, resulting in tissue damage (Kirsner and Vivas, 2015). Specifically, the formation of an ulcer follows sustaining high pressure and shear forces in aged skin. Aged skin contains reduced levels of elastin, with friction leading to superficial erosions and blisters and resulting in hypoxia and tissue necrosis. When the wound opens, it becomes exposed to damaged tissue, excess moisture from urine and sweat, and even fecal matter (Kirsner and Vivas, 2015). Patients with diabetes mellitus have an increased risk of developing pressure ulcers. Other risk factors include paralysis, cardiovascular diseases, unconsciousness, malnutrition, and fecal incontinence. Pressure ulcers are often colonized by many bacteria species (Ammons et al., 2015; de Wert et al., 2020; Shibata et al., 2021). Common bacteria include *Streptococcus*, *Peptoniphilus*, *Anaeroccocus*, *Gemella*, *Staphylococcus*, *Corynebacterium*, Enterobacteraceae, *Anaerococcus* (Ammons et al., 2015), *Acinetobacter*, *Finegoldia*, *Micrococcs*, *Enhydrobacter*, *Prevotella*, Lachnospiraceae, *Blautia*, *Lactobacillus*, and *Enterococcus*.

21.5.2.3 Diabetic foot ulcers

Diabetic ulcers are commonly found in the foot and are high-risk cutaneous wounds for patients with diabetes mellitus, resulting in a high incidence of hospitalization and increased risk for amputation if the antibiotic treatments are not effective in controlling infections. These ulcers are the result of nonhealing lesions due to neurological disorders, diabetic neuropathy, and peripheral vascular disease of the lower limbs (Kirsner and Vivas, 2015). The presence of peripheral neuropathy significantly increases the risk of developing diabetic foot ulcers because lacking proper sensory greatly reduces the patient's awareness of a mechanical trauma in the foot (Ammons et al., 2015). Patients cannot feel the pressure, pain, and discomfort that an injury normally elicits; when the injury is not addressed, ulceration is the result (Kirsner and Vivas, 2015). Common causes for diabetic foot ulcers are neuropathy and ischemia, but complications from the pathology of diabetes itself, including altered regulation in blood flow and impaired oxygenation of extremities, greatly increase the risk of developing diabetic foot ulcers. Most patients with diabetic foot ulcers show symptoms of infection and may require hospitalization. Diabetic foot ulcers are colonized by a variety of bacteria species. These include *Acinetobacter* (Dowd et al., 2008b; Gjødsbøl et al., 2006, 2012; James et al., 2008; Wolcott et al., 2016), *Bacillus* (Dowd et al., 2008a,b; Gontcharova, 2010; Wolcott et al., 2016), *Enterobacter* (Dowd et al., 2008b; Gjødsbøl et al., 2006, 2012; James et al., 2008; Price et al., 2009; Wolcott et al., 2016), *Escherichia* (Dowd et al., 2008b; Gontcharova, 2010; Gjødsbøl et al., 2006, 2012; James et al., 2008; Scales and Huffnagle, 2013), *Propionibacterium* (Dowd et al., 2008a; Gontcharova, 2010; Wolcott et al., 2016), *Enterococcus* (Dowd et al., 2008a,b; James et al., 2008; Price et al., 2009; Scales and Huffnagle, 2013; Wolcott et al., 2016), *Pseudomonas* (Dowd et al., 2008a; Gontcharova, 2010; Gjødsbøl et al., 2012; James et al., 2008; Price et al., 2009; Scales and Huffnagle, 2013; Wolcott et al., 2016), and *Staphylococcus* (Dowd et al., 2008a,b; Gontcharova, 2010; James et al., 2008; Price et al., 2009; Wolcott et al., 2016).

21.6 Common techniques for surveying the skin microbiota

To begin understanding the nature of the skin microbiota and its effect on skin function, new methods and technologies were developed to survey the microbiota. Methods to identify primary infectious microbes previously relied on culture-dependent methods. This led the identification of important wound pathogens that commonly colonize wounds with impaired healing. These culture-dependent techniques, while still commonly used today in health clinics, depend on the use of specific media that cater to the growth needs of the microbe. Therefore, this method cannot be reliably used to evaluate complex multimetabolism-dependent microbial communities, which need very specific culture conditions to grow. For example, strictly anaerobic microbes cannot grow in conditions in which oxygen is present in high concentrations, and thus require a strictly anoxic environment or a very low oxygen concentration in the growth chambers to facilitate growth. Other microbes may require specific unknown metabolites to grow and would consequently never grow artificially in media. Accordingly, microbes that require the complex microenvironment of human skin will not consistently grow in simplified artificial media and will be missed in these culture-dependent surveys.

To overcome the inability to evaluate complex microbiomes, culture-independent methods were developed that do not require specific culture conditions to grow different bacterial taxa (Temmerman et al., 2004). Culture-independent metagenomic techniques have been used to describe the microbial diversity of complex communities, such as the wound-associated microbiome (Hodkinson and Grice, 2015; Ye et al., 2008). One of the most common methods is high-throughput sequencing of bacterial 16S ribosomal RNA (16S rRNA) genes (Hannigan et al., 2014; Hodkinson and Grice, 2015; Misic et al., 2014). Fungal species can be identified by sequencing the 18S rRNA. Results from these methods showed that the microbiome of wounds is complex and crosses both bacterial and eukaryotic domains. Dynamic changes in relative abundances between these microbes over time are shown to correspond to healing and impaired wounds.

Other techniques of culture-independent methods, such as sequencing the internal transcribed space (ITS) and shotgun metagenomic sequencing, are now being used to survey complex wound microbiomes across domains with dynamic interactions between microbes over time corresponding to healing and nonhealing wounds (Hannigan and Grice, 2013; Kalan et al., 2016; Loesche et al., 2017; Misic et al., 2014). With the ITS method, the genomic ITS rRNA gene is targeted over traditional 16S rRNA sequencing to obtain taxonomical information about the bacterial community. The ITS region offers species- and subspecies-level resolution because this region could harbor higher sequence variations (Ruegger et al., 2014). While sequencing the 16S rRNA revealed the complicated nature of microbiomes, the data were commonly represented at the genus taxonomical level, which made interpretation of the data difficult because bacteria act as either pathogens or commensal bacteria at the species level. Other bacteria need to be identified at the subspecies or strain levels to identify their status of pathogenicity (Kalan et al., 2019).

21.7 Manipulation of the skin microbiota to prevent or treat skin-associated ailments

The composition of commensal bacterial species is relatively stable over time but can be perturbed during infection or disease. Recently, interest in manipulating the microbiota has come into play in studying potential therapeutic strategies for the treatment and study of diseases, and improving the health of patients affected by skin ailments. The goal is to limit the growth of a few pathogenic species and support the growth of commensal or beneficial bacterial species to improve overall health. Thus, manipulations of the microbiota may directly promote healing by reducing the presence of pathogens and enhancing the abundance of beneficial species in the affected area. Probiotics, microbiota transplants, and even prebiotics have been heavily researched in the gut microbiota and have been used to treat many gut-related conditions and diseases. While recent research has shown a strong relationship between skin health and probiotics, the mechanisms by which probiotics facilitate improved skin health remains unclear.

21.7.1 Probiotics

Live microbial supplements may be beneficial to health by improving the microbiota community of the skin (Ouwehand et al., 2003). Providing beneficial skin probiotics may promote homeostasis by influencing both the skin tissue and cells and the bacteria that reside on the skin (Yu et al., 2020). Because the skin is a complex ecosystem with many interactions between microbes and skin cells, probiotics may be beneficial in treating skin diseases or ailments that are caused by an imbalance of microorganisms (Knackstedt et al., 2020). Both diseased and healthy skin may be affected by probiotic treatment. Research on skin health benefits of probiotics indicates that oral consumption might reduce skin sensitivity and support skin immune function. Oral ingestion of several probiotic *Lactobacillus* species, including *Lactobacillus johnsonii*, *L. paracasei*, *L. casei*, and *L. bulgaris*, improve the health of the skin. Benefits include improving the skin immune system, decreasing the neurosensitivity of reactive skin, and reducing transepidermal water loss (Roudsari et al., 2015). Lactobacilli bacteria are a major group of probiotics that can modulate the host immune response in the gut by binding to pattern recognition receptors that are expressed on a number of immune cells and other tissues, including the intestinal epithelium. Research has also shown that probiotics can help to maintain skin homeostasis by protecting against UV-induced skin damage (Kober and Bowe, 2015). How probiotics can affect other skin ailments, such as psoriasis and rosacea, have yet to be explored (Knackstedt et al., 2020).

21.7.2 Skin microbiota transplant

While success has been found in the gut, the most prominent example of this procedure has been shown with the treatment of the antibiotic-resistant bacteria *Clostridium difficile* within the gut microbiome with the help of fecal transplantation (Schneider and Nelson, 2019). Following this successful treatment, many projects have been focused on developing treatments for gut diseases through the use of microbiome transplants. Efforts to manipulate the skin microbiome have been made by using microbiome transplants as a novel therapeutic approach for skin diseases and conditions. *C. acnes* and its strain diversity represent a major part of the human skin microbiome, and specific bacterial strains are associated with acne vulgaris. An effort has been made to determine whether this subpopulation of species can be modulated at the strain level. It has been shown that the human skin microbiome composition can be modulated through approaches similar to those used in fecal transplantation of the gut microbiome. Probiotic solutions from donor microbiomes have been applied to the skin of healthy volunteers, whose skin microbiome was monitored during and after the treatment (Paetzold et al., 2019). The dose of bacteria to be applied played an important role in the capacity of modulation. Indeed, temporary modulation of the *C. acnes* population at the strain level was feasible without a negative reaction of the host. It is important to note that all subjects underwent dermatological inspection to establish that no adverse effects were detected.

21.7.3 Prebiotics

Prebiotics are chemicals or metabolites that are beneficial to the commensal bacteria of the gut and skin in order to improve the health of the gut and skin (Krutmann, 2009; Bockmühl et al., 2007). This means that prebiotics need to be specific or targeted enough to affect only commensal bacteria, as opposed to the pathogenic bacteria, in order to have a positive effect on the organ or tissue. This is because the beneficial or commensal bacteria should be selectively stimulated to grow over the bacteria that can potentially cause disease or infection. Prebiotics can be as simple as simple sugars or molecules but can be as complex as oligomers such as galactooligosaccharides. Examples of prebiotics include glycerin, which supports *C. acnes*. The cosmetics industry has long utilized extracts from natural sources, such as plants and yeasts, to stimulate cellular processes in the skin to enhance the appearance of skin (Bockmühl et al., 2007; Krutmann, 2009). Extracts from plants such as ginseng, black currant, and pine can be formulated into cosmetic products to be applied to the skin. Research has been conducted to begin elucidating the effects of plant extracts on skin microbiota. Daily application of such products for 3 weeks showed inhibition of the growth of *P. acnes*, while the growth of coagulase-negative staphylococci bacteria that are beneficial to the skin was not affected (Bockmühl et al., 2007).

21.8 Perspectives

Thus far, studies have focused on understanding how different cell types, extracellular matrix components, and signaling molecules play a role in maintaining healthy skin and how they contribute to cutaneous wound healing. However, research into the contributions of the skin microbiota to skin health, disease, and impaired wounds has lagged behind, even though microbes are acknowledged to be an important factor in these processes. Whereas key infectious microbes were identified through culture-dependent methods, identifying the myriad of microbes residing in the skin and wounds has been a challenge because different microbes have specific and often unknown growing conditions that have not yet been replicated in culture. In the past decade, as a result of appreciation of the complexity of the gut microbiota and its role in gut health and disease, there has been a strong push to describe the skin microbiota and to define the microbiota that are present in healing and in chronic wounds. With the development and advancement of high-throughput sequencing platforms and the support of large databases, physicians and scientists have been able to show how complicated and complex the microbial community of the skin is.

Once microbes that significantly contribute to wound healing or chronic wound development are known, experiments that elucidate the functional role of the microbes must be conducted. With shotgun metagenomic sequencing platforms, and with bioinformatics and statistics support becoming more prevalent, the information from these sequencing experiments provide insight beyond taxonomical identification. Gene-level metagenomic profiles can characterize not just the identification of the microbe but also the genetic profile of the microbiota. Genes that are important for virulence, biofilm formation, quorum sensing, antibiotic resistance, and host-defense evasion can be identified as potential targets for new drugs and antibiotics. Several publicly available databases that include bacterial genes can help to identify key processes and signaling pathways in order to generate new hypotheses that can be explored both in vitro and in vivo using animal models to study mechanisms.

Other large data-driven experiments, such as transcriptomics experiments, can identify genes that are activated and transcribed during infection or toxic stress in response to drug therapeutics. Proteomic and metabolomic studies can provide information about how microbes are able to evade host innate and adaptive immunity and provide insight into which proteins or metabolites should be targeted in future studies.

Altogether, the future understanding of how microbes contribute to healthy skin and skin-based ailments will result from hypothesis-driven experiments researching the interactions between the skin of the host and the microbial community that resides on the skin. With the discovery and treatment with antibiotics, many bacterial infections that previously had high infection, morbidity, and mortality rates can be treated. As the battle against primary pathogens continues, the number of antibiotic-resistant pathogens is on the rise. In addition, the number of cases involving opportunistic pathogens in patients with compromised innate and adaptive immune systems will continue to increase dramatically. It will be necessary to study how the health of patients, who may or may not exhibit symptoms of a disease or condition, will interact with pathogenic members of the microbial community to result in infection.

References

Ammons, M.C.B., et al., 2015. Biochemical association of metabolic profile and microbiome in chronic pressure ulcer wounds. PLoS One 10, e0126735.

Ashrafi, M., et al., 2020. A microbiome and metabolomic signature of phases of cutaneous healing identified by profiling sequential acute wounds of human skin: an exploratory study. PLoS One 15. e0229545.

Baker, L.B., 2019. Physiology of sweat gland function: the roles of sweating and sweat composition in human health. Temperature 6, 211–259.

Barrientos, S., Stojadinovic, O., Golinko, M.S., Brem, H., Tomic-Canic, M., 2008. Perspective article: growth factors and cytokines in wound healing. Wound Repair. Regen. 16, 585–601.

Batista, D., et al., 2002. Comparative integrated omics: identification of key functionalities in microbial community-wide metabolic networks. NPJ Biofilms Microbiomes 57, 10–13.

Bockmühl, D., et al., 2007. Prebiotic cosmetics: an alternative to antibacterial products. Int. J. Cosmet. Sci. 29, 63–64.

Büchau, A.S., et al., 2007. S100A15, an antimicrobial protein of the skin: regulation by E. coli through toll-like receptor 4. J. Invest. Dermatol. 127, 2596–2604.

Byrd, A.L., Belkaid, Y., Segre, J.A., 2018. The human skin microbiome. Nat. Rev. Microbiol. 16, 143–155.

Chen, Y.E., Tsao, H., 2013. The skin microbiome: current perspectives and future challenges. J. Am. Acad. Dermatol. 69, 143–155. e3.

Coates, M., Lee, M.J., Norton, D., MacLeod, A.S., 2019. The skin and intestinal microbiota and their specific innate immune systems. Front. Immunol. 10, 1–11.

Costerton, W., et al., 2007. The application of biofilm science to the study and control of chronic bacterial infections. J. Clin. Invest. 117, 278.

Culp, B., Scheinfeld, N., 2009. Rosacea: a review. P T 34, 38–45.

Daou, H., Paradiso, M., Hennessy, K., Seminario-Vidal, L., 2021. Rosacea and the microbiome: a systematic review. Dermatol. Ther. (Heidelb). 11, 1–12.

de Wert, L.A., et al., 2020. The cutaneous microbiome in hospitalized patients with pressure ulcers. Sci. Rep. 10, 5963.

Demidova-Rice, T.N., Hamblin, M.R., Herman, I.M., 2012. Acute and impaired wound healing. Adv. Skin. Wound Care 25, 304–314.

Dhall, S., et al., 2014. Generating and reversing chronic wounds in diabetic mice by manipulating wound redox parameters. J. Diabetes Res. 2014, 1–18.

Dowd, S.E., et al., 2008a. Polymicrobial nature of chronic diabetic foot ulcer biofilm infections determined using bacterial tag encoded FLX amplicon pyrosequencing (bTEFAP). PLoS One 3, e3326.

Dowd, S.E., et al., 2008b. Survey of bacterial diversity in chronic wounds using Pyrosequencing, DGGE, and full ribosome shotgun sequencing. BMC Microbiol. 8, 43.

Dréno, B., et al., 2018. Cutibacterium acnes (Propionibacterium acnes) and acne vulgaris: a brief look at the latest updates. J. Eur. Acad. Dermatol. Venereol. 32, 5–14.

Egert, M., Simmering, R., 2016. Microbiota Hum. Skin. 902, 61–81.

Egert, M., Simmering, R., Riedel, C.U., 2017. The association of the skin microbiota with health, immunity, and disease. Clin. Pharmacol. Ther. 102, 62–69.

Egozi, E.I., Ferreira, A.M., Burns, A.L., Gamelli, R.L., Dipietro, L.A., 2003. Mast cells modulate the inflammatory but not the proliferative response in healing wounds. Wound Repair. Regen. 11, 46–54.

Eming, S.A., Krieg, T., Davidson, J.M., 2007. Inflammation in wound repair: molecular and cellular mechanisms. J. Invest. Dermatol. 127, 514–525.

Fitz-Gibbon, S., et al., 2013. Propionibacterium acnes strain populations in the human skin microbiome associated with acne. J. Invest. Dermatol. 133, 2152–2160.

Fux, C.A., Costerton, J.W., Stewart, P.S., Stoodley, P., 2005. Survival strategies of infectious biofilms. Trends Microbiol. 13, 34–40.

Fyhrquist, N., Salava, A., Auvinen, P., Lauerma, A., 2016. Skin biomes. Curr. Allergy Asthma Rep. 16, 40.

Gallo, R.L., et al., 1994. Syndecans, cell surface heparan sulfate proteoglycans, are induced by a proline-rich antimicrobial peptide from wounds. Proc. Natl. Acad. Sci. 91, 11035–11039.

Gao, Z., Tseng, C.H., Pei, Z., Blaser, M.J., 2007. Molecular analysis of human forearm superficial skin bacterial biota. Proc. Natl. Acad. Sci. U. S. A. 104, 2927–2932.

Gilaberte, Y., Prieto-Torres, L., Pastushenko, I., Juarranz, Á., 2016. Anatomy and function of the skin. Nanoscience in Dermatology. Elsevier, pp. 1–14. Available from: https://doi.org/10.1016/B978-0-12-802926-8.00001-X.

Gilbert, J.A., et al., 2018. Current understanding of the human microbiome. Nat. Med. 24, 392–400.

Gjødsbøl, K., et al., 2006. Multiple bacterial species reside in chronic wounds: a longitudinal study. Int. Wound J. 3, 225–231.

Gjødsbøl, K., et al., 2012. No need for biopsies: comparison of three sample techniques for wound microbiota determination. Int. Wound J. 9, 295–302.

Gontcharova, V., 2010. A Comparison of bacterial composition in diabetic ulcers and contralateral intact skin. Open. Microbiol. J. 4, 8–19.

Gordon, S., 2003. Alternative activation of macrophages. Nat. Rev. Immunol. 3, 23–35.

Grey, J.E., Enoch, S., Harding, K.G., 2006. ABC of wound healing: venous and arterial leg ulcers. BMJ 332, 0604140.

Grice, E.A., Segre, J.A., 2011. The skin microbiome. Nat. Rev. Microbiol. 9, 244.

Grice, E.A., et al., 2009. Topographical and temporal diversity of the human skin microbiome. Science (80-.) 324, 1190–1192.

Groscurth, P., 2002. Anatomy of sweat glands. Curr. Probl. Dermatol. 30, 1–9.

Hannigan, G.D., Grice, E.A., 2013. Microbial ecology of the skin in the era of metagenomics and molecular microbiology. Cold Spring Harb. Perspect. Med. 3, 1–15.

Hannigan, G.D., et al., 2014. Culture-independent pilot study of microbiota colonizing open fractures and association with severity, mechanism, location, and complication from presentation to early outpatient follow-up. J. Orthop. Res. 32, 597–605.

Hodkinson, B.P., Grice, E.A., 2015. Next-generation sequencing: a review of technologies and tools for wound microbiome research. Adv. Wound Care 4, 50–58.

Huang, C.M., et al., 2005. Prospective highlights of functional skin proteomics. Mass. Spectrom. Rev. 24, 647–660.

James, G.A., et al., 2008. Biofilms in chronic wounds. Wound Repair. Regen. 16, 37–44.

Johnson, T.R., et al., 2018. The cutaneous microbiome and wounds: new molecular targets to promote wound healing. Int. J. Mol. Sci. 19.

Kalan, L., et al., 2016. Redefining the chronic-wound microbiome: fungal communities are prevalent, dynamic, and associated with delayed healing. MBio 7, 1–12.

Kalan, L.R., et al., 2019. Strain- and species-level variation in the microbiome of diabetic wounds is associated with clinical outcomes and therapeutic efficacy. Cell Host Microbe 25, 641–655. e5.

Kali, A., 2015. Human microbiome engineering: the future and beyond. J. Clin. Diagn. Res. 9, 1–4.

Karin, M., Lawrence, T., Nizet, V., 2006. Innate immunity gone awry: linking microbial infections to chronic inflammation and cancer. Cell 124, 823–835.

Khavkin, J., Ellis, D.A.F., 2011. Aging skin: histology, physiology, and pathology. Facial Plast. Surg. Clin. North. Am. 19, 229–234.

Kim, J.H.J.H., et al., 2019. High levels of oxidative stress and skin microbiome are critical for initiation and development of chronic wounds in diabetic mice. Sci. Rep. 9, 19318.

Kim, J.H., et al., 2020. High levels of oxidative stress create a microenvironment that significantly decreases the diversity of the microbiota in diabetic chronic wounds and promotes biofilm formation. Front. Cell. Infect. Microbiol. 10, 1–20.

Kirsner, R.S., Vivas, A.C., 2015. Lower-extremity ulcers: diagnosis and management. Br. J. Dermatol. 173, 379–390.

Knackstedt, R., Knackstedt, T., Gatherwright, J., 2020. The role of topical probiotics in skin conditions: a systematic review of animal and human studies and implications for future therapies. Exp. Dermatol. 29, 15–21.

Kober, M.-M., Bowe, W.P., 2015. The effect of probiotics on immune regulation, acne, and photoaging. Int. J. Women's Dermatol. 1, 85–89.

Kolarsick, P.A.J., Kolarsick, M.A., Goodwin, C., 2011. Anatomy and physiology of the skin. J. Dermatol. Nurses. Assoc. 3, 203–213.

Kolluru, G.K., Bir, S.C., Kevil, C.G., 2012. Endothelial dysfunction and diabetes: effects on angiogenesis, vascular remodeling, and wound healing. Int. J. Vasc. Med. 2012, 1–30.

Kong, H.H., 2011. Skin microbiome: genomics-based insights into the diversity and role of skin microbes. Trends Mol. Med. 17, 320–328.

Kong, H.H., Segre, J.A., 2012. Skin microbiome: looking back to move forward. J. Invest. Dermatol. 132, 933–939.

Krishna, S., Miller, L.S., 2012. Host–pathogen interactions between the skin and Staphylococcus aureus. Curr. Opin. Microbiol. 15, 28–35.

Krutmann, J., 2009. Pre- and probiotics for human skin. J. Dermatol. Sci. 54, 1–5.

Lai, Y., Gallo, R.L., 2009. AMPed up immunity: how antimicrobial peptides have multiple roles in immune defense. Trends Immunol. 30, 131–141.

Lai, Y., et al., 2010. Activation of TLR2 by a small molecule produced by Staphylococcus epidermidis increases antimicrobial defense against bacterial skin infections. J. Invest. Dermatol. 130, 2211–2221.

Lai-Cheong, J.E., McGrath, J.A., 2017. Structure and function of skin, hair and nails. Medicine (Baltimore) 45, 347–351.

Li, Y.-H., Tian, X., 2012. Quorum sensing and bacterial social interactions in biofilms. Sensors 12, 2519–2538.

Loesche, M., et al., 2017. Temporal stability in chronic wound microbiota is associated with poor healing. J. Invest. Dermatol. 137, 237–244.

Loots, M.A.M., et al., 1998. Differences in cellular infiltrate and extracellular matrix of chronic diabetic and venous ulcers vs acute wounds. J. Invest. Dermatol. 111, 850–857.

Lunjani, N., Hlela, C., O'Mahony, L., 2019. Microbiome and skin biology. Curr. Opin. Allergy Clin. Immunol. 19, 328–333.

MacLeod, A.S., Mansbridge, J.N., 2016. The innate immune system in acute and chronic wounds. Adv. Wound Care 5, 65–78.

Margalit, A., Kowalczyk, M.J., Żaba, R., Kavanagh, K., 2016. The role of altered cutaneous immune responses in the induction and persistence of rosacea. J. Dermatol. Sci. 82, 3–8.

Martins-Green, M., 1997. Dynamics of cell-ECM interactions with implications for tissue enginnering. Princ. Tissue Eng. 23–46.

Mcmurdie, P.J., et al., 2015. Advancing our understanding of the human microbiome using QIIME. Methods Enzymol. 531.

Meisel, J.S., et al., 2018. Commensal microbiota modulate gene expression in the skin. Microbiome 6, 1–15.

Menon, G.K., 2015. Skin basics: structure and function. In: Pappas, A. (Ed.), Lipids and Skin Health. Springer International Publishing, pp. 9–23. Available from: https://doi.org/10.1007/978-3-319-09943-9_2.

Misic, A.M., Gardner, S.E., Grice, E.A., 2014. The wound microbiome: modern approaches to examining the role of microorganisms in impaired chronic wound healing. Adv. Wound Care 3, 502–510.

Morton, L.M., Phillips, T.J., 2016. Wound healing and treating wounds differential diagnosis and evaluation of chronic wounds. J. Am. Acad. Dermatol. 74, 589–605.

Nakatsuji, T., et al., 2013. The microbiome extends to subepidermal compartments of normal skin. Nat. Commun. 4, 1–8.

Nouvong, A., Ambrus, A.M., Zhang, E.R., Hultman, L., Coller, H.A., 2016. Reactive oxygen species and bacterial biofilms in diabetic wound healing. Physiol. Genomics 48, 889–896.

Ouwehand, A.C., Batsman, A., Salminen, S., 2003. Probiotics for the skin: a new area of potential application? Lett. Appl. Microbiol. 36, 327–331.

Paetzold, B., et al., 2019. Skin microbiome modulation induced by probiotic solutions. Microbiome 7, 95.

Pang, M., et al., 2020. Changes in foot skin microbiome of patients with diabetes mellitus using high-throughput 16s rRNA gene sequencing: a case control study from a single center. Med. Sci. Monit. 26, 1–11.

Price, L.B., et al., 2009. Community analysis of chronic wound bacteria using 16S rRNA gene-based pyrosequencing: impact of diabetes and antibiotics on chronic wound microbiota. PLoS One 4, e6462.

Raghav, A., et al., 2018. Financial burden of diabetic foot ulcers to world: a progressive topic to discuss always. Ther. Adv. Endocrinol. Metab. 9, 29–31.

Roudsari, M.R., Karimi, R., Sohrabvandi, S., Mortazavian, A.M., 2015. Health effects of probiotics on the skin. Crit. Rev. Food Sci. Nutr. 55, 1219–1240.

Ruegger, P.M., Clark, R.T., Weger, J.R., Braun, J., Borneman, J., 2014. Improved resolution of bacteria by high throughput sequence analysis of the rRNA internal transcribed spacer. J. Microbiol. Methods 105, 82–87.

Salava, A., et al., 2016. Skin microbiome in melanomas and melanocytic nevi. Eur. J. Dermatol. 26, 49–55.

Salvo, P., Dini, V., Di Francesco, F., Romanelli, M., 2015. The role of biomedical sensors in wound healing. Wound Med. 8, 15–18.

Sanmiguel, A., Grice, E.A., 2015. Interactions between host factors and the skin microbiome. Cell. Mol. Life Sci. 72, 1499–1515.

Scales, B.S., Huffnagle, G.B., 2013. The microbiome in wound repair and tissue fibrosis. J. Pathol. 229, 323–331.

Schäfer, M., Werner, S., 2008. Oxidative stress in normal and impaired wound repair. Pharmacol. Res. 58, 165–171.

Schmidt, C., 2020. Out of your skin. Nat. Biotechnol. 38, 392–397.

Schneider, L.A., Korber, A., Grabbe, S., Dissemond, J., 2007. Influence of pH on wound-healing: a new perspective for wound-therapy? Arch. Dermatol. Res. 298, 413–420.

Schneider, A.M., Nelson, A.M., 2019. Skin microbiota: friend or foe in pediatric skin health and skin disease. Pediatr. Dermatol. 36, 815–822.

Schommer, N.N., Gallo, R.L., 2013. Structure and function of the human skin microbiome. Trends Microbiol. 21, 660–668.

Shibata, K., et al., 2021. Skin physiology and its microbiome as factors associated with the recurrence of pressure injuries. Biol. Res. Nurs. 23, 75–81.

Squarzanti, D.F., et al., 2020. Non-melanoma skin cancer: news from microbiota research. Crit. Rev. Microbiol. 46, 433–449.

Stadelmann, W.K., Digenis, A.G., Tobin, G.R., 1998. Physiology and healing dynamics of chronic cutaneous wounds. Am. J. Surg. 176, 26S–38S.

Stehlikova, Z., et al., 2019. Dysbiosis of skin microbiota in psoriatic patients: co-occurrence of fungal and bacterial communities. Front. Microbiol. 10.

Stewart, P.S., Franklin, M.J., 2008. Physiological heterogeneity in biofilms. Nat. Rev. Microbiol. 6, 199–210.

Sumikawa, Y., et al., 2006. Induction of β-defensin 3 in keratinocytes stimulated by bacterial lipopeptides through toll-like receptor 2. Microbes Infect. 8, 1513–1521.

Temmerman, R., Huys, G., Swings, J., 2004. Identification of lactic acid bacteria: culture-dependent and culture-independent methods. Trends Food Sci. Technol. 15, 348–359.

Tett, A., et al., 2017. Unexplored diversity and strain-level structure of the skin microbiome associated with psoriasis. NPJ Biofilms Microbiomes 3, 14.

Jagadeesh, A.T, Prakash, P.Y., Rao, N.K., Ramya, V., 2017. Culture characterization of the skin microbiome in Type 2 diabetes mellitus: A focus on the role of innate immunity. Diabetes Res Clin Pract 1872-8227, 134, 1–7. Available from: https://doi.org/10.1016/j.diabres.2017.09.007.

Tsilimigras, M.C.B., Fodor, A.A., 2016. Compositional data analysis of the microbiome: fundamentals, tools, and challenges. Ann. Epidemiol. 26, 330–335.

Verbanic, S., Shen, Y., Lee, J., Deacon, J.M., Chen, I.A., 2020. Microbial predictors of healing and short-term effect of debridement on the microbiome of chronic wounds. NPJ Biofilms Microbiomes 6, 1–11.

Williams, M.R., Gallo, R.L., 2015. The role of the skin microbiome in atopic dermatitis. Curr. Allergy Asthma Rep. 15.

Williams, H., et al., 2017. Cutaneous Nod2 expression regulates the skin microbiome and wound healing in a murine model. J. Invest. Dermatol. 137, 2427–2436.

Wlaschek, M., Peus, D., Achterberg, V., Meyer-Ingold, W., Scharffetter-Kochanek, K., 1997. Protease inhibitors protect growth factor activity in chronic wounds. Br. J. Dermatol. 137, 646–663.

Wlaschek, M., Scharffetter-Kochanek, K., 2005. Oxidative stress in chronic venous leg ulcers. Wound Repair. Regen. 13, 452–461.

Wolcott, R.D.R.A.R.D.R.A., et al., 2016. Analysis of the chronic wound microbiota of 2,963 patients by 16S rDNA pyrosequencing. Wound Repair. Regen. 24, 163–174.

Yan, S., Wu, G., 2019. Can biofilm be reversed through quorum sensing in pseudomonas aeruginosa? Front. Microbiol. 10.

Ye, J., et al., 2008. Bacteria and bacterial rRNA genes associated with the development of colitis in IL-10-/- mice. Inflamm. Bowel Dis. 14, 1041–1050.

Yu, Y., Champer, J., Beynet, D., Kim, J., Friedman, A.J., 2015. The role of the cutaneous microbiome in skin cancer: lessons learned from the gut. J. Drugs Dermatol. 14, 461–465.

Yu, Y., Dunaway, S., Champer, J., Kim, J., Alikhan, A., 2020. Changing our microbiome: probiotics in dermatology. Br. J. Dermatol. 182, 39–46.

Yurchenco, P.D., 2011. Basement membranes: cell scaffoldings and signaling platforms. Cold Spring Harb. Perspect. Biol. 3, a004911.

Zhao, G., et al., 2013. Biofilms and inflammation in chronic wounds. Adv. Wound Care 2, 389–399.

Section VII

Microbiome and Cancer

Chapter 22

Healthy gut microbiome in the prevention of colorectal cancer

Marufa Rumman[1], Misako Nagasaka[2], Bayan Al-Share[2], Asfar S. Azmi[2] and Md. Hafiz Uddin[2]

[1]*Department of Microbiology, National Center for Toxicological Research, US Food and Drug Administration, Redfield, AR, United States,*
[2]*Department of Oncology, Karmanos Cancer Institute, Wayne State University School of Medicine, Detroit, MI, United States*

22.1 Introduction

In 2020, about 1.9 million people (one in ten cancer cases) were newly diagnosed with colorectal cancer (CRC), which was responsible for 935,000 deaths (9.4%) worldwide with a slightly higher incidence in males (Sung et al., 2021). CRC remains the third leading cause of cancer but is now the second leading cause of cancer-associated deaths, just after lung cancer. The CRC incidence rate is about four times higher in transitioned or developed countries when compared with transitioning or developing countries and up to nine times higher in some world regions, including Europe, Australia and New Zealand, and Northern America (Sung et al., 2021). A uniform increase in the incidence of CRC with the Human Development Index of transitioning countries suggests CRC as a marker of socioeconomic development (F., 2014; Fidler et al., 2016). By the year 2030, CRC is expected to increase to 2.2 million cases (Arnold et al., 2017). In low- and middle-income countries such as Brazil and Russia, the increased incidence of CRC correlates with the recent transition from a healthier lifestyle to a Western lifestyle (Center et al., 2009). On the other hand, the incidence rate of CRC in high-income countries such as the United States and Australia is declining, most likely as a result of better screening (i.e., colonoscopy) and early interventions (i.e., polypectomy). Colorectal polyps are precursor lesions for CRC and have been found to result from a Western lifestyle (Arnold et al., 2017).

Unfortunately, the incidence of CRC is steadily rising among people of 50 years of age, and lifestyle factors such as diet and obesity act as main contributing factors. Studies have shown that genetic alterations or mutations in the germline are responsible for a fraction of case as compared to lifestyle factors in that age group (Lui et al., 2019). Unhealthy dietary choices are probably linked to the development or progression of CRC and other cancers, as is evident from recent metaanalysis (Grosso et al., 2017). A multicenter case-control study that was conducted in Spain compared a Western diet and a Mediterranean diet in terms of the risk of developing CRC. The Mediterranean diet, that is rich in legumes, vegetables, olives, olive oil, fish, and fruits, seems to play a protective role in comparison to the Western diet, which is abundant in sugar, processed meat, and fats (Castello et al., 2019; Randi et al., 2010). A report that was published in 2018 in collaboration with the World Cancer Research Fund and the American Institute for Cancer Research recognizes red and processed meats, obesity, and consumption of alcohol as major risk factors for CRC. The same report identifies enhanced physical activity and consumption of whole grains and dietary fiber as major protective factors (Research, 2018). Because diet contributes directly to the gut flora, an increasing number of studies are investigating the relationship among diet, the gut microbiome, and CRC. In this chapter we will summarize the evidence that connects the gut microbiome to CRC and will discuss a number of preventive approaches.

22.2 The gut microbiome in the progression of colorectal cancer

The composition of the gut microbiome is usually assessed by two main approaches: whole-genome shotgun sequencing and 16S ribosomal RNA amplicon sequencing. Both these approaches use next-generation sequencing technologies to sequence microbes in a given sample and align the results with known sequence databases for the identification and abundance of particular taxa (Saus et al., 2019). There is a growing amount of research investigating the mechanism of

Microbiome, Immunity, Digestive Health and Nutrition. DOI: https://doi.org/10.1016/B978-0-12-822238-6.00005-4

colorectal carcinogenesis in relation to diet or gut microbiota (Niederreiter et al., 2018; Rothschild et al., 2018). The human intestine is populated by trillions of microbes (Niederreiter et al., 2018) that encode about a hundred times more unique genes than our own genome (Rothschild et al., 2018). The most dominant bacterial population belongs to the Firmicutes and Bacteroidetes groups. It is evident from different studies that an alteration in the balance of microbial populations in the gut may results in inflammatory conditions or even cancer (Niederreiter et al., 2018; Telle-Hansen et al., 2018). The gut flora largely influenced by dietary pattern. A comparative study of rural African and urban Italian children observed significant difference in gut microbiomes and postulated different dietary habits as a reason (De Filippo et al., 2010).

The association of colorectal carcinogenesis and the gut microbiome is evident from several preclinical and epidemiological studies (Wong et al., 2017). It has been demonstrated that the introduction of a colorectal cancer patient's stool into the gut of mice causes hyperplastic and dysplastic changes in their intestine (Wong et al., 2017). The genetic background of the recipient mouse plays an important role in promoting intestinal carcinogenesis when challenged with the stool from colorectal cancer patients, which is evident from a APC Min/+ mouse models (Tomkovich et al., 2017; Burns et al., 2018). A molecular epidemiological analysis showed an alteration of human gut microbiota in CRC patients compared to healthy controls. In a cohort of Japanese colorectal carcinoma in adenoma patients, a significant change in the composition of the gut microbial population has been demonstrated (Kasai et al., 2016). Intake of antibiotics eliminates different populations of bacteria and causes microbial imbalance in the gut. Such disturbances have been associated with a reduction in mucus-producing goblet cells, polyp formation, and progression of CRC in a mouse model (Kaur et al., 2018). Prolonged antibiotic use also increases the risk of developing cancers, including CRC in humans (Kilkkinen et al., 2008).

In 2019, a metaanalysis of fecal metagenomes revealed specific global microbial signatures for CRC. The study identified a number of bacteria that are significantly associated with CRC. The most important species were *Fusobacterium nucleatum*, *Dialister* sp., *Solobacterium moorei*, *Gemella morbillorum*, *Parvimonas micra*, and *Peptostreptoccus stomatis* (Wirbel et al., 2019). Thomas et al. observed similar trends and found significant increases in *Clostridium symbiosum* and *Streptococcus gallolyticus* populations in CRC patients (Thomas et al., 2019). Other studies demonstrated the abundance of *S. gallolyticus* (*S. bovis*), *F. nucleatum*, *Escherichia coli*, *B. fragilis*, and *E. faecalis* and depletions in *Roseburia*, *Clostridium*, *Faecalibacterium*, and *Bifidobacterium* in CRC patients (Feng et al., 2015; Gagniere et al., 2016; Gao et al., 2015; Shang and Liu, 2018; Wang et al., 2012; Yu et al., 2017; Zhang et al., 2019b). A gut microbial analysis of patients with familial adenomatous polyposis or CRC shows many bacterial species in abundance, such as *Bacteroides fragilis* and *Akkermansia muciniphila*, which could be associated with the development of CRC (Dejea et al., 2018; Wang et al., 2017). *B. fragilis* and *E. faecalis* were shown to secrete enterotoxins such as fragylisin along with the generation of reactive oxygen species (ROS) responsible for the damage of gut epithelium via inflammation and oxidative DNA damage (Goodwin et al., 2011; Wu et al., 2009). *B. fragilis* also stimulates the proliferation of the epithelial cells by activating beta-catenin nuclear signaling (Wu et al., 2003).

The gut flora can take part in the promotion of carcinogenesis by altering various metabolic pathways, inflammatory pathways, and immune responses. For example, *F. nucleatum* induces inflammation and carcinogenesis by activating the NF-κB pathway via binding of FadA to E-cadherin and is well known for CRC progression (Allali et al., 2018). Such interaction potentiates beta-catenin and Wnt signaling, promoting cellular proliferation (Kostic et al., 2013; Rubinstein et al., 2013) as well as suppression of immune surveillance (Kostic et al., 2013; Park et al., 2017; Ye et al., 2017). The constant activation of NF-κB pathway could be achieved by the binding of flagellin to toll-like receptors. The bacterium *B. fragilis* can induce inflammation by secreting enterotoxin that leads to subsequent activation of STAT3 and enhanced Th17 response in the gut. Such activation of inflammation-associated pathways and immune response in the gut epithelial cells can lead to CRC (Allali et al., 2018). Moreover, IL17-induced inflammation can be trigged as a result of the accumulation of regulatory T cells (Treg), as shown in APC^{min} mice post BFT (metalloproteinase enterotoxin) colonization (Geis et al., 2015). Chronic inflammation is regulated by ROS and plays an important role in cancer progression (Aggarwal et al., 2019). It has been demonstrated that by acting on toll-like receptors, the bacterium *Peptidostreptococcus anaerobius* causes an increase in ROS production (Tsoi et al., 2017).

Alteration in the cellular metabolism is considered a hallmark of cancer (Ward and Thompson, 2012). CRC samples have been found to be overexpressed with enzymes such as aldehyde dehydrogenase and acyl-CoA synthetase, which are involved in glycolysis and fatty acid metabolism, respectively. It can be assumed that some CRC-associated bacteria overexpress certain metabolic pathways, which produce metabolites that can trigger and promote CRC. Acetaldehyde is a known carcinogen that is produced during these metabolic processes. The activation of the pentose phosphate pathway leads to an increase in production of nicotinamide adenine dinucleotide phosphate, which is essential for cellular proliferation via the promotion of nucleic acid synthesis (Allali et al., 2018).

It is suggested from metabolic analyses that a shift in the gut microbiome from carbohydrate-consuming bacteria to amino acid–consuming bacteria, primarily *Clostridiales*, could be related to consumption of diets that are high in meat and fats and low in fiber. A metabolism of amino acids results in the production of polyamines that can contribute to carcinogenesis (Wirbel et al., 2019; Thomas et al., 2019).

Genetic modulation of gut bacterial species has been detected in cancer. It has been demonstrated in CRC samples, particularly in *Clostridium* spp., that bai operon activity becomes enhanced. This operon is associated with the conversion of primary bile acids to secondary bile acids. As secondary bile acids have oncogenic potentials, they can play a significant role in the promotion of CRC (Wirbel et al., 2019). Another cancer-promoting metabolite that has been observed in the CRC patients is trimethylamine (TMA). In stool samples from CRC patients, Thomas and coworkers observed an elevated expression of *cutC* gene that encodes choline TMA-lyase. This gene is associated with the production of TMA through the enzymatic degradation of choline (Thomas et al., 2019). *F. nucleatum* inhibits proliferation and promotes cell death of T cells, making them incapable of promoting a robust immune response. As a result, tumor cells are less likely to be attacked by immune cells (Mima et al., 2015; Nosho et al., 2016).

22.3 Factors affecting the gut microbiome

Dietary fiber is fermented by colonic bacteria to produce short-chain fatty acids (SCFAs) as their energy source (Encarnacao et al., 2015). Alternatively, a diet that is low in dietary fiber causes gut bacteria to degrade and utilize the glycoprotein in the mucus layer. This leads to a disruption of the protective mucosal barrier, exposing the epithelial lining to various inflammatory and toxic agents (Desai et al., 2016).

22.3.1 Dietary fats

A diet that is high in fat may promote colorectal cancer by a variety of mechanisms (Beyaz et al., 2016; Owczarek et al., 2016; Romagnolo et al., 2019; Zhang et al., 2019a). It can modify the gut microbiota and support oncogenic changes. High-fat diets have been shown to increase the percentage of Firmicutes to Bacteroidetes in rats (Hamilton et al., 2015). Mice fed with saturated fat had stimulated growth of bacteria such as *Bilophilia wadsworthia* that release hydrogen sulfide, which is associated with CRC (Lam et al., 2015). Similar findings were observed in clinical studies (David et al., 2014; Printz, 2017). In a randomized controlled trial involving 217 healthy adults, high- and low-fat diet groups were shown to have increased pathogenic (*Alistipes* and *Bacteroides*) and beneficial SCFA-producing bacteria (*Blautia* and *Faecalibacteria*), respectively (Wan et al., 2019).

In several in vitro and in vivo studies, secondary bile acids have been found to be carcinogenic. Association of CRC with bile acids is evident from a metaanalysis (Tong et al., 2008). Intestinal bacteria such as *Clostridia* convert excess primary bile acids such as deoxycholic acid (DCA) into secondary bile acids when the subject is fed with high-fat diet. This secondary bile acid is involved in enhanced inflammatory response, development of intestinal carcinoma from adenoma, and depletion of the SCFA-producing bacteria (Cao et al., 2017; Islam et al., 2011; Wang et al., 2019).

Some fatty acids in the category of n-3 polyunsaturated fatty acids (PUFA), such as alpha-linolenic acid, eicosapentaenoic acid (EPA), and docosahexaenoic acid (DHA), have been shown to prevent CRC by exerting anticancer properties such as antioxidant effects, reduction in inflammation, proliferation, angiogenesis, and the induction of apoptosis (Fazio et al., 2016; Fini et al., 2010; Liu et al., 2016; Moradi Sarabi et al., 2018; Ortea et al., 2018; Piazzi et al., 2014). PUFAs have been demonstrated to influence the microbiome. Consumption of EPA/DHA in a dose of 4 g/day for 8 weeks was shown to stimulate the growth of *Bifidobacterium* (Watson et al., 2018); n-3 PUFA was also shown to increase the number bacteria in the gut that produce SCFAs (Noriega et al., 2016; Robertson et al., 2017). In addition, they may help to make checkpoint inhibitors more effective by preventing the tumor from escaping immune surveillance (Ilag, 2018).

22.3.2 Red and processed meats

Red meat and processed meat are probable causes of CRC (Research, 2018). There are several roles of gut microbiota in such carcinogenic process: (1) bacteria that ferment protein, generating metabolites for example indoles, phenols, amines, ammonia, and hydrogen sulfide, which are known to induce inflammation and damage to the DNA; (2) bacteria that ferment amino acids having sulfur involving *E. coli*, *Klebsiella*, *Enterobacteriaceae*, *Clostridium*, and so on; (3) bacteria that ferment amino acids having aromatic structure, such as *Bacteroides* and *Clostridium* (Yao et al., 2016); and (4) bacteria that deconjugate taurocholic acid to taurine (and DCA), which in turn is converted into hydrogen sulfide by bacteria such as *Bilophila wadsworphia* (Ridlon et al., 2016). Additionally, some bacteria are able to metabolize L-carnitine from red meat to generate the carcinogenic compound trimethylamine N-oxide (Xu et al., 2015).

Heme, which is found in meat, is another possible causative agent of CRC that is known to cause lipid peroxidation. It generates of reactive oxygen species (ROS) that induce cytotoxic stress leading to the damage of colonic superficial epithelial cells. Prolonged exposure to heme can modulate cell cycle genes and increase the risk of cancer. The above roles of heme are facilitated by several groups of gut bacteria, including *Proteobacteria*, *Verrucomicrobia*, *Bacteroides*, and Enterobacteriaceae. Heme forms a toxic complex with free radicals that can reach to the colonic epithelium in the presence of mucin-consuming bacteria such as *Akkermansia muciniphilia*, *Prevotella*, *Ruminococcus*, and *Desulfovibrio* (Constante et al., 2017; Ijssennagger et al., 2015). By contrast, some bacteria, such as *Lactobacillus reuteri*, could convert heterocyclic amines into less mutagenic substances (Zhang et al., 2017).

22.3.3 Alcohol consumption

Alcohol consumption, even in small quantities, increases the risk for CRC. Alcohol consumption promotes several groups of harmful bacteria and suppresses several groups of beneficial bacteria (Mutlu et al., 2012; Schnabl and Brenner, 2014). An increase in gram-negative bacteria causes the release of higher level of endotoxins, resulting in activation of proteins as well as immune cells that induce inflammation (Bishehsari et al., 2017; Rao et al., 2004). Disruption of the balance of colonic bacteria leads to dysregulated gut microbiome metabolism (Bjorkhaug et al., 2017). This occurs as a result of chronic overconsumption of alcohol. The gut of such patients showed overgrowth of inflammatory gram-negative bacteria, including *Clostridium*, *Holdemania* (Firmicutes), and *Sutterella* (Proteobacteria) (Bjorkhaug et al., 2017; Seesaha et al., 2020). Overconsumption also causes the depletion of beneficial SCFA-producing bacteria as well as bacterial populations of the genus *Faecalibacterium* (Bultman, 2017; Vipperla and O'Keefe, 2016). Several studies have revealed that alcohol can modulate the metabolism of bile acids, resulting in a decrease in bile acid production and overgrowth of harmful bacteria. However, it has been observed that treatment of certain bile acids can reverse such bacterial overgrowth in a rat model (Lorenzo-Zuniga et al., 2003; Schnabl and Brenner, 2014; Xie et al., 2013). Furthermore, there are gut bacterial groups that can metabolize alcohol directly and produce excessive acetaldehyde in the colon, which is probably responsible for the inflammation-induced carcinogenesis (Zhong and Zhou, 2014).

22.3.4 Smoking

Although the latency period is long, smoking has been linked to increased risk of CRC. Even after quitting smoking, individuals remained at higher risk for about 25 years as compared to never-smokers (Gong et al., 2012). A metagenomics study revealed a negative correlation between smoking and gut microbial diversity (Zhernakova et al., 2016). *Bacteroides* spp. and *Prevotella* spp. were found to be increased in the fecal samples of smokers (Benjamin et al., 2012) and were associated with the degradation of the mucin layer and inflammation (Allais et al., 2016). Preclinical studies demonstrated that cigarette smoke can reduce *Bifidobacterium* and hence the production of SCFAs (Tomoda et al., 2011; Hu et al., 2015). Diversity of the gut microbiota was partially restored after quitting smoking shown by the decrease in Bacteroidetes and Proteobacteria and the increase in Firmicutes and Actinobacteria (Biedermann et al., 2014).

22.3.5 Obesity

The relationship between obesity and the gut microbiota is bidirectional and complex (Song and Chan, 2019), and the age of individuals further adds complexity in this relationship (Li et al., 2018). In an in vivo mouse model, it was demonstrated that young obese mice favored long-chain fatty acid oxidation that increased proliferation of intestinal stem cells, whereas aged obese mice showed a downregulation of tumor suppressors as well as proapoptotic signaling (Li et al., 2018). These observations may explain the increase in CRC incidence in young adults.

Obesity-associated modulation of gut microbes can occur through inflammation via proinflammatory molecules such as lipopolysaccharide and metabolites such as elevated acetate and reduced butyrate, leading to the disruption of gut barrier (Cani and Jordan, 2018). Epigenetic changes dysregulate the expression of genes in colonic epithelial cells (Li et al., 2018, 2014; Qin et al., 2018). Obesity induced by high-fat diet causes remodeling of acetylation and subsequent transcriptional changes associated with the initiation and progression of CRC (Li et al., 2014). Besides acetylation, obesity can promote DNA methylation as well (Li et al., 2018). An alteration of gut microbiome–associated metabolites, such as DCA and lipoteichoic acid, results in DNA damage (Loo et al., 2017; Ohtani et al., 2014;

Yoshimoto et al., 2013). The lipoteichoic acid upregulates prostaglandin E2 via prostaglandin-endoperoxide synthase 2 and promotes CRC (Loo et al., 2017). The role of obesity in prostaglandin E2− and DCA-mediated CRC progression (Loo et al., 2017) requires further investigation.

22.3.6 Physical activity

There is increasing preclinical evidence suggesting that the gut flora can be altered by exercise along with an increase in SCFA production (Evans et al., 2014; Matsumoto et al., 2008; Petriz et al., 2014; Queipo-Ortuno et al., 2013). The ratio of Bacteroidetes to Firmicutes is increased in obese animals that undergo exercise (Petriz et al., 2014) in a manner that is roughly proportional to the intensity of the exercise (Evans et al., 2014). This finding also reflected in the abundance of metabolites. Both the concentration of intestinal butyrate and butyrate-producing bacteria are increased with exercise (Matsumoto et al., 2008; Evans et al., 2014). In a small cohort of athletes ($n = 23$), diverse gut microbes were observed in comparison to matched controls (Clarke et al., 2014). There was observed increase in floral diversity, an increase in different beneficial metabolites such as SCFAs and an increase in amino acid and antibiotic biosynthesis and carbohydrate metabolism etc (Barton et al., 2018). These encouraging results warrant further investigation in human studies with larger sample sizes.

22.4 The gut microbiome in the protection against colorectal cancer

In contrast to the topic of previous section, bacteria can play an important role in cancer protection including CRC (Seesaha et al., 2020) (Fig. 22.1). Bacteria such as *Lachnospiraceae* spp., *Bifidobacterium animalis*, and *Streptococcus thermophilus* are found to be depleted in patients with CRC (Dai et al., 2018; Feng et al., 2015) and are suggested to have a protective role in experimental models (Fong et al., 2020). The *Gordonibacter pamelae* and *Bifidobacterium catenulatum*

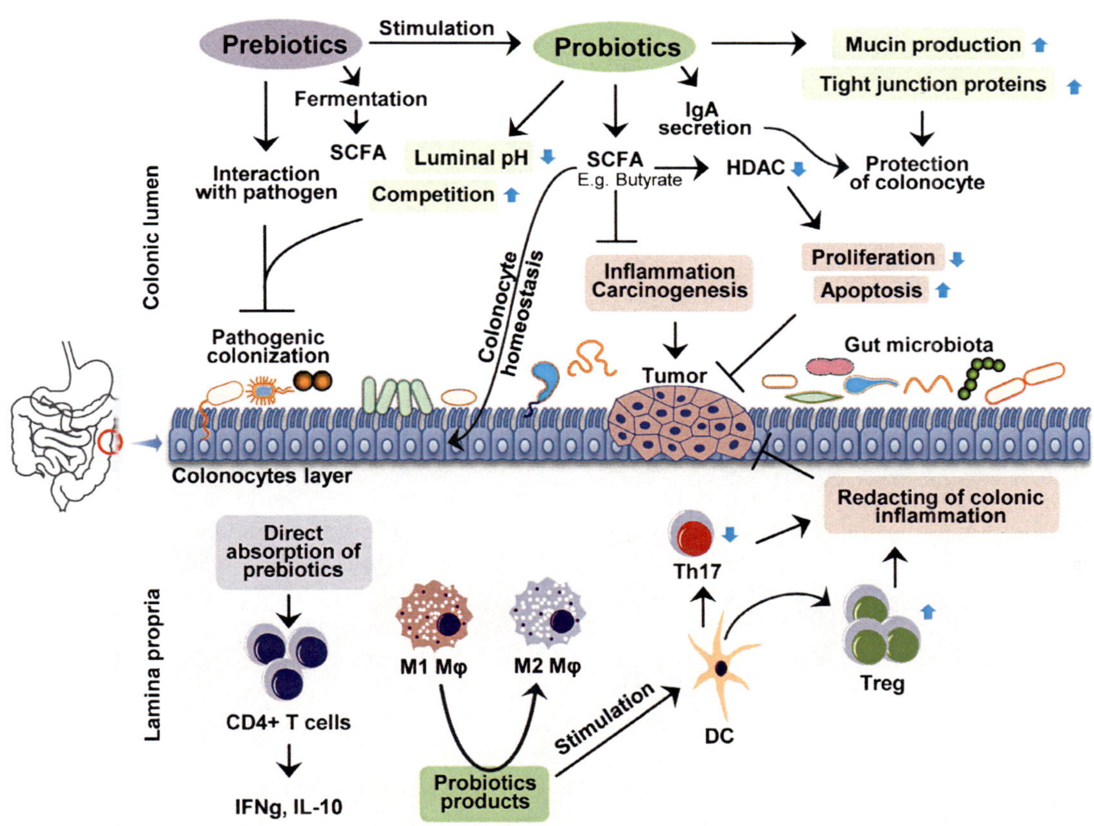

FIGURE 22.1 **Schematic diagram illustrating the role of gut microbiota in the prevention of colorectal cancer.** *SCFA*, short chain fatty acid; *HDAC*, histone deacetylases; *Mφ*, macrophage; *DC*, dendritic cells; *Treg*, regulatory T cells; *Th17*, T helper cells 17; *IgA*, immunoglobulin type A; INFg: interferon gamma.

populations were significantly elevated in the control group compared to CRC patients in a metaanalysis performed by Thomas et al. (2019), suggesting a protective role by these bacterial species. Maturation and activation of dendritic cells and releasing IFN-γ could be the mechanisms by which *Bifidobacterium* spp. induce anticancer benefits (Yi et al., 2018). Among other gut microbiota, *Akkermansia muciniphilia* could enhance the effectiveness of immune checkpoint inhibitors by promoting the recruitment of CD^{4+} T cells (Routy et al., 2018). The gut microbial population that decreases in CRC patients includes *Clostridium butyricum*, *Streptococcus salivanes*, *Streptococcus thermophilus*, *Carnobacterium maltaromaticum*, and *Lactobacillus gallinarum*. These bacteria may make significant contributions to the maintenance of optimal gut health. Reductions in these bacterial populations may directly or indirectly influence the growth of CRC-promoting microbes (Dai et al., 2018). It is likely that the introduction of such a bacterial population in CRC patients may provide protection against the progression of the malignant disease. Further research is needed in order to shed light on how different gut floras interact with one another and contribute to protection against malignant disease.

22.5 Prevention of colorectal cancer by enhancing healthy gut flora

A healthy diet stimulates the gut flora in a proper way and helps to maintain a balance among numerous populations of the microbiota and settle a healthy gut. There are several ways to keep this balance intact, by aiding the microbes already in the gut to grow by providing certain foods (prebiotics), by introducing living microbes in to the gut (probiotics), and by introducing a synergistic combination of prebiotics and probiotics (synbiotics) (Fig. 22.1).

22.5.1 Prebiotics

According to the International Scientific Association for Prebiotics and Probiotics, prebiotic food is "a substrate that is selectively utilized by host microorganisms conferring health benefit." It includes oligosaccharides (fructooligosaccharides, galactooligosaccharides, mannanoligosaccharides, xylooligosaccharides, and inulin), phenolics, phytochemicals, and some dietary fibers (Gibson et al., 2017). The oligofructose and inulin induce the formation of SCFA by influencing the proliferation of *Bifidobacterium* spp. and subsequent fermentation. The SCFAs, namely, butyrate, acetate, and propionate, have been shown to provide protection against CRC (Encarnacao et al., 2015). Colonic cells utilize butyrate as main energy source, whereas liver and muscle cells use propionate and acetate for the production of glucose and energy, respectively (Wong et al., 2006). Moreover, as an inhibitor of HDAC, SCFA can exerts suppressive effect on the oxidative effect, NF-κB signaling, and Treg regulation (Encarnacao et al., 2015).

Prebiotics were shown to reduce the prostaglandin level (by inhibiting COX-2) and the proliferation of cancer cells and to increase apoptosis in different model systems (Belobrajdic et al., 2012; Campbell et al., 1997; Chen and Vitetta, 2018; van der Beek et al., 2017). Prebiotic inulin can enhance the growth of *Lactobacillus* spp. and can prevent the formation of aberrant crypt foci and subsequent colorectal carcinogenesis (Pattananandecha et al., 2016). The finding was evaluated in different mouse models (Verghese et al., 2002a,b). The proliferation of *Bifidobacterium* spp. due to the consumption of wheat bran containing arabinoxylan-oligosaccharides was evident from a randomized controlled trial in healthy subjects (Windey et al., 2015). Few clinical trials demonstrated that in individuals who have been treated with prebiotics there is enhancement of the bacterial populations of several genera, including *Faecalibacterium*, *Akkermansia*, *Ruminococcus*, and *Rosebura* (Azcarate-Peril et al., 2017; Dewulf et al., 2013; Maier et al., 2017). Prebiotic oligosaccharides have been shown to prevent pathogens colonization (Monteagudo-Mera et al., 2019; Shoaf et al., 2006), owing to their similarity to glycoconjugates of microvilli and subsequent interaction with bacterial receptors (Forchielli and Walker, 2005). In addition, research has shown that consumption of high amounts of dietary fiber facilitates the activity of prebiotics (Healey et al., 2018).

Polyphenolic prebiotics found in green tea, specifically catechins, have been shown to promote the growth of *Bifidobacteria* spp. and other SCFA-producing bacteria in mouse models (Jin et al., 2012). A similar finding has been demonstrated in humans by Yuan et al. (2018) in contrast to a cohort study conducted on a Japanese population (Wada et al., 2019). Prebiotic anthocyanins are phenolic compounds present in blueberries that can induce apoptosis as well as growth inhibition of cancer cells, as is evident from both in vitro and in vivo studies (Fernandez et al., 2018; Lippert et al., 2017; Pan et al., 2017, 2018; Yi et al., 2005). It has also been shown that polyphenols present in potatoes can activate antioxidant systems when metabolized by colonic bacteria (Khairallah et al., 2018). Walnuts can also induce SCFA-forming bacteria and can contribute to anticancer activities (Nakanishi et al., 2016).

In contrast to the beneficial roles of prebiotics there are reports of a tumor-promoting role of prebiotics. The prebiotic product butyrate can exert its oncogenic role by enhancing cellular proliferation and polyp formation (Belcheva et al., 2014).

Long-term consumption of inulin-containing foods may promote the development of hepatocellular carcinoma (Singh et al., 2018). Such discrepancies might be due to the host's genetic background (Bultman and Jobin, 2014).

22.5.2 Probiotics

A probiotic is a live microorganism that is administered in adequate amounts to benefit the host in accordance with the FAO/WHO guidelines (Reid, 2016). Common probiotics include *Lactobacillus* spp., *Bifidobacterium* spp., and *Propionibacterium* spp. and yeasts such as *Saccharomyces*, found in yogurt, cheese, kefir, sour cream, and so on (Fontana et al., 2013).

Several research groups have shown that certain probiotic strains are able to prevent the colonization of pathogenic bacteria such as *Clostridium difficile* (Mills et al., 2018) and *Staphylococcus aureus* (Piewngam et al., 2018) by adhering to colonic epithelia or mucus (Tuomola et al., 1999) or by competing for nutrients (Kamada et al., 2012). Many probiotics plays a role in reducing luminal pH by secreting lactic acid, acetic acid, or bacteriocins to interfere with the growth of pathogens (Fayol-Messaoudi et al., 2005) as well as by exerting antimicrobial action (Gillor et al., 2008). It has been demonstrated that the probiotic *Bacillus* spp. secretes lipopeptide fengycins, inhibiting the colonization of *Staphylococcus aureus* through disruption of quorum sensing, the bacterial signaling system (Kamada et al., 2012).

A recent metaanalysis observed an inverse association between consumption of fermented dairy products and the risk of CRC (Aryana and Olson, 2017). The Continuous Update Project report published in 2018 also showed a decreased risk of CRC with increased consumption of dairy products (Research, 2018). The mechanism of probiotic-mediated CRC protection involves SCFA production, direct binding to carcinogen and reducing absorption, providing a suitable environment for beneficial bacteria. In addition, they enhance the immune system's ability to fight tumors by promoting dendritic cells, T cells, and B lymphocytes to differentiate and mature (Molska and Regula, 2019). The inhibition of cancer cell proliferation and induction of apoptosis by probiotic bacteria are evident from a number of in vitro and in vivo studies. Mechanistically, they modulate apoptotic pathway–associated proteins such as TNF-related apoptosis, inducing ligand (TRAIL), BAX, and BCL-2 (Agah et al., 2019; Bohlul et al., 2019; Chen et al., 2017; Elmadfa et al., 2010; Gamallat et al., 2016; Irecta-Najera et al., 2017; Olivares et al., 2006; Song et al., 2018). Probiotics have also been shown to act on key genes such as KRAS and APC, probably through the expression of micro-RNA (Heydari et al., 2019; Konishi et al., 2016; Sakatani et al., 2016). It has been demonstrated in animal studies that probiotics exerts their antioxidant properties through increased levels of antioxidant enzymes, such as SOD and catalase (Walia et al., 2018). In addition to providing protection through bacterial exopolysaccharides (Di et al., 2017), probiotics appear to provide protection via other mechanisms (Di et al., 2018). For example, ferrichrome produced by *Lactobacillus casei* plays a proapoptotic role using the JNK pathway in malignant cells compared to normal intestinal epithelial cells (Konishi et al., 2016). Also, the P40 and P75 soluble proteins from *L. rhamnosus* have been shown to reduce inflammation in the colon epithelial cells and protect it from being oncogenic (Yan et al., 2007). Moreover, production of polyphosphate by *L. brevis* plays an important growth inhibitory role in cancer cells via the modulation of ERK signaling (Sakatani et al., 2016). The probiotics also contribute to the maintenance of intestinal mucous barrier integrity and provide protection to the epithelial cells of the underlying mucosa from various carcinogens (Molska and Regula, 2019).

The immunomodulatory role of probiotics has been well recognized (Klaenhammer et al., 2012). Two species of *Bifidobacterium*, *B. infantis* (Konieczna et al., 2012) and *B. breve* (Jeon et al., 2012), have been shown to activate FOXP3 + dendritic cells that lead to the expression of Treg via toll-like receptors and the metabolism of retinoid acid. Several species of *Lactobacillus*, such as *L. rhamnosus* GG and *L. acidophilus*, can suppress IL17 and IL23 via STAT3 and NF-κB inhibition (Chen et al., 2015; Ghadimi et al., 2012). Furthermore, phenotypic switching of macrophages can be facilitated by *Lactobacillus* spp. (Sichetti et al., 2018). Suppression of tumor growth is also evident from the activity of *L. acidophilus* NCFM in CT-26-implanted animals, possibly through the downregulation of CXCR4 (Chen et al., 2012).

The permeability of the tight junction is a common problem in CRC that is associated with epithelial-to-mesenchymal transition and metastasis (Ahmad et al., 2017; Soler et al., 1999). *Lactobacillus rhamnosus*, *L. plantarum*, and *Escherichia coli* Nissle 1917 are reported to restore tight junction proteins such as claudin-1, occludin, ZO-1, and ZO-2 (Alvarez et al., 2016; Zyrek et al., 2007) and hence, play an important role in CRC prevention.

In addition to bacteria, yeast is an effective probiotic. The role of probiotic fermented milk products such as kefir and koumiss and cheese containing *Debaryomyces hansenii*, *Kluyveromyces lactis*, and *Yarrowia lipolytica* are well documented (Kumura et al., 2004). Yeasts can stimulate SCFA production, apoptosis, innate and adaptive immunity, and the suppression of inflammation in the mucosa of the colon (Moslehi-Jenabian et al., 2010; Saber et al., 2017).

22.5.3 Synbiotics

Recent cancer research has focused on combinations of prebiotics and probiotics that are termed symbiotic. The prebiotics oligofructose and maltodextrin along with *Lactobacillus acidophilus*, *Bifidobacterium bifidum*, and *B. infantum* were found to be effective in protection against inflammation-mediated carcinogenesis in in vivo model systems (Kuugbee et al., 2016). Synbiotics can reduce the oxidative effect and proliferation, can enhance apoptosis in the cancer cells, and may prevent CRC effectively compared to single-component treatments (Lin et al., 2019). By contrast, some studies did not find any significant changes in boosting the immune system with synbiotic compared to individual compounds (Ohara and Suzutani, 2018; Roller et al., 2004). However, the findings warrant further investigation.

22.6 Conclusion and future perspectives

The occurrence of CRC is on the rise, and different prevention strategies are needed to reduce the risk of developing CRC. Modification of the diet might be an appealing strategy, considering the interaction among diet, the gut microbiota, and CRC for protection against the disease. In fact, it has been demonstrated that shifting to an African diet (high in fiber, low in fat) from an American diet (low in fiber, high in fat) decreases inflammation and microbial proliferation (O'Keefe et al., 2015). It is possible that CRC can be prevented with such dietary interventions, but several questions and uncertainties remain.

In recent years, probiotics, prebiotics, and synbiotics have received a lot of attention. Engineering of probiotics can be performed via genetic modulation or protein change for a transformation of a proinflammatory microbe to an antiinflammatory species. It is evident that lipoteichoic acid–associated protein deletion in *L. acidophilus* transforms it into an antiinflammatory agent and reduces inflammation and polyp formation in the colon (Khazaie et al., 2012; Mohamadzadeh et al., 2011). The use of these products has shown benefits for improving the effectiveness of checkpoint inhibitors (Rossi et al., 2018); however, they have yet to be evaluated for safety. Long-term probiotic use can increase the opportunity of horizontal gene transfer, enabling relocation of antibiotic-resistant genes among different species (Fong et al., 2020). In the past, probiotics have been reported to cause bacteremia (Lewis et al., 2014). Once safety profiles of such products get unraveled, a broader application may be considered for the prevention of CRC or even as an agent for combination therapy. Most of the studies involving diet, the gut microbiome, and CRC have been performed in vitro or in vivo. More human cohort studies are of the utmost importance for the verification of preclinical findings.

The gut microbiome can differ from person to person and may respond differently to different dietary interventions (Healey et al., 2017). Consequently, changing dietary lifestyle in different individuals will not yield the same outcome. As a result of whole-genome sequencing and metabolomics analysis, more and more bacterial species are being found to be associated with CRC. However, it is necessary to determine how different genera interact with each other to induce and promote carcinogenesis. Furthermore, the gut microbiome can be considered to provide potential diagnostic biomarkers and for the prediction of the therapeutic efficacy of anticancer drugs in treating CRC. Despite the uncertainties and controversies that may exist, there is no harm in promoting a healthy lifestyle for a healthy gut flora.

Conflict of interest

The authors declare no conflict of interest.

References

Agah, S., et al., 2019. More protection of Lactobacillus acidophilus than Bifidobacterium bifidum probiotics on azoxymethane-induced mouse colon cancer. Probiotics Antimicrob. Proteins 11 (3), 857–864.

Aggarwal, V., et al., 2019. Role of reactive oxygen species in cancer progression: molecular mechanisms and recent advancements. Biomolecules 9 (11).

Ahmad, R., et al., 2017. Loss of claudin-3 expression induces IL6/gp130/Stat3 signaling to promote colon cancer malignancy by hyperactivating Wnt/beta-catenin signaling. Oncogene 36 (47), 6592–6604.

Allais, L., et al., 2016. Chronic cigarette smoke exposure induces microbial and inflammatory shifts and mucin changes in the murine gut. Env. Microbiol. 18 (5), 1352–1363.

Allali, I., et al., 2018. Gut microbiome of Moroccan colorectal cancer patients. Med. Microbiol. Immunol. 207 (3–4), 211–225.

Alvarez, C.S., et al., 2016. Outer membrane vesicles and soluble factors released by probiotic Escherichia coli Nissle 1917 and commensal ECOR63 enhance barrier function by regulating expression of tight junction proteins in intestinal epithelial cells. Front. Microbiol. 7, 1981.

Arnold, M., et al., 2017. Global patterns and trends in colorectal cancer incidence and mortality. Gut 66 (4), 683–691.

Aryana, K.J., Olson, D.W. 2017. A 100-year review: yogurt and other cultured dairy products. J. Dairy. Sci. 100 (12), 9987–10013.

Azcarate-Peril, M.A., et al., 2017. Impact of short-chain galactooligosaccharides on the gut microbiome of lactose-intolerant individuals. Proc. Natl Acad. Sci. U. S. A. 114 (3), E367–E375.

Barton, W., et al., 2018. The microbiome of professional athletes differs from that of more sedentary subjects in composition and particularly at the functional metabolic level. Gut 67 (4), 625–633.

Belcheva, A., et al., 2014. Gut microbial metabolism drives transformation of MSH2-deficient colon epithelial cells. Cell 158 (2), 288–299.

Belobrajdic, D.P., et al., 2012. Fructan extracts from wheat stem and barley grain stimulate large bowel fermentation in rats. Nutr. Res. 32 (8), 599–606.

Benjamin, J.L., et al., 2012. Smokers with active Crohn's disease have a clinically relevant dysbiosis of the gastrointestinal microbiota. Inflamm. Bowel Dis. 18 (6), 1092–1100.

Beyaz, S., et al., 2016. High-fat diet enhances stemness and tumorigenicity of intestinal progenitors. Nature 531 (7592), 53–58.

Biedermann, L., et al., 2014. Smoking cessation alters intestinal microbiota: insights from quantitative investigations on human fecal samples using FISH. Inflamm. Bowel Dis. 20 (9), 1496–1501.

Bishehsari, F., et al., 2017. Alcohol and gut-derived inflammation. Alcohol. Res. 38 (2), 163–171.

Bjorkhaug, S.T., et al., 2017. Chronic alcohol overconsumption may alter gut microbial metabolism: a retrospective study of 719 (13)C-D-xylose breath test results. Microb. Ecol. Health Dis. 28 (1), 1301725.

Bohlul, E., et al., 2019. TRAIL-expressing recombinant Lactococcus lactis induces apoptosis in human colon adenocarcinoma SW480 and HCT116 cells. J. Appl. Microbiol. 126 (5), 1558–1567.

Bultman, S.J., 2017. Interplay between diet, gut microbiota, epigenetic events, and colorectal cancer. Mol. Nutr. Food Res. 61 (1).

Bultman, S.J., Jobin, C., 2014. Microbial-derived butyrate: an oncometabolite or tumor-suppressive metabolite? Cell Host Microbe 16 (2), 143–145.

Burns, M.B., et al., 2018. Colorectal cancer mutational profiles correlate with defined microbial communities in the tumor microenvironment. PLoS Genet. 14 (6), e1007376.

Campbell, J.M., et al., 1997. Selected indigestible oligosaccharides affect large bowel mass, cecal and fecal short-chain fatty acids, pH and microflora in rats. J. Nutr. 127 (1), 130–136.

Cani, P.D., Jordan, B.F., 2018. Gut microbiota-mediated inflammation in obesity: a link with gastrointestinal cancer. Nat. Rev. Gastroenterol. Hepatol. 15 (11), 671–682.

Cao, H., et al., 2017. Secondary bile acid-induced dysbiosis promotes intestinal carcinogenesis. Int. J. Cancer 140 (11), 2545–2556.

Castello, A., et al., 2019. Low adherence to the western and high adherence to the mediterranean dietary patterns could prevent colorectal cancer. Eur. J. Nutr. 58 (4), 1495–1505.

Center, M.M., et al., 2009. International trends in colorectal cancer incidence rates. Cancer Epidemiol. Biomark. Prev. 18 (6), 1688–1694.

Chen, J., Vitetta, L., 2018. Inflammation-modulating effect of butyrate in the prevention of colon cancer by dietary fiber. Clin. Colorectal Cancer 17 (3), e541–e544.

Chen, C.C., et al., 2012. Oral inoculation of probiotics Lactobacillus acidophilus NCFM suppresses tumour growth both in segmental orthotopic colon cancer and extra-intestinal tissue. Br. J. Nutr. 107 (11), 1623–1634.

Chen, L., et al., 2015. Lactobacillus acidophilus suppresses colitis-associated activation of the IL-23/Th17 axis. J. Immunol. Res. 2015, 909514.

Chen, Z.Y., et al., 2017. Inhibitory effects of probiotic lactobacillus on the growth of human colonic carcinoma cell line HT-29. Molecules 22 (1).

Clarke, S.F., et al., 2014. Exercise and associated dietary extremes impact on gut microbial diversity. Gut 63 (12), 1913–1920.

Constante, M., et al., 2017. Dietary heme induces gut dysbiosis, aggravates colitis, and potentiates the development of adenomas in mice. Front. Microbiol. 8, 1809.

Dai, Z., et al., 2018. Multi-cohort analysis of colorectal cancer metagenome identified altered bacteria across populations and universal bacterial markers. Microbiome 6 (1), 70.

David, L.A., et al., 2014. Diet rapidly and reproducibly alters the human gut microbiome. Nature 505 (7484), 559–563.

De Filippo, C., et al., 2010. Impact of diet in shaping gut microbiota revealed by a comparative study in children from Europe and rural Africa. Proc. Natl Acad. Sci. U. S. A. 107 (33), 14691–14696.

Dejea, C.M., et al., 2018. Patients with familial adenomatous polyposis harbor colonic biofilms containing tumorigenic bacteria. Science 359 (6375), 592–597.

Desai, M.S., et al., 2016. A dietary fiber-deprived gut microbiota degrades the colonic mucus barrier and enhances pathogen susceptibility. Cell 167 (5), 1339–1353 e1321.

Dewulf, E.M., et al., 2013. Insight into the prebiotic concept: lessons from an exploratory, double blind intervention study with inulin-type fructans in obese women. Gut 62 (8), 1112–1121.

Di, W., et al., 2017. Physicochemical characterization and antitumour activity of exopolysaccharides produced by Lactobacillus casei SB27 from yak milk. Carbohydr. Polym. 171, 307–315.

Di, W., et al., 2018. Exopolysaccharides produced by Lactobacillus strains suppress HT-29 cell growth via induction of G0/G1 cell cycle arrest and apoptosis. Oncol. Lett. 16 (3), 3577–3586.

Elmadfa, I., et al., 2010. Immune-stimulating effects of lactic acid bacteria in vivo and in vitro. Proc. Nutr. Soc. 69 (3), 416–420.

Encarnacao, J.C., et al., 2015. Revisit dietary fiber on colorectal cancer: butyrate and its role on prevention and treatment. Cancer Metastasis Rev. 34 (3), 465–478.

Evans, C.C., et al., 2014. Exercise prevents weight gain and alters the gut microbiota in a mouse model of high fat diet-induced obesity. PLoS One 9 (3), e92193.

F., B., 2014. Transitions in human development and the global cancer burden. In: Stewart, B.W., Wild, C.P. (Eds.), World Cancer Report 2014, 2014. *WHO Press*, p. 14.

Fayol-Messaoudi, D., et al., 2005. pH-, Lactic acid-, and non-lactic acid-dependent activities of probiotic lactobacilli against salmonella enterica serovar typhimurium. Appl. Env. Microbiol. 71 (10), 6008–6013.

Fazio, C., et al., 2016. Inflammation increases NOTCH1 activity via MMP9 and is counteracted by eicosapentaenoic acid-free fatty acid in colon cancer cells. Sci. Rep. 6, 20670.

Feng, Q., et al., 2015. Gut microbiome development along the colorectal adenoma-carcinoma sequence. Nat. Commun. 6, 6528.

Fernandez, J., et al., 2018. Functional anthocyanin-rich sausages diminish colorectal cancer in an animal model and reduce pro-inflammatory bacteria in the intestinal microbiota. Genes. (Basel) 9 (3).

Fidler, M.M., et al., 2016. A global view on cancer incidence and national levels of the human development index. Int. J. Cancer 139 (11), 2436–2446.

Fini, L., et al., 2010. Highly purified eicosapentaenoic acid as free fatty acids strongly suppresses polyps in Apc(Min/ +) mice. Clin. Cancer Res. 16 (23), 5703–5711.

Fong, W., et al., 2020. Gut microbiota modulation: a novel strategy for prevention and treatment of colorectal cancer. Oncogene 39 (26), 4925–4943.

Fontana, L., et al., 2013. Sources, isolation, characterisation and evaluation of probiotics. Br. J. Nutr. 109 (Suppl 2), S35–S50.

Forchielli, M.L., Walker, W.A., 2005. The role of gut-associated lymphoid tissues and mucosal defence. Br. J. Nutr. 93 (Suppl 1), S41–S48.

Gagniere, J., et al., 2016. Gut microbiota imbalance and colorectal cancer. World J. Gastroenterol. 22 (2), 501–518.

Gamallat, Y., et al., 2016. Lactobacillus rhamnosus induced epithelial cell apoptosis, ameliorates inflammation and prevents colon cancer development in an animal model. Biomed. Pharmacother. 83, 536–541.

Gao, Z., et al., 2015. Microbiota disbiosis is associated with colorectal cancer. Front. Microbiol. 6, 20.

Geis, A.L., et al., 2015. Regulatory T-cell response to enterotoxigenic bacteroides fragilis colonization triggers IL17-dependent colon carcinogenesis. Cancer Discov. 5 (10), 1098–1109.

Ghadimi, D., et al., 2012. Epigenetic imprinting by commensal probiotics inhibits the IL-23/IL-17 axis in an in vitro model of the intestinal mucosal immune system. J. Leukoc. Biol. 92 (4), 895–911.

Gibson, G.R., et al., 2017. Expert consensus document: the International Scientific Association for Probiotics and Prebiotics (ISAPP) consensus statement on the definition and scope of prebiotics. Nat. Rev. Gastroenterol. Hepatol. 14 (8), 491–502.

Gillor, O., et al., 2008. The dual role of bacteriocins as anti- and probiotics. Appl. Microbiol. Biotechnol. 81 (4), 591–606.

Gong, J., et al., 2012. A pooled analysis of smoking and colorectal cancer: timing of exposure and interactions with environmental factors. Cancer Epidemiol. Biomark. Prev. 21 (11), 1974–1985.

Goodwin, A.C., et al., 2011. Polyamine catabolism contributes to enterotoxigenic Bacteroides fragilis-induced colon tumorigenesis. Proc. Natl Acad. Sci. U. S. A. 108 (37), 15354–15359.

Grosso, G., et al., 2017. Possible role of diet in cancer: systematic review and multiple *meta*-analyses of dietary patterns, lifestyle factors, and cancer risk. Nutr. Rev. 75 (6), 405–419.

Hamilton, M.K., et al., 2015. Changes in intestinal barrier function and gut microbiota in high-fat diet-fed rats are dynamic and region dependent. Am. J. Physiol. Gastrointest. Liver Physiol 308 (10), G840–G851.

Healey, G.R., et al., 2017. Interindividual variability in gut microbiota and host response to dietary interventions. Nutr. Rev. 75 (12), 1059–1080.

Healey, G., et al., 2018. Habitual dietary fibre intake influences gut microbiota response to an inulin-type fructan prebiotic: a randomised, double-blind, placebo-controlled, cross-over, human intervention study. Br. J. Nutr. 119 (2), 176–189.

Heydari, Z., et al., 2019. Effects of Lactobacillus acidophilus and Bifidobacterium bifidum probiotics on the expression of MicroRNAs 135b, 26b, 18a and 155, and their involving genes in mice colon cancer. Probiotics Antimicrob. Proteins 11 (4), 1155–1162.

Hu, J., et al., 2015. Effects of cigarette smoke condensate on the production and characterization of exopolysaccharides by Bifidobacterium. An. Acad. Bras. Cienc. 87 (2), 997–1005.

Ijssennagger, N., et al., 2015. Gut microbiota facilitates dietary heme-induced epithelial hyperproliferation by opening the mucus barrier in colon. Proc. Natl Acad. Sci. U. S. A. 112 (32), 10038–10043.

Ilag, L.L., 2018. Are long-chain polyunsaturated fatty acids the link between the immune system and the microbiome towards modulating cancer? Med. (Basel) 5 (3).

Irecta-Najera, C.A., et al., 2017. Protective effect of lactobacillus casei on DMH-induced colon carcinogenesis in mice. Probiotics Antimicrob. Proteins 9 (2), 163–171.

Islam, K.B., et al., 2011. Bile acid is a host factor that regulates the composition of the cecal microbiota in rats. Gastroenterology 141 (5), 1773–1781.

Jeon, S.G., et al., 2012. Probiotic Bifidobacterium breve induces IL-10-producing Tr1 cells in the colon. PLoS Pathog. 8 (5), e1002714.

Jin, J.S., et al., 2012. Effects of green tea consumption on human fecal microbiota with special reference to Bifidobacterium species. Microbiol. Immunol. 56 (11), 729–739.

Kamada, N., et al., 2012. Regulated virulence controls the ability of a pathogen to compete with the gut microbiota. Science 336 (6086), 1325–1329.

Kasai, C., et al., 2016. Comparison of human gut microbiota in control subjects and patients with colorectal carcinoma in adenoma: terminal restriction fragment length polymorphism and next-generation sequencing analyses. Oncol. Rep. 35 (1), 325–333.

Kaur, K., et al., 2018. Antibiotic-mediated bacteriome depletion in Apc(Min/ +) mice is associated with reduction in mucus-producing goblet cells and increased colorectal cancer progression. Cancer Med. 7 (5), 2003–2012.

Khairallah, J., et al., 2018. Microbial biotransformation of a polyphenol-rich potato extract affects antioxidant capacity in a simulated gastrointestinal model. Antioxid. (Basel) 7 (3).

Khazaie, K., et al., 2012. Abating colon cancer polyposis by Lactobacillus acidophilus deficient in lipoteichoic acid. Proc. Natl Acad. Sci. U. S. A. 109 (26), 10462–10467.

Kilkkinen, A., et al., 2008. Antibiotic use predicts an increased risk of cancer. Int. J. Cancer 123 (9), 2152–2155.

Klaenhammer, T.R., et al., 2012. The impact of probiotics and prebiotics on the immune system. Nat. Rev. Immunol. 12 (10), 728–734.

Konieczna, P., et al., 2012. Bifidobacterium infantis 35624 administration induces Foxp3 T regulatory cells in human peripheral blood: potential role for myeloid and plasmacytoid dendritic cells. Gut 61 (3), 354–366.

Konishi, H., et al., 2016. Probiotic-derived ferrichrome inhibits colon cancer progression via JNK-mediated apoptosis. Nat. Commun. 7, 12365.

Kostic, A.D., et al., 2013. Fusobacterium nucleatum potentiates intestinal tumorigenesis and modulates the tumor-immune microenvironment. Cell Host Microbe 14 (2), 207–215.

Kumura, H., et al., 2004. Screening of dairy yeast strains for probiotic applications. J. Dairy. Sci. 87 (12), 4050–4056.

Kuugbee, E.D., et al., 2016. Structural change in microbiota by a probiotic cocktail enhances the gut barrier and reduces cancer via TLR2 signaling in a rat model of colon cancer. Dig. Dis. Sci. 61 (10), 2908–2920.

Lam, Y.Y., et al., 2015. Effects of dietary fat profile on gut permeability and microbiota and their relationships with metabolic changes in mice. Obesity (Silver Spring) 23 (7), 1429–1439.

Lewis, J.D., et al., 2014. Nutrition, gut microbiota and immunity: therapeutic targets for IBD. Concluding remarks. Nestle Nutr. Inst. Workshop Ser. 79, 161–162.

Li, R., et al., 2014. Obesity, rather than diet, drives epigenomic alterations in colonic epithelium resembling cancer progression. Cell Metab. 19 (4), 702–711.

Li, R., et al., 2018. Transcriptome and DNA methylome analysis in a mouse model of diet-induced obesity predicts increased risk of colorectal cancer. Cell Rep. 22 (3), 624–637.

Lin, P.Y., et al., 2019. Germinated brown rice combined with Lactobacillus acidophilus and Bifidobacterium animalis subsp. lactis inhibits colorectal carcinogenesis in rats. Food Sci. Nutr. 7 (1), 216–224.

Lippert, E., et al., 2017. Anthocyanins prevent colorectal cancer development in a mouse model. Digestion 95 (4), 275–280.

Liu, M., et al., 2016. Elevation of n-3/n-6 PUFAs ratio suppresses mTORC1 and prevents colorectal carcinogenesis associated with APC mutation. Oncotarget 7 (47), 76944–76954.

Loo, T.M., et al., 2017. Gut microbiota promotes obesity-associated liver cancer through PGE2-mediated suppression of antitumor immunity. Cancer Discov. 7 (5), 522–538.

Lorenzo-Zuniga, V., et al., 2003. Oral bile acids reduce bacterial overgrowth, bacterial translocation, and endotoxemia in cirrhotic rats. Hepatology 37 (3), 551–557.

Lui, R.N., et al., 2019. Global increasing incidence of young-onset colorectal cancer across 5 continents: a joinpoint regression analysis of 1,922,167 cases. Cancer Epidemiol. Biomark. Prev. 28 (8), 1275–1282.

Maier, T.V., et al., 2017. Impact of dietary resistant starch on the human gut microbiome, metaproteome, and metabolome. mBio 8 (5).

Matsumoto, M., et al., 2008. Voluntary running exercise alters microbiota composition and increases n-butyrate concentration in the rat cecum. Biosci. Biotechnol. Biochem. 72 (2), 572–576.

Mills, J.P., et al., 2018. Probiotics for prevention of Clostridium difficile infection. Curr. Opin. Gastroenterol. 34 (1), 3–10.

Mima, K., et al., 2015. Fusobacterium nucleatum and T cells in colorectal carcinoma. JAMA Oncol. 1 (5), 653–661.

Mohamadzadeh, M., et al., 2011. Regulation of induced colonic inflammation by Lactobacillus acidophilus deficient in lipoteichoic acid. Proc. Natl Acad. Sci. U S A 108 (Suppl 1), 4623–4630.

Molska, M., Regula, J., 2019. Potential mechanisms of probiotics action in the prevention and treatment of colorectal cancer. Nutrients 11 (10).

Monteagudo-Mera, A., et al., 2019. Adhesion mechanisms mediated by probiotics and prebiotics and their potential impact on human health. Appl. Microbiol. Biotechnol. 103 (16), 6463–6472.

Moradi Sarabi, M., et al., 2018. The effects of dietary polyunsaturated fatty acids on miR-126 promoter DNA methylation status and VEGF protein expression in the colorectal cancer cells. Genes. Nutr. 13, 32.

Moslehi-Jenabian, S., et al., 2010. Beneficial effects of probiotic and food borne yeasts on human health. Nutrients 2 (4), 449–473.

Mutlu, E.A., et al., 2012. Colonic microbiome is altered in alcoholism. Am. J. Physiol. Gastrointest. Liver Physiol 302 (9), G966–G978.

Nakanishi, M., et al., 2016. Effects of walnut consumption on colon carcinogenesis and microbial community structure. Cancer Prev. Res. (Phila.) 9 (8), 692–703.

Niederreiter, L., et al., 2018. Food, microbiome and colorectal cancer. Dig. Liver Dis. 50 (7), 647–652.

Noriega, B.S., et al., 2016. Understanding the impact of omega-3 rich diet on the gut microbiota. Case Rep. Med. 2016, 3089303.

Nosho, K., et al., 2016. Association of Fusobacterium nucleatum with immunity and molecular alterations in colorectal cancer. World J. Gastroenterol. 22 (2), 557–566.

Ohara, T., Suzutani, T., 2018. Intake of Bifidobacterium longum and Fructo-oligosaccharides prevents colorectal carcinogenesis. Euroasian J. Hepatogastroenterol 8 (1), 11–17.

Ohtani, N., et al., 2014. Obesity and cancer: a gut microbial connection. Cancer Res. 74 (7), 1885–1889.

O'Keefe, S.J., et al., 2015. Fat, fibre and cancer risk in African Americans and rural Africans. Nat. Commun. 6, 6342.

Olivares, M., et al., 2006. Dietary deprivation of fermented foods causes a fall in innate immune response. Lactic acid bacteria can counteract the immunological effect of this deprivation. J. Dairy. Res. 73 (4), 492–498.

Ortea, I., et al., 2018. Proteomics study reveals that docosahexaenoic and arachidonic acids exert different in vitro anticancer activities in colorectal cancer cells. J. Agric. Food Chem. 66 (24), 6003–6012.

Owczarek, D., et al., 2016. Diet and nutritional factors in inflammatory bowel diseases. World J. Gastroenterol. 22 (3), 895–905.

Pan, P., et al., 2017. Black raspberries and their anthocyanin and fiber fractions alter the composition and diversity of gut microbiota in F-344 rats. Nutr. Cancer 69 (6), 943–951.

Pan, P., et al., 2018. Gut bacteria are required for the benefits of black raspberries in Apc (Min/ +) mice. J. Berry Res. 8 (4), 239–249.

Park, H.E., et al., 2017. Intratumoral Fusobacterium nucleatum abundance correlates with macrophage infiltration and CDKN2A methylation in microsatellite-unstable colorectal carcinoma. Virchows Arch. 471 (3), 329–336.

Pattananandecha, T., et al., 2016. Hydrolysed inulin alleviates the azoxymethane-induced preneoplastic aberrant crypt foci by altering selected intestinal microbiota in Sprague-Dawley rats. Pharm. Biol. 54 (9), 1596–1605.

Petriz, B.A., et al., 2014. Exercise induction of gut microbiota modifications in obese, non-obese and hypertensive rats. BMC Genomics 15, 511.

Piazzi, G., et al., 2014. Eicosapentaenoic acid free fatty acid prevents and suppresses colonic neoplasia in colitis-associated colorectal cancer acting on Notch signaling and gut microbiota. Int. J. Cancer 135 (9), 2004–2013.

Piewngam, P., et al., 2018. Pathogen elimination by probiotic Bacillus via signalling interference. Nature 562 (7728), 532–537.

Printz, C., 2017. Sulfide-producing bacteria linked to higher colon cancer risk in African Americans. Cancer 123 (15), 2795.

Qin, Y., et al., 2018. An obesity-associated gut microbiome reprograms the intestinal epigenome and leads to altered colonic gene expression. Genome Biol. 19 (1), 7.

Queipo-Ortuno, M.I., et al., 2013. Gut microbiota composition in male rat models under different nutritional status and physical activity and its association with serum leptin and ghrelin levels. PLoS One 8 (5), e65465.

Randi, G., et al., 2010. Dietary patterns and the risk of colorectal cancer and adenomas. Nutr. Rev. 68 (7), 389–408.

Rao, R.K., et al., 2004. Recent advances in alcoholic liver disease I. Role of intestinal permeability and endotoxemia in alcoholic liver disease. Am. J. Physiol. Gastrointest. Liver Physiol 286 (6), G881–G884.

Reid, G., 2016. Probiotics: definition, scope and mechanisms of action. Best. Pract. Res. Clin. Gastroenterol. 30 (1), 17–25.

Research., W.C.R.F.A.I.f.C., 2018 Diet, nutrition, physical activity and colorectal cancer. *Continuous Update Project Expert Report 2018*. dietandcancerreport.org

Ridlon, J.M., et al., 2016. Taurocholic acid metabolism by gut microbes and colon cancer. Gut Microbes 7 (3), 201–215.

Robertson, R.C., et al., 2017. Deficiency of essential dietary n-3 PUFA disrupts the caecal microbiome and metabolome in mice. Br. J. Nutr. 118 (11), 959–970.

Roller, M., et al., 2004. Intestinal immunity of rats with colon cancer is modulated by oligofructose-enriched inulin combined with Lactobacillus rhamnosus and Bifidobacterium lactis. Br. J. Nutr. 92 (6), 931–938.

Romagnolo, D.F., et al., 2019. n-6 Linoleic acid induces epigenetics alterations associated with colonic inflammation and cancer. Nutrients 11 (1).

Rossi, M., et al., 2018. Nutraceuticals in colorectal cancer: a mechanistic approach. Eur. J. Pharmacol. 833, 396–402.

Rothschild, D., et al., 2018. Environment dominates over host genetics in shaping human gut microbiota. Nature 555 (7695), 210–215.

Routy, B., et al., 2018. Gut microbiome influences efficacy of PD-1-based immunotherapy against epithelial tumors. Science 359 (6371), 91–97.

Rubinstein, M.R., et al., 2013. Fusobacterium nucleatum promotes colorectal carcinogenesis by modulating E-cadherin/beta-catenin signaling via its FadA adhesin. Cell Host Microbe 14 (2), 195–206.

Saber, A., et al., 2017. Secretion metabolites of probiotic yeast, Pichia kudriavzevii AS-12, induces apoptosis pathways in human colorectal cancer cell lines. Nutr. Res. 41, 36–46.

Sakatani, A., et al., 2016. Polyphosphate derived from Lactobacillus brevis inhibits colon cancer progression through induction of cell apoptosis. Anticancer. Res. 36 (2), 591–598.

Saus, E., et al., 2019. Microbiome and colorectal cancer: roles in carcinogenesis and clinical potential. Mol. Asp. Med. 69, 93–106.

Schnabl, B., Brenner, D.A., 2014. Interactions between the intestinal microbiome and liver diseases. Gastroenterology 146 (6), 1513–1524.

Seesaha, P.K., et al., 2020. The interplay between dietary factors, gut microbiome and colorectal cancer: a new era of colorectal cancer prevention. Future Oncol. 16 (7), 293–306.

Shang, F.M., Liu, H.L., 2018. Fusobacterium nucleatum and colorectal cancer: a review. World J. Gastrointest. Oncol. 10 (3), 71–81.

Shoaf, K., et al., 2006. Prebiotic galactooligosaccharides reduce adherence of enteropathogenic Escherichia coli to tissue culture cells. Infect. Immun. 74 (12), 6920–6928.

Sichetti, M., et al., 2018. Anti-inflammatory effect of multistrain probiotic formulation (L. rhamnosus, B. lactis, and B. longum). Nutrition 53, 95–102.

Singh, V., et al., 2018. Dysregulated microbial fermentation of soluble fiber induces cholestatic liver cancer. Cell 175 (3), 679–694 e622.

Soler, A.P., et al., 1999. Increased tight junctional permeability is associated with the development of colon cancer. Carcinogenesis 20 (8), 1425–1431.

Song, M., Chan, A.T., 2019. Environmental factors, gut microbiota, and colorectal cancer prevention. Clin. Gastroenterol. Hepatol. 17 (2), 275–289.

Song, H., et al., 2018. Pretreatment with probiotic Bifico ameliorates colitis-associated cancer in mice: transcriptome and gut flora profiling. Cancer Sci. 109 (3), 666–677.

Sung, H., et al., 2021. Global cancer statistics 2020: globocan estimates of incidence and mortality worldwide for 36 cancers in 185 countries. CA Cancer J. Clin. 71 (3), 209–249.

Telle-Hansen, V.H., et al., 2018. Impact of a healthy dietary pattern on gut microbiota and systemic inflammation in humans. Nutrients 10 (11).

Thomas, A.M., et al., 2019. Metagenomic analysis of colorectal cancer datasets identifies cross-cohort microbial diagnostic signatures and a link with choline degradation. Nat. Med. 25 (4), 667–678.

Tomkovich, S., et al., 2017. Locoregional effects of microbiota in a preclinical model of colon carcinogenesis. Cancer Res. 77 (10), 2620–2632.

Tomoda, K., et al., 2011. Cigarette smoke decreases organic acids levels and population of bifidobacterium in the caecum of rats. J. Toxicol. Sci. 36 (3), 261–266.

Tong, J.L., et al., 2008. Association between fecal bile acids and colorectal cancer: a *meta*-analysis of observational studies. Yonsei Med. J. 49 (5), 792–803.

Tsoi, H., et al., 2017. Peptostreptococcus anaerobius induces intracellular cholesterol biosynthesis in colon cells to induce proliferation and causes dysplasia in mice. Gastroenterology 152 (6), 1419–1433 e1415.

Tuomola, E.M., et al., 1999. The effect of probiotic bacteria on the adhesion of pathogens to human intestinal mucus. FEMS Immunol. Med. Microbiol. 26 (2), 137–142.

van der Beek, C.M., et al., 2017. Role of short-chain fatty acids in colonic inflammation, carcinogenesis, and mucosal protection and healing. Nutr. Rev. 75 (4), 286–305.

Verghese, M., et al., 2002a. Dietary inulin suppresses azoxymethane-induced aberrant crypt foci and colon tumors at the promotion stage in young Fisher 344 rats. J. Nutr. 132 (9), 2809–2813.

Verghese, M., et al., 2002b. Dietary inulin suppresses azoxymethane-induced preneoplastic aberrant crypt foci in mature Fisher 344 rats. J. Nutr. 132 (9), 2804–2808.

Vipperla, K., O'Keefe, S.J., 2016. Diet, microbiota, and dysbiosis: a 'recipe' for colorectal cancer. Food Funct. 7 (4), 1731–1740.

Wada, K., et al., 2019. Green tea intake and colorectal cancer risk in Japan: the Takayama study. Jpn. J. Clin. Oncol. 49 (6), 515–520.

Walia, S., et al., 2018. Chemoprevention by probiotics during 1,2-dimethylhydrazine-induced colon carcinogenesis in rats. Dig. Dis. Sci. 63 (4), 900–909.

Wan, Y., et al., 2019. Effects of dietary fat on gut microbiota and faecal metabolites, and their relationship with cardiometabolic risk factors: a 6-month randomised controlled-feeding trial. Gut 68 (8), 1417–1429.

Wang, T., et al., 2012. Structural segregation of gut microbiota between colorectal cancer patients and healthy volunteers. ISME J. 6 (2), 320–329.

Wang, X., et al., 2017. Gut flora profiling and fecal metabolite composition of colorectal cancer patients and healthy individuals. Exp. Ther. Med. 13 (6), 2848–2854.

Wang, S., et al., 2019. Interplay between bile acids and the gut microbiota promotes intestinal carcinogenesis. Mol. Carcinog. 58 (7), 1155–1167.

Ward, P.S., Thompson, C.B., 2012. Metabolic reprogramming: a cancer hallmark even warburg did not anticipate. Cancer Cell 21 (3), 297–308.

Watson, H., et al., 2018. A randomised trial of the effect of omega-3 polyunsaturated fatty acid supplements on the human intestinal microbiota. Gut 67 (11), 1974–1983.

Windey, K., et al., 2015. Wheat bran extract alters colonic fermentation and microbial composition, but does not affect faecal water toxicity: a randomised controlled trial in healthy subjects. Br. J. Nutr. 113 (2), 225–238.

Wirbel, J., et al., 2019. *Meta*-analysis of fecal metagenomes reveals global microbial signatures that are specific for colorectal cancer. Nat. Med. 25 (4), 679–689.

Wong, J.M., et al., 2006. Colonic health: fermentation and short chain fatty acids. J. Clin. Gastroenterol. 40 (3), 235–243.

Wong, S.H., et al., 2017. Gavage of fecal samples from patients with colorectal cancer promotes intestinal carcinogenesis in germ-free and conventional mice. Gastroenterology 153 (6), 1621–1633 e1626.

Wu, S., et al., 2003. Bacteroides fragilis enterotoxin induces c-Myc expression and cellular proliferation. Gastroenterology 124 (2), 392–400.

Wu, S., et al., 2009. A human colonic commensal promotes colon tumorigenesis via activation of T helper type 17 T cell responses. Nat. Med. 15 (9), 1016–1022.

Xie, G., et al., 2013. Alteration of bile acid metabolism in the rat induced by chronic ethanol consumption. FASEB J. 27 (9), 3583–3593.

Xu, R., et al., 2015. A genome-wide systems analysis reveals strong link between colorectal cancer and trimethylamine N-oxide (TMAO), a gut microbial metabolite of dietary meat and fat. BMC Genomics 16 (Suppl 7), S4.

Yan, F., et al., 2007. Soluble proteins produced by probiotic bacteria regulate intestinal epithelial cell survival and growth. Gastroenterology 132 (2), 562–575.

Yao, C.K., et al., 2016. Review article: insights into colonic protein fermentation, its modulation and potential health implications. Aliment. Pharmacol. Ther. 43 (2), 181–196.

Ye, X., et al., 2017. Fusobacterium nucleatum subspecies animalis influences proinflammatory cytokine expression and monocyte activation in human colorectal tumors. Cancer Prev. Res. (Phila.) 10 (7), 398–409.

Yi, W., et al., 2005. Phenolic compounds from blueberries can inhibit colon cancer cell proliferation and induce apoptosis. J. Agric. Food Chem. 53 (18), 7320–7329.

Yi, M., et al., 2018. Gut microbiome modulates efficacy of immune checkpoint inhibitors. J. Hematol. Oncol. 11 (1), 47.

Yoshimoto, S., et al., 2013. Obesity-induced gut microbial metabolite promotes liver cancer through senescence secretome. Nature 499 (7456), 97–101.

Yu, J., et al., 2017. Metagenomic analysis of faecal microbiome as a tool towards targeted non-invasive biomarkers for colorectal cancer. Gut 66 (1), 70–78.

Yuan, X., et al., 2018. Green tea liquid consumption alters the human intestinal and oral microbiome. Mol. Nutr. Food Res. 62 (12), e1800178.

Zhang, J., et al., 2017. Gut microbial transformation of the dietary imidazoquinoxaline mutagen MelQx reduces its cytotoxic and mutagenic potency. Toxicol. Sci. 159 (1), 266–276.

Zhang, S., et al., 2019a. Metabolic reprogramming links chronic intestinal inflammation and the oncogenic transformation in colorectal tumorigenesis. Cancer Lett. 450, 123–131.

Zhang, X., et al., 2019b. Fecal Fusobacterium nucleatum for the diagnosis of colorectal tumor: a systematic review and *meta*-analysis. Cancer Med. 8 (2), 480–491.

Zhernakova, A., et al., 2016. Population-based metagenomics analysis reveals markers for gut microbiome composition and diversity. Science 352 (6285), 565–569.

Zhong, W., Zhou, Z., 2014. Alterations of the gut microbiome and metabolome in alcoholic liver disease. World J. Gastrointest. Pathophysiol. 5 (4), 514–522.

Zyrek, A.A., et al., 2007. Molecular mechanisms underlying the probiotic effects of Escherichia coli Nissle 1917 involve ZO-2 and PKCzeta redistribution resulting in tight junction and epithelial barrier repair. Cell Microbiol. 9 (3), 804–816.

Section VIII

Microbiome, Arthritis and Multiple Sclerosis

Chapter 23

Regulatory roles of the microbiome in arthritis, fibromyalgia, and multiple sclerosis

Liz J. Vázquez-Chinea[1], Savannah N. Johnson[1] and Elizabeth A. Lipski[2]

[1]*Department of Clinical Nutrition, Maryland University of Integrative Health, Laurel, MD, United States,* [2]*Faculty Advisor, Department of Clinical Nutrition, Maryland University of Integrative Health, Laurel, MD, United States*

23.1 Introduction

The human gastrointestinal system houses trillions of microbes from over 1000 bacterial species. The colonization of the gut microbiome begins while an infant is being born, starting with exposure to the mother's microbiota as the infant goes through the birth canal. Additional environmental exposures and dietary changes throughout life further develop and diversify the microbiota. Conversely, exposures to antimicrobials such as antibiotic medications may decrease diversity and colonization. Although an individual's microbiome is unique and flexible over time, diversity is essential to a balanced microbiome and health promotion.

The microbiota is well known to play important roles digesting food and producing beneficial byproducts such as vitamins and short-chain fatty acids, but the benefits are not limited to the gastrointestinal tract. The microbiome is proving to be a pillar of immune health, given its impact on the integrity of the intestinal epithelium and modulation of inflammation. The gastrointestinal lining serves as a barrier, microbiota harbor, surface for nutrient uptake, and site of essential interactions between the endocrine, immune, and nervous systems (Buscarinu et al., 2018). While these interactions can be bidirectional and multifactorial, an unbalanced microbiome, also known as dysbiosis, can compromise the epithelial lining, causing increased permeability, or "leaky gut." The impaired intestinal barrier function may lead to bacterial translocation and systemic inflammatory signaling. The resulting immune activation can potentially manifest as autoimmunity and pain.

About 25 diverse health conditions are currently associated with alteration in the gut microbiota composition in which much attention has been paid to inflammatory bowel disease, obesity, allergy, and diabetes (van Hemert et al., 2014). Since the gut microbiome has already been found to be an independent factor for systemic disease and can potentially modulate pain as discussed above, there is reason to suspect commonalities in the microbiome profiles of those experiencing pain. In this chapter, chronic pain conditions, such as rheumatoid arthritis, fibromyalgia, and multiple sclerosis, will be examined in relation to the microbiome, and the potential use of probiotics in such disorders will be discussed.

23.2 Rheumatoid arthritis

Rheumatoid arthritis (RA) is a chronic inflammatory disease that targets the joints of the body. Because it is an autoimmune condition, it is evident that there are environmental and genetic factors involved in its development. Given the role that the microbiome plays in immune modulation, current research has begun to investigate whether the microbiome plays a major role in RA etiopathogenesis. Studies that have found associations between the microbiome profile and RA have shown significant differences compared to healthy controls. Assuming this relationship, it is hypothesized that probiotic supplementation could be utilized as a potential treatment for RA. However, there have not been many randomized controlled trials (RCTs) that currently have tested this treatment.

Microbiome, Immunity, Digestive Health and Nutrition. DOI: https://doi.org/10.1016/B978-0-12-822238-6.00028-5

23.2.1 Microbiome profile of individuals with rheumatoid arthritis

When the microbiomes of individuals with RA are compared with those of healthy controls, there are significant differences: primarily low diversity (De Filippo et al., 2019). One study utilized 16S rRNA hybridization and DNA staining to compare the microbiomes of 51 patients with RA and 50 patients with fibromyalgia (Vaahtovuo et al., 2008). The researchers found that bifidobacterial levels were significantly decreased in individuals with RA. Bacteria in the following groups were also low: the *Bacteroides-Porphyromonas-Prevotella* group, the *Bacteroides fragilis* subgroup, and the *Eubacterium rectale—Clostridium coccoides* group. Other researchers found similar results utilizing a small sample size: 25 patients with early RA and 23 healthy controls (Toivanen et al., 2002). Not only did patients with RA have lower levels of the *Bacteroides-Prevotella-Porphyromonas* genera, but the average was only half the abundance in the healthy controls.

Overabundance of certain microbes has also been associated with RA. An observational study specifically analyzed the presence of *Lactobacillus* in RA patients and found significantly more *Lactobacillus* in their fecal microbiota compared to that of the healthy controls (Liu et al., 2013). The researchers concluded that there may be a potential relationship between *Lactobacillus* abundance and RA progression.

Although lab-induced RA cannot fully demonstrate the pathogenesis of RA in humans, evidence shows a consistent relationship between the microbiota and RA. In microbiota-humanized mice with early RA, researchers not only found increased *Prevotella copri* levels, but also found that its abundance was correlated with RA severity, suggesting that RA may be associated with dysbiosis (Maeda et al., 2016). This was found to be true in humans as well (Scher et al., 2015).

A cross-sectional study evaluated the microbiota profile of four groups: 28 healthy controls, 26 individuals with RA (treated), 44 individuals with RA (nontreated), and 16 individuals with psoriatic arthritis (Scher et al., 2015). After 16S ribosomal DNA was utilized to determine the microbiota profile of each group, it was evident that the individuals with RA had increased *P. copri* abundance (75%) compared with the controls (21.4%). There was an inverse correlation between the abundance of *P. copri* and beneficial bacteria groups, primarily *Bacteroides*. Thus controlling the abundance of *P. copri* may be a new approach to preventing RA.

Similar results in relation to the increased abundance of *P. copri* in participants with RA were found in another study (Zhong et al., 2018). The researchers found gram-positive microbes seemed to dominate the microbiome of the RA individuals. *Clostridium asparagiforme*, *Gordonibacter pameleae*, *Eggerthella lenta*, and *Lachnospiraceae bacterium* were some of the more prevalent microbes, while gram-negative bacteria, specifically *Proteobacteria* and *Firmicutes* (Veillonellaceae family), were depleted.

As was shown by the previously mentioned studies, it appears that the typical microbiome of individuals with RA has overall low diversity. While the prevalence of *Prevotella* species were low in the participants with RA in one study, multiple other studies have shown that *P. copri* levels are high in individuals with RA. Therefore it is important to note the specific strain. Overall, individuals with RA tend to have dysbiosis. These similarities support the theory that the microbiota may be associated with the etiopathology of RA.

23.2.2 Probiotic supplementation in individuals with rheumatoid arthritis

Animal models have shown probiotic supplementation have favorable effects on RA, specifically by modulating inflammation. *Lactobacillus casei* supplementation in rats resulted in increased antiinflammatory cytokines (interleukin 10 (IL-10) and transforming growth factor) and decreased inflammatory cytokines (IL-1, IL-2, IL-6, IL-12, interferon, tumor necrosis factor) (So et al., 2008). Similar results found specifically that *L. casei* inhibited COX-2 and improved joint pain (Amdekar et al., 2011). *Lactobacillus rhamnosus* GG not only improved arthritis but prevented it (Baharav et al., 2004). *L. delbrueckii* administration has also been shown to significantly reduce the development of lab-induced arthritis in mice (Kano et al., 2002). Similarly to *L. rhamnosus* GG, *Prevotella histicola* supplementation reduces arthritis incidence and severity primarily by mucosal modulation. By generating regulatory T cells (Treg) and other immune-regulating cells, *P. histicola* reduces intestinal permeability (Marietta et al., 2016). Thus there is sufficient evidence that probiotics as a treatment for RA may have potential in humans.

Probiotic treatment in animal models has had favorable results, but research in humans has been limited. Currently, there have been only four RCTs that have investigated the efficacy of probiotics as a treatment for RA. The primary strains that were used were *L. casei* and *L. rhamnosus*. Markers used to analyze outcomes include total antioxidant capacity (TAC), Health Assessment Questionnaire (HAQ), global health assessment (GHA), erythrocyte sedimentation rate (ESR), malondialdehyde (MDA), homeostatic model assessment of beta cell function (HOMA-B), erythrocyte superoxide dismutase, glutathione peroxidase, and catalase, disease activity score (DAS), swollen joint count (SJC), tender joint count (TJC), tumor necrosis factor alpha (TNF-α), interleukins (IL), and high-sensitivity C-reactive protein (hs-CRP). Some studies utilized a combination of these known as the ACR20, which is a method endorsed by the American College of Rheumatology and the European League Against Rheumatism to measure multiple markers of improvement in individuals with RA (Felson, 2007; Felson and LaValley, 2014). Improved ACR20 would mean that the number of affected joints was reduced by 20% along with a 20% improvement in three of the other markers: GHA, HAQ, pain scale, ESR, and CRP (Felson and LaValley, 2014).

In a randomized control trial, 60 individuals ages 25–70 were split into a treatment group and a control group (Zamani et al., 2017). All participants had moderate to severe RA. The treatment group was given a synbiotic supplement composed of a prebiotic, inulin, and three species of probiotics: *Lactobacillus acidophilus*, *L. casei*, and *Bifidobacterium bifidum*. The probiotic group took two capsules that contained 10^9 CFU/g of each probiotic every day for 8 weeks. Inflammatory biomarkers such as CRP were observed along with TJC-28, SJC-28, DAS, MDA, TAC, and clinical symptoms. After only 8 weeks, the treatment group showed improved DAS and significantly decreased insulin levels, HOMA-B, and HAQ compared to placebo. It was concluded that probiotic supplementation has favorable effects in individuals with RA (Zamani et al., 2017). The multispecies probiotics paired with the prebiotic in this study might have had a synergistic effect.

Other researchers conducted a similar study but utilized only the single strain *L. casei* 01 in the treatment group (Alipour et al., 2014). This double-blind, placebo-controlled study had sample sizes of 46 individuals. All participants were women ages 20–80 years who had been on medication for 3 months before starting the study. The treatment groups were given a 10^8 CFU/capsule once daily for 8 weeks. Alipour et al. (2014) looked at DAS, GHA, SJC-28, TJC-28, and the inflammatory markers TNF-α, IL-1, IL-6, IL-10, and IL-12. The outcomes consisted of a significant difference in IL-10, IL-12, and TNF-α in comparing the treatment group to placebo (Alipour et al., 2014). Therefore probiotics may aid in dampening the inflammatory cytokines that are involved in RA, which is the primary mechanism of many RA medications.

A Canadian study looked at probiotic supplementation effects in severe cases of RA in 26 individuals ages 18–80, most of whom were women (Pineda Mde et al., 2011). The treatment group (14) received a probiotic twice a day with a 10^9 CFU/capsule of *L. rhamnosus* GR-1 and *L. reuteri* RC-14. The intervention duration was 3 months, and the researchers evaluated SIC-28, RIC, ESR CRP, HAQ, IL-1, IL-6, IL-10, IL-12p70, and TNF-α. While more individuals in the treatment group (three) reached an ACR20 response compared to the placebo group (one), there were no significant differences in the individual markers that comprise the ACR20 criteria. In fact, cytokine levels improved more in the placebo group. The treatment group had a significant improvement in the HAQ but not when compared to the placebo. The probiotic group did show improved functionality, although this result was not statistically significant. A limitation of this study is the use of probiotics as an adjunct to the participants' current treatments.

Researchers evaluated the effects of *L. rhamnosus* GG in 21 individuals ages 18–64 with mild RA (Hatakka et al., 2003). Although this pilot study had the smallest sample size, it had the longest intervention period at 12 months. Biomarkers that were observed included SJC-28, CRP, TJC-28, ESR, HAQ, TNF-a, IL-1B, IL-6, IL-10, and IL-12. SJC and TJC decreased from a total measure of 8.3 to 4.6 in the treatment group, along with the GHA (70%). The placebo group also had clinical parameter improvements but not to the same extent. The treatment group had a slight increase in inflammatory cytokines, yet more participants reported a better sense of well-being in this group. Both groups experienced some improvements, but when compared to each other, there were no significant differences. Thus the benefits of supplementation with *L. rhamnosus* GG as a potential approach to managing chronic pain experienced by those with RA are still unknown.

The primary limitation of the studies on probiotics in RA is a lack in the number of trials. Also, when the currently available kinds of literature are analyzed, there are significant differences in the methods and variables, including the use of various immune-suppressing medications. Thus with the wide range of probiotic dosages, durations, strains, and combinations, it is difficult to compare outcomes and extrapolate clinically applicable conclusions.

Overall, it seems that probiotics do not have consistent effects on RA symptoms and biomarkers in humans. Interestingly, multiple current studies testing the efficacy of probiotic supplementation in RA revealed improvements in both the treatment and placebo groups, suggesting that there may be a mental or emotional component to RA and/or perception of pain. Although improved functionality and well-being were observed in a couple of the studies, the effects were not considered significant. Only two of the four RCTs found significant effects of probiotics in treating RA (Alipour et al., 2014; Zamani et al., 2017). Both studies used the probiotic species *L. casei*, although one in isolation and one as a synbiotic. These were also two of the studies with the shortest duration. One administered the lowest in dose of CFUs and utilized a single strain (Alipour et al., 2014). Therefore duration and dose may not be major influencing factors, but the species may be: *L. casei* in particular. To determine this, future studies should attempt to conduct RCTs with multiple treatment groups that are given various isolated species and, ideally, strains.

23.3 Fibromyalgia

Fibromyalgia (FM) is a chronic widespread pain disease which is typically associated with symptoms such as fatigue, depression, morning stiffness, and sleeping disorders (Roman et al., 2018a). The etiology of this chronic disorder is still unknown. It is thought that genetic disposition, infections, neurotransmitter abnormalities, and central sensitization could play a role in the pathogenesis of the disease, researchers state. The prevalence of FM is thought to range from 0.2 to 6.6% (Erdrich et al., 2020), and appears to be more prevalent among women (Kim & Chang, 2012).

FM seems to be associated with gastrointestinal disorders (Erdrich et al., 2020). FM patients usually show a high incidence of gastrointestinal symptoms with associated diarrhea, although constipation is also reported. Around 32% of patients with FM are labeled as having irritable bowel syndrome (IBS), and 81% of them express having irregular bowel patterns (Pimentel et al., 2004). The patients with both FM and IBS report an increase in the severity of symptoms and poorer quality of life (Erdrich et al., 2020).

23.3.1 Pain and fibromyalgia

Fibromyalgia is diagnosed according to two scales from the American College of Rheumatology: the Widespread Pain Index and the Symptom Severity Scale (Kim & Chang, 2012). Both are used for diagnostic criteria based on the duration of symptoms, including whether there is an absence of another disease to explain the pain shown and whether there is a persistent comparable level of symptoms for at least 3 months (Kim & Chang, 2012).

The medications that have been approved in the United States for the treatment of FM are pregabalin, duloxetine, and milnacipran, but these have limited to moderate effects in reducing chronic pain among this population (Supraha et al., 2013). Diet and probiotics, as well as exercise, have been suggested to be beneficial for individuals with FM. Probiotics could help to reverse bacterial overgrowth, strengthening the gut barrier function, modifying the bowel inflammatory response, and impeding pathogens that could be affecting the person with FM.

23.3.2 Gut microbiota and fibromyalgia

Research studies examining the differences in the gut microbiome among patients with FM have been developed. Gut bacteria appears to play a role in numerous medical conditions, including FM. The gut bacterial composition and their metabolic activity appeared to be related to the expression of symptoms in individuals with FM (Minerbi and Fitzcharles, 2020). For instance, *Bacteroides* spp. appeared to be positively correlated with the total symptoms score on the Fibromyalgia Impact Questionnaire (Erdrich et al., 2020). In addition, the presence of SCFAs, neurotransmitters, bacterial antigens, and bile acids in the host circulation has been suggested as a mechanism in which these bacterial species can cause FM symptoms, which are expressed as fatigue, mood, and/or pain (Minerbi and Fitzcharles, 2020).

Researchers have investigated whether increased intestinal permeability is present in FM patients and whether it is correlated with the intensity of pain symptoms (Goebel et al., 2008). The sample consisted of 40 patients with FM. Among the sample, 62.5% of patients described gastrointestinal symptoms (e.g., heartburn, bloating, constipation/diarrhea). Gastroduodenal permeability was increased in 32.5% of patients with FM compared to one healthy volunteer. The small bowel permeability index was increased in 37.5% of patients with FM and in none of the healthy volunteers. Researchers found that increased intestinal permeability and a leaky gut barrier are present among the population affected by the disease compared to healthy controls. No correlation between intestinal permeability values and pain intensity or abdominal symptoms was found. Among the limitations of the study was that small intestinal bacterial overgrowth (SIBO) was not measured, suggesting the need for further studies to measure lactulose breath testing and SIBO to determine a possible association between the two.

In a prospective, double-blind study, researchers tested lactulose breath test and its association with IBS and FM and a possible link between pain and FM (Pimentel et al., 2004). The study included 42 participants with FM (mostly women) and 111 participants with IBS. The researchers were able to find abnormal lactulose breath tests among both FM participants (100%) and IBS subjects (84%) in comparison to 20% of normal controls, which suggested the presence of bacterial overgrowth. They also identified increased hydrogen production in the FM group, indicating a larger quantity of small intestinal bacteria among them, which seemed to correlate with somatic pain.

In addition to gastrointestinal symptoms, mood and emotional problems are common in patients with FM, possibly related to the gut-brain axis connection. Memory loss, confusion, and forgetfulness have been reported in individuals with FM (Roman et al., 2018a,b). Researchers have pointed to reduced values of *Bifidobacterium* and increased levels of *Enterococcus* spp. being associated with cognitive and neurological deficits in patients with FM (Roman et al., 2018a). In their RCT the authors tested whether multispecies probiotics could improve emotional symptoms, cognition, and functional state in individuals with FM. The sample consisted of 40 subjects (20 in the intervention/probiotic group and 20 in the placebo group). The study lasted 8 weeks. The intervention group received four bottles of Ergyphilus Plus (which contains *L. casei*, *L. Rhamnosus* GG, *L. acidophilus*, and *Bifidobacteria bifidus*). The researchers found that the intervention group had improved cognition status; no other benefits were seen.

An investigation from 2018 showed beneficial effects from probiotics, specifically *Lactobacillus plantarum* and *Pediococcus acidilactici*, in IBS, which is seen in individuals with FM (Roman et al., 2018a,b). These probiotic strains appeared to be beneficial in improving IBS-related quality of life in people with IBS and diarrhea. Also, prebiotics have shown a significant effect ($p < 0.05$) in anxiety scores but not in depression levels. In addition, *L. rhamnosus* GG appeared to significantly lower the severity of abdominal pain in children with IBS. The use of probiotics also appeared to reduce pain levels in patients with RA, which may suggest a beneficial effect in patients with FM (Roman et al., 2018a,b).

Further research evaluating the gut microbiome in FM patients is warranted. However, the existing research may suggest a possible association between the gastrointestinal microbiota composition, metabolism, and FM. In addition, there is ongoing research evaluating the use of fecal microbiota transplants with FM patients as an emerging treatment option (Choi & Cho, 2016). The expected results could be promising.

23.4 Multiple sclerosis

Multiple sclerosis (MS) is a chronic autoimmune, proinflammatory disease of the central nervous system (CNS). Around 2.5 million people worldwide are affected by the disorder, whose frequent symptoms include fatigue, depression, vision loss, numbness, bladder and bowel dysfunction, and pain, presumably affected by abnormal T cell responses in the CNS (van den Hoogen et al., 2017). The exact etiology of MS is still unknown, but it is thought that genetic, herpetic, and environmental factors play a significant role. Genetic components account for 30% of the disease risk; among the nongenetic components, possible contributors include smoking, low levels of vitamin D, exposure to Epstein-Barr virus (a herpesvirus), and, more recently encountered, the gut microbiota (Freedman et al., 2018).

23.4.1 Gut microbiota and multiple sclerosis

The gut microbiota serves as an intestinal barrier that is influenced by food metabolism, immune responses, and nutrient uptake. Researchers evaluating the microbiome profiling for different autoimmune and autoinflammatory conditions (e.g., type 1 diabetes mellitus, celiac disease, RA, inflammatory bowel disease, and MS, including individuals with the active or remission phases of relapsing-remitting MS [RRMS]) have found that these patients show microbial dysbiosis (Freedman et al., 2018). For instance, patients with MS may demonstrate increased intestinal permeability, changes in bile acid metabolism, increased constipation, and fecal incontinence as well as higher chances of IBD. This suggests a connection between the gut and the CNS, known as the gut-brain axis (Chen et al., 2016a,b). Treatment of patients with such gastrointestinal presentations has included gut-focused behavioral therapy and biofeedback (Ghareghani et al., 2018).

Alterations in the gut microbiota are influenced by diet, either short- or long-term dietary habits, genes, antibiotics, and probiotics among colonization and ecological processes. In healthy individuals the gut bacterial flora mainly comprise bacteria from the phyla Bacteroides and Firmicutes, which aid in metabolism and nutrient availability from fermentation processes (van den Hoogen et al., 2017). Human studies evaluating the microbiota from seven patients with RRMS, found reductions in *Bacteroidetes*, *Firmicutes*, and *Proteobacteria* species (de Oliveira et al., 2017).

Research evidence from murine models and human studies suggests that the gut microbiota contributes to the pathogenesis of MS (Liu et al., 2018). In addition, research studies report that alteration in the gut microbiome may influence inflammatory responses in animals and individuals with MS (Morshedi et al., 2019). Much of the research that has found a correlation between MS and the gut microbiota has involved studies using experimental autoimmune encephalomyelitis (EAE) mouse models. Research from de Oliveira et al. (2017), showed that animals that were colonized with *B. fragilis* and polysaccharides. appeared to be protected from the disease. By contrast, animals that were treated with antibiotics showed reductions in their clinical scores, suggesting a role of the microbiota in the inflammatory cells of such models (de Oliveira et al., 2017).

An investigation from 2018 showed that transgenic mice that were prone to spontaneous brain autoimmunity and received fecal microbiota from subjects with MS developed severe disease in contrast to mice that acquired fecal microbiota from healthy matching twins (Camara-Lemarroy et al., 2018). These researchers also pointed to differences in the microbiome composition of patients with MS versus healthy controls, as those with RRMS showed increases in the bacterial families *Faecalibacterium*, *Anaerostipes*, *Pseudomonas*, *Mycoplasma*, *Haemophilus*, *Blautia*, and *Dorea* species, whereas they tended to show decreases in *Bacteroides*, *Prevotella*, *Adlercreutzia*, and *Parabcteroides*.

A research study pointed to a decreased abundance of the families *Lactobacilli* in human studies of MS compared to healthy controls (Liu et al., 2018). In another study, higher serum reactivity to *Clostridium perfringens* type B in patients with MS was found when compared to controls, which led to the hypothesis of prior exposure to this bacterium by patients with the disease (Pröbstel and Baranzini, 2018). These researchers also pointed to decreased levels of alpha1,3-galactose, a highly immunogenic sugar moiety produced by the gut microbiota, in patients with MS, suggesting another modification in the gut microbiota. They also described an increased proportion of the genus *Akkermansia* (which converts mucin to SCFA and has proinflammatory and regulatory properties) in MS patients compared to controls.

A research review determined that there is a reduced abundance of SCFA producers in the gut of patients with MS; these include *Prevotella*, *Lachnospiraceae*, *Faecalibacterium prausnitizii*, and *Butyricimonas* (van den Hoogen et al., 2017). It is thought that since SCFA have appeared to decrease the disease score in EAE models, decreased SCFA production could have an impact on the disease pathology. The authors also mentioned two articles that found *Enterobacteriaceae*, which could be either pathogenic or nonpathogenic species, to be increased in MS patients compared to healthy individuals. In a cohort of 18 MS pediatric patients pointed out by the authors, *B. fragilis* was reduced in comparison to controls.

In a 2018 study, researchers investigated whether dysfunction in gut permeability contributes to the immune pathogenesis of MS with 22 patients, five twin pairs, and 18 healthy donors (Buscarinu et al., 2018). Authors found a significant difference in the proportion of subjects with increased intestinal permeability (73% versus 28% patients). Such a difference was more significant when twin pairs were not considered, which suggested a possible genetic influence on intestinal permeability determinants (81% patients versus 28% healthy donors). Their findings showed that intestinal permeability was frequently seen in patients with MS in which genetic influence could be a determinant.

23.4.2 Multiple sclerosis and modifiable factors

Modifiable factors of the gut microbiota include probiotics, as was previously mentioned. More research in this specific area is needed to elucidate which particular strains may show a beneficial effect in patients with MS, as much of the research has been conducted in animals. Research evaluating the modulation of MS by food and gut microbiota in its animal EAE model pointed out the following findings:

- Prophylactic therapy with *Bifidobacterium animalis* reduced the duration of EAE symptoms.
- Commercially available probiotic drinks that contain *L. casei* lessened the EAE disorder score in Lewis rats, although this was not observed in other models.
- The use of *Lactobacilli* monostrains as prophylaxis diminished autoreactive T cells and prevented EAE.
- Treatment with heat-killed *P. acidilactici* as prophylaxis reduced the disease score in EAE.
- Prophylactic therapy with *Streptococcus thermophilus*, three *Lactobacillus* subspecies, and *B. bifidum* suppressed the incidence and severity of EAE in MOG mice (van den Hoogen et al., 2017).

A randomized, double-blind placebo-controlled trial was conducted to assess the effects of probiotics on metabolic condition, disability, and mental health in patients with MS (Kouchaki et al., 2017). A probiotic capsule (*L. acidophilus*, *B. bifidum*, *L. casei*, and *Lactobacillus fermentum*) taken for 12 weeks had positive effects on mental health, insulin resistance, high-density lipoprotein cholesterol, inflammatory markers, and on Expanded Disability status scale compared to placebo. It was also found that probiotic supplementation could modify hs-CRP levels compared to the placebo group, and that B cell function was significantly reduced after 12 weeks of receiving previously mentioned probiotics.

A research review showed that in the EAE model, there was evidence of fecal microbial dysbiosis and reductions of *Prevotella* during the disease (Liu et al., 2018). At the same time, *L. reuteri* appeared to improve the severity of EAE and decreased Th1 and Th17 cytokine levels in the serum and gut. In a human study, *L. reuteri* was suggested to expand the quality of life in patients with MS, thus providing insightful evidence of the connection between the gut-brain axis and potential impact and modifying factors in the disease.

One systematic review described the role of probiotic supplementation in inflammation in individuals with MS. *L. casei*, *L. plantarum*, *L. fermentum*, *B. infantis*, and *B. lactis* appeared to decrease IL-6 and hs-CRP levels, compared to those in the control group who received maltodextrin capsules (Morshedi et al., 2019). No significant modification in TNF-α was found. In their review, consumption of probiotics had beneficial effects on systemic, central inflammation, and immune control in EAE models and individuals with MS (Morshedi et al., 2019). However, further research is needed to explain the exact mechanism of action by which probiotics may exert such benefits.

More research investigating the role of the gut microbiota in MS is warranted to understand better whether and how diet triggers or perpetuates MS and whether prior animal results are transferable to humans. However, studies so far have shown differences in the gut environment of patients with MS versus control groups, which indicates a possible relationship. Gut dysbiosis is commonly present in autoimmune and inflammatory disorders that disrupt the gut barrier, causing systemic inflammation. As probiotics play an essential role in modulating barrier function, appropriate formulation, dose, and study design are needed for additional and ongoing trials evaluating their role and/or effectiveness in MS.

Along with fecal transplant and well-designed clinical trials with probiotics, bile acid supplementation in patients with MS shows encouraging results. Bile acids can modulate intestinal barrier function and can modify the intestinal microbiota community, and the microbiome can, in turn, alter bile acid metabolism (Camara-Lemarroy et al., 2018).

TABLE 23.1 Microbiota profile of individuals with chronic pain conditions.

Condition	Decreased microbes	Increased microbes
Rheumatoid arthritis	• Gram-negative (Zhong et al., 2018) • *Bacteroides* (Scher et al., 2015) • Bifidobacteria (Vaahtovuo et al., 2008) • *Bacteroides-Porphyromonas-Prevotella* (Vaahtovuo et al., 2008; Toivanen et al., 2002) • *Bacteroides fragilis* subgroup (Vaahtovuo et al., 2008) • *Eubacterium rectale–Clostridium coccoides* group (Vaahtovuo et al., 2008) • *Proteobacteria* (Zhong et al., 2018) • *Firmicutes* (Zhong et al., 2018)	• Gram-positive (Zhong et al., 2018) • *Prevotella copri* (Maeda et al., 2016; Scher et al., 2015; Zhong et al., 2018) • *Clostridium perfringens* • *Lactobacillus* (Liu et al., 2018) • *Clostridium asparagiforme* (Zhong et al., 2018) • *Gordonibacter pamelaeae* (Zhong et al., 2018) • *Eggerthella lenta* (Zhong et al., 2018) • *Lachnospiraceae* (Zhong et al., 2018)
Multiple sclerosis	• *Bacteroides* • *Firmicutes* • *Lachnospiraceae, Faecalibacterium prausnitizii, Butyricimonas, B. fragilis* (van den Hoogen et al., 2017) • *Proteobacteria* (de Oliveira et al., 2017) • *Parabcteroides* • *Prevotella* • *Adlercreutzia* (Camara-Lemarroy et al., 2018) • *Lactobacilli* (Liu et al., 2018)	• *Faecalibacterium* • *Anaerostipes* • *Pseudomonas* • *Mycoplasma* • *Haemophilus* • *Blautia* • *Dorea* (Camara-Lemarroy et al., 2018) • *C. perfringens* type B, • *Akkermansia* (Pröbstel and Baranzini, 2018)
Fibromyalgia	• *Bifidobacterium* (Roman et al., 2018a)	• *Enterococcus* spp. (Roman et al., 2018a)

According to the authors, metabolomic research has discovered alterations in bile acid in patients with MS compared to healthy control groups. More investigation could demonstrate encouraging results.

23.5 Discussion

In this review, the commonalities of gut microbiota profiles of chronic pain conditions were explored, as well as how they differ from healthy control groups, as seen in Table 23.1. Improving the integrity of the intestinal barrier and decreasing the overall immune response are potential mechanisms by which the microbiome aids in pain modulation unrelated to gastrointestinal conditions. Previous evidence has revealed that similar rheumatic manifestations have appeared in patients with celiac disease, a condition that is known to increase intestinal permeability and dysbiosis (Lerner and Matthias, 2015). Gastrointestinal manifestations that are characteristic of celiac disease have also appeared in RA (Lerner and Matthias, 2015). Inflammation, pain, and food sensitivities are evident in both celiac disease and RA. Thus the inflammatory cascade and immune activation that is largely influenced by the gut microbiome of both conditions likely overlap in etiology.

Probiotics are living microorganisms that are beneficial for the host's health, and these may enhance the integrity of the intestinal epithelial barrier (Dai et al., 2017). Since dysbiosis was present in chronic pain conditions, as seen in Table 23.2, the use of probiotics to improve either gut health or pain stimulus appears to be a potential alternative for

TABLE 23.2 Probiotic supplementation and chronic pain conditions.

Condition	Probiotic/strain/combo	Outcome/suggested effect
Rheumatoid arthritis	Synbiotic including: *Lactobacillus acidophilus*, *Lactobacillus philus*, *Lactobacillus casei*, and *Bifidobacterium bifidum*	Improved DAS. Significantly decreased insulin levels, HOMA-B, and HAQ (Zamani et al., 2016).
	L. casei 01	Significant difference in IL-10, IL-12, and TNF-α (Alipour et al., 2014).
	Lactobacillus rhamnosus GR-1 and *Lactobacillus reuteri* RC-14	Improved HAQ and functionality cytokine levels improved more in the placebo (Pineda Mde et al., 2011).
	L. rhamnosus GG	Improved SJC, TJC, and GH. Slight increased inflammatory cytokines Better sense of well-being (Hatakka et al., 2003).
Multiple sclerosis	*Bifidobacterium animalis* *Lactobacillus casei* *Lactobacilli monostrains* Heat-killed *Pediococcus acidilactici* Prophylactic therapy with *Streptococcus thermophilus*, three *Lactobacillus*	Prophylactic therapy appeared to reduce the duration of EAE symptoms. Appeared to lessen the EAE disorder score in Lewis rats, although this was not observed in other models As prophylaxis, shown to reduce autoreactive

(Continued)

TABLE 23.2 (Continued)

Condition	Probiotic/strain/combo	Outcome/suggested effect
	subspecies, and *B. bifidum*	T cells and prevented EAE. As prophylaxis, showed to reduce the disease score in EAE. Appeared to suppress the incidence and severity of EAE in MOG mice (van den Hoogen et al., 2017).
	Lactobacillus reuteri	improved the severity of EAE and decreased Th1 and Th17 cytokine levels in the serum and gut; in humans it appeared to help improved quality of life in MS patients (Liu et al., 2018).
	L. acidophilus, *B. bifidum*, *L. casei*, and *Lactobacillus fermentum*	Showed positive effects on mental health, insulin resistance, high-density lipoprotein cholesterol, inflammatory markers, and on Expanded Disability Status Scale compared to placebo. These appeared to modify hs-CRP levels compared to the placebo group. Reduced B cell function (Kouchaki et al., 2017).
	L. casei, *Lactobacillus plantarum*, *L. fermentum*, *Bifidobacterium infantis*, and *Bifidobacterium lactis*	Appeared to decrease IL-6 and hs-CRP levels compared to controls. Beneficial effects on systemic, central inflammation, and immune control in EAE models and individuals with MS were seen (Morshedi et al., 2019).
Fibromyalgia	*L. casei*, *Lactobacillus rhamnosus* GG, *Lactobacillus acidophilus*, and *Bifidobacteria bifidus*	Shown to improve cognitive status, yet no other benefits were seen (Roman et al., 2018a).

treatment. Probiotics have shown favorable results in those with chronic pain conditions such as decreased inflammatory markers (hs-CRP) in patients with MS (Kouchaki et al., 2017) and improved cognition in individuals with FM. However, using larger-scale randomized control trials are needed to determine how and which probiotic strains can modulate the pain response among the populations affected by these chronic pain conditions so results could be applied in clinical practice. As was previously discussed, specific durations, dose, and strains are pertinent factors to be considered and investigated.

Although there is no clear commonality in the microbial profile among all of these conditions, it is apparent that they share in having intestinal permeability. Dysbiosis may be only one of many factors contributing to a compromised intestinal barrier. Thus probiotic supplementation alone might not be enough. A multifactorial approach that addresses the other causes, such as stress, diet, medications, and nutrient deficiencies, may be necessary to restore the integrity of the intestinal epithelium of individuals with these conditions.

23.6 Conclusion

The microbiome profiles were observed in chronic pain conditions, including rheumatoid arthritis, fibromyalgia, and multiple sclerosis. Our findings suggest that dysbiosis plays a crucial role in the pathogenesis of these inflammatory conditions as compared to healthy control groups. The research discussed in this chapter supports the suspicion that dysbiosis, and hence increased intestinal permeability, is present in chronic pain disorders. The resulting disrupted gut barrier is likely the mechanism, if not one of many, that leads to systemic inflammation and pain. The use of probiotics may serve as a potential therapeutic approach, given their ability to improve gut function for those suffering from such chronic pain conditions, but further research is warranted.

References

Alipour, B., Homayouni-Rad, A., Vaghef-Mehrabany, E., Sharif, S.K., Vaghef-Mehrabany, L., Asghari-Jafarabadi, M., et al., 2014. Effects of Lactobacillus casei supplementation on disease activity and inflammatory cytokines in rheumatoid arthritis patients: a randomized double-blind clinical trial. Int. J. Rheum. Dis 17 (5), 519–527. Available from: https://doi.org/10.1111/1756-185X.12333.

Amdekar, S., Singh, V., Singh, R., Sharma, P., Keshav, P., Kumar, A., 2011. Lactobacillus casei reduces the inflammatory joint damage associated with collagen-induced arthritis (CIA) by reducing the pro-inflammatory cytokines. J. Clin. Immunol 31 (2), 147–154.

Baharav, E., Mor, F., Halpern, M., Weinberger, A., 2004. Lactobacillus GG bacteria ameliorate arthritis in Lewis rats. J. Nutr 134 (8), 1964–1969.

Buscarinu, M.C., Romano, S., Mechelli, R., Umeton, R.P., Ferraldeschi, M., Fornasiero, A., et al., 2018. Intestinal permeability in relapsing-remitting multiple sclerosis. Neurotherapeutics 15 (1), 68–74.

Camara-Lemarroy, C.R., Metz, L.M., Yong, V.W., 2018. Focus on the gut-brain axis: multiple sclerosis, the intestinal barrier and the microbiome. World J. Gastroenterol 24 (37), 4217–4223. Available from: https://doi.org/10.3748/wjg.v24.i37.4217.

Chen, J., Chia, N., Kalari, K.R., et al., 2016a. Multiple sclerosis patients have a distinct gut microbiota compared to healthy controls. Sci. Rep. 6, 28484. Available from: https://doi.org/10.1038/srep28484. Published 2016 Jun 27.

Chen, J., Wright, K., Davis, J.M., et al., 2016b. An expansion of rare lineage intestinal microbes characterizes rheumatoid arthritis. Genome Med. 8 (1), 43. Available from: https://doi.org/10.1186/s13073-016-0299-7. Published 2016 Apr 21.

Choi, H.H., Cho, Y.S., 2016. Fecal microbiota transplantation: current applications, effectiveness, and future perspectives. Clin. Endosc 49 (3), 257–265. Available from: https://doi.org/10.5946/ce.2015.117.

De Filippo, C., Di Paola, M., Giani, T., Tirelli, F., Cimaz, R., 2019. Gut microbiota in children and altered profiles in juvenile idiopathic arthritis. J. Autoimmun 98, 1–12. Available from: https://doi.org/10.1016/J.JAUT.2019.01.001.

de Oliveira, G.L.V., Leite, A.Z., Higuchi, B.S., Gonzaga, M.I., Mariano, V.S., 2017. Intestinal dysbiosis and probiotic applications in autoimmune diseases. Immunology 152 (1), 1–12. Available from: https://doi.org/10.1111/imm.12765.

Erdrich, S., Hawrelak, J.A., Myers, S.P., Harnett, J.E., 2020. Determining the association between fibromyalgia, the gut microbiome and its biomarkers: a systematic review. BMC Musculoskelet. Disord. 21 (1), 1–12. Available from: https://doi.org/10.1186/s12891-020-03201-9.

Felson, D., 2007. A proposed revision to the ACR20: the hybrid measure of American College of Rheumatology response. Arthritis Rheumatism 57 (2), 193–202. Available from: https://doi.org/10.1002/art.22552.

Felson, D.T., LaValley, M.P., 2014. The ACR20 and defining a threshold for response in rheumatic diseases: too much of a good thing. Arthritis Res. Ther 16 (1), 101. Available from: https://doi.org/10.1186/ar4428.

Freedman, S.N., Shahi, S.K., Mangalam, A.K., 2018. The "Gut Feeling": breaking down the role of gut microbiome in multiple sclerosis. Neurotherapeutics 15 (1), 109–125. Available from: https://doi.org/10.1007/s13311-017-0588-x.

Ghareghani, M., Reiter, R.J., Zibara, K., Farhadi, N., 2018. Latitude, vitamin D, melatonin, and gut microbiota act in concert to initiate multiple sclerosis: a new mechanistic pathway. Front. Immunol 9.

Goebel, A., Sprotte, G., Lochs, H., Schedel, R., Buhner, S., 2008. Altered intestinal permeability in patients with primary fibromyalgia and in patients with complex regional pain syndrome. Rheumatology 47 (8), 1223–1227. Available from: https://doi.org/10.1093/rheumatology/ken140.

Hatakka, K., Martio, J., Korpela, M., Herranen, M., Poussa, T., Laasanen, T., et al., 2003. Effects of probiotic therapy on the activity and activation of mild rheumatoid arthritis—a pilot study. Scand. J. Rheumatol 32 (4), 211—215.

Kano, H., Kaneko, T., Kaminogawa, S., 2002. Oral intake of Lactobacillus delbrueckii subsp. bulgaricus OLL1073R-1 prevents collagen-induced arthritis in mice. J. Food Prot 65 (1), 153—160. Available from: http://www.ncbi.nlm.nih.gov/pubmed/11808787.

Kim, S.E., Chang, L., 2012. Overlap between functional GI disorders and other functional syndromes: what are the underlying mechanisms? Neurogastroenterol. Motil. Off. J. Eur. Gastrointest. Motil. Society. 24 (10), 895—913. Available from: https://doi.org/10.1111/j.1365-2982.2012.01993.x.

Kouchaki, E., Tamtaji, O.R., Salami, M., Bahmani, F., Daneshvar Kakhaki, R., Akbari, E., et al., 2017. Clinical and metabolic response to probiotic supplementation in patients with multiple sclerosis: a randomized, double-blind, placebo-controlled trial. Clin. Nutr 36 (5), 1245—1249. Available from: https://doi.org/10.1016/j.clnu.2016.08.015.

Lerner, A., Matthias, T., 2015. Rheumatoid arthritis—celiac disease relationship: joints get that gut feeling. Autoimmun. Rev 14 (11), 1038—1047.

Liu, X., Zou, Q., Zeng, B., Fang, Y., Wei, H., 2013. Analysis of fecal lactobacillus community structure in patients with early rheumatoid arthritis. Curr. Microbiol 67 (2), 170—176. Available from: https://doi.org/10.1007/s00284-013-0338-1.

Liu, Y., Alookaran, J.J., Rhoads, J.M., 2018. Probiotics in autoimmune and inflammatory disorders. Nutrients 10 (10), 1537. Available from: https://doi.org/10.3390/nu10101537.

Maeda, Y., Kurakawa, T. Umemoto, E., Motooka, D., Ito, Y., Gotoh, K., et al., 2016. Dysbiosis contributes to arthritis development via activation of autoreactive T cells in the intestine. Arthritis Rheumatol 68 (11), 2646—2661.

Marietta, E.V., Murray, J.A., Luckey, D.H., et al., 2016. Suppression of inflammatory arthritis by human gut-derived prevotella histicola in humanized mice. Arthritis Rheumatol 68 (12), 2878—2888. Available from: https://doi.org/10.1002/art.39785.

Minerbi, A., Fitzcharles, M.A., 2020. Gut microbiome: pertinence in fibromyalgia. Clin. Exp. Rheumatol. 38 Suppl 123 (1), 99—104.

Morshedi, M., Hashemi, R., Moazzen, S., Sahebkar, A., Hosseinifard, E.S., 2019. Immunomodulatory and anti-inflammatory effects of probiotics in multiple sclerosis: a systematic review. J. Neuroinflamm. 16 (1), 18—21. Available from: https://doi.org/10.1186/s12974-019-1611-4.

Pimentel, M., Wallace, D., Hallegua, D., et al., 2004. A link between irritable bowel syndrome and fibromyalgia may be related to findings on lactulose breath testing. Ann. Rheum. Dis 63 (4), 450—452. Available from: https://doi.org/10.1136/ard.2003.011502.

Pineda Mde, L., Thompson, S.F., Summers, K., de Leon, F., Pope, J., Reid, G., 2011. A randomized, double-blinded, placebo-controlled pilot study of probiotics in active rheumatoid arthritis. Med. Sci. Monit. 17 (6), CR347—CR354. Available from: https://doi.org/10.12659/msm.881808. Published 2011 Jun 1.

Pröbstel, A.K., Baranzini, S.E., 2018. The role of the gut microbiome in multiple sclerosis risk and progression: towards characterization of the "MS Microbiome.". Neurotherapeutics 15 (1), 126—134. Available from: https://doi.org/10.1007/s13311-017-0587-y.

Roman, P., Estévez, A.F., Miras, A., et al., 2018a. A pilot randomized controlled trial to explore cognitive and emotional effects of probiotics in fibromyalgia. Sci. Rep. 8 (1), 10965. Available from: https://doi.org/10.1038/s41598-018-29388-5. Published 2018 Jul 19.

Roman, P., Carrillo-Trabalon, F., Sanchez-Labraca, N., Canadas, F., Estevez, A., Cardona, D., 2018b. Are probiotic treatments useful on fibromyalgia syndrome or chronic fatigue syndrome patients? A systematic review. Benef. Microbes 9 (4), 603—611.

Scher, J.U., Ubeda, C., Artacho, A., et al., 2015. Decreased bacterial diversity characterizes the altered gut microbiota in patients with psoriatic arthritis, resembling dysbiosis in inflammatory bowel disease. Arthritis Rheumatol 67 (1), 128—139. Available from: https://doi.org/10.1002/art.38892.

So, J.S., Lee, C.G., Kwon, H., et al., 2008. Lactobacillus casei potentiates induction of oral tolerance in experimental arthritis. Mol. Immunol 46 (1), 172—180. Available from: https://doi.org/10.1016/j.molimm.2008.07.038.

Supraha, V., Francis, D.K., Utrobicic, A., Choy, E.H., Tenzera, D., Kordic, A., 2013. Probiotics for fibromyalgia (Protocol). Cochrane Database Syst. Rev (3). Available from: https://doi.org/10.1002/14651858.CD010451.

Toivanen, P., Vartiainen, S., Jalava, J., et al., 2002. Intestinal anaerobic bacteria in early rheumatoid arthritis (RA). Arthritis Res 4 (Suppl 1), 5. Available from: https://doi.org/10.1186/ar492.

Vaahtovuo, J., Munukka, E., Korkeamäki, M., Luukkainen, R., Toivanen, P., 2008. Fecal microbiota in early rheumatoid arthritis. J. Rheumatol 35 (8), 1500—1505.

van den Hoogen, W.J., Laman, J.D., 't Hart, B.A., 2017. Modulation of multiple sclerosis and its animal model experimental autoimmune encephalomyelitis by food and gut microbiota. Front. Immunol. 8, 1081. Available from: https://doi.org/10.3389/fimmu.2017.01081. Published 2017 Sep 5.

van Hemert, S., Breedveld, A.C., Rovers, J.M., Vermeiden, J.P., Witteman, B.J., Smits, M.G., de Roos, N.M., 2014. Migraine associated with gastrointestinal disorders: review of the literature and clinical implications. Front. Neurol. 5, 241. Available from: https://doi.org/10.3389/fneur.2014.00241.

Zamani, B., Golkar, H.R., Farshbaf, S., et al., 2016. Clinical and metabolic response to probiotic supplementation in patients with rheumatoid arthritis: a randomized, double-blind, placebo-controlled trial. Int. J. Rheumatic Dis 19 (9), 869—879. Available from: https://doi.org/10.1111/1756-185X.12888.

Zamani, B., Farshbaf, S., Golkar, H.R., Bahmani, F., Asemi, Z., 2017. Synbiotic supplementation and the effects on clinical and metabolic responses in patients with rheumatoid arthritis: a randomised, double-blind, placebo-controlled trial. Br. J. Nutr 117 (08), 1095—1102. Available from: https://doi.org/10.1017/s000711451700085x.

Zhong, D., Wu, C., Zeng, X., Wang, Q., 2018. The role of gut microbiota in the pathogenesis of rheumatic diseases. Clin. Rheumatol 37 (1), 25—34.

Section IX

Environmental Pollutants and Gut Microbiome

Chapter 24

Microplastic toxicity and the gut microbiome

Muhammad Manjurul Karim[1], Pooja Shivappa[2], Nadiya Dileep[3], Tania Akter Jhuma[1] and Ashfaque Hossain[3]

[1]*Department of Microbiology, University of Dhaka, Dhaka, Bangladesh,* [2]*Department of Basic Sciences, RAK Medical and Health Sciences University, Ras Al Khaimah, United Arab Emirates,* [3]*Department of Medical Microbiology and Immunology, RAK Medical and Health Sciences University, Ras Al Khaimah, United Arab Emirates*

24.1 Introduction

Minor pieces of plastic, predominantly those that are less than 5 mm in size, are commonly called microplastics (MPs) (Akdogan and Guven, 2019). Therefore MPs are solid synthetic or polymeric particles, irregular or regular in shape with sizes varying from 1 μm to 5 mm (Fig. 24.1). MPs have both primary and secondary sources of origin. Primary MPs are manufactured for commercial use, while secondary MPs are particles that result from the breakdown of larger plastic items. They are mostly insoluble in water and pervade both the marine and terrestrial environments (de Souza Machado et al., 2018; Galloway et al., 2017). It is now established that humans are consuming MPs in various ways: ingestion (through contaminated food or food transport), inhalation, or skin contact (Campanale et al., 2020). MPs with a diameter of less than 20 μm can penetrate the organs, while those of about 5 μm in size can penetrate all the organs, cut the cell membrane, and cross the blood-brain barrier as well as the placenta (Boyle and Örmeci, 2020). How toxic chemicals are adsorbed onto MP particles is still unclear; however, factors such as pH variations, hydrophobic interactions, aging of particles, and polymer composition are reported to contribute to their toxicity (Hwang et al., 2020; Joo et al., 2021).

24.2 History of microplastics

Plastic production, use, and consumption have risen dramatically since the 1950s in tandem with the rising human population and economic growth. Since then, the demand for plastic has grown exponentially from half a million metric tons to 359 million metric tons in 2018, outpacing all other manufactured materials (Lambert et al., 2014; Wu et al.,

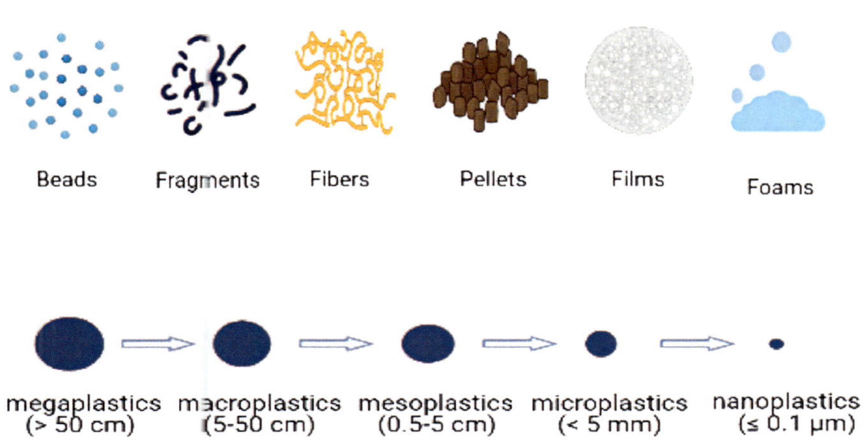

FIGURE 24.1 Plastic particles in different shapes and sizes.

Microbiome, Immunity, Digestive Health and Nutrition. DOI: https://doi.org/10.1016/B978-0-12-822238-6.00019-4

2019). The production and usage of plastics have increased steadily, and one-third of it is discarded within 3 years of manufacture (Paul et al., 2020). According to Geyer et al. (2017), approximately 6300 million metric tons of plastic waste had been produced as of 2015, which is expected to rise to 12,000 million metric tons by 2050 if current trends continue (Geyer et al., 2017), largely as a result of its nonbiodegradability. Although many attempts have been made to recycle it, a considerable amount of trash has accumulated in the environment. Ultimately, the greatest invention becomes a curse for the ecosystem and is recognized as a serious global concern (Paul et al., 2020).

Owing to their excellent physical and chemical properties and extensive use in manufacturing, agriculture, and daily life, demand for plastics has risen day by day, which ultimately makes a world without plastic unimaginable (Lu et al., 2018; Yong et al., 2020). Most of the plastics break down into smaller fragments known as MPs. These miniaturized forms with a diameter of less than 5 mm have a high fragmentation rate and break down further, resulting in nanoplastics (<0.1 μm in diameter, Fig. 24.1) (Banerjee and Shelver, 2021; de Souza Machado et al., 2018; Jin et al., 2019a,b; Paul et al., 2020). The most common types of MPs include polyvinyl chloride, nylons, polyethylene terephthalate, polyethylene, polypropylene, polystyrene, polyvinyl alcohol, and polyamide, which have become prevalent as a result of their immense usage in our daily life (Auta et al., 2017; Mammo et al., 2020; Prakash et al., 2020; Tanaka and Takada, 2016).

Plastics break down into MPs in a variety of ways, including extended exposure to ultraviolet (UV) light, physical abrasion, biological degradation, and so on (Banerjee and Shelver, 2021; Wan et al., 2019). Furthermore, excessive usage and improper waste management systems are to be blamed for the pervasive contamination by MPs. MPs are introduced into the environment in a number of ways, including domestic sewage, vehicle tire abrasions, fabric fiber, fertilizers, effluent from the wastewater treatment plants, vinyl used in agricultural purposes, litter from ships and recreational events, wind erosion, landfills from urban and industrial areas, and surface currents, to mention a few (Chae and An, 2018; de Souza Machado et al., 2018; Mammo et al., 2020). Nowadays, MPs are manufactured for use in cosmetics, cleansers, paints, drug delivery formulations, and so on, and their ultimate destination is deposition in the environment (Banerjee and Shelver, 2021; Hirt and Body-Malapel, 2020; Lambert and Wagner, 2018).

MPs of various forms are reported, such as beads, spheres, fragments, fibers, nurdles, pellets, foams, and films (Hirt and Body-Malapel, 2020; Qiao et al., 2019) (Fig. 24.1). MPs are divided into two categories based on the sources of origin: primary and secondary (Auta et al., 2017; Da Costa et al., 2019; Hirt and Body-Malapel, 2020; Horton et al., 2017; Lambert and Wagner, 2018; Prakash et al., 2020; Wu et al., 2019). Primary MPs are predominantly manufactured products. They are industrially produced as resin pellets to produce large items or produced directly for cleaning products, cosmetic products (facial scrub), oral products (toothpaste), and so on. (Auta et al., 2017; Hirt and Body-Malapel, 2020; Lambert and Wagner, 2018). Primary MPs are likely to flow into wastewater treatment plants (WWTPs) from industrial or domestic drainage systems. Although WWTPs removes only a portion of MPs, some sewage treatment can remove up to 99.9% of MP particles from wastewater. However, there is still room for improvement in this treatment technology (Hernandez et al., 2017; Horton et al., 2017; Mintenig et al., 2017). These particles might be followed by fibers emitted from the waste of washing machines, which is estimated to consist of 1900 polyacrylic fibers per item. These are secondary MPs that are released into the aquatic and terrestrial environment from washing machines (Browne et al., 2011; Da Costa et al., 2019).

Secondary MPs are formed by the degradation of plastic waste through physical, chemical, and biological processes which include UV radiation, weathering, photolysis, physical abrasion, oxidation, hydrolytic degradation, biodegradation, etc. (Auta et al., 2017; Banerjee and Shelver, 2021; Hirt and Body-Malapel, 2020; Lambert and Wagner, 2018). This degradation leads to different sizes of particles: megaplastics (>50 cm), macroplastics (5–50 cm), mesoplastics (0.5–5 cm), MPs (<5 mm) and nanoplastics (≤ 0.1 μm) (Lebreton et al., 2018) (Fig. 24.1).

24.3 Microplastics pollution and the threat to habitats

The widespread use of MPs has led to serious environmental pollution, which is considered to be one of the prime global environmental concerns, owing to its abundance and persistence in the environment (Deng et al., 2020; Luo et al., 2019a,b; Zhao et al., 2020). Understandably, it is an important research topic in the field of international ecology and environmental science (Luo et al., 2019a,b). These tiny ubiquitous particles have become the leading contributors to the pollution of terrestrial and aquatic ecosystems. Although there are many reports on the impacts of MPs in the marine environment, the terrestrial ecosystem remains mostly unexplored in this regard. Freshwater and terrestrial environments are recognized as origins and transporters for the accumulation of MPs in the ocean, and it is reported that MP contamination on land might be 4- to 23-fold larger than that of the ocean (de Souza Machado et al., 2018; Horton et al., 2017).

24.4 Gut microbiome

The gut microbiota, a diverse assemblage of microorganisms in the gastrointestinal tract (GIT), has been one of the most dynamic research areas in recent years. The GIT is home to a complex and diverse microbiota population that has coevolved with its host over millennia to create a symbiotic relationship and supports its host in a variety of ways (Heintz-Buschart and Wilmes, 2018; Hossain et al., 2015; Schmidt et al., 2018; Wu and Wu, 2012). The GIT consists of an integrated, versatile ecosystem with a robust community of microorganisms known as the "gut microbiome" which has an interface of 250–400 m^2 (Heintz-Buschart and Wilmes, 2018; Hossain et al., 2020; Wu and Wu, 2012). The molecular biologist, Joshua Lederberg outlined "gut microbiome" as the integrality of microorganisms and their collective microbial genome present in the GIT (Cresci and Bawden, 2015).

The gut microbiome is an ecosystem of complex symbiotic, commensal and pathogenic microorganisms living in a particular niche with stable and favorable conditions. The number of genes of the combined microbial genome outnumbers the genes of their hosts by a factor of 150–500 (Cresci and Bawden, 2015; Lane and Yadav, 2020; Lazar et al., 2018). There are at least 10^{14} microorganisms in the GIT, representing over 2000 species and 12 different phyla (Lazar et al., 2018; Shi et al., 2017). *Proteobacteria*, *Firmicutes*, *Actinobacteria*, and *Bacteroidetes* are the four major phyla that account for 93.5% of the overall species in the gut microbiota (Belizário and Napolitano, 2015; Thursby and Juge, 2017). According to 16S rRNA sequencing data, Firmicutes (predominantly gram-positive) and Bacteroidetes (predominantly gram-negative) are the most predominant phyla, representing approximately 92% of the human microbiome, followed by *Actinobacteria*, *Proteobacteria*, and *Verrucomicrobia*, which might vary among individuals (Lobionda et al., 2019; Shi et al., 2017). From the stomach (10^1–10^3 CFU/mL) to the jejunum (10^4–10^7 CFU/mL), and to the colon (10^9–10^{12} CFU/mL), the taxonomic diversity and bacterial load of the microbiota increase steadily. The bacterial diversity also varies depending on the pH of the environment (Hossain et al., 2020; Mailhe et al., 2018) (Fig. 24.2).

A number of international projects are underway to better understand the human microbiota and unfold previously unknown information. The baseline structure and function of the microbiome were established by the contributions of two major collaborative efforts of Human Microbiome Project funded by the National Institute of Health, and Metagenomics of the Human Intestinal Tract financed by the European Commission (Belizário and Napolitano, 2015; Cresci and Bawden, 2015; Schmidt et al., 2018). These two studies accumulated vast amounts of metagenomics data using next-generation sequencing (NGS) on the human microbiome adopting the pan-genome approach, which helped to get insights into the diversity and complexity of the human microbiota. Now, metagenome-wide association studies

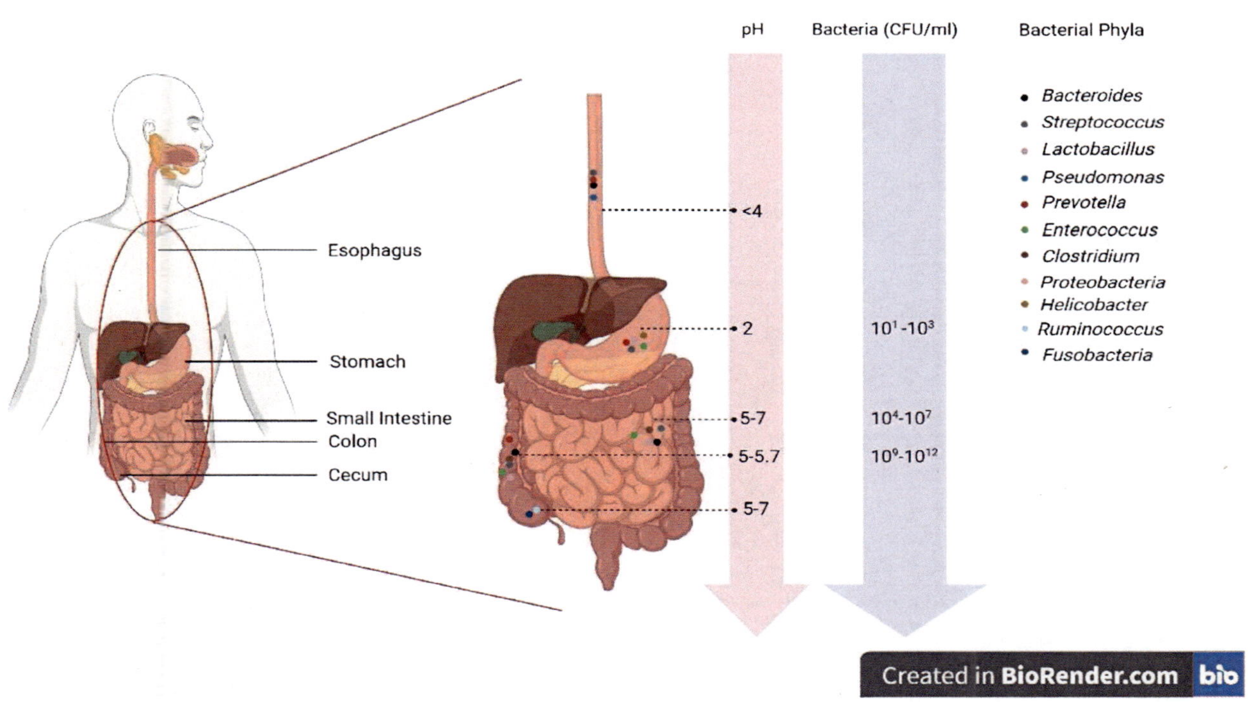

FIGURE 24.2 Human gut microbiome.

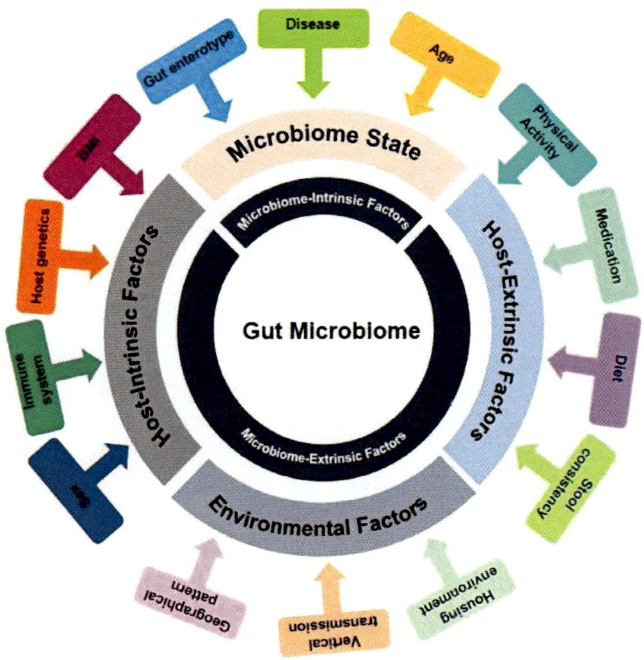

FIGURE 24.3 Factors influencing gut microbiome composition.

in analogy to genome-wide association studies in addition to metabolomic studies are being carried out to provide an integrated view of the relative impact of various host and environmental factors on microbiome composition (Schmidt et al., 2018; Zhu et al., 2020).

As a result of both intrinsic and extrinsic factors, the taxonomic composition of the gut microbiome differs significantly among individuals. Microbiome-extrinsic factors can be divided into three categories, namely, host-intrinsic, host-extrinsic, and environmental (Schmidt et al., 2018) (Fig. 24.3), and influence the composition of the gut microbiota. To verify the changes in microbiota composition, the fecal microbiota of different groups were examined, and three groups were classified in the human gut flora known as enterotypes. Enterotype 1 has a complex carbohydrate-degrading group with Bacteroidetes predominance; enterotype 2 has a proteolytic group with *Prevotella* predominance; and enterotype 3 has a mixed group with *Ruminococcus* predominance (Arumugam et al., 2011).

24.4.1 Dysbiosis and eubiosis

The human gut harbors more than 100 trillion microorganisms, which thrive in its host, maintaining a balanced composition (Cresci and Bawden, 2015; Hossain et al., 2020). The microbiota and the host have a mutually beneficial symbiotic relationship in which the host provides habitat and nutrients for the microbiome and the gut microbiota regulates several important functions, including host metabolism, immunity, and intestinal barrier function (Lobionda et al., 2019; Shi et al., 2017; Hossain et al., 2016).

Human gut microbiomes with high diversity are considered stable, and those with low diversity are considered unstable. Here, the state of dysbiosis or eubiosis plays a significant role in modulating an individual's health. Eubiosis refers to a balanced microbial ecosystem that denotes a healthy microbiome, and dysbiosis represents an environment that is devoid of mutual harmony among bacteria and the host. The latter is an imbalanced situation in the microbiome structure caused by an abnormal ratio of commensal and pathogenic microorganisms (Belizário and Napolitano, 2015; Hossain et al., 2020; Iebba et al., 2016; Lobionda et al., 2019). The unnatural shift of microorganisms, resulting from an increase in the Firmicutes/Bacteroidetes ratio, causes several disorders, such as type 1 diabetes (Hossain et al., 2020; Lazar et al., 2018; Richardson et al., 2018), rheumatoid arthritis (Lazar et al., 2018), obesity (Claus et al., 2016; Hossain et al., 2020; Shi et al., 2017), cardiovascular diseases, hypertension (Shi et al., 2017), cancer (Hossain et al., 2020; Lazar et al., 2018), inflammatory bowel disease (Belizário and Napolitano, 2015; Lazar et al., 2018; Shi et al., 2017), Crohn's disease (Belizário and Napolitano, 2015), autoimmune disease (Lazar et al., 2018), neurological inflammatory disease (Lazar et al., 2018), and so on. Potentially beneficial microbes, mostly from the bacterial phyla Firmicutes and Bacteroidetes, predominate in the eubiotic state of the gut, while potentially pathogenic organisms, such as those from the phyla

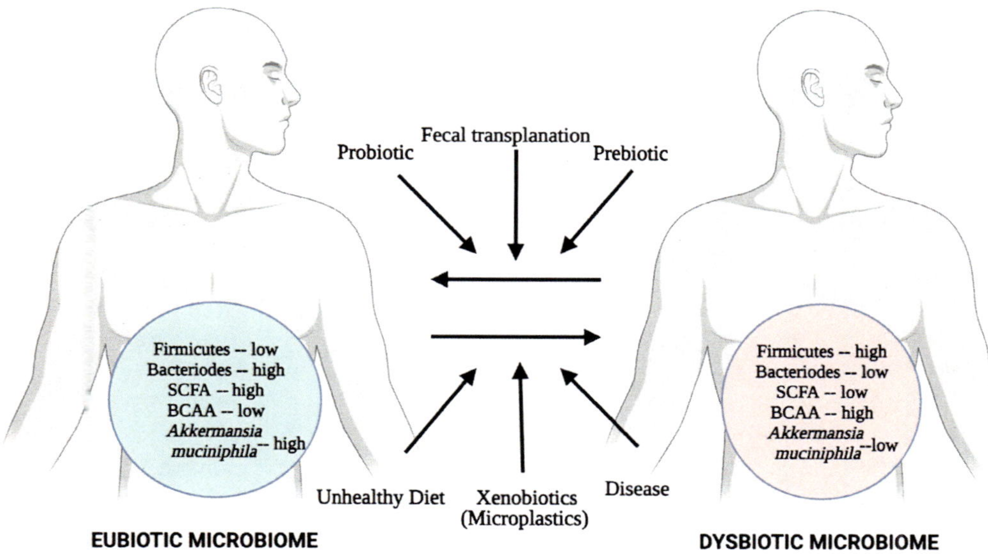

FIGURE 24.4 Factors influencing reversible transition of dysbiotic and eubiotic states of microbiome.

Proteobacteria, are present in small numbers. When dysbiosis occurs, the "good bacteria" lose control and the "bad bacteria" take over (Hossain et al., 2015; Iebba et al., 2016; Zhang et al., 2015). Both germ-free animals and gnotobiotic animal models have been critical in generating valuable data regarding the role of the gut microbiome in different diseases and health conditions (Franklin and Ericsson, 2017; Kennedy et al., 2018). Various factors that mediate the transition from the eubiotic gut microbiome to the dysbiotic gut and vice versa are shown in Fig. 24.4.

24.4.2 Short-chain fatty acids and branched-chain fatty acids

Short-chain fatty acids (SCFAs) are saturated aliphatic organic acids that are 1–6 carbons long (Fig. 24.5), arising from anaerobic intestinal microbial fermentation of dietary fibers, resistant starch, and proteins in the colon that are not digested by humans (Macfarlane and Macfarlane, 2011). Among the many health benefits offered by a healthy gut microbiome, the production of SCFAs ranks high, as these organic acids have both local and systemic effects on the regulation of gene expression. The main SCFAs are acetate (C2), propionate (C3), and butyrate (C4) (Fig. 24.5), which are produced primarily in the proximal and transverse colon at a rate of 150–600 mmol/day, depending on the type of diet. (Cummings et al., 1987, Cummings and Macfarlane, 1991). Approximately 95% of the SCFAs that are produced in the gut are absorbed by the colon, providing approximately 500 kcal of energy per day, and exert multiple beneficial effects on mammalian energy metabolism (Besten et al., 2013). The absorbed SCFAs are used as an energy source by the colonocytes (butyrate being the preferred substrate) and by the other tissues, including the liver and the muscles. In fact, they are the indicator of a healthy gut system because of their elevated production by the eubiotic microbiome (Fig. 24.4) (enterotype 1 as defined in Table 24.1), and destabilization or decreased production of SCFA triggers the gut microbiome toward dysbiosis. Excess consumption of protein and fat (described as a Western diet) is an unhealthy practice. Continued consumption of such a diet leads to a dysbiotic microbiome (enterotype 2 as defined in Table 24.1), which is characterized by two events: enrichment of proteolytic bacterial species (Fig. 24.4) and accumulation of the branched chain amino acids (BCAA) (Fig. 24.6). The former event leads to degradation of gut mucin, resulting in impairment of the gut permeability barrier. The resulting enhanced translocation of bacteria, bacterial products, and bacterial components (e.g., LPS, lipoteichoic acid) into systemic circulation may cause inflammation and eventually metabolic syndrome (Besten et al., 2013). The other event, that is, accumulation of excess BCAA, has been reported to correlate with insulin resistance and development of type 2 diabetes (Yoon, 2016).

24.4.3 Virome and microplastics

Viruses are the most abundant biological entity on earth with an estimated number of 10^{31} particles (Rohwer et al., 2009; Suttle, 2007) and can be found in every ecosystem, including the human body. The totality of viruses present in and on our body is known as the virome. (Güemes et al., 2016). All possible types of viruses are present in gut virome,

SCFAs

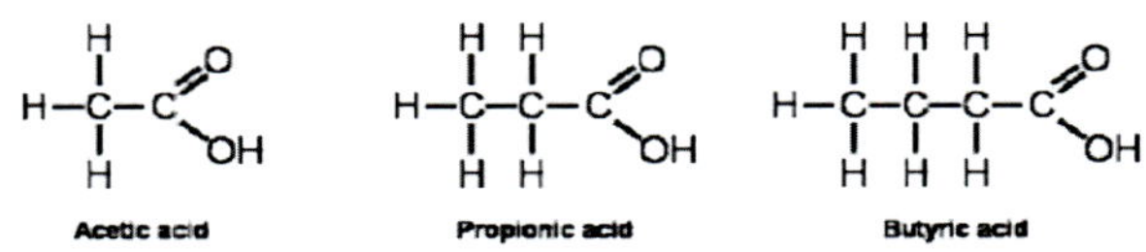

FIGURE 24.5 Short-chain fatty acids.

TABLE 24.1 Terminology.

Akkermansia muciniphila: A mucin-degrading bacterial member of the gut microbiome that is considered to be an indicator bacterium of the eubiotic microbiome.

Biofilm: Community of microorganism attached to a surface (biotic or abiotic) enclosed in a self-produced polymeric material composed of protein, carbohydrates, and nucleic acids. The bacteria in biofilm usually exhibit increased resistance to antibiotics and disinfectants.

Branched chain amino acids (BCAAs): Valine, isoleucine, and leucine.

Dysbiosis: Transition of a healthy microbiome (eubiotic microbiome) to an unhealthy microbiome (dysbiotic microbiome); imbalance in the intestinal microbiota, usually representing a disease state or health condition.

Enterotype: Classification of the human gut microbiome according to its characteristics. Microbiomes of enterotype 1, enterotype 2 and enterotype 3 represents eubiotic, dysbiotic, and mixed-type microbiomes, respectively (Arumugam et al., 2011).

Eubiosis: Transition of an unhealthy, dysbiotic microbiome to a healthy microbiome; the opposite of dysbiosis.

Fecal microbial transplantation (FMT): A procedure that involves the transfer of fecal material from a healthy person with a eubiotic microbiome (presumably containing gut health bacteria) to the gastrointestinal tract of a person with a dysbiotic microbiome with the aim of inducing eubiosis in the recipient.

Firmicutes/Bacteroidetes ratio: Firmicutes and Bacteroidetes are two major phyla of bacteria that are present in the human gut microbiome. Together, they represent 90% of the total microbiome. An increase in the Firmicutes/Bacteroidetes ratio indicates the transition of the gut microbiome toward dysbiosis and has been linked to a variety of disease conditions.

Germ-free animals: Animals that have no microbiome, that is, no microorganisms living in or on them. Such animals are raised in sterile conditions and provided with sterile food. Immediately after birth, various methods are employed to control their exposure to viral, bacterial, or parasitic agents. Such animals are crucial in microbiome studies.

Gnotobiotic animals: Animals in which only known microorganisms are present. Such animals are developed by introducing specific known organisms to germ-free animals. They are very useful in determining the role of different members of the microbiome singly and in combination.

Inflammasome: Cytosolic multiprotein complexes of the innate immune system, which induce activation of the inflammatory responses, resulting in proinflammatory form of programmed cell death, termed pyroptosis.

Metabolic syndrome: A combination of three health conditions: obesity, diabetes, and hypertension (high blood pressure).

Metabolome: Total metabolites found in a tissue or a sample. This profile represents a snapshot of what metabolites are present in a sample at a given time.

Metagenomics: Genomic analyses of DNA extracted directly from a clinical or environmental sample that help to assess community diversity (through 16S rRNA metagenomics), or fuller complexity (through whole-genome shotgun metagenomics).

Microbiome: The combined genetic material of the microorganisms that are present in a particular environment.

Microbiota: Microorganisms such as bacteria, fungi, archaea, protozoa, and viruses that are harbored by normal, healthy individuals.

Next-generation sequencing (NGS): Advanced DNA sequencing technology that enables rapid sequencing of DNA fragment of large sizes (in excess of 1 million base pairs); even the entire human genome (3 billion bases) in a single day.

Pan-genome The collection of the entire set of genes found within a species.

Plasticizer: A substance that is added to a material to make it more flexible and softer, to decrease its viscosity, and to increase its plasticity so that there is less friction in handling during its manufacture.

Plastisphere: Microbiome on microplastics.

Prebiotics: Food ingredients that are nondigestible by humans but promote the growth of beneficial microorganisms (probiotics) in the intestine.

Probiotics: Living microorganisms (bacteria and yeasts) that can be consumed by humans for health benefits.

Short-chain fatty acids (SCFAs): Acetic acid, propionic acid, and butyric acid.

Trophic level: A position in the food chain in an ecosystem. The organisms belonging to a food chain are classified into trophic levels on the basis of their feeding behavior; the first level comprises the producers, the green plants, while the second and third levels contain the herbivores and the carnivores, respectively.

Trophic transfer: The transfer of elements, including contaminants, from one trophic level to another along the food chain.

Virome: The totality of viruses (both prokaryotic and eukaryotic) that are present in and on the body.

Xenobiotic: The term that is used to describe the presence of two types of chemical substances in an organism (1) that are not naturally produced by that organism and (2) that are present in much higher concentration than usual.

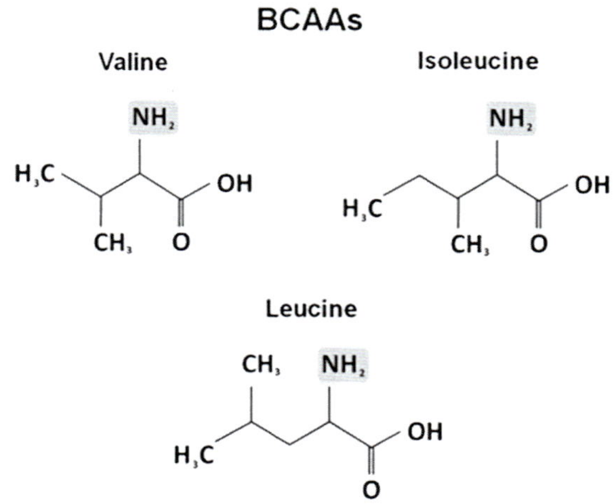

FIGURE 24.6 Branched-chain amino acids.

including bacteriophages infecting members of the gut microbiome, nonpathogenic and pathogenic DNA and RNA viruses infecting eukaryotic host cells, prophages, giant viruses, and a few types of plant viruses (Raghavendra and Pullaiah, 2018). The virome, as an intrinsic component of the gut microbiome, is being analyzed by using NGS in the context of the pan-genome (with the host genome and bacterial genome) for gaining useful insights into the ever-evolving interkingdom interactions (Garmaeva et al., 2019; Wylie et al., 2012). As the gut virome infects members of the gut microbiome, it indirectly influences our health and disease states by modulating the structure and function of the gut microbiota (Mukhopadhya et al., 2019). It can be speculated that the ingested plastisphere is also affected by the gut virome. As MP particles were found to serve as fomites for both pathogenic and nonpathogenic microbes, its potential relevance in the transmission of SARS-CoV-2 is speculated. Virus-laden bioaerosols can adhere to MP particles and thus can be directly transported to airways and via inhalations. Moreover, such MPs can serve as a vehicle for long-distance travel of the virus particles (VishnuRadhan et al., 2020).

24.4.4 Akkermansia muciniphila

Akkermansia muciniphila is considered an indicator organism of the eubiotic microbiome. It is an obligate anaerobic, nonmotile, non-spore-forming, gram-negative bacteria colonizing the intestinal mucosal layers of humans and other animals. It has the unique ability to grow by utilizing mucin as its sole nutritional source (both nitrogen and carbon source). (Derrien et al., 2004; de Vos, 2017; Liu et al., 2021a,b). One of the most abundant bacteria in the healthy gut, *A. muciniphila* represents 3%–5% of the total bacterial population. Studies on the probiotic characteristics of this microbiome member have demonstrated that it is linked with a variety of diseases and health conditions (generalized inflammation, type 2 diabetes, obesity, ulcerative colitis, and many others) in a positive way; that is, a decrease in its abundance led to increasing disease burden (Caesar et al., 2015; Earley et al., 2019; Everard et al., 2013; Liu et al., 2021a,b). In preclinical studies, *A. muciniphila* has been reported to respond to immunotherapy for cancer. Considered to be the next-generation probiotic (Naito et al., 2018; Zhang, et al., 2019), this gut microbiome member can target specific diseases and hence can be used as a revolutionary class of medicines for therapeutic purposes. *A. muciniphila* is commercially available as an over-the-counter health supplement, marketed by major health supplement brands.

24.5 Toxic effects of microplastics on human systems

There are different ways in which MPs find their way into the human body. The three best-studied portals of entry are ingestion of contaminated food, inhalation of MP particles present in the atmosphere, and direct skin contact (Rahman et al., 2021) (Fig. 24.7). The most common entry point is through ingestion of contaminated food, which includes marine species, plants, salts, sugar, and bottled water. (Waring et al., 2018). Most undigested MP particles are defecated in large quantities, but a fraction of these particles enters the bloodstream. The ingested MPs first encounter the intestinal epithelium. The particles that are of irregular shape and carry higher concentrations of environmental debris and toxicants are more likely to cause inflammation at the intestinal lining (Wright and Kelly, 2017). However, the effect of the presence of MPs, which are not efficiently eliminated by the gut, remains unknown (Yee et al., 2021).

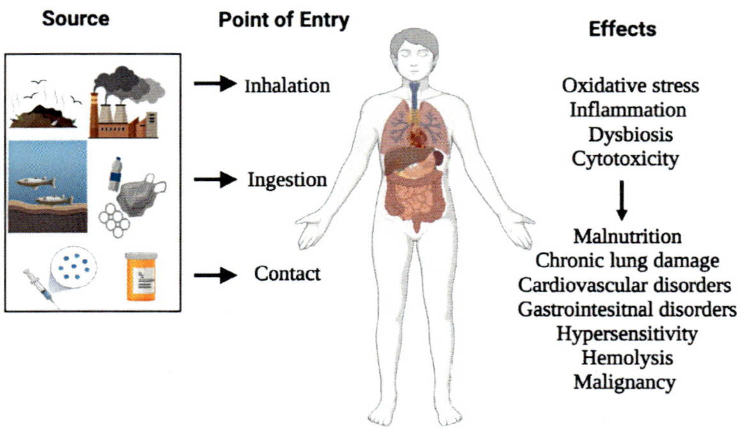

FIGURE 24.7 Microplastics and human health.

Intestinal diseases caused by MPs have been demonstrated in fish and mouse models (Deng et al., 2018; Mattsson et al., 2017). Experiments with rats have provided some clear illustrations of the consequences of MP toxicity. MPs can cause damage to the endovascular septum and enter the liver through the portal vein (Deng et al., 2017, 2018). Prolonged accumulation of MPs in liver tissues and the resultant chronic inflammation can lead to metabolic syndrome (Yee et al., 2021; Yong et al., 2020).

24.5.1 Microplastic modulation of cellular pathways

Mechanisms underlying MP-induced chronic or acute toxicity are not clear. However, it has been observed that high-level accumulation of MPs induces a cytotoxic effect (Campanale et al., 2020). Cell death could result via necrotic cell membrane burst or apoptosis. At high concentrations, MPs are usually disruptive to plasma membranes, while at a moderate level, they can damage vital cellular and extracellular components such as proteoglycans, extracellular ligands, and membrane receptors. Destabilization of cell physiology ensues (Liu et al., 2021a,b; Wang et al., 2020).

MP particles can enter cells, depending upon the type of cell, through endocytosis (Oh and Park, 2014). Endocytosed MPs create major issues for many reasons. First, MPs at high concentrations damage cell membrane permeability. Second, once inside the cell, MPs could possibly interact with vital organelles of the cytoplasm, for instance, the nucleus or mitochondria, and interfere with cellular mechanisms such as the cell division process, mitotic spindle development, chromosomal relocation, and energy generation. Third, MPs can potentially interfere with the carrier proteins that are involved in membrane trafficking (Horstmann et al., 2002; Liu et al., 2021a,b; Treyer et al., 2016), thereby affecting the cellular transportation process across the cell membrane. Fourth, MPs can disturb endosomal membrane passage, for instance, membrane protein turnover and signaling pathways. It may be mentioned that if the MP particles could end up in the lysosome, they are not likely to be digested by lysosomal enzymes (Wang et al., 2020; Yee et al., 2021). The accumulation of MPs in lysosomes or endosomes would disturb the degradation process of organelles and notably damage the important cell membrane turnover course of macroautophagy (Lim et al., 2019). This could initiate a positive feedback mechanism that eventually terminates in programmed cell death. Even moderate levels of MPs could influence autophagy. At higher concentrations, MPs could create cellular stress (Yee et al., 2021), for instance, damaging the plasma membrane. The endolysosomes would trigger further stress responses. Earlier studies in animal models revealed that exposure to MPs could affect cellular development and reproduction (Besseling et al., 2014), and intriguingly, they elevate 5' adenosine monophosphate (AMP) and 5'-AMP-activated protein kinase, leading to a cellular stress response (Liu et al., 2018).

24.5.2 Microplastics-induced reactive oxygen species

The production of reactive oxygen species (ROS) is a common phenomenon related to the cellular stress response, and it was recently identified as a molecular initiating event (Jeong and Choi, 2019). ROS are generated from diverse sources, including mitochondrial respiratory chain, enzymatic activation of cytochrome p450, and nicotinamide adenine dinucleotide phosphate oxidases (Bedard and Krause, 2007; Kany et al., 2019). Increased mitochondrial ROS can lead to functional impairment of cells. Every cell contains an evolutionarily protected innate immune system, which usually acts against the invasion of pathogens or exposure to vital xenobiotic molecules (Yee et al., 2021; Yong et al., 2020).

Nevertheless, cells may respond to a group of endogenous or secretory molecules known as toll-like receptors or to signaling induced by damage-related molecular patterns derived from degrading host cell components. This is called inflammatory inflammation, that is, inflammatory reactions in the absence of pathogenic infection (De Lorenzo et al., 2018; Shen et al., 2013). Inside the human body the cytokines that are released from this local inflammation can attract immune cells into the circulation, which can worsen the local inflammation and lead to cell and tissue death (Kany et al., 2019; Yee et al., 2021).

The accumulation of MPs in lung cells can lead to chronic lung damage involving the inflammasome pathway, leading to pyroptosis. Long-term exposures to MPs, especially occupational exposures, have been found to cause pulmonary fibrosis and granulomatous reactions (Lim et al., 2019; Liu et al., 2021a,b). Chronic inflammation and lung damage due to inhalation of MPs have been linked to the overproduction of ROS, which can induce cell proliferation and secondary genotoxicity (Amato-Lourenço et al., 2020). The formation of malignant lesions can be explained because of an interplay involving dust overload, which can impair the clearance mechanism and induce oxidative stress and cytotoxicity (Prata, 2018).

24.5.3 Accumulation of microplastic particles in different tissues

Translocation is the process by which inhaled MPs are deposited in distant organs and the bloodstream. A study in a human placental perfusion model demonstrated the translocation of polystyrene particles for transplacental transfer (Wick et al., 2010). The presence of MPs in brain cells was also documented in fishes (Mattsson et al., 2017). However, tissue accumulation, diseases, and metabolic disturbances caused by MPs have not yet been clearly demonstrated in humans.

MP-induced intestinal dysbiosis was evidenced in a mouse model (Luo et al., 2019a,b). Here, changes in the gut microbiome caused digestive system dysfunction, eventually upsetting the physiological homeostasis. Importantly, alterations of the gut microbiota have been associated with chronic diseases of other organs, including kidney disease (Plata et al., 2019) neurological disorders (Limonta et al., 2019), cardiovascular disorders (Jin et al., 2019a,b), and even cancer (Francescone et al., 2014). Albumin and globulin blood proteins have also been reported to interact with MPs to form plastic-protein complexes (Mohr et al., 2014). If there are large quantities of such complex formations, the blood vessels may become blocked. Nevertheless, it is complicated to imagine a significant accumulation of MPs in the human bloodstream under natural conditions (Yong et al., 2020). The presence of polypropylene particles exhibited different and harmful effects on different cell lines, depending on their sizes (25−200 μm) and the densities used. Therefore the presence of such particles in the human system can cause cytotoxicity, hypersensitivity, unwanted immune responses, and acute reactions, such as hemolysis, thus posing a risk to human health (Campanale et al., 2020). As was mentioned earlier in the chapter, we are yet to pinpoint our understanding of the effects of MPs on human health; nevertheless, it has become apparent that physical and chemical properties of the particles, biofilm growth on MPs, concentration, and duration of exposure (Lim et al., 2019; Limonta et al., 2019) are the likely predisposing factors of MP-induced toxicity.

24.6 Plastisphere: biofilm on microplastics

Biofilm is a community of microorganisms attached to a surface (biotic or abiotic, both natural and artificial) at the solid-liquid interface enclosed in a self-produced highly hydrated extracellular polymeric material (EPM) composed of protein, carbohydrates, and nucleic acids. Bacterial biofilms have been found in every sphere of the environment. The microbial cells residing inside the EPM exist in a quiescent state (Jhuma et al., 2021) and are protected from relatively higher concentrations of antibiotics and disinfectants or heavy metals often in the range of 10X to 100X in comparison to their free-floating counterparts (Mah, 2012, Sultana et al., 2020). Although the transfer of antibiotic resistance genes (ARGs) is usually considered to occur in clinical settings, the exchange of genetic materials among pathogens in natural settings is increasingly being recognized as a matter of deep concern. Unlike chemical pollutants, ARGs can be multiplied and transferred among the bacterial population and can persist in the presence of very low or no selective pressure. MPs can serve as an ideal platform for biofilm formation by the adhered bacteria, and thereby facilitate transfer and amplification of ARGs, which can be carried to different geographic locales by flowing water (Abe et al., 2020; Gorokhova et al., 2015). Rapid aggregation of biofilm-covered MPs with marine biogenic particles and spatial and seasonal variation in diversity and structure of microbial biofilms on marine plastics were observed in Northern European waters (Besemer, 2015; Oberbeckmann et al., 2014). Plastic materials such as polyethylene and polyethylene terephthalate serve as excellent surfaces for colonization and biofilm formation by different types of bacteria, where they survive longer than nonadhered bacteria (Lobelle and Cunliffe, 2011; Oberbeckmann, et al., 2016; Webb et al., 2009).

24.7 Microplastics as pathogen carriers

The carriage of bacteria on MPs has emerged as an important area of investigation, as MPs serve as vehicles of transmission of pathogens that can be taken up by humans through inhalation, infection, or dermal contact. Bodies of water are another means of MP-mediated bacterial transmission, including potential pathogens, demonstrated by metagenomics and bioinformatics studies (Brandt and House, 2016; Oberbeckmann et al., 2016). They persist in water much longer than other floating natural substances, such as wood and feathers. The buoyancy of MPs helps them to float, providing a liquid-solid interface, which is ideal for the formation of biofilm. Pathogens such as *Vibrio cholerae* and *V. fluvialis*, which cause gastroenteritis, were found to be carried by MPs, highlighting the fact that pathogens with epidemic and even pandemic potential can be dispersed by plastic particles floating in water bodies (Bowley et al., 2021; Rogers et al., 2020).

24.8 Conclusion and perspectives

The microbiome influences our health and well-being in many different ways, and we are witnessing remarkable progress in delineating its role in various disease processes. Realization of the health significance of eubiotic and dysbiotic states of the gut microbiota and the factors influencing reversible transition between them has prompted scientists to explore ways to engineer the microbiome for our health benefit at both generalized and personalized levels. Because of their ubiquitous presence and their apparently unavoidable ingestion, or trophic-level transfer, MP particles are now recognized as an important environmental pollutant with a harmful effect on shaping the microbiome composition of all living animals. At present, the major concerns are the acute toxicity or long-term exposure in the human system, which might lead to significant increases in mortality and morbidity. Nevertheless, it is not clearly understood how MPs from the environment, seafood, or bottled water affect human health. There are also many unanswered questions, such as the following: What is the critical dose of MP particles needed to initiate and sustain the process of dysbiosis? How does MP-induced dysbiosis differ from dysbiosis induced by a fat-rich-diet or dysbiosis induced by antibiotics? What microbial taxa are enriched or decreased when the MP-induced dysbiosis sets in? Further studies are needed to answer these clinically relevant questions. Collaborative efforts are needed among scientists from different disciplines to evaluate the bioaccumulation of MPs in humans across different geographic locations.

References

Abe, K., Nomura, N., Suzuki, S., 2020. Biofilms: hot spots of horizontal gene transfer (HGT) in aquatic environments, with a focus on a new HGT mechanism. FEMS Microbiol. Ecol. 96. Available from: https://doi.org/10.1093/femsec/fiaa031.

Akdogan, Z., Guven, B., 2019. Microplastics in the environment: a critical review of current understanding and identification of future research needs. Environ. Pollut. 254, 113011.

Amato-Lourenço, L.F., Galvao, L.D.S., deWeger, L.A., Hiemstra, P.S., Vijver, M.G., Mauad, T., 2020. An emerging class of air pollutants: potential effect of microplastics to respiratory human health. Sci. Total. Environ. 749, 141676.

Arumugam, M., Raes, J., Pelletier, E., et al., 2011. Enterotypes of the human gut microbiome. Nature. 473, 174–180.

Auta, H.S., Emenike, C.U., Fauziah, S.H., 2017. Distribution and importance of microplastics in the marine environment: a review of the sources, fate, effects, and potential solutions. Environ. Int. 102, 165–176. Available from: https://doi.org/10.1016/j.envint.2017.02.013.

Banerjee, A., Shelver, W.L., 2021. Micro- and nanoplastic induced cellular toxicity in mammals: a review. Sci. Total. Environ. 755, 142518. Available from: https://doi.org/10.1016/j.scitotenv.2020.142518.

Bedard, K., Krause, K.-H., 2007. The NOX family of ROS-generating NADPH oxidases: physiology and pathophysiology. Physiol. Rev. 87, 245–313.

Belizário, J.E., Napolitano, M., 2015. Human microbiomes and their roles in dysbiosis, common diseases, and novel therapeutic approaches. Front. Microbiol 6, 1–16. Available from: https://doi.org/10.3389/fmicb.2015.01050.

Besemer, K., 2015. Biodiversity, community structure and function of biofilms in stream ecosystems. Res. Microbiol 166, 774–781. Available from: https://doi.org/10.1016/j.resmic.2015.05.006.

Besseling, E., Wang, B., Lürling, M., Koelmans, A.A., 2014. Nanoplastic affects growth of *S. obliquus* and reproduction of *D. magna*. Environ. Sci. Technol. 48, 12336–12343.

Besten, G., Eunen, K., Groen, A.K., Venema, K., Reijngoud, D., Bakker, B.M., 2013. The role of short-chain fatty acids in the interplay between diet, gut microbiota, and host energy metabolism. J. Lipid Res. 54, 2325–2340. Available from: https://doi.org/10.1194/jlr.R036012.

Bowley, J., Baker-Austin, C., Porter, A., Hartnell, R., Lewis, C., 2021. Oceanic hitchhikers - assessing pathogen risks from marine microplastic. Trends Microbiol 29, 107–116.

Boyle, K., Örmeci, B., 2020. Microplastics and nanoplastics in the freshwater and terrestrial environment: a review. Water. 12, 2633. Available from: https://doi.org/10.3390/w12092633.

Brandt, L.D., House, C.H., 2016. Marine subsurface microbial community shifts across a hydrothermal gradient in Okinawa trough sediments. Archaea . Available from: https://doi.org/10.1155/2016/2690329.

Browne, M.A., Crump, P., Niven, S.J., Teuten, E., Tonkin, A., Galloway, T., et al., 2011. Accumulation of microplastic on shorelines woldwide: sources and sinks. Environ. Sci. Technol. 45, 9175–9179.

Caesar, R., Tremaroli, V., Kovatcheva-Datchary, P., 2015. Crosstalk between gut microbiota and dietary lipids aggravates WAT inflammation through TLR signaling. Cell Metab. 22, 658–668.

Campanale, C., Massarell , C., Savino, I., Locaputo, V., Uricchio, V.F., 2020. A detailed review study on potential effects of microplastics and additives of concern on human health. Int. J. Environ. Res. Public. Health 17, 1212.

Chae, Y., An, Y.J., 2018. Current research trends on plastic pollution and ecological impacts on the soil ecosystem: a review. Environ. Pollut. 240, 387–395. Available from: https://doi.org/10.1016/j.envpol.2018.05.008.

Claus, S.P., Guillou, H., Ellero-Simatos, S., 2016. The gut microbiota: a major player in the toxicity of environmental pollutants? Biofilms Microbiomes 2, 1–12. Available from: https://doi.org/10.1038/npjbiofilms.2016.3.

Cresci, G.A., Bawden, E., 2015. Gut microbiome: what we do and don't know. Nutr. Clin. Pract.: Off. Publ. Am. Soc. Parenter. Enter. Nutr. 30, 734–746. Available from: https://doi.org/10.1177/0884533615609899.

Cummings, J.H., Macfarlane, G.T., 1991. The control and consequences of bacterial fermentation in the human colon. J. Appl. Bacteriol. 70, 443–459.

Cummings, J.H., Pomare, E.W., Branch, W.J., Naylor, C.P., Macfarlane, G.T., 1987. Short chain fatty acids in human large intestine, portal, hepatic and venous blood. Gut. 28, 1221–1227.

Da Costa, J.P., Paço, A., Santos, P.S.M., Duarte, A.C., Rocha-Santos, T., 2019. Microplastics in soils: assessment, analytics and risks. Environ. Chem. 16, 18–30. Available from: https://doi.org/10.1071/EN18150.

De Lorenzo, G., Ferrari, S., Cervone, F., Okun, E., 2018. Extracellular DAMPs in plants and mammals: immunity, tissue damage and repair. Trends Immunol 39, 937–950.

de Souza Machado, A.A., Kloas, W., Zarfl, C., Hempel, S., Rillig, M.C., 2018. Microplastics as an emerging threat to terrestrial ecosystems. Glob. Change Biol. 24, 1405–1416.

de Vos, W.M., 2017. Microbe profile: *Akkermansia muciniphila*: a conserved intestinal symbiont that acts as the gatekeeper of our mucosa. Microbiology 1635, 646–648.

Deng, Y., Zhang, Y., Lemos, B., Ren, H., 2017. Tissue accumulation of microplastics in mice and biomarker responses suggest widespread health risks of exposure. Sci. Rep. 7, 1–10.

Deng, Y., Zhang, Y., Qiao, R., Bonilla, M.M., Yang, X., Ren, H., et al., 2018. Evidence that microplastics aggravate the toxicity of organophosphorus flame retardants in mice (Mus musculus). J. Hazard. Mater. 357, 348–354.

Deng, Y., Yan, Z., Shen R., et al., 2020. Microplastics release phthalate esters and cause aggravated adverse effects in the mouse gut. Environ. Int. 143, 105916. Available from: https://doi.org/10.1016/j.envint.2020.105916.

Derrien, M., Vaughan, E.E., Plugge, C.M., de Vos, W.M., 2004. *Akkermansia muciniphila* gen. nov., sp. nov., a human intestinal mucin-degrading bacterium. Int. J. Syst. Evolut. Microbiol 54, 1469–1476.

Earley, H., Lennon, G., Balfe, A., et al., 2019. The abundance of *Akkermansia muciniphila* and its relationship with sulphated colonic mucins in health and ulcerative colitis. Sci. Report. 9, 15683. Available from: https://doi.org/10.1038/s41598-019-51878-3.

Everard, A., Belzer, C., Geurts, L., et al., 2013. Cross-talk between *Akkermansia muciniphila* and intestinal epithelium controls diet-induced obesity. Proc. Natl Acad. Sci. U. S. A 110, 9066–9071.

Francescone, R., Hou, V , Grivennikov, S.I., 2014. Microbiome, inflammation and cancer. Cancer J. 20, 181–189.

Franklin, C.L., Ericsson, A.C., 2017. Microbiota and reproducibility of rodent models. Lab. Anim. 46, 114–122.

Galloway, T.S., Cole, M , Lewis, C., 2017. Interactions of microplastic debris throughout the marine ecosystem. Nat. Ecol. Evol 1, 1–8.

Garmaeva, S., Sinha, T. Kurilshikov, A., et al., 2019. Studying the gut virome in the metagenomic era: challenges and perspectives. BMC Biol. 17, 84. Available from: https://doi.org/10.1186/s12915-019-0704-y.

Geyer, R., Jambeck, J.R., Law, K.L., 2017. Production, use, and fate of all plastics ever made. Sci. Adv. 3, 25–29. Available from: https://doi.org/10.1126/sciadv.1700782.

Gorokhova, E, Asa, M., El-Shehawy Rehab, E., 2021. Understanding biofilm formation in ecotoxicological assays with natural and anthropogenic particulates. Front. Microbiol. 12, 632947. 10.3389/fmicb.2021.632947.

Güemes, C.A.G., Youle, M., Cantu, V.A., Felts, B., Nulton, J., Rohwer, F., 2016. Viruses as winners the game of life. Annu. Rev. Virol. 3, 197–214.

Heintz-Buschart, A., W lmes, P., 2018. Human gut microbiome: function matters. Trends Microbiol 26, 563–574. Available from: https://doi.org/10.1016/j.tim.2017. 1.002.

Hernandez, E., Nowack B., Mitrano, D.M., 2017. Polyester textiles as a source of microplastics from households: a mechanistic study to understand microfiber release during washing. Environ. Sci. Technol. 51, 7036–7046. Available from: https://doi.org/10.1021/acs.est.7b01750.

Hirt, N., Body-Malapel, M., 2020. Immunotoxicity and intestinal effects of nano- and microplastics: a review of the literature. Part. Fibre Toxicol. 17, 1–22. Available from: https://doi.org/10.1186/s12989-020-00387-7.

Horstmann, H., Ng, C.P., Tang, B.L., Hong, W., 2002. Ultrastructural characterization of endoplasmic reticulum—Golgi transport containers (EGTC). J. Cell Sci. 115, 4263–4273.

Horton, A.A., Walton, A., Spurgeon, D.J., Lahive, E., Svendsen, C., 2017. Microplastics in freshwater and terrestrial environments: evaluating the current understanding to identify the knowledge gaps and future research priorities. Sci. Total. Environ. 586, 127–141. Available from: https://doi.org/10.1016/j.scitotenv.2017.01.190.

Hossain, A., Akhter, S., Kabir, Y., 2015. Diet, microbiome and human health. In: Tamang, J.P. (Ed.), Health Benefits of Fermented Foods and Beverages. Taylor & Francis Publications, Boca Raton, Florida, pp. 198–220.

Hossain, A., Menezes, G., Al-Mogbel, M., Ashankyty, I., 2016. Modulation of environmental toxicants and therapeutic agents by gut microbiome. In: Bagchi, D. (Ed.), Food Toxicology. Taylor & Francis Publications Boca Raton, Florida, pp. 492–518.

Hossain, A., Karim, M.M., Jhuma, T.A., Menezes, G., 2020. Gut microbiome and heavy metals. In: Bagchi, D., Bagchi, M. (Eds.), Metal Toxicology Handbook. CRC Press. Taylor & Francis Group, Boca Raton, Florida, USA, pp. 81–98.

Hwang, J., Choi, D., Han, S., et al., 2020. Potential toxicity of polystyrene microplastic particles. Sci. Rep. 10, 7391. Available from: https://doi.org/10.1038/s41598-020-64464-9.

Iebba, V., Totino, V., Gagliardi, A., et al., 2016. Eubiosis and dysbiosis: the two sides of the microbiota. N. Microbiol 39, 1–12.

Jeong, J., Choi, J., 2019. Adverse outcome pathways potentially related to hazard identification of microplastics based on toxicity mechanisms. Chemosphere. 231, 249–255.

Jhuma, T.A., Rafeya, J., Sultana, S., Rahman, M.T., Karim, M.M., 2021. Isolation of endophytic salt-tolerant plant growth-promoting rhizobacteria from *Oryza sativa* and evaluation of their plant growth-promoting traits under salinity stress condition. Front. Sustain. Food Syst. 5, 250. Available from: https://www.frontiersin.org/articles/10.3389/fsufs.2021.687531/full.

Jin, M., Qian, Z., Yin, J., Xu, W., Zhou, X., 2019a. The role of intestinal microbiota in cardiovascular disease. J. Cell. Mol. Med. 23, 2343–2350.

Jin, Y., Lu, L., Tu, W., Luo, T., Fu, Z., 2019b. Impacts of polystyrene microplastic on the gut barrier, microbiota and metabolism of mice. Sci. Total. Environ. 649, 308–317. Available from: https://doi.org/10.1016/j.scitotenv.2018.08.353.

Joo, S.H., Liang, Y., Kim, M., Byun, J., Choi, H., 2021. Microplastics with adsorbed contaminants: mechanisms and treatment. Environ. Chall. 3 (100042), 100042.

Kany, S., Vollrath, J.T., Relja, B., 2019. Cytokines in Inflammatory Disease. Int. J. Mol. Sci. 20, 008. Available from: https://doi.org/10.3390/ijms20236008.

Kennedy, E.A., King, K.Y., Baldridge, M.T., 2018. Mouse Microbiota models: comparing germ-free mice and antibiotics treatment as tools for modifying gut bacteria. Front. Physiol. 9, 1534.

Lambert, S., Wagner, M., 2018. Microplastics are contaminants of emerging concern in freshwater environments: an overview, The Handbook of Environmental Chemistry, 58. *Springer International Publishing, Cham*, pp. 1–23, 10.1007/978-3-319–61615-5_1.

Lambert, S., Sinclair, C., Boxall, A., 2014. Occurrence, degradation, and effect of polymer-based materials in the environment. Rev. Environ. Contamin. Toxicol. 227, 1–53. Available from: https://doi.org/10.1007/978-3-319-01327-5_1.

Lane, M., Yadav, V., 2020. 199-Multiple sclerosis, Textbook of Natural Medicine, 1587–1599. Elsevier, p. e3. Available from: https://doi.org/10.1016/B978-0-323-43044-9.00199-0.

Lazar, V., Ditu, L.M., Pircalabioru, G.G., et al., 2018. Aspects of gut microbiota and immune system interactions in infectious diseases, immunopathology, and cancer. Front. Immunol 9, 1–18. Available from: https://doi.org/10.3389/fimmu.2018.01830.

Lebreton, L., Slat, B., Ferrari, F., et al., 2018. Evidence that the great pacific garbage patch is rapidly accumulating plastic. Sci. Rep. 8, 1–15.

Lim, S.L., Ng, C.T., Zou, L., et al., 2019. Targeted metabolomics reveals differential biological effects of nanoplastics and nanoZnO in human lung cells. Nanotoxicology. 13, 1117–1132.

Limonta, G., Mancia, A., Benkhalqui, A., Bertolucci, C., Abelli, L., Fossi, M.C., et al., 2019. Microplastics induce transcriptional changes, immune response and behavioral alterations in adult zebrafish. Sci. Rep. 9, 1–11.

Liu, Z., Cai, M., Yu, P., Chen, M., Wu, D., Zhang, M., et al., 2018. Age-dependent survival, stress defense, and AMPK in *Daphnia pulex* after short-term exposure to a polystyrene nanoplastic. Aquat. Toxicol. 204, 1–8.

Liu, X., Zhao, F., Liu, H., Xie, Y., Zhao, D., Li, C., 2021a. Transcriptomics and metabolomics reveal the adaption of *Akkermansia muciniphila* to high mucin by regulating energy homeostasis. Sci. Rep. 11. Available from: https://doi.org/10.1038/s41598-021-88397-z.

Liu, Z., Hu, G., Li, Z., et al., 2021b. Study of growth, metabolism, and morphology of *Akkermansia muciniphila* with an in vitro advanced bionic intestinal reactor. BMC Microbiol 21, 61.

Lobelle, D., Cunliffe, M., 2011. Early microbial biofilm formation on marine plastic 208 debris. Mar. Pollut. Bull. 62, 197–200.

Lobionda, S., Sittipo, P., Kwon, H.Y., Lee, Y.K., 2019. The role of gut microbiota in intestinal inflammation with respect to diet and extrinsic stressors. Microorganisms 7, 271. Available from: https://doi.org/10.3390/microorganisms7080271.

Lu, L., Wan, Z., Luo, T., Fu, Z., Jin, Y., 2018. Polystyrene microplastics induce gut microbiota dysbiosis and hepatic lipid metabolism disorder in mice. Sci. Total. Environ. 631–632, 449–458. Available from: https://doi.org/10.1016/j.scitotenv.2018.03.051.

Luo, T., Wang, C., Pan, Z., Jin, C., Fu, Z., Jin, Y., 2019a. Maternal polystyrene microplastic exposure during gestation and lactation altered metabolic homeostasis in the dams and their F1 and F2 offspring. Environ. Sci. Technol. 53, 10978–10992.

Luo, T., Zhang, Y., Wang, C., et al., 2019b. Maternal exposure to different sizes of polystyrene microplastics during gestation causes metabolic disorders in their offspring. Environ. Pollut. 255, 113122. Available from: https://doi.org/10.1016/j.envpol.2019.113122.

Macfarlane, G.T., Macfarlane, S., 2011. Fermentation in the human large intestine its physiologic consequences and the potential contribution. J. Clin. Gastroenterol. 45, S120–S127.

Mah, T.-F., 2012. Biofilm-specific antibiotic resistance. Future Microbiol 7, 1061–1072.

Mailhe, M., Ricaboni, D., Vitton, V., et al., 2018. Repertoire of the gut microbiota from stomach to colon using culturomics and next-generation sequencing. BMC Microbiol 18, 1–11. Available from: https://doi.org/10.1186/s12866-018-1304-7.

Mammo, F.K., Amoah, I.D., Gani, K.M., et al., 2020. Microplastics in the environment: interactions with microbes and chemical contaminants. Sci. Total. Environ. 743, 140518.

Mattsson, K., Johnson, E.V., Malmendal, A., Linse, S., Hansson, L.-A., Cedervall, T., 2017. Brain damage and behavioral disorders in fish induced by plastic nanoparticles delivered through the food chain. Sci. Rep. 7, 1–7.

Mintenig, S.M., Int-Veen, I., Löder, M.G.J., Primpke, S., Gerdts, G., 2017. Identification of microplastic in effluents of waste water treatment plants using focal plane array-based micro-Fourier-transform infrared imaging. Water Res. 108, 365–372. Available from: https://doi.org/10.1016/j.watres.2016.11.015.

Mohr, K., Sommer, M., Baier, G., et al., 2014. Aggregation behavior of polystyrene-nanoparticles in human blood serum and its impact on the in vivo distribution in mice. J. Nanomed. Nanotechnol. 5 (2). Available from: https://doi.org/10.4172/2157-7439.1000193.

Mukhopadhya, I., Segal, J.P., Carding, S.R., 2019. The gut virome: the 'missing link' between gut bacteria and host immunity? Therapeutic Adv. Gastroenterol. 12, 1756284819836620.

Naito, Y., Uchiyama, K., Takagi, T., 2018. A next-generation beneficial microbe: *Akkermansia muciniphila*. J. Clin. Biochem. Nutr. 63, 33–35.

Oberbeckmann, S., Loeder, M.G., Gerdts, G., Osborn, A.M., 2014. Spatial and seasonal variation in diversity and structure of microbial biofilms on marine plastics in Northern European waters. FEMS Microbiol. Ecol. 90, 478–492. Available from: https://doi.org/10.1111/1574-6941.12409.

Oberbeckmann, S., Osborn, A.M., Duhaime, M.B., 2016. Microbes on a bottle: substrate, season and geography influence community composition of microbes colonizing marine plastic debris. PLoS One 11 (8), e0159289. Available from: https://doi.org/10.1371/journal.pone.0159289.

Oh, N., Park, J.-H., 2014. Endocytosis and exocytosis of nanoparticles in mammalian cells. Int. J. Nanomed. 9, 51–63.

Paul, M.B., Stock, V., Cara-Carmona, J., et al., 2020. Micro and nanoplastics-current state of knowledge with the focus on oral uptake and toxicity. Nanoscale Adv. 2, 4350–4367. Available from: https://doi.org/10.1039/d0na00539h.

Plata, C., Cruz, C., Cervantes, L.G., Ramírez, V., 2019. The gut microbiota and its relationship with chronic kidney disease. Int. Urol. Nephrol 51, 2209–2226.

Prakash, V., Dwivedi, S. Gautam, K., Seth, M., Anbumani, S., 2020. Occurrence and ecotoxicological effects of microplastics on aquatic and terrestrial ecosystems. Handb. Environ. Chem. 95, 223–243. Available from: https://doi.org/10.1007/698_2020_456.

Prata, J.C., 2018. Airborne microplastics: consequences to human health? Environ. Pollut. 234, 115–126.

Qiao, R., Deng, Y., Zhang, S., et al., 2019. Accumulation of different shapes of microplastics initiates intestinal injury and gut microbiota dysbiosis in the gut of zebrafish. Chemosphere 236, 124334. Available from: https://doi.org/10.1016/j.chemosphere.2019.07.065.

Raghavendra, P., Pullaiah, T., 2018. Pathogen identification using novel sequencing methods. Advances in Cell and Molecular Diagnostics. Elsevier, pp. 161–202.

Rahman, A., Sarker, A., Yadav, O.P., Achari, G., Slobodnik, J., 2021. Potential human health risks due to environmental exposure to nano- and microplastics and knowledge gaps: a scoping review. Sci. Total. Environ. 757, 143872. Available from: https://doi.org/10.1016/j.scitotenv.2020.143872.

Richardson, J.B., Dancy, B.C.R., Horton, C.L., et al., 2018. Exposure to toxic metals triggers unique responses from the rat gut microbiota. Sci. Rep. 8, 1–12. Available from: https://doi.org/10.1038/s41598-018-2493.

Rogers, K.L., Carreres-Calabulg, J.A., Gorokova, E., Posth, N.R., 2020. Micro-by-micro interactions: how microorganisms influence the fate of marine microplastics. Limnol. Oceanogr. Lett. 5, 18–36.

Rohwer, F., Prangishvili, D., Lindell, D., 2009. Roles of viruses in the environment. Environ. Microbiol 11, 2771–2774.

Schmidt, T.S.B., Raes, J., Bork, P., 2018. The human gut microbiome: from association to modulation. Cell 172, 1198–1215. Available from: https://doi.org/10.1016/j.cell.2018.02.044.

Shen, H., Kreisel, D., Goldstein, D.R., 2013. Processes of sterile inflammation. J. Immunol 191, 2857–2863.

Shi, N., Li, N., Duan, X., Niu, H., 2017. Interaction between the gut microbiome and mucosal immune system. Mil. Med. Res. 4, 1–7. Available from: https://doi.org/10.1186/s40779-017-0122-9.

Sultana, S., Paul, S.C., Parveen, S., Alam, S., Rahman, N., Jannat, B., et al., 2020. Isolation and identification of salt-tolerant plant growth-promoting rhizobacteria and its application for rice cultivation under salt stress. Can. J. Microbiol. 66 (2), 144–160. Available from: https://doi.org/10.1139/cjm-2019-0323.

Suttle, C.A., 2007. Marine viruses: major players in global ecosystem. Nat. Rev. Microbiol 5, 801–812.

Tanaka, K., Takada, H., 2016. Microplastic fragments and microbeads in digestive tracts of planktivorous fish from urban coastal waters. Sci. Rep. 6, 1–8. Available from: https://doi.org/10.1038/srep34351.

Thursby, E., Juge, N., 2017. Introduction to the human gut microbiota. Biochem. J. 474, 1823–1836. Available from: https://doi.org/10.1042/BCJ20160510.

Treyer, A., Pujato, M., Pechuan, X., Müsch, A., 2016. Iterative sorting of apical and basolateral cargo in Madin–Darby canine kidney cells. Mol. Biol. Cell 27, 2259–2271.

VishnuRadhan, R., David, D., Eldho T and Bhagat, J., 2020. Do atmospheric plastics act as fomites for novel viruses? Earth Arxiv (unpublished data). Available from: https://doi.org/10.31223/osf.io/2vj8m.

Wan, Z., Wang, C., Zhou, J., et al., 2019. Effects of polystyrene microplastics on the composition of the microbiome and metabolism in larval zebrafish. Chemosphere 217, 646–658. Available from: https://doi.org/10.1016/j.chemosphere.2018.11.070.

Wang, Y.-L., Lee, Y., Chiu, I., Lin, Y., Chiu, H., 2020. Potent impact of plastic nanomaterials and micromaterials on the food chain and human health. Int. J. Mol. Sci. 21 1727. Available from: https://doi.org/10.3390/ijms21051727.

Waring, R.H., Harris, R.M., Mitchell, S.C., 2018. Plastic contamination of the food chain: a threat to human health? Maturitas. 115, 64–68.

Webb, H.K., Crawford, R.J., Sawabe, T., Ivanova, E.P., 2009. Poly (ethylene 321 terephthalate) Polymer surfaces as a substrate for bacterial attachment and biofilm 322 formation. Microbes Environ. 24, 39–42.

Wick, P., Malek, A., Manser, P., et al., 2010. Barrier capacity of human placenta for nanosized materials. Environ. Health Perspect. 118, 432–436.

Wright, S.L., Kelly, F.J., 2017. Plastic and human health: a micro issue? Environ. Sci. Technol. 51, 6634–6647.

Wu, H.J., Wu, E., 2012. The role of gut microbiota in immune homeostasis and autoimmunity. Gut Microbes 3, 4–14. Available from: https://doi.org/10.4161/gmic.19320.

Wu, P., Huang, J., Zheng, Y., et al., 2019. Environmental occurrences, fate, and impacts of microplastics. Ecotoxicol. Environ. Saf. 184, 109612. Available from: https://doi.org/10.1016/j.ecoenv.2019.109612.

Wylie, K.M., Weinstock, G.M., Storch, G.A., 2012. Emerging view of the human virome. Transl. Res. 160, 283–290.

Yee, M.S.-L., et al., 2021. Impact of microplastics and nanoplastics on human health. Nanomaterials (Basel, Switz.) 11 (2), 496. Available from: https://doi.org/10.3390/nano11020496.

Yong, C.Q.Y., Valiyaveetill, S., Tang, B.L., 2020. Toxicity of microplastics and nanoplastics in mammalian systems. Int. J. Environ. Res. Public. Health 17, 1509. Available from: https://doi.org/10.3390/ijerph17051509.

Yoon, M.-S., 2016. The emerging role of branched-chain amino acids in insulin resistance and metabolism. Nutrients 8 (7), 405. Available from: https://doi.org/10.3390/nu8070405.

Zhang, Y.J., Li, S., Gan, R.Y., et al., 2015. Impacts of gut bacteria on human health and diseases. Int. J. Mol. Sci. 16, 7493–7519. Available from: https://doi.org/10.3390/ijms16047493.

Zhang, T., Li, Q., Lei, C., Buch, H., Zhang, F., 2019. Akkermansia muciniphila is a promising probiotic. Microb. Biotechnol. 12, 1109–1125.

Zhao, H.J., Xu, J.K., Yan, Z.H., Ren, H.Q., Zhang, Y., 2020. Microplastics enhance the developmental toxicity of synthetic phenolic antioxidants by disturbing the thyroid function and metabolism in developing zebrafish. Environ. Int. 140, 105750. Available from: https://doi.org/10.1016/j.envint.2020.105750.

Zhu, F., Ju, Y., Wang, W., et al., 2020. Metagenome-wide association of gut microbiome features for schizophrenia. Nat. Commun. 11, 1–10. Available from: https://doi.org/10.1038/s41467-020-15457-9.

Section X

Molecular and Immunological Mechanisms Associated with Healthy Gut Microbiome Functions

Chapter 25

Role of the gut microbiome and probiotics for prevention and management of tuberculosis

Tejaswini Baral[1], Shilia Jacob Kurian[1], Sonal Sekhar M.[1], Murali Munisamy[2], Chandrashekar Udyavara Kudru[3], Bidita Khandelwal[4], Mithu Banerjee[5], Chiranjay Mukhopadhyay[6], Kavitha Saravu[7], Jitendra Singh[2], Sarman Singh[8] and Mahadev Rao[9]

[1]Department of Pharmacy Practice, Manipal College of Pharmaceutical Sciences, Manipal Academy of Higher Education, Manipal, Karnataka, India, [2]Department of Translational Medicine, All India Institute of Medical Sciences, Bhopal, Madhya Pradesh, India, [3]Department of Medicine, Kasturba Medical College and Hospital, Manipal Academy of Higher Education, Manipal, India, [4]Department of General Medicine, Sikkim Manipal Institute of Medical Sciences, Sikkim Manipal University, Gangtok, Sikkim, India, [5]Department of Biochemistry, All India Institute of Medical Sciences, Jodhpur, Rajasthan, India, [6]Department of Microbiology, Coordinator, Center for Emerging and Tropical Diseases, Kasturba Medical College and Hospital, Manipal Academy of Higher Education, Manipal, Karnataka, India, [7]Department of Infectious Diseases, Coordinator, Manipal Center for Infectious Diseases, Kasturba Medical College and Hospital, Manipal Academy of Higher Education, Manipal, Karnataka, India, [8]All India Institute of Medical Sciences, Bhopal, Madhya Pradesh, India, [9]Department of Pharmacy Practice, Coordinator, Center for Translational Research, Manipal College of Pharmaceutical Sciences, Manipal Academy of Higher Education, Manipal, Karnataka, India

25.1 Introduction

Tuberculosis (TB) is considered to be one of the topmost killer diseases worldwide. According to the World Health Organization (WHO), in 2018, the estimated global incidence was ten million cases. TB primarily affects the lungs, termed pulmonary TB. However, being a multisystemic disease, TB also affects other organs, including the pleura, lymph nodes, skeleton, and meninges (Kang et al., 2020). Although there have been advances in TB treatment and care, challenges remain (Fig. 25.1). For example, the emergence of resistant strains of TB such as multidrug-resistant and extensively drug-resistant TB, latent TB, side effects of anti-TB drugs, comorbidities such as HIV, diabetes mellitus (DM) (Al-Humadi et al., 2017; Forget and Menzies, 2006). The adverse effects of anti-TB medications include drug-induced hepatotoxicity, gastrointestinal (GI) side effects, neurological disorders, dermatological reactions, hematological reactions, and nephrotoxicity. These effects cause patient noncompliance with the therapy and lead to medication discontinuation and unfavorable treatment outcomes (Forget and Menzies, 2006; Lv et al., 2013). During the first phase of treatment, adverse effects related to the liver and GI tract are reportedly high. The GI reactions often include nausea and vomiting, abdominal pain, and loss of appetite. These are often the reasons behind patients' irregular behavior toward intake of medication (Lin et al., 2020). A prospective population-based cohort study by Lv et al. (2013) profiled the adverse effects of directly observed treatment, short-course in people with TB in China (Lv et al., 2013). They reported that approximately 15% of patients had adverse drug reactions, which affected control of their TB. In the study of 766 patients, liver dysfunction occurred in 273 patients (35.6%), and 161 patients (21%) had adverse GI effects. They also found that 32 (11.7%) patients who developed liver dysfunction and 15 (9.3%) patients who suffered from GI effects refused to continue the TB treatment because of these adverse effects.

Current robust evidence shows that the human microbiome and microbiota have great potential in prevention and treatment of TB. The term "microbiome" refers to the gene level, and the microbiota is at the organism level. The microbiome is the collection of all genes that contribute to the microbiota. The gut microbiome (GM) is considered the center of microbial activity, exceeding the human genome by 100 times (Zhou et al., 2020). The human gut microbial ecosystem consists predominantly of bacteria along with other microbes such as yeasts and archaea. The GM plays a

Microbiome, Immunity, Digestive Health and Nutrition. DOI: https://doi.org/10.1016/B978-0-12-822238-6.00036-4

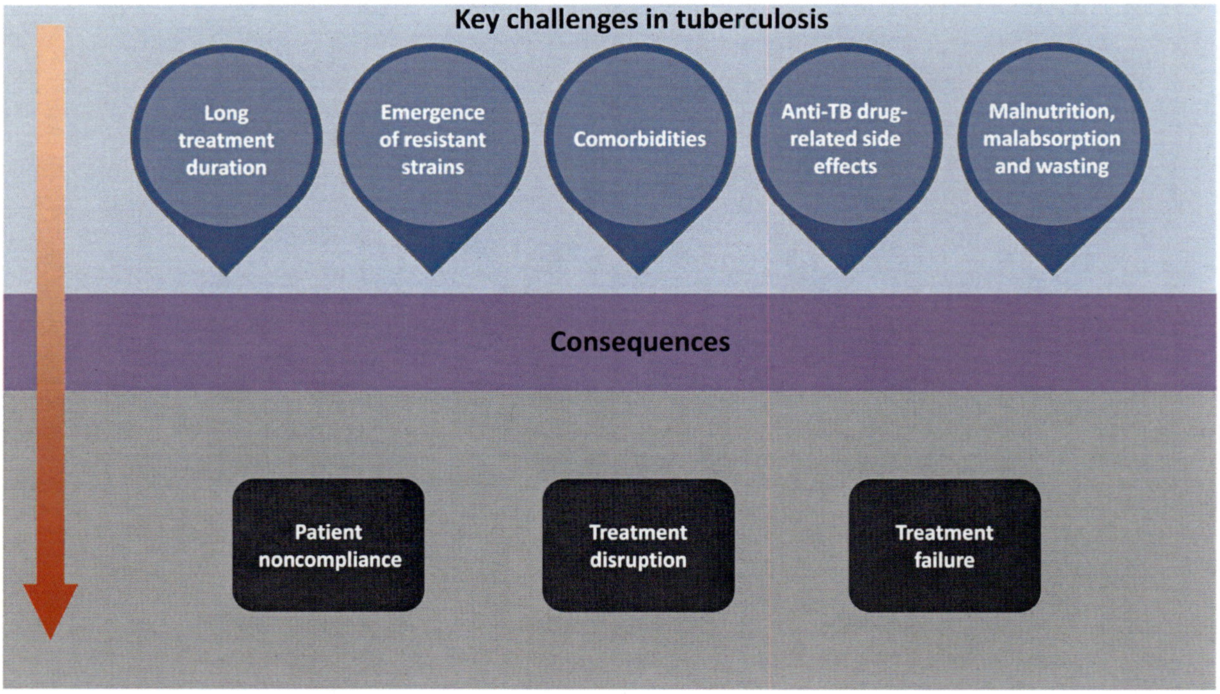

FIGURE 25.1 Challenges in the treatment of tuberculosis.

vital role in digestion, metabolism, homeostasis, immunity, and so on (Wood et al., 2017). The GM is believed to be interconnected with the lungs, liver, brain, and immune system via the gut-lung axis, the gut-liver axis, the gut-brain axis, and the gut–immune system axis, respectively.

25.2 Gut-lung axis

Healthy human lungs are not sterile as was earlier assumed. Microbial colonization in the respiratory system has been studied with the evolution of microbiological tests for microbiome profiling. Compared to the gut microbiota, the lung microbiota has less microbial mass. In a healthy gut, the major bacteria phyla are Firmicutes and Bacteroidetes followed by the *Proteobacteria*, *Actinobacteria*, *Fusobacteria*, and *Verrucomicrobia*. In the GI tract, the site-specific variation in the abundance and diversity of the microbiota might be due to pH, properties of mucin, bile acid concentration, and host defense factors. In contrast, the diversity and abundance of the lung microbiota depend on the salivary microinhalations, mucociliary clearance, the immune system, microbial proliferation, pH, and oxygen concentration. In the lungs, on the phylum level, most bacteria are similar to those in the gut (Budden et al., 2017; Enaud et al., 2020).

The gut-lung axis implies bidirectional microbial communication between the gut and the lungs (Enaud et al., 2020; Zhou et al., 2020). Although the exact mechanism of the pathways is unknown, it is believed that the cytokines, endotoxins, and metabolites that are produced by the gut microbiota connect the tissue of the intestine with that of lungs through the blood circulation, endotoxins, and metabolites produced by gut microbiota. Therefore dysbiosis of the gut microbiota is associated with an increased risk of lung infection and disorders with alteration of the immune response. Similarly, dysbiosis of the lung microbiota associated with asthma, chronic obstructive pulmonary disease, and pneumonia alters intestinal immunity (Zhou et al., 2020). Ultimately, dysbiosis of the GM may cause a decrease in resistance to colonization and containment of pathogens and consequently may increase susceptibility to TB (Naidoo et al., 2019).

25.3 Gut-liver axis

The gut-liver axis may be considered the model of anatomical and metabolic interaction. The human gut provides 70% of blood and gut microbe–derived components to the liver. The gut epithelium provides barriers to pathogen-associated molecular patterns (PAMPs) entering the portal circulation, and these PAMPs are related to the pathogenesis of liver diseases (Milosevic et al., 2019). The GM plays a role in bile acid metabolism via bile acid salt hydrolase activity, thus altering bile acid levels during gut dysbiosis and consequently causing health issues in the host (Mancini et al., 2018). It is presumed that the severity of liver diseases is associated with the degree of gut dysbiosis. The short-chain fatty acid (SCFA)-producing bacteria are responsible for bile acid metabolism as well as the production of immunoglobulin A (Mancini et al., 2018). The gut-liver axis is vital to consider in TB because the anti-TB drugs have profound hepatotoxicity and subsequently hinder the desired treatment outcome.

25.4 Gut–immune system axis

The GM plays a crucial role in immune homeostasis by involving mainly SCFA-producing microorganisms. This helps to release the proinflammatory and antiinflammatory cytokines and controls the mononuclear phagocytes (such as dendritic cells, or DCs) functions (Negi et al., 2019). By accumulating innate and adaptive immune cells, gut-associated lymphoid tissue provides an active immune response against the infiltration of pathogenic microbes (He et al., 2017). A few researchers have suggested that the common mucosal immune system is responsible for the immunological communication between the gut and the lungs (He et al., 2017). The commensal microbes that line the gut mucosa influences these responses. The main SCFA-producing bacteria are *Faecalibacterium*, *Ruminococcus*, *Bacteroides*, *Alistipes*, *Prevotella*, and *Bifidobacterium* (Ramakrishna, 2013).

A large amount of gut-derived metabolites are responsible for the microbiome-cytokine interaction and modulation. A clinical study by Schirmer et al. (2016) established the link between the GM and its ability to produce cytokines. They reported that the metabolic processes of cytokine production include the metabolism of microbial tryptophan and palmitoleic acid. Tryptophol, the metabolite of tryptophan, is believed to have strong tumor necrosis factor alpha (TNF-α) inhibitory effects, whereas the palmitoleic acid showed a specific effect on monocyte-derived cytokines, that is, TNF-α, interleukin 1 beta (IL-1β), and IL-6. Palmitoleic acid inhibits IL-1β and interferon gamma (IFN-γ) induced apoptosis. The production capacity of TNF-α and IFN-γ and, to some extent, IL-1β, IL-6, IL-17, and IL-22 depends on the GM (Schirmer et al., 2016). This explains the impaired immune function during gut dysbiosis.

25.5 Role of the human gut microbiome in tuberculosis

Since GM is fundamental to the entire microbiome, any dysbiosis will increase the chances of invasive infections (Dubourg et al., 2013). Evidence from preclinical and clinical studies highlights the role of the GM in tuberculosis (Table 25.1). Alteration of the GM during TB infection has been reported in these studies.

The diversity in GM perturbation in TB infection varies according to several factors, such as genetics, diet, the severity of infection, the duration of infection, and the medication that is used during the infection treatment. There is no margin scale of comparison of which particular phylum or order of bacteria alters during TB infection. The microbiota plays an indirect factor in preventing disease progression and provides resistance to initial infection via enhancement of innate immunity from commensal bacteria (Namasivayam et al., 2018). This may explain why only 10%–15% of exposed patients who are exposed to TB develop active TB disease. The GM perturbation promotes the survival of *Mycobacteria*, leading to an increased bacterial burden on the lungs (Negi et al., 2019), early lung colonization (Dumas et al., 2018), and inflammatory and immune dysfunction (Khaliq et al., 2021). An in vitro study by Winglee et al. (2014) reported a rapid reduction in gut bacterial composition within 6 days after TB infection, suggesting the modulation of gut microbiota during infection (Winglee et al., 2014). As an impact of broad-spectrum antibiotics, GM perturbation increases the susceptibility to TB infection by modulating the frequency of T cells (Khan et al., 2016; Winglee et al., 2014). It is presumed that GM dysbiosis after TB treatment may be associated with post-TB sequelae and even relapse (Naidoo et al., 2019) (Fig. 25.2). Wipperman et al. (2017) reported long-lasting microbiome dysbiosis even 1.2 years after completion of TB therapy (Wipperman et al., 2017). Negi et al. (2020) showed decreased *Mycobacterium* clearance by isoniazid, aggravated tissue pathology, and reduced CD4 + cells in mice with gut dysbiosis (Negi et al., 2020).

25.6 Role of probiotics in tuberculosis

The role of probiotics in TB is a relatively new and exciting area of research involving the clinical health benefits of probiotics. According to the WHO and the Food and Agriculture Organization, probiotics are "live microorganisms which, when administered in adequate amounts confer a health benefit on the host" (Hill et al., 2014). Each strain of probiotic has been designed to have a specific therapeutic purpose. The organisms that have been most studied as probiotics are *Lactobacillus*, *Bifidobacterium*, *Lactococcus*, *Streptococcus*, *Escherichia*, *Bacillus*, *Bacteroides*,

TABLE 25.1 Studies on gut microbiota taxonomy profiling in treating tuberculosis infection.

Study	Type of study	Enrichment in Taxa (Phyla)	Depletion in Taxa (Phyla)
Winglee et al. (2014)	Preclinical	*Lachnospiraceae, Ruminococcaceae*	*Clostridiales, Bacteroidales*
Khan et al. (2016)	Preclinical	–	*Lactobacillus, Campylobacter, Bacteroides, Enterococcus*
Namasivayam et al. (2017)	Preclinical	–	*Clostridiales, Porphyromonadaceae*
Dumas et al. (2018)	Preclinical	*Proteobacteria*	*Bacteroidetes, Firmicutes*
Negi et al. (2020), India	Preclinical	*Bacteroidetes, Enterococcus*	*Lactobacillus, Bifidobacterium, Campylobacter*
Khan et al. (2019), Canada	Preclinical	*Bacteroidete, Verrucomicrobia, Clostridiaceae*	*some families of Firmicutes*
Maji et al. (2017)	Clinical	*Faecaliacterium, Coprococcus, Phascolarcobacterium, Pseudobutyrivibrio, Roseburia*	*Prevotella, Bifidobacterium*
Wipperman et al. (2017)	Clinical	*Erysipelatoclostridium, Fusobacterium, Prevotella*	*Bifidobacterium, Ruminococcus, Coprococcus, Blautia, Lactobacillus*
Khaliq et al. (2021)	Clinical	*Fusobacteria, Actinobacteria, Firmicutes*	*Bacteroidetes*
Luo et al. (2017)	Clinical	*Proteobacteria*	*Bacteroidetes, Firmicutes*
Hu et al. (2019), China	Clinical	*Bacteroidetes, Firmicutes*	*Clostridium*

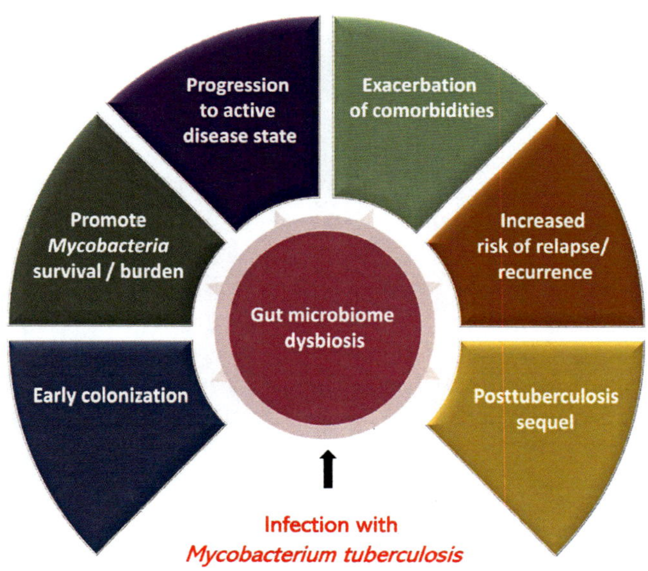

FIGURE 25.2 Consequences of gut microbiome dysbiosis due to tuberculosis infection.

Enterococcus, *Escherichia*, *Faecalibacterium*, *Propionibacterium*, and *Saccharomyces*. The probiotics have been isolated as pure organisms and can be used in single-strain or multistrain formulations to yield the desired health benefits (Preidis and Versalovic, 2009). However, probiotics require stability against stomach acids, enzymes, bile acids, and so on so that these can easily pass through the GI tract to exert their therapeutic action. The ideal probiotics should

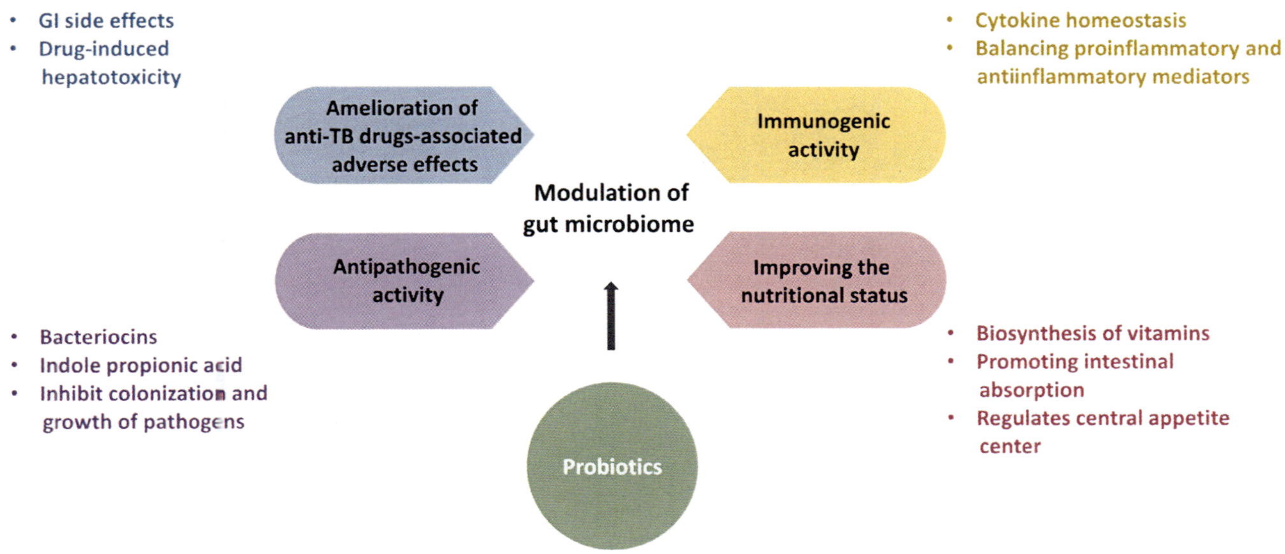

FIGURE 25.3 Mechanisms involved for beneficial effects of probiotics in tuberculosis.

demonstrate these attributes to enhance prevention and management of disease (Halloran and Underwood, 2019). Apart from the organisms mentioned earlier, other strains of commensal microorganisms are also being studied to implement probiotics. The adherence of probiotics to the intestinal mucosa is important to exert its therapeutic action (Cremonini et al., 2002). By adhering to the intestinal lining, the probiotics inhibit the pathogenic microbes' growth and restore the depleted function of the commensal microbes due to gut dysbiosis. After reaching the target site and its direct effects, probiotics exhibit therapeutic action via their byproducts. For example, the indole propionic acid (IPA) is a metabolite that is reported to exhibit antimycobacterial property (Negatu et al., 2020), and the bacteriocins are peptides that possess antipathogenic action (Karpiński and Szkaradkiewicz, 2016). These mechanisms explain the efficacy and clinical benefits of probiotics, including in tuberculosis (Fig. 25.3).

25.6.1 Probiotics as immunomodulator

The probiotic strains adhere to the intestinal epithelium cells and, via toll-like receptors (TLRs), mediate immune stimulation (Halloran and Underwood, 2019; Maldonado Galdeano et al., 2019). The TLRs, which are the key receptors in innate immunity, that are present in the membrane proteins can recognize the microbial molecule. Exopolysaccharides (EPS), lipopolysaccharides (LPS), lipoteichoic acid (LTA), lipoproteins, peptidoglycan, glycolipids, some heat shock proteins, and flagellin are TLR ligands. Most of the probiotics strain are gram-positive bacteria, and their cell wall constituents polysaccharide and glycolipids are the primary targets as immunomodulatory molecules. Via TLR2, the LTA acts as an immunostimulator. EPS produced by *Lactobacillus* and *Bifidobacteria* possess selective immunomodulation. LPS in the gram-negative bacteria cell wall acts via TLR4 and is involved in the progression of inflammation, obesity, and type 2 DM by high-fat content (Fedorova and Danilenko, 2014). It is assumed that the leakage of LPS into systemic circulation leads to chronic subclinical inflammation in the peripheral adipose tissue and thereby alters it metabolic function. d'Hennezel et al. (2017) studied the role of LPS in immunity. Their findings suggested the overall TLR4 silencing capacity of LPS and exhibited the immunoinhibitory action. Although the free form of LPS that is generated from the gut lumen tends to prevent inflammation, a microbiome shift with a decrease in Bacteroidetes species relative to an increase in Firmicutes species gram-positive bacteria, has been shown to be associated with altering metabolic function. The exact role of LPS in peripheral inflammation in relation to GM composition has yet to be elucidated (d'Hennezel et al., 2017). The DNA of various *Lactobacillus* strains is effective in Type 1T helper (Th1)/Th2 balance maintenance in the peripheral blood mononuclear cells and induces B cell proliferation and activation of DCs. Strains of *Bifidobacterium* contain the cytosine-phosphate-guanosine-oligodeoxynucleotides (CpG-ODNs), which are PAMPs, induce the production of TNF-α via TLR9 (Fedorova and Danilenko, 2014). In an in vitro model, the *Bifidobacterium* strain was reported to possess the ability to enhance the polarization of T cell, maturation of DCs, production of IL-10, TNF-α and IFN-γ, and activation of $CD8^+$ cell (Ruiz et al., 2017). The strain initiates the adaptive immunity by activating the DCs. The DCs are considered the main link in initiating the specific microbe-associated molecular patterns, which play a role in immune regulation. *Bifidobacterium* activated DCs model has been found to have antiinflammatory potential. The elucidation of bacterial DNA as a strain-specific immune response can be correlated with cytokine profile.

25.6.2 Probiotics as an antiinfective mediator

Probiotics synthesize the bacteriocins, which are antimicrobial substances. The bacteriocins are peptides that do not harm the producing strain. Bacteriocins are mostly produced by the *Propionibacterium*, *Lactobacillus*, *Enterococcus*, *Escherichia coli*, and so on. Bacteriocins possess both killing and growth suppression capacity by inhibiting cell wall components' biosynthesis, developing bacterial spores, forming pores in the cell membrane, and activating autolytic enzymes. *Lactobacillus* species such as *L. bulgaricus*, *L. acidophilus*, *L. casei*, *L. helveticus*, *L. lactis*, *L. salivarius*, and *L. plantarum* inhibit the growth of pathogenic bacteria in the GI tract by creating an unfavorable environment and synthesizing pH-lowering components. Probiotic bacteria maintain the microbiological equilibrium in the GI tract. *Bifidobacterium* has been well-known in treating infections of the urogenital tract. *E. coli* prevents salmonella and shigella infection by producing microcins and colicins (Karpiński and Szkaradkiewicz, 2016). The bacteriocins that are produced from the *Lactobacillus* are more natural and safe because of their presence in food items. *Lactobacillus* bacteriocins include mainly lantibiotics and nisin, and these possess antimycobacterial properties. The bacteriocins possess high anionic lipid content and form pores on pathogen bacterial cell walls, including *Mycobacterium*, and inhibit cell wall biosynthesis. In vitro *Lactobacillus*-derived bacteriocins showed activity against a broad range of *Mycobacteria*, including *Mycobacterium smegmatis*, *Mycobacterium avium*, *Paratuberculosis*, *Mycobacterium tuberculosis*, and *Mycobacterium kansasii* (Sivaraj et al., 2018). *Clostridia* and *Peptostreptococcus anaerobius* are responsible for the production of IPA (Negatu et al., 2020). With antiinflammatory and antioxidant properties, the IPA shows a broad-spectrum antimycobacterial property. It is effective against the multiresistant *Mycobacteria tuberculosis* and *Mycobacteria avium*. The IPA, the analog of tryptophan, is an antimycobacterial metabolite that is generated by the gut bacteria. Tryptophan biosynthesis is the important pathway for the *M. tuberculosis* growth and viability both in vitro and in vivo (Negatu et al., 2019). This pathway protects the *Mycobacteria* in the phagosomes and helps in disease progression (Lott, 2020). The IPA is considered one of the missing functional links between GM and TB (Negatu et al., 2020).

Negatu et al. (2019) demonstrated that IPA, acting as an allosteric inhibitor, blocks tryptophan biosynthesis. IPA binds to the allosteric binding site of tryptophan and thereby shuts off the enzymatic action in *Mycobacteria* (Negatu et al., 2019). A small indole-based scaffold IPA has been detected in the human bloodstream and even in cerebrospinal fluid. The ability of this substance to cross the blood-brain barrier could help in acting against TB meningitis. Moreover, IPA possesses neuroprotective effects. With the development of host-directed microbiota therapy, IPA could be used as adjunctive therapy for TB (Negatu et al., 2020).

25.6.3 Preclinical and clinical evidence of the effects of probiotics in managing tuberculosis

Microbiological assays have shown that the probiotics strains are stable against anti-TB drugs, which implies that the probiotics will not be having any drug interactions when used along with the anti-TB drugs. Gavrilova et al. (2014) studied the efficacy of a probiotics strain on *Mycobacterium* B5 and reported that lactic acid bacteria (which were isolated from healthy animals) were steady against many of the antibiotics that are used in treatment of TB (Gavrilova et al., 2014).

The macrophage-inducible C-type lectin (Mincle), also known as pattern recognition receptors expressed on immune cells, is important for recognizing microbial stimuli and binding, which initiate an innate and adaptive immune response to infections. Negi et al. (2019) studied the stimulation of this Mincle by *Lactobacillus* against TB (Negi et al., 2019). They found that *Mycobacterial* survival in the lungs in antibiotic-induced gut dysbiosis is restricted by *Lactobacillus* incorporation in mice via Mincle stimulation. The Mincle stimulation generated the protective cytokine response, decreasing the expression of regulatory T cells (Treg) and increasing IL-6, IL-12, IL-17, IFN-γ, and phagocytosis. The Treg are known to decrease the helper T cells, and thereby the *Lactobacillus* induces the Th1 immune response (Negi et al., 2019). The study found that the probiotic strain causing suppression of Treg expression could have beneficial effects in fighting against TB.

Dumas et al. (2018) studied the impact of fecal transplantation (FT) from healthy mice into antibiotic-treated mice (Dumas et al., 2018). The FT has been used in treating many GI diseases to restore the normal gut flora. In this study, the researchers reported the impact of FT on *Mycobacteria* lung colonization. The FT restored the Bacteroidetes and

Firmicutes that had been depleted by the antibiotic treatment and prevented early lung colonization by *Mycobacterium tuberculosis*. Furthermore, the probiotics strain could help in preventing the progression of TB by restricting the *Mycobacteria* lung colonization. After the entry of the *Mycobacteria* into the body, when lung macrophages fail in its elimination, it starts colonizing the lungs and progresses to an active TB state. This usually happens as a result of impaired immunity, which is presumed to be a part of gut dysbiosis. The study reported the microbiota dysbiosis leads to early lung colonization of *M. tuberculosis* (Dumas et al., 2018).

A quasiexperiment with randomization study by Setiyaningrum et al. (2016) reported the efficacy of the combined use of probiotics and zinc on the immunity of pulmonary TB patients (Setiyaningrum et al., 2016). They divided the 54 pulmonary TB patients into two groups. One group was treated with daily probiotic and zinc supplementation for 4 weeks, and the other group was treated with placebo for 4 weeks; both groups received the standard TB therapy. The researchers discovered an increased level of lymphocytes and decreased neutrophil-lymphocyte ratio (NLR) in the intervention group. Lymphocytes and neutrophils are important in specific and nonspecific defense mechanisms. The NLR levels are presumed to be a stronger predictor of bacteremia. Through competitive attachment to enterocytes, the probiotics inhibit the pathogenic bacterial growth in the intestinal and activate macrophages, promoting the production of ILs (Setiyaningrum et al., 2016).

25.7 Role of probiotics in preventing anti-TB drug-induced adverse effects

25.7.1 Amelioration of adverse gastrointestinal effects

The most common anti-TB drug-induced GI side effects are diarrhea, nausea and vomiting, dyspepsia, and abdominal pain. The probiotics have well-established clinical benefits in alleviating GI ailments such as diarrhea, constipation, and abdominal pain. Their application is well-known in treating antibiotic-associated diarrhea. Species of *Lactobacillus*, *Bifidobacterium*, *Enterococcus*, *Saccharomyces*, and so on have been found to be effective against GI ailments (Cremonini et al., 2002).

Human GM regulates the appetite by directly acting on the central appetite pathways, the intestinal release of peptide tyrosine and ghrelin (the hormone that is responsible for satiety), and GI motility by regulating the metabolism of bile acids and serotonin. Therefore gut dysbiosis associated with anti-TB drugs may cause vomiting and loss of appetite. An open-label dose-response randomized controlled trial (RCT) by Lin et al. (2020) with 429 TB patients studied the efficacy of *Lactobacillus casei* on anti-TB drug-related adverse GI effects, including nausea, vomiting, diarrhea, flatulence, loss of appetite, and constipation, during the intensive phase of treatment. This RCT involved three groups (low-dose, high-dose, and control groups consisting of 142, 143, and 144 subjects, respectively). The intervention groups were asked to consume the probiotic for 2 months. The low-dose group consumed one bottle of *L. casei* per day, and the high-dose group consumed two bottles. Every bottle of *L. casei* contained 10 billion CFUs. The researchers observed a dose-related response; that is, the high *L. casei* dose resulted in a lower incidence of vomiting and appetite loss (Lin et al., 2020).

Intestinal motility has been known to improve with the use of strains of *Lactobacillus*, *Bifidobacterium*, *Clostridium*, *Streptococcus*, and so on (Halloran and Underwood, 2019; Mekonnen et al., 2020). These are well-known probiotics for the treatment of constipation by exhibiting laxatives effects. The fermented products of our gut commensals affect gut motility. The chemotactic peptides such as formyl-methionyl-leucyl-phenylalanine stimulate the enteric nervous system and primary afferent nerves, which helps in the propulsive movement of the small intestine. The SCFAs produced by bacteria also affect gut motility by improving solute and water absorption (Dimidi et al., 2017).

25.7.2 Amelioration of drug-induced hepatotoxicity

According to American Thoracic Society guidelines, drug-induced hepatotoxicity is defined as follows: (1) five times the upper limit of normal (ULN) level of serum aspartate aminotransferase (AST) and/or alanine aminotransferase (ALT) or (2) AST and/or ALT levels higher than three times the ULN, combined with symptoms such as easy fatigability, nausea, vomiting, abdominal pain, and/or insufficient oral intake, and (3) at least a 50% improvement in the liver enzyme elevations after discontinuation of anti-TB drugs (Song et al., 2019). The anti-TB drugs isoniazid, rifampicin, and pyrazinamide cause drug-induced hepatotoxicity (also termed idiosyncratic hepatotoxicity) (Ramappa and Aithal, 2013). Song, et al. (2019) reported a lower TB treatment success rate in groups of TB patients with drug-induced hepatotoxicity than in groups with non-drug-induced hepatotoxicity (Song et al., 2019). Drug-induced hepatotoxicity is one of the most common side effects of anti-TB drugs, usually occurring within the first 2 months of therapy. Drug-induced hepatotoxicity requires drug interruption or modification, and these changes in treatment affect the success of TB eradication (Song et al., 2019). Patients are usually advised to stop taking the medication when their ALT levels reach three times the ULN with symptoms attributable to hepatotoxicity or detection of ALT levels that are five times the ULN. Once liver tests normalize, reintroduction can be attempted according to the patients' clinical needs. In severe illness in which the patient is smear-positive or when discontinuation of therapy is unsafe, nonhepatotoxic medication such as ethambutol, cycloserine, or fluoroquinolone can be considered (Ramappa and Aithal, 2013). Any treatment modification during the intensive phase of therapy lead to risks of a subtherapeutic effect, potentially resulting in a risk of TB relapse or development of resistance in the future (Song et al., 2019). An in vitro study by Jeon et al. (2020) reported the hepatoprotective effects of *Streptococcus thermophillus* in preventing liver damage by acting against oxidative stress and inflammatory response (Jeon et al., 2020). A metaanalysis of 17 studies by Khalesi et al. (2018) has concluded that more than 8 weeks' consumption of probiotics may help to improve serum levels of liver enzymes, primarily by reducing serum levels of ALT, AST, and gamma-glutamyl transferase (Khalesi et al., 2018). The outcome is better if synbiotics are used along with probiotics because synbiotics help in sustaining and enhancing the effects of probiotics. A metaanalysis of four RCTs by Ma et al. (2013) concluded that GM modulation by probiotics strains (*Lactobacillus*, *Bifidobacterium*, and *Streptococcus*) might help in reducing the AST, total cholesterol, inflammatory mediators, and insulin resistance in nonalcoholic fatty liver disease (Ma et al., 2013). Shang et al. (2011) reported anti-TB drug-induced liver injury in 106 of 4304 patients in a prospective population-based cohort study. They concluded that there would be a decrease in 16.58% in unsuccessful treatment by preventing the occurrence of anti-TB drug-induced liver injury (Shang et al., 2011).

25.8 Probiotics in the improvement of nutritional status in tuberculosis

Usually, patient with TB are malnourished. The reasons behind this are many, including a decrease in appetite as a side effect of anti-TB drugs, GM dysbiosis leading to compromise in central appetite pathways, and a decrease in nutrient absorption and vitamin synthesis by the gut microbes. Probiotics have yielded clinical benefits in many GI disorders, such as diarrhea, irritable bowel syndrome, inflammatory bowel disease, colorectal cancer, and lactose intolerance. Usually, in these disorders, nutrient absorption decreases, whereas probiotics confer better absorption (Brown and Valiere, 2004). The small intestinal barrier consists of commensal bacteria, antimicrobial peptides (AMPs), a mucus layer, junctional complexes, and so on. Nutrient absorption occurs mainly in the small intestine region. Therefore the disorders that affect the GI tract interrupt absorption by disturbing the normal microbial flora that lines the intestine. Consequently, the structure and permeability of the small intestinal barrier are altered (Judkins et al., 2020). Normal secretion of AMPs is one of the potential factors in the maintenance of a healthy intestinal barrier and permeability function by enhancing the mucin expression, synthesis of tight junction proteins, and intestinal repair (Judkins et al., 2020; Wang et al., 2015). In addition to maintaining intestinal absorption, probiotics contribute to the production of SCFAs (acetate, propionate, and butyrate), which help in the digestion and metabolism of carbohydrates, proteins, and lipids (Ríos-Covián et al., 2016).

A broad range of *Lactobacillus*, *Enterococcus*, *Bifidobacterium*, and *Sphingobacterium* species have the potential to synthesize and supply vitamin K and vitamin B in the human body. Therefore probiotic bacteria and vitamin synthesis also provide vitamin-mediated immunity (Gu and Li, 2018; Yoshii et al., 2019). Hence the use of probiotics will help to combat malnutrition in patients with TB.

25.9 Dose and dosage of probiotic formulations for dietary supplements

Commercially, probiotics are available as capsules, tablets, powders, and liquids or are incorporated into foods containing a wide variety of strains and doses. These products often contain mixed cultures of live microorganisms rather than single strains (Kligler and Cohrssen, 2008). Of the many promising species that are used as probiotics microorganisms, some of the main species that have been used in pharmaceutical preparations and application in in vivo and in vitro studies are *Lactobacillus* species (*L. acidophilus*, *L. amylovorus*, *L. casei*, *L. crispatus*, *L. bulgarius*, *L. paracasei*, *L. plantarum*, *L. rhamnosus*, *L. gasseri*, *L. gallinarum*, *L. johnsonii*, etc.), *Bifidobacterium* species (*B. adolescentis*, *B. animalis*, *B. bifidum*, *B. breve*, *B. infantis*, *B. lactis*, *B. longum* etc.), and others species (*Enterococcus faecalis*, *E. faecium*, *Lactococcus lactis*, *Leuconstoc mesenteroides*, *Pediococcus acidilactici*, *Sporolactobacillus inulinus*, *Streptococcus thermophilus*, *Propionibacterium freudenreichii*, *Saccharomyces cerevisiae*, *Saccharomyces boulardii*, etc.) (Holzapfel et al., 2001).

Probiotics are measured in colony-forming units (CFU), which indicates the number of viable cells or microorganisms. Many probiotic supplements contain 1–10 billion CFU per dose, and some products contain up to 50 billion CFU or more. The effective dose that has been documented consists of 5–10 billion CFUs per day for children and 10–20 billion CFUs per day for adults. A wide range of dosages have been studied in clinical trials, such as 250–500 mg *S. boulardii* per day, 100 million to 1.8 trillion CFUs *Lactobacillus* per day, and so on. No evidence has shown that a higher dose can be unsafe, but it may be unnecessary. Hence according to clinical trial data and formulation efficacy, the dose or CFU per dose can be determined or adjusted (Kligler and Cohrssen, 2008).

25.10 Conclusion

Profound gut dysbiosis during TB and persistent alteration by anti-TB drugs have been well studied in both preclinical and clinical studies. The decline in normal gut microbial function contributes to disease progression in TB. The remodeling of the microbial ecosystem of the gut could provide a promising improvement in the resilience of the GM to challenges in case of infection or concomitant antimicrobial use. If we can predict how probiotics can modulate or restore the perturbed gut flora, we can better design them for effective clinical use. Well-controlled clinical trials are much needed to validate the novel therapeutic strategies of probiotics for TB treatment and prevention.

References

Al-Humadi, H.W., Al-Saigh, R.J., Al-Humadi, A.W., 2017. Addressing the challenges of tuberculosis: a brief historical account. Front. Pharmacol. 8, 689.

Brown, A.C., Valiere, A., 2004. Probiotics and medical nutrition therapy. Nutr. Clin. Care 7 (2), 56–68.

Budden, K.F., Gellatly, S.L., Wood, D.L., Cooper, M.A., Morrison, M., Hugenholtz, P., et al., 2017. Emerging pathogenic links between microbiota and the gut-lung axis. Nat. Rev. Microbiol. 15 (1), 55–63.

Cremonini, F., Di Caro, S., Santarelli, L., Gabrielli, M., Candelli, M., Nista, E.C., et al., 2002. Probiotics in antibiotic-associated diarrhoea. Dig. Liver Dis. 34 (Suppl 2), S78–S80.

d'Hennezel, E., Abubucker, S., Murphy, L.O., Cullen, T.W., 2017. Total lipopolysaccharide from the human gut microbiome silences toll-like receptor signaling. mSystems 2 (6), e00046-17.

Dimidi, E., Christodoulides, S., Scott, S.M., Whelan, K., 2017. Mechanisms of action of probiotics and the gastrointestinal microbiota on gut motility and constipation. Adv. Nutr. 8 (3), 484–494.

Dubourg, G., Lagier, J.C., Armougom, F., Robert, C., Hamad, I., Brouqui, P., et al., 2013. The gut microbiota of a patient with resistant tuberculosis is more comprehensively studied by culturomics than by metagenomics. Eur. J. Clin. Microbiol. Infect. Dis. 32 (5), 637–645.

Dumas, A., Corral, D., Colom, A., Levillain, F., Peixoto, A., Hudrisier, D., et al., 2018. The host microbiota contributes to early protection against lung colonization by *Mycobacterium tuberculosis*. Front. Immunol. 9, 2656.

Enaud, R., Prevel, R., Ciarlo, E., Beaufils, F., Wieërs, G., Guery, B., et al., 2020. The gut-lung axis in health and respiratory diseases: a place for inter-organ and inter-kingdom crosstalks. Front. Cell Infect. Microbiol. 10, 9.

Fedorova, I.A., Danilenko, V.N., 2014. Immunogenic properties of a probiotic component of the human gastrointestinal tract microbiota. Biol. Bull. Rev. 4 (6), 457–465.

Forget, E., Menzies, D., 2006. Adverse reactions to first-line antituberculosis drugs. Expert. Opin. Drug. Saf. 5 (2), 231–249.

Gavrilova, N.N., Ratnikova, I.A., Sadanov, A.K., Bayakisheva, K., Tourlibaeva, Z.J., Belikova, O.A., 2014. Application of probiotics in complex treatment of tuberculosis. Int. J. Eng. Res. Appl. 4 (11), 13–18.

Gu Q, Li P., 2018. Biosynthesis of vitamins by probiotic bacteria.

Halloran, K., Underwood, M.A., 2019. Probiotic mechanisms of action. Early Hum. Dev. 135, 58–65.

He, Y., Wen, Q., Yao F., Xu, D., Huang, Y., Wang, J., 2017. Gut-lung axis: the microbial contributions and clinical implications. Crit. Rev. Microbiol. 43 (1), 81–95.

Hill, C., Guarner, F., Reid, G., Gibson, G.R., Merenstein, D.J., Pot, B., et al., 2014. Expert consensus document. the international scientific association for probiotics and prebiotics consensus statement on the scope and appropriate use of the term probiotic. Nat. Rev. Gastroenterol. Hepatol. 11 (8), 506–514.

Holzapfel, W.H., Haberer, P., Geisen, R., Björkroth, J., Schillinger, U., 2001. Taxonomy and important features of probiotic microorganisms in food and nutrition. Am. J. Clin. Nutr. 73 (2 Suppl), 365S–373S.

Hu, Y., Yang, Q., Liu, B., Dong, J., Sun, L., Zhu, Y., et al., 2019. Gut microbiota associated with pulmonary tuberculosis and dysbiosis caused by anti-tuberculosis drugs. J. Infect. 78 (4), 317–322.

Jeon, M.G., Kim, T.R., Lee, J.Y., Kim, H.S., Ji, Y., Holzapfel, W.H., et al., 2020. Hepatoprotective effects of *Streptococcus thermophilus* LM1012 in mice exposed to air pollutants. J. Med. Food 23 (8), 852–861.

Judkins, T.C., Archer, D.L., Kramer, D.C., Solch, R.J., 2020. Probiotics, nutrition, and the small intestine. Curr. Gastroenterol. Rep. 22 (1), 2.

Kang, W., Yu, J., Du, J., Yang, S., Chen, H., Liu, J., et al., 2020. The epidemiology of extrapulmonary tuberculosis in China: a large-scale multi-center observational study. PLoS One 15 (8), e0237753.

Karpiński, T., Szkaradkiewicz, A., 2016. Bacteriocins. Encycl. Food Health 312–319.

Khalesi, S., Johnson, D.W., Campbell, K., Williams, S., Fenning, A., Saluja, S., et al., 2018. Effect of probiotics and synbiotics consumption on serum concentrations of liver function test enzymes: a systematic review and *meta*-analysis. Eur. J. Nutr. 57 (6), 2037–2053.

Khaliq, A., Ravindran, R., Afzal, S., Jena, P.K., Akhtar, M.W., Ambreen, A., et al., 2021. Gut microbiome dysbiosis and correlation with blood biomarkers in active-tuberculosis in endemic setting. PLoS One 16 (1), e0245534.

Khan, N., Vidyarthi, A., Nadeem, S., Negi, S., Nair, G., Agrewala, J.N., 2016. Alteration in the gut microbiota provokes susceptibility to tuberculosis. Front. Immunol. 7, 529.

Khan, N., Mendonca, L., Dhariwal, A., Fontes, G., Menzies, D., Xia, J., et al., 2019. Intestinal dysbiosis compromises alveolar macrophage immunity to Mycobacterium tuberculosis. Mucosal Immunol. 12 (3), 772–783.

Kligler, B., Cohrssen, A., 2008. Probiotics. Am. Fam. Phys 78 (9), 1073–1078.

Lin, S., Zhao, S., Liu, J., Zhang, J., Zhang, C., Hao, H., et al., 2020. Efficacy of proprietary *Lactobacillus casei* for anti-tuberculosis associated gastrointestinal adverse reactions in adult patients: a randomized, open-label, dose–response trial. Food Funct. 11 (1), 370–377.

Lott, J.S., 2020. The tryptophan biosynthetic pathway is essential for *Mycobacterium tuberculosis* to cause disease. Biochem. Soc. Trans. 48 (5), 2029–2037.

Luo, M., Liu, Y., Wu, P., Luo, D.X., Sun, Q., Zheng, H., et al., 2017. Alternation of gut microbiota in patients with pulmonary tuberculosis. Front. Physiol. 8, 822.

Lv, X., Tang, S., Xia, Y., Wang, X., Yuan, Y., Hu, D., et al., 2013. Adverse reactions due to directly observed treatment strategy therapy in Chinese tuberculosis patients: a prospective study. PLoS One 8 (6), e65037.

Ma, Y.Y., Li, L., Yu, C.H., Shen, Z., Chen, L.H., Li, Y.M., 2013. Effects of probiotics on nonalcoholic fatty liver disease: a *meta*-analysis. World J. Gastroenterol. 19 (40), 6911–6918.

Maji, A., Misra, R., Dhakan, D., Gupta, V., Mahato, N., Saxena, R., et al., 2017. Gut microbiome contributes to impairment of immunity in pulmonary tuberculosis patients by alteration of butyrate and propionate producers. Env. Microbiol. 20 (1), 402–419.

Maldonado Galdeano, C., Cazorla, S.I., Lemme Dumit, J.M., Vélez, E., Perdigón, G., 2019. Beneficial effects of probiotic consumption on the immune system. Ann. Nutr. Metab. 74 (2), 115–124.

Mancini, A., Campagna, F., Amodio, P., Tuohy, K.M., 2018. Gut: liver: brain axis: the microbial challenge in the hepatic encephalopathy. Food Funct. 9 (3), 1373–1388.

Mekonnen, S.A., Merenstein, D., Fraser, C.M., Marco, M.L., 2020. Molecular mechanisms of probiotic prevention of antibiotic-associated diarrhea. Curr. Opin. Biotechnol. 61, 226–234.

Milosevic, I., Vujovic, A., Barac, A., Djelic, M., Korac, M., Radovanovic Spurnic, A., et al., 2019. Gut-liver axis, gut microbiota, and its modulation in the management of liver diseases: a review of the literature. Int. J. Mol. Sci. 20 (2), 395.

Naidoo, C.C., Nyawo, G.R., Wu, B.G., Walzl, G., Warren, R.M., Segal, L.N., et al., 2019. The microbiome and tuberculosis: state of the art, potential applications, and defining the clinical research agenda. Lancet Respir. Med. 7 (10), 892–906.

Namasivayam, S., Maiga, M., Yuan, W., Thovarai, V., Costa, D.L., Mittereder, L.R., et al., 2017. Longitudinal profiling reveals a persistent intestinal dysbiosis triggered by conventional anti-tuberculosis therapy. Microbiome. 5 (1), 71.

Namasivayam, S., Sher, A., Glickman, M.S., Wipperman, M.F., 2018. The microbiome and tuberculosis: early evidence for cross talk. mBio. 9 (5), e01420-18.

Negatu, D.A., Yamada, Y., Xi, Y., Go, M.L., Zimmerman, M., Ganapathy, U., et al., 2019. Gut microbiota metabolite indole propionic acid targets tryptophan biosynthesis in *Mycobacterium tuberculosis*. mBio. 10 (2), e02781-18.

Negatu, D.A., Gengenbacher, M., Dartois, V., Dick, T., 2020. Indole propionic acid, an unusual antibiotic produced by the gut microbiota, with anti-inflammatory and antioxidant properties. Front. Microbiol. 11, 575586.

Negi, S., Pahari, S., Bashir, H., Agrewala, J.N., 2019. Gut microbiota regulates mincle mediated activation of lung dendritic cells to protect against *Mycobacterium tuberculosis*. Front. Immunol. 10, 1142.

Negi, S., Pahari, S., Bashir, H., Agrewala, J.N., 2020. Intestinal microbiota disruption limits the isoniazid mediated clearance of *Mycobacterium tuberculosis* in mice. Eur. J. Immunol. 50 (12), 1976–1987.

Preidis, G.A., Versalovic, J., 2009. Targeting the human microbiome with antibiotics, probiotics, and prebiotics: gastroenterology enters the metagenomics era. Gastroenterology. 136 (6), 2015–2031.

Ramakrishna, B.S., 2013. Role of the gut microbiota in human nutrition and metabolism. J. Gastroenterol. Hepatol. 28 (Suppl 4), 9–17.

Ramappa, V., Aithal, G.P., 2013. Hepatotoxicity related to anti-tuberculosis drugs: mechanisms and management. J. Clin. Exp. Hepatol. 3 (1), 37–49.

Ríos-Covián, D., Ruas-Madiedo, P., Margolles, A., Gueimonde, M., de Los Reyes-Gavilán, C.G., Salazar, N., 2016. Intestinal short chain fatty acids and their link with diet and human health. Front. Microbiol. 7, 185.

Ruiz, L., Delgado, S., Ruas-Madiedo, P., Sánchez, B., Margolles, A., 2017. Bifidobacteria and their molecular communication with the immune system. Front. Microbiol. 8, 2345.

Schirmer, M., Smeekens, S.P., Vlamakis, H., Jaeger, M., Oosting, M., Franzosa, E.A., et al., 2016. Linking the human gut microbiome to inflammatory cytokine production capacity. Cell. 167 (4), 1125–36.e8.

Setiyaningrum, Z., Darmono, S., Sofro, M., Dharmana, E., Widyastiti, N., 2016. Effect of combined probiotics an zinc supplementation on immune status of pulmonary tuberculosis patients. Pak. J. Nutr. 15 (7), 680–685.

Shang, P., Xia, Y., Liu, F., Wang, X., Yuan, Y., Hu, D., et al., 2011. Incidence, clinical features and impact on anti-tuberculosis treatment of anti-tuberculosis drug induced liver injury (ATLI) in China. PLoS One 6 (7), e21836.

Sivaraj, A., Sundar, R., Manikkam, R., Parthasarathy, K., Rani, U., Kumar, V., 2018. Potential applications of lactic acid bacteria and bacteriocins in anti-mycobacterial therapy. Asian Pac. J. Trop. Med. 11 (8), 453.

Song, J.H., Yoon, S.Y., Park, T.Y., Heo, E.Y., Kim, D.K., Chung, H.S., et al., 2019. The clinical impact of drug-induced hepatotoxicity on anti-tuberculosis therapy: a case control study. Respir. Res. 20 (1), 283.

Wang, S., Thacker, P.A. Watford, M., Qiao, S., 2015. Functions of antimicrobial peptides in gut homeostasis. Curr. Protein Pept. Sci. 16 (7), 582–591.

Winglee, K., Eloe-Fadrosh, E., Gupta, S., Guo, H., Fraser, C., Bishai, W., 2014. Aerosol *Mycobacterium tuberculosis* infection causes rapid loss of diversity in gut microbiota. PLoS One 9 (5), e97048.

Wipperman, M.F., Fitzgerald, D.W., Juste, M.A.J., Taur, Y., Namasivayam, S., Sher, A., et al., 2017. Antibiotic treatment for Tuberculosis induces a profound dysbiosis of the microbiome that persists long after therapy is completed. Sci. Rep. 7 (1), 10767.

Wood, M.R., Yu, E.A., Mehta, S., 2017. The human microbiome in the fight against tuberculosis. Am. J. Trop. Med. Hyg. 96 (6), 1274–1284.

Yoshii, K., Hosomi, K., Sawane, K., Kunisawa, J., 2019. Metabolism of dietary and microbial vitamin B family in the regulation of host immunity. Front. Nutr. 6, 48.

Zhou, A., Lei, Y., Tang, L., Hu, S., Yang, M., Wu, L., et al., 2020. Gut microbiota: the emerging link to lung homeostasis and disease. J. Bacteriol. 203 (4), e00454-20.

Chapter 26

Lactic acid bacteria–based beverages in the promotion of gastrointestinal tract health

Luís Cláudio Lima de Jesus[1], Tales Fernando da Silva[1], Rafael de Assis Glória[1], Andria dos Santos Freitas[1], Monique Ferrary Américo[1], Lucas Jorge da Silva Fernandes[1], Gabriela Munis Campos[1], Gabriel Camargos Gomes[1], Rhayane Cristina Viegas Santos[1], Rodrigo Dias de Oliveira Carvalho[1], Debmalya Barh[1,2] and Vasco Azevedo[1]

[1]Laboratório de Genética Celular e Molecular (LGCM), Instituto de Ciências Biológicas, Departamento de Genética, Ecologia e Evolução, Universidade Federal de Minas Gerais (UFMG), Belo Horizonte, Minas Gerais, Brazil, [2]Centre for Genomics and Applied Gene Technology, Institute of Integrative Omics and Applied Biotechnology (IIOAB), Nonakuri, West Bengal, India

26.1 Introduction

Humanity has known fermented beverages since the beginning of civilizations, and they are recognized as beneficial for therapeutic use. It was an innovation that was introduced in prehistoric agricultural societies, with evidence of fermented beverage production found in China 7000 B.C. and in Mesopotamia 5000 B.C. Fermented milk has been used in the Middle East, Egypt, Greece, and Italy since 7000–5000 B.C., while in Asia, at approximately the same time, the primary consumption was products that were fermented from rice (Gasbarrini et al., 2016). Since then, modern technology has been employed to investigate the health effects that are provided by fermented products, with increasing research in discovering microorganism strains able to confer additional health improvement in daily consumption (Gasbarrini et al., 2016).

Fermented beverages must contain a sugar source and a natural microbiota or surrounding microorganisms that are responsible for the natural fermentation process (Otles and Ozyurt, 2019). *Lactobacillus* and *Bifidobacterium* comprise most species present in dairy fermented beverages and food (Tamang et al., 2016a,b). However, other bacterial genera (*Enterococcus*, *Lactococcus*, *Pediococcus*, *Bacillus*, *Brevibacterium*, *Leuconostoc*, and *Propionibacterium*) and yeasts, such as *Saccharomyces,* were also isolated and are used in fermented beverages worldwide (Hittinger et al., 2018; Tamang et al., 2016a, 2016b).

The health benefits for the gastrointestinal tract (GIT) of fermented beverages were recognized even before the existence of microorganisms was discovered. Fermented beverages provide vital nutrients to a well-balanced diet: calcium, proteins, potassium, phosphorus, vitamins, and lipids. Also, the acid that is produced slowly helps digestion. Besides nutrients, fermented beverages contain microorganisms that transform the raw material properties that enhance sensorial quality, degrade toxic components and nonnutritive factors, fortify and deliver bioactive compounds, and produce antioxidant and antimicrobial components (Hittinger et al., 2018). The consumption of dairy products was associated with a reduction in the risk of type 2 diabetes, better digestion, attenuation of irritable bowel syndrome, fatty acid liver disease through immunomodulation, and even a decrease in cholesterol levels (Chandan et al., 2017). Many of these effects are derived from Lactic Acid Bacteria (LAB), the mainly microorganism group used in the production of fermented products.

26.2 Lactic Acid Bacteria

Lactic Acid Bacteria are a heterogeneous group of gram-positive, facultative anaerobic microorganisms (Pachla et al., 2018), generally nonmotile, nonsporulating, catalase-negative, and able to produce lactic acid as the primary metabolic end product of carbohydrate fermentation (Bintsis, 2018). They can adapt to different environments, which could

Microbiome, Immunity, Digestive Health and Nutrition. DOI: https://doi.org/10.1016/B978-0-12-822238-6.00008-X

explain their wide use in the fermentation of different food products (Hayek et al., 2019). The main species belong to the genera *Carnobacterium*, *Enterococcus*, formerly *Lactobacillus*, *Lactococcus*, *Leuconostoc*, *Oenococcus*, *Pediococcus*, *Streptococcus*, *Sporolactobacillus*, *Tetragenococcus*, *Vagococcus*, and *Weissella* (Burgain et al., 2014).

Most of the LAB strains are considered to be GRAS (Generally Recognized as Safe) by the U.S. Food and Drug Administration, playing an essential role in health maintenance and the prevention and treatment of different human pathological conditions, such as inflammatory bowel diseases (Martín and Langella, 2019). In this context, several bacterial strains belonging to this group have been characterized and demonstrate a probiotic profile.

Probiotics are defined as "live microorganisms that, when administered in adequate amounts, confer health benefits on the host" (Hill et al., 2014). The mechanisms of action proposed for the beneficial effects of probiotic are diverse and strain-specific (Plaza-Diaz et al., 2019). They include improvements in the epithelial barrier through the maintenance of tight junction proteins and mucin production by goblet cells, adhesion to the intestinal mucosa and concomitant inhibition of pathogen adhesion, competitive exclusion of pathogenic microorganisms through the production of antimicrobial substances, and modulation of the mucosa immune system (Batista et al., 2020; Pachla et al., 2018).

From a biotechnological aspect, LAB is essential in the food industry, since these microorganisms are used in the production and preservation of fermented products, such as yogurt and some types of cheeses and wines. Owing to the ability to acidify the medium, they improve the organoleptic characteristics (e.g., texture, flavor, and aroma) and nutritional value of these products (Tamang et al., 2016a,b). LAB metabolizes carbohydrates into different useful compounds (mainly lactic acid) through a process known as fermentation, which is defined as sugar metabolism in which energy is derived from the partial oxidation of an organic compound (Hayek and Ibrahim, 2013). In addition, this bacteria group has proteolytic activities of great importance in the maturation and enzymatic modification of food products, once they serve as means of manufacturing proteins, peptides, and amino acids that are available for bacterial growth, and organoleptic properties of fermented foods (Hayek and Ibrahim, 2013).

Many food products such as meat, vegetables, and especially dairy products use LAB as fermenters. These bacteria use the proteins and sugars present in milk and end up metabolizing lactic acid, and others compounds, such as bacteriocins, peptides, organic acids, and short-chain fatty acids (SCFAs). These metabolites produced by LAB microorganisms are highly studied for their beneficial potential for health (Pessione, 2012).

26.3 LAB-derived bioactive compounds

Lactic acid production is the primary metabolite produced by LAB during carbohydrate fermentation (Wang et al., 2016). Owing to its acidic nature, this product lowers the pH of the environment in which these bacteria are inserted, making it unfavorable for the growth of other organisms. Therefore this antimicrobial factor is highly studied for its probiotic potential (Wang et al., 2016). Furthermore, these bacteria evolved by acquiring resistance to the low pH of the stomach, bile acids, and other stresses promoted by the human and animals GIT, which allows them to reach their target site of action in viable amounts to promote their beneficial effects.

Bacteriocins are other essential compounds that are produced by LAB. These molecules act as natural bactericides, increasing the shelf life of LAB-derived products, thus making artificial preservatives unnecessary for the food industry. Moreover, they can prevent the infection of potentially pathogenic organisms colonizing the intestinal mucosa (Batista et al., 2020). These proteins are divided into four classes, which act in different ways in their antimicrobial effect. Class I bacteriocins (or lantibiotics) are modified peptides that consist mainly of amino acids such as lanthionine and beta-methylthionine synthesized on the ribosome. Some common examples are nisin, lacticin 481, and cytolysin (McAuliffe et al., 2001). Class II bacteriocins are known as those that do not have lanthionine. They are thermostable molecules that have high specificity (Moll et al., 1999). Class III bacteriocins, or bacteriolysins, are divided into bacteriolytic and nonlytic and contain some molecules, such as lysostaphin, that generally act on the walls of gram-positive bacteria, thus causing their lysis (Moll et al., 1999). Class IV bacteriocins are still considered complex molecules, being composed of combinations of lipids and carbohydrates, and their actions are still being studied (Moll et al., 1999).

Other metabolites that are produced by LAB are fatty acids. Bacteria like *Lactiplantibacillus plantarum* are responsible for metabolizing polyunsaturated fatty acids that are present in the diet, thus generating intermediate molecules such as SCFAs, conjugates, and partially saturated trans-fatty acids, among others (Kishino et al., 2013). The variety of these intermediates has been studied since they can facilitate the absorption of lipids by the body. These SCFAs are end products of glycolytic and phosphoketolases and may contribute to the regulation of epithelial cell proliferation and differentiation. In addition, they provide extra energy for their host, which can cause weight gain (Pessione, 2012).

Organic acids produced by these bacteria are already known to prevent the spoilage of foods. For example, some LAB strains have shown the ability to inhibit both fungal growth and mycotoxin production (Guimarães et al., 2018).

As was mentioned earlier in the chapter, these characteristics make LAB the most used organisms in food fermentation processes.

26.4 The LAB fermentation process

Fermented foods are obtained through the action of microorganisms such as dairy bacteria and yeasts (Hotessa and Robe, 2020). During the process, carbohydrates are partially oxidized, and energy is released in the absence of an external electron acceptor, generating metabolites such as alcohol and acids (Guergoletto et al., 2019; Melini et al., 2019).

The fermentation process that is carried out by microorganisms occurs from the catabolism of carbohydrates and leads to the formation of two main products, alcohol and organic acids (Ayivi et al., 2020; Sieuwerts, 2016). Several food products, such as milk, wines, cheeses, meats, and vegetables, undergo fermentation by microorganisms. This process allows food preservation, increases its shelf life, and promotes organoleptic characteristics, such as flavor, aroma, and food texture (Ayivi et al., 2020; Mokoena, 2017; Sieuwerts, 2016).

According to the fermentation capacity of carbohydrates, LAB is divided into two categories: homofermentative and heterofermentative. Homofermentative LAB (members of the Enterococcaceae, Streptococcaceae, and several Lactobacillaceae families) metabolize glucose, producing lactic acid as the main product. This process occurs via the Embden-Meyerhof (glycolytic) pathway (Bintsis, 2018; Bosma et al., 2017; Hatti-Kaul et al., 2018). They metabolize a sugar molecule (such as glucose) to two lactic acid molecules and two ATP molecules, resulting in more than 85% lactic acid from a glucose molecule (Hayek and Ibrahim, 2013).

Heterofermentative LAB (Leuconostocaceae and Lactobacillaceae species) also metabolize pentoses, so this fermentation can use two pathways, namely the phosphoketolase (PK) and pentose phosphate (PP)/glycolytic pathways. Thus, these microorganisms are able to produce as much lactic acid, as well as alcohol, acetic acid, and carbon dioxide. However, homofermentative strains can also produce ethanol and acetate during mixed acid fermentation using the glycolytic pathway, but this process depends on certain environmental conditions, such as a change in oxygen concentration or carbon limitation (Bosma et al., 2017; Hatti-Kaul et al., 2018). Therefore the first step of glucose degradation in heterofermentative LAB is called the pentose phosphate pathway. Participating in this pathway are glyceraldehyde 3-phosphate, acetyl-phosphate, and CO_2. Next, glyceraldehyde 3-phosphate enters in the glycolytic glycolysis pathway, being transformed into lactic acid through pyruvate, while acetyl phosphate is converted into acetic acid and/or ethanol (Bintsis, 2018).

Some LAB species are classified as facultative heterofermentative, which carry out both fermentations, consuming hexoses via the homolactic route and pentoses via the heterolactic route (Bintsis, 2018).

26.5 Organoleptic characteristics of LAB-fermented foods

The fermentation process and the characteristics of the fermented product depend on the material to be fermented and the initial culture, which consists of one or a consortium of microorganisms added to a mixture of organic compounds to produce the fermented food. In general, LAB cultures are used in milk fermentation, where differents carbohydrates or proteins can be metabolized, resulting in lactic acid and other compouds (Ayivi et al., 2020) that participate in acidification, flavor and texture of fermented products (Ayivi et al., 2020; Ghosh et al., 2019). Thus the choice of strain is important to obtain the desired properties (Alkema et al., 2016).

For example, some *Lactococcus lactis* and *Leuconostoc* strains produce diacetyl, a compound that guarantees a characteristic buttery flavor in dairy foods such as cheeses and butter. This compound is produced as a result of spontaneous oxidative decarboxylation of alpha- acetolactate, formed through the condensation of two pyruvates or citric acid. Another product that guarantees aroma in dairy products is acetaldehyde; in general, this molecule is formed as an intermediate product of ethanol production in heterofermentative LABs (Hatti-Kaul et al., 2018). In addition, some compounds such as exopolysaccharides can influence the texture of yogurts, improving them (Alkema et al., 2016). Likewise, the production of carbon dioxide by heterofermentative LAB can determine the consistency of some dairy products, such as some types of cheese that have holes in their shapes (Ayivi et al., 2020).

In particular, LAB is often used in fermentation that uses a consortium of microorganisms that interact with each other and affect each other, such as in yogurts that use *Streptococcus thermophilus* and *Lactobacillus delbrueckii* subsp. *bulgaricus* in the fermentation of bovine milk. In this product, the two species work mutually. However, there are cases of strains that compete and release toxic substances, such as bacteriocins, which can lead to the exclusion of one of the strains. Furthermore, in the cooperative work of these strains, one stimulates the growth of the other through the exchange of metabolites.

In this sense, the yogurt fermentation phases are divided into exponential growth phases and lower growth phases, in which each species during its exponential growth phase produces compounds that are necessary for the growth of the other. For example, the production of formic acid and folic acid by *S. thermophilus* contributes in the synthesis of purines from *L. delbrueckii*. However, the change in interactions between these species affects the product characteristics. The count reduction of *L. bulgaricus* ensures a smoother yogurt, as the species is related to a lower pH. Furthermore, *S. thermophilus* contributes mainly with diacetyl, acetoin and acetaldehyde, and exopolysaccharides. These components determine the product viscosity and flavor and, consequently, provide changes in the properties of the food. For this reason, yogurt producers try to create combinations of these strains in which *S. thermophilus* has a high cell count and *L. bulgaricus* has a low cell count (Sieuwerts, 2016).

26.6 Commercial products from LAB fermented foods

LAB plays an important role in the fermented food industry, giving unique sensory characteristics to food products (Mathur et al., 2020). LAB-fermented products include dairy products, fermented meats, cereals, and fermented vegetables (Ayivi et al., 2020). Fermentation can increase the nutritional value of foods and beverages, which are incorporated beneficial metabolites such as polyphenols, SCFAs, and products that improve the bioavailability of micronutrients and absorption of food (Annunziata et al., 2020; Tamang et al., 2016a,b). In addition, LAB-driven fermentations usually generate bioproducts with bioactivity and a wide range of health-promoting effects, including protection against infectious agents, immunomodulatory effects, antiallergic effects, antiobesity effects, antioxidant effects, increasing the bioavailability of vitamins/minerals and antianxiety effects, among others (Ayivi et al., 2020; Chen et al., 2019).

Most products that are developed with the use of LAB also provide superior health benefits to the consumer, which is critical to maintaining a healthy gastrointestinal system (Smerud et al., 2009). The bacterial genera that are most associated with fermented dairy products are *Lactobacillus, Lactococcus*, *Propionibacterium*, and *Bifidobacterium* strains (Varzakas, 2020). Some of the fermented food products of LAB include kefir, cheese, butter, yogurt, fermented vegetables, soy curd, koumiss, tofu, fermented meat juices, and lactic beverages, which can be seen in Table 26.1, addressing lactic products fermented by LAB along with the main activities associated with microorganisms that confer health benefits.

According to Aspri et al. (2020), several probiotic cereal-based foods have appeared on the market in the past years. The fermentation of oats by *L. plantarum* 299V was the first non-dairy probiotic drink on the market in 2006; today LAB is associated with products such as cereal bars, probiotic flakes, and breakfast cereals in countries such as France, the United Kingdom, and the United States (Table 26.1) (Aspri et al., 2020; Speranza et al., 2020).

In recent decades, the increasing interest in probiotic foods has been driven by the beneficial properties of food with regard to health (Melini et al., 2019). Besides, the knowledge about health-promoting components and their activity in fermented foods have drawn the attention of the food industry and accelerated the search for strains with superior characteristics and properties to improve sensory quality (Annunziata et al., 2020; Dimidi et al., 2019). In this context, the patents presented in Table 26.2 provide a summary of the publications related to the health benefits caused by LAB.

Fermented foods have other benefits for human health and can also be used in the prevention of chronic diseases, for improvements in gastrointestinal functions, for reduction of serum cholesterol levels, and for their antimutagenic activities (Muncey and Hekmat, 2021; Smerud et al., 2009). Food industry fermentation can be taken into account as a strategy for the production of functional foods and beverages with higher nutritional value and health-promoting compounds. With the rise of vegetarianism and veganism, probiotic drinks based on fruits, vegetables, and cereals have received attention from the scientific community and the food industry.

26.7 Application of LAB-fermented beverages in gastrointestinal tract–related diseases

LABs have been used in a variety of processes, mainly in the fermentation and preservation of functional foods used for human consumption (Rakhmanova et al., 2018; Tamang et al., 2016a,b). Besides promoting organoleptic characteristics to fermented products, these microorganisms provide essential micronutrients (vitamins) and other bioactive compounds such as SCFAs, vitamins, and peptides with potential health benefits for the host organism (Ebringer et al., 2008; Ghosh et al., 2019).

Several studies have demonstrated the association between probiotics-based beverages consumption with a broad spectrum of beneficial effects mainly focused on GIT, as its effectiveness in antibiotic-associated diarrhea (Dietrich, 2014), constipation (Khuropakhonphong et al., 2021), treatment of inflammatory bowel diseases (IBDs) (Levit et al., 2017),

TABLE 26.1 Dairy and nondairy probiotics beverages by lactic acid bacteria and associated mechanisms.

Dairy and nondairy probiotic beverages	Type	Microrganisms	Associated action	References	Company
Biola	Fruit juice	*Lacticaseibacillus rhamnosus* GG *Lactobacillus acidophilus* LA-5 and *Bifidobacterium Bb-12*	Improves intestinal microbiota, immunity, and intestinal transit	(Smerud et al., 2009)	Tine BA.
Bioprofit	Fruit juice	*L. rhamnosus* GG, *Probionibacterium freudenreichii* subsp. *Shermanii JS*	Bacteriocin production, antibacterial activity	(Ayivi et al., 2020)	Valio Ltd.
Bravo Friscus	Fruit juice	*Lactiplantibacillus plantarum* HEAL9, *Lactobacillus paracasei* 8700:2	Antioxidant activity	(Kaźmierczak-Siedlecka et al., 2020)	Skanemajerier
Dahi	Fermented milk	*L. acidophilus*	Antibacterial compounds production	(Mathur et al., 2020)	
Gefilus	Grains, beans, and seeds	*L. rhamnosus* GG	Improvement of the immune system by the increase of IgA and TCD^{4+} cells	(Ayivi et al., 2020)	Valio Ltd.
GoodBelly drink	Fruit juice	*L. plantarum* 299v	Influences innate immunity of bowel-associated lymphoid tissue	(Mathur et al., 2020)	NextFoods
Kefir	Probiotc liquid	*Lentilactobacillus kefir*, *L. kefiranofaciens*, *L. kefirgranum*	Bacteriocin production increases antibacterial activity	(Dias et al., 2020)	Life way
Kefir Soy	Soybeans	*Kluyveromyces marxianus*, *K. lactis*, *Levilactobacillus brevis*, *L. kefir*, *Leuconostoc mesenteroides*, *Lactobacillus helveticus*	Improves intestinal microbiota, immunity, and intestinal transit	(Ayivi et al., 2020)	Life way
KeVita	Sparkling lemon and ginger drink	*Bacillus coagulans* GBI-306086, *Lacticaseibacillus paracasei* 8700:2 *L. plantarum* HEAL 9	Antioxidant and antiinflammatory properties	(Smerud et al., 2009)	H-E-B.
Korean Kimchi	Fermented vegetables	*L. plantarum*	Antimicrobial properties	(Park et al., 2014)	Choi's Kinchi Co.
Koumiss	Fermented vegetables	*Lactobacillus* sp.	Antimicrobial properties	(Mathur et al., 2020)	Kumiss LLC.

(*Continued*)

TABLE 26.1 (Continued)

Proviva	Fermented fruit drink with oatmeal	*L. plantarum* 299 v	Produces EPS, beta-galactosidase and bacteriocins	(Kaźmierczak-Siedlecka et al., 2020)	Skane Dairy
Swiss cheese	Fermented milk	*L. helveticus* R389	Improvement of the immune system by the increase of IgA and TCD^{4+} cells	(Ayivi et al., 2020)	Life way
Tofu	Fermented soy milk	*L. plantarum*	Antioxidant activity	(Mathur et al., 2020)	Live way
Yakult	Fermented milk	*Lacticaseibacillus casei* Shirota	Promote a significant production of SCFAs	(Chen et al., 2019)	Yakult S/A
Natural yogurt	Fermented milk	*L. acidophilus*, *L. delbrueckii*	Improve intestinal microbiota and intestinal transit	(Chen et al., 2019)	Life way

intestinal mucositis (De Jesus et al., 2019), colorectal cancer (del Carmen et al., 2016), and enteric infections (Acurcio et al., 2017). The main probiotic-based beverages related to beneficial effects on the GIT include fermented milks and yogurts (Sakandar and Zhang, 2021).

26.7.1 Antibiotic-associated diarrhea

Antibiotic-associated diarrhea (AAD) is a common adverse effect during antibiotic therapy for the treatment of infectious diseases, leading to the overgrowth of opportunistic and pathogenic bacteria, such as *Clostridium difficile*, leading to disruption of the homeostatic intestinal microbiota (Mullish and Williams, 2018; Sun et al., 2019).

Studies of probiotic beverages have demonstrated a positive association between their consumption and antibiotic-associated diarrhea reduction. For example, a commercial probiotic milk drink (Biola TINE BA, Oslo) fermented with *Lactobacillus acidophilus* La-5 (10^7 CFU/mL), *Lacticaseibacillus rhamnosus* GG, and *Bifidobacterium* Bb-12 (10^8 CFU/mL) reduced by 79% the risk for development of AAD by adult hospitalized patients (Wenus et al., 2008).

A commercially available probiotic drink containing *Lacticaseibacillus casei* DN 114001 (Actimel) (10^{10} CFU) cost-efficiently reduced the prevalence and duration of AAD during antibiotic treatment in patients with infections in a primary care hospital in Germany. The greatest reduction was observed in patients who were treated with ampicillin, sulbactam, ceftriaxone, and metronidazole (Dietrich, 2014). The same effect was reported by Alberda et al. (2018), who demonstrated that the administration of a probiotic yogurt drink (Danactive) containing this strain provided a noticeable preventive effect for AAD as well as *C. difficile*–associated diarrhea in an intensive care unit.

It was reported that milk fermented by *L. acidophilus*, *Bifidobacterium animalis*, and *S. thermophilus* (10^6 CFU/g), supplemented with an apple-pear fruit preparation, significantly shortened the duration of antibiotics-associated diarrhea and reduced the duration and severity of abdominal pain, flatulence, passing of gas, nausea, and passage of mucus with stool. In addition, increased fecal excretion of *B. animalis* and *L. acidophilus* and *C. difficile* suppression were observed (de Vrese et al., 2011).

Another study demonstrated that consumption of yogurt containing a combination of *L. rhamnosus* GG (LGG) *Bifidobacterium lactis* (Bb-12), and *L. acidophilus* (La-5) reduced the incidence of antibiotic-associated gastrointestinal disturbance in a clinical trial conducted in children aged 1–12 (Fox et al., 2015).

26.7.2 Constipation

Constipation most commonly occurs when bowel movements become less frequent and stools become difficult to pass through the digestive tract or cannot be eliminated effectively from the rectum (Tamura et al., 2016). The positive effect of the consumption of probiotics in constipation treatment has been extensively demonstrated in studies using animal models and clinical trials.

TABLE 26.2 Patents related to the health benefits of associated strains and fermented lactic acid bacteria (LAB).

Patent inventors	Title	Patent number	Associated Strains and Fermented	Health effects
Andrea Budelli, Francesca Romano Fasano, Miryam Terzano, and Lorenzo Bramati (HJ Heinz Company LLC)	Probiotic compositions and methods.	EP 3 366 143 A1	*Lacticaseibacillus paracasei* CBA L74 forms replicant or nonreplicant. Metabolites (including acids) of *Lb. paracasei* CBA L74 strain in a replicating or nonreplicating form in fermented foods.	Modulate the immune system of the mucosa and improve intestinal disorders. Mucosal-induced immune system, excluded pathogens, improved enteral nutrition.
Francesca Romana Fasano and Andrea Luigi Budelli (HJ Heinz Company Brands LLC)	Gluten-related disorders.	EP 3 351 554 A1	*L. paracasei* CBA L74 metabolites	Improved gluten-associated disorders. Induced reduced influence of gliadin peptides on cells.
Francesca Romana Fasano and Maria Rescigno (HJ Heinz Company Brands LLC)	Probiotics and methods of use.	EP 3 470 074 A1	Foods fermented with *L. paracasei* CBA L74 in a form of replication or not (test-based results in mice)	Reduced severity of infection in descendants of lactating mice. Effective against various pathogens including *Listeria monocytogenes*, *Salmonella typhimurium*, and *Cronobacter sakazakii*.
Gongnian Xiao, Haina Yuan, Jinyan Gong, Shujun Xiao, Hui Zhang, Guangsheng Zhao, Bingquan Chu Ruosi Fang, and Lin Li (University of Zhejiang Science and Technology)	Method for improving coagulability of yogurt using glutamine transaminase.	NL 202 577 9 A;NL 202 577 9B1	*Lactobacillus acidophilus*, *L. delbrueckii*, and *Streptococcus thermophilus*	A method for improving coagulability of yogurt using glutamine transaminase, which belongs to the technical field of processing of dairy products.
Kobayashi Tatsuya, Saito Junki, Suzuki Takao, and Kamikawa Tomohiro (Yakult Honsha Kk /Yakult Mat Co Ltd.)	Method for producing lactic acid bacteria–fermented food.	JP 202 016 247 7A	*Lacticaseibacillus casei* Shirota	To provide a lactic acid bacteria–fermented food production method capable of obtaining an effect of stimulating the growth of lactic acid bacteria at the time of production and a survival improving effect during storage.
Li Guoyong, Wang Fengxin, Gong Zhiqing, Wu Xuezhi, and Zhang Pinnan (Jiangxi New Born Dairy Nutrition Co Ltd.)	Composition beneficial to intestinal tract regulation, digestion promotion, and constipation improvement and preparation method thereof.	CN 111 771 97 1A	Food containing or treated only with, respectively, lactic acid producing bacteria, *Bifidobacterium* ssp. or propionic LAB, as well as methods for their preparation.	The invention discloses a composition that is beneficial to intestinal tract regulation, digestion promotion, and constipation improvement and a preparation method thereof; relates to the technical field of food, and can effectively maintain human body and intestinal tract health.

(Continued)

TABLE 26.2 (Continued)

Patent inventors	Title	Patent number	Associated Strains and Fermented	Health effects
Liu Xiaojun, Ma Jie, Zhang Lingyu, Zhang Jianjun, and Qimugesudu Yang Huijuan (Beijing Scitop Bio Tech Shareholding Co Ltd.)	*Lactobacillus casei* Zhang and application thereof in preventing and treating chronic gastroenteritis.	CN 111 424 00 3A	*L. casei* in a form of replication or not (test-based results in vivo).	LAB separated from traditional fermented dairy products. Has the functions of regulating intestinal flora; improving body immunity; preventing, treating, and repairing chronic gastroenteritis; and relieving diarrhea and abdominal discomfort.
Zhao Linsong, Jiang Haihua, and Lyu Guang (Hebei Brothers Ilong Food Tech Co Ltd.)	Lactic acid bacteria beverage with clean label as well as preparation method and stability judging method of lactic acid bacteria beverage.	CN 111 771 97 2A	LAB beverage stability assessment method	Method is also provided and can rapidly, efficiently, and accurately judge the amount of water separation and precipitation in the shelf life of clean lactic acid bacteria beverage.

For example, continuous intake of Bulgarian yogurt (containing *S. thermophilus* and *Lactobacillus bulgaricus*) relieved constipation symptoms (rectal pain, separate hard or lumpy stool, and inability to pass stool) in healthy constipated Thai women. This beneficial effect was associated with increased SCFAs (butyrate, propionate, and acetate) and possible stimulation of the gut-brain axis (Khuropakhonphong et al., 2021).

It has also been reported that consumption of a fermented milk product containing dairy starters (*S. thermophilus* CNCM I-163, *L. delbrueckii* subsp. *bulgaricus* CNCM I-1632 and CNCM I-1519, *L. lactis* CNCM I-1631) (1.2×10^9), and *Bifidobacterium animalis* I-2494 (1.25×10^{10}) improved abdominal distension and accelerated orocecal and colonic transit times in subjects suffering from irritable bowel syndrome with constipation. This probiotic beverage also potentiated colonic SCFA production and decreased the abundance of *Bilophila wadsworthia* pathobiont (Veiga et al., 2015).

Another study showed that daily intake of fermented milk containing *L acticaseibacillus casei* Shirota (LcS) (108 CFU/mL) strongly prevented the incidence of constipation and diarrhea in young nutrient-deprived Vietnamese children (Mai et al., 2021) by improving their nutritional status.

26.7.3 Inflammatory Bowel Diseases

The IBDs are described as disorders that involve chronic inflammation of the GIT. They primarily include ulcerative colitis, which affects mainly the colon, and Crohn's disease, which causes inflammation anywhere along the lining of the GIT (Nemati and Teimourian, 2017). Several studies have demonstrated the efficiency of probiotic beverages in treatment of IBDs, mainly colitis.

Zhang and colleagues reported positive results with the fermented milk containing *Bacillus subtilis* JNFE0126 to prevent and treat epithelial damage in a murine model of dextran sodium sulfate (DSS)-induced IBD. It was observed that oral administration of this probiotic beverage formulation prevented colon shortening, reduced the inflammatory response of the intestinal mucosa by inhibiting TNF-α and neutrophils infiltration, and increasing IL-10 levels. Additionally, it was observed preservation of epithelial architecture of colon and small intestine, induction of proliferation of the intestinal stem cell, reconstruction of the mucosal barrier by increasing tight junction zonulin 1 and mucin 2 levels, and microbiota modulation by reducing Lachnospiraceae and Bacteroidaceae abundance (Zhang et al., 2021).

Another study showed that the administration of soy milk fermented by riboflavin-producing *L. plantarum* CRL2130, ($5 \pm 3 \times 10^8$ CFU/mL) was also able to prevent TNBS-induced colitis in a murine model, similar to the results found by using commercial riboflavin. The anti-inflammatory effect observed with this probiotic formulation was associated with a reduction in the production of inducible nitric oxide synthase and proinflammatory cytokines (TNF-α, IL17, IFN-γ, and IL6), and an increase in IL-10 cytokine (Levit et al., 2017).

The beneficial properties of yogurt, Kombucha and Pulpe consumption in IBD has also been evaluated. A study was conducted in Poland with the strains *Lactobacillus bulgaricus* 151 and *S. termophilus* MK10 administered in an in vivo model for DSS-induced colitis showed improvement in the treated animals' clinical aspects, including weight variation and intestinal compliance (Wasilewska et al., 2019). Histological analysis of the intestinal epithelium revealed a reduction in mucosal inflammation and myeloperoxidase activity. The modulation of proinflammatory and anti-inflammatory cytokines was also evaluated. It was demonstrated that the therapeutic effect on DSS-induced colitis in mice by these probiotics was mediated by modulation of the activities of the subset of T helper cells (Th1, Th2, Th17) and Treg response, improving colon healing and downregulation of CD3 + , CD4 + , CD25 + , Foxp3 + , and T cells in the gut (Wasilewska et al., 2019).

Human clinical trials have revealed the probiotic potential of yogurt supplementation in patients with IBD, demonstrating an improvement in intestinal function by modulating the microbiota (Shadnoush et al., 2015) and formation of a desirable anti-inflammatory environment in the peripheral blood of these patients with the increasing populations of CD4 + , CD25 + , and Treg cells (Lorea Baroja et al., 2007).

A study conducted at the Alborz University of Medical Sciences concluded that oral administration of kombucha in mice improved symptoms of DSS-induced colitis (Pakravan et al., 2019). Among the aspects that were analyzed, an increase in the expression of zonula occludens 1 and 2 proteins in animals treated with this probiotic beverage, revealing its essential role in maintaining the epithelial barrier. It was also observed that in addition to reduced intestinal permeability, animals that were treated with kombucha showed improvement in body weight loss, bleeding, diarrhea, and survival rate after colitis induction, as well as upregulated the gene expression of TLR-2, TLR-4, MYD88, demonstrating this ability to induce an immune response (Pakravan et al., 2019).

Regarding pulque, a traditional Mexican alcoholic fermented beverage, a study conducted by French and Mexican scientists showed that *Luteimonas composti*, *Fructilactobacillus sanfransciensis*, and *L. plantarum* strains isolated from pulque demonstrated that this strain reduced body weight loss, intestinal permeability, and proinflammatory cytokines (IFN-γ, IL-17A, IL-5, IL-22) in a murine model of DSS-induced colitis (Torres-Maravilla et al., 2016).

Another study evaluated the probiotic potential of a bacterial isolate of pulque and aguamiel, verifying its antibacterial activity against pathogenic organisms. After isolating and confirming the LAB strains with probiotic potential, the action of these bacteria on *Helicobacter pylori* strains isolated from a gastric biopsy of bacteria was evaluated. All probiotic strains that were isolated from pulque and aguamiel inhibited the growth of the standard *H. pylori* strain ATCC 43504, and two strains from LAB were able to inhibit the growth of one of the *H. pylori* strains isolated from the biopsy. However, all LAB inhibited the *H. pylori* urease enzyme (Cervantes-Elizarrarás et al., 2019). The results demonstrate the potential of pulque as a probiotic drink with a beneficial action on the intestinal microbiota.

Few studies are reporting the beneficial effects of probiotics beverages on Crohn's disease. However, kefir consumption had a positive impact on the biochemical parameters of patients, such as hemoglobin, erythrocyte sedimentation rate and C-reactive protein, and modulated gut microbiota (increase fecal *Lactobacillus* abundance). Kefir consumption improved the quality of life, abdominal pain, and bloating of patients with Crohn's disease (Yilmaz et al., 2020).

26.7.4 Intestinal mucositis

Mucositis is a common cytotoxic side effect of cancer chemotherapy. This clinically important gastrointestinal disorder is characterized by inflammation and ulceration of the intestinal epithelium (Sonis, 2004). Some studies conducted with probiotic-based beverages have also demonstrated beneficial effects for the prevention of intestinal mucositis induced by chemotherapeutic agents, such as 5-fluorouracil, oxaliplatin, irinotecan, and methotrexate.

In this context, preliminary evidence has shown that sheep's milk yogurt containing *Lactobacillus bulcaricus* and *S. thermophilus* (10^7 CFU) ameliorated the deleterious effects of intestinal mucositis induced by methotrexate chemotherapy (2.5 mg/kg/3 days). This probiotic beverage maintained the mucosal barrier and improved intestinal permeability, brush border, sucrase and lactase activity, and tissue architecture following MTX-induced small intestinal mucositis (Southcott et al., 2008).

Positive effects of fermented products on intestinal mucositis induced by antineoplastic drugs have also been reported by Cordeiro et al. (2018), who studied the role of fermented milk using the immunomodulatory *L. casei* BL23 or *Propionibacterium freudenreichii* CB129 supplemented with 30% (w/v) of whey protein. The researchers demonstrated that this probiotic beverage increases the survival rate of both strains when challenged with acid, bile salts, high temperature, and cold storage stresses and prevented weight loss, goblet cell degeneration, and villus shortening in mice that received 5-FU (300 mg/kg) (Cordeiro et al., 2018).

Another study demonstrated that fermented milk by *L. delbrueckii* subsp. *lactis* CIDCA 133 (7.5×10^7 CFU/mL) protected the intestinal mucosa of mice from inflammatory and histopathological damage caused by 5-FU chemotherapy (300 mg/kg). Promising results were observed in the recovery of epithelial architecture to prevent goblet cell degeneration, IgA secretion and polymorphonuclear cell infiltrate reduction, and intestinal permeability decrease (De Jesus et al., 2019).

26.7.5 Colorectal cancer

Some scientific evidence from preclinical studies suggests that the consumption of probiotic beverages can contribute to the inhibition of the growth of colorectal cancer (CRC), one of the most common cancers (Méndez Utz et al., 2017).

For example, the effect of fermented milk containing *L. acidophilus* CL1285, *Lacticaseibacillus casei* LBC80R and *Lacticaseibacillus rhamnosus* CLR2 (10^9 CFU) reduced the initial stage of CRC development by increasing detoxifying enzymes and decreasing aberrant crypt foci and beta-glucuronidase activity in a rat model of dimethylhydrazine (DMH)-induced colorectal cancer (Desrouillères et al., 2015).

Regular consumption of kefir was also able to suppress the development of colorectal cancer that was chemically induced by DMH. The anticancer effects were promoted by increasing the production of SCFAs (mainly propionate), immunomodulation (TNF-α, and IL-1β increase), antioxidant property (an increase of catalase activity), and reducing intestinal permeability and incidence of aberrant crypt foci (dos Reis et al., 2019).

Another study showed that yogurt consumption containing *S. thermophilus* CRL807 and *L. delbrueckii* subsp. *bulgaricus* CRL864 reduced DMH-induced colon cancer development by decreasing severe loss of mucosal architecture, cellular inflammation, and goblet cell depletion; reducing crypt abscess formation and levels of MCP-1; and increasing IL-10 cytokine production (del Carmen et al., 2016). Narushima et al. (2010) reported the anticancer effect that demonstrated the inhibitory effects of yogurt consisting of milk fermented by *L. delbrueckii* 2038 and *S. salivarius* 1131 on the formation of colonic aberrant crypt foci and reduced numbers of tumors induced by DMH in rats (Narushima et al., 2010).

26.7.6 Enteric infections

Some probiotic beverages have shown positive results in treating enteric infections caused by contaminated consumption of food or water with pathogenic microorganisms, such as *Salmonella*, *E. coli*, and *H. pylori*.

Acurcio et al. (2017) showed that *L. plantarum* B7 strain fermented milk (10^8 CFU/mL) prevents a murine salmonellosis model through inhibition of bacterial translocation, antibacterial activity against the *Salmonella* population, downregulating proinflammatory cytokines (IFN-γ and IL-6), and upregulating TGF-β gene expression (Acurcio et al., 2017). Similar effects were also observed in a fermented milk formulation using the *L. paracasei* NCC2461 strain. Its beneficial effect was on decreasing gene expression of proinflammatory cytokines (IFN-γ, IL-6, TNF-α, and IL-17), reducing *S. typhimurium* translocation for spleen and liver, and increasing preservation of intestinal mucosa (Acurcio et al., 2020).

It was also reported that fermented milk containing *L. rhamnosus* S1K3 (MTCC5957) protects mice against *Salmonella* infection–induced epithelial damage by reducing TGF-β and enhancing IgA secretion and IL-4 cytokine levels in the intestinal fluid. In addition, the researchers observed preservation of intestinal mucosa architecture, increases in TLR gene expression at the Peyer's patches, and lower *Salmonella* count in the feces, spleen, liver, and intestine tissues (Kemgang et al., 2016).

The protective effect of probiotic beverages was related to *H. pylori* inhibition. Yoon et al. (2019) showed that fermented milk by *L. paracasei* (HP7) and herbal extract *Glycyrrhiza glabra* had a beneficial effect in patients with *H. pylori* infection by reducing *H. pylori* urease activity, reducing *H. pylori* bacteria load, and improving chronic inflammation and gastrointestinal symptoms, such as abdominal pain (Yoon et al., 2019).

Another study showed that a probiotic beverage formulation containing *L. acidophilus* NCDC-13 ($10^{11}-10^{12}$ CFU/mL) supplemented with whey-skim milk (60:40, v/v), germinated pearl millet flour (4.73%, w/v), and liquid barley malt extract (3.27%, w/v) acted against *Shigella dysenteriae* GIT pathogenicity by reducing translocation of the pathogen in the intestine, spleen, and liver, and increasing the IgA level in intestinal fluid (Ganguly et al., 2019).

26.8 Final considerations

The marketing of dairy and non-dairy probiotic beverages has increased significantly, mainly owing to their reported beneficial effects on host health. Fermented beverages containing microorganisms become suitable matrixes as probiotic carriers, allowing them to resist GIT stressors and perform better at their health-promoting site of action during fermentation and marketing. The results mentioned in this chapter show that probiotic beverages are a promising therapeutic alternative for treating and preventing different diseases associated with GIT. Many of these effects are related to specific properties of the microorganisms, such as immunomodulation, inhibition of pathogens, production of bioactive metabolites with beneficial effects on health, and protection of the epithelial barrier by the expression of mucins and tight junctions' proteins. These preclinical studies and clinical trials highlight the need for new formulations of probiotic drinks to be evaluated in terms of their organoleptic characteristics, their beneficial effects on health, and their safety for human consumption.

References

Acurcio, L.B., Bastos, R.W., Sandes, S.H., de, C., Guimarães, A.C., de, C., et al., 2017. Protective effects of milk fermented by *Lactobacillus plantarum* B7 from Brazilian artisanal cheese on a *Salmonella enterica* serovar Typhimurium infection in BALB/c mice. J. Funct. Foods 33, 436–445.

Acurcio, L.B., Wuyts, S., de Cicco Sandes, S.H., Sant'anna, F.M., Pedroso, S.H.S.P., Bastos, R.W., et al., 2020. Milk fermented by Lactobacillus paracasei NCC 2461 (ST11) modulates the immune response and microbiota to exert its protective effects against Salmonella typhimurium infection in mice. Probiotics Antimicrob. Proteins 12 (4), 1398–1408.

Alberda, C., Marcushamer, S., Hewer, T., Journault, N., Kutsogiannis, D., 2018. Feasibility of a *Lactobacillus casei* drink in the intensive care unit for prevention of antibiotic associated diarrhea and *Clostridium difficile*. Nutrients 10 (5), 539.

Alkema, W., Boekhorst, J., Wels, M., van Hijum, S.A.F.T., 2016. Microbial bioinformatics for food safety and production. Brief. Bioinforma. 17 (2), 283–292.

Annunziata, G., Arnone, A., Ciampaglia, R., Tenore, G.C., Novellino, E., 2020. Fermentation of foods and beverages as a tool for increasing availability of bioactive compounds. focus on short-chain fatty acids. Foods 9 (8).

Aspri, M., Papademas, P., Tsaltas, D., 2020. Review on non-dairy probiotics and their use in non-dairy based products. Fermentation 6 (1), 30.

Ayivi, R.D., Gyawali, R., Krastanov, A., Aljaloud, S.O., Worku, M., Tahergorabi, R., et al., 2020. Lactic acid bacteria: food safety and human health applications. Dairy 1 (3), 202–232.

Batista, V.L., da Silva, T.F., de Jesus, L.C.L., Coelho-Rocha, N.D., Barroso, F.A.L., Tavares, L.M., et al., 2020. Probiotics, prebiotics, synbiotics, and paraprobiotics as a therapeutic alternative for intestinal mucositis. Front. Microbiol. 11, 2246.

Bintsis, T., 2018. Lactic acid bacteria as starter cultures: an update in their metabolism and genetics. AIMS Microbiol. 4 (4), 665–684.

Bosma, E.F., Forster, J., Nielsen, A.T., 2017. Lactobacilli and pediococci as versatile cell factories—evaluation of strain properties and genetic tools. Biotechnol. Adv. 35 (4), 419–442.

Burgain, J., Scher, J., Francius, G., Borges, F., Corgneau, M., Revol-Junelles, A.M., et al., 2014. Lactic acid bacteria in dairy food: surface characterization and interactions with food matrix components. Adv. Colloid Interface Sci. 213, 21–35.

Cervantes-Elizarrarás, A., Cruz-Cansino, N., Ramírez-Moreno, E., Vega-Sánchez, V., Velázquez-Guadarrama, N., Zafra-Rojas, Q., et al., 2019. In vitro probiotic potential of lactic acid bacteria isolated from aguamiel and pulque and antibacterial activity against pathogens. Appl. Sci. 9 (3), 601.

Chandan, R.C., Gandhi, A., Shah, N.P., 2017. Yogurt: historical background, health benefits, and global trade. In: Nagendra, P.S. (Ed.), Yogurt in Health and Disease Prevention. 1–29.

Chen, M., Ye, X., Shen, D., Ma, C., 2019. Modulatory effects of gut microbiota on constipation: the commercial beverage yakult shapes stool consistency. J. Neurogastroenterol. Motil. 25 (3), 475–477.

Cordeiro, B.F., Oliveira, E.R., da Silva, S.H., Savassi, B.M., Acurcio, L.B., Lemos, L., et al., 2018. Whey protein isolate-supplemented beverage, fermented by Lactobacillus casei BL23 and Propionibacterium freudenreichii 138, in the prevention of mucositis in Mice. Front. Microbiol. 9.

del Carmen, S., de Moreno de LeBlanc, A., LeBlanc, J.G., 2016. Development of a potential probiotic yoghurt using selected anti-inflammatory lactic acid bacteria for prevention of colitis and carcinogenesis in mice. J. Appl. Microbiol. 121 (3), 821–830.

De Jesus, L.C.L., Drumond, M.M., de Carvalho, A., Santos, S.S., Martins, F.S., Ferreira, Ê., et al., 2019. Protective effect of *Lactobacillus* delbrueckii subsp. *Lactis* CIDCA 133 in a model of 5 fluorouracil-induced intestinal mucositis. J. Funct. Foods 53, 197–207.

dos Reis, S.A., da Conceição, L.L., e Dias, M.M., Siqueira, N.P., Rosa, D.D., de Oliveira, L.L., et al., 2019. Kefir reduces the incidence of pre-neoplastic lesions in an animal model for colorectal cancer. J. Funct. Foods 53, 1–6.

Desrouillères, K., Millette, M., Vu, K.D., Touja, R., Lacroix, M., 2015. Cancer preventive effects of a specific probiotic fermented milk containing *Lactobacillus acidophilus* CL1285, *L. casei* LBC80R and *L. rhamnosus* CLR2 on male F344 rats treated with 1,2-dimethylhydrazine. J. Funct. Foods 17, 816–827.

de Vrese, M., Kristen, H., Rautenberg, P., Laue, C., Schrezenmeir, J., 2011. Probiotic lactobacilli and bifidobacteria in a fermented milk product with added fruit preparation reduce antibiotic associated diarrhea and Helicobacter pylori activity. J. Dairy. Res. 78 (4), 396–403.

Dias, W.C., Dornelas, A., Martins, O., Moreira Júnior, S., 2020. KEFIR: CARACTERÍSTICAS E BENEFÍCIOS. Alimentos: Ciência, Tecnologia e Meio Ambiente 1 (7).

Dietrich, C.G., 2014. Commercially available probiotic drinks containing *Lactobacillus casei* DN-114001 reduce antibiotic-associated diarrhea. World J. Gastroenterol. 20 (42), 15837.

Dimidi, E., Cox, S.R., Rossi, M., Whelan, K., 2019. Fermented foods: definitions and characteristics, impact on the gut microbiota and effects on gastrointestinal health and disease. Nutrients 11 (8), 1806.

Ebringer, L., Ferenčík, M., Krajčovič, J., 2008. Beneficial health effects of milk and fermented dairy products—review. Folia Microbiol. 53 (5), 378–394.

Fox, M.J., Ahuja, K.D.K., Robertson, I.K., Ball, M.J., Eri, R.D., 2015. Can probiotic yogurt prevent diarrhoea in children on antibiotics? A double-blind, randomised, placebo-controlled study. BMJ Open. 5 (1), e006474–e006474.

Ganguly, S., Sabikhi, L., Singh, A.K., 2019. Effect of whey-pearl millet-barley based probiotic beverage on Shigella-induced pathogenicity in murine model. J. Funct. Foods 54, 498–505.

Gasbarrini, G., Bonvicini, F., Gramenzi, A., 2016. Probiotics history. J. Clin. Gastroenterol 50, S116–S119.

Ghosh, T., Beniwal, A., Semwal, A., Navani, N.K., 2019. Mechanistic insights into probiotic properties of lactic acid bacteria associated with ethnic fermented dairy products. Front. Microbiol. 10.

Guergoletto, K.B., Farinazzo, F.S., Mauro, C.S.I., Fernandes, M.T.C., Alves, G., Prudencio, S.H., et al., 2019. Non-dairy probiotic and prebiotic beverages: applications, nutrients, benefits, and challenges. Nutrients in Beverages 12, 277–314.

Guimarães, A., Venancio, A., Abrunhosa, L., 2018. Antifungal effect of organic acids from lactic acid bacteria on Penicillium nordicum. Food Addit. Contamin. Part. A 35 (9), 1803–1818.

Hatti-Kaul, R., Chen, L., Dishisha, T., Enshasy, H.El, 2018. Lactic acid bacteria: from starter cultures to producers of chemicals. FEMS Microbiol. Lett. 365 (20).

Hayek, S.A., Gyawali, R., Aljaloud, S.O., Krastanov, A., Ibrahim, S.A., 2019. Cultivation media for lactic acid bacteria used in dairy products. J. Dairy. Res. 86 (4), 490–502.

Hayek, S.A., Ibrahim, S.A., 2013. Current limitations and challenges with lactic acid bacteria: a review. Food Nutr. Sci. 04 (11), 73–87.

Hill, C., Guarner, F., Reid, G., Gibson, G.R., Merenstein, D.J., Pot, B., et al., 2014. The international scientific association for probiotics and prebiotics consensus statement on the scope and appropriate use of the term probiotic. Nat. Rev. Gastroenterol. Hepatol. 11 (8), 506–514.

Hittinger, C.T., Steele, J.L., Ryder, D.S., 2018. Diverse yeasts for diverse fermented beverages and foods. Curr. Opin. Biotechnol. 49, 199–206.

Hotessa, N., Robe, J., 2020. Ethiopian indigenous traditional fermented beverage: the role of the microorganisms toward nutritional and safety value of fermented beverage. Int. J. Microbiol. 2020, 1–11.

Kaźmierczak-Siedlecka, K., Daca, A., Folwarski, M., Witkowski, J.M., Bryl, E., Makarewicz, W., 2020. The role of *Lactobacillus plantarum* 299v in supporting treatment of selected diseases. Cent. Eur. J. Immunol. 45 (4), 488–493.

Kemgang, T.S., Kapila, S., Shanmugam, V.P., Reddi, S., Kapila, R., 2016. Fermented milk with probiotic *Lactobacillus rhamnosus* S1K3 (MTCC5957) protects mice from Salmonella by enhancing immune and nonimmune protection mechanisms at intestinal mucosal level. J. Nutr. Biochem. 30, 62–73.

Khuropakhonphong, R., Whanmek, K., Purttiponthanee, S., Chathiran, W., Srichamnong, W., Santivarangkna, C., et al., 2021. Bulgarian yogurt relieved symptoms and distress and increased fecal short-chain fatty acids in healthy constipated women: a randomized, blinded crossover controlled trial. NFS J. 22, 20–31.

Kishino, S., Takeuchi, M., Park, S.-B., Hirata, A., Kitamura, N., Kunisawa, J., et al., 2013. Polyunsaturated fatty acid saturation by gut lactic acid bacteria affecting host lipid composition. Proc. Natl Acad. Sci. 110 (44), 17808–17813.

Levit, R., de Giori, G.S., de Moreno de LeBlanc, A., LeBlanc, J.G., 2017. Evaluation of the effect of soymilk fermented by a riboflavin-producing *Lactobacillus plantarum* strain in a murine model of colitis. Beneficial Microbes 8 (1), 65–72.

Lorea Baroja, M., Kirjavainen, P.V., Hekmat, S., Reid, G., 2007. Anti-inflammatory effects of probiotic yogurt in inflammatory bowel disease patients. Clin. Exp. Immunol. 149 (3), 470–479.

Mai, T.T., Thi Thu, P., Thi Hang, H., Trang, T.T.T., Yui, S., Shigehisa, A., et al., 2021. Efficacy of probiotics on digestive disorders and acute respiratory infections: a controlled clinical trial in young Vietnamese children. Eur. J. Clin. Nutr. 75 (3), 513–520.

Martín, R., Langella, P., 2019. Emerging health concepts in the probiotics field: streamlining the definitions. Front. Microbiol. 10.

Mathur, H., Beresford, T.P., Cotter, P.D., 2020. Health benefits of lactic acid bacteria (Lab) fermentates. Nutrients 12 (6), 1–16.

McAuliffe, O., Ross, R.P., Hill, C., 2001. Lantibiotics: structure, biosynthesis and mode of action. FEMS Microbiol. Rev. 25 (3), 285–308.

Melini, F., Melini, V., Luziatelli, F., Ficca, A.G., Ruzzi, M., 2019. Health-promoting components in fermented foods: an up-to-date systematic review. Nutrients 11 (5).

Mokoena, M.P., 2017. Lactic acid bacteria and their bacteriocins: classification, biosynthesis and applications against uropathogens: a mini-review. Molecules 22 (8), 1255.

Moll, G.N., Konings, W.N., Driessen, A.J., 1999. Bacteriocins: mechanism of membrane insertion and pore formation. Antonie van. Leeuwenhoek 76 (1–4), 185–198.

Mullish, B.H., Williams, H.R., 2018. *Clostridium difficile* infection and antibiotic-associated diarrhoea. Clin. Med. 18 (3), 237–241.

Muncey, L., Hekmat, S., 2021. Development of probiotic almond beverage using *Lacticaseibacillus rhamnosus* GR-1 fortified with short-chain and long-chain inulin fibre. Fermentation 7 (2), 90.

Méndez Utz, V.E., Perdigón, G., de Moreno de LeBlanc, A., 2017. Fermented milks and cancer. Dairy. Hum. Health Dis. Across Lifesp. 343–351.

Narushima, S., Sakata, T. Hioki, K., Itoh, T., Nomura, T., Itoh, K., 2010. Inhibitory effect of yogurt on aberrant crypt foci formation in the rat colon and colorectal tumorigenesis in RasH2 Mice. Exp. Anim. 59 (4), 487–494.

Nemati, S., Teimourian, S., 2017. An overview of inflammatory bowel disease: general consideration and genetic screening approach in diagnosis of early onset subsets. Middle East. J. Dig. Dis. 9 (2), 69–80.

Otles, S., Ozyurt, V.H., 2019. Probiotic and prebiotic beverages. In: Grumezescu, A.A., Holban, M. (Eds.), Functional and Medicinal Beverages, Volume 11. The Science of Beverages. Elsevier, pp. 447–458.

Pachla, A., Wicha, M., Ptaszyńska, A.A., Borsuk, G., Trokenheim, Ł.Ł., Małek, W., 2018. The molecular and phenotypic characterization of fructophilic lactic acid bacteria isolated from the guts of Apis mellifera L. derived from a Polish apiary. J. Appl. Genet. 59 (4), 503–514.

Pakravan, N., Kermanian, F., Mahmoudi, E., 2019. Filtered Kombucha tea ameliorates the leaky gut syndrome in young and old mice model of colitis. Iran. J. Basic. Med. Sci. 22 (10), 1158–1165.

Park, K.Y., Jeong, J.K., Lee, Y.E., Daily, J.W., 2014. Health benefits of kimchi (korean fermented vegetables) as a probiotic food. J. Med. Food 17 (1), 6–20.

Pessione, E., 2012. Lactic acid bacteria contribution to gut microbiota complexity: lights and shadows. Front. Cell. Infect. Microbiol. 2.

Plaza-Diaz, J., Ruiz-Ojeda, F.J., Gil-Campos, M., Gil, A., 2019. Mechanisms of action of probiotics. Adv. Nutr. 10 (suppl_1), S49–S66.

Rakhmanova, A., Khan, Z.A., Shah, K., 2018. A mini review fermentation and preservation: role of lactic acid bacteria. MOJ Food Process. Technol. 6 (5), 414–417.

Sakandar, H.A., Zhang, H., 2021. Trends in probiotic(s)-fermented milks and their in vivo functionality: a review. Trends Food Sci. Technol. 110, 55–65.

Shadnoush, M., Hosseini, R.S., Khalilnezhad, A., Navai, L., Goudarzi, H., Vaezjalali, M., 2015. Effects of probiotics on gut microbiota in patients with inflammatory bowel disease: a double-blind, placebo-controlled clinical trial. Korean. J. Gastroenterol. 65 (4), 215.

Sieuwerts, S., 2016. Microbial interactions in the yoghurt consortium: current status and product implications. SOJ Microbiol. Infect. Dis. 4 (2), 01–05.

Smerud, H.K., Kleiveland, C.R., Mosland, A.R., Grave, G., Birkeland, S.-E., Kloster Smerud, H., et al., 2009. Effect of a probiotic milk product on gastrointestinal and respiratory infections in children attending day-care. Microb. Ecol. Health Dis. 20 (2).

Sonis, S.T., 2004. The pathobiology of mucositis. Nat. Rev. Cancer 4 (4), 277–284.

Southcott, E., Tooley, K.L., Howarth, G.S., Davidson, G.P., Butler, R.N., 2008. Yoghurts containing probiotics reduce disruption of the small intestinal barrier in methotrexate-treated rats. Digestive Dis. Sci. 53 (7), 1837–1841. Available from: https://doi.org/10.1007/s10620-008-0275-1.

Speranza, B., Campaniello, D., Petruzzi, L., Altieri, C., Sinigaglia, M., Bevilacqua, A., et al., 2020. The inoculation of probiotics in vivo is a challenge: strategies to improve their survival, to avoid unpleasant changes, or to enhance their performances in beverages. Beverages 6 (2), 1–18.

Sun, L., Zhang, X., Zhang, Y., Zheng, K., Xiang, Q., Chen, N., et al., 2019. Antibiotic-induced disruption of gut microbiota alters local metabolomes and immune responses. Front. Cell. Infect. Microbiol. 9.

Tamang, J.P., Shin, D.-H., Jung, S.-J., Chae, S.-W., 2016a. Functional properties of microorganisms in fermented foods. Front. Microbiol. 7.

Tamang, J.P., Watanabe, K., Holzapfel, W.H., 2016b. Review: diversity of microorganisms in global fermented foods and beverages. Front. Microbiol. 7, 377.

Tamura, A., Tomita, T., Oshima, T., Toyoshima, F., Yamasaki, T., Okugawa, T., et al., 2016. Prevalence and self-recognition of chronic constipation: results of an internet survey. J. Neurogastroenterol. Motil. 22 (4), 677–685.

Torres-Maravilla, E., Lenoir, M., Mayorga-Reyes, L., Allain, T., Sokol, H., Langella, P., et al., 2016. Identification of novel anti-inflammatory probiotic strains isolated from pulque. Appl. Microbiol. Biotechnol. 100 (1), 385–396.

Varzakas, T., 2020. Microbiology of Fermented. Foods Beverages. Foods, 9(11), 1660.

Veiga, P., Pons, N., Agrawal, A., Oozeer, R., Guyonnet, D., Brazeilles, R., et al., 2015. Changes of the human gut microbiome induced by a fermented milk product. Sci. Rep. 4 (1), 6328.

Wang, M., Gao, Z., Zhang, Y., Pan, L., 2016. Lactic acid bacteria as mucosal delivery vehicles: a realistic therapeutic option. Appl. Microbiol. Biotechnol. 100 (13), 5691–5701.

Wasilewska, E., Zlotkowska, D., Wroblewska, B., 2019. Yogurt starter cultures of *Streptococcus thermophilus* and *Lactobacillus bulgaricus* ameliorate symptoms and modulate the immune response in a mouse model of dextran sulfate sodium-induced colitis. J. Dairy. Sci. 102 (1), 37–53.

Wenus, C., Goll, R., Loken, E.B., Biong, A.S., Halvorsen, D.S., Florholmen, J., 2008. Prevention of antibiotic-associated diarrhoea by a fermented probiotic milk drink. Eur. J. Clin. Nutr. 62 (2), 299–301.

Yilmaz, I., Dolar, M.E., Ozpinar, H., 2020. Effect of administering kefir on the changes in fecal microbiota and symptoms of inflammatory bowel disease: a randomized controlled trial. Turkish J. Gastroenterol. 30 (3), 242–253.

Yoon, J.Y., Cha, J.M. Hong, S.S., Kim, H.K., Kwak, M.S., Jeon, J.W., et al., 2019. Fermented milk containing *Lactobacillus paracasei* and Glycyrrhiza glabra has a beneficial effect in patients with *Helicobacter pylori* infection. Medicine 98 (35), e16601.

Zhang, X., Tong, Y., Lyu, X., Wang, J., Wang, Y., Yang, R., 2021. Prevention and alleviation of dextran sulfate sodium salt-induced inflammatory bowel disease in mice with *Bacillus subtilis*-fermented milk via inhibition of the inflammatory responses and regulation of the intestinal flora. Front. Microbiol. 11.

Chapter 27

Prospective role of prebiotics and probiotics in gut immunity

Saptadip Samanta

Department of Physiology, Midnapore College, Midnapore, West Bengal, India

27.1 Introduction

The human gastrointestinal (GI) tract is directly connected with the environment and is always facing challenges from different microbes and food-borne antigens. The GI tract is also the prime residential site of a large number of commensal microorganisms. The characteristic features of colonization of microorganisms are variable, owing to the diverse biochemical environments of the GI tract. Acid-tolerant microorganisms (e.g., *Helicobacter pylori*) can reside in the stomach. However, the variation in microbial population has been started from the small intestine, where pH tends to be neutral or moves toward an alkaline medium. Finally, the population of microorganisms reaches a maximum level in the colon. The variation in the microbial population depends on several factors, including genetic variations, diet, age, recurring oral antibiotics treatment, and stress (environmental and psychological) (Osadchiy et al., 2019).

The gut microbiota is beneficial for both the host and the microbes. The homeothermic environment and supply of nutrients promote the growth of the microbial population, while the gut flora takes part in the fermentation of nondigestible complex carbohydrates and the production of B vitamins, short-chain fatty acids (acetate, propionate, and butyrate), and others. The gut microflora covers the mucosal surface of the colonic area, which restricts the colonization and growth of the pathogenic microorganisms (Buffie and Pamer, 2013). The gut microbiota also prevents metabolic disorders, neutralizes bacterial toxins, maintains intestinal barrier functions, promotes SCFA synthesis from prebiotics, and exerts oncosuppressive effects in the GI and extraintestinal areas (Scott et al., 2018). Alteration in the gut microbial population promotes different types of diseases. Dysbiosis causes gastritis, diarrhea, inflammatory bowel diseases (IBD), and GI cancer (stomach, colon, hepatobiliary) (Shanahan, 2013; Tanaka and Nakayama, 2017). The gut microbiome essentially regulates systemic immunity. The commensal organisms stimulate myelopoiesis, influencing the growth and development of granulocyte-macrophage progenitors in the bone marrow (Gorjifard and Goldszmid, 2016). They have a positive role in the maturation and activation of dendritic cells (DCs), macrophages, and neutrophils in the peripheral lymphoid organs. Commensal organisms profoundly shape the gut immune system by increasing the size and number of germinal centers in Peyer's patches, IgA-secreting plasma cells, $CD4^+$ T cells in the lamina propria, and $\alpha\beta$ T cell receptor−expressing intraepithelial $CD8\alpha\beta^+$ T cells (Hooper and Macpherson, 2010). The microflora forms a mutualistic environment in the gut to tolerate beneficial bacteria and common food-borne antigens. Short-chain fatty acids (SCFAs) from microbial metabolism involved in the immunomodulatory functions and regulate the inflammatory response. The immunomodulatory effects are stimulated by bacteria-derived molecules, microbe- or pathogen-associated molecular patterns (MAMPs or PAMPs). The components are recognized by pattern recognition receptors (PRRs) of the immune cells to elicit an innate response through MyD88-dependent signaling (Fessler et al., 2019). The gut microflora strictly maintains the balance on the maturation and activation of DCs, Th1, Th2, Th17, and Treg cells for regulation of immune functions and inflammatory response.

Prebiotics (inulin, fructooligosaccharides, galactooligosaccharides, resistant starch, pectin) are the nondigestible components that selectively enhance the growth of beneficial bacteria in the gut. Probiotics are live microorganisms, such as *Lactobacillus* spp., *Enterococcus* spp., *Bifidobacterium* spp., and *Lactococcus* spp. They utilize nondigestible components, producing different types of biomolecules that are beneficial for the host. Currently, prebiotics, probiotics, and synbiotics (a mixture of prebiotics and probiotics) have been included in therapeutic regimens for the treatment of

Microbiome, Immunity, Digestive Health and Nutrition. DOI: https://doi.org/10.1016/B978-0-12-822238-6.00014-5

dysbiosis-related disorders (IBD, ulcerative colitis, and antibiotics-induced diarrhea) and colorectal cancers. People consume a certain amount of prebiotics and probiotics in a natural diet that include fruits, vegetables, fermented foods, and fermented drinks.

27.2 The gut microbiota and its beneficial effects

27.2.1 Shape of the gut microbiota

The human gut is the greatest harbor of the microbial community. Colonization of the microbial population starts in the gestational period when the fetus ingests some amount of amniotic fluid that brings the microbes to the fetal intestine. The constitution of the gut microbiota is complete by the age of 3 years. Different factors, such as anatomical and physiological variations in the GI tract, site-specific preferences, and selection pressure promote the colonization process (Fessler et al., 2019). Although the colonization step is initiated during fetal life, there is no identical picture in the composition of gut microbiota. It varies among the individuals. Various contributing factors, such as the delivery process of the fetus (normal vaginal or cesarean delivery), exposure to antibiotics and other medicines, genetic factors, feeding practices, childcare practices, and environmental pressures affect the variation of the microbial population from the early years of life (LeBlanc et al., 2011; Tanaka and Nakayama, 2017).

The gut microbial community comprises bacteria, archaea, viruses, fungi, protozoa, and helminths (Garrett, 2019; Kong and Cai, 2019). Among the different microbial groups, the bacterial population is almost the major contributing portion. The human gut contains approximately 3×10^{13} resident bacteria (Wong et al., 2019). The total genome content of the microorganisms is about 150 times higher than that of the host (Qin et al., 2010). There is a large diversity in bacterial species, and about 1000–1150 types of bacterial species make a heterogeneous population in the human gut (Sommer and Bäckhed, 2013). These bacterial populations belongs to five phyla: Firmicutes, Bacteroidetes, Actinobacteria, Proteobacteria, and Fusobacteria (Table 27.1). However, organisms in the Firmicutes and Bacteroidetes phyla predominate (Samanta, 2022).

27.2.2 Role of the microbiota in the synthesis of essential metabolites and bioactive components

The gut microbiota has several beneficial effects on human health (Fig. 27.1). The metabolites of gut organisms amplify immune surveillance. The gut microbiome ferments prebiotic components, such as different oligosaccharides, dietary fiber, and resistant starch, for the production of SCFAs such as acetate, propionate, butyrate, hexanoate, and valerate. SCFAs enhance the proliferation and phagocytic capacity of DCs, modulating the inflammatory response. SCFAs are protective against ulcerative colitis (UC), Crohn's disease, and other inflammatory diseases.

Gut bacteria are involved in vitamin synthesis. They provide vitamin B1, vitamin B2, vitamin B6, biotin, and folic acid. *Propionibacterium freudenreichii, Salmonella enterica*, *Listeria innocua*, and *Lactobacillus reuteri* provide vitamin B12. *Enterobacter* sp., *Eubacterium lentum*, *Veillonella* sp., and *Bacteroides* sp convert vitamin K1 (phylloquinone) to vitamin K2 (menaquinone) (Samanta, 2022). Bioactive compounds are synthesized from dietary components. Tryptophan is abundantly present in milk, eggs, red meat, vegetables, and other foods. *Lactobacilli* convert tryptophan to indole-3-aldehyde, which is a ligand of aryl hydrocarbon receptor for modulation of the inflammatory response (Zelante et al., 2013). Commensal microbes transform the primary bile acid into secondary bile acids (deoxycholic acid, lithocholic acid), which have diverse functions.

Gut microorganisms produce different indole derivatives (indole, indoxyl sulfate, and indole-3-propionic acid (IPA)). IPA has antioxidant properties, which restrict oxidative stress (OS), block amyloid beta synthesis, and decrease the risk of neurodegeneration. IPA also activates pregnane X receptor for the regulation of the inflammatory response and tumor necrosis factor alpha (TNF-α) release. It also upregulates junctional protein expression (Kho and Lal, 2018).

Numerous phenolic compounds, including 4-OH phenylacetic acid, equol, urolithins, enterolactone, enterodiol, 8-prenylnaringenin, 2-(3,4-dihydroxy phenyl) acetic acid, 3-(4-hydroxyphenyl) propionic acid, and 5-(3,4-dihydroxyphenyl) valeric acid are the metabolic products of gut bacteria. The antimicrobial properties of the phenolic derivatives prevent the growth of the pathogenic bacteria (Selma et al., 2009). They also reduce OS, platelet aggregation, and circulatory defects.

The human colon contains polyamines (putrescine, spermidine, and spermine) resulting from microbial metabolism. These significantly preserve the high proliferation rate of intestinal epithelial cells. Polyamines also induce the expression of the intercellular junctional proteins occludin, zonula occludens-1, and E-cadherin. These proteins effectively

TABLE 27.1 Major groups of human gut microbiota and their importance (prepared from Samanta, 2022).

Major phyla	Important genera of gut flora under different phyla	Major functional importance
Firmicutes (spore-forming, Gram-positive, anaerobic bacteria)	*Lactobacillus* *Clostridium* *Eubacterium* *Ruminococcus* *Peptococcus* *Enterococcus* *Streptococcus* *Peptostreptococcus* *Acetobacterium* *Succiniclasticum*	• Produce SCFAs, lactate, and succinate • Decrease risk of pathogenesis and metabolic diseases • *Lactobacillus* spp. are the most beneficial probiotic bacteria • *Clostridium* spp. produce beta-glucuronidase • Some are pathogenic. • *Enterococcus faecalis* produce superoxide molecule to increase OS.
Bacteroidetes (Gram-negative)	*Bacteroides* *Prevotella* *Butyricimonas*	• Take part in fermentation of nondigestible sugars • Opportunistic pathogens • Induce the occurrence of IBD • *Bacteroides fragilis* promotes gamma-secretase-dependent cleavage of E-cadherin. • It also releases metalloproteinase that destabilizes the tight junction and influences the loss of intestinal epithelial cells.
Actinobacteria (Gram-positive bacteria)	*Bifidobacterium* *Corynebacterium*	• Metabolize dietary fibers and proteins • Produce antimicrobial agents • *Bifidobacteria* spp. are the major probiotic organisms
Proteobacteria (Gram-negative, *Gammaproteobacteria*, *Epsilonproteobacteria*)	*Escherichia coli* *Pseudomonas* *Vibrio cholera* *Yersinia pestis* *Salmonella* *Enterobacter* *Proteus* *Shigella* *Desulfovibrio* *Helicobacter* *Campylobacter* (*C. jejuni*)	• Mostly pathogenic and show opportunistic activity • Produce virulent factors and cytotoxic agent • *Salmonella enterica* secretes avirulence protein A (AvrA). It has intrinsic deubiquitinase activity • *Shigella flexneri* provides inositol phosphate phosphatase D (IpgD) and cysteine protease-like virulence gene A (VirA). These agents interfere with the p53-mediated DNA repair system. • *E. coli* is the major source of colibactin, and cytolethal distending toxin (CDT). These factors break DNA strands and promote mutation. • *Helicobacter pylori* produces cytotoxin-associated gene A, which disrupts tight junction and also activates the Ras/MAPK pathway.
Fusobacteria	*Fusobacterium nucleatum* subsp. *animalis* *F. russii*	• Mostly prevent pathogenesis and inflammatory response • Sometimes, *F. nucleatum* provides adhesin A (FadA), which suppresses E-cadherin expression and disrupts the tight junction.
Fungi	*Saccharomyces* *Aspergillus* *Penicillium*	• Ferment carbohydrates and maintain intestinal health and systemic immunity

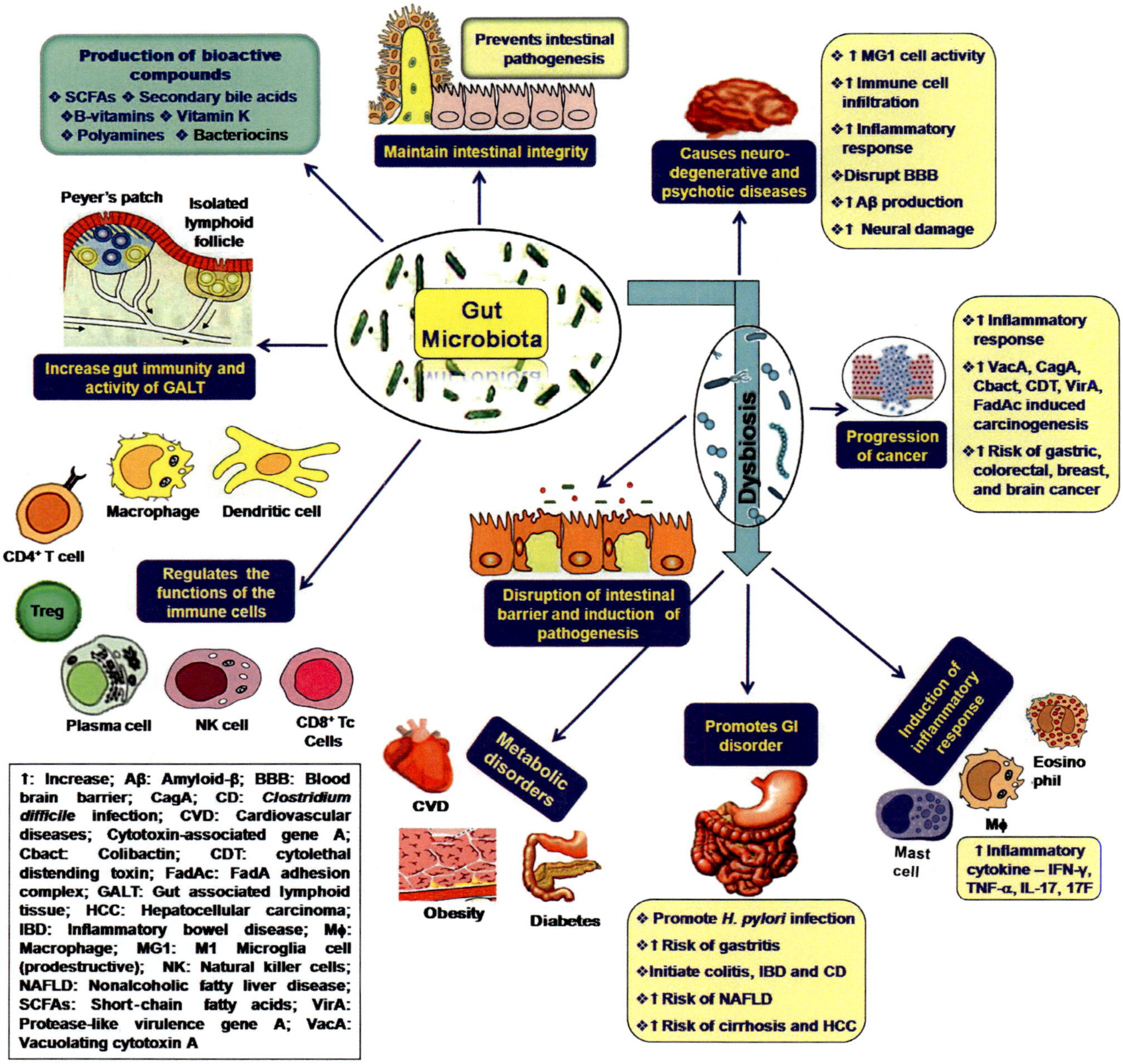

FIGURE 27.1 Beneficial effects of the gut microbiota and impacts of dysbiosis.

increase cell junctional strength and protect the intestinal barrier. Moreover, polyamines, particularly sepermine, inhibit M1 macrophage-mediated proinflammatory activity (Rooks and Garrett, 2016).

27.3 Impacts of dysbiosis

Alteration of the distribution pattern of the gut microflora causes dysbiosis. The improper expression of surface markers on the intestinal epithelial cells due to silence genetic defects, impaired lymphoid tissue along with a small number of myeloid or lymphoid cells in the intestinal wall, recurring GI tract infection, multiple antibiotic treatments, aging, exposure of xenobiotic compounds, and inappropriate dietary habits increase the risk of dysbiosis (Zitvogel et al., 2015). An imbalance in the gut microbiota aggravates various diseases.

27.3.1 Inflammatory bowel diseases

Ulcerative colitis (UC) and Crohn's disease (CD) are the most common form of IBD. These diseases start with inflammation of the mucosal membrane. Dysbiosis is associated with IBD. Increased populations of Neisseriaceae, Fusobacteriaceae, Enterobacteriaceae, Veillonellaceae, Gemellaceae, and Pasteurellaceae and lower number of Bacteroidales, Erysipelotrichaceae, Clostridiales, and Bifidobacteriaceae were associated with IBD (Qin et al., 2010). CD is not site specific. It can occur in any part of the GI tract. However, the occurrence of UC is restricted to the colon. The common symptoms of CD and UC are recurring diarrhea, fever, and abdominal pain. Several antibodies (anti-*Saccharomyces cerevisiae*, anti-OmpC, perinuclear antineutrophil cytoplasmic antibody) against the host's microbial factors and autoantigens are the infective agents in IBD (Landers et al., 2002). *Faecalibacterium prausnitzii* and *Roseburia* sp. diminish proinflammatory cytokines interleukin 12 (IL-12) and interferon gamma (IFN-γ) and elevate the antiinflammatory agent IL-10 (Sokol et al., 2008; Machiels et al., 2014). GI inflammation reduces the secretion of antimicrobial peptides (defensins), resulting in the progression of the diseases. *Bacteroides fragilis* contains endotoxic lipopolysaccharide (LPS) on their surfaces, which promotes the formation of colits. The prebiotic component oligofructose-enriched inulin decreases the mucolytic *Ruminococcus gnavus* population and promotes the growth of *Bifidobacterium longum*, which neutralizes reactive oxygen species (ROS) and reactive nitrogen species and decreases OS and inflammatory effects (Joossens et al., 2012).

27.3.2 *Clostridium difficile* infection

Dysbiosis increases the population of *Clostridium difficile*, which secretes toxins A and B. These catalytic toxins damage the cytoskeleton and the colonic epithelial barrier, followed by induction of inflammatory response and cell death (Pruitt et al., 2012). *C. difficile* infection causes severe diarrhea, pseudomembranous colitis, and sepsis. Fecal microbiota transplant from a healthy subject decreases the cytokines levels of TNF-α, IL-1β, IL-6, IL-8, and IL-12. Moreover, this treatment increases the beneficial bacteria in the phyla Lactobacillaceae, Ruminococcaceae, Desulfovibrionaceae, Sutterellaceae, and Porphyromonadaceae and the human antimicrobial peptide LL-37 (Konturek et al., 2016). Another dysbiosis-mediated illness is celiac disease. Lerner et al. (2017a) reported that gut bacteria synthesize microbial transglutaminases. This enzyme forms stable neocompounds from gliadin through the process called post-translational modification of proteins (PTMP). Supplementation of *Bifidobacterium* and *Lactobacillus* strains protects against gliadin-induced toxic effects and restricts the levels of proinflammatory cytokine in celiac disease (Kho and Lal, 2018).

27.3.3 Metabolic disorders

Imbalances in the population of butyrate-producing Firmicutes and Bacteroidetes have been associated with obesity (Ley et al., 2006) and subsequent type 2 diabetes. Gut microbiota is involved in glucose homeostasis and lipogenesis. These organisms regulate starch metabolism and SCFA production, leading to an increased supply of lipogenic substances (monosaccharides, SCFAs) to the liver. Administration of *Lactobacillus rhamnosus* and bifidogenic prebiotics decreases plasma cholesterol, triacylglyceride levels, and adipogenic activity (Lee et al., 2006). Dysbiosis increases the permeability of the intestinal barrier, enhancing the passage of dietary and microbial antigens to the immunogenic environment. The subsequent effects are activation of immune cells and triggering of prolonged inflammation, causing damage to the pancreatic beta cells, followed by hyperglycemic effects (Endesfelder et al., 2016). The probiotic organisms *Lactobacillus* and *Bifidobacterium* can improve glucose homeostasis.

Nonalcoholic fatty liver disease (NAFLD) is also associated with metabolic disorders. The high population of Firmicutes increases SCFA production. SCFAs induce lipogenesis and gluconeogenesis in the liver. Chronic inflammation and alteration in bile acid metabolism change the gut permeability and potentiate NAFLD. Gut microbiota synthesizes trimethylamine and trimethylamine-N-oxide (TMAO) from choline and L-carnitine, which are abundantly present in red meat. TMAO suppresses bile acid synthesis and increases blood cholesterol levels, which potentiate cardiovascular disease (Griffin et al., 2015).

27.3.4 Role in cancer progression

Different bacteria are involved in GI cancer. *B. fragilis* and *Fusobacterium nucleatum* induce chronic inflammation by producing proinflammatory substances. Chronic inflammation plays a prime role in carcinogenesis (Kho and Lal, 2018; Vivarelli et al., 2019). The inflammatory response promotes the release of different cytokines that alter intracellular signaling pathways, activating oncogenic promoter. The inflammatory response also stimulates OS, followed by DNA damage and mutagenesis. Moreover, chronic inflammation induces the release of metabolic carcinogens such as colibactin from *Escherichia coli* (Tomkovich et al., 2017) and cytolethal distending toxin from *Campylobacter jejuni* (He et al., 2019). Both these components exhibit carcinogenic effects.

27.3.5 Gastric cancer

The stomach and colon are the most susceptible sites for dysbiosis-induced carcinogenesis. *H. pylori* advances gastric carcinoma. Initially, *H. pylori* induces gastritis, which leads to gastric atrophy, intestinal metaplasia, and dysplasia. This organism is recognized as a Group I carcinogen by the International Agency on Research on Cancer (IARC) (Wong et al., 2019). *H. pylori* synthesizes two major compounds: vacuolating cytotoxin A (VacA) and cytotoxin-associated gen A (CagA). VacA and CagA induce the Ras-dependent extracellular signal-regulated kinase pathway, resulting in activation of phosphatidyl-inositol 3-kinase. The subsequent effects are cell proliferation, cell survival, cell growth, and prevention of apoptosis. Additionally, *H. pylori* induces OS, and ROS generation. ROS promotes lipid peroxidation, DNA strand breakage, and genomic instability. All these are the progressive factors for carcinogenesis. Besides *H. pylori*, *Peptostreptococcus stomatis*, *Parvimonas micra*, and *F. nucleatum* also contribute to gastric carcinoma.

27.3.6 Colorectal cancer

Colorectal cancer is a serious problem throughout the world. It is a type of adenocarcinoma. Several factors, including heredity, lifestyle, food habits, and composition of colonic microbiota, have an impact on carcinogenesis of colorectal cancer (Garrett, 2019). Dysbiosis is directly associated with colon carcinoma. Microbial antigens are recognized by several PRRs, such as toll-like receptors (TLRs), nucleotide-binding oligomerization-like receptors, and RIG-I-like receptors of the immune cells (Wong and Yu, 2019). This activation increases the expression of chemokines CXCL9, CXCL10, CCL 17, and CCL20. CXCL9 and CXCL10 recruit T helper cell 1 (Th1), while CCL 17 and CCL20 activate Th17 cells (Cremonesi et al., 2018). Th17 cells are potential producers of IL-17, a prime cytokine of the inflammatory response.

A group of microorganisms acts as cancer-inducing microbiota. Enterotoxigenic *B. fragilis*, genotoxic *E. coli*, *Enterococcus faecalis*, *F. nucleatum*, and *Peptostreptococcus* can activate Th17 cells for the secretion of IL-17, which sustains chronic inflammation (Wu et al., 2009; Park et al., 2018). In addition to the bacteria, cytomegalovirus, John Cunningham virus, human papillomavirus, and fungus (*Malassezia* and others) are also involved in tumorigenesis (Wong and Yu, 2019). *F. nucleatum* shows FadA adhesion complex (FadAc), which binds with E-cadherin of the colonic cells to activate the Wntβ-catenin pathway (Rubinstein et al., 2013). The cell surface marker galactose *N*-acetyl-D-galactosamine (Gal-GalNAc) of the tumor cells binds with another adhesion molecule Fap2 to create a link between colonic cells and the organisms (Brennan and Garrett, 2019). *F. nucleatum* activates TLR4 for the expression of microRNAs, making the cells chemoresistant (Yu et al., 2017). *E. faecalis* induces genotoxicity in the colonic epithelial cells for the progression of cancer (Wu et al., 2009). *B. fragilis* (ETBF) release toxins to activate immune cells and NF-κB-mediated chronic inflammatory response (Garrett, 2019). Polyketide synthase–expressing *E. coli* produces genotoxic colibactin, which forms mutagenic DNA adducts after alkylation (Garrett, 2019). *Helicobacter hepaticus* decreases the stability of DNA, leading to cancer progression in the colon (Vivarelli et al., 2019).

A high-fat diet increases primary bile acid synthesis in the liver. Colonic bacteria convert primary bile acids to secondary bile acids, which increase the risk of cancer through Wnt/β-catenin signaling (Bernstein et al., 2009). Deoxycholic acid acts as an oxidant that causes DNA damage (Bernstein et al., 2011).

27.3.7 Hepatocellular carcinoma

The gut microbiota has a link with hepatocellular carcinoma (HCC). The microbial LPS, bacterial DNA, and the metabolites of peptidoglycans are transported to the liver through enterohepatic circulation. LPS and others activate NF-κB-mediated chronic inflammation for the advancement of HCC (Fatima et al., 2017). *H. hepaticus* also advance the virus-induced HCC (Wong et al., 2019).

27.3.8 Other cancers

Wong et al. (2019) reported that oral microorganisms are involved in periodontal disease, followed by pancreatic, colorectal, and other extraintestinal cancers. The gut microorganisms are also associated with breast cancer and brain cancer. Breast tissue is estrogen sensitive. Colonic bacteria synthesize hydroxylated estrogen, decreasing the circulating estrogen levels. They also convert phytosterol to phytoestrogen, which lowers the efficacy of the estrogen receptors. Thus the gut microbiota decreases the risk of breast cancer. Dysbiosis increases the circulating estrogen levels by expressing beta-glucuronidase for deconjugation of conjugated estrogen, leading to increases of estrogen to the estrogen-sensitive breast cancer cells (Samanta, 2022). Gliomagenesis is closely associated with the disruption of the blood-brain barrier and infiltration of immune cells. The glioma cells and macrophages stimulate microglia cells for the release of cytokines and growth factors related to tumorigenesis in the brain (Samanta, 2022).

27.4 Prebiotics

Classically, there are several definitions of prebiotic. A prebiotic is "a nondigestible food ingredient that beneficially affects the host by selectively stimulating the growth and/or activity of one or a limited number of bacteria in the colon (Gibson and Roberfroid, 1995)." The International Scientific Association for Probiotics and Prebiotics (ISAPP) defines a prebiotic as "a substrate that is selectively utilized by host microorganisms conferring health benefit" (Gibson et al., 2017). The term "selectivity" indicates the growth of the particular beneficial microorganisms. Different prebiotic components (fructooligosaccharides, galactooligosaccharides, and resistant starch) specifically advance the growth of beneficial microbes such as *Lactobacillus* and *Bifidobacterium*.

Prebiotics are oligosaccharides or polysaccharides. They are present as nondigestible dietary fiber. Fructans (inulin and fructooligosaccharides) are well-class prebiotics. Other oligosaccharides, such as galactooligosaccharides (GOS), isomalto-oligosaccharides, xylooligosaccharides, soybean oligosaccharides, resistant starch, pectin, and lactulose (synthetic disaccharide), are also known as prebiotic components. Noncarbohydrate cocoa-derived flavanols show prebiotic activity (Tzounis et al., 2011). Fruits and vegetables are the natural sources of prebiotics. Consumption of different food items, including honey, watermelon, pears, blueberries, leeks, Jerusalem artichokes, chicory, onions, garlic, bananas, asparagus, rye, milk, barley, and salsify, provides prebiotic ingredients within the range of 0.3%−6% of their fresh weight (Khangwal and Shukla, 2019). Sometimes the concentration of prebiotic components in dietary sources is insufficient to elicit satisfactory effects (Manning and Gibson, 2004).

Inulin and fructooligosaccharides (FOS) are the polymers of D-fructose. A group of vegetables and cereals, including onions (25%−40%), garlic, honey, bananas, tomatoes, asparagus, Jerusalem artichokes, sugar beets, wheat, rye, barley, and mushrooms, contain FOS. However, aloe vera extract and steviol glycosides of stevia leaves are the most common sources of inulin and FOS. The human intestine does not contain any fructopolymer-degrading enzyme. These substances reach the colon without any digestion and absorption in the small intestine. Colonic bacteria use them as fermentable sugars, producing lactate and SCFAs from FOS (Bornet et al., 2002). FOS improves gut immunity. They can increase the expression of IL-4, activation of $CD282^+/TLR2^+$ dendritic cells (Clarke et al., 2016), secretion of IgA, and proliferation of NK cells (Lomax et al., 2015). GOS consist of oligogalactose with some lactose and glucose. Milk is the richest source of GOS. The human intestine cannot hydrolyze GOS, owing to deficiencies in beta-galactosidase and beta-glucosidase. Colonic bacteria use GOS for the production of SCFAs. They have bifidogenic activity. The elevated levels of serum IL-8, IL-10, and C-reactive protein are restored by GOS. They can increase NK cell activity (Samanta, 2022; Vulevic et al., 2015). GOS of breast milk protects infants from GI pathogenesis (Fanaro et al., 2005). The synthetic disaccharide lactulose has bifidogenic activity, acting as a growth inhibitor agent of *Clostridium perfringens* and *Bacteroidaceae*. It increases fecal nitrogen elimination by decreasing urinary nitrogen excretion (Samanta, 2022). Irrespective of its prebiotic effects, lactulose is treated as a laxative agent in the clinical system. The pentose sugar xylose is the monomeric unit of xylooligosaccharides. Oats, corncobs, and honey are good sources of xylan. Pectic oligosaccharides are acidic polysaccharides. They are composed of galacturonic acid. Homogalacturonan, rhamnogalacturonan I, rhamnogalacturonan II, and arabinopectin are collectively known as pectic substances (Samanta, 2019). Citrus fruits, apples, and vegetables provide pectin through the diet. Pectin has bifidogenic activity. Various foods, such as peas, lentils, beans, mustard, and chickpeas, contain raffinose oligosaccharides that have prebiotic properties. The polysaccharides and oligosaccharides of laminarin, red seaweed (*Gelidium sesquipedale*), and brown seaweed (*Osmundea pinnatifida*, and *Ecklonia radiate*) are beneficial for the gut microflora (Okolie et al., 2017). Dark-roasted coffee beans and brewed coffee contain a wide range of oligosaccharides that have prebiotic effects (Tian et al., 2017).

Starch contains amylose (15%) and amylopectin (85%). Highly branched amylopectin has a large surface area that exposes the susceptible bonds for pancreatic alpha-amylase to attack. Amylose has a limited surface area, owing to its straight-chain configuration. High amylase-containing starches are resistant to digestion in the intestinal lumen; they are called resistant starch (RS). Oats, maize, potatoes, and green bananas supply the RS after consumption. The colonic microbiota ferments the RS for the synthesis of butyrate. RS shows bifidogenic activity. The large number of starch-binding cell surface proteins on the surface of *Bifidobacterium* spp. increases the availability of starch for metabolism, influencing the growth of the bacteria (Birt et al., 2013). RS also increases the *Lactobacilli* population manyfold (Silvi et al., 1999). Every day, approximately 20 g of RS is metabolized in the gut (Topping and Clifton, 2001). *Bifidobacterium* spp. and *Clostridium butyricum* utilize RS for the production of byproducts.

27.5 Probiotics

The concept of probiotics is not new. The idea was conceived when Henry Tissier had isolated *Bifidobacteria* from the stools of breast-fed infants in 1899. Later, *Bifidobacteria* supplementation was recommended for the treatment of infantile diarrhea. Metchnikoff and his coworkers reported that intestinal microorganisms are beneficial for human health. In 1907, Metchnikoff documented the importance of the gut flora in his famous book *The Prolongation of Life*. The term "probiotic" was first used by Kollath in 1953 to describe the beneficial effects of the intestinal bacteria. The word "probiotic" has two parts; in Latin the preposition "pro" means "for" or "in support," and the term "biotic" had taken from the Greek "bios," which means "life." The overall meaning of "probiotic" is "for life" or "in support of life." The Food and Agricultural Organization and the World Health Organization jointly proposed a definition, which is stated as "live microorganisms which when administered in adequate amounts confer a health benefit on the host" (Food and Agriculture Organization of the United Nations and World Health Organization, 2001). ISAPP approves of this definition. Simply, it can be stated that probiotics are live microbial food supplements that are taken to improve the host's intestinal functions.

The most common probiotic organisms belong to the genera *Lactobacilli* and *Bifidobacteria* (Adnan and Pramaningtyas, 2020). The probiotic organisms in the genus *Lactobacillus* are *L. acidophilus*, *L. casei*, *L. rhamnosus*, *L. helveticus*, *L. gasseri*, *L. reuteri*, *L. crispatus*, *L. salivarius*, *L. ruminis*, *L. fermentum*, *L. johnsonii*, *L. plantarum*, *L. brevis*, *L. delbrueckii*, *L. curvatus*, and *L. sakei* (Lebeer et al., 2008). *B. longum*, *B. adolescentis*, *B. lactis*, *B. animalis*, *B. breve*, and *B. infantis* are probiotic organisms in the genus *Bifidobacteria*. Other bacteria, such as *Lactococci* and *Streptococci*, and a few species of *Bacillus*, *Escherichia*, *Propionibacterium*, and *Saccharomyces* are also in the list of probiotic organisms. Fermented foods are the natural sources of probiotics. Several medicinal preparations are also available that contain a combination of different probiotic organisms. Probiotic treatment is effective against various GI pathogens, including *E. coli*, *Bacillus*, *Salmonella*, *Shigella*, *Vibrio cholera*, *Klebsiella*, and *Pseudomonas* (Mohanty and Ray, 2016). The combination of prebiotics and probiotics is more effective for beneficial effects. This combination is denoted by a new term: synbiotics. Prebiotics increase the growth of the selective beneficial population, such as *Bifidobacteria* and *Lactobacilli*, as they can efficiently ferment different FOS, RS, GOS, pectin, and others (Davani-Davari et al., 2019). Probiotic organisms express different enzymes (beta-fructofuranosidase, fructokinase, alpha-glucosidase, pectinase, and xylanase) for the breakdown of prebiotics. They produce acidic metabolites and prevent the growth of pathogenic bacteria (Samanta, 2022).

27.6 Role of prebiotics and probiotics in immunomodulation

27.6.1 Prebiotics and immune-boosting effects

Prebiotic components act as immunosaccharides. They can improve functions of neutrophils, phagocytic activity, lysosomal enzyme activity (Fig. 27.2), and activation of the alternative pathway of the complement system (Khangwal and Shukla, 2019). The proteoglycans and polysaccharides of mushrooms show immunomodulatory effects. These components boost up both adaptive and innate immunity by stimulating macrophages, splenocytes, and thymocytes (Rodriguez-Lagunas et al., 2017).

27.6.2 Production of short-chain fatty acids and their immunomodulatory effects

The colonic environment of the gut is less oxygenic, which favors the fermentation process. The anaerobic or microaerophilic gut bacteria potentially enable fermentation of the nondigestible carbohydrates for the production of secondary metabolites. Fermentable carbohydrates are initially hydrolyzed to simple sugars (glucose, fructose, galactose, etc.), which are further metabolized to pyruvate, lactate, acetyl-CoA, and succinate. Acetate-producing bacteria *Akkermansia mucinophila*, *Bacteroides* spp., *Bifidobacterium* spp., *Prevotella* spp., and *Ruminococcus* spp. make acetate from acetyl-CoA. *Clostridium* spp. and *Streptococcus* spp. use the Wood-Ljungdahl pathway for acetate production. *Bacteroides*

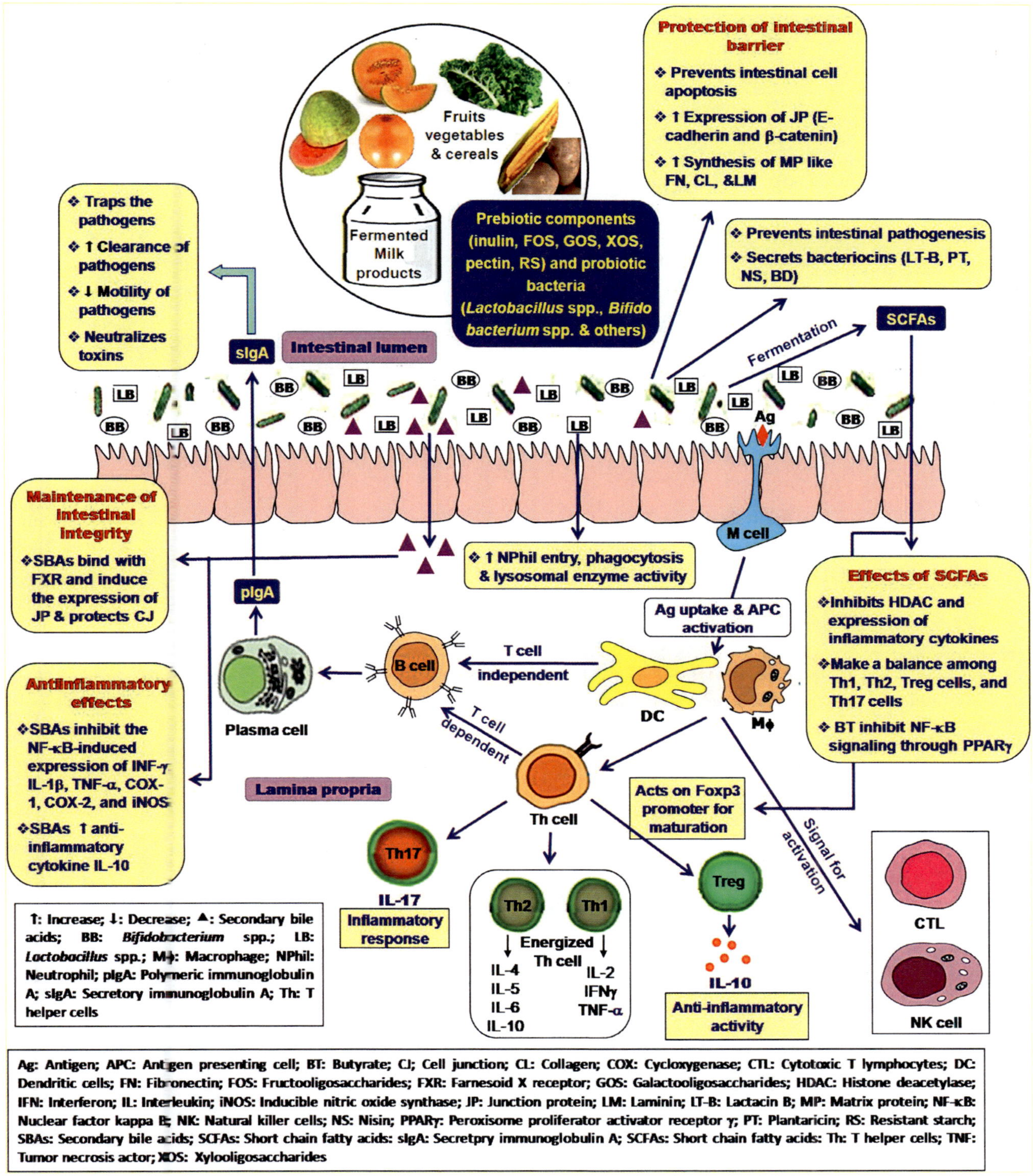

FIGURE 27.2 Immunomodulatory effects of short-chain fatty acids and probiotic bacteria in the gut.

spp., *Phascolarctobacterium succinatutens*, *Dialister* spp., and *Veillonella* spp. mainly adopt the propionate-succinate pathway for the production of propionate. *Roseburia inulinivorans* and *Ruminococcus obeum* specifically use the propanediol pathway for propionate synthesis. The formation of butyrate is most important. The gut microbiota expresses the genes encoding butyrate kinase (*buk*) and butyryl-CoA:acetate-CoA transferase (*ptb*) (Louis et al., 2004). The enzyme butyrate kinase converts the acetyl CoA to butyrate (Zhu et al., 2005). Some organisms (*Eubacterium rectale* (A),

Eubacterium hallii (A, L), *F. prausnitzii* (A), and *Roseburia* spp. (A)) express butyryl-CoA:acetate-CoA transferase, which uses acetate for butyrate production (Koh et al., 2016). Butyrate regulates the proliferation and barrier function of colonocytes and decreases oxidative stress–mediated DNA damage (Gibson et al., 1999). The action of SCFAs is mediated by three receptors: GPR41, GPR109A, and GPR43. Not all these receptors are exclusively present on the immune cells. GPR41 is present in adipocytes and peripheral mononuclear cells. Immune cells exclusively express the GPR43 receptor (Maslowski et al., 2009).

Butyrate protects the integrity of the colonic epithelium by upregulating the expression of junctional protein (Wang et al., 2012) (Fig. 27.2). Butyrate inhibits histone deacetylases to repress the expression of proinflammatory cytokine genes (Chang et al., 2014) (Fig. 27.2). SCFAs induce the maturation of FoxP3$^+$ Treg cells. Macia et al. (2015) reported that SCFA-induced inflammasome-mediated activity increased the production of IL-18 for the protection of the mucosal layer during the inflammatory response. SCFAs prevent the formation of Th1 and Th17 cells to diminish the secretion of inflammatory cytokines. In IBD experimental model, butyrate increases Treg cell population and reduces the expression of IFN-γ (Klampfer et al., 2003). SCFAs maintain a balance among the Th1, Th2, Treg, and Th17 cells along with the secretion of proinflammatory and antiinflammatory cytokine for the potential regulation of colitis, IBD, and other inflammatory-related diseases. Acetate influences the chemotactic activity of neutrophils. It also advances the maturation of Treg cells from naïve CD4$^+$ T cells. Propionate binds with GPR43 of FoxP3$^+$ cells for differentiation of Treg cell and induction of IL-10 (antiinflammatory cytokine) production.

27.6.3 Protective role against inflammation

The inflammatory response is the mediator of many diseases. Butyrate potentially inhibits the inflammatory response through activation of peroxisome proliferator activator receptor gamma (PPARγ), which blocks STAT, AP-1, and NF-κB signaling (Vieira et al., 2013) (Fig. 27.2). SCFAs reduce colonic inflammation and IBD. Patients with IBD show a lower number of SCFA-producing bacteria (Huda-Faujan et al., 2010). Treg cells exhibit GPR43 on their surface that binds with SCFAs, particularly butyrate for their maturation and production of antiinflammatory cytokine IL-10. The reduction of IL-12 and IL-6 and the expression of RelB gene are done through the butyrate-dependent inhibition of histone deacetylase. Chang et al. (2014) reported that butyrate also diminishes IL-6, NOS2, and IL-12 release from macrophages. Vitamin B3 (niacin) is also the ligand of GPR109A. Both butyrate and niacin induce the release of IL-18 from epithelial cells and IL-10 from DCs and macrophages to maintain intestinal integrity and suppression of inflammation (Singh et al., 2014). Thus SCFAs, particularly butyrate, have a vital role in the antiinflammatory response in the colon.

27.6.4 Role of bile acids

Commensal bacteria are directly associated with bile acid metabolism. Bile salts appear in the duodenum through the biliary secretion and are extensively reabsorbed through enterohepatic circulation. Unabsorbed bile salts reach the colon, and commensal microorganisms convert the primary bile acids to secondary bile acids with the help of bile salt hydrolases. Both primary and secondary bile acids bind with the nuclear receptor known as the farnesoid X receptor (FXR). The FXR maintains the expression of junctional protein as well as the integrity of the epithelial cells (Inagaki et al., 2006) (Fig. 27.2). Secondary bile acids bind with TGR5, which is a G-protein-coupled receptor in the rhodopsin subfamily. TGR5-mediated activation decreases macrophage activity, inflammation, and lipid metabolism (Pols et al., 2011). Vavassori et al. (2009) reported that secondary bile acids regulate the expression of NF-κB-mediated genes such as IL-1β, TNF-α, cycloxygenase 1 (COX-1), cycloxygenase 2 (COX-2), and inducible nitric oxide synthase (Fig. 27.2). The bile acid metabolite taurine can activate the nucleotide-binding domain and leucine-rich repeat containing gene family 6 (Nlrp6) signal dependent IL-18 secretion and antimicrobial peptide production. Histamine and spermine inhibit Nlrp6 activation to stop the overexpression (Levy et al., 2015).

27.6.5 Effect on development of gut-associated lymphoid tissue and expansion of gut immunity

The human gut contains a large number of immune cells. Peyer's patches and isolated lymphoid follicles are part of the gut-associated lymphoid tissue (GALT). Lymphoid tissue inducer (LTi) cells promote the development of lymphoid organs after colonization of the intestinal bacteria during fetal life resulting from the consumption of amniotic fluid (Argenio, 2018). LTi cells (CD45intCD4-CD3-CD127hiLin-cells) (Eberl and Littman, 2003; Cupedo et al., 2009) and the mesenchymal organizer cells interact with each other with the help of lymphotoxin $\alpha1\beta2$ for the formation of lymphoid tissue (Jetten, 2009; Cupedo and Mebius, 2005). Antigen-presenting cells and innate lymphoid cells, CD4$^+$ and

$CD8^+$ T cells, and B cells are present in GALT. The lamina propria also contains dendritic cells and macrophages. Thus the development of lymphoid tissue and recruitment of immune cells start before birth. Normally, intestinal M cells (specialized epithelial cells) are capable of uptake of antigens directly from the intestinal lumen. These cells provide antigens to the dendritic cells of the lamina propria. Activated DCs stimulate $CD8^+$ T cells, Th17 cells, and Treg cells. $CD8^+$ cells and Th17 cells are involved in the immune reaction, whereas Treg cells provide balance in the immune response. Commensal microbes induce Treg cell maturation and regulate the activity of aberrant effector T cells during inflammation, leading to progression of the tolerogenic capacity of the host defenses against nonhost gut microbes. This effect prevents unwanted inflammation in the GI tract. The importance of the gut microbiota has been established from the study of germ-free (GF) animals. Experimental evidence from GF animals has shown that the normal microbial population of the gut facilitates the maturation of myeloid and lymphoid progenitor cells for the progression of innate immunity and adaptive immunity (Khosravi et al., 2014).

Probiotics act through MAMPs and PAMPs to exert signaling related to epithelial cell protection, proliferation, and barrier function (Eberl and Boneca, 2010). TLRs and NOD-like receptors (NLRs) help in the interaction with immune cells and the gut microbiota. During infection, TLRs stimulate the secretion of proinflammatory mediators, while SCFAs regulate the inflammatory response via binding with NLRs (Vieira et al., 2013). Mice treated with *S. cerevisiae* strain UFMG 905 show modulation of signaling pathways such as mitogen-activated protein kinase (p38 and JNK), NF-kB, and AP-1 to reduce the transcriptional activity of inflammatory mediators in the intestine (Martins et al., 2011). *Bifidobacterium animalis* var. *lactis* increases intestinal immunoglobulin A (IgA) and antiinflammatory cytokine IL-10 levels along with elevated levels of serum IgA and IgM (Martins et al., 2010).

27.6.6 Protective role of probiotic organisms against pathogenesis

Lactic acid bacteria and bifidobacteria are the most common types of probiotic organisms. *Lactobacilli* are facultative anaerobic, Gram-positive, rod-shaped, spore-forming bacteria. They are a good producer of lactic acid. *Bifidobacteria* are obligate anaerobe, pleomorphic rod-shaped, Gram-positive, nonsporing organisms. They provide both lactic acid and acetic acid. A medicine such as VSL#3 is a mixture of *Lactobacillus casei*, *L. plantarum*, *L. bulgaricus*, *L. acidophilus*, *B. longum*, *B. breve*, *B. infantis*, and *S. thermophilus*. It is applicable for the treatment of ulcerative colitis, irritable bowel syndrome, and diarrhea (Dong et al., 2016).

Adherence of pathogens on the surface of the epithelium followed by disruption of the epithelial barrier initiates the process of pathogenesis. *Lactobacillus* increases the expression of tight junctional protein E-cadherin and beta-catenin to improve the strength of the epithelial barrier (Fig. 27.2). The glycocalyx is the outermost covering of the intestinal epithelial cells, which is a viscous mucilage layer. The glycocalyx is made by glycolipids and glycoproteins (mucin). This layer prevents colonization of pathogenic bacteria and protects the intestinal cells' integrity (Monteagudo-Mera et al., 2019). Probiotic bacteria also protect intestinal cells against cytokine-induced apoptosis.

Probiotic bacteria recognize the sugar residues of the glycoprotein mucin for their adherence and colonization. *Lactobacillus* and *Bifidobacterium* express different factors, such as lipoteichoic acid, surface layer—associated proteins, and mucin-binding proteins (Mubs) on their surfaces for interaction with the mucin layer (Lebeer et al., 2010). Mubs contain Leu-Pro-any amino acid-Thr-Gly (LPxTG) motif at their C terminal. Specific mucin-binding domain facilitates the adherence process of the probiotic bacteria (van Tassell and Miller, 2011). Probiotic organisms induce the expression of *MUC2* and *MUC3* genes to maintain the levels of mucin (Samanta, 2022). They also induce the synthesis of extracellular matrix substances such as fibronectin, collagen, and laminin to increase the strength of the intestinal barrier (Monteagudo-Mera et al., 2019). Pathogenic organisms also bear mucin-binding motifs for their attachment (Popowska et al., 2017). Moreover, alteration of the sugar residues in the mucin layer causes binding of harmful bacteria during dysbiosis (Larsson et al., 2011; Sommer et al., 2014). Therefore improper diet, antibiotic treatment, recurring infection, and low numbers of beneficial organisms influence intestinal pathogenesis.

Additionally, type IV pili of the *Bifidobacterium* and SpaCBA pili of the *Lactobacillus* also help in binding with the surface epithelium (Piepenbrink and Sundberg, 2016; Reunanen et al., 2012). SpaCBA pili help in the binding with macrophages. This association induces the expression of the antiinflammatory cytokine IL-10, while represses IL-6 production (Vargas García et al., 2015). Probiotic bacteria express fibronectin-binding proteins for binding with extracellular matrix glycoprotein fibronectin. *Lactobacillus salivarius* REN shows a specific protein, choline-binding protein A for their binding with epithelial cell layer (Wang et al., 2017). Thus probiotic organisms cover the extracellular matrix layer to decrease the accessibility of the attachment site for the pathogens (Hymes et al., 2016).

Probiotic bacteria produce different types of bacteriocins as antimicrobial agents. *Lactococcus acidophilus* provides lactacin B. *L. plantarum* and *L. lactis* provide plantaricin and nisin, respectively. Another organism, *Bifidobacterium bifidum* NCFB 1454, releases bifidocin B (Nielsen et al., 2010) (Fig. 27.2). Bacteriocins act as a bactericidal agent.

Commensal organisms always fight against pathogenesis and exaggeration of inflammatory response. Segmented filamentous bacteria exert defensive action against pathogens by activating Th17 cells and release inflammatory cytokines IL-17 and IL-22 (Ivanov et al., 2009). Probiotic organisms improve the immune defense. Probiotic bacteria or their antigenic fragments interact with the intestinal M cells, gut epithelial cells, and interdigitant DCs for association to the immune environment of the lamina propria. This interaction induces the expression of cytokines (TNF-α, IFN-γ). Later, cytokines make a network among the different immune cells and epithelial cells, resulting in the release of other cytokines, such as IL-4, IL-6, and IL-10 (Fig. 27.2). Moreover, probiotic bacteria influence the clonal selection of IgA-producing B cells for IgA synthesis.

27.6.7 Role in innate immunity

Probiotic bacteria can tolerate gastric HCl, bile acids, intestinal pH profile, proteolytic enzymes, antimicrobial peptides, and intestinal peristalsis (Galdeano et al., 2007). The viable bacteria or their cellular fragments stimulate innate and adaptive immune responses by activating the macrophages and DCs of the Peyer's patches and the lamina propria. The lipoteichoic acid of *Bifidobacteria* and *Lactobacilli* trigger nitric oxide (NO)-dependent pathogen-infected cell death by increasing TNF-α secretion from macrophages. NO also induces the expression of phagocytosis receptors FcγRIII and TLR (Delcenserie et al., 2008). Experimental evidence indicated that strains of *Lactobacillus paracasei* and *L. acidophilus* promote TNF-α secretion and recruit polymorphonuclear cells in the events of phagocytosis (Kourelis et al., 2010).

27.6.8 Probiotic bacteria and modulation of cytokine profile

Probiotic bacteria can regulate the cytokine profile. These bacteria prevent the degradation of inhibitor of kappa B (IκB). They also modulate the translocation of the RelA subunit of NF-κB by controlling the PPAR-γ-dependent system (Kelly et al., 2004). Thus, gut microbiota makes a protective defense against pathogenesis. Probiotic bacteria stimulate Th2 cells and macrophages for the production of IL-10 to balance the immune response and maintenance of gut homeostasis (Sichetti et al., 2018). In the same event, SCFAs increase the number of Treg cells for the production of IL-10. McCarthy et al. (2003) indicated that *L. salivarius* and *Bifidobacterium infantis* had ability to reduce INF-γ and TNF-α even in IL-10-deficient mice. The balancing property between inflammatory and antiinflammatory activity creates a healthy situation in the colonic environment. Other organisms, such as *B. fragilis* and Firmicutes, increase the numbers of FoxP3$^+$ Treg cells and IL-10 levels (El Aidy et al., 2012).

Probiotic organisms may act as immunostimulatory and immunoregulatory agents. Immunostimulatory bacteria activate Th1 and NK cells and increase the production of proinflammatory cytokines. *Lactobacillus* and *Bifidobacterium* act as immunobiotic agents. Lactic acid from these bacteria binds with TLR2, inducing a Th1 response for the production of IFN-β during GI infection (Zhang et al., 2018). Alternatively, these bacteria diminish the activity of Th2 cells and release IL-4, IL-13, and IgE (Rosa et al., 2017). The initial inflammatory response and cytokine levels help the probiotic bacteria for the expression of human beta 2 defensin (Fig. 27.2).

Lactobacillus and *Bifidobacterium* modulate cytokine levels in a reversible fashion. They induce the expression of IL-10 but downregulate IL-6 and TNF-α expression (Fig. 27.2). *Lactobacillus plantarum* L67 significantly decreased allergic activity by lowering the circulatory IgE (Takeda et al., 2014).

27.6.9 Role in enhanced secretion of IgA

Probiotic bacteria stimulate the production of IgA from the plasma cells. IgA-expressing B cells enter the circulation through the mesenteric lymph node and may return to the gut after recirculation from the other tissues. IgA is the crucial factor in systemic immunity. Probiotic bacteria increase the production of secretory IgA (sIgA) from the plasma cells of the lamina propria. Intestinal epithelial cells secrete this antibody to the intestinal lumen, where the sIgA binds with the mucus layer and advances the clearance of pathogens and toxins without activating the inflammatory response, the condition known as immune exclusion (Xiong and Hu, 2015) (Fig. 27.2). Cytokines IL-10, IL-6, IL-4, IL-2, TGF-β, and vasoactive intestinal peptide exhibit T-independent activation of IgA production. These cytokines promote the switching of IgA synthesis over IgM synthesis (Galdeano et al., 2007). T-independent activity produces low-affinity innate-type IgA for nonspecific antigen binding. The T-dependent response occurs through B cell receptor and CD40, resulting in B cell activation and production of high-affinity antigen-specific IgA (Pabst, 2012). The plasma cells of the lamina propria secrete polymeric IgA (pIgA) comprising two monomers of IgA linked by the joining (J) chain. The polymeric immunoglobulin receptor (pIgR) binds with pIgA for its transportation. At the luminal face of the epithelial cells, the pIgA-pIgR complex provides sIgA. The sIgA interacts with self-antigens, food components, commensal organisms, pathogenic bacteria, and toxins and traps the antigen on the mucus layer, blocking the accessibility of the antigen to the cell surface receptors. sIgA acts as a first-line barrier (Pabst, 2012).

27.6.10 Role in immunosurveillance-mediated cancer prevention

Some probiotic organisms are protective against cancer. Lee et al. (2008) reported that *L. reuteri* inhibits NF-κB-mediated expression of COX-2, cyclin D1, and Bcl-2 to restrict the inflammatory response, cell proliferation, and antiapoptotic activity, respectively. *L. casei* strain Shirota promotes T-cell-induced cytotoxicity in colon cancer cells (Yamazaki et al., 2000). *Bifidobacterium* activates DCs and tumor-specific $CD8^+$ T cells to restrict the cancerous growth (Routy et al., 2018). Another organism, *Bacillus polyfermenticus*, increased the number of $CD4^+$ and $CD8^+$ cells and increased circulatory IgG levels in patients with colon cancer during clinical trials (Rossi et al., 2018). Purified exopolysaccharide LA-EPS-20079 of *L. acidophilus* 20079 is also protective in human colon cancer (El-Deeb et al., 2018). *Bifidobacterium*, *Collinsella aerofaciens*, *Enterococcus faecium*, *Bacteroides thetaiotaomicron*, and *B. fragilis* block PD-1 (programmed cell death protein 1) and CTLA-4 (cytotoxic T-lymphocytes associated protein 4) activity to resist the immune tolerance capacity of the tumor cells (Matson et al., 2018; Vétizou et al., 2015).

27.7 Additional benefits of probiotics

Gut microbiota has a role in the maturation and development of the human central nervous system. The gut-brain axis makes an association with brain tissue, the autonomic nervous system, the hypothalamic pituitary adrenal axis, and the enteric nervous system (Cussotto et al., 2018). Gut bacteria have an impact on the maturation and functions of the microglia in the central nervous system (Erny et al., 2015).

A large number of women have suffered from bacterial vaginosis (BV), urinary tract infection (UTI), and yeast infections. Probiotic bacteria such as *Lactobacillus brevis*, *L. casei*, *L. vaginalis*, *L. delbrueckii*, *L. salivarius*, *L. reuteri*, and *L. rhamnosus* are common residents of the urinary tract. Any imbalance in the microbial population advances the state of BV and UTI (Waigankar and Patel, 2011).

27.8 Concluding remarks

Commensal organisms are the most mostly beneficial, but dysbiosis can promote the activity of opportunistic bacteria and advances several adverse effects. Dietary ingredients (cereal products, fruits, and vegetables) and fermented foods contribute prebiotic and probiotic components that preserve a healthy microbial community within the gut. The synergistic effects of prebiotic and probiotic components increase the production of SCFAs and other essential bioactive components, which have various immunomodulatory effects, including the development of GALT, protection of gut integrity, prevention of pathogenesis, maintenance of cytokine levels, and regulation of the inflammatory response. The multidimensional aspects of immune defense preserve the normal homeostasis of the gut. Thus prebiotics and probiotics can improve and maintain the normal health of the individual. Probiotics-mediated treatment regimens are also popular for the treatment of GI infections, gastritis, diarrhea, IBD, colorectal cancer, and others. Recently, probiotics have been used as adjuvants in clinical trials for the chemotherapeutic treatment of some cancers. It can be concluded that regular intake of prebiotics and probiotics have health promotional effects.

References

Adnan, M.L., Pramaningtyas, M.D., 2020. Probiotics as promising immunomodulatory agents to prevent covid-19. infection: a narrative review. Int. J. Med. Stud. 8 (2), 121–125. Available from: https://doi.org/10.5195/ijms/2020.486.

Argenio, V.D., 2018. The prenatal microbiome: a new player for human health. High-Throughput 7, 38. Available from: https://doi.org/10.3390/ht7040038.

Bernstein, C., et al., 2011. Carcinogenicity of deoxycholate, a secondary bile acid. Arch. Toxicol. 85, 863–871.

Bernstein, H., Bernstein, C., Payne, C.M., Dvorak, K., 2009. Bile acids as endogenous etiologic agents in gastrointestinal cancer. World J. Gastroenterol. 15, 3329–3340.

Birt, D.F., Boylston, T., Hendrich, S., et al., 2013. Resistant Starch: promise for improving human health. Adv. Nutr. 4, 587–601. Available from: https://doi.org/10.3945/an.113.004325.

Bornet, F.R., Brouns, F., Tashiro, Y., Duvillier, V., 2002. Nutritional aspects of short-chain fructooligosaccharides: natural occurrence, chemistry, physiology and health implications. Dig. Liver Dis. 34 (Suppl 2), S111–S120.

Brennan, C.A., Garrett, W.S., 2019. *Fusobacterium nucleatum*—symbiont, opportunist and oncobacterium. Nat. Rev. Microbiol. 17 (3), 156–166. Available from: https://doi.org/10.1038/s41579-018-0129-6.

Buffie, C.G., Pamer, E.G., 2013. Microbiota-mediated colonization resistance against intestinal pathogens. Nat. Rev. Immunol. 3, 790–801.

Chang, P.V., Hao, L., Offermanns, S., Medzhitov, R., 2014. The microbial metabolite butyrate regulates intestinal macrophage function via histone deacetylase inhibition. Proc. Natl Acad. Sci. U. S. A. 111, 2247–2252.

Clarke, S., Green-Johnson, J., Brooks, S., et al., 2016. B2-1 fructan supplementation alters host immune responses in a manner consistent with increased exposure to microbial components: results from a double-blinded, randomised, cross-over study in healthy adults. Br. J. Nutr. 115, 1748–1759.

Cremonesi, E., et al., 2018. Gut microbiota modulate T cell trafficking into human colorectal cancer. Gut 67, 1984–1994.

Cupedo, T., Mebius, R.E., 2005. Cellular interactions in lymph node development. J. Immunol. 174, 21–25.

Cupedo, T., Crellin, N.K., Papazian, N., et al., 2009. Human fetal lymphoid tissue-inducer cells are interleukin 17-producing precursors to RORC + CD127 + natural killer-like cells. Nat. Immunol. 10, 66–74.

Cussotto, S., Sandhu, K.V., Dinan, T.G., Cryan, J.F., 2018. The neuroendocrinology of the microbiota-gut-brain axis: a behavioural perspective. Front. Neuroendocrinol. 51, 80–101. Available from: https://doi.org/10.1016/j.yfrne.2018.04.002.

Davani-Davari, D., Negahdaripour, M., Karimzadeh, I., et al., 2019. Prebiotics: definition, types, sources, mechanisms, and clinical applications. Foods 8, 92. Available from: https://doi.org/10.3390/foods8030092.

Delcenserie, V., et al., 2008. Immunomodulatory effects of probiotics in the intestinal tract. Curr. Iss. Mol. Biol. 10 (1–2), 37–54.

Dong, J., Teng, G., Wei, T., Gao, W., et al., 2016. Methodological quality assessment of *meta*-analyses and systematic reviews of probiotics in inflammatory bowel disease and pouchitis. PLoS One 11 (12), Article ID e0168785.

Eberl, G., Boneca, I.G., 2010. Bacteria and MAMP-induced morphogenesis of the immune system. Curr. Opin. Immunol. 22, 448–454. Available from: https://doi.org/10.1016/j.coi.2010.06.002.

Eberl, G., Littman, D.R., 2003. The role of the nuclear hormone receptor RORgammat in the development of lymph nodes and Peyer's patches. Immunol. Rev. 195, 81–90.

El Aidy, S., van Baarlen, P., Derrien, M., et al., 2012. Temporal and spatial interplay of microbiota and intestinal mucosa drive establishment of immune homeostasis in conventionalized mice. Mucosal Immunol. 5, 567–579. Available from: https://doi.org/10.1038/mi.2012.32.

El-Deeb, N.M., Yassin, A.M., Al-Madboly, L.A., El-Hawiet, A., 2018. A novel purified *Lactobacillus acidophilus* 20079 exopolysaccharide, LA-EPS-20079, molecularly regulates both apoptotic and NF-KB inflammatory pathways in human colon cancer. Microb. Cell Fact. 17 (1), Article 29.

Endesfelder, D., Engel, M., Castell, W.Z., 2016. Gut immunity and type 1 diabetes: a melange of microbes, diet, and host interactions? Curr. Diab. Rep. 16 (7), 60. Available from: https://doi.org/10.1007/s11892-016-0753-3.

Erny, D., Hrabe de Angelis, A.L., Jaitin, D., et al., 2015. Host microbiota constantly control maturation and function of microglia in the CNS. Nat. Neurosci. 18 (7), 965–977. Available from: https://doi.org/10.1038/nn.4030.

Fanaro, S., Boehm, G., Garssen, J., et al., 2005. Galacto-oligosaccharides and long-chain fructo-oligosaccharides as prebiotics in infant formulas: a review. Acta Paediatr. Suppl. 94, 22–26.

Fatima, N., Akhtar, T., Sheikh, N., 2017. Prebiotics: a novel approach to treat hepatocellular carcinoma. Can. J. Gastroenterol. Hepatol. Available from: https://doi.org/10.1155/2017/6238106Article ID 6238106.

Fessler, F., Matson, V., Gajewski, T.F., 2019. Exploring the emerging role of the microbiome in cancer immunotherapy. J. Immunother. Cancer 7, 108. Available from: https://doi.org/10.1186/s40425-019-0574-4.

Food and Agriculture Organization of the United Nations and World Health Organization, 2001. Health and nutritional properties of probiotics in food including powder milk with live lactic acid bacteria. World Health Organization. Available from: http://www.who.int/foodsafety/publications/fs_management/en/probiotics.pdf

Galdeano, C.M., Leblanc DeMorenoDe, A., Vinderola, G., Bonet Bibas, M.E., Perdigón, G., 2007. Proposed model: mechanisms of immunomodulation induced by probiotic bacteria. Clin. Vaccine Immunol. 14 (5), 485–492.

Garrett, W.S., 2019. The gut microbiota and colon cancer. Science 364 (6446), 1133–1135. Available from: https://doi.org/10.1126/science.aaw2367.

Gibson, G.R., Roberfroid, M.B., 1995. Dietary modulation of the human colonic microbiota. Introducing the concept of prebiotics. J. Nutr. 125, 1401–1412. Available from: https://doi.org/10.1093/jn/125.6.1401.

Gibson, P.R., Rosella, O., Wilson, A.J., et al., 1999. Colonic epithelial cell activation and the paradoxical effects of butyrate. Carcinogenesis 20, 539–544. Available from: https://doi.org/10.1093/carcin/20.4.539.

Gibson, G.R., Hutkins, R., Sanders, M.E., et al., 2017. The international scientific association for probiotics and prebiotics (ISAPP) consensus statement on the definition and scope of prebiotics. Nat. Rev. Gastroenterol. Hepatol. 14, 491–502. Available from: https://doi.org/10.1038/nrgastro.2017.75.

Gorjifard, S., Goldszmid, R.S., 2016. Microbiota—myeloid cell crosstalk beyond the gut. J. Leukoc. Biol. 100 (5), 865–879. Available from: https://doi.org/10.1189/jlb.3RI0516-222R.

Griffin, J.L., Wang, X.Z., Stanley, E., 2015. Does our gut microbiome predict cardiovascular risk? A review of the evidence from metabolomics. Circ. Cardiovasc. Genet. 8, 187–191. Available from: https://doi.org/10.1161/CIRCGENETICS.114.000219.

He, Z., et al., 2019. *Campylobacter jejuni* promotes colorectal tumorigenesis through the action of cytolethal distending toxin. Gut 68, 289–300. Available from: https://doi.org/10.1136/gutjnl-2018-317200.

Hooper, L.V., Macpherson, A.J., 2010. Immune adaptations that maintain homeostasis with the intestinal microbiota. Nat. Rev. Immunol. 10, 159–169. Available from: https://doi.org/10.1038/nri2710.

Huda-Faujan, N., Abdulamir, A.S., Fatimah, A.B., et al., 2010. The impact of the level of the intestinal short chain fatty acids in inflammatory bowel disease patients vs healthy subjects. Open. Biochem. J. 4, 53–58. Available from: https://doi.org/10.2174/1874091X01004010053.

Hymes, J.P., Johnson, B.R., Barrangou, R., Klaenhammer, T.R., 2016. Functional analysis of an S-layer-associated fibronectin-binding protein in *Lactobacillus acidophilus* NCFM. Appl. Env. Microbiol. 82, 2676–2685. Available from: https://doi.org/10.1128/AEM.00024-16.

Inagaki, T., Moschetta, A., Lee, Y.-K., Peng, L., et al., 2006. Regulation of antibacterial defense in the small intestine by the nuclear bile acid receptor. Proc. Natl. Acad. Sci. U. S. A. 103, 3920–3925. 182.

Ivanov, I.I., Atarashi, K. Manel, N., Brodie, E.L., et al., 2009. Induction of intestinal Th17 cells by segmented filamentous bacteria. Cell 139, 485–498. Available from: https://doi.org/10.1016/j.cell.2009.09.033.

Jetten, A.M., 2009. Retinoid-related orphan receptors (RORs): critical roles in development, immunity, circadian rhythm, and cellular metabolism. Nucl. Recept. Signal. 7, e003. Available from: https://doi.org/10.1621/nrs.07003.

Joossens, M., De Preter, V., Ballet, V., et al., 2012. Effect of oligofructose-enriched inulin (OF-IN) on bcterial composition and disease activity of patients with Crohn's disease: results from a double-blinded randomised controlled trial. Gut 61, 958. Available from: https://doi.org/10.1136/gutjnl-2011-300413.

Kelly, D., Campbell, J.I., King, T.P., et al., 2004. Commensal anaerobic gut bacteria attenuate inflammation by regulating nuclear-cytoplasmic shuttling of PPAR-gamma and RelA. Nat. Immunol. 5, 104–112.

Khangwal, I., Shukla, P., 2019. Potential prebiotics and their transmission mechanisms: recent approaches. J. Food Drug. Anal. 27, 649–656. Available from: https://doi.org/10.1016/j.jfda.2019.02.003.

Kho, Z.Y., Lal, S.K., 2018. The human gut microbiome – a potential controller of wellness and disease. Front. Microbiol. 9, 1835. Available from: https://doi.org/10.3389/fmicb.2018.01835.

Khosravi, A., Yáñez, A., Price, J.G., et al., 2014. Gut microbiota promote hematopoiesis to control bacterial infection. Cell Host Microbe 15, 374–381. Available from: https://doi.org/10.1016/j.chom.2014.02.006.

Klampfer, L., Huang, J., Sasazuki, T., Shirasawa, S., Augenlicht, L., 2003. Inhibition of interferon gamma signaling by the short chain fatty acid butyrate. Mol. Cancer Res. 1, 855–862.

Koh, A., De Vadder, F., Kovatcheva-Datchary, P., Backhed, F., 2016. From dietary fiber to host physiology: short-chain fatty acids as key bacterial metabolites. Cell 165, 1332–1345. Available from: https://doi.org/10.1016/j.cell.2016.05.041.

Kong, F., Cai, Y., 2019. Study insights into gastrointestinal cancer through the gut microbiota. BioMed. Res. Int. Available from: https://doi.org/10.1155/2019/8721503Article ID 8721503.

Konturek, P.C., Koziel, J., Dieterich, W., et al., 2016. Successful therapy of *Clostridium difficile* infection with fecal microbiota transplantation. J. Physiol. Pharmacol. 67, 859–866.

Kourelis, A., Zinonos, I., Kakagianni, M., et al., 2010. Validation of the dorsal air pouch model to predict and examine immunostimulatory responses in the gut. J. Appl. Microbiol. 108 (1), 274–284.

Landers, C.J., Cohavy, C., Misra, R., et al., 2002. Selected loss of tolerance evidenced by Crohn's disease–associated immune responses to auto- and microbial antigens. Gastroenterol. 123, 689–699. Available from: https://doi.org/10.1053/gast.2002.35379.

Larsson, J.M.H., Karlsson, H., Crespo, J.G., et al., 2011. Altered O-glycosylation profile of MUC2 mucin occurs in active ulcerative colitis and is associated with increased inflammation. Inflamm. Bowel Dis. 17, 2299–2307. Available from: https://doi.org/10.1002/ibd.21625.

Lebeer, S., Vanderleyden, J., De Keersmaecker, S.C.J., 2008. Genes and molecules of lactobacilli supporting probiotic action. Microbiol. Mol. Biol. Rev. 72 (4), 728–764. Available from: https://doi.org/10.1128/MMBR.00017-08.

Lebeer, S., Vanderleyden, J., De Keersmaecker, S.C.J., 2010. Host interactions of probiotic bacterial surface molecules: comparisonwith commensals and pathogens. Nat. Rev. Microbiol. 8 (3), 171–184.

LeBlanc, J.G., Laiño, J.E., del Valle, M.J., et al., 2011. B-Group vitamin production by lactic acid bacteria–current knowledge and potential applications. J. Appl. Microbiol. 111, 1297–1309. Available from: https://doi.org/10.1111/j.1365-2672.2011.05157.x.

Lee, H.Y., Park, J.H., Seok, S.-H., Baek, M.-W., et al., 2006. Human originated bacteria, *Lactobacillus rhamnosus* PL60, produce conjugated linoleic acid and show anti-obesity effects in diet-induced obese mice. Biochim. Biophys. Acta 1761, 736–744. Available from: https://doi.org/10.1016/j.bbalip.2006.05.007.

Lee, D.K., Jang, S., Kim, M.J., et al., 2008. Anti-proliferative effects of *Bifidobacterium adolescentis* SPM0212 extract on human colon cancer cell lines. BMC Cancer 8, 310. Available from: https://doi.org/10.1186/1471-2407-8-310.

Lerner, A., Aminov, R., Matthias, T., 2017a. Transglutaminases in dysbiosis as potential environmental drivers of autoimmunity. Front. Microbiol. 8, 66. Available from: https://doi.org/10.3389/fmicb.2017.00066.

Levy, M., Thaiss, C.A., Zeevi, D., Dohnalová, L., et al., 2015. Microbiotamodulated metabolites shape the intestinal microenvironment by regulating NLRP6 inflammasome signaling. Cell 163, 1428–1443.

Ley, R.E., Turnbaugh, P.J., Klein, S., Gordon, J.I., 2006. Microbial ecology:human gut microbes associated with obesity. Nature 444, 1022–1023. Available from: https://doi.org/10.1038/4441022a.

Lomax, A.R., Cheung, L.V., Noakes, P.S., Miles, E.A., Calder, P.C., 2015. Inulin-type β-2-1 fructans have some effect on the antibody response to seasonal influenza vaccination in healthy middle-aged humans. Front. Immunol. 6, 490. Available from: https://doi.org/10.3389/fimmu.2015.00490.

Louis, P., Duncan, S.H., McCrae, S.I., et al., 2004. Restricted distribution of the butyrate kinase pathway among butyrate-producing bacteria from the human colon. J. Bacteriol. 186, 2099–2106.

Machiels, K., Joossens, M., Sabino, J., 2014. De et al. A decrease of the butyrate-producing species *Roseburia hominis* and *Faecalibacterium prausnitzii* defines dysbiosis in patients with ulcerative colitis. Gut 63, 1275–1283. Available from: https://doi.org/10.1136/gutjnl-2013-304833.

Macia, L., et al., 2015. Metabolite-sensing receptors GPR43 and GPR109A facilitate dietary fibre-induced gut homeostasis through regulation of the inflammasome. Nat. Commun. 6, 6734. Available from: https://doi.org/10.1038/ncomms7734.

Manning, T.S., Gibson, G.R., 2004. Prebiotics. Best Pract. Res. Clin. Gastroenterol. 18 (2), 287–298. Available from: https://doi.org/10.1016/j.bpg.2003.10.008.

Martins, A.K., Martins, F.S., Gomes, D.A., et al., 2010. Evaluation of in vitro antagonism and of in vivo immune modulation and protection against pathogenic experimental challenge of two probiotic strains of *Bifidobacterium animalis* var. *lactis*. Arch. Microbiol. 192, 995–1003. Available from: https://doi.org/10.1007/s00203-010-0626-0.

Martins, F.S., Elian, S.D., Vieira, A.T., Tiago, F.C., et al., 2011. Oral treatmentwith *Saccharomycescerevisiae* strain UFMG 905 modulates immune responses and interferes with signal pathways involved in the activation of inflammation in a murine model of typhoid fever. Int. J. Med. Microbiol. 301, 359–364. Available from: https://doi.org/10.1016/j.ijmm.2010.11.002.

Maslowski, K.M., Vieira, A.T., Ng, A., Kranich, J., et al., 2009. Regulation of inflammatory responses by gut microbiota and chemoattractant receptor GPR43. Nature 461, 1282–1286. Available from: https://doi.org/10.1038/nature08530.

Matson, V., Fessler, J., Bao, R., et al., 2018. The commensal microbiome is associated with anti–PD-1 efficacy in metastatic melanoma patients. Science 359, 104–108. Available from: https://doi.org/10.1126/science.aao3290.

McCarthy, J., O'Mahony, L., O'Callaghan, L., et al., 2003. Double blind, placebo controlled trial of two probiotic strains in interleukin 10 knockout mice and mechanistic link with cytokine balance. Gut 52 (7), 975–980.

Mohanty, D., Ray, P., 2016. Evaluation of probiotic and antimicrobial properties of lactobacillus strains isolated from dairy products. Int. J. Pharm. Sci. 8 (11), 230–234.

Monteagudo-Mera, A., Rastall, R.A., Gibson, G.R., et al., 2019. Adhesion mechanisms mediated by probiotics and prebiotics and their potential impact on human health. Appl. Microbiol. Biotechnol. 103, 6463–6472. Available from: https://doi.org/10.1007/s00253-019-09978-7.

Nielsen, D.S., Cho, G.-S., Hanak, A., Huch, M., Franz, C.M.A.P., Arneborg, N., 2010. The effect of bacteriocin-producing *Lactobacillus plantarum* strains on the intracellular pH of sessile and planktonic *Listeria monocytogenes* single cells. Int. J. Food Microbiol. 141, S53–S59.

Okolie, C.L., Rajendran, S.R.C.K., Udenigwem, C.C., Aryee, A.N.A., Mason, B., 2017. Prospects of brown seaweed polysaccharides (BSP) as prebiotics and potential immunomodulators. J. Food Biochem. 41 (5), e12392. Available from: https://doi.org/10.1111/jfbc.12392.

Osadchiy, V., Martin, C.R., Mayer, E.A., 2019. The gut–brain axis and the microbiome: mechanisms and clinical implications. Clin. Gastroenterol. Hepatol. 17 (2), 322–332.

Pabst, O., 2012. New concepts in the generation and functions of IgA. Nat. Rev. Immunol. 12, 821–832. Available from: https://doi.org/10.1038/nri3322.

Park, C.H., Eun, C.S., Han, D.S., 2018. Intestinal microbiota, chronic inflammation, and colorectal cancer. Intest. Res. 16 (3), 338–345.

Piepenbrink, K.H., Sundberg, E.J., 2016. Motility and adhesion through type IV pili in gram-positive bacteria. Biochem. Soc. Trans. 44, 1659–1666. Available from: https://doi.org/10.1042/BST20160221.

Pols, T.W.H., Nomura, M., Harach, T., Lo Sasso, G.M., et al., 2011. TGR5 activation inhibits atherosclerosis by reducing macrophage inflammation and lipid loading. Cell Metab. 14, 747–757.

Popowska, M., Krawczyk-Balska, A., Ostrowski, R., Desvaux, M., 2017. InlL from *Listeria monocytogenes* is involved in biofilm formation and adhesion to mucin. Front. Microbiol . Available from: https://doi.org/10.3389/fmicb.2017.00660.

Pruitt, R.N., Chumbler, N.M., Rutherford, S.A., et al., 2012. Structural determinants of *Clostridium difficile* toxin a glucosyltransferase activity. J. Biol. Chem. 287, 8013–8020. Available from: https://doi.org/10.1074/jbc.M111.298414.

Qin, J., Li, R., Raes, J., et al., 2010. Ahumangutmicrobial gene catalogue established by metagenomic sequencing. Nature 464 (7285), 59–65.

Reunanen, J., von Ossowski, I., Hendrickx, A.P.A., Palva, A., de Vosa, W.M., 2012. Characterization of the SpaCBA pilus fibers in the probiotic *Lactobacillus rhamnosus* GG. Appl. Env. Microbiol. 78, 2337–2344. Available from: https://doi.org/10.1128/AEM.07047-11.

Rodriguez-Lagunas, M.J., Azagra-Boronat, I., Saldaña-Ruíz, S., Massot-Cladera, M., et al., 2017. Immunomodulatory role of probiotics in early life. In: Muñoz-Torrero, D., Riu, M., Feliu 2, C. (Eds.), Recent Adv. Pharm. Sci. In: VII. Research Signpost Trivandrum, Kerala, India, pp. 19–34.

Rooks, M.G., Garrett, W.S., 2016. Gut microbiota, metabolites and host immunity. Nat. Rev. Immunol. 16, 341–352. Available from: https://doi.org/10.1038/nri.2016.42.

Rosa, D.D., Dias, M.M.S., Grześkowiak, Ł.M., Reis, S.A., Conceição, L.L., Peluzio, M.D.C.G., 2017. Milk kefir: nutritional, microbiological and health benefits. Nutr. Res. Rev. 30 (1), 82–96.

Rossi, M., Mirbagheri, S.S., Keshavarzian, A., Bishehsari, F., 2018. Nutraceuticals in colorectal cancer: a mechanistic approach. Eur. J. Pharmacol. 833, 396–402.

Routy, B., et al., 2018. The gut microbiota influences anticancer immunosurveillance and general health. Nat. Rev. Clin. Oncol. 15, 382–396. Available from: https://doi.org/10.1038/s41571-018-0006-2.

Rubinstein, M.R., Wang, X., Liu, W., et al., 2013. *Fusobacterium nucleatum* promotes colorectal carcinogenesis by modulating E-Cadherin/β-catenin signaling via its FadA adhesion. Cell Host Microbe 14, 195–206.

Samanta, S., 2019. Microbial pectinases: a review on molecular and biotechnological perspectives. J. Microbiol. Biotechnol. Food Sci. 9 (2), 248–266. Available from: https://doi.org/10.15414/jmbfs.2019.9.2.248-266.

Samanta, S., 2022. Potential impacts of prebiotics and probiotics on cancer prevention. Anti-Cancer Agents Med. Chem. 22 (4), 605–628. Available from: https://doi.org/10.2174/1871520621999201210220442.

Scott, A.J., Merrifield, C.A., Younes, J.A., Pekelharing, E.P., 2018. Pre-, pro- and synbiotics in cancer prevention and treatment—a review of basic and clinical research. Ecancer 12, 869. Available from: https://doi.org/10.3332/ecancer.2018.869.

Selma, M.V., Espin, J.C., Tomas-Barberan, F.A., 2009. Interaction between phenolics and gut microbiota: role in human health. J. Agric. Food Chem. 57, 6485–6501. Available from: https://doi.org/10.1021/jf902107d.

Shanahan, F., 2013. The colonic microbiota in health and disease. Curr. Opin. Gastroenterol. 29 (1), 49–54.

Sichetti, M., De Marco, S., Pagiotti, R., et al., 2018. Antiinflammatory effect of multistrain probiotic formulation (*L. rhamnosus, B. lactis*, and *B. longum*). Nutrition 53, 95–102.

Silvi, S., Rumney, C.J., Cresci, A., Rowland, I.R., 1999. Resistant starch modifies gut microflora and microbial metabolism in human flora-associated rats inoculated with faeces from Italian and UK donors. J. Appl. Microbiol. 86, 521–530.

Singh, N., Gurav, A., Sivaprakasam, S., Brady, E., et al., 2014. Activation of Gpr109a, receptor for niacin and the commensal metabolite butyrate, suppresses colonic inflammation and carcinogenesis. Immunity 40, 128–139.

Sokol, H., Pigneur, B., Watterlot, L., et al., 2008. Faecalibacterium prausnitzii is an anti-inflammatory commensal bacterium identified by gut microbiota analysis of Crohn's disease patients. Proc. Natl. Acad. Sci. U. S. A. 105, 16731–16736. Available from: https://doi.org/10.1073/pnas.0804812105.

Sommer, F., Adam, N., Johansson, M.E.V., et al., 2014. Altered mucus glycosylation in core 1 O-glycan-deficient mice affects microbiota composition and intestinal architecture. PLoS One 9, e85254. Available from: https://doi.org/10.1371/journal.pone.0085254.

Sommer, F., Bäckhed, F., 2013. The gut microbiota-masters of host development and physiology. Nat. Rev. Microbiol. 11, 227–238.

Takeda, S., Hidaka, M., Yoshida, H., et al., 2014. Antiallergic activity of probiotics from Mongolian dairy products on type I allergy in mice and mode of antiallergic action. J. Funct. Foods 9, 60–69.

Tanaka, M., Nakayama, J., 2017. Development of the gut microbiota in infancy and its impact on health in later life. Allergol. Int. 66 (4), 515–522. Available from: https://doi.org/10.1016/j.alit.2017.07.010.

Tian, T., Freeman, S., Corey, M., German, J.B., Barile, D., 2017. Chemical characterization of potentially prebiotic oligosaccharides in brewed coffee and spent coffee grounds. J. Agric. Food Chem. 65, 2784e92.

Tomkovich, S., et al., 2017. Locoregional effects of microbiota in a preclinical model of colon carcinogenesis. Cancer Res. 77, 2620–2632.

Topping, D.L., Clifton, P.M., 2001. Short-chain fatty acids and human colonic function: roles of resistant starch and nonstarch polysaccharides. Physiol. Rev. 81, 1031–1064.

Tzounis, X., Rodriguez-Mateos, A., Vulevic, J., et al., 2011. Prebiotic evaluation of cocoa-derived flavanols in healthy humans by using a randomized, controlled, double-blind, crossover intervention study. Am. J. Clin. Nutr. 93, 62–72.

van Tassell, M.L., Miller, M.J., 2011. *Lactobacillus* adhesion to mucus. Nutrients 3 (5), 613–636.

Vargas García, C.E., Petrova, M., Claes, I.J.J., et al., 2015. Piliation of *Lactobacillus rhamnosus* GG promotes adhesion, phagocytosis, and cytokine modulation in macrophages. Appl. Env. Microbiol. 81, 2050–2062. Available from: https://doi.org/10.1128/AEM.03949-14.

Vavassori, P., Mencarelli, A., Renga, B., Distrutti, E., Fiorucci, S., 2009. The bile acid receptor FXR is a modulator of intestinal innate immunity. J. Immunol. 183, 6251–6261.

Vétizou, M., Pitt, J.M., Daillère, R., et al., 2015. Anticancer immunotherapy by CTLA-4 blockade relies on the gut microbiota. Science 350, 1079–1084. Available from: https://doi.org/10.1126/science.aad1329.

Vieira, A.T., Teixeira, M.M., Martins, F.S., 2013. The role of probiotics and prebiotics in inducing gut immunity. Front. Immunol. 4, 445. Available from: https://doi.org/10.3389/fimmu.2013.00445. Article.

Vivarelli, S., Salemi, R. Candido, S., et al., 2019. Gut microbiota and cancer: from pathogenesis to therapy. Cancers 11, 38. Available from: https://doi.org/10.3390/cancers11010038.

Vulevic, J., Juric, A., Walton, G.E., et al., 2015. Influence of galacto-oligosaccharide mixture (β-gos) on gut microbiota, immune parameters and metabonomics in elderly persons. Br. J. Nutr. 114, 586–595.

Waigankar, S.S., Patel, V., 2011. Role of probiotics in urogenital healthcare. J. Midlife Health 2 (1), 5–10. Available from: https://doi.org/10.4103/0976-7800.83253.

Wang, H.B., Wang, P.Y., Wang, X., Wan, Y.L., Liu, Y.C., 2012. Butyrate enhances intestinal epithelial barrier function via up-regulation of tight junction protein Claudin-1 transcription. Dig. Dis. Sci. 57, 3126–3335. Available from: https://doi.org/10.1007/s10620-012-2259-4.

Wang, R., Jiang, L., Zhang, M., et al., 2017. The adhesion of *Lactobacillus salivarius* REN to a human intestinal epithelial cell line requires S-layer proteins. Sci. Rep. 7, 44029. Available from: https://doi.org/10.1038/srep44029.

Wong, S.H., Yu, J., 2019. Gut microbiota in colorectal cancer: mechanisms of action and clinical applications. Nat. Rev. Gastroenterol. Hepatol. 16, 690–704. Available from: https://doi.org/10.1038/s41575-019-0209-8.

Wong, S.H., Kwong, T.N.Y., Wu, C.-Y., Yu, J., 2019. Clinical applications of gut microbiota in cancer biology. Semin. Cancer Biol. 55, 28–36. Available from: https://doi.org/10.1016/j.semcancer.2018.05.003.

Wu, S., et al., 2009. A human colonic commensal promotes colon tumorigenesis via activation of T helper type 17 T cell responses. Nat. Med. 15, 1016–1022.

Xiong, N., Hu, S., 2015. Regulation of intestinal IgA responses. Cell Mol. Life Sci. 72, 2645–2655.

Yamazaki, K., Tsunoda, A., Sibusawa, M., et al., 2000. The effect of an oral administration of Lactobacillus casei strain shirota on azoxymethane-induced colonic aberrant crypt foci and colon cancer in the rat. Oncol. Rep. 7 (5), 977–982. Available from: https://doi.org/10.3892/or.7.5.977.

Yu, T.C., Guo, F., Yu, Y., Sun, T., et al., 2017. *Fusobacterium nucleatum* promotes chemoresistance to colorectal cancer by modulating autophagy. Cell 170, 548–563e16. Available from: https://doi.org/10.1016/j.cell.2017.07.008.

Zelante, T., Iannitti, R.G., Cunha, C., et al., 2013. Tryptophan catabolites from microbiota engage aryl hydrocarbon receptor and balance mucosal reactivity via interleukin-22. Immunity 39, 372–385.

Zhang, H., Yeh, C., Jin, Z., et al., 2018. Prospective study of probiotic supplementation results in immune stimulation and improvement of upper respiratory infection rate. Synth. Syst. Biotechnol. 3 (2), 113–120. Available from: https://doi.org/10.1016/j.synbio.2018.03.001.

Zhu, Y., Liu, X.G., Yang, S.T., 2005. Construction and characterization of pta gene-deleted mutant of *Clostridium tyrobutyricum* for enhanced butyric acid fermentation. Biotechnol. Bioeng. 90, 154–166.

Zitvogel, L., Galluzzi, L., Viaud, S., et al., 2015. Cancer and the gut microbiota: an unexpected link. Sci. Transl. Med. 7 (271), 271ps1. Available from: https://doi.org/10.1126/scitranslmed.3010473.

Chapter 28

Gut–brain communication: a novel application of probiotics

Sujit Das[1], Trupti J. Undhad[2] and Subrota Hati[2]

[1]*Department of Rural Development and Agricultural Production, North-Eastern Hill University, Tura, Meghalaya, India,* [2]*Department of Dairy Microbiology, SMC College of Dairy Science, Kamdhenu University, Anand, Gujarat, India*

28.1 Introduction

The human body is a complicated biological system with approximately 100 trillion bacteria that have ten times the number of human body cells, 150 times the number of human genes, 1000 species, and above 7000 strains, with over 90% of microbiota cells and ten million microbiota genes (Kelly et al., 2016). The intestinal tract is the primary section in the upper digestive system, with bacterial densities varying from 10–1000 bacteria per gram in the upper digestive tract to 1011–1012 bacteria per gram in the colon, comprising the principal Firmicutes and Bacteroides groups (Kelly et al., 2016; Savignac et al., 2015). In comparison, the fetal intestine is sterile, the maternal microbiota is the original bacterial colonization at birth, and the intestinal microbiome of an adult individual consists of about 1 kg of bacteria, viruses, protozoa, fungi, and archaea (Latalova et al., 2017). The gut-brain axis (GBA) is thought to be connected to neuroendocrine-immune pathways, which produce the intestinal microbiota. An army surgeon's study tracking gastric juices released through an intragastric fistula provided the first proof of the GBA, and Mangiola et al. (2016) revealed that mood changes were connected with intestinal activities. The intestinal microbiota can also create metabolic chemicals, including bile acids, choline, and short-chain fatty acids (SCFAs), which can help the host's metabolism. The intestinal microbiota can digest and ferment complex carbohydrates such as dietary fiber into the SCFAs butyrate, acetate, and propionate, all of which have neuroprotective properties and can penetrate the bloodstream and play a significant function in the brain via two types of distributed 7-trans membrane G-protein-coupled receptors, including the free fatty acid receptors (Dinan et al., 2015).

Only a few clinical trials examining the effect of probiotic intake on behavior, mood, and cognition in the general population are available; however, data from well-defined clinical trials is required to accept probiotics as a viable alternative in the management of depression or other neuropsychiatric disorders. In a recent metaanalysis of ten randomized controlled trials, Ng et al. (2018) found that probiotic supplementation had no overall effect on mood, with only mild effects on individuals with preexisting mood disorders and no effect on healthy people in the group. According to this metaanalysis, more research is needed to prove the usefulness of probiotics in improving depression and anxiety symptoms, as well as standard of living and inflammatory biomarkers. The intestinal microbiota thus plays a crucial role in depressive disorders via the GBA. This chapter addresses the GBA and its relationship to mood disorders.

28.2 The gut-brain axis: a concept

The concept of the GBA, a bidirectional transmission medium between the "big brain" in the cranium and the "little brain" (i.e., the enterogenic nervous system), was created to be able comprehend the central and peripheral effects of gut-brain peptides such as cholecystokinin and bombesin. It was soon implicated in pathologies such as anorexia nervosa (Banks, 1980). The GBA, or bidirectional coordination between the brain and the stomach, has long been recognized. The brain influences the gastrointestinal (GI) system by modulating movement, production, intake, and blood flow; at the same time, the stomach can influence brain function and conduct (Grenham et al., 2011). The GI tract, central nervous system (CNS), autonomic nervous system (ANS), enteric nervous system (ENS), neuroendocrine system, and

Microbiome, Immunity, Digestive Health and Nutrition. DOI: https://doi.org/10.1016/B978-0-12-822238-6.00017-0

immune response are included in the GBA scaffolding (Grenham et al., 2011). Previous findings have shown that the gut microbiota is implicated in the formation of neurons and various brain functions by controlling the GBA (Erny et al., 2015).

The GBA (Fig. 28.1) is a bidirectional signaling axis that is critical for metabolic homeostasis. The GI tract converts sensory data into neuronal, hormonal, and immunological responses, which are then transferred to the CNS. While new evidence links alterations in intestinal immune signaling to changes in gut-mediated energy homeostasis (Mayer, 2011), the majority of existing impacts of the GBA on the energy equation are the result of neural and hormonal gut-derived signals. Individualized sensory pathways that are distinct to nutrients can generate signals in the GI tract at various locations to connect with the brain about not only the calorific value of a meal, but also perhaps about the specific macronutrient makeup of calories taken (Hamr et al., 2015). Gut-derived signals are then transmitted to a number of brain locations to sustain energy homeostasis during both feeding and fasting in order to elicit reactions that eventually lead to both acute and chronic alterations in energy intake and utilization.

In reaction to preabsorptive nutrients, gut peptides are liberated from enteroendocrine cells and serve to transmit information to the brain about incoming energy. Gut peptides might act specifically on gut peptide receptors expressed on vagal or spinal afferent nerve terminals innervating the intestine if discharged into the subcellular space to drive gut-brain neuronal signaling. Gut peptides could also act indirectly to relay neuronal signaling to afferent nerves through receptors on intrinsic enteric nervous system neurons. In comparison, gut peptides can spread through the systemic or lymphatic circulation to ultimately enter the brain and function in an endocrine fashion on central receptors (Rhee et al., 2009). Studies into gut-brain crosstalk have shown a complex interaction process that not only ensures normal GI homeostasis, but also is believed to have many effects on influence, desire, and higher mental processes. The intricacy of these experiences is included in the GBA denomination. Its purpose is to regulate and integrate intestinal functions and to connect the brain's affective and psychological centers to peripheral intestine functions and processes, including autoimmune response, intestinal absorption, colonic reflex, and enteroendocrine transmission (Rhee et al., 2009).

Neuroimmunoendocrine mediators include the pathways underlying GBA communications. This bidirectional communication network is made up of the CNS, which includes the brain and spinal cord, the ANS, the ENS, and the hypothalamus pituitary adrenal axis (HPA). The sympathetic and parasympathetic branches of the autonomic system transmit incoming signals from the lumen to the CNS via intestinal, spinal, and vagal pathways, as well as efferent signals from the CNS to the intestinal wall. The HPA axis is the body's main stress efferent axis, and it coordinates the body's adaptive responses to all types of stressors (Tsigos and Chrousos, 2002). Nausea, diarrhea, and stomach cramps are common comorbidities (the presence of one or more disorders in addition to the primary condition) in many neurological conditions (Westfall et al., 2017). Furthermore, recent advances in metagenomic sequencing have revealed that disruption in the makeup of the gut microbiota (gut dysbiosis) is evident in a number of neurological illnesses. The importance of maintaining a healthy microbiota population (gut symbiosis) in the control of the GBA cannot be

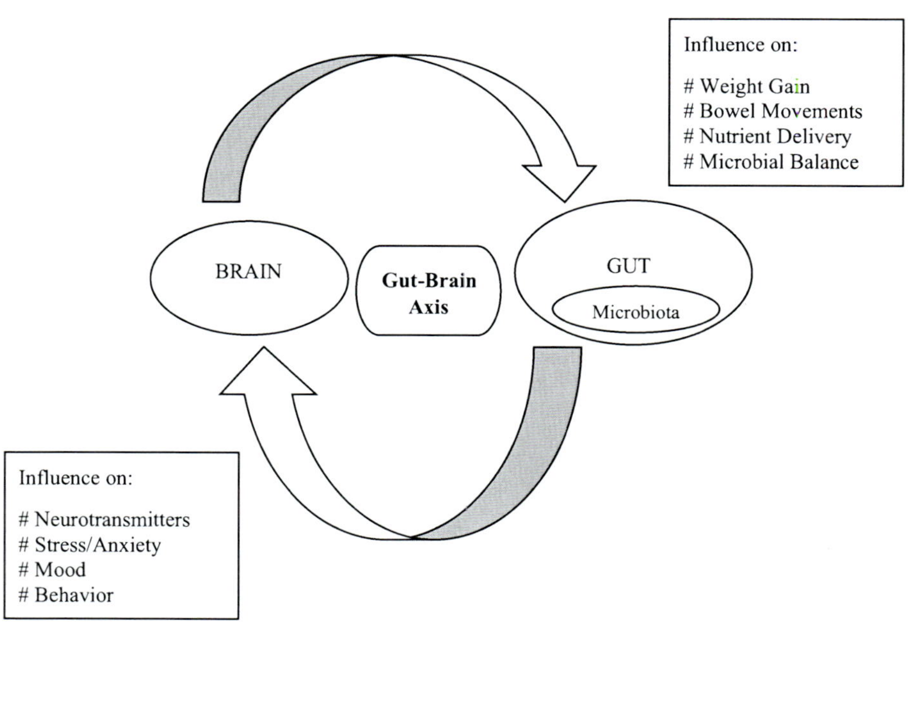

FIGURE 28.1 A schematic representation of the microbiota-gut-brain axis.

understated. The phrase microbiota-GBA was coined to emphasize the role of the microbiota in the GBA. Between the brain, the intestinal tract, and the gut, the GBA facilitates the coordination of the hormonal and immunological systems that are involved in gut function maintenance (Bienenstock and Collins, 2010).

28.3 Factors regulating the microbiota-gut-brain axis

28.3.1 Genetics and epigenetics

The link between host genetics and microbiota makeup is an important and underrated subject of research, particularly in terms of brain health. A number of studies are looking into the connections between human and mouse host genetics as well as the microbiota's diversity and individual constituent taxa (Bonder et al., 2016). For example, taxa abundance was significantly more associated in monozygotic twins than in dizygotic twins, with Christensenellaceae being the most highly heritable family, and also related to a minimal body mass index, indicating that the gut flora is a key contributor of host genetic and phenotype interaction (Rothschild et al., 2018). It is possible that neuronal transcription and host activity are influenced by gut microbial products. Via gene-environment interactions, the microbiota could be an essential host genome modulator and might eventually be recognized as a distinct epigenetic entity (Stilling et al., 2018).

28.3.2 Mode of delivery at birth

The mode of delivery has a significant impact on the formation of an infant's microbiota, as birth provides the first opportunity for large-scale bacterial colonization. The early implantation of microbiota during vaginal delivery (i.e., natural birth) occurs during parturition, when the baby travels via the birth canal and is in contact with the mother's vaginal microbiota. Vertical bacterial transfer from the mother to the newborn was discovered in a seminal study, with the microflora of vaginally delivered newborns closely matching their mothers' vaginal microbiome (Dominguez-Bello et al., 2010).

28.3.3 Fats

Diets that are high in saturated fat and trans fat are linked to a higher prevalence of cardiovascular diseases and an elevation in total serum low-density lipoprotein cholesterol. High-fat diet administration to healthy patients, for instance, resulted in a drop in Bacteroidetes levels and an increase in Firmicutes and Proteobacteria levels in the gut microbiota (Zhang et al., 2012). Mice that are fed a high-fat and high-sucrose diet have shown similar phylum changes (Parks et al., 2013).

28.3.4 Medications and the microbiome

Given the growing attention to the microbiota-gut-brain axis, it is not surprising that a lot of research has gone into understanding the impact of psychiatric medicines in recent years. Antidepressant serotonin reuptake inhibitors (SSRIs), such as sertraline, paroxetine, and fluoxetine, have antimicrobial action against gram-positive bacteria such as *Staphylococcus* and *Enterococcus* (Coban and Durupinar, 2009). Furthermore, the synergistic effects of some antibiotic-associated SSRIs and their effects on some antibiotic-resistant bacteria have been used to verify the antimicrobial activity of specific antidepressants (Bohnert et al., 2011).

In one study, prolonged fluoxetine administration resulted in lower levels of *Prevotella* and *Succinivibrio* in the cecum (Cussotto et al., 2019), and another study found lower levels of *Lactobacillus johnsonii* and *Bacteroidales* S24-7, both of which correspond to a phylum related with body mass regulation (Lyte et al., 2019). Another category of antidepressants, tricyclic antidepressants, has been proven to prevent the spread of gut infectious agents such as *Escherichia coli*, *Enterocolitica*, *Yersinia*, and the parasite *Giardia lamblia* (Weinbach et al., 1992).

28.4 Interactions within the gut-brain axis

The gut and the brain communicate with each other in a bidirectional manner by way of various mechanisms and pathways, including neuronal, endocrine, humoral, and immunological signals. For a complete understanding of the GBA, the main signaling pathways have been elucidated.

28.4.1 Neuronal pathway

The ENS performs most of the neural connection in the GI system. The ENS is a branch of the ANS that regulates GI functions such as motility, mucus secretion, and blood circulation (Browning et al., 2017). The ENS connects with the ANS and CNS via neurotransmitters (adrenaline, noradrenaline, and acetylcholine) and also sensory-motor neurons, all which convey impulses from the gut to the brain. Electrophysiological thresholds of ENS neurons are controlled by the gut microbiome (Sarkar et al., 2016). The metabolic compounds of *Bifidobacterium longum* NCC3001 evoke a decrease in action potential spikes in myenteric neurons following electrical stimulation (Bercik et al., 2011a). Myenteric sensory neurons displayed reduced excitability in germ-free (GF) mice relative to normal mice (McVey Neufeld et al., 2013). In addition, after hyperpolarization, these neurons showed enhanced excitability in GF mice (McVey Neufeld et al., 2013). In contrast to their activities on myenteric neurons, gut bacteria serve an important role in the initial colonization and homeostasis of glial cells in the gut.

Another neural channel in gut-brain transmission is the vagus nerve, which has both efferent and afferent functions (Cryan and Dinan, 2012). Approximately 80% of vagal fibers are visual, transmitting information to the CNS on the status of the body's organs (Thayer and Sternberg, 2009). The vagus nerve controls several critical activities, notably bronchial constriction, pulse rate, and GI motility. Several of the cognitive benefits of the gut microbiota or probiotic strains depend on the vagus nerve's (sensory information) activity (Bercik et al., 2011a; Bravo et al., 2011). Chronic treatment with *Lactobacillus rhamnosus* JB-1, for example, resulted in an increase in gamma-aminobutyric acid (GABA) mRNA expression within certain brain areas in addition to lowering stress-induced corticosterone and anxiety- and depression-related activities (Bravo et al., 2011). Investigations in vagotomized animals confirmed the vagus nerve as a crucial modulator in the gut-brain transmission route (Bravo et al., 2011). However, vagus nerve–independent pathways persist in gut-brain communication, as antimicrobial treatments have not demonstrated gut-brain reliance on vagus nerve activity (Bercik et al., 2011b).

28.4.2 Endocrine pathway

The gut microbiota alters the supply of nutrients and thereby affects the release from enteroendocrine cells of biologically active peptides, which may influence the GBA. The neuropeptide galanin, for example, is believed to be associated with many important neurobiological functions, such as nociception, control of the sleep-wake sequence, nourishing, mood, regulation of blood pressure, parenteral actions, and neurotrophic activities functions. Galanin stimulates the core branch of the HPA axis, regulating the release of corticotropin-releasing factor and adrenocorticotropic hormone, and thereby enhances glucocorticoid release from the adrenal cortex (Azzam et al., 2017). It may also directly induce the secretion of cortisol from the adrenal cortex and the release of norepinephrine from the adrenal medulla, illustrating its participation in the response of the HPA axis to stress (Picciotto, 2008).

28.4.3 Humoral/metabolic pathway

Bacterial metabolites are powerful humoral influences (most notably SCFAs formed by the bacterial fermentation of dietary carbohydrates). They also exhibit hormone-like activity and immunomodulatory capabilities, and they connect with nerve cells through the sympathetic branch of the ANS, which is well recognized for controlling enterocyte feeding. Furthermore, microbiota-derived SCFAs have been demonstrated to influence microglia homeostasis, which is important for correct neurodevelopment and brain tissue homeostasis, as well as being implicated in behavior regulation (Mayer et al., 2015).

Changes in SCFA metabolism have been linked to the onset of autism through impairing microglial communication and function (Frye et al., 2016; Rogers et al., 2016). SCFAs also regulate the discharge of intestinal peptides from enteroendocrine cells, which has been demonstrated to govern the generation of gut-derived serotonin by enterochromaffin cells, all of which influence hormonal gut-brain interactions (Wang and Kasper, 2014). The gut contains about 95% of the total serotonin in the body, much of which resides in plasma. Although serotonin has an innate function in the gut and peripheral processes, it can also stimulate afferent nerve endings in the brain (MacFabe, 2013). While most therapists are aware of fluoxetine's ability to prevent intestinal serotonin from reaching the bloodstream, many are unaware that higher plasma serotonin levels have been reported in autistic children (Marler et al., 2016). High plasma serotonin levels and poor serotonergic neurotransmission have been found to have an inverse relationship in young male adults with autistic disorders (Mayer et al., 2014).

28.4.4 Immune pathway

During situations of dysbiosis the gut microbiome affects the metabolism of inflammation within the GI system, principally through the immune systems' secretion of cytokines (interleukin [IL]-10 and IL-4) and other mediators of cellular communication, such as interferon-gamma. As an instance, irregular microbiome stimulates the mucosal innate immune responses in irritable bowel syndrome (IBS), which enhances gut epithelial permeability, stimulates sensory pathways of gut pain, and dysregulates the ENS (Mayer et al., 2014; Dupont, 2014); both brain and gut brain disorders occur, the former being predominant (Koloski et al., 2012). Intestinal motility and secretion are impaired by disturbances of the GBA related to visceral hypersensitivity and contribute to cellular changes in the enteroendocrine and immune systems (Mayer et al., 2014).

28.5 Function of the microbiota in the gut-brain-axis

The enteric microbiota appears to have a significant impact on GBA, communicating with intestinal cells and ENS not only locally, but also explicitly with the CNS through neuroendocrine and metabolic pathways, according to both therapeutic and experimental proof (Marilia et al., 2015). In humans, more than 20 years ago, the most credible finding of GI microbe–brain interaction arose from the improvements that were seen in patients with hepatic encephalopathy following ingestion of oral antibiotics (Morgan, 1991). In the meantime, new evidence and, more significantly, autism dysbiosis support the function of microbiota in insecurity and depressive-like behaviors (Foster and Neufeld 2013; Naseribafrouei et al., 2014). In addition, the microbiota of autistic patients varies depending on the severity of the condition (Mayer et al., 2014).

28.5.1 From the gut microbiota to the brain

There has been a proliferation of experimental work in recent years, carried out primarily on animals, aimed at investigating the impact of the microbiota on GBA modulation. Various technical techniques involving the use of GF animals, probiotics, antibiotics and infection studies have been used (Bravo et al., 2012). Results of experiments on GF animals showed that intestinal bacterial colonization is key to the growth and maturation of both the ENS and the CNS (Stilling et al., 2014). Upon animal colonization, all of these abnormalities are recovered in a bacterial species–specific manner. Furthermore, studies on the effects of probiotics and/or antibiotics on the intestinal microbiota have verified the impact of the microbiota on the GBA. These findings also confirm that by affecting brain neurochemistry, the microbiota influences anxiety and the HPA system. Chronic treatment with *L. rhamnosus* JB-1 resulted in region-dependent changes in GABA mRNA in the brain (Saulnier et al., 2013).

$GABA_B$ was elevated in cortical cingulate and prelimbic areas relative to mice with regulated diets, whereas in the hippocampus, amygdala, and locus coeruleus, it decreased concomitantly. In contrast, the concentrations of $GABA_{A\alpha 2}$ mRNA were lowered in the prefrontal cortex and amygdala but expanded in the hippocampus. The probiotics also reduced stress-triggered cortisol release as well as anxiety and depression-based activities (Bravo et al., 2011). Likewise, in distinct pathogen-free mice, transient changes in microbiota composition caused by oral antimicrobial agents (neomycin, bacitracin, etc.) raised exploratory activity and hippocampal release of brain-derived neurotrotropin (Bercik et al., 2011b). Moreover, improvement in the composition of the microbiota with the interaction of probiotics VSL #3 (medical food used as a probiotic) contributed to an improvement in neurotrophic factor expression derived from the brain, modulation of age-related hippocampal alterations, and reversion of visceral hypersensitivity caused by neonatal maternal separation in a rat model of IBS (Distrutti et al., 2013). In the latter model of stress, a shift in the activity of subsets of genes engaged in pain propagation and inflammation was detected, which was resolved by implementation of probiotics in early life.

Evidence suggests that the vagus nerve requires microbiota contact with the brain, which transmits information to the CNS from the luminal setting. Neurochemical and behavioral consequences were absent in vagotomized mice, indicating that the vagus is the central modulatory constitutive connectivity route between the microbiota and the brain (Bravo et al., 2011). The anxiolytic effect that was obtained with *B. longum* therapy in a chronic colitis model correlated with anxiety-like behavior was missing in mice that were vagotomized prior to initiation of colitis (Bercik et al., 2011a). Through various mechanisms, the microbiota may associate with the GBA; the most important such mechanism is likely to be intestinal barrier modulation, the disturbance of which may affect all the underlying compartments. Probiotic species-specific central effects are also aligned with retaining tight junction integrity and maintaining the intestinal barrier, as was recently seen in an animal model of water avoidance stress (Ait-Belgnaoui et al., 2014).

Pretreatment of animals with the coupled probiotic formulations of *Lactobacillus helveticus* R0052 and *B. longum* R0175 reinstated the integrity of the tight junction barrier and attenuated HPA axis and the activities of the ANS, assessed by measurements of plasma cortisol and catecholamine. Probiotics have also been shown to inhibit alterations in hippocampal neurogenesis and the expression of hypothalamic genes associated with synaptic plasticity. Microbiota can also communicate with GBA through the activation of afferent sensory nerves, as confirmed for *Lactobacillus reuteri*, which modulates gut motility and pain perception by improving their excitability by hindering the emergence of calcium-dependent potassium channels (Kunze et al., 2009). Furthermore, the microbiota can affect the function of the ENS and generate a biologically active type of catecholamines in the gut lumen by producing molecules that can serve as local neurotransmitters such as GABA, serotonin, melatonin, histamine, and acetylcholine (Asano et al., 2012). Lactobacilli utilize nitrate and nitrite to create nitric oxide and hydrogen sulfide (Sobko et al., 2006), which regulate gut motility by binding to the vanilloid receptor (possessing a vanillyl group) on capsaicin-sensitive nerve fibers (Schicho et al., 2006).

28.5.2 From the brain to the gut microbiota

Different types of psychological stressors, regardless of time, affect the composition and overall biomass of the enteric microbiota. In particular, the usage of short stressors also affects the microbiota, being substantially capable of modifying the population profile and reducing the relative ratios of the primary microbiota phyla for only 2 hours (Galley et al., 2014). These impacts may be regulated by the efferent systems of parallel neuroendocrine development (i.e., ANS and HPA), both directly and indirectly by host-enteric microbiota signaling. The so-called emotional motor system is made up of efferent neural pathways linked to endogenous pain-modulator pathways (Rhee et al., 2009).

The direct effect is regulated by the release of molecules for signaling via neurons, immune cells and enterochromaffin cells under brain control, which may influence the microbiota. Contact between effectors of the CNS and bacteria depends on bacteria for the existence of neurotransmitter receptors. A few reports have shown that bacteria contain binding sites for host-produced enteric neurotransmitters, which can influence the function of microbiota components, may lead to increased predisposition to stimuli of inflammation and infection (Hughes and Sperandio, 2008). *Pseudomonas fluorescens* has been found to have a high preference for the GABA mechanism, with binding characteristics close to those of a brain receptor (Guthrie and Nicholson-Guthrie, 1989). The epinephrine/norepinephrine receptor in *E. coli* O157:H7 is generated from the host and can be inhibited directly by adrenergic antagonists (Clarke et al., 2006).

The brain is involved in modulating gut functions such as motility, acid secretion, bicarbonates and mucus, intestinal fluid management, and mucosal immune reflex, all of which are essential for the preservation of the mucus layer and biofilm, where particular bacteria groups grow in a variety of microhabitats and metabolic niches related to the mucosa (Macfarlane and Dillon, 2007). Through disruption of the normal mucosal habitat, GBA dysregulation can affect the gut microbiota. Stress allows the size and consistency of mucus secretion to differ (Rubio and Huang, 1992). Mental stress increases the intensity of the cecocolonic spike burst activation that is caused by central stimulation of corticotropin-releasing factor (Gue et al., 1991). Modifications in GI transit, both regional and global, can have a major impact on the processing of vital nutrients, mainly prebiotics and dietary fibers, to the enteric microbiota.

The brain can influence microbiota composition and function by altering intestinal absorption, enabling bacterial antigens to infiltrate the epithelium and trigger an immune response in the mucosa. Acute stress—increased colonic paracellular permeability could result in an increase in interferon output as well as a decrease in ZO-2 mRNA expression and occlusion (Demaude et al., 2006). The ANS can also influence immune function in the brain. In muscle dysfunction caused by stress, the sympathetic branch regulates the number, degranulation, and behavior of mast cells, resulting in an imbalance in tryptase and histamine release (Santos et al., 1998). In turn, other mast cell products, such as corticotropin-releasing factor, may increase epithelial bacterial permeability, which enhances their access to lamina propria immune cells (Rhee et al., 2009). Corticotropin-producing hormone receptors also are engaged in colonic barrier dysfunction that relates to depression and increased vulnerability to colitis in adult rats in relation to mild stress in neonatal maternal separation (Soderholm et al., 2002).

Bilateral olfactory bulbectomy caused depression-like behavior in mice, which was correlated with increased central corticotropin-releasing factor expression and serotonin levels as well as improvements in colonic motility and intestinal microbial profile (Park et al., 2013). The increase in the release of alpha-defensin from Paneth cells is another potential disruption in the microbiota habitat triggered by stress (Alonso et al., 2008). Finally, it is important to remember that the expression of virulent bacteria is facilitated by gut alterations associated with stress. The expression of *Pseudomonas aeruginosa*, which can contribute to gut sepsis, is caused by norepinephrine released during surgery

(Alverdy et al., 2000). Norepinephrine can also promote the proliferation of many enteric pathogens strains and maximize the virulent characteristics of *Campylobacter jejuni* and may favor the overgrowth of *E. coli* as well as *E. coli* 0157:H7:33 pathogenic isolates (Freestone et al., 2003).

28.6 The gut-brain axis in the regulation of food intake

Once fasted rats have been refed, food consumption drops within minutes of refeeding and declines gradually over the course of a meal, implying that negative feedback transmissions are quickly transmitted to the brain after food enters the GI tract to stop an abundance of arriving nutrients (Davis and Smith, 1990). One of the first signals that is produced is that bolus (food) reaches the stomach for mechanoreception. Saline processing reduces food intake as rapidly as within 3 minutes, depending on the volume of the stomach, when a gastric cuff is used to close the pyloric sphincter and enable the stomach to fill without emptying (Phillips and Powley, 1996). The vagal and spinal afferents that innervate the stomach express stretch-receptive calcium channels, and gastric distention causes vagal afferent firing and stimulation in the neurons of the hindbrain, facilitating a neuronal stomach-brain axis that reduces food intake (Raybould et al., 1999). Nutrients are discharged into the small intestine by the pyloric sphincter, and the frequency of gastric emptying begins to decline as nutrients enter the small intestine, increasing stomach distention after food ingestion. The release of CCK and GLP-1 slows the emptying of the stomach (Talsania et al., 2005) and vagal stimulation, as subdiaphragmatic vagotomy, a procedure involving the separation of the afferent vagus fibers at the brain stem, modulates the influence of CCK or GLP-1 on gastric discharge (Wickbom et al., 2008).

28.7 Behavior and the microbiota gut-brain axis

Even though the underlying mechanisms of the symbiotic association between host and microbe have yet to be fully established, especially in terms of neural well-being, a variety of animal and human studies (Table 28.1) have been beneficial in allowing the scientific community to continuously narrow gaps in our understanding of the microbiota-gut-brain axis.

28.7.1 Animal studies

A variety of probiotics have been tested in animal models of neurological diseases. *Bifidobacterium* and *Lactobacillus* are the two main genera that have shown potential benefits for treating neurological disorders (e.g., agitation, depression, fatigue, and autism spectrum disorder). A maternal separation model was used in recent research to induce early-life stress in infant rats. The maternally separated rat pups performed poorly in the forced swimming task, with enhanced inflammation and corticotrophin release factor mRNA activation in the amygdala area of the brain. In this animal model, probiotic *B. infantis* treatment reversed behavioral deficiencies and inflammation (indexed by the release of interleukin-6) and displayed enhancement of serotonergic precursor plasma tryptophan (Desbonnet et al., 2010).

A mouse model of autism spectrum disorder (ASD) with maternal immune activation demonstrated GI membrane deficiencies and impaired microbial content, which are all recognized symptoms of ASD. The anomalies in gut permeability and microbial composition as well as ASD-related behaviors such as cognitive deficits, depression-like behaviors, and sensorimotor behaviors were corrected by oral therapy with the human commensal *Bacteroides fragilis*. Upon treatment with *B. fragilis*, normal levels of certain metabolites were also regained. A shift in 4-ethylphenyl sulfate is enough among those restored metabolites to induce anxiety-like behavior (Hsiao et al., 2013).

The latest research has presented clear proof of the regulation of neurotransmitter concentrations by the use of magnetic resonance spectroscopy in the treatment by probiotics. Mice that were fed with *L. rhamnosus* JB-1 depicted raised concentrations of glutamate, N-acetyl aspartate, and GABA in the brain, implying that probiotics might influence brain function through metabolic pathways and indicating the potential of clinical translation of this probiotic method through neurological disorder therapies (Janik et al., 2016).

Several other studies have been performed to illustrate the mechanism through which probiotics act on the GBA. The relief of inflammation by probiotics was addressed in one of the approaches. It demonstrated that *B. infantis* 35624 and *L. rhamnosus* improve the plasma concentration of IL-10, an antiinflammatory cytokine, while decreasing interferon gamma (IFN-γ) and tumor necrosis factor alpha (TNF-α) producing cells (Donato et al., 2010). Increased blood barrier integrity and concomitant suppression of neuroinflammation have been linked to probiotic inflammation control in patients with neurological conditions (Felger and Lotrich, 2013).

TABLE 28.1 Effects of probiotics on brain function and behavior.

Probiotic strain	Model	Duration	Effects	References
Bifidobacterium infantis	Rodent (maternal separation model of depression)	45 days	Normalized immune response, reduced behavioral deficits	Desbonnet et al. (2010)
Bacteroides fragilis	Rodent (maternal immune activation)	6 days	Reversed ASD-related behavior, Improved gut barrier integrity	Hsiao et al. (2013)
A mixture of *Lactobacillus helveticus* R0052 and *Bifidobacterium longum* R0175	Rodent (healthy)	2 weeks	Reduced anxiety-like behavior	Messaoudi et al. (2011)
B. longum 1714, *B. breve* 1205	Rodent (innately anxious)	6 weeks	Reduced anxiety-like behavior	Savignac et al. (2015)
A mixture of *Bifidobacterium animalis* subsp. *lactis*, *Streptococcus thermophilus*, *Lactobacillus bulgaricus*, and *Lactococcus lactis* subsp. *lactis*	Human (healthy)	4 weeks	Influenced brain activity in emotional centers	Tillisch et al. (2013)
L. helveticus R0052 and *B. longum* R0175	Human (healthy)	30 days	Less psychological distress	Messaoudi et al. (2011)
Lactobacillus casei strain Shirota	Human (healthy, exposed to academic stress)	11 weeks	Reduced stress-related gastrointestinal symptoms	Kato-Kataoka et al. (2016)
L. casei strain Shirota	Human (healthy)	3 weeks	Improved mood	Benton et al. (2007)
A mixture of *Bifidobacterium bifidum* W23, *B. lactis* W52, *Lactobacillus acidophilus* W37, *L. brevis* W63, *L. casei* W56, *L. salivarius* W24, and *L. lactis* (W19 and W58)	Human (healthy)	4 weeks	Reduced rumination and aggressive cognition	Steenbergen et al. (2015)
L. acidophilus	Human (autistic children)	2 months	Reduced D-arabinitol level in urine, improvement in ability to concentrate and carry out orders	Kałużna-Czaplińska and Błaszczyk (2012)
VSL#3 (VSL Pharmaceuticals Inc.), a multistrain mixture of 10 probiotics	Human (child with autism spectrum disorder, case report)	4 weeks	Improved autistic core symptoms, relieved gastrointestinal symptoms	Grossi et al. (2016)
L. casei strain Shirota	Human (PD: Parkinson's disease)	5 weeks	Reduced constipation	Cassani et al. (2015)
L. acidophilus, L. casei, B. bifidum, and *Limosilactobacillus fermentum*	Human (AD: Alzheimer disease)	12 weeks	Improved cognition	Akbari et al. (2016)

28.7.2 Human studies

Although there have been fewer human clinical trials than animal research studies, there is increasing research evidence that probiotics can help with depression and anxiety disorders. Over the course of 30 days, oral delivery of a probiotic formulation containing *L. helveticus* R0052 and *B. longum* R0175 has been shown to improve mood in otherwise healthy volunteers (Messaoudi et al., 2011). As measured by the global Hospital Anxiety and Depression scale scores and the Hopkins symptoms checklist's global severity ranking, probiotic-treated volunteers had less anxiety and depressive symptoms than the placebo group. In participants who were treated with the probiotic formulations, urinary free cortisol was also reduced, indicating a decreased degree of stress (Messaoudi et al., 2011). These results show that probiotic therapy has a beneficial impact on anxiety and depression-related behaviors in human volunteers (Liu et al., 2015).

Under academic pressure, the stress-relieving properties of *L. casei* strain Shirota have also been investigated. Medical students were given either *L. casei* fermented milk or placebo milk for 8 weeks before taking an academic success test. Those who were given *L. casei strain* Shirota showed substantially decreased salivary cortisol levels and alleviated GI symptoms that were alleviated compared to the levels prior to research. In addition, the probiotics group had slightly more intestinal microbial species in their feces than the placebo group, and the proportion of Bacteroidaceae was significantly lower in the probiotics group (Kato-Kataoka et al., 2016).

Taken collectively, these animal and human studies suggest that probiotics may serve a crucial function in the bidirectional association between the gut and the brain as well as modulating brain function. Although several research teams have attempted to illustrate a framework for probiotics' control of the GBA in both healthy and diseased states, the underlying extent has yet to be established. As a result, elucidating their masked interactions in the GBA will be critical for the effective utilization of probiotics in treating neurological disorders.

28.8 Conclusions and future perspectives

Over the past decade, understanding how gut microbes affect the connectivity of the GBA has been the focus of critical research. At all nodes of the axis, there is an increasing attempt to dissect the dynamics of this contact. The gut microbiota is now widely acknowledged as being important for the proper development and regulation of brain activity. In addition, evidence from animal and clinical studies concerning the microbiota in a variety of psychological, neurological, and neurodegenerative disturbances is accumulating, as was mentioned earlier in the chapter. However, since this field is still in its initial stages, it is important not to overinterpret such findings.

Modifications in the microbial content of the gut as a result of probiotic bacteria absorption or changes in the composition of the gut microbiota in many other ways may cause a variety of effects. These include effects on the host's immune, nervous, and endocrine systems, all of which affect one another, implying that gut-host cross-talk is essential. Conduct, mood, and stress response may all be influenced by ingestion of probiotic bacteria. Such effects are very important and have drawn public and scientific interest. There is currently a large difference between the experimental results and the clinical evidence. Now, the goal is to turn these animal studies into clinical use. The structure, abundance, and function of particular probiotics, combined with a better understanding of the gut microbiota composition, could contribute to more effective diet and drug therapies in the future.

References

Ait-Belgnaoui, A., Colom, A., Braniste, V., Ramalho, L., Marrot, A., Cartier, C., et al., 2014. Probiotic gut effect prevents the chronic psychological stress-induced brain activity abnormality in mice. Neurogastroenterol. Motil. 26 (4), 510–520.

Akbari, E., Asemi, Z., Daneshvar Kakhaki, R., Bahmani, F., Kouchaki, E., Tamtaji, O.R., et al., 2016. Effect of probiotic supplementation on cognitive function and metabolic status in Alzheimer's disease: a randomized, double-blind and controlled trial. Front. Aging Neurosci. 8, 256.

Alonso, C., Guilarte, M. Vicario, M., Ramos, L., Ramadan, Z., Antolín, M., et al., 2008. Maladaptive intestinal epithelial responses to life stress may predispose healthy women to gut mucosal inflammation. Gastroenterology 135 (1), 163–172.

Alverdy, J., Holbrook, C., Rocha, F., Seiden, L., Licheng, R., 2000. Gut-derived sepsis occurs when the right pathogen with the right virulence genes meets the right host: evidence for in vivo virulence expression in Pseudomonas aeruginosa. Ann. Surg. 232 (4), 480.

Asano, Y., Hiramoto, T. Nishino, R., Aiba, Y., Kimura, T., Yoshihara, K., et al., 2012. Critical role of gut microbiota in the production of biologically active, free catecholamines in the gut lumen of mice. Am. J. Physiol.-Gastrointestinal Liver Physiol. 303 (11), G1288–G1295.

Azzam, I., Gilad, S., Limor, R., Stern, N., Greenman, Y., 2017. Ghrelin stimulation by hypothalamic–pituitary–adrenal axis activation depends on increasing cortisol levels. Endocr. Connect. 6 (8), 847–855.

Banks, W.A., 1980. Evidence for a cholecystokinin gut-brain axis with modulation by bombesin. Peptides 1 (4), 347–351.

Benton, D., Williams, C., Brown, A., 2007. Impact of consuming a milk drink containing a probiotic on mood and cognition. Eur. J. Clin. Nutr. 61 (3), 355–361.

Bercik, P., Park, A.J., Sinclair, D., Khoshdel, A., Lu, J., Huang, X., et al., 2011a. The anxiolytic effect of Bifidobacterium longum NCC3001 involves vagal pathways for gut–brain communication. Neurogastroenterol. Motil. 23 (12), 1132–1139.

Bercik, P., Denou, E., Collins, J., Jackson, W., Lu, J., Jury, J., et al., 2011b. The intestinal microbiota affect central levels of brain-derived neurotropic factor and behavior in mice. Gastroenterology 141 (2), 599–609.

Bienenstock, J., Collins, S., 2010. 99th Dahlem conference on infection, inflammation and chronic inflammatory disorders: psycho-neuroimmunology and the intestinal microbiota: clinical observations and basic mechanisms. Clin. Exp. Immunol., 160 (1), 85–91.

Bohnert, J.A., Szymaniak-Vits, M., Schuster, S., Kern, W.V., 2011. Efflux inhibition by selective serotonin reuptake inhibitors in Escherichia coli. J. Antimicrobial Chemother. 66 (9), 2057–2060.

Bonder, M.J., Kurilshikov, A., Tigchelaar, E.F., Mujagic, Z., Imhann, F., Vila, A.V., et al., 2016. The effect of host genetics on the gut microbiome. Nat. Genet. 48 (11), 1407–1412.

Bravo, J.A., Julio-Pieper, M., Forsythe, P., Kunze, W., Dinan, T.G., Bienenstock, J., et al., 2012. Communication between gastrointestinal bacteria and the nervous system. Curr. Opin. Pharmacol. 12 (6), 667–672.

Bravo, J.A., Forsythe, P., Chew, M.V., Escaravage, E., Savignac, H.M., Dinan, T.G., et al., 2011. Ingestion of Lactobacillus strain regulates emotional behavior and central GABA receptor expression in a mouse via the vagus nerve. Proc. Natl Acad. Sci. 108 (38), 16050–16055.

Browning, K.N., Verheijden, S., Boeckxstaens, G.E., 2017. The vagus nerve in appetite regulation, mood, and intestinal inflammation. Gastroenterology 152 (4), 730–744.

Cassani, E., Barichella, M., Cancello, R., Cavanna, F., Iorio, L., Cereda, E., et al., 2015. Increased urinary indoxyl sulfate (indican): new insights into gut dysbiosis in Parkinson's disease. Park. Relat. Disord. 21 (4), 389–393.

Clarke, M.B., Hughes, D.T., Zhu, C., Boedeker, E.C., Sperandio, V., 2006. The QseC sensor kinase: a bacterial adrenergic receptor. Proc. Natl Acad. Sci. 103 (27), 10420–10425.

Coban, A.Y., Durupinar, B., 2009. Investigation of antibacterial activity of sertralin. Mikrobiyoloji Bulteni 43 (4), 651–656.

Cryan, J.F., Dinan, T.G., 2012. Mind-altering microorganisms: the impact of the gut microbiota on brain and behaviour. Nat. Rev. Neurosci. 13 (10), 701–712.

Cussotto, S., Strain, C.R., Fouhy, F., Strain, R.G., Peterson, V.L., Clarke, G., et al., 2019. Differential effects of psychotropic drugs on microbiome composition and gastrointestinal function. Psychopharmacol. (Berl.) 236, 1671–1685.

Davis, J.D., Smith, G.P., 1990. Learning to sham feed: behavioral adjustments to loss of physiological postingestional stimuli. Am. J. Physiol. Reg. Integr. Comp. Physiol. 259 (6), R1228–R1235.

Demaude, J., Salvador-Cartier, C., Fioramonti, J., Ferrier, L., Bueno, L., 2006. Phenotypic changes in colonocytes following acute stress or activation of mast cells in mice: implications for delayed epithelial barrier dysfunction. Gut 55 (5), 655–661.

Desbonnet, L., Garrett, L., Clarke, G., Kiely, B., Cryan, J.F., Dinan, T.G., 2010. Effects of the probiotic Bifidobacterium infantis in the maternal separation model of depression. Neuroscience 170 (4), 1179–1188.

Dinan, T.G., Stilling, R.M., Stanton, C., Cryan, J.F., 2015. Collective unconscious: how gut microbes shape human behavior. J. Psychiatr. Res. 63, 1–9.

Distrutti, E., Cipriani, S., Mencarelli, A., Renga, B., Fiorucci, S., 2013. Probiotics VSL# 3 protect against development of visceral pain in murine model of irritable bowel syndrome. PLoS One 8 (5), e63893.

Dominguez-Bello, M.G., Costello, E.K., Contreras, M., Magris, M., Hidalgo, G., Fierer, N., et al., 2010. Delivery mode shapes the acquisition and structure of the initial microbiota across multiple body habitats in newborns. Proc. Natl Acad. Sci. 107 (26), 11971–11975.

Donato, K.A., Gareau, M.G., Wang, Y.J.J., Sherman, P.M., 2010. Lactobacillus rhamnosus GG attenuates interferon-γ and tumour necrosis factor-α-induced barrier dysfunction and pro-inflammatory signalling. Microbiology 156 (11), 3288–3297.

Dupont, H.L., 2014. Evidence for the role of gut microbiota in irritable bowel syndrome and its potential influence on therapeutic targets. Aliment. Pharmacol. Ther. 39 (10), 1033–1042.

Erny, D., de Angelis, A.L.H., Jaitin, D., Wieghofer, P., Staszewski, O., David, E., et al., 2015. Host microbiota constantly control maturation and function of microglia in the CNS. Nat. Neurosci. 18 (7), 965–977.

Felger, J.C., Lotrich, F.E., 2013. Inflammatory cytokines in depression: neurobiological mechanisms and therapeutic implications. Neuroscience 246, 199–229.

Foster, J.A., Neufeld, K.A.M., 2013. Gut–brain axis: how the microbiome influences anxiety and depression. Trends Neurosci. 36 (5), 305–312.

Freestone, P.P., Haigh, R.D., Williams, P.H., Lyte, M., 2003. Involvement of enterobactin in norepinephrine-mediated iron supply from transferrin to enterohaemorrhagic Escherichia coli. FEMS Microbiol. Lett. 222 (1), 39–43.

Frye, R.E., Rose, S., Chacko, J., Wynne, R., Bennuri, S.C., Slattery, J.C., et al., 2016. Modulation of mitochondrial function by the microbiome metabolite propionic acid in autism and control cell lines. Transl. Psychiat. 6 (10), e927. e927.

Galley, J.D., Nelson, M.C., Yu, Z., Dowd, S.E., Walter, J., Kumar, P.S., et al., 2014. Exposure to a social stressor disrupts the community structure of the colonic mucosa-associated microbiota. BMC Microbiol. 14 (1), 1–13.

Grenham, S., Clarke, G., Cryan, J.F., Dinan, T.G., 2011. Brain–gut–microbe communication in health and disease. Front. Physiol. 2, 94.

Grossi, E., Melli, S., Dunca, D., Terruzzi, V., 2016. Unexpected improvement in core autism spectrum disorder symptoms after long-term treatment with probiotics. SAGE Open. Med. Case Rep. 4, 2050313X16666231.

Gue, M., Junien, J.L., Bueno, L., 1991. Conditioned emotional response in rats enhances colonic motility through the central release of corticotropin-releasing factor. Gastroenterology 100 (4), 964–970.

Guthrie, G.D., Nicholson-Guthrie, C.S., 1989. gamma-Aminobutyric acid uptake by a bacterial system with neurotransmitter binding characteristics. Proc. Natl. Acad. Sci. 86 (19), 7378–7381.

Hamr, S.C., Wang, B., Swartz, T.D., Duca, F.A., 2015. Does nutrient sensing determine how we "see" food? Curr. Diabetes Rep. 15 (6), 38.

Hsiao, E.Y., McBride, S.W., Hsien, S., Sharon, G., Hyde, E.R., McCue, T., et al., 2013. Microbiota modulate behavioral and physiological abnormalities associated with neurodevelopmental disorders. Cell 155 (7), 1451–1463.

Hughes, D.T., Sperandio, V., 2008. Inter-kingdom signalling: communication between bacteria and their hosts. Nat. Rev. Microbiol. 6 (2), 111–120.

Janik, R., Thomason, L.A., Stanisz, A.M., Forsythe, P., Bienenstock, J., Stanisz, G.J., 2016. Magnetic resonance spectroscopy reveals oral Lactobacillus promotion of increases in brain GABA, N-acetyl aspartate and glutamate. Neuroimage 125, 988–995.

Kałużna-Czaplińska, J., Błaszczyk, S., 2012. The level of arabinitol in autistic children after probiotic therapy. Nutrition 28 (2), 124–126.

Kato-Kataoka, A., Nishida, K., Takada, M., Kawai, M., Kikuchi-Hayakawa, H., Suda, K., et al., 2016. Fermented milk containing Lactobacillus casei strain Shirota preserves the diversity of the gut microbiota and relieves abdominal dysfunction in healthy medical students exposed to academic stress. Appl. Environ. Microbiol. 82 (12), 3649–3658.

Kelly, J.R., Clarke, G., Cryan, J.F., Dinan, T.G., 2016. Brain-gut-microbiota axis:challenges for translation in psychiatry. Ann. Epidemiol. 26, 366–372.

Koloski, N.A., Jones, M., Kalantar, J., Weltman, M., Zaguirre, J., Talley, N.J., 2012. The brain–gut pathway in functional gastrointestinal disorders is bidirectional: a 12-year prospective population-based study. Gut 61 (9), 1284–1290.

Kunze, W.A., Mao, Y.K., Wang, B., Huizinga, J.D., Ma, X., Forsythe, P., et al., 2009. Lactobacillus reuteri enhances excitability of colonic AH neurons by inhibiting calcium-dependent potassium channel opening. J. Cell. Mol. Med. 13 (8b), 2261–2270.

Latalova, K., Hajda, M., Prasko, J., 2017. Can gut microbes play a role in mental disorders and their treatment? Psychiatria Danubina 29 (1), 28–30.

Liu, X., Cao, S., Zhang, X., 2015. Modulation of gut microbiota–brain axis by probiotics, prebiotics, and diet. J. Agric. Food Chem. 63 (36), 7885–7895.

Lyte, M., Daniels, K.M., Schmitz-Esser, S., 2019. Fluoxetine-induced alteration of murine gut microbial community structure: evidence for a microbial endocrinology-based mechanism of action responsible for fluoxetine-induced side effects. PeerJ 7, e6199.

MacFabe, D., 2013. Autism: metabolism, mitochondria, and the microbiome. Glob. Adv. Health Med. 2 (6), 52–66.

Macfarlane, S., Dillon, J.F., 2007. Microbial biofilms in the human gastrointestinal tract. J. Appl. Microbiol. 102 (5), 1187–1196.

Mangiola, F., Ianiro, G., Franceschi, F., Fagiuoli, S., Gasbarrini, G., Gasbarrini, A., 2016. Gut microbiota in autism and mood disorders. World J. Gastroenterol. 22 (1) 361.

Marilia, C., Annunziata, S., Maria, A.M., Carola, S., 2015. The gut-brain axis: interactions between enteric microbiota, central and enteric nervous systems. Ann. Gastroenterol. 28, 203–209.

Marler, S., Ferguson, B.J., Lee, E.B., Peters, B., Williams, K.C., McDonnell, E., et al., 2016. Brief report: whole blood serotonin levels and gastrointestinal symptoms in autism spectrum disorder. J. Autism Dev. Disord. 46 (3), 1124–1130.

Mayer, E.A., 2011. Gut feelings: the emerging biology of gut–brain communication. Nat. Rev. Neurosci. 12 (8), 453–466.

Mayer, E.A., Savidge, T., Shulman, R.J., 2014. Brain–gut microbiome interactions and functional bowel disorders. Gastroenterology 146 (6), 1500–1512.

Mayer, E.A., Tillisch, K. Gupta, A., 2015. Gut/brain axis and the microbiota. J. Clin. Invest. 125 (3), 926–938.

McVey Neufeld, K.A., Mao, Y.K., Bienenstock, J., Foster, J.A., Kunze, W.A., 2013. The microbiome is essential for normal gut intrinsic primary afferent neuron excitability in the mouse. Neurogastroenterol. Motil. 25 (2), e183–e188.

Messaoudi, M., Lalonde, R., Violle, N., Javelot, H., Desor, D., Nejdi, A., et al., 2011. Assessment of psychotropic-like properties of a probioticformulation (Lactobacillus helveticus R0052 and Bifidobacteriumlongum R0175) in rats and human subjects. Br. J. Nutr. 105, 755–764.

Morgan, M.Y., 1991. The treatment of chronic hepatic encephalopathy. Hepato-gastroenterology 38 (5), 377–387.

Naseribafrouei, A., Hestad, K., Avershina, E., Sekelja, M., Linløkken, A., Wilson, R., et al., 2014. Correlation between the human fecal microbiota and depression. Neurogastroenterol. Motil. 26 (8), 1155–1162.

Ng, Q.X., Peters, C., Ho, C.Y.X., Lim, D.Y., Yeo, W.S., 2018. A meta-analysis of the use of probiotics to alleviate depressive symptoms. J. Affect. Disord. 228, 13–19

Park, A.J., Collins, J., Blennerhassett, P.A., Ghia, J.E., Verdu, E.F., Bercik, P., et al., 2013. Altered colonic function and microbiota profile in a mouse model of chronic depression. Neurogastroenterol. Motil. 25 (9), 733–e575.

Parks, B.W., Nam, E., Org, E., Kostem, E., Norheim, F., Hui, S.T., et al., 2013. Genetic control of obesity and gut microbiota composition in response to high-fat, high-sucrose diet in mice. Cell Metab. 17 (1), 141–152.

Phillips, R.J., Powley, T.L., 1996. Gastric volume rather than nutrient content inhibits food intake. Am. J. Physiol. Regul. Integr. Comp. Physiol. 271 (3), R766–R769.

Picciotto, M.R., 2008. Galanin–25 years with a multitalented neuropeptide. Cell. Mol. Life Sci. 65 (12), 1872–1879.

Raybould, H.E., Gschossman, J.M., Ennes, H., Lembo, T., Mayer, E.A., 1999. Involvement of stretch-sensitive calcium flux in mechanical transduction in visceral afferents. J. Autonomic Nerv. Syst. 75 (1), 1–6.

Rhee, S.H., Pothoulakis, C., Mayer, E.A., 2009. Principles and clinical implications of the brain–gut–enteric microbiota axis. Nat. Rev. Gastroenterol. Hepatol. 6 (5), 306–314.

Rogers, G.B., Keating, D.J., Young, R.L., Wong, M.L., Licinio, J., Wesselingh, S., 2016. From gut dysbiosis to altered brain function and mental illness: mechanisms and pathways. Mol. Psychiatry 21 (6), 738–748.

Rothschild, D., Weissbrod, O., Barkan, E., Kurilshikov, A., Korem, T., Zeevi, D., et al., 2018. Environment dominates over host genetics in shaping human gut microbiota. Nature 555 (7695), 210–215.

Rubio, C.A., Huang, C.B., 1992. Quantification of the sulphomucin-producing cell population of the colonic mucosa during protracted stress in rats. Vivo (Athens, Greece) 6 (1), 81–84.

Santos, J., Saperas, E., Nogueiras, C., Mourelle, M., Antolín, M., Cadahia, A., et al., 1998. Release of mast cell mediators into the jejunum by cold pain stress in humans. Gastroenterology 114 (4), 640–648.

Sarkar, A., Lehto, S.M., Harty, S., Dinan, T.G., Cryan, J.F., Burnet, P.W., 2016. Psychobiotics and the manipulation of bacteria–gut–brain signals. Trends Neurosci. 39 (11), 763–781.

Saulnier, D.M., Ringel, Y., Heyman, M.B., Foster, J.A., Bercik, P., Shulman, R.J., et al., 2013. The intestinal microbiome, probiotics and prebiotics in neurogastroenterology. Gut. Microbes 4 (1), 17–27.

Savignac, H.M., Tramullas, M., Kiely, B., Dinan, T.G., Cryan, J.F., 2015. Bifidobacteria modulate cognitive processes in an anxious mouse strain. Behav. Brain Res. 287, 59–72.

Schicho, R., Krueger, D., Zeller, F., Von Weyhern, C.W.H., Frieling, T., Kimura, H., et al., 2006. Hydrogen sulfide is a novel prosecretory neuromodulator in the Guinea-pig and human colon. Gastroenterology 131 (5), 1542–1552.

Sobko, T., Huang, L., Midtvedt, T., Norin, E., Gustafsson, L.E., Norman, M., et al., 2006. Generation of NO by probiotic bacteria in the gastrointestinal tract. Free. Radic. Biol. Med. 41 (6), 985–991.

Soderholm, J.D., Yates, D.A., Gareau, M.G., Yang, P.C., MacQueen, G., Perdue, M.H., 2002. Neonatal maternal separation predisposes adult rats to colonic barrier dysfunction in response to mild stress. Am. J. Physiol. -Gastrointest. Liver Physiol. 283 (6), G1257–G1263.

Steenbergen, L., Sellaro, R., van Hemert, S., Bosch, J.A., Colzato, L.S., 2015. A randomized controlled trial to test the effect of multispecies probiotics on cognitive reactivity to sad mood. Brain, Behav. Immun. 48, 258–264.

Stilling, R.M., Dinan, T.G., Cryan, J.F., 2014. Microbial genes, brain & behaviour–epigenetic regulation of the gut–brain axis. Genes, Brain Behav. 13 (1), 69–86.

Stilling, R.M., Moloney, G.M., Ryan, F.J., Hoban, A.E., Bastiaanssen, T.F., Shanahan, F., et al., 2018. Social interaction-induced activation of RNA splicing in the amygdala of microbiome-deficient mice. eLife 7, e33070.

Talsania, T., Anini, Y., Siu, S., Drucker, D.J., Brubaker, P.L., 2005. Peripheral exendin-4 and peptide YY3–36 synergistically reduce food intake through different mechanisms in mice. Endocrinology 146 (9), 3748–3756.

Thayer, J.F., Sternberg, E.M., 2009. Neural concomitants of immunity—focus on the vagus nerve. NeuroImage 47 (3), 908–910.

Tillisch, K., Labus, J., Kilpatrick, L., Jiang, Z., Stains, J., Ebrat, B., et al., 2013. Consumption of fermented milk product with probiotic modulates brain activity. Gastroenterology 144 (7), 1394–1401.

Tsigos, C., Chrousos, G.P., 2002. Hypothalamic–pituitary–adrenal axis, neuroendocrine factors and stress. J. Psychosom. Res. 53 (4), 865–871.

Wang, Y., Kasper, L.H., 2014. The role of microbiome in central nervous system disorders. Brain, Behav. Immun. 38, 1–12.

Weinbach, E.C., Levenbook, L., Alling, D.W., 1992. Binding of tricyclic antidepressant drugs to trophozoites of Giardia lamblia. Comp. Biochem. Physiol. C, Comp. Pharmacol. Toxicol. 102 (3), 391–396.

Westfall, S., Lomis, N., Kahouli, I., Dia, S.Y., Singh, S.P., Prakash, S., 2017. Microbiome, probiotics and neurodegenerative diseases: deciphering the gut brain axis. Cell. Mol. Life Sci. 74 (20), 3769–3787.

Wickbom, J., Herrington, M.K., Permert, J., Jansson, A., Arnelo, U., 2008. Gastric emptying in response to IAPP and CCK in rats with subdiaphragmatic afferent vagotomy. Regul. Pept. 148 (1–3), 21–25.

Zhang, C., Zhang, M., Pang, X., Zhao, Y., Wang, L., Zhao, L., 2012. Structural resilience of the gut microbiota in adult mice under high-fat dietary perturbations. ISME J. 6 (10), 1848–1857.

Chapter 29

Flaxseed has a pronounced effect on gut microbiota

M.N. Sharath Kumar and S. Devaraja

Department of Studies and Research in Biochemistry, Tumkur University, Tumkur, Karnataka, India

29.1 Introduction

The gut microbiota, which is composed of thousands of different microbial species and more than 15,000 kinds of bacteria and weighs around 1 kg, represents the first protection system of the gastrointestinal (GI) tract. The presence of the microbiota varies within the GI tract, from a few microorganisms in the stomach and small intestine up to a concentration of approximately 10^{12} bacteria in the colon, mostly represented by the Firmicutes and Bacteriodetes phyla (Sartor, 2008; Schmidt and Stallmach, 2005). Within the species that compose the microbiota, it is also possible to recognize the kingdom of Archaea, eukaryotes, and many viruses and bacteriophages (Breitbart et al., 2003; Eckburg et al., 2003). Finally, there are several families of fungi, whose physiological role in the GI system is still unclear. The gut microbiota is involved in the maturation of the immune system. It stimulates innate immunity in the early years of life, leading to the maturation of the, and acquired immunity through stimulation of local and systemic immune responses (Nell et al , 2010). is the gut microbiota also has a role in the synthesis and metabolism of certain nutrients, hormones, and vitamins and clearance of drugs and toxic. The hypothesis that the gut microbiota has an important role in maintaining the physiological state of the GI system is supported by several studies that have shown qualitative and quantitative alterations of the intestinal flora in a number of gastrointestinal and extragastrointestinal diseases. The trillions of bacteria that are housed in the GI tract play an essential role in host homeostasis. This bacterial population was shown to be able to modulate the host's energy metabolism (Tremaroli and Backhed, 2012), inflammation (Cani and Delzenne, 2011; Cani et al., 2012), and immunity (Maynard et al., 2012). The interactions between bacteria and their host are mediated, among others, by metabolites produced by the bacteria (Nicholson et al., 2012). Owing to the huge size of the microbiome, a very high metabolic potential of the gut microbiota could be envisaged. The contribution of bacteria to fatty acid metabolism has been mostly studied in ruminants (Chilliard et al., 2007) (Fig. 29.1).

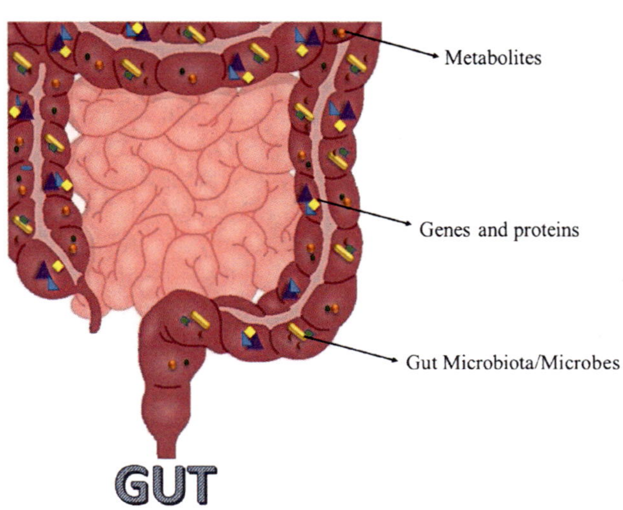

FIGURE 29.1 Gut microbiota.

Microbiome, Immunity, Digestive Health and Nutrition. DOI: https://doi.org/10.1016/B978-0-12-822238-6.00011-X

29.2 Flaxseed oil prevents alcoholic liver disease by modulating the gut microbiota through its antiinflammatory property

Excess intake of alcohol leads to be liver damage as it generates alcoholic liver disease (ALD). ALD a chronic wide-spectrum liver injury caused by consistently excessive alcohol intake. It ranking major causes of morbidity and mortality worldwide among people who abuse alcohol (Rehm et al., 2009). ALD includes a histological spectrum of liver injure ranging from simple steatosis to hepatitis characterized by inflammation, with potential progression to fibrosis and cirrhosis. Hepatitis, with an occurrence of approximately 10%−35% in chronic drinkers, is responsible for more than one-third of the significant morbidity and mortality rate through worldwide, thus it play a crucial role in reversible pathological process of ALD (Di, 2000). The gut microbiota plays a crucial role in the progression and pathogenesis of ALD. Accumulating evidence has revealed that the gut microbiota is closely associated with the liver in ALD in the gut-liver axis (Kirpich et al., 2016). Impairment of gut microbiota homeostasis in ALD induces proliferation of gram-negative pathogenic bacteria, which generate lipopolysaccharide (LPS) and translocate to the liver tissue as a trigger for hepatitis by binding to toll-like receptor 4 (TLR-4) on macrophages and neutrophils. Moreover, Campos Canesso et al. showed that the administration of alcohol to germ-free mice is associated with the absence of liver inflammation and injury, indicating that alcohol alone is not sufficient for the development of liver disease and that the presence of microbiota alterations is also necessary (Allampati and Mullen, 2016). Thus modulation of gut microbiota dysbiosis could attenuate hepatic injury in ALD (Altamirano and Bataller, 2011). Flaxseed oil (FO) is rich in plant-derived omega-3 (ω-3) polyunsaturated fatty acids (PUFAs), mainly alpha-linolenic acid (ALA, 18:3 ω-3). Clinical studies have reported that a low levels of ω-3 PUFAs in serum and liver tissue is a common characteristic of patients ALD (Szabo, 2015). Dietary FO prevented acute alcoholic hepatic steatosis via ameliorating lipid homeostasis at adipose tissue-liver axis in mice (Szabo and Bala, 2010). However, the impact of dietary FO on inflammation and gut microbiota in chronic ALD remains unknown. According to the study by Zhang et al., team members FO showed more effective in reduction of hepatic damage associated with altered gut microbiota and the decrease of liver inflammation. Numerous studies have indicated that alcohol exposure significantly reduces final body weight in chronic ALD (Canesso et al., 2014). FO may positively affect nutrients absorption and efficiency of calorie utilization in the GI tract in ALD. Significant reductions in plasma alanine aminotransferase (ALT) and aspartate aminotransferase (AST) in AF/FO group revealed that supplementary FO alleviated liver damage caused by chronic ethanol feeding by significantly decreasing the plasma ALT (alanine aminotransferase) and AST in ALD. It is well known that LPS induces hepatic inflammation in ALD by translocating to the liver via the portal vein and binds to TLR-4 of antigen-presenting cells to induce an inflammatory immune response and finally cause chronic hepatitis (Chen et al., 2016). FO may decrease gut permeability and reduce LPS translocation from the intestines to the liver and systematic circulation in ALD, which contributes to the reduction in inflammatory response in the liver. This attenuation may be associated with the intestinal innate immune system, and the underlying mechanism needs to be further researched (Shapiro et al., 2011). Activation of Kupffer cells and neutrophils induces oxidative stress and produces inflammatory cytokines, such as tumor necrosis factor alpha (TNF-α), interleukin 1 beta (IL-1β) and IL-6, which cause apoptosis and necrosis of hepatocytes and result in liver injury (Thaiss et al., 2016). FO alleviated hepatic inflammation via antiinflammatory cytokines. IL-10 is an antiinflammatory cytokine that is released by Kupffer cells and monocytes (Gao, 2012). Additionally, regulatory immune cells, especially regulatory T lymphocytes (Tregs) (Neuman et al., 2015), which play a critical role in regulation of proinflammation to keep maintain the immune balance in ALD (Byun et al., 2013), need to be investigated further study. Gut microbiota dysbiosis is thought to play a crucial role in the pathogenesis of ALD (de Waal et al., 1991). Bacteriodetes and Firmicutes were the most dominant in all four groups, and the proportion of Firmicutes was notably increased in alcohol feeding groups. Decreased levels of Bacteriodetes and higher levels of Proteobacteria in the alcohol intake group are the main criteria for gut dysbiosis, as recently described in human and animal studies (Hawrylowicz and O'Garra, 2005). Moreover, dietary FO notably reduced the proportion of Proteobacteria in chronic alcohol consumption, revealing that dietary FO may attenuate gut dysbiosis, presumably by modulating gut Proteobacteria. Exact mechanism(s) underlying these effects remain to be determined. At the genus level, decreased gut Porphyromonadaceae and inversely elevated Parabacteroides were found in chronic alcohol administration. Porphyromonadaceae was negatively correlated with TNF-α expression in the liver in ALD (Chen et al., 2016). Elevated levels of Parabacteroides were also involved in the prevention of hepatic inflammation in ALD (Kasztelan-Szczerbinska et al., 2015). FO ameliorates alcoholic liver disease via antiinflammation and modulating the gut microbiota, thus helping in the prevention and treatment of ALD (Fig. 29.2).

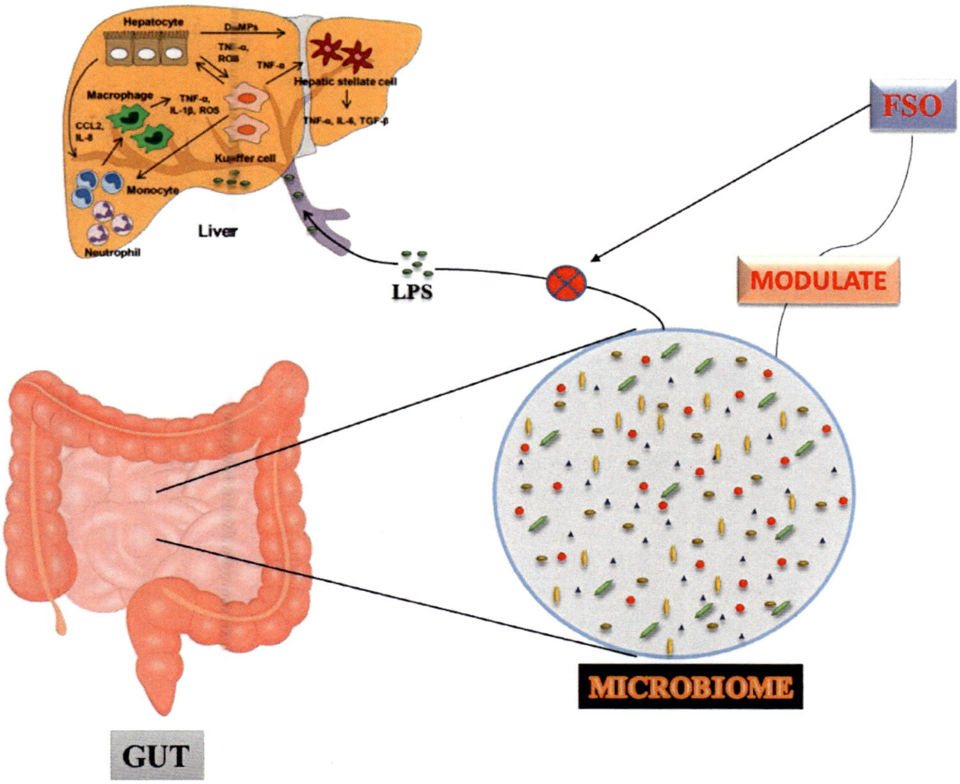

FIGURE 29.2 Prevention and treatment of ALD.

29.3 Flaxseed oil reduces the severity of type 2 diabetes mellitus by modulating the gut microbiota via its antiinflammation effect

Type 2 diabetes mellitus (T2DM) is a major disorder that leads to several diseases. The prevalence of T2DM presents a global public health problem. According to statistical data of the International Diabetes Federation, more than 450 million people suffered from diabetes worldwide in 2017, and this number is expected to hit 629 million by 2045 (Larsen et al., 2010). The gut microbiota represents a complicated community involving bacteria in the GI tract, which usually maintains a mutually beneficial relationship with the host. Numerous studies have demonstrated that the pathogenesis of T2DM is closely related to the gut microbiota (Zhang et al., 2018). It has been proven that diabetic individuals possess a high quantity of gram-negative bacteria, especially those belonging to the phylum Proteobacteria (Bouter et al., 2017). In a rat model of T2DM, the proportion of Firmicutes was increased, while the abundance of Bacteroides was decreased (Delzenne et al., 2015). Regulation of the gut microbiota may improve metabolism in T2DM (Moura-Assis et al., 2018). In T2DM, gut microbiota dysbiosis promotes the generation of LPS triggering chronic low-degree inflammation by binding to TLR-4 (Nie et al., 2019). Excessive translocation of gut-derived LPS to the liver via the portal circulation can subsequently promote the inflammatory cascade, leading to the release of inflammatory cytokines involving IL-1β, IL6, IL-17A, and TNF-α (Conlon and Bird, 2014). In addition to LPS, other metabolites (such as short-chain fatty acids, SCFAs) of the gut microbiota have been demonstrated to target multiple pathways in the intestine, liver, pancreas, and peripheral blood circulation, resulting in improvement in gut homeostasis, glycemic control, lipid profile, insulin resistance, and inflammation (Hu et al., 2018). Patients with diabetes exhibited lower levels of SCFAs and a smaller proportion of SCFA-producing gut bacterial communities, including *Bifidobacterium*, *Bacteroidetes*, and *Lactobacillus* (Ulven and Christiansen, 2015). At present, SCFAs have been recognized as potential mediators in intestinal immune function. Over the past few decades, humans' lifestyle has changed greatly, with increasing consumption of animal products and dietary fat. The incidence of T2DM has been positively related to these unhealthy lifestyle patterns, indicating that dietary factors play a crucial role in the onset and progression of T2DM (White and Marette, 2014). One of the most common dietary approaches against T2DM is the increase in the consumption of ω-3 PUFAs (Bang et al., 1980). A survey has suggested that the low incidence of diabetes in Greenland Eskimos may partly due to high consumption of fish oil, which is rich in ω-3 PUFAs, especially eicosapentaenoic acid

(EPA) and docosahexaenoic acid (DHA) (Soleimani et al., 2017). FO, a plant source of ω-3 PUFAs, has shown benefit in chronic metabolic disorders, such as diabetes (Liu et al., 2012). The advantage of FO over fish oil is that the food sources of FO are less expensive and are free of heavy metal contamination. In addition, FO has been demonstrated to suppress the production of TNF-α, IL-6, and IL-1β (Gaspar et al., 2019). Accumulating studies have shown that FO protects against inflammation and oxidative stress (Jangale et al., 2016) in metabolic disease models.

Intraperitoneal injection of streptozotocin (STZ) is mainly attributed to overproduction of reactive oxygen species, leading to toxicity of pancreatic cells (Masiello et al., 1998), which regulate the synthesis and release of insulin via the glucose transporter and alkylation of DNA. In contrast, nicotinamide (NA) prevents excess damage to pancreatic beta cells during the induction of T2DM. A T2DM rat model has been widely used to assess the effects and mechanisms of intervention (Yuan et al., 2019). FO was shown to be more effective in alleviation of T2DM, which may be the result of the suppression of inflammation and restoration of gut dysbiosis, potentially providing a theoretical foundation as an inexpensive intervention for prevention and treatment of T2DM. Consumption of ω-3 PUFAs is beneficial for chronic metabolic diseases, including T2DM, via antiinflammation and suppressing oxidative stress in clinical and experimental studies (Zhang et al., 2017). FO as a source of a plant-derived ω-3 PUFAs (rich in ALA) has been extensively investigated and has shown antiinflammatory properties throughout the body (Zhao, 2013). Wang et al. showed that supplementation with perilla oil, which is rich in ALA, improved hypertriglyceridemia and gut dysbiosis in diabetic KKAy mice. A randomized double-blind placebo-controlled clinical trial revealed that FO supplementation significantly improved levels of gene expression related to insulin, lipids, and inflammation in diabetic patients with coronary heart disease (Hashemzadeh et al., 2017). FO may positively affect nutrient absorption and efficiency of calorie utilization in the GI tract in T2DM. Yang et al. (2011) found that an increase in ω-3 PUFA levels and the concomitant decrease in the ω-6/ω-3 PUFA level ratio were likely to be involved in the beneficial changes to the metabolic indicators. FO treatment effectively alleviated routine parameters, including fasting blood glucose, glycated hemoglobin, liver hematoxylin eosin, and metabolic indicators (total cholesterol (TC), triglycerides (TG), low-density lipoproteins and high-density lipoproteins), which revealed that FO can regulate the metabolic disorders in a manner that is consistent with recent studies. Oxidative stress is considered to be one of the chief factors responsible for the development and progression of T2DM. Haliga et al. (2009) showed that dietary flaxseed reduced renal oxidative stress by improving the activity of serum superoxide dismutase (SOD) and decreasing thiobarbituric acid reactive substances in a STZ-induced diabetic hamster model. FO induced an increase in the serum oxidative indicator SOD as well as a decrease in malondialdehyde. Collectively, it has been suggested that oxidative stress may be linked to the etiology of T2DM and can be ameliorated by a healthy diet. LPS derived from pathogenic bacteria, a causal link between the gut microbiota and systemic low-grade inflammation, is translocated to the liver and binds to TLR-4 to induce an inflammatory cascade reaction, ultimately leading to diabetes mellitus (Badawy et al., 2015). LPS levels were notably decreased after FO intervention, indirectly demonstrating that the integrity and permeability of the intestinal mucosa barriers were improved to alleviate peripheral endotoxemia. The damage of intestinal tight junction proteins such as Zo-1 and occludin is essential and critical for LPS translocation. However, the direct evidence of gut permeability and integrity still needs further research to measure and assess the efforts of the intestinal mucosal barrier, including physical tight junctions (Zo-1, occludin) and other components of the gut barriers involving intestinal mucosal immunity (γδ T cells, intraepithelial lymphocytes, Tregs, natural killer T cells). Diabetes is a chronic metabolic disease involving a low-grade inflammation (Burcelin et al., 2012). Studies have demonstrated that activation of neutrophils and macrophages results in oxidative stress as well as the release of inflammatory cytokines, including IL-1β, IL-6, and TNF-α, leading to a decrease in insulin secretion, hepatic steatosis, and a series of metabolic disorders. Hotamisligil et al. (1993) first showed that the proinflammatory cytokine TNF-α was able to induce insulin resistance. Accelerating studies have demonstrated that ω-3 fatty acids are considered to possess antiinflammatory abilities in chronic metabolic diseases. In T2DM, inflammation and glucose are interrelated, with reciprocal causation (Saisho, 2015). A high glucose concentration leads to protein glycation and the production of advanced glycose end products (AGEs). The accumulation of AGEs activates NF-κB to induce TNF-α expression which lead to chronic inflammation. Intake of dietary ω-3 fatty acids is associated with reduced lymphocyte proliferation and Th1 cell development (Morey et al., 2019); lower circulating levels of leptin, C-reactive protein, and other proinflammatory cytokines; and a lower risk of infection (Zhang et al., 2005). It has been shown that supplementation with FO for 8 weeks caused a significant decrease in the levels of glucose tolerance, IL-1β, and TNF-α in obese and diabetic subjects (Merchant et al., 2005). Significant augmentation in a series of proinflammatory cytokines, including IL-1β, TNF-α, IL-6, and IL-17A is seen in STZ-NA-induced diabetic rats. However, the levels of those proinflammatory cytokines were dramatically reduced after FO treatment, which was in agreement with a recent study in which dietary FO modulated the expression of inflammatory genes with alleviation of the protein glycation status in STZ-NA-induced diabetic rats (Jangale et al., 2013). The possible mechanism of antiinflammation is that ALA in the

FO diet mediated G-protein-coupled receptor 120 (GPR120) activation and beta-arrestin 2 recruitment in the inhibition of TLR-4 and TNF-α downstream signaling. But the exact evidence needs to be further researched. IL-10 is an antiinflammatory cytokine that is released by Kupffer cells and monocytes. The role of IL-10 and Treg cells in underlying mechanisms of FO-treated T2DM needs to be further investigated. Growing evidence has demonstrated that the gut microbiota plays a critical role in the development of T2DM (Van der Poll et al., 1994). Gut dysbiosis can facilitate LPS entry into systemic circulation through increasing gut permeability, which leads to inflammation and metabolic dysfunction. Firmicutes plays a major role in absorbing calories from the diet and storing fat in gut cells. A decrease in Firmicutes after FO treatment revealed that FO may attenuate diabetes by reducing Firmicutes involved in energy absorption. In addition, the ratio of Firmicutes-Bacteroidetes was increased in diabetic group, suggesting a characteristics of micro-ecological disorder in intestinal microbes (Brun et al., 2007). Dietary FO intervention restored this ratio, demonstrating that dietary FO can positively shape the host microbial ecosystem. A positive correlations of Firmicutes with proinflammatory IL-1β, TNF-α, IL-6, IL-17A, and LPS demonstrated that both proinflammatory indicators and Firmicutes contribute to pathogenesis of diabetes. In contrast, negative correlations of Bacteroidetes with proinflammatory indicators may indicate protection in diabetes. However, the details in the relationship between Firmicutes-Bacteroidetes and inflammation need to be further investigated. At the genus level, *Blautia* is a gram-positive, anaerobic bacterium belonging to the family Lachnospiraceae that is thought to take part in the development of glucose metabolism disturbances (Martinez et al., 2013). A recent study showed that *Blautia*, especially *Blautia coccoides*, may activate the secretion of TNF-α to an even greater extent than the LPS (Lambeth et al., 2015). *Blautia* has been positively correlated with inflammatory indicators (IL-1β, TNF-α, IL-6, and LPS) in diabetes. Meanwhile, abnormal elevated levels of *Blautia* in diabetes were restored to normal by FO intervention. Similarly, supplementation with ALA that is rich in perilla oil resulted in decreased abundance of *Blautia* in diabetic KKAy mice (Kashtanova et al., 2018). In addition, FO treatment significantly regulated the abundance of *Alistipes*, which belongs to family Rikenellaceae. A recent study showed that *Alistipes* was consistently decreased in gestational diabetes. Furthermore, *Alistipes* was negatively correlated with TNF-α and LPS, demonstrating that *Alistipes* may suppress diabetic inflammation. These findings revealed that *Alistipes* may be involved in the improvement of diabetes. More and more studies have suggested that SCFAs, one of the critical gut microbiota metabolites, as a link between the microbiota and host homeostasis, play an important role in the regulation of inflammation and glucose and lipid metabolism. Acetic acid, propionic acid, and butyric acid account for 90%–95% of the total SCFAs (Egshatyan et al., 2016). Acetate and propionate are the main products of Bacteroidetes, and butyrate is produced mainly by Firmicutes. A study showed that ω-3 PUFA exerted a positive action by reverting microbiota-derived SCFAs (Kuang et al., 2017). FO may alleviate diabetes via modulating gut microbiota acetic acid. In addition, increased Bacteroidetes, decreased Firmicutes, and a reduced ratio of Firmicutes-Bacteroidetes after dietary FO treatment demonstrated that Bacteroidetes acted as the main source of increased acetate. Intriguingly, dietary FO supplementation also significantly increased propionic acid and butyric acid, which were decreased in diabetes group but without significant difference, suggesting that dietary FO has the ability to increase SCFAs within or without diabetes. But a study reported that oral butyrate administration significantly alleviated diabetic endotoxemia with improved gut integrity and a decreased Firmicutes-Bacteroidetes ratio (Cummings et al., 1987). The exact role of increased propionic acid and butyric acid after dietary FO intervention needs to be further investigated. FO may regulate the gut microbiota metabolites SCFAs, especially acetate, with beneficial effects on diabetes (Fig. 29.3).

29.4 By altering the composition of the gut microbiota, flaxseed oil has beneficial immunological and metabolic effects

To be healthy and to lead a healthy life style, the metabolic process in the body play a major role. Metabolic health can be described as the body's ability to successfully balance fuel oxidation and fuel availability in an endocrine-influenced environment (Aparicio-Soto et al., 2016). In the current obesogenic climate, attaining this metabolic balance is becoming ever more difficult to achieve; as a consequence, the burden of lifestyle-related metabolic diseases is rapidly increasing worldwide. Low-grade, sustained systemic inflammation, considered to lie at the heart of metabolic dysfunction, is strongly linked to the consumption of obesogenic diets that are high in proinflammatory saturated fats (Kajla et al., 2015). Chronic intake of diets that are high in saturated fats not only leads to inevitable weight gain but can also compromise intestinal barrier function, causing metabolic end toxemia, inducing systemic inflammation partly through TLR signaling (Oliveira et al., 2015). The mucosal immune system employs a number of mechanisms to assist in the protection of the gut barrier and preservation of host defenses. Antimicrobial peptides are one such example. Primarily produced by Paneth cells, however, also by enterocytes in the intestine, they create a physical barrier, separating gut microbiota from intestinal cells, employing a variety of bacteriostatic and bactericidal mechanisms (Cicerale et al., 2012). For example,

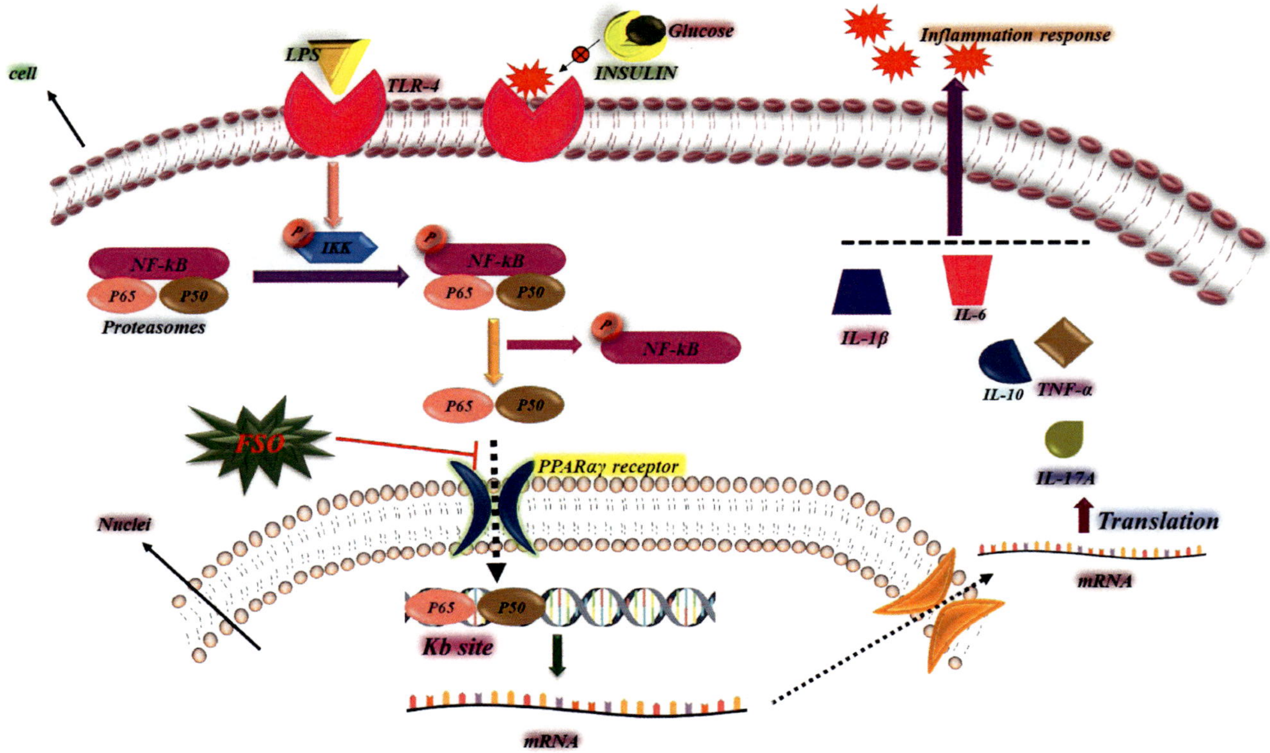

FIGURE 29.3 Regulation of gut microbiota.

RegIIIγ, a representative antimicrobial peptide of the lectin family, uses a nonenzymatic attack to destroy pathogenic bacteria through the TLR–myeloid differentiation primary response 88 (MyD88)-signaling pathway via recognition of microbiota-associated molecular patterns, eventually promoting its transcription and subsequent production (Vazquez-Baeza et al., 2013). The adaptive immune system also plays a fundamental role in modulating the local gut environment and composition of microbiota, with key players such as intestinal Treg cells, aiding mucosal tolerance, and enforcing commensalism (Langille et al., 2013). It has been expeditiously recognized that the gut microbiota is crucial in both mediating host inflammation and influencing host metabolic health, with diet as a critical factor in determining the diversity and function of the microbial community (Nishitsuji et al., 2017). A diet that is high in saturated fatty acids and low in dietary fibers is considered to be one of the main factors driving dysbiosis of the gut microbiota, decreasing microbial diversity and beneficial species of bacteria as well as increasing the abundance of certain pathogenic species of bacteria, overall negatively affecting metabolic health in both humans and rodents (Duncan et al., 2007). The potential impact of specific plant-derived dietary fats and oils on the composition of the gut microbiota and host metabolic health has been somewhat elucidated. FO contain a wide variety of compounds that are purported to have antiinflammatory properties and beneficial effects on markers of metabolic health, including blood glucose and lipids, body weight, and inflammation (Kubeck et al., 2016). FO is an abundant source of ALA, an essential ω-3 PUFA, but contains only a small quantity of phenolic compounds (Pellizzon and Ricci, 2018). FO exerts superior effects on the composition and diversity of the gut microbiota, intestinal immune function, and metabolic health in comparison. FO contains ω-9 and ω-3 fatty acids, which have been shown to increase glucagon-like peptide 1 in the intestine, reverse insulin resistance, and exert antiinflammatory activities through GPR120 and GPR40 pathways in the liver, skeletal muscle, and adipose tissue (Tirosh et al., 2019). Higher circulating levels of propionate are implicated in exaggerated gluconeogenesis and insulin resistance in both mice and humans (Sonnenburg and Backhed, 2016). Individuals with low bacterial richness show more insulin resistance and other metabolic abnormalities in comparison with individuals with high bacterial richness (Menni et al., 2017). The presence of the ω-3 fatty acid ALA in FO and its ability to be converted into the more potent ω-3 fatty acid DHA may play a role in increasing microbiota diversity (Al-Assal et al., 2018). Although it is well known that the majority of dietary fats are digested and absorbed in the small intestine, there is evidence that dietary fats, in particular, monounsaturated fatty acids and PUFAs, can enter the large intestine, potentially exerting effects on gut tissue as well as interacting with the resident microbiota (Loy et al., 2017). A recent study by Campbell et al. demonstrated that when the border-dwelling bacteria *Mucispirillum* was depleted in Treg-deficient mice, a subsequent increase in type 2 responses occurred, negatively altering epithelial homeostasis (Campbell et al., 2018). They also noted that the growth and expansion of *Mucispirillum* were in turn enhanced by Treg by promoting immune tolerance to this bacterium, subsequently shaping the microbiota. Bacteria belonging to unclassified *Clostridia* express bile acid-inducible genes and generate secondary bile acids, which exert widespread effects on host health, including antimicrobial effects (Gefen and Geva-Zatorsky, 2018). Although the mechanisms remain to be elucidated, an increase in these genera of bacteria and subsequent enrichment in genes involved in pathways of bile acid production would be a key impact factor of FO consumption on the gut microbiota. Specifically, these species of bacteria deliver prothermogenic metabolites such as LPS because of a defective intestinal barrier, causing chronic low-grade inflammation and atherosclerosis (Round et al., 2011). FO can beneficially affect the composition of the gut microbiota, increase microbial diversity, and influence the production of microbial-derived metabolites, contributing to the activation of the mucosal immune system (Fu et al., 2015), enhancement of markers of gut barrier integrity, and ultimately support of metabolic health (Vital et al., 2019). Further studies are warranted to test this hypothesis in human clinic-based dietary interventions (Fig. 29.4).

29.5 Flaxseed polysaccharide modulates the composition and structure of the gut microbiota by altering the firmicutes-bacteroidetes ratio

The gut microbiota, consisting of numerous microbial species, is considered a vital organ that plays a pivotal role in host health (Xu et al., 2014). Numerous studies have indicated that human gut microbiota can exert numerous benefits to the host health, including nutrient absorption, glucose tolerance, and pathogen defense (Al-Lahham et al., 2010). Growing evidence has demonstrated that plant polysaccharides cannot be digested in the upper digestive tract but can reach the distal intestine and be degraded by the gut microbiota (Canani et al., 2011). The fermentation of polysaccharides and their interaction with the gut microbiota are closely related to human health. For example, polysaccharides extracted from soybean were found to promote the production of acetic acid and the relative richness of *Bifidobacterium*. In addition, polysaccharide fermentation can further improve the intestinal microecosystem by inhibiting certain harmful bacteria in the host (Turnbaugh et al., 2009). Recently, scientists have proposed using polysaccharides as a new therapeutic strategy to regulate the gut microbiota for improving human health and preventing diseases (Fan et al., 2014). Bioactive polysaccharide intake regulates the composition and metabolites of the gut microbiota (Luo et al., 2018). Most polysaccharides modulate the diversity of the gut microbiota by decreasing the ratio of Firmicutes to Bacteroidetes, which might be beneficial for regulating host lipid metabolism (Jones et al., 2014). Polysaccharides could also protect

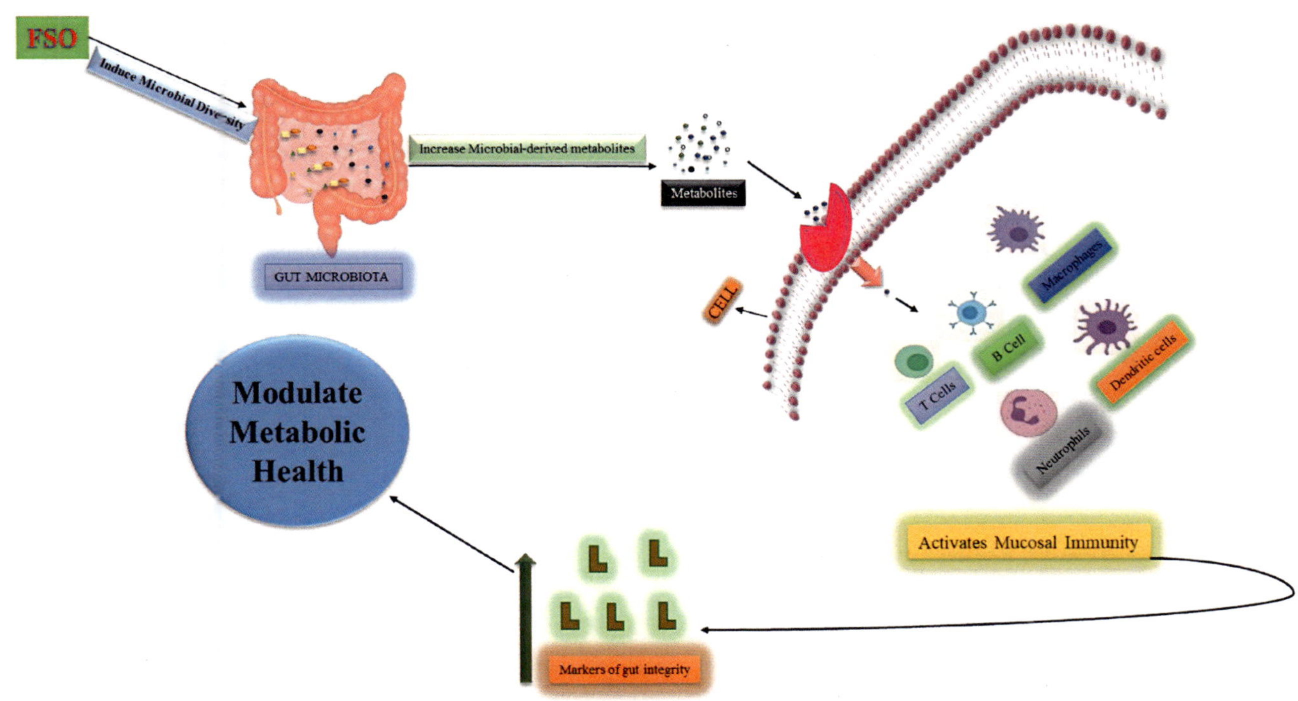

FIGURE 29.4 Modulation of metabolic health.

against dysbiosis of the gut microbiota and inflammatory bowel disease by modulating the multiformity of the intestinal flora (Hiippala et al., 2016). Moreover, SFCAs, the major metabolites arising from the microbial fermentation of dietary fiber, play a vital role in providing energy for epithelial cells (Zhang et al., 2015) and maintaining the epithelial barrier. SCFAs not only can modulate the immune response to prevent colorectal cancer, but also can regulate the characteristics of metabolic syndrome (Li et al., 2015). Because flaxseed (*Linum usitatissimum* L.) contains various biological compounds such as ω-3 fatty acids, good-quality dietary fiber, and bioactive lignans, it is considered a good functional food ingredient (Roulard et al., 2016). The main soluble dietary fiber that can be extracted from flaxseed is a heterogeneous polysaccharide that consists of neutral arabinoxylan and acidic rhamnose (Fu et al., 2018). Flaxseed polysaccharide (FSP) exhibited good antioxidant activities in scavenging radicals in vitro (Tedeschi et al., 2009). In addition, FSP intake can reduce the elevated blood glucose and body weight gain that are induced by a high-fat diet in rats (Ding et al., 2019). Our previous work demonstrated that FSP alleviates metabolic syndrome by modulating the gut microbiota and increasing the content of SCFAs in mice that were fed a high-fat diet (Li and Klaus, 1989). The ratio of Firmicutes to Bacteroidetes was significantly decreased by FSP fermentation. The Firmicutes-Bacteroidetes ratio is widely considered to be associated with energy harvest and can characterize the risk of obesity (Toma and Leung, 1987). The decreased Firmicutes-Bacteroidetes ratio caused by FSP fermentation suggested that FSP might have a beneficial effect on host weight loss. It has been demonstrated that FSP decreased the Firmicutes-Bacteroidetes ratio and inhibited obesity in mice that were fed a high-fat diet (Zhao et al., 2006). Additionally, at the genus level, fermentation with FSP exhibited a higher number of *Phascolarctobacterium* but lower numbers of *Bifidobacterium* and *Faecalibacterium*. *Prevotella* is the principal genus in the gut microbiota that degrades and uses polysaccharides. *Prevotella* consists of beneficial microbes that can improve glucose metabolism (Yang et al., 2019). *Phascolarctobacterium* was proved to be associated with propionate production through the succinate cycle (Wang et al., 2018). FSP displayed a different pattern of regulation of the gut microbiota compared with inulin (Huang et al., 2016). Altogether, the fermentation of FSP modulated the microbial community structure and increased the microbial diversity, suggesting that FSP has potential prebiotic potency. In conclusion as FSP cannot be degraded by simulated saliva or gastric or small intestine conditions, so it reaches the large intestine intact. FSP was gradually hydrolyzed and used by human fecal microbiota during fermentation. FSP fermentation exhibited a modulatory effect on the gut microbial community structure by reducing the ratio of Firmicutes to Bacteroidetes. FSP stimulated the growth of beneficial and SCFA-producing bacteria such as *Prevotella* and *Phascolarctobacterium*. Furthermore, the content of SCFAs, especially acetic and propionic acids, was significantly increased during FSP fermentation. Therefore FSP might contribute to gut health and could be a candidate for use as a prebiotic in functional foods (Fig. 29.5).

29.6 Flaxseed oil ameliorated polycystic ovary syndrome through the sex steroid hormone–microbiota-inflammation axis

Polycystic ovary syndrome (PCOS) is a common endocrine-metabolic disorder among women of reproductive age, with a worldwide prevalence of 4%–21% (Larsen et al., 2010). Patients with PCOS largely present with metabolic and other clinical features, including hyperandrogenemia, hyperinsulinemia, obesity, dyslipidemia, insulin resistance, ovulatory dysfunction, menstrual irregularity, and polycystic ovaries. Moreover, PCOS may cause an increased risk for metabolic disorders, such as T2DM, cardiovascular disease, metabolic syndrome, endometrial cancer, infertility, obstetrical complications, psychiatric conditions, and other complications (Zhao et al., 2006). Genetics, lifestyle, and environmental triggers have been considered to be the main drivers of PCOS, but the exact pathogenesis remains largely unclear. In view of the limitations of current therapeutics in treating the metabolic symptoms of PCOS, novel and more practical strategies for the control of PCOS are needed. Considerable evidence has demonstrated that abnormal sex steroid hormones are crucial parameters in PCOS manifestation and development (Toma and Leung, 1987). One of the core pathophysiological features of PCOS is hyperandrogenism, which is caused by elevated testosterone, inhibiting estrogen production. Subsequently, decreased estrogen is associated with increased susceptibility to PCOS. The gut microbiota and the vaginal microbiota (VMB) are inextricably linked to the occurrence of PCOS (Pellizzon and Ricci, 2018). Gut dysbiosis in PCOS embodies a decrease in alpha diversity and *Bifidobacterium* as well as an increase in *Bacteroides*. Intriguingly, remediation of gut dysbiosis by dietary prebiotic inulin supplementation contributes to the amelioration of PCOS. Regarding the VMB, in approximately 70% of healthy women of reproductive age, predominant *Lactobacillus* has beneficial effects on the vaginal ecosystem via generating acidic fermentation products (primarily lactic acid) (Canani et al., 2011). However, the roles of the gut microbiota and the VMB in the pathogenesis of PCOS need to be fully elucidated. Chronic low-grade inflammation is thought to be a key contributor in the pathogenesis of PCOS, with

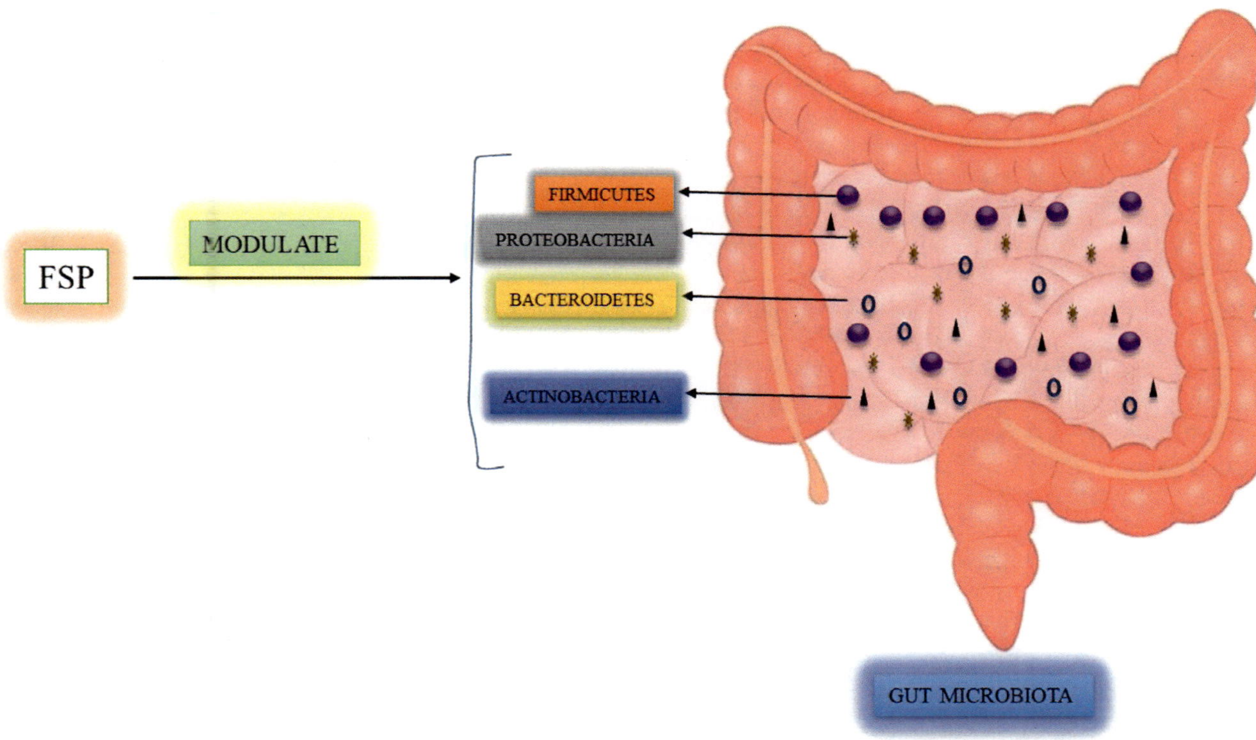

FIGURE 29.5 Role of FSP on gut microbiota.

the main performance of a higher concentration of interleukin (IL)- 6 and TNF-α. Regulatory element-binding protein 1 may promote TLR-4-induced proinflammatory responses by reprogramming fatty acid metabolism. Accumulating studies have demonstrated that sex steroid hormones are closely associated with dysbacteriosis and inflammation in various metabolic disturbances (Xu et al., 2014). High-throughput sequencing has shown that sex steroid hormones modulate the microbiota composition of mammals. In turn, host hormones can affect bacterial gene expression, virulence, and growth, with consequences on the host physiology (Jones et al., 2014). In PCOS, dysbacteriosis results in gut mucosal permeability, with a resultant increased translocation of LPS from the pathogenic into the systemic circulation. Moreover, the translocation of gut-derived LPS to the liver via the portal circulation subsequently gets it bound to TLR-4 on macrophages, which is responsible for driving the production of proinflammatory cytokines. The amount of gut-derived LPS that is translocated to the ovary can subsequently promote the inflammatory cascade, leading to ovarian inflammatory changes. In addition to LPS, other metabolites of the gut microbiota (e.g., SCFAs) have been demonstrated to regulate gut homeostasis, the lipid profile, insulin resistance, and inflammation (Menni et al., 2017). Thus chronic inflammation might be reduced by improving the gut-vaginal microbiota dysbiosis in PCOS. Nutritional intervention represents a promising strategy for the treatment of PCOS. Omega-3 PUFAs were found to provide health benefits such as antiinflammatory effects, insulin sensitivity, cellular differentiation, and ovulation (Huang et al., 2016). FO, which is obtained from the seeds of the flax plant, can provide plant-derived ω-3 PUFAs (ALA, 18:3 ω-3) to non–fish eaters. Dietary FO supplementation has the ability to attenuate chronic metabolic diseases. Furthermore, FO has been reported to enhance the insulin level, lower the blood glucose level, and restore the enzymatic antioxidant barrier in diabetic rats by blocking insulin resistance. The effect of FO on PCOS may be due to modulating the hormone–inflammation–gut/vaginal microbiota axis, potentially providing an inexpensive intervention for the control of PCOS. The letrozole-induced PCOS model has been widely utilized to assess the effects and the mechanisms of interventions on precautions and pathogenesis in PCOS (Yang et al., 2019). Letrozole, an aromatase inhibitor, induces hyperandrogenemia by reducing the conversion of testosterone to estrogen. Accumulated human and animal studies have demonstrated that the ω-3 PUFAs DHA and EPA may ameliorate the progression of PCOS. Anovulation was dramatically improved by DHA supplementation (Sonnenburg and Backhed, 2016). FO has the ability to ameliorate PCOS, indicating that inexpensive FO exhibits effects similar to those of EPA and DHA. We speculate that in vivo interconversion among FO, EPA, and DHA may partially contribute to understanding this effect of ω-3 PUFAs. However, the dynamic interconversion after FO supplementation needs to be further determined. PCOS represents a chronic endocrine metabolic disease with hyperandrogen and low estrogen (Round et al., 2011). The luteinizing hormone (LH)/follicle stimulating hormone (FSH) ratio has been regarded as a main biomarker of diagnosis in PCOS. The notably increased testosterone level is usually considered a marker of hyperandrogenism in PCOS. During the development and the progression of PCOS, abnormally increased testosterone secretion suppresses the production of estrogen, sex hormone-binding globulin (SHBG), and FSH (Pellizzon and Ricci, 2018). Furthermore, the suppressed FSH results in the arrest of ovarian folliculogenesis and corpora lutea under the synergistic effect of LH, thereby decreasing the progesterone that is produced by the corpus luteum. SHBG produced by the liver, a transporter of sex hormones with a high affinity for T but a low affinity for E2, can be used to evaluate the severity of hyperandrogenism and the therapeutic efficacy. FO supplementation has been found to be capable of improving the homeostasis of sex steroid hormones in PCOS. A study has indicated that an intake of dietary EPA exerts beneficial effects on androgens in PCOS. Moreover, EPA treatment increases the number of follicular cell layers and decreases the levels of LH and T in PCOS (Al-Lahham et al., 2010).

Recent studies have shown that patients with PCOS mainly presented with dyslipidemia and insulin resistance, which might be caused by hyperandrogenism (Huang et al., 2016). Omega-3 fatty acids may be recommended for the treatment of PCOS patients with insulin resistance as well as high TC and TG levels. Dietary FO improved dyslipidemia in patients with PCOS. FO alleviated systematic and ovarian inflammation via suppressing the proinflammatory cytokines (TNF-α, IL-1β, IL-6, IL-17A, and MCP-1) and elevating the antiinflammatory IL-10, suggesting the antiinflammation role of inexpensive dietary FO administration in patients with PCOS. Cosupplementation with ω-3 PUFAs and vitamin E downregulated IL-8 and TNF-α expression in PCOS patients (Fu et al., 2015). Dietary ALA-rich FO inhibited the production of TNF-α, IL-1β, and IL-6 in T2DM and ALD. EPA treatment can decrease IL-1β and TNF-α but increases IL-10 in PCOS. Clinical studies have shown that the supplementation of ω-3 PUFAs is beneficial in decreasing the MCP-1 expression in macrophages and the levels of endothelial chemokine (Li and Klaus, 1989). Owing to accumulating evidence of the critical role of macrophages in the pathogenesis of PCOS, we speculate that the antiinflammatory effect of dietary FO may contribute to inhibition of the activation of macrophages and their polarization. The exact role of the macrophage in the underlying mechanism of FO-treated PCOS needs to be further investigated. Growing evidence has demonstrated that the gut microbiota and its metabolites are closely associated with the occurrence and the development of PCOS (Wang et al., 2018). Liu et al. found a decrease in Ruminococcaceae and an increase in gram-negative bacteria, including *Bacteroides* and *Desulfovibrio*, in women with PCOS, which was consistent with our research. Regarding to the phylum level of the gut microbiota, we found that the Firmicutes, Bacteroidetes, and Proteobacteria were the most dominant in all groups. An increase in. Firmicutes-Bacteroidetes ratio is closely related to obesity (Li and Klaus, 1989). An increased Firmicutes-Bacteroidetes ratio in PCOS was rectified by administration of dietary FO. Moreover, at the genus level, FO restored gut dysbiosis in PCOS by upregulating *Lactobacillus*, *Allobaculum*, *Desulfovibrio*, and *Bifidobacterium* and downregulating *Actinobacteria* and *Bacteroides*, showing that dietary FO attenuated PCOS via restoring the gut dysbiosis. Similarly, dietary prebiotic inulin supplementation ameliorated PCOS in mice via improving the gut microbiota (Li and Klaus, 1989). A recent clinical study found that *Bifidobacterium lactis* V9 was significantly decreased in patients with PCOS. This probiotic regulates the secretion of sex hormones in PCOS patients through the gut-brain axis. Microbial LPS and SCFAs are thought to play a pivotal role in regulating inflammation and metabolism in PCOS (Hiippala et al., 2016). LPS, derived from gram-negative bacteria, activates the inflammation of the peripheral circulation and the ovary through the TLR-4-NF-κB-signaling pathway in macrophages to release inflammatory cytokines (Tirosh et al., 2019). Abnormal plasma LPS was restored by dietary FO treatment, demonstrating that FO improved gut dysbiosis and the gut barrier, decreasing the intestinal permeability of LPS translocation from the intestines to the systematic circulation in PCOS, finally contributing to the suppression of systematic and ovarian inflammation. In addition, the microbial fermentation end products, namely, SCFAs, are pivotal in the regulation of intestinal permeability and inflammation in the gut and the systemic circulation (Ding et al., 2019). The antiinflammatory and immunomodulatory effects of SCFAs might be due to the activation of the specific cell receptors GPR109a, GPR41, and GPR43 and the intracellular target via inhibiting histone deacetylase activity (Fu et al., 2018). Elevated SCFA levels, including levels of acetic acid, propionic acid, butyric acid, and valeric acid, after dietary FO treatment may enhance the integrity of the intestinal mucosal barrier as well as inhibiting enteral and parenteral inflammation. The alteration of SCFAs may regulate inflammation through the SCFAs−GPR−inflammatory cell−signaling pathway. Other microbial metabolites may also be involved in the effects of dietary FO intervention on PCOS and need to be further investigated using metabonomics methodology. Studies have indicated that the VMB plays a crucial role in the development of reproductive system diseases and HPV/HIV (Roulard et al., 2016). However, the effects of the proportions of VMB on PCOS are rarely reported. The richness and the diversity of the VMB significantly affect the host's reproductive ability and metabolic function as well as the defense ability of the immune system (Vazquez-Baeza et al., 2013). A reduction in VMB diversity led to a lack of estrogen-metabolizing bacteria and thereby decreasing the circulating estrogen, which may induce a hypoestrogen-related disease. Moreover, FO altered VMB via upregulating *Lactobacillus*, *Faecalibacterium*, and *Parabacteroides* as well as downregulating *Proteobacteria* and *Streptococcus*. *Lactobacillus* in VMB can generate acidic fermentation products (primarily lactic acid) which serve to create an acidic environment to restrict the growth of most pathogens in the vagina. The ability of *Lactobacillus* to inhibit infection without inducing inflammation may maximize fecundity and successful pregnancy outcome in women. Additionally, *Faecalibacterium* is a commensal bacterium, the absence of which may be associated with lipid metabolism and inflammation in Crohn's disease (Kashtanova et al., 2018). Studies on the effects of the VMB in reproductive and gynecological diseases are still limited, and the effects are largely unknown; this topic needs to be further studied. Sex steroid hormones, The gut and vaginal microbiota, and inflammation appear to be closely correlated. The abundances of beneficial bacteria (*Lactobacillus*, *Firmicutes*, *Butyrivibrio*, *Desulfovibrio*, and *Bifidobacterium*) were positively correlated with SCFAs and E2, while these were negatively associated with LPS, testosterone, and proinflammatory indicators. By contrast, *Actinobacteria*, *Bacteroides*, and *Streptococcus* were negatively correlated with SCFAs and estrogen, whereas these were positively correlated with LPS, testosterone, and proinflammatory indicators. Recent studies indicated the bidirectional regulation of the gut microbiota and circulating estrogen levels (Badawy et al., 2015). The estrogen level may improve bacterial virulence by inhibiting quorum-sensing signaling. Progesterone has been proven to promote the growth of *Bacteroides* and *Prevotella* (Langille et al., 2013). The gut microbiome promotes the enterohepatic circulation of estrogenic metabolites in the host through the secretion of beta-glucuronidase enzymes to hydrolyze estrogenic glucuronides, increasing the reabsorption of estrogens into the blood and reducing their elimination from the body (Wang et al., 2018). Additionally, butyrate, a microbial fermentation end product, can suppress the levels of proinflammation cytokines and LPS and enhance the integrity of the gut mucosa by upregulating the expression of tight junction proteins and the production of retinoic acid (Menni et al., 2017). In conclusion, FO ameliorated PCOS via the sex steroid hormone–gut/vaginal microbiota–inflammation axis, which may potentially serve as an inexpensive intervention for the control of PCOS, especially in patients who are vegetarians.

References

Al-Assal, K., Martinez, A.C., Torrinhas, R.S., Cardinelli, C., Waitzberg, D., 2018. Gut microbiota and obesity. Clin. Nutr. Exp. 20, 60–64.

Al-Lahham, S.H., Peppelenbosch, M.P., Roelofsen, H., Vonk, R.J., Venema, K., 2010. Biological effects of propionic acid in humans; metabolism, potential applications and underlying mechanisms. Biochim. Biophys. Acta 1801, 1175–1183.

Allampati, S., Mullen, K.D., 2016. Long-term management of alcoholic liver disease. Clin. Liver Dis. 20, 551–562.

Altamirano, J., Bataller, R., 2011. Alcoholic liver disease: pathogenesis and new targets for therapy. Nat. Rev. Gastroenterol. Hepatol. 8, 491–501.

Aparicio-Soto, M., Sanchez-Hidalgo, M., Rosillo, M.A., Castejon, M.L., Alarcon-de-la-Lastra, C., 2016. Extra virgin olive oil: a key functional food for prevention of immune-infammatory diseases. Food Funct. 7 (11), 4492–4505. Available from: https://doi.org/10.1039/c6fo01094f.

Badawy, E.A., Rasheed, W.I., Elias, T.R., Hussein, J., Harvi, M., Morsy, S., et al., 2015. Flaxseed oil reduces oxidative stress and enhances brain monoamines release in streptozotocin-induced diabetic rats. Hum. Exp. Toxicol. 34 (11), 1133–1138.

Bang, H.O., Dyerberg, J., Sinclair, H.M., 1980. The composition of the Eskimo food in north western Greenland. Am. J. Clin. Nutr. 33 (12), 2657–2661.

Bouter, K.E., van Raalte, D.H., Groen, A.K., Nieuwdorp, M., 2017. Role of the gut microbiome in the pathogenesis of obesity and obesity-related metabolic dysfunction. Gastroenterology. 152 (7), 1671–1678.

Breitbart, M., Hewson, I., Felts, B., Mahaffy, J.M., Nulton, J., Salamon, P., et al., 2003. Metagenomic analyses of an uncultured viral community from human feces. J. Bacteriol. 185, 6220–6223 [PMID: 14526037].

Brun, P., Castagliuolo, I., Di Leo, V., Buda, A., Pinzani, M., Palu, G., et al., 2007. Increased intestinal permeability in obese mice: new evidence in the pathogenesis of nonalcoholic steatohepatitis. Am. J. Physiol. Gastrointest. Liver Physiol 292 (2), G518–G525.

Burcelin, R., Garidou, L., Pomie, C., 2012. Immuno-microbiota cross and talk: the new paradigm of metabolic diseases. Semin. Immunol. 24 (1), 67–74.

Byun, J.S., Suh, Y.G., Yi, H.S., Lee, Y.S., Jeong, W.I., 2013. Activation of toll-like receptor 3 attenuates alcoholic liver injury by stimulating Kupffer cells and stellate cells to produce interleukin-10 in mice. J. Hepatol. 58, 342–349.

Campbell, C., Dikiy, S., Bhattarai, S.K., Chinen, T., Matheis, F., Calafore, M., et al., 2018. Extrathymically generated regulatory T cells establish a niche for intestinal border-dwelling bacteria and afect physiologic metabolite balance. Immunity 48 (6), 1245–1257.

Canani, R.B., Costanzo, M.D., Leone, L., Pedata, M., Meli, R., Calignano, A., 2011. Potential beneficial effects of butyrate in intestinal and extraintestinal diseases. World J. Gastroenterol. 17, 1519–1528.

Canesso, M.C.C., Lacerda, N.L., Ferreira, C.M., Gonçalves, J.L., Almeida, D., Gamba, C., et al., 2014. Comparing the effects of acute alcohol consumption in germ-free and conventional mice: the role of the gut microbiota, BMC Microbiol, 14. p. 240.

Cani, P.D., Delzenne, N.M., 2011. The gut microbiome as therapeutic target. Pharmacol. Ther. 130, 202–212.

Cani, P.D., Osto, M., Geurts, L., Everard, A., 2012. Involvement of gut microbiota in the development of low-grade inflammation and type 2 diabetes associated with obesity. Gut Microbes 3, 279–288.

Chen, J.R., Chen, Y.L., Peng, H.C., Lu, Y.A., Chuang, H.L., Chang, H.Y., et al., 2016. Fish oil reduces hepatic injury by maintaining normal intestinal permeability and microbiota in chronic ethanol-fed rats. Gastroenterol. Res. Pract. 2016, 4694726.

Chen, R.C., Xu, L.M., Du, S.J., Huang, S.S., Wu, H., Dong, J.J., et al., 2016. Lactobacillus rhamnosus GG supernatant promotes intestinal barrier function, balances Treg and TH17 cells and ameliorates hepatic injury in a mouse model of chronic-binge alcohol feeding. Toxicol. Lett. 241, 103–110.

Chilliard, Y., Glasser, F., Ferlay, A., Bernard, L., Rouel, J., et al., 2007. Diet, rumen biohydrogenation and nutritional quality of cow and goat milk fat. Eur. J. Lipid Sci. Technol. 109, 828–855.

Cicerale, S., Lucas, L.J., Keast, R.S., 2012. Antimicrobial, antioxidant and anti-infammatory phenolic activities in extra virgin olive oil. Curr. Opin. Biotechnol. 23 (2), 129–135. Available from: https://doi.org/10.1016/j.copbio.2011.09.006.

Conlon, M.A., Bird, A.R., 2014. The impact of diet and lifestyle on gut microbiota and human health. Nutrients. 7 (1), 17–44.

Cummings, J.H., Pomare, E.W., Branch, W.J., Naylor, C.P., Macfarlane, G.T., 1987. Short chain fatty acids in human large intestine, portal, hepatic and venous blood. Gut. 28 (10), 1221–1227.

Delzenne, N.M., Cani, P.D., Everard, A., Neyrinck, A.M., Bindels, L.B., 2015. Gut microorganisms as promising targets for the management of type 2 diabetes. Diabetologia. 58 (10), 2206–2217.

de Waal, M.R., Abrams, J., Bennett, B., Figdor, C.G., de Vries, J.E., 1991. Interleukin 10(IL10) inhibits cytokine synthesis by human monocytes: an autoregulatory role of IL-10 produced by monocytes. J. Exp. Med. 174, 1209–1220.

Di, M.A., 2000. Therapy of digestive disorders: a companion to sleisenger and Fordtran's gastrointestinal and liver disease. Gastroenterology. 118, 1275–1276.

Ding, Y., Yan, Y., Peng, Y., Chen, D., Mi, J., Lu, L., et al., 2019. In vitro digestion under simulated saliva, gastric and small intestinal conditions and fermentation by human gut microbiota of polysaccharides from the fruits of Lycium barbarum. Int. J. Biol. Macromol. 125, 751–760.

Duncan, S.H., Belenguer, A., Holtrop, G., Johnstone, A.M., Flint, H.J., Lobley, G.E., 2007. Reduced dietary intake of carbohydrates by obese subjects results in decreased concentrations of butyrate and butyrate-producing bacteria in feces. Appl. Env. Microbiol. 73 (4), 1073–1078. Available from: https://doi.org/10.1128/aem.02340-06.

Eckburg, P.B., Lepp, P.W., Relman, D.A., 2003. Archaea and their potential role in human disease. Infect. Immun. 71, 591–596 [PMID: 12540534].

Egshatyan, L., Kashtanova, D., Popenko, A., Tkacheva, O., Tyakht, A., Alexeev, D., et al., 2016. Gut microbiota and diet in patients with different glucose tolerance. Endocr. Connect. 5 (1), 1–9.

Fan, W., Huo, G., Li, X., Yang, L., Duan, C., 2014. Impact of diet in shaping gut microbiota revealed by a comparative study in infants during the six months of life. J. Microbiol. Biotechnol. 24, 133–143.

Fu, J., Bonder, M.J., Cenit, M.C., Tigchelaar, E.F., Maatman, A., Dekens, J.A., et al., 2015. The gut microbiome contributes to a substantial proportion of the variation in blood lipids. Circ. Res. 117 (9), 817–824.

Fu, X., Cao, C., Ren, B., Zhang, B., Huang, Q., Li, C., 2018. Structural characterization and in vitro fermentation of a novel polysaccharide from Sargassum thunbergii and its impact on gut microbiota. Carbohydr. Polym. 183, 230–239.

Gao, B., 2012. Hepatoprotective and anti-inflammatory cytokines in alcoholic liver disease. J. Gastroenterol. Hepatol. 27 (Suppl 2), 89–93.

Gaspar, R.C., Veiga, C.B., Bessi, M.P., Datilo, M.N., Sant'Ana, M.R., Rodrigues, P.B., et al., 2019. Unsaturated fatty acids from flaxseed oil and exercise modulate GPR120 but not GPR40 in the liver of obese mice: a new anti-inflammatory approach. J. Nutr. Biochem. 66, 52–62.

Gefen, T., Geva-Zatorsky, N., 2018. What came first: the microbiota or the Tr(egg) cells? Immunity 48 (6), 1072–1074. Available from: https://doi.org/10.1016/j.immuni.2018.060.005.

Haliga, R., Mocanu, V., Paduraru, I., Stoica, B., Oboroceanu, T., Luca, V., 2009. Effects of dietary flaxseed supplementation on renal oxidative stress in experimental diabetes. Rev. Med. Chir. Soc. Med Nat. Iasi 113 (4), 1200–1204.

Hashemzadeh, A.A., Nasoohi, N., Raygan, F., Aghadavod, E., Akbari, E., Taghizadeh, M., et al., 2017. Flaxseed oil supplementation improve gene expression levels of PPAR-gamma, LP(a), IL-1 and TNF-alpha in type 2 diabetic patients with coronary heart disease. LIPIDS. 52 (11), 907–915.

Hawrylowicz, C.M., O'Garra, A., 2005. Potential role of interleukin-10-secreting regulatory T cells in allergy and asthma. Nat. Rev. Immunol. 5, 271–283.

Hiippala, K., Kainulainen, V., Kalliomaki, M., Arkkila, P., Satokari, R., 2016. Mucosal prevalence and interactions with the epithelium indicate commensalism of sutterella spp. Front. Microbiol. 7, 1706.

Hotamisligil, G.S., Shargill, N.S., Spiegelman, B.M., 1993. Adipose expression of tumor necrosis factor-alpha: direct role in obesity-linked insulin resistance. Science. 259 (5091), 87–91.

Huang, F., Zhang, R., Liu, Y., Xiao, J., Su, D., Yi, Y., et al., 2016. Characterization and mesenteric lymph node cells-mediated immunomodulatory activity of litchi pulp polysaccharide fractions. Carbohydr. Polym. 152, 496–503.

Hu, J.L., Nie, S.P., Xie, M.Y., 2018. Antidiabetic mechanism of dietary polysaccharides based on their gastrointestinal functions. J. Agric. Food Chem. 66 (19), 4781–4786.

Jangale, N.M., Devarshi, P.P., Bansode, S.B., Kulkarni, M.J., Harsulkar, A.M., 2016. Dietary flaxseed oil and fish oil ameliorates renal oxidative stress, protein glycation, and inflammation in streptozotocin-nicotinamide-induced diabetic rats. J. Physiol. Biochem. 72 (2), 327–336.

Jangale, N.M., Devarshi, P.P., Dubal, A.A., Ghule, A.E., Koppikar, S.J., Bodhankar, S.L., et al., 2013. Dietary flaxseed oil and fish oil modulates expression of antioxidant and inflammatory genes with alleviation of protein glycation status and inflammation in liver of streptozotocin-nicotinamide induced diabetic rats. Food Chem. 141 (1), 187–195.

Jones, M.L., Ganopolsky, J.G., Martoni, C.J., Labbe, A., Prakash, S., 2014. Emerging science of the human microbiome. Gut Microbes 5, 446–457.

Kajla, P., Sharma, A., Sood, D.R., 2015. Flaxseed-a potential functional food source. J. Food Sci. Technol. 52 (4), 1857–1871. Available from: https://doi.org/10.1007/s13197-014-1293-y.

Kashtanova, D.A., Tkacheva, O.N., Doudinskaya, E.N., Strazhesko, I.D., Kotovskaya, Y.V., Popenko, A.S., et al., 2018. Gut microbiota in patients with different metabolic statuses: Moscow study. Microorganisms. 6, 4.

Kasztelan-Szczerbinska, B., Surdacka, A., Celinski, K., Rolinski, J., Zwolak, A., Miacz, S., et al., 2015. Prognostic significance of the systemic inflammatory and immune balance in alcoholic liver disease with a focus on gender-related differences. PLoS One 10, e0128347.

Kirpich, I.A., Petrosino, J., Ajami, N., Feng, W., Wang, Y., Liu, Y., et al., 2016. Saturated and unsaturated dietary fats differentially modulate ethanol-induced changes in Gut microbiome and metabolome in a mouse model of alcoholic liver disease. Am. J. Pathol. 186, 765–776.

Kuang, Y.S., Lu, J.H., Li, S.H., Li, J.H., Yuan, M.Y., He, J.R., et al., 2017. Connections between the human gut microbiome and gestational diabetes mellitus. Gigascience. 5 (8), 1–12.

Kubeck, R., Bonet-Ripoll, C., Hofmann, C., Walker, A., Muller, V.M., Schuppel, V.L., et al., 2016. Dietary fat and gut microbiota interactions determine diet induced obesity in mice. Mol. Meta b5 (12), 1162–1174.

Lambeth, S.M., Carson, T., Lowe, J., Ramaraj, T., Leff, J.W., Luo, L., et al., 2015. Composition, diversity and abundance of gut microbiome in Prediabetes and type 2 diabetes. J. Diabetes Obes. 2 (3), 1–7.

Langille, M.G., Zaneveld, J., Caporaso, J.G., McDonald, D., Knights, D., Reyes, J.A., et al., 2013. Predictive functional profiling of microbial communities using 16S rRNA marker gene sequences. Nat. Biotechnol. 31 (9), 814–821. Available from: https://doi.org/10.1038/nbt0.2676.

Larsen, N., Vogensen, F.K., van den Berg, F.W., Nielsen, D.S., Andreasen, A.S., Pedersen, B.K., et al., 2010. Gut microbiota in human adults with type 2 diabetes differs from non-diabetic adults. PLoS One 5 (2), e9085.

Liu, Y., Chen, F., Odle, J., Lin, X., Jacobi, S.K., Zhu, H., et al., 2012. Fish oil enhances intestinal integrity and inhibits TLR4 and NOD2 signaling pathways in weaned pigs after LPS challenge. J. Nutr. 142 (11), 2017–2024.

Li, X., Klaus, B., 1989. A novel carboxylesterase from Aspergillus niger and its hydrolysis of succinimide esters. Carlsberg. Res. Commun. 54, 241–249.

Li, T., Long, M., Gatesoupe, F.J., Zhang, Q., Li, A., Gong, X., 2015. Comparative analysis of the intestinal bacterial communities in different species of carp by pyrosequencing. Microb. Ecol. 69, 25–36.

Loy, A., Pfann, C., Steinberger, M., Hanson, B., Herp, S., Brugiroux, S., et al., 2017. Lifestyle and horizontal gene transfer-mediated evolution of mucispirillum schaedleri, a core member of the murine gut microbiota. mSystems.

Luo, J., Li, Y., Mai, Y., Gao, L., Ou, S., Wang, Y., et al., 2018. Flaxseed gum reduces body weight by regulating gut microbiota. J. Funct. Foods 47, 136–142 [CrossRef] 40. Wojciechowicz, M.; Heinrichova, K.; Ziolecki, A. A polygalacturonate lyase produced by Lachnospira multiparus isolated from the bovine rumen. J. Gen. Microbiol. 1980, 117, 193–199.

Martinez, I., Muller, C.E., Walter, J., 2013. Long-term temporal analysis of the human fecal microbiota revealed a stable core of dominant bacterial species. PLoS One 8 (7), e69621.

Masiello, P., Broca, C.R., Roye, M., Manteghetti, M., Hillaire-Buys, D., Novelli, M., et al., 1998. Experimental NIDDM: development of a new model in adult rats administered streptozotocin and nicotinamide. Diabetes. 47 (2), 224–229.

Maynard, C.L., Elson, C.O., Hatton, R.D., Weaver, C.T., 2012. Reciprocal interactions of the intestinal microbiota and immune system. Nature 489, 231–241.

Menni, C., Jackson, M.A., Pallister, T., Steves, C.J., Spector, T.D., Valdes, A.M., 2017. Gut microbiome diversity and high-fibre intake are related to lower long-term weight gain. Int. J. Obes. (Lond.) 41 (7), 1099–1105. Available from: https://doi.org/10.1038/ijo.2017.66.

Merchant, A.T., Curhan, G.C., Rimm, E.B., Willett, W.C., Fawzi, W.W., 2005. Intake of n-6 and n − 3 fatty acids and fish and risk of community-acquired pneumonia in US men. Am. J. Clin. Nutr. 82 (3), 668–674.

Morey, M., O'Gaora, P., Pandit, A., Helary, C., 2019. Hyperglycemia acts in synergy with hypoxia to maintain the pro-inflammatory phenotype of macrophages. PLoS One 14 (8), e220577.

Moura-Assis, A., Afonso, M.S., de Oliveira, V., Morari, J., Dos, S.G., Koike, M., et al., 2018. Flaxseed oil rich in omega-3 protects aorta against inflammation and endoplasmic reticulum stress partially mediated by GPR120 receptor in obese, diabetic and dyslipidemic mice models. J. Nutr. Biochem. 53, 9–19.

Nell, S., Suerbaum, S., Josenhans, C., 2010. The impact of the microbiota on the pathogenesis of IBD: lessons from mouse infection models. Nat. Rev. Microbiol. 8, 564–577. Available from: https://doi.org/10.1038/nrmicro2403. PMID: 20622892.

Neuman, M.G., Maor, Y., Nanau, R.M., Melzer, E., Mell, H., Opris, M., et al., 2015. Alcoholic liver disease: role of cytokines. Biomolecules. 5, 2023–2034.

Nicholson, J.K., Holmes, E., Kinross, J., Burcelin, R., Gibson, G., et al., 2012. Host-gut microbiota metabolic interactions. Science 336, 1262–1267.

Nie, Q., Chen, H., Hu, J. Fan, S., Nie, S., 2019. Dietary compounds and traditional Chinese medicine ameliorate type 2 diabetes by modulating gut microbiota. Crit. Rev Food Sci. Nutr. 59 (6), 848–863.

Nishitsuji, K., Xiao, J., Nagatomo, R., Umemoto, H., Morimoto, Y., Akatsu, H., et al., 2017. Analysis of the gut microbiome and plasma short-chain fatty acid profles in a spontaneous mouse model of metabolic syndrome. Sci. Rep. 7 (1), 15876. Available from: https://doi.org/10.1038/s41598-017-16189-5.

Oliveira, V., Marinho, R., Vitorino, D., Santos, G.A., Moraes, J.C., Dragano, N., et al., 2015. Diets containing alpha-linolenic (omega3) or Oleic (omega9) fatty acids rescues obese mice from insulin resistance. Endocrinology 156 (11), 4033–4046. Available from: https://doi.org/10.1210/en.2014-1880.

Pellizzon, M.A., Ricci, M.R., 2018. The common use of improper control diets in diet-induced metabolic disease research confounds data interpretation: the fber factor. Nutr. Metab. (Lond.) 15, 3. Available from: https://doi.org/10.1186/s12986-018-0243-5.

Rehm, J., Mathers, C., Popova, S., Thavorncharoensap, M., Teerawattananon, Y., Patra, J., 2009. Global burden of disease and injury and economic cost attributable to alcohol use and alcohol-use disorders. Lancet. 373, 2223–2233.

Roulard, R., Petit, E., Mesnard, F., Rhazi, L., 2016. Molecular investigations of flaxseed mucilage polysaccharides. Int. J. Biol. Macromol. 86, 840–847.

Round, J.L., Lee, S.M., Li, J., Tran, G., Jabri, B., Chatila, T.A., et al., 2011. The Toll-like receptor 2 pathway establishes colonization by a commensal of the human microbiota. Science 332 (6032), 974–977. Available from: https://doi.org/10.1126/science.1206095.

Saisho, Y., 2015. Metformin and inflammation: its potential beyond glucoselowering effect. Endocr. Metab. Immune Disord. Drug. Targets 15 (3), 196–205.

Sartor, R.B., 2008. Microbial influences in inflammatory bowel diseases. Gastroenterology 134, 577–594. Available from: https://doi.org/10.1053/j.gastro.2007.11.059. PMID: 18242222.

Schmidt, C., Stallmach, A., 2005. Etiology and pathogenesis of inflammatory bowel disease. Minerva Gastroenterol. Dietol. 51, 127–145 [PMID: 15990703].

Shapiro, H., Tehilla, M., Attal-Singer, J., Bruck, R., Luzzatti, R., Singer, P., 2011. The therapeutic potential of long-chain omega-3 fatty acids in non-alcoholic fatty liver disease. Clin. Nutr. 30, 6–19.

Soleimani, Z., Hashemdokht, F., Bahmani, F., Taghizadeh, M., Memarzadeh, M.R., Asemi, Z., 2017. Clinical and metabolic response to flaxseed oil omega-3 fatty acids supplementation in patients with diabetic foot ulcer: a randomized, double-blind, placebo-controlled trial. J. Diabetes Complicat. 31 (9), 1394–1400.

Sonnenburg, J.L., Backhed, F., 2016. Diet-microbiota interactions as moderators of human metabolism. Nature 535 (7610), 56–64. Available from: https://doi.org/10.1038/nature18846.

Szabo, G., 2015. Gut-liver axis in alcoholic liver disease. Gastroenterology. 148, 30–36.

Szabo, G., Bala, S., 2010. Alcoholic liver disease and the gut-liver axis. World J. Gastroenterol. 16, 1321–1329.

Tedeschi, C., Clement, V., Rouvet, M., Valles-Pamies, B., 2009. Dissolution tests as a tool for predicting bioaccessibility of nutrients during digestion. Food Hydrocoll. 23, 1228–1235.

Thaiss, C.A., Zmora, N., Levy, M., Elinav, E., 2016. The microbiome and innate immunity. Nature. 535, 65–74.

Tirosh, A., Calay, E.S., Tuncman, G., Claiborn, K.C., Inouye, K.E., Eguchi, K., et al., 2019. The short-chain fatty acid propionate increases glucagon and FABP4 production, impairing insulin action in mice and humans. Sci. Transl. Med.

Toma, R.B., Leung, H.K., 1987. Determination of reducing Sugars in French fried potatoes by 3,5-dinitrosalicylic acid. Food Chem. 23, 29–33.

Tremaroli, V., Backhed, F., 2012. Functional interactions between the gut microbiota and host metabolism. Nature 489 (242–249), 2.

Turnbaugh, P.J., Hamady, M., Yatsunenko, T., Cantarel, B.L., Duncan, A., Ley, R.E., et al., 2009. A core gut microbiome in obese and lean twins. Nature 457, 480–484.

Ulven, T., Christiansen, E., 2015. Dietary fatty acids and their potential for controlling metabolic diseases through activation of FFA4/GPR120. Annu. Rev. Nutr. 35, 239–263.

Van der Poll, T., Jansen, J., Levi, M., Ten, C.H., Ten, C.J., van Deventer, S.J., 1994. Regulation of interleukin 10 release by tumor necrosis factor in humans and chimpanzees. J. Exp. Med. 180 (5), 1985–1988.

Vazquez-Baeza, Y., Pirrung, M., Gonzalez, A., Knight, R., 2013. EMPeror: a tool for visualizing high-throughput microbial community data. Gigascience 2 (1), 16. Available from: https://doi.org/10.1186/2047-217x-2-16.

Vital, M., Rud, T., Rath, S., Pieper, D.H., Schlüter, D., 2019. Diversity of bacteria exhibiting bile acid-inducible 7α-dehydroxylation genes in the human gut. Comput. Struct. Biotechnol. J. 17, 1016–1019. Available from: https://doi.org/10.1016/j.csbj.2019.070.012.

Wang, C., Li, W., Chen, Z., Gao, X., Yuan, G., Pan, Y., et al., 2018. Effects of simulated gastrointestinal digestion in vitro on the chemical properties, antioxidant activity, alpha-amylase and alpha-glucosidase inhibitory activity of polysaccharides from Inonotus obliquus. Food Res. Int. 103, 280–288.

White, P.J., Marette, A., 2014. Potential role of omega-3-derived resolution mediators in metabolic inflammation. Immunol. Cell Biol. 92 (4), 324–330.

Xu, P., Wu, J., Zhang, Y., Chen, H., Wang, Y., 2014. Physicochemical characterization of puerh tea polysaccharides and their antioxidant and α-glycosidase inhibition. J. Funct. Foods 6, 545–554.

Yang, C., Deng, Q., Xu, J., Wang, X., Hu, C., Tang, H., et al., 2019. Sinapic acid and resveratrol alleviate oxidative stress with modulation of gut microbiota in high-fat diet-fed rats. Food Res. Int. 116, 1202–1211.

Yang, Z.H., Miyahara, H., Takemura, S., Hatanaka, A., 2011. Dietary saury oil reduces hyperglycemia and hyperlipidemia in diabetic KKAy mice and in dietinduced obese C57BL/6J mice by altering gene expression. LIPIDS. 46 (5), 425–434.

Yuan, T., Chu, C., Shi, R., Cui, T., Zhang, X., Zhao, Y., et al., 2019. ApoE-dependent protective effects of Sesamol on high-fat dietinduced behavioral disorders: regulation of the microbiome-gut-brain Axis. J. Agric. Food Chem. 67 (22), 6190–6201.

Zhang, P., Smith, R., Chapkin, R.S., McMurray, D.N., 2005. Dietary (n-3) polyunsaturated fatty acids modulate murine Th1/Th2 balance toward the Th2 pole by suppression of Th1 development. J. Nutr. 135 (7), 1745–1751.

Zhang, L., Wang, Y., Xiayu, X., Shi, C., Chen, W., Song, N., et al., 2017. Altered gut microbiota in a mouse model of Alzheimer's disease. J. Alzheimers Dis. 60 (4), 1241–1257.

Zhang, Q., Yu, H., Xiao, X., Hu, L., Xin, F., Yu. X., 2018. Inulin-type fructan improves diabetic phenotype and gut microbiota profiles in rats. PEERJ. 6, e4446.

Zhang, X., Zhao, Y., Xu, J., Xue, Z., Zhang, M., Pang, X., et al., 2015. Modulation of gut microbiota by berberine and metformin during the treatment of high-fat diet-induced obesity in rats. Sci. Rep. 5, 14405.

Zhao, L., 2013. The gut microbiota and obesity: from correlation to causality. Nat. Rev. Microbiol. 11 (9), 639–647.

Zhao, G., Nyman, M., Jonsson, J.A., 2006. Rapid determination of short-chain fatty acids in colonic contents and faeces of humans and rats by acidified water-extraction and direct-injection gas chromatography. Biomed. Chromatogr. 20, 674–682.

Chapter 30

Role of mushroom polysaccharides in improving gut health and associated diseases

Divya Yadav[1,2] and Pradeep Singh Negi[1,2]

[1]*Department of Fruit and Vegetables Technology, CSIR-Central Food Technological Research Institute, Mysuru, India,* [2]*Academy of Scientific and Innovative Research (AcSIR), Ghaziabad, India*

30.1 Introduction

Extensive research targeting the effective health-promoting impact and vast nutritional value of mushrooms has been expanding dynamically over the past few decades. Mushrooms have long been recognized for their vital role in curing and treating various human diseases, such as asthma, cancer, hypercholesterolemia, hyperlipidemia, hypertension, immunodeficiency, and obesity, owing to their higher content of proteins, polysaccharides, unsaturated fatty acids, minerals , triterpenes, sterols, and secondary metabolites (Ma et al., 2018; Yadav and Negi, 2021). Mushrooms have a great potential to improve nutrition and provide health benefits (Fig. 30.1).

Globally, the production of cultivated mushrooms is increasing greatly. Mushroom consumption has increased 21-fold in the last 56 years (Food and Agriculture Organization Statistical. 2017. FAOSTAT— Data-crops-visualized-

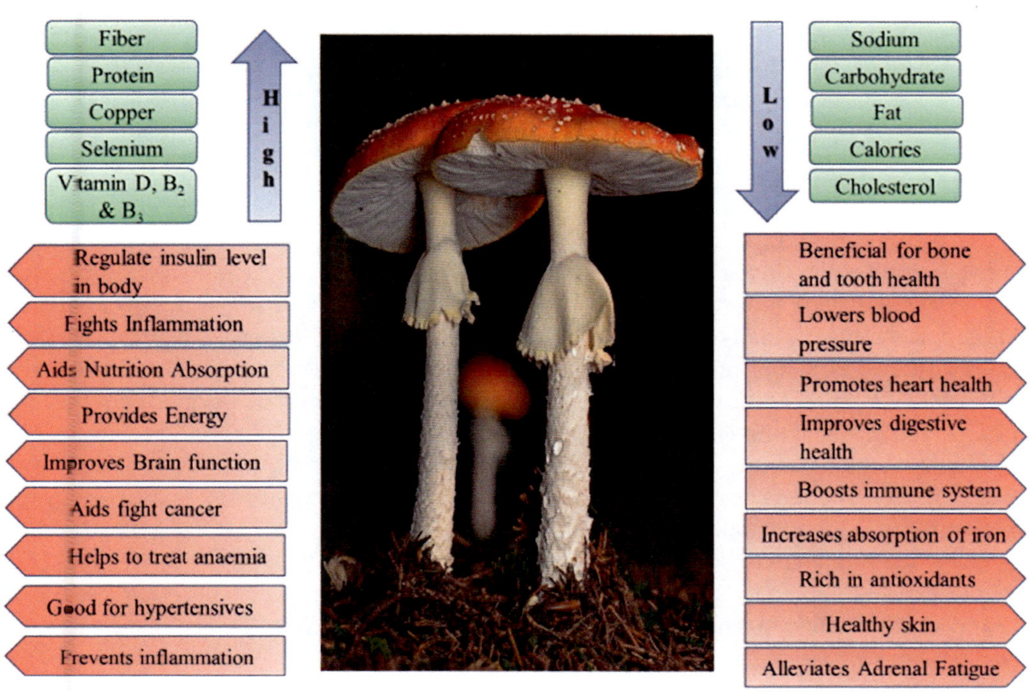

FIGURE 30.1 Health benefits of mushrooms.

Microbiome, Immunity, Digestive Health and Nutrition. DOI: https://doi.org/10.1016/B978-0-12-822238-6.00013-3

mushrooms and truffles). In 2016 the total global production of cultivated edible mushroom species reached 11 billion metric tons, with China (72.3%), Italy (6.34%), the United States (3.89%), the Netherlands (2.78%), and Poland (2.41%) contributing significantly (Kalac, 2019). The introduction of newer varieties of mushrooms for commercialization has flourished in the worldwide mushroom industry during the last few decades (Gupta et al., 2018).

Edible mushrooms are considered an abundant source of natural nutraceuticals and are finding application in advanced functional foods. Mushrooms have an ancient legacy of therapeutic use, with benefits ranging including delaying aging, curing asthma and cough, encouraging diuresis, and revitalizing the spirit by boosting cognitive performance (Xiao et al., 2020). Polysaccharides, protein and /peptides, glycoproteins, unsaturated fatty acids, phenolic compounds, and terpenes are some promising nutraceutical compounds with remarkable biological properties that are found in mushrooms (Ma et al., 2021). The polysaccharide fractions have been reported to hold the majority of the bioactive components, and mushroom polysaccharides offer multiple health-promoting benefits, including antidiabetes, antiinflammation, antiobesity, antioxidation, antitumor, lipid absorption, and metabolism modulation (Yin et al., 2020). Currently, mushroom polysaccharides are gaining a great deal of interest for their ability to regulate the gut microbiota by lowering pathogen levels and promoting the growth of beneficial microbes. They are therefore being explored for their utilization in functional food and nutraceutical industries (Jin et al., 2017).

30.2 The gut microbiome

The gastrointestinal tract comprises the largest human-associated microbial population, which has a significant impact on human health. The microbial population in the human intestine is a diverse ecosystem, with over ten trillion microbes consisting of around three million genes, which is about 500-fold the number of human genes. These microorganisms (bacteria, viruses, fungi, and archaea) interact with the host to ensure metabolic homeostasis (Xu et al., 2013).

Studies have revealed that various factors regulate the composition, diversity, and functional role of the gut microbiota, including the host's age, diet, surrounding atmosphere, and sex, as well as pharmacological treatments and pathological conditions (Nie et al., 2019). Actinobacteria, Bacteroidetes, Firmicutes, Fusobacteria, Proteobacteria, and Verrucomicrobia are the most prominent phylum in the gastrointestinal system. These have been associated with physiological processes, including food digestion, absorption, metabolism, and immune health. It has been discovered that the complex microbial community in the gastrointestinal tract has several favorable health benefits attributable to the unique metabolic features induced by the millions of genes present in them (Ma et al., 2021).

A growing body of research suggests that the gut microbiota interacts with the intestine's mucosal barrier via a range of immunological mechanisms (Daoust et al., 2021). Dysbiosis of the intestinal microbiota has been associated with the emergence of multiple diseases, including obesity, type 1 and type 2 diabetes, cancer, hepatic disorders, inflammatory bowel disease, and sometimes even multiple organ failure (Yin et al., 2020). Modulated microbiota populations have been reported in the majority of disorders, causing pathophysiologies in vital human organs (Yin et al., 2020). The gut microbiota has a symbiotic association with its host, resulting in the production of secondary metabolites and maintenance of gut homeostasis.

30.3 Mushroom polysaccharides

Mushrooms are hypocaloric and have a low glycogen carbohydrate content, making them an excellent source of dietary fiber (Cateni et al., 2021). Mushroom dietary fiber aids in the prevention of constipation, hemorrhoids, colon disorders, diabetes, and cardiovascular disease. Polysaccharides derived from mushrooms have recently gained a lot of recognition because of their broad range of biological properties, including antioxidative, antitumor, immunomodulatory, and antiinflammatory potential (Fig. 30.2) (El-Enshasy et al., 2018). The extraction procedures and the source that is utilized, such as the fruiting body, mycelia, fermentation supernatants, or even residues of edible or medicinal fungus, affect the structural characteristics and bioactivities of mushroom polysaccharides (Yin et al., 2020).

Polysaccharides are a vital component in mushrooms. Their structural complexity is due to the result of the ether linkage between two sugar units in a multitude of ways. The biological functions of these polysaccharides are based on the molecular weight, monosaccharide composition, stereochemistry of simple sugars, the glycosidic linkage having alpha-, beta- or both the configurations, degree of branching, and length of the side chain of the molecule. Mushrooms contain a range of polysaccharides, including glucans, heteroglycans, and proteoglycans (Table 30.1). However, the physicochemical properties of various mushroom polysaccharides differ, including their size and structure, molecular weight, water activity, and bioactivity (Cateni et al., 2021; Gong et al., 2020). Various polysaccharides that are present in mushroom are described in this section.

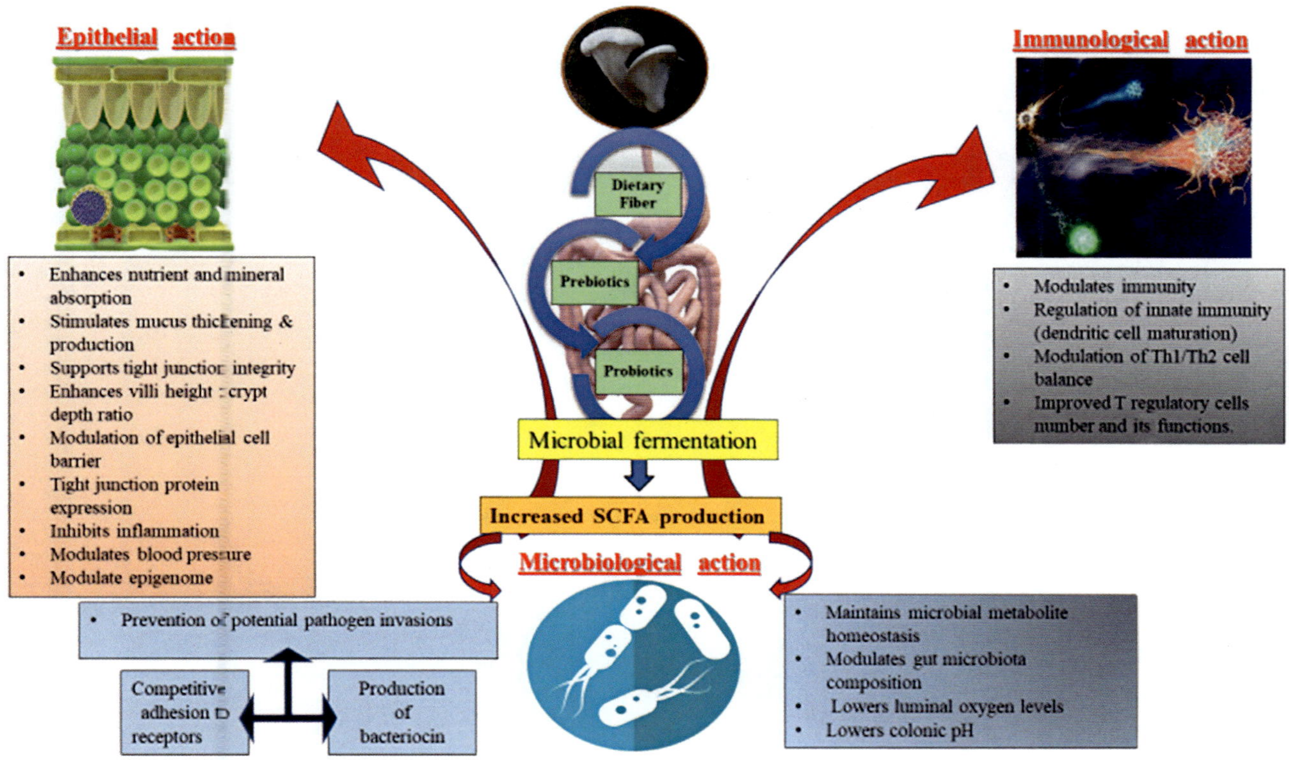

FIGURE 30.2 Mechanisms of action of mushroom dietary fibers.

30.3.1 D-glucan

D-glucans from mushrooms are acknowledged as dietary fibers and bioactive polysaccharides and are recognized for their therapeutic properties. D-glucans from mushrooms have a diverse range of molecular structures with varied linkage patterns (α-D-glucans, β-D-glucans, and α/β-D-glucans), the degree of solubility, and physicochemical characteristics; and have variable prebiotic potential. Mushroom D-glucans function as dietary fibers, as they are neither digested nor absorbed by human digestive enzymes in the digestive tract, making them a significant carbon and energy source. They are also referred to as biological response modifiers, owing to their potential to regulate the gut microbiota, which further promotes host health (Ruthes et al., 2021). D-glucans enhance the population of beneficial microbes as a result of their interaction with the gut microbiome, leading to increased levels of short-chain fatty acids (SCFAs) (acetate, butyrate, and propionate) and gases (H_2, CO_2, and CH_4) (Boulaka et al., 2020).

Mushroom D-glucans have a unique combination of physicochemical and structural properties that can modulate human gut microbiota and enhance the abundance of health-beneficial bacteria (Table 30.2). However, mushroom comprises mainly the beta form of D-glucan polysaccharide.

30.3.1.1 β-glucan

β-glucans are polysaccharides made up of D-glucose monomers linked together by β-glycosidic linkages. β-D-glucans function as a biological response modifier, stimulating the production of macrophages, NK cells, T cells, and cytokines (Zhang et al., 2015).

1. β-1,3 glucan

   The β-glucan molecule binds effectively with the innate immune cells such as NK cells and macrophages effectively. It has been observed that 0.1%–0.5% of β-1,3 glucan in the diet stimulates the synthesis of interleukin 4 (IL-4), IL-12, and interferon (IFN). It is also utilized as an antiallergic agent (Bhakta and Kumar, 2013).

2. β-1,4 glucan

   Chitin is a natural amino polysaccharide with a long-chain polymer of N-acetylglucosamine linked by a β-1,4 bond. Chitin has broad applicability, owing to its antimicrobial effect, biocompatibility, and biodegradability. It improves gastrointestinal health and limits pathogen abundance as a result of its prebiotic potential. Chitin

TABLE 30.1 Polysaccharides that are present in various mushrooms.

Mushroom	Polysaccharide	Structure
B-Glucan		
Several mushrooms	Chitin	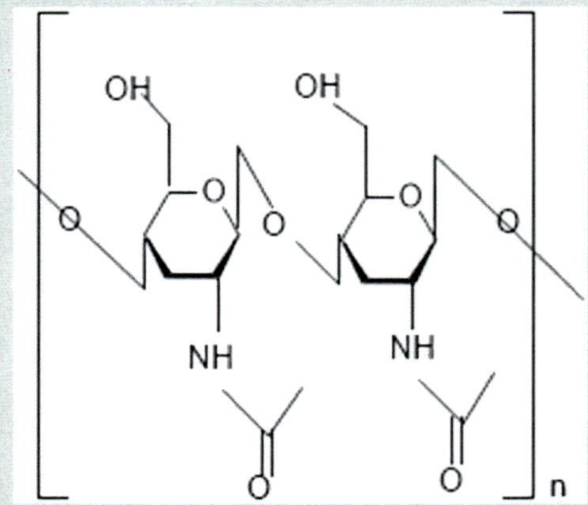
Cordyceps sinensis	Cordyglucan	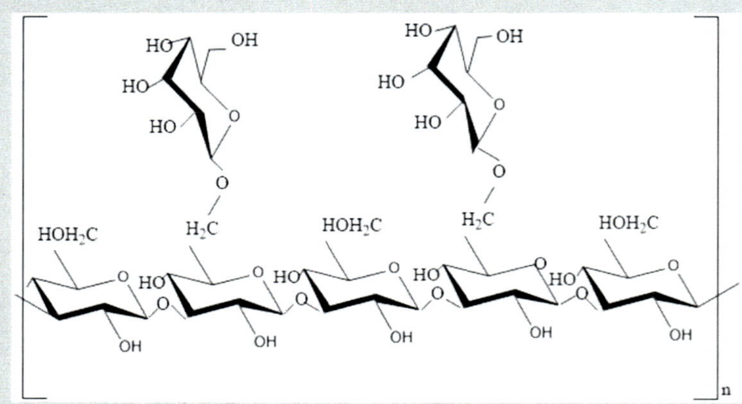
Pleurotus ostreatus	Pleuran	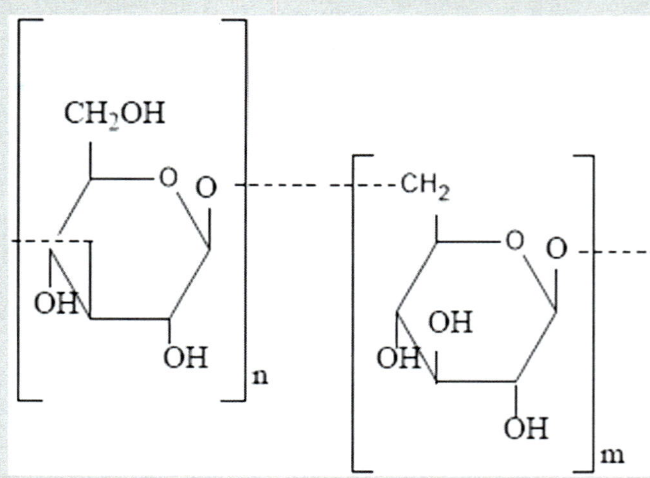

(Continued)

TABLE 30.1 (Continued)

Mushroom	Polysaccharide	Structure
Schizophyllum commune	Schizophyllan	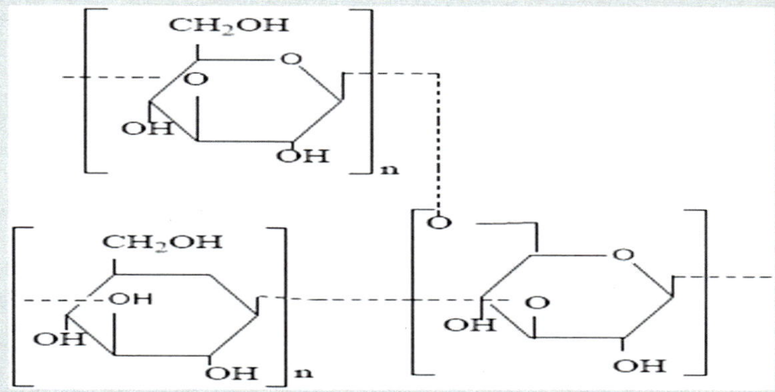
Grifola frondosa	Grifolan	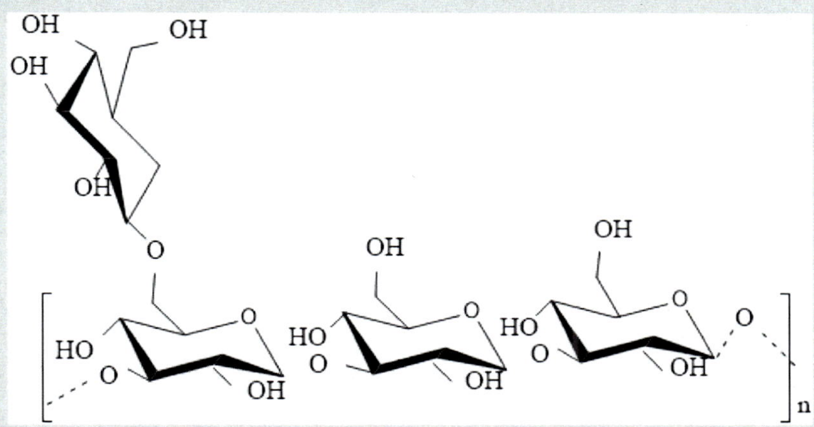
Proteoglycan		
Trametes versicolor	Krestin	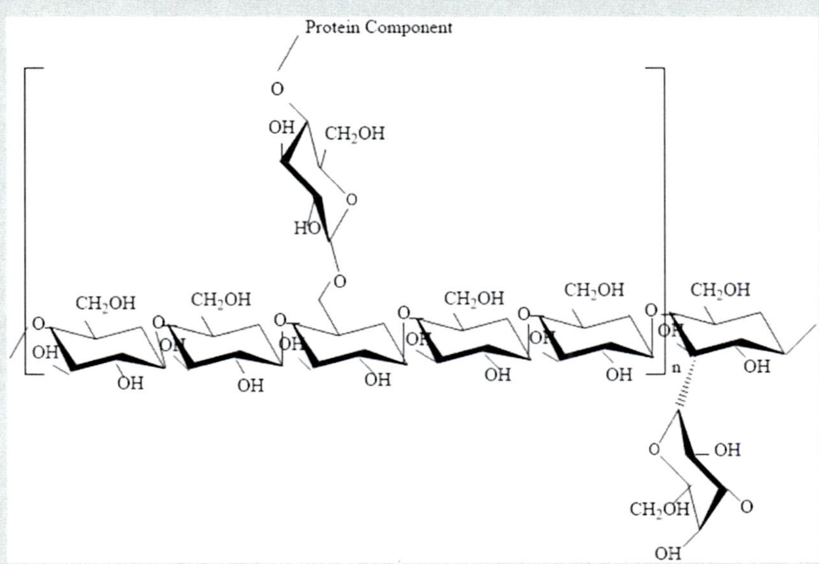

TABLE 30.2 Prebiotic properties of mushroom D-glucans.

Mushroom	Structural Features	Solubility	Model of Study	Importance	Ref.
Wolfiporia cocos	$(1\rightarrow3)$-linked β-D-glucan	Water- insoluble	In vivo (alcoholic hepatic steatosis mouse model)	Increased: • Firmicutes to Proteobacteria ratio • Lachnospiraceae	Sun et al. (2020)
Schizophyllum spp.	$(1\rightarrow6)$-linked β-D-glucan		In vivo (low-fiber diet mice model)	Increased: • Firmicutes to Proteobacteria ratio	Muthuramalingam et al. (2019)
	$(1\rightarrow3),(1\rightarrow6)$-linked β-D-glucan	Not specified	In vivo (ob/ob mice model)	Increased: • Bacteroidetes • *Clostridium* IV • *Clostridium* XIVa • Lachnospiraceae Decreased: • *Megamonas* • *Proteus*	Shan-Shan et al. (2019)
Lentinula edodes	$(1\rightarrow3),(1\rightarrow6)$-linked β-D-glucan	Soluble	In vivo (piglet model)	Increased: • Faecalibacterium Ruminococcaceae Decreased: • *Actinobacillus* • *Coprococcus* • *Oscillospira* • *Phascolarctobacterium* • *Solobacterium*	Wang et al. (2019)
Ganoderma lucidum	$(1\rightarrow4),(1\rightarrow6)$-linked β-D-glucan	Not specified	In vivo (dextran sodium sulfate–induced colitis rat model)	Increased: • Firmicutes • *Ruminococcus* Decreased: • *Escherichia coli* • *Shigella*	Xie et al. (2019)
	$(1\rightarrow3)$-linked β-D-glucan		In vivo (dextran sodium sulfate–induced colorectal cancer mouse model)	In CRC mice, decreased: • *Desulfovibrionaceae* • *Oscillospira*	Luo et al. (2018)
Cookeina speciosa	$(1\rightarrow3),(1\rightarrow6)$-linked β-D-glucan (glucan A)	Insoluble	In vitro (fecal batch fermentation)	Increased: • Fermentation profile	Cantu-Jungles et al. (2018)
	$(1\rightarrow3)$-linked β-D-glucan (glucan B)			Increased: • *Clostridium* cluster	
Pleurotus tuber-regium	$(1\rightarrow3),(1\rightarrow4)$-linked β-D-glucan	Alkali soluble	In vitro (fecal batch fermentation)	Increased: • *Lactobacillus*	Lam et al. (2018)
Schizophyllum commune	β-D-glucan and oligo-β-D- glucan	Soluble	In vitro (fecal batch fermentation)	Increased:(β-D-glucan) • *Bacteroides* • *Clostridium* (Oligo-β-D-glucan) • *Bacteroides* • *Clostridium* • *Eubacteria* spp.	Chaikliang et al. (2015)

(*Continued*)

TABLE 30.2 (Continued)

Mushroom	Structural Features	Solubility	Model of Study	Importance	Ref.
Auricularia auricula-judae	β-D-glucan and oligo-β-D-glucan	Soluble	In vitro (fecal batch fermentation)	Increased:(β-D-glucan) • *Bifidobacterium* • *Lactobacillus* (Oligo-β-D-glucan) • *Bacteroides* • *Bifidobacterium* • *Clostridia* • *Lactobacillus*	

consumption also improves glucose intolerance, increases insulin secretion, alleviates dyslipidemia, and maintains intestinal integrity as well as the gut microbiome (Lopez-Santamarina et al., 2020).

3. β-1,3 glucan and β-1,6 glucan

Cordyglucans obtained from *Cordyceps sinensis* are consists of a β-1,3-D-glucan backbone with β-1,6 linked branches. Cordyglucans extracted from the cell wall were reported to have high anticancer potential. Its antitumor efficacy is dosage-dependent, tending to increase with the increase in the dose (Yalin et al., 2005).

Pleuran is a 762-kDa water-soluble polysaccharide [β-(1,3/1,6)-D-glucan] isolated from *Pleurotus ostreatus*. The compound displays antineoplastic effects against various cell lines, such as HT-29 colorectal cancer cells, PC-3 prostate cancer cells, and MCF-7 breast cancer cells. It also exhibits antioxidant and antiviral effects (Golak-Siwulska et al., 2018).

Lentinan is a β-glucan derived from *Lentinus edodes*, which exhibits in vitro cytotoxicity toward cancer cells but serves primarily as a biological response modifier. Owing to their broad array of therapeutic potential, minimal toxicity, and relatively low cost, glucan-based drugs such as lentinan are promising and excellent immunoregulating medications when paired with chemotherapeutics (Zhang et al., 2019).

Schizophyllan is an extracellular polysaccharide derived from *Schizophyllum commune*. It has been extensively studied for its anticancer, antiviral, and other health implications, contributing to its use in pharmaceutical applications as an adjuvant cancer therapy alongside standard medications (Zhang et al., 2013).

Grifolan is a β-glucan polysaccharide extracted from the mycelium or the fruiting body of *Grifola frondosa*. Grifolan is a macrophage activator that enhances cytokine production and elevates macrophage mRNA levels of IL-1, IL-6, and tumor necrosis factor alpha (TNF-α). It also exhibited immunomodulatory properties (Sonawane et al., 2014).

Ganoderan is a hypoglycemic polysaccharide derived from *G lucidum*. Its antiangiogenic and immunomodulation properties are primarily responsible for *G. lucidum* anticancer activity, inhibiting cell proliferation, metastasis, and invasion while enhancing apoptosis. Ganoderan have antiinflammatory, antioxidant, and immunostimulating properties (Venturella et al., 2021).

30.4 Proteoglycan

The mushroom-derived proteoglycans or polysaccharide peptides are sequences of polypeptides or small protein units linked with β-D-glucan units. Proteoglycans are cell surface or extracellular matrix macromolecules with one or multiple glycosaminoglycan units covalently attached to a membrane protein. Aspartic, glutamic, and other acidic amino acids are abundant in the polypeptide component of the protein (Sonawane et al., 2014).

In vivo and in vitro aqueous extracts of *Phellinus linteus* containing proteoglycans from the fruiting body or mycelia of *P. linteus* have been shown to have an anticancer and immunomodulatory effect. *P. linteus* has potent antitumor effects on human hepatocellular carcinoma cells (HepG2). In *P. linteus*—treated HepG2 cells, the expression of calreticulin, cyclin D1, CDK2, and cyclin E was reduced, and the expression of cyclin A and P27kip1 increased (Sonawane et al., 2014).

Protein-bound polysaccharide K (PSK) derived from *Trametes versicolor* is a protein-bound β-glucan. It was reported to reduce plasma triglycerides and free fatty acids, further downregulating the proinflammatory cytokines

IFN-γ, IL-1, and IL-6, and upregulating the expression of the antiinflammatory cytokine (IL-10). Therefore PSK could be an effective adjuvant in preventing hyperlipidemia-related cardiovascular risk (Venturella et al., 2021).

Polysaccharide peptide can be obtained from the *T. versicolor* strain. PSP has immunomodulatory effects and has the potential to increase an innate immune response (Rodríguez-Valentín et al., 2018). Numerous in vitro and in vivo models and even some clinical trials have found that PSP has immunomodulating, antitumor, antiinflammatory, and antiviral properties. PSP has also exhibited antiaging, antiulcer, learning and memory-enhancing effects, and it limits adverse outcomes due to chemotherapy and radiotherapy treatments (Venturella et al., 2021).

G. frondosa proteoglycan (GFP) is a β-glucan proteoglycan compound derived from *G. frondosa*. In vitro, GFPs show remarkable hypoglycemic effects due to increased insulin resistance induced by reactivation of the insulin receptor. It also lowers fasting serum insulin and fasting serum glucose levels (Venturella et al., 2021).

30.4.1 Glycoprotein

G. frondosa glycoprotein (GFG) is a novel glycoprotein isolated from *G. frondosa*—fermented mycelium. It affects the stress response, caspase-8/-3-dependent, p53-dependent mitochondrial-mediated, and PI3k/Akt pathways to promote apoptosis in human gastric cancer cells (Cui et al., 2016).

30.5 Prebiotic efficiency of mushroom polysaccharides

Prebiotics are nondigestible food components that aid in promoting the proliferation of beneficial bacteria in the gastrointestinal tract. Dietary fibers and oligosaccharides constitute the majority of prebiotic compounds present in mushrooms. They have a favorable impact on health, such as maintaining gut health, cancer inhibition, cholesterol limitation, immunopotentiation, and obesity prevention. Mushroom polysaccharides are one of the potential source of prebiotics. Mushroom dietary fiber aids in the prevention of constipation, hemorrhoids, colon disorders, diabetes, and cardiovascular disease by improving gastrointestinal tract function and cholesterol as well as insulin metabolism (Cateni et al., 2021).

Mushroom polysaccharides have been appreciated for their potential to regulate the gut microbiota by decreasing pathogen levels and encouraging the development of health-promoting microorganisms, thus presenting opportunities for their application in the functional and nutraceutical food industries (Yin et al., 2020). Most polysaccharides are β-glucans, which are indigestible and resistant to the stomach's acidic environment as well as digestive enzymes in the gastrointestinal tract (Ma et al., 2021). Human digestive enzymes are ineffective toward various edible mushroom polysaccharides, and these intact polysaccharides interact with the gut microbiota in several ways, indicating that they can be utilized as prebiotics and dietary fiber (Ma et al., 2021). Most of the mushroom polysaccharides have shown no significant alteration in functional group or variations in molecular weight without synthesis of free monosaccharides throughout the simulated digestion system, indicating that intact polysaccharides can enter the large intestine, where they can be cleaved and utilized by gut microbes. Polysaccharides extracted from several edible mushroom species have the potential to regulate the gut microbiota (Ma et al., 2021). The consumption of *Ganoderma lucidum* polysaccharide substantially increased the species of *Lactobacillus* and Lachnospiraceae (Khan et al., 2018). Also, *G. lucidum* extracts increased abundance of *Bifidobacteria* (Chang et al., 2015).

Polysaccharides from *Flammulina velutipes* could modulate the gut microbiota structure in vivo by lowering the Firmicutes-Bacteroidetes ratio and increasing the abundance of *Lactobacillus* and *Akkermansia* species (Zhao et al., 2019b). Supplementation with polysaccharides isolated from *Auricularia auricular* also lowered the Firmicutes-Bacteroidetes ratio and improved the diversity of the fecal microbiota (Zhao et al., 2019a). Hence supplementing the diet with mushroom polysaccharides will potentially raise the abundance of various gut microbiota species, resulting in several health benefits. Furthermore, polysaccharides may impede the proliferation of health-threatening gut microbiota species (Ma et al., 2021). Fermentation of mushroom polysaccharides by the gut microbiota produced SCFAs as end products, which reduce the pH of the gastrointestinal tract and stimulate the proliferation of health-promoting gut bacteria (Yin et al., 2020). These SCFAs could be utilized as carbon and energy sources by gut microbiota, favoring the proliferation of healthy microbiota in the gastrointestinal tract (Cockburn and Koropatkin, 2016). SCFAs are further absorbed by gastrointestinal tissues and provide a range of nutritional health benefits (Ma et al., 2021).

The spatial organization of the gut microbiota in mice is modified by *Lentinula edodes*—derived polysaccharide, as it modulates the richness and evenness of the gut microbiota (Xu et al., 2015). *Pleurotus abalonus*, *P. sajor-caju*, and *P. ostreatus* exhibit the bifidogenic effect by stimulating the growth of *Bifidobacterium animalis*, *B. bifidum*, *B. breve*, and *B. longum* (Saman et al., 2016). These findings confirm that edible mushrooms should be investigated further for their potential to aid in the improvement of host health via improvement in the host microbiome through their prebiotic characteristics.

30.6 Hydrolysis of mushroom polysaccharides by intestinal bacterial enzymes

Humans catabolize complex carbohydrates such as β-glucan, cellulose, and lignin inefficiently, owing to the lack of carbohydrate-active enzymes in their genomes (CAZymes). Hence most polysaccharides that are extracted from mushrooms are in β-glucan forms and can eventually reach to the large intestine, where they can be broken down by the gut microbiota (Yin et al., 2020). Carbohydrate esterases, glycoside hydrolases, glycosyltransferases, and polysaccharide lyases are some of the CAZymes that are produced by the gut microbiota to catabolize mushroom polysaccharides. Firmicutes tend to metabolize specific polysaccharides, whereas Bacteroidetes can metabolize a broad spectrum of polysaccharides. Degradation of polysaccharides by the gut microbiota is carried out by several mechanisms, including the ATP-binding cassette (ABC) transport system and multienzyme complexes system (Yin et al., 2020).

Firmicutes have been found to have fewer CAZymes for degrading extracellular polysaccharides. Still, they can utilize various transporters to import smaller sugars considering intracellular processing, including ABC transporters, glycosidepentoside-hexuronide transporters, major facilitator superfamily transporters, and phosphoenolpyruvate:carbohydrate phosphotransferase system transporters. Starch degradation in *Eubacterium rectale* via ABC transport system has a well-defined mechanism. This mechanism includes an extracellular substrate-binding protein for ATP hydrolyzed sugar importation. Long-chain starch is degraded into short-chain maltooligosaccharides by glycoside hydrolases, which are then detected by two substrate-binding proteins from the ABC transport system and further transferred to the cytoplasm (Cockburn and Koropatkin, 2016; Cockburn et al., 2015).

In Bacteroidetes, the Sus-like transport system is considered to be the crucial mechanism to degrade polysaccharide. Polysaccharide utilization loci are distinct gene clusters that code for detecting proteins and enzymes and capturing, disintegrating, transporting, and regulating transcription. However, numerous distinctive polysaccharide utilization loci are target-specific polysaccharides, such as fructans, glycans, mannans, xylans, xyloglucans, and marine polysaccharides (Sonnenburg et al., 2005).

Ruminococcus champanellensis isolated from human feces samples utilize a multienzyme complexes system that breaks down cellulose and starch. Cellulosomes are multienzyme complexes that interact with the domains of dockerins and cohesins to combine substrates, carbohydrate-binding proteins, and enzymes (Ben David et al., 2015; Cockburn and Koropatkin, 2016; Zhang et al., 2018).

30.7 Impact of mushroom polysaccharides on the gut microbiome

Diet plays a vital role in directly affecting the composition, diversity, and richness of the gut microbiota (Makki et al., 2018). The abundance of fiber degrading bacteria such as *Bifidobacterium*, *Clostridium*, *Eubacterium*, and *Ruminoccocus* is significantly lowered when carbohydrates are eliminated from the diet (Mardinoglu et al., 2018). Polysaccharides can be exploited as an energy source by colonized microbiota in the gastrointestinal tract, resulting in the production of bioactive metabolites (SCFAs) that are crucial for expressing the gut functional genes, including apoptosis, chemotaxis, proliferation, and differentiation of immune-related cells. Furthermore, polysaccharides use may impede the proliferation of specific opportunistic pathogens in the gut microbiota with health-threatening effects (Ma et al., 2021).

The potential benefits of mushroom polysaccharides from various edible mushroom species for the gut microbiota have been well proven, and the species richness of *Lactobacillus* and Lachnospiraceae rose significantly after consumption of *G. lucidum* polysaccharide (Khan et al., 2018). Polysaccharides from *F. velutipes* could influence the organization of gut microbiota in vivo by lowering the Firmicutes-Bacteroidetes ratio and the relative abundances of *Akkermansia* and *Lactobacillus* (Zhao et al., 2019b). Additionally, recent research has revealed that supplementing with *A. auricula* polysaccharides reduced the Firmicutes-Bacteroidetes ratio and enhanced the diversity of the fecal microbiota (Zhao et al., 2019a).

Recent findings on interactions between mushroom polysaccharides and the gut microbiota have gained tremendous interest because of their protective health-related effects. The gut bacteria can metabolize mushroom polysaccharides and transform them into bioactive compounds, such as SCFAs (Ma et al., 2021). SCFAs reduce gastrointestinal pH by modulating the composition of the gut microbiota and microbial metabolite homeostasis (Yin et al., 2020). In an in vitro assay, polysaccharides extracted from *F. velutipes* decreased the pH from 6.80 to 6.10 after 6 hours of fermentation (Su et al., 2019). Similarly, fermentation of polysaccharides that were extracted from *Ganoderma atrum* and *Lentinula edodes* reduced pH from 6.3 to 4.9 and from 7.75 to 7.0, respectively, in 6 hours (Ding et al., 2017; Xue et al., 2020). In vivo studies confirmed these findings, with a decrease in pH values in the cecum and colon of mice that were fed with polysaccharides extracted from *A. auricula*, *F. velutipes*, and *Hericium erinaceus* (Wang et al., 2018 Zhao et al., 2019a,b). The utilization of polysaccharides by gut bacteria results in SCFAs as end products, which lower the pH of the gastrointestinal tract and encourage an environment that is conducive to the proliferation of a health-beneficial gut microbiota (Jayachandran et al., 2018; Yin et al., 2020).

Various mushroom polysaccharides may have varying effects on the production of SCFAs. Many researchers have found that mushroom-derived polysaccharides aided the growth of Bacteroidetes, the phylum that produces the majority of acetate and propionate in the lower intestine (Ríos-Covián et al., 2016). In vitro fermentation studies reveal that fibers from *Lentinula edodes* enhanced total SCFAs and propionic and butyric acids; polysaccharides from *Ganoderma atrum* increased acetic, butyric, and propionic acids; and the β-glucan from *Schizophylum commune* increased total SCFAs and butyric and acetic acids (Chaikliang et al., 2015; Ding et al., 2017; Xue et al., 2020).

The structure of mushroom polysaccharides may have an impact on SCFA synthesis. In vivo studies on polysaccharides from *F. velutipes* showed enhancement in total SCFA content as the higher content of some monosaccharides (glucose, galactose, mannose) increased the levels of butyric and isobutyric acids in rats, whereas others (xylose, arabinose, fucose) increased the levels of acetic, n-butyric, and propanoic acids (Ye et al., 2020; Zhao et al., 2019b). Administration of *A. auricula* polysaccharides was reported to significantly enhance total SCFA concentrations in the gastrointestinal tract in an in vivo animal model (Zhao et al., 2019a). Numerous studies have linked the generation of SCFAs by *Pleurotus eryngii*, *F. velutipes*, and *Cordyceps militaris* to a myriad of health benefits, including improved cognitive impairment, improvements in colitis symptoms, and immunological modulation (Ji et al., 2020; Ma et al., 2017; Zhao et al., 2019b).

Studies that focused on the interaction between the structural properties of edible mushroom polysaccharides and their vital regulatory SCFAs and intestinal flora species laid the groundwork for elucidating the mechanisms of the health-promoting benefits of mushroom polysaccharides. Such research will also have a significant impact on the field of functional components and gut bacteria.

30.8 Prevention of various diseases by prebiotic mushroom polysaccharides through modulation of the gut microbiota

The gut microbiota of the fetus has little or no colonization; hence during and after birth, the gut microbiota is rapidly established through contact with the mother and the environment. The emergence of the unique core microbiome is initiated soon after birth. It is associated with various maternal pregnancy-related aspects and early childhood factors involving mode of delivery, feeding, gestational age, childhood maltreatment, antibiotic intake, and environmental factors. The microbiome that is established during the first three years of life remains generally constant until late adulthood. Habitat, food, and the degree of persistence of the core microbiome are characteristics that are associated with many of these age-related changes (Sarkar et al., 2021).

Edible mushroom polysaccharides have been widely studied because of their health improvement potential, as they modulate the gut microbiota in the host significantly (Table 30.3). Gut dysbiosis occurs as a result of an imbalance in the gut microbiome that is primarily due to the lack of commensals, decreased richness, or increasing virulent species (Sarkar et al., 2021). Alterations in microbiota architecture and SCFA levels have been observed in the majority of diseases, causing pathophysiologies in major human organs (Fig. 30.3). The significance of the gut microbiota in human health has prompted researchers to look for new mechanisms to manipulate the composition of the gut composition for the treatment of chronic diseases and symptoms (Khan et al., 2018).

30.8.1 Inflammatory bowel disease

Inflammatory bowel disease (IBD), which includes ulcerative colitis and Crohn's disease, is a chronic inflammation of the gastrointestinal tract. The proinflammatory state in IBD is related to an impaired host's inflammatory response to the gut microbiome in genetically vulnerable individuals. Patients with IBD have a higher population of proinflammatory microbes (*Proteobacteria*, *Fusobacterium* spp., *Escherichia* spp.) and a lesser population of beneficial microbes (Bacteroidetes, butyrate-producing species, Firmicutes) (Lau et al., 2021). The structure of IBD patients' intestinal flora changes, and this is reflected in the decline in the abundance of bacteria that strengthen antiinflammatory activity

TABLE 30.3 Effect of mushroom polysaccharide on metabolic disorders.

Mushroom	Model (High-Fat Diet–Induced Rodent)	Increased Gut Microbiota	Decreased Gut Microbiota	Effect on Host	Ref.
Ganoderma lucidum (MY)	Obese C57BL/6NCrlBltwmice	*Anaerotruncus colihominis*, *Bacteroides* spp., *Parabacteroides goldsteinii*	Ratio of Firmicutes to Bacteroidetes, *Proteobacteria*	Prevents obesity, inflammation, and insulin resistance; maintains gut barrier integrity, and improves gut dysbiosis.	Chang et al. (2015)
	Obese male C57BL/6Jmice	*Enterococcus*	—	Reduces insulin concentration and insulin-regulated lipogenesis in the liver, ameliorating insulin sensitivity by decreasing inflammation and inflammation-induced ectopic lipotoxicity.	Xu et al. (2017)
	Type 2 diabetic male SD rats	*Adlecreuzia*, *Bacteroides*, *Parabacteroides*, *Rothia*	*Aerococcus*, *Coprococcus*, *Proteus*, *Ruminococcus*	Reduces serum IL-1, IL-6, insulin, HOMA-IR, LDL-C, TC, TG, levels; regulates the gut microbiota.	Chen et al. (2020c)
Ganoderma lucidum (FB)	Male Syrian golden hamsters	*Alistipes*, *Alloprevotella*, *Bifidobacterium*, *Oscillibacter*, *Paraprevotella*, *Prevotella*, *Ruminococcus*	*Clostridium* IV, *Clostridium* XIV, *Desulfovermiculus*, *Desulfovibrio*, *Intestinimonas*, *Pseudoflavonifractor*, *Schwartzia*	Reduces serum AST, LDL-C, TC, and TG levels, alters gut microbial composition.	Tong et al. (2020)
	Male Wistar rats	*Aerococcus*, *Alistipes*, *Allopravotella*, *Anaerovorax*, *Bacteroides*, *Barnesiella*, *Ignavigranum*, *Jeotgalicoccus*, *Parabacteroides*, *Parasutterella*, *Psychrobacter*, *Staphylococcus*	*Blautia*, *Enterorhabdus*, *Roseburia*	Improves intestinal integrity and hepatic ectopic fat deposition by lessening lipid metabolism disorders.	Lv et al. (2019)
Grifola frondosa (FB)	Type 2 diabetic male ICR mice	*Akkermansia muciniphila*, *Bacteroides acidifaciens*, *Lactobacillus acidophilus*, *Porphyromonas gingivalis*, *Roseburia intestinalis*, *Tannerella forsythia*	Ratio of Firmicutes to Bacteroidetes	Modulates various diabetic markers.	Chen et al. (2019)

(*Continued*)

TABLE 30.3 (Continued)

Mushroom	Model (High-Fat Diet–Induced Rodent)	Increased Gut Microbiota	Decreased Gut Microbiota	Effect on Host	Ref.
	Male Wistar rats	*Barnesiella, Flavonifracter*	*Butyricicoccus, Clostridium* XVI	Improves hepatic ectopic fat deposition by increasing fecal total bile acids.	Li et al. (2019a)
	Male Wistar rats	*Allobaculum, Bacteroides, Bifidobacterium*	Ratio of Firmicutes to Bacteroidetes, *Acetatifactor, Alistipes, Flavonifractor, Paraprevotella, Oscillibacter*	Improves adiposity, liver function, and lipid levels in NAFLD rats by modulating ACC, CYP4A1, CYP7A1, SOCS2, and TNF-α expression levels.	Li et al. (2019b)
	Male Wistar rats	*Bacteroides, Barnesiella, Defluvitalea, Oscillibacter, Proteobacteria*	Ratio of Firmicutes to Bacteroidetes	Reduces body weight, cholesterol, and serum glucose levels and liver tissue damage; upregulates AMPK, CYP7A1, GK, and PPAR levels; downregulates ACC, FAS, PEPCK and SREBP-1c levels.	Pan et al. (2020)
	Male Kunming mice	*Alistipes*	*Aerococcus, Enterococcus, Staphylococcus*	Reduces serum levels of FBG, LDL-C, OGT, TC, and TG; modulates gut microbiome; improve bile acid production.	Guo et al. (2020)
Phellinus linteus (MY)	Type 2 diabetic male SD rats	*Blautia, Lachnospiraceae-NK4A136, Lachnospiraceae-UCG-006, Prevotella, Roseburia, Ruminiclostridium*	*Akkermansia, Bacteroidales, Clostridium, Escherichia, Shigella*	Modulates various diabetic markers.	Liu et al. (2020)
Phellinus linteus (FB)	Obese male C57BL/6J mice	*Porphyromonas*		Improves insulin resistance; lowers fasting blood glucose levels and alleviates glucose intolerance.	Feng et al. (2018)
Cordyceps sinensis (Nm)	C57BL/6J mice	*Actinobacteria, Acidobacteria, Olsenella*	*Bacteroidetes*	Improves inflammation, lipid metabolism markers, alanine aminotransferase and insulin resistance but exacerbates liver fibrosis and steatosis.	Chen et al. (2020b)

(*Continued*)

TABLE 30.3 (Continued)

Mushroom	Model (High-Fat Diet–Induced Rodent)	Increased Gut Microbiota	Decreased Gut Microbiota	Effect on Host	Ref.
Hirsutella sinensis (MY)	Obese male C57BL/6J mice	*Clostridium cocleatum, Intestinimonas butyriciproducens, Parabacteroides goldsteinii*	*Escherichia coli, Pseudomonas aeruginosa, Shewanella algae*	Cures obesity-induced gut dysbiosis and leaky gut; improves inflammation, metabolic endotoxemia, insulin resistance, and dyslipidemia.	Wu et al. (2019)
Pleurotus eryngii (FB)	Obese male C57BL/6J mice	*Anaerostipes, Clostridium, Parabacteroides*	*Lactobacillus, Roseburia*	Reduces total cholesterol and LDL cholesterol levels in serum; enhances glucose tolerance; increases total bile acids in feces.	Nakahara et al. (2019)
Poria cocos (Sclerotium)	ob/ob mice	*Alloprevotella, Bacteroides, Clostridium IV, Lachnospiracea, Parabacteroides, Ruminococcus*	Ratio of Firmicutes to Bacteroidetes, *Megamonas, Proteus*	Activates the intestinal PPAR pathway; enhances glucose and lipid metabolism; alleviates hepatic steatosis.	Shan-Shan et al. (2019)
Sarcodon aspratus (FB)	Obese male C57BL/6J mice	*Akkermansia, Bacteroides Lactobacillus*	Ratios of Firmicutes to Bacteroidetes	Ameliorates adipocyte differentiation, dyslipidemia, glucose intolerance, inflammation, and hepatic steatosis.	Chen et al. (2020a)
Schizophyllum spp.	Low-fiber diet–induced male C57BL/6J mice	*Anaerostipes, Clostridiales*	*Lachnospiracea, Oscillospira, Ruminococcus gnavus*	Increases the density of goblet cells and the thickness of the mucosal layer.	Muthuramalingam et al. (2019)

FB, fruiting body; *MY*, mycelium.

(e.g., *Bifidobacteria* and *Lactobacillus*) and an enhancement in the abundance of bacteria that are detrimental to health (e.g., *Clostridium* and *Staphylococcus*) (Wlodarska et al., 2015).

Recent research has confirmed that edible mushroom polysaccharides can prevent IBD via regulating the gut microbiota (Xu et al., 2013). The effects of mushroom polysaccharides on the remodeling of gut microbiota and/or production of microbiota-derived metabolites (SCFAs) contribute to its antiinflammatory potential (Ma et al., 2021). The polysaccharide that is derived from *A. auricula-judae* exhibited significant preventive effects on dextran sulfate sodium (DSS)-induced colitis by reducing the elevated abundances of Deferribacteres and Actinobacteria in IBD mice in a DSS-induced colitis animal model. Further findings revealed that *Dictyophora indusiata* polysaccharide (DIP) could reduce IBD symptoms by reducing gut microbiota abundances, mainly Enterobacteriales and Gammaproteobacteria in the DIP-fed mice compared to their increased abundance in model mice (Kanwal et al., 2020).

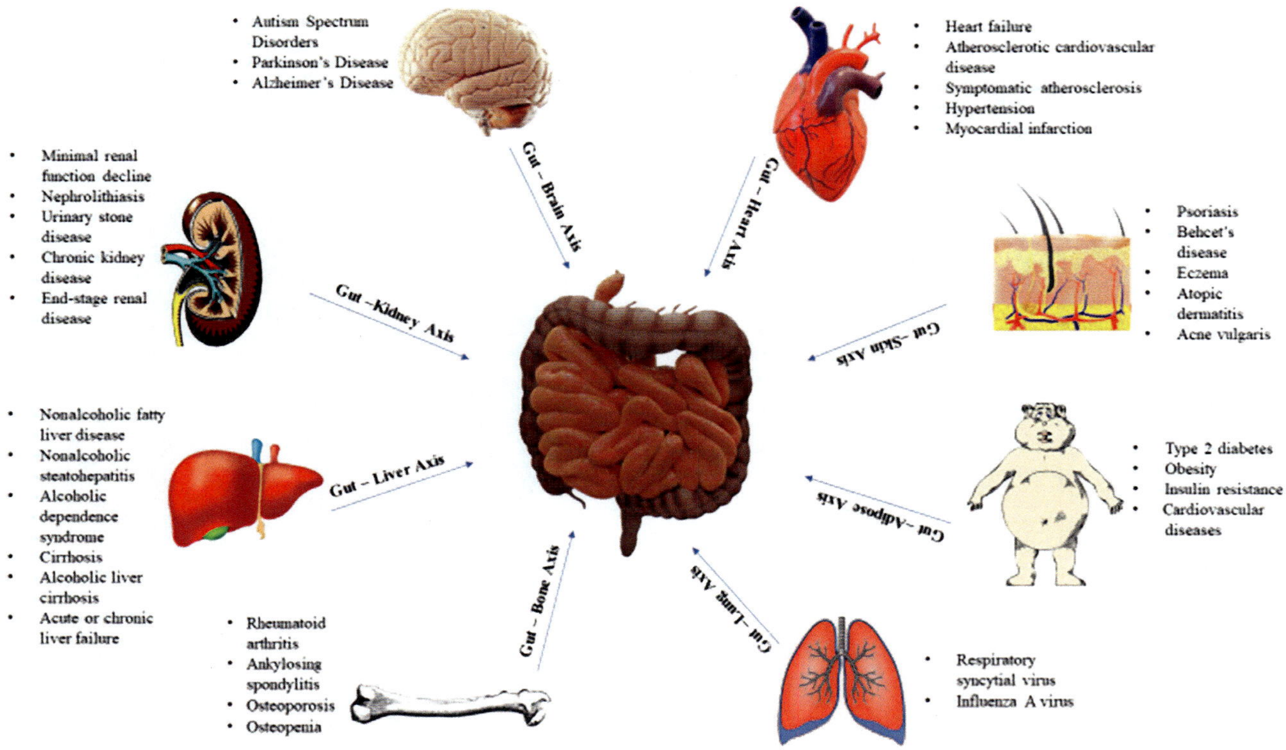

FIGURE 30.3 Diseases associated with the gut-organ axes.

The administration of *Hericium erinaceus* polysaccharide led to a significant alteration in the composition of the gut microbiota, including the restoration of *Akkermansia*, *Clostridiales*, and *Desulfovibrio* relative abundances, and also the prevention of colitis manifestations (Ren et al., 2018). Treatment with *F. velutipes* polysaccharide can cure colonic microbial disturbances and improve SCFA concentrations, contributing to a downregulation of the TLR4\NF-β signal pathway and alleviation from colitis (Zhao et al., 2020c). The effectiveness of beta-type glycosidic polysaccharides derived from *F. velutipes* in curing colitis was correlated with the modulation in the abundance of various SCFA-producing species such as *Clostridium, Bifidobacterium*, and *Lactobacillus* (Zhao et al., 2020c). The concentrations of SCFAs in the cecum rose significantly in association with the polysaccharide fractions (Hu et al., 2019).

30.8.2 Antitumor effect

Intestinal flora and its derived metabolites, such as azomethane and lipopolysaccharides, can induce inflammatory and immunological responses in the gut, leading to the transformation of normal cells into tumor cells (Lofgren et al., 2011). Colorectal cancer (CRC) is among the most confronted cancers concerning the gut microbiota (Chen and Li, 2020). High-fat and low-fiber diets have been found to increase the prevalence of CRC. The principal mechanism is primarily due to the modulation in specific gut microbiota composition and related metabolites (Ajouz et al., 2014). Furthermore, correlations between the gut microbiota and the onset of breast cancer, gastric cancer, intestinal adenocarcinoma, and liver cancer have been reported (Feng et al., 2015; Goedert et al., 2015; Wu et al., 2015). Edible mushroom polysaccharides could be a viable and coherent approach for cancer prevention and treatment. Acid-soluble polysaccharides derived from *G. frondosa* can significantly reduce hepatocellular carcinoma cell proliferation by increasing mitochondrial membrane permeability in a caspase-dependent manner (Yu et al., 2020). Depending on the suppression of apoptosis by reverting microtubule polymerization, *G. lucidum* polysaccharide could alleviate paclitaxel-induced intestinal barrier impairment (Li et al., 2020). *G. lucidum* polysaccharide, one of the most explored and prominent polysaccharides, is considered a prebiotic alternative and is known for its antitumor properties (Ahmad, 2018). Administration of polysaccharide isolated from *G. lucidum* could significantly ameliorate CRC symptoms in mice, relying on enhancing the relative abundances of Bacteroides and Enterobacteriaceae along with the diminution of *Clostridiales*, *Desulfovibrionaceae*, *Oscillospira*, and *Ruminococcus* (Luo et al., 2018). Furthermore, in comparison to guar gum, *G. lucidum* polysaccharide intake has been reported to increase the abundance of *Akkermansia*. *G. lucidum* polysaccharide was more efficient in ameliorating CRC symptoms (Luo et al., 2020). Blending *G. lucidum* polysaccharide and jiaogulan saponins could significantly improve CRC-related symptoms such as an inflamed gut barrier, tumor cell multiplication, and oncogenic signaling molecule production with an increased relative abundance of Bacteroidetes and other bacteria that produce SCFA (Khan et al., 2019). Furthermore, polysaccharide isolated from *G. lucidum* improved CRC via gut microbiota modulation and significantly relieved breast cancer symptoms, owing to gut microbiota–associated regulatory effects (Su et al., 2018). Edible mushroom polysaccharides have the ability to suppress tumor proliferation and treat related symptoms in cancer patients.

30.8.3 Immunomodulatory effect

The intestine is an intricate ecosystem with several species of microbes that are differentiated from the host by a mucus membrane secreted by the goblet cells. A monolayer of intestinal epithelial cells exists under the mucus membrane, creating a barrier for pathogenic microbe invasion (Tang et al., 2019). The undigested dietary components in the gut undergo anaerobic fermentation by the intestinal microbiota, producing various metabolites. Metabolites derived from the gut microbiota, namely, LPS and SCFAs, can penetrate the intestinal barrier, favoring host cell interaction and leading to an immunomodulatory effect (Yin et al., 2020). (Li et al., 2016) reported that in mice with chronic pancreatitis, polysaccharides from *G. lucidum* improved the population of several health-promoting bacteria in the feces. These modifications in fecal bacteria were followed by an increased rate of reactive oxygen species detoxification in the pancreas with a decrease in IFN-γ and TNF-α expression levels in serum and led to improved pancreatic secretion. Mushroom-derived polysaccharides have been proven to be an effective bioactive component for immunomodulation by stimulating the production of SCFAs by the gut microbiota (Yin et al., 2020).

30.9 Conclusion and future perspectives

Mushrooms have been considered a valuable source of bioactive polysaccharides. Their biological activities are inherently connected to their interactions with the gut microbiota, their composition and regulation, and related stimulation of biological functions. Mushroom polysaccharides are resistant to hydrolysis by human digestive enzymes and act as prebiotics by entering the large intestine, where they are effectively fermented by the indigenous microbiome. Mushroom polysaccharide fermentation by the gut microbiota might lower the risk of several diseases by strengthening the gastrointestinal tract environment and minimizing disease symptoms, specifically in microbiota-related diseases. SCFAs are responsible for the effective interaction of polysaccharides with the gut microbiota and the associated health benefits. SCFAs can stimulate specific cell-signaling pathways by targeting host tissues that are involved in metabolic regulation and immunological function. The diversity of the gut microbiota is attributed to interindividual differences, such as lifestyle, dietary habits, climate, surroundings, and aging. Extensive research is required for a better understanding of structural variability in mushroom polysaccharides and their biological effect; as a result, explicit inferences about the prebiotic potential of mushroom polysaccharides still need in-depth findings. In addition, research on interindividual variability should be factored and recognized to explore the underlying links between mushroom polysaccharides and the gut microbiota. The significance of mushroom polysaccharides in promoting human health through modulation in the gut microbiota can be further explored by using effective techniques such as metabolomics, metagenomics, and proteomics.

References

Ahmad, M.F., 2018. *Ganoderma lucidum*: persuasive biologically active constituents and their health endorsement. Biomed. Pharmacother. 107, 507–519.

Ajouz, H., Mukherji, D., Shamseddine, A., 2014. Secondary bile acids: an underrecognized cause of colon cancer. World J. Surg. Oncol. 12 (1), 1–5.

Ben David, Y., Dassa, B., Borovok, I., Lamed, R., Koropatkin, N.M., Martens, E.C., et al., 2015. *Ruminococcal cellulosome* systems from rumen to human. Environ. Microbiol. 17 (9), 3407–3426.

Bhakta, M., Kumar, P., 2013. Mushroom polysaccharides as a potential prebiotics. Int. J. Health Sci. Res. 3 (8), 77–84.

Boulaka, A., Christodoulou, P., Vlassopoulou, M., Koutrotsios, G., Bekiaris, G., Zervakis, G.I., et al., 2020. Genoprotective properties and metabolites of β-glucan-rich edible mushrooms following their *in-vitro* fermentation by human faecal microbiota. Molecules 25 (15), 3554.

Cantu-Jungles, T.M., Ruthes, A.C., El-Hindawy, M., Moreno, R.B., Zhang, X., Cordeiro, L.M., et al., 2018. *In-vitro* fermentation of *Cookeina speciosa* glucans stimulates the growth of the butyrogenic *Clostridium* cluster XIVa in a targeted way. Carbohydr. Polym. 183, 219–229.

Cateni, F., Gargano, M.L., Procida, G., Venturella, G., Cirlincione, F., Ferraro, V., 2021. Mycochemicals in wild and cultivated mushrooms: nutrition and health. Phytochem. Rev. 1–45.

Chaikliang, C., Wichienchot, S., Youravoug, W., Graidist, P., 2015. Evaluation on prebiotic properties of β-glucan and oligo-β-glucan from mushrooms by human fecal microbiota in fecal batch culture. Funct. Foods Health Dis. 5 (11), 395–405.

Chang, C.J., Lin, C.S., Lu, C.C., Martel, J., Ko, Y.F., Ojcius, D.M., et al., 2015. *Ganoderma lucidum* reduces obesity in mice by modulating the composition of the gut microbiota. Nat. Commun. 6 (1), 1–19.

Chen, C., and Li, H. (2020). The inhibitory effect of gut microbiota and its metabolites on colorectal cancer. Available from: https://doi.org/10.4014/jmb.2002.02032.

Chen, Y., Liu, D., Wang, D., Lai, S., Zhong, R., Liu, Y., et al., 2019. Hypoglycemic activity and gut microbiota regulation of a novel polysaccharide from *Grifola frondosa* in type 2 diabetic mice. Food Chem. Toxicol. 126, 295–302.

Chen, J., Liu, J., Yan, C., Zhang, C., Pan, W., Zhang, W., et al., 2020a. *Sarcodon aspratus* polysaccharides ameliorated obesity-induced metabolic disorders and modulated gut microbiota dysbiosis in mice fed a high-fat diet. Food Funct. 11 (3), 2588–2602.

Chen, L., Zhang, L., Wang, W., Qiu, W., Liu, L., Ning, A., et al., 2020b. Polysaccharides isolated from *Cordyceps Sinensis* contribute to the progression of NASH by modifying the gut microbiota in mice fed a high-fat diet. PLoS One 15 (6), e0232972. Available from: https://doi.org/10.1371/journal.pone.0232972.

Chen, M., Xiao, D., Liu, W., Song, Y., Zou, B., Li, L., et al., 2020c. Intake of *Ganoderma lucidum* polysaccharides reverses the disturbed gut microbiota and metabolism in type 2 diabetic rats. Int. J. Biol. Macromol. 155, 890–902.

Cockburn, D.W., Koropatkin, N.M., 2016. Polysaccharide degradation by the intestinal microbiota and its influence on human health and disease. J. Mol. Biol. 428 (16), 3230–3252.

Cockburn, D.W., Orlovsky, N.I., Foley, M.H., Kwiatkowski, K.J., Bahr, C.M., Maynard, M., et al., 2015. Molecular details of a starch utilization pathway in the human gut symbiont *Eubacterium rectale*. Mol. Microbiol. 95 (2), 209–230.

Cui, F., Zan, X., Li, Y., Sun, W., Yang, Y., Ping, L., 2016. *Grifola frondosa* glycoprotein GFG-3a arrests S phase, alters proteome, and induces apoptosis in human gastric cancer cells. Nutr. Cancer 68 (2), 267–279.

Daoust, L., Pilon, G., Marette, A., 2021. Perspective: nutritional strategies targeting the gut microbiome to mitigate COVID-19 outcomes. Adv. Nutr. Available from: https://doi.org/10.1093/advances/nmab031.

Ding, Q., Nie, S., Hu, J., Zong, X., Li, Q., Xie, M., 2017. *In-vitro* and *in-vivo* gastrointestinal digestion and fermentation of the polysaccharide from *Ganoderma atrum*. Food Hydrocoll. 63, 646–655.

El-Enshasy, H.A., Selvamani, S., Dailin, D.J., Abd Malek, R., Hanapi, S.Z., Ambehabati, K.K., et al., 2018. Antioxidant compounds of the edible mushroom *Pleurotus ostreatus*. Int. J. Biotechnol. Wellness Ind. 7 (1), 1–14.

Feng, Q., Liang, S., Jia, H., Stadlmayr, A., Tang, L., Lan, Z., et al., 2015. Gut microbiome development along the colorectal adenoma–carcinoma sequence. Nat. Commun. 6 (1), 1–13.

Feng, H., Zhang, S., Wan, J.M.F., Gui, L., Ruan, M., Li, N., et al., 2018. Polysaccharides extracted from *Phellinus linteus* ameliorate high-fat high-fructose diet induced insulin resistance in mice. Carbohydr. Polym. 200, 144–153.

Food and Agriculture Organization Statistical. 2017. FAOSTAT— Data-crops-visualized-mushrooms and truffles. (Accessed 9 April 2022).

Goedert, J.J., Jones, G., Hua, X., Xu, X., Yu, G., Flores, R., et al., 2015. Investigation of the association between the fecal microbiota and breast cancer in postmenopausal women: a population-based case-control pilot study. JNCI: J. Natl Cancer Inst. 107 (8). Available from: https://doi.org/10.1093/jnci/djv147.

Golak-Siwulska, I., Kałużewicz, A., Spiżewski, T., Siwulski, M., Sobieralski, K., 2018. Bioactive compounds and medicinal properties of Oyster mushrooms (*Pleurotus* sp.). Folia Hortic. 30 (2), 191–201.

Gong, P., Wang, S., Liu, M., Chen, F., Yang, W., Chang, X., et al., 2020. Extraction methods, chemical characterizations and biological activities of mushroom polysaccharides: A mini-review. Carbohydr. Res. 494, 108037.

Guo, W.L., Deng, J.C., Pan, Y.Y., Xu, J.X., Hong, J.L., Shi, F.F., et al., 2020. Hypoglycemic and hypolipidemic activities of *Grifola frondosa* polysaccharides and their relationships with the modulation of intestinal microflora in diabetic mice induced by high-fat diet and streptozotocin. Int. J. Biol. Macromol. 153, 1231–1240.

Gupta, S., Summuna, B., Gupta, M., Annepu, S.K., 2018. Edible mushrooms: cultivation, bioactive molecules, and health benefits. Bioact. Mol. Food 1–33. Available from: https://doi.org/10.1007/978-3-319-54528-8_86-1.

Hu, Q., Yuan, B., Wu, X., Du, H., Gu, M., Han, Y., et al., 2019. Dietary intake of *Pleurotus eryngii* ameliorated dextran-sodium-sulfate-induced colitis in mice. Mol. Nutr. Food Res. 63 (17), 1801265.

Jayachandran, M., Chen, J., Chung, S.S.M., Xu, B., 2018. A critical review on the impacts of β-glucans on gut microbiota and human health. J. Nutr. Biochem. 61, 101–110.

Ji, Y., Su, A., Ma, G., Tao, T., Fang, D., Zhao, L., et al., 2020. Comparison of bioactive constituents and effects on gut microbiota by *in-vitro* fermentation between *Ophicordyceps sinensis* and *Cordyceps militaris*. J. Funct. Foods 68, 103901.

Jin, M., Zhu, Y., Shao, D., Zhao, K., Xu, C., Li, Q., et al., 2017. Effects of polysaccharide from mycelia of *Ganoderma lucidum* on intestinal barrier functions of rats. Int. J. Biol. Macromol. 94, 1–9.

Kalac, P., 2019. Mineral composition and radioactivity of edible mushrooms. Academic Press.

Kanwal, S., Joseph, T.P., Aliya, S., Song, S., Saleem, M.Z., Nisar, M.A., et al., 2020. Attenuation of DSS induced colitis by *Dictyophora indusiata* polysaccharide (DIP) via modulation of gut microbiota and inflammatory related signaling pathways. J. Funct. Foods 64, 103641.

Khan, I., Huang, G., Li, X., Leong, W., Xia, W., Hsiao, W.W., 2018. Mushroom polysaccharides from *Ganoderma lucidum* and *Poria cocos* reveal prebiotic functions. J. Funct. Foods 41, 191–201.

Khan, I., Huang, G., Li, X.A., Liao, W., Leong, W.K., Xia, W., et al., 2019. Mushroom polysaccharides and jiaogulan saponins exert cancer preventive effects by shaping the gut microbiota and microenvironment in ApcMin/+ mice. Pharmacol. Res. 148, 104448.

Lam, K.L., Keung, H.Y., Ko, K.C., Kwan, H.S., Cheung, P.C.K., 2018. In vitro fermentation of beta-glucans and other selected carbohydrates by infant fecal inoculum: An evaluation of their potential as prebiotics in infant formula. Bioact. Carbohydr. Diet. Fibre 14, 20–24.

Lau, A.W.Y., Tan, L.T.H., Ab Mutalib, N.S., Wong, S.H., Letchumanan, V., Lee, L.H., 2021. The chemistry of gut microbiome in health and diseases. Prog. Microbes Mol. Biol. 4 (1). Available from: https://doi.org/10.36877/pmmb.a0000175.

Li, L., Guo, W.L., Zhang, W., Xu, J.X., Qian, M., Bai, W.D., et al., 2019a. *Grifola frondosa* polysaccharides ameliorate lipid metabolic disorders and gut microbiota dysbiosis in high-fat diet fed rats. Food Funct. 10 (5), 2560–2572.

Li, X., Zeng, F., Huang, Y., Liu, B., 2019b. The positive effects of *Grifola frondosa* heteropolysaccharide on NAFLD and regulation of the gut microbiota. Int. J. Mol. Sci. 20 (21), 5302.

Li, D., Gao, L., Li, M., Luo, Y., Xie, Y., Luo, T., et al., 2020. Polysaccharide from spore of *Ganoderma lucidum* ameliorates paclitaxel-induced intestinal barrier injury: Apoptosis inhibition by reversing microtubule polymerization. Biomed. Pharmacother. 130, 110539.

Li, K., Zhuo, C., Teng, C. Yu, S., Wang, X., Hu, Y., et al., 2016. Effects of *Ganoderma lucidum* polysaccharides on chronic pancreatitis and intestinal microbiota in mice. Int. J. Biol. Macromol. 93, 904–912.

Liu, Y., Wang, C., Li, J., Li, T., Zhang, Y., Liang, Y., et al., 2020. *Phellinus linteus* polysaccharide extract improves insulin resistance by regulating gut microbiota composition. FASEB J. 34 (1), 1065–1078.

Lofgren, J.L., Whary, M.T., Ge, Z., Muthupalani, S., Taylor, N.S., Mobley, M., et al., 2011. Lack of commensal flora in *Helicobacter pylori*–infected INS-GAS mice reduces gastritis and delays intraepithelial neoplasia. Gastroenterology 140 (1), 210–220.

Lopez-Santamarina, A., Mondragon, A.D.C., Lamas, A., Miranda, J.M., Franco, C.M., Cepeda, A., 2020. Animal-origin prebiotics based on chitin: an alternative for the future? A critical review. Foods 9 (6), 782.

Luo, J., Zhang, C., Liu, R., Gao, L., Ou, S., Liu, L., et al., 2018. *Ganoderma lucidum* polysaccharide alleviating colorectal cancer by alteration of special gut bacteria and regulation of gene expression of colonic epithelial cells. J. Funct. Foods 47, 127–135.

Luo, J., Li, T., Xie, J., Guo, H., Liu, L., Zhang, G., et al., 2020. Guar gum different from *Ganoderma lucidum* polysaccharide in alleviating colorectal cancer based on omics analysis. Food Funct. 11 (1), 572–584.

Lv, X.C., Guo, W.L., Li, L., Yu, X.D., Liu, B., 2019. Polysaccharide peptides from *Ganoderma lucidum* ameliorate lipid metabolic disorders and gut microbiota dysbiosis in high-fat diet-fed rats. J. Funct. Foods 57, 48–58.

Ma, G., Kimatu, B.M., Zhao, L., Yang, W., Pei, F., Hu, Q., 2017. *In-vivo* fermentation of a *Pleurotus eryngii* polysaccharide and its effects on fecal microbiota composition and immune response. Food Funct. 8 (5), 1810–1821.

Ma, G., Yang, W., Zhao, L., Pei, F., Fang, D., Hu, Q., 2018. A critical review on the health promoting effects of mushrooms nutraceuticals. Food Sci. Hum. Wellness 7 (2), 125–133.

Ma, G., Du, H., Hu, Q., Yang, W., Pei, F., Xiao, H., 2021. Health benefits of edible mushroom polysaccharides and associated gut microbiota regulation. Crit. Rev. Food Sci. Nutr. 1–18. Available from: https://doi.org/10.1080/10408398.2021.1903385.

Makki, K., Deehan, E.C., Walter, J., Bäckhed, F., 2018. The impact of dietary fiber on gut microbiota in host health and disease. Cell Host Microbe 23 (6), 705–715.

Mardinoglu, A., Wu, H. Bjornson, E., Zhang, C., Hakkarainen, A., Räsänen, S.M., et al., 2018. An integrated understanding of the rapid metabolic benefits of a carbohydrate-restricted diet on hepatic steatosis in humans. Cell Metab. 27 (3), 559–571.

Muthuramalingam, K., Singh, V., Choi, C., Choi, S.I., Park, S., Kim, Y.M., et al., 2019. Effect of mushroom (*Schizophyllum* spp.) derived β-glucan on low-fiber diet induced gut dysbiosis. J. Appl. Biol. Chem. 62 (2), 211–217.

Nakahara, D., Nan, C., Mori, K., Hanayama, M., Kikuchi, H., Hirai, S., et al., 2019. Effect of mushroom polysaccharides from *Pleurotus eryngii* on obesity and gut microbiota in mice fed a high-fat diet. Eur. J. Nutr. 1–14. Available from: https://doi.org/10.1007/s00394-019-02162-7.

Nie, Q., Chen, H., Hu, J., Fan, S., Nie, S., 2019. Dietary compounds and traditional Chinese medicine ameliorate type 2 diabetes by modulating gut microbiota. Crit. Rev. Food Sci. Nutr. 59 (6), 848–863.

Pan, Y., Wan, X., Zeng, F., Zhong, R., Guo, W., Lv, X.C., et al., 2020. Regulatory effect of *Grifola frondosa* extract rich in polysaccharides and organic acids on glycolipid metabolism and gut microbiota in rats. Int. J. Biol. Macromol. 155, 1030–1039.

Ren, Y., Geng, Y., Du, Y., Li, W., Lu, Z.M., Xu, H.Y., et al., 2018. Polysaccharide of *Hericium erinaceus* attenuates colitis in C57BL/6 mice via regulation of oxidative stress, inflammation-related signaling pathways and modulating the composition of the gut microbiota. J. Nutr. Biochem. 57, 67–76.

Rodríguez-Valentín, M., López, S., Rivera, M., Ríos-Olivares, E., Cubano, L., Boukli, N.M., 2018. Naturally derived anti-HIV polysaccharide peptide (PSP) triggers a toll-like receptor 4-dependent antiviral immune response. J. Immunol. Res . Available from: https://doi.org/10.1155/2018/8741698.

Ruthes, A.C., Cantu-Jungles, T.M., Cordeiro, L.M., Iacomini, M., 2021. Prebiotic potential of mushroom D-glucans: implications of physicochemical properties and structural features. Carbohydr. Polym. 117940. Available from: https://doi.org/10.1016/j.carbpol.2021.117940.

Ríos-Covián, D., Ruas-Madiedo, P., Margolles, A., Gueimonde, M., De Los Reyes-gavilán, C.G., Salazar, N., 2016. Intestinal short chain fatty acids and their link with diet and human health. Front. Microbiol. 7, 185.

Saman, P., Chaiongkarn, A., Moonmangmee, S., Sukcharoen, J., Kuancha, C., Fungsin, B., 2016. Evaluation of prebiotic property in edible mushrooms. Biol. Chem Res. 3, 75–85.

Sarkar, A., Yoo, J.Y., Valeria Ozorio Dutra, S., Morgan, K.H., Groer, M., 2021. The association between early-life gut microbiota and long-term health and diseases. J. Clin. Med. 10 (3), 459.

Shan-Shan, S.U.N., Ka , W.A.N.G., Ke, M.A., Li, B.A.O., Hong-Wei, L.I.U., 2019. An insoluble polysaccharide from the sclerotium of *Poria cocos* improves hyperglycemia, hyperlipidemia and hepatic steatosis in ob/ob mice via modulation of gut microbiota. Chin. J. Nat. Med. 17 (1), 3–14.

Sonawane, H., Bhosle, S., Bapat, G., Vikram, G., 2014. Pharmaceutical metabolites with potent bioactivity from mushrooms. J. Pharm. Res. 8 (7), 969–972.

Sonnenburg, J.L., Xu, J., Leip, D.D., Chen, C.H., Westover, B.P., Weatherford, J., et al., 2005. Glycan foraging *in-vivo* by an intestine-adapted bacterial symbiont. Science 307 (5717), 1955–1959.

Su, J., Su, L., Li, D., Shuai, O., Zhang, Y., Liang, H., et al., 2018. Antitumor activity of extract from the sporoderm-breaking spore of *Ganoderma lucidum*: restoration on exhausted cytotoxic T cell with gut microbiota remodeling. Front. Immunol. 9, 1765.

Su, A., Ma, G., Xie, M., Ji, Y., Li, X., Zhao, L., et al., 2019. Characteristic of polysaccharides from *Flammulina velutipes in-vitro* digestion under salivary, simulated gastric and small intestinal conditions and fermentation by human gut microbiota. Int. J. Food Sci. Technol. 54 (6), 2277–2287.

Sun, S., Wang, K., Sun, L., Cheng, B., Qiao, S., Dai, H., et al., 2020. Therapeutic manipulation of gut microbiota by polysaccharides of *Wolfiporia cocos* reveals the contribution of the gut fungi-induced PGE2 to alcoholic hepatic steatosis. Gut Microbes 12 (1), 1830693.

Tang, C., Ding, R., Sun, J., Liu, J., Kan, J., Jin, C., 2019. The impacts of natural polysaccharides on intestinal microbiota and immune responses—a review. Food Funct. 10 (5), 2290–2312.

Tong, A.J., Hu, R.K., Wu, L.X., Lv, X.C., Li, X., Zhao, L.N., et al., 2020. Ganoderma polysaccharide and chitosan synergistically ameliorate lipid metabolic disorders and modulate gut microbiota composition in high fat diet-fed golden hamsters. J. Food Biochem. 44 (1), e13109.

Venturella, G., Ferraro, V., Cirlincione, F., Gargano, M.L., 2021. Medicinal mushrooms: bioactive compounds, use, and clinical trials. Int. J. Mol. Sci. 22 (2), 634.

Wang, X.Y., Yin, J.Y., Nie, S.P., Xie, M.Y., 2018. Isolation, purification and physicochemical properties of polysaccharide from fruiting body of *Hericium erinaceus* and its effect on colonic health of mice. Int. J. Biol. Macromol. 107, 1310–1319.

Wang, X., Wang, W., Wang, L., Yu, C., Zhang, G., Zhu, H., et al., 2019. Lentinan modulates intestinal microbiota and enhances barrier integrity in a piglet model challenged with lipopolysaccharide. Food Funct. 10 (1), 479–489.

Wlodarska, M., Kostic, A.D., Xavier, R.J., 2015. An integrative view of microbiome-host interactions in inflammatory bowel diseases. Cell Host Microbe 17 (5), 577–591.

Wu, S., Xu, W., Wang, F.R., Yang, X.W., 2015. Study of the biotransformation of tongmai formula by human intestinal flora and its intestinal permeability across the *Caco-2* cell monolayer. Molecules 20 (10), 18704–18716.

Wu, T.R., Lin, C.S., Chang, C.J., Lin, T.L., Martel, J., Ko, Y.F., et al., 2019. Gut commensal *Parabacteroides goldsteinii* plays a predominant role in the anti-obesity effects of polysaccharides isolated from *Hirsutella sinensis*. Gut 68 (2), 248–262.

Xiao, Z., Zhou, W., Zhang, Y., 2020. Fungal polysaccharides. Adv. Pharmacol. 87, 277–299.

Xie, J., Liu, Y., Chen, B., Zhang, G., Ou, S., Luo, J., et al., 2019. *Ganoderma lucidum* polysaccharide improves rat DSS-induced colitis by altering cecal microbiota and gene expression of colonic epithelial cells. Food Nutr. Res. 63. Available from: https://doi.org/10.29219/fnr.v63.1559.

Xu, S., Dou, Y., Ye, B., Wu, Q., Wang, Y., Hu, M., et al., 2017. *Ganoderma lucidum* polysaccharides improve insulin sensitivity by regulating inflammatory cytokines and gut microbiota composition in mice. J. Funct. Foods 38, 545–552.

Xu, X., Xu, P., Ma, C., Tang, J., Zhang, X., 2013. Gut microbiota, host health, and polysaccharides. Biotechnol. Adv. 31 (2), 318–337.

Xu, X., Yang, J., Ning, Z., Zhang, X., 2015. *Lentinula edodes*-derived polysaccharide rejuvenates mice in terms of immune responses and gut microbiota. Food Funct. 6 (8), 2653–2663.

Xue, Z., Ma, Q., Chen, Y., Lu, Y., Wang, Y., Jia, Y., et al., 2020. Structure characterization of soluble dietary fiber fractions from mushroom *Lentinula edodes* (Berk.) Pegler and the effects on fermentation and human gut microbiota *in-vitro*. Food Res. Int. 129, 108870. Available from: https://doi.org/10.1016/j.foodres.2019.108870.

Yadav, D., Negi, P.S., 2021. Bioactive components of mushrooms: processing effects and health benefits. Food Res. Int. 110599.

Yalin, W., Ishurd, O., Cuirong, S., Yuanjiang, P., 2005. Structure analysis and antitumor activity of (1→ 3)-β-D-glucans (cordyglucans) from the mycelia of *Cordyceps sinensis*. Planta Medica 71 (04), 381–384.

Ye, J., Wang, X., Wang, K., Deng, Y., Yang, Y., Ali, R., et al., 2020. A novel polysaccharide isolated from *Flammulina velutipes*, characterization, macrophage immunomodulatory activities and its impact on gut microbiota in rats. J. Anim. Physiol. Anim. Nutr. 104 (2), 735–748.

Yin, C., Noratto, G.D., Fan, X., Chen, Z., Yao, F., Shi, D., et al., 2020. The impact of mushroom polysaccharides on gut microbiota and its beneficial effects to host: a review. Carbohydr. Polym. 116942. Available from: https://doi.org/10.1016/j.carbpol.2020.116942.

Yu, J., Liu, C., Ji, H.Y., Liu, A.J., 2020. The caspases-dependent apoptosis of hepatoma cells induced by an acid-soluble polysaccharide from *Grifola frondosa*. Int. J. Biol. Macromol. 159, 364–372.

Zhang, Y., Kong, H., Fang, Y., Nishinari, K., Phillips, G.O., 2013. Schizophyllan: a review on its structure, properties, bioactivities and recent developments. Bioact. Carbohydr. Diet. Fibre 1 (1), 53–71.

Zhang, Y., Li, Q., Shu, Y., Wang, H., Zheng, Z., Wang, J., et al., 2015. Induction of apoptosis in S180 tumour bearing mice by polysaccharide from *Lentinus edodes* via mitochondria apoptotic pathway. J. Funct. Foods 15, 151–159.

Zhang, T., Yang, Y., Liang, Y., Jiao, X., Zhao, C., 2018. Beneficial effect of intestinal fermentation of natural polysaccharides. Nutrients 10 (8), 1055.

Zhang, M., Zhang, Y., Zhang, L., Tian, Q., 2019. Mushroom polysaccharide lentinan for treating different types of cancers: A review of 12 years clinical studies in China. Prog. Mol. Biol. Transl. Sci. 163, 297–328.

Zhao, R., Cheng, N., Nakata, P.A., Zhao, L., Hu, Q., 2019a. Consumption of polysaccharides from *Auricularia auricular* modulates the intestinal microbiota in mice. Food Res. Int. 123, 383–392.

Zhao, R., Hu, Q., Ma, G., Su, A., Xie, M., Li, X., et al., 2019b. Effects of *Flammulina velutipes* polysaccharide on immune response and intestinal microbiota in mice. J. Funct. Foods 56, 255–264.

Zhao, R., Ji, Y., Chen, X., Su, A., Ma, G., Chen, G., et al., 2020c. Effects of a β-type glycosidic polysaccharide from *Flammulina velutipes* on anti-inflammation and gut microbiota modulation in colitis mice. Food Funct. 11 (5), 4259–4274.

Section XI

Microbiome and Immunomodulatory Peptides

Chapter 31

The role of the microbiome in sports nutrition

Estela González-Rodríguez, Laura Albero Montava, Isabel Sospedra López and José Miguel Martínez-Sanz

Nursing Department, Faculty of Health Sciences, University of Alicante, Alicante, Spain

31.1 Introduction

The human intestinal microbiota is made up of thousands of bacterial taxa, with *Lactobacillus, Bifidobacterium, Proteobacteria, Akkermansia, Streptococcus, Clostridium, Turicibacter*, and *Rikenellaceae* are commonly present among athletes (Hughes, 2020), and their composition varies depending on diet, habitat, and other factors such as physical exercise. These microorganisms have important functions such as the absorption of nutrients, the synthesis of vitamins, the production of energy, the modulation of inflammation, and participation in the immune response of the individual. In addition, the microbiota acts as an endocrine organ and the variations produced in its state due to physical exercise are related to modifications in the physiology, metabolism, immunity, and behavior of the host, although these changes depend on the intensity and duration of the exercise (Zeppa et al., 2020). Therefore the intestinal microbiota indirectly influences various parameters of physical performance, recovery, and disease patterns (e.g., by signaling through myokines and other cytokines); modulates the activation of the hypothalamic−pituitary−adrenal axis; and also affects the metabolic pathways associated with performance (Mohr et al., 2020).

The quality of the diet influences the composition of the intestinal microbiota, since those diets with a low fiber content and rich in refined carbohydrates and fats will lead to a reduction in the microbial community diversity, with alterations in its structure and activity (Zeppa et al., 2020).

Probiotics are live microorganisms that are administered orally for a few weeks to increase the amount of beneficial bacteria in our intestines. They are related to a wide variety of benefits for intestinal health and to the modulation of the immune function, through the maintenance of the intestinal barrier and the production of vitamins, short-chain fatty acids (SCFA), and molecules that act as neurotransmitters (Jäger et al., 2019). To promote the improvement of the composition of the microbiota, supplementation with probiotics is recommended for athletes (Maughan et al., 2018).

In the colon, microorganisms are responsible for digesting and fermenting complex carbohydrates to yield SCFA such as n-butyrate, acetate, and propionate. These last two are transported to the bloodstream, in which they are distributed to the different organs so that they can be used as substrates in energy metabolism, specifically in hepatocytes, which use propionate for gluconeogenesis (Mach and Fuster-Botella, 2017).

Therefore diet is also a key factor to take into account since it is capable of having beneficial effects on the intestinal microbiota, increasing the microbial diversity and the abundance of genera associated with health. Such effects aid the digestion of nutrients and the provision of energy during exercise, which could provide metabolic benefits for an athlete during high intensity exercise and recovery (Marttinen et al., 2020).

31.2 The microbiome and sport

This growing interest in the microbiota also encompasses the field of sports, and recent studies have related the composition of the intestinal microbiota in professional athletes to dietary factors and extreme exercise (Clark and Mach, 2016; Aya et al., 2021). The athlete's nutrition or diet can modify the intestinal microbiota in a way that has a beneficial or counterproductive effect in relation to their sports performance (Aya et al., 2021). It has been seen that there is

Microbiome, Immunity, Digestive Health and Nutrition. DOI: https://doi.org/10.1016/B978-0-12-822238-6.00039-X

a positive correlation between greater diversity of the microbiota due to a high-protein diet and high performance in physical training (Clarke et al., 2014).

It has been shown that professional athletes, or people who practice sports regularly, have a more diverse intestinal microbiota than that found in healthy individuals who do not practice any sport, which could indicate that physical activity is a factor promoting greater bacterial diversity at the intestinal level (Aya et al., 2021).

The microbiotas of athletes differ from those of healthy sedentary individuals in various taxa. It is not clear, however, to what extent changes in the microbiota can be attributed to observed changes in athletic performance. The gut microbial communities of athletes differ from those of sedentary persons in both diversity and the presence of certain taxa. However, the degree to which the gut microbial community differs between elite athletes and nonelite athletes is unclear, and it remains to be seen if we can effectively monitor the potential of athletes based on the microbiota (Clark and Mach, 2016; Aya et al., 2021).

Three genera that are increased in abundance after physical exercise and that can influence sports performance are highlighted.

The first genus to highlight is *Veillonella*. The articles of Grosicki et al. (2019) and Scheiman et al. (2019) clearly show an increase in its abundance after the performance of acute physical exercise, especially in marathon runners, and this may be related to an increase in sports performance. To evaluate this hypothesis, an experiment was carried out with mice and it was observed that, in comparison with the control group (to which *Lactobacillus bulgaricus* was introduced by catheter), the treatment group (*Veillonella* by catheter), 5 hours after running to exhaustion, had peak run times that were significantly longer (statistically). Although this experiment had limitations as it was not carried out in humans, the data could be extrapolated, since everything indicates that *Veillonella* could have beneficial effects on the performance of athletes, by conserving the metabolic enzymes that help in the decomposition of lactate to propionate and acetate. Furthermore, a replicate experiment was performed in an independent cohort of human athletes and it was verified that there was high production of acetate and propionate in these individuals. This suggests that the lactate produced during exercise reached the intestine and was metabolized by *Veillonella*.

The second genus to highlight is *Prevotella*. In the study of Petersen et al. (2017), the abundance of *Prevotella* was related significantly and positively to the time spent exercising during a 1-week period, which indicates that exercise increases the presence of this genus. Furthermore, the presence of *Prevotella* is positively correlated with a number of amino acid metabolism pathways such as lysine biosynthesis, alanine, aspartate, and glutamate metabolism, and D-glutamine and D-glutamate metabolism. In contrast, it is negatively related to certain pathways of amino acid metabolism and carbohydrate metabolism, including the metabolism of SCFA. Furthermore, as observed in Murtaza et al. (2019), *Prevotella* is favored when athletes eat a diet rich in carbohydrates/fiber and is negatively correlated with the amounts of sucrose consumed (Kulecka et al., 2020). Regarding performance improvements, changes in the abundance of this fermenter of complex carbohydrates suggest alterations at the metabolic level, but more studies are needed to reinforce and affirm this hypothesis.

Finally, we highlight the genus *Akkermansia*, which Clarke et al. (2014) and Petersen et al. (2017) reported as being significantly increased in athletes. *Akkermansia* is responsible for degrading the mucin located in the mucus layer and its increase has been shown to be inversely correlated with obesity and metabolic disorders. In addition, it has been observed that athletes, with higher levels of *Akkermansia* than the nonathlete controls, had lower values of metabolic markers associated with inflammation, since it helps to maintain intestinal permeability during the state of inflammation associated with disease. This genus also showed a negative correlation with oxidative stress (Huang et al., 2020a), so the health of athletes would be improved, and this could lead to better sports performance, although more exhaustive studies are needed to affirm this hypothesis.

31.3 Research related to the microbiome and athletic performance

The following section describes different investigations that relate physical exercise to the microbiome in athletes (Table 31.1).

The gut microbiota varies depending on the sport performed or its intensity (O'Donovan et al., 2020). In addition, it has been found that the diversity of this microbiota is greater for athletes than for sedentary individuals (Liang et al., 2019; Kulecka et al., 2020; Clarke et al., 2014). Among the bacteria that produce fecal SCFA, *Clostridiales*, *Ruminococcaceae*, and *Faecalibacterium* are dominant in the microbial community of elite athletes (Han et al., 2020).

According to Scheiman et al. (2019), the abundance of the genus *Veillonella*, present above all in marathon runners, is enriched in these athletes after exercise. This is associated with an enhancement of sports performance, since there is

TABLE 31.1 Research concerning physical exercise and the microbiome in athletes.

Author	Research aim	Type of study	Sample	Microbiome	Test	Results
Clarke et al. (2014)	To explore the impact of exercise and diet on the gut microbiota	Observational	40 elite rugby players and 46 healthy controls	*Akkermansia, Prevotella, Ruminococcaceae*	Genetic analysis of stool samples. Blood samples. Measurements of creatine kinase (CK) (marker of extreme exercise). Food consumption frequency questionnaire.	The diversity of the microorganisms of the athletes was greater than that of the healthy controls. This change was positively correlated with athletic performance and with protein and CK consumption.
Petersen et al. (2017)	To compare the characteristics of the gut microbiomes of professional and amateur cyclists	Observational	33 cyclists (22 professionals and 11 amateurs)	*Bacteroides, Prevotella, Eubacterium, Rumiinococcus, Akkermansia*		The greater number of hours of training of the professionals increased the amounts of *Prevotella* and *Methanobrevibacter smithii*.
Scheiman et al. (2019)	To observe the relationship between the amount of *Veillonella* and the improvement in sports performance through lactate metabolism	Observational cohorts	87 ultramarathonerunners and Olympic rowers	*Veillonella atypica*	Genetic analysis of stool samples. Samples were collected 1 week before and 1 week after the day of the competition.	The genus *Veillonella* was increased in athletes after exercise, as was the metabolic pathway that *Veillonella* species use for lactate metabolism. This result raised the possibility that systemic lactate resulting from muscle activity during exercise could enter the gastrointestinal lumen and be metabolized by *Veillonella*.
Liang et al. (2019)	To investigate whether the gut microbiome differs significantly between upper and lower level athletes	Observational	12 professional martial arts athletes at the top level and 16 at the bottom level	*Parabacteroides, Phascolarctobacterium, Oscillibascter, Bilophila, Megasphaera*	Genetic analysis of stool samples. Questionnaire on frequency of food consumption during the last 3 months. Duration of the workout and the average number of hours of exercise per week (specifying the exercise load).	The results indicated that athletes had a greater diversity in their gut microbiota than sedentary individuals.

(Continued)

TABLE 31.1 (Continued)

Author	Research aim	Type of study	Sample	Microbiome	Test	Results
Grosicki et al. (2019)	To examine the gut microbiome of a world-class ultramarathon runner before and after competing in the Western States Endurance Run (WSER), a 163 km foot-mountain race	Observational case study	1 world-class ultramarathon runner	*Veillonella, Streptococcus, Alloprevotella, Subdolingranulum*	Daily training records recorded via a Suunto Ambit3 Peak GPS watch. Gut microbiome analyses were performed after collection of stool at four time points (21 weeks before WSER, 2 weeks before WSER, 2 h after WSER, and 10 days after WSER).	The Firmicutes/ Bacteroidetes ratio was almost tripled 2 h after the race due to the loss of Bacteroidetes. The proportion of Proteobacteria had increased by more than five times 2 h after the race due to the 29-fold increase in *Haemophilus*. This global value also includes the proliferation of *Veillonella* (+ 14,229%) and Streptococcus (+ 438%) and reductions in *Alloprevotella* (−79%) and *Subdolingranulum* (−50%)
Murtaza et al. (2019)	To investigate the effects of extreme changes in diet patterns on the gut microbiota of elite race walkers undergoing intensified training and their possible links to athletic performance	Nonrandomized clinical trial (RCT)	29 men between 20 and 35 years of age	*Faecalibacterium, Dorea, Bascteroides, Prevotella, Clostridiaceae, Lachnospiraceae, Ruminococcaceae, Akkermansia*	Genetic analysis of stool samples. Physiological measurements of VO_2, running time, and respiratory exchange rate were performed.	The low-carbohydrate, high-fat diet increased Bacteroides and Dorea and reduced *Faecalibacterium*. Runners with a high-carbohydrate diet had an improved performance.
O'Donovan et al. (2020)	To determine whether the gut microbiome and metabolome differ between sport classification groups among elite Irish athletes, many of whom were participating in the 2016 Olympics	Observational	37 elite Irish athletes, who competed in 16 different sports	*Eubacterium, Polynucleobacter, Faecalibacterium, Bacteroides, Gordonibacter*	Genetic analysis of stool samples. Urine analysis. Blood pressure was measured for static sports, and oxygen volume (VO2 max) for dynamic sports.	The microbiomes of the athletes participating in sports with a high static component had associated species such as *Streptococcussuis, Clostridiumbolteae, Lactobacillus phage Lfelnf*, and *Anaerostipeshadrus*, while those of the athletes participating in more dynamic sports

						had associated species such as *Bifidobacterium animalis, Lactobacillus acidophilus, Prevotella intermedia* and *Faecalibacterium prausnitzii.*
Han et al. (2020)	To compare the gut microbiomes of the different cohorts, and to investigate their associations with dietary factors, physical characteristics, and sports performance.	Observational	19 female rowers	*Firmicutes, Bacteroidetes, Proteobacteria, and Actinobacteria*	306 fecal samples were collected (on average, 16 samples per person)	The mean relative abundances of *Firmicutes* and *Proteobacteria* of adult elite athletes (76.27% and 8.73%) and youth elite athletes (64.7% and 10.69%) were higher than those of elite youth athletes (58.12% and 8.01%,); also, the average relative abundance of *Bacteroidetes* in nonelite youth athletes (26.19%) was significantly higher than that in adult elite athletes (11.41%).
Kulecka et al. (2020)	To investigate differences in the gut microbiota between healthy controls and endurance athletes	Observational cases/controls	14 marathon runners, 11 cross-country skiers, and 46 sedentary healthy controls	*Prevotella, Phascolarctobascterium, Christensenellaceae, Agathobacter*	Stool samples were collected and the diet was evaluated through a 24-h recall.	Both the athletes and the sedentary controls had a low abundance of Bacteroidetes and a high abundance of *Prevotella*. The microbiome was more diverse in the cross-country skiers than in the sedentary controls, due to their training load. *Prevotella* showed an inverse correlation with the intake of sucrose. *Phascolarctobacterium* was inversely correlated with the amount of polyunsaturated fatty acids in the diet. There was a direct positive

(Continued)

TABLE 31.1 (Continued)

Author	Research aim	Type of study	Sample	Microbiome	Test	Results
						correlation of *Christensenellaceae* with the intake of folic acid and of *Agathobacter* with the amount of fiber in the diet.
Huang et al. (2020)	To investigate how probiotics affect the microbiota of athletes and how this translates into functional activities	RCT	20 male triathlon athletes	*Anaerotruncus, Caproiciproducens, Coprobacillus, Desulfovibrio, Dielma, Family_XIII, Holdemania, Oxalobacter, Akkermansia, Bifidobacterium, Butyricimonas, Lactobacillus*	-Genetic analysis of the microbiota. -Blood samples and complete blood count analysis. Body composition. Biochemical variables. Dietary records were analyzed for individual nutritional compositions and calories, as a reference.	PS128 supplementation was associated with an improvement in endurance running performance through alterations in the microbiota and related metabolites, but not in the maximal oxygen uptake.
Tabone et al. (2021)	To determine exercise-driven changes in the gut microbiota, the serum, and the fecal metabolome using 16S RNA genetic analysis and uncoded metabolomics	Non-RCT	40 male endurance runners, aged 18–50	*Romboustsia, Ruminococcocaceae UCG-005, Escherichia coli TOP498, Blautia, Ruminoclostridium 9, Clostridium phoceensis*	The pH and fecal ammonia were determined in samples collected before and after the exercise session.	The exercise session increased the abundance of the genus *Romboutsia, Ruminococcocaceae UCG-005, E. coli TOP498*, and the *genus Blautia*, and decreased the abundance of *Ruminiclostridium 9* and *Clostridiumphoceensis*.
Dong et al. (2021)	To explore the efficacy of probiotics in improving athletes' state of anxiety and their athletic performance in stressful situations	Non-RCT	21 male divers	*Parasutterella, Lachnospiraceae, Propionibacteriaceae*	The anxiety state of the competition situation was measured with the Mental Preparation Form-3 before, during, and after the exercise. Fecal samples were also analyzed.	The intervention did not change the intestinal microbial composition in terms of diversity, but it did increase the abundance of *Bifidobacteriaceae*. In the 8th week, the performance of the athletes s stress and anxiety improved significantly.

an improvement in execution time as this microorganism is capable of metabolizing the lactate synthesized during exercise, yielding propionate.

Another genus that stands out for its ability to improve performance is *Prevotella*, since it is positively correlated with the average time of physical exercise performed during 1 week, that is, the greater the amount of exercise performed, the greater the abundance of *Prevotella*. In addition, a greater abundance of *Prevotella* coincides with the appearance of amino acid and carbohydrate metabolism pathways, including those of branched chain amino acids (BCAAs), which are important in muscle recovery. BCAAs decrease exercise-induced fatigue and promote muscle protein synthesis. A positive correlation between yield and the presence of the genus *Akkermansia* has also been observed. This is associated with better immunity and intestinal barrier function, and with a healthier metabolic profile, as this is negatively correlated with obesity (Murtaza et al., 2019).

The species *Methanobrevibacter smithii* is present at a higher proportion in professional cyclists, compared to amateur cyclists, and positively regulates methane production. This implies an improvement in the metabolism of energy and carbohydrates, and thus an enhanced athletic performance. *M. smithii* is positively associated with high-carbohydrate diets (Petersen et al., 2017).

The findings of Murtaza et al. (2019) indicate that the composition of the microbiota can be influenced by the type of diet and that changes can modify the performance of an athlete. In this case, an increase in the amount of carbohydrates ingested was correlated with improvements in the performance of athletes, compared to athletes who followed a diet low in carbohydrates and high in fat.

In addition, there are data that suggest that the immune system, oxidative stress, metabolism, and cognitive capacity of athletes can be positively regulated by probiotic supplementation.

Among the existing probiotics, *L-Plantarum PS128* has been shown to be capable of reducing inflammation and oxidative stress in athletes by modifying the composition of their microbiota, with an increase in genera such as *Akkermansia*; as a consequence, performance is improved (Huang et al., 2020b). The research of Dong et al. (2021) verified that supplementation with *Bifidobacterium animalis subsp. lactis BB-12* can also produce an improvement in the performance of athletes under stress situations. Although it did not change the composition of their microbiota, it did produce an increase in *Bifidobacteriaceae* and this resulted in certain improvements—such as increased maximal oxygen absorption, reduction of gastrointestinal symptoms and plasma endotoxin levels after competition, better performance and muscle tone, and increased proliferation of beneficial metabolites.

31.4 Conclusion

The genus *Veillonella* increases in abundance after physical exercise and this could produce an improvement in execution time since it is able to metabolize the lactate synthesized during the exercise, converting it into propionate.

The abundance of the genus *Prevotella* is correlated with the exercise time, which is why it is increased in professional athletes. This could positively influence muscle recovery due to a decrease in the fatigue caused by exercise and the promotion of muscle protein synthesis.

The genus *Akkermansia* is associated with better immunity, an improved intestinal barrier function, and a healthier metabolic profile, which is negatively correlated with obesity. This genus is more abundant in athletes.

References

Aya, V., Flórez, A., Perez, L., Ramírez, J.D., 2021. Association between physical activity and changes in intestinal microbiota composition: a systematic review. PLoS One 16. Available from: https://doi.org/10.1371/journal.pone.0247039.

Clark, A., Mach, N., 2016. Exercise-induced stress behavior, gut-microbiota-brain axis and diet: a systematic review for athletes. J. Int. Soc. Sports Nutr. 13, 43. Available from: https://doi.org/10.1186/s12970-016-0155-6.

Clarke, S.F., Murphy, E.F., O'Sullivan, O., Lucey, A.J., Humphreys, M., Hogan, A., et al., 2014. Exercise and associated dietary extremes impact on gut microbial diversity. Gut 63, 1913–1920. Available from: https://doi.org/10.1136/gutjnl-2013-306541.

Dong, W., Wang, Y., Liao, S., Tang, W., Peng, L., Song, G., 2021. *Bifidobacterium animalis* subsp. lactis BB-12 improves the state anxiety and sports performance of young divers under stress situations: a single-arm, prospective proof-of-concept study. Frontiers in Psychology 11. Available from: https://doi.org/10.3389/fpsyg.2020.570298.

Grosicki, G.J., Durk, R.P., Bagley, J.R., 2019. Rapid gut microbiome changes in a world-class ultramarathon runner. Physiological Reports 7. Available from: https://doi.org/10.14814/phy2.14313.

Han, M., Yang, K., Yang, P., Zhong, C., Chen, C., Wang, S., et al., 2020. Stratification of athletes' gut microbiota: the multifaceted hubs associated with dietary factors, physical characteristics and performance. Gut Microbes 12, 1–18. Available from: https://doi.org/10.1080/19490976.2020.1842991.

Huang, W.C., Pan, C.H., Wei, C.C., Huang, H.Y., 2020a. *Lactobacillus plantarum* ps128 improves physiological adaptation and performance in triathletes through gut microbiota modulation. Nutrients 12, 1–15. Available from: https://doi.org/10.3390/nu12082315.

Huang, W.C., Pan, C.H., Wei, C.C., Huang, H.Y., 2020b. *Lactobacillus plantarum* ps128 improves physiological adaptation and performance in triathletes through gut microbiota modulation. Nutrients 12, 1–15. Available from: https://doi.org/10.3390/nu12082315.

Hughes, R.L., 2020. A review of the role of the gut microbiome in personalized sports nutrition. Frontiers in Nutrition . Available from: https://doi.org/10.3389/fnut.2019.00191.

Jäger, R., Mohr, A.E., Carpenter, K.C., Kerksick, C.M., Purpura, M., Moussa, A., et al., 2019. International Society of Sports Nutrition Position Stand: probiotics. J. Int. Soc. Sports Nutr. Available from: https://doi.org/10.1186/s12970-019-0329-0.

Kulecka, M., Fraczek, B., Mikula, M., Zeber-Lubecka, N., Karczmarski, J., Paziewska, A., et al., 2020. The composition and richness of the gut microbiota differentiate the top Polish endurance athletes from sedentary controls. Gut Microbes 11, 1374–1384. Available from: https://doi.org/10.1080/19490976.2020.1758009.

Liang, R., Zhang, S., Peng, X., Yang, W., Xu, Y., Wu, P., et al., 2019. Characteristics of the gut microbiota in professional martial arts athletes: a comparison between different competition levels. PLoS One 14. Available from: https://doi.org/10.1371/journal.pone.0226240.

Mach, N., Fuster-Botella, D., 2017. Endurance exercise and gut microbiota: a review. Journal of Sport and Health Science . Available from: https://doi.org/10.1016/j.jshs.2016.05.001.

Marttinen, M., Ala-Jaakkola, R., Laitila, A., Lehtinen, M.J., 2020. Gut microbiota, probiotics and physical performance in athletes and physically active individuals. Nutrients 12, 1–39. Available from: https://doi.org/10.3390/nu12102936.

Maughan, R.J., Burke, L.M., Dvorak, J., Larson-Meyer, D.E., Peeling, P., Phillips, S.M., et al., 2018. IOC consensus statement: dietary supplements and the high-performance athlete. Br. J. Sports Med. 52, 439–455. Available from: https://doi.org/10.1136/bjsports-2018-099027.

Mohr, A.E., Jäger, R., Carpenter, K.C., Kerksick, C.M., Purpura, M., Townsend, J.R., et al., 2020. The athletic gut microbiota. J. Int. Soc. Sports Nutr. Available from: https://doi.org/10.1186/s12970-020-00353-w.

Murtaza, N., Burke, L.M., Vlahovich, N., Charlesson, B., O' Neill, H., Ross, M.L., et al., 2019. The effects of dietary pattern during intensified training on stool microbiota of elite race walkers. Nutrients 11. Available from: https://doi.org/10.3390/nu11020261.

O'Donovan, C.M., Madigan, S.M., Garcia-Perez, I., Rankin, A., O' Sullivan, O., Cotter, P.D., 2020. Distinct microbiome composition and metabolome exists across subgroups of elite Irish athletes. Journal of Science and Medicine in Sport 23, 63–68. Available from: https://doi.org/10.1016/j.jsams.2019.08.290.

Petersen, L.M., Bautista, E.J., Nguyen, H., Hanson, B.M., Chen, L., Lek, S.H., et al., 2017. Community characteristics of the gut microbiomes of competitive cyclists. Microbiome 5. Available from: https://doi.org/10.1186/s40168-017-0320-4.

Scheiman, J., Luber, J.M., Chavkin, T.A., MacDonald, T., Tung, A., Pham, L.D., et al., 2019. Meta-omics analysis of elite athletes identifies a performance-enhancing microbe that functions via lactate metabolism. Nat. Med. 25, 1104–1109. Available from: https://doi.org/10.1038/s41591-019-0485-4.

Tabone, M., Bressa, C., García-Merino, J.A., Moreno-Pérez, D., Van, E.C., Castelli, F.A., et al., 2021. The effect of acute moderate-intensity exercise on the serum and fecal metabolomes and the gut microbiota of cross-country endurance athletes. Scientific reports 11 (1), 3558. Available from: https://doi.org/10.1038/s41598-021-82947-1.

Zeppa, S.D., Agostini, D., Gervasi, M., Annibalini, G., Amatori, S., Ferrini, F., et al., 2020. Mutual interactions among exercise, sport supplements and microbiota. Nutrients . Available from: https://doi.org/10.3390/nu12010017.

Section XII

Study Design and Statistical Interpretation

Chapter 32

Food-derived immunomodulatory peptides: science to applications

Chopada Kevalkumar Dineshbhai[1], Pratik Shukla[1], Amar Sakure[2], Brij Pal Singh[3], Sujit Das[4], Sreeja V.[1] and Subrota Hati[1]

[1]Department of Dairy Microbiology, SMC College of Dairy Science, Kamdhenu University, Anand, Gujarat, India, [2]Deparment of Agriculture Biotechnology, Anand Agricultural University, Anand, Gujarat, India, [3]Deparment of Microbiology, Central University of Haryana, Mahendragarh, Haryana, India, [4]Department of Rural Development and Agricultural Production, North-Eastern Hill University, Tura, Meghalaya, India

32.1 Introduction

Protein plays an important role in the diet. As a source of physiologically active constituents, nutritional proteins are produced that have a favorable impact. Specific protein fragments are recognized as bioactive peptides that have an excellent influence on the characteristics of the body and can eventually alter the fitness of health following gastrointestinal digestion (Kitts and Weiler, 2003). Such, components, termed bioactive peptides, have the ability to affect various biological characteristics (Hartmann and Meisel, 2007).

Bioactive peptides have shorter-chain amino acid sequences that comprise a molecular sequencing of 2–20 amino acids per molecule (Sanchez and Vazquez, 2017). Food-derived bioactive peptides are originally in an inactive form in the precursor molecules sequence but in three ways that could be liberated: (1) hydrolysis with digestive enzymes such as alcalase, chymotrypsin, pepsin, trypsin, and many more; (2) proteolytic action of starter cultures used for milk digestion; and (3) hydrolysis with milk enzymes derived from proteolytic organisms. When released, bioactive peptides can affect a number of physiological reactions, such as cardiovascular, endocrine, digestive, immunological, and neurological activities; weight control; and improvements in bone health. The bioactive peptides that can be obtained from food have attracted the interest of many researchers to develop numerous possible nutraceutical supplements to promote healthy functional meals as well as other pharmaceutical products (Korhonen and Pihlanto, 2003).

Food proteins, are the precursors for bioactive peptides (Reyes-Diaz et al., 2018). Antioxidative, antimicrobial, antihypertensive, antithrombotic, and immunomodulative actions are linked to valuable health implications of bioactive peptides. The health implications may allow these bioactive peptides to be used in the food system, in nutritional products, and in natural medicaments (Haque et al., 2008).

Newborn mammals are immunologically immature because of impairments in some activities of the T and B lymphocytes. It is generally known that milk includes several host defense proteins and different proteins that improve the immature system cell function (Hata et al., 1998). The immune system is a self-sufficient defense system that consists of several organic structures and techniques within an organism to save it and treat certain infectious illnesses. Many immunomodulatory medications are inappropriate for chronic or preventive use, so there is an increasing demand for herbal and nonpharmaceutical immunomodulators (Xu et al., 2019).

One of the main types of bioactive peptides is immunomodulatory peptide. Since oxidative stress is strictly associated with immunological function, several immunomodulatory peptides correspondingly demonstrate antioxidant or anti-inflammatory activity (Dadar et al., 2019). These peptides have been shown to improve inflammation by inhibition of cytokine secretions or abatement of proinflammatory expressions (Chalamaiah et al., 2018).

There was a lengthy acknowledgment of the relationship between diet and immunity. In both in vitro and in vivo investigations, bioactive peptides generated through large number of protein sources were seen to possess immunomodulatory activity. Nevertheless, the majority of research analyzed the impact of peptides and protein hydrolysates upon particular immune responses, while one small unit of research evaluated their influence upon the innate immune response (Shahidi and Zhong, 2008).

Microbiome, Immunity, Digestive Health and Nutrition. DOI: https://doi.org/10.1016/B978-0-12-822238-6.00007-8

Bioactive peptides derived from fermented milk can give certain health advantages, which include improving the functioning of the innate immune response by activating the macrophages. Nevertheless, further knowledge of the role of these bioactive peptides as inducers of immunological responses is essential. The immunomodulatory activity is also known for the bioactive peptides that are produced from caseins and whey proteins. The peptides may modify human lymphocyte proliferation, reduce cytokine generation, and boost macrophage phagocytic activity. As an outcome, the growth of the immune response within infant could be regulated (Korhonen and Pihlanto, 2006).

In vitro designing of in vivo gastrointestinal processed milk has been the main reason for the study using immunologically active peptides (Svedberg et al., 1985). Thereafter, the authentication of immunostimulatory effects was also gained with in vitro research on real impacts of certain peptides on immune response cells along with in vivo investigations of the impact of peptides produced via dietetic methods (Schlimme and Meisel, 1995).

32.2 Immunomodulation and immunomodulatory peptides

Immunomodulation is a promising technique because it has properties that are similar to those of antibiotic medications but can prevent or minimize drug resistance issues (Finlay and Hancock, 2004). Proper functioning of the immune system is essential for infection interruption and the immune response seems to be the first layer of pathogenic defense, which offers conservation against impaired body functioning. Several distinct immune cells seem to be doing monitoring. For example, the nonspecific immune system has been predominantly characterized as dendritic cells, macrophages, and natural killer (NK) cells, which are components of natural immunological activity. The antigens are taken by macrophages and dendritic cells and handled by the T cells of the immune response (Hou et al., 2012).

Cytotoxic activities are controlled by NK cells and react on microbe—contaminated cells together with tumors cells. Its revival encourages cytokine generation (Clare and Swaisgood, 2000). The major cells of the effector in cell-mediated immunity are the T ($CD4^+$ and $CD8^+$) lymphocytes. B lymphocytes are important for the humor reaction and are responsible for producing antigen elimination antibodies (Abid et al., 2012).

As the system defends the organism, the immune system comprises specialized cells, antibodies, and a lymphatic circulation system. Peptides obtained from caseins and major whey proteins from milk-protein hydrolysates can increase immune cell activity, evaluated as a proliferation of lymphocytes and antibody formation as well as cytokine modulation (Gill et al., 2000). The physical features that are associated with these dietary peptides show a basic mechanism based on target enzyme inhibition, which is somehow implicated in important processes, including blood coagulation, phagocytosis, and pathogenic infections (Tidona et al., 2009).

The development of new immunomodulatory peptides in dietary proteins is becoming more and more interesting. Peptides are thus recognized that upgrading proinflammatory and anti-inflammatory cytokines could either improve or impair immunity. These promote the synthesis of certain antibodies and can also help to proliferate or eliminate T and B lymphocytes as well as influence the phagocytic activity of macrophages and NK cells (Wagar et al., 2009).

Because of its diverse consequences, including frequently imbricate methods of action, immunomodulating peptides have been difficult to describe. Peptides influence an organism's immunological response within the existence of an antigen. That impact might have been beginning or defusing a nonspecific or acquired immune response. It is also difficult to distinguish between the impacts of nonspecific (innate) and acquired (adaptive) immune activity, as nonspecific immunological activity influences acquired immunological response via interactions with the cell and the reaction on cytokine outlines (Parish and O'Neill, 1997).

Meisel (1997) describes immunomodulatory peptides as those that are "stimulating the proliferation of human lymphocytes and the phagocytic activities of macrophages." Udenigwe and Aluko (2012) explained immunomodulatory peptides that "act by enhancing the functions of the immune system, including regulation of cytokine expression, antibody production, and ROS (reactive oxygen species) induced immune functions."

Antiproliferative peptides are classified as immunomodulatory by Schanbacher et al. (1998), but they are classified as a different group in other studies (Hernandez-Ledesma et al., 2014). The immune system has an important role to play for avoiding as well as stimulating neoplasia (Chalamaiah et al., 2014), and numerous publications have demonstrated that there is an association between immunosuppression and growth of cancer (Cata et al., 2013). This association arises from variations in the concentration and activity of various important components, including interleukins and growth factor of humoral and cellular immunity (Waugh and Wilson, 2008). Comparison, among the irritation, carcinogenesis as well as increased proinflammatory cytokines throughout the early phases of tumor developmen are the sign of lethality (Grivennikov and Karin, 2011). It was decided that the anti-inflammatory peptides can also be regarded as immunostimulatory peptides. The anti-inflammation peptides support this process, which inhibits the aggregation of neutrophil and regulates generation of the adhesive particles of the leukocyte surface (Zouki et al., 2000).

On the other hand, immunomodulatory peptides help to defend the host but do not directly interact with the pathogen. Implication of the gastrointestinal system in the immunity of the organism will improve the significance of dietary peptides in immunomodulation (Wittkopf et al., 2014). Numerous peptides have been shown to either promote or suppress immunological responses (Agyei and Danquah, 2012); the best-recognized of these are the breakdown products of different milk proteins (Gill et al., 2000). Immunomodulatory peptides promote immune cell expansion along with maturation; accelerate NK cell activity as well as macrophage phagocytosis; increase antibody, cytokine, and chemokine production; and inhibit inflammatory chemicals. Glycine, along with other residues that are found in peptides, promote the growth of human lymphocytes in the surrounding blood (Kayser and Meisel, 1996).

32.3 Mechanism of immunomodulatory peptide

Immunity is the body's natural defense mechanism against a wide range of infectious illnesses. Previous infection, immunization, and numerous environmental stimuli are all factors that cause immunity. Furthermore, the immune system may differentiate between the body's proteins or cells and alien substances. The immune response begins immediately after the external element has been detected, and it consists of the collective and synchronized reaction on certain cells with mediators to unfamiliar constituents (Baxter, 2007). The immune system has been divided into two major groups based on function: innate immune response and adaptive immune response (Vesely et al., 2011) (Fig. 32.1).

The most important mediators in the immune system, which provide immediate defense, are the cytokines, monocytes, macrophages, complements, acute-phase proteins, and neutrophils; microbial, chemical, and physical barriers are also occasionally involved in the nonspecific immune system. Pathogen-associated molecular patterns (PAMPs) are unique moieties produced from pathogens that are recognized by the host to identify the presence of a pathogen. PAMPs are recognized by pattern recognition receptors, which are germline-encoded and evolutionarily maintained host sensors. When the PRRs identify the PAMPs, the activation of various interferons, chemokines, and cytokines prompts a variety of immunological responses. PRR families such as DNA receptors, NOD-like receptors, RIG-I-like receptors, and toll-like receptors (TLRs) play a vital role in the host's defense (Parkin and Cohen, 2001).

Peptides that have immunomodulatory effects may act through directly affecting receptors (e.g., TLRs) or through passive diffusion in cells, peptide transporter (PepT1 situated in gastrointestinal epithelial cells transporting tiny peptides, e.g., dipeptides and tripeptides), endocytosis of the liquid stage, and subsequent interfering inflammation signaling paths, namely, NFκB PI3K, and MAPK (Horner et al., 2016) (Fig. 32.2). Host defense peptides (HDPs) and milk-based peptides are shown in terms of their immunomodulatory effects. Natural peptides and modulation of peptide levels are obviously prospective therapeutic techniques for the obstruction of inflammatory markers, infection therapy, and treatment of inflammatory illnesses. The immunomodulatory peptides that are generated from milk may also be considered health promotional food complements (Gronberg et al., 2014).

The HDPs are generated primarily in response to injury and infection by several types of cells, both epithelial and immunological cells (e.g., neutrophils). Peptides resulting from milk are hydrolysates of milk protein. They are absorbed after digestion in the gastrointestinal system. Both HDPs and innate defense regulator peptides (IDRs) with milk-derived peptides have straight antimicrobial effects in contrast to attacking pathogenic microorganisms. Milk-derived peptides as well as HDP and IDRs can directly or indirectly support the recruitment of immune cells (neutrophils and monocytes) by suppressing their production of proinflammatory cytokines (tumor necrosis factor alpha (TNF-α), interleukin 6 (IL-6), and IL-1β) as well as increasing the level of anti-inflammatory cytokines (IL-10). It may trigger the synthesis of chemokines (CCL-2/ MCP1), stimulate the activation, distinguishing of the macrophages and dendritic cells. They can also affect adaptive immunity by recruitment of T-cells. Moreover, HDPs/IDRs might control cell-specific processes such as generation and autophagy of neutrophil-derived extracellular traps (Cai et al., 2020) (Fig. 32.3).

32.4 Sources of some immunomodulatory peptides

Val-Glu-Pro-Ile-Pro-Tyr fraction (54–59 sequences of peptide) isolated from tryptic hydrolyzed β-caseins was the first to exhibit immune-stimulating activities, (Parker et al., 1984). Since then, many additional peptides have been discovered, including peptides derived through bovine beta-casein such as fraction 63–68 and fraction 191–193 along with single peptides derived from bovine α_{s1}-casein, which is fraction 194–199 (Migliore-Samour and Jolles, 1988).

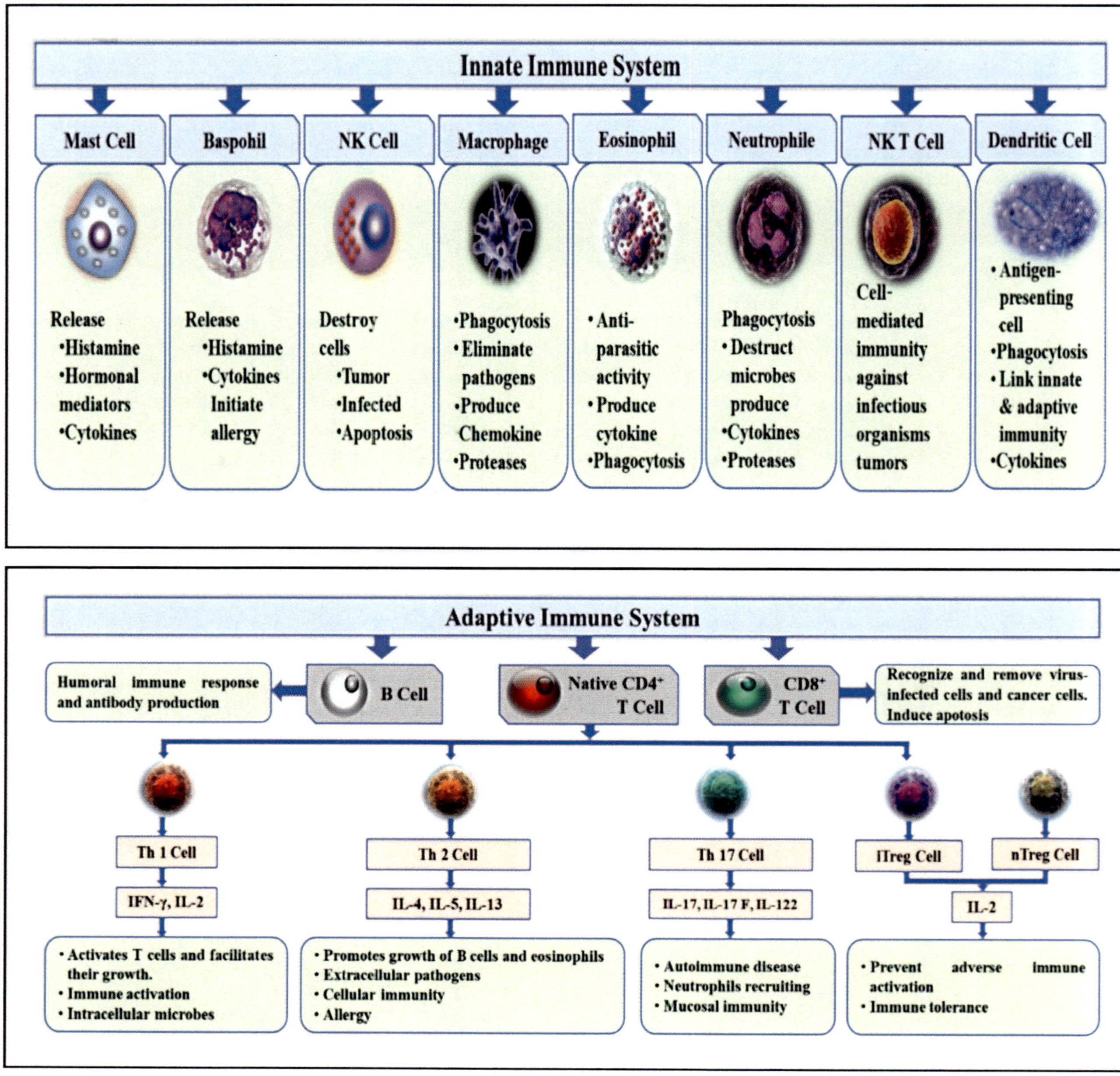

FIGURE 32.1 Innate (nonspecific) and adaptive (specific or acquired) immunological activities of cells (Jantan et al., 2015).

32.4.1 Milk and milk products

Milk is a rich source of energy and contains a range of important biomolecules, such lactose, lipids, vitamins, minerals, and proteins. It is one of the richest sources of bioactive peptides. Owing to the availability of a diverse variety of nutrients, milk is frequently referred to as a "nutritious diet" (Pereira, 2014). Despite allegations that milk lipids are harmful to the cardiovascular system, milk is a rich source of nutritional benefits. Milk and other dairy products have been investigated as potential sources of bioactive peptides. Apart from nutritional benefits, milk has crucial characteristics for immune system development (Givens, 2018).

The number of immunomodulated peptides with action on particular types of cells in the immune system were discovered in bovine milk. A range of milk constituents, namely, alpha, beta, and kappa casein, as well as whey protein along with lactoferrin have been demonstrated to influence lymphocyte proliferation (Wong et al., 1996). Furthermore, lactic acid fermented beta-casein has an immunomodulating effect on the mononuclear and T supporter cells, primarily Th1-cells (Laffineur et al., 1996). Synthesis peptides consisting of bovine beta-casomorphin-7, beta-casomorphine-10,

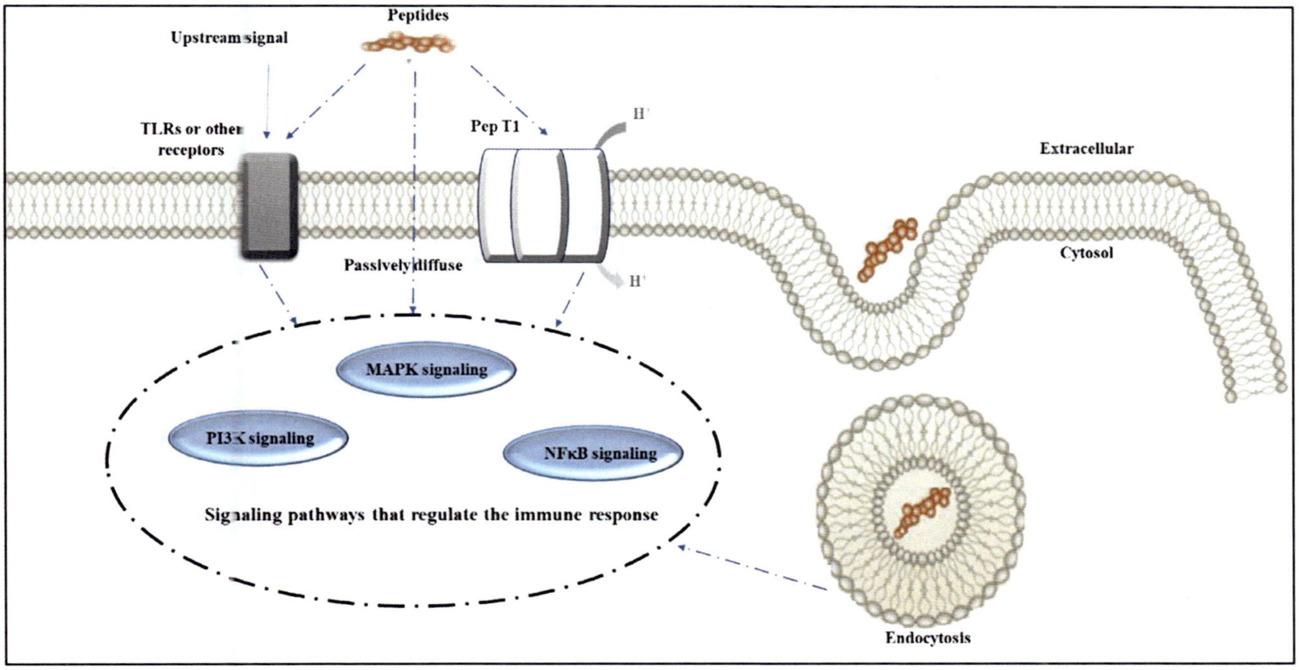

FIGURE 32.2 Mechanisms of immunomodulatory peptides (Cai et al., 2020).

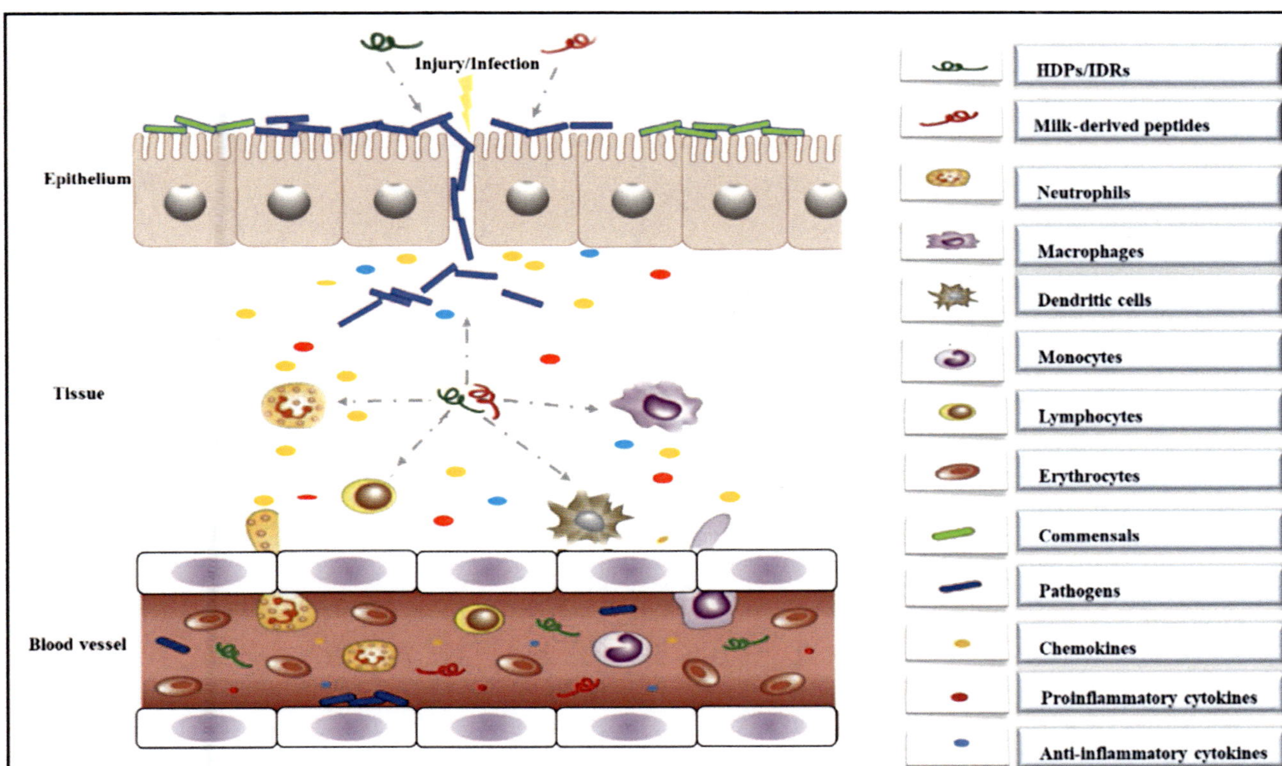

FIGURE 32.3 Immunomodulation activity of host defense peptides and innate defense regulator peptides along with peptides derived from milk (Cai et al., 2020).

and kappa-casein fragments significantly increased the dose-dependent expansion on human lymphocytic cells (Kayser and Meisel, 1996).

Bioactive peptides in yogurt have decreased tumor cell growth, which might partly explain how yogurt intake is linked to a diminished rate of colon cancer. Peptides generated by lactic bacteria during milk fermentation regulate the growth of human lymphocytes through modulation of cytokine production and enhance macrophage phagocytic activity (Matar et al., 2003) (Table 32.1). Immunomodulatory milk peptides have also been shown to diminish atopic allergy reactions in humans and boost the immunity of the gastrointestinal tract mucosa (Korhonen and Pihlanto, 2003).

Masotti et al. (2011) studied the potential impact of soy and milk immunomodulation activity in association with *Streptococcus thermophilus* ST5 and *Bifidobacterium longum* R0175 or *Lactobacillus helveticus* R0052 on the human monocyte system (U937). The combined impact from soy and milk on transforming growth factor beta-1 cell generation was studied, either in the monocyte phase or in the macrophage, via U937 cells confronted with proinflammatory cytokine TNF-α. Li and Mine (2004) showed that the immunoenhancing effects upon this cell's proliferative reaction on human macrophage cells such as U937 were similarly noted for bovine glycomacropeptide (GMP) and its derivatives. GMP has been discovered to be an effective low level immune enhancer that meaningfully improves the proliferation

TABLE 32.1 Immunomodulating impact of peptides derived from milk protein (Maestri et al., 2016).

Protein	Enzyme	Amino acid sequences	Immunological impact	Reference
α-Lactalbumin	Trypsin	GLF LLY	Increase phagocytic activity, increase lymphocyte proliferation	Kayser and Meisel (1996)
Lactoferrin	Pepsin	FKCRRWQWR NKKLGAPSITCVRRAF	Increase phagocytic activity, increase proliferation of splenocytes	Miyauchi et al. (1997)
α_{s1}-Casein	Chymosin	RPKHPIKHQGLP QEVLNENLLRF	Increase phagocytic activity	Lahov and Regelson (1996)
α_{s1}-Casein	Trypsin	QMEAESISSSEE IVPNSVEQK	Increase immunoglobulins production (IgG, IgM, IgA)	Hata et al. (1998)
β-Casein	Trypsin	VEPIPY	Increase phagocytic activity	Parker et al. (1984)
β-Casein	Not reported	RELEELNVPGEIVE SLSSSEESITRINK	Increase growth of monocytes, T cells, and B cells	Kawahara and Otani (2004)
β-Casein	Not reported	Casein phosphopeptide (1–28)	Increase generation of Ig-A cells	Kitamura and Otani (2002)
κ-Casein	Chymosin	YG	Increase lymphocytic effects	Kayser and Meisel (1996)
κ-Casein	Chymosin	Glycomacropeptides (106–169)	Increase phagocytosis effects Increase articulation of tumor necrosis factor, interleukin-1β, and interleukin -8	Requena et al. (2009)
Lactoferrin	Pepsin	FKCRRWQWR NKKLGAPSI TCVRRAF	Increase phagocytosis effects	Miyauchi et al. (1997)

Source: Reyes-Diaz, A., González-Córdova, A.F., Hernández-Mendoza, A., Reyes-Díaz, R. and Vallejo-Cordoba, B., 2018. Immunomodulation by hydrolysates and peptides derived from milk proteins. Int. J. Dairy. Technol. 71 (1), 1–9.

and phagocytic activity of U937 cells. The GMP dosage response was shown to be an increased effect when dose given from 1 to maximal 10 μg/mL for the proliferation of U947 cells, however activity reduced quickly to 100 μg/mL, indicating the GMP seem to be a greater immune enhancement when its in vitro concentration becomes very low. Kayser and Meisel (1996) observed a much higher proliferative effects on peripheral human blood cells in vivo in Tyr-Gly and Tyr-Gly-Gly derived from bovine kappa-casein and alpha-lactalbumin correspondingly, for example, certain peptides have been used to suppress the occurrence of acute illnesses among individuals with pre-AIDS HIV in order to treat human immune deficiency virus infections (Hadden, 1991).

When beta-casein is hydrolyzed, Pro-Gly-Pro-Ile-Pro-Asn and Leu-Leu-Tyr peptide sequences are formed. It has been found that both peptides boost the production of antibodies and increase in vitro phagocytic effects through mice peritoneal macrophages, while Leu-Leu-Tyr also stimulates T cell proliferation based on antigen (Migliore-Samour and Jolles, 1988). Antimicrobial and immunomodulatory actions are integrated into lactoferrin, an iron-binding glycoprotein. Lactoferrin and its derivatives also have an influence on the cytokine production of the body in inflammatory and immunological processes (El-Salam and El-Shibiny, 2013). Lactoferrin-produced peptides have also exhibited immunomodulation by reducing the generation of granulopoiesis and antibodies and by controlling NK cells (Crouch et al., 1992). Furthermore, ACE inhibitor peptides have been involved in immune system activation, since ACE inhibition raises the concentrations of bradykinin, which then stimulates macrophages and improves lymphocyte motility and lymphokine production (Paegelow and Werner, 1986).

During milk fermentation by lactic acid bacteria, casein-derived peptides are generated. Owing to their immune cell actions, these peptides are becoming particularly relevant for the food and food-processing industries. The immunomodulatory peptides were demonstrated to influence human lymphocytes proliferation, to increase macrophage phagocytic activity, and to decrease control of the regulation over specific cytokines (Matar et al., 2003). The immunomodulation action of beta-casomorphin-7 has been induced by the spreading with human colonic lamina propria lymphocytic cells in which the antiproliferative impact of doses of micromolar is altered via opiate receptor naloxone opponent (Elitsur and Luk, 1991). Glutamine peptides could be used as a free amino acid "glutamine" when glutamine has been advisable to lymphocytic growth, and immunocompetent cells are used for immunomodulatory action at high rates (Calder et al., 1994). Likewise, beta-casokinine-10 and beta-casomorphine-7 (various beta-casein segments) cause a proliferative response to rat lymphocytes, depending upon the peptide concentration, with suppression and stimulation (Meisel and Bockelmann, 1999).

32.4.1.1 *Cheese*

Four pressed curd cheeses have been explored for their immunomodulation potential for human T cells in vitro. Optimized 5-bromo-20-deoxyuridine and 3-(4.5-dimethylthiazol-2-yl)-2,5-diphenyl tetrazolium bromide have been assessed for T lymphocyte proliferation and metabolic activities. The existence of immunomodulating substances in some of these cheeses was shown by an approximately 4.6 times augmentation of T lymphocyte cells along with metabolism via water-soluble extracts from Abondance and Tomme de Savoie at lag stage. Since many milk protection peptides show an immunomodulator function, the nitrogen fractions of the cheese extract seem to have been identified through a particular focus on peptides (Durrieu et al., 2006).

Apart from the bacterial cell components, antibacterial elements such as peptides might have the immunomodulatory effects of cheese. Peptides in a lyophilized extract from gouda cheese inhibited the growth of cultured human peripheral blood lymphocytes in vitro and caused apoptosis in human promyelocytic leukemia cells (IL-60) (Meisel and Gunther, 1998).

32.4.2 Fish

Yang et al. (2009) reported that marine oligopeptide obtained from chum salmon (*Oncorhynchus keta*) in a preparation with the enzyme hydrolysis method could increase mitogen concanavalin A–induced lymphocyte proliferation, plaque-forming cells, activity of NK cells, $CD4^+$ T helping percentages of spleen cells, and the production of Th1 (IL-2, interferon gamma (IFN-γ)) as well as Th2 (IL-5, IL-6) type cytokines. Duarte et al. (2006) have shown that the amount of IGA^+ cells in the small intestinal lamina propria from mice and the number of S-IgA cells inside the intestinal lumen are augmented by the oral administration of fish protein concentrate (FPC) from a fermentation process. In the lamina propria from the small intestine in animals that were collected from FPC, IL-4, IL-6, and IL-10 were all considerably elevated. Some inflammatory cytokines, such as IFN-γ and TNF-α, also spiked simultaneously.

Chalamaiah et al. (2014) found that the immunological mechanism of mice has been improved via stimulation of NK cells, macrophages, T cell subclass ($CD4^+$ & $CD8^+$), and mucosal immune system (S-IgA) from rohu egg protein hydrolysates (REPHs). The analysis reveals that REPHs contain strong, nonallergic immune-enhancing peptides. He et al. (2015) discovered that peptide fragments isolated from the enzymatic hydrolysis of *Paphia undulata* (shellfish) after two-step isolation consisting of peptide sequences such as Pro-His-Thr-Cys, Val-Gly-Try-Thr, Glu-Phe, Leu-Phe, and Glu-Gly-Ala-Lys, Trp-Ile, or Trp-Leu, might greatly improve lymphocyte proliferation activity. Hou et al. (2012) produced hydrolysates for the immunomodulatory peptides of Alaska pollock frame. Three immunomodulating peptides were been produced, and the peptide chains were Asn-Gly-Met-Thr-Tyr, Asn-Gly-Leu-Ala-Pro, and Trp-Thr. Pollock hydrolysate has been selected as the potential source of immunostimulatory peptides containing significant lymphocytosis proliferative properties. Mallet et al. (2014) studied the breakdown of shark proteins by enzymes, resulting in shark protein hydrolysate peptides, which promoted the functioning of the intestinal barrier by improving immunoglobulin A cell production and by the generation of the intestinal cytokines IL-6 and TNF-α. The TGF-β and IL-10 increases help to accomplish uncontrolled inflammatory responses generated by *Escherichia coli* infection.

32.4.3 Marine organisms

Marine organisms have been shown to provide a great supply from bioactive elements along with immunomodulating peptides (Harnedy and Fitz-Gerald, 2011). Morris et al. (2007) investigated the impact on undernourished mice of green microalgae enzymatic protein hydrolysate *Chlorella vulgaris* (Cv-PH). The increase in peripheral blood carbon clearance suggested stimulation of the mononuclear phagocytosis process. Furthermore, Cv-PH positively promoted cell-moderated immunological activities, such as a direct reaction from T-dependent antibody as well as the reconstitution from delayed-type hypersensitivity response.

32.4.4 Eggs

He et al. (2021) isolated ovalbumin immunomodulatory peptides from white duck egg. The hydrolysates catalyzed by five proteases with the greatest levels of hydrolyzes and most immunomodulative action was seen in papain hydrolysate. Using TLR-2 and TLR-4, papain hydrolysate may greatly improve phagocytic activity and stimulate nitric oxide, tumor necrosis factor alpha, and IL-6 formation of RAW 264.7 cells. The researchers identified nine different immunomodulatory peptides: FDKLPGF, ISSTVSLKM, ISFQTAADQAR, WTSSTMMEER, WVESQTNGIIK, KNILQPSSVDSQ, IDKVVHFDKLPG, TQIDKVVHFDKLPGFG, and SEKMKILELPFASGMM. Ova transferrin is one such egg white protein immunomodulatory peptide that has proven effects in granulocytes and macrophages (Sanchez and Vazquez, 2017).

32.4.5 Plants

The major importance of bioactive peptides derived from plants contained in food is its potential pharmacological and/or nutraceutical advantage (Pang et al., 2012). This is because the peptides were either purposely added to the product or have emerged as enriched at some stage in its processing (Hayes and Tiwari, 2015). A few of the bioactive peptides in plant-based foods are protein hydrolases, which are pharmaceutically inactive in their natural state and are frequently referred to as "cryptides." They operate either as a defense against pathogens or as signaling molecules (Marmiroli and Maestri, 2014) (Table 32.2).

Mercier et al. (2004) identified the sequences of immunomodulatory peptides His-Cys-Gln-Arg-Pro-Arg, and Gln-Arg-Pro-Arg derived from soy protein hydrolysates showing a stimulating phagocytic action. For stimulation of immunomodulatory action, peptides having low molecular weight and positive charge may be important. Lymphocyte proliferation was induced at a significantly lower concentration by positive charge of peptides extracted from whey protein hydrolysate than the total hydrolysates. Kong et al. (2008) discovered that the greatest immunomodulatory effect was the augmentation of mouse spleen lymphocytic cells as well as the phagocytosis impact of peritoneal macrophages from hydrolysates of soy protein, made from insoluble soy protein with alcalase. The peptide generated from soybean soymetide is through beta-conglycinin. They come into contact with macrophage and neutrophil surface receptors, which emit the signal of the immune system similarly obtained by a bacteriological disease. Several soybean varieties have contrasting soymetide compositions (Guijarro-Díez et al., 2013).

The 43-residue peptide lunasin, which is derived from soybean, has a number of beneficial immunomodulatory effects (Singh et al., 2014), and equivalent compounds have been discovered in the top smaller granules of cereals

TABLE 32.2 In vitro studies of plant-derived immunomodulatory peptides.

Source	Plant protein	Peptide sequence	Experimental methods	Reference
Soybean (*Glycine max*)	Soy protein, alcalase hydrolysates	WGAPSL	Immunomodulatory activity	Zhong et al. (2007)
	Peptidase R enzymes for protein digestion	GGRKQGQHQQEE	Immunomodulatory activity	
	Trypsin digest of β-conglycinin α' subunit (soymetide)	Soymetide-13: MITLAIPVNKPGR	Stimulation of phagocytosis	Kuba et al. (2005)
		Soymetide 9: MITLAIPVN Soymetide-4: MITL	Enhancement of phagocytic activity in vitro; most active in stimulating phagocytosis	Yoshikawa et al. (2000)
	Pepsin hydrolysate	Val-Glu-Pro-Ile-Pro-Tyr and Gly-Leu-Phe	Phagocytosis of macrophages	Chen et al. (1995)
Rice	Albumin digestion by tryptic	GYPMYPLPR	Immunomodulation effects	Takahashi et al. (1994)
Rice protein hydrolysates	Trypsin hydrolysis	YGIYPR	Macrophage proliferation	Xu et al. (2019)
Selenium-enriched rice protein hydrolysates	Enzymatic hydrolysis	SeMDPGQQ TSeMMM	Immunostimulatory effects	Fang et al. (2019)
Wheat (*Triticum aestivum* L.)	α-Gliadin	PPYCTIVPFGIFGTN YR	Immunomodulatory peptides	Lammers et al. (2011)
	Alcalase hydrolysate	ECFSTA	Immunostimulatory activity	Wu et al. (2017)
Corn (zein protein hydrolysate)	Thermolysin hydrolysis	FLPFNQL PFNQL SQLALTNPT PFNQLAG FLPPVT	Production of interleukin 6 in human macrophage-like U937 cells	Liu et al. (2020)

Source: Maestri, E., Marmiroli, M., Marmiroli, N., 2016. Bioactive peptides in plant-derived foodstuffs. J. Proteom. 147, 140–155.

(Ortiz-Martinez et al., 2014). Rice and soybean immunomodulatory peptides promote the generation of ROS, which aids the host's protection against various environmental factors (Wang and De Mejia, 2005) (Table 32.2).

Chickpeas are a primary staple food that is an integral part of the diets in certain Mediterranean nations such as Central America, South America, and India. Chickpeas are an excellent source of lysine and thus become a good alternative in treating lysine deficits. Dominguez-Vega et al. (2011) discovered certain immunomodulating activities from chickpea peptides produced by microbiological proteases that have a particular binding site on human blood phagocytic cells and encouraged rate of microbial killing via phagocytic activity. Met-Ile-Thr-Leu-Ala-Ile-Phe-Val-Asn-Lys-Phe-Gly-Arg was described as the bioactive peptide sequence. Similar research described by Girón-Calle et al. (2008) on the hydrolysis of chickpeas using microbial proteases led to the stimulation of the growth of human monocyte THP-1 cells, which are derived from chickpea peptides. Following that, the authors examined the effect on the THP-1 and Caco-2 cells growth from pepsin and pancreatin hydrolysates (Giron-Calle et al., 2010). An immunomodulatory peptide that prevents cancer chemotherapy–induced alopecia are extracted from soybean protein and is digested by enzymes (Tsuruki and Yoshikawa, 2004).

32.4.5.1 Rice

Xu et al. (2019) have found peptides having immunomodulatory effects in rice protein hydrolysate peptides (RPHs). Enzymatic hydrolysis has been diagnosed with trypsin. A high-performance liquid chromatography electrospray-ionization quadrupole time-of-flight mass spectrometry revealed the sequence of peptide as Tyr-Gly-Ile-Tyr-Pro-Arg (YGIYPR). The assessment of its immunomodulation effects has shown by YGIYPR has increased macrophage RAW 264.7 cells over 12.5–100 μg/mL and a significant proliferation is accomplished even with a lowest dosage. Fang et al. (2019) used for screening the immunostimulatory selenium comprising peptides derived from rice protein hydrolysates supplemented with selenium (SPHs) in RAW 264.7 cell. TSeMMM peptides were shown to have good immunomodulatory action at a concentration of 80 g/mL, out of two chemically synthesized peptides SeMDPGQQ and TSeMMM, respectively. The immunomodulating activity of SPHs is found to correlate with the sequence of SeMet (selenomethionin) in the structure of peptides containing selenium. Additionally, TSeMMM showed potential for promoting human health through a greater immunomodulatory activity as a functional food additive. The Gly-Tyr-Pro-Met-Tyr-Pro-Pro-Pro-Arg sequence demonstrated resistant towards digestion by pancreas and pepsin. These amino acid sequence have shown substantial action that promotes phagocytosis activity (Takahashi et al., 1994).

32.4.5.2 Peas

Ndiaye et al. (2012) demonstrated the need for intestinal mucosal immunostimulation in vivo of pea-protein hydrolysates by using thermolysin as a key enzyme. Enhanced IL-6 intestinal epithelial cell (IEC) production contributed to understanding the enhancement of the IgA^+ cell number observed in the intestinal lamina propria in this investigation. It is obvious that PPHs can provide IECs with immunological signals to regulate the production of IL-6 in such a way that B cells can be differentiated into plasmocytes without achieving inflammatory levels that might cause harm to the intestines.

32.4.5.3 Wheat

Wu et al. (2017) isolated the innovative immunostimulatory amino acid sequence from wheat germ globulin, which is stated as Glu-Cys-Phe-Ser-Thr-Ala (ECFSTA). Peptides have already been shown to have excellent immunomodulatory activity, and the results suggest that ECFSTA may improve the phagocytic effects for RAW 264.7 cells together with boost greatly of its nitric oxide, IL-6, TNF-α and ROS. ECFSTA stimulates predominantly macrophages by using PRRs from TLR-2 and TLR-4 and at the same time boosting macrophages for TNF-α to be produced by ROS generation. Gluten is a protein that contains over 80% of its total proteins in wheat and other cereal flours. In its sequence of amino acids, gluten contains glutamine, which is essential in the gastrointestinal and immune system cell cycle process. Gluten-derived peptides might become reservoirs of immunostimulants and cause NK cells to be activated in people (Horiguchi et al., 2005).

32.4.5.4 Corn (zein protein)

Liu et al. (2020) found zein protein hydrolysates with immunomodulatory peptides by thermolysin digestion. The zein protein hydrolysate–generated peptides have been shown to produce substantial outcomes in decreasing IL-6 in human macrophages, for example, U937 cells also used as the marker. The immunomodulatory peptides that are produced from zein protein hydrolysate are FLPFNQL, PFNQL, SQLALTNPT, PFNQLAG, and FLPPVT. Vo et al. (2013) identified bioactive peptides sequenced LDAVNR (686 Da) and MMLDF (655 Da) from plants. Both low–molecular-weight peptides were found to have good hydrophobicity and to display excellent immunomodulatory activity. The peptides P1 and P2 found from *Spirulina maxima* were able to prevent the secretion and synthesis of histamine from mast cells and endothelial IL-8. These peptides have facilitated the inhibitory activity through the abolition of intracellular ROS generation in cells.

32.5 Bioavailability of bioactive peptides in foods

A major objective in determining a link between in vitro and in vivo activities is the bioavailability of bioactive peptides. The probability of some bioactive peptide generated through fermenting intervention depends upon its peptides' potential to access a relevant target site. Thus peptides might have to be impervious to additional breakdown in the digestive tract by proteolytic and peptidolytic enzymes. Afterwards, peptides must be ingested and reach the systemic circulation. One of the major variables in the bioavailability of bioactive peptides is their susceptibility to hydrolysis (Hernandez-Ledesma et al., 2004).

In addition to biological activity, recent research into peptides has focused on their physiological stability and intestinal absorption in order to fully understand the bioaccessibility and bioavailability of biologically active peptides (Matsui, 2018). Active transcellular transport or passive methods may be used for the transportation of peptides. Even though significant amino acid absorption takes place on the apical side of enterocytes in the form of dipeptides and tripeptides, the release of integral peptides into the general circulation via basolateral membrane seems insignificant (Daniel, 2004). In vitro experiments on the intestinal absorption of peptides and simulated intestinal epithelium were carried out by using a monolayer of intestinal cell lines, and peptides and its derivatives were subsequently analyzed in animal and clinical research in blood samples (Pihlanto, 2013).

Determining the number and type of peptides found in feeding hydrolysates from a certain protein involves a degree of uncertainty. Basing studies on a purified peptide may seem straightforward, but the gastrointestinal tract may also handle such molecules. A novel experimental technique is provided through reproduction of the human gastrointestinal system (Vu et al., 2014). During experiments, bacteria originated from the human bowel have been subjected to plant proteins to examine whether or not the intestinal flora may lead to formation of bioactive peptides, in either in the presence or absence of probiotic supplements (Pang et al., 2012). The mechanisms of absorption, distribution, metabolism, and excretion (ADME) continue to be unclear, which inhibits substitution of synthetic substances as pharmaceuticals by peptides in their natural form (Li-Chan, 2015). Research into the fate of dietary antigens can help to clarify certain elements of ADME in dietary peptides (Menard et al., 2010).

32.5.1 Absorption

During the digestive processes in the gastrointestinal system, bioactive peptides are sometime produced through dietary ingredients. For other circumstances in which bioactive peptides are collected in foods prior to consumption, their bioactivity can be produced only on the basis of aging or fermenting and processing if they stay unmodified through the gastrointestinal tract. In concept, short peptides are least likely to be degraded by the enzymes (e.g., proteases, peptidases) that are found in the gastrointestinal tract and separated through the pancreas and confined inside the brush border membrane of the gut. In stomach short peptides are present, especially those that carry a smaller peptide (proline-proline) on their C end (Panchaud et al., 2012). Larger peptides (six or more residues having mass more than 700 Da) have bioactivities without penetrating an intestinal epithelial cell, which would be significant for gastrointestinal function and food allergic peptides (Menard et al., 2010).

The 33mer α_2 and 26mer γ-gliadin are very resistant to protease disintegration and have significant glutamine and proline content throughout the situation with immunomodulatory peptides produced by degradation in cereal gluten. Very short peptides (having two to six residues) seem to be carried through the intestinal epithelium more easily than unbound amino acids (Brandsch et al., 2008). Renukuntla et al. (2013) determined that paracellular and transcellular endocytosis as well as transportation through plasma membrane are potential input pathways accessible. PepT1 is an H^+/peptide cotransporter, which is located in an apical duodenum and jejunum enterocyte membrane that is able to transport dipeptides and tripeptides in a manner that depends upon the membrane potential and pH. PepT1 seems to transport all probable dipeptides and tripeptides derived from protein digestion and free amino acids (Gilbert et al., 2008); however, it prefers peptides with large hydrophobic surfaces (Saaby et al., 2013). Dipeptide (e.g., valine-tyrosine) absorption happens quickly upon absorption (Matsui et al., 2002). Enterocytes have strong peptidase action, and peptides that become resilient with proteolysis cross inside the transcellular mode by intestine epithelium using the peptide transporter system, which is located in the basolateral membrane (Brandsch et al., 2008). Endocytosis is capable of the M (microfold) cells that cover the Peyer's patches, which permits the peptides to be delivered to the underlying lymphatic system related with tissue. Hence small peptides may reach the bloodstream via the thoracic duct via the lymphatic system (Wasan, 2002) and therefore bypass the liver. The colon can potentially represent an additional site of absorption, as its activity of peptidase is free and has a lengthy residence period (Renukuntla et al., 2013).

Monolayers containing cultivated Caco-2 cells obtained via a human colorectal carcinoma are used as imitations of human enterocytes from the small intestines to assess the absorption and transport of peptides. Vij et al. (2015) have shown in a recent application that the cell was able to transport a seven-residue (VLPVPQK) peptide within minutes, but at the same time it was hydrolyzed at a cell's apical area, since the VLPVPQK peptide was retrieved on the basolateral side. The paracellular and transcellular pathways have also been proven not to be involved in their transport, opening up the opportunity of using oligopeptide transporters. A full description of different methods in vitro as well as in vivo for investigation of peptide bowel absorption is presented by Antunes et al. (2013).

32.5.2 Distribution

Owing to their low concentrations (picograms per milliliter), identifying peptides within plasma seems technologically complex. When study participants were given peptide-supplemented beverages, peptides with five to six residues could be found in the blood (Van Platerink et al., 2006). In plasma, as with immunomodulatory lunasin following soybean ingestion, larger peptides were intact (Dia et al., 2009). Animal models were utilized to show that this 43-residue peptide may access various tissues to pass through the blood-brain barrier (Hsieh et al., 2010). The tripeptide may be recognized in picomolar quantities within 15 minutes of the intake of an IPP tripeptide-enhanced yogurt drink (Foltz et al., 2007). Its concentration returns after 2 hours to baseline level, demonstrating that tiny oligopeptides may be delivered to the blood without destruction from the gastrointestinal system. Additional dipeptides and tripeptides were found in plasma in modest amounts, independent of the dietary intake, thereby showing the metabolic processing of the proteins. The VY dipeptide blood levels likewise increased according to the dose after an enriched drink was taken (Matsui et al., 2002). The dipeptide (VY) half-life the has been estimated as being greater than that of the tripeptide (IPP) half-life.

The simultaneous supplement to a meal significantly improves the half-life, raising the potential of a contribution to peptidase inhibitors in food (Foltz et al., 2007). The allergic reactivity to some IgE-mediated food proteins demonstrates that dietary peptides can enter the blood. However, both endothelium and plasma lead in peptides' half-life estimated in minutes, owing to peptidase activity. As a result, peptides brought into the body are believed to be of limited biological effectiveness (Foltz et al., 2010). Lunasin is highly bioavailable, since protease inhibitors have simultaneously existed such that 30% of the peptide may reach the target tissues (Hsieh et al., 2010).

32.5.3 Metabolism

Even though precise data are not available, it is widely considered that peptides are catabolized and can be employed as building blocks for protein synthesis as a result of free amino acids. Researchers have previously observed that the brush border, enterocyte, plasma, endothelium, and target cells are present with peptidases. Peptides could be absorbed and metabolized in the lysosome by endocytosis. For immunogenic gluten peptides a distinct metabolic pathway occurs because the residues of glutamines are either deaminated or transamidated and then converted into glutamic acid, a step that is needed to bind determinants to HLA-DQ2 or DQ8 as well as for the buildup of peptides inside the lamina propria, including its striated-border epithelium, accompanied by antibody identification (Sollid, 2002).

32.5.4 Elimination

Peptides existing in the blood may have been detected in urine (Sigdel et al., 2014), and commercially developed peptide medicines are known to accumulate at relatively high levels in the liver and kidneys. Peptides can be filtered into glomerular zone by shape and size (mass $<$ 25 kDa) and reabsorbed through renal tubules once they are present in the filtrate, as with endogenous peptides such as interleukins, and some are hydrolyzed in the tubular epithelium (Shi and Derendorf, 2010). Small peptides are carried into renal epithelial cells by PepT2 and PepT1; subsequently the peptides are hydrolyzed in the cytoplasm (Brandsch et al., 2008).

32.6 Methods used to assay immunomodulatory activity

Peptides and other substances may be evaluated by several approaches for their immunomodulatory effects, emphasizing certain types of cell and activities. A large series of studies supports the nonspecific effect on immunostimulatory peptides, addressing inhibition as well as activation of the activities of the immune system (Becker et al., 2014). The majority of experiments use human volunteers and animal models for investigation of basic activities for immune cells, including proliferation, phagocytic effects, differentiation, and cytokine generation. Table 32.3 summarizes various techniques that are employed for investigation of the immunomodulating activities of peptides. The exposure to cells as well as animals with recognized stimulants for immune cells with pretreatment of peptides and moderating consequences in treated samples is used to evaluate immunosuppressive and anti-inflammatory activities (Chalamaiah et al., 2014).

The phagocytic ability of treated mice is normally tested ex vivo, along with the group of peritoneum-derived macrophages and the suspension of heat killing of *Candida albicans* or *Saccharomyces cerevisiae* (Duarte et al., 2006). The macrophage phagocytic activity is assessed according to frequency of activated macrophages comprising at least one yeast cell (Chalamaiah et al., 2014). NK cells are accountable and vulnerable to immune function—interfering chemicals for nonspecific defense responses. Mouse splenocytes have been treated with sensitive cancer cells in the normal ex vivo test for NK activity (Yuan et al., 2006), and cancer cells' vitality has been assessed by using the MTT/formazan test. The activity level of NK cells has been determined by comparing test samples against controls (having an absence of splenocytes). The flow cytometry following interaction involving specific antibodies to surface antigens is measured by the quantity of macrophages, NK cells, plasma, T-helper cells, and T-cytotoxic lymphocytic cells (Lammers et al., 2011). The sampling from peripheral blood mononuclear cells is being characterized through T cells, monocytes, and B cells, using antibodies attached to magnetic beads. ELISA may also be used to test mouse sera with Ig-E, Ig-A, Ig-M, and Ig-G and the sample cells that secrete IgA from the small bowel. Multiple tests on peptides having

TABLE 32.3 Various in vitro and in vivo assay methods to study immunomodulating effects from peptides.

Endpoint	Analytical methodology	Cells to be targeted	Outcomes	References
Proliferation of immune cells	WST:1 test, Spectrophotometry reading to check absorbance of formazan, sulforhodamine B method	Neutrophils, peripheral blood mononuclear, lymphocytes, macrophage cell line	Stimulation index for proliferation, increase or decrease of proliferation, interact to recognized activators	Kong et al. (2008)
Mixed lymphocyte reaction, proliferation of T lymphocytes cells	Liquid scintillation spectroscopy methods used for assessments of incorporated thymidine by tritiated assay	Dendritic cells	Peptides may modify dendritic cell response into heterologous immune cells	Giordani et al. (2014)
Phagocytosis activity	For neutral red absorbance taken spectrophotometer, triggered macrophages harboring at least one yeast cell using microscope, fluorescence of internalized cells	Peritoneal macrophages, macrophage cell lines, for example, RAW 264.7	Increase in phagocytic activity (immunostimulation)	Duarte et al. (2006), Kong et al. (2008), Ndiaye et al. (2012), Chalamaiah et al. (2014)
Natural killer cell activity	Absorbance of formazan with spectrophotometry reading	Activity of mouse spleen cell (splenocytes)	Natural killer cell activity being compared either cytotoxicity against target cells mostly with standards	Yuan et al. (2006), Chalamaiah et al. (2014)
Phenotype of immune cell (such as T cell population)	Flow cytometry, subsequent staining using certain antibodies for cell surface antigen	Blood donors' peripheral blood mononuclear cells, spleen cells from mouse	Changes the lymphocyte cells: • $CD4^+$ (T helper) • $CD8^+$ (T cytotoxic) • $CD11b^+$ (macrophages) • Cd49b* (NK cells)	Chalamaiah et al. (2014)
B cell population in intestine	Quantity of IgA^+ and IgG^+ cells recognized by immunofluorescence having specific antibody against IgA and IgG	B cells in intestine sections from mice	Increased immune monitoring of cells with in gut generating antibodies	Duarte et al. (2006), Ndiaye et al. (2012)
Upregulation on cytokine production activation, downregulation of immunosuppressor cytokine	Sandwich ELISA medium used to assessment of cytokine concentration	Cytokines • Macrophage • Neutrophil • Chemokines	Peptides could promote or suppress the secretion causes from immunostimulants	
Nitric oxide	Cell culture supernatant tested using Griess reagent, compared to a typical sodium-nitrite standard curve, Western blotting used for assessments of iNOS protein	Nitrite, indicative of nitric oxide generation, iNOS (inducible nitric oxide synthase)	Increase or decrease in nitric oxide production or in inducible nitric oxide synthase expression encouraged from LPS (anti-inflammatory)	Ndiaye et al. (2012), Cardeno et al. (2014)

Source: Maestri, E., Marmiroli, M., Marmiroli, N., 2016. Bioactive peptides in plant-derived foodstuffs. J. Proteom. 147, 140–155.

immunomodulatory activity are based upon the assessment of cytokine production and the inflammatory mediators that activate the immune response; since the immune response is released early, their quantification could convey signals regarding the effects of peptides' immunostimulation or immunosuppression.

The spectrophotometric measurements of nitrite levels generated by sulfanilamide and N-1-naphthylethylenediamine quantify the generation of NO, which is an essential part of the inflammatory action (Coneski and Schoenfisch, 2012). Inflammation modulation can be evaluated by monitoring enzyme activity involving inflammation (particularly phospholipase A2, cyclooxygenase, and thrombin) or by evaluating the amount of gene expression for the respective genes (Cardeno et al., 2014).

32.7 Clinical studies

32.7.1 Related to COVID-19

CIGB-258 (formerly APL1 or CIGB-814 and currently known as CIGB-258) is an immunoregulation peptide that is obtained from the human heat shock protein 60 (HSP60). HSP60-derived peptides can also stimulate the regulatory functioning of T cells. Therefore different research groups have given HSP 60 to study the role of a regulatory immune response (Van Eden et al., 2005). CIGB-258 enhances regulatory T cell frequency as well as their ability to inhibit $CD4^+$ T cells in an antigen reaction (Prada et al., 2018).

Hernandez-Cedeno et al. (2021) investigated, in a group of COVID-19 patients who were given CIGB-258 peptides, certain biomarkers related to hyperinflammation and several cytokines associated with a cytokine storm granzyme B as well as perforin. All severely sick patients got 1 or 2 mg of CIGB-258 intravenously every 12 hours under invasive mechanical ventilation. Extremely sick patients received 1 mg CIGB-258 every 12 hours and recovered; each patient was in an extreme situation with oxygen treatment. Biomarkers, such as IL-6, IL-10, TNF-α, and granzyme B along with perforin, were reduced substantially in levels of hyperinflammation throughout therapy.

32.7.2 Impact of probiotics strains on T cell response

The use of adjuvant characteristics of probiotics was described by pro-Th1 cytokines initiating as well as strengthening Th1 type immunological resection. *Lactococcus lactis*– and *Lactobacillus plantarum*–derived splenocytes promoted the generation of IL-12 and IFN-γ when lactic acid bacteria and allergens were intranasally coadministered to mice (Repa et al., 2003). *Lactobacillus fermentum* strain CECT5716 improved the Th1 influenza-generated response and increased antibody neutralization of the virus (Olivares et al., 2007).

Eleven distinct strains of probiotic were examined in human peripheral cells in the generation of cytokines and TNF-α, and some bacterial strains also resulted in IL-12 and IFN-γ generated by each test (Kekkonen et al., 2008). Earlier investigations in gnotobiotic pigs demonstrated that the action of both the Th1 (IL-12 and IFN-γ) and Th2 (IL-4 and IL-10) cytokines for virulent human rhinovirus (HRV) infection was improved by mixing of *Lactobacillus acidophilus* NCFM and *Lactobacillus reuteri* strains (Azevedo et al., 2012). The *L. acidophilus* NCFM strain improved by generating $CD8^+$ T cells in gnotobiotic pigs from the HRV-specific IFN-γ rotavirus vaccination, demonstrating *L. acidophilus* strain adjuvanticity (Zhang et al., 2008).

In animal trials by various research organizations, dose effects can explain some of the conflicts that arise from the same strain of probiotics. $CD4^+$ T cells, upregulated IL-10, and transforming growth factor beta concentration were inhibited in mice by the treatment of *Lactobacillus casei* proinflammatory cytokine expressions (So et al., 2008a,b). However additional research indicated that *L. casei* has been a pure enhancer of Th1 in mice. The dosage of *L. casei* for the various experiments varied greatly, in addition to varying species of animal, with considerably higher dosages employed throughout the study of mice (So et al., 2008a,b). The dose of *L. casei* that was administered was 5×10^9 or 2×10^{10} CFU/dose per mouse, three times per week for 11–12 weeks. In another study, the dose of *L. casei* was 2×10^8 CFU/dose per mouse twice per week for 8 weeks (Van Overtvelt et al., 2010).

32.8 Applications

With the advances in understanding of the physiological actions of bioactive peptides, there have been increasing industrial demands to utilize them as essential components of foodstuffs as well as drugs. Technological innovations involving such peptides are in progress for industrial manufacturing industries (Korhonen et al., 1998). Commercially available peptides are generated from fish proteins and are said to be advantageous for a number of medical concerns, including their impacts on the immune system. PeptiBal, an ultrafiltrated hydrolyzed fish protein hydrolysate, has been found to have immunomodulatory effects in human clinical studies. Randomized double-blind placebo-controlled research showed a statistically significant increase in IgAs production (Boutin et al., 2012).

Fitz-Gerald et al. (2005) proposed that the immunomodulating action of fish protein hydrolysates may have valuable therapeutic effects in the intestines, including its effectiveness in utilizing several models of epithelial and human colon. Seacure is a commercial supplement of human protein produced via fermentation by a specific yeast strain of fish protein. The immunomodulatory effects of such a fish powder concentrate, which was designed for human dietary supplements, were assessed by a murine model earlier used in the immunomodulating activity of, cheese (Medici et al., 2004), kefir (Vinderola et al., 2005), and other fermented milks (Matar et al., 2001). A caseinophosphopeptide CPP-III, which is available commercially and comprises primarily fraction 1–32 from bovine α_{s2}-casein as well as fraction 1–28 from beta-casein, stimulates proliferation reactions and generation of immunoglobulin in mouse spleen cell cultures. An o-phospho-l-serine residue was associated with this immunostimulating activity, Therefore the bioactivity in the intestinal tract is rather constant for proteinase action (Hata, 1999).

32.9 Conclusion

Bioactive peptides may be produced from dietary protein as global biomolecules. Thus the number of peptides that are accessible from a single dietary protein cannot be limited. Protein-rich nutrients are a component of our everyday diet, precursors of biologically effective peptides. In the presence of bacteria, such as lactic acid bacteria, the immunological modulatory ability of hydrolysates produced from milk proteins and the appropriate impacts on stimulation of immunological response offer benefits. These protein-generated foods might be used for the creation of immunomodulatory functional meals as novel materials. Food protein hydrolysates and peptides have been shown to have immunomodulatory properties, mainly in vitro (in cultured cell) and in animal studies. Therefore further clinical trials are necessary for greater understanding to determine the gastrointestinal sustainability, bioavailability, and security of such protein hydrolysates or peptides before they can be used as functional foods.

References

Abid, S., Khajuria, A., Parvaiz, Q., Sidiq, T., Bhatia, A., Singh, S., et al., 2012. Immunomodulatory studies of a bioactive fraction from the fruit of Prunus cerasus in BALB/c mice. Int. Immunopharmacol. 12 (4), 626–634.

Agyei, D., Danquah, M.K., 2012. Rethinking food-derived bioactive peptides for antimicrobial and immunomodulatory activities. Trends Food Sci. Technol. 23 (2), 62–69

Antunes, F., Andrade, F., Ferreira, D., Morck Nielsen, H., Sarmento, B., 2013. Models to predict intestinal absorption of therapeutic peptides and proteins. Curr. Drug Metab. 14 (1), 4–20.

Azevedo, M.S.P., Zhang, W., Wen, K., Gonzalez, A.M., Saif, L.J., Yousef, A.E., et al., 2012. Lactobacillus acidophilus and L. reuteri modulate cytokine responses in gnotobiotic pigs infected with human rotavirus. Benef. Microbes 3 (1), 33.

Baxter, D., 2007. Active and passive immunity, vaccine types, excipients and licensing. Occup. Med. 57 (8), 552–556.

Becker, K., Schroecksnadel, S., Gostner, J., Zaknun, C., Schennach, H., Überall, F., et al., 2014. Comparison of in vitro tests for antioxidant and immunomodulatory capacities of compounds. Phytomedicine 21 (2), 164–171.

Boutin, Y., Paradis, M.E., Couture, P., Lamarche, B., 2012. Immunological effect of fish protein supplementation on healthy adults. J. Nat. Prod. 5, 37–44.

Brandsch, M., Knütter, I., Bosse-Doenecke, E., 2008. Pharmaceutical and pharmacological importance of peptide tragonsporters. J. Pharm. Pharmacol. 60 (5), 543–585.

Cai, J., Li, X., Du, H., Jiang, C., Xu, S., Cao, Y., 2020. Immunomodulatory significance of natural peptides in mammalians: promising agents for medical application. Immunobiology 225 (3), 151936.

Calder, P.C., Sherrington, E.J., Askanazi, J., Newsholme, E.A., 1994. Inhibition of lymphocyte proliferation in vitro by two lipid emulsions with different fatty acid compositions. Clin. Nutr. 13 (2), 69–74.

Cardeno, A., Sanchez-Hidalgo, M., Aparicio-Soto, M., Alarcón-de-la-Lastra, C., 2014. Unsaponifiable fraction from extra virgin olive oil inhibits the inflammatory response in LPS-activated murine macrophages. Food Chem. 147, 117–123.

Cata, J.P., Wang, H., Gottumukkala, V., Reuben, J., Sessler, D.I., 2013. Inflammatory response, immunosuppression, and cancer recurrence after perioperative blood transfusions. Br. J. Anaesth. 110 (5), 690–701.

Chalamaiah, M., Hemalatha, R., Jyothirmayi, T., Diwan, P.V., Kumar, P.U., Nimgulkar, C., et al., 2014. Immunomodulatory effects of protein hydrolysates from rohu (Labeo rohita) egg (roe) in BALB/c mice. Food Res. Int. 62, 1054–1061.

Chalamaiah, M., Yu, W., Wu, J., 2018. Immunomodulatory and anticancer protein hydrolysates (peptides) from food proteins: a review. Food Chem. 245, 205–222.

Chen, J., Suetsuna, K., Yamauchi, F., 1995. Isolation and characterization of immunostimulative peptides from soybean. J. Nutr. Biochem. 6 (6), 310–313.

Clare, D.A., Swaisgood, H.E., 2000. Bioactive milk peptides: a prospectus. J. Dairy Sci. 83 (6), 1187–1195.

Coneski, P.N., Schoenfisch, M.H., 2012. Nitric oxide release: part III. Measurement and reporting. Chem. Soc. Rev. 41 (10), 3753–3758.

Crouch, S.P., Slater, K.J., Fletcher, J., 1992. Regulation of cytokine release from mononuclear cells by the iron-binding protein lactoferrin. Am. Soc. Hematol. 80, 235–240.

Dadar, M., Shahali, Y., Chakraborty, S., Prasad, M., Tahoori, F., Tiwari, R., et al., 2019. Antiinflammatory peptides: current knowledge and promising prospects. Inflamm. Res. 68 (2), 125–145.

Daniel, H., 2004. Molecular and integrative physiology of intestinal peptide transport. Annu. Rev. Physiol. 66, 361–384.

Dia, V.P., Torres, S., De Lumen, B.O., Erdman Jr, J.W., Gonzalez De Mejia, E., 2009. Presence of lunasin in plasma of men after soy protein consumption. J. Agric. Food Chem. 57 (4), 1260–1266.

Dominguez-Vega, E., Kotkowska, O., Garcia, M.C., Crego, A.L., Marina, M.L., 2011. Fast determination of the functional peptide soymetide in different soybean derived foods by capillary-high performance liquid chromatography. J. Chromatogr. A1218 (30), 4928–4933.

Duarte, J., Vinderola, G., Ritz, B., Perdigón, G., Matar, C., 2006. Immunomodulating capacity of commercial fish protein hydrolysate for diet supplementation. Immunobiology 211 (5), 341–350.

Durrieu, C., Degraeve, P., Chappaz, S., Martial-Gros, A., 2006. Immunomodulating effects of water-soluble extracts of traditional French Alps cheeses on a human T-lymphocyte cell line. Int. Dairy J. 16 (12), 1505–1514.

Elitsur, Y., Luk, G.D., 1991. Beta-casomorphin (BCM) and human colonic lamina propria lymphocyte proliferation. Clin. Exp. Immunol. 85 (3), 493–497.

El-Salam, M.A., El-Shibiny, S., 2013. Bioactive peptides of buffalo, camel, goat, sheep, mare, and yak milks and milk products. Food Rev. Int. 29 (1), 1–23.

Fang, Y., Pan, X., Zhao, E., Shi, Y., Shen, X., Wu, J., et al., 2019. Isolation and identification of immunomodulatory selenium-containing peptides from selenium-enriched rice protein hydrolysates. Food Chem. 275, 696–702.

Finlay, B.B., Hancock, R.E., 2004. Can innate immunity be enhanced to treat microbial infections? Nat. Rev. Microbiol. 2 (6), 497–504.

Fitz-Gerald, A.J., Rai, P.S., Marchbank, T., Taylor, G.W., Ghosh, S., Ritz, B.W., et al., 2005. Reparative properties of a commercial fish protein hydrolysate preparation. Gut 54 (6), 775–781.

Foltz, M., Meynen, E.E., Bianco, V., van Platerink, C., Koning, T.M., Kloek, J., 2007. Angiotensin converting enzyme inhibitory peptides from a lactotripeptide-enriched milk beverage are absorbed intact into the circulation. J. Nutr. 137 (4), 953–958.

Foltz, M., van der Pijl, P.C., Duchateau, G.S., 2010. Current in vitro testing of bioactive peptides is not valuable. J. Nutr. 140 (1), 117–118.

Gilbert, E.R., Wong, E.A., Webb Jr, K.E., 2008. Board-invited review: peptide absorption and utilization: implications for animal nutrition and health. J. Anim. Sci. 86 (9), 2135–2155.

Gill, H.S., Doull, F., Rutherfurd, K.J., Cross, M.L., 2000. Immunoregulatory peptides in bovine milk. Br. J. Nutr. 84 (S1), 111–117.

Giordani, L., Del Pinto, T., Vincentini, O., Felli, C., Silano, M., Viora, M., 2014. Two wheat decapeptides prevent gliadin-dependent maturation of human dendritic cells. Exp. Cell Res. 321 (2), 248–254.

Girón-Calle, J., Vioque, J., Pedroche, J., Alaiz, M., Yust, M.M., Megías, C., et al., 2008. Chickpea protein hydrolysate as a substitute for serum in cell culture. Cytotechnology 57 (3), 263–272.

Giron-Calle, J., Alaiz, M., Vioque, J., 2010. Effect of chickpea protein hydrolysates on cell proliferation and in vitro bioavailability. Food Res. Int. 43 (5), 1365–1370.

Givens, D.I., 2018. Milk and dairy foods: implications for cardiometabolic health. Cardiovasc. Endocrinol. Metab. 7 (3), 56.

Grivennikov, S.I., Karin, M., 2011. Inflammatory cytokines in cancer: tumour necrosis factor and interleukin 6 take the stage. Ann. Rheumatic Dis. 70 (Suppl 1), i104–i108.

Gronberg, A., Mahlapuu, M., Ståhle, M., Whately-Smith, C., Rollman, O., 2014. Treatment with LL-37 is safe and effective in enhancing healing of hard-to-heal venous leg ulcers: a randomized, placebo-controlled clinical trial. Wound Repair. Regen. 22 (5), 613–621.

Guijarro-Díez, M., García, M.C., Marina, M.L., Crego, A.L., 2013. LC-ESI-TOF MS method for the evaluation of the immunostimulating activity of soybeans via the determination of the functional peptide soymetide. J. Agric. Food Chem. 61 (15), 3611–3618.

Hadden, J.W., 1991. Immunotherapy of human immunodeficiency virus infection. Trends Pharmacol. Sci. 12, 107–111.

Haque, E., Chand, R., Kapila, S., 2008. Biofunctional properties of bioactive peptides of milk origin. Food Rev. Int. 25 (1), 28–43.

Harnedy, P.A., Fitz-Gerald, R.J., 2011. Bioactive proteins, peptides, and amino acids from macroalgae 1. J. Phycol. 47 (2), 218–232.

Hartmann, R., Meisel, H., 2007. Food-derived peptides with biological activity: from research to food applications. Curr. Opin. Biotechnol. 18 (2), 163–169.

Hata, I., 1999. Immunostimulatory action of a commercially available casein phosphopeptide preparation, CPP-III, in cell cultures. Milchwissenshaft 54, 3–7.

Hata, I., Higashiyama, S., Otani, H., 1998. Identification of a phosphopeptide in bovine αs1-casein digest as a factor influencing proliferation and immunoglobulin production in lymphocyte cultures. J. Dairy Res. 65 (4), 569–578.

Hayes, M., Tiwari, B.K., 2015. Bioactive carbohydrates and peptides in foods: an overview of sources, downstream processing steps and associated bioactivities. Int. J. Mol. Sci. 16 (9), 22485–22508.

He, X.Q., Cao, W.H., Pan, G.K., Yang, L., Zhang, C.H., 2015. Enzymatic hydrolysis optimization of Paphia undulata and lymphocyte proliferation activity of the isolated peptide fractions. J. Sci. Food Agric. 95 (7), 1544–1553.

He, P., Wang, Q., Zhan, Q., Pan, L., Xin, X., Wu, H., et al., 2021. Purification and characterization of immunomodulatory peptides from enzymatic hydrolysates of duck egg ovalbumin. Food Funct. 12 (2), 668–681.

Hernandez-Cedeno, M., Venegas-Rodriguez, R., Peña-Ruiz, R., Bequet-Romero, M., Santana-Sanchez, R., Penton-Arias, E., et al., 2021. CIGB-258, a peptide derived from human heat-shock protein 60, decreases hyperinflammation in COVID-19 patients. Cell Stress. Chaperones 26 (3), 515–525.

Hernandez-Ledesma, B., Amigo, L., Ramos, M., Recio, I., 2004. Application of high-performance liquid chromatography–tandem mass spectrometry to the identification of biologically active peptides produced by milk fermentation and simulated gastrointestinal digestion. J. Chromatogr. A 1049 (1–2), 107–114.

Hernandez-Ledesma, B., García-Nebot, M.J., Fernández-Tomé, S., Amigo, L., Recio, I., 2014. Dairy protein hydrolysates: peptides for health benefits. Int. Dairy. J. 38 (2), 82–100.

Horiguchi, N., Horiguchi, H., Suzuki, Y., 2005. Effect of wheat gluten hydrolysate on the immune system in healthy human subjects. Biosci. Biotechnol. Biochem. 69 (12), 2445–2449.

Horner, K., Drummond, E., Brennan, L., 2016. Bioavailability of milk protein-derived bioactive peptides: a glycaemic management perspective. Nutr. Res. Rev. 29 (1), 91–101.

Hou, H., Fan, Y., Li, B., Xue, C., Yu, G., Zhang, Z., et al., 2012. Purification and identification of immunomodulating peptides from enzymatic hydrolysates of Alaska pollock frame. Food Chem. 134 (2), 821–828.

Hsieh, C.C., Hernandez-Ledesma, B., Jeong, H.J., Park, J.H., Ben, O., 2010. Complementary roles in cancer prevention: protease inhibitor makes the cancer preventive peptide lunasin bioavailable. PLoS One 5 (1), e8890.

Jantan, I., Ahmad, W., Bukhari, S.N.A., 2015. Plant-derived immunomodulators: an insight on their preclinical evaluation and clinical trials. Front. Plant. Sci. 6, 655.

Kawahara, T., Otani, H., 2004. Stimulatory effects of casein phosphopeptide (CPP-III) on mRNA expression of cytokines in Caco-2 cells. Biosci. Biotechnol. Biochem. 68 (8), 1779–1781.

Kayser, H., Meisel, H., 1996. Stimulation of human peripheral blood lymphocytes by bioactive peptides derived from bovine milk proteins. FEBS Lett. 383 (1–2), 18–20.

Kekkonen, R.A., Kajasto, E., Miettinen, M., Veckman, V., Korpela, R., Julkunen, I., 2008. Probiotic Leuconostoc mesenteroides ssp. cremoris and Streptococcus thermophilus induce IL-12 and IFN-γ production. World J. Gastroenterol. (8), 1192.

Kitamura, H., Otani, H., 2002. Fecal IgA levels in healthy persons who ingested cakes with or without bovine casein phosphopeptides. Milchwissenshaft 57, 611–614.

Kitts, D.D., Weiler, K., 2003. Bioactive proteins and peptides from food sources. Applications of bioprocesses used in isolation and recovery. Curr. Pharm. Des. 9 (16), 1309–1323.

Kong, X., Guo, M., Hua, Y., Cao, D., Zhang, C., 2008. Enzymatic preparation of immunomodulating hydrolysates from soy proteins. Bioresour. Technol. 99 (18), 8873–8879.

Korhonen, H., Pihlanto, A., 2003. Food-derived bioactive peptides-opportunities for designing future foods. Curr. Pharm. Des. 9 (16), 1297–1308.

Korhonen, H., Pihlanto, A., 2006. Bioactive peptides: production and functionality. Int. Dairy J. 16 (9), 945–960.

Korhonen, H., Pihlanto-Leppala, A., Rantamaki, P., Tupasela, T., 1998. Impact of processing on bioactive proteins and peptides. Trends Food Sci. Technol. 9 (8–9), 307–319.

Kuba, M., Tana, C., Tawata, S., Yasuda, M., 2005. Production of angiotensin I-converting enzyme inhibitory peptides from soybean protein with Monascus purpureus acid proteinase. Process. Biochem. 40 (6), 2191–2196.

Laffineur, E., Genetet, N., Leonil, J., 1996. Immunomodulatory activity of β-casein permeates medium fermented by lactic acid bacteria. J. Dairy Sci. 79 (12), 2112–2120.

Lahov, E., Regelson, W., 1996. Antibacterial and immunostimulating casein-derived substances from milk: casecidin, isracidin peptides. Food Chem. Toxicol. 34 (1), 131–145.

Lammers, K.M., Khandelwal, S., Chaudhry, F., Kryszak, D., Puppa, E.L., Casolaro, V., et al., 2011. Identification of a novel immunomodulatory gliadin peptide that causes interleukin-8 release in a chemokine receptor CXCR3-dependent manner only in patients with coeliac disease. Immunology 132 (3), 432–440.

Li, E.W., Mine, Y., 2004. Immunoenhancing effects of bovine glycomacropeptide and its derivatives on the proliferative response and phagocytic activities of human macrophage like cells, U937. J. Agric. Food Chem. 52 (9), 2704–2708.

Li-Chan, E.C., 2015. Bioactive peptides and protein hydrolysates: research trends and challenges for application as nutraceuticals and functional food ingredients. Curr. Opin. Food Sci. 1, 28–37.

Liu, P., Liao, W., Qi, X., Yu, W., Wu, J., 2020. Identification of immunomodulatory peptides from zein hydrolysates. Eur. Food Res. Technol. 246, 931–937.

Maestri, E., Marmiroli, M., Marmiroli, N., 2016. Bioactive peptides in plant-derived foodstuffs. J. Proteom. 147, 140–155.

Mallet, J.F., Duarte, J., Vinderola, G., Anguenot, R., Beaulieu, M., Matar, C., 2014. The immunopotentiating effects of shark-derived protein hydrolysate. Nutrition 30 (6), 706–712.

Marmiroli, N., Maestri, E., 2014. Plant peptides in defense and signaling. Peptides 56, 30–44.

Masotti, A.I., Buckley, N., Champagne, C.P., Green-Johnson, J., 2011. Immunomodulatory bioactivity of soy and milk ferments on monocyte and macrophage models. Food Res. Int. 44 (8), 2475–2481.

Matar, C., Valdez, J.C., Medina, M., Rachid, M., Perdigon, G., 2001. Immunomodulating effects of milks fermented by Lactobacillus helveticus and its non-proteolytic variant. J. Dairy Res. 68 (4), 601–609.

Matar, C., LeBlanc, J.G., Martin, L., Perdigon, G., 2003. Biologically active peptides released in fermented milk: role and functions. In: Farnworth, E. R. (Ed.), Handbook of Fermented Functional Foods. CRC Press, pp. 177–201.

Matsui, T., 2018. Are peptides absorbable compounds? J. Agric. Food Chem. 66 (2), 393.

Matsui, T., Tamaya, K., Seki, E., Osajima, K., Matsumoto, K., Kawasaki, T., 2002. Absorption of Val–Tyr with in vitro angiotensin I-converting enzyme inhibitory activity into the circulating blood system of mild hypertensive subjects. Biol. Pharm. Bull. 25 (9), 1228–1230.

Medici, M., Vinderola, C.G., Perdigon, G., 2004. Gut mucosal immunomodulation by probiotic fresh cheese. Int. Dairy J. 14 (7), 611–618.

Meisel, H., 1997. Biochemical properties of regulatory peptides derived from mil proteins. Peptide Sci. 43 (2), 119–128.

Meisel, H., Bockelmann, W., 1999. Bioactive peptides encrypted in milk proteins: proteolytic activation and thropho-functional properties. Antonie Van. Leeuwenhoek 76 (1), 207–215.

Meisel, H., Gunther, S., 1998. Food proteins as precursors of peptides modulating human cell activity. Nahrung (Weinheim), 42(3–4), 175–176.

Menard, S., Cerf-Bensussan, N., Heyman, M., 2010. Multiple facets of intestinal permeability and epithelial handling of dietary antigens. Mucosal Immunol. 3 (3), 247–259.

Mercier, A., Gauthier, S.F., Fliss, I., 2004. Immunomodulating effects of whey proteins and their enzymatic digests. Int. Dairy J. 14 (3), 175–183.

Migliore-Samour, D., Jolles, P., 1988. Casein, a prohormone with an immunomodulating role for the newborn? Experientia 44 (3), 188–193.

Miyauchi, H., Kaino, A., Shinoda, I., Fukuwatari, Y., Hayasawa, H., 1997. Immunomodulatory effect of bovine lactoferrin pepsin hydrolysate on murine splenocytes and Peyer's patch cells. J. Dairy Sci. 80 (10), 2330–2339.

Morris, H.J., Carrillo, O., Almarales, A., Bermudez, R.C., Lebeque, Y., Fontaine, R., et al., 2007. Immunostimulant activity of an enzymatic protein hydrolysate from green microalga Chlorella vulgaris on undernourished mice. Enzyme Microb. Technol. 40 (3), 456–460.

Ndiaye, F., Vuong, T., Duarte, J., Aluko, R.E., Matar, C., 2012. Anti-oxidant, anti-inflammatory and immunomodulating properties of an enzymatic protein hydrolysate from yellow field pea seeds. Eur. J. Nutr. 51 (1), 29–37.

Olivares, M., Díaz-Ropero, M.P., Sierra, S., Lara-Villoslada, F., Fonollá, J., Navas, M., et al., 2007. Oral intake of Lactobacillus fermentum CECT5716 enhances the effects of influenza vaccination. Nutrition 23 (3), 254–260.

Ortiz-Martinez, M., Winkler, R., García-Lara, S., 2014. Preventive and therapeutic potential of peptides from cereals against cancer. J. Proteom. 111, 165–183.

Paegelow, I., Werner, H., 1986. Immunomodulation by some oligopeptides. Methods Find. Exp. Clin. Pharmacol. 8 (2), 91–95.

Panchaud, A., Affolter, M., Kussmann, M., 2012. Mass spectrometry for nutritional peptidomics: how to analyze food bioactives and their health effects. J. Proteom. 75 (12), 3546–3559.

Pang, G., Xie, J., Chen, Q., Hu, Z., 2012. How functional foods play critical roles in human health. Food Sci. Hum. Wellness 1 (1), 26–60.

Parish, C.R., O'Neill, E.R., 1997. Dependence of the adaptive immune response on innate immunity: some questions answered but new paradoxes emerge. Immunol. Cell Biol. 75 (6), 523–527.

Parker, F., Migliore-Samour, D., Floch, F., Zerial, A., Werner, G.H., Jolles, J., et al., 1984. Immunostimulating hexapeptide from human casein: amino acid sequence, synthesis and biological properties. Eur. J. Biochem. 145 (3), 677–682.

Parkin, J., Cohen, B., 2001. An overview of the immune system. Lancet 357 (9270), 1777–1789.

Pereira, P.C., 2014. Milk nutritional composition and its role in human health. Nutrition 30 (6), 619–627.

Pihlanto, A., 2013. Lactic fermentation and bioactive peptides. In: Marcelino Kongo, J. (Ed.), Lactic Acid Bacteria–R and D for Food, Health and Livestock Purposes. InTech, Rijeka, Croatia, pp. 310–331.

Prada, D., Gómez, J., Lorenzo, N., Corrales, O., Lopez, A., Gonzalez, E., 2018. Phase I clinical trial with a novel altered peptide ligand derived from human heat-shock protein 60 for treatment of rheumatoid arthritis: safety, pharmacokinetics and preliminary therapeutic effects. J. Clin. Trials 8, 2167-0870.

Renukuntla, J., Vadlapudi, A.D., Patel, A., Boddu, S.H., Mitra, A.K., 2013. Approaches for enhancing oral bioavailability of peptides and proteins. Int. J. Pharm. 447 (1–2), 75–93.

Repa, A., Grangette, C., Daniel, C., Hochreiter, R., Hoffmann-Sommergruber, K., Thalhamer, J., et al., 2003. Mucosal co-application of lactic acid bacteria and allergen induces counter-regulatory immune responses in a murine model of birch pollen allergy. Vaccine 22 (1), 87–95.

Requena, P., Daddaoua, A., Guadix, E., Zarzuelo, A., Suárez, M.D., De Medina, F.S., et al., 2009. Bovine glycomacropeptide induces cytokine production in human monocytes through the stimulation of the MAPK and the NF-κB signal transduction pathways. Br. J. Pharmacol. 157 (7), 1232–1240.

Reyes-Diaz, A., González-Córdova, A.F., Hernández-Mendoza, A., Reyes-Díaz, R., Vallejo-Cordoba, B., 2018. Immunomodulation by hydrolysates and peptides derived from milk proteins. Int. J. Dairy Technol. 71 (1), 1–9.

Saaby, L., Nielsen, C.U., Steffansen, B., Larsen, S.B., Brodin, B., 2013. Current status of rational design of prodrugs targeting the intestinal di/tri-peptide transporter hPEPT1 (SLC15A1). J. Drug. Deliv. Sci. Technol. 23 (4), 307–314.

Sanchez, A., Vazquez, A., 2017. Bioactive peptides: a review. Food Qual. Saf. 1 (1), 29–46.

Schanbacher, F.L., Talhouk, R.S., Murray, F.A., Gherman, L.I., Willett, L.B., 1998. Milk-borne bioactive peptides. Int. Dairy J. 8 (5–6), 393–403.

Schlimme, E., Meisel, H., 1995. Bioactive peptides derived from milk proteins. Structural, physiological and analytical aspects. Food/Nahrung 39 (1), 1–20.

Shahidi, F., Zhong, Y., 2008. Bioactive peptides. J. AOAC Int. 91 (4), 914–931.

Shi, R., Derendorf, H., 2010. Pediatric dosing and body size in biotherapeutics. Pharmaceutics 2 (4), 389–418.

Sigdel, T.K., Nicora, C.D., Hsieh, S.C., Dai, H., Qian, W.J., Camp, D.G., et al., 2014. Optimization for peptide sample preparation for urine peptidomics. Clin. Proteom. 11 (1), 1–8.

Singh, B.P., Vij, S., Hati, S., 2014. Functional significance of bioactive peptides derived from soybean. Peptides 54, 171–179.

So, J.S., Kwon, H.K., Lee, C.G., Yi, H.J., Park, J.A., Lim, S.Y., et al., 2008a. Lactobacillus casei suppresses experimental arthritis by down-regulating T helper 1 effector functions. Mol. Immunol. 45 (9), 2690–2699.

So, J.S., Lee, C.G., Kwon, H.K., Yi, H.J., Chae, C.S., Park, J.A., et al., 2008b. Lactobacillus casei potentiates induction of oral tolerance in experimental arthritis. Mol. Immunol. 46 (1), 172–180.

Sollid, L.M., 2002. Coeliac disease: dissecting a complex inflammatory disorder. Nat. Rev. Immunol. 2 (9), 647–655.

Svedberg, J., de Haas, J., Leimenstoll, G., Paul, F., Teschemacher, H., 1985. Demonstration of β-casomorphin immunoreactive materials in in vitro digests of bovine milk and in small intestine contents after bovine milk ingestion in adult humans. Peptides 6 (5), 825–830.

Takahashi, M., Moriguchi, S., Yoshikawa, M., Sasaki, R., 1994. Isolation and characterization of oryzatensin: a novel bioactive peptide with ileum-contracting and immunomodulating activities derived from rice albumin. Biochem. Mol. Biol. Int. 33 (6), 1151–1158.

Tidona, F., Criscione, A., Guastella, A.M., Zuccaro, A., Bordonaro, S., Marletta, D., 2009. Bioactive peptides in dairy products. Ital. J. Anim. Sci. 8 (3), 315–340.

Tsuruki, T., Yoshikawa, M., 2004. Design of soymetide-4 derivatives to potentiate the anti-alopecia effect. Biosci. Biotechnol. Biochem. 68 (5), 1139–1141.

Udenigwe, C.C., Aluko, R.E., 2012. Food protein-derived bioactive peptides: production, processing, and potential health benefits. J. Food Sci. 77 (1), R11–R24.

Van Eden, W., Van der Zee, R., Prakken, B., 2005. Heat-shock proteins induce T-cell regulation of chronic inflammation. Nat. Rev. Immunol. 5 (4), 318–330.

Van Overtvelt, L., Moussu, H., Horiot, S., Samson, S., Lombardi, V., Mascarell, L., et al., 2010. Lactic acid bacteria as adjuvants for sublingual allergy vaccines. Vaccine 28 (17), 2986–2992.

Van Platerink, C.J., Janssen, H.G.M., Horsten, R., Haverkamp, J., 2006. Quantification of ACE inhibiting peptides in human plasma using high performance liquid chromatography–mass spectrometry. J. Chromatogr. B 830 (1), 151–157.

Vesely, M.D., Kershaw, M.H., Schreiber, R.D., Smyth, M.J., 2011. Natural innate and adaptive immunity to cancer. Annu. Rev. Immunol. 29, 235–271.

Vij, R., Reddi, S., Kapila, S., Kapila, R., 2015. Transepithelial transport of milk derived bioactive peptide VLPVPQK. Food Chem. 190, 681–688.

Vinderola, C.G., Duarte, J., Thangavel, D., Perdigón, G., Farnworth, E., Matar, C., 2005. Immunomodulating capacity of kefir. J. Dairy Res. 72 (2), 195–202.

Vo, T.S., Ryu, B., Kim, S.K., 2013. Purification of novel anti-inflammatory peptides from enzymatic hydrolysate of the edible microalgal Spirulina maxima. J. Funct. Foods 5 (3), 1336–1346.

Vu, L.T., Less, R.R., Rajagopalan, P., 2014. The promise of organotypic hepatic and gastrointestinal models. Trends Biotechnol. 32 (8), 406–413.

Wagar, L.E., Champagne, C.P., Buckley, N.D., Raymond, Y., Green-Johnson, J.M., 2009. Immunomodulatory properties of fermented soy and dairy milks prepared with lactic acid bacteria. J. Food Sci. 74 (8), M423–M430.

Wang, W., De Mejia, E.G., 2005. A new frontier in soy bioactive peptides that may prevent age-related chronic diseases. Compr. Rev. Food Sci. Food Saf. 4 (4), 63–78.

Wasan, K.M., 2002. The role of lymphatic transport in enhancing oral protein and peptide drug delivery. Drug. Dev. Ind. Pharm. 28 (9), 1047–1058.

Waugh, D.J., Wilson, C., 2008. The interleukin-8 pathway in cancer. Clin. Cancer Res. 14 (21), 6735–6741.

Wittkopf, N., Neurath, M.F., Becker, C., 2014. Immune-epithelial crosstalk at the intestinal surface. J. Gastroenterol. 49 (3), 375–387.

Wong, C.W., Seow, H.F., Liu, A.H., Husband, A.J., Smithers, G.W., Watson, D.L., 1996. Modulation of immune responses by bovine β-casein. Immunol. Cell Biol. 74 (4), 323–329.

Wu, W., Zhang, M., Ren, Y., Cai, X., Yin, Z., Zhang, X., et al., 2017. Characterization and immunomodulatory activity of a novel peptide, ECFSTA, from wheat germ globulin. J. Agric. Food Chem. 65 (27), 5561–5569.

Xu, Z., Mao, T.M., Huang, L., Yu, Z.C., Yin, B., Chen, M.L., et al., 2019. Purification and identification immunomodulatory peptide from rice protein hydrolysates. Food Agric. Immunol. 30 (1), 150–162.

Yang, R., Zhang, Z., Pei, X., Han, X., Wang, J., Wang, L., et al., 2009. Immunomodulatory effects of marine oligopeptide preparation from Chum Salmon (Oncorhynchus keta) in mice. Food Chem. 113 (2), 464–470.

Yoshikawa, M., Fujita, H., Matoba, N., Takenaka, Y., Yamamoto, T., Yamauchi, R., et al., 2000. Bioactive peptides derived from food proteins preventing lifestyle-related diseases. Biofactors 12 (1-4), 143–146.

Yuan, H., Song, J., Li, X., Li, N., Dai, J., 2006. Immunomodulation and antitumor activity of κ-carrageenan oligosaccharides. Cancer Lett. 243 (2), 228–234.

Zhang, W., Azevedo, M.S., Wen, K., Gonzalez, A., Saif, L.J., Li, G., et al., 2008. Probiotic Lactobacillus acidophilus enhances the immunogenicity of an oral rotavirus vaccine in gnotobiotic pigs. Vaccine 26 (29–30), 3655–3661.

Zhong, F., Zhang, X., Ma, J., Shoemaker, C.F., 2007. Fractionation and identification of a novel hypocholesterolemic peptide derived from soy protein Alcalase hydrolysates. Food Res. Int. 40 (6), 756–762.

Zouki, C., Ouellet, S., Filep, J.G., 2000. The anti-inflammatory peptides, antiflammins, regulate the expression of adhesion molecules on human leukocytes and prevent neutrophil adhesion to endothelial cells. FASEB J. 14 (3), 572–580.

Chapter 33

Clinical intervention, study design, and statistical methodology: statistical approaches for pre-post studies

Kohsuke Hayamizu
Yokohama University of Pharmacy, Yokohama, Japan

33.1 Introduction

A clinical study examines the effects of an intervention in humans, whereas an intervention study aims to verify the effect of an experimental intervention. A clinical trial in drug development is equivalent to an intervention study. Phase III clinical trials generally test the hypothesis that a candidate therapy is useful for treating a given disease. Meanwhile, human intervention research is important not only in drug research, but also in research on functional foods and nutrients such as vitamins. The strength of the evidence that is obtained by such studies is the same regardless of whether the research target is a drug or food ingredient. Randomized controlled trials (RCTs) are ideal for this purpose. Therefore many studies employ RCTs to verify the effectiveness of functional foods. In addition, the regulations regarding the operation of clinical trials for functional foods are basically the same as those for drugs, including ethical considerations for volunteers and the registration system for trial protocols. Thus clinical trials are similar in both areas of study.

Nevertheless, there are some differences between clinical trials for functional foods and drugs. For example, studies of the effectiveness of functional foods often include subjects who are healthy but are on the borderline of having health conditions. In addition, the expected effect size of the intervention is small. Furthermore, it is often financially unfeasible to utilize very large sample sizes. However, if the number of subjects is small, such as in a pilot study, the power of a statistical test to determine the intervention effect becomes weak, which might result in a failure to validate the intervention effect. Another difference between clinical trials for functional foods and drugs is that after subjects are allocated to the treatment or control group, there is a risk of imbalances in background factors, which might affect the efficacy outcome analysis. These issues can be resolved by devising appropriate allocation plans and data analysis methods. This chapter focuses on the risk of imbalances in background factors and introduces a commonly used data analysis method for continuous variable data obtained from an RCT both before and after intervention. In addition, this chapter introduces example applications of a new statistical model called the change point regression model (CPRM), which is useful for analyzing the effectiveness of functional foods.

33.2 Pre-post design

Covariance is a subject-specific factor that affects the intervention effect. When the covariance is unbalanced between the treatment and control groups, it is called a confounder or concomitant variable. If the evaluation of an intervention effect does not consider confounders, then bias will be introduced (Chow and Liu, 2013). In parallel design studies, subjects who meet the inclusion criteria are randomly assigned to each treatment group. Because the subject population is assumed to be homogeneous, the subjects' characteristics (e.g., age, sex, body weight) are expected to be comparable between treatment groups. In clinical trials with very large sample sizes, random assignment into two groups results in almost equal allocation of known and unknown confounders, so there is little risk of bias. However, in trials with small sample sizes, there are risks of imbalance and bias even with random assignment. If the subject population is known to be heterogeneous with respect to certain characteristics, then stratified randomization according to these variables should be applied.

Microbiome, Immunity, Digestive Health and Nutrition. DOI: https://doi.org/10.1016/B978-0-12-822238-6.00037-6

In intervention studies, to verify the intervention effect, the subjects are divided into groups, and the preintervention value ("pre-value" hereafter) and postintervention value ("post-value" hereafter) of the efficacy index are obtained from a given individual. Next, the existence of the intervention effect is verified by data analysis. While multiple post-values can be acquired during follow-up, this chapter considers only the final values for postintervention data. Therefore repeated measurements are limited to two points before and after the intervention. This is called a pre-post design. Pretests and posttests are conducted to adjust for the effects of confounding factors and to evaluate intervention tests in small sample sizes.

33.3 Outcome measure

In studies with a pre-post design, the post-value, pre-post value (i.e., the Δ value), post-value/pre-value ratio (i.e., the rate of change), and so on can be considered outcome measures. This chapter considers two cases of post-values and pre-post values. In addition, because the pre-value can greatly affect the intervention effect, an analysis method that adjusts for the effect of the pre-value is considered; that is, the pre-value is treated as covariance. Moreover, the data are assumed to have a normal distribution. Table 33.1 shows the statistical analysis methods that are used according to the outcome measurement and the presence or absence of covariance.

33.4 Sample data: hydroxycitric acid study

(−)-Hydroxycitric acid (HCA), which is abundant in the rind of *Garcinia cambogia*, a fruit that is native to Southeast Asia, inhibits ATP-citrate lyase and blocks the conversion of citric acid to acetyl-CoA. This inhibitory activity of HCA is reported to have an antiobesity effect by blocking the first step in fatty acid synthesis (Hayamizu et al., 2001). To verify the antiobesity effect of HCA, a double-blind, randomized, parallel-group comparative study was conducted in 40 preobese to class 1 obese people (body mass index: 25−35 kg/m^2) with placebo given as a control. The primary endpoint was the visceral fat area (VFA), which was calculated by image analysis from computed tomography images at the umbilicus level, which were obtained before and 8 weeks after HCA administration. The antiobesity effect was evaluated as a decrease in VFA. Because the individual differences in preintervention VFA among the subjects were large, they were expected to affect postintervention VFA levels. Herein, data that were randomly extracted from 15 patients in each group are used. Table 33.2 shows some of the data from the HCA study.

33.5 Method for analyzing the intervention effect without adjusting for the pre-value effect

Let x_{ij} and y_{ij} be the pre-value and post-value, respectively, for the ith subject in the jth treatment group, where $i = 1, 2, \ldots, n$ and $j = 1, 2, \ldots, k$. It is assumed that x_{ij} and y_{ij} follow a normal distribution with mean μ_{jx} and variance $\sigma_x{}^2$ and μ_{jy} and $\sigma_y{}^2$, respectively. Of note, $\sigma_x{}^2$ and $\sigma_y{}^2$ do not have to be equal. The pre- and post-values of subject i are data obtained from the same individual, and their correlation is ρ ($\neq 0$). The results of the HCA study are shown in Table 33.3 and Fig. 33.1.

From Table 33.3 the VFAs in the HCA and placebo groups were 70.5 and 79.7 cm^2, respectively, and the effect size of HCA administration is calculated as follows:

$$D = \mu_{\mathrm{Hy}} - \mu_{\mathrm{Py}} = 70.5 - 79.7 = -9.2$$

TABLE 33.1 Classification of statistical analysis model according to covariance and types of endpoints in a pre-post design (Kakuma and Hayamizu, 2010).

Covariance	Outcome measure	Statistical analysis
Not adjust with pre-value	Post-value	ANOVA (*t*-test)
	Pretest-posttest value	ANOVA (*t*-test)
Adjust with pre-value	Post-value	ANCOVA
		CPRM
	Pretest-posttest value	ANCOVA
		CPRM

CPRM, Change point regression model.
ANOVA: Analysis of variance. The *t*-test of the two groups can be defined as a one-way ANOVA model, $Y_{ij} = \theta_j + \varepsilon_{ij}$ group factor effect comparison.

TABLE 33.2 Sample data of the (–)-hydroxycitric acid study.

ID	Group	Pre VFA	Post VFA	ΔVFA
1	HCA	38.8	59.5	20.7
2	Placebo	59.5	73	13.5
3	HCA	72.1	84.5	12.4
4	Placebo	97.5	107.0	9.5
5	HCA	34.2	43.2	9.0
		· · · · Data omitted · · · ·		
28	Placebo	70	54.2	− 15.8
29	HCA	125	108.8	− 16.2
30	HCA	99.7	82.2	− 17.5

HCA, (–)-hydroxycitric acid; *VFA*, visceral fat area.

TABLE 33.3 Summary of the visceral fat area in the (–)-hydroxycitric acid study.

Group	Measurement period	mean	SD
HCA ($n = 15$)	Pre (0 week)	78.9	41.7
	Post (8 weeks)	70.5	31.2
Placebo ($n = 15$)	Pre (0 week)	81.9	35.5
	Post (8 weeks)	79.7	36.0

HCA, (–)-hydroxycitric acid.

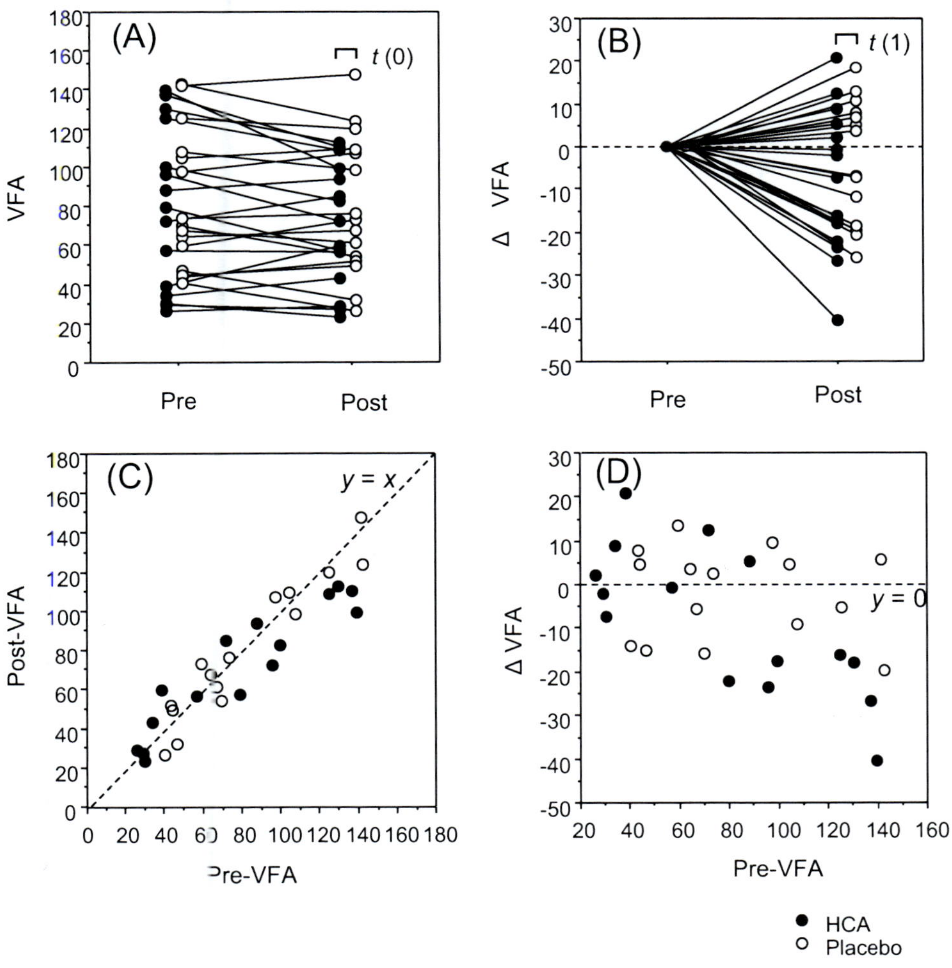

FIGURE 33.1 Postintervention changes in visceral fat area in the (–)-hydroxycitric acid study. (A, B) Parallel plots. (C, D) Scatter plots.

The effect size due to ΔVFA is calculated as follows:

$$\delta D = \left(\mu_{Hy} - \mu_{Hx}\right) - \left(\mu_{Py} - \mu_{Px}\right)$$
$$= (70.5 - 78.9) - (79.7 - 81.9) = -8.4 - (-2.2) = -6.2$$

Thus HCA administration appears to exhibit efficacy in both outcome measures.

The intervention effect of the pretest-posttest design is modeled as follows:

$$\tau(b) = \left(\mu_{Hy} - \mu_{Py}\right) - b\left(\mu_{Hx} - \mu_{Px}\right) \tag{33.1}$$

where b is a constant and takes a value of 0 or 1. When $b = 0$,

$$\tau(0) = \left(\mu_{Hy} - \mu_{Py}\right)$$

which corresponds to the case in which the post-value is used as the outcome measure.

Meanwhile, when $b = 1$,

$$\tau(1) = \left(\mu_{Hy} - \mu_{Py}\right) - \left(\mu_{Hx} - \mu_{Px}\right) = \left(\mu_{Hy} - \mu_{Hx}\right) - \left(\mu_{Py} - \mu_{Px}\right)$$

which corresponds to the case in which the pretest-posttest value is used as the outcome measure.

In testing the difference of outcome measures, the null hypothesis (H) and the alternative hypothesis (K) are as follows (Kakuma and Hayamizu, 2010):

When the post-value is used as the outcome measure,

$$H:\quad \tau(0) = \left(\mu_{Hy} - \mu_{Py}\right) = 0$$
$$K:\quad \tau(0) = \left(\mu_{Hy} - \mu_{Py}\right) \neq 0$$

When the pretest-posttest value is used as the outcome measure,

$$H:\quad \tau(1) = \left(\mu_{Hy} - \mu_{Hx}\right) - \left(\mu_{Py} - \mu_{Px}\right) = 0$$
$$K:\quad \tau(1) = \left(\mu_{Hy} - \mu_{Hx}\right) - \left(\mu_{Py} - \mu_{Px}\right) \neq 0$$

The test statistic, $t(b)$, for these hypotheses is as follows:

$$t(b) = \frac{\hat{\tau}(b)}{\hat{\mathrm{SE}}(\hat{\tau}(b))} = \frac{\left(\mu_{Hy} - \mu_{Py}\right) - b\left(\mu_{Hx} - \mu_{Px}\right)}{\hat{\mathrm{SE}}\left[\left(\mu_{Hy} - \mu_{Py}\right) - b\left(\mu_{Hx} - \mu_{Px}\right)\right]} \tag{33.2}$$

When the post-value is used as the outcome measure,

$$t(0) = \frac{\left(\mu_{Hy} - \mu_{Py}\right)}{\sqrt{\left(\frac{1}{n_H} + \frac{1}{n_P}\right)\left(\frac{SS_y}{n_H + n_P - 2}\right)}} \tag{33.3}$$

When the pretest-posttest value is used as the outcome measure,

$$t(1) = \frac{\left(\mu_{Hy} - \mu_{Hx}\right) - \left(\mu_{Py} - \mu_{Px}\right)}{\sqrt{\left(\frac{1}{n_H} + \frac{1}{n_P}\right)\left\{\left(\frac{SS_y}{n_H + n_P - 2}\right) + \left(\frac{1}{n_H} + \frac{1}{n_P}\right)\left(\frac{SS_x}{n_H + n_P - 2}\right) - 2\left(\frac{SS_{xy}}{n_H + n_P - 2}\right)\right\}}} \tag{33.4}$$

Here, SS_y, SS_x, and SS_{xy} represent the sum of squares defined by the following formulas:

$$SS_y = \sum_{j} \sum_{i\,=\,1}^{n_j} \left(y_{ij} - \mu_{jy}\right)^2,$$

TABLE 33.4 Analysis of intervention effect without adjusting the effect of the pre-value in the (–)-hydroxycitric acid study.

Outcome measure	(–)-Hydroxycitric acid	Placebo	*T* value	Difference between groups (95% CI)	*P* value
Post-value	70.53	79.73	– 0.7476 ($t(0)$)	– 9.20 (–34.41:16.01)	0.4609
Pre-post value	– 8.35	– 2.18	– 1.1972 ($t(1)$)	– 6.17 (–16.72: 4.38)	0.2413

$$SS_x = \sum_j \sum_{i=1}^{n_j} \left(x_{ij} - \mu_{jx}\right)^2,$$

$$SS_{xy} = \sum_j \sum_{i=1}^{n_j} \left(x_{ij} - \mu_{jx}\right)\left(y_{ij} - \mu_{jy}\right)$$

The test statistics $t(0)$ and $t(1)$ follow the t-distribution with degrees of freedom ($n_H + n_P - 2$) under the null hypothesis.

Table 33.4 shows the test results. When the post-value is used as the outcome measure, the t-value is -0.7476 and the P-value is 0.4609. The null hypothesis cannot be rejected, and no significant difference is observed. When the pretest-posttest value is the outcome measure, the t-value is -1.1972 and the P-value is 0.2413. This is better than that when the post-value is the outcome measure because the intervention effect of subject i is treated as the magnitude of change from the subject's pre-value, and the interindividual error is decreasing to some extent. However, the null hypothesis cannot be rejected even if the pretest-posttest value is used as the outcome measure, and no significant VFA reduction effect is observed.

Now the test data are plotted, and the entire dataset is examined. Visually expressing the state of raw data is useful, and a more accurate presentation is important. The distribution of raw data is generally presented as mean $\pm$ SD (or $\pm$ SE) or in a box plot. However, note that because these expression methods do not clearly present the information from the raw data, some information is obfuscated or lost.

Fig. 33.1A and B are parallel plots, which are typical data plots that express the results of pre-post design tests. The pre- and post-values of a given subject i are related and connected by a line. Fig. 33.1A shows the variation before and after the intervention for each individual, comparing both the HCA and placebo groups, with the post-value as the outcome measure shown at $t(0)$. Meanwhile, Fig. 33.1B shows the comparison of both groups with the pretest-posttest value as the outcome measure at $t(1)$. However, it seems necessary to plot the data differently in order to better understand how VFA changed before and after the intervention.

Let us express the state of raw data using a scatter plot like those in Fig. 33.1C and D. The pre-values, post-values, or pre-post values can be plotted on either axis the scatter plot. In a scatter plot, the data of one subject can be presented as a single point with two coordinates. When the x-axis is the pre-value and the y-axis is the post-value (Fig. 33.1C) or pre-post value (Fig. 33.1D), if there is no change in VFA after the intervention, the data will fall on the line $y = x$; if there is a change in VFA, the data will fall on the line $y = 0$. Therefore if there is indeed an intervention effect, the plot will depart from $y = x$ or $y = 0$. Thus scatter plots are useful for visualizing test data repeatedly measured from a given subject. Visual inspection of Fig. 33.1C and D suggests that the pre-values are likely correlated with the post-values and pretest-posttest values. That is, the pre-value is the covariance of the post-value or pre-post value.

Next, an analysis of intervention effects adjusting for the effect of the pre-value will be introduced.

33.6 Method for analyzing an intervention effect, adjusting for the effect of the pre-value: analysis of covariance

As was described previously, Fig. 33.1C suggests a (positive) correlation between the pre-values and post-values. Indeed, the correlation coefficient between them is $r = 0.929$, confirming a strong correlation. Therefore to analyze the intervention effect while considering the influence of the pre-value, analysis of covariance (ANCOVA) can be used. ANCOVA is a linear model that applies the principle of regression analysis to analysis of variance (ANOVA) and considers the influence of other data to comprehensively compare target data between groups.

A linear model is a model in which the relationship between the objective and explanatory variables is linear. Other linear models besides ANCOVA include ANOVA, *t*-tests, multivariate analysis of variance (MANOVA), and multiple regression analysis. ANCOVA encompasses several models. First, is a model in which the influence of the pre-value on the post-value is constant, termed "ANCOVA1" herein. ANCOVA1 is represented by a linear model as follows:

$$y_{ij} = \theta_j + \beta x_{ij} + \varepsilon_{ij} \tag{33.5}$$

Here, y_{ij} and x_{ij} are the post-value and pre-value for the *i*th subject in the *j*th treatment group, respectively. It is assumed that θ_j is the effect of the *j* group, β is the effect of the pre-value, ε_{ij} is an independent random error for all values of *i* and *j*, the mean is 0, and the variance follows a normal distribution with σ_ε^2. As the intervention effect of ANCOVA1 is the difference in the group effect θ_j, it is defined as follows:

$$\Delta = \theta_H - \theta_P \tag{33.6}$$

The concept of ANCOVA1 is shown in Fig. 33.2A. The assumption that the influence of the pre-value on the post-value is constant means that β is a coefficient that is common to both groups. In ANCOVA1 the null hypothesis (*H*) and the alternative hypothesis (*K*) in testing the difference of the outcome measure are as follows: *H*: $\Delta = 0$, *K*: $\Delta \neq 0$.

The results of ANCOVA1 applied to the HCA study data are shown in Fig. 33.2B and Table 33.5. The result of the linear model is expressed as an ANOVA table like Table 33.5. The estimated value of each level is shown in Table 33.6, and the model formula is as follows:

$$y = 13.2407 + \begin{pmatrix} -3.3689 \\ 3.3689 \end{pmatrix} + 0.8117x$$

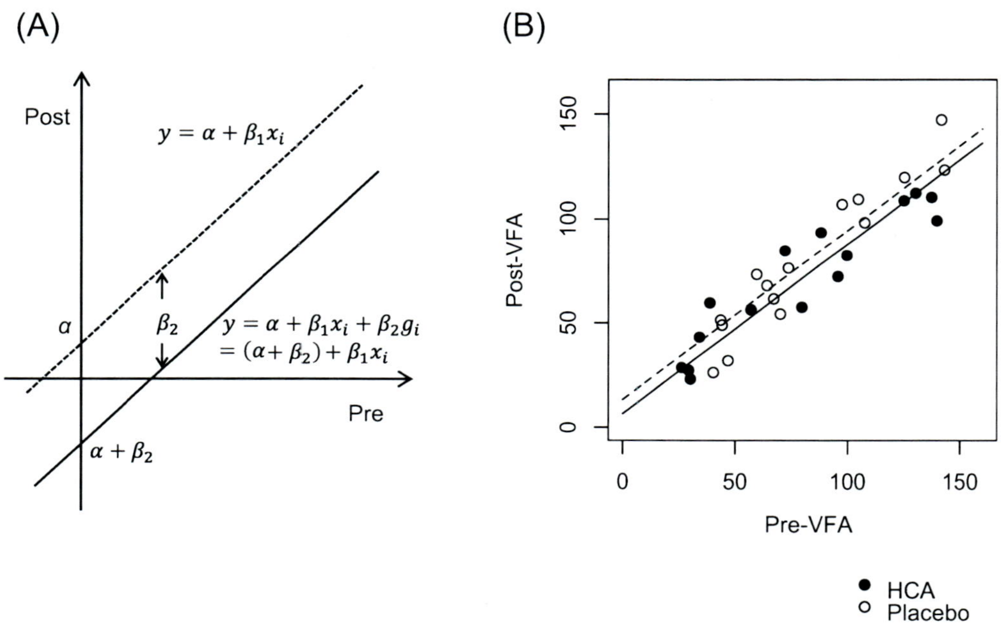

FIGURE 33.2 The concept of ANCOVA1 and its application to (−)-hydroxycitric acid study. (A) Concept. (B) Application to (−)-hydroxycitric acid study.

TABLE 33.5 Analysis of the (−)-hydroxycitric acid study by ANCOVA 1.

Outcome measure		df	Sum Sq	Mean Sq	*F* value	*P* value
Post-value	Pre (x)	1	28019	28019	185.411	1.31e-13
	Group (θ)	1	340	340	2.249	0.1453
	Residuals	27	4080	151		
Pre-post value	Pre (x)	1	1436	1436.4	9.505	0.00468
	Group (θ)	1	340	339.9	2.249	0.1453
	Residuals	27	4080	151.1		

TABLE 33.6 Estimated coefficients of ANCOVA 1 in the (−)-hydroxycitric acid study.

Outcome measure	Parameter	Coefficient	SE	*T* value	*P* value
Post-value	Intercept	13.2407	5.8459	2.265	0.0318
	HCA	− 3.3689	2.2462	− 1.500	0.1453
	Placebo	3.3689			
	Slope	0.8117	0.0599	13.545	<0.0001
Pre-post value	Intercept	13.2407	5.8459	2.265	0.0318
	HCA	− 3.3689	2.2462	− 1.500	0.1453
	Placebo	3.3689			
	Slope	− 0.1883	0.0599	− 3.141	0.0041

where 13.2407 is an intercept that is common to both groups and the difference between the HCA and placebo groups—the intervention effect—is $\Delta = 3.3689 \times 2 = 6.7378$. The result of the intervention effect is $p = 0.1453$, indicating no significant difference.

If the y_{ij} value of ANCOVA1 is changed to the pretest-posttest value, the pre-value now influences the change due to treatment in the model:

$$\Delta y_{ij} = \theta_j + \beta x_{ij} + \varepsilon_{ij} \tag{33.7}$$

where $\Delta y_i = y_i - x_i$.

Even if the outcome measure is changed to the pretest-posttest value, the test result does not change ($p = 0.1453$). As indicated in Fig. 33.1C and D, Eq. (33.7) equals Eq. (33.5) rotated by $-\pi/4$. In other words, the regression line of Eq. (33.7) does not change the y-intercept (θ_H and θ_P) because Eq. (33.5) only rotates.

In ANCOVA1, although it is assumed that the influence of the pre-value on the post-value is constant, the effect of the pre-value may change depending on the group. Herein, this case is referred to as ANCOVA2 and is expressed as follows:

$$y_i = \alpha + \beta_1 x_i + \beta_2 g_i + \beta_3 x_i * g_i + \varepsilon_i \tag{33.8}$$

Unlike ANCOVA1, the interaction (β_3) of the pre-value effect and the group effect are included in the model in ANCOVA2. In ANCOVA2, α and β_2 are the common constant term and slope of both groups, respectively. When 1 (i.e., the intervention group) is substituted into g_i, Eq. (33.8) becomes the following:

$$\begin{aligned} y_i &= \alpha + \beta_1 x_i + \beta_2 g_i + \beta_3 x_i + \varepsilon_i \\ &= (\alpha + \beta_2) + (\beta_1 + \beta_3) x_i + \varepsilon_i \end{aligned}$$

Accordingly, the constant term and slope change by β_2 and β_3, respectively.

The general concept of ANCOVA2 is shown in Fig. 33.3A. In this model, the post-value is affected by the interaction between the pre-value and group effect. Moreover, the null hypothesis (H) and the alternative hypothesis (K) in testing the difference of the outcome measure are as follows: H: $\beta_3 = 0$, K: $\beta_3 \neq 0$.

The results of applying ANCOVA2 to the HCA study data are shown in Fig. 33.3B and Table 33.7. Looking at the results for the interaction term [Pre(x) × Group(θ)], $p = 0.0222$, indicating a significant difference. In Fig. 33.3B the slope of the HCA group is smaller than that of the placebo group. Thus subjects with greater VFA pre-values exhibited greater decreases in VFA due to the HCA intervention. The estimated values for each level are shown in Table 33.8. The regression line for both groups is

$$y = 0.3477 + 0.9691x$$

that the regression line for the HCA group alone is

$$\begin{aligned} y &= (0.3477 + 7.581) + (0.9691 - -.136)x \\ &= 15.5097 + 0.8331x \end{aligned}$$

In ANCOVA2, β_2 may be meaningless because the intervention effect is judged according to the resultant β_3. The estimated model formula is as follows:

$$y = 0.3477 + \begin{pmatrix} 7.581 \\ -7.581 \end{pmatrix} + 0.9691x + \begin{pmatrix} -0.136 \\ 0.136 \end{pmatrix} x$$

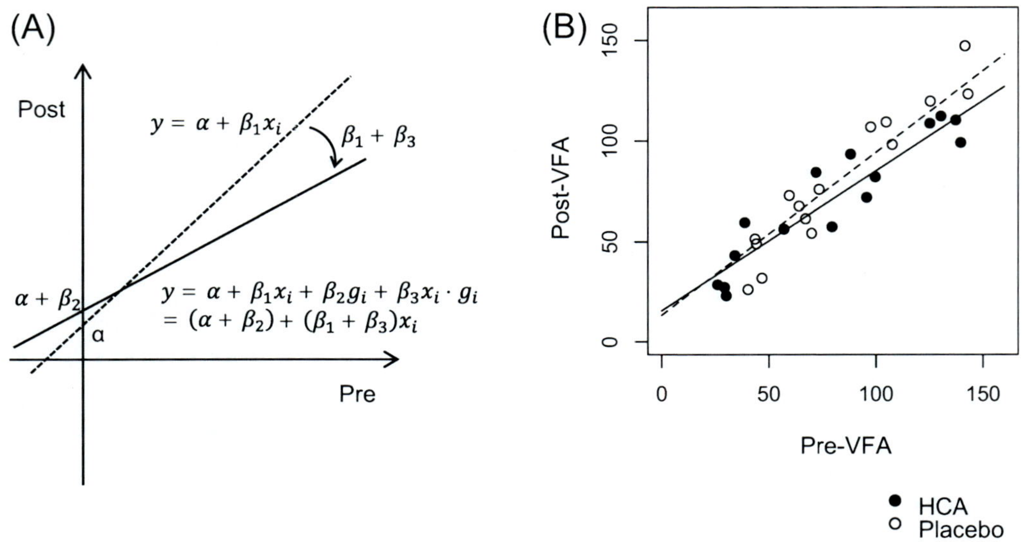

FIGURE 33.3 The concept of ANCOVA2 and its application to (−)-hydroxycitric acid study. (A) Concept. (B) Application to (−)-hydroxycitric acid study.

TABLE 33.7 Analysis of the (−)-hydroxycitric acid study by ANCOVA 2.

Outcome measure		df	Sum Sq	Mean Sq	*F* value	*P* value
Post-value	Pre (*x*)	1	28019	28019	219.160	< 0.0001
	Group (θ)	1	340	340	2.659	0.1150
	Pre (*x*)*Group (θ)	1	756	756	5.915	0.0222
	Residuals	26	3324	128		
Pre-post value	Pre (*x*)	1	1436	1436.4	11.236	0.0025
	Group (θ)	1	340	339.9	2.659	0.1150
	Pre (*x*)*Group (θ)	1	756	756.2	5.915	0.0222
	Residuals	26	3324	127.8		

TABLE 33.8 Estimated coefficients of ANCOVA 2 in the (−)-hydroxycitric acid study.

Outcome measure	Parameter	Coefficient	SE	*T* value	*P* value
Post-value	Intercept	0.3477	7.559	0.046	0.9636
	HCA	7.581	4.954	1.530	0.1380
	Placebo	− 7.582			
	Slope	0.9691	0.085	11.400	< 0.0001
	HCA	− 0.136	0.056	− 2.432	0.0222
	Placebo	0.136			
Pre-post value	Intercept	0.3477	7.559	0.046	0.9636
	HCA	7.581	4.954	1.530	0.1380
	Placebo	− 7.582			
	Slope	− 0.0309	0.085	− 0.363	0.7196
	HCA	− 0.136	0.056	− 2.432	0.0222
	Placebo	0.136			

HCA, (−)-hydroxycitric acid.

TABLE 33.9 Estimated coefficients of the *T*-test in the (−)-hydroxycitric acid study.

Outcome measure	Parameter	Coefficient	SE	*T* value	*P* value
Post-value	Intercept	79.733	8.702	9.163	<0.0001
	HCA	−4.600	6.153	−0.748	0.461
	Placebo	4.600			
Pre-post value	Intercept	−2.180	3.642	−0.599	0.554
	HCA	−0.308	2.576	−1.197	0.241
	Placebo	0.308			

If the y_{ij} of ANCOVA2 is changed to the pretest-posttest value, then the change due to the intervention is influenced by the interaction between the pre-value and the effect of group as follows:

$$\Delta y_i = \alpha + \beta_1 x_i + \beta_2 g_i + \beta_3 x_i * g_i + \varepsilon_i \tag{33.9}$$

Furthermore, in this case, the *P*-value of the interaction term does not change, and a significant difference is recognized. Like an ANCOVA1, Eq. (33.9) is obtained by rotating Eq. (33.8) by $-\pi/4$.

Thus by using ANCOVA2, which includes the interaction effect in the model, the results of the HCA study demonstrate that HCA decreases VFA. Therefore it is important to select an appropriate statistical analysis method. A typical commercial statistical analysis program implements ANCOVA, which includes interactions in the model. Accordingly, several clinical trials of functional foods report analysis results using ANCOVA (Table 33.9).

Of note, the *t*-test that was described in the previous section is expressed in a linear model as follows:

$$y_i = \alpha + g_i + \varepsilon_i \tag{33.10}$$

$$\Delta y_i = \alpha + g_i + \varepsilon_i \tag{33.11}$$

The null hypothesis (*H*) and the alternative hypothesis (*K*) in testing the difference of the outcome measure are as follows: *H*: $g_i = 0$, *K*: $g_i \neq 0$. When this linear model is analyzed, the *P*-values of Eqs. (33.10) and (33.11) are 0.4609 and 0.2413, respectively, which are concordant with the *t*-test result.

33.7 Change point regression model

Randomized controlled trials have recently been conducted to evaluate the effectiveness of functional foods. However, as was mentioned at the beginning of the chapter, there are some differences between RCTs for functional foods and those for clinical trials for drug development. Specifically, RCTs for functional foods often target people who are healthy, are on the borderline of illness, or have only mild illness. Therefore the expected effect size might be smaller than the clinical effect size. For example, consider a clinical trial of a functional food that is expected to improve hyperglycemia. The inclusion criteria for the study subjects are healthy people with slightly elevated blood glucose levels. Therefore even if the functional food truly has a hypoglycemic effect, such an effect cannot be expected in a person whose blood glucose level is near normal or slightly elevated at baseline. In this example the blood glucose level before the intervention affects the blood glucose level after the intervention. However, in an analysis of the outcome measure, the *t*-test and ANCOVA introduced earlier in the chapter cannot distinguish subjects who can expect an effect from those who cannot. Therefore the overall effect size becomes small. To resolve this issue, the subjects can be grouped into those who can expect an effect and those who cannot in advance, followed by subgroup analysis.

There are three main challenges in this approach. First, the cutoff for subgroups must be known in advance. Second, there must be a sufficient number of subjects within each subgroup, which increases the cost and time of the study. The third challenge is the problem of multiple comparisons. ANCOVA2, which evaluates the interaction between the pre-value and group, can be used to address these challenges. However, ANCOVA2 assumes that there is a certain effect even for subjects in whom no effect is expected, hampering its use as a model for evaluating the effectiveness of functional foods. To overcome this challenge, the CPRM has been proposed (Hayamizu et al., 2009).

The CPRM is a statistical model that can search for a threshold value that produces an effect with the pre-value as a boundary and distinguish between areas in which the effect appears or does not appear (Fig. 33.4). There are

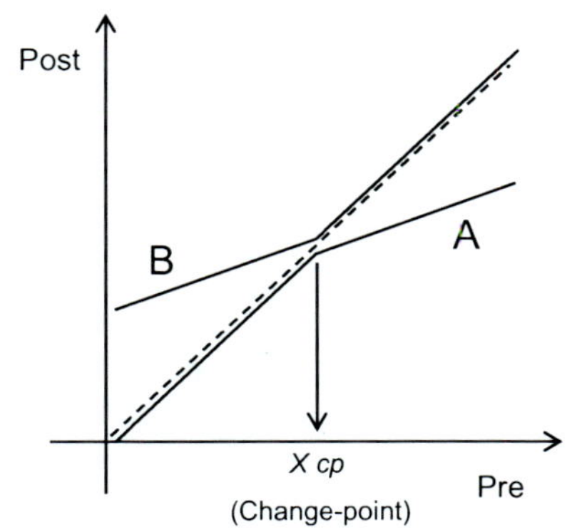

FIGURE 33.4 The concept of change point regression model $A{:}y_i = \alpha + \beta_1 x_i + I(x_i > x_{\text{cp}})(x_i - x_{\text{cp}})g_i + \varepsilon_i$ $B{:}y_i = \alpha + \beta_1 x_i + I(x_i < x_{cp})(x_i\text{-}x_{cp})g_i + \varepsilon_i$.

theoretically two cases for the model assumed by CPRM: (1) The intervention reduces the outcome measure (e.g., anti-obesity effects, hyperglycemia, and hyperlipidemia), or (2) the intervention increases the outcome measure (e.g., bone density, muscle mass, and blood ferritin level). In both cases, if there is a threshold or change point (i.e., the point at which the effect appears), it can differentiate a type A or type B pattern (Fig. 33.4). The CPRMs of type A (33.12) and type B (33.13) are defined as follows:

$$y_i = \alpha + \beta_1 x_i + \beta_2 I(x_i > x_{\text{cp}})(x_i - x_{\text{cp}})g_i + \varepsilon_i \tag{33.12}$$

$$y_i = \alpha + \beta_1 x_i + \beta_2 I(x_i < x_{\text{cp}})(x_i - x_{\text{cp}})g_i + \varepsilon_i \tag{33.13}$$

Here, x_{cp} is the effect change point, and $I(\bullet)$ is an indicator function. For example, in the case of type A, $I(\bullet)$ is 1 if $x_i > x_{\text{cp}}$ is true and 0 if it is false. When the pre-value is smaller than x_{cp}, Eq. (33.12) yields $y_i = \alpha + \beta_1 x_i + \varepsilon_i$ from $I(x_i > x_{\text{cp}}) = 0$. That is, if the pre-value is smaller than x_{cp}, then both the intervention and control groups move in the same manner. On the other hand, if the pre-value is larger than x_{cp} because $I(x_i > x_{\text{cp}}) = 1$, Eq. (33.13) yields $y_i = \alpha + \beta_1 x_i + \beta_2(x_i - x_{\text{cp}}) + \varepsilon_i = (\alpha + \beta_2 x_{\text{cp}}) + (\beta_1 + \beta_2)x_i + \varepsilon_i$. In other words, the coefficient of the intervention group in the region in which x_i is greater than x_{cp} changes to $\beta_1 + \beta_2$ and deviates from the control group. Thus as $g_i = 0$ in the control group, $y_i = \alpha + \beta_1 x_i + \varepsilon_i$ remains. Meanwhile, type B patterns utilize an indicator function that yields 1 if $x_i < x_{cp}$ is true and 0 if it is false. The CPRM may also use the pretest-posttest value as the outcome measure:

Type A:

$$\Delta y_i = \alpha + \beta_1 x_i + \beta_2 I(x_i > x_{\text{cp}})(x_i - x_{\text{cp}})g_i + \varepsilon_i \tag{33.14}$$

Type B:

$$\Delta y_i = \alpha + \beta_1 x_i + \beta_2 I(x_i < x_{\text{cp}})(x_i - x_{\text{cp}})g_i + \varepsilon_i \tag{33.15}$$

In the CPRM the null hypothesis (H) and the alternative hypothesis (K) when testing the difference of the outcome measure are as follows: H: $\beta_2 = 0$, K: $\beta_2 \neq 0$.

The results of applying the CPRM to the HCA study data are shown in Fig. 33.5 and Table 33.10. The P-value of β_2 is 0.0034, indicating a significant difference between groups. The slope of the HCA group changes after x_{cp} (=70.94) and deviates from the control group (Fig. 33.5A). In other words, the VFA-decreasing effect of HCA begins to appear in obese individuals with a baseline VFA ≥ 70.94 cm^2. The slope of the regression line of the HCA group after x_{cp} is 0.9404 − 0.3695 = 0.5709, which indicates a decrease of approximately 57% compared to the pre-value. The behavior of the placebo group after the intervention is almost unchanged from $\beta_1 = 0.9404$. Moreover, even when y_i is set to the pretest-posttest value, the P-value of β_2 does not change, again indicating a significant difference. This is because Eq. (33.14) is a model that is obtained by rotating Eq. (33.12) by − π/4 as in ANCOVA. When the HCA study data are used as an example, the ANCOVA2 model is most successful in detecting differences between groups using the conventional analysis method. Meanwhile, the CPRM can detect the difference between groups with much greater power. However, this is not the only benefit of the CPRM. The change point estimated by the CPRM is a threshold for the appearance of effects.

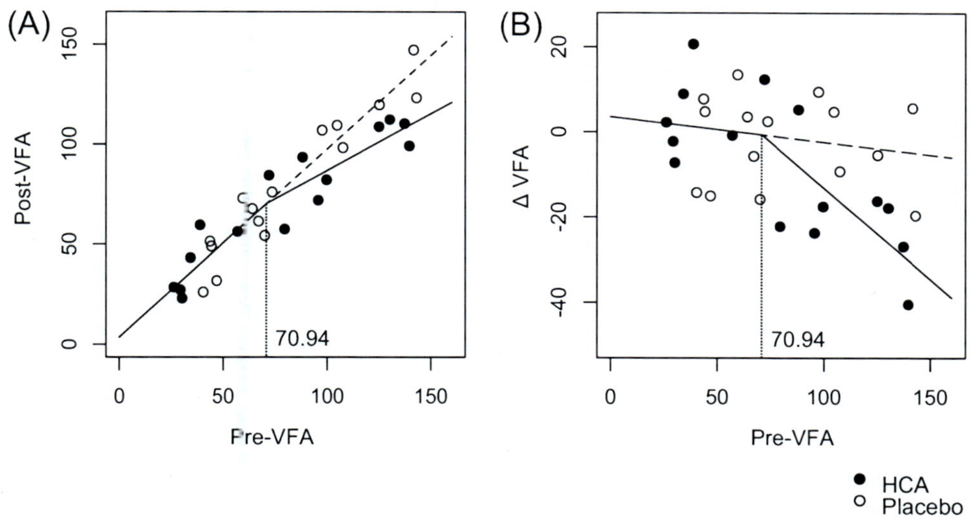

FIGURE 33.5 Application of change point regression model to the (−)-hydroxycitric acid study. (A) The outcome measure is the post-value. (B) The outcome measure is the pretest-posttest value.

TABLE 33.10 Analysis of the (−)-hydroxycitric acid study by change point regression model.

Outcome measure	Parameter	Estimate	SE	*T* value	*P* value
Post-value (x_{cp} = 70.94)	Intercept (α)	3.5813	5.0614	0.708	0.4853
	β_1	0.9404	0.0658	14.293	< 0.0001
	β_2	− 0.3695	0.1152	− 3.209	0.0034
Pre-post value (x_{cp} = 70.94)	Intercept (α)	3.5813	5.0614	0.708	0.4853
	β_1	− 0.0596	0.0658	− 0.906	0.3729
	β_2	− 0.3696	0.1152	− 3.209	0.0034

Hence if the HCA study was a pilot study, it could provide critical information for protocol development in planning large trials for verification. In addition, by comparing the value of the inflection point at which the effect appears with a clinically meaningful threshold, the result can provide stronger evidence. Thus the CPRM is not only an analysis method for evaluating differences between groups, but also a convenient method that provides key information.

In this chapter the intervention effect in a pretest-posttest study design was evaluated by using ANOVA (i.e., *t*-test), ANCOVA, and the CPRM, and the results were compared by using the sample data of the HCA study. The analysis aimed to examine the intervention effect and showed that the results were the same regardless of whether the outcome measure was the post-value or the pretest-posttest value in ANCOVA or the CPRM.

However, in the exploratory stages of clinical trials, it is necessary to incorporate factors that are considered important (i.e., covariates) into the model and examine their effects on the outcome measure. Thus the goodness of fit of the model is different from the outcome measure that is used, making it appropriate to determine the factors on the basis of a model with the best fit; this is used in the search for x_{cp} in the CPRM. That is, in CPRMs such as Eqs. (33.12)–(33.15), it is the same question as finding the value of x_{cp} that yields the best fit.

While there are several goodness-of-fit indicators, the Akaike information criterion (AIC), a representative indicator, will be used here. The AIC is used to "balance model complexity against data fit" (Burnham and Anderson, 2002). As an example, consider the creation of a model that statistically describes some measurement data. The more parameters or the higher the order, the better the goodness of fit with the measurement data. On the other hand, unintentional (i.e., unrelated to the structure of the measurement target) fluctuations such as noise preclude the fitting of the same kind of data; this is termed the overfitting problem. To avoid this problem, it is necessary to reduce the number of modeling parameters. However, it is difficult to decide how many should be used. The AIC is one solution. Specifically, selecting the model with the smallest AIC is often adequate. The AIC is defined as follows:

$$\mathrm{AIC} = -2\ln L + 2k$$

TABLE 33.11 Comparison of AIC in each analysis model.

Outcome measure	Model	AIC	Change-point	*P* value
Post-value	ANOVA (*t*-test) (30.10)	300.1211	–	.4609
	ANCOVA1 (30.5)	240.5174	–	.1453
	ANCOVA2 (30.8)	236.3684	–	.0222
	CPRM (30.12)	233.2235	70.94	.0034
Pre-post value	ANOVA (*t*-test) (30.11)	247.8621	–	.2413
	ANCOVA1 (30.7)	240.5174	–	.1453
	ANCOVA2 (30.9)	236.3684	–	.0222
	CPRM (30.14)	233.2235	70.94	.0034

AIC, Akaike information criterion; *CPRM*, change point regression model.

Here, L is the maximum likelihood, and k is the number of parameters. If the number of parameters is increased, L will be larger and the AIC will be smaller. However, because the second term on the right side is $+2k$, the cost of adding a parameter is given as a penalty; that is, when the effect of the parameter added to the right side of the equation exceeds the penalty and the maximum likelihood is improved (i.e., when the AIC is lowered), the incorporated parameter is suitable for explaining the response variable on the left side.

A comparison of AICs and the test results for each test method are shown in Table 33.11. Accordingly, when the AIC increases, the *P*-value of the test of the intervention effect decreases. Among the analysis methods described in the HCA study, the CPRM is the most fitted. It can be interpreted as showing that 70.94 cm^2 of the pre-value is the effect appearance point (i.e., change point), indicating its superiority for reflecting the intervention effect of HCA administration.

In this chapter, the results of the HCA study have been presented as an example application of the CPRM. As another example application, Mastuse et al. performed an RCT to verify the effect of neuromuscular electrical stimulation during walking in obese women and analyzed the efficacy, using the CPRM (Matsuse et al., 2020). The CPRM has other applications besides evaluating efficacy in RCTs, such as analyzing alternating beats of the right ventricular contraction pattern in children (Hashimoto and Watanabe, 2014), analyzing echocardiography in newborns (Nagasawa et al., 2018), examining the relationship between deep trunk muscle thickness and 100-m sprint time in sprinters (Fujita et al., 2019), and estimating protein requirements in the elderly (Bandegan et al., 2017; Hayamizu et al., 2011; Ogawa et al., 2011; Rafii et al., 2015). All of these studies aimed to find change points. It is hoped that the application of the CPRM in RCTs will continue to increase.

References

Bandegan, A., Courtney-Martin, G., Rafii, M., Pencharz, P.B., Lemon, P.W., 2017. Indicator amino acid-derived estimate of dietary protein requirement for male bodybuilders on a nontraining day is several-fold greater than the current recommended dietary allowance. J. Nutr. 147, 850–857.

Burnham, K.H., Anderson, D.R., 2002. Model Selection and Multimodel Inference, second ed. Springer-Verlag, New York, p. 2002.

Chow, S.H., Liu, J., 2013. Design and Analysis of Clinical Trials, third ed. John Wiley & Sons, New Jersey.

Fujita, S., Kusano, S., Sugiura, Y., Sakuraba, K., Kubota, A., Sakuma, K., et al., 2019. A 100-m sprint time is associated with deep trunk muscle thickness in collegiate male sprinters. Front. Sports Active Living 1, 32.

Hashimoto, I., Watanabe, K., 2014. Alternation of right ventricular contraction pattern in healthy children-shift from radial to longitudinal direction at approximately 15 mm of tricuspid annular plane systolic excursion. Circ. J. 78, 1967–1973.

Hayamizu, K., Ishii, Y., Kaneko, I., Shen, M., Sakaguchi, H., Okuhara, Y., et al., 2001. Effect of long-term administration of *Garcinia cambogia* extracts on visceral at accumulation in human: A placebo-controlled double blind trial. J. Oleo Sci. 50, 805–812.

Hayamizu, K., Kato, M., Hattori, S., 2011. Determining amino acid requirements from repeated observations on indicator amino acid oxidation method by mixed-effect change-point regression models. J. Clin. Biochem. Nutr. 49, 115–120.

Hayamizu, K., Yamashita, N., Hattori, S., Kakuma, T., 2009. A change-point regression approach for efficacy evaluation of dietary supplement. J. Clin. Biochem. Nutr. 44, 285–290.

Kakuma, T., Hayamizu, K., 2010. Statistical analysis method for evaluating the effectiveness of functional foods. J. Jpn. Soc. Fd. Eng. 30, 39–43 (in Japanese).

Matsuse, H., Segal, N.A., Rabe, K.G., Shiba, N., 2020. The effect of neuromuscular electrical stimulation during walking on muscle strength and knee pain in obese women with knee pain: a randomized controlled trial. Am. J. Phys. Med. Rehabil. 99, 56–64.

Nagasawa, H., Motz, R., Hamada, C., Kohno, Y., Yamada, Y., 2018. An interpretation of the neonatal period definition obtained with echocardiographic examination by using change point regression analysis. J. Neonatal-Perinatal Med. 11, 387–392.

Ogawa, A., Naruse, Y., Shigemura, Y., Kobayashi, Y., Suzuki, I., Wada, S., et al., 2011. An evaluation of protein intake for metabolic demands and the quality of dietary protein in rats using an indicator amino acid oxidation method. J. Nutr. Sci. Vitaminol. 57, 418–425.

Rafii, M., Chapman, K., Owens, J., Elango, R., Campbell, W.W., Ball, R.O., et al., 2015. Dietary protein requirement of female adults >65 years determined by the indicator amino acid oxidation technique is higher than current recommendations. J. Nutr. 145, 18–24.

Commentary from the Editor's Desk

Commentary from the Editor's Desk

A treatise on a healthy microbiome: contribution to human health and disease prevention

Debasis Bagchi[1,2,3], Samudra Prosad Banik[4] and Bernard William Downs[5]

[1]College of Pharmacy and Health Sciences, Texas Southern University, Houston, TX, United States, [2]Department of Biology, Adelphi University, Garden City, NY, United States, [3]Department of R&D, Victory Nutrition International, Inc., Bonita Springs, FL, United States, [4]Department of Microbiology, Maulana Azad College, Kolkata, West Bengal, India, [5]Division of Precision Nutrition, Victory Nutrition International, Inc., Bonita Springs, FL, United States

A.1 Microbiome terminology: the origin

The term "microbiome" was coined by J.L. Mohr in 1952 (Davenport et al., 2017; Ursell et al., 2012). It refers to the collective genome of the microbiota of a living organism and led to the creation of several terms, such as transcriptome, genome, proteome, metabolome, holobiome. In 1988, J.M. Whipps and his collaborators redefined this term as "a distinctive microbial community regulating a reasonably explicit habitat with recognizable physico-chemical characteristics" (Berg et al., 2020; Davenport et al., 2017; Ursell et al., 2012). In 2020, Berg and colleagues provided a more discrete, clear, and comprehensive terminology with extended explanation. Berg and colleagues distinctly segregated these two terms "microbiome" and "microbiota" and provided six discrete strategic perspectives (1) the microbiota and its composition; (2) the heterogeneity and dynamics of microbiomes; (3) microbial networks, their stability, and their resilience; (4) the core microbiome and its definition; (5) microbiome-host interactions; and (6) microbiome interspecies interactions. These distinct perspectives should enhance the standardization of future microbiome studies and their implementation to improve human health (Berg et al., 2020).

A.2 Microbiota, microorganisms, and microbiome

Trillions of microbes that exist in the gut and are referred to as the gut microbiota play a vital role in human biology, survival and existence, and disease (Barko et al., 2018; Razavi et al., 2019; Tetro and Allen-Vercoe, 2016). The term "biome" refers to a novel ecosystem that is reasonably well-defined with distinct biophysicochemical properties and comprises natural flora and fauna, while the prefix "micro," which implies that something is invisible to the human eye, refers to the premise that the human body is naturally equipped with a huge collection of microorganisms, mostly bacteria, but also archaea, algae, protists, yeast, viruses, and fungi, which manage and maintain the bacterial ecosystem that resides mainly in the human gut and attempts to equilibrate the environment's ecology and to maintain the stability of the underlying human-microbiome interactions.

Two major categories of microbes survive in the gut microbiota: autochthonous microbes (on the epithelium of the colonic mucosa) and allochthonous microbes (most enteric and pathogenic microorganisms) (Barko et al., 2018; Moludia et al., 2019; Nagano et al., 2019; Razavi et al., 2019; Tetro and Allen-Vercoe, 2016). Note that the progression of cirrhosis can be determined from the ratio of autochthonous to nonautochthonous microbes.

The microbiota consists of multiple microorganisms, including prokaryotes (bacteria), eukaryotes (such as protozoa, algae, and fungi), archaea, and viruses living in harmony with the cells of the human body, while an array of internal ecosystem components and residents are regulated by microbial elements, including proteins, peptides, lipids, polysaccharides, DNA, and RNA, and are influenced or affected by genetic elements including phages, transposons, viruses, and relic DNA that exist in the environmental conditions of the habitat (Crakes and Jiang, 2019; Nichols et al., 2019;

Microbiome, Immunity, Digestive Health and Nutrition. DOI: https://doi.org/10.1016/B978-0-12-822238-6.00032-7

Nieman and Wentz, 2019). Finally, a plethora of microbial metabolites, including diverse signaling molecules, toxic metabolites and organic molecules, are produced in the body under the changing conditions of the internal environment (Nichols et al., 2019; Razavi et al., 2019).

The term "microbiome" refers to a characteristic microbial community defining a reasonably well-characterized habitat with distinctive physiochemical and genetic properties. Basically, the microbiome, or the genetic material of all microbes belonging to a specific ecological niche, either free or living in symbiotic space with an organism, involves a distinctive and diverse theater of activity and functions, specific to the habitat that they share (Barko et al., 2018; Razavi et al., 2019; Tetro and Allen-Vercoe, 2016). The microbiome is a powerful and bilateral microecosystem that changes with time and exposure conditions and is further integrated in macroecosystems, including eukaryotic hosts and their functioning, which in ideal bioenvironmental conditions ultimately contributes to human health (Crakes and Jiang, 2019; Dabke et al., 2019; Nagano et al., 2019; Nichols et al., 2019; Nieman and Wentz, 2019; Ussar et al., 2016).

The microbiome survives as a community within living hosts and establishes commensal, amensal, mutualistic, parasitic, and/or pathogenic relationships. This collection of microorganisms is known as the microbiome or microbiota (Barko et al., 2018; Razavi et al., 2019; Tetro and Allen-Vercoe, 2016). Digestion of food, immune system regulation, protection against diverse disease-causing bacteria, and synthesis of multiple vitamins, including vitamins B12, thiamine and riboflavin, vitamin K (a blood coagulant), and short-chain fatty acids (SCFAs), are regulated by the bacteria in the microbiome (Nichols et al., 2019; Tetro and Allen-Vercoe, 2016). Thus in addition to the activity of microorganisms, the microbiome also encompasses their theater of activity, leading to the equilibration of specific ecological niches. In a nutshell, the microbiome produces an aggressive and reciprocal microecosystem that changes with time, exposure conditions, and propensity, is further consolidated into macroecosystems involving eukaryotic hosts, and is intricately associated with physiological performance, overall health, and well-being (Nieman and Wentz, 2019; Tetro and Allen-Vercoe, 2016).

It is important to emphasize that secondary metabolites have a pivotal role in arbitrating complex interspecies interactions that enable survival in competitive environments. Small molecules initiate quorum sensing, which allows the bacteria to initiate, perform, and regulate cooperative activities and facilitate the adaptation of their phenotypes to the biotic surroundings, leading to cell-cell adhesion or biofilm formation. For most anaerobic ecosystems, direct interspecies electron transfer is a crucial mechanism (Miller and Bassler, 2001; Moludia et al., 2019; Nagano et al., 2019; Nichols et al., 2019; Tetro and Allen-Vercoe, 2016). It has been demonstrated that volatile molecules can serve as long-term messengers for distant cross-kingdom communications (Tetro and Allen-Vercoe, 2016; Ussar et al., 2016).

A.3 Microbiome: human health and disease pathology

Inside the human body, as well as on the skin, diverse microbial communities such as commensal, symbiotic, and pathogenic communities survive and can modulate host energy metabolism and immune competence, regulate several diseases and disorders, and improve the host's health and well-being. Fig. A.1 depicts the diverse activity continuum of the microbiota. Scientists are continuing to conduct research in this area to understand how the microbiome influences

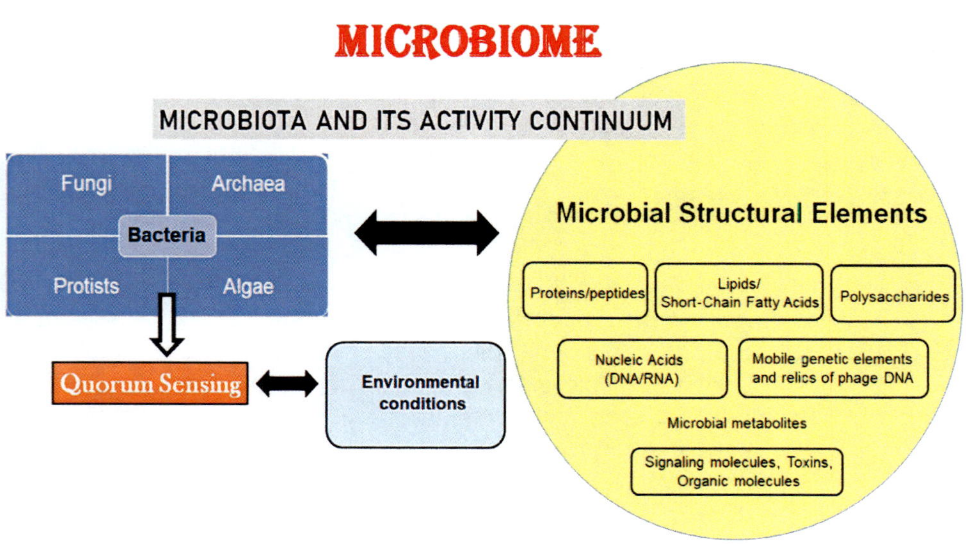

FIGURE A.1 The microbiota and its activity continuum.

human health. This book has focused on revealing and delving into knowledge gaps and future research opportunities (Barko et al., 2018; Nieman and Wentz, 2019; Razavi et al., 2019; Tetro and Allen-Vercoe, 2016). Unveiling and identifying discrete unique microbiota may help to improve human health and therapeutic strategies. For example, the composition of the lung microbiome is altered during chronic pulmonary diseases and disorders, which can dramatically influence host defense and immune competence, while an infection can greatly influence the microbiome of the blood or pulmonary tissues (Moludia et al., 2019; Nagano et al., 2019; Nichols et al., 2019; Nieman and Wentz, 2019; Razavi et al., 2019; Tetro and Allen-Vercoe, 2016; Ussar et al., 2016).

Ongoing research is attempting to reveal and articulate intricate aspects of human physiology and microbiome standardization and their correlation to boost human health by corrective manipulation of the microbiome. Understanding the microbiota community structure and its influence on preventing human diseases and improving human health can be done by analyzing the abundance in the body of both healthy microbes and opportunistic pathogens, which compete for resources (Barko et al., 2018; Miller and Bassler, 2001; Razavi et al., 2019; Tetro and Allen-Vercoe, 2016). Novel microbiota investigations may help to improve health, prevent disease, and augment the development of novel therapeutics.

The human microbiome, especially the gut microbiome, is involved in coordinating several essential activities of the host, such as digestive health, proper nutrition, metabolism and detoxification, host defense, maturation and improving of the host immune defense, and disease initiation or prevention (Barko et al., 2018; Razavi et al., 2019; Tetro and Allen-Vercoe, 2016). It is interesting to note that a mammalian gut contains diverse numbers and types of microbes, mostly Firmicutes and Bacteroidetes. Interestingly, European and Chinese individuals with type 2 diabetes exhibit distinctly dissimilar gut microbiome compositions, which may be to the result of environmental and/or genetic differences (Barko et al., 2018; Moludia et al., 2019; Nagano et al., 2019; Nichols et al., 2019; Tetro and Allen-Vercoe, 2016).

In mammals, host diet and phylogeny extensively modulate and modify the gut microbial community (Crakes and Jiang, 2019; Tetro and Allen-Vercoe, 2016). Alterations in the host diet significantly modify the composition of three representative human gut bacteria, namely, *Bacteroides thetaiotaomicron*, *Eubacterium rectale*, and *Methanobrevibacter smithii*. Furthermore, different compositions of these three gut bacteria greatly influence host metabolism and allied disorders and diseases. For example, infectious mycobacterial and staphylococcal organisms are closely associated with inflammatory bowel disease. Another microbial product, lipopolysaccharide (LPS), which is biosynthesized by gram-negative bacteria and carried by chylomicrons, triggers innate immunity in organisms (Moludia et al., 2019; Razavi et al., 2019; Tetro and Allen-Vercoe, 2016). Incidentally, the level of LPS is a critical survival indicator for peritoneal dialysis patients. In particular, it has been established that plasma LPS and bacterial DNA are biomarkers for systemic inflammation and for prognosis. Trimethylamine, another microbial product, is oxidized to trimethylamine N-oxide by the liver enzymes, flavine monooxygenases, which have a pronounced effect on patient morbidity (Barko et al., 2018; Crakes and Jiang, 2019; Moludia et al., 2019; Nagano et al., 2019; Nichols et al., 2019; Razavi et al., 2019; Tetro and Allen-Vercoe, 2016).

It is well known that the SCFAs acetate, propionate, and butyrate are major fermentation products of the gut microbiome and are generated during metabolism (Nichols et al., 2019; Tetro and Allen-Vercoe, 2016) (Fig. 32.1). Moreover, several factors, including host lifestyle, dietary habits, drug usage, genetics, and immune competence, could influence the composition of the microbiome, which will greatly influence host physiology and overall health. In addition, other risk factors may greatly affect the interaction between host physiology and the microbiota (Tetro and Allen-Vercoe, 2016; Ussar et al., 2016).

A.4 Microbiome and metabolic syndrome

Impairment of the microbiome, especially the gut microbiome, has an adverse influence on metabolic homeostasis and significantly contributes to multiple intricate mechanisms of metabolic syndrome, including aberrant blood pressure, type 2 diabetes mellitus, insulin resistance, a dysregulated renin-angiotensin system, cardiovascular dysfunction, hyperlipidemia, hypercholesterolemia, hyperglycemia, hypertriglyceridemia, hyperlipoproteinemia, adiposity, and visceral fat accumulation (Mazidi et al., 2016). Mazidi et al. (2016) reported that the gut microbiome comprises about 2 kg of the entire body weight. An impaired gut microbiome also disrupts energy homeostasis and immune competence and triggers inflammatory signaling pathways. A recent collaborative study by researchers at Vanderbilt University Medical Center and the University of California at Davis demonstrated that a link exists between heart disease and consumption of a high-fat diet in a mouse model (Gut Microbiome Link Between High-Fat Diet and Heart Disease Identified. Clinical Omics, 2021). These researchers demonstrated that disruption of the gut microbiome can occur as a result of frequent consumption of high-fat foods through enhanced production of toxic metabolites, which can lead to diverse cardiovascular diseases and dysfunctions. Probiotics, prebiotics, synbiotics, specific nutraceuticals, SCFAs, and botanical extracts exhibit significant benefits in appropriately modulating the gut microbiome to ameliorate symptoms of metabolic syndrome and restore health.

A.5 Microbiome and obesity

The trillions of bacteria of the gut microbiome play a vital role in human metabolism. Research studies have demonstrated that microbial imbalance in both humans and animals leads to an upsurge in obesity and metabolic syndrome (Barko et al., 2018; Razavi et al., 2019; Tetro and Allen-Vercoe, 2016). John and Muallin (2016) reported that bacteria can induce an increase in weight, change the expression of related genes that are associated with lipid and carbohydrate metabolism, and increase energy homeostasis from a healthy diet. It has also been demonstrated that obesity can develop as a result of changes in the ratio of Bacteroidetes to Firmicutes. These researchers indicated that gut microbiome dysbiosis contributes significantly to obesity and is more complicated than just an alteration in the ratio of Bacteroidetes to Firmicutes. Appropriate changes in the gut microbiome are possible through modification of dietary habits and supplementation of prebiotics (i.e., nondigestible beneficial fibers), probiotics, and digestive enzymes to combat the obesity epidemic (John and Muallin, 2016; Tetro and Allen-Vercoe, 2016).

A.6 Microbiome and diabetes

Dysbiosis is a consequence of compositional and structural changes in the gut microbiota that modify metabolic activities, causing diverse metabolic disorders leading to diabetes and obesity among other pathologies. Some of the latest state-of-the-art technologies, including next-generation sequencing of metagenomic DNA, polymerase chain reaction—denaturing gradient gel electrophoresis, microarrays, and fluorescence in situ hybridization have given us advanced knowledge and greater technological expertise to explore the gut microbiome (Barko et al., 2018; Razavi et al., 2019; Tetro and Allen-Vercoe, 2016; Vallianou et al., 2018). Dysbiosis leads to elevated plasma concentrations of lipopolysaccharides (LPSs) and has demonstrated an increased incidence of cytokine surge and consequential increased inflammatory response. Studies based on metabolomics demonstrated that butyrate and propionate inhibit weight gain in mice supplemented with high fat diet, while acetate was shown to lower food consumption in healthy mice (Vallianou et al., 2018). Extensive research of newer technologies using prebiotics, probiotics, synbiotics, genetically modified bacteria, and fecal microbiota transplantation offer new opportunities to correct dysbiosis and consequential pathologies, including amelioration of type 2 diabetes.

A.7 Microbiome and cancer

It has been well demonstrated that a healthy microbiota functions synergistically to confer benefits on the host. However, changes in the microbiome can occur as a result of diverse environmental stressors, including food, infection, and/or unhealthy lifestyle, which may perturb the healthy symbiotic relationships of microbes and cause diverse degenerative diseases (Barko et al., 2018; Razavi et al., 2019; Tetro and Allen-Vercoe, 2016). Recent studies have demonstrated an important role for the bacterial microbiota in the induction of carcinogenesis that is mediated through immune dysfunction, dysbiosis, and metabolic disruption. Researchers are unveiling novel therapeutic strategies to make a healthy microbiome to prevent and combat cancer (Goodman and Gardner, 2018; Rajagopala et al., 2017; Schwabe and Jobin, 2013).

A.8 Microbiome, brain health, and advancing age

Researchers have recently come to a more detailed understanding of the intimate link between gut health, brain health, and aging (Nagpal et al., 2018). These researchers have also demonstrated that restoring a healthy microbiome can reverse progressive age-associated cognitive decline and dysfunctions. Recent published research has shown growing evidence that gut health is necessary for optimal brain health as well as overall physical health (Carabotti et al., 2015; Clapp et al., 2017).

A.9 Summary and conclusions

Researchers have proposed that the establishment of a healthy bacterial ecosystem in a human body is based on an individualized microbiome with a composition of high specificity. However, ecosystem generalizations are daunting, owing to extensive variations that exist from one individual to another. Also, researchers have established the significance of the microbiome in blood circulation and have shown that macrovesicles equipped with exogenous microbial RNA can serve as signaling molecules in human plasma. With advancing age, the ratio of Firmicutes to Bacteroidetes alters, which is associated with an age-related reduction in genetic integrity and a concomitant increase in numerous disease pathologies. For example, inflammatory bowel and hepatic diseases are associated with an increase in Bacteroidetes and a reduction in Firmicutes. Significant research studies have demonstrated that administration of select prebiotics, probiotics, and synbiotics, singly and in combination, promote the growth of beneficial microorganisms in the intestine, which ultimately boost physical well-being and immune competence. Epigenetic factors combine the impact of lifestyle influences on gene expression. With a gradual and progressive reduction in advancing age-related genetic integrity, it is crucial to increase and maintain adherence to a better quality of lifestyle behaviors, such as food, exercise, and sleep, and reduce the negative influence of such factors as excessive stress. Thus in the very near future, exploring and monitoring the personalized genome, transcriptome, and microbiome may reveal new health assessment opportunities that will enable more therapeutic corrective interventions to arrest a wide range of chronic degenerative diseases.

References

Barko, P.C., McMichael, M.A., Swanson, K.S., Williams, D.A., 2018. The gastrointestinal microbiome: a review. J. Vet. Intern. Med. 32 (1), 9–25. Available from: https://doi.org/10.1111/jvim.14875.

Berg, G., Rybakova, D., Fischer, D., Cernava, T., Vergès, M.C., Charles, T., et al., 2020. Microbiome definition re-visited: old concepts and new challenges. Microbiome 8 (1), 103. Available from: https://doi.org/10.1186/s40168-020-00875-0.

Carabotti, M., Scirocco, A., Maselli, M.A., Severia, C., 2015. The gut-brain axis: interactions between enteric microbiota, central and enteric nervous systems. Ann. Gastroenterol. 28 (2), 203–209. PMCID: PMC4367209 PMID: 25830558.

Clapp, M., Aurora, N., Herrera, L., Bhatia, M., Wilen, E., Wakefield, S., 2017. Gut microbiota's effect on mental health: the gut-brain axis. Clin. Pract. 7 (4), 987. Available from: https://doi.org/10.4081/cp.2017.987.

Crakes, K.R., Jiang, G., 2019. Gut microbiome alterations during HIV/SIV infection: implications for HIV cure. Front. Microbiol. 10, 1104. Available from: https://doi.org/10.3389/fmicb.2019.01104.

Dabke, K., Hendrick, G., Devkota, S., 2019. The gut microbiome and metabolic syndrome. J. Clin. Invest. 129 (10), 4050–4057. Available from: https://doi.org/10.1172/JCI129194.

Davenport, E.R., Sanders, J.G., Song, S.J., Amato, K.R., Clark, A.G., Knight, R., 2017. The human microbiome in evolution. BMC Biol. 15, 127. Available from: https://doi.org/10.1186/s12915-017-0454-7.

Goodman, B., Gardner, H., 2018. The microbiome and cancer. J. Pathol. 244, 667–676. Available from: https://doi.org/10.1002/path.5047.

Gut Microbiome Link Between High-Fat Diet and Heart Disease Identified. Clinical Omics. https://www.clinicalomics.com/topics/patient-care/cardiovascular-disease/gut-microbiome-link-between-high-fat-diet-and-heart-disease-identified/?utm_medium = newsletter&utm_source = Clinical + OMICs + Update&utm_content = 01&utm_campaign = Clinical + OMICs + Update_20210813&oly_enc_id = 8797B0386467B6R (accessed August 13, 2021).

John, G.K., Muallin, G.E., 2016. The gut microbiome and obesity. Curr. Oncol. Rep. 18, 45. Available from: https://doi.org/10.1007/s11912-016-0528-7.

Mazidi, M., Rezaie, P., Kengne, A.P., Ghayour, M., Mobarhan, M.G., Ferns, G.A., 2016. Gut microbiome and metabolic syndrome. Diabetes Metab. Syndrome: Clin. Res. Rev. 10 (2), S150–S157. Available from: https://doi.org/10.1016/j.dsx.2016.01.024.

Miller, M.B., Bassler, E.L., 2001. Quorum sensing in bacteria. Annu. Rev. Microbiol. 55, 165–199. Available from: https://doi.org/10.1146/annurev.micro.55.1.165.

Moludia, J., Alizadeh, M., Davari, M., Golmohammadi, A., Maleki, V., 2019. The efficacy and safety of probiotics intervention in attenuating cardiac remodeling following myocardial infraction: literature review and study protocol for a randomized, double-blinded, placebo-controlled trial. Contemp. Clin. Trials Comm. 15, 100364. Available from: https://doi.org/10.1016/j.conctc.2019.100364.

Nagano, T., Otoshi, T., Hazama, D., Kiriu, T., Umezawa, K., Katsurada, N., et al., 2019. Novel cancer therapy targeting microbiome. OncoTargets Ther. 12, 3619–3624. Available from: https://doi.org/10.2147/OTT.S207546.

Nagpal, R., Mainali, R., Ahmadia, S., Wang, S., Singh, R., Kavanagh, K., et al., 2018. Gut microbiome and aging: physiological and mechanistic insights. Nutr. Healthy Aging 4, 267–285. Available from: https://doi.org/10.3233/NHA-170030.

Nichols, R.G., Peters, J.M., Patterson, A.D., 2019. Interplay between the host, the human microbiome, and drug metabolism. Hum. Genomics 13, 27. Available from: https://doi.org/10.1186/s40246-019-0211-9.

Nieman, D.C., Wentz, L.M., 2019. The compelling link between physical activity and the body's defense system. J. Sport. Health Sci. 8, 201–217. Available from: https://doi.org/10.1016/j.jshs.2018.09.009.

Rajagopala, S.V., Vashee, S., Oldfield, L.M., Suzuki, Y., Venter, J.C., Telenti, A., et al., 2017. The human microbiome and cancer. Cancer Prev. Res. (Phila.) 10 (4), 226–234. Available from: https://doi.org/10.1158/1940-6207.CAPR-16-0249.

Razavi, A.C., Potts, K.S., Kelly, T.K., Bazzano, L.A., 2019. Sex, gut microbiome, and cardiovascular disease risk. Biol. Sex. Differ. 10, 29. Available from: https://doi.org/10.1186/s13293-019-0240-z.

Schwabe, R.F., Jobin, C., 2013. The microbiome and cancer. Nat. Rev. Cancer 13 (11), 800–812. Available from: https://doi.org/10.1038/nrc3610.

Tetro, J.A., Allen-Vercoe, E. (Eds.), 2016. The Human Microbiome Handbook. Destech Publications Inc., Lancaster, PA.

Ursell, L.K., Metcalf, J.L., Parfrey, L.W., Knight, R., 2012. Defining the human microbiome. Nutr. Rev. 70 (Suppl. 1), S38–S44. Available from: https://doi.org/10.1111/j.1753-4887.2012.00493.x.

Ussar, S., Fujisaka, S., Kahn, C.R., 2016. Interactions between host genetics and gut microbiome in diabetes and metabolic syndrome. Mol. Metab. 5 (9), 795–803. Available from: https://doi.org/10.1016/j.molmet.2016.07.004.

Vallianou, N.G., Stratigou, T., Tsagarakis, S., 2018. Microbiome and diabetes: where are we now? Diabetes Res. Clin. Pract. 146, 111–118. Available from: https://doi.org/10.1016/j.diabres.2018.10.008.

Index

Note: Page numbers followed by "*f*" and "*t*" refer to figures and tables, respectively.

A

Acetate, 95–96
Acidaminococcaceae, 198
Active living, 179
Aerobic glycolysis, 208
Akkermansia, 118–119, 198, 452, 457
Alcoholic liver disease, 4[illegible]8
Allergic diseases, 80
Alzheimer's disease (AD), 5–6
American diet. *See* Western diet
Antibiotic-associated diarrhea, 367, 378
Antioxidant therapy, 206–207
Athletic performance, microbiome and, 452–457
Autoimmune disease, 75

B

Bacteriophages, 297
Bacteroidales S24-7 group, 198
Bacteroidetes, 209
Bifidobacteria, 44–45
Bifidobacterium, 100, 118–119, 367
Bifidobacterium lactis BB12, 99–100
Bile acids, 95–96, 144–145
Bilophila, 198
Bioactive peptides in foods
  absorption, 471
  elimination, 472
  metabolism, 472
  research into, 470
Body's normal physiological homeostasis
  de novo para-aminobenzoic acid (PABA), 250
  gut microbiota in, 249–251
  inflammatory responses, 251
  maintenance and development of, 250
  major role in, 250
  neutrophil aging, 251
Brain-derived neurotrophic factor, 30–31
Brain functioning, 266–267
Butyrate, 95–96
Butyric acid, 213

C

Cancer progression, 391–392
Carcinogen, 316
Cardiovascular disease (CVD), 6
Cecum, 41
Celiac disease, 79
Change point regression model (CPRM), 481, 490*f*
  Akaike information criterion (AIC), 491–492, 492*t*
  challenges approach, 489
  hydroxycitric acid study by, 491*f*, 491*t*
  statistical model, 489–490
Christensenellaceae, 198
Chronic inflammatory
  allergic diseases, 80
  and autoimmune diseases, 79–82
  cancer, 81–82
  celiac disease, 79
  inflammatory bowel disease, 80
  neurological inflammatory diseases, 80
  periodontal disease, 80–81
  rheumatoid arthritis, 79
  skin-related autoimmune pathology, 80
  systemic lupus erythematosus, 80
  type 1 diabetes, 79
Clinical intervention, study design, and statistical methodology
  change point regression model, 489–492
  outcome measure, 482
  pre-value effect, 482–489
Clinical study, immunoregulation peptide
  probiotics strains on T cell response, 474
  related to COVID-19, 474
Clostridium difficile infection, 97, 391
Colonization resistance, 30
Colorectal cancer, 97–99, 382, 392
  by enhancing healthy gut flora
    prebiotics, 320–321
    prevention of, 320–322
    probiotics, 321
    synbiotics, 322
  progression, 315–317
  protection against, 319–320
Commendamide, 145
Commensals, 109
Constipation, 378–380
Conventionalized (CONV) mice, 195
Coriobacteriaceae, 198
Curli possesses amyloidogenic behavior, 31–32
Cutaneous wound healing
  bacteria in chronic ulcers, 305*t*
  chronic wounds and wound microbiome, 304–306
    diabetic foot ulcers, 306
    pressure ulcers, 305
    venous ulcers, 305
  inflammatory cells, 302
  inflammatory phase of, 302
  microbiome and chronic development, 302–306
  M1 macrophages secrete factors, 302
  remodeling, 303
  venous and arterial ulcers, 304
Cytokines, 362

D

Depression, 34
Desaminotyrosine (DAT), 144
Desulfovibrionaceae, 198
Diabetes mellitus (DM), 6–7
Diet and cognition, 268
Dietary components in boosting a healthy microbiome
  carbohydrates, 158
  fats, 159
  fatty acids, 159–160
  healthy diet and poor-quality diet, 159*f*
  polyphenols, 160
  prebiotics, 160–161
  probiotics, 161
  proteins, 158–159
  selective diets, 161
  spices, 160
  syndrome risk, 157–161
Dietary components on metabolic endotoxemia
  effects of, 172
  exercise and, 172
  fasting on, 172
  next generation of bacteria, 172
Dietary interventions on the microbiota
  diet and gut, 129*t*
  effects of, 127–133
Dietary modulation of the gut microbiota, 100–102
Diet, gut microbes, and cognitive decline
  diet and cognition, 268
  microbiome and cognitive decline, 268
  microbiome composition and evolution, 265–266
  microbiome-gut-brain axis, 267–268
  microbiota and brain functioning, 266–267
Digestion, 206

Diseases of the gut
central to gut, 276
contribution of human intestinal microbes, 96–99
Clostridium difficile infection, 97
colorectal cancer, 97–98
esophageal cancer, 98–99
inflammatory bowel diseases (IBD), 97
obesity, 98
rheumatoid arthritis, 98
Parkinson's disease, 279*f*
psychiatric and neurological disorders, 277–280
major depressive disorder (MDD), 278*f*
neurological disorders, 278–280
psychiatric disorders, 277–278
toxins and Alzheimer's disease, 280*f*
Diverse pathological conditions
gastrointestinal microbiota, 261
innate and adaptive immune system, 260
maternal immune activation (MIA) mouse model, 261
microbiota and its implications in, 259–262
Parkinson's disease, 260
Serotonin's vasoconstrictive effects, 260–261
DNA methylation in Rett syndrome, 282
Dopamine homeostasis to combat brain-gut functional impairment
brain reward cascade, 230*f*
deficiency syndrome as a behavioral octopus, 230*f*
dopamine release, 231*f*
endorphin deficiency syndrome, 232–233
gut-hypothalamic homeostatic energy regulation, 234–237
obesity as a hypodopaminergia, 233–234
opioid deficiency syndrome, 232–233
Dysbiosis as obesogenic signature
decreased diversity, 168
firmicutes/bacteroidetes ratio, 168
gut barrier permeability, 169
biome-focused lab testing, 169
oligosaccharides, 169
omega-3 index, 169
polyphenols, 170
postbiotics, 170
prebiotics, 169
probiotics, 170
resistant starch, 169–170
stool microbiome testing, 169
testing, 169
keystone bacteria, 168
Dysbiosis of normal flora
gut microbiome in human ailments, 32–34
cancer, 33
depression, 34
diabetes, 33
gout, 33–34
hypertension, 33
infant health, 34
inflammatory bowel disease, 33
obesity, 32–33
rheumatoid arthritis, 34
human aliments and the therapeutic probiotic, 35*t*
in human diseases and intervention strategy, 32–35
prebiotics, probiotics, postbiotics, and synbiotics, 34–35

E

Eating, gut biome balance
bioactive lipids, 177
category foods, 177
clinical protocol for microbiome-centered, 178
clinical protocol for physical activity, 180
exercise *versus* nonexercise, 179
and fasting, 176–179
intermittent, 178–179
versus mimicking, 178–179
mimicking time-restricted eating, 179
fats, 177
microbiota-accessible carbohydrates, 177
micronutrients, 177
phytonutrients, 178
prebiotics, 178
prebiotic whole foods, 177
probiotics, 178
protein, 177
sedentary *versus* sitting disease, 179–180
vitamins, 177
Emotional motor system, 410
Endocrine pathway, 408
Endorphin deficiency syndrome
epigenetic induction of opioid peptide, 233*f*
and opioid, 232–233
Endotoxins, 170, 196, 362
Enteric infections, 382
Enterohemorrhagic *E. coli*, 218–219
Enterotypes, 348
Epigenetic influences, 209
Epithelial cells, 20–21
Escherichia coli, 100
Esophageal cancer, 98–99
Ethylphenylsulfate, 143–144

F

Farnesoid X receptor (FXR), 396
Farnesol X, 154
Fasting-inducible adipose factor, 174
Fermentation, 374
Fermented beverages, 373
Fibromyalgia
gut microbiota and, 334–335
pain and, 334
Fight-or-flight response, 253
Flavonoids, 145
Flaxseed pronounced effect on gut microbiota
alcoholic liver disease by, 418
antiinflammatory property, 418
firmicutes-bacteroidetes ratio, 423–424
immunological and metabolic effects, 421–423
sex steroid hormone-microbiota-inflammation axis, 424–427
type 2 diabetes mellitus, 419–421
Food-derived immunomodulatory peptides
bioactive, 461
bioavailability of, 470–472
clinical study, 474
immunomodulation and, 462–463
mechanism of, 463
methods used to assay, 472–473
sources of, 463–470
Forgotten organ, 137

G

Gastric cancer, 99–100, 392
Gastrointestinal tract (GIT), 93
development of gut flora, 94–95
dietary modulation of the gut microbiota, 100–102
microbes in, 93–94
causing diseases, 96–99
microorganism that can prevent human disease, 99–100
colorectal cancer, 99
gastric cancer, 99–100
hepatic encephalopathy, 100
inflammatory bowel diseases, 100
obesity, 100
physiological functions of gut microbes, 95–96
central nervous system development, 96
immune system development, 96
metabolism, 95–96
Gastrointestinal tract-related diseases
antibiotic-associated diarrhea, 378
colorectal cancer, 382
constipation, 378–380
enteric infections, 382
inflammatory bowel diseases, 380–381
intestinal mucositis, 381–382
laboratory-fermented beverages in, 376–382
Gliadin, 75
Gluten, 75
Gout, 33–34
G protein-coupled membrane receptor, 154
Granulation tissue, 302
Gut-associated lymphoid tissue, 114–115, 396–397
Gut-brain axis
animal study, 411–412
behavior and, 411–413
bidirectional signaling, 406
brain function and behavior, 412*t*
concept of, 405–406
human study, 413
interactions within, 407–409
endocrine pathway, 408
humoral/metabolic pathway, 408
immune pathway, 409
neuronal pathway, 408
microbiota in, 409–411
to brain, 409–410
from brain to, 410–411
in regulation of food intake, 411
schematic representation of, 406*f*

Gut-brain communication
- axis, 405–407, 406*f*
- behavior and microbiota, 411–413
- function of microbiota in, 409–411
- interactions within, 407–409
- regulating the microbiota, 407
- regulation of food intake, 411

Gut dysbiosis and its role in chronic disease
- autoimmune, 79–82
- chronic inflammatory, 79–82
- diet and, 72
- fasting and, 72
- gliadin, 75
- gluten, 75
- healthy diets for, 73–74
- inflammation, 74–75
- inner tube of, 77–78
- irritants to, 76–77
- leaky, 76
- short-chain fatty acids, 72
- unhealthy diets on, 72
- unraveling the mystery of treatment, 82–85

Gut flora, development of, 94–95

Gut-hypothalamic homeostatic energy regulation
- feeding behavior regulation by dopaminergic systems, 235–236
- food and drug addiction identification, 236–237
- homeostatic systems, 235
- set point dilemma, 235

Gut-immune system axis 363

Gut-lung axis, 362

Gut microbes, physiological functions of, 95–96
- central nervous system development, 96
- immune system development, 96
- metabolism, 95–96

Gut microbial flora in body's serotonin turnover and diseases
- axis and serotonin turnover, 257–259
- diverse pathological conditions, 259–262
- gastrointestinal tract (GIT), 246
- gut microbiota, 246–249
- microbiota-gut-brain axis, 246
- neurotransmitter-modulating activity, 246
- normal physiological homeostasis, 249–251
- population through gut-brain axis, 251–257

Gut microbial metabolites
- and cardiovascular disease, 138–145
- of clinical significance, 143–145
  - bile acids, 144–145
  - desaminotyrosine, 144
  - ethylphenylsulfate 143–144
  - flavonoids, 145
  - imidazole propionate, 144
  - long-chain fatty acids, 143
  - N-acyl amides, 145
  - polyamines, 144
  - retinoic acid, 144
  - taurine, 144
- preferences, 139*f*
- short-chain fatty acids, 139–143
  - cardiovascular disease pathology implications, 141
  - indoxyl sulfate, 140–141
  - role of, 139–140
- and trimethylamine-N-oxide, 142
  - and cardiovascular disease, 142–143

Gut microbiome, 185–188, 273, 347–351, 432
- absorption, distribution, metabolism and excretion, 274
- ADME, representation of, 275*f*
- *Akkermansia muciniphila*, 351
- altering medication efficacy, 274–275
- branched-chain amino acids, 351*f*
- branched-chain fatty acids, 349, 350*f*
- digestion and, 273–274
- dysbiosis and eubiosis, 348–349, 349*f*
- effect of various dietary types on, 101*t*
- factors affecting
  - alcohol consumption, 318
  - dietary fats, 317
  - obesity, 318–319
  - physical activity, 319
  - red and processed meats, 317–318
  - smoking, 318
- factors influencing gut, 348*f*
- human, 347*f*
- human nutrition and immunity
  - in immune system development, 46–47
  - role of microbiome, 44–46
  - studying, 42–44
- improving
  - avoid artificial sweeteners, 47
  - breastfeed for six months, 47
  - consume polyphenol-containing food, 48
  - consume prebiotic foods, 47
  - eating diverse diet, 47
  - fermented foods, 47
  - plant-based foods, 48
  - take antibiotics when required, 48
  - ways to improve gut microbiome, 47–48
  - whole grains, 48
- introduction to, 273–276
- and its importance
  - in body, 43*f*
  - studying, 42–44
  - types of bacteria in, 43*f*
- medication-induced gut dysbiosis, 276
- probiotics and management of tuberculosis
  - amelioration of gastrointestinal, 367
  - amelioration of hepatotoxicity, 367–368
  - as an antiinfective mediator, 366
  - anti-TB drug-induced adverse effects, 367–368
  - beneficial effects of, 365*f*
  - challenges in treatment of, 362*f*
  - dose and dosage of, 368–369
  - dysbiosis due to, 364*f*
  - human role in, 363
  - -immune system axis, 363
  - as immunomodulator, 365–366
  - improvement of nutritional status in, 368
  - -liver axis, 362–363
  - -lung axis, 362
  - microbiota taxonomy profiling in, 364*t*
  - preclinical and clinical evidence of, 366–367
  - role of probiotics in, 363–367
- in protection against colorectal cancer prevention of, 319*f*
- resiliency, 167
- Rett syndrome, 280–284
  - diseases of, 276–280
  - diversity and dysbiosis, 285–286
  - intestinal inflammation in, 289
  - introduction to, 273–276
  - loss of function mutations, 290
  - neurological and psychiatric disorders, 287–288
  - recent and past study, 286–287
  - relationship between, 284–290
  - therapeutics and, 289–290
- short-chain fatty acids (SCFAs), 274*f*, 349
- taxonomic composition of, 348
- terminology, 350*t*
- virome and microplastics, 349–351

Gut microbiome on human health and diseases
- dysbiosis of normal flora, 32–35
- microbiome, 25
- physiological roles of, 27–32
- significant niche of, 26–27
- uneven distribution of microbiota, 25–26

Gut microbiota
- and beneficial effects
  - groups of human, 389*t*
  - and impacts of dysbiosis, 390*f*
  - shape of, 388
- and cardiovascular disease, 113–114
- and diabetes, 112
- dietary modulation of, 100–102
- distribution of human, 248*f*
- and dyslipidemia, 112–113
- function of, 249*f*
- and hyperglycemia, 112
- and hypertension, 113
- and metabolic disorders, 187
- and metabolic syndrome, 110–114
- and MetS, 114
- microbiome and energy homeostasis, 187
- obesity, 111–112
- and obesity
  - among children, 195
  - microbial richness to, 196
  - multifactorial disease, 195
  - preclinical and human data reporting, 197*t*
  - resveratrol, 196–199
- in protein metabolizing machinery, 249
- taxonomic heterogeneity, 247
- and their distribution, 247*t*
- Type 2 diabetes, 112

H

Healthy diets for gut
- benefit of, 73–74
- ketogenic, 74
- low-carbohydrate, 74
- Mediterranean, 74
- paleolithic, 74

Healthy gut microbiome
of colorectal cancer
by enhancing, 320–322
factors affecting, 317–319
progression of, 315–317
protection against, 319–320
composition of, 150–153
in metabolic syndrome
atherosclerosis, 158*f*
beneficial role of, 154–157
in cardiovascular diseases, 156–157
composition of, 150–153
dietary components in, 151*t*, 157–161
dysbiosis, 110*f*
gut, 149
gut microbiota and, 110–114
health and diseases, related to, 154
metabolic syndrome, 150
and metabolic syndrome, 154–155
micronutrients, prebiotics, and probiotics, 157*f*
at molecular level, 114–117
and obesity, 156*f*
in preventing metabolic syndrome, 154–157
risk factor of, 149–150
16S rRNA gene sequencing, 111*f*
therapy by managing gut microbiota, 117–119
to treat obesity, 155–156
trimethylamine-N-oxide, 158*f*
taxonomic, 153*t*
Heat-killed probiotics, 190
Helicobacter pylori, 93–94, 99–100
Hepatic encephalopathy, 100
Hepatocellular carcinoma, 392
Human intestinal microbes, 96–97
Clostridium difficile infection, 97
colorectal cancer, 97–98
esophageal cancer, 98–99
inflammatory bowel diseases (IBD), 97
obesity, 98
rheumatoid arthritis, 98
Humoral/metabolic pathway, 408
Hydrolysis of mushroom polysaccharides
by intestinal bacterial enzymes, 439
Ruminococcus champanellensis, 439
Sus-like transport system, 439
Hydroxycitric acid study, 482, 483*t*
analysis of, 486*t*
by ANCOVA 2, 488*t*
by change point regression model, 491*t*
coefficients of ANCOVA 1 in, 487*t*, 488*f*
concept of ANCOVA1, 486*f*
estimated coefficients of ANCOVA 2, 488*t*
intervention effect, 485*t*
point regression model to, 491*f*
postintervention changes in, 483*f*
T-test in, 489*t*
visceral fat area, 483*t*
Hygiene hypothesis, 42

I

Imidazole propionate, 144
Immune cell subsets in early life
epithelial cells, 20–21
and function of, 13–21
in germ-free animals, 18*f*
lymphoid, 14–19
B cells, 19
$CD4^+$ T cells, 14–15
$CD8^+$ T cells, 16–17
germ-free mice with *Bacteroides fragilis*, 16*f*
innate-like lymphocytes, 17–18
innate lymphoid cells, 18–19
invariant natural killer T cells, 17–18
NK cells, 19
regulating serum IgE levels, 20*f*
T cell repertoire, 14
Th1 and Th2 Cells, 15
Th17 cells, 15
T regulatory cells, 15–16
microbiota on development, 13–21
myeloid cells, 13–14
erythrocytes in the spleen, 14
myeloid-derived suppressor cells, 14
Immune exclusion, 398–399
Immune pathway, 409
Immune reactions control
Human Microbiome Project (HMP), 217
Lactobacillus salivarius into the lung, 221–225
microbiota detection tools, 218–225
control mice, 221*f*
germ-free IgA knockout and B6 mice, 220*f*, 222*t*
lactobacilli strain, 220*f*
lactobacilli strains, 223*f*
mice models, 218–221
Schaedler's *E. coli*, 220*f*
surfactant protein D, 221–225
technological approaches, 218
total IgM production, 224*f*
translocation of lactobacilli, 221*f*
wild-type and IgA knockout mice, 223*f*
nutrition and energy mechanisms, 218
Immunity in early life
and functional tuning, 12
germ-free mice, characteristics of, 12*f*
microbiota shapes the development of, 11–13
Immunomodulatory activity
methods used to assay, 472–473
in vitro and in vivo, 473*t*
Immunomodulatory peptide mechanism, 465*f*
innate and adaptive, 464*f*
and innate defense regulator, 465*f*
mediators in immune system, 463
Immunomodulatory peptides
antiproliferative, 462
cytotoxic activity, 462
development of new, 462
immunomodulation and, 462–463
Immunosurveillance-mediated cancer prevention, 399
Impacts of dysbiosis
Clostridium difficile infection, 391
colorectal cancer, 392
gastric cancer, 392
hepatocellular carcinoma, 392
inflammatory bowel diseases, 391
metabolic disorders, 391
other cancers, 392–393
role in cancer progression, 391–392
Individual microbiota correction and human health
brief methodology, 55
and cardiovascular diseases, 55
classical concept of, 53
early formation of, 54
endocannabinoid axis, 54
formation and early colonization, 54
healthy, 53–54
results and discussion, 55–65
Infant health, 34
Inflammaging, 266–267
Inflammation
acute, 75
chronic, 75
Inflammation, 188
Inflammatory bowel disease, 33, 80, 97, 100, 380–381, 391
Inflammatory events, 210
Inflammatory inflammation, 352–353
Innate immunity, 398
Innate-like lymphocytes, 17–18
Innate lymphoid cells, 18–19
Inner tube of gut
active microbial community, 77
cross-talk between microbes, 78
microbiome and infections, 78
mucosal immune system, 77
SCFAs, 77
toll-like receptors (TLRs), 78
Insulin-dependent diabetes, 6
Intervention effect without adjusting pre-value effect
ANCOVA 1 in the (-)-hydroxycitric acid study, 487*t*
ANCOVA 2 in the (-)-hydroxycitric acid study, 488*t*
application to(-)-hydroxycitric acid study, 486*f*
application to(?)-hydroxycitric acid study, 488*f*
hydroxycitric acid study by ANCOVA 1, 486*t*
hydroxycitric acid study by ANCOVA 2, 488*t*
method for, 485–489
T-test in the (-)-hydroxycitric acid study, 489*t*
Intestinal microbiome, 265–266
Intestinal microbiome in overweight and obesity
according to weight status, 126–127
causes of, 125
dietary interventions on, 127–133
microbiota and, 125–126
Intestinal mucositis, 381–382
Intestinal transplantation, 188–189
Invariant natural killer T cells, 17–18
Isoquinolines, 231–232

K

Kernel, 176

L

Lactobacillus acidophilus, 99–100
Laboratory-derived bioactive compounds, 374–375
Lactic acid bacteria (LAB)
  based beverages
    commercial products from, 376
    fermentation process, 375
    laboratory-derived bioactive compounds, 374–375
    laboratory-fermented beverages in, 376–382
    lactic acid bacteria, 373–374
    organoleptic characteristics of lab-fermented foods, 375–376
  fermentation process, 375
  fermented foods
    commercial products from, 376
    dairy and nondairy probiotics beverages, 377*t*
    healthy gastrointestinal system, 376
    strains and lactic acid bacteria (LAB), 379*t*
Lactobacillus, 367
Lactobacillus reuteri, 388
Leaky gut, 76
Listeria innocua, 388
Long-chain fatty acids, 143
L-Plantarum PS128, 457
Lymphoid cells, 14–19

M

Medication-induced gut microbiome dysbiosis, 276
Metabolic disorders, 391
Metabolic syndrome
  and cardiovascular diseases, 55
  flowchart depicting factors, 153*f*
  medical conditions, 150
  person-oriented treatment of type 2 diabetes, 55
  therapy
    diet influence, 117–118
    exercise influence, 119
    by managing gut microbiota, 117–119
    prebiotics influence, 118
    probiotics influence, 118–119
Metabolites, 362
Methanobrevibacter smithii, 457
Microbial dysbiosis in skin-related diseases
  acne vulgaris, 300–301
  atopic dermatitis, 300
  bacteria in, 300*t*
  psoriasis, 301
  rosacea, 301
  skin cancers, 302
  skin integrity and health, 301*f*
Microbial metabolites in cardiovascular and human health
  bioactive in, 138*f*
  gut, 138–145
Microbial regulation of host energy balance
  associated harvest, 173
  fasting and autophagy, 174
  gut-adipose-liver axis and energy storage, 173–174
  gut biome and circadian rhythm, 174
  gut-brain axis of eating behavior, 173
  gut microbiota and the host metabolism, 173
  gut-muscle axis and energy output, 173
Microbial strains, 212
Microbiome functional competence, 209
Microbiome-gut-brain axis, 267–268
  and serotonin turnover
    association between, 257–259
    brain-gut microbiome interaction, 260*f*
    enzyme tryptophan hydroxylase 2 (TpH2), 258
    intestinal serotonin activity, 258
    neurons in CNS, 257
    serum 5-HT levels and urinary 5-HIAA excretion, 259*t*
Microbiome in arthritis
  with chronic pain conditions, 337*t*
  commonalities of gut, 338
  fibromyalgia, 334–335
  multiple sclerosis, 335–338
  probiotic supplementation, 338*t*
  rheumatoid, 331–333
Microbiome in immune system development
  healthy probiotic bacteria, 46
  irritable bowel syndrome, 47
  role of, 46–47
Microbiome in nutrition
  boosting of cardiovascular health, 45
  brain health, promotion of, 45
  breast milk, digestion of, 45
  diabetes risk, 45
  fiber, digestion of, 45
  immune competence, enhancement of, 45
  immune system, and diet, 44*f*
  insulin sensitization, potentiation of, 45
  prebiotic foods and dietary fiber, 45–46
  probiotic foods, 46
  role of, 44–46
Microbiome in sports nutrition, 451–457
Microbiomes to combat metabolic syndrome and disorders, 187
  dysbiosis, 187–188
  gut, 185–187
  harnessing the gut to treat disorders, 188
  heat-killed probiotics, 190
  and inflammation, 188
  intestinal transplantation, 188–189
  oral butyrate, 189
  postbiotics, 190
  prebiotics, 190
  synbiotics, 190
Microbiota according to weight status
  bacterial composition of, 126–127
  high (HGC) and low (LGC) microbial gene count, 128*t*
  lean and obese individuals, 127*t*
Microbiota and obesity
  composition of, 126
  dietary habits, 126
  Human Microbiome Project and MetaHit, 125
  in overweight, 126
Microbiota-gut-brain axis
  factors regulating, 407
  fats, 407
  genetics and epigenetics, 407
  medications and microbiome, 407
  mode of delivery at birth, 407
Microcephaly, 281
Microplastics on human systems, 352*f*
  induced reactive oxygen species, 352–353
  modulation of cellular pathways, 352
  particles in different tissues, 353
  toxic effects of, 351–353
Microplastic toxicity
  gut microbiome, 347–351
  history of, 345–346
  as pathogen carriers, 354
  plastic particles in sizes and chemistry, 345*f*
  plastisphere, 353–354
  pollution and the threat to habitats, 346
Modulation of cytokine profile, 398
Molecular level mechanism
  energy metabolites, 115–117
    bile salt hydrolase, 116
    modulation of cell signaling and immune response, 116
    putative molecular basis, 117*f*
    short-chain fatty acids, 115–116
    trimethylamine oxide, 116–117
  gut barrier, 114–115, 115*f*
Mucin, 93–94
Mucosal layer, 83
Multifaceted systems biology intervention
  antioxidant therapy, 206–207
  butyric acid, 213
  conceptual overview, 210–211
  digestion, 206
  epigenetic influences, 209
  ideal biological environment, 205–206
  inflammatory events, 210
  microbial dysbiosis, 211–212
  microbial restoration, 212
  microbial strains, 211*t*, 212
  microbiome functional competence, 209
  microbiome, role of, 207–209
  optimal homeostatic equilibrium, 205–206
Multiple sclerosis
  gut microbiota and, 335–336
  and modifiable factors, 336–338
Muscle-gut or gut-muscle axis, 173
Mushroom polysaccharides, 432–437
  action of dietary fibers, 433*f*
  D-glucan, 433–437, 436*t*
    prebiotic properties of, 436*t*
  dietary fibers, action of, 433*f*
  gut microbiome, 432
  on gutmicrobiome, 439–440
  and gut microbiota have interactions, 439–440
  hydrolysis of, 439
  impact of, 439–440
  potential benefits of, 439

Mushroom polysaccharides (*Continued*)
prebiotic efficiency of, 438–439
present in mushrooms, 434*t*
prevention of diseases by, 440–445
proteoglycan, 437–438
structural property of, 440
structure of, 440
in various mushrooms, 434*t*
Myeloid cells, 13–14

N

N-acyl amides, 145
Neurological disorders, 278–280
Neurological inflammatory diseases, 80
Neuronal pathway, 408
Neurotransmitters on the gut microbiome population
of biogenic amines, 252–256
catecholamines, 253
histidine derivative on, 255–256
norepinephrine, 253
brain-gut axis and serotonergic metabolism, 255*f*
communication route, 252*f*
through gut-brain axis, 251–257
5-hydroxytryptamine, 253–255
miscellaneous neurotransmitters, 257
gamma-aminobutyric acid derivatives in, 257
producing bacteria, 254*t*
role of esters in, 256–257
glutamic acid in, 256–257
quaternary ammonium esters in, 256
Newborn immune system
microbiota on the development and function of, 13–21
microbiota shapes, 11–13
Niche of the human microbiome
by fetus and newborn, 27*f*
of gut microbiome, 26
gut shaping, 27
heritable component of gut, 26–27
significant, 26–27
Nucleotide-binding oligomerization domain–like (NOD-like) receptors, 96

O

Obesity as a hypodopaminergia
comprehensive systems biology approach, 234
digestive malfunction, 234
dopamine and neurotransmitters, 234
strategy to overcome, 234
understanding, 233–234
Obesogenic features of the gut microbiome
transferrable, 168
mother-to-child transmission, 168
Obesogenic metabolic endotoxemia
diet-induced lipopolysaccharide endotoxemia, 170–171
immune activation, 172
and insulin resistance, 170–172
insulin resistance and inflammation, 172
stress and metabolic endotoxemia, 172
Oil pulling, 84
Optimal homeostatic equilibrium, 205–206
Oral butyrate, probiotics, 189
Oral cancer, 4–5
Oral microbiome
cancer, 4–5
caries, 3–4
and disease, 3–5
periodontal disease, 4
and systemic disease, 5–7
Alzheimer's, 5–6
cardiovascular, 6
diabetes, 6–7
Organoleptic characteristics of lab-fermented foods, 375–376
Outcome measure, pretest-posttest design statistical analysis model, 482*t*
Oxidative respiration, 208
Oxygen utilization, 207*f*

P

Patients with metabolic syndrome
anaerobic microorganisms, 65
chromatography of alycha, 62*f*
databases, 58–59
data from human trials, 56–58
by novel pharmabiotics, 57–58
personalized nutrition patients with type 2 diabetes, 56–57
Escherichia coli 058
Enterococcus faecalis, and *Bacillus subtilis 090*, 61*t*
gastrointestinal microbiome for, 55–56
health programming, 62–65
IT tools, 59
lactic acid bacteria in dynamics, 60*t*
mechanisms involved and prospective research, 59
number of microorganisms, 64*f*
obesity, type 2 diabetes, and cardiovascular diseases, 55–56
pathogenic microbiota, 64
3P medicine and individual nutrition, IT tools, 58
as potential modern prebiotics, 59–62
pregnancy and childbirth in women, 63*t*
proinflammatory (IL-17) and antiinflammatory (IL-10) cytokines and SIgA, 65*f*
PD dementia, 279–280
Pectic substances, 393
Pentose phosphate pathway, 375
Periodontal disease, 4, 80–81
Personal oral hygiene protocol, 84–85
Pharmacodynamics, 274–275
Physiological roles of gut microbiome
associated lymphoid tissue, 29–30
in boosting immune response, 28–29
cognitive development, role in, 30–32
defense mechanism against pathogens, 30
functions, 31*f*
as a human organ, 32
microbiota in, 27–28
phylum *Firmicutes*, 28
propionate and acetate, 28
short-chain fatty acids, 29*t*
Plastisphere, 353–354
Polyamines, 144
Porphyromonas gingivalis, 4
Prebiotics, 34, 83, 160–161, 190
components, 393
definitions of, 393
mushroom polysaccharides
antitumor effect, 444–445
from *Flammulina velutipes*, 438
of gut microbiota, 440–445
gut-organ axes, 444*f*
Hericium erinaceus, 444
immunomodulatory effect, 445
inflammatory bowel disease, 440–444
by *Lentinula edodes*, 438–439
on metabolic disorders, 441*t*
Pretest-posttest design, 481–482
Prevotella, 452, 457
Probiotics, 451
additional benefits of, 399
bacteria, 44–45, 398
bile acids, 396
concept of, 394
enhanced secretion of IgA, 398–399
gut-associated lymphoid tissue, development of, 396–397
gut immunity expansion, 396–397
in immunomodulation, 394–399
bacteria in gut, 395*f*
boosting effects, 394
short-chain fatty acids, 395*f*
immunomodulatory effects, 394–396
immunosurveillance-mediated cancer prevention, 399
innate immunity, role in, 398
and modulation of cytokine profile, 398
organisms, 394
organisms against pathogenesis, 397–398
prebiotics and in gut immunity
benefits of, 399
dysbiosis impacts, 390–393
in immunomodulation, 394–399
and its beneficial effects, 388–390
prebiotics, 393–394
protective role against inflammation, 396
role of, 394–399
short-chain fatty acids, 394–396
Progression of colorectal cancer
cancer progression role, 316
cellular metabolism, 316
colorectal carcinogenesis, 316
gut microbiome in, 315–317
metaanalysis of fecal metagenomes, 316
Propionate, 95–96
Propionibacterium freudenreichii, 388
Proteoglycan
glycoprotein, 438
Phellinus linteus, 437
Pseudo-organ, 137

Psychiatric and neurological disorders, 277–280
Psychiatric disorders, 277–278

R

Red complex, 4
Resistant starch (RS), 393–394
Restoring gut biome balance for weight loss
dietary components on metabolic endotoxemia, 172
dysbiosis as obesogenic signature, 168–170
eating and fasting, 176–179
gut microbiome resiliency, 167
microbial regulation of host energy balance, 173–174
obesogenic features of, 168
obesogenic metabolic endotoxemia and insulin resistance, 170–172
physical activity *versus* physical inactivity, 179–180
wake-sleep circadian rhythm, 180–181
western lifestyle, 174–176
Resveratrol-mediated gut microbiot
and obesity, 194–199
QRC, EGCG, and resveratrol(RSV), 194*f*
and type 2 diabetes mellitus, 199–200
Retinoic acid, 144
Rett syndrome, 280–284
background and history of, 280–281
biogenic amine systems in, 282
DNA methylation in, 282
excitatory and inhibitory receptor systems, 283
genetic links, 281
gut microbiome and, 280–284
in males, 282–283
pathology, symptoms, and progression, 281
treatment paradigm, 283–284
Rheumatoid arthritis, 79, 98
microbiome profile of, 332
probiotic supplementation in, 332–333
Ruminococcaceae, 198

S

Salmonella enterica, 388
Secondary bile acids, 116
Sequence databases, 315–316
Sex steroid hormone-microbiota-inflammation axis, 424–427
Short-chain fatty acids (SCFAs), 72, 95–96, 394–396
Sitting disease, 179
Skin microbiome diversity
body parts, 299–300
dermatological disorders, 299
Propionibacterium acnes, 299
Proteobacteria and Flavobacteriales, 299–300
Staphylococcus species. *P. acnes*, 299
Skin microbiota manipulation
associated ailments, 307–308
microbiota transplant, 307
prebiotics, 307–308
probiotics, 307
Skin-related autoimmune pathology, 80
Skin structure
and function, 298–299
proteins and lipids, 298
sweat glands, 298–299
Sources of immunomodulatory peptides
eggs, 468
fish, 467–468
marine organisms, 468
milk and milk products, 464–467, 466*t*
cheese, 467
plant
corn, 470
-derived, 469*t*
peas, 470
rice, 469–470
wheat, 470
plants, 468–470
Spatial data, 59
Spore-based bacillus probiotics, 83
Sport
microbiome and, 451–452
research related to the microbiome and athletic performance, 452–457
Standard American diet (SAD), 167, 174
Stratum corneum, 298
Stratum granulosum, 298
Stratum spinosum, 298
Streptococcus, 367
Surveying the skin microbiota techniques, 306
Synbiotics, 35, 118–119, 190
Systemic lupus erythematosus, 80

T

Tannerella forsythia, 4
Taurine, 144
T cell repertoire, 14
TePe EasyPick, 84
Toxin producer, 287–288
Treatise on a healthy microbiome
brain health and advancing age, 500
and cancer, 500
and diabetes, 500
human health and disease pathology, 498–499
and metabolic syndrome, 499–500
microbiota
and its activity, 498*f*
microorganisms, 497–498
and obesity, 500
terminology, 497
Treponema denticola, 4
Type 2 diabetes mellitus
gut microbiota and, 199–200
metabolic endotoxemia, 199–200
resveratrol, 200
T2DM development, 199

U

Uncultivated phylotypes, 3
Unraveling the mystery of treatment
for active infections, 82
importance of mouth, 84
mucosal layer, 83
personal oral hygiene protocol, 84–85
prebiotics, 83
repairing and restoring the gut, 82–83
spore-based bacillus probiotics, 83
understanding periodontal disease, 82
Uremic toxin, 143–144

V

Veillonella, 452–457
Virome, 109
Vitamin K_2, 95–96

W

Wake-sleep for circadian rhythm balance, 180–181
Western diet, 72
Western lifestyle as menace to gut biome
eating habits, 174
NOVA classification, 174–176
Wound healing and chronic wound development
and cutaneous wound healing, 302–306
microbial dysbiosis in skin-related diseases, 300–302
skin cancers, 302
skin integrity and health, 301*f*
skin microbiome diversity, 299–300
skin structure and function, 298–299
surveying the skin microbiota, 306
treat skin-associated ailments, 307–308

X

X-chromosome inactivation, 282–283